Advanced Level Physics

Grenville Jones

Physics

Fifth edition

Books by M. Nelkon

Published by Heinemann

ADVANCED LEVEL PRACTICAL PHYSICS (*with J. Ogborn*)
SCHOLARSHIP PHYSICS
OPTICS, WAVES, SOUND
MECHANICS AND PROPERTIES OF MATTER
PRINCIPLES OF ATOMIC PHYSICS AND ELECTRONICS
REVISION NOTES IN PHYSICS
Book I. Mechanics, Electricity, Atomic Physics
Book II. Optics, Waves, Sound, Heat, Properties of Matter
GRADED EXERCISES AND WORKED EXAMPLES IN PHYSICS
(with Multiple Choice Questions)
NEW TEST PAPERS IN PHYSICS
REVISION BOOK IN ORDINARY LEVEL PHYSICS
ELEMENTARY PHYSICS, Book I and II (*with A. F. Abbott*)
MATHEMATICS OF PHYSICS (*with J. H. Avery*)
ELECTRICAL PRINCIPLES (*with H. I. Humphreys*)
ELECTRONICS AND RADIO PRINCIPLES (*with H. I. Humphreys*)
SOLUTIONS TO ADVANCED LEVEL PHYSICS QUESTIONS
SOLUTIONS TO ORDINARY LEVEL PHYSICS QUESTIONS
BASIC MATHEMATICS FOR SCIENCE

Published by Hart-Davis Educational

PRINCIPLES OF PHYSICS
EXERCISES IN ORDINARY LEVEL PHYSICS
C.S.E. PHYSICS

Published by Edward Arnold

ELECTRICITY

Published by Blackie

HEAT

Books by P. Parker

Published by Heinemann

HEAT

Published by Arnold

ELECTRONICS

Cover Photographs

(Front) A 'cubic' soap bubble trapped in a wire cube. (*Courtesy of Dr. C. Isenberg, Physics Laboratory, University of Kent*)
(Back) Two photographs of a decorated plate, one taken by reflected light at room temperature (left), the other by its own emitted light when incandescent at about 1100 K (right). (*Courtesy of the Worcester Royal Porcelain Company Limited and Tom Biro, FRSA, FSIAD*)

Advanced Level Physics

Fifth edition

M. NELKON, M.Sc. (Lond.), F.Inst.P., A.K.C.
formerly Head of the Science Department,
William Ellis School, London

P. PARKER, M.Sc., F.Inst.P., A.M.I.E.E.
Late Senior Lecturer in Physics,
The City University, London

Heinemann Educational Books
London

Heinemann Educational Books Ltd
22 Bedford Square, London WC1B 3HH

London   Edinburgh   Melbourne   Auckland
Hong Kong   Singapore   Kuala Lumpur   New Delhi
Ibadan   Nairobi   Johannesburg
Exeter (NH)   Kingston   Port of Spain

ISBN 0 435 68666 6
© M. Nelkon and Mrs. P. Parker 1958, 1964, 1970, 1977, 1982
First published as one volume 1958
Reprinted four times
Second edition 1964
Reprinted three times
Third edition (SI) 1970
Reprinted four times
Fourth edition 1977
Reprinted three times
Fifth edition 1982
Reprinted 1983, 1984 (twice)

British Library Cataloguing in Publication Data

Nelkon, M.
  Advanced Level physics.—5th ed.
  1. Physics
  I. Title   II. Parker, P.
530   QC23

ISBN 0-435-68666-6

Illustrations drawn by
George Hartfield Illustrators, Carshalton, Surrey

Set in 10/11 Monophoto Times by
Northumberland Press Ltd, Gateshead, Tyne and Wear, England
Printed and bound in Great Britain by
William Clowes Limited,
Beccles, Suffolk

Preface to Fifth Edition

In this edition the text has been revised and updated to take account of the Advanced level syllabuses of the major Examining Boards, such as the London Advanced level syllabus and that of the Joint Matriculation Board. To assist the student more worked examples in all branches of physics have been added in illustration of the subject matter and more straightforward questions on basic principles have been added to many exercises. The examination questions in exercises have also been updated and graded and two new multiple-choice papers are provided.

Some of the more important text changes are briefly as follows:

(1) In *Mechanics*, the conservation of linear momentum has been given further prominence, the main dynamics sections are now followed by exercises, the rotational dynamics text has been revised, and statics and fluids are reduced in content but there are more exercises on Bernoulli's principle.

The *Properties and Matter* sections have been revised and reordered.

(2) In *Heat*, the section on real gases has been merged with that on vapours and some older matter deleted.

(3) *Geometrical Optics* has been reduced in content in accordance with syllabuses and the telescope section has been rewritten.

(4) *Waves and Optics* has been extended, with additions in the air wedge and diffraction grating topics and an account of the Michelson interferometer.

(5) *Electricity and Atomic Physics*. The sections on electric circuits, magnetic fields and force on conductor, electromagnetic induction, photoelectricity and parts of nuclear energy have been rewritten, and magnetic materials reduced in content.

I am very grateful to the following for their considerable assistance with preparation of the new edition: J. H. Avery, formerly Stockport Grammar School; M. V. Detheridge, William Ellis School, London; S. S. Alexander, formerly Woodhouse School and The Mount School for Girls, Mill Hill, London; and Dr. M. Crimes, Woodhouse School, London.

I am also considerably indebted to the following for their generous assistance with the Fourth Edition on which the new edition is based: Mrs. J. Pope, formerly Middlesex Polytechnic; C. F. Tolman, Whitgift School, Croydon; D. Deutsch, formerly Clare College, Cambridge; P. Betts, The Crossley and Porter School, Halifax; R. D. Harris, Ardingly College, Sussex; R. Croft, The City University; N. Phillips, Loughborough University; and Dr. L. S. Julien, University of Surrey. I am also indebted to Richard Gale and Trevor Hook of the publishers for their unfailing courtesy and expert advice.

Note to 1983 and 1984 Reprints

Further text has been added on the resolving power of a diffraction grating, on the transistor switch with a sine wave input and use as an amplifier, on the motion of a rider round a curve and on the helical path of electrons. I acknowledge with thanks permission by The Associated Examining Board to reproduce questions (AEB) in past Advanced level examinations, and notice of amendments by M. P. Preston, Lewes, and R. D. Harris, Ardingly College.

Preface to Fourth Edition

In this edition we have taken account of the revised syllabuses of the Examining Boards, including the new London Advanced level syllabus. Briefly, the main changes in the text are as follows:

Mechanics and Properties of Matter
(i) Gravitation now includes an account of the potential and kinetic energy of satellites. (ii) In molecular theory, the variation of potential energy with molecular separation has been amplified, and properties of solids deduced. (iii) An account of the various types of bonds, mainly in solids, has been added. (iv) A discussion of dislocation and slip has been included in elasticity.

Heat
(i) The first law of thermodynamics in the form of $\Delta Q = \Delta U + p.\Delta V$ has been applied to ideal gases. (ii) In the kinetic theory of gases, there is now an account of the maxwellian distribution of molecular speeds, mean free path, and viscosity and thermal conductivity. (iii) Electrical methods have been given prominence in specific heat capacity and specific latent heat, and a mechanical method added for specific heat capacity. (iv) In radiation, the non-equilibrium case has been considered. (v) The triple point and its determination have been discussed in thermometry. (vi) In accordance with the new syllabuses, the section on the thermal expansion of solids and liquids has been revised.

Geometrical Optics
There is now a more concise account of mirrors, a more direct treatment of lenses, and early consideration of refractor and reflector telescopes. Except for basic definitions, photometry has been omitted.

Waves
This section has been expanded in accordance with the new syllabuses. It contains a general treatment of (i) mechanical and electromagnetic waves; (ii) progressive and stationary waves; (iii) reflection, refraction, interference, diffraction and polarization of waves. Sound waves, and the measurement of their velocity in air, have been fully discussed.

Wave Optics
(i) A qualitative account of the effect of lenses and mirrors on waves has been given. (ii) There are now separate chapters on interference, diffraction and polarization. The principles of holography and of the radio-telescope resolving power have been added, and the section on polarization has been expanded.

Sound
This section has been revised and now includes the case of the reflector in the Doppler effect and a brief account of high-fidelity reproduction.

Electricity
The electrostatics text now contains applications of the electrometer and d.c. amplifier, and the charge and discharge of a capacitor through a resistor. In current electricity: (i) a comparison of ohmic and non-ohmic conductors has been given; (ii) the classical electron theory has been extended; (iii) a fuller account of the e.m.f. of a thermocouple by the potentiometer, and of the absolute

method of measuring resistance, have been added; (iv) the chapters on electrolysis and magnetic materials have been revised.

Atomic Physics
(i) The chapter on Electrons includes the cathode ray oscilloscope, and the triode valve is only utilised as an introduction to the field effect transistor. (ii) Junction diodes and Transistors now discusses rectifiers, amplifiers, oscillators, switches, logic gates, the multivibrator and the field effect transistor. (iii) In the section on Energy Levels, there is an account of spontaneous and stimulated emission and its application to the laser. (iv) The chapter on Radioactivity and the Nucleus contains a discussion of absorption by metals, more details of the Rutherford scattering law and the nuclear reactor, and an account of the neutron–proton ratio aspect of unstable nuclei.

Preface to First Edition

This text-book is designed for Advanced level students of Physics, and covers Mechanics and Properties of Matter, Heat, Optics, and Sound. Electricity and Atomic Physics to that standard. It is based on the experience gained over many years of teaching and lecturing to a wide variety of students in schools and polytechnics.

In the treatment, an Ordinary level knowledge of the subject is assumed. We have aimed at presenting the physical aspect of topics as much as possible, and then at providing the mathematical arguments and formulae necessary for a thorough understanding. Historical details have also been given to provide a balanced perspective of the subject. As a help to the student, numerous worked examples from past examination papers have been included in the text.

It is possible here to mention only a few points borne in mind by the authors. In Mechanics and Properties of Matter, the theory of dimensions has been utilized where the mathematics is difficult, as in the subject of viscosity, and the 'excess pressure' formula has been extensively used in the treatment of surface tension. In Heat, the kinetic theory of gases has been fully discussed, and the experiments of Joule and Andrews have been presented in detail. The constant value of $n \sin i$ has been emphasised in refraction at plane surfaces in Optics, there is a full treatment of optical treatment of optical instruments, and accounts of interference, diffraction and polarization. In Sound, the physical principles of stationary waves, and their application to pipes and strings, have been given prominence. Finally, in Electricity the electron and ion have been used extensively to produce explanations of phenomena in electrostatics, electro-magnetism, electrolysis and atomic physics; the concept of e.m.f. has been linked at the outset with energy; and there are accounts of measurements and instruments.

Publisher's Note

Since the first publication of *Advanced Level Physics*, the revisions for reprints and new editions have been undertaken by Mr. Nelkon owing to the death of Mr. Parker.

ACKNOWLEDGEMENTS

Thanks are due to the following Examining Boards for their kind permission to reprint past questions:

London University School Examinations (*L.*)
Oxford and Cambridge Schools Examination Board (*O. & C.*)
Joint Matriculation Board (*N.*)
Cambridge Local Examinations Syndicate (*C.*)
Oxford Delegacy of Local Examinations (*O.*)

I am indebted to the following for kindly supplying photographs and permission to reprint them:

The late Lord Blackett and the Science Museum, Fig. 41.19; Head of Physics Department, The City University, London, Figs. 23.13, 24.2, 25.5; Dr. B. H. Crawford, National Physical Laboratory, Fig. 21.20; R. Croft, The City University, Figs, 23.6, 27.4, 24.17; Hilger and Watts Limited, Figs. 23.10, 23.14, 24.9, 24.12; National Chemical Laboratory, Fig. 40.22(i); N. Phillips, Loughborough University, Fig. 24.20; late Sir G. P. Thomson and the Science Museum, Fig. 40.22(ii); late Sir J. J. Thomson, Fig. 41.25; United Kingdom Atomic Energy Authority, Figs. 28.10, 41.30; The Worcester Royal Porcelain Company Limited and Tom Biro, Fig. 13.25 and back cover.

Contents

Part One: Mechanics and Properties of Matter

1	Dynamics	3
2	Circular motion. Gravitation, S.H.M.	38
3	Rotational Dynamics	77
4	Static Bodies. Fluids	94
5	Elasticity. Molecules and Matter	115
6	Surface Tension	138
7	Solid Friction. Viscosity	160

Part Two: Heat

8	Introduction: Temperature, Heat, Energy	173
9	Heat Capacity. Latent Heat	179
10	Gases: Gas Laws and Heat Capacities	194
11	Kinetic Theory of Gases	219
12	Changes of State. Vapours. Real Gases	236
13	Transfer of Heat: Conduction and Radiation	257
14	Thermal Expansion of Solids and Liquids	290
15	Thermometers.	298

Part Three: Geometrical Optics

16	Introduction: Reflection at Plane Surfaces and Curved Mirrors	311
17	Refraction at Plane Surfaces	326
18	Refraction Through Prisms	339
19	Lenses and Defects. Spectra	349
20	Optical Instruments	382

Part Four: Waves, Wave Optics, Sound

21	Oscillations and Waves	407
22	Wave Theory of Light. Velocity of Light	440
23	Interference of Light Waves	456
24	Diffraction of Light Waves	480
25	Polarization of Light Waves	499
26	Characteristics of Sound Waves	511
27	Waves in Pipes, Strings, Rods	530

Part Five: Electricity and Atomic Physics

28	Electrostatics	562
29	Capacitors	593
30	Current Electricity	619
31	Measurements by Potentiometer and Wheatstone Bridge	656
32	Chemical Effect of Current	674

33 Magnetic Fields and Force on Conductor   690
34 Magnetic Fields of Current-Carrying Conductors   709
35 Electromagnetic Induction   726
36 Magnetic Properties of Materials   763
37 A.C. Circuits   772
38 Electrons. Motion in Fields. Electron Tubes   793
39 Junction Diode, Transistor and Applications   822
40 Photoelectricity. Energy Levels. X-Rays   843
41 Radioactivity. The Nucleus   874
    Revision Papers   911
    Multiple Choice Papers   917
    Answers to Exercises and Papers   929
    Index   937

Part One

Mechanics and Properties of Matter

1 Dynamics

Motion in a Straight Line. Velocity

If a car travels in a constant direction and covers a distance s in a time t, then its *mean* or *average velocity* in that direction is defined as s/t. It therefore follows that

$$distance\ s = average\ velocity \times t.$$

The term 'displacement' is given to the distance moved in a constant direction, for example, from L to C in Fig. 1.1 (i). Velocity may therefore be defined as the *rate of change of displacement*.

Velocity can be expressed in *metre per second* (m s^{-1}) or in *kilometre per hour* (km h^{-1}). By calculation, 36 km h^{-1} = 10 m s^{-1}.

If an object moving in a straight line travels equal distances in equal times, no matter how small these distances may be, the object is said to be moving with *uniform* velocity. The velocity of a falling stone increases continuously, and so is a *non-uniform* velocity.

If, at any point of a journey, Δs is the small change in displacement in a small time Δt, the velocity v is given by $v = \Delta s/\Delta t$. In the limit, using calculus notation,

$$v = \frac{ds}{dt}.$$

We call ds/dt the *instantaneous velocity* at the time or place concerned. The term 'mean velocity' refers to finite times and finite distances.

Vectors and Scalars

Displacement and *velocity* are examples of a class of quantities called *vectors* which have both magnitude and direction. They may therefore be represented to scale by a line drawn in a particular direction. For example, Cambridge is 80 km from London in a direction 20° E. of N. We can therefore represent the displacement between the cities in magnitude and direction by a straight line LC 4 cm long 20° E. of N., where 1 cm represents 20 km, Fig. 1.1. (i). Similarly, we can represent the velocity u of a ball initially thrown at an angle of 30° to the horizontal by a straight line OD drawn to scale in the direction of the velocity u, the arrow on the line showing the direction, Fig. 1.1. (ii).

Unlike vectors, *scalars* are quantities which have magnitude but no direction. A car moving along a winding road or a circular track at 80 km h^{-1} is said to have a *speed* of 80 km h^{-1}. 'Speed' is a quantity which has no direction

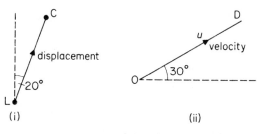

Fig. 1.1 Vectors

but only magnitude, like 'mass' or 'density' or 'temperature'. These quantities are examples of scalars.

The distinction between speed and velocity can be made clear by reference to a car moving round a circular track at say 80 km h^{-1}, Fig. 1.2. At every point on the track the *speed* is the same—it is 80 km h^{-1}. At every point, however, the *velocity* is different. At A, B or C, for example, the velocity is in the direction of the corresponding tangent AP, BQ or CR. So even though they have the same magnitude, the three velocities are all different because they point in different directions.

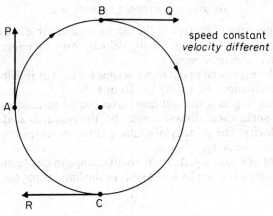

speed constant
velocity different

Fig. 1.2 Velocity and speed

Distance-Time Graphs

When the distance, s of a car moving in a constant direction from some fixed point is plotted against the time t, a *distance-time (s-t) graph* of the motion is obtained. The velocity of the car at any instant is given by the change in distance per second at that instant. At E in Fig. 1.3, for example, if the change in distance s is Δs and this change is made in a time Δt,

$$\text{velocity at E} = \frac{\Delta s}{\Delta t}.$$

In the limit, then, when Δt approaches zero, the velocity at E becomes equal to the *gradient of the tangent to the curve* at E. Using calculus notation, $\Delta s/\Delta t$

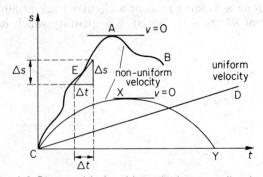

Fig. 1.3 Distance (s)–time (t) graphs (constant direction)

then becomes equal to ds/dt (p. 3). So the gradient of the tangent at E is the instantaneous velocity at E.

If the distance-time graph is a straight line CD, the gradient is constant at all points; it therefore follows that the car is moving with a *uniform* velocity, Fig. 1.3. If the distance-time graph is a curve CAB, the gradient varies at different points. The car then moves with non-uniform velocity. At the instant corresponding to A the velocity is zero, since the gradient at A of the curve CAB is zero.

When a ball is thrown upwards, the graph of the height s reached at any instant t is represented by the parabolic curve CXY in Fig. 1.3. The gradient at X is zero, illustrating that the velocity of the ball at its maximum height is zero.

Velocity-Time Graphs. Acceleration

When the velocity of a moving train is plotted against the time, a 'velocity-time (v-t) graph' is obtained. Useful information can be deduced from this graph, as we shall see shortly. If the velocity is uniform, the velocity-time graph is a straight line parallel to the time-axis, as shown by line (1) in Fig. 1.4. If the train increases in velocity steadily from rest, the velocity-time graph is a straight line, line (2), inclined to the time-axis. If the velocity change is not steady, the velocity-time graph is curved. In Fig. 1.4, for example, the velocity-time graph OAB represents the velocity of a train starting from rest which reaches a maximum velocity at A, and then comes to rest at the time corresponding to B.

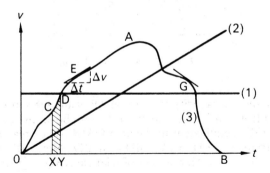

Fig. 1.4 Velocity (v)–time (t) curves

Acceleration is the 'rate of change of velocity', that is, the change of velocity per second. We can see that *the acceleration of the train at any instant is given by the gradient to the velocity-time graph* at that instant, as at E. At the peak point A of the curve OAB the gradient is zero, that is, the acceleration is then zero. At any point, such as G, between A, B the gradient to the curve is negative because the graph slopes downwards. Here the train has a *deceleration* or decrease in velocity with time. Like velocity, acceleration is a vector.

The gradient to the curve at any point such as E is given by:

$$\frac{\text{velocity change}}{\text{time}} = \frac{\Delta v}{\Delta t}$$

where Δv represents a small change in v in a small time Δt. In the limit, the ratio $\Delta v/\Delta t$ becomes dv/dt, using calculus notation.

Area between Velocity-Time Graph and Time-Axis

Consider again the velocity-time graph OAB in Fig. 1.4, and suppose the velocity increases in a very small time-interval XY from a value represented by XC to a value represented by YD. Since the small distance travelled = average velocity × time XY, the distance travelled is represented by the *area* between the curve CD and the time-axis, shown shaded in Fig. 1.4. By considering every small time-interval between OB in the same way, it follows that *the total distance travelled by the train in the time OB is given by the area between the velocity-time graph and the time-axis.* This result applies to any velocity-time graph, whatever its shape.

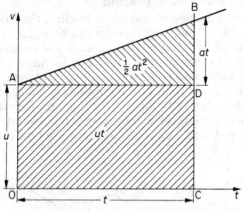

Fig. 1.5 Uniform acceleration

Figure 1.5 shows the velocity-time graph AB of an object moving with uniform acceleration a from an initial velocity u. From above, the distance s travelled in a time t or OC is equivalent to the area OABC. The area OADC $= u.t$. The area of the triangle ABD $= \frac{1}{2}$AD.BD $= \frac{1}{2}t$.BD. Now BD $=$ the increase in velocity in a time $t = at$. Hence area of triangle ABD $= \frac{1}{2}t.at = \frac{1}{2}at^2$

$$\therefore \text{total area OABC} = s = ut + \tfrac{1}{2}at^2.$$

This result is also deduced on p. 7.

Acceleration

As we have seen, the *acceleration* of a moving object at an instant is the *rate of change of its velocity* at that instant. In the case of a train accelerating steadily from 10 m s^{-1} to 15 m s^{-1} in 10 seconds, the steady or uniform acceleration

$$= (15 - 10) \text{ m s}^{-1} \div 10 \text{ seconds} = 0.5 \text{ m s}^{-1} \text{ per second.}$$

Since the time element (second) is repeated twice, the acceleration is written 0.5 m s^{-2}.

In terms of the calculus, the acceleration a of a moving object is given by

$$a = \frac{dv}{dt}$$

where dv/dt is the rate of change of velocity or the velocity change per second.

Distance travelled with Uniform Acceleration. Equations of Motion

If the velocity changes by equal amounts in equal times, no matter how small the time-intervals may be, the acceleration is said to be *uniform*. Suppose that the velocity of an object moving in a straight line with uniform acceleration a increases from a value u to a value v in a time t. Then, from the definition of acceleration,

$$a = \frac{v - u}{t},$$

from which

$$\mathbf{v} = \mathbf{u} + \mathbf{at} \qquad . \qquad . \qquad . \qquad . \qquad . \qquad (1)$$

Suppose an object with a velocity u accelerates with a uniform acceleration a for a time t and attains a velocity v. The distance s travelled by the object in the time t is given by

$$s = \text{average velocity} \times t$$

$$= \tfrac{1}{2}(u + v) \times t$$

But

$$v = u + at$$

$$\therefore s = \tfrac{1}{2}(u + u + at)t$$

$$\therefore \mathbf{s} = \mathbf{ut} + \tfrac{1}{2}\mathbf{at}^2 \qquad . \qquad . \qquad . \qquad . \qquad . \qquad (2)$$

If we eliminate t by substituting $t = (v - u)/a$ from (1) in (2), we obtain, on simplifying,

$$\mathbf{v}^2 = \mathbf{u}^2 + 2\mathbf{as} \qquad . \qquad . \qquad . \qquad . \qquad . \qquad (3)$$

Equations (1), (2), (3) are the equations of motion of an object moving in a straight line with uniform acceleration. When an object undergoes a uniform *deceleration*, for example when brakes are applied to a car, a has a *negative* value.

Examples

1. A car moving with a velocity of 54 km h^{-1} accelerates uniformly at the rate of 2 m s^{-2}. Calculate the distance travelled from the place where acceleration began to that where the velocity reaches 72 km h^{-1}, and the time taken to cover this distance.

(i) 54 km $h^{-1} = 15$ m s^{-1}, 72 km $h^{-1} = 20$ m s^{-1}, acceleration $a = 2$ m s^{-2}.

Using

$$v^2 = u^2 + 2as,$$

$$\therefore 20^2 = 15^2 + 2 \times 2 \times s$$

$$\therefore s = \frac{20^2 - 15^2}{2 \times 2} = 43\tfrac{3}{4} \text{ m.}$$

(ii) Using

$$v = u + at$$

$$\therefore 20 = 15 + 2t$$

$$\therefore t = \frac{20 - 15}{2} = 2 \cdot 5 \text{ s.}$$

2. A train travelling at 72 km h^{-1} undergoes a uniform deceleration of 2 m s^{-2} when brakes are applied. Find the time taken to come to rest and the distance travelled from the place where the brakes were applied.

(i) 72 km $h^{-1} = 20$ m s^{-1}, and $a = -2$ m s^{-2}, $v = 0$.

Using $$v = u + at$$

$$\therefore 0 = 20 - 2t$$

$$\therefore t = 10 \text{ s}$$

(ii) The distance, s, $= ut + \frac{1}{2}at^2$.

$$\therefore s = 20 \times 10 - \frac{1}{2} \times 2 \times 10^2 = 100 \text{ m}.$$

Motion Under Gravity

When an object falls to the ground under the action of gravity, experiment shows that the object has a constant or uniform acceleration of about $9 \cdot 8$ m s^{-2} or 10 m s^{-2} approximately, while it is falling. The numerical value of this acceleration is usually denoted by the symbol g. Drawn as a vector quantity, g is represented by a straight vertical line with an arrow on the line pointing dowards.

Suppose that an object is dropped from a height of 20 m above the ground. Then the initial velocity $u = 0$, and the acceleration $a = g = 10$ m s^{-2} (approx.). When the object reaches the ground, $s = 20$ m. Substituting in $s = ut + \frac{1}{2}at^2$, then

$$s = \tfrac{1}{2}gt^2 = 5t^2.$$

$$\therefore 20 = 5t^2, \quad \text{or} \quad t = 2 \text{ s}.$$

Thus the objects takes 2 seconds to reach the ground.

If a cricket-ball is thrown vertically upwards, it slows down owing to the attraction of the earth. The magnitude of the deceleration is $9 \cdot 8$ m s^{-2}, or g. Mathematically, a deceleration can be regarded as a negative acceleration in the direction along which the object is moving; and so $a = -9 \cdot 8$ m s^{-2} in this case.

Suppose the ball was thrown vertically upwards with an initial velocity, u, of 30 m s^{-1}. The time taken to reach the top of its motion can be obtained from the equation $v = u + at$. The velocity, v, at the top is zero; and since $u = 30$ m and $a = -9 \cdot 8$ or 10 m s^{-2} (approx.), we have

$$0 = 30 - 10t.$$

$$\therefore t = \frac{30}{10} = 3 \text{ s}.$$

The highest distance reached is thus given by

$$s = ut + \tfrac{1}{2}at^2$$

$$= 30 \times 3 - 5 \times 3^2 = 45 \text{ m}.$$

Resultant. Components

If a boy is running along the deck of a ship in a direction OA, and the ship is moving in a different direction OB, the boy will move relatively to the sea along a direction OC, between OA and OB, Fig. 1.6 (i). Now in one second the boat moves from O to B, where OB represents the velocity of the boat, a vector quantity, in magnitude and direction. The boy moves from O to A in the same time, where OA represents the velocity of the boy in magnitude and direction. Thus in one second the boy moves from O to C relative to the sea.

It can now be seen that if lines OA, OB are drawn to represent in magnitude

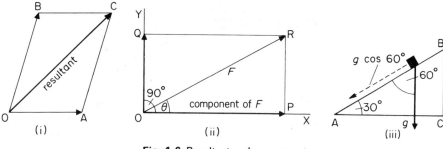

Fig. 1.6 Resultant and component

and direction the respective velocities of the boy and the ship, the magnitude and direction of the *resultant* velocity of the boy is represented by the diagonal OC of the completed parallelogram having OA, OB as two of its sides. OACB is known as a *parallelogram of velocities*.

Although velocities have been considered, the diagonal OC of the parallelogram represents the resultant or addition of *any* two vectors represented by OA and OB. Conversely, a velocity represented completely by OC can be regarded as having an 'effective part', or *component* represented by OA, and another component represented by OB.

In practice, we often require to find the component of a vector quantity in a certain direction. Suppose OR represents the vector *F*, and OX is the direction, Fig. 1.6 (ii). If we complete the parallelogram OQRP by drawing a perpendicular RP from R to OX, and a perpendicular RQ from R to OY, where OY is perpendicular to OX, we can see that OP, OQ represent the components of *F* along OX, OY respectively. Now the component OQ has no effect in a perpendicular direction; consequently OP represents the total effect of *F* along the direction OX. OP is called the 'resolved component' in this direction. If θ is the angle ROX, then, since triangle OPR has a right angle at P,

$$OP = OR \cos \theta = F \cos \theta \qquad . \qquad . \qquad . \qquad (4)$$

Since the angle between *F* and OY is $(90° - \theta)$, the component of *F* in the direction OY

$$= F \cos (90° - \theta) = F \sin \theta$$

The acceleration due to gravity, *g*, acts vertically downwards. In free fall, an object has an acceleration *g*. An object sliding freely down an inclined plane ABC, however, has an acceleration due to gravity equal to the component of *g* down the plane, Fig. 1.6 (iii). If the plane is inclined at 60° to the vertical, the acceleration down the plane is then $g \cos 60°$ or $9.8 \cos 60°$ m s^{-2}, which is 4.9 m s^{-2}.

Since $\cos 60° = \sin 30°$, we can say that the acceleration down the plane is also given by $g \sin 30°$, where the angle made by the plane with the horizontal is 30°.

Projectiles

Consider an object O thrown forward from the top of a cliff OA with a horizontal velocity *u* of 15 m s^{-1}, Fig. 1.7. Since *u* is horizontal, it has no component in a *vertical* direction. Similarly, since *g* acts vertically, it has no component in a *horizontal* direction.

We may thus treat vertical and horizontal motion independently. Consider

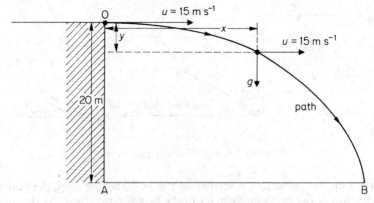

Fig. 1.7 Motion under gravity

the vertical motion from O. If OA is 20 m, the ball has an initial vertical velocity of zero and a vertical acceleration of g, which is 9·8 m s^{-2} (10 m s^{-2} approximately). Thus, from $s = ut + \frac{1}{2}at^2$, *the time t* to reach the bottom of the cliff is given, using $g = 10$ m s^{-2}, by

$$20 = \tfrac{1}{2}.10.t^2 = 5t^2, \quad \text{or} \quad t = 2 \text{ s}.$$

So far as the horizontal motion is concerned, the ball continues to move forward with a constant horizontal velocity of 15 m s^{-1} since g has no component horizontally. In 2 seconds, therefore,

horizontal distance AB = distance from cliff = $15 \times 2 = 30$ m.

Generally, in a time t the ball falls a vertical distance, y say, from O given by $y = \frac{1}{2}gt^2$. In the same time the ball travels a horizontal distance, x say, from O given by $x = ut$, where u is the velocity of 15 m s^{-1}. If t is eliminated by using $t = x/u$ in $y = \frac{1}{2}gt^2$, we obtain $y = gx^2/2u^2$. This is the equation of a *parabola*. It is the path OB in Fig. 1.7. In our discussion air resistance has been ignored.

Horizontal Range of Projectiles

In Fig. 1.8, a projectile at O on the ground is thrown with a velocity u at an angle α to the horizontal. We consider the vertical and horizontal motion *separately* in motion of this kind.

Vertical motion. The vertical component of u is $u \cos (90° - \alpha)$ or $u \sin \alpha$; the acceleration $a = -g$. When the projectile reaches the ground at B, the *vertical distance s travelled is zero*. So, from $s = ut + \frac{1}{2}at^2$, we have

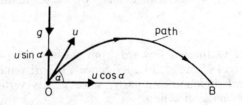

Fig. 1.8

$$0 = u \sin \alpha . t - \tfrac{1}{2}gt^2.$$

Thus $$t = \frac{2u \sin \alpha}{g} \qquad . \qquad . \qquad . \qquad . \qquad . \qquad (1)$$

Horizontal motion. Since g acts vertically, it has no component in a horizontal direction. So the projectile moves in a horizontal direction with an unchanged or constant velocity $u \cos \alpha$ because this is the component of u horizontally. So

$$\text{Range } R = \text{OB} = \text{velocity} \times \text{time}$$

$$= u \cos \alpha \times \frac{2u \sin \alpha}{g}$$

$$= \frac{2u^2 \sin \alpha \cos \alpha}{g} = \frac{u^2 \sin 2\alpha}{g}.$$

The *maximum range* is obtained when $\sin 2\alpha = 1$, or $2\alpha = 90°$. So $\alpha = 45°$ for maximum range with a given velocity of throw u. In this case, the range is u^2/g.

At the maximum height A of the path, the *vertical* velocity of the projectile is zero. So, applying $v = u + at$ in a vertical direction, the time t to reach A is given by

$$0 = u \sin \alpha - gt, \quad \text{or} \quad t = u \sin \alpha/g.$$

From (1), we see that this is half the time to reach B.

Example

A small ball A, suspended from a string OA, is set into oscillation, Fig. 1.9. When the ball passes through the lowest point of the motion, the string is cut. If the ball is then moving with the velocity 0·8 m s^{-1} at a height 5 m above the ground, find the horizontal distance travelled by the ball. (Assume $g = 10$ m s^{-2}.)

When A is at the lowest point of the oscillation, it is moving horizontally with velocity 0·8 m s^{-1}. The ball lands at B on the ground.

To find the time of travel, consider the vertical motion. In this case the vertical distance s travelled is 5 m; the initial vertical velocity $u = 0$; and $a = g = 10$ m s^{-2}. From $s = ut + \tfrac{1}{2}at^2$, we have

$$5 = \tfrac{1}{2} \times 10 \times t^2$$

So $$t^2 = 1, \quad \text{or} \quad t = 1 \text{ s.}$$

To find the horizontal distance travelled, consider the horizontal motion. The velocity is 0·8 m s^{-1} and this is constant. So

$$\text{horizontal distance to B} = 0\!\cdot\!8 \text{ m s}^{-1} \times 1 \text{ s} = 0\!\cdot\!8 \text{ m.}$$

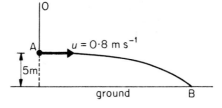

Fig. 1.9 Projectile example

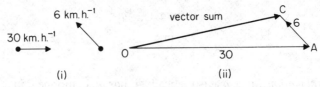

Fig. 1.10 Addition of vectors

Addition of Vectors

Suppose a ship is travelling due east at 30 km h^{-1} and a boy runs across the deck in a north-west direction at 6 km h^{-1}, Fig. 1.10 (i). We can find the velocity and direction of the boy relative to the sea by adding the two velocities. Since velocity is a vector quantity, we draw a line OA to represent 30 km h^{-1} in magnitude and direction, and then, from the end of A, draw a line AC to represent 6 km h^{-1} in magnitude and direction, Fig. 1.10 (ii). The sum, or resultant, of the velocities is now represented by the line OC in magnitude and direction, because a distance moved in one second by the ship (represented by OA) together with a distance moved in one second by the boy (represented by AC) is equivalent to a movement of the boy from O to C relative to the sea.

Subtraction of Vectors

We now consider the *subtraction* of vectors. If a car A travelling at 50 km h^{-1} is moving in the same direction as another car B travelling at 60 km h^{-1}, the *relative velocity* of B to A = difference in velocities = $60 - 50 = 10$ km h^{-1}. If however, the cars are travelling in opposite directions, the relative velocity of B to A = $60 - (-50) = 110$ km h^{-1}, since the velocity of A is opposite (negative) compared to B.

Suppose that a car X is travelling with a velocity v along a road 30° east of north, and a car Y is travelling with a velocity u along a road due east, Fig. 1.11 (i). Since 'velocity' has direction as well as magnitude, that is, 'velocity' is a vector quantity (p. 4), we cannot subtract u and v numerically to find the relative velocity. We must adopt a method which takes into account the direction as well as the magnitude of the velocities, that is, a vector subtraction is required.

The velocity of X relative to Y = difference in velocities = $\vec{v} - \vec{u} = \vec{v} + (-\vec{u})$. Suppose OA represents the velocity, v, of X in magnitude and direction, Fig. 1.11 (ii). Since Y is travelling due east, a velocity AB numerically equal to u but in the due *west* direction represents the vector $(-\vec{u})$. The vector sum of OA and AB is OB from above, which therefore represents in magnitude

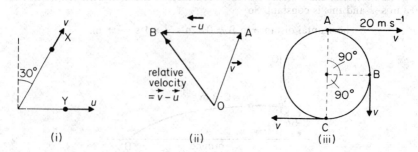

Fig. 1.11 Subtraction of velocities

and direction the velocity of X minus that of Y. By drawing an accurate diagram of the two velocities, OB can be found.

Example

A car is moving round a circular track with a constant speed v of 20 m s^{-1}, Fig. 1.11 (iii). At different times the car is at A, B and C respectively. Find the velocity change (*a*) from A to C, (*b*) from A to B.

(*a*) Velocity change from A to C $= \vec{v_C} - \vec{v_A} = (+20) - (-20)$

$$= 40 \text{ m s}^{-1} \text{ in the direction of C.}$$

(*b*) Velocity change from A to B $= \vec{v_B} - \vec{v_A} = \vec{v_B} + (-\vec{v_A})$.

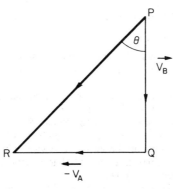

Fig. 1.12

In Fig 1.12, PQ represents the vector $\vec{v_B}$ or 20 m s^{-1} and QR represents $-\vec{v_A}$ or 20 m s^{-1}.

So $PR = \vec{v_B} - \vec{v_A} = \sqrt{20^2 + 20^2} = 28 \text{ m s}^{-1}$ (approx.), and its direction θ

relative to v_B is 45°.

Exercises 1A

Linear motion. Projectiles. Relative velocity

(Assume $g = 10 \text{ m s}^{-2}$ or 10 N kg^{-1} unless otherwise given)

1. A car moving with a velocity of 10 m s^{-1} accelerates uniformly at 1 m s^{-2} until it reaches a velocity of 15 m s^{-1}. Calculate (i) the time taken, (ii) the distance travelled during the acceleration, (iii) the velocity reached 100 m from the place where the acceleration began.

2. A ball is thrown vertically upwards with an initial speed of 20 m s^{-1}. Calculate (i) the time taken to return to the thrower, (ii) the maximum height reached.

3. A ball is thrown vertically upwards and caught by the thrower on its return. Sketch a graph of *velocity* (taking the upward direction as positive) against *time* for the whole of its motion, neglecting air resistance. How, from such a graph, would you obtain an estimate of the height reached by the ball? (*L.*)

4. A ball is dropped from a height of 20 m and rebounds with a velocity which is 3/4 of the velocity with which it hit the ground. What is the time interval between the first and second bounces?

5. A ball is thrown forward horizontally from the top of a cliff with a velocity of 10 m s^{-1}. The height of the cliff above the ground is 45 m. Calculate (i) the time to reach the ground, (ii) the distance from the cliff of the ball on hitting the ground, (iii) the direction of the ball to the horizontal just before it hits the ground.

6. In question **5**, prove that the path of the ball in flight is a parabola.

7. A projectile is fired with a velocity of 320 m s^{-1} at an angle of 30° to the horizontal. Find (i) the time to reach its greatest height, (ii) its horizontal range.

With the same velocity, what is the maximum possible range?

8. A small smooth object slides from rest down a smooth inclined plane inclined at 30° to the horizontal. What is (i) the acceleration down the plane, (ii) the time to reach the bottom if the plane is 5 m long?

The object is now thrown up the plane with an initial velocity of 15 m s^{-1}. (iii) How long does the object take to come to rest? (iv) How far up the plane has the object then travelled?

9. A stone attached to a string is whirled round in a horizontal circle with a constant speed of 10 m s^{-1}. Calculate the difference in the *velocity* when the stone is (i) at opposite ends of a diameter, (ii) in two positions A and B, where angle AOB is 90° and O is the centre of the circle.

10. Two ships A and B are 4 km apart. A is due west of B. If A moves with a uniform velocity of 8 km h^{-1} due east and B moves with a uniform velocity of 6 km h^{-1} due south, calculate (i) the magnitude of the velocity of A relative to B, (ii) the closest distance apart of A and B.

11. Define *uniform acceleration*. State, for each case, *one* set of conditions sufficient for a body to describe (*a*) a parabola, (*b*) a circle.

A projectile is fired from ground level, with velocity 500 m s^{-1} at 30° to the horizontal. Find its horizontal range, the greatest vertical height to which it rises, and the time to reach the greatest height. What is the least speed with which it could be projected in order to achieve the same horizontal range? (The resistance of the air to the motion of the projectile may be neglected.) (*O*.)

12. A lunar landing module is descending to the Moon's surface at a steady velocity of 10 m s^{-1}. At a height of 120 m, a small object falls from its landing gear. Taking the Moon's gravitational acceleration as 1·6 m s^{-2}, at what speed, in m s^{-1}, does the object strike the Moon?

A 202 B 22 C 19·6 D 16·8 E 10 (*AEB*, 1980)

Laws of Motion. Force and Momentum

Newton's Laws of Motion

In 1687 SIR ISAAC NEWTON published a work called *Principia*, in which he set out clearly the Laws of Mechanics. He gave three 'laws of motion':

Law I. *Every body continues in its state of rest or uniform motion in a straight line, unless it is acted on by a resultant force.*

Law II. *The change of momentum per unit time is proportional to the impressed force, and takes place in the direction of the straight line along which the force acts.*

Law III. *Action and reaction are always equal and opposite.*

These laws cannot be proved in a formal way; we believe they are correct because all the theoretical results obtained by assuming their truth agree with the experimental observations, as for example in astronomy (p. 47).

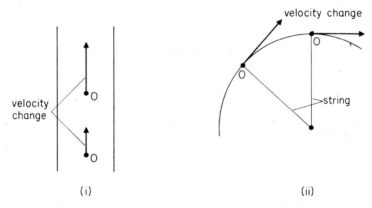

Fig. 1.13 Velocity changes: (i) magnitude (ii) direction

Inertia. Mass

Newton's first law expresses the idea of *inertia*. The inertia of a body is its reluctance to start moving, and its reluctance to stop once it has begun moving. Thus an object at rest begins to move only when it is pushed or pulled, i.e., when a *force* acts on it. An object O moving in a straight line with constant velocity will change its direction or move faster only if a new force acts on it. Figure 1.13 (i). This can be demonstrated by a puck moving on a cushion of gas on a smooth level sheet of glass. As the puck slides over the glass, photographs taken at successive equal times by a stroboscope method show that the motion is practically that of uniform velocity. Passengers in a bus or car are jerked forward when the vehicle stops suddenly. They continue in their state of motion until brought to rest by friction or collision. The use of safety belts reduces the shock.

Figure 1.13 (ii) illustrates a velocity change when an object O is whirled at constant speed by a string. This time the magnitude of the velocity v is constant but its direction changes. So a force acts on the object O.

'Mass' is a measure of the inertia of a body. If an object changes its direction or its velocity slightly when a large force acts on it, its inertial mass is high. The mass of an object is constant all over the world; it is the same on the earth as on the moon. Mass is measured in kilogram (kg) by means of a chemical balance, where it is compared with standard masses based on the International Prototype Kilogram.

Force. The Newton

When an object X is moving it is said to have an amount of *momentum* given, by definition, by

$$momentum = mass\ of\ X \times velocity \qquad . \qquad . \qquad . \qquad (1)$$

Thus an object of mass 20 kg moving with a velocity of 10 m s^{-1} has a momentum of 200 kg m s^{-1}. If another object collides with X its velocity alters, and so the momentum of X alters.

From Newton's second law, a *force F* acts on X which is equal to the change in momentum per second. Using (1), it follows that if m is the mass of X,

$$F \propto m \times \text{change of velocity per second}$$

But the change in velocity per second is the *acceleration a* produced by the force.

$$\therefore F \propto ma,$$

so
$$F = kma \qquad . \qquad . \qquad . \qquad . \qquad . \qquad (i)$$

where k is a constant.

With SI units, the **newton** (N) is the unit of force. It is defined as the force which gives a mass of 1 kg an acceleration of 1 m s^{-2}. Substituting $F = 1$ N, $m = 1$ kg and $a = 1$ m s^{-2} in the expression for F in (i), we obtain $k = 1$. Hence, with units as stated, $k = 1$.

$$\therefore F = ma, \qquad . \qquad . \qquad . \qquad . \qquad (2)$$

which is a standard equation in dynamics. Thus if a mass of 0·2 kg is acted upon by a force F which produces an acceleration a of 4 m s^{-2}, then, since $m = 0·2$ kg,

$$F = ma = 0·2\ (\text{kg}) \times 4\ (\text{m s}^{-2}) = 0·8\ \text{N}.$$

Weight and Mass

The *weight* of an object is defined as the *force* acting on it due to gravitational attraction, or gravity. So the weight of an object can be measured by attaching it to a spring-balance and noting the extension, as the extension is proportional to the force on it (p. 115).

Suppose the weight of an object of mass m is denoted by W. If the object is released so that it falls to the ground, its acceleration is g. Now $F = ma$. Consequently the force acting on it, that is its weight, is given by

$$W = mg.$$

If the mass is 1 kg, then, since $g = 9·8$ m s^{-2}, the weight $W = 1 \times 9·8 = 9·8$ N. The weight of a 5 kg mass is thus $5 \times 9·8$ N or 49 N. Note that the weight of a 100 g (0·1 kg) mass is about 1 N; the weight of an average-sized apple is about 1 N.

We can see that on the surface of the earth, the value of g may be expressed as about '9·8 N kg^{-1}'. The *force per unit mass* in a gravitational field is called the field *gravitational intensity*. On the moon's surface the gravitational intensity is about 1.6 N kg^{-1}.

The reader should note carefully that the *mass* of an object is constant all over the universe, whereas its *weight* is a *force* whose magnitude depends on the value of g. The acceleration due to gravity, g, depends on the distance of the place considered from the centre of the earth; it is slightly greater at the poles than at the equator, since the earth is not a perfectly spherical shape. It therefore follows that the weight of an object differs in different parts of the world. On the moon, which is smaller than the earth and has a smaller density, an object would have the same mass as on the earth but it would weigh about one-sixth of its weight on the earth, as the acceleration of free fall on the moon is about $g/6$.

Experimental Investigation of *F* = *ma*

An experimental investigation of $F = ma$ can be carried out by accelerating a trolley, mass m, down a friction-compensated inclined plane by a constant force F due to a stretched piece of elastic, and measuring the acceleration a with a ticker tape. Details of the experiment can be obtained from O-level texts, such as the author's *Principles of Physics* (Hart-Davis Educational, London). We assume the reader is familiar with the method. Figure 1.14 shows the results obtained. When the force is increased in the ratio 1:2:3, experiment shows that the acceleration increases in the ratio 2·4:4·8:7·3, which is approximately 1:2:3. Thus $a \propto F$ with constant mass.

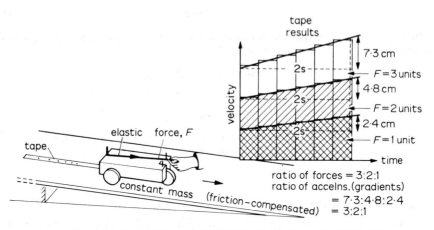

Fig. 1.14 Investigation of acceleration and force (mass constant)

The mass m can also be varied by placing similar trolleys on top of each other and the force F can be kept constant. One experiment shows that with 1, 2 and 3 trolleys the accelerations decrease in the ratio 7·5:4·9:2·5, which is 3:2:1 approximately. Thus $a \propto 1/m$ when F is constant.

Applications of *F* = *ma*

The following examples illustrate the application of $F = ma$. It should be carefully noted that (i) F represents the *resultant* force on the object of mass m, (ii) F must be expressed in the appropriate unit of 'force', the newton (N), and m in the corresponding unit of 'mass', the kilogram (kg).

Examples

1. A force of 200 N pulls a sledge of mass 50 kg and overcomes a constant frictional force of 40 N. What is the acceleration of the sledge?

$$\text{Resultant force, } F = 200 - 40 = 160 \text{ N}$$

From $F = ma$,

$$\therefore \ 160 = 50 \times a$$

$$\therefore \ a = 3\cdot2 \text{ m s}^{-2}.$$

2. An object of mass 2·00 kg is attached to the hook of a spring-balance, and the latter is suspended vertically from the roof of a lift. What is the reading on the spring-balance when the lift is (i) ascending with an acceleration of 0·2 m s^{-2}, (ii) descending with an acceleration of 0·1 m s^{-2}, (iii) ascending with a uniform velocity of 015 m s^{-1} ($g = 10$ m s^{-2} or 10 N kg^{-1}).

Suppose T is the tension (force) in the spring-balance in N.

(i) The object is acted on by two forces: (a) The tension T in the spring-balance, which acts upwards, (b) its weight, 20 N, which acts downwards. Since the object moves upwards, T is greater than 20 N. Hence the net force, F, acting on the object $= T - 20$ N. Now

$$F = ma,$$

where a is the acceleration in m s^{-2}, 0·2 m s^{-2}.

$$\therefore \ T - 20 = 2 \times a = 2 \times 0\cdot2$$

$$\therefore \ T = 20\cdot4 \text{ N} \quad . \qquad . \qquad . \qquad . \qquad . \qquad (1)$$

(ii) When the lift descends with an acceleration of 0·1 m s^{-2}, the weight, 20 N, is now greater than T_1, the tension in the spring-balance.

$$\therefore \ \text{resultant force} = 20 - T_1$$

$$\therefore \ F = 20 - T_1 = ma2 \times 0\cdot1$$

$$\therefore \ T_1 = 20 - 0\cdot2 = 19\cdot8 \text{ N}.$$

(iii) When the lift moves with constant velocity, the acceleration is zero. In this case the reading on the spring-balance is exactly equal to the weight, 20 N.

Linear Momentum

Newton defined the force acting on an object as the rate of change of its momentum, the momentum being the product of its mass and velocity (p. 15). *Momentum is thus a vector quantity*; its direction is that of the velocity. Suppose that the mass of an object is m, its initial velocity due to a force F acting on it for a time t is v. Then

$$\text{change of momentum} = mv - mu,$$

and hence

$$F = \frac{mv - mu}{t}$$

$$\therefore \ Ft = mv - mu = \textit{momentum change} \quad . \qquad . \qquad . \qquad (1)$$

The quantity Ft (force $\times$ time) is known as the *impulse* of the force on the object, and from (1) it follows that the units of momentum are the same as those of

Ft, that is, *newton second* (N s). From 'mass × velocity', alternative units are 'kg m s^{-1}'.

Force and Momentum Change

A person of mass 50 kg who is jumping from a height of 5 metres will land on the ground with a velocity $= \sqrt{2gh} = \sqrt{2 \times 10 \times 5} = 10$ m s^{-1}, assuming $g = 10$ m s^{-2} approx. If he does not flex his knees on landing, he will be brought to rest very quickly, say in $\frac{1}{10}$th second. The force F acting is then given by

$$F = \frac{\text{momentum change}}{\text{time}}$$

$$= \frac{50 \times 10}{\frac{1}{10}} = 5000 \text{ N}.$$

This is a force of about 10 times the person's weight and the large force has a severe effect on the body.

Suppose, however, that the person flexes his knees and is brought to rest much more slowly on landing, say in 1 second. Then, from above, the force F now acting is 10 times less than before, or 500 N. Consequently, much less damage is done to the person on landing.

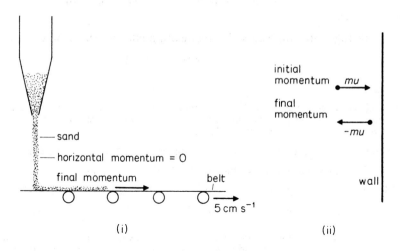

Fig. 1.15 Linear momentum changes

Suppose sand is allowed to fall vertically at a steady rate of 100 g s^{-1} on to a horizontal conveyor belt moving at a steady velocity of 5 cm s^{-1}, Fig. 1.15 (i). The initial horizontal velocity of the sand is zero. The final horizontal velocity is 5 cm s^{-1}. Now in one second in a horizontal direction,

mass $= 100$ g $= 0.1$ kg, velocity gained $= 5$ cm s^{-1} $= 5 \times 10^{-2}$ m s^{-1}

∴ momentum change per second horizontally $= 0.1 \times 5 \times 10^{-2}$

$$= 5 \times 10^{-3} \text{ newton}$$

$$= \text{force on belt}.$$

Observe that this is a case where the *mass* changes with time and the velocity gained is constant. In terms of the calculus, force is the rate of change of momentum mv, which is $v \times dm/dt$, and dm/dt is 100 g s^{-1} in this numerical example.

Consider a molecule of mass m in a gas, which strikes the wall of a vessel repeatedly with a velocity u and rebounds with a velocity $-u$, Fig. 1.15 (ii). Since momentum is a vector quantity, the momentum change = final momentum $-$ initial momentum $= mu - (-mu) = 2mu$. If the containing vessel is a cube of side l, the molecule repeatedly takes a time $2l/u$ to make a collision with the same side as it moves to-and-fro across the vessel. So

$$\text{number of collisions per second, } n = \frac{1}{2l/u} = \frac{u}{2l}.$$

The average force on wall = $n \times$ one momentum change

$$= \frac{u}{2l} \times 2mu = \frac{mu^2}{l}.$$

The total gas pressure is the average force per unit area on the walls of the container due to all the numerous gas molecules and is discussed in *Heat*.

Suppose a ball of mass 0.1 kg hits a smooth wall normally with a velocity of 10 m s^{-1} four times per second, rebounding each time with a velocity of 10 m s^{-1}. Then

$$\text{average force on wall} = 4 \times \text{momentum change per second}$$

$$= 4 \times 2 \times 0.1 \times 10 = 8 \text{ N}.$$

Examples

1. A hose ejects water at a speed of 20 cm s^{-1} through a hole of area 100 cm^2. If the water strikes a wall normally, calculate the force on the wall in newtons, assuming the velocity of the water normal to the wall is zero after collision.

The volume of water per second striking the wall $= 100 \times 20 = 2000$ cm^3.

$$\therefore \text{ mass per second striking wall} = 2000 \text{ g s}^{-1} = 2 \text{ kg s}^{-1}.$$

Velocity change of water on striking wall $= 20 - 0 = 20$ cm s^{-1} $= 0.2$ m s^{-1}.

$$\therefore \text{ momentum change per second} = 2 \text{ (kg s}^{-1}) \times 0.2 \text{ (m s}^{-1}) = 0.4 \text{ N}.$$

2. Sand drops vertically at the rate of 2 kg s^{-1} on to a conveyor belt moving horizontally with a velocity of 0·1 m s^{-1}. Calculate (i) the extra power needed to keep the belt moving, (ii) the rate of change of kinetic energy of the sand. Why is the power twice as great as the rate of change of kinetic energy?

(i) Force required to keep belt moving = rate of increase of horizontal momentum of sand = mass per second $(dm/dt) \times$ velocity change $= 2 \times 0.1 = 0.2$ newton.

$$\therefore \text{ power} = \text{work done per second} = \text{force} \times \text{rate of displacement}$$

$$= \text{force} \times \text{velocity} = 0.2 \times 0.1 = 0.02 \text{ watt (p. 28)}.$$

(ii) Kinetic energy of sand $= \frac{1}{2}mv^2$.

$$\therefore \text{ rate of change of energy} = \frac{1}{2}v^2 \times \frac{dm}{dt}, \text{ since } v \text{ is constant,}$$

$$= \frac{1}{2} \times 0.1^2 \times 2 = 0.01 \text{ watt.}$$

Thus the power supplied is twice as great as the rate of change of kinetic energy. The extra power is due to the fact that the sand does not immediately assume the velocity of the belt, so that the belt at first moves relative to the sand. The extra power is needed to overcome the friction between the sand and belt.

Newton's Third Law

Newton's third law—action and reaction are equal and opposite—means that if a body A exerts a force (action) on a body B, then B will exert an equal and opposite force (reaction) on A.

These forces are produced between objects by direct contact when they touch, or by gravitational forces, for example, when they are apart. Thus if a ball is kicked upwards, the force on the ball by the kicker is equal and opposite to the force on the kicker by the ball. The initial upward acceleration of the ball is usually very much greater than the downward acceleration of the kicker because the mass of the ball is much less than that of the kicker.

As the ball falls downwards towards the ground, the force of attraction on the ball by the Earth is equal and opposite to the force of attraction on the Earth by the ball. The upward acceleration of the Earth is not noticeable since its mass is so large.

In the case of a rocket, the downward force on the burning gases from the exhaust is equal to the upward force on the rocket. This is one application of the reaction force. Another is a water-sprinkler which spins backwards as the water is thrown forwards.

Action-reaction forces therefore always occur in pairs. It should be noted that the two forces act on *different* bodies. So only *one* of the forces is used in discussing the motion of one of the two bodies. In the case of a man standing in a lift moving upwards, for example, the upward reaction of the floor *on the man* is the force we need to take into account in applying $F = ma$ to the motion of the *man*. The equal downward force on the floor is *not* required.

Example

Figure 1.16 shows a truck A of mass 1000 kg pulling a trailer of mass 3000 kg. The frictional force on A is 1000 N, on B it is 2000 N, and the truck engine exerts a force of 8000 N.

Calculate (i) the acceleration of the truck and trailer, (ii) the tension T in the tow-bar connecting A and B.

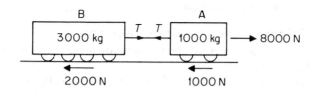

Fig. 1.16 Example

For B only   From $F = ma$, where F is the resultant force,

$$T - 2000 = 3000\,a \qquad . \qquad . \qquad . \qquad . \qquad . \qquad . \qquad (1)$$

For A only $8000 - 1000 - T = 1000\,a \qquad . \qquad . \qquad . \qquad . \qquad . \qquad . \qquad (2)$

Adding (1) and (2) to eliminate T, then

$$8000 - 1000 - 2000 = 4000\,a.$$

So $5000 = 4000\,a,$ and $a = 1\cdot25$ m s^{-2}

From (1), $T = 3000 \times 1\cdot25 + 2000 = 5750$ N.

Conservation of Linear Momentum

We now consider what happens to the linear momentum of objects which *collide* with each other.

Experimentally, this can be investigated by several methods:

1. Trolleys in collision, with ticker-tapes attached to measure velocities.
2. Linear air-track, using perspex models in collision and stroboscopic photo-graphy for measuring velocities.

As an illustration of the experimental results, the following measurements were taken in trolley collisions (Fig. 1.17):

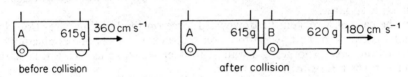

Fig. 1.17 Linear momentum experiment

Before collision
   Mass of trolley A = 615 g; initial velocity = 360 cm s^{-1}.

After collision
   A and B coalesced and both moved with velocity of 180 cm s^{-1}.

Thus the total linear momentum of A and B before collision = 0·615 (kg) × 3·6 (m s^{-1}) + 0 = 2·20 kg m s^{-1} (approx.). The total momentum of A and B after collision = 1·235 × 1·8 = 2·20 kg m s^{-1} (approx.).

Within the limits of experimental accuracy, it follows that *the total momentum of A and B before collision = the total momentum after collision*. Similar results are obtained if A and B are moving with different speeds after collision, or in opposite directions before collision.

Principle of Conservation of Linear Momentum

These experimental results can be shown to follow from Newton's second and third laws of motion (p. 15).

Suppose that a moving object A, of mass m_1 and velocity u_1, collides with another object B, of mass m_2 and velocity u_2, moving in the same direction, Fig. 1.18. By Newton's law of action and reaction, the force F exerted by A on B is equal and opposite to that exerted by B on A. Moreover, the time t during which the force acted on B is equal to the time during which the force of reaction acted on A. Thus the magnitude of the impulse, Ft, on B is equal and opposite to the magnitude of the impulse on A. From equation (1), p. 18, the impulse is equal to the change of momentum. It therefore follows that the change in the total momentum of the two objects is *zero*, i.e., the total momentum of the two objects is constant although a collision had occurred. Thus if A moves with a reduced velocity v_1 after collision, and B then moves with an increased velocity v_2,

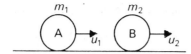

Fig. 1.18 Conservation of linear momentum

$$m_1 u_1 + m_2 u_2 = m_1 v_1 + m_2 v_2.$$

The *principle of the conservation of linear momentum* states that, *if no external forces act on a system of colliding objects, the total momentum of the objects in a given direction remains constant.*

Examples

1. An object A of mass 2 kg is moving with a velocity of 3 m s^{-1} and collides head on with an object B of mass 1 kg moving in the opposite direction with a velocity of 4 m s^{-1}, Fig. 1.18. After collision both objects coalesce, so that they move with a common velocity v. Calculate v.

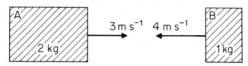

Fig. 1.19 Example

Total momentum before collision of A and B in the direction of A

$$= 2 \times 3 - 1 \times 4 = 2 \text{ kg m s}^{-1}.$$

Note that momentum is a vector and the momentum of B is of opposite sign to A. After collision, momentum of A and B in the direction of A $= 2v + 1v = 3v$.

$$\therefore 3v = 2$$

$$\therefore v = \tfrac{2}{3} \text{ m s}^{-1}.$$

2. What is understood by (*a*) the principle of the *conservation of energy*, (*b*) the principle of the *conservation of momentum*?

A bullet of mass 20 g travelling horizontally at 100 m s^{-1}, embeds itself in the centre of a block of wood mass 1 kg which is suspended by light vertical strings 1 m in length. Calculate the maximum inclination of the strings to the vertical.

Describe in detail how the experiment might be carried out and used to determine the velocity of the bullet just before the impact of the block. (*N.*)

Suppose A is the bullet, B is the block suspended from a point O, and θ is the maximum inclination to the vertical, Fig. 1.20. If v m s^{-1} is the common velocity of block and bullet when the latter is brought to rest relative to the block, then, from the principle of the conservation of momentum, since 20 g $= 0.02$ kg,

$$(1 + 0.02)v = 0.02 \times 100$$

$$\therefore v = \frac{2}{1.02} = \frac{100}{51} \text{ m s}^{-1}.$$

The vertical height risen by block and bullet is given by $v^2 = 2gh$, where $g = 9.8$ m s^{-2} and $h = l - l \cos \theta = l(1 - \cos \theta)$.

$$\therefore v^2 = 2gl(1 - \cos \theta).$$

$$\therefore \left(\frac{100}{51}\right)^2 = 2 \times 9.8 \times 1(1 - \cos \theta).$$

$$\therefore 1 - \cos \theta = \left(\frac{100}{51}\right)^2 \times \frac{1}{2 \times 9.8} = 0.1962.$$

$$\therefore \cos \theta = 0.8038, \text{ or } \theta = 37° \text{ (approx.).}$$

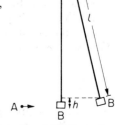

Fig. 1.20 Example

The velocity, v, of the bullet can be determined by applying the conservation of momentum principle.

Thus $mv = (m + M)V$, where m is the mass of the bullet, M is the mass of the block, and V is the common velocity. then $v = (m + M)V/m$. The quantities m and M can be found by weighing. V is calculated from the horizontal displacement a of the block, since (i) $V^2 = 2gh$ and (ii) $h(2l - h) = a^2$ from the geometry of the circle, so that, to a good approximation, $2h = a^2/l$.

Inelastic and Elastic Collisions

In collisions, the total momentum of the colliding objects is always conserved. Usually, however, their total kinetic energy is not conserved. Some of it is changed to heat or sound energy, which is not recoverable. Such collisions are said to be *inelastic*. For example, when a lump of putty falls to the ground, the total momentum of the putty and earth is conserved, that is, the putty loses momentum and the earth gains an equal amount of momentum. But all the kinetic energy of the putty is changed to heat and sound on collision.

If the total kinetic energy is conserved, the collision is said to be *elastic*. Gas molecules make elastic collisions. The collision between two smooth billiard balls is approximately elastic. Electrons may make elastic or inelastic collisions with atoms of a gas. As proved on p. 28, the kinetic energy of a mass m moving with a velocity v has kinetic energy equal to $\frac{1}{2}mv^2$.

As an illustration of the mechanics associated with elastic collisions, consider a sphere A of mass m and velocity v incident on a stationary sphere B of equal mass m, Fig. 1.21 (i). Suppose the collision is elastic, and after collision let A move with a velocity v_1 at an angle of 60° to its original direction and B move with a velocity v_2 at an angle θ to the direction of v.

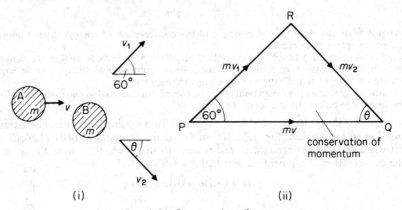

Fig. 1.21 Conservation of momentum

Since momentum is a vector (p. 18), we may represent the momentum mv of A by the line PQ drawn in the direction of v, Fig. 1.21 (ii). Likewise, PR represents the momentum mv_1 of A after collision. *Since momentum is conserved, the vector RQ must represent the momentum mv_2 of B after collision*, that is,

$$\overrightarrow{mv} = \overrightarrow{mv_1} + \overrightarrow{mv_2}.$$

Hence $$\overrightarrow{v} = \overrightarrow{v_1} + \overrightarrow{v_2},$$

or PQ represents v in magnitude, PR represents v_1 and RQ represents v_2. But if the collision is elastic,

$$\tfrac{1}{2}mv^2 = \tfrac{1}{2}mv_1^2 + \tfrac{1}{2}mv_2^2$$

$$\therefore\ v^2 = v_1^2 + v_2^2.$$

Consequently, triangle PRQ is a right-angled triangle with angle R equal to 90°.

$$\therefore\ v_1 = v \cos 60° = \frac{v}{2}.$$

Also, $\theta = 90° - 60° = 30°$, and $v_2 = v \cos 30° = \dfrac{\sqrt{3}\,v}{2}$.

Momentum and Explosive Forces

There are numerous cases where momentum changes are produced by *explosive* forces. An example is a bullet of mass $m = 50$ g say, fired from a rifle of mass $M = 2$ kg with a velocity v of 100 m s^{-1}. Initially, the total momentum of the bullet and rifle is zero. From the principle of the conservation of linear momentum, when the bullet is fired the total momentum of bullet and rifle is still zero, since no external force has acted on them. Thus if V is the velocity of the rifle,

$$mv\ \text{(bullet)} + MV\ \text{(rifle)} = 0$$

$$\therefore\ MV = -mv, \quad \text{or} \quad V = -\frac{m}{M}v.$$

The momentum of the rifle is thus *equal and opposite* to that of the bullet. Further, $V/v = -m/M$. Since $m/M = 50/2000 = 1/40$, it follows that $V = -v/40 = 2\cdot5$ m s^{-1}. This means that the rifle moves back or *recoils* with a velocity only about $\frac{1}{40}$th that of the bullet.

If it is preferred, one may also say that the explosive force produces the same numerical momentum change in the bullet as in the rifle. Thus $mv = MV$, where V is the velocity of the rifle in the *opposite* direction to that of the bullet.

The joule (J) is the unit of energy (p. 27).

The kinetic energy, E_1, of the bullet $= \tfrac{1}{2}mv^2 = \tfrac{1}{2}.0\cdot05.100^2 = 250$ J.

The kinetic energy, E_2, of the rifle $= \tfrac{1}{2}MV^2 = \tfrac{1}{2}.2.2\cdot5^2 = 6\cdot25$ J.

Thus the total kinetic energy produced by the explosion $= 256\cdot25$ J. The kinetic energy E_1 of the bullet is thus 250/256·25, or about 98%, of the total energy. This is explained by the fact that the kinetic energy depends on the *square* of the velocity. The high velocity of the bullet thus more than compensates for its small mass relative to that of the rifle. See also p. 29.

Exercises 1B

Force and Momentum

(Assume g = 10 m s^{-2} or 10 N kg^{-1} unless otherwise stated)

1. A car of mass 1000 kg is accelerating at 2 m s^{-2}. What resultant force acts on the car? If the resistance to the motion is 1000 N, what is the force due to the engine?

2. A box of mass 50 kg is pulled up from the hold of a ship with an acceleration of 1 m s^{-2} by a vertical rope attached to it. Find the tension in the rope.

What is the tension in the rope when the box moves up with a uniform velocity of 1 m s^{-1}?

3. A lift moves (i) up and (ii) down with an acceleration of 2 m s^{-2}. In each case, calculate the reaction of the floor on a man of mass 50 kg standing in the lift.

4. A ball of mass 0·2 kg falls from a height of 45 m. On striking the ground it rebounds in 0·1 s with two-thirds of the velocity with which it struck the ground. Calculate (i) the momentum change on hitting the ground, (ii) the force on the ball due to the impact.

5. A ball of mass 0·05 kg strikes a smooth wall normally four times in 2 seconds with a velocity of 10 m s^{-1}. Each time the ball rebounds with the same speed of 10 m s^{-1}. Calculate the average force on the wall.

Draw a sketch showing how the momentum varies with time over the 2 seconds.

6. The mass of gas emitted from the rear of a toy rocket is initially 0·1 kg s^{-1}. If the speed of the gas relative to the rocket is 50 m s^{-1}, and the mass of the rocket is 2 kg, what is the initial acceleration of the rocket?

7. A ball A of mass 0·1 kg, moving with a velocity of 6 m s^{-1}, collides directly with a ball B of mass 0·2 kg at rest. Calculate their common velocity if both balls move off together.

If A had rebounded with a velocity of 2 m s^{-1} in the opposite direction after collision, what would be the new velocity of B?

8. A bullet of mass 20 g is fired horizontally into a suspended stationary wooden block of mass 380 g with a velocity of 200 m s^{-1}. What is the common velocity of the bullet and block if the bullet is embedded (stays inside) the block?

If the block and bullet experience a constant opposing force of 2 N, find the time taken by them to come to rest.

9. A hose directs a horizontal jet of water, moving with a velocity of 20 m s^{-1}, on to a vertical wall. The cross-sectional area of the jet is 5 × 10^{-4} m^2. If the density of water is 1000 kg m^{-3}, calculate the force on the wall assuming the water is brought to rest there.

10. In a nuclear collision, an alpha-particle A of mass 4 units is incident with a velocity v on a stationary helium nucleus B of 4 mass units, Fig. 1A. After collision, A moves in the direction BC with a velocity $v/2$, where BC makes an angle of 60° with the initial direction AB, and the helium nucleus moves along BD.

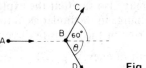

Fig. 1A

Calculate the velocity of rebound of the helium nucleus along BD and the angle θ made with the direction AB. (A solution by drawing is acceptable.)

11. Define linear momentum. State the law of conservation of linear momentum and describe in detail an experiment to demonstrate this law.

Rain falls vertically on to a plane roof, 1·5 m square, which is inclined to the horizontal at an angle of 30°. The raindrops strike the roof with a vertical velocity of 3 m s^{-1}, and a volume of 2·5 × 10^{-2} m^3 of water is collected from the roof in one minute. Assuming that the conditions are steady and that the velocity of the raindrops after impact is zero, calculate (*a*) the vertical force exerted on the roof by the impact of the falling rain, and (*b*) the pressure normal to the roof due to the impact of the rain.

If, instead, the roof were subject to a rain of hard spheres, which collided elastically, with the same mass per unit time and the same vertical velocity, what would the normal pressure on the roof then be? (Density of water = 10^3 kg m^{-3}.) (*O. & C.*)

12. State the *law of conservation of linear momentum*. Before the start of a race, each runner has momentum which differs from that which he has during the race. Does this represent a violation of the law? Give reasons for your answer.

A uranium atom (U), travelling with a velocity of 5·00 × 10^5 m s^{-1} relative to the containing tube, breaks up into krypton (Kr) and barium (Ba). The krypton atom is ejected directly backwards at a velocity of 2.35×10^6 m s^{-1} relative to the barium after separation. With what velocity does the barium atom move forward relative to the tube? You may assume that no other particles are produced, and that relativistic corrections are small.

What is the velocity of the krypton atom relative to the containing tube? (Kr = 95·0; Ba = 140; U = 235 u.) (*C.*)

13. State Newton's Laws of Motion and deduce from them the relation between the distance travelled and the time for the case of a body acted upon by a constant force. Explain the units in which the various quantities are measured.

A fire engine pumps water at such a rate that the velocity of the water leaving the nozzle is 15 m s^{-1}. If the jet be directed perpendicularly on to a wall and the rebound of the water be neglected, calculate the pressure on the wall (1 m^3 water has mass 1000 kg). (*O. & C.*)

14. Answer the following questions making particular reference to the physical principles concerned: (*a*) explain why the load on the back wheels of a motor car increases when the vehicle is accelerating; (*b*) the diagram, Fig. 1B, shows a painter in a crate which hangs alongside a building. When the painter who weighs 1000 N pulls on the rope the force he exerts on the floor of the crate is 450 N. If the crate weighs 250 N find the acceleration. (*N.*)

Fig. 1B

15. In an elastic head-on collision, a ball of mass 1·0 kg moving at 4·0 m s^{-1} collides with a stationary ball of mass 2·0 kg. Calculate the velocities of the balls after the collision indicating the directions in which they are then travelling. (*AEB*, 1980)

16. A stone is projected vertically upwards and eventually returns to the point of projection. Ignoring any effects due to air resistance draw sketch graphs to show the variation with time of the following properties of the stone: (i) velocity, (ii) kinetic energy, (iii) potential energy, (iv) momentum, (v) distance from point of projection, (vi) speed. (*AEB*, 1982)

17. Explain what is meant by a *force*. Define the SI unit in which it is measured.

Distinguish carefully the conditions under which (*a*) linear momentum is conserved and (*b*) kinetic energy is conserved.

A gun fires a shell with the horizontal component of its velocity equal to 200 m s^{-1}. At the highest point in its flight, the shell explodes into three fragments. Two of these fragments, which have equal mass, fly off with equal speeds of 300 m s^{-1} relative to the ground, one along the flight direction of the shell at the instant of fragmentation and the other perpendicular to it and in a horizontal plane. Find the magnitude and direction of the velocity of the third fragment immediately after the explosion, assuming its mass is three times that of each of the other two fragments. Neglect air resistance.

Describe and explain qualitatively the subsequent motion of the three fragments. (*O. & C.*)

Work, Energy, Power

Work

When an engine pulls a train with a constant force of 50 units through a distance of 20 units in its own direction, the engine is said by definition to do an amount of *work* equal to 50 × 20 or 1000 units, the product of the force and the distance. Thus if W is the amount of work,

$$W = \text{force} \times \text{distance moved in direction of force.}$$

Work is a *scalar* quantity; it has no property of direction but only magnitude. When the force is one newton and the distance moved is one metre, then the work done is one *joule*. Thus a force of 50 N moving through a distance of 10 m does 50 × 10 or 500 joule of work. Note this is also a measure of the *energy* transferred to the object.

The force to raise steadily a mass of 1 kg is equal to its weight, which is about 10 N (see p. 16). Thus if the mass of 1 kg is raised vertically through 1 m, then, approximately, work done = 10 (N) × 1 (m) = 10 joule.

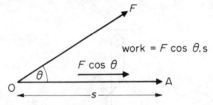

Fig. 1.22 Work and displacement

Before leaving the topic of 'work', the reader should note carefully that we have assumed the force to move an object in its own direction. Suppose, however, that a force F pulls an object a distance s along a line OA acting at an angle θ to it, Fig. 1.22. The component of F along OA is $F \cos \theta$ (p. 9), and this is the effective part of F pulling along the direction OA. The component of F along a direction perpendicular to OA has no effect along OA. Consequently

$$\text{work done} = F \cos \theta \times s.$$

In general, the work done by a force is equal to the product of the force and the displacement of its point of application in the direction of the force.

Power

When an engine does work quickly, it is said to be operating at a high *power*; if it does work slowly it is said to be operating at a lower power. 'Power' is defined as the *work done per second*, or

$$\text{power} = \frac{\text{work done}}{\text{time taken}}.$$

The practical unit of power, the SI unit, is 'joule per second' or *watt* (W); the watt is defined as the rate of working at 1 joule per second.

$$1 \text{ horse-power (hp)} = 746 \text{ W} = \tfrac{3}{4} \text{ kW (approx.)},$$

where 1 kW = 1 kilowatt or 1000 watt. Thus a small motor of $\tfrac{1}{6}$ hp in a vacuum carpet cleaner has a power of about 125 W.

Kinetic Energy

An object is said to possess *energy* if it can do work. When an object possesses energy because it is moving, the energy is said to be *kinetic*, e.g., a flying stone can break a window. Suppose that an object of mass m is moving with a velocity u, and is gradually brought to rest in a distance s by a constant force F acting against it. The kinetic energy originally possessed by the object is equal to the work done against F, and hence

$$\text{kinetic energy} = F \times s.$$

But $F = ma$, where a is the deceleration of the object. Hence $F \times s = mas$. From $v^2 = u^2 + 2as$ (see p. 7), we have, since $v = 0$ and a is negative in this case,

$$0 = u^2 - 2as, \text{ or } as = \frac{u^2}{2}.$$

$$\therefore \textit{kinetic energy} = mas = \tfrac{1}{2}mu^2.$$

When m is in kg and u is in m s^{-1}, then $\frac{1}{2}mu^2$ is in *joule*. Thus a car of mass 1000 kg, moving with a velocity of 36 km h^{-1} or 10 m s^{-1}, has an amount W of kinetic energy given by

$$W = \tfrac{1}{2}mu^2 = \tfrac{1}{2} \times 1000 \times 10^2 = 50\,000 \text{ J.}$$

Kinetic Energies due to Explosive Forces

Suppose that, due to an explosion or nuclear reaction, a particle of mass m breaks away from the total mass concerned and moves with velocity v, and the mass M left moves with velocity V in the opposite direction.

The kinetic energy E_1 of the mass m is given by

$$E_1 = \tfrac{1}{2}mv^2 = \frac{(mv)^2}{2m} = \frac{p^2}{2m} \qquad . \qquad . \qquad . \qquad (1)$$

where p is the momentum mv of the mass. Similarly, the kinetic energy E_2 of the mass M is given by

$$E_2 = \tfrac{1}{2}MV^2 = \frac{p^2}{2M} \qquad . \qquad . \qquad . \qquad (2)$$

because the momentum $MV = mv = p$ from the conservation of momentum. Dividing (1) by (2), we see that

$$\frac{E_1}{E_2} = \frac{1/m}{1/M} = \frac{M}{m}.$$

Hence the energy is *inversely*-proportional to the masses of the particles, that is, the smaller mass, m say, has the larger energy. Thus if E is the total energy of the two masses, the energy of the smaller mass $= ME/(M + m)$.

Suppose a bullet of mass $m = 50$ g is fired from a rifle of mass $M = 2$ kg $= 2000$ g and that the total kinetic energy produced by the explosion is 2050 J. Since the energy is shared inversely as the masses,

$$\text{kinetic energy of bullet} = \frac{2000}{2000 + 50} \times 2050 \text{ J}$$

$$= \frac{2000}{2050} \times 2050 \text{ J} = 2000 \text{ J.}$$

So $\qquad$ kinetic energy of rifle $= 50$ J.

An α-particle has a mass of 4 units and a radium nucleus a mass of 228 units. If disintegration of a thorium nucleus, mass 232, produces an α-particle and radium nucleus, and a release of energy of 4·05 MeV, where 1 MeV = 1·6 × 10^{-13} J, then

$$\text{energy of } \alpha\text{-particle} = \frac{228}{(4 + 228)} \times 4\cdot05 = 3\cdot98 \text{ MeV}.$$

The α-particle thus travels a relatively long distance before coming to rest compared with the comparatively heavy radium nucleus, which moves back or recoils a small distance.

Potential Energy

A mass held stationary above the ground has energy, because, when released, it can raise another object attached to it by a rope passing over a pulley, for example. A coiled spring also has energy, which is released gradually as the spring uncoils. The energy of the weight or spring is called *potential energy*, because it arises from the position or arrangement of the body and not from its motion. In the case of the weight, the energy given to it is equal to the work done by the person or machine which raises it steadily to that position against the force of attraction of the earth. So this is *gravitational potential energy*. In the case of the spring, the energy is equal to the work done in displacing the molecules from their normal equilibrium positions against the forces of attraction of the surrounding molecules. So this is *molecular potential energy*.

If the mass of an object is m, and the object is held stationary at a height h above the ground, the energy released when the object falls to the ground is equal to the work done

$$= \text{force} \times \text{distance} = \text{weight of object} \times h.$$

Suppose the mass m is 5 kg, so that the weight is 5 × 9·8 N or 50 N approx., and h is 4 metre. Then

gravitational potential energy, p.e. = 50 (N) × 4(m) = 200 J
(more accurately, p.e. = 196 J).

Generally, at a height of h,

gravitational potential energy = mgh,

where m is in kg, h is in metre, $g = 9\cdot8$ N kg^{-1}.

Conservative Forces

If a ball of weight W is raised steadily from the ground to a point X at a height h above the ground, the work done is $W.h$. The potential energy, p.e., of the ball relative to the ground is thus $W.h$. Now whatever route is taken from ground level to X, the work done is the same—if a *longer* path is chosen, for example, the component of the weight in the particular direction must then be overcome and so the force required to move the ball is correspondingly smaller. The p.e. of the ball at X is thus independent of the route to X. This implies that if the ball is taken in a closed path round to X again, *the total work done is zero*. Work or energy has been expended on one part of the closed path, and regained on the remaining part.

When the work done in moving round a closed path in a field to the original point is zero, the forces in the field are called *conservative forces*. The earth's

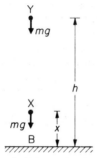

Fig. 1.23 Mechanical energy in a gravitational field

gravitational field is an example of a field containing conservative forces, as we now show.

Suppose the ball falls from a place Y at a height h to another X at a height of x above the ground, Fig. 1.23. Then, if W is the weight of the ball and m its mass,

$$\text{p.e. at X} = Wx = mgx$$

and $$\text{k.e. at X} = \tfrac{1}{2}mv^2 = \tfrac{1}{2}m.2g(h - x) = mg(h - x),$$

using $v^2 = 2as = 2g(h - x)$. Hence

$$\text{p.e.} + \text{k.e.} = mgx + mg(h - x) = mgh.$$

Thus at any point such as X, the total mechanical energy of the falling ball is equal to the original energy at Y. The mechanical energy is hence constant or conserved. This is the case for a conservative field.

Non-Conservative Forces. Principle of Conservation of Energy

The work done in taking a mass m round a closed path in the conservative earth's gravitational field is zero. Figure 1.24 (i). If the work done in taking an object round a closed path to its original position is not zero, the forces in the field are said to be *non-conservative*. This is the case, for example, when a wooden block B is pushed round a closed path on a rough table to its initial position O. Work is therefore done against friction, both as A moves away from O and as it returns. In a conservative field, however, work is done during part of the path and regained for the remaining part.

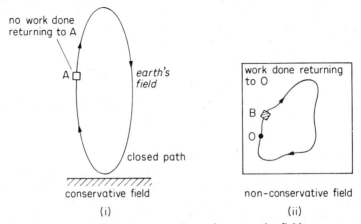

Fig. 1.24 Non-conservative and conservative fields

When a body falls in the earth's gravitational field, a small part of the energy is used up in overcoming the resistance of the air. This energy is dissipated or lost as heat—it is not regained in moving the body back to its original position. This resistance is another example of a non-conservative force.

Although energy may be transformed from one form to another, as in the last example from mechanical energy to heat, *the total energy in a closed system is always constant*. If an electric motor is supplied with 1000 J of energy, 850 J of mechanical energy, 140 J of heat energy and 10 J of sound energy may be produced. This is called the *Principle of the Conservation of Energy* and is one of the key principles in science.

Momentum and Energy in Gravitational Attraction

Consider a ball B held stationary at a height above the earth E. Relative to each other, the momentum and kinetic energy of B and E are both zero.

Suppose the ball is now released. The gravitational force of E on B accelerates the ball. So its momentum increases as it falls. From the law of conservation of momentum, the equal and opposite gravitational force of B on E produces an equal but opposite momentum on the earth. The earth is so heavy, however, that its velocity V towards B is extremely small. For example, suppose the mass m is 0·2 kg and the mass M of E is 10^{25} kg, and the velocity v of the ball B is 10 m s^{-1} at an instant. Then, from the conservation of momentum,

$$MV = mv$$

or $\qquad V = \dfrac{m}{M} \times v = \dfrac{0\cdot2}{10^{25}} \times 10 = 2 \times 10^{-25} \text{ m s}^{-1}.$

So the earth's velocity V is extremely small.

The total energy of B and E remains constant while the ball B is falling, since there are no external forces acting on the system. When B hits the ground the force (action) of B on E is equal and opposite to the force (reaction) of E on B, and the forces act for the same short time. So the total momentum of B and E is conserved. Thus when the ball rebounds, B moves upward with a momentum change equal and opposite to that of E. As B continues to rise its velocity and momentum decreases. So the momentum of E decreases. When B reaches its maximum height its momentum is zero. So the momentum of E is then zero.

Although momentum is conserved on collision with the earth, some mechanical energy is transformed to heat and sound. Hence the total mechanical energy of B and E is less after collision. As we showed on page 29, the ratio of the kinetic energy of the ball to the earth after collision is *inversely-proportional* to their masses. So with the above figures,

$$\frac{\text{kinetic energy of earth}}{\text{kinetic energy of ball}} = \frac{0\cdot2}{10^{25}} = 2 \times 10^{-26}.$$

Hence the kinetic energy of the earth after collision is extremely small. Practically all the kinetic energy is transferred to the ball.

Dimensions

By the *dimensions* of a physical quantity we mean the way it is related to the fundamental quantities mass, length and time; these are usually denoted by M, L and T respectively. An area, length × breadth, has dimensions L × L or L^2; a volume has dimensions L^3; density, which is mass/volume, has dimensions M/L^3 or ML^{-3}; an angle has no dimensions, since it is the ratio of two lengths.

As an area has dimensions L^2, the *unit* may be written in terms of the metre as 'm^2'. Similarly, the dimensions of a volume are L^3 and hence the unit is 'm^3'. Density has dimensions ML^{-3}. The density of mercury is thus written as '13 600 kg m^{-3}'. If some physical quantity has dimensions $ML^{-1}T^{-1}$, its unit may be written as 'kg m^{-1} s^{-1}'.

The following are the dimensions of some quantities in Mechanics:

Velocity. Since velocity $= \dfrac{\text{displacement}}{\text{time}}$, its dimensions are L/T or LT^{-1}.

Acceleration. The dimensions are those of velocity/time, i.e., L/T^2 or LT^{-2}.

Force. Since force $=$ mass $\times$ acceleration, its dimensions are MLT^{-2}.

Work or Energy. Since work $=$ force $\times$ distance, its dimensions are ML^2T^{-2}.

Example

In the gas equation $\left(p + \dfrac{a}{V^2}\right)(V - b) = RT$, what are the dimensions of the constants *a* and *b*?

p represents pressure, V represents volume. The quantity a/V^2 must represent a pressure since it is added to p. The dimensions of $p = [\text{force}]/[\text{area}] = MLT^{-2}/L^2 = ML^{-1}T^{-2}$; the dimensions of $V = L^3$. Hence

$$\frac{[a]}{L^6} = ML^{-1}T^{-2}, \text{ or } [a] = ML^5T^{-2}.$$

The constant b must represent a volume since it is subtracted from V. Hence

$$[b] = L^3.$$

Application of Dimensions. Simple Pendulum

We can often use dimensions to solve problems. As an example, suppose a small mass is suspended from a long thread so as to form a simple pendulum. We may reasonably suppose that the period, T, of the oscillations depends only on the mass m, the length l of the thread, and the acceleration, g, of free fall at the place concerned. Suppose then that

$$T = km^x l^y g^z \qquad . \qquad . \qquad . \qquad . \qquad . \qquad \text{(i)}$$

where x, y, z, k are unknown numbers. The dimensions of g are LT^{-2} from p. 32. Now the dimensions of both sides of (i) must be the same.

$$\therefore T \equiv M^x L^y (LT^{-2})^z.$$

Equating the indices of M, L, T on both sides, we have

$$x = 0,$$
$$y + z = 0,$$

and
$$-2z = 1.$$
$$\therefore z = -\tfrac{1}{2}, \quad y = \tfrac{1}{2}, \quad x = 0.$$

Thus, from (i), the period T is given by

$$T = kl^{\frac{1}{2}}g^{-\frac{1}{2}},$$

or
$$T = k\sqrt{\frac{l}{g}}.$$

We cannot find the magnitude of k by the method of dimensions, since it is a number. A complete mathematical investigation shows that $k = 2\pi$ in this case, and hence $T = 2\pi\sqrt{l/g}$. (See also p. 72.)

Velocity of Transverse Wave in a String

As another illustration of the use of dimensions, consider a wave set up in a stretched string by plucking it. The velocity, V, of the wave depends on the tension, F, in the string, its length l, and its mass m, and we can therefore suppose that

$$V = kF^x l^y m^z, \qquad \qquad \text{(i)}$$

where x, y, z are numbers we hope to find by dimensions and k is a constant.

The dimensions of velocity, V, are LT^{-1}, the dimensions of tension, F, are MLT^{-2}, the dimensions of length, l, is L, and the dimension of mass, m, is M. From (i), it follows that

$$LT^{-1} \equiv (MLT^{-2})^x \times L^y \times M^z.$$

Equating powers of M, L, and T on both sides,

$$\therefore 0 = x + z, \qquad \qquad \text{(i)}$$
$$1 = x + y, \qquad \qquad \text{(ii)}$$
and
$$-1 = -2x, \qquad \qquad \text{(iii)}$$
$$\therefore x = \tfrac{1}{2}, \quad z = -\tfrac{1}{2}, \quad y = \tfrac{1}{2}.$$
$$\therefore V = k.F^{\frac{1}{2}} l^{\frac{1}{2}} m^{-\frac{1}{2}},$$

or
$$V = k\sqrt{\frac{Fl}{m}} = k\sqrt{\frac{F}{m/l}} = k\sqrt{\frac{\text{Tension}}{\text{mass per unit length}}}$$

A complete mathematical investigation shows that $k = 1$.

The method of dimensions can thus be used to find the relation between quantities when the mathematics is too difficult. It has been extensively used in hydrodynamics, for example.

Exercises 1 C

Energy. Power. Dimensions

(*Assume g = 10 m s^{-2} or 10 N kg^{-1} unless otherwise stated.*)

1. An object A of mass 10 kg is moving with a velocity of 6 m s^{-1}. Calculate its kinetic energy and its momentum.

If a constant opposing force of 20 N suddenly acts on A, find the time it takes to come to rest and the distance through which it moves.

2. An object A moving horizontally with kinetic energy of 800 J experiences a constant horizontal opposing force of 100 N while moving from a place X to a place Y, where XY is 2 m. What is the energy of A at Y?

In what further distance will A come to rest if this opposing force continues to act on it?

3. A ball of mass 0·1 kg is thrown vertically upwards with an initial speed of 20 m s^{-1}. Calculate (i) the time taken to return to the thrower, (ii) the maximum height reached, (iii) the kinetic and potential energies of the ball half-way up.

4. Calculate the energy of (i) a 2 kg object moving horizontally with a velocity of 10 m s^{-1}, (ii) a 10 kg object held stationary 5 m above the ground.

5. A 4 kg ball moving with a velocity of 10·0 m s^{-1} collides with a 16 kg ball moving with a velocity of 4·0 m s^{-1} (i) in the same direction, (ii) in the opposite direction. Calculate the velocity of the balls in each case if they coalesce on impact, and the loss of energy resulting from the impact. State the principle used to calculate the velocity.

6. A ball of mass 0·1 kg is thrown vertically upwards with a velocity of 20 m s^{-1}. What is the potential energy at the maximum height? What is the potential energy of the ball when it reaches three-quarters of the maximum height while moving upwards?

7. A stationary mass explodes into two parts of mass 4 units and 40 units respectively. If the larger mass has an initial kinetic energy of 100 J, what is the initial kinetic energy of the smaller mass? Explain your calculation.

8. What is an *elastic* and an *inelastic* collision? Give one example of each type. A bullet of mass 10 g is fired vertically with a velocity of 100 m s^{-1} into a block of wood of mass 190 g suspended by a long string above the gun. If the bullet comes to rest in the block, through what height does the block move?

9. A car of mass 1000 kg moves at a constant speed of 20 m s^{-1} along a horizontal road where the frictional force is 200 N. Calculate the power developed by the engine.

If the car now moves up an incline at the same constant speed, calculate the new power developed by the engine. Assume that the frictional force is still 200 N and that sin θ = 1/20, where θ is the angle of the incline to the horizontal.

10. Which of the following are (i) scalars, (ii) vectors? Obtain the dimensions of each.

A momentum	B work	C speed	D force
E energy	F weight	G mass	H acceleration

11. The velocity v of a wave along a plucked string is given by $v = \sqrt{Tl/M}$, where T is the tension in the string, l is its length and M is its mass. Show that this formula is dimensionally correct.

12. The volume per second of a liquid flowing through a horizontal pipe of length l is given by $kpa^x/l\eta$, where k is a constant, p is the excess pressure (force per unit area), a is the radius of the pipe and η is a frictional quantity of dimensions $ML^{-1}T^{-1}$. By dimensions, find the number x.

13. The period of vibration t of a liquid drop is given by $t = ka^x\rho^y\gamma^z$, where k is a constant, a is the radius of the drop, ρ is the density of the liquid and γ is the surface tension of dimensions MT^{-2}.

By dimensions, find the values of the indices x, y, z and the relation for t.

14. Explain what is meant by *kinetic energy*, and show that for a particle of mass m moving with velocity v, the kinetic energy is $\frac{1}{2}mv^2$.

A steel ball is (*a*) projected horizontally with velocity v, at a height h above the ground, (*b*) dropped from a height h and bounces on a fixed horizontal steel plate. Neglecting air resistance and using suitable sketch graphs, explain how the kinetic energy of the ball varies in (*a*) with its height above the ground, and in (*b*) with its height above the plate. (*N.*)

15. Sand falls at a rate of 0·15 kg s^{-1} on to a conveyor belt moving horizontally at a constant speed of 2 m s^{-1}. Calculate (*a*) the extra force necessary to maintain this speed, (*b*) the rate at which work is done by this force, (*c*) the change in kinetic energy per second of the sand on the belt. Account for the difference between your answers to (*b*) and (*c*). (*N.*)

16. A railway truck of mass 4×10^4 kg moving at a velocity of 3 m s^{-1} collides with another truck of mass 2×10^4 kg which is at rest. The couplings join and the trucks move off together. What fraction of the first truck's initial kinetic energy remains as kinetic energy of the two trucks after the collision? Is energy conserved in a collision such as this? Explain your answer briefly. (*L.*)

17. A body moving through air at a high speed v experiences a retarding force F given by $F = kA\rho v^x$, where A is the surface area of the body, ρ is the density of the air and k is a numerical constant. Deduce the value of x.

A sphere of radius 50 mm and mass 1·0 kg falling vertically through air of density 1·2 kg m^{-3} attains a steady velocity of 11·0 m s^{-1}. If the above equation then applies to its fall what is the value of k in this case? (*L.*)

18. A ball falls freely to Earth from a height H and rebounds to a height $h(< H)$. Discuss the linear momentum and energy changes that occur during (i) the fall, (ii) the rebound, with reference to the principles of conservation of momentum and energy. The mass of the Earth, though very large compared with that of the ball, should not be taken as infinite. (Detailed mathematical treatment is not required.)

In a pile-driver a mass m falls freely from height H on to a vertical post of mass M and does not rebound. The ground exerts a constant force F opposing the motion of the post into the ground. The post is driven in a distance d. Find an expression for F (the motion of the earth due to the impact may be neglected). (*O. & C.*)

19. The nucleus of a radioactive isotope of thorium decays by the emission of an α-particle ($A = 4$) to an isotope of radium ($A = 226$). Calculate the ratio of the speeds of the α-particle and radium nucleus and hence find the recoil kinetic energy in MeV of the radium nucleus if the ejected α-particle has a kinetic energy of 4·61 MeV. (Assume the thorium nucleus was at rest.) (*L.*)

20. (a) A particle of mass m, initially at rest, is acted upon by a constant force until its velocity is v. Show that the kinetic energy of the particle is $\frac{1}{2}mv^2$.

(b) A train of mass $2·0 \times 10^5$ kg moves at a constant speed of 72 km h^{-1} up a straight incline against a frictional force of $1·28 \times 10^4$ N. The incline is such that the train rises vertically 1·0 m for every 100 m travelled along the incline. Calculate (i) the rate of increase per second of the potential energy of the train, (ii) the necessary power developed by the train. (*N.*)

21. Explain what is meant by *energy* and distinguish between *potential energy* and *kinetic energy*.

Define *linear momentum* and state the law of conservation of linear momentum. Describe an experiment to verify this law.

A rocket of mass M is moving at constant speed in free space and initially has kinetic energy E. An explosive charge of negligible mass divides it into three parts of equal mass $M/3$ in such a way that one part moves in the same direction as the parent rocket with kinetic energy $E/3$, and the other two portions move off at an angle $60°$ to this direction. Determine the total energy W imparted to the parts of the rocket in the explosion. (*O. & C.*)

22. The diagram, Fig. 1C, shows two simple pendulums each 0·8 m long with identical bobs hanging side by side. Bob A is raised with the string taut to the horizontal position A' and released.

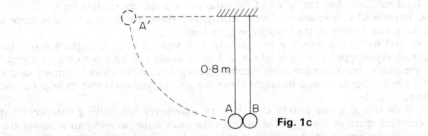

Fig. 1c

(a) Calculate the velocity with which A strikes B. (b) Calculate the velocities of A and B just after A makes a perfectly elastic collision with B. (c) A and B are replaced by a similar arrangement with five identical bobs A, B, C, D, E, hanging side by side but not in contact, each being a simple pendulum 0·8 m long. If A is taken to A' as before and released, discuss the early stages of the subsequent motion of the bobs. (*N.*)

23. Define *linear momentum*.

Describe an experiment which can be performed to investigate inelastic collisions between two bodies moving in one dimension. Explain how the velocities of the bodies can be measured. Summarise the results which would be obtained in terms of kinetic energies and momenta of the bodies before and after impact. How would these summarised results differ if the collision were perfectly elastic?

Two identical steel balls B and C lie in a smooth horizontal straight groove so that they are touching. A third identical ball A moves at a speed v along the groove and collides with B. Assuming that the collisions are all perfectly elastic explain why it is impossible for

(a) A to stop while B and C move off together at speed $v/2$,

(b) A to stop while B and C move off together at speed $v/\sqrt{2}$. (L.)

Fig. 1D

24. As shown in the diagram, two trolleys P and Q of masses 0·50 kg and 0·30 kg respectively are held together on a horizontal track against a spring which is in a state of compression. When the spring is released the trolleys separate freely and P moves

Fig. 1E

to the left with an initial velocity of 6 m s^{-1}. Calculate (a) the initial velocity of Q, (b) the initial total kinetic energy of the system. Calculate also the initial velocity of Q if trolley P is held still when the spring under the same compression as before is released. (N.)

25. State the principle of the conservation of linear momentum and show how it follows from Newton's laws of motion.

A stationary radioactive nucleus of mass 210 units disintegrates into an alpha particle of mass 4 units and a residual nucleus of mass 206 units. If the kinetic energy of the alpha particle is E, calculate the kinetic energy of the residual nucleus. (N.)

26. Define linear momentum and state the principle of conservation of linear momentum. Explain briefly how you would attempt to verify this principle by experiment.

Sand is deposited at a uniform rate of 20 kilogram per second and with negligible kinetic energy on to an empty conveyor belt moving horizontally at a constant speed of 10 metre per minute. Find (a) the force required to maintain constant velocity, (b) the power required to maintain constant velocity, and (c) the rate of change of kinetic energy of the moving sand. Why are the latter two quantities unequal? (O. & C.)

27. What do you understand by the *conservation of energy*? Illustrate your answer by reference to the energy changes occurring (a) in a body whilst falling to and on reaching the ground, (b) in an X-ray tube.

The constant force resisting the motion of a car, of mass 1500 kg, is equal to one-fifteenth of its weight. If, when travelling at 48 km per hour, the car is brought to rest in a distance of 50 m by applying the brakes, find the additional retarding force due to the brakes (assumed constant) and the heat developed in the brakes. (N.)

28. Define *momentum* and state the *law of conservation of linear momentum*.

Discuss the conservation of linear momentum in the following cases (a) a freely falling body strikes the ground without rebounding, (b) during free flight an explosive charge separates an earth satellite from its propulsion unit, (c) a billiard ball bounces off the perfectly elastic cushion of a billiard table.

A bullet of mass 10 g travelling horizontally with a velocity of 300 m s^{-1} strikes a block of wood of mass 290 g which rests on a rough horizontal floor. After impact the block and bullet move together and come to rest when the block has travelled a distance of 15 m. Calculate the coefficient of sliding friction between the block and the floor. (O. & C.)

29. Derive an expression for the kinetic energy of a moving body.

A vehicle of mass 2000 kg travelling at 10 m s^{-1} on a horizontal surface is brought to rest in a distance of 12·5 m by the action of its brakes. Calculate the average retarding force What power must the engine develop in order to take the vehicle up an incline of 1 in 10 at a constant speed of 10 m s^{-1} if the frictional resistance is equal to 200 N? (L.)

2 Circular motion. Gravitation S.H.M.

Circular Motion

Angular Velocity

In the previous chapter we discussed the motion of an object moving in a straight line. There are numerous cases of objects moving in a *curve* about some fixed point. The earth and the moon revolve continuously round the sun, for example, and the rim of the balance-wheel of a watch moves to-and-fro in a circular path about the fixed axis of the wheel. In this chapter we shall study the motion of an object moving in a circle with a *uniform speed* round a fixed point O as centre, Fig. 2.1.

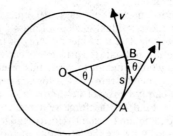

Fig. 2.1 Circular motion

If the object moves from A to B so that the radius OA moves through an angle θ, its *angular velocity*, ω, about O is defined as the *change of the angle per second*. Thus if t is the time taken by the object to move from A to B,

$$\omega = \frac{\theta}{t}. \qquad \qquad \qquad (1)$$

Angular velocity is usually expressed in 'radian per second' (rad s^{-1}). From (1),

$$\theta = \omega t \qquad \qquad \qquad (2)$$

which is analogous to the formula 'distance = uniform velocity × time' for motion in a straight line. It will be noted that the time T to describe the circle once, known as the *period* of the motion, is given by

$$T = \frac{2\pi}{\omega}, \qquad \qquad \qquad (3)$$

since 2π radians is the angle in 1 revolution.

If s is the length of the arc AB, then $s/r = \theta$, by definition of an angle in radians.

$$\therefore s = r\theta.$$

Dividing by t, the time taken to move from A to B,

$$\therefore \frac{s}{t} = r\frac{\theta}{t}.$$

But s/t = the *speed*, v, of the rotating object, and θ/t is the angular velocity.

$$\therefore v = r\omega \quad . \qquad . \qquad . \qquad . \qquad . \qquad (4)$$

Example

A model car moves round a circular track of radius 0·3 m at 2 revolutions per second. What is (a) the angular velocity ω, (b) the period T, (c) the speed v of the car? Find also (d) the angular velocity of the car if it moves with a uniform speed of 2 m s^{-1} in a circle of radius 0·4 m?

(a) For 1 revolution, angle turned $\theta = 2\pi$ rad (360°). So

$$\omega = 2 \times 2\pi = 4\pi \text{ rad s}^{-1}.$$

(b) Period T = time for 1 rev $= \dfrac{2\pi}{\omega} = \dfrac{2\pi}{4\pi} = 0\cdot5$ s. (Or, $T = 1$ s/2 rev $= 0\cdot5$ s.)

(c) Velocity $v = r\omega = 0\cdot3 \times 4\pi = 1\cdot2\pi = 3\cdot8$ m s^{-1}.
(d) From $v = r\omega$,

$$\omega = \frac{v}{r} = \frac{2 \text{ m s}^{-1}}{0\cdot4 \text{ m}} = 5 \text{ rad s}^{-1}.$$

Acceleration in a Circle

When a stone is attached to a string and whirled round at constant speed in a circle, one can feel the force (pull) in the string needed to keep the stone moving in its circular path. Although the stone is moving with a constant speed, the presence of the force implies that the stone has an *acceleration*.

The force on the stone acts *towards the centre* of the circle. We call it a *centripetal force*. The direction of the acceleration is in the same direction as the force, that is, towards the centre. We now show that if v is the uniform speed in the circle and r is the radius of the circle,

$$\text{acceleration towards centre} = \frac{v^2}{r} \qquad . \qquad . \qquad . \qquad . \qquad . \qquad (1)$$

or, since $v = r\omega$,

$$\text{acceleration towards centre} = \frac{r^2\omega^2}{r} = r\omega^2 \qquad . \qquad . \qquad . \qquad (2)$$

The dimensions of v are LT^{-1} and of r is L. So v^2/r has the dimensions LT^{-2}, which is an acceleration. Also, the dimension of ω is T^{-1}, so $r\omega^2$ has the dimensions LT^{-2}, which is an acceleration.

Proof of v^2/r or $r\omega^2$

To obtain an expression for the acceleration towards the centre, consider an object moving with a constant speed v round a circle of radius r, Fig. 2.2 (i). At A, its *velocity* v_A is in the direction of the tangent AC; a short time Δt later at B, its velocity v_B is in the direction of the tangent BD. Since their directions are different, the velocity v_B is different from the velocity v_A, although their magnitudes are both equal to v. Thus a velocity change or acceleration has occurred from A to B.

The velocity change from A to B $= \vec{v_B} - \vec{v_A} = \vec{v_B} + (-\vec{v_A})$. The arrows denote vector quantities. In Fig. 2.2 (ii), PQ is drawn to represent $\vec{v_B}$ in magnitude (v)

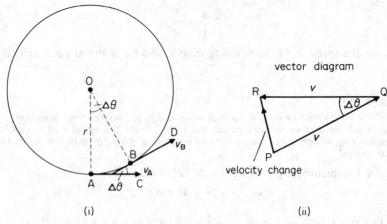

Fig. 2.2 Acceleration in a circle

and direction (BD); QR is drawn to represent $(-\overrightarrow{v_A})$ in magnitude (v) and direction (CA). Then, as shown on p. 12,

$$\text{velocity change} = \overrightarrow{v_B} + (-\overrightarrow{v_A}) = \text{PR}.$$

When Δt is small, the angle AOB or $\Delta\theta$ is small. Thus angle PQR, equal to $\Delta\theta$, is small. PR then points towards O, the centre of the circle. *The velocity change or acceleration is thus directed towards the centre.*

The magnitude of the acceleration, a, is given by

$$a = \frac{\text{velocity change}}{\text{time}} = \frac{\text{PR}}{\Delta t}$$

$$= \frac{v.\Delta\theta}{\Delta t}$$

since $\text{PR} = v.\Delta\theta$. In the limit, when Δt approaches zero, $\Delta\theta/\Delta t = d\theta/dt = \omega$, the angular velocity. But $v = r\omega$ (p. 39). Hence, since $a = v\omega$,

$$a = \frac{v^2}{r} \quad \text{or} \quad r\omega^2 \qquad . \qquad . \qquad . \qquad . \qquad (5)$$

Thus an object moving in a circle of radius r with a constant speed v has an acceleration towards the centre equal to v^2/r or $r\omega^2$.

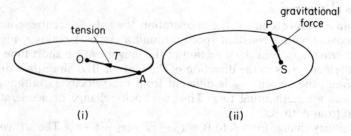

Fig. 2.3 Centripetal forces

Centripetal Forces

The centripetal force F required to keep an object of mass m moving in a circle of radius $r = ma = mv^2/r$. As already stated, it acts towards the centre of the circle. When a stone A is whirled in a horizontal circle of centre O by means of a string, the tension T provides the centripetal force, Fig. 2.3 (i). For a racing car moving round a circular track, the friction at the wheels provides the centripetal force. Planets such as P, moving in a circular orbit round the sun S, have a centripetal force due to gravitational attraction between S and P (p. 47), Fig. 2.3 (ii).

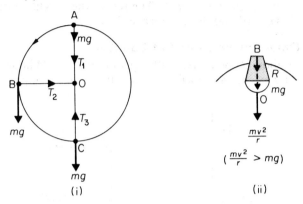

Fig. 2.4 Circular motion

Fig. 2.4 (i) shows an object of mass m whirled with constant speed v in a *vertical* circle of centre O by a string of length r. At A, the top of the motion, suppose T_1 is the tension (force) in the string. Then, since the weight mg acts downwards towards the centre O,

$$\text{force towards centre, } F = T_1 + mg = \frac{mv^2}{r}.$$

So
$$T_1 = \frac{mv^2}{r} - mg \qquad . \qquad . \qquad . \qquad (1)$$

At the point B, where OB is horizontal, suppose T_2 is the tension in the string. The weight mg acts vertically downwards and has no component in the horizontal direction BO. So

$$\text{force towards centre, } F = T_2 = \frac{mv^2}{r} \qquad . \qquad . \qquad . \qquad . \qquad (2)$$

At C, the lowest point of the motion, the weight mg acts in the *opposite* direction to the tension T_3 in the string. So

$$\text{force towards centre, } F = T_3 - mg = \frac{mv^2}{r}.$$

So
$$T_3 = \frac{mv^2}{r} + mg \qquad . \qquad . \qquad . \qquad . \qquad (3)$$

From (1), (2) and (3), we see that (a) the *maximum* tension is given by (3) and (b) this occurs at the bottom C of the circle. Here the tension T_3 must be greater than mg by mv^2/r to make the object keep moving in a circular path. The *minimum* tension is given in (1) and this occurs at A, the top of the motion. Here part of the required centripetal force is provided by the weight *mg* and the rest by T_1.

If some water is placed in a bucket B attached to the end of a string, the bucket can be whirled in a vertical plane without any water falling out. When the bucket is vertically above the point of support O, the weight *mg* of the water is less than the required force mv^2/r towards the centre and so the water stays in, Fig. 2.4 (ii). The reaction *R* of the bucket base on the water provides the rest of the force mv^2/r. If the bucket is whirled slowly and $mg > mv^2/r$, part of the weight provides the force mv^2/r. The rest of the weight causes the water to accelerate downward and hence to leave the bucket.

Weight of Object at Poles and at Equator

The earth turns about its polar axis with a period of about 24 hours. Its angular velocity $\omega = 2\pi/T = 2\pi/(24 \times 3600 \text{ s}) = 7\cdot3 \times 10^{-5}$ rad s^{-1} (approx).

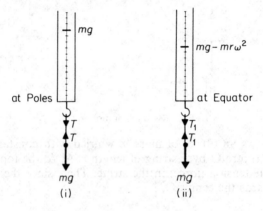

Fig. 2.5 Weight of object and rotation of earth

Poles. Suppose an object of mass *m* is attached to a spring balance by a string at the north or south pole, Fig. 2.5 (i). Here the object has no rotation about the polar axis. So the tension *T* in the spring $= mg$. The spring balance always reads the tension in the string. So the balance reading gives the weight *mg* of the object.

Equator. Suppose the same object is now taken to the equator and attached by a string to the spring balance, Fig. 2.5 (ii). The object now has an angular velocity ω about the centre of the earth as the earth rotates about its polar axis and an acceleration $r\omega^2$ towards the centre of the earth. So if T_1 is the tension in the string, which acts upwards *on the mass*,

$$\text{centripetal force} = \text{force towards centre}$$

$$= mg - T_1 = mr\omega^2.$$

So $T_1 = mg - mr\omega^2.$

The spring balance reads T_1. So the balance records an 'apparent weight' $m'g$ say, which is *less* than the true weight *mg* by $mr\omega^2$. The fractional difference is $mr\omega^2/mg$ or $r\omega^2/g$. Now $r = 6\cdot37 \times 10^6$ m, $\omega = 7\cdot3 \times 10^{-5}$ rad s^{-1} and $g = 9\cdot78$ m s^{-2}. So

$$\text{percentage difference} = \frac{r\omega^2}{g} \times 100\% = \frac{6.37 \times 10^6 \times (7.3 \times 10^{-5})^2}{9.78} \times 100\%$$

$$= 0.35\%.$$

Motion of Car (or Train) Round Banked Track

Suppose a car (or train) is moving round a banked track in a circular path of horizontal radius r, Fig. 2.6. If the only forces at the wheels A, B are the normal reactions R_1, R_2 respectively, that is, there is no side-slip or strain at the wheels, the force towards the centre of the track is $(R_1 + R_2) \sin \theta$, where θ is the angle of inclination of the plane to the horizontal.

$$\therefore (R_1 + R_2) \sin \theta = \frac{mv^2}{r} \qquad . \qquad . \qquad . \qquad . \qquad \text{(i)}$$

The car does not move in a vertical direction. So, for vertical equilibrium,

$$(R_1 + R_2) \cos \theta = mg \qquad . \qquad . \qquad . \qquad . \qquad \text{(ii)}$$

Dividing (i) by (ii), $\qquad\qquad \therefore \tan \theta = \frac{v^2}{rg} \qquad . \qquad . \qquad . \qquad . \qquad \text{(iii)}$

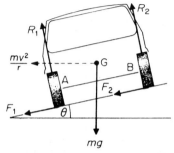

Fig. 2.6 Car on banked track

Thus for a given velocity v and radius r, the angle of inclination of the track for no side-slip must be $\tan^{-1} (v^2/rg)$. As the speed v increases, the angle θ increases, from (iii). A racing-track is made saucer-shaped because at high speeds the cars can move towards a part of the track which is steeper and sufficient to prevent side-slip. The outer rail of a curved railway track is raised about the inner rail so that the force towards the centre is largely provided by the component of the reaction at the wheels. It is desirable to bank a road at corners for the same reason as a racing track is banked.

Conical Pendulum

Suppose a small object A of mass m is tied to a string OA of length l and then whirled round in a *horizontal* circle of radius r, with O fixed directly above the

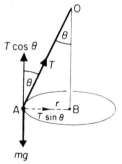

Fig. 2.7 Conical pendulum

centre B of the circle, Fig. 2.7. If the circular speed of A is constant, the string turns at a constant angle θ to the vertical. This is called a conical pendulum.

Since A moves with a constant speed v in a circle of radius r, there must be a centripetal force mv^2/r acting towards the centre B. The horizontal component, $T \sin \theta$, of the tension T in the string provides this force along AB. So

$$T \sin \theta = \frac{mv^2}{r} \qquad \qquad (1)$$

Also, since the mass does not move vertically, its weight mg must be counterbalanced by the vertical component $T \cos \theta$ of the tension. So

$$T \cos \theta = mg \qquad \qquad (2)$$

Dividing (1) by (2), then $\qquad \tan \theta = \dfrac{v^2}{rg}.$

A similar formula for θ was obtained for the angle of banking of a track which prevented side-slip.

If $v = 2$ m s^{-1}, $r = 0.5$ m and $g = 10$ m s^{-2}, then

$$\tan \theta = \frac{v^2}{rg} = \frac{2^2}{0.5 \times 10} = 0.8.$$

So $\qquad\qquad\qquad \theta = 39°.$

If $m = 2.0$ kg, it follows from (2) that

$$T = \frac{mg}{\cos \theta} = \frac{2 \times 10}{\cos 39°} = 25.7 \text{ N.}$$

A pendulum suspended from the ceiling of a train does not remain vertical while the train goes round a circular track. Its bob moves *outwards* away from the centre and the string becomes inclined at an angle θ to the vertical, as shown in Fig. 2.7. In this case the centripetal force is provided by the horizontal component of the tension in the string, as we have already explained.

Motion of Bicycle Rider round Circular Track

When a person on a bicycle rides round a circular racing track, the necessary centripetal force mv^2/r is actually provided by the frictional force F at the ground. Fig. 2.8. The force F has a moment about the centre of gravity G equal to $F.h$ which tends to turn the rider outwards. When the rider leans inwards as

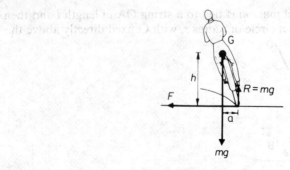

Fig. 2.8 Rider on a circular track

shown, this is counterbalanced by the moment $R.a$ about G. $R = mg$ since there is no vertical motion, so the moment is $mg.a$. Thus, provided no slipping occurs, $F.h = mg.a$.

$$\therefore \frac{a}{h} = \tan \theta = \frac{F}{mg},$$

where θ is the angle of inclination to the vertical. Now $F = mv^2/r$.

$$\therefore \tan \theta = \frac{v^2}{rg}.$$

When F is greater than the limiting friction, skidding occurs. In this case $F > \mu mg$, or $mg \tan \theta > \mu mg$. Thus $\tan \theta > \mu$ is the condition for skidding.

Exercises 2A

Circular Motion

(Assume $g = 10 \text{ m s}^{-2}$ or 10 N kg^{-1} unless otherwise given)

1. An object of mass 4 kg moves round a circle of radius 6 m with a constant speed of 12 m s^{-1}. Calculate (i) the angular velocity, (ii) the force towards the centre.

2. An object of mass 10 kg is whirled round a horizontal circle of radius 4 m by a revolving string inclined to the vertical. If the uniform speed of the object is 5 m s^{-1}, calculate (i) the tension in the string, (ii) the angle of inclination of the string to the vertical.

3. A racing-car of 1000 kg moves round a banked track at a constant speed of 108 km h^{-1}. Assuming the total reaction at the wheels is normal to the track, and the horizontal radius of the track is 100 m, calculate the angle of inclination of the track to the horizontal and the reaction at the wheels.

4. An object of mass 8·0 kg is whirled round in a vertical circle of radius 2 m with a constant speed of 6 m s^{-1}. Calculate the maximum and minimum tensions in the string.

5. Calculate the force necessary to keep a mass of 0·2 kg moving in a horizontal circle of radius 0·5 m with a period of 0·5 s. What is the direction of the force?

6. Calculate the mean angular velocity of the Earth assuming it takes 24·0 h to rotate about its axis.

An object of mass 2·00 kg is (i) at the Poles, (ii) at the Equator. Assuming the Earth is a perfect sphere of radius 6·4 × 10^6 m, calculate the change in weight of the mass when taken from the Poles to the Equator. Explain your calculation with the aid of a diagram.

7. A stone is rotated steadily in a horizontal circle with a period T by a string of length l. If the tension in the string is constant and l increases by 1%, find the percentage change in T.

8. A mass of 0·2 kg is whirled in a horizontal circle of radius 0·5 m by a string inclined at 30° to the vertical. Calculate (i) the tension in the string, (ii) the speed of the mass in the horizontal circle.

9. An object of mass 0·5 kg is rotated in a horizontal circle by a string 1 m long. The maximum tension in the string before it breaks is 50 N. What is the greatest number of revolutions per second of the object?

10. A mass of 0·4 kg is rotated by a string at a constant speed v in a vertical circle of radius 1 m. If the minimum tension of the string is 3 N, calculate (i) v, (ii) the maximum tension, (iii) the tension when the string is just horizontal.

11. What force is necessary to keep a mass of 0·8 kg revolving in a horizontal circle of radius 0·7 m with a period of 0·5 s? What is the direction of this force? (Assume that $\pi^2 = 10$.) (*L.*)

12. A spaceman in training is rotated in a seat at the end of a horizontal rotating arm of length 5 m. If he can withstand accelerations up to $9g$, what is the maximum number

of revolutions per second permissible? The acceleration of free fall (g) may be taken as
10 m s^{-2}. (*L.*)

13. Define the terms (*a*) *acceleration*, and (*b*) *force*. Show that the acceleration of a body
moving in a circular path of radius r with uniform speed v is v^2/r, and draw a diagram
to show the direction of the acceleration.

A small body of mass m is attached to one end of a light inelastic string of length l.
The other end of the string is fixed. The string is initially held taut and horizontal, and
the body is then released. Find the values of the following quantities when the string
reaches the vertical position: (*a*) the kinetic energy of the body, (*b*) the velocity of the
body, (*c*) the acceleration of the body, and (*d*) the tension in the string. (*O. & C.*)

14. Explain what is meant by *angular velocity*. Derive an expression for the force required
to make a particle of mass m move in a circle of radius r with uniform angular velocity ω.

A stone of mass 500 g is attached to a string of length 50 cm which will break if the
tension in it exceeds 20 N. The stone is whirled in a vertical circle, the axis of rotation
being at a height of 100 cm above the ground. The angular speed is very slowly increased
until the string breaks. In what position is this break most likely to occur, and at what
angular speed? Where will the stone hit the ground? (*C.*)

15. A special prototype model aeroplane of mass 400 g has a control wire 8 m long
attached to its body. The other end of the control line is attached to a fixed point. When
the aeroplane flies with its wings horizontal in a horizontal circle, making one revolution
every 4 s, the control wire is elevated 30° above the horizontal. Draw a diagram showing
the forces exerted on the plane and determine (*a*) the tension in the control wire, (*b*) the
lift on the plane. (Assume acceleration of free fall, $g = 10$ m s^{-2} and $\pi^2 = 10$.) (*AEB*)

Gravitation

Kepler's Laws

Kepler (1571–1630) had studied for many years the records of observations on the planets made by TYCHO BRAHE, and he discovered three laws now known by his name. *Kepler's laws* state:

(1) The planets describe ellipses about the sun as one focus.
(2) The line joining the sun and the planet sweeps out equal areas in equal times.
(3) The squares of the periods of revolution of the planets are proportional to the cubes of their mean distances from the sun.

Newton's Investigation on Planetary Motion

About 1666, at the early age of 24, Newton investigated the motion of a planet moving in a circle round the sun S as centre, Fig. 2.9 (i). The force acting on the planet of mass m is $mr\omega^2$, where r is the radius of the circle and ω is the angular velocity of the motion (p. 39). Since $\omega = 2\pi/T$, where T is the period of the motion,

$$\text{force on planet} = mr\left(\frac{2\pi}{T}\right)^2 = \frac{4\pi^2 mr}{T^2}$$

This is equal to the force of attraction of the sun on the planet. *Assuming an inverse-square law*, then, if k is a constant,

$$\text{force on planet} = \frac{km}{r^2}$$

$$\therefore \frac{km}{r^2} = \frac{4\pi^2 mr}{T^2}$$

$$\therefore T^2 = \frac{4\pi^2}{k} r^3$$

$$\therefore T^2 \propto r^3,$$

since k, π are constants.

Now Kepler had announced that the squares of the periods of revolution of the planets are proportional to the cubes of their mean distances from the sun

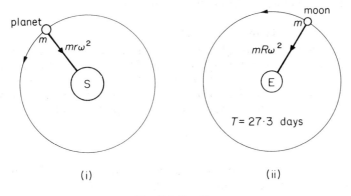

(i)   (ii)

Fig. 2.9 Satellites

(see above). Newton thus suspected that *the force between the sun and the planet was inversely proportional to the square of the distance between them.*

Motion of Moon round Earth

Newton now tested the inverse-square law by applying it to the case of the moon's motion round the earth. Fig. 2.9 (ii). The moon has a period of revolution, T, about the earth of approximately 27·3 days, and the force on it $= mR\omega^2$, where R is the radius of the moon's orbit and m is its mass.

$$\therefore \text{force} = mR\left(\frac{2\pi}{T}\right)^2 = \frac{4\pi^2 mR}{T^2}.$$

If the planet were at the earth's surface, the force of attraction on it due to the earth would be mg, where g is the acceleration due to gravity. Figure 2.9 (ii). Assuming that the force of attraction varies as the inverse square of the distance between the earth and the moon, then, by ratio,

$$= \frac{4\pi^2 mr}{T^2} : mg = \frac{1}{R^2} : \frac{1}{r_E^2},$$

where r_E is the radius of the earth.

$$\therefore \frac{4\pi^2 R}{T^2 g} = \frac{r_E^2}{R^2},$$

$$\therefore g = \frac{4\pi^2 R^3}{r_E^2 T^2} \qquad . \qquad . \qquad . \qquad . \qquad (1)$$

Newton substituted the then known values of R, r_E, and T, but was disappointed to find that the answer for g was not near to the observed value, 9·8 m s^{-2}. Some years later, he heard of a new estimate of the radius of the earth, and we now know that r_E is about $6·4 \times 10^6$ m. The radius R of the moon's orbit is about $60·1 r_E$ and the period T of the moon is about 27·3 days or $27·3 \times 24 \times 3600$ s. So

$$g = \frac{4\pi^2 R^3}{r_E^2 T^2} = \frac{4\pi^2 \times (60·1\, r_E)^3}{r_E^2 T^2} = \frac{4\pi^2 \times 60·1^3\, r_E}{T^2}$$

$$= \frac{4\pi^2 \times 60·1^3 \times 6·4 \times 10^6}{(27·3 \times 24 \times 3600)^2} = 9·9 \text{ m s}^{-2}.$$

The result is very close to the measured value of g.

Newton's Law of Gravitation. G

Newton saw that a universal law could be stated for the attraction between any two particles of matter. He suggested that: *The force of the attraction between two given particles is inversely proportional to the square of their distance apart.*

From this law it follows that the force of attraction, F, between two particles of masses m and M respectively, at a distance r apart, is given by

$$F = G\frac{mM}{r^2}, \qquad . \qquad . \qquad . \qquad . \qquad (2)$$

where G is a universal constant known as the *gravitational constant*. This expression for F is *Newton's law of gravitation*. It is a universal law.

From (2), $G = Fr^2/mM$. So G can be expressed in N m^2 kg^{-2}. Careful measurement shows that $G = 6.67 \times 10^{-11}$ N m^2 kg^{-2}. The dimensions of G are

$$[G] = \frac{MLT^{-2} \times L^2}{M^2} = L^3 M^{-1} T^{-2}.$$

So the unit of G may also be expressed as m^3 kg^{-1} s^{-2}.

A celebrated experiment to measure G was carried out by C. V. BOYS in 1895, using a method similar to one of the earliest determinations of G by CAVENDISH in 1798. Two identical balls, a, b, of gold, 5 mm in diameter, were suspended by a long and a short fine quartz fibre respectively from the ends, C, D, of a highly-polished bar CD, Fig. 2.10. Two large identical lead spheres, A, B, 115 mm in diameter, were brought into position near a, b respectively. As a result of the attraction between the masses, two equal but opposite forces acted on CD. The bar was thus deflected, and the angle of deflection, θ, was measured by a lamp and scale method by light reflected from CD. The high sensitivity of the quartz fibres enabled the small deflection to be big enough to be measured accurately. The small size of the apparatus allowed it to be screened considerably from air convection currents.

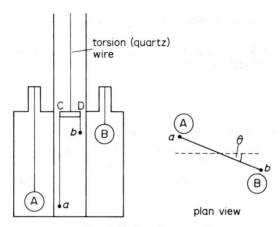

Fig. 2.10 Experiment on G

Calculation for G

Suppose d is the distance between a, A, or b, B, when the deflection is θ. Then if m, M are the respective masses of a, A,

$$\text{torque of couple on CD} = G\frac{mM}{d^2} \times \text{CD}.$$

But $\text{torque} = c\theta,$

where c is the torque in the torsion wire per unit radian of twist (p. 125).

$$\therefore G\frac{mM}{d^2} \times \text{CD} = c\theta.$$

$$\therefore G = \frac{c\theta d^2}{mM \times \text{CD}} \qquad . \qquad . \qquad . \qquad . \quad (1)$$

The constant c was determined by allowing CD to oscillate through a small angle and then observing its period of oscillation, T, which was of the order of 3 minutes. If I is the known moment of inertia of the system about the torsion wire, then (see p. 91),

$$T = 2\pi \sqrt{\frac{I}{c}}.$$

Gravitational Force on Masses on Earth and Outside Earth

On the earth's surface, an object of mass m has a gravitational force of mg on it, where g is the acceleration of free fall. So a mass of 1 kg has a weight of $1g$ or 10 N, assuming g is 10 m s^{-2} at the earth's surface.

To find the gravitational force on masses on the earth or outside it, it is legitimate to consider that the whole mass M_E of the earth is concentrated at its centre. Assuming the earth is a sphere of radius r_E, a mass m on the surface is at a distance r_E from the mass M_E. If the same mass is taken above the earth to a distance $2r_E$ from the centre, the force between M_E and m is reduced to $1/2^2$ or 1/4, since the force between given masses is inversely-proportional to the square of their distance apart. So now

$$\text{gravitational force} = \frac{1}{4} \times 10 \text{ N} = 2\cdot5 \text{ N}.$$

At a higher height of $2r_E$ above the earth, which is a distance $3r_E$ from the centre, the gravitational force F is now reduced to

$$F = \frac{1}{3^2} \times 10 \text{ N} = 1\cdot1 \text{ N}.$$

Variation of Acceleration of Free Fall

For points *outside* the earth, the gravitational force obeys an inverse-square law. So the acceleration of free fall, g', $\propto l/r^2$, where r is the distance to the centre of the earth, Fig. 2.11. The maximum value of g' is obtained at the earth's surface, where $r = r_E$.

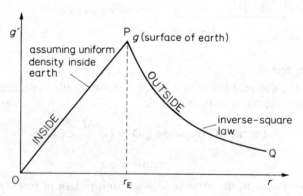

Fig. 2.11 Variation of g', acceleration of free fall

Inside the earth, the value of g' is *not* inversely-proportional to the square of the distance from the centre. Assuming a uniform earth density, which is not true in practice, theory shows that g' varies linearly with the distance from the centre, as shown in Fig. 2.11.

Since the gravitational force F on a mass m is given generally by $F = mg'$, then $g' = F/m$. We see that g' can be expressed in 'newtons per kilogram'

(N kg^{-1}). The *force per unit mass* in the gravitational field of the earth is called its *gravitational intensity*. We see that, on the earth, $g = 9\cdot8$ m s$^{-2} = 9\cdot8$ N kg^{-1}.

Example

A man can jump 1·5 m on earth. Calculate the approximate height he might be able to jump on a planet whose density is one-quarter that of the earth and whose radius is one-third that of the earth.

Suppose the man of mass m leaps a height h on the earth and a height h_1 on the planet. Assuming he can give himself the same initial kinetic energy on the two planets, the potential energy gained is the same at the maximum height. So

$$mg_1h_1 = mgh,$$

where g_1 and g are the respective gravitational intensities on the planet and earth. So

$$h_1 = \frac{g}{g_1} \times h \qquad . \qquad . \qquad . \qquad . \qquad . \qquad (1)$$

But for the earth, $g = GM/r_E{}^2$ (p. 52) $= G.\frac{4}{3}\pi r_E{}^3 \rho_E/r_E{}^2 = G.\frac{4}{3}\pi r_E \rho_E$, where ρ_E is the density of the earth. Similarly, $g_1 = G.\frac{4}{3}\pi r_1 \rho_1$, where r_1, ρ_1 are the respective radius and density of the planet. So

$$\frac{g}{g_1} = \frac{r_E \rho_E}{r_1 \rho_1} = 4 \times 3 = 12.$$

From (1), we have

$$h_1 = 12 \times 1\cdot5 \text{ m} = 18 \text{ m}.$$

Force on Astronaut. Weightlessness

When a rocket is fired to launch a spacecraft and astronaut into orbit round the earth, the initial acceleration must be very high owing to the large initial thrust required. This acceleration, a, is of the order of $15g$, where g is the gravitational acceleration at the earth's surface.

Suppose S is the reaction of the couch to which the astronaut is initially strapped, Fig. 2.12 (i). Then, from $F = ma$, $S - mg = ma = m.15g$, where m is the mass of the astronaut. Thus $S = 16mg$. This force is 16 times the weight of the astronaut and thus, initially, he experiences a large force.

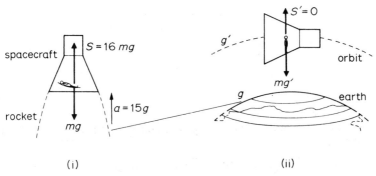

Fig. 2.12 Weight and weightlessness

In orbit, however, the state of affairs is different. This time the acceleration of the spacecraft and astronaut are both g' in magnitude, where g' is the acceleration due to gravity at the particular height of the orbit, Fig. 2.12 (ii). If S' is the reaction of the surface of the spacecraft in contact with the astronaut, then, for circular motion,

$$F = mg' - S' = ma = mg'.$$

Thus $S' = 0$. The astronaut now experiences no reaction at the floor when he walks about, for example, and so he experiences the sensation of being 'weightless' although he has a gravitational force mg' acting on him.

At the earth's surface we feel the reaction at the ground and are thus conscious of our weight. Inside a lift which is falling fast, the reaction at our feet diminishes. If the lift falls freely, the acceleration of objects inside is the same as that outside and hence the reaction on them is zero. This produces the sensation of 'weightlessness'. In orbit, as in Fig. 2.12 (ii), objects inside a spacecraft are also in 'free fall' because they have the same acceleration g' as outside the spacecraft.

Earth Satellites

Satellites can be launched from the earth's surface to circle the earth. They are kept in their orbit by the gravitational attraction of the earth. Consider a satellite of mass m which just circles the earth of mass M close to its surface in an orbit 1, Fig. 2.13. Then, if r_E is the radius of the earth,

$$\frac{mv^2}{r_E} = G\frac{Mm}{r_E^2} = mg,$$

where g is the acceleration due to gravity at the earth's surface and v is the

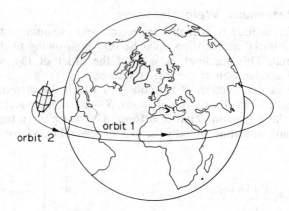

Fig. 2.13 Orbits round earth

velocity of m in its orbit. Thus $v^2 = r_E g$, and hence, using $r_E = 6.4 \times 10^6$ m and $g = 9.8$ m s^{-2},

$$v = \sqrt{r_E g} = \sqrt{6.4 \times 10^6 \times 9.8} = 8 \times 10^3 \text{ m s}^{-1} \text{ (approx.)},$$
$$= 8 \text{ km s}^{-1}.$$

The velocity v in the orbit is thus about 8 km s^{-1}. In practice, the satellite is carried by a rocket to the height of the orbit and then given an impulse, by firing jets, to deflect it in a direction parallel to the tangent of the orbit (see p. 54). Its velocity is boosted to 8 km s^{-1} so that it stays in the orbit. The period in orbit

$$= \frac{\text{circumference of earth}}{v} = \frac{2\pi \times 6 \cdot 4 \times 10^6 \text{ m}}{8 \times 10^3 \text{ m s}^{-1}}$$

$$= 5000 \text{ seconds (approx.)} = 83 \text{ min.}$$

Parking Orbits

Consider now a satellite of mass m circling the earth in the plane of the equator in orbit 2 concentric with the earth, Fig. 2.13. Suppose the direction of rotation as the same as the earth and the orbit is at a distance R from the centre of the earth. Then if v is the velocity in orbit,

$$\frac{mv^2}{R} = \frac{GMm}{R^2}.$$

But $\qquad GM = gr_E^2$, where r_E is the radius of the earth.

$$\therefore \frac{mv^2}{R} = \frac{mgr_E^2}{R^2}$$

$$\therefore v^2 = \frac{gr_E^2}{R}.$$

If T is the period of the satellite in its orbit, then $v = 2\pi R/T$.

$$\therefore \frac{4\pi^2 R^2}{T^2} = \frac{gr_E^2}{R}$$

$$\therefore T^2 = \frac{4\pi^2 R^3}{gr_E^2} \qquad . \qquad . \qquad . \qquad . \qquad (i)$$

If the period of the satellite in its orbit is exactly equal to the period of the earth as it turns about its axis, which is 24 hours, *the satellite will stay over the same place on the earth* while the earth rotates. This is sometimes called a 'parking orbit'. Relay satellites can be placed in parking orbits, so that television programmes can be transmitted continuously from one part of the world to another.

Since $T = 24$ hours, the radius R can be found from (i). Thus from

$$R = \sqrt[3]{\frac{T^2 gr_E^2}{4\pi^2}} \quad \text{and} \quad g = 9 \cdot 8 \text{ m s}^{-2}, \ r_E = 6 \cdot 4 \times 10^6 \text{ m},$$

$$\therefore R = \sqrt[3]{\frac{(24 \times 3600)^2 \times 9 \cdot 8 \times (6 \cdot 4 \times 10^6)^2}{4\pi^2}} = 42\,400 \text{ km.}$$

The height above the earth's surface of the parking orbit

$$= R - r_E = 42\,400 - 6400 = 36\,000 \text{ km.}$$

In the orbit, assuming it is circular the velocity of the satellite

$$= \frac{2\pi R}{T} = \frac{2\pi \times 42\,400 \text{ km}}{24 \times 3600 \text{ s}} = 3 \cdot 1 \text{ km s}^{-1}.$$

The satellite, with the necessary electronic equipment inside, rises vertically from the equator when it is fired. At a particular height the satellite is given a horizontal momentum by firing rockets on its surface and the satellite then turns into the required orbit. This is illustrated in the next example.

Example

A satellite is to be put into orbit 500 km above the earth's surface. If its vertical velocity after launching is 2000 m s^{-1} at this height, calculate the magnitude and direction of the impulse required to put the satellite directly into orbit, if its mass is 50 kg. Assume $g = 10$ m s^{-2}, radius of earth, $r_E = 6400$ km.

Suppose u is the velocity required for orbit, radius r. Then, with usual notation,

$$\frac{mu^2}{R} = \frac{GmM}{R^2} = \frac{gr_E^2 m}{R^2}, \text{ as } \frac{GM}{r_E^2} = g.$$

$$\therefore u^2 = \frac{gr_E^2}{R}.$$

Now $r_E = 6400$ km, $R = 6900$ km, $g = 10$ m s^{-2}.

$$\therefore u^2 = \frac{10 \times (6400 \times 10^3)^2}{6900 \times 10^3}.$$

$$\therefore u = 7700 \text{ m s}^{-1} \text{ (approx.)}.$$

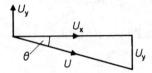

Fig. 2.14 Example

At this height, vertical momentum

$$U_y = mv = 50 \times 2000 = 100\,000 \text{ kg m s}^{-1}.$$

Horizontal momentum required $U_x = mu = 50 \times 7700 = 385\,000$ kg m s^{-1}.

$$\therefore \textit{impulse needed, } U, = \sqrt{U_y^2 + U_x^2} = \sqrt{100\,000^2 + 385\,000^2} \quad \text{(Fig. 2.14)}$$

$$= 4\cdot0 \times 10^5 \text{ kg m s}^{-1} \qquad . \qquad . \qquad . \qquad . \qquad . \qquad (1)$$

Direction. The angle θ made by the total impulse with the horizontal or orbit tangent is given by $\tan \theta = U_y/U_x = 100\,000/385\,000 = 0\cdot260$. Thus $\theta = 14\cdot6°$.

Mass and Density of Earth

At the earth's surface the force of attraction on a mass m is mg, where g is the acceleration due to gravity. Now it can be shown in this case that we can assume that the mass, M, of the earth is concentrated at its centre, if it is a sphere (p. 50). Assuming that the earth is spherical of radius r_E, it then follows that the force of attraction of the earth on the mass m is GmM/r_E^2. So

$$G\frac{mM}{r_E^2} = mg.$$

$$\therefore g = \frac{GM}{r_E^2}.$$

$$\therefore M = \frac{gr_E^2}{G}.$$

Now, $g = 9.8$ m s^{-1}, $r_E = 6.4 \times 10^6$ m, $G = 6.7 \times 10^{-11}$ N m^2 kg^{-2}.

$$\therefore M = \frac{9.8 \times (6.4 \times 10^6)^2}{6.7 \times 10^{-11}} = 6.0 \times 10^{24} \text{ kg.}$$

The volume of a sphere is $4\pi r^3/3$, where r is its radius. So the mean density, ρ, of the earth is approximately given by

$$\rho = \frac{M}{V} = \frac{g r_E^2}{4\pi r_E^3 G/3} = \frac{3g}{4\pi r_E G}.$$

By substituting known values of g, G and r_E, the mean density of the earth is found to be about 5500 kg m^{-3}. The density of the earth is actually non-uniform and may approach a value of 10 000 kg m^{-3} towards the interior.

Mass of Sun

The mass M_S of the sun can be found from the period of a satellite and its distance from the sun. Consider the case of the earth. Its period T is about 365 days or $365 \times 24 \times 3600$ seconds. Its distance r_S from the centre of the sun is about 1.5×10^{11} m. If the mass of the earth is m, then, for circular motion round the sun,

$$\frac{GM_S m}{r_S^2} = m r_S \omega^2 = \frac{m r_S 4\pi^2}{T^2},$$

$$\therefore M_S = \frac{4\pi^2 r_S^3}{GT^2} = \frac{4\pi^2 \times (1.5 \times 10^{11})^3}{6.7 \times 10^{-11} \times (365 \times 24 \times 3600)^2} = 2 \times 10^{30} \text{ kg.}$$

In the equation $GM_S m/r_S^2 = m r_S \omega^2$ above, we see that the mass m of the satellite cancels on both sides and does not appear in the final equation for ω. So ω, the angular velocity in the orbit, is *independent* of the mass of the satellite. The angular velocity ω (and the period) depends only on the value of r_S, the orbit distance from the sun. This is true for all planets, that is, their angular velocity is independent of the mass of the planet and depends only on the radius of the orbit.

Potential

The *potential*, V, at a point due to the gravitational field of the earth is defined as numerically equal to the work done in taking a unit mass from infinity to that point. This is analogous to 'electric potential'. The potential at infinity is conventionally taken as *zero*.

For a point outside the earth, assumed spherical, we can imagine the whole mass M of the earth concentrated at its centre. The force of attraction on a unit mass outside the earth is thus GM/r^2, where r is the distance from the centre. The work done by the gravitational force in moving a distance Δr towards the earth = force × distance = $GM . \Delta r/r^2$. Hence the potential at a point distant a from the centre greater than r is given by

$$V_a = \int_\infty^a \frac{GM}{r^2} dr = -\frac{GM}{a} \qquad . \qquad . \qquad . \qquad (1)$$

if the potential at infinity is taken as zero by convention. The negative sign indicates that the potential at infinity (zero) is *higher* than the potential close to the earth.

On the earth's surface, of radius r_E, we therefore obtain

$$V = -\frac{GM}{r_E}. \qquad . \qquad . \qquad . \qquad . \qquad . \qquad (2)$$

For large distances from the earth, for example, when a rocket travels from the earth to the moon, the change in potential energy of a mass can only be calculated by using $mass \times (GM/a - GM/b)$, where b and a are the distances from the centre of the earth. For small distances above the earth, however, the gravitational force on a mass is fairly constant. So the change in potential energy in this case can be calculated using $force \times distance$ or mgh.

Velocity of Escape

Suppose a rocket of mass m is fired from the earth's surface Q so that it just escapes from the gravitational influence of the earth. Then work done $=$ $m \times$ potential difference between infinity and Q

$$= m \times \frac{GM}{r_E}.$$

$\therefore$ kinetic energy of rocket $= \frac{1}{2}mv^2 = m \times \dfrac{GM}{r_E}.$

$$\therefore v = \sqrt{\frac{2GM}{r_E}} = \text{velocity of escape.}$$

Now $GM/r_E^2 = g.$

$$\therefore v = \sqrt{2gr_E}.$$

$$\therefore v = \sqrt{2 \times 9\cdot8 \times 6\cdot4 \times 10^6} = 11 \times 10^3 \text{ m s}^{-1} = 11 \text{ km s}^{-1} \text{ (approx.).}$$

With an initial velocity, then, of about 11 km s^{-1}, a rocket will completely escape from the gravitational attraction of the earth. It can be made to travel towards the moon, for example, so that eventually it comes under the gravitational attraction of this planet. At present, 'soft' landings on the moon have been made by firing retarding retro rockets.

Summarising, with a velocity of about 8 km s^{-1}, a satellite can describe a circular orbit close to the earth's surface (p. 52). With a velocity greater than

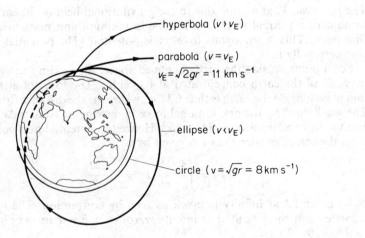

Fig. 2.15 Orbits

8 km s^{-1} but less than 11 km s^{-1}, a satellite describes an elliptical orbit round the earth. Its maximum and minimum height in the orbit depends on its particular velocity. Figure 2.15 illustrates the possible orbits of a satellite launched from the earth.

The molecules of air at normal temperatures and pressures have an average velocity of the order of 480 m s^{-1} or 0.48 km s^{-1} which is much less than the velocity of escape. Many molecules move with higher velocity than 0.48 km s^{-1} but gravitational attraction keeps the atmosphere round the earth. The gravitational attraction of the moon is much less than that of the earth and this accounts for the lack of atmosphere round the moon.

P.E. and K.E. of Satellite

A satellite of mass m in orbit round the earth has both kinetic energy, k.e., and potential energy, p.e. The k.e. $= \frac{1}{2}mv^2$, where v is the velocity in the orbit. Now for circular motion in an orbit of radius r_0, if M is the mass of the earth,

$$\text{force towards centre} = \frac{mv^2}{r_0} = G\frac{Mm}{r_0^2}$$

$$\therefore \text{ k.e.} = \tfrac{1}{2}mv^2 = G\frac{Mm}{2r_0} \qquad . \qquad . \qquad . \qquad . \qquad (1)$$

Assuming the zero of potential energy in the earth's field is at infinity (p. 55),

$$\text{p.e. of mass in orbit} = -G\frac{Mm}{r_0} \qquad . \qquad . \qquad . \qquad (2)$$

So, from (1), the potential energy of the mass in orbit is numerically *twice* its kinetic energy and of opposite sign.

From (1) and (2),

$$\text{total energy in orbit} = -\frac{GMm}{r_0} + \frac{GMm}{2r_0}$$

$$= -\frac{GMm}{2r_0} \qquad . \qquad . \qquad . \qquad . \qquad (3)$$

Owing to friction in the earth's atmosphere, the satellite energy diminishes and the radius of the orbit decreases to r_1 say. The total energy in this orbit, from above, is $-GMm/2r_1$. Since this is *less* than the initial energy in (3), it follows that

$$\frac{GMm}{2r_1} > \frac{GMm}{2r_0}.$$

From (1), these two quantities are the kinetic energy values in the respective orbits of radius r_1 and r_0. Hence the kinetic energy of the satellite *increases* when it falls to an orbit of smaller radius, that is, the satellite speeds up. This apparent anomaly is explained by the fact that the potential energy decreases by twice as much as the kinetic energy increases, from (2). Thus on the whole there *is* a loss of energy, as we expect.

Example

A satellite of mass 1000 kg moves in a circular orbit of radius 7000 km round the earth, assumed to be a sphere of radius 6400 km. Calculate the total energy needed to place the satellite in orbit from the earth, assuming $g = 10$ N kg^{-1} at the earth's surface.

To launch the satellite, mass m, from the earth's surface of radius r_E into an orbit of radius r_0,

energy needed W = increase in potential energy and kinetic energy

$$= \frac{GMm}{r_E} - \frac{GMm}{r_0} + \tfrac{1}{2}mv^2$$

$$= \frac{GMm}{r_E} - \frac{GMm}{2r_0}$$

from equation (3) of the previous section. But $GM/r_E{}^2 = g$, or $GM/r_E = gr_E$.

So
$$W = mgr_E - \frac{mgr_E{}^2}{2r_0} = mg\left(r_E - \frac{r_E{}^2}{2r_0}\right)$$

$$= 1000 \times 10\left(6{\cdot}4 \times 10^6 - \frac{6{\cdot}4^2 \times 10^{12}}{2 \times 7 \times 10^6}\right)$$

$$= 3{\cdot}5 \times 10^{10} \text{ J}.$$

Exercises 2B

(Assume $g = 10$ N kg^{-1} unless otherwise stated)

Gravitation

1. The gravitational force on a mass of 1 kg at the earth's surface is 10 N. Assuming the earth is a sphere of radius R, calculate the gravitational force on a satellite of mass 100 kg in a circular orbit of radius $2R$ from the centre of the earth.

2. Assuming the earth is a uniform sphere of mass M and radius R, show that the acceleration of free fall at the earth's surface is given by $g = GM/R^2$.

What is the acceleration of a satellite moving in a circular orbit round the earth of radius $2R$?

3. A planet of mass m moves round the sun of mass M in a circular orbit of radius r with an angular velocity ω. Show (i) that ω is independent of the mass m of the planet, (ii) that in a circular orbit of radius $4r$ round the sun, the angular velocity decreases to $\omega/8$.

4. Obtain the dimensions of G.

The period of vibration T of a star under its own gravitational attraction is given by $T = 2\pi/\sqrt{G\rho}$, where ρ is the mean density of the star. Show that this relation is dimensionally correct.

5. A satellite X moves round the earth in a circular orbit of radius R. Another satellite Y of the same mass moves round the earth in a circular orbit of radius $4R$. Show that (i) the speed of X is twice that of Y, (ii) the kinetic energy of X is greater than that of Y, (iii) the potential energy of X is less than that of Y.

Has X or Y the greater total energy (kinetic plus potential energy)?

6. Find the period of revolution of a satellite moving in a circular orbit round the earth at a height of $3{\cdot}6 \times 10^6$ m above the earth's surface. Assume the earth is a uniform sphere of radius $6{\cdot}4 \times 10^6$ m, the earth's mass is 6×10^{24} kg and G is $6{\cdot}7 \times 10^{-11}$ N m^2 kg^{-1}.

7. If the acceleration of free fall at the earth's surface is 9·8 m s^{-2}, and the radius of the earth is 6400 km, calculate a value for the mass of the earth. ($G = 6{\cdot}7 \times 10^{-11}$ N m^2 kg^{-2}). Give the theory.

8. Assuming the mean density of the earth is 5500 kg m^{-3}, that G is $6{\cdot}7 \times 10^{-11}$ N m^2 kg^{-2}, and that the earth's radius is 6400 km, find a value for the acceleration of free fall at the earth's surface. Derive the formula used.

9. Two binary stars, masses 10^{20} kg and 2×10^{20} kg respectively, rotate about their common centre of mass with an angular velocity ω. Assuming that the only force on a star is the mutual gravitational force between them, calculate ω. Assume that the distance between the stars is 10^6 km and that G is 6.7×10^{-11} N m^2 kg^{-2}.

10. A preliminary stage of spacecraft *Apollo 11*'s journey to the moon was to place it in an earth parking orbit. This orbit was circular, maintaining an almost constant distance of 189 km from the earth's surface. Assuming the gravitational field strength in this orbit is 9.4 N kg^{-1}, calculate (*a*) the speed of the spacecraft in this orbit and (*b*) the time to complete one orbit. (Radius of the earth = 6370 km.) (*L*.)

11. Explorer 38, a radio-astronomy research satellite of mass 200 kg, circles the earth in an orbit of average radius $3R/2$ where R is the radius of the earth. Assuming the gravitational pull on a mass of 1 kg at the earth's surface to be 10 N, calculate the pull on the satellite. (*L*.)

12. The earth is elliptical with polar and equatorial radii equal to 6.357×10^6 m and 6.378×10^6 m respectively. Determine the difference, Δg, in the values of the acceleration of free fall at a pole and at the equator due to this difference in radii. What other factors affect Δg? (Mass of earth = 5.957×10^{24} kg. Gravitational constant, $G = 6.670 \times 10^{-11}$ N m^2 kg^{-2}.) (*L*.)

13. (*a*) What simple assumptions are made and what additional data are required to deduce from the gravitational constant G an estimate for the mass of the earth? Show how the mass is calculated from the data.

(*b*) What are the dimensions of G? A large fluid star oscillates in shape under the influence of its own gravitational attraction. Use the method of dimensions to show how the period of oscillation T is related to the mean radius R, the mean density D of the fluid, and G.

(*c*) Two uniform spheres revolve in space about their common centre of gravity under the influence of their mutual gravitational attraction. Find an expression for the period T in terms of the masses M_1 and M_2 of the spheres, the distance D between their centres, and G. (*O. & C.*)

14. Explain what is meant by the *constant of gravitation*. Describe a laboratory experiment to determine it, showing how the result is obtained from the observations.

A proposed communication satellite would revolve round the earth in a circular orbit in the equatorial plane, at a height of 35 880 km above the earth's surface. Find the period of revolution of the satellite in hours, and comment on the result. (Radius of earth = 6370 km, mass of earth = 5.98×10^{24} kg, constant of gravitation = 6.66×10^{-11} N m^2 kg^{-2}.) (*N*.)

15. Assuming that the planets are moving in circular orbits, apply Kepler's laws to show that the acceleration of a planet is inversely proportional to the square of its distance from the sun. Explain the significance of this and show clearly how it leads to Newton's law of universal gravitation.

Obtain the value of g from the motion of the moon, assuming that its period of rotation round the earth is 27 days 8 hours and that the radius of its orbit is 60.1 times the radius of the earth. (Radius of earth = 6.36×10^6 m.) (*N*.)

16. (*a*) Describe how the SI unit of force is defined from Newton's Laws of Motion. Why is it necessary to introduce the dimensional constant G in Newton's Law of Gravitation? Find the dimensions of G in terms of mass M, length L and time T.

(*b*) Derive an expression for the acceleration g due to gravity on the surface of the earth in terms of G, the radius of the earth R and its density ρ.

The maximum vertical distance through which a fully-dressed astronaut can jump on the earth is 0.5 m. Estimate the maximum vertical distance through which he can jump on the moon, which has a mean density two-thirds that of the earth and a radius one-quarter that of the earth, stating any assumptions made. Determine the ratio of the time duration of his jump on the moon to that of his jump on the earth. (*O. & C.*)

17. Explain what is meant by the universal constant G.

Derive the relationship between G and the acceleration of free fall, g, at the surface of the earth (neglecting rotation of the earth and assuming that it is spherical).

Explain why the rotation of the earth about its axis affects the value of g at the equator.

Calculate the percentage change in g between the poles and the equator (again assuming that the earth is spherical).

The orbit of the moon is approximately a circle of radius 60 times the equatorial radius of the earth. Calculate the time taken for the moon to complete one orbit, neglecting the rotation of the earth.

(Acceleration of free fall at the poles of the earth = 9·8 m s^{-2}. Equatorial radius of the earth = 6·4 × 10^6 m. 1 day = 8·6 × 10^4 seconds.) (*L.*)

18. How do you account for the sensation of 'weightlessness' experienced by the occupant of a space capsule (*a*) in a circular orbit round the earth, (*b*) in outer space? Give one other instance in which an object would be 'weightless'. (*N.*)

19. Describe, briefly, a method for the measurement of the gravitational constant *G*.

(*a*) Express the acceleration due to gravity, *g*, at the surface of the earth, in terms of *G*, the mass *M* of the earth and the radius *R* of the earth, assuming the earth is a uniform sphere. The effect of the earth's rotation may be neglected.

(*b*) Express the period *T* of a satellite in a circular orbit round the earth, in terms of the radius *r* of the orbit, *g* at the surface of the earth and *R*.

(*c*) For communication purposes it is desirable to have a satellite which stays vertically above one point on the earth's surface. Explain why the orbit of such a satellite (i) must be circular, and (ii) must lie in the plane of the equator. Find the radius of this orbit. (Radius of earth = 6400 km.) (*O. & C.*)

20. What do you understand by the *intensity of gravity* (*gravitational field strength*) and the *gravitational potential* at a point in the earth's gravitational field? How are they related?

Taking the earth to be uniform sphere of radius 6400 km, and the value of *g* at the surface to be 10 m s^{-2}, calculate the total energy needed to raise a satellite of mass 2000 kg to a height of 800 km above the ground and to set it into circular orbit at that altitude.

Explain briefly how the satellite is set into orbit once the intended altitude has been reached, and also what would happen if this procedure failed to come into action. (*O.*)

Simple Harmonic Motion

When the bob of a pendulum moves to-and-fro through a small angle, the bob is said to be moving with *simple harmonic motion*. The prongs of a sounding tuning fork, and the layers of air near it, are moving to-and-fro with simple harmonic motion. Light waves can be considered due to simple harmonic variations of electric and magnetic forces.

Simple harmonic motion is closely associated with circular motion. An example is shown in Fig. 2.16. This illustrates an arrangement used to convert the

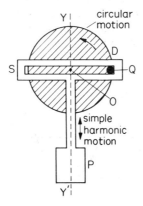

Fig. 2.16 Simple harmonic motion

circular motion of a disc D into the to-and-fro or simple harmonic motion of a piston P. The disc is driven about its axle O by a peg Q fixed near its rim. The vertical motion drives P up and down. Any horizontal component of the motion merely causes Q to move along the slot S. Thus the simple harmonic motion of P is the *projection* on the vertical line YY' of the circular motion of Q.

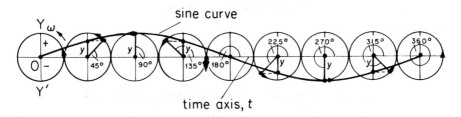

Fig. 2.17 Simple harmonic and circular motion. The diagram shows eight positions of a particle moving round a circle through 360° at constant angular velocity. The distances y from O of the foot of the projection on YOY' all lie on a sine curve as shown

The projection of Q on YY' is the *foot* of the perpendicular from Q to the diameter passing through YY'. Figure 2.17 shows how the distance y from O of the projection varies as Q moves round the circular disc D with constant angular velocity ω. In this rough sketch the horizontal axis represents angle of rotation or time, as the angle turned is proportional to the time. On one side of O, y has positive values; on the other side of O it has negative values. The graph of y against time t is a *simple harmonic* curve or *sine* (*sinusoidal*) *curve* as we see shortly. The maximum value of y is called the *amplitude*. One complete set of values of y is called one *cycle* because the graph repeats itself after one cycle.

Formulae in Simple Harmonic Motion

Consider an object moving round a circle of radius r and centre Z with a uniform angular velocity ω, Fig. 2.18. As we have just seen, if CZF is a fixed diameter, the *foot* of the perpendicular from the moving object to this diameter moves from Z to C, back to Z and across to F, and then returns to Z, while the object

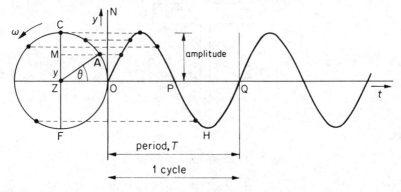

Fig. 2.18 Simple harmonic curve

moves once round the circle from O in an anti-clockwise direction. The to-and-fro motion along CZF of the foot of the perpendicular may be defined as *simple harmonic motion*.

Suppose the object moving round the circle is at A at some instant, where angle OZA $= \theta$, and suppose the foot of the perpendicular from A to CZ is M. The acceleration of the object at A is $\omega^2 r$, and this acceleration is directed along the radius AZ (see p. 40). Hence the acceleration of M towards Z

$$= \omega^2 r \cos AZC = \omega^2 r \sin \theta.$$

But $r \sin \theta = MZ = y$ say.

$$\therefore \text{ acceleration of M towards Z} = \omega^2 y.$$

Now ω^2 is a constant.

$$\therefore \textit{acceleration of M towards Z} \propto \textit{distance of M from Z.}$$

If we wish to express mathematically that the acceleration is always directed towards Z, we must say

$$\text{acceleration towards Z} = -\omega^2 y \qquad . \qquad . \qquad . \qquad (1)$$

The minus indicates, of course, that the object begins to decelerate as it passes the centre, Z, of its motion. As we see later in discussing cases of simple harmonic motion, this is due to an opposing force. If the minus were omitted from equation (1) the latter would imply that the acceleration increases as y increases, and the object would then never return to its original position.

We can now form a definition of simple harmonic motion. It is the motion of a particle *whose acceleration is always* (i) *directed towards a fixed point,* (ii) *directly proportional to its distance from that point.*

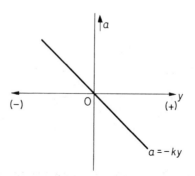

Fig. 2.19 Graph of acceleration against displacement

The straight-line graph in Fig. 2.19 shows how the acceleration a varies with displacement y from a fixed point for simple harmonic motion. The line has a *negative* gradient, since $a = -ky$, where k is a positive constant.

Period, Amplitude. Sine Curve

The time taken for the foot of the perpendicular to move from C to F and back to C is known as the *period* (T) of the simple harmonic motion. In this time, the object moving round the circle goes exactly once round the circle from C; and since ω is the angular velocity and 2π radians (360°) is the angle described, the period T is given by

$$T = \frac{2\pi}{\omega} \qquad . \qquad . \qquad . \qquad . \qquad . \qquad (1)$$

The distance ZC, or ZF, is the maximum distance from Z of the foot of the perpendicular, and is known as the *amplitude* of the motion. It is equal to r, the radius of the circle.

We have now to consider the variation with time, t, of the distance, y, from Z of the foot of the perpendicular. The distance $y = \text{ZM} = r \sin \theta$. But $\theta = \omega t$, where ω is the angular velocity.

$$\therefore \ y = r \sin \omega t \qquad . \qquad . \qquad . \qquad . \qquad (2)$$

The graph of y against t is shown in Fig. 2.20; ON represents the y-axis and OQ the t-axis. Since the angular velocity of the object moving round the circle is constant, θ is proportional to the time t. So at X, the angle θ or ωt is equal to 90° or $\pi/4$ in radians; at P, the angle θ is 180° or π radians; and at Q, the angle θ is 360° or 2π radians. The simple harmonic graph is therefore a sine (sinusoidal) curve. See also Fig. 2.18.

A cosine curve such as $y = r \cos \omega t$, has the same waveform as a sine curve. So this also represents simple harmonic motion. But as $y = r$ when θ or ωt is

zero, the cosine curve starts at a maximum value instead of zero as in a sine
curve.

The complete set of values of y from O to Q is known as a cycle. The number
of cycles per second is called the *frequency*. The unit '1 cycle per second'
is called '1 *hertz* (*Hz*)'. The mains frequency in Great Britain is 50 Hz or 50
cycles per second.

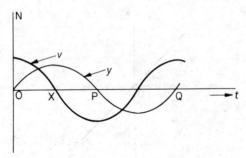

Fig. 2.20 Graph of velocity v against time t

Velocity in Simple Harmonic Motion

If y is the displacement at an instant, then the velocity v at this instant is dy/dt,
the rate of change of displacement (p. 4). Now $y = r \sin \omega t$. Figure 2.20. To
find v or dy/dt from this graph we take the *gradient* value of the curve at the
time t considered. Figure 2.20 shows how v varies with time, t.

The velocity–time (v–t) graph is a cosine curve. At $t = 0$, v has a maximum
value. So $v = A \cos \omega t$ where A is the amplitude or maximum value of y. Now
A is the gradient of the y–t graph at $t = 0$. We can see by drawing different
graphs of y against t that A depends on both r, the maximum value of
y, and ω, the angular velocity (or number of cycles per second). Since
$dy/dt = \omega r \cos \omega t = v$, we see that $A = \omega r$. So the velocity v is given by

$$v = \omega r \cos \omega t. \qquad . \qquad . \qquad . \qquad (1)$$

We can also express the velocity v in terms of y and r. From $y = r \sin \omega t$
and $v = r\omega \cos \omega t$, we have $\sin \omega t = y/r$ and $\cos \omega t = v/r\omega$. Now $\sin^2 \omega t +$
$\cos^2 \omega t = 1$, from trigonometry. So

$$\frac{v^2}{r^2\omega^2} + \frac{y^2}{r^2} = 1$$

Simplifying $\qquad\qquad\qquad v = \omega\sqrt{r^2 - y^2} \qquad . \qquad . \qquad . \qquad . \qquad (2)$

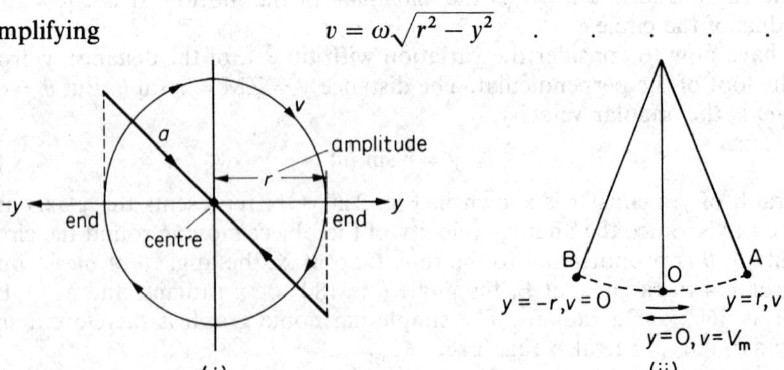

Fig. 2.21 Simple harmonic motion

Figure 2.21 (i) shows the variation of v with *displacement* y. It is an ellipse. We can understand why this graph is obtained by considering the motion of a bob at the end of an oscillating simple pendulum, Fig. 2.21 (ii). At the centre O ($y = 0$), the velocity v is a maximum. At the end A of the oscillation ($y = r$), $v = 0$. At the other end B ($y = -r$), $v = 0$. Note that v has an opposite direction on each half of the cycle. From (2), it follows that the maximum velocity v_m, when $y = 0$, is given by

$$v_m = \omega r \qquad . \qquad . \qquad . \qquad . \qquad . \qquad (3)$$

S.H.M. Equations—Alternative Derivation

As we shall now show, all the equations used in s.h.m. can be derived by calculus without using the circle. With the usual notation,

$$\text{acceleration, } a = \frac{dv}{dt} = \frac{dy}{dt} \cdot \frac{dv}{dy} = v \frac{dv}{dy}$$

Now by definition of s.h.m., $a = -\omega^2 y$ (p. 62).

$$\therefore v \frac{dv}{dy} = -\omega^2 y$$

Integrating,
$$\therefore \frac{v^2}{2} = -\omega^2 \frac{y^2}{2} + c \qquad . \qquad . \qquad . \qquad (1)$$

where c is a constant. Now $v = 0$ when $y = r$, the amplitude. So $c = \omega^2 r^2 / 2$, from (1). Substituting for c in (1) and simplifying,

$$\therefore v = \omega \sqrt{r^2 - y^2}$$

$$\therefore \frac{dy}{dt} = \omega \sqrt{r^2 - y^2} \qquad . \qquad . \qquad . \qquad . \qquad (2)$$

$$\therefore \frac{1}{\omega} \int \frac{dy}{\sqrt{r^2 - y^2}} = \int dt + C$$

$$\therefore \frac{1}{\omega} \sin^{-1} \left(\frac{y}{r} \right) = t + C \qquad . \qquad . \qquad . \qquad (3)$$

When $t = 0$, then $y = 0$; so $C = 0$, from (3).

$$\therefore \frac{1}{\omega} \sin^{-1} \left(\frac{y}{r} \right) = t$$

$$\therefore y = r \sin \omega t \qquad . \qquad . \qquad . \qquad . \qquad (4)$$

When t increases to $t + 2\pi/\omega$, $y = r \sin (\omega t + 2\pi) = r \sin \omega t$, which is the same displacement value as at t. Hence the *period* T of the motion $= 2\pi/\omega$.

Summarising our Results:

(1) If the acceleration a of an object $= -\omega^2 y$, where y is the distance or displacement of the object from a fixed point, the motion is simple harmonic motion. The graph of a against y is a straight line through the origin with a negative gradient.

(2) The *period*, T, of the motion $= 2\pi/\omega$, where T is the time to make a complete to-and-fro movement or cycle. The *frequency*, f, $= 1/T$ and its unit is 'Hz'. Note that $\omega = 2\pi/T = 2\pi f$.

(3) The amplitude, r, of the motion is the maximum distance on either side of the centre of oscillation.

(4) The velocity at any instant, $v, = \omega\sqrt{r^2 - y^2}$; the maximum velocity $= \omega r$. The graph of the variation of v with displacement y is an ellipse.

Example

A steel strip, clamped at one end, vibrates with a frequency of 20 Hz and an amplitude of 5 mm at the free end, where a small mass of 2 g is positioned. Find (a) the velocity of the end when passing through the zero position, (b) the acceleration at maximum displacement, (c) the maximum kinetic and potential energy of the mass.

Suppose $y = r \sin \omega t$ represents the vibration of the strip where r is the amplitude.

(a) The velocity, $v, = \omega\sqrt{r^2 - y^2}$ (p. 64). When the end of the strip passes through the zero position $y = 0$; and the maximum speed, v , is given by

$$v_m = \omega r.$$

Now $\omega = 2\pi f = 2\pi \times 20$, and $r = 0{\cdot}005$ m.

$$\therefore v_m = 2\pi \times 20 \times 0{\cdot}005 = 0{\cdot}628 \text{ m s}^{-1}.$$

(b) The acceleration $= -\omega^2 y = -\omega^2 r$ at the maximum displacement.

$$\therefore \text{ acceleration} = (2\pi \times 20)^2 \times 0{\cdot}005$$
$$= 79 \text{ m s}^{-2}.$$

(c) $m = 2$ g $= 2 \times 10^{-3}$ kg, $v_m = 0{\cdot}628$ m s^{-1}.

$$\therefore \text{ maximum k.e.} = \tfrac{1}{2}mv_m{}^2 = \tfrac{1}{2} \times (2 \times 10^{-3}) \times 0{\cdot}628^2 = 3{\cdot}9 \times 10^{-4} \text{ J (approx.).}$$

$$\text{Maximum p.e. } (v = 0) = \text{maximum k.e.} = 3{\cdot}9 \times 10^{-4} \text{ J.}$$

S.H.M. and g

If a small coin is placed on a horizontal platform connected to a vibrator, and the amplitude is kept constant as the frequency is increased from zero, the coin will be heard 'chattering' at a particular frequency f_0. At this stage the reaction of the table with the coin becomes zero at some part of every cycle, so that it loses contact periodically with the surface, Fig. 2.22.

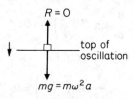

Fig. 2.22

The maximum acceleration in s.h.m. occurs at the end of the oscillation because the acceleration is directly proportional to the displacement. Thus maximum acceleration $= \omega^2 a$, where a is the amplitude and ω is $2\pi f_0$.

The coin will lose contact with the table when it is moving *down* with acceleration g, Fig. 2.22. Suppose the amplitude a is $0{\cdot}08$ m. Then

$$(2\pi f_0)^2 a = g$$

$$\therefore 4\pi^2 f_0{}^2 \times 0{\cdot}08 = 9{\cdot}8$$

$$\therefore f_0 = \sqrt{\frac{9{\cdot}8}{4\pi^2 \times 0{\cdot}08}} = 1{\cdot}8 \text{ Hz.}$$

Oscillating System—Spring and Mass

Suppose that one end of a spring S of negligible mass is attached to a smooth object A, and that S and A are laid on a horizontal smooth table, Fig. 2.23. If the free end of S is attached to the table and A is pulled slightly to extend the spring and then released, the system vibrates with simple harmonic motion. The centre of oscillation O is the position of A at the end of the spring corresponding to its natural length, that is, when the spring is neither extended nor compressed.

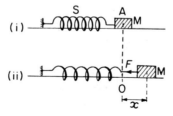

Fig. 2.23 Oscillating spring and mass

Suppose the extension x of the spring is directly proportional to the force F in the spring (Hooke's law). F acts in the opposite direction to x, so $F = -kx$, where k is known as the *force constant* of the spring or 'force per unit extension'. If m is the mass of A, the acceleration a is given by $F = ma$. So

$$ma = -kx$$

Thus

$$a = -\frac{k}{m}x = -\omega^2 x,$$

where $\omega^2 = k/m$. So the motion of A is simple harmonic and the period T is given by

$$T = \frac{2\pi}{\omega} = 2\pi\sqrt{\frac{m}{k}}.$$

Potential and Kinetic Energy Exchanges in Oscillating Systems

The energy of the stretched spring is *potential energy*, p.e.—its molecules are continually displaced or compressed relative to their normal distance apart. The p.e. for an extension $x = \int F.dx = \int kx.dx = \frac{1}{2}kx^2$.

The energy of the mass is *kinetic energy*, k.e., or $\frac{1}{2}mv^2$, where v is the velocity. Now from $x = r \sin \omega t$, $v = dx/dt = \omega r \cos \omega t$

$\therefore$ total energy of spring plus mass $= \frac{1}{2}kx^2 + \frac{1}{2}mv^2$

$$= \frac{1}{2}kr^2 \sin^2 \omega t + \frac{1}{2}m\omega^2 r^2 \cos^2 \omega t.$$

But $\omega^2 = k/m$, or $k = m\omega^2$.

$\therefore$ total energy $= \frac{1}{2}m\omega^2 r^2 (\sin^2 \omega t + \cos^2 \omega t) = \frac{1}{2}m\omega^2 r^2 = constant.$

Thus the total energy of the vibrating mass and spring is constant. When the k.e. of the mass is a maximum (energy $= \frac{1}{2}m\omega^2 r^2$ and mass passing through the centre of oscillation), the p.e. of the spring is then zero ($x = 0$). Conversely, when the p.e. of the spring is a maximum (energy $= \frac{1}{2}kr^2 = \frac{1}{2}m\omega^2 r^2$ and mass at the end of the oscillation), the k.e. of the mass is zero ($v = 0$). Figure 2.24 shows the variation of p.e. and k.e. with displacement x; the force F extending the spring, also shown, is directly proportional to the displacement from the centre of oscillation.

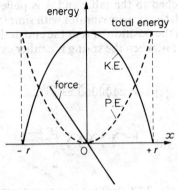

Fig. 2.24 Energy in S.H.M.

The constant interchange of energy between potential and kinetic energies is essential for producing and maintaining oscillations, whatever their nature. In the case of the oscillating bob of a simple pendulum, for example, the bob loses kinetic energy after passing through the middle of the swing, and then stores the energy as potential energy as it rises to the top of the swing. The reverse occurs as it swings back. In the case of oscillating layers of air when a sound wave passes, kinetic energy of the moving air molecules is converted to potential energy when the air is compressed. In the case of electrical oscillations, a coil L and a capacitor C in the circuit constantly exchange energy; this is stored alternately in the magnetic field of L and the electric field of C.

Oscillation of Mass Suspended from Helical Spring

Consider a helical spring or an elastic thread PA suspended from a fixed point P, Fig. 2.25. When a mass m is placed on it, the spring stretches to O by a length e given by

$$mg = ke, \qquad \qquad \qquad \qquad \text{(i)}$$

where k is the force constant (force per unit extension) of the spring, since the tension in the spring is then mg. If the mass is pulled down a little and then

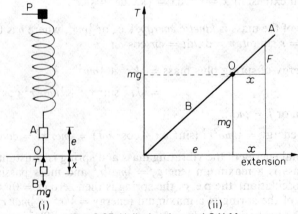

Fig. 2.25 Helical spring and S.H.M.

released, it vibrates up-and-down above and below O. Suppose at an instant that B is at a distance x below O. The tension T of the spring at B is then equal to $k(e + x)$. Hence the resultant force F towards $O = k(e + x) - mg = ke - kx - mg = -kx$, since $ke = mg$ from (i). From $F = ma$,

$$\therefore \quad -kx = ma,$$

$$\therefore \quad a = -\frac{k}{m}x = -\omega^2 x,$$

where $\omega^2 = k/m$. Thus the motion is simple harmonic about O, and the period T is given by

$$T = \frac{2\pi}{\omega} = 2\pi\sqrt{\frac{m}{k}} \quad . \qquad . \qquad . \qquad . \qquad . \qquad (1)$$

Also, since $mg = ke$, it follows that $m/k = e/g$.

$$\therefore \ T = 2\pi\sqrt{\frac{e}{g}} \quad . \qquad . \qquad . \qquad . \qquad . \qquad (2)$$

Figure 2.25(ii) shows the straight-line variation of the tension T in the spring with the extension, assuming Hooke's law (p. 115). The point O on the line corresponds to the extension e when the weight mg is on the spring and $T = mg$. When the mass is pulled down and released as in Fig. 2.25(i), the tension values vary along the straight line AOB. So at a displacement x from O, the *resultant* force $F(T - mg)$ on m is proportional to x. From $F = ma$, the acceleration a of m is proportional to x. So the motion is simple harmonic about O. Also, from Fig. 2.25(ii), $F/x = mg/e$. Hence $F/m = a = gx/e$. So $\omega^2 = g/e$ and the period $= 2\pi/\omega = 2\pi\sqrt{e/g}$, as deduced in (2).

From (1), it follows that $T^2 = 4\pi^2 m/k$. Consequently a graph of T^2 against m should be a straight line through the origin. In practice, when the load m is varied and the corresponding period T is measured, a straight line graph is obtained when T^2 is plotted against m, thus verifying indirectly that the motion of the load was simple harmonic. The graph does not pass through the origin, however, owing to the mass and the movement of the various parts of the spring. This has not been taken into account in the foregoing theory.

From (1), the period of oscillation T depends on the mass m and the force constant k of the spring. Since m and k are constants, it follows that if the same mass and spring are taken to the moon, the period of oscillation would be the same. The period of oscillation T of a simple pendulum of length l would change, however, if it were taken to the moon, as $T = 2\pi\sqrt{l/g}$ and the moon's gravitational intensity is about $g/6$.

Springs in Series and Parallel

Consider a helical spring of force constant k where $F = k \times$ extension. A mass m of weight mg then extends the spring by a length x given by $mg = kx$, and the period of oscillation of the mass is $T = 2\pi\sqrt{m/k}$ as previously obtained.

Suppose two identical helical springs are connected in *series*, each of force constant k, Fig. 2.26 (i). The same weight mg will extend the springs twice as much as for a single spring since the total length is twice as much. So $mg = F = k_1.2x$, where k_1 is the force constant of the two springs in series. But $mg = kx$. So $2k_1 = k$, or $k_1 = k/2$. The period of oscillation of the mass m is now given by

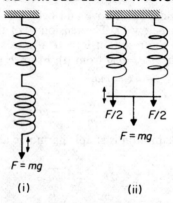

$F = mg$

(i)   (ii)

Fig. 2.26 Springs in series and parallel

$$T_1 = 2\pi\sqrt{\frac{m}{k_1}} = 2\pi\sqrt{\frac{m}{k/2}} = 2\pi\sqrt{\frac{2m}{k}}.$$

So $T_1 = \sqrt{2}T.$

The mass therefore oscillates with a longer period at the end of the two springs.

Now suppose the two springs are placed in *parallel* and the mass m is attached at the middle of a short horizontal connecting bar, Fig. 2.26 (ii). This time the force on each spring is $mg/2$. So the extension is half as much as for a single spring, or $x/2$. The force on the system of parallel springs is mg. So $mg = F = k_2 . x/2$, where k_2 is the force constant of the system. Since $mg = kx$ for a single spring, it follows that $k_2/2 = k$ or $k_2 = 2k$. So the period T_2 of the system is given by

$$T_2 = 2\pi\sqrt{\frac{m}{k_2}} = 2\pi\sqrt{\frac{m}{2k}} = \frac{1}{\sqrt{2}}T.$$

The period of the parallel system is therefore less than for a single spring. Also, from above,

$$\frac{T_1}{T_2} = \frac{\sqrt{2}\,T}{T/\sqrt{2}} = \sqrt{4} = 2.$$

The same results are obtained from the formula $T = 2\pi\sqrt{e/g}$, where e is the spring extension for a mass m.

Example

A small mass of 0·2 kg is attached to one end of a helical spring and produces an extension of 15 mm or 0·015 m. The mass is now set into vertical oscillation of amplitude 10 mm. What is (*a*) the period of oscillation, (*b*) the maximum kinetic energy of the mass, (*c*) the potential energy of the spring when the mass is 5 mm below the centre of oscillation? ($g = 9\cdot8$ m s^{-2})

(*a*) The force constant k of the spring in N m^{-1} is given by

$$k = \frac{mg}{e} = \frac{0\cdot2 \times 9\cdot8}{0\cdot015}.$$

As we have previously shown,

$$T = 2\pi\sqrt{\frac{m}{k}} = 2\pi\sqrt{\frac{0\cdot2 \times 0\cdot015}{0\cdot2 \times 9\cdot8}}$$

$$= 2\pi\sqrt{\frac{0 \cdot 015}{9 \cdot 8}} = 0 \cdot 25 \text{ s.}$$

(b) The maximum k.e. $= \frac{1}{2}mv_m{}^2$, where v_m is the maximum velocity. Now for simple harmonic motion, $v_m = r\omega$ where r = amplitude = 10 mm = 0·01 m. So

$$\text{maximum k.e.} = \frac{1}{2} \times 0 \cdot 2 \times r^2\omega^2$$

$$= \frac{1}{2} \times 0 \cdot 2 \times 0 \cdot 01^2 \times \frac{9 \cdot 8}{0 \cdot 015}$$

$$= 6 \cdot 5 \times 10^{-3} \text{ J.}$$

(c) The potential energy of the spring is given generally by $\frac{1}{2}kx^2$, where k is the force constant and x is the extension from its *original* length. The centre of oscillation is 15 mm below the unstretched length, so 5 mm below the centre of oscillation corresponds to an extension x of 20 mm or 0·02 m. Also, from p. 70,

$$k = mg/0 \cdot 015.$$

So
$$\text{potential energy of spring} = \frac{1}{2}kx^2 = \frac{\frac{1}{2} \times 0 \cdot 2 \times 9 \cdot 8}{0 \cdot 015} \times 0 \cdot 02^2$$

$$= 2 \cdot 6 \times 10^{-2} \text{ J.}$$

Simple Pendulum

We shall now study another case of simple harmonic motion. Consider a *simple pendulum*, which consists of a small mass m attached to the end of a length l of wire, Fig. 2.27. If the other end of the wire is attached to a fixed point P and the mass is displaced slightly, it oscillates to-and-fro along the arc of a circle of centre P. We shall now show that the motion of the mass about its original position O is simple harmonic motion.

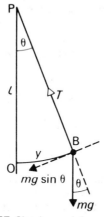

Fig. 2.27 Simple pendulum

Suppose that the vibrating mass is at B at some instant, where OB = y and angle OPB = θ. At B, the force pulling the mass towards O is directed along the tangent at B, and is equal to $mg \sin \theta$. The tension, T, in the wire has no component in this direction, since PB is perpendicular to the tangent at B. Thus, since force = mass × acceleration,

$$-mg \sin \theta = ma,$$

where a is the acceleration along the arc OB; the minus indicates that the force is towards O, while the displacement, y, is measured along the arc from O in the opposite direction. *When θ is small*, $\sin \theta = \theta$ in radians; also $\theta = y/l$. Hence,

$$-mg\theta = -mg\frac{y}{l} = ma$$

$$\therefore a = -\frac{g}{l}y = -\omega^2 y,$$

where $\omega^2 = g/l$. Since the acceleration is proportional to the distance y from a fixed point, the motion of the vibrating mass is simple harmonic motion (p. 63). Further, the period $T = 2\pi/\omega$.

$$\therefore T = \frac{2\pi}{\sqrt{g/l}} = 2\pi\sqrt{\frac{l}{g}} \quad . \quad . \quad . \quad . \quad (1)$$

At a given place on the earth, where g is constant, the formula shows that the period T depends only on the length, l, of the pendulum. Moreover, the period remains constant even when the amplitude of the vibration diminishes owing to the resistance of the air. This result was first obtained by Galileo, who noticed a swinging lantern, and timed the oscillations by his pulse as clocks had not yet been invented. He found that the period remained constant although the swings gradually diminished in amplitude.

On the moon, g is about one-sixth that on the earth. From (1), we see that a pendulum of given length on the moon would have a period over twice as long as on the earth.

Example

A small bob of mass 20 g oscillates as a simple pendulum, with amplitude 5 cm and period 2 seconds. Find the velocity of the bob and the tension in the supporting thread, when the velocity of the bob is a maximum.

The velocity, v, of the bob is a maximum when it passes through its original position. With the usual notation (see p. 66), the maximum velocity v_m is given by

$$v_m = \omega r,$$

where r is the amplitude of 0·05 m. Since $T = 2\pi/\omega$,

$$\therefore \omega = \frac{2\pi}{T} = \frac{2\pi}{2} = \pi \quad . \quad . \quad . \quad . \quad (1)$$

$$\therefore v_m = \omega r = \pi \times 0{\cdot}05 = 0{\cdot}16 \text{ m s}^{-1}.$$

Suppose F is the tension in the thread. The net force towards the centre of the circle along which the bob moves is then given by $(F - mg)$. The acceleration towards the centre of the circle, which is the point of suspension, is v_m^2/l, where l is the length of the pendulum.

$$\therefore F - mg = \frac{mv_m^2}{l}$$

$$\therefore F = mg + \frac{mv_m^2}{l} \quad . \quad . \quad . \quad . \quad (2)$$

From $T = 2\pi\sqrt{l/g}$, $l = gT^2/4\pi^2 = 9{\cdot}8 \times 2^2/4\pi^2$. Also, $m = 20$ g $= 0{\cdot}02$ kg. So, from (2),

$$F = 0{\cdot}02 \times 9{\cdot}8 + \frac{0{\cdot}02 \times (0{\cdot}05\pi)^2 \times \pi^2}{9{\cdot}8}$$

$$= 19{\cdot}65 \times 10^{-2} \text{ N}.$$

Oscillations of a Liquid in a U-Tube

If the liquid on one side of a U-tube T is depressed by blowing gently down that side, the levels of the liquid will oscillate for a short time about their respective initial positions O, C, before finally coming to rest, Fig. 2.28.

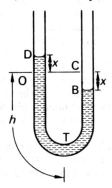

Fig. 2.28 S.H.M. of liquid

At some instant, suppose that the level of the liquid on the left side of T is at D, at a height x above its original (undisturbed) position O. The level B of the liquid on the other side is then at a depth x below its original position C. So the excess pressure on the whole liquid, as shown on p. 102,

$$= \text{excess height} \times \text{liquid density} \times g = 2x\rho g.$$

Since pressure = force per unit area,

force on liquid = pressure × area of cross-section of the tube = $2x\rho g \times A$,

where A is the cross-sectional area of the tube. The mass of liquid in the U-tube = volume × density = $2hA\rho$, where $2h$ is the total length of the liquid in T. So, from $F = ma$ the acceleration, a, towards O or C is given by

$$-2x\rho g A = 2hA\rho a.$$

The minus indicates that the force towards O is opposite to the displacement measured from O at that instant.

$$\therefore a = -\frac{g}{h}x = -\omega^2 x,$$

where $\omega^2 = g/h$. So the motion of the liquid about O (or C) is simple harmonic, and the period T is given by

$$T = \frac{2\pi}{\omega} = 2\pi\sqrt{\frac{h}{g}}.$$

In practice the oscillations are heavily damped owing to friction, which we have ignored.

Exercises 2C

Simple Harmonic Motion

$$(Assume \ g = 10 \ m \ s^{-2} \ or \ 10 \ N \ kg^{-1})$$

1. An object moving with simple harmonic motion has an amplitude of 0·02 m and a frequency of 20 Hz. Calculate (i) the period of oscillation, (ii) the acceleration at the middle and end of an oscillation, (iii) the velocities at the corresponding instants.

2. A body of mass 0·2 kg is executing simple harmonic motion with an amplitude of 20 mm. The maximum force which acts upon it is 0·064 N. Calculate (*a*) its maximum velocity, and (*b*) its period of oscillation. (*L.*)

3. A steel strip, clamped at one end, vibrates with a frequency of 50 Hz and an amplitude of 8 mm at the free end. Find (*a*) the velocity of the end when passing through the zero position, (*b*) the acceleration at the maximum displacement.

Draw a sketch showing how the velocity and the acceleration vary with the displacement of the free end.

4. Some of the following graphs refer to simple harmonic motion, where v is the velocity, a is the acceleration, E_k is the kinetic energy, E is the total energy and x is the displacement from the mean (zero) position. Which graphs are correct?

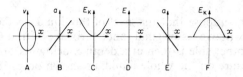

Fig. 2A

5. A spring of force constant k of 5N m^{-1} $(F = -kx)$ is placed horizontally on a smooth table. One end of the spring is fixed and a mass X of 0·20 kg is attached to the free end. X is displaced a distance of 4 mm along the table and then released. Show that the motion of X is simple harmonic, and calculate (i) the period, (ii) the maximum acceleration, (iii) the maximum kinetic energy, (iv) the maximum potential energy of the spring.

6. A simple pendulum has a period of 4·2 s. When the pendulum is shortened by 1 m, the period is 3·7 s. From these measurements, calculate the acceleration of free fall g and the original length of the pendulum.

If the pendulum is taken from the earth to the moon where the acceleration of free fall is $g/6$, what relative change, if any, occurs in the period T?

7. (*a*) Define *simple harmonic motion*. Give *three* examples of systems which vibrate with approximately simple harmonic motion. How does the displacement of a simple harmonic motion vary with time?

What is meant by the *phase difference* between two simple harmonic motions of the same frequency? Illustrate your answer graphically, by considering the variation of the displacement with time of two motions vibrating with simple harmonic motion of the same frequency but which have phase differences of (i) 90° and (ii) 180°.

(*b*) At what points in a simple harmonic motion are (i) the acceleration, (ii) the kinetic energy and (iii) the potential energy of the system each at (1) a maximum, and (2) a minimum?

Sketch the graphs showing how (iv) the kinetic energy, (v) the potential energy, and (vi) the sum of the kinetic and potential energies for a simple harmonic oscillator each vary with displacement.

(*c*) Calculate the period of oscillation of a simple pendulum of length 1·8 m, with a bob of mass 2·2 kg. What assumption is made in this calculation? ($g = 9·8$ m s^{-2})

If the bob of this pendulum is pulled aside a horizontal distance of 20 cm and released, what will be the values of (i) the kinetic energy and (ii) the velocity of the bob at the lowest point of the swing? (*L.*)

8. Define *simple harmonic motion*. State one condition for simple harmonic motion.

A spring is extended 10 mm when a small weight is attached to its free end. The

weight is now pulled down slightly and released. Show that its motion is simple harmonic
and calculate the period.

9. (*a*) The displacement y of a mass vibrating with simple harmonic motion is given by
$y = 20 \sin 10\pi t$, where y is in millimetres and t is in seconds. What is (i) the amplitude,
(ii) the period, (iii) the velocity at $t = 0$?

(*b*) A mass of 0·1 kg oscillates in simple harmonic motion with an amplitude of 0·2 m
and a period of 1·0 s. Calculate its maximum kinetic energy. Draw a sketch showing
how the kinetic energy varies with (i) the displacement, (ii) the time.

10. A mass X of 0·1 kg is attached to the free end of a vertical helical spring whose
upper end is fixed and the spring extends by 0·04 m. X is now pulled down a small
distance 0·02 m and then released. Find (i) its period, (ii) the maximum force acting on
it during the oscillations, (iii) its kinetic energy when X passes through its mean position.

11. The displacement of a particle vibrating with simple harmonic motion of angular
speed ω is given by $y = a \sin \omega t$ is the time. What does a represent? Sketch a graph of
the *velocity* of the particle as a function of time starting from $t = 0$ s.

A particle of mass 0·25 kg vibrates with a period of 2·0 s. If its greatest displacement
is 0·4 m what is its maximum kinetic energy? (*L.*)

12. Explain what is meant by *simple harmonic motion*.

Show that the vertical oscillations of a mass suspended by a light helical spring are
simple harmonic and describe an experiment with the spring to determine the acceleration
due to gravity.

A small mass rests on a horizontal platform which vibrates vertically in simple harmonic
motion with a period of 0·50 second. Find the maximum amplitude of the motion which
will allow the mass to remain in contact with the platform throughout the motion. (*L.*)

13. (*a*) (i) Define *simple harmonic motion*. (ii) Show that the equation $y = a \sin (\omega t + \varepsilon)$
represents such a motion and explain the meaning of the symbols y, a, ω and ε. (iii) Draw
with respect to a common time axis graphs showing the variation with time t the
displacement, velocity and kinetic energy of a heavy particle that is describing such a
motion.

(*b*) When a metal cylinder of mass 0·2 kg is attached to the lower end of a light helical
spring the upper end of which is fixed, the spring extends by 0·16 m. The metal cylinder
is then pulled down a further 0·08 m. (i) Find the force that must be exerted to keep
it there if Hooke's law is obeyed. (ii) The cylinder is then released. Find the period
of vertical oscillations, and the kinetic energy the cylinder possesses when it passes
through its mean position. (*O.*)

14. Give *two* practical examples of oscillatory motion which approximate to simple
harmonic motion. What conditions must be satisfied if the approximations are to be
good ones?

A point mass moves with simple harmonic motion. Draw on the same axes sketch
graphs to show the variation with position of (*a*) the potential energy, (*b*) the kinetic
energy, and (*c*) the total energy of the particle.

A particle rests on a horizontal platform which is moving vertically in simple harmonic
motion with an amplitude of 10 cm. Above a certain frequency, the thrust between the
particle and the platform would become zero at some point in the motion. What is this
frequency, and at what point in the motion does the thrust become zero at this frequency?
(*C.*)

15. Define *simple harmonic motion*. Show that a heavy body supported by a light helical
spring executes simple harmonic motion when displaced vertically from its equilibrium
position by an amount which does not exceed a certain value and then released. How
would you determine experimentally the maximum amplitude for simple harmonic
motion?

A helical spring gives a displacement of 5 cm for a load of 500 g. Find the maximum
displacement produced when a mass of 80 g is dropped from a height of 10 cm on to
a light pan attached to the spring. (*N.*)

16. Define *simple harmonic motion*. Explain what is meant by the *amplitude*, the *period*
and the *phase* of such a motion.

A simple pendulum of length 1·5 m has a bob of mass 2·0 kg. (*a*) State the formula
for the period of small oscillations and evaluate it in this case. (*b*) If, with the string taut,

the bob is pulled aside a horizontal distance of 0·15 m from the mean position and then released from rest, find the kinetic energy and the speed with which it passes through the mean position. (c) After 50 complete swings, the maximum horizontal displacement of the bob has become only 0·10 m. What fraction of the initial energy has been lost? (d) Estimate the maximum horizontal displacement of the bob after a further 50 complete swings. (Take g to be 10 m s^{-2}.) (O.)

17. (a) Define *simple harmonic motion*. (b) A light helical spring, for which the force necessary to produce unit extension is k, hangs vertically from a fixed support and carries a mass M at its lower end. Assuming that Hooke's law is obeyed and that there is no damping, show that if the mass is displaced in a vertical direction from its equilibrium position and released, the subsequent motion is simple harmonic. Derive an expression for the time period in terms of M and k. (c) If $M = 0.30$ kg, $k = 30$ N m^{-1} and the initial displacement of the mass is 0·015 m, calculate (i) the maximum kinetic energy of the mass, (ii) the maximum and minimum values of the tension in the spring during the motion. (d) Sketch graphs showing how (i) the kinetic energy of the mass, (ii) the tension in the spring vary with displacement from the equilibrium position. (e) If the same spring with the same mass attached were taken to the moon, what would be the effect, if any, on the time period of the oscillations? Explain your answer. (N.)

18. (a) Write down, and explain the meaning of, the equation which defines simple harmonic motion. Explain the meanings of the following terms used in connection with such a motion: (i) amplitude, (ii) period, (iii) phase.

(b) A point P moves in a circular path of radius r with a speed v. Show that the motion of the foot of the perpendicular, drawn from P onto the diameter of the circle, is simple harmonic. What are the amplitude and period of the motion?

(c) An object is hung on the free end of a helical spring which is suspended from a rigid support. When displaced vertically the object executes simple harmonic motion. Explain what measurements you would make to obtain a value for the acceleration of free fall. How would you calculate your result from your measurements? (AEB, 1982)

3 Rotational Dynamics

So far in this book we have considered the equations of linear motion and other dynamical formulae associated with a particle. We now consider the dynamics of large rotating objects, which consist of millions of particles, each at different places.

Moment of Inertia, I

Suppose a rigid object is rotating about a fixed axis O, and a particle A of the object makes an angle θ with a fixed line OY in space at some instant, Fig. 3.1. The angular velocity, $d\theta/dt$ or ω, of every particle about O is the same, since we

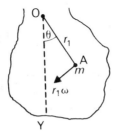

Fig. 3.1 Rotating rigid body

are dealing with a rigid body, and the velocity v_1 of A at this instant is given by $r_1\omega$, where $r_1 = $ OA. Thus the kinetic energy of A $= \frac{1}{2}m_1v_1{}^2 = \frac{1}{2}m_1r_1{}^2\omega^2$. Similarly, the kinetic energy of another particle of the body $= \frac{1}{2}m_2r_2{}^2\omega^2$, where r_2 is its distance from O and m_2 is its mass. In this way we see that the kinetic energy, k.e., of the whole object is given by

$$\text{k.e.} = \frac{1}{2}m_1r_1{}^2\omega^2 + \frac{1}{2}m_2r_2{}^2\omega^2 + \frac{1}{2}m_3r_3{}^2\omega^2 + \ldots$$
$$= \frac{1}{2}\omega^2(m_1r_1{}^2 + m_2r_2{}^2 + m_3r_3{}^2 + \ldots)$$
$$= \frac{1}{2}\omega^2(\Sigma mr^2),$$

where Σmr^2 represents the sum of the magnitudes of 'mr^2' for all the particles of the object. We shall see shortly how the quantity Σmr^2 can be calculated for a particular object. The magnitude of Σmr^2 is known as the *moment of inertia* of the object about the axis concerned, and we shall denote it by the symbol I (see also p. 82). Thus

$$\text{kinetic energy, k.e.} = \frac{1}{2}I\omega^2 \qquad . \qquad . \qquad . \qquad . \qquad (1)$$

The unit of I is kg m^2. The unit of ω is rad s^{-1}. Thus if $I = 2$ kg m^2 and $\omega = 3$ rad s^{-1}, then

$$\text{k.e.} = \frac{1}{2}I\omega^2 = \frac{1}{2} \times 2 \times 3^2 \text{ joule} = 9 \text{ J}.$$

The kinetic energy of a particle of mass m moving with a velocity v is $\frac{1}{2}mv^2$. It will thus be noted that the formula for the kinetic energy of a rotating object is similar to that of a moving particle, the mass m being replaced by the moment of inertia I and the linear velocity v being replaced by the angular velocity ω.

Moment of Inertia of Uniform Rod

(1) *About axis through middle.* The moment of inertia of a small element Δx about an axis PQ through its centre O perpendicular to the length $= \left(\dfrac{\Delta x}{l}M\right)x^2$, where l is the length of the rod, M is its mass, and x is the distance of the small element from O, Fig. 3.2.

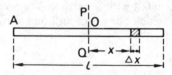

Fig. 3.2 Moment of inertia—uniform rod

$$\therefore \text{ moment of inertia, } I = 2\int_0^{l/2}\left(\frac{dx}{l}M\right)x^2$$

$$= \frac{2M}{l}\int_0^{l/2} x^2\,dx = \frac{Ml^2}{12} \qquad . \qquad (1)$$

Thus if the mass of the rod is 60 g and its length is 20 cm, $M = 6 \times 10^{-2}$ kg, $l = 0.2$ m, and $I = 6 \times 10^{-2} \times 0.2^2/12 = 2 \times 10^{-4}$ kg m^2.

(2) *About the axis through one end, A.* In this case, measuring distances x from A instead of O,

$$\text{moment of inertia, } I = \int_0^l\left(\frac{dx}{l}M\right)\times x^2 = \frac{Ml^2}{3} \qquad . \qquad . \qquad (2)$$

Moment of Inertia of Ring and Other Objects

Every element of a ring is the same distance from the centre. Hence the moment of inertia about an axis through the centre perpendicular to the plane of the ring $= Ma^2$, where M is the mass of the ring and a is its radius.

The moment of inertia of an object depends on the axis of rotation. The values of I for some circular shaped objects are as follows:

(1) *Circular disc.* About an axis through its centre perpendicular to its plane, $I = Ma^2/2$, where M is the mass and a is the radius.

(2) *Cylinder.* About its axis of symmetry, (a) a solid cylinder has a moment of inertia $I = Ma^2/2$, (b) a hollow cylinder open at both ends has a moment of inertia $I = Ma^2$, where M is the mass and a is the radius.

(3) *Sphere.* About a diameter, a solid sphere has a moment of inertia $I = 2Ma^2/5$, where M is the mass and a is the radius.

Radius of Gyration

The moment of inertia of an object about an axis, Σmr^2, is sometimes written as Mk^2, where M is the mass of the object and k is a quantity called the *radius of gyration* about the axis. For example, the moment of inertia of a uniform rod about an axis through one end $= Ml^2/3 = M(l/\sqrt{3})^2$. Thus the radius of gyration, $k = l/\sqrt{3} = 0.58l$. The moment of inertia of a sphere about its centre $= \frac{2}{5}Ma^2 = M \times (\sqrt{\frac{2}{5}}a)^2$. Thus the radius of gyration, $k = \sqrt{\frac{2}{5}}a = 0.63a$ in this case.

Relation between Moment of Inertia about C.G. and Parallel Axis

Suppose I is the moment of inertia of a body about an axis CD and I_G is the moment of inertia about a parallel axis PQ through the centre of gravity, G, distant h from the axis CD, Fig. 3.3. If A is a particle of mass m whose distance from PQ is x, its moment of inertia about CD $= m(h - x)^2$

$$\therefore I = \Sigma m(h - x)^2 = \Sigma mh^2 + \Sigma mx^2 - \Sigma 2mhx.$$

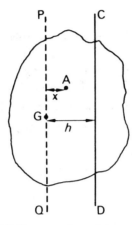

Fig. 3.3 Theorem of parallel axes

Now $\Sigma mh^2 = h^2 \times \Sigma m = Mh^2$, where M is the total mass of the object, and $\Sigma mx^2 = I_G$, the moment of inertia through the centre of gravity.

Also, $$\Sigma 2mhx = 2h\Sigma mx = 0,$$

since Σmx, the sum of the moments about the centre of gravity, is zero; this follows because the moment of the resultant (the weight) about G is zero.

$$\therefore I = I_G + Mh^2 \qquad . \qquad . \qquad . \qquad . \qquad (1)$$

From this result, it follows that the moment of inertia, I, of a disc of radius a and mass M about an axis perpendicular to its plane through a point on its circumference $= I_G + Ma^2$, since $h = a =$ radius of disc in this case. But $I_G =$ moment of inertia about the centre $= Ma^2/2$ (p. 78).

$$\therefore \text{ moment of inertia, } I = \frac{Ma^2}{2} + Ma^2 = \frac{3Ma^2}{2}.$$

Relation between Moments of Inertia about Perpendicular Axes

Suppose OX, OY are any two perpendicular axes and OZ is an axis perpendicular to OX and OY, Fig. 3.4 (i). The moment of inertia, I, of a lamina in the plane YOX about the axis OZ $= \Sigma mr^2$, where r is the distance of a particle A from OZ and m is its mass. But $r^2 = x^2 + y^2$, where x, y are the distances of A from the axis OY, OX respectively.

$$\therefore I = \Sigma m(x^2 + y^2) = \Sigma mx^2 + \Sigma my^2$$

$$\therefore I = I_y + I_x \qquad . \qquad . \qquad . \qquad . \qquad . \qquad . \qquad (1)$$

where I_y, I_x are the moments of inertia about OX, OY respectively.

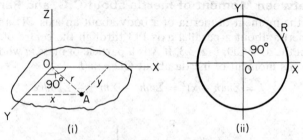

Fig. 3.4 Theorem of perpendicular axes

As a simple application, consider a ring R and two perpendicular axes OX, OY in its plane, Fig. 3.4 (ii). Then from the above result,

$$I_y + I_x = I = \text{moment of inertia through O perpendicular to ring.}$$

$$\therefore I_y + I_x = Ma^2.$$

But $I_y = I_x$, by symmetry.

$$\therefore I_x + I_x = Ma^2,$$

$$\therefore I_x = \frac{Ma^2}{2}.$$

This is the moment of inertia of the ring about any diameter in its plane.

Torque on Rotating Body

Consider a rigid body rotating about a fixed axis O with an angular velocity ω at some instant. Fig. 3.5.

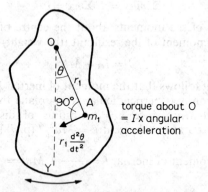

torque about O
= I x angular
acceleration

Fig. 3.5 Torque of rotating rigid body

The force acting on the particle $A = m_1 \times \text{acceleration} = m_1 \times \dfrac{d}{dt}(r_1\omega) =$

$m_1 \times r_1\dfrac{d\omega}{dt} = m_1 r_1\dfrac{d^2\theta}{dt^2}$, since $\omega = \dfrac{d\theta}{dt}$. The moment of this force about the axis

$O = \text{force} \times \text{perpendicular distance from O} = m_1 r_1\dfrac{d^2\theta}{dt^2} \times r_1$, since the force

acts perpendicularly to the line OA.

$$\therefore \text{ moment or } torque = m_1 r_1{}^2 \frac{d^2\theta}{dt^2}.$$

$\therefore$ total moment of all forces on body about O, or *total torque*,

$$= m_1 r_1{}^2 \frac{d^2\theta}{dt^2} + m_2 r_2{}^2 \frac{d^2\theta}{dt^2} + m_3 r_3{}^2 \frac{d^2\theta}{dt^2} + \ldots$$

$$= (\Sigma m r^2) \times \frac{d^2\theta}{dt^2},$$

since the angular acceleration, $d^2\theta/dt^2$, about O is the same for all particles.

$$\therefore \text{ total torque about O, } T = I \frac{d^2\theta}{dt^2}, \qquad . \qquad . \qquad . \qquad (1)$$

where $I = \Sigma m r^2$ = moment of inertia about O. In general, for any rotating rigid body,

$$Torque, \ T = I \frac{d^2\theta}{dt^2} = I\alpha,$$

where α is the *angular acceleration*, $d^2\theta/dt^2$ or $d\omega/dt$.

Torque and Angular Acceleration

A *force F* gives a mass m a linear acceleration a, and $F = ma$. In contrast, a *torque T* produces a rotation of a body on which it acts and gives it an angular acceleration discussed shortly. A torque is the turning-effect or moment of forces. For example, if F is the tension in a belt round a wheel of radius r which can turn about its centre, Fig. 3.6 (i), the torque T on the wheel $= F \times r$. If a large weight Mg is hung from one end of a rope round a pulley, and a small weight mg is hung from the other end, Fig. 3.6 (ii), then

$$\text{torque } T \text{ on pulley} = Mgr - mgr.$$

The unit of torque is N m (newton metre).

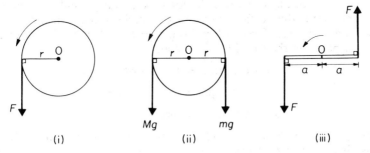

Fig. 3.6 Examples of torque

In a moving-coil ammeter, a torque is produced on the coil by two equal and opposite separated forces F, Fig. 3.6 (iii). These two forces are together called a *couple*. Their torque about O, an axis through the middle of the coil, is given by

$$T = F.a + F.a = F \times 2a$$

$$= F \times \text{ perpendicular distance between forces.}$$

A torque T acting on an object produces an *angular acceleration* α given by

$$T = I\alpha,$$

where I is the moment of inertia about the axis of rotation, as we have just shown on p. 81.

This result is analogous to the case of a particle of mass m which has an acceleration a when a force F acts on it. Here $F = ma$. In place of F we have a *torque* T for a rotating rigid object; in place of m we have the *moment of inertia* I; and in place of linear acceleration a, we have *angular acceleration* α. Thus the moment of inertia I may also be defined as the ratio *torque/angular acceleration* or T/α.

Angular velocity ω has a unit 'rad s^{-1}'. Angular acceleration α is the rate of change of ω, or $d\omega/dt$, and has a unit 'rad s^{-2}'. Suppose an object is rotating with an angular velocity ω_0 and a torque then gives it an angular acceleration α. In time t, the new angular velocity is given by

$$\omega = \omega_0 + \alpha t.$$

This is analogous to the linear motion formula $v = u + at$.

As an illustration, suppose a wheel has a moment of inertia I of 2 kg m^2 about its centre and is rotating about the centre with an angular velocity of 1 rad s^{-1}. If a torque T of 1 N m acts on the wheel and gives it an angular acceleration α about the centre, then, from $T = I\alpha$,

$$\alpha = \frac{T}{I} = \frac{1}{2} = 0{\cdot}5 \text{ rad s}^{-2}.$$

So 4 s later the angular velocity has increased to

$$\omega = \omega_0 + \alpha t = 1 + 0{\cdot}5 \times 4 = 3 \text{ rad s}^{-1}.$$

If a rotating wheel is slowed down by an applied torque, then α is an angular *deceleration* and is given a negative value in $\omega = \omega_0 + \alpha t$.

Work Done by a Couple

Suppose two equal and opposite forces F act tangentially to a wheel W, and rotate it through an angle θ while the forces keep tangentially to the wheel, Fig. 3.7. The torque due to the two forces or couple is then constant.

The work done by each force $= F \times$ distance $= F \times r\theta$, since $r\theta$ is the distance moved by a point on the rim if θ is in radians.

$$\therefore \text{ total work done} = Fr\theta + Fr\theta = 2Fr\theta.$$

But   torque $= F \times 2r = 2Fr$

$$\therefore \text{ work done by couple} = \text{torque} \times \theta.$$

Although we have chosen a simple case, the result for the work done by a couple is always given by *torque* $\times$ *angle of rotation*. In the formula, it should be carefully noted that θ is in radians. Thus suppose $F = 20$ N, $r = 4$ cm $= 0{\cdot}04$ metre, and the wheel makes 5 revolutions while the torque is kept constant. Then, from above,

$$\text{torque} = 20 \times 0{\cdot}08 \text{ N m,}$$

and   angle of rotation $= 2\pi \times 5$ radian.

$$\therefore \text{ work done} = 20 \times 0{\cdot}08 \times 2\pi \times 5 = 50 \text{ J (approx.)}$$

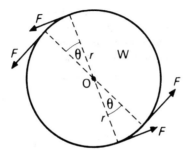

Fig. 3.7 Work done by couple

Examples

1. A heavy flywheel of moment of inertia 0·3 kg m² is mounted on a horizontal axle of radius 0·01 m and negligible mass compared with the flywheel. Neglecting friction, find (i) the angular acceleration if a force of 40 N is applied tangentially to the axle, (ii) the angular velocity of the flywheel after 10 seconds from rest.

(i) Torque $T = 40$ (N) $\times 0·01$ (m) $= 0·4$ N m.

From
$$T = I\alpha,$$

$$\text{angular acceleration } \alpha = \frac{T}{I} = \frac{0·4}{0·3} = 1·3 \text{ rad s}^{-2}.$$

(ii) After 10 seconds, angular velocity $\omega = \alpha t$

$$= 1·3 \times 10 = 13 \text{ rad s}^{-1}.$$

2. The moment of inertia of a solid flywheel about its axis is 0·1 kg m². It is set in rotation by applying a tangential force of 20 N with a rope wound round the circumference, the radius of the wheel being 0·1 m. Calculate the angular acceleration of the flywheel. What would be the angular acceleration if a mass of 2 kg were hung from the end of the rope? (*O. & C.*)

$$\text{Torque } T = I\alpha, \text{ and } T = 20 \times 0·1 \text{ N m.}$$

So
$$\text{angular acceleration } \alpha = \frac{T}{I} = \frac{20 \times 0·1}{0·1} = 20 \text{ rad s}^{-2}.$$

If a mass m of 2 kg, or weight 20 N assuming $g = 10$ m s^{-2}, is hung from the end of the rope, it moves down with an acceleration a, Fig. 3.8. In this case, if F is the tension in the rope,

$$mg - F = ma \qquad . \qquad . \qquad . \qquad . \qquad . \qquad (1)$$

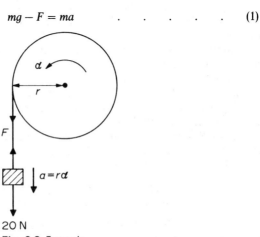

20 N

Fig. 3.8 Example

For the flywheel, $\qquad\qquad\qquad F.r = \text{torque} = I\alpha$ . . . . . (2)

where r is the radius of the wheel and α the angular acceleration about the centre. Now the mass descends a distance given by $r\theta$, where θ is the angle the flywheel has turned. Hence the acceleration $a = r\alpha$. Substituting in (1),

$$\therefore mg - F = mr\alpha$$

Multiplying by r,

$$\therefore mgr - F.r = mr^2\alpha \qquad . \qquad . \qquad . \qquad . \qquad (3)$$

Adding (2) and (3),

$$\therefore mgr = (I + mr^2)\alpha$$

$$\therefore \alpha = \frac{mgr}{I + mr^2} = \frac{2 \times 10 \times 0.1}{0.1 + 2 \times 0.1^2}$$

$$= 16.7 \text{ rad s}^{-2},$$

Angular Momentum and its Conservation

In linear or straight-line motion, an important property of a moving object is its linear momentum (p. 18). When an object spins or rotates about an axis, its *angular momentum* plays an important part in its motion.

Consider a particle A of a rigid object rotating about an axis O, Fig. 3.9 (i). The momentum of A = mass × velocity = $m_1 v = m_1 r_1 \omega$. The 'angular momentum' of A about O is defined as the *moment of the momentum* about O. Its magnitude is thus $m_1 v \times p$, where p is the perpendicular distance from O to the direction of v. Thus angular momentum of A = $m_1 v p = m_1 r_1 \omega \times r_1 = m_1 r_1^2 \omega$.

$$\therefore \text{ total angular momentum of whole body} = \Sigma m_1 r_1^2 \omega = \omega \Sigma m_1 r_1^2$$

$$= I\omega,$$

where I is the moment of inertia of the body about O.

Angular momentum is analogous to 'linear momentum', mv, in the dynamics of a moving particle. In place of m we have I, the moment of inertia; in place of v we have ω, the angular velocity.

Further, the *conservation of angular momentum*, which corresponds to the conservation of linear momentum, states that *the angular momentum about an axis of a given rotating body or system of bodies is constant, if no external torque acts about that axis*. Thus when a high diver jumps from a diving board, his moment of inertia, I, can be decreased by curling his body more, in which case his angular velocity ω is increased, Fig. 3.9 (ii). He may then be able to turn more somersaults before striking the water. Similarly, a dancer on skates can spin faster by folding her arms.

The earth rotates about an axis passing through its geographic north and south poles with a period of 1 day. If it is struck by meteorites, then, since action and reaction are equal, no external couple acts on the earth and meteorites. Their total angular momentum is thus conserved. Neglecting the angular momentum of the meteorites about the earth's axis before collision compared with that of the earth, then

angular momentum of *earth plus meteorites* after collision =
angular momentum of *earth* before collision.

Since the effective mass of the earth has increased after collision the moment of inertia has increased. So the earth will slow down slightly. Similarly, if a

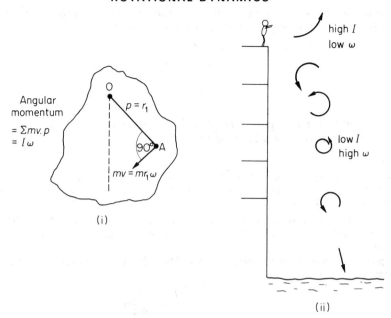

Fig. 3.9 Angular momentum

mass is dropped gently on to a turntable rotating freely at a steady speed, the conservation of angular momentum leads to a reduction in the speed of the table.

Angular momentum, and the principle of the conservation of angular momentum, have wide applications in physics. They are used in connection with enormous rotating masses such as the earth, as well as minute spinning particles such as electrons, neutrons and protons found inside atoms.

Experiment on Conservation of Angular Momentum

A simple experiment on the principle of the conservation of angular momentum is illustrated below.

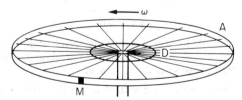

Fig. 3.10 Conservation of angular momentum

Briefly, in Fig. 3.10 a bicycle wheel A, without a tyre, is set rotating in a horizontal plane and the time for three complete revolutions is obtained with the aid of a white tape marker M on the rim. A ring D of known moment of inertia, I, is then gently placed on the wheel concentric with it, by 'dropping' it from a small height. The time for the next three revolutions is then determined. This is repeated with several more rings of greater known moment of inertia.

If the principle of conservation of angular momentum is true, then $I_0\omega_0 = (I_0 + I_1)\omega_1$, where I_0 is the moment of inertia of the wheel alone, ω_0 is the angular frequency of the wheel alone, and ω_1 is the angular frequency with a ring. Thus if t_0, t_1 are the respective times for three revolutions,

$$\frac{I_0 + I_1}{t_1} = \frac{I_0}{t_0}$$

$$\therefore \frac{I_1}{I_0} + 1 = \frac{t_1}{t_0}.$$

Thus a graph of t_1/t_0 against I_1 should be a straight line. Within the limits of experimental error, this is found to be the case.

Examples

1. A ballet dancer spins about a vertical axis at 1 revolution per second with arms outstretched. With her arms folded, her moment of inertia about the vertical axis decreases by 60%. Calculate the new rate of revolution.

Suppose I is the initial moment of inertia about the vertical axis and ω is the initial angular velocity corresponding to 1 rev s^{-1}. The new moment of inertia $I_1 = 40\%$ of $I = 0.4\ I$.

Suppose the new angular velocity is ω_1. Then, from the conservation of angular momentum,

$$I_1\omega_1 = I\omega.$$

So
$$\omega_1 = \frac{I}{I_1}\omega = \frac{1}{0.4}\omega.$$

Since angular velocity $\propto$ number of revs per second, the new number n of revs per second is given by

$$n = \frac{1}{0.4} \times 1 \text{ rev s}^{-1} = 2.5 \text{ rev s}^{-1}$$

2. A disc of moment of inertia 5×10^{-4} kg m² is rotating freely about axis O through its centre at 40 r.p.m., Fig. 3.11. Calculate the new revolutions per minute (r.p.m.) if some wax of mass 0.02 kg is dropped gently on to the disc 0.08 m from its axis.

Initial angular momentum of disc $= I\omega = 5 \times 10^{-4}\omega$,

where ω is the angular velocity corresponding to 40 r.p.m.

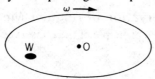

Fig. 3.11 Example

When the wax of mass 0.02 kg is dropped gently on to the disc at a distance r of 0.08 m from the centre O, the disc slows down. Suppose the angular velocity is now ω_1. The total angular momentum about O of disc plus wax

$$= I\omega_1 + mr^2\omega_1 = 5 \times 10^{-4}\omega_1 + 0.02 \times 0.08^2 . \omega_1$$

$$= 6.28 \times 10^{-4}\omega_1.$$

From the conservation of angular momentum for the disc and wax about O

$$6.28 \times 10^{-4}\omega_1 = 5 \times 10^{-4}\omega.$$

$$\therefore \frac{\omega_1}{\omega} = \frac{500}{628} = \frac{n}{40},$$

where n is the r.p.m. of the disc, because the angular velocity is proportional to the r.p.m.

$$\therefore n = \frac{500}{628} \times 40 = 32 \text{ (approx.)}.$$

Kepler's Law and Angular Momentum

Consider a planet moving in an orbit round the sun S, Fig. 3.12. At an instant when the planet is at O, its velocity v is along the tangent to the orbit at O.

Suppose the planet moves a very small distance Δs from O to B in a small time Δt, so that the velocity $v = \Delta s/\Delta t$ and its direction is practically along OB. Then, if the conservation of angular momentum is obeyed,

$$mv \times p = \text{constant},$$

where m is the mass of the planet and p is the perpendicular from S to OB produced.

$$\therefore \ \frac{m.\Delta s.p}{\Delta t} = \text{constant}.$$

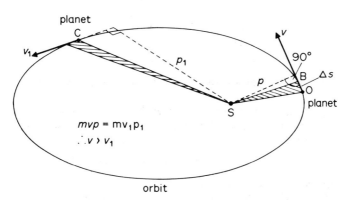

Fig. 3.12 Angular momentum and planetary motion

But the area ΔA of the triangle SBO $= \frac{1}{2}$ base $\times$ height $= \Delta s \times p/2$.

So, from above,
$$m.2\frac{\Delta A}{\Delta t} = \text{constant}$$

$$\therefore \ \frac{\Delta A}{\Delta t} = \text{constant},$$

since $2m$ is constant. So if the conservation of angular momentum is true, the area swept out per second by the radius SO is constant while the planet O moves in its orbit. In other words, equal areas are swept out in equal times. *But this is Kepler's second law*, which has been observed to be true for centuries (see p. 47). Consequently, the principle of the conservation of angular momentum has stood the test of time. Since the angular momentum about S at O and at C are equal, and p is less than p_1, it follows that v is greater than v_1. So the planet speeds up on approaching S.

The force on O is always one of attraction towards S. It is described as a *central force*. The force has no moment about S and so the angular momentum of the planet about S remains constant.

Kinetic Energy of a Rolling Object

When an object such as a cylinder or ball rolls on a plane, the object is rotating as well as moving down the plane. So it has rotational energy in addition to translational energy.

Consider a uniform cylinder C rolling along a plane without slipping, Fig. 3.13. The forces on C are (a) its weight Mg acting at its central axis O, (b) the frictional force F at the plane which prevents slipping. The force which produces linear acceleration and translational kinetic energy down the plane $= Mg \sin \alpha - F$. The *torque* about O which produces angular acceleration and rotational kinetic energy $= F.r$, where r is the radius of the cylinder.

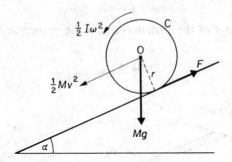

Fig. 3.13 Energy and acceleration of object rolling down plane

Since energy is a scalar quantity (one with no direction), we can add the translational and rotational kinetic energies to obtain the total energy of the cylinder. So at a given instant,

$$\text{total kinetic energy} = \tfrac{1}{2}Mv^2 + \tfrac{1}{2}I\omega^2$$

where I is the moment of inertia about the axis O, ω is the angular velocity about O and v is the translational velocity down the plane. If the cylinder does not slip, then $v = r\omega$. So

$$\text{total kinetic energy} = \tfrac{1}{2}Mv^2 + \tfrac{1}{2}I\left(\frac{v}{r}\right)^2$$

$$= \tfrac{1}{2}v^2\left(M + \frac{I^2}{r^2}\right).$$

Suppose the cylinder rolls from rest through a distance s along the plane. The loss of potential energy $= Mgs \sin \alpha =$ gain in kinetic energy $= \tfrac{1}{2}v^2(M + I/r^2)$ from above. So

$$v^2 = \frac{2Mgs \sin \alpha}{M + (I/r^2)}.$$

But $v^2 = 2as$ where a is the *linear acceleration* down the plane. So

$$2as = \frac{2 Mgs \sin \alpha}{M + (I/r^2)}$$

Thus
$$a = \frac{Mg \sin \alpha}{M + (I/r^2)}. \qquad (1)$$

A uniform *solid* cylinder of mass M and radius r has a moment of inertia $I = Mr^2/2$ about its axis (p. 78). So $I/r^2 = M/2$. Substituting in (1), we find that the acceleration down the plane $a = 2g \sin \alpha/3$. A uniform *hollow* cylinder open at both ends has a moment of inertia about its axis given by $I = Mr^2$, where M is the mass and r is the radius. From (1), we find that the acceleration down the plane, a, $= g \sin \alpha/2$. So the solid cylinder would have a greater acceleration down the plane than a hollow cylinder of the same mass. If no other tests were available, we could distinguish between a solid cylinder and a hollow cylinder closed at both ends, both of the same mass, by allowing them to roll from rest down an inclined plane. Starting from the same place the cylinder which reaches the bottom first would be the solid cylinder.

Measurement of Moment of Inertia of Flywheel

The moment of inertia of a flywheel W about a horizontal axle A can be determined by passing one end of some string through a hole in the axle, winding the string round the axle, and attaching a mass M to the other end of the string, Fig. 3.14. The length of string is such that M reaches the floor, when released, at the same instant as the string is completely unwound from the axle.

M is released, and the number of revolutions, n, made by the wheel W up to the occasion when M strikes the ground is noted. The further number of revolutions n_1 made by W until it comes finally to rest, and the time t taken, are also observed by means of a chalk-mark on W.

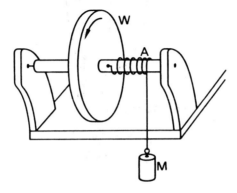

Fig. 3.14 Moment of inertia of flywheel

Now the loss in potential energy of M = gain in kinetic energy of M + gain in kinetic energy of flywheel + work done against friction.

$$\therefore Mgh = \tfrac{1}{2}Mr^2\omega^2 + \tfrac{1}{2}I\omega^2 + nf \qquad . \qquad . \qquad . \qquad . \qquad \text{(i)}$$

where h is the distance M has fallen, r is the radius of the axle, ω is the angular velocity, I is the moment of inertia, and f is the energy per turn expended against friction. Since the energy of rotation of the flywheel when the mass M reaches the ground = work done against friction in n_1 revolutions, then

$$\tfrac{1}{2}I\omega^2 = n_1 f.$$

$$\therefore f = \tfrac{1}{2}\frac{I\omega^2}{n_1}.$$

Substituting for f in (i),

$$\therefore Mgh = \tfrac{1}{2}Mr^2\omega^2 + \tfrac{1}{2}I\omega^2\left(1 + \frac{n}{n_1}\right) \qquad . \qquad . \qquad . \qquad \text{(ii)}$$

Since the angular velocity of the wheel when M reaches the ground is ω, and the final angular velocity of the wheel is zero after a time t, the average angular velocity $= \omega/2 = 2\pi n_1/t$. Thus $\omega = 4\pi n_1/t$. Knowing ω and the magnitude of the other quantities in (ii), the moment of inertia I of the flywheel can be calculated.

Period of Oscillation of Rigid Body

Consider a rigid object oscillating about a fixed horizontal axis O near one end for example. Suppose h is the distance OG where G is the centre of gravity, and θ is the angle with the vertical made by OG at an instant. The torque on the object is then $I\alpha$ where α is the angular acceleration $d^2\theta/dt^2$ and the opposing torque when θ is small is $mgh.\theta$, since the perpendicular distance from G to the vertical through O is $h\sin\theta$ and $\sin\theta = \theta$ when the angle is small. So

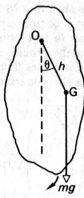

Fig. 3.15 Oscillation of rigid body

$$I\alpha = -mgh.\theta$$

and
$$\alpha = -\frac{mgh}{I}.\theta.$$

This is an equation of simple harmonic motion, where, with the usual notation, $\omega^2 = mgh/I$. So the period of oscillation T is given by

$$T = 2\pi/\omega = 2\pi\sqrt{\frac{I}{mgh}} \qquad . \qquad . \qquad . \qquad (1)$$

The oscillating rigid body is called a 'compound pendulum'. I is the moment of inertia about the axis O.

Consider a rigid body oscillating in a horizontal direction at the lower end of a vertical elastic or so-called torsion wire. In a moving-coil galvanometer a rigid coil is suspended from the end of a vertical torsion wire (p. 699). This time the opposing torque is $c\theta$, where θ is the small angle of deflection from the equilibrium position of the coil and c is the opposing torque per radian deflection of the torsion wire, known as its 'elastic constant' (p. 125). So if I is the moment of inertia of the coil about the vertical and α is the angular acceleration, then

$$I\alpha = -c\theta$$

and
$$\alpha = -\frac{c}{I}\theta.$$

This is a simple harmonic motion equation where $\omega^2 = c/I$. The period of oscillation T is given by

$$T = 2\pi/\omega = 2\pi\sqrt{\frac{I}{c}} \qquad . \qquad . \qquad . \qquad . \qquad (2)$$

Summary

We conclude with a summary showing a comparison between formulae in rotational and linear motion.

Linear Motion	Rotational Motion
1. Velocity, v	Angular velocity, $\omega = v/r$
2. Momentum $= mv$	Angular momentum $= I\omega$
3. Energy $= \frac{1}{2}mv^2$	Rotational energy $= \frac{1}{2}I\omega^2$
4. Force, F, $= ma$	Torque, T, $= I\alpha$
5. Simple pendulum: $T = 2\pi\sqrt{\dfrac{l}{g}}$	Rigid body: $T = 2\pi\sqrt{\dfrac{I}{mgh}}$
6. Motion down inclined plane – energy equation: $\frac{1}{2}mv^2 = mgh \sin\theta$	Rotating without slipping down inclined plane – energy equation: $\frac{1}{2}Mv^2 + \frac{1}{2}I\omega^2 = Mgh \sin\theta$
7. Conservation of linear momentum on collision, if no external forces	Conservation of angular momentum on collision, if no external torque.

Example

What is meant by the *moment of inertia* of an object about an axis?

Describe and give the theory of an experiment to determine the moment of inertia of a flywheel mounted on a horizontal axle.

A uniform circular disc of mass 20 kg and radius 0·15 m is mounted on a horizontal cylindrical axle of radius 0·015 m and negligible mass. Neglecting frictional losses in the bearings, calculate (a) the angular velocity acquired from rest by the application for 12 seconds of a force of 20 N tangential to the axle, (b) the kinetic energy of the disc at the end of this period, (c) the time required to bring the disc to rest if a braking force of 1 N were applied tangentially to its rim. (L.)

Moment of inertia of disc, I, $= \frac{1}{2}Ma^2 = \frac{1}{2} \times 20 \times 0\cdot15^2 = 0\cdot225$ kg m^2.

(a) Torque due to 20 N tangential to axle

$$= 20 \times 0\cdot015 = 0\cdot3 \text{ N m.}$$

$$\therefore \text{ angular acceleration} = \frac{\text{torque}}{I} = \frac{0\cdot3}{0\cdot225} \text{ rad s}^{-2}.$$

$$\therefore \text{ after 12 seconds, angular velocity} = \frac{12 \times 0\cdot3}{0\cdot225} = 16 \text{ rad s}^{-1}.$$

(b) K.E. of disc after 12 seconds $= \frac{1}{2}I\omega^2$

$$= \frac{1}{2} \times 0\cdot225 \times 16^2 = 28\cdot8 \text{ J.}$$

(c) Decelerating torque $= 1 \times 0\cdot15$.

$$\therefore \text{ angular deceleration} = \frac{\text{torque}}{I} = \frac{0\cdot15}{0\cdot225} \text{ rad s}^{-2}.$$

$$\therefore \text{ time to bring disc to rest} = \frac{\text{initial angular velocity}}{\text{angular deceleration}}$$

$$= \frac{16 \times 0\cdot225}{0\cdot15} = 24 \text{ s.}$$

Exercises 3

1. A disc of moment of inertia 10 kg m^2 about its centre rotates steadily about the centre with an angular velocity of 20 rad s^{-1}. Calculate (i) its rotational energy, (ii) its angular momentum about the centre, (iii) the number of revolutions per second of the disc.

2. A constant torque of 200 N m turns a wheel about its centre. The moment of inertia about this axis is 100 kg m^2. Find (i) the angular velocity gained in 4 s, (ii) the kinetic energy gained after 20 revs.

3. A flywheel has a kinetic energy of 200 J. Calculate the number of revolutions it makes before coming to rest if a constant opposing couple of 5 N m is applied to the flywheel.

If the moment of inertia of the flywheel about its centre is 4 kg m^2, how long does it take to come to rest?

4. A constant torque of 500 N m turns a wheel which has a moment of inertia 20 kg m^2 about its centre. Find the angular velocity gained in 2 s and the kinetic energy gained.

5. A ballet dancer spins with 2·4 rev s^{-1} with her arms outstretched, when the moment of inertia about the axis of rotation is I. With her arms folded, the moment of inertia about the same axis becomes 0·6 I. Calculate the new rate of spin.

State the principle used in your calculation.

6. A disc rolling along a horizontal plane has a moment of inertia 2·5 kg m^2 about its centre and a mass of 5 kg. The velocity along the plane is 2 m s^{-1}.

If the radius of the disc is 1 m, find (i) the angular velocity, (ii) the total energy (rotational and translation) of the disc.

7. A wheel of moment of inertia 20 kg m^2 about its axis is rotated from rest about its centre by a constant torque T and the energy gained in 10 s is 360 J. Calculate (i) the angular velocity at the end of 10 s, (ii) T, (iii) the number of revolutions made by the wheel before coming to rest if T is removed at 10 s and a constant opposing torque of 4 N m^{-1} is then applied to the wheel.

8. A uniform rod of length 3 m is suspended at one end so that it can move about an axis perpendicular to its length. The moment of inertia about the end is 6 kg m^2 and the mass of the rod is 2 kg. If the rod is initially horizontal and then released, find the angular velocity of the rod when (i) it is inclined at 30° to the horizontal, (ii) reaches the vertical.

9. A recording disc rotates steadily at 45 rev min^{-1} on a table. When a small mass of 0·02 kg is dropped gently on the disc at a distance of 0·04 m from its axis and sticks to the disc, the rate of revolution falls to 36 rev min^{-1}. Calculate the moment of inertia of the disc about its centre.

Write down the principle used in your calculation.

10. A disc of moment of inertia 0·1 kg m^2 about its centre and radius 0·2 m is released from rest on a plane inclined at 30° to the horizontal. Calculate the angular velocity after it has rolled 2 m down the plane if its mass is 5 kg.

11. A flywheel with an axle 1·0 cm in diameter is mounted in frictionless bearings and set in motion by applying a steady tension of 2 N to a thin thread wound tightly round the axle. The moment of inertia of the system about its axis of rotation is 5·0 × 10^{-4} kg m^2. Calculate (a) the angular acceleration of the flywheel when 1 m of thread has been pulled off the axle, (b) the constant retarding couple which must then be applied to bring the flywheel to rest in one complete turn, the tension in the thread having been completely removed. (N.)

12. Define the moment of inertia of a body about a given axis. Describe how the moment of inertia of a flywheel can be determined experimentally.

A horizontal disc rotating freely about a vertical axis makes 100 r.p.m. A small piece of wax of mass 10 g falls vertically on to the disc and adheres to it at a distance of 9 cm from the axis. If the number of revolutions per minute is thereby reduced to 90, calculate the moment of inertia of the disc. (N.)

13. Write down an expression for the angular momentum of a point mass m moving in a circular path of radius r with constant angular velocity ω. Extend this result to a system of several point masses each rotating about a common axis with the same angular velocity and show how this leads to the concept of the moment of inertia I of a rigid body.

Show that the quantity $\frac{1}{2}I\omega^2$ is the kinetic energy of rotation of a rigid body rotating about an axis with angular velocity ω.

Describe how you would determine by experiment the moment of inertia of a flywheel.

The atoms in the oxygen molecule O_2 may be considered to be point masses separated by a distance of $1\cdot2 \times 10^{-10}$ m. The molecular speed of an oxygen molecule at s.t.p. is 460 m s^{-1}. Given that the rotational kinetic energy of the molecule is two-thirds of its translational kinetic energy, calculate its angular velocity at s.t.p. assuming that molecular rotation takes place about an axis through the centre of, and perpendicular to, the line joining the atoms. (*O. & C.*)

14. (*a*) For a rigid body rotating about a fixed axis, explain with the aid of a suitable diagram what is meant by *angular velocity*, *kinetic energy* and *moment of inertia*.

(*b*) In the design of a passenger bus, it is proposed to derive the motive power from the energy stored in a flywheel. The flywheel, which has a moment of inertia of $4\cdot0 \times 10^2$ kg m^2, is accelerated to its maximum rate of rotation $3\cdot0 \times 10^3$ revolutions per minute by electric motors at stations along the bus route.

(i) Calculate the maximum kinetic energy which can be stored in the flywheel. (ii) If, at an average speed of 36 kilometres per hour, the power required by the bus is 20 kW, what will be the maximum possible distance between stations on the level? (*N.*)

15. (*a*) Explain what is meant by (i) a *couple*, (ii) the *moment of a couple*. Show that a force acting along a given line can always be replaced by a force of the same magnitude acting along a parallel line, together with a couple.

(*b*) A flywheel of moment of inertia $0\cdot32$ kg m^2 is rotated steadily at 120 rad s^{-1} by a 50 W electric motor. (i) Find the kinetic energy and angular momentum of the flywheel. (ii) Calculate the value of the frictional couple opposing the rotation. (iii) Find the time taken for the wheel to come to rest after the motor has been switched off. (*O.*)

16. Define angular momentum. A point mass m is attached to one end of a light rigid rod of length r. The rod rotates in the horizontal plane with uniform angular velocity ω about a vertical axis through its other end. Write down an expression for the angular momentum of the mass about the axis. Describe qualitatively, with the aid of a sketch, the variation of the angular momentum of the mass about a fixed point on its path.

State the law of conservation of angular momentum, and describe an experiment to verify the law.

Derive expressions for the final angular velocity of the point mass m above in the following circumstances: (*a*) m remains unchanged and the rod contracts to half its length; (*b*) r remains unchanged and a second identical mass moves from the axis to a point halfway along the rod. (*O. & C.*)

17. Explain the meaning of the term *moment of inertia*. Describe in detail how you would find experimentally the moment of inertia of a bicycle wheel about the central line of its hub.

A uniform cylinder 20 cm long, suspended by a steel wire attached to its mid-point so that its long axis is horizontal, is found to oscillate with a period of 2 seconds when the wire is twisted and released. When a small thin disc, of mass 10 g, is attached to each end the period is found to be $2\cdot3$ seconds. Calculate the moment of inertia of the cylinder about the axis of oscillation. (*N.*)

18. A flywheel rotates about a horizontal axis fitted into friction free bearings. A light string, one end of which is looped over a pin on the axle, is wrapped ten times round the axle and has a mass of $1\cdot5$ kg attached to its free end. Discuss the energy changes as the mass falls. If the moment of inertia of the wheel and axle is $0\cdot10$ kg m^2 and the diameter of the axle $5\cdot0$ cm, calculate the angular velocity of the flywheel at the instant when the string detaches itself from the axle after ten revolutions. (*AEB*, 1982)

4 Static Bodies. Fluids

Static Bodies

Statics

1. Statics is a subject which concerns the *equilibrium* of forces, such as the forces which act on a bridge. In Fig. 4.1 (i), for example, the joint O of a light bridge is in equilibrium under the action of the two forces P, Q acting in the girders meeting at O and the reaction S of the masonry at O.

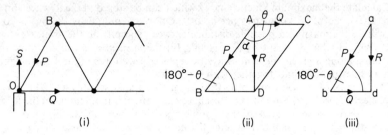

Fig. 4.1 Equilibrium of forces

Parallelogram of Forces

A force is a vector quantity, i.e., it can be represented in magnitude and direction by a straight line (p. 3). If AB, AC represent the forces P, Q respectively at the joint O, their *resultant*, R, is represented in magnitude and direction by the diagonal AD of the parallelogram ABDC which has AB, AC as two of its adjacent sides, Fig. 4.1 (ii). This is known as the *parallelogram of forces*, and is exactly analogous to the parallelogram of velocities discussed on p. 9. Alternatively, a line ab may be drawn to represent the vector P and bd to represent Q, in which case ad represents the resultant R.

By trigonometry for triangle ABD, we have

$$AD^2 = BA^2 + BD^2 - 2BA \cdot BD \cos ABD.$$

$$\therefore R^2 = P^2 + Q^2 + 2PQ \cos \theta,$$

where θ = angle BAC; the angle between the forces P, Q = $180°$ − angle ABD. This formula enables R to be calculated when P, Q and the angle between them are known. The angle BAD, or α say, between the resultant R and the force P can then be found from the relation

$$\frac{R}{\sin \theta} = \frac{Q}{\sin \alpha},$$

applying the sine rule to triangle ABD and noting that angle ABD = $180° - \theta$.

Resolved component. On p. 9 we saw that the effective part, or resolved component, of a vector quantity X in a direction θ inclined to it is given by $X \cos \theta$. Thus the resolved component of a force P in a direction making an angle of $30°$ with it is $P \cos 30°$; in a perpendicular direction to the latter the resolved component is $P \cos 60°$, or $P \sin 30°$. In Fig. 4.1 (i), the downward component of the force P on the joint of O is given by $P \cos$ BOS.

Forces in Equilibrium. Triangle of Forces

Since the joint O is in equilibrium, Fig. 4.1 (i), the resultant of the forces P, Q in the rods meeting at this joint is equal and opposite to the reaction S at O. Now the diagonal AD of the parallelogram ABDC in Fig. 4.1 (ii) represents the resultant R of P, Q since ABDC is the parallelogram of forces for P, Q; and hence DA represents the force S. Consequently the sides of the triangle ABD represents the three forces at O in magnitude and direction. This result can be generalised as follows. *If three forces are in equilibrium, they can be represented in magnitude and direction by the three sides of a triangle taken in order.* This theorem in Statics is known as the *triangle of forces*. In Fig. 4.1 (ii), AB, BD, DA, in this order, represent, P, Q, S respectively in Fig. 4.1 (i).

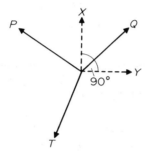

Fig. 4.2 Resolution of forces

We can derive another relation between forces in equilibrium. Suppose X, Y are the respective algebraic sums of the resolved components in two perpendicular directions of three forces P, Q, T in equilibrium, Fig. 4.2. Then, since X, Y can each be represented by the sides of a *rectangle* drawn to scale, their resultant R is given by

$$R^2 = X^2 + Y^2 \qquad . \qquad . \qquad . \qquad . \qquad \text{(i)}$$

Now if the forces are in equilibrium, R is zero. It then follows from (i) that X must be zero and Y must be zero. Thus *if forces are in equilibrium the algebraic sum of their resolved components in any two perpendicular directions is respectively zero*. This result applies to any number of forces in equilibrium.

Equilibrium of Three Coplanar Forces

If any object is in equilibrium under the action of *three* forces, the resultant of two of the forces must be equal and opposite to the third force. Thus the line of action of the third force must pass through the point of intersection of the lines of action of the other two forces.

As an example of calculating unknown forces in this case, suppose that a 12 m ladder of 200 N is placed at an angle of 60° to the horizontal, with one end B leaning against a smooth wall and the other end A on the ground, Fig. 4.3. The force R at B on the ladder is called the *reaction* of the wall, and if the latter is smooth, R acts perpendicularly to the wall. Assuming the weight, W, of the ladder acts at its mid-point G, the forces W and R meet at O, as shown. Consequently the force F at A passes through O.

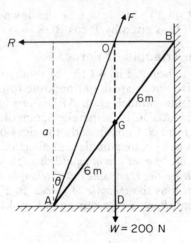

Fig. 4.3 Triangle of forces

The *triangle of forces* can be used to find the unknown forces R, F. Since DA is parallel to R, AO is parallel to F, and OD is parallel to W, the triangle of forces is represented by AOD. By means of a scale drawing R and F can be found, since

$$\frac{W(200)}{OD} = \frac{F}{AO} = \frac{R}{DA}.$$

A quicker method is to take moments about A for all the forces. The algebraic sum of the moments is zero about any point since the object is in equilibrium, and hence

$$R.a - W.AD = 0,$$

where a is the perpendicular from A to R. (F has zero moment about A.) But $a = 12 \sin 60°$, and $AD = 6 \cos 60°$.

$$\therefore R \times 12 \sin 60° - 200 \times 6 \cos 60° = 0.$$

$$\therefore R = 100 \frac{\cos 60°}{\sin 60°} = 58 \text{ N (approx.)}$$

Suppose θ is the angle F makes with the vertical.

Resolving the forces vertically, $F \cos \theta = W = 200$ N.

Resolving horizontally, $F \sin \theta = R = 58$ N.

$$\therefore F^2 \cos^2 \theta + F^2 \sin^2 \theta = F^2 = 200^2 + 58^2.$$

$$\therefore F = \sqrt{200^2 + 58^2} = 208 \text{ N (approx.)}$$

Moments

When the steering-wheel of a car is turned, the applied force is said to exert a *moment*, or turning-effect, about the axle attached to the wheel. The magnitude of the moment of a force P about a point O is defined as *the product of the force P and the perpendicular distance OA from O to the line of action of P,* Fig. 4.4 (i). Thus

$$\text{moment} = P \times \text{AO}.$$

The magnitude of the moment is expressed in *newton metre* (N m) when P is in newton and AO is in metre. We shall take an anticlockwise moment as positive in sign and a clockwise moment as negative in sign.

Parallel Forces and Equilibrium

If a rod carries loads of 10, 20, 30, 15, and 25 N at O, A, B, C, D respectively, Fig. 4.4 (ii), the resultant R of the weights, which are parallel forces, is given by

$$R = 10 + 20 + 30 + 15 + 25 = 100 \text{ N}.$$

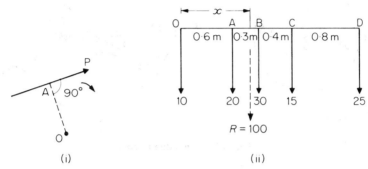

Fig. 4.4 (i) Moment (ii) Parallel forces

Experiment and theory show that *the moment of the resultant of a number of forces about any point is equal to the algebraic sum of the moments of the individual forces about the same point.* This result enables us to find where the resultant R acts. Taking moments about O for all the forces in Fig. 4.4 (ii), we have

$$(20 \times 0 \cdot 6) + (30 \times 0 \cdot 9) + (15 \times 1 \cdot 3) + (25 \times 2 \cdot 1),$$

because the distances between the forces are 0·6 m, 0·3 m, 0·4 m, 0·8 m, as shown. If x is the distance of the line of action of R from O, the moment of R about O $= R \times x = 100 \times x$.

$$\therefore \ 100x = (20 \times 0 \cdot 6) + (30 \times 0 \cdot 9) + (15 \times 1 \cdot 3) + (25 \times 2 \cdot 1),$$

from which $\qquad\qquad\qquad x = 1 \cdot 1 \text{ m}.$

The resultant of a number of forces *in equilibrium* is zero; and the moment of the resultant about any point is hence zero. It therefore follows that the algebraic sum of the moments of all the forces about any point is zero when those forces are in equilibrium. This means that the total clockwise moment of the forces about any point = the total anticlockwise moment of the remaining forces about the same point.

Torque due to Couple

There are many examples in practice where two forces, acting together, exert a moment or turning-effect on some object. As a very simple case, suppose two strings are tied to a wheel at X, Y, and *two equal and opposite forces, F*, are exerted tangentially to the wheel, Fig. 4.5 (i). If the wheel is pivoted at its centre, O, it begins to rotate about O in an anticlockwise direction.

Two equal and opposite forces whose lines of action do not coincide are said to form a *couple*. The two forces always have a turning-effect, or moment, called a *torque*, which is defined by (see p. 81)

$$torque = one force \times perpendicular\ distance\ between\ forces \quad . \quad (1)$$

Since XY is perpendicular to each of the forces F in Fig. 4.5 (i), the torque on the wheel $= F \times XY = F \times$ diameter of wheel. Thus if $F = 10$ newton and the diameter is 2 metre, the torque $= 20$ N m.

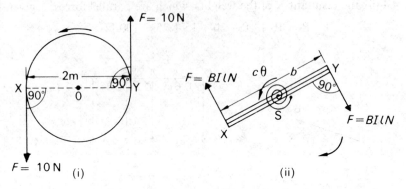

Fig. 4.5 Torque due to couple

In the theory of the *moving-coil electrical instrument*, we meet a case where a coil rotates when a current I is passed into it and comes to rest after deflection through an angle θ, Fig. 4.5 (ii). The forces F on the two sides X and Y of the coil are both equal to $BIlN$, where B is the strength of the magnetic field, l is the length of the coil and N is the number of turns (see Electricity section, chapter 33). Thus the coil is deflected by a couple. The torque of the deflecting couple $= F \times b$, where $b = XY =$ breadth of coil. Hence

$$torque = BIlN \times b = BANI,$$

where $A = lb =$ area of coil. The opposing torque, due to the spring S, is $c\theta$, where c is its elastic constant (p. 125). So, for equilibrium, $BANI = c\theta$.

Centre of Gravity

Every particle is attracted towards the centre of the earth by the force of gravity, and the *centre of gravity* of a body is the point where the *resultant* force of attraction or *weight* of the body acts. In the simple case of a ruler, the centre of gravity is the point of support when the ruler is balanced. A similar method can be used to find roughly the centre of gravity of a flat plate. A more accurate method consists of suspending the object in turn from two points on it, so that it hangs freely in each case, and finding the point of intersection of a plumb-line, suspended in turn from each point of suspension. This experiment is described in elementary books.

An object can be considered to consist of many small particles. The forces on the particles due to the attraction of the earth are all parallel since they act vertically, and hence their resultant is the sum of all the forces. The resultant is the *weight* of the whole object, of course. In the case of a rod of uniform cross-sectional area, the weight of a particle A at one end, and that of a corresponding particle A′ at the other end, have a resultant which acts at the mid-point O of the rod, Fig. 4.6 (i). Similarly, the resultant of the weight of a particle B, and that of a corresponding particle B', have a resultant acting at O. In this way, i.e., by symmetry, it follows that the resultant of the weights of all the particles of the rod acts at O. Hence the centre of gravity of a uniform rod is at its mid-point.

The centre of gravity, c.g., of the curved surface of a hollow cylinder acts at the mid-point of the cylinder axis. This is also the position of the c.g. of a uniform solid cylinder. The c.g. of a triangular plate or lamina is two-thirds of the distance along a median from corresponding point of the triangle. The c.g. of a uniform right solid cone is three-quarters along the axis from the apex.

Centre of Mass

Consider a smooth uniform rod on a horizontal surface with negligible friction such as ice. If the rod is struck by a force near one end, the rod will rotate as it accelerates. If it is struck at the centre, it will accelerate without rotation. The *centre of mass* of an object may be defined as the point at which an applied force produces acceleration but no rotation.

With two separated masses m_1 and m_2 connected by a rigid rod of negligible mass, the centre of mass is at a distance x_1 from m_1 and a distance x_2 from m_2 given numerically by $m_1x_1 = m_2x_2$. So if the mass m_1 is 1 kg and the mass m_2 is 2 kg, and the length of the rod is 3 m, the centre of mass is 2 m from the 1 kg mass and 1 m from the 2 kg mass.

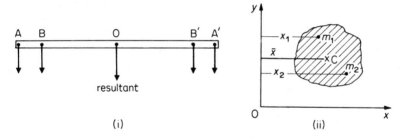

Fig. 4.6 Centre of gravity and mass

Figure 4.6 (ii) shows the particles of masses m_1, m_2,... which together form the object of total mass M. If x_1, x_2,... are the respective x-co-ordinates of the particles relative to axes Ox, Oy, then generally the co-ordinate $\bar{x}$ of the centre of mass C is defined by

$$\bar{x} = \frac{m_1x_1 + m_2x_2 + ...}{m_1 + m_2 + ...} = \frac{\Sigma mx}{M}.$$

Similarly, the distance $\bar{y}$ of the centre of mass C from Ox is given by

$$\bar{y} = \frac{\Sigma my}{M}.$$

In a molecule of sodium chloride, the sodium atom has a relative atomic mass of about 23·0 and the chlorine atom one of about 35·5. If the separation of the atoms is a, the centre of mass has a distance $\bar{x}$ from the sodium atom given from above by

$$\bar{x} = \frac{(23 \times 0) + (35·5 \times a)}{23 + 35·5} = \frac{35·5a}{58·5} = 0·6a \text{ (approx.).}$$

If the earth's field is uniform at all parts of an object, then the *weight* of a small mass m of it is typically mg. Thus, by moments, the distance of the centre of gravity from an axis Oy is given by

$$\frac{\Sigma mg \times x}{\Sigma mg} = \frac{\Sigma mx}{\Sigma m} = \frac{\Sigma mx}{M}.$$

The gravitational field intensity, g, cancels in numerator and denominator. It therefore follows that the centre of mass *coincides* with the centre of gravity. However, if the earth's field is *not* uniform at all parts of the object, the weight of a small mass m_1 of it is then $m_1 g_1$ say and the weight of a small mass m_2 at another part is $m_2 g_2$. Clearly, the centre of gravity does not now coincide with the centre of mass. A very long or very large object has different values of g at various parts of it.

Fluids

Pressure

Liquids and gases are called *fluids*. Unlike solid objects, fluids can flow.

If a piece of cork is pushed below the surface of a pool of water and then released, the cork rises to the surface again. The liquid thus exerts an upward force on the cork and this is due to the *pressure* exerted on the cork by the surrounding liquid. Gases also exert pressures. For example, when a thin closed metal can is evacuated, it usually collapses with a loud bang. The surrounding air now exerts a pressure on the outside which is no longer counterbalanced by the pressure inside, and hence there is a resultant force.

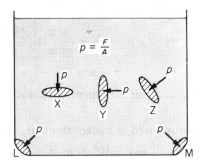

Fig. 4.7 Pressure in liquid

Pressure is defined as the *average force per unit area* at the particular region of liquid or gas. In Fig. 4.7, for example, X represents a small horizontal area, Y a small vertical area, and Z a small inclined area, all inside a vessel containing a liquid. The pressure p acts normally to the planes of X, Y or Z. In each case

$$\text{average pressure, } p = \frac{F}{A},$$

where F is the normal force due to the liquid on one side of an area A of X, Y or Z. Similarly, the pressure p on the sides L or M of the curved vessel acts normally to L and M and has magnitude F/A. In the limit, when the area is very small, $p = dF/dA$.

At a given point in a liquid, the pressure can act in any direction. *Pressure is a scalar*, not a vector. The direction of the force on a particular surface is normal to the surface.

Formula for Pressure

Observation shows that the pressure increases with the depth, h, below the liquid surface and with its density ρ.

To obtain a formula for the pressure, p, suppose that a horizontal plate X of area A is placed at a depth h below the liquid surface, Fig. 4.8. By drawing vertical lines from points on the perimeter of X, we can see that the force on X due to the liquid is equal to the weight of liquid of height h and uniform cross-section A. Since the volume of this liquid is Ah, the mass of the liquid $= Ah \times \rho$.

$$\therefore \text{ weight} = Ah\rho g \text{ newton,}$$

where g is 9·8, h is in m, A is in m^2, and ρ is in kg m^{-3}.

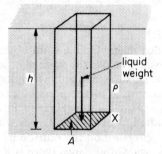

Fig. 4.8 Pressure and depth

$$\therefore \text{ pressure, } p, \text{ on } X = \frac{\text{force}}{\text{area}} = \frac{Ah\rho g}{A}$$

$$\therefore p = h\rho g \qquad . \qquad . \qquad . \qquad . \qquad . \qquad (1)$$

When h, ρ, g have the units already mentioned, the pressure p is in *newton metre^{-2}* (N m^{-2}).

The *bar* is a unit of pressure used in meteorology. By definition,

$$1 \text{ bar} = 10^5 \text{ N m}^{-2} \qquad . \qquad . \qquad . \qquad . \qquad (2)$$

The *pascal* (Pa) is the name given to a pressure of 1 N m^{-2}. Thus

$$1 \text{ bar} = 10^5 \text{ Pa} \qquad . \qquad . \qquad . \qquad . \qquad (3)$$

Pressure is often expressed in terms of that due to a height of mercury (Hg). One unit is the *torr* (after Torricelli):

$$1 \text{ torr} = 1 \text{ mmHg} = 133\cdot3 \text{ N m}^{-2} \text{ (approx.)}.$$

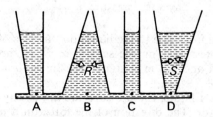

Fig. 4.9 Pressure and cross-section

From $p = h\rho g$ it follows that *the pressure in a liquid is the same at all points on the same horizontal level in it*. Experiment also gives the same result. Thus a liquid filling the vessel shown in Fig. 4.9 rises to the same height in each section if ABCD is horizontal. The cross-sectional area of B is greater than that of D; but the force on B is the sum of the weight of water above it together with the downward component of reaction R of the sides of the vessel, whereas the force on D is the weight of water above it *minus* the upward component of the reaction S of the sides of the vessel. It will thus be noted that the pressure in a vessel is independent of the cross-sectional area of the vessel.

Atmospheric Pressure

A *barometer* is an instrument for measuring the pressure of the atmosphere, which is required in weather-forecasting, for example. An accurate form of barometer consists basically of a vertical barometer tube about a metre long containing mercury, with a vacuum at the closed top, Fig. 4.10 (i). The other end of the tube is below the surface of mercury contained in a vessel B.

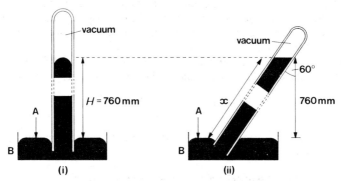

Fig. 4.10 Barometer (i) vertical and (ii) inclined

The pressure on the surface of the mercury in B is atmospheric pressure, A; and since the pressure is transmitted through the liquid, the atmospheric pressure supports the column of mercury in the tube. Suppose the column is a vertical height H above the level of the mercury in B. Now the pressure, p, at the bottom of a column of liquid of vertical height H and density ρ is given by $p = H\rho g$ (p. 102). Thus if $H = 760$ mm $= 0.76$ m and $\rho = 13\,600$ kg m^{-3},

$$p = H\rho g = 0.76 \times 13\,600 \times 9.8 = 1.013 \times 10^5 \text{ N m}^{-2}.$$

The pressure at the bottom of a column of mercury 760 mm high for a particular mercury density and value of g is known as *standard pressure* or *one atmosphere*. By definition, 1 atmosphere $= 1.01325 \times 10^5$ N m^{-2}. *Standard temperature and pressure* (s.t.p.) is 0°C and 760 mmHg pressure.

The *bar* is 10^5 N m^{-2} and is thus very nearly equal to one atmosphere.

It should be noted that the pressure p at a place X below the surface of a liquid is given by $p = H\rho g$, where H is the *vertical* distance of X below the surface. In Fig. 4.10 (ii), a very long barometer tube is inclined at an angle of 60° to the vertical. The length of the mercury along the slanted side of the tube is x mm say. If the atmospheric pressure here is the same as in Fig. 4.10 (i), this means that the *vertical* height to the mercury surface is still 760 mm. So

$$x \cos 60° = 760,$$

and
$$x = \frac{760}{\cos 60°} = \frac{760}{0.5} = 1520 \text{ mm.}$$

'Correction' to the Barometric Height

For comparison purposes, the pressure read on a barometer is often 'reduced' or 'corrected' to the magnitude the pressure would have at 0°C and at sea-level, latitude 45°. Suppose the 'reduced' pressure is H_o cm of mercury, and the observed pressure is H_t cm of mercury, corresponding to a temperature of t°C. Then, since pressure $= h\rho g$ (p. 102),

$$H_o \rho_o g = H_t \rho_t g',$$

where g is the acceleration due to gravity at sea-level, latitude 45°, and g' is the acceleration at the latitude of the place where the barometer was read.

$$\therefore \; H_o = H_t \times \frac{\rho_t}{\rho_o} \times \frac{g'}{g}.$$

The magnitude of g'/g can be obtained from standard tables. The ratio ρ_t/ρ_o of the densities $= 1/(1 + \gamma t)$, where γ is the absolute cubic expansivity of mercury. Further, the observed height H_t on the brass scale requires correction for the expansion of brass from the temperature at which it was correctly calibrated. If the latter is 0°C, then the corrected height is $H_t(1 + \alpha t)$, where α is the mean linear expansivity of brass. Thus, finally, the 'corrected' height H_o is given by

$$H_o = H_t \cdot \frac{1 + \alpha t}{1 + \gamma t} \cdot \frac{g'}{g}.$$

For further accuracy, a correction must be made for the surface tension of mercury (p. 142).

Variation of Atmospheric Pressure with Height

The density of a liquid varies very slightly with pressure. The density of a gas, however, varies appreciably with pressure. Thus at sea-level the density of the atmosphere is about $1 \cdot 2$ kg m^{-3}; at 1000 m above sea-level the density is about $1 \cdot 1$ kg m^{-3}; and at 5000 m above sea-level it is about $0 \cdot 7$ kg m^{-3}. Standard atmospheric pressure is the pressure at the base of a column of mercury 760 mm high, a liquid which has a density about 13 600 kg m^{-3}. Suppose air has a constant density of about $1 \cdot 2$ kg m^{-3}. Then the height of an air column of this density which has a pressure equal to standard atmospheric pressure

$$= \frac{760}{1000} \times \frac{13\,600}{1 \cdot 2} \text{m} = 8 \cdot 6 \text{ km.}$$

In fact, the air 'thins' the higher one goes, as explained above. The height of the air is thus much greater than 8·6 km.

Density

As we have seen, the pressure in a fluid depends on the density of the fluid.

The *density* of a substance is defined as its *mass per unit volume*. Thus

$$\text{density, } \rho = \frac{\text{mass of substance}}{\text{volume of substance}} \qquad . \qquad . \qquad . \qquad (1)$$

The density of copper is about $9 \cdot 0$ g cm^{-3} or $9 \cdot 0 \times 10^3$ kg m^{-3}; the density of aluminium is $2 \cdot 7$ g cm^{-3} or $2 \cdot 7 \times 10^3$ kg m^{-3}; the density of water at 4°C is 1 g cm^{-3} or 1000 kg m^{-3}.

Substances which float on water have a density less than 1000 kg m^{-3} (p. 106). For example, ice has a density of about 900 kg m^{-3}; cork has a density of about 250 kg m^{-3}. Steel, of density 7800 kg m^{-3}, will float on mercury, whose density is about 13 600 kg m^{-3} at 0°C.

Archimedes' Principle

An object immersed in a fluid experiences a resultant upward force owing to the pressure of fluid on it. This upward force is called the *upthrust* of the fluid on the object. ARCHIMEDES stated that *the upthrust is equal to the weight of fluid displaced by the object*, and this is known as *Archimedes' Principle*. Thus if an iron cube of volume 400 cm^3 is totally immersed in water of density 1 g cm^{-3}, the upthrust on the cube = weight of 400 × 1 g = 4 N. If it is totally immersed in oil of density 0·8 g cm^{-3}, the upthrust on it = weight of 400 × 0·8 g = 3·2 N.

Figure 4.11 shows why Archimedes' Principle is true. If S is a solid immersed in a liquid, the pressure on the lower surface C is greater than on the upper surface B, since the pressure at the greater depth h_2 is more than that at h_1. The pressure on the remaining surfaces D and E act as shown. The *force* on each of the four surfaces is calculated by summing the values of *pressure × area* over every part, remembering that vector addition is needed to sum forces. With a simple *rectangular-shaped solid* and the sides, D, E vertical, it can be seen that (i) the resultant horizontal force is zero, (ii) the upward force on C = pressure × area $A = h_2 \rho g A$, where ρ is the liquid density, and the downward force on B = pressure × area $A = h_1 \rho g A$. Thus

resultant force on solid = upward force (upthrust) = $(h_2 - h_1)\rho g A$.

But $(h_2 - h_1)A$ = volume of solid, V,

∴ upthrust = $V \rho g = mg$, where $m = V\rho$.

∴ upthrust = weight of liquid displaced.

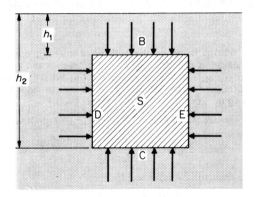

Fig. 4.11 Archimedes' principle

With a solid of irregular shape, taking into account horizontal and vertical components of forces, the same result is obtained. The upthrust is the weight of *liquid* displaced whatever the nature of the object immersed, or whether it is hollow or not. This is due primarily to the fact that the pressure on the object depends on the liquid in which it is placed.

Density measurement by Archimedes' Principle

The upthrust on an object immersed in water, for example, is the difference between (i) its weight in air when attached to a spring-balance and (ii) the reduced reading on the spring-balance or 'weight' when it is totally immersed in the liquid. Suppose the upthrust is found to be 1·0 N. Then, from Archimedes' Principle, the object displaces 100 g of water. But the density of water is 1 g cm^{-3}. Hence the volume of the object = 100 cm^3. The volume of any other object can be deduced from the difference in weighings in (i) and (ii).

The density of a *solid* such as brass or iron can thus be determined by (1) weighing it in air, (2) weighing it when it is totally immersed in water.

The volume V is then found as explained, and the density ρ is calculated from $\rho = m/V$ where m is the mass.

Flotation

When an object *floats* in a fluid, the upthrust = the weight of the object, for equilibrium.

In *air*, for example, a balloon of constant volume 5000 m^3 and mass 4750 kg rises to an altitude where the upthrust is 4750 g newtons, where g is the acceleration due to gravity at this height. From Archimedes' Principle, the upthrust = the weight of air displaced = 5000 ρg newtons, where ρ is the density of air at this height. Thus

$$5000\ \rho g = 4750\ g$$

$$\therefore\ \rho = 0.95\ \text{kg m}^{-3}.$$

If a block of ice of volume 1 m^3 and mass 900 kg floats in *water* of density 1000 kg m^{-3}, the mass of water displaced is 900 kg, from Archimedes' Principle. Thus the volume of water displaced by the ice is 0.9 m^3. So the block floats with 0.1 m^3 above the water surface. If the ice, mass 900 kg, all melts, the water formed has a volume of 0.9 m^3. So all the melted ice takes up *exactly* the whole of the space which the solid ice had originally occupied below the water.

Example

An ice cube of mass 50.0 g floats on the surface of a strong brine solution of volume 200.0 cm^3 inside a measuring cylinder. Calculate the level of the liquid in the measuring cylinder (i) before and (ii) after all the ice is melted. (iii) What happens to the level if the brine is replaced by 200.0 cm^3 water and 50.0 g of ice is again added? (Assume density of ice, brine = 900, 1100 kg m^{-3} or 0.9, 1.1 g cm^{-3} respectively.)

(i) Floating ice displaces 50 g of brine since upthrust equals weight of ice.

$$\therefore\ \text{volume displaced} = \frac{\text{mass}}{\text{density}} = \frac{50}{1.1} = 45.5\ \text{cm}^3.$$

$$\therefore\ \text{level on measuring cylinder} = 245.5\ \text{cm}^3.$$

(ii) 50 g of ice forms 50 g of water when all of it is melted.

$$\therefore\ \text{level on measuring cylinder } \textit{rises} \text{ to } 250.0\ \text{cm}^3.$$

(iii) *Water.* Initially, volume of water displaced = 50 cm^3, since upthrust = 50 g.

$$\therefore\ \text{level on cylinder} = 250.0\ \text{cm}^3.$$

If 1 g of ice melts, volume displaced is 1 cm^3 less. But volume of water formed is 1 cm^3. Thus the net change in water level is zero. Hence the water level remains unchanged as the ice melts.

Fluids in Motion. Streamlines and Velocity

A stream or river flows slowly when it runs through open country and faster through narrow openings or constrictions. As shown shortly, this is due to the fact that water is practically an incompressible fluid, that is, changes of pressure cause practically no change in fluid density at various parts.

Figure 4.12 shows a tube of water flowing steadily between X and Y, where X has a bigger cross-sectional area A_1 than the part Y, of cross-sectional area A_2. The *streamlines* of the flow represent the directions of the velocities of the particles of the fluid and the flow is uniform or laminar (p. 143). Assuming the

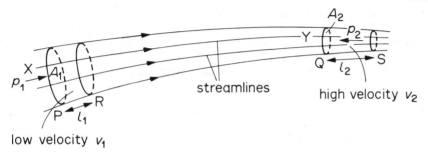

Fig. 4.12 Bernoulli's theorem

liquid is incompressible, then, if it moves from PQ to RS, the volume of liquid between P and R is equal to the volume between Q and S. Thus $A_1 l_1 = A_2 l_2$, where l_1 is PR and l_2 is QS, or $l_2/l_1 = A_1/A_2$. Hence l_2 is greater than l_1. Consequently the *velocity* of the liquid at the narrow part of the tube, where, it should be noted, the streamlines are closer together, is greater than at the wider part Y, where the streamlines are further apart. For the same reason, slow-running water from a tap can be made into a fast jet by placing a finger over the tap to narrow the exit.

Pressure and Velocity. Bernoulli's Principle

About 1740, Bernoulli obtained a relation between the pressure and velocity at different parts of a moving incompressible fluid. If the viscosity is negligibly small, there are no frictional forces to overcome (p. 163). In this case the work done by the pressure difference per unit volume of a fluid flowing along a pipe steadily is equal to the gain in kinetic energy per unit volume plus the gain in potential energy per unit volume.

Now the work done by a pressure in moving a fluid through a distance = force × distance moved = (pressure × area) × distance moved = pressure × volume moved, assuming the area is constant at a particular place for a short time of flow. At the beginning of the pipe where the pressure is p_1, the work done per unit volume on the fluid is thus p_1; at the other end, the work done per unit volume by the fluid is likewise p_2. Hence the net work done *on* the fluid per unit volume = $p_1 - p_2$. The kinetic energy per unit volume = $\frac{1}{2}$ mass per unit volume × velocity^2 = $\frac{1}{2}\rho$ × velocity^2, where ρ is the density of the fluid. Thus if v_2 and v_1 are the final and initial velocities respectively at the end and the beginning of the pipe, the kinetic energy gained per unit volume = $\frac{1}{2}\rho(v_2{}^2 - v_1{}^2)$. Further, if h_2 and h_1 are the respective heights measured from a fixed level at the end and beginning of the pipe, the potential energy gained per unit volume = mass per unit volume × g × $(h_2 - h_1)$ = $\rho g(h_2 - h_1)$.

Thus, from the conservation of energy,

$$p_1 - p_2 = \tfrac{1}{2}\rho(v_2{}^2 - v_1{}^2) + \rho g(h_2 - h_1)$$

$$\therefore\ p_1 + \tfrac{1}{2}\rho v_1{}^2 + \rho g h_1 = p_2 + \tfrac{1}{2}\rho v_2{}^2 + \rho g h_2$$

$$\therefore\ p + \tfrac{1}{2}\rho v^2 + \rho g h = \text{constant},$$

where p is the pressure at any part and v is the velocity there. Hence it can be said that, for streamline motion of an incompressible non-viscous fluid,

the sum of the pressure at any part plus the kinetic energy per unit volume plus the potential energy per unit volume there is always constant.

This is known as *Bernoulli's Principle*.

Bernoulli's Principle shows that at points in a moving fluid where the potential energy change ρgh is very small, or zero as in flow through a horizontal pipe, the pressure is low where the velocity is high; conversely, the pressure is high where the velocity is low. The principle has wide applications.

Example

As a numerical illustration of the previous analysis, suppose the area of cross-section A_1 of X in Fig. 4.12 is 4 cm², the area A_2 of Y is 1 cm², and water flows past each section in laminar flow at the rate of 400 cm³ s⁻¹. Then

$$\text{at X, speed } v_1 \text{ of water} = \frac{\text{vol. per second}}{\text{area}} = 100 \text{ cm s}^{-1} = 1 \text{ m s}^{-1};$$

$$\text{at Y, speed } v_2 \text{ of water} = 400 \text{ cm s}^{-1} = 4 \text{ m s}^{-1}.$$

The density of water, $\rho = 1000$ kg m⁻³. So, if p is the pressure difference,

$$\therefore p = \tfrac{1}{2}\rho(v_2^2 - v_1^2) = \tfrac{1}{2} \times 1000 \times (4^2 - 1^2) = 7\cdot5 \times 10^3 \text{ N m}^{-2}.$$

If h is in metres, $\rho = 1000$ kg m⁻³ for water, $g = 9\cdot8$ m s⁻², then, from $p = h\rho g$,

$$h = \frac{7\cdot5 \times 10^3}{1000 \times 9\cdot8} = 0\cdot77 \text{ m (approx.)}.$$

The pressure head h is thus equivalent to $0\cdot77$ m of water.

Applications of Bernoulli's Principle

1. A suction effect is experienced by a person standing close to the platform at a station when a fast train passes. The fast-moving air between the person and train produces a decrease in pressure and the excess air pressure on the other side pushes the person towards the train.

2. *Filter pump*. A filter pump has a narrow section in the middle, so that a jet of water from the tap flows faster here, Fig. 4.13 (i). This causes a drop in pressure near it and air therefore flows in from the side tube to which a vessel is connected. The air and water together are expelled through the bottom of the filter pump.

3. *Aerofoil lift*. The curved shape of an aerofoil creates a faster flow of air over its top surface than the lower one, Fig. 4.13 (ii). This is shown by the closeness of the streamlines above the aerofoil compared with those below. From Bernoulli's Principle, the pressure of the air below is greater than that above, and this produces the lift on the aerofoil.

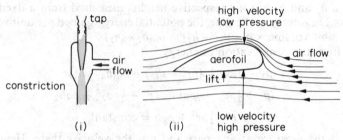

Fig. 4.13 Fluid velocity and pressure

4. *Flow of liquid from wide tank*. Suppose a liquid flows through a hole H at the bottom of a wide tank, as shown in Fig. 4.14. Assuming negligible viscosity

and streamline flow at a small distance from the hole, which is an approximation, Bernoulli's theorem can be applied. At the top X of the liquid in the tank, the pressure is atmospheric, say B, the height measured from a fixed level such as the hole H is h, and the kinetic energy is negligible if the tank is wide so that the level falls very slowly. At the bottom, Y, near H, the pressure is again B, the height above H is now zero, and the kinetic energy is $\frac{1}{2}\rho v^2$, where ρ is the density and v is the velocity of emergence of the liquid. Thus, from Bernoulli's Principle,

$$B + \rho h g = B + \tfrac{1}{2}\rho v^2$$

$$\therefore v^2 = 2gh.$$

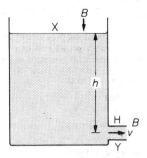

Fig. 4.14 Torricelli's theorem

Thus the velocity of the emerging liquid is the same as that which would be obtained if it fell through a height h, and this is known as *Torricelli's theorem*. In practice the velocity is less than that given by $\sqrt{2gh}$ owing to viscous forces, and the lack of streamline flow must also be taken into account.

Measurement of Fluid Velocity. Pitot-static Tube

The velocity at a point in a fluid flowing through a horizontal tube can be measured by the application of the Bernoulli equation on p. 107. In this case h is zero and so

$$p + \tfrac{1}{2}\rho v^2 = \text{constant}.$$

Here p is the static pressure at a point in the fluid, that is, the pressure unaffected by its velocity. The pressure $p + \frac{1}{2}\rho v^2$ is the total or dynamic pressure, that is, the pressure which the fluid would exert if it is brought to rest by striking a surface placed normally to the velocity at the point concerned.

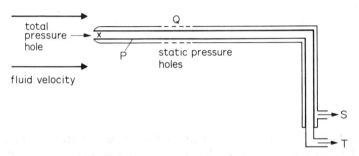

Fig. 4.15 Principle of Pitot-static tube

Figure 4.15 illustrates the principle of a *Pitot-static tube*. The inner or Pitot tube P, named after its inventor, has an opening X at one end *normal* to the fluid velocity. A manometer connected to T would measure the total pressure $p + \frac{1}{2}\rho v^2$. The outer or static tube has holes Q in its side which are *parallel* to the fluid velocity. A manometer connected to S would measure the static pressure p. The difference in pressure, $h\rho'g$, in the two sides of a single manometer joined respectively to T and S would hence be equal to $\frac{1}{2}\rho v^2$. Thus v can be calculated from $v = \sqrt{2h\rho'g/\rho}$. In practice, corrections are applied to the manometer readings to take account of differences from the simple theory outlined.

Example

(i) Water flows steadily along a horizontal pipe at a volume rate of 8×10^{-3} m³ s⁻¹. If the area of cross-section of the pipe is 40 cm² (40×10^{-4} m²), calculate the flow velocity of the water.

(ii) Find the total pressure in the pipe if the static pressure in the horizontal pipe is $3 \cdot 0 \times 10^4$ Pa, assuming the water is incompressible, non-viscous and its density is 1000 kg m⁻³.

(iii) What is the new flow velocity if the total pressure is $3 \cdot 6 \times 10^4$ Pa?

(i) Velocity of water $= \dfrac{\text{volume per second}}{\text{area}}$

$$= \frac{8 \times 10^{-3}}{40 \times 10^{-4}} = 2 \text{ m s}^{-1}$$

(ii) Total pressure = static pressure $+ \frac{1}{2}\rho v^2$

$$= 3 \cdot 0 \times 10^4 + \frac{1000 \times 2^2}{2}$$

$$= 3 \cdot 0 \times 10^4 + 0 \cdot 2 \times 10^4 = 3 \cdot 2 \times 10^4 \text{ Pa}$$

(iii) $\frac{1}{2}\rho v^2 = $ total pressure $-$ static pressure

So $\frac{1}{2} \times 1000 \times v^2 = 3 \cdot 6 \times 10^4 - 3 \cdot 0 \times 10^4 = 0 \cdot 6 \times 10^4$

$$v = \sqrt{\frac{0 \cdot 6 \times 10^4}{500}} = 3 \cdot 5 \text{ m s}^{-1}.$$

Exercises 4

Statics

1. Figure 4A shows three forces acting at a point. The lines drawn represent the forces roughly in magnitude and direction. Which diagram best represents equilibrium?

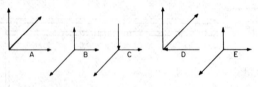

Fig. 4A

2. In Fig. 4B (i) and (ii), calculate the torque acting on the rod AB, 0·4 m long.

In Fig. 4B (i), what work would be done if the torque remained constant and AB is rotated through 2 revolutions?

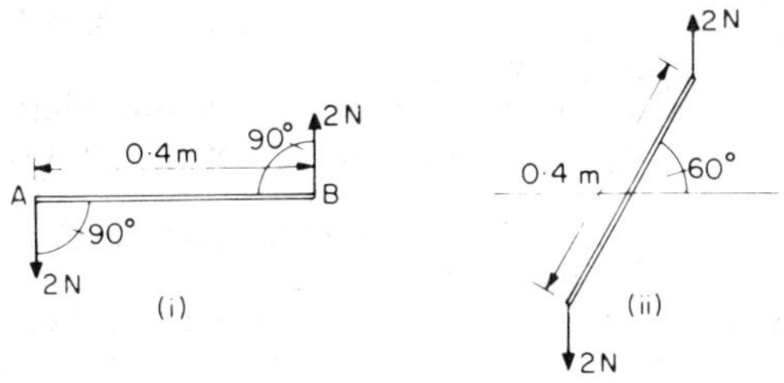

Fig. 4B

3. The foot of a uniform ladder is on a rough horizontal ground, and the top rests against a smooth vertical wall. The weight of the ladder is 400 N, and a man weighing 800 N stands on the ladder one-quarter of its length from the bottom. If the inclination of the ladder to the horizontal is 30°, find the reaction at the wall and the total force at the ground.

4. A rectangular plate ABCD has two forces of 100 N acting along AB and DC in opposite directions. If AB = 3 m, BC = 5 m, what is the moment of the couple or torque acting on the plate? What forces acting along BC and AD respectively are required to keep the plate in equilibrium?

5. A flat plate is cut in the shape of a square of side 20·0 cm, with an equilateral triangle of side 20·0 cm adjacent to the square. Calculate the distance of the centre of mass from the apex of the triangle.

6. A hollow metal cylinder 2 m tall has a base of diameter 35 cm and is filled with water to a height of (i) 1 m, (ii) 50 cm. Calculate the distance of the centre of gravity in metre from the base in each case if the cylinder has no top.

(Metal weighs 20 kg m^{-2} of surface. Assume $\pi = 22/7$.)

7. A trap-door 120 cm by 120 cm is kept horizontal by a string attached to the mid-point of the side opposite to that containing the hinge. The other end of the string is tied to a point 90 cm vertically above the hinge. If the trap-door weight is 50 N, calculate the tension in the string and the reaction at the hinge.

8. Three forces in one plane act on a rigid body. What are the conditions for equilibrium?

The plane of a kite of mass 6 kg is inclined to the horizon at 60°. The thrust of the air acting normally on the kite acts at a point 25 cm above its centre of gravity, and the string is attached at a point 30 cm above the centre of gravity. Find the thrust of the air on the kite, and the tension in the string. (C.)

9. State and explain the conditions under which a rigid body remains in equilibrium under the action of a set of coplanar forces. Describe an experiment to determine the position of the centre of gravity of a flat piece of cardboard cut into the shape of a triangle. Use the conditions of equilibrium you have stated to justify your practical method.

A simple model of the ammonia molecule (NH_3) consists of a pyramid with the hydrogen atoms at the vertices of the equilateral base and the nitrogen atom at the apex. The N–H distance is 0·10 nm and the angle between two N–H bonds is 108°. Find the centre of gravity of the molecule. (Atomic weights: H = 1·0; N = 14; 1 nm = 1 × 10^{-9} m.) (O.)

10. Summarise the various conditions which are being satisfied when a body remains in equilibrium under the action of three non-parallel forces.

A wireless aerial attached to the top of a mast 20 m high exerts a horizontal force upon it of 600 N. The mast is supported by a stay-wire running to the ground from a point 6 m below the top of the mast, and inclined at 60° to the horizontal. Assuming that the action of the ground on the mast can be regarded as a single force, draw a diagram of the forces acting on the mast, and determine by measurement or by calculation the force in the stay-wire. (C.)

Fluids

11. An alloy of mass 588 g and volume 100 cm^3 is made of iron of density 8·0 g cm^{-3}

and aluminium of density $2\cdot7$ g cm^{-3}. Calculate the proportion (i) by volume, (ii) by mass of the constituents of the alloy.

12. A string supports a solid iron object of mass 180 g totally immersed in a liquid of density 800 kg m^{-3}. Calculate the tension in the string if the density of iron is 8000 kg m^{-3}.

13. A hydrometer floats in water with $6\cdot0$ cm of its graduated stem unimmersed, and in oil of density $0\cdot8$ g cm^{-3} with $4\cdot0$ cm of the stem unimmersed. What is the length of stem unimmersed when the hydrometer is placed in a liquid of density $0\cdot9$ g cm^{-3}?

14. A uniform capillary tube contains air trapped by a mercury thread 40 mm long. When the tube is placed horizontally as in Fig. 4c (i), the length of the air column is 36 mm. When placed vertically, with the open end of the tube downwards, the length of the air column is now x mm, Fig. 4c (ii). Calculate x if the atmospheric pressure is 760 mmHg, assuming that the air obeys Boyle's law, $pV = $ constant.

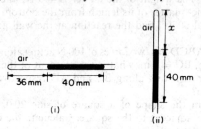

Fig. 4c

15. A barometer tube, 960 mm long above the mercury in the reservoir, contains a little air above the mercury column inside it. When vertical, Fig. 4D (i), the mercury column is 710 mm above the mercury in the reservoir. When inclined at 30° to the horizontal, Fig. 4D (ii), the mercury column is now 910 mm along the barometer tube.

Assuming the air obeys Boyle's law, $pV = $ constant, calculate the atmospheric pressure.

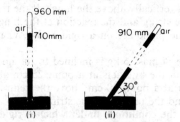

Fig. 4D

16. State the principle of Archimedes and use it to derive an expression for the resultant force experienced by a body of weight W and density σ when it is totally immersed in a fluid of density ρ.

A solid weighs $237\cdot5$ g in air and $12\cdot5$ g when totally immersed in a liquid of density $0\cdot9$ g cm^{-3}. Calculate (a) the density of the solid, (b) the density of a liquid in which the solid would float with one-fifth of its volume exposed above the liquid surface. (L.)

17. State (a) the laws of fluid pressure and (b) the principle of Archimedes. Show how (b) is a consequence of (a). Describe a simple experiment which verifies Archimedes' Principle.

The volume of a hot-air balloon is 600 m^3 and the density of the surrounding air is $1\cdot25$ kg m^{-3}. The balloon just hovers clear of the ground when the burner has heated the air inside to a temperature at which its density is $0\cdot80$ kg m^{-3}. (a) What is the total mass of the balloon, including the hot air inside it? (b) What is the total mass of the envelope of the balloon and its load? (c) Find the acceleration with which the balloon will start to rise when the density of the air inside is reduced to $0\cdot75$ kg m^{-3}. (Take g to be 10 m s^{-2}.) (O.)

Fluid Motion

18. An open tank holds water 1·25 m deep. If a small hole of cross-section area 3 cm² is made at the bottom of the tank, calculate the mass of water per second initially flowing out of the hole. ($g = 10$ m s^{-2}, density of water $= 1000$ kg m^{-3}.)

19. A lawn sprinkler has 20 holes each of cross-section area $2·0 \times 10^{-2}$ cm² and is connected to a hose-pipe of cross-section area 2·4 cm². If the speed of the water in the hose-pipe is 1·5 m s^{-1}, estimate the speed of the water as it emerges from the holes.

20. Show that the term $\frac{1}{2}\rho v^2$ which enters into the Bernoulli equation has the same dimensions as pressure p.

A fluid flows through a horizontal pipe of varying cross-section. Assuming the flow is streamline and applying the Bernoulli equation $p + \frac{1}{2}\rho v^2 = $ constant, show that the pressure in the pipe is greatest where the cross-section area is greatest.

21. Water flows along a horizontal pipe of cross-section area 48 cm² which has a constriction of cross-section area 12 cm² at one place. If the speed of the water at the constriction is 4 m s^{-1}, calculate the speed in the wider section.

The pressure in the wider section is $1·0 \times 10^5$ Pa. Calculate the pressure at the constriction. (Density of water $= 1000$ kg m^{-3}.)

22. Water flows steadily along a uniform flow tube of cross-section 30 cm². The static pressure is $1·20 \times 10^5$ Pa and the total pressure is $1·28 \times 10^5$ Pa.

Calculate the flow velocity and the mass of water per second flowing past a section of the tube. (Density of water $= 1000$ kg m^{-3}.)

23. (a) Distinguish between *static pressure*, *dynamic pressure* and *total pressure* when applied to streamline (laminar) fluid flow and write down expressions the fluid density applied to streamline (laminar) fluid flow and write down expressions for these three pressures at a point in the fluid in terms of the fluid velocity v, the fluid density ρ, pressure p, and the height h, of the point with respect to a datum.

(b) Describe, with the aid of a labelled diagram, the Pitot-static tube and explain how it may be used to determine the flow velocity of an incompressible, non-viscous fluid.

(c) The static pressure in a horizontal pipeline is $4·3 \times 10^4$ Pa, the total pressure is $4·7 \times 10^4$ Pa, and the area of cross-section is 20 cm². The fluid may be considered to be incompressible and non-viscous and has a density of 10^3 kg m^{-3}. Calculate (i) the flow velocity in the pipeline, (ii) the volume flow rate in the pipeline. (*N*.)

5 Elasticity. Molecules and Matter

Elasticity

Elasticity of Metals

A bridge used by traffic is subjected to loads or forces of varying amounts. Before a steel bridge is constructed, therefore, samples of the steel are sent to a research laboratory. Here they undergo tests to find out whether the steel can withstand the loads likely to be put on them.

Figure 5.1 illustrates a simple laboratory method of investigating the property of steel we are discussing. Two long thin steel wires, P, Q, are suspended beside each other from a rigid support B, such as a girder at the top of the ceiling. The wire P is kept taut by a weight A attached to its end and carries a scale M graduated in millimetres. The wire Q carries a vernier scale V which is alongside the scale M.

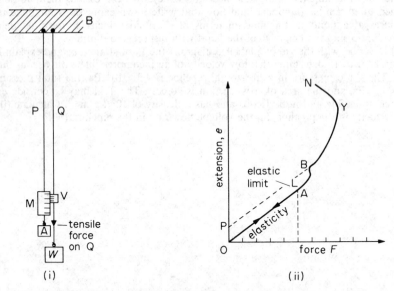

Fig. 5.1 (i) Elasticity experiment and (ii) result—extension against load

When a load W such as 10 N is attached to the end of Q, the wire increases in length by a small amount which can be read from the change in the reading on the vernier V. If the load is taken off and the reading on V returns to its original value, the wire is said to be *elastic* for loads from zero to 10 N, a term adopted by analogy with an elastic thread. When the load W is increased to 20 N the extension (increase in length) is obtained from V again; and if the reading on V returns to its original value when the load is removed the wire is said to be elastic at least for loads from zero to 20 N. The load or force which stretches a wire is called a *tensile force*.

Proportional and Elastic Limits

The extension of a thin wire such as Q for increasing loads or forces F may be found by experiment to be as follows:

Force (N)	0	10	20	30	40	50	60	70	80
Extension (mm)	0	0·14	0·20	0·42	0·56	0·70	0·85	1·01	1·19

When the extension, e, is plotted against the force F in the wire, a graph is obtained which is a *straight line* OA, followed by a curve ABY rising slowly at first and then very sharply, Fig. 5.1 (ii). Up to A, about 50 N, the results show that the extension increased by 0·014 mm per N added to the wire. A, then, is the *proportional limit*.

Along OA, and up to L just beyond A, the wire returned to its original length when the load was removed. The force at L is called the *elastic limit*. Along OL the metal is said to undergo changes called *elastic deformation*.

Beyond the elastic limit L, however, the wire has a permanent extension such as OP when the force is removed at B, for example, Fig. 5.1 (ii). Beyond L, therefore, the wire is no longer elastic. The extension increases rapidly along the curve ABY as the force on the wire is further increased and at N the wire thins and breaks. Molecular theory, discussed on p. 125, explains why this occurs.

Hooke's Law

From the straight line graph OA, we deduce that *the extension is proportional to the force or tension in a wire if the proportional limit is not exceeded.* This is known as *Hooke's law*, after ROBERT HOOKE, founder of the Royal Society, who discovered the relation in 1676.

The extension of a wire is due to the displacement of its molecules from their mean (average) positions. So the law shows that when a molecule of the metal is slightly displaced from its mean position the restoring force is proportional to its displacement (see p. 132). One may therefore conclude that the molecules of a solid metal are undergoing simple harmonic motion (p. 63). Up to the elastic limit the energy gained or stored by a stretched wire is molecular potential energy, which is recovered when the load is removed.

The measurements also show that it would be dangerous to load the wire with weights greater than the magnitude of the elastic limit, because the wire then suffers a permanent strain. Similar experiments in the research laboratory enable scientists to find the maximum load which a steel bridge, for example, should carry for safety. Rubber samples are also subjected to similar experiments, to find the maximum safe tension in rubber belts used in machinery. See Fig. 5A, p. 116.

Yield Point. Ductile and Brittle Substances. Breaking Stress

Careful experiments show that, for mild steel and iron for example, the molecules of the wire begin to 'slide' across each other soon after the load exceeds the elastic limit, that is, the material becomes *plastic*. This is indicated by the slight 'kink' at B beyond L in Fig. 5.1 (ii), and it is called the *yield point* of the wire. The change from an elastic to a plastic stage is shown by a sudden increase in the extension. In the plastic stage, the energy gained by the stretched wire is dissipated as heat and unlike the elastic stage, the energy is not recovered when the load is removed.

As the load is increased further the extension increases rapidly along the curve YN and the wire then becomes narrower and finally breaks. The *breaking stress* of the wire is the corresponding force per unit area of the narrowest cross-section of the wire.

Substances such as those just described, which lengthen considerably and undergo plastic deformation until they break, are known as *ductile* substances. Lead, copper and wrought iron are ductile. Other substances, however, break just after the elastic limit is reached; they are known as *brittle* substances. Glass and high carbon steels are brittle.

Brass, bronze, and many alloys appear to have no yield point. These materials increase in length beyond the elastic limit as the load is increased without the appearance of a plastic stage.

The strength and ductility of a metal, its ability to flow, depend on defects in the metal crystal lattice. This is discussed later (p. 125).

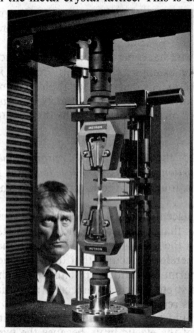

Fig. 5A The photograph shows a metal sample at the point of failure following a tensile test on an Instron Model 1185 Universal Materials Testing Machine. For precise measurement of extension, the system is fitted with an Automatic Extensometer. The load range is 0·1 N to 100 kN and the machine is used for a wide range of materials. (*Courtesy of Instron Limited*)

Tensile Stress and Tensile Strain. Young Modulus

We have now to consider the technical terms used in the subject of elasticity of wires. When a force or tension F is applied to the end of a wire of cross-sectional area A, Fig. 5.2 (i),

$$\text{the } tensile\ stress = force\ per\ unit\ area = \frac{F}{A} \qquad . \qquad . \qquad . \qquad (1)$$

If the extension of the wire is e, and its original length is l,

$$\text{the } tensile\ strain = extension\ per\ unit\ length = \frac{e}{l} \qquad . \qquad . \qquad (2)$$

Suppose 2 kg is attached to the end of a vertical wire of length 2 m and diameter 0·64 mm, and the extension is 0·60 mm. Then

$$F = 2 \times 9 \cdot 8 \text{ N}, \quad A = \pi \times 0 \cdot 032^2 \text{ cm}^2 = \pi \times 0 \cdot 032^2 \times 10^{-4} \text{ m}^2.$$

$$\therefore \text{ tensile stress} = \frac{2 \times 9 \cdot 8}{\pi \times 0 \cdot 032^2 \times 10^{-4}} \text{N m}^{-2},$$

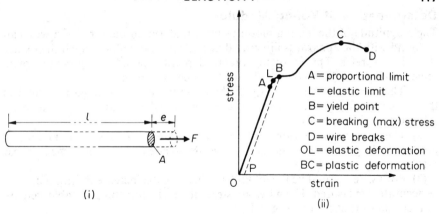

Fig. 5.2 (i) Tensile stress and tensile strain (ii) Stress against strain, ductile material

and $$\text{tensile strain} = \frac{0\cdot6 \times 10^{-3} \text{ metre}}{2 \text{ metre}} = 0\cdot3 \times 10^{-3}.$$

It will be noted that 'stress' has units such as 'N m^{-2}'; 'strain' has no units because it is the ratio of two lengths. Figure 5.2 (ii) shows the general stress–strain graph for a ductile material.

Under elastic conditions, a *modulus of elasticity* of the wire, called the **Young modulus (E)**, is defined as the ratio

$$E = \frac{\text{tensile stress}}{\text{tensile strain}}. \qquad\qquad . \qquad . \qquad . \qquad . \qquad (3)$$

Thus $$E = \frac{F/A}{e/l}.$$

Using the above figures, when the elastic limit is not exceeded,

$$E = \frac{2 \times 9\cdot8/(\pi \times 0\cdot032^2 \times 10^{-4})}{0\cdot3 \times 10^{-3}},$$

$$= \frac{2 \times 9\cdot8}{\pi \times 0\cdot032^2 \times 10^{-4} \times 0\cdot3 \times 10^{-3}},$$

$$= 2\cdot0 \times 10^{11} \text{ N m}^{-2}.$$

Dimensions of Young Modulus

As stated before, the 'strain' of a wire has no dimensions of mass, length, or time, since, by definition, it is the ratio of two lengths. Now

$$\text{dimensions of stress} = \frac{\text{dimensions of force}}{\text{dimensions of area}}$$

$$= \frac{MLT^{-2}}{L^2} = ML^{-1}T^{-2}.$$

∴ dimensions of the Young modulus, E,

$$= \frac{\text{dimensions of stress}}{\text{dimensions of strain}} = ML^{-1}T^{-2}.$$

Determination of Young Modulus

The magnitude of the Young modulus for a material in the form of a wire can be found with the apparatus illustrated in Fig. 5.1 (i), p. 114, to which the reader should now refer. The following practical points should be specially noted, remembering that the elastic limit must not be exceeded:

(1) The use of two wires, P, Q, of the same material and length, eliminates the correction for (i) the yielding of the support when loads are added to Q, (ii) changes of temperature.

(2) The wire is made *thin* so that a moderate load of several kilograms produces a large tensile stress. The wire is also made *long* so that a measurable extension is produced.

(3) Both wires should be free of kinks, otherwise the increase in length cannot be accurately measured. The wires are straightened by attaching suitable weights to their ends, as shown in Fig. 5.1 (i).

(4) A vernier scale is necessary to measure the extension of the wire since this is always small. The 'original length' of the wire is measured from the top B *to the vernier V* by a ruler, since an error of 1 millimetre is negligible compared with an original length of several metres. For very accurate work, the extension can be measured by using a spirit level between the two wires, and adjusting a vernier screw to restore the spirit level to its original reading after a load is added.

(5) The diameter of the wire must be found by a micrometer screw gauge at several places, and the average value then calculated. The area of cross-section, $A, = \pi r^2$, where r is the radius.

(6) The readings on the vernier are also taken when the load is gradually removed in steps of 1 kilogram; they should be very nearly the same as the readings on the vernier when the weights were added, showing that the elastic limit was not exceeded.

Calculation and Magnitude of Young Modulus

From the measurements, a graph can be plotted of the force F in newtons against the average extension e in metres. A straight line graph AB passing through the origin is drawn through all the points, Fig. 5.3.

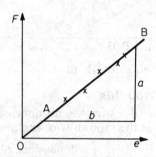

Fig. 5.3 Calculation of E

Now
$$E = \frac{F/A}{e/l} = \frac{F}{e} \times \frac{l}{A}$$

with the usual notation. The value of F/e is the gradient, a/b, of the straight line AB and this can be found. So knowing F/e, the original length l of the wire and the cross-section area A ($\pi d^2/4$, where d is the diameter of the wire), E can be calculated.

Mild steel ($0\cdot2\%$ carbon) has a Young modulus value of about $2\cdot0 \times 10^{11}$ N m^{-2}, copper has a value about $1\cdot2 \times 10^{11}$ N m^{-2}; and brass a value about $1\cdot0 \times 10^{11}$ N m^{-2}.

The breaking stress (tenacity) of cast-iron is about $1\cdot5 \times 10^{8}$ N m^{-2}; the breaking stress of mild steel is about $4\cdot5 \times 10^{8}$ N m^{-2}.

At Royal Ordnance and other Ministry of Supply factories, tensile testing is carried out by placing a sample of the material in a machine known as an *extensiometer*, which applies stresses of increasing value along the length of the sample and automatically measures the slight increase in length. When the elastic limit is reached, the pointer on the dial of the machine flickers, and soon after the yield point is reached the sample becomes thin at some point and then breaks. A graph showing the load against extension is recorded automatically by a moving pen while the sample is undergoing test.

Example

Find the maximum load which may be placed on a steel wire of diameter $1\cdot0$ mm if the permitted strain must not exceed $\frac{1}{1000}$ and the Young modulus for steel is $2\cdot0 \times 10^{11}$ N m^{-2}.

We have $\dfrac{\text{max. stress}}{\text{max. strain}} = 2 \times 10^{11}$.

$\therefore$ max. stress $= \frac{1}{1000} \times 2 \times 10^{11} = 2 \times 10^{8}$ N m^{-2}.

Now area of cross-section in m$^2 = \dfrac{\pi d^2}{4} = \dfrac{\pi \times 1\cdot0^2 \times 10^{-6}}{4}$

and $\qquad$ stress $= \dfrac{\text{load } F}{\text{area}}$

$\therefore F = \text{stress} \times \text{area} = 2 \times 10^{8} \times \dfrac{\pi \times 1\cdot0^2 \times 10^{-6}}{4}$ N

$= 157$ N

Force in Bar due to Contraction or Expansion

When a bar is heated, and then prevented from contracting as it cools, a considerable force is exerted at the ends of the bar. We can derive a formula for the force if we consider a bar of Young modulus E, a cross-sectional area A, a linear expansivity of magnitude α, and a decrease in temperature of $\theta°$C. Then, if the original length of the bar is l, the decrease in length e if the bar were free to contract $= \alpha l\theta$ since, by definition, α is the change in length per unit length per degree temperature change.

Now $$E = \frac{F/A}{e/l}.$$

$$\therefore F = \frac{EAe}{l} = \frac{EA\alpha l\theta}{l}.$$

$$\therefore F = EA\alpha\theta.$$

As an illustration, suppose a steel rod of cross-sectional area $2\cdot0$ cm^2 is heated to $100°$C, and then prevented from contracting when it is cooled to $10°$C. The linear expansivity of steel $= 12 \times 10^{-6}$ K^{-1} and Young modulus $= 2\cdot0 \times 10^{11}$ N m^{-2}. Then

$$A = 2 \text{ cm}^2 = 2 \times 10^{-4} \text{ m}^2, \quad \theta = 90°\text{C}.$$

$$\therefore F = EA\alpha\theta = 2 \times 10^{11} \times 2 \times 10^{-4} \times 12 \times 10^{-6} \times 90 \text{ N}$$

$$= 43\,200 \text{ N}.$$

Energy Stored in a Wire

Suppose that a wire has an original length l and is stretched by a length e when a force F is applied at one end. If the elastic limit is not exceeded, the extension is directly proportional to the applied load (p. 115). Consequently the force *in the wire* has increased uniformly in magnitude from zero to F, and so the average force in the wire while stretching was $F/2$. Now

$$\text{work done} = \text{force} \times \text{distance}.$$

$$\therefore \ \text{work} = \text{average force} \times \text{extension}$$

$$= \tfrac{1}{2}Fe \qquad \cdot \qquad \cdot \qquad \cdot \qquad \cdot \qquad (1)$$

This is the amount of energy stored in the wire. It is the gain in molecular potential energy of the molecules due to their displacement from their mean positions. The formula $\tfrac{1}{2}Fe$ gives the energy in *joules* when F is in newtons and e is in metres.

Further, since $F = EAe/l$,

$$\text{energy } W = \tfrac{1}{2}EA\frac{e^2}{l}.$$

Suppose that a vertical wire, suspended from one end, is stretched by attaching a weight of 20 N to the lower end. If the weight extends the wire by 1 mm or 1×10^{-3} m, then

$$\text{energy gained by wire} = \tfrac{1}{2}Fe = \tfrac{1}{2} \times 20 \times 1 \times 10^{-3}$$

$$= 10^{-2} \text{ J} = 0.01 \text{ J}.$$

The gravitational potential energy (mgh) lost by the weight in dropping a distance of 1 mm $= 20 \times 1 \times 10^{-3}$ J $= 0.02$ J. Half of this energy, 0.01 J, is the molecular energy gained by the wire; the remainder is the energy dissipated as heat when the weight comes to rest after vibrating at the end of the wire.

Graph of F against e and Energy Measurement

The energy in the wire when it is stretched can also be found from the graph of F against e. Figure 5.4. Suppose the wire extension is e_1 when a force F_1 is applied.

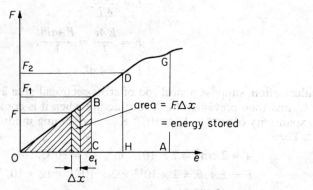

Fig. 5.4 Energy in stretched wire

At some stage before the extension e_1 is reached suppose that the force in the wire is F and that the wire now extends by a very small amount Δx, as shown. Then over this small extension,

$$\text{energy in wire} = \text{work done} = F.\Delta x.$$

Now $F.\Delta x$ is represented by the small *area* between the axis of e and the graph, shown shaded in Fig. 5.4. So the total work done between zero extension and e_1 is the *area OBC between the graph and the axis of e*. The area of the triangle $\text{OBC} = \frac{1}{2}$ base × height $= \frac{1}{2}F_1e_1$, which is in agreement with our formula on p. 120 for the energy stored in the wire.

The area result is a general one. It can be used for both the linear (elastic) and the non-linear (non-elastic) parts of the force F against extension e graph. So in Fig. 5.4, the work done when the force F_1 (extension e_1) is increased to F_2 (extension e_2) is the area of the trapezium BDCH. If the extension occurs from O to A, which is beyond the elastic limit, the work done is still equal to the area of OGA.

It should be noted that the energy in the wire is equal to the area between the graph and the e-axis because F is plotted vertically and e is plotted horizontally. If e is plotted vertically and F is plotted horizontally, the energy in the wire would then be the area between the graph and the *vertical* or e-axis.

Energy per unit volume of wire

The energy per unit volume of a stretched wire is given by a simple formula, as we now see.

The energy stored $= \frac{1}{2}Fe$ and the volume of the wire $= Al$, where A is the cross-section area and l is the length of the wire. So

$$\text{energy per unit volume} = \frac{1}{2}\frac{F.e}{A.l} = \frac{1}{2} \times \left(\frac{F}{A}\right) \times \left(\frac{e}{l}\right)$$

$$= \frac{1}{2} \text{ stress} \times \text{strain}$$

So if the stress in a wire is 2×10^7 N m^{-2} and the strain is 10^{-2}, then

$$\text{energy per unit volume} = \frac{1}{2} \times 2 \times 10^7 \times 10^{-2}$$

$$= 10^5 \text{ J m}^{-3}$$

Examples

1. A uniform steel wire of length 4 m and area of cross-section 3×10^{-6} m² is extended 1 mm. Calculate the energy stored in the wire. (Young modulus $= 2\cdot0 \times 10^{11}$ N m^{-2}.)

$$\text{Stretching force } F = EA\frac{e}{l}$$

So $$\text{energy stored} = \tfrac{1}{2}Fe = \tfrac{1}{2}\frac{EAe^2}{l}$$

$$= \tfrac{1}{2} \times \frac{2 \times 10^{11} \times 3 \times 10^{-6} \times (1 \times 10^{-3})^2}{4} \text{ J},$$

$$= 0\cdot075 \text{ J}.$$

2. Define *stress* and *strain*. Describe the behaviour of a copper wire when it is subjected to an increasing longitudinal stress. Draw a stress–strain diagram and mark on it the elastic region, yield point and breaking stress.

A wire of length 5 m, of uniform circular cross-section of radius 1 mm is extended by 1·5 mm when subjected to a uniform tension of 100 newton. Calculate from first principles the strain energy per unit volume assuming that deformation obeys Hooke's law.

Show how the stress–strain diagram may be used to calculate the work done in producing a given strain, when the material is stretched beyond the Hooke's law region. (*O. & C.*)

$$\text{Strain energy} = \tfrac{1}{2} \text{ tension} \times \text{extension}$$

Tension = 100 N. Extension = $1\cdot5 \times 10^{-3}$ m.

$$\therefore \text{ energy} = \tfrac{1}{2} \times 100 \times 1\cdot5 \times 10^{-3} = 0\cdot075 \text{ J.}$$

Volume of wire = length × area = $5 \times \pi \times 1 \times 10^{-6}$ m³.

$$\therefore \text{ energy per unit volume} = \frac{0\cdot075}{5 \times \pi \times 1 \times 10^{-6}} = 4\cdot7 \times 10^3 \text{ J m}^{-3} \text{ (approx.).}$$

In the stress–strain diagram, the area between the graph and the strain-axis is the work done per unit volume, even though the material is stretched beyond the Hooke's law region. The whole work done = volume of wire × area obtained from graph.

Bulk Modulus

When a gas or a liquid is subjected to an increased pressure the substance contracts. A change in bulk thus occurs, and the *bulk strain* is defined by:

$$\text{strain} = \frac{\text{change in volume}}{\text{original volume}}.$$

The *bulk stress* on the substance is the increased force per unit area, by definition, and the bulk modulus, K, is given by:

$$K = \frac{\text{bulk stress}}{\text{bulk strain}}$$

$$= \frac{\text{increase in force per unit area}}{\text{change in volume/original volume}}.$$

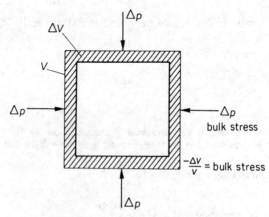

Fig. 5.5 Bulk stress and bulk strain

If the original volume of the substance is V, the change in volume may be denoted by $-\Delta V$ when the pressure increases by a small amount Δp; the minus $(-)$ indicates that the volume decreases. Thus (Fig. 5.5)

$$K = -\frac{\Delta p}{\Delta V/V}.$$

When Δp and ΔV become very small, then, in the limit,

$$K = -V\frac{dp}{dV} \qquad . \qquad . \qquad . \qquad . \qquad . \qquad (1)$$

The bulk modulus of water is about 2×10^9 N m^{-2} for pressures in the range 1–25 atmospheres; the bulk modulus of mercury is about 27×10^9 N m^{-2}. The bulk modulus of gases depends on the pressure, as now explained. Generally, since the volume change is relatively large, the bulk modulus of a gas is low compared with that of a liquid.

Bulk Modulus of a Gas

If the pressure, p, and volume, V, of a gas change under conditions such that

$$pV = \text{constant},$$

which is Boyle's law, the changes are said to be *isothermal* ones. In this case, by differentiating the product pV with respect to V, we have

$$p + V\frac{dp}{dV} = 0.$$

$$\therefore p = -V\frac{dp}{dV}.$$

But the bulk modulus, K, of the gas is equal to $-V\frac{dp}{dV}$ by definition (see above).

$$\therefore K = p. \qquad . \qquad . \qquad . \qquad . \qquad (2)$$

Thus the *isothermal bulk modulus is equal to the pressure.*

When the pressure, p, and volume V, of a gas change under conditions such that

$$pV^\gamma = \text{constant},$$

where $\gamma = C_p/C_V =$ the ratio of the molar heat capacities of the gas, the changes are said to be *adiabatic* (p. 213). This equation is the one obeyed by local values of pressure and volume in air when a sound wave travels through it. Differentiating both sides with respect to V,

$$\therefore p \times \gamma V^{\gamma-1} + V^\gamma\frac{dp}{dV} = 0,$$

$$\therefore \gamma p = -V\frac{dp}{dV},$$

$$\therefore \text{adiabatic bulk modulus} = \gamma p \qquad . \qquad . \qquad . \qquad (3)$$

For air at normal pressure, $K = 10^5$ N m^{-2} isothermally and $1 \cdot 4 \times 10^5$ N m^{-2} adiabatically. The values of K are of the order 10^5 times smaller than liquids as gases are much more compressible.

Velocity of Sound

The velocity of sound waves through any material depends on (i) its density, ρ, (ii) its modulus of elasticity, E. Thus if v is the velocity, we may say that

$$v = kE^x\rho^y \qquad . \qquad . \qquad . \qquad . \qquad \text{(i)}$$

where k is a constant and x, y are indices we can find by the method of dimensions (p. 33).

The units of velocity, v, are LT^{-1}; the units of density, ρ, are ML^{-3}; and the units of modulus of elasticity, E, are $ML^{-1}T^{-2}$ (see p. 117). Equating the dimensions on both sides of (i),

$$\therefore LT^{-1} = (ML^{-1}T^{-2})^x \times (ML^{-3})^y.$$

Equating the indices of M, L, T on both sides, we have

$$0 = x + y,$$
$$1 = -x - 3y,$$
$$-1 = -2x.$$

Solving, we find $x = \frac{1}{2}$, $y = -\frac{1}{2}$. Thus $v = kE^{\frac{1}{2}}\rho^{\frac{1}{2}}$. A rigid investigation shows $k = 1$, and thus

$$v = E^{\frac{1}{2}}\rho^{-\frac{1}{2}} = \sqrt{\frac{E}{\rho}}.$$

In the case of a solid, E is the Young modulus. In the case of air and other gases, and of liquids, E is replaced by the bulk modulus K. Laplace showed that the adiabatic bulk modulus must be used in the case of a gas, and since this is γp, the velocity of sound in a gas is given by the expression

$$v = \sqrt{\frac{\gamma p}{\rho}}.$$

Modulus of Rigidity or Shear Modulus. Torsion Wire

So far we have considered the strain in one direction, or tensile strain, to which the Young modulus is applicable and the strain in bulk or volume, to which the bulk modulus is applicable.

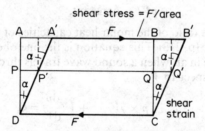

Fig. 5.6 Shear stress and shear strain

Consider a block of material ABCD such as rubber for convenience, Fig. 5.6. Suppose the lower plane CD is fixed, and a stress parallel to CD is applied by a force F to the upper side AB. The block then changes its shape and takes up a position A'B'CD. It can now be seen that planes in the material parallel to DC are displaced relative to each other. The plane AB, for example, which was originally directly opposite the plane PQ, is displaced to A'B' and PQ is displaced to P'Q'. The *angular displacement* α is defined as the *shear strain*. α is the angular displacement between any two planes, for example, between CD and P'Q'.

No volume change occurs in Fig. 5.6. Further, since the force along CD is F in magnitude, it forms a *couple* with the force F applied to the upper side AB. The *shear stress* is defined as the 'shear force per unit area' on the face AB (or CD), as in the Young modulus or the bulk modulus. Unlike the case for these moduli, however, the shear stress has a turning or 'displacement' effect owing to the couple present. The solid does not collapse because in a strained equilibrium position such as A′B′CD in Fig. 5.6, the external couple acting on the solid due to the forces F is balanced by an opposing couple due to stresses inside the material

If the elastic limit is not exceeded when a shear stress is applied, that is, the solid recovers its original shape when the stress is removed, the *modulus of rigidity* or *shear modulus*, G, is defined by:

$$G = \frac{shear\ stress\ (force\ per\ unit\ area)}{shear\ strain\ (angular\ displacement,\ \alpha)}.$$

Shear strain has no units; shear stress has units of N m^{-2}. The modulus of rigidity of copper is $4{\cdot}8 \times 10^{10}$ N m^{-2}; for phosphor-bronze it is $4{\cdot}4 \times 10^{10}$ N m^{-2}, and for quartz fibre it is $3{\cdot}0 \times 10^{10}$ N m^{-2}.

If a helical spring is stretched, all parts of the spring become twisted. The applied force has thus developed a 'torsional' or shear strain. The extension of the spring hence depends on its shear modulus, in addition to its dimensions.

In sensitive current-measuring instruments, a very weak control is needed for the rotation of the instrument coil. This may be provided by using a long elastic or *torsion wire* of phosphor bronze in place of a spring. The coil is suspended from the lower end of the wire and when it rotates through an angle θ, the wire sets up a weak opposing torque equal to $c\theta$, where c is the elastic constant of the wire. Calculation shows that $c = \pi G a^4 / 2l$, where a is the radius of the wire and l its length. Quartz fibres are very fine but comparatively strong, and have elastic properties used for sensitive control of rotation.

Theory of Plastic Deformation

We conclude with a brief account of the theory underlying the deformation of metals due to applied stress.

X-ray diffraction patterns show that metals are *crystalline* (p. 862). Thus a metal consists of an arrangement of regularly spaced atoms. Normally, bonds between the atoms keep them stacked in a particular structure or crystal lattice. Figure 5.7 (i) illustrates a perfect crystal. Certain planes through the crystal, called *crystal planes*, are rich in atoms.

As discussed earlier, the atoms are slightly displaced at low tensile stress and return to their normal positions when the stress is released if this is less than the elastic limit of the metal. If the metal is stretched beyond the elastic limit, it does not return to its original length. The metal is then said to have undergone *plastic deformation*. This is caused by movement of crystal planes called *slip*. The origin of slip lies in crystal *dislocations*, which we now discuss.

Dislocation and Slip

In a real crystal, the arrays of atoms are not perfectly regular. They have imperfections caused during formation or manufacture, such as missing atoms, extra or foreign (impurity) atoms, and *dislocations*.

Figure 5.7 (ii) shows one type of dislocation, called an *edge dislocation*. As illustrated, it introduces a lack of symmetry in the crystal. We can see this by considering atoms in a normal crystal, Fig. 5.7 (i). Here a path ABCD, followed

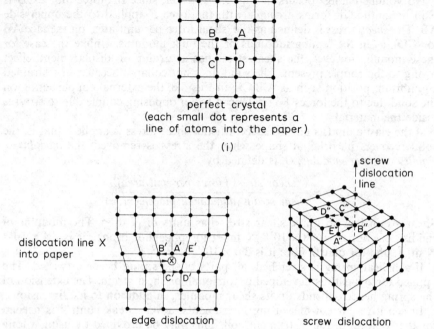

Fig. 5.7 (i) Perfect crystal (ii, iii) Dislocations

from A with displacements successively west (AB), south (BC), east (CD) and north (DA), is always a 'closed' path. A dislocation line, however, is a set of points for which a similar path does *not* close. For example, in the edge dislocation in Fig. 5.7 (ii), the similar path starting at A' ends at E' and not A'.

Figure 5.7 (iii) shows another type of dislocation called a *screw dislocation*. Here a similar path to ABCD in Fig. 5.7 (i) starts at A" and ends at E", following a helical path A"B"C"D"E". It should be noted that in a screw dislocation, the extra displacement A"E" from A" is *along* the dislocation line, whereas in an edge dislocation the extra displacement A'E', Fig. 5.7 (ii), is *perpendicular* to the dislocation line.

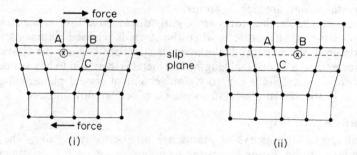

Fig. 5.8 Slip in crystal

When a metal is stressed so that plastic deformation occurs, the crystal planes slip by movement of dislocations. Figure 5.8 illustrates how this happens. The dislocation line at X in Fig. 5.8 (i) moves from A to B, Fig. 5.8 (ii), when the bond CB is broken and replaced by the bond CA. In this way the dislocation moves from left to right through the whole crystal. The process is analogous to the movement of a ruck from one side of a carpet to the other. Further, the movement of the dislocation through the crystal requires a very much lower stress than one involving the movement of whole planes of atoms over each other, which is in agreement with experiment. The movement of whole planes of atoms is analogous to moving a carpet bodily. Summarising, *slip is due to movement of dislocations.*

A raft of equally-sized bubbles on the surface of a liquid is an extremely good model of a metal crystal. The *bubble raft* was first used by Sir Lawrence Bragg. The bubbles attract each other with close-range forces similar to the van der Waals forces which hold atoms together and they tend to form close-packed structures. Dislocations can be seen at places in the bubbles. By touching these places lightly, the motion of the dislocations and their interaction with each other can be observed.

Work Hardening

If a metal is repeatedly deformed, its resistance to plastic deformation increases. This is called 'work hardening'. A thick copper wire, deformed by bending it repeatedly, becomes harder and brittle—this is 'cold' work hardening.

Work hardening can be explained from the movement of dislocations. As we have seen, if a sample of crystalline solid such as a metal undergoes plastic deformation, the dislocation lines move through it. When, however, the dislocations encounter other types of imperfections, such as an impurity atom, they tend to become *pinned* and tangled. The more the solid is deformed, the greater is the amount of entanglement. The movement of the dislocation thus becomes more difficult and so the strength of the metal increases. This accounts for work hardening.

Recently it has been shown that single crystal 'whiskers' can be made practically dislocation-free. Such crystals are brittle but very strong, since plastic deformation can only occur by the movement of whole planes of atoms. When set in a suitable resin they can be used as the basis of a strong composite material.

Non-Metals. Rubber and Glass

Rubber. A rubber band can be stretched to many times its length. Rubber stretches very easily compared with copper, that is, it is much less stiff than the metal. Figure 5.9 is an exaggerated sketch. Before it snaps, a length of rubber returns to its original length and shape when released. So rubber is elastic at high strains, such as 700%, whereas copper is elastic at relatively small strains, such as 0·1%.

The reason for the large stretching of rubber without breaking lies in its molecular structure. Unstretched rubber has coiled molecules. When stretched the molecules unwind and become straight. Thus, as X-ray diffraction shows, stretched rubber has a more ordered structure than unstretched rubber. After rubber is stretched and all the molecules are straight, it is much harder to stretch the rubber further. See Fig. 5B overleaf.

When a metal is slightly stretched, the interatomic spacing increases. The spacing returns to its original value when the force is removed (see p. 115). When the metal is stretched beyond the elastic limit, slip occurs as explained previously and the metal undergoes plastic deformation. In contrast, the stretch-

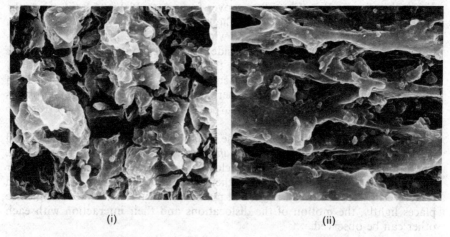

(i)   (ii)

Fig. 5B Scanning electron micrographs of (i) unstretched rubber chains—coiled molecules, (ii) stretched rubber chains—uncoiled molecules. (*Courtesy of Biophoto Associates*)

ing of rubber occurs by the straightening of its molecules, even for large strains. So unlike metals there is no plastic deformation when rubber is stretched.

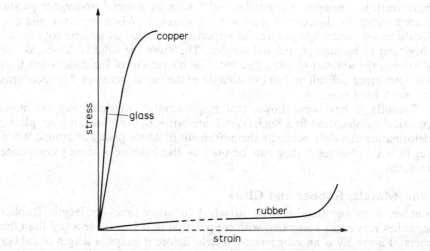

Fig. 5.9 Stress-strain curves (*not to scale*)

Glass. Glass is very stiff at room temperature; its stiffness is then greater than that of steel. It has only a small elastic region, however, and is brittle. So it has no plastic region and fractures easily, Fig. 5.9. This behaviour of glass is due to the existence of cracks in its surface. Unlike metals, which can yield easily to high stresses appearing across cracks and thus relieve the stress, glass does not yield. So the high concentration of stress at the crack makes the glass break. Glass is a super-cooled liquid. At high temperatures it begins to melt and flow easily.

Further information about non-metals is outside the scope of this book and may be found in *The New Science of Strong Materials* by J. E. Gordon (Penguin).

Molecules and Matter

Particle Nature of Matter

Matter is made up of many millions of molecules, which are particles whose dimensions are about 3×10^{-10} m. Evidence of the existence of molecules is given by experiments demonstrating *Brownian motion*, with which we assume the reader is familiar. One example is the irregular motion of smoke particles in air, which can be observed by means of a microscope. This is due to continuous bombardment of a tiny smoke particle by numerous air molecules all round it. The air molecules move with different velocities in different directions. The resultant force on the smoke particle is therefore unbalanced, and irregular in magnitude and direction. Larger particles do not show Brownian motion when struck on all sides by air molecules. The resultant force is then relatively negligible.

More evidence of the existence of molecules is supplied by the successful predictions made by the *kinetic theory of gases*. This theory assumes that a gas consists of millions of separate particles or molecules moving about in all directions (p. 219). *X-ray diffraction patterns* of crystals also provide evidence for the particle nature of matter (p. 862). The symmetrical patterns of spots obtained are those which one would expect from a three-dimensional grating or lattice formed from particles. A smooth continuous medium would not give a diffraction pattern of spots.

Size and Separation of Molecules

The size of atoms and molecules can be estimated in several different ways. By allowing an oil drop to spread on water, for example, an upper limit of about 5×10^{-7} cm is obtained for the size of an oil molecule. X-ray diffraction experiments enable the interatomic spacing between atoms in a crystal to be accurately found. The results are of the order of a few angstrom units, such as $3 \, \text{Å}$ or 3×10^{-10} m or 0.3 nm (nanometre).

A simple calculation shows the order of magnitude of the enormous number of molecules present in a small volume. One gram of water occupies 1 cm^3. One mole has a mass of 18 g, and thus occupies a volume of 18 cm^3 or 18×10^{-6} m^3. Assuming the diameter of a molecule is 3×10^{-10} m, its volume is roughly $(3 \times 10^{-10})^3$ or 27×10^{-30} m^3. Hence the number of molecules in one mole $= 18 \times 10^{-6}/(27 \times 10^{-30}) = 7 \times 10^{23}$ approximately.

The *Avogadro constant*, N_A, is the number of molecules in one mole of a substance. Accurate values show that $N_A = 6.02 \times 10^{23}$ mol^{-1}, or 6.02×10^{26} kmol^{-1}, where 'kmol' represents a kilomole, 1000 moles.

The order of separation of molecules in liquids is about the same as in solids. We can calculate the separation of gas molecules at standard pressure from the fact that a mole of any gas occupies about 22·4 litres at s.t.p., or 22.4×10^{-3} m^3. Since one mole contains about 6×10^{23} molecules, then, roughly, taking the cube root of the volume per molecule,

$$\text{average separation} = \sqrt[3]{\frac{22.4 \times 10^{-3}}{6 \times 10^{23}}} \ \text{m}$$

$$= 33 \times 10^{-10} \ \text{m (approx.)}$$

This is about 10 times the separation of molecules in solids or liquids.

The lightest atom is hydrogen. Since about 6×10^{23} hydrogen molecules have a mass of 2 g or 2×10^{-3} kg, and each hydrogen molecule consists of two atoms, then

$$\text{mass of hydrogen atom} = \frac{2 \times 10^{-3}}{2 \times 6 \times 10^{23}} = 1.7 \times 10^{-27} \text{ kg (approx.).}$$

Heavier atoms have masses in proportion to their relative atomic masses.

Example

Estimate the order of separation of atoms in aluminium metal, given the density is 2700 kg m^{-3}, the relative atomic mass is 27 and the Avogadro constant is 6×10^{23} mol^{-1}.

1 mole of aluminium has a mass of 27 g. From the density value,

$$1 \text{ m}^3 \text{ of aluminium has } \frac{2700 \times 10^3}{27} \text{ or } 10^5 \text{ moles.}$$

Now 1 mole contains 6×10^{23} molecules or atoms of aluminium

So $6 \times 10^{23} \times 10^5$ atoms occupy a volume of 1 m^3.

Hence $\text{volume per atom} = \dfrac{1}{6 \times 10^{28}} = 1.7 \times 10^{-29} \text{ m}^3.$

The volume occupied per atom is of the order d^3, where d is the separation of the atoms. So, approximately,

$$d = \sqrt[3]{1.7 \times 10^{-29}} \text{ m}$$
$$= 2.6 \times 10^{-10} \text{ m.}$$

Intermolecular forces

The forces which exist between molecules can explain many of the bulk properties of solids, liquids and gases. These intermolecular forces arise from two main causes:

(1) The *potential energy* of the molecules, which is due to interactions with surrounding molecules (this principally electrical in origin).

(2) The *thermal energy* of the molecules—this is the kinetic energy of the molecules and it depends on the temperature of the substance concerned.

We shall see later that the particular state or phase in which matter appears—that is, solid, liquid or gas—and the properties it then has, are determined by the relative magnitudes of these two energies.

Potential Energy and Force

In bulk, matter consists of numerous molecules. To simplify the situation, Fig. 5.10 shows the variation of the mutual potential energy V between two molecules at a distance r apart.

Along the part BCD of the curve, the potential energy V is negative. Along the part AB, the potential energy V is positive. Generally, V can be written approximately as

$$V = \frac{a}{r^p} - \frac{b}{r^q} \qquad . \qquad . \qquad . \qquad . \qquad . \qquad (1)$$

where p and q are powers of r, and a and b are constants. The positive term with the constant a indicates a repulsive force and the negative term with the constant b an attractive force, as discussed shortly.

There are different kinds of bonds or forces between atoms and molecules in solids, depending on the nature of the solid. In an *ionic solid*, for example, sodium chloride, V can be approximated by

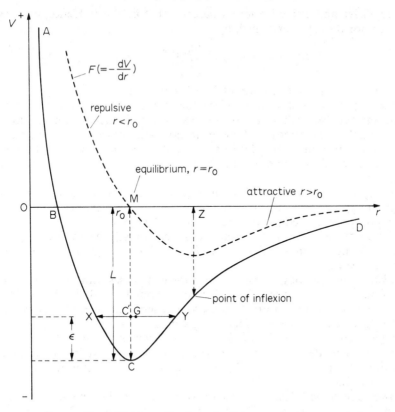

Fig. 5.10 Intermolecular potential energy and force

$$V = \frac{a}{r^9} - \frac{b}{r} \qquad \qquad \text{(2)}$$

The force F between molecules is generally given by $F = -dV/dr$, the negative *potential gradient* in the field (see p. 584). From (2), it follows that, for the two ions,

$$F = \frac{9a}{r^{10}} - \frac{b}{r^2} \qquad \qquad \text{(3)}$$

The $+$ve term in (3) indicates a *repulsive* force since the force acts in the direction of increasing r. This is the force along ABC in Fig. 5.10. The $-$ve term in (3) indicates an *attractive* force since the force acts oppositely to the direction of increasing r. This force acts along CD in Fig. 5.10; as shown, it decreases with increasing separation, r, of the two molecules.

Properties of Solids from Molecular Theory

Several properties of a model solid can be deduced or calculated from the potential–separation $(V - r)$ graph or the force–separation $(F - r)$ graph.

Equilibrium spacing of molecules. The value of r when the potential energy V is a minimum corresponds to the stable or equilibrium spacing between the molecules. At the absolute zero, where the thermal energy is zero, this corresponds to C in Fig. 5.10, or a separation $r = r_0$. At this separation or spacing, the repulsive and attractive forces balance, that is, $F = 0$. Hence, from (3), r_0 is given, for the ions concerned, by

$$r_0 = \left(\frac{9a}{b}\right)^{1/8}.$$

The value of r_0 for solids is about 2 to 5×10^{-10} m.

If the separation r of the molecules is slightly increased from r_0, the attractive force between them will restore the molecules to their equilibrium position after the external force is removed. If the separation is decreased from r_0, the repulsive force will restore the molecules to their equilibrium position after the external force is removed. So the molecules of a solid *oscillate* about their equilibrium or mean position.

Elasticity and Hooke's law. Near the equilibrium position r_0, the graph of F against r approximates to a straight line, Fig. 5.10. This means that the extension is proportional to the applied force (Hooke's law, p. 115).

The 'force constant', k, between the molecules is given by $F = -k(r - r_0)$, where r is slightly greater than r_0. So $k = -dF/dr = -$gradient of tangent to curve at $r = r_0$.

Breaking strain. So long as the restoring force increases with increasing separation from $r = r_0$, the molecules will remain bound together. This is the case from $r = r_0$ to $r = OZ$ in Fig. 5.10. Beyond a separation $r = OZ$, however, the restoring force *decreases* with increasing separation. OZ is therefore the separation between the molecules at the *breaking point* of the solid (see p. 119). It corresponds to the value of r for which $dF/dr = 0$, that is, to the point of inflexion of the $V - r$ curve. The *breaking strain* = extension/r_0 = $(x - r_0)/r_0$, where $x = OZ$.

Thermal expansion. Molecules remain stationary at absolute zero, since their thermal energy is then zero. This corresponds to the point C of the energy curve in Fig. 5.10. At a higher temperature, the molecules have some energy, ε, above the minimum value, as shown. Hence they oscillate between points such as X and Y. Since the $V - r$ curve is not symmetrical, the mean position G of the oscillation is on the right of C', as shown. This corresponds to a greater separation than r_0. Thus the solid *expands* when its thermal energy is increased.

At a slightly higher temperature, the mean position moves further to the right of G and so the solid expands further. When the energy equals CM (latent heat, L), the energy enables the molecules to break completely the bonds of attraction which keep them in a bound state. The molecules then have little or no interaction and now form a *gas*.

Latent Heat of Vaporisation

Inside a liquid, molecules continually break and reform bonds with neighbours. The 'latent heat of vaporisation L' of a liquid is the energy to break all the bonds between its molecules.

Suppose ε is the energy to separate a particular molecule X from its nearest neighbour, that is, the energy per pair of molecules. If there are n nearest neighbours per molecule, and we neglect the effect of the other molecules, then the energy to break the bonds between X and its neighbours is $n\varepsilon$.

With a mole of liquid, there are N_A molecules inside it, where N_A is the Avogadro constant. The number of pairs of molecules is $\frac{1}{2}N_A$. So the energy required to break the bonds of all the molecules at the boiling point is roughly $\frac{1}{2}N_A n\varepsilon$. Thus the latent heat of vaporisation per mol, $L = \frac{1}{2}N_A n\varepsilon$.

Bonds between Atoms and Molecules

The atoms and molecules in solids, liquids and gases are held together by so-called *bonds* between them. There are different types of bonds. All are due to electrostatic forces which arise from the +ve charge on the nucleus of an atom and its surrounding electrons which carry −ve charges.

(*a*) *Ionic bond.* As explained on p. 678, sodium chloride in the solid state consists of +ve sodium ions and −ve chlorine ions; the ions are formed because an electron is readily given up by a sodium atom to a chlorine atom. The ions in the crystal are held together by the electrostatic attraction between the opposite charges, Fig. 5.11. This is called an 'ionic bond', and is a strong bond.

(*b*) *Covalent bond.* In a hydrogen molecule, H_2, the electron in one atom wanders to the other. Thus at any instant the atom A from which the electron departed is left with a +ve charge and the atom B to which the electron moved has a −ve charge, Fig. 5.11. Thus the atoms attract each other. The same result is obtained when an electron moves from B to A.

The bond between atoms due to shared electrons is called a 'covalent bond'. Note there is no transfer of electrons as in the case of the ionic bond. Diamond is a solid kept in this state by covalent bonds between neighbouring carbon atoms. As shown by the stability of the hydrogen molecule and the high melting point of a diamond, the covalent bond is very strong.

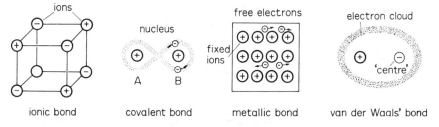

Fig. 5.11 Types of bonds (diagrammatic)

(*c*) *Metallic bond.* In solid metals such as sodium or copper, one or more electrons in the outermost part of the atom may leave and occupy the orbit of another atom. These so-called 'free' electrons wander through the metal crystal structure, which consists of fixed +ve ions. The metallic bond is similar to a covalent bond except that electrons are not attached to any particular atoms; it keeps the metal in its solid state. The metallic bond is not as strong as the ionic and covalent bonds.

(*d*) *Van der Waals' bond.* Over a long time-interval, the 'centre' of an electron cloud round the nucleus is at the nucleus itself. At any instant, however, more electrons may appear on one side of the nucleus than the other. In this case the 'centre' of the electron cloud, or −ve charge, is slightly displaced from the +ve charge on the nucleus. The two charges now form an 'electric dipole' (see p. 603). A dipole attracts the electrons in neighbouring atoms, forming other dipoles.

The electric dipoles have weak forces between them, called 'van der Waals' forces because similar attractive forces were predicted by van der Waals in connection with the molecules of gases (see p. 251). Solid neon, an inert element, is kept in this state by these bonds; the low melting point of solid neon shows that the bonds are weak.

Finally, it may be noted that the ionic bond is 'non-saturated'—the attractive force between two ions is not affected by the presence of any attached or surrounding ions. It is also 'non-directional'—the force due to an ion is exerted equally all round the ion. In contrast, the covalent bond can be saturated and may have direction. The metallic bond is an unsaturated covalent bond; hence close-packing of atoms occurs in metals.

Summarising. The strong ionic bond is due to the electrostatic force between ions. The strong covalent bond is due to shared electrons between atoms. The fairly strong metallic bond is similar to a covalent bond except that electrons are not attached to any particular atoms. The weak van der Waals' bond is due to the force between electric dipoles formed by the nucleus and the electron cloud.

Exercises 5

(Assume $g = 10$ N kg^{-1} unless otherwise given)

Elasticity

1. A wire 2 m long and cross-sectional are a 10^{-6} m^2 is stretched 1 mm by a force of 50 N in the elastic region. Calculate (i) the strain, (ii) the Young modulus, (iii) the energy stored in the wire.

2. Figure 5A shows the variation of F, the load applied to two wires X and Y, and their extension e. The wires are both iron and have the same length.

Fig. 5A

(i) Which wire has the smaller cross-section? (ii) Explain how you would use the graph for X to obtain a value for the Young modulus of iron, listing the additional measurements needed.

3. Define *tensile stress, tensile strain, Young modulus.* What are the units and dimensions of each?

A force of 20 N is applied to the ends of a wire 4 m long, and produces an extension of 0·24 mm. If the diameter of the wire is 2 mm, calculate the stress on the wire, its strain, and the value of the Young modulus.

4. What force must be applied to a steel wire 6 m long and diameter 1·6 mm to produce an extension of 1 mm? (Young modulus for steel $= 2·0 \times 10^{11}$ N m^{-2}.)

5. Find the extension produced in a copper wire of length 2 m and diameter 3 mm when a force of 30 N is applied. (Young modulus for copper $= 1·1 \times 10^{11}$ N m^{-2}.)

6. A spring is extended by 30 mm when a force of 1·5 N is applied to it. Calculate the energy stored in the spring when hanging vertically supporting a mass of 0·20 kg if the spring was unstretched before applying the mass. Calculate the loss in potential energy of the mass. Explain why these values differ. (*L.*)

7. Define *elastic limit* and *Young modulus* and describe how you would find the values for a copper wire.

What stress would cause a wire to increase in length by one-tenth of one per cent if the Young modulus for the wire is 12×10^{10} N m^{-2}? What force would produce this stress if the diameter of the wire is 0·56 mm? (*L.*)

8. In an experiment to measure the Young modulus for steel a wire is suspended vertically and loaded at the free end. In such an experiment, (*a*) why is the wire long and thin, (*b*) why is a second steel wire suspended adjacent to the first?

Sketch the graph you would expect to obtain in such an experiment showing the relation between the applied load and the extension of the wire. Show how it is possible to use the graph to determine (a) Young modulus for the wire, (b) the work done in stretching the wire.

If the Young modulus for steel is 2.00×10^{11} N m^{-2}, calculate the work done in stretching a steel wire 100 cm in length and of cross-sectional area 0.030 cm^2 when a load of 100 N is slowly applied without the elastic limit being reached. (N.)

9. Define the terms *tensile stress* and *tensile strain* and explain why these quantities are more useful than *force* and *extension* for a description of the elastic properties of matter.

Describe the apparatus you would use and the measurements you would perform to investigate the relation between the tensile stress applied to a wire and the strain it produces.

A cylindrical copper wire and a cylindrical steel wire, each of length 1.5 m and diameter 2 mm, are joined at one end to form a composite wire 3 m long. The wire is loaded until its length becomes 3.003 m. Calculate the strains in the copper and steel wires and the force applied to the wire.

(Young modulus for copper = 1.2×10^{11} N m^{-2}; for steel 2.0×10^{11} N m^{-2}.) (O. & C.)

10. State Hooke's law and describe, with the help of a rough graph, the behaviour of a copper wire which hangs vertically and is loaded with a gradually increasing load until it finally breaks. Describe the effect of gradually reducing the load to zero (a) before, (b) after the elastic limit has been reached.

A uniform steel wire of density 7800 kg m^{-3} weighs 16 g and is 250 cm long. It lengthens by 1.2 mm when stretched by a force of 80 N. Calculate (a) the value of the Young modulus for the steel, (b) the energy stored in the wire. (N.)

11. Explain the terms *stress, strain, modulus of elasticity* and *elastic limit*. Derive an expression in terms of the tensile force and extension for the energy stored in a stretched rubber cord which obeys Hooke's law.

The rubber cord of a catapult has a cross-sectional area 1.0 mm^2 and a total unstretched length 10.0 cm. It is stretched to 12.0 cm and then released to project a missile of mass 5.0 g. From energy considerations, or otherwise, calculate the velocity of projection, taking the Young modulus for the rubber as 5.0×10^8 N m^{-2}. State the assumptions made in your calculation. (L.)

12. (a) Define the *Young modulus*. Describe how you would make an accurate determination of this quantity for material in the form of a wire.

(b) A wire of length 3.0 m and cross-sectional area 1.0×10^{-6} m^2 has a mass of 15 kg hung on it. What is the stress produced in the wire? ($g = 9.8$ m s^{-2}.)

If the Young modulus for the material is 2.0×10^{11} N m^{-2}, what is the extension, x, produced? When extended how much energy is stored in the wire?

If the mass of 15 kg were allowed to fall through a distance x, what would be the change in its gravitational potential energy? Why is this change not equal to the final energy stored in the wire when extended by the mass? (L.)

13. (a) A wire when stretched behaves at first elastically and then plastically until it finally breaks. Describe carefully what happens to the energy used to stretch the wire at each stage.

(b) A rod of mild steel, of uniform cross-section 0.030 m^2 and length 1.0 m, is stretched

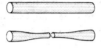

Fig. 5B

steadily by a machine. At first it stretches uniformly but only by a very small amount; then a neck is produced in the rod, and finally it breaks as shown in Fig. 5B. Sketch a force–extension graph to represent what takes place.

In such an experiment a series of readings of the force and extension are recorded and a graph is plotted which is found to be a straight line through the origin up to a value of the stretching force of 120 kN. If the strain at that point is 1/2500, what is the value of the Young modulus for mild steel?

The rod is found to break at a stretching force of 240 kN. Explain why the stress at the section of the rod where the break occurs is likely to be much greater than 8000 kN m^{-2}.

Immediately before the rod breaks, its extension is about 4 cm. Make a rough estimate of the work done in stretching it this amount. By referring to your force–extension graph, or otherwise, suggest why your value of work done is likely to be too small. (*L.*)

14. Define *stress* and *strain*, and explain why these quantities are useful in studying the elastic behaviour of a material.

State one advantage and one disadvantage in using a long wire rather than a short stout bar when measuring the Young modulus by direct stretching.

Calculate the minimum tension with which platinum wire of diameter 0·1 mm must be mounted between two points in a stout invar frame if the wire is to remain taut when the temperature rises 100 K. Platinum has linear expansivity 9×10^{-6} K^{-1} and Young modulus 17×10^{10} N m^{-2}. The thermal expansion of invar may be neglected. (*O. & C.*)

15. Describe the changes which take place when a wire is subjected to a steadily increasing tension. Include in your description a sketch graph of tension against extension for (*a*) a ductile material such as drawn copper and (*b*) a brittle one such as cast iron.

Show that the energy stored in a rod of length L when it is extended by a length l is $\frac{1}{2}El^2/L^2$ per unit volume where E is Young modulus of the material.

A railway track uses long welded steel rails which are prevented from expanding by friction in the clamps. If the cross-sectional area of each rail is 75 cm^2 what is the elastic energy stored per kilometre of track when its temperature is raised by 10°C? (Linear expansivity of steel = $1·2 \times 10^{-5}$ K^{-1}; Young modulus for steel = 2×10^{11} N m^{-2}.) (*O. & C.*)

16. What is meant by saying that a substance is 'elastic'?

A vertical brass rod of circular section is loaded by placing a 5 kg weight on top of it. If its length is 50 cm, its radius of cross-section 1 cm, and the Young modulus of the material $3·5 \times 10^{10}$ N m^{-2}, find (*a*) the contraction of the rod, (*b*) the energy stored in it. (*C.*)

17. (*a*) Define *stress, strain, Young modulus*. (*b*) The formula for the velocity v of compressional waves travelling along a rod made of material of Young modulus E and density ρ is $v = (E/\rho)^{\frac{1}{2}}$. Show that this formula is dimensionally consistent. (*c*) A uniform wire of unstretched length 2·49 m is attached to two points A and B which are 2·0 m apart and in the same horizontal line. When a 5 kg mass is attached to the midpoint C of the wire, the equilibrium position of C is 0·75 m below the line AB. Neglecting the weight of the wire and taking the Young modulus for its material to be 2×10^{11} N m^{-2}, find (i) the strain in the wire, (ii) the stress in the wire, (iii) the energy stored in the wire. (*O.*)

18. Define the Young modulus of elasticity. Describe an accurate method of determining it. The rubber cord of a catapult is pulled back until its original length has been doubled. Assuming that the cross-section of the cord is 2 mm square, and that Young modulus for rubber is 10^7 N m^{-2} calculate the tension in the cord. If the two arms of the catapult are 6 cm apart, and the unstretched length of the cord is 8 cm what is the stretching force? (*O. & C.*)

19. State Hooke's law, and describe in detail how it may be verified experimentally for copper wire. A copper wire, 200 cm long and 1·22 mm diameter, is fixed horizontally to two rigid supports 200 cm long. Find the mass in grams of the load which, when suspended at the mid-point of the wire, produces a sag of 2 cm at that point. Young modulus for copper = $12·3 \times 10^{10}$ N m^{-2}. (*L.*)

Molecular Forces and Energy

20. Figure 5c shows (i) the variation PADE of potential energy V between two molecules with their separation r, and (ii) the variation QBLC of the force F between the molecules with their separation r.

(*a*) Explain how the $F - r$ curve is obtained from the $V - r$ curve, (*b*) State which part of the $F - r$ curve corresponds to a repulsive force and which part to an attractive force. Describe briefly one experiment which shows the repulsive force and one experiment which shows the attractive force.

21. Explain how you would use the $V - r$ curve in Fig. 5c (i) (a) to obtain the equilibrium separation of the molecules, (b) to find the energy needed to completely separate two molecules initially at the equilibrium separation, (c) to show that a solid usually expands when its thermal energy is increased.

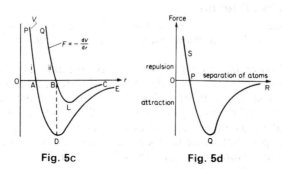

Fig. 5c   Fig. 5d

22. Using the $F - r$ curve in Fig. 5c (ii), explain how you would (a) find the force constant $k (F = -kx$, where x is the extension of a wire in the elastic region) (b) account for Hooke's law of elasticity, (c) obtain a value for the breaking force of a solid.

23. The force between two molecules may be regarded as an attractive force which increases as their separation decreases and a repulsive force which is only important at small separations and which there varies very rapidly. Draw sketch graphs (a) for force-separation, (b) for potential energy-separation. On each graph mark the equilibrium distance and on (b) indicate the energy which would be needed to separate two molecules initially at the equilibrium distance.

With the help of your graphs discuss briefly the resulting motion if the molecules are displaced from the equilibrium position (N).

24. The graph (Fig. 5D) represents the relationship between the interatomic forces which exist in a material and the separation of the atoms. What point on the graph corresponds to the separation when the material is not subjected to any stress? Use the graph to explain (i) why energy is stored in a material when it is compressed and when it is extended, and (ii) why, and over what region, the material can be expected to obey Hooke's law. (*AEB*, 1982)

25. An elastic modulus is the ratio of stress to strain. Explain carefully the precise nature of the stress and strain in each case for (i) the bulk modulus, (ii) the shear modulus, (iii) the Young modulus of an elastic solid. Which, if any, of these moduli can be used to describe the properties of (a) a liquid, (b) a gas?

Sketch the form of the force–distance curve between atoms of a solid and discuss, with reference to your curve, why most solids appear to obey Hooke's law.

Show, for the case of bulk modulus, that the strain energy per unit volume of a material obeying Hooke's law is $\frac{1}{2} \times$ stress $\times$ strain. The bulk modulus of lead is $4 \cdot 5 \times 10^{10}$ N m^{-2}. Calculate the pressure which must be applied to lead in order to produce a strain energy of 100 J m^{-3}. (*O. & C.*)

6 Surface Tension

Surface Tension Phenomena

We now consider in detail a phenomenon of a liquid surface called *surface tension*. As we shall soon show, surface tension is due to intermolecular attraction.

It is a well-known fact that some insects, for example a 'pond skater', are able to walk across a water surface; that a drop of water may remain suspended for some time from a tap before falling, as if the water particles were held together in a bag; that mercury gathers into small droplets when split; and that a dry steel needle may be made, with care, to float on water, Fig. 6.1. These observations suggest that *the surface of a liquid acts like an elastic skin covering the liquid or is in a state of tension.* Thus forces S in the liquid support the weight W of the needle, as shown in Fig. 6.1.

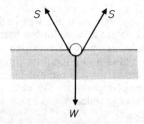

Fig. 6.1 Needle floating on water

Energy of Liquid Surface. Molecular Theory

The fact that a liquid surface is in a state of tension can be explained by the intermolecular forces discussed on p. 130. In the bulk of the liquid, which begins only a few molecular diameters downwards from the surface, a particular molecule such as A is surrounded by an equal number of molecules on all sides. This can be seen by drawing a sphere round A, Fig. 6.2. The average distance apart of the molecules is such that the attractive forces balance the repulsive forces (p. 132). Thus the average intermolecular force between A and the surrounding molecules is zero, Fig. 6.2.

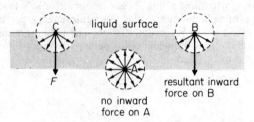

Fig. 6.2 Molecular forces in liquid

Consider now a molecule such as C or B in the surface of the liquid. There are very few molecules on the vapour side above C or B compared with the liquid below, as shown by drawing a sphere round C or B. Thus if C is displaced very slightly upward, a resultant attractive force F on C, due to the large number of molecules below C, now has to be overcome. It follows that if all the molecules in the surface were removed to infinity, a definite amount of work would be needed. *Consequently molecules in the surface have potential energy.*

A molecule in the bulk of the liquid forms bonds with more neighbours than one in the surface. Thus bonds must be broken, i.e. work must be done, to bring a molecule into the surface. So molecules in the surface of the liquid have more potential energy than those in the bulk.

Surface Area. Shape of Drop

The potential energy of any system in stable equilibrium is a minimum. Thus under surface tension forces, the area of a liquid surface will have the least number of molecules in it, that is, *the surface area of a given volume of liquid is a minimum.* Mathematically, it can be shown that the shape of a given volume of liquid with a minimum surface area is a *sphere.*

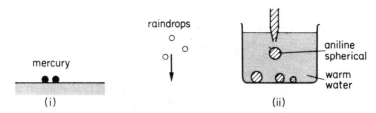

Fig. 6.3 Liquid drops

This is why raindrops, and small droplets of mercury, are approximately spherical in shape, Fig. 6.3 (i). To eliminate completely the effect of gravitational forces, Plateau placed a drop of oil in a mixture of alcohol and water of the same density. In this case the weight of the drop is counterbalanced by the upthrust of the surrounding liquid. He then observed that the drop was a perfect sphere. Plateau's 'spherule' experiment can be carried out by warming water in a beaker and then carefully introducing aniline with the aid of a pipette, Fig. 6.3 (ii). At room temperature the density of aniline is slightly greater than water. At a higher temperature the densities of the two liquids are roughly the same and the aniline is then seen to form *spheres*, which rise and fall in the liquid.

A soap bubble is spherical because the gravitational force on its weight is extremely small and the liquid shape is then mainly due to surface tension forces. Although the density of mercury is high, small drops of mercury are spherical. The ratio of surface area ($4\pi r^2$) to weight (or volume, $4\pi r^3/3$) of a sphere is proportional to the ratio r^2/r^3, or to $1/r$. Thus the smaller the radius, the greater is the influence of surface tension forces compared with the weight. Large mercury drops, however, are flattened on top. This time the gravitational force is relatively greater. The shape of the drop must agree with the principle that the sum of the gravitational potential energy and the surface energy is a minimum, and so the centre of gravity moves down as low as possible.

Lead shot is manufactured by spraying lead from the top of a tall tower. As they fall, the small drops form spheres under the action of surface tension forces.

Surface Tension Definition. Units, Dimensions

Since the surface of a liquid acts like an elastic skin, the surface is in a state of tension. A blown-up football bladder has a surface in a state of tension but the surface tension of a bladder increases as the surface area increases, whereas the surface tension of a liquid is independent of surface area. Any line in the bladder surface is acted on by two equal and opposite forces, and if the bladder is cut with a knife the rubber is drawn away from the cut by the two forces present and the bladder bursts open.

The *surface tension*, γ, of a liquid, sometimes called the *coefficient of surface tension*, is defined as *the force per unit length acting in the surface at right angles to one side of a line drawn in the surface*. In Fig. 6.4 AB represents a line 1 m long. The unit of γ is *newton metre*$^{-1}$ (N m^{-1}).

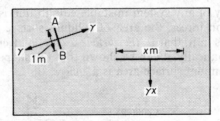

Fig. 6.4 Surface tension

The magnitude of γ depends on the temperature of the liquid and on the medium on the other side of the surface. For water at 20°C in contact with air, $\gamma = 7 \cdot 26 \times 10^{-2}$ newton metre^{-1}. For mercury at 20°C in contact with air, $\gamma = 46 \cdot 5 \times 10^{-2}$ N m^{-1}. The surface tension of a water-oil (olive-oil) boundary is $2 \cdot 06 \times 10^{-2}$ N m^{-1}, and for a mercury-water boundary it is $42 \cdot 7 \times 10^{-2}$ N m^{-1}.

Since surface tension γ is a 'force per unit length', the dimensions of

$$\text{surface tension} = \frac{\text{dimensions of force}}{\text{dimensions of length}} = \frac{MLT^{-2}}{L}$$

$$= MT^{-2}.$$

In addition to defining surface tension as a force per unit length, we shall see later that surface tension can also be defined in terms of surface energy (p. 154).

Some Surface Tension Phenomena

The effect of surface tension forces in a soap film can be demonstrated by placing a thread B carefully on a soap film formed in a metal ring A, Fig. 6.5 (i). The surface tension forces on both sides of the thread counterbalance, as shown in Fig. 6.5 (i). If the film enclosed by the thread is pierced, however, the

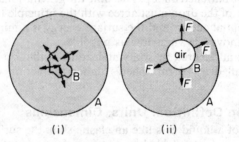

Fig. 6.5 Contraction of surface

thread is pulled out into a circle by the surface tension forces F which now act on one side only of the thread, Fig. 6.5 (ii). Observe that the film has contracted to a minimum area.

Another demonstration of surface tension forces can be made by sprinkling light dust or lycopodium powder over the surface of water contained in a dish. If the middle of the water is touched with the end of a glass rod previously dipped into soap solution, the powder is carried away to the sides by the water. The explanation lies in the fact that the surface tension of pure water is greater than that of soapy (impure) water (p. 142). The resultant force at the place where the soapy rod touched the water is hence *away from* the rod towards the surrounding water. So the powder moves away from the centre towards the sides of the vessel.

A toy duck moves by itself across the surface of water when a small bag of camphor is tied to its base. The camphor lowers the surface tension of the water in contact with it, and the duck is driven across the water by the resultant force on it.

Capillarity

When a capillary tube is immersed in water, and then placed vertically with one end in the liquid, observation shows that the water rises in the tube to a height above the surface. As we see later, the effect is due to surface tension. The narrower the tube, the greater is the height to which the water rises, Fig. 6.6 (i). This phenomenon is known as *capillarity*, and it occurs when blotting-paper is used to dry ink. The liquid rises up the pores of the paper when it is pressed on the ink.

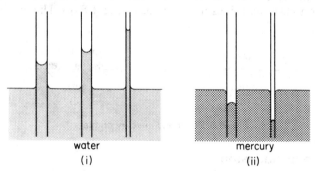

water   mercury
(i)   (ii)

Fig. 6.6 Capillary rise and fall

When a capillary tube is placed inside mercury, however, the liquid is depressed *below* the outside level, Fig. 6.6 (ii). The depression increases as the diameter of the capillary tube decreases. See also p. 150.

Angle of Contact

In the case of water in a glass capillary tube, if the glass is clean observation of the meniscus shows that it is hemispherical, that is, the glass surface is tangential to the meniscus where the water touches it. In other cases where liquids rise in a capillary tube, the tangent BN to the liquid surface where it touches the glass may make an acute angle θ with the glass, Fig. 6.7 (i). The angle θ is known as the *angle of contact* between the liquid and the glass, and is always measured *through the liquid*. The angle of contact between two given surfaces varies largely with their freshness and cleanliness. The angle of contact

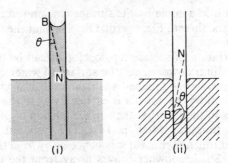

Fig. 6.7 Angle of contact

between water and very clean glass is zero, but when the glass is not clean the angle of contact may be about 8° for example. The angle of contact between alcohol and clean glass is zero.

When a capillary tube is placed inside mercury, observation shows that the surface of the liquid is depressed in the tube and is convex upwards, Fig. 6.7 (ii). The tangent BN to the mercury at the point B where the liquid touches the glass thus makes an obtuse angle, θ, with the glass when measured through the liquid. We shall see later (p. 148) that a liquid will rise in a capillary tube if the angle of contact is acute, and that a liquid will be depressed in the tube if the angle of contact is obtuse. For the same reason, clean water spreads over, or 'wets', a clean glass surface when spilt on it, Fig. 6.8 (i); the angle of contact is zero. On the other hand, mercury gathers itself into small pools or globules when spilt on glass, and does not 'wet' glass, Fig. 6.8 (ii). The angle of contact is obtuse.

Fig. 6.8 Water and mercury on glass

Cohesion and Adhesion

The difference in behaviour of water and mercury on clean glass can be explained in terms of the attraction between the molecules of these substances. It appears that the force of *cohesion* between two molecules of water is less than the force of *adhesion* between a molecule of water and a molecule of glass; and thus water spreads over glass. On the other hand, the force of cohesion between two molecules of mercury is greater than the force of adhesion between a molecule of mercury and a molecule of glass; and so mercury gathers in pools when spilt on glass.

A drop of olive-oil spreads into a thin film of large area when placed on clean water (p. 129). This is due to the fact that one end of the oil molecule has a greater adhesive force for a neighbouring water molecule than its cohesive force for a neighbouring oil molecule.

As we have already seen, a solution of soap in water has a smaller surface tension than pure water. Molecules of soap between water molecules may reduce the cohesive force between the water molecules.

Before washing a greasy plate with detergent solutions, the angle of contact between the plate and the grease is zero and between the plate and water is 180°. In dilute detergent solutions, this is reversed. So the grease now forms globules which are easily removed. In rainproofing materials, the fibres are coated with a wax which increases the angle of contact of water to more than 90°. Drops of water are therefore formed and so the water does not penetrate the material.

Angle of Contact Measurement

The angle of contact can be found by means of the method outlined in Fig. 6.9 (i), (ii).

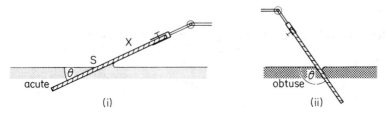

Fig. 6.9 Angle of contact measurement

A plate X of the solid is placed at varying angles to liquid until the surface S appears to be *plane* at X. The angle θ made with the liquid surface is then the angle of contact. For an obtuse angle of contact, a similar method can be adopted. In the case of mercury and glass, a thin plane mirror enables the liquid surface to be seen by reflection. For a freshly-formed mercury drop in contact with a clean glass plate, the angle of contact is 137°.

Measurement of Surface Tension by Capillary Tube Method

Theory. Suppose γ is the magnitude of the surface tension of a liquid such as water, which rises up a clean glass capillary tube and has an angle of contact zero. Figure 6.10 shows a section of the meniscus M at B, which is a hemisphere. Since the glass AB is a tangent to the liquid at its meniscus, the surface tension forces, which act along the boundary of the liquid with the air, act vertically downwards on the glass. By the law of action and reaction, the glass exerts an upward force on the liquid meniscus. Now the length of liquid in contact with the glass is $2\pi r$, where r is the radius of the capillary and γ is the force per unit length acting in the surface of the liquid.

$$\therefore \ 2\pi r \times \gamma = \text{upward force on liquid} \qquad . \qquad . \qquad (1)$$

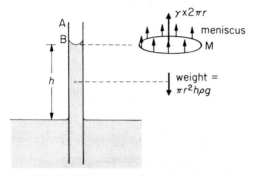

Fig. 6.10 Rise in capillary tube—theory

If γ is in newton metre^{-1} and r is in metre, then the upward force is in *newton*.

This force supports the weight of the column of liquid of height h above the outside level. The volume of the liquid $= \pi r^2 h$. So the mass, m, of the liquid column $=$ volume $\times$ density $= \pi r^2 h \rho$, where ρ is the density. The *weight* of the liquid $= mg = \pi r^2 h \rho g$.

If ρ is in kg m^{-3}, r and h in metres, and $g = 9.8$ m s^{-2}, then $\pi r^2 h \rho g$ is in *newton*. From (1), it now follows that

$$\therefore \ 2\pi r \gamma = \pi r^2 h \rho g$$

$$\therefore \ \gamma = \frac{rh\rho g}{2} \qquad . \qquad . \qquad . \qquad (2)$$

This formula will be proved later by a different (and better) method. See p. 149.

If $r = 0.2$ mm $= 0.2 \times 10^{-3}$ m, $h = 6.6$ cm for water $= 6.6 \times 10^{-2}$ m, and $\rho = 1000$ kg m^{-3}, then

$$\gamma = \frac{0.2 \times 10^{-3} \times 6.6 \times 10^{-2} \times 1000 \times 9.8}{2} = 6.5 \times 10^{-2} \text{ N m}^{-1}.$$

In deriving this formula for γ it should be noted that we have (i) assumed the angle of contact to be zero, (ii) neglected the weight of the small amount of liquid above the bottom of the meniscus at B, Fig. 6.10.

Experiment. In the experiment, the cleaned capillary tube C is supported in a beaker filled with clean water. A pin P, bent at right angles at two places, is tied to C by a rubber band, Fig. 6.11. P is adjusted until its point just touches the horizontal level of the liquid in the beaker. A travelling microscope X is now focused on the meniscus M in C, and then it is focused on to the point of P after the beaker is taken away. In this way the height h of M above the liquid level in the beaker is determined. The radius of the capillary at M can be found by cutting the tube at this place and measuring the diameter by the travelling microscope; or if the capillary tube is a uniform one, by measuring the length, l, and mass, m, of a mercury thread drawn into the tube, and calculating the radius, r, from the relation $r = \sqrt{m/\pi l \rho}$, where ρ is the density of mercury. The surface tension γ is then calculated from the formula $\gamma = rh\rho g/2$. Its magnitude for water at 15°C is 7.33×10^{-2} N m^{-1}.

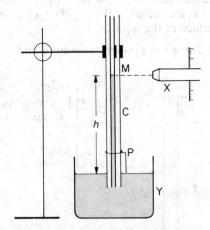

Fig. 6.11 Surface tension by capillary rise

Measurement of Surface Tension by Microscope Slide

Besides the capillary tube method, the surface tension of water can be measured by weighing a microscope slide in air, and then lowering it until it just meets the surface of water, Fig. 6.12. The surface tension force acts vertically downward round the boundary of the slide as the angle of contact is zero, and so the slide is pulled down by the water clinging to it (see inset fig.). If a and b are the respective length and thickness of the slide, then $(2a + 2b)$ is the length of the liquid in contact with the slide. Since γ is the force per unit length in the liquid surface the downward force $= \gamma(2a + 2b)$. If the mass required to counterbalance the force is m, then

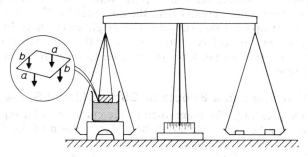

Fig. 6.12 Surface tension by microscope slide

$$\gamma(2a + 2b) = mg,$$

$$\therefore \gamma = \frac{mg}{2a + 2b}.$$

If $m = 0.88$ g $= 0.88 \times 10^{-3}$ kg, $a = 6.0$ cm, $b = 0.2$ cm, then:

$$\gamma = \frac{0.88 \times 10^{-3} \text{ (kg)} \times 9.8 \text{ (m s}^{-2})}{2 \times (6 + 0.2) \times 10^{-2} \text{ (m)}} = 7.0 \times 10^{-2} \text{ N m}^{-1}.$$

Surface Tension of a Soap Solution

The surface tension of a soap solution can be found by a similar method. A soap-film is formed in a three-sided metal frame ABCD, and the apparent weight is found, Fig. 6.13. When the film is broken by piercing it, the decrease in the apparent weight, mg, is equal to the surface tension force acting downwards when the film existed. This is equal to $2\gamma b$, where $b = BC$, since the film has *two* sides.

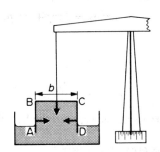

Fig. 6.13 Surface tension of soap film

$$\therefore 2\gamma b = mg,$$

$$\therefore \gamma = \frac{mg}{2b}.$$

It will be noted that the surface tension forces on the sides AB, CD of the frame act horizontally as shown, and their resultant is zero.

A soap film can be supported in a vertical rectangular frame but a film of water can not. This is due to the fact that the soap drains downward in a vertical film, so that the top of the film has a lower concentration of soap than the bottom. The surface tension at the top is thus *greater* than at the bottom (soap diminishes the surface tension of pure water). The upward pull on the film by the top bar is hence greater than the downward pull on the film by the lower bar. The net upward pull supports the weight of the film. In the case of pure water, however, the surface tension would be the same at the top and the bottom, and so there is no net force in this case to support a water film in a rectangular frame.

Pressure Difference in a Bubble or Curved Liquid Surface

As we shall see presently, the magnitude of the curvature of a liquid, or of a bubble formed in a liquid, is related to the surface tension of the liquid.

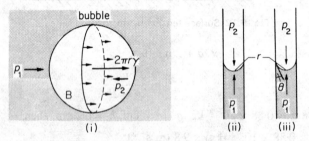

Fig. 6.14 Excess pressure in bubble

Consider a bubble formed inside a liquid, Fig. 6.14 (i). If we consider the equilibrium of *one half*, B, of the bubble, we can see that the surface tension force on B plus the force on B due to external pressure p_1 = the force on B due to the internal pressure p_2 inside the bubble. The force on B due to the pressure p_1 is given by $\pi r^2 \times p_1$, since πr^2 is the area of the vertical circular face of B and pressure is 'force per unit area'. The force on B due to the pressure p_2 is given similarly by $\pi r^2 \times p_2$. The surface tension force acts around the *circumference* of the bubble, which has a length $2\pi r$; so the force is $2\pi r\gamma$. It follows that, for equilibrium of B,

$$\pi r^2 p_2 = 2\pi r\gamma + \pi r^2 p_1.$$

Simplifying,

$$\therefore r(p_2 - p_1) = 2\gamma,$$

or

$$p_2 - p_1 = \frac{2\gamma}{r}.$$

Now $(p_2 - p_1)$ is the excess pressure, p, in the bubble over the outside pressure.

$$\therefore \textit{excess pressure, } p, = \frac{2\gamma}{r} \qquad \qquad . \qquad . \qquad . \qquad . \qquad (1)$$

Although we considered a bubble, the same formula for the excess pressure holds for any curved surface or meniscus, where r is its radius of curvature and γ is its surface tension, provided the angle of contact is zero, Fig. 6.14 (ii). If the angle of contact is θ, the formula is modified by replacing γ by $\gamma \cos \theta$, Fig. 6.14 (iii). Thus, in general,

$$\text{excess pressure, } p, = \frac{2\gamma \cos \theta}{r} \qquad . \qquad . \qquad . \qquad (2)$$

Excess Pressure in Soap Bubble

A soap bubble has two liquid surfaces in contact with air, one inside the bubble and the other outside the bubble. The force on one half, B, of the bubble due

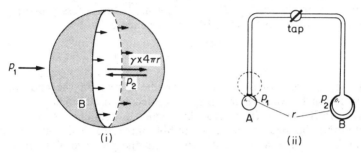

Fig. 6.15 (i) Excess pressure in soap bubble. (ii) Collapse of bubble A

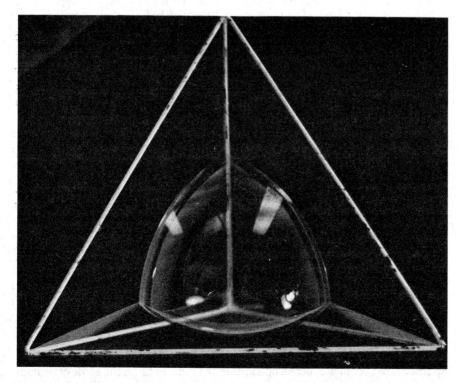

Fig. 6A Soap bubble trapped in tetrahedron (*Courtesy of Dr C. Isenberg, University of Kent*)

to surface tension forces is thus $\gamma \times 2\pi r \times 2$, i.e., $\gamma \times 4\pi r$, Fig. 6.15 (i). For the equilibrium of B, it follows that

$$\pi r^2 p_2 = 4\pi r\gamma + \pi r^2 p_1$$

where p_2, p_1 are the pressures inside and outside the bubble respectively.

Simplifying, $\qquad\qquad \therefore\ p_2 - p_1 = \dfrac{4\gamma}{r},$

$$\therefore\ excess\ pressure,\ p, = \frac{4\gamma}{r} \qquad . \qquad . \qquad . \qquad . \qquad (3)$$

This result for excess pressure should be compared with the result obtained for a bubble formed inside a liquid, equation (1).

If γ for a soap solution is 25×10^{-3} N m^{-1}, the excess pressure inside a bubble of radius 0·5 cm or $0·5 \times 10^{-2}$ m is hence given by:

$$p = \frac{4 \times 25 \times 10^{-3}}{0·5 \times 10^{-2}} = 20\ \text{N m}^{-2}.$$

Two soap-bubbles of unequal size can be blown on the ends of a tube, communication between them being prevented by a closed tap in the middle (Fig. 6.15 (ii). If the tap is opened, the *smaller* bubble is observed to collapse gradually and the size of the larger bubble increases. This can be explained from our formula $p = 4\gamma/r$, which shows that the pressure of air p_1 inside the smaller bubble is *greater* than that p_2 inside the larger bubble. Consequently air flows from the smaller to the larger bubble when communication is made between the bubbles, and the smaller bubble A thus gradually collapses to a soap film of radius r equal to the radius of the new bubble B on the other side, Fig. 6.15 (ii). No more air then flows as the air pressure is the same on both sides.

Rise or Fall of Liquids in Capillary Tubes

From our knowledge of the angle of contact and the excess pressure on one side of a curved liquid surface, we can deduce that some liquids will rise in a capillary tube, whereas others will be depressed.

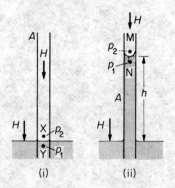

Fig. 6.16 Capillary rise by excess pressure

Suppose the tube A is placed in water, for example, Fig. 6.16 (i). At first the liquid surface becomes concave upwards in the tube, because the angle of contact with the glass is zero. Consequently the pressure on the air side, X, of the curved surface is greater than the pressure on the liquid side Y by $2\gamma/r$, where γ is the surface tension and r is the radius of curvature of the tube. But the pressure at X is atmospheric, H. Hence the pressure at Y must be less than atmospheric by $2\gamma/r$. Figure 6.16 (i) is therefore impossible because it shows the pressure at Y equal to the atmospheric pressure. Thus, as shown in Fig. 6.16 (ii), the liquid rises up the tube to such a height h that the pressure at N is less than at M by $2\gamma/r$. A similar argument now shows that a liquid falls in a capillary tube when the angle of contact is obtuse.

The angle of contact between mercury and glass is obtuse (p. 142). Thus when a capillary tube is placed in mercury the liquid first curves downwards. The pressure inside the liquid just below the curved surface is now greater than the pressure on the other side, which is atmospheric, and the mercury therefore moves down the tube until the excess pressure $= 2\gamma \cos \theta/r$, with the usual notation. A liquid thus falls in a capillary tube if the angle of contact is obtuse.

Capillary Rise and Fall by Pressure Method

We shall now calculate the capillary rise of water by the excess pressure formula $p = 2\gamma/r$.

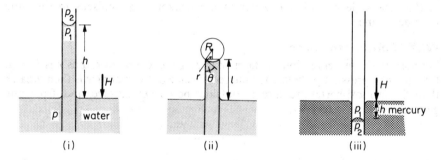

Fig. 6.17 Excess pressure application

In the case of a capillary tube dipping into water, the angle of contact is practically zero, Fig. 6.17 (i). Thus if p_2 is the pressure of the atmosphere, and p_1 is the pressure in the liquid, we have

$$p_2 - p_1 = \frac{2\gamma}{r}.$$

Now if H is the atmospheric pressure, h is the height of the liquid in the tube and ρ its density,

$$p_2 = H \quad \text{and} \quad p_1 = H - h\rho g,$$

$$\therefore H - (H - h\rho g) = \frac{2\gamma}{r},$$

$$\therefore h\rho g = \frac{2\gamma}{r},$$

$$\therefore h = \frac{2\gamma}{r\rho g} \qquad . \qquad . \qquad . \qquad . \qquad \text{(i)}$$

The formula shows that h increases as r decreases, that is, the narrower the tube, the greater is the height to which the water rises (see Fig. 6.6 (i), p. 141).

If the height l of the tube above the water is *less* than the calculated value of h in the above formula, the water surface at the top of the tube now meets it at an *acute angle of contact* θ, Fig. 6.17 (ii). The radius R of the meniscus is greater than the capillary tube radius r. By geometry, we see that $\cos \theta = r/R$. But from the excess pressure formula for the meniscus, $p_2 - p_1 = 2\gamma/R$ and so $H - (H - l\rho g) = 2\gamma/R = l\rho g$. Hence

$$l = \frac{2\gamma}{R\rho g} \qquad \qquad . \qquad . \qquad . \qquad . \qquad . \qquad \text{(ii)}$$

Dividing (ii) by (i), it follows that $l/h = r/R$. So

$$\cos \theta = \frac{r}{R} = \frac{l}{h}.$$

Thus suppose water rises to a height of 10 cm in a capillary tube when it is placed in a beaker of water. If the tube is pushed down until the top is only 5 cm above the outside water surface, then $\cos \theta = \frac{5}{10} = 0.5$. Thus $\theta = 60°$. The meniscus now makes an angle of contact of 60° with the glass. As the tube is pushed down further, the angle of contact increases beyond 60°. When the top of the tube is level with the water in the beaker, the meniscus in the tube becomes plane.

With Mercury in Glass

Suppose that the depression of the mercury inside a tube of radius r is h, Fig. 6.17 (ii). The pressure p_2 below the curved surface of the mercury is then greater than the (atmospheric) pressure p_1 outside the curved surface; and, from our general result on page 147,

$$p_2 - p_1 = \frac{2\gamma \cos \theta}{r},$$

where θ is the supplement of the obtuse angle of contact of mercury with glass, that is, θ is an acute angle and its cosine is positive. But $p_1 = H$ and $p_2 = H + h\rho g$, where H is the atmospheric pressure.

$$\therefore (H + h\rho g) - H = \frac{2\gamma \cos \theta}{r}.$$

$$\therefore h\rho g = \frac{2\gamma \cos \theta}{r}.$$

$$\therefore h = \frac{2\gamma \cos \theta}{r\rho g}. \qquad \qquad . \qquad . \qquad . \qquad \text{(1)}$$

The height of depression, h, thus increases as the radius r of the tube decreases. See Fig. 6.6 (ii), p. 141.

Examples

1. Water rises to a height h inside a clean glass capillary tube of radius 0·2 mm when the tube is placed vertically inside a beaker of water. Calculate h if the surface tension of water is 7.0×10^{-2} N m^{-1} and the angle of contact is zero.

    The tube is now pushed into the water until 4·0 cm of its length is above the surface. Describe and explain what happens. (Density of water = 1000 kg m^{-3}, g = 10 m s^{-2}.)

(i) The height h is given, with the usual notation, by

$$\gamma = \frac{rh\rho g}{2},$$

so
$$h = \frac{2\gamma}{r\rho g} = \frac{2 \times 7 \times 10^{-2}}{0 \cdot 2 \times 10^{-3} \times 1000 \times 10}$$

$$= 7 \times 10^{-2} \text{ m.}$$

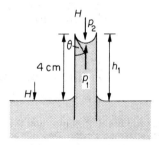

Fig. 6.18

(ii) When the tube is pushed down so that the height h_1 in Fig. 6.18 is 4 cm, the angle of contact at the top of the tube changes from zero to a value θ as shown. Now $\cos \theta = r/R$, where R is the radius of the meniscus (p. 150). From the excess pressure formula

$$p_2 - p_1 = \frac{2\gamma}{R}$$

where $p_2 = $ atmospheric pressure H and $p_1 = H - h_1\rho g$.

So
$$h_1\rho g = \frac{2\gamma}{R}.$$

But from above,
$$h\rho g = \frac{2\gamma}{r}.$$

Dividing, then
$$\frac{h_1}{h} = \frac{r}{R} = \cos \theta.$$

So
$$\cos \theta = 4/7, \text{ and } \theta = 55°.$$

2. A soap bubble X of radius 0·03 m and another bubble Y of radius 0·04 m are brought together so that the combined bubble has a common interface S of radius r, Fig. 6.19. Calculate r.

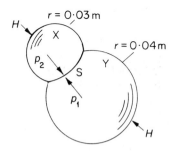

Fig. 6.19

Suppose the pressure inside X is p_2 and that inside Y is p_1. The excess pressure in the soap bubble X $= p_2 - H = 4\gamma/0 \cdot 03$, where H is the atmospheric pressure. Similarly, $p_1 - H = 4\gamma/0 \cdot 04$ for the bubble Y. We see that p_2 is greater than p_1. So the interface S is *convex* towards Y as shown.

To find r, use the excess pressure formula. Then, for S,

$$p_2 - p_1 = \frac{4\gamma}{r}.$$

But from above, by subtraction,

$$p_2 - p_1 = \frac{4\gamma}{0\cdot03} - \frac{4\gamma}{0\cdot04}.$$

So, cancelling 4γ,

$$\frac{1}{r} = \frac{1}{0\cdot03} - \frac{1}{0\cdot04} = \frac{0\cdot01}{0\cdot03 \times 0\cdot04}.$$

Thus $r = 3 \times 0\cdot04 = 0\cdot12$ m.

3. 'The surface tension of water is $7\cdot5 \times 10^{-2}$ N m^{-1} and the angle of contact of water with glass is zero.' Explain what these statements mean. Describe an experiment to determine *either* (a) the surface tension of water, *or* (b) the angle of contact between paraffin wax and water.

A glass U-tube is inverted with the open ends of the straight limbs, of diameters respectively 0·500 mm and 1·00 mm, below the surface of water in a beaker. The air pressure in the upper part is increased until the meniscus in one limb is level with the water outside. Find the height of water in the other limb. (The density of water may be taken as 1000 kg m^{-3}.) (*L.*)

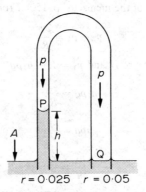

$r = 0\cdot025$ $r = 0\cdot05$

Fig. 6.20 Example (radii in cm)

Suppose p is the air pressure inside the U-tube when the meniscus Q is level with the water outside and P is the other meniscus at a height h, Fig. 6.20. Let A be the atmospheric pressure. Then, if r_1 is the radius at P, 0.25×10^{-3} m,

$$p - (A - h\rho g) = \frac{2\gamma}{r_1} \qquad\qquad\qquad (i)$$

since the pressure in the liquid below P is $(A - h\rho g)$.

The pressure in the liquid below Q $= A$. Hence, for Q,

$$p - A = \frac{2\gamma}{r_2} \qquad\qquad\qquad (ii)$$

where r_2 is the radius $0\cdot25 \times 10^{-3}$ m,

From (i) and (ii), it follows that

$$h\rho g = \frac{2\gamma}{r_1} - \frac{2\gamma}{r_2}$$

$$\therefore h = \frac{1}{\rho g}\left[\frac{2\gamma}{r_1} - \frac{2\gamma}{r_2}\right]$$

$$= \frac{1}{9800}\left[\frac{2 \times 0\cdot075}{0\cdot25 \times 10^{-3}} - \frac{2 \times 0\cdot075}{0\cdot5 \times 10^{-3}} \right]$$

$$= 3\cdot1 \times 10^{-2} \text{ m (approx.)}.$$

4. A small drop of clean water is squeezed between two clean glass plates so that a very thin layer of large area is formed. Explain why it is difficult to pull the plates apart, assuming the thickness of the layer is 10^{-6} m, the area of it is 40 cm², and the surface tension of water is 7×10^{-2} N m⁻¹.

Suppose X, Y are the two plates, A is the external atmospheric pressure and p is the pressure inside the very thin water layer of thickness d, Fig. 6.21. Assuming zero angle of contact, the meniscus shown has a curvature $1/r$ or $1/(d/2)$ which is very much greater than the curvature parallel to the plates. If we neglect the curvature parallel to the plates, the excess pressure is given by γ/r in place of $2\gamma/r$ obtained with a *spherical* meniscus. So

$$A - p = \frac{\gamma}{r} = \frac{\gamma}{d/2} = \frac{2\gamma}{d}.$$

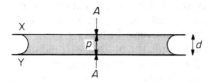

Fig. 6.21

From Fig. 6.21, we see that the force pushing the two plates X,Y together is given by

$$F = \text{excess pressure} \times \text{area of layer}$$

$$= (A - p) \times \text{area} = \frac{2\gamma}{d} \times \text{area}$$

$$= \frac{2 \times 7 \times 10^{-2}}{10^{-6}} \times 40 \times 10^{-4}$$

$$= 560 \text{ N}.$$

So a force of 560 N is needed to pull the plates apart. With a thick layer of water d is bigger and F becomes smaller, so the plates are now easier to pull apart.

Variation of Surface Tension with Temperature. Jaeger's Method

By forming a bubble inside a liquid, and measuring the excess pressure, JAEGER was able to find how the surface tension of a liquid varied with temperature. One form of the apparatus is shown in Fig. 6.22 (i). A capillary or drawn-out tubing A is connected to a vessel W containing a funnel C, so that air is driven slowly through A when water enters W through C. The capillary A is placed inside a beaker containing the liquid L. When air is passed through it at a slow rate a bubble forms slowly at the end of A.

Figure 6.22 (ii) shows the bubble at three possible stages of growth. The radius grows from that a to a hemispherical shape at b. Here its pressure is larger since the radius is smaller. If we consider the bubble growing to c, the radius of c would be *greater* than that of b and hence it cannot contain the increasing pressure. The downward force of the bubble due to the pressure, in fact, would be greater than the upward force due to surface tension. Hence *the bubble becomes unstable and breaks away from A when its radius is the same as*

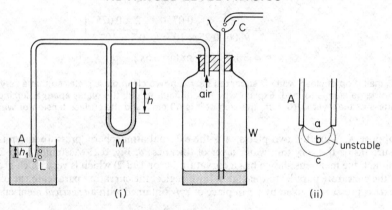

Fig. 6.22 Jaeger's method

that of A. Thus as the bubble grows the pressure in it increases to a maximum, and then decreases as the bubble breaks away. The maximum pressure is observed from a manometer M containing a light oil of density ρ, and a series of observations are taken as several bubbles grow.

The maximum pressure inside the bubble $= H + h\rho g$ where h is the maximum difference in levels in the manometer M, and H is the atmospheric pressure. The pressure outside the bubble $= H + h_1\rho_1 g$, where h_1 is the depth of the orifice of A below the level of the liquid L, and ρ_1 is the liquid density.

$$\therefore \text{ excess pressure} = (H + h\rho g) - (H + h_1\rho_1 g) = h\rho g - h_1\rho_1 g.$$

But
$$\text{excess pressure} = \frac{2\gamma}{r},$$

where r is the radius of the orifice of A (p. 146).

$$\therefore \frac{2\gamma}{r} = h\rho g - h_1\rho_1 g,$$

$$\therefore \gamma = \frac{rg}{2}(h\rho - h_1\rho_1).$$

By adding warm liquid to the vessel containing L, the variation of the surface tension with temperature can be determined. Experiment shows that the surface tension of liquids, and water in particular, *decreases* with increasing temperature along a fairly smooth curve. Various formulae relating the surface tension to temperature have been proposed, but none has been found to be completely satisfactory. The decrease of surface tension with temperature may be attributed to the greater average separation of the molecules at higher temperature. The force of attraction between molecules is then reduced, and so the surface energy is reduced, as can be seen from the potential energy curve on p. 131.

Surface Tension and Surface Energy

We now consider the surface energy of a liquid and its relation to its surface tension γ. Consider a film of liquid stretched across a horizontal frame ABCD, Fig. 6.23. Since γ is the force per unit length, the force on the rod BC of length $l = \gamma \times 2l$, because there are two surfaces to the film.

Suppose the rod is now moved a distance b from BC to B'C' against the surface tension forces, so that the surface area of the film increases. The tempera-

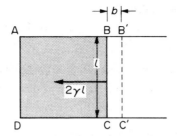

Fig. 6.23 Surface energy and work

ture of the film then usually decreases, in which case the surface tension alters (p. 154). If the surface area increases under *isothermal* (constant temperature) conditions, however, the surface tension is constant; and we can then say that, if γ is the surface tension at that temperature,

work done in enlarging surface area = force × distance,

$$= 2\gamma l \times b = \gamma \times 2lb.$$

But $2lb$ is the total increase in surface area of the film.

∴ work done per unit area in enlarging area = γ.

Thus the surface tension, γ, can be defined as *the work done per unit area in increasing the surface area of a liquid under isothermal conditions.* This is also called the *free surface energy* because the mechanical work done can be released when the surface contracts.

Examples

1. A soap bubble in a vacuum has a radius of 3 cm and another soap bubble in the vacuum has a radius of 6 cm. If the two bubbles coalesce under isothermal conditions, calculate the radius of the bubble formed.

Since the bubbles coalesce under isothermal conditions, the surface tension γ is constant. Suppose R is the radius in cm, $R \times 10^{-2}$ m, of the bubble formed.

Then   work done $= \gamma \times$ surface area $= \gamma \times 8\pi R^2 \times 10^{-4}$

But   original work done $= (\gamma \times 8\pi.3^2 + \gamma \times 8\pi.6^2) \times 10^{-4}$

∴ $\gamma \times 8\pi R^2 = \gamma \times 8\pi.3^2 + \gamma.8\pi.6^2$.

∴ $R^2 = 3^2 + 6^2$.

∴ $R = \sqrt{3^2 + 6^2} = 6 \cdot 7$ cm.

2. (i) Calculate the work done against tension surface forces in blowing a soap bubble of 1 cm diameter if the surface tension of soap solution is $2 \cdot 5 \times 10^{-2}$ N m⁻¹. (ii) Find the work required to break up a drop of water of radius 0·5 cm into drops of water each of radii 1 mm assuming isothermal conditions. (Surface tension of water = 7×10^{-2} N m⁻¹.)

(i) The original surface area of the bubble is zero, and the final surface area $= 2 \times 4\pi r^2$ (two surfaces of bubble) $= (2 \times 4\pi \times 0 \cdot 5^2) \times 10^{-4} = 2\pi \times 10^{-4}$ m².

∴ work done $= \gamma \times$ increase in surface area.

$$= 2 \cdot 5 \times 10^{-2} \times 2\pi \times 10^{-4} = 1 \cdot 57 \times 10^{-5} \text{ J}.$$

(ii) Since volume of a drop $= \frac{4}{3}\pi r^3$,

$$\text{number of drops formed} = \frac{\frac{4}{3}\pi \times 0 \cdot 5^3}{\frac{4}{3}\pi \times 0 \cdot 1^3} = 125.$$

∴ final total surface area of drops

$$= 125 \times 4\pi r^2 = 125 \times 4\pi \times 0.1^2 \times 10^{-4},$$
$$= 5\pi \times 10^{-4} \text{ m}^2.$$

But original surface area of drop $= 4\pi \times 0.5^2 \times 10^{-4} = \pi \times 10^{-4}$ m².

∴ work done $= \gamma \times$ change in surface area

$$= 7 \times 10^{-2} \times (5\pi - \pi) \times 10^{-4} = 8.8 \times 10^{-5} \text{ J}.$$

Molecular Bonds and Surface Tension

Molecules which reach the surface from the interior break and reform bonds with neighbours all round them as they rise. At the surface, however, *half* the bonds have not reformed. The bond energy thus released results in greater energy in the surface than in the bulk of the liquid.

We can find an expression for γ from molecular theory. Suppose n is the number of nearest neighbours per molecule in the liquid and ε is the energy per molecule pair, that is, the energy to break permanently the bonds between a pair of neighbouring molecules in the liquid. Then $n\varepsilon$ is the energy to break the bonds between a particular molecule and all its neighbours. For a molecule in the surface, about *half* the number of bonds have not reformed, as stated above. So $\frac{1}{2}n\varepsilon$ is roughly the energy released when a molecule reaches the surface.

If there are N molecules per unit area in the surface, the number of pairs is $\frac{1}{2}N$. So the energy per unit area increase of the surface is $\frac{1}{2}N \times \frac{1}{2}n\varepsilon = \frac{1}{4}Nn\varepsilon$. So

$$\gamma = \tfrac{1}{4}Nn\varepsilon.$$

Surface Energy and Latent Heat

Inside a liquid molecules move about in all directions, continually breaking and reforming bonds with neighbours. If a molecule in the surface passes into the vapour outside, a definite amount of energy is needed to permanently break the bonds with its neighbouring molecules in the liquid. We have seen that the surface tension is the energy per unit area needed to break the bonds with half the molecules all round the molecules. So the energy needed to evaporate a liquid is related to its free surface energy or surface tension. The latent heat of vaporisation, l, is the energy needed to change liquid to vapour at the boiling point and is therefore related to the free surface energy. See p. 132.

Total Surface energy

As we have seen, the surface energy is increased when the surface area of a liquid is increased. The molecules which then reach the surface are slowed up by the force pulling them inwards, so the average translational kinetic energy of all the liquid molecules is reduced. On this account the liquid cools while the surface is increased, and heat flows in from the surroundings to restore the temperature.

The increase in the *total surface energy per unit area*, E, is thus given by

$$E = \gamma + H \qquad \qquad \qquad \qquad (1)$$

where H is the heat per unit area absorbed by the surface from the surroundings. E should be distinguished from the *free surface energy* per unit area, which is γ.

The variation of E with temperature is shown in Fig. 6.24, together with the similar variation of L, the latent heat of vaporisation (see p. 245). Both vanish at the critical temperature, since no liquid exists above the critical temperature whatever the pressure (p. 248).

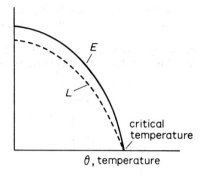

Fig. 6.24 Variation of E and L with temperature

Exercises 6

(Assume $g = 10\ m\ s^{-2}$ or $10\ N\ kg^{-1}$ unless otherwise given.)

1. Define *surface tension*. A rectangular plate of dimensions 6 cm by 4 cm and thickness 2 mm is placed with its largest face flat on the surface of water. Calculate the downward force on the plate due to surface tension assuming zero angle of contact.

What is the downward force if the plate is placed vertical so that its longest side just touches the water? (Surface tension of water = $7\cdot0 \times 10^{-2}\ N\ m^{-1}$.)

2. An air bubble inside a liquid of surface tension $1\cdot0 \times 10^{-3}\ N\ m^{-1}$ grows from a radius of $1\cdot0 \times 10^{-5}$ m to $1\cdot0 \times 10^{-4}$ m in 6 μs (6×10^{-6} s). Calculate the average rate of change of the pressure inside the bubble.

3. Define surface tension in terms of *surface energy*. Calculate the change in surface energy of a soap bubble when its radius decreases from 5 cm to 1 cm. (Surface tension of soap solution = $2\cdot0 \times 10^{-2}\ N\ m^{-1}$.)

State any assumption made in your calculation.

4. What are the *dimensions* of surface tension? A capillary tube of 0·4 mm diameter is placed vertically inside (i) water of surface tension $6\cdot5 \times 10^{-2}\ N\ m^{-1}$ and zero angle of contact, (ii) a liquid of density 800 kg m^{-3}, surface tension $5\cdot0 \times 10^{-2}\ N\ m^{-1}$ and angle of contact 30°. Calculate the height to which the liquid rises in the capillary in each case.

5. The water rises to a height of 8 cm above the outside level when a long clean capillary tube is dipped into a beaker of clean water and then withdrawn.

Explain what happens when a capillary tube of the same diameter but length 4 cm is dipped into the water.

6. (i) A soap-bubble has a diameter of 4 mm. Calculate the pressure inside it if the atmospheric pressure is $10^5\ N\ m^{-2}$. (Surface tension of soap solution = $2\cdot8 \times 10^{-2}\ N\ m^{-1}$.) (ii) Estimate the total surface energy of a million drops of water each of radius $1\cdot0 \times 10^{-4}$ m, if the surface tension of water is $7 \times 10^{-2}\ N\ m^{-1}$. State any assumptions made.

7. Explain briefly, with the aid of a diagram, what you would expect to happen to a nearly spherical water droplet resting on a clean horizontal surface if a tiny amount of detergent were added to it. How do you account for the change that might occur? (*L.*)

8. Explain briefly (*a*) the approximately spherical shape of a rain drop, (*b*) the movement of tiny particles of camphor on water, (*c*) the possibility of floating a needle on water, (*d*) why a column of water will remain in an open vertical capillary tube after the lower end has been dipped in water and withdrawn. (*N.*)

9. Surface tension may be defined in terms of force per unit length or of energy per unit area. Show, by considering an increase in surface area of a liquid, that these definitions are equivalent. State any necessary condition.

Derive the equation $\gamma = hr\rho g/2$, where γ is the surface tension of a liquid which wets glass, h is its observed rise in a capillary tube of radius r and ρ is its density. Describe the experiment to determine γ for a liquid which is based on this formula. (*L.*)

10. By considering the work done per unit in increasing the surface area of a bubble blown in a liquid, or otherwise, derive an expression for the excess pressure p inside a bubble of radius r.

The diagrams below represent glass capillary tubes dipping into a liquid. Explain why the situation represented by (i) is unstable while that in (ii) is stable.

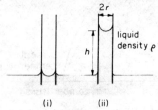

2r

liquid
density ρ

h

(i)   (ii)

Fig. 6A

Use the data given in diagram (ii) to derive an expression for the height h to which the liquid rises, given that the angle of contact between the liquid and glass is zero.

By considering intermolecular forces explain why the surface of a liquid is different from the bulk of the liquid.

Suggest why there might be a connection between the surface energies (surface tensions) of liquids and their normal boiling points. (*L.*)

11. A clean glass capillary tube, of internal diameter 0·04 cm, is held vertically with its lower end below the surface of clean water in a beaker, and with 10 cm of the tube above the surface. To what height will the water rise in the tube? What will happen if the tube is now depressed until only 5 cm of its length is above the surface? The surface tension of water is $7·2 \times 10^{-2}$ N m^{-1}.

Describe, and give the theory of some method, other than that of the rise in a capillary tube, of measuring surface tension. (*O. & C.*)

12. (*a*) An air bubble is blown in water. Deduce an expression for the excess pressure inside the bubble. (*b*) The bubbles in a bubble chamber are formed with an average radius $1·0 \times 10^{-6}$ m. They grow to a radius of $1·0 \times 10^{-5}$ m in 2·0 μs. Calculate the mean rate at which the pressure in the bubbles is changing during their growth. (Surface tension of liquid in chamber $= 8·0 \times 10^{-3}$ N m^{-1}.)

(*c*) Describe an experiment to measure the surface tension of a liquid which is based on measuring the excess pressure in a bubble. How may this experiment be developed to investigate the way in which the surface tension of the liquid varies with temperature?

Explain why such an experiment is likely to be more suitable in measuring the ratio of surface tensions of a liquid at various temperatures than in measuring the value of its surface tension at a particular temperature. Explain qualitatively why it is likely that surface tension will decrease with temperature rise. (*L.*)

13. Define the terms surface tension, angle of contact. Describe a method for measuring the surface tension of a liquid which wets glass. List the principal sources of error and state what steps you would take to minimise them.

A glass tube whose inside diameter is 1 mm is dipped vertically into a vessel containing mercury with its lower end 1 cm below the surface. To what height will the mercury rise in the tube if the air pressure inside it is 3×10^3 N m^{-2} below atmospheric pressure? Describe the effect of allowing the pressure in the tube to increase gradually to atmospheric pressure. (Surface tension of mercury $= 0·5$ N m^{-1}, angle of contact with glass $= 180°$, density of mercury $= 13\,600$ kg m^{-3}, $g = 9·81$ m s^{-2}.) (*O. & C.*)

14. Explain what is meant by surface tension, and show how its existence is accounted for by molecular theory.

Find an expression for the excess pressure inside a soap-bubble of radius R and surface tension T. Hence find the work done by the pressure in increasing the radius of the bubble from a to b. find also the increase in surface area of the bubble, and in the light of this discuss the significance of your result. (*C.*)

15. Define surface tension and use your definition to obtain an expression for the work done per unit area in changing the area of a surface.

Explain the following phenomena: (*a*) Water forms globules on a greasy plate but not on a clean one, (b) two flat glass blocks with a thin layer of water between them are difficult to pull apart, (c) small bubbles on the surface of water cluster together.

Two soap bubbles, one of radius 50 mm and the other of radius 80 mm, are brought together so that they have a common interface. Calculate the radius of curvature of this interface and explain whether it is convex towards the larger or smaller bubble. (L.)

16. Define *surface tension* and state the effect on the surface tension of water of raising its temperature.

Describe an experiment to measure the surface tension of water over the range of temperatures from 20°C to 70°C. Why is the usual capillary rise method unsuitable for this purpose?

Two unequal soap bubbles are formed one on each end of a tube closed in the middle by a tap. State and explain what happens when the tap is opened to put the two bubbles into connection. Give a diagram showing the bubbles when equilibrium has been reached. (L.)

17. The lower end of a vertical clean glass capillary tube is just immersed in water. Why does water rise up the tube?

A vertical capillary tube of internal radius r m has its lower end dipping in water of surface tension T newton m^{-1}. Assuming the angle of contact between water and glass to be zero, obtain from first principles an expression for the pressure excess which must be applied to the upper end of the tube in order just to keep the water levels inside and outside the tube the same.

A capillary of internal diameter 0·7 mm is set upright in a beaker of water with one end below the surface; air is forced slowly through the tube from the upper end, which is also connected to a U-tube manometer containing a liquid of density 800 kg m^{-3}. The difference in levels on the manometer is found to build up to 9·1 cm, drop to 4·0 cm, build up to 9·1 cm again, and so on. Estimate (a) the depth of the open end of the capillary below the free surface of the water in the beaker, (b) the surface tension of water. [State clearly any assumptions you have made in arriving at these estimates.] (O.)

18. (a) Outline briefly a method for measuring the surface tension of water, mentioning the precautions you would take to obtain an accurate result. (Derivation of the equation used to obtain the result is *not* required.)

(b) A molecule in the surface of a liquid has about twice the potential energy of a molecule in the body of that liquid. Explain why when the area of the free surface of a liquid is suddenly increased, the temperature of the liquid falls. What would happen if the increase in the area took place very slowly?

(c) An open rectangular tank has internal dimensions 3·0 m by 1·0 m and contains 500 kg of water.

(i) Estimate the number of water molecules, each of diameter $3·0 \times 10^{-10}$ m, in the free water surface.

(ii) The surface tension of water is 70·0 mN m^{-1}, which can be regarded as stating that the *extra* potential energy possessed by the molecules in the liquid surface is 70·0 mJ m^{-2}. Calculate the extra potential energy possessed by a water molecule in the free surface, compared with a molecule in the body of the liquid.

(iii) If the density of water is 10^3 kg m^{-3}, estimate the total number of water molecules in the tank.

(iv) Use the information given in (b) to help estimate the total potential energy possessed by all the water molecules in the tank.

(v) To which of the following physical quantities would you think the value calculated in (iv) should correspond? Specific heat capacity. Specific latent heat of fusion. Specific latent heat of vaporisation. Give a reason for your choice. (L.)

7 Solid Friction. Viscosity

Solid Friction

Static Friction

When a person walks along a road, he or she is prevented from slipping by the force of friction at the ground. In the absence of friction, for example on an icy surface, the person's shoe would slip when placed on the ground. The frictional force always *opposes* the motion of the shoe.

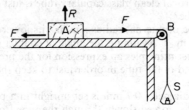

Fig. 7.1 Solid friction

The frictional force between the surface of a table and a block of wood A can be investigated by attaching one end of a string to A and the other to a scale-pan S, Fig. 7.1. The string passes over a fixed grooved wheel B. When small weights are added to S, the block does not move. The frictional force between the block and table is thus equal to the total weight on S together with the weight of S. When more weights are added, A still does not move, showing that the frictional force has increased. But as the weight is increased further, A suddenly begins to slip. The maximum frictional force is now present between the surfaces. It is called the *limiting frictional force*, and we are said to have reached *limiting friction*.

Coefficient of Static Friction

The normal reaction, R, of the table on A is equal to the weight of A. By placing various weights on A to alter the magnitude of R, we can find how the limiting frictional force F varies with R by the experiment just described. The results show that, approximately,

$$\frac{\text{limiting frictional force } (F)}{\text{normal reaction } (R)} = \mu, \text{ a constant,}$$

and μ is known as the *coefficient of static friction* between the two surfaces. The magnitude of μ depends on the nature of the two surfaces; for example, it is about 0·2 to 0·5 for wood on wood, and about 0·2 to 0·6 for wood on metals. Experiment also shows that the limiting frictional force is the same if the block A in Fig. 7.1 is turned on one side so that its surface area of contact with the table decreases, and thus the limiting frictional force is independent of the area of contact when the normal reaction is the same.

The coefficient of static friction, μ, can also be found by placing the block A on the surface S, and then gently tilting S until A is on the point of slipping down the plane, Fig. 7.2. The static frictional force F is then equal to $mg \sin$

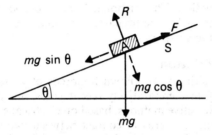

Fig. 7.2 Coefficient of friction by inclined plane

θ, where θ is the angle of inclination of the plane to the horizontal; the normal reaction R is equal to $mg \cos \theta$.

$$\therefore \mu = \frac{F}{R} = \frac{mg \sin \theta}{mg \cos \theta} = \tan \theta,$$

and hence μ can be found by measuring θ.

Kinetic Friction. Coefficient of Kinetic (Dynamic) Friction

When brakes are applies to a bicycle, a frictional force is exerted between the moving wheels and brake blocks. In contrast to the case of static friction, when one of the objects is just on the point of slipping, the frictional force between the moving wheel and brake blocks is called a *kinetic* (or *dynamic*) *frictional force*. Kinetic friction thus occurs between two surfaces which have relative motion.

The *coefficient of kinetic (dynamic) friction*, μ', between two surfaces is defined by the relation

$$\mu' = \frac{F'}{R}$$

where F' is the frictional force when the object moves with a uniform velocity and R is the normal reaction between the surfaces. The coefficient of kinetic friction between a block A and a table can be found by the apparatus shown in Fig. 7.1. Weights are added to the scale-pan, and each time A is given a slight push. At one stage A continues to move with a constant velocity, and the kinetic frictional force F' is then equal to the total weight in the scale-pan together with the latter's weight. On dividing F' by the weight of A, the coefficient can be calculated. Experiment shows that, when weights are placed on A to vary the normal reaction R, the magnitude of the ratio F'/R is approximately constant. Results also show that the coefficient of kinetic friction between two given surfaces is less than the coefficient of static friction between the same surfaces, and that the coefficient of kinetic friction between two given surfaces is approximately independent of their relative velocity.

Laws of Solid Friction

Experimental result on solid friction are summarised in the *laws of friction*, which state:

(1) The frictional force between two surfaces opposes their relative motion.
(2) The frictional force is independent of the area of contact of the given surfaces when the normal reaction is constant.
(3) The limiting frictional force is proportional to the normal reaction for

the case of static friction. The frictional force is proportional to the normal reaction for the case of kinetic (dynamic) friction, and is independent of the relative velocity of the surfaces.

Theory of Solid Friction

The laws of solid friction were known hundreds of years ago, but they have been explained only in comparatively recent years, mainly by F. P. Bowden and collaborators. Sensitive methods, based on electrical conductivity measurements, reveal that the true area of contact between two surfaces is extremely small, perhaps one ten-thousandth of the area actually placed together for steel surfaces. This is explained by photographs which show that some of the atoms of a metal project slightly above the surface, making a number of crests or 'humps'. As Bowden has stated: 'The finest mirror, which is flat to a millionth of a centimetre, would to anyone of atomic size look rather like the South Downs—valley and rolling hills a hundred or more atoms high.' Two metal surfaces thus rest on each other's projections when placed one on the other.

Fig. 7A Micrograph of the smooth surface of a special steel, magnification × 500. It shows the molecular projections which account for solid friction between two smooth metal surfaces. (*Courtesy of Triton Instruments Limited*)

Since the area of actual contact is extremely small, the pressures at the points of contact are very high, perhaps $10\,000$ million N m^{-2} for steel surfaces. The projections merge a little under the high pressure, producing adhesion or 'welding' at the points, and a force which opposes motion is therefore obtained. This explains Law 1 of the laws of solid friction. When one of the objects is turned over, so that a smaller or larger surface is presented to the other object, measurements show that the small area of actual contact remains constant. Thus the frictional force is independent of the area of the surfaces, which explains Law 2. When the load increases the tiny projections are further squeezed by the enormous pressures until the new area of contact becomes big enough to support the load. The greater the load, the greater is the area of actual contact, and the frictional force is thus approximately proportional to the load, which explains Law 3.

Viscosity or Fluid Friction

If we move through a pool of water we experience a resistance to our motion. This shows that there is a *frictional force* in liquids. We say this is due to the **viscosity** of the liquid. If the frictional force is comparatively low, as in water, the viscosity of the liquid is low; if the frictional force is large, as in glue or glycerine, the viscosity of the liquid is high. We can compare roughly the viscosity of two liquids by filling two measuring cylinders with each of them, and allowing identical small steel ball-bearings to fall through each liquid. The sphere falls more slowly through the liquid of higher viscosity.

As we shall see later, the viscosity of a lubricating oil is one of the factors which decide whether it is suitable for use in an engine. The Ministry of Aircraft Production, for example, listed viscosity values to which lubricating oils for aero-engines must conform. The subject of viscosity has thus considerable practical importance.

Newton's Formula. Coefficient of Viscosity

When water flows slowly and steadily through a pipe, the layer A of the liquid in contact with the pipe is practically stationary, but the central part C of the water is moving relatively fast, Fig. 7.3. At other layers between A and C, such as B, the water has a velocity less than at C, the magnitude of the velocities being represented by the length of the arrowed lines in Fig. 7.3. Now as in the case of two solid surfaces moving over each other, a frictional force is exerted between two liquid layers when they move over each other. Thus because the velocities of neighbouring layers are different, as shown in Fig. 7.3, a frictional force occurs between the various layers of a liquid when flowing through a pipe.

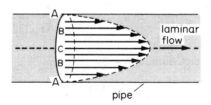

Fig. 7.3 Laminar (uniform) flow through pipe

The basic formula for the frictional force, F, in a liquid was first suggested by NEWTON. He saw that the larger the *area* of the surface of liquid considered, the greater was the frictional force F. He also stated that F was directly proportional to the *velocity gradient* at the part of the liquid considered. This is the case for most common liquids, called *Newtonian liquids*. If v_1, v_2 are the velocities of C, B respectively in Fig. 7.3, and h is their distance apart, the velocity gradient between the liquids is defined as $(v_1 - v_2)/h$. The velocity gradient can thus be expressed in (m/s)/m, or as 's^{-1}'.

Thus if A is the area of the liquid surface considered, the frictional force F on the surface is given by

$$F \propto A \times \text{velocity gradient,}$$

or $\qquad\qquad F = \eta A \times \text{velocity gradient} \qquad . \qquad . \qquad . \qquad . \qquad (1)$

where η is a constant of the liquid known as the *coefficient of viscosity*. This expression for the frictional force in a liquid should be contrasted with the case of solid friction, in which the frictional force is independent of the area of contact and of the relative velocity between the solid surfaces concerned (p. 161).

Definition, Units, and Dimensions of Coefficient of Viscosity

The magnitude of η is given by

$$\eta = \frac{F}{A \times \text{velocity gradient}}.$$

The unit of F is a newton, the unit of A is m², and the unit of velocity gradient is 1 m/s per m. Thus η may be defined as *the frictional force per unit area of a liquid when it is in a region of unit velocity gradient.*

The 'unit velocity gradient' = 1 m s⁻¹ change per m. Since the 'm' cancels, the 'unit velocity gradient' = 1 per second. From $\eta = F/(A \times \text{velocity gradient})$, it follows that η may be expressed in units of N s m⁻².

The coefficient of viscosity of water at 10°C is $1\cdot3 \times 10^{-3}$ N s m⁻². Since $F = \eta A \times$ velocity gradient, the frictional force on an area of 10 cm² in water at 10°C, between two layers of water 1 mm apart, which move with a relative velocity of 2 cm s⁻¹, is found as follows:

coefficient of viscosity $\eta = 1\cdot3 \times 10^{-3}$ N s m⁻², $A = 10 \times 10^{-4}$ m², velocity gradient $= 2 \times 10^{-2}$ m s⁻¹ $\div (1 \times 10^{-3})$ m $= 20$ s⁻¹.

$$\therefore F = 1\cdot3 \times 10^{-3} \times 10 \times 10^{-4} \times 20 = 2\cdot6 \times 10^{-5} \text{ N}.$$

Dimensions. The dimensions of a force, F, (= mass × acceleration = mass × velocity change/time) are MLT⁻². See p. 32. The dimensions of an area, A, are L². The dimensions of velocity gradient

$$= \frac{\text{velocity change}}{\text{distance}} = \frac{\text{L}}{\text{T}} \div \text{L} = \frac{1}{\text{T}}.$$

Now

$$\eta = \frac{F}{A \times \text{velocity gradient}},$$

$$\therefore \text{ dimensions of } \eta = \frac{\text{MLT}^{-2}}{\text{L}^2 \times 1/\text{T}},$$

$$= \text{ML}^{-1}\text{T}^{-1}.$$

Thus η may be expressed in units 'kg m⁻¹ s⁻¹'.

Experiment shows that the viscosity coefficient of a liquid diminishes as its temperature rises. Thus for water, η at 15°C is $1\cdot1 \times 10^{-3}$ N s m⁻², at 30°C it is $0\cdot8 \times 10^{-3}$ N s m⁻² and at 50°C it is $0\cdot6 \times 10^{-3}$ N s m⁻². Lubricating oils for motor engines which have the same coefficient of viscosity in summer and winter are known as 'viscostatic' oils.

Steady Flow of Liquid through Pipe. Poiseuille's Formula

The steady flow of liquid through a pipe was first investigated thoroughly by POISEUILLE in 1844, who derived an expression for the volume of liquid issuing per second from the pipe. We can derive most of the formula by the *method of dimensions* (p. 32).

The volume of liquid issuing per second from the pipe depends on (i) the coefficient of viscosity, η, (ii) the radius, a, of the pipe, (iii) the *pressure gradient*, g, set up along the pipe. The pressure gradient $= p/l$, where p is the pressure difference between the ends of the pipe and l is its length. Thus x, y, z being indices which require to be found, suppose

$$\text{volume per second} = k\eta^x a^y g^z \qquad . \qquad . \qquad . \qquad (1)$$

Now the dimensions of volume per second are L^3T^{-1}; the dimensions of η are $ML^{-1}T^{-1}$; the dimensions of a is L; and the dimensions of g are

$$\frac{[\text{pressure}]}{[\text{length}]}, \quad \text{or} \quad \frac{[\text{force}]}{[\text{area}][\text{length}]}, \quad \text{or} \quad \frac{MLT^{-2}}{L^2 \times L}, \text{ which is } ML^{-2}T^{-2}.$$

Thus from (1), equating dimensions on both sides,

$$L^3T^{-1} \equiv (ML^{-1}T^{-1})^x L^y (ML^{-2}T^{-2})^z.$$

Equating the respective indices of M, L, T on both sides, we have

$$x + z = 0,$$

$$-x + y - 2z = 3,$$

$$x + 2z = 1.$$

Solving, we obtain $x = -1$, $z = 1$, $y = 4$. Hence, from (1),

$$\text{volume per second} = k\frac{a^4 g}{\eta} = k\frac{pa^4}{l\eta}.$$

We cannot obtain the numerical factor k from the method of dimensions. Mathematical analysis shows that the factor $\pi/8$ enters into the formula, which is:

$$\textit{Volume per second} = \frac{\pi pa^4}{8\eta l} \qquad . \qquad . \qquad . \qquad . \qquad (2)$$

Poiseuille's formula holds as long as the velocity of each layer of the liquid is parallel to the axis of the pipe and the flow pattern has been developed. This is *orderly (laminar) flow*. As the pressure difference between the ends of the pipe is increased, a critical velocity is reached at some stage, and the motion of the liquid changes from an orderly to a *turbulent* one. Poiseuille's formula does not apply to turbulent motion.

Turbulent Flow

The onset of turbulence was first demonstrated by O. REYNOLDS in 1883, and was shown by placing a horizontal tube T, about 0·5 cm in diameter, at the bottom of a tank W of water, Fig. 7.4 (i). The flow of water along T is controlled by a clip C on rubber tubing connected to T. A drawn-out glass jet B, attached to a reservoir A containing coloured water, is placed at one end of T, and at low velocities of flow a thin coloured stream of water is observed flowing along

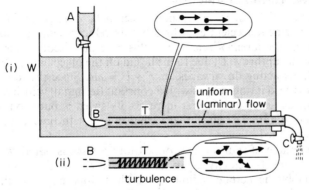

Fig. 7.4 Laminar and turbulent flow

the middle of T. As the rate of flow of the water along T is increased, a stage
is reached when the colouring in T begins to spread out and fill the whole of
the tube, Fig. 7.4 (ii). The critical velocity has now been exceeded, and turbulence
has begun.

Figure 7.4 shows diagrammatically in inset: (i) laminar or uniform flow—here
particles of liquid at the same distance from the axis always have equal velocities
directed parallel to the axis, (ii) turbulence—here particles at the same distance
from the axis have different velocities, and these vary in magnitude and direction
with time.

Determination of Viscosity by Poiseuille's Formula

The viscosity of a liquid such as water can be measured by connecting one end
of a capillary tube T to a constant pressure apparatus A, which provides a *steady*
flow of liquid, Fig. 7.5. By means of a beaker B and a stop-clock, the volume
of water per second flowing through the tube can be measured. The pressure
difference between the ends of T is $h\rho g$, where h is the pressure head, ρ is the
density of liquid, and g is $9 \cdot 8$ m s^{-2}. Provided the flow is laminar or uniform,
Poiseuille showed that

$$\text{volume per second} = \frac{\pi p a^4}{8\eta l} = \frac{\pi h \rho g a^4}{8\eta l},$$

where l is the length of T and a is its radius. The radius of the tube can be
measured by means of a mercury thread or by a microscope. The coefficient
of viscosity η can then be calculated, since all the other quantities in the above
equation are known.

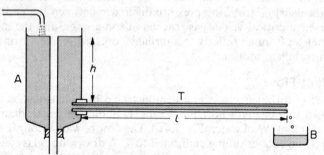

Fig. 7.5 Absolute measurement of viscosity

Stokes' Law. Terminal Velocity

When a small object, such as a small steel ball, is released in a viscous
liquid like glycerine it accelerates at first, but its velocity soon reaches a steady
value known as the *terminal velocity*. In this case the viscous force, F, acting
upwards, and the upthrust, U, due to the liquid on the object, are together equal
to its weight, mg acting downwards, or $F + U = mg$. Since the resultant force
on the object is zero it now moves with a constant (terminal) velocity. An object
dropped from an aeroplane at first increases its speed v, but soon reaches its
terminal speed. Figure 7.6 shows that variation of v with time as the terminal
velocity v_0 is reached

Suppose a sphere of radius a is dropped into a viscous liquid of coefficient
of viscosity η, and its velocity at an instant is v. The frictional force, F, can
be partly found by the method of dimensions. Thus suppose $F = k a^x \eta^y v^z$, where k
is a constant. The dimensions of F are MLT^{-2}; the dimensions of a is L; the
dimensions of η are $ML^{-1}T^{-1}$; and the dimensions of v are LT^{-1}.

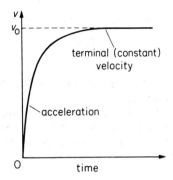

Fig. 7.6 Motion of falling sphere

$$\therefore \text{MLT}^{-2} \equiv \text{L}^x \times (\text{ML}^{-1}\text{T}^{-1})^y \times (\text{LT}^{-1})^z.$$

Equating indices of M, L, T on both sides,

$$\therefore y = 1,$$

$$x - y + z = 1,$$

$$-y - z = -2.$$

Hence $z = 1$, $x = 1$, $y = 1$. Consequently $F = k\eta av$. In 1850 STOKES showed mathematically that the constant k was 6π, and he arrived at the formula

$$F = 6\pi\eta av \qquad . \qquad . \qquad . \qquad . \qquad . \qquad (1)$$

Measurement of Viscosity of Viscous Liquid

Stokes' formula can be used to measure the coefficients of viscosity of very viscous liquids such as glycerine or treacle. A tall glass vessel G is filled with the liquid, and a small steel ball P is dropped gently into the liquid so that it falls along the axis of G, Fig. 7.7. Towards the middle of the liquid P reaches its terminal velocity v_0, which is measured by timing its fall through a distance AB or BC.

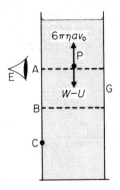

Fig. 7.7 Measurement of viscosity

The upthrust, U, on P due to the liquid $= 4\pi a^3 \sigma g/3$, where a is the radius of P and σ is the density of the liquid. The weight, W, of P is $4\pi a^3 \rho g/3$, where ρ is density of the bearing's material. The net downward force is thus $4\pi a^3 g(\rho - \sigma)/3$. When the opposing frictional force grows to this magnitude, the resultant force on the bearing is zero. Thus for the terminal velocity v_0, we have

$$6\pi\eta a v_0 = \frac{4}{3}\pi a^3 g(\rho - \sigma),$$

$$\therefore \eta = \frac{2ga^2(\rho - \sigma)}{9v_0} \qquad \qquad \text{. . . . (i)}$$

The coefficient of viscosity of the liquid can be calculated from (i) when v_0 and the other quantities are known.

In Millikan's experiment to determine the charge on an electron (p. 793), an oil drop reaches its terminal velocity v_0 while falling in air of viscosity η. The radius a of the drop is needed in the experiment and this is found from equation (i) when the value of η is known.

Example

A small oil drop falls with a terminal velocity of $4{\cdot}0 \times 10^{-4}$ m s^{-1} through air. Calculate the radius of the drop.

What is the new terminal velocity for an oil drop of half this radius? (Viscosity of air $= 1{\cdot}8 \times 10^{-5}$ N s m^{-2}, density of oil $= 900$ kg m^{-3}, $g = 10$ m s^{-2}; neglect density of air.)

At terminal (steady) velocity,

$$\text{frictional force on drop} = \text{weight} - \text{upthrust}.$$

With the usual notation,

$$6\pi\eta a v_0 = \frac{4}{3}\pi a^3 \rho g - \frac{4}{3}\pi a^3 \sigma g$$

where ρ is the oil density and σ that of the air. Neglecting σ, and simplifying,

$$a = \left(\frac{9v_0\eta}{2\rho g}\right)^{\frac{1}{2}}$$

$$= \left(\frac{9 \times 4 \times 10^{-4} \times 1{\cdot}8 \times 10^{-5}}{2 \times 900 \times 10}\right)^{\frac{1}{2}}$$

$$= 1{\cdot}9 \times 10^{-6} \text{ m}.$$

The terminal velocity $v_0 \propto a^2$ from above. So when the radius a is decreased to one-half,

$$\text{new terminal velocity} = \tfrac{1}{4} \times 4{\cdot}0 \times 10^{-4} = 1{\cdot}0 \times 10^{-4} \text{ m s}^{-1}.$$

Molecular Theory of Viscosity

Viscous forces are detected in gases as well as in liquids. Thus if a disc is spun round in a gas close to a suspended stationary disc, the latter rotates in the same direction. The gas hence transmits frictional forces. The flow of gas through pipes, particularly in long pipes as in transmission of natural gas from the North Sea area, is affected by the viscosity of the gas.

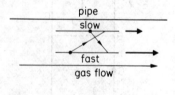

Fig. 7.8 Viscosity of gas due to momentum transfer

The viscosity of *gases* is explained by the transfer of momentum which takes place between neighbouring layers of the gas as it flows in a particular direction. Fast-moving molecules in a layer X cross with their own velocity to a layer Y say where molecules are moving with a slower velocity, Fig. 7.8. Molecules in

Y likewise move to X. The net effect is an increase in momentum in Y and a corresponding decrease in X, although on the average the total number of molecules in the two layers is unchanged. Thus the layer Y speeds up and the layer X slows down, that is, a *force* acts on the layers of the gas while they move. This is the viscous force. We consider the movement of the molecules in more detail later.

Although there is transfer of momentum as in the gas, the viscosity of a *liquid* is mainly due to the molecular attraction between molecules in neighbouring layers. Energy is needed to drag one layer over the other against the force of attraction. Thus a shear stress is required to make the liquid move in laminar flow.

Exercises 7

(Assume $g = 10$ m s^{-2} or 10 N kg^{-1} unless otherwise given)

Solid Friction

1. State the laws of solid friction. Describe an experiment to determine the coefficient of dynamic (or sliding) friction between two surfaces.

A horizontal circular turntable rotates about its centre at the uniform rate of 120 revolutions per minute. Find the greatest distance from the centre at which a small body will remain stationary relative to the turntable, if the coefficient of static friction between the turntable and the body is 0·80. (*L.*)

2. State (*a*) the laws of solid friction, (*b*) the triangle law for forces in equilibrium. Describe an experiment to determine the coefficient of sliding (dynamic) friction between two wooden surfaces.

A block of wood of mass 150 g rests on an inclined plane. If the coefficient of static friction between the surfaces in contact is 0·30, find (*a*) the greatest angle to which the plane may be tilted without the block slipping, (*b*) the force parallel to the plane necessary to prevent slipping when the angle of the plane with the horizontal is 30°, showing that this direction of the force is the one for which the force required to prevent slipping is a minimum. (*L.*)

3. Distinguish between *static* and *sliding* (kinetic) friction and define the *coefficient of sliding friction*.

How would you investigate the laws of sliding friction between wood and iron?

An iron block, of mass 10 kg, rests on a wooden plane inclined at 30° to the horizontal. It is found that the least force parallel to the plane which causes the block to slide *up* the plane is 100 N. Calculate the coefficient of sliding friction between wood and iron. (*N.*)

4. State the laws of solid friction.

Describe experiments to verify these laws, and to determine the coefficient of static friction, for two wooden surfaces.

A small coin is placed on a gramophone turntable at a distance of 7·0 cm from the axis of rotation. When the rate of rotation is gradually increased from zero the coin begins to slide outwards when the rate reaches 60 revolutions per minute. Calculate the rate of rotation for which sliding would commence if (*a*) the coin were placed 12·0 cm from the axis, (*b*) the coin were placed in the original position with another similar coin stuck on top of it. (*L.*)

Viscosity

5. Draw a velocity-time graph for a body released from rest and falling through air from a considerable height. Explain the shape of the graph. (*L.*)

6. The viscous force on a sphere, of radius r, moving through a fluid with velocity v can be expressed as $6\pi\eta rv$, where η is the coefficient of viscosity of the fluid. What is the limitation on the use of this expression?

A small sphere, of radius r and density σ, is released from the bottom of a column of liquid of density ρ, which is slightly larger than σ. Deduce expressions for (a) the initial acceleration of the sphere, and (b) its terminal velocity. (L.)

7. Explain what is meant by laminar flow and define the coefficient of viscosity.

A flat plate of area $0 \cdot 1$ m^2 is placed on a flat surface and is separated from it by a film of oil 10^{-5} m thick whose coefficient of viscosity is $1 \cdot 5$ N s m^{-2}. Calculate the force required to cause the plate to slide on the surface at a constant speed of 1 mm s^{-1}.

(Assume that the flow is laminar and that the oil adjacent to each surface moves with that surface.) (O. &. C.)

8. Explain briefly what is meant by saying that a fluid is *viscous*?

Discuss the factors which affect the volume rate of orderly flow of liquid through a tube. What is meant by the condition that the flow must be orderly? How would you demonstrate the the flow through a tube was of this nature?

By considering the forces acting on a sphere falling through a viscous medium, explain why it eventually reaches a terminal velocity. Explain whether or not it would be possible for these forces to bring the sphere to rest.

Two spherical raindrops of equal size are falling vertically through air with a terminal velocity of $0 \cdot 150$ m s^{-1}. What would be the terminal velocity if these two drops were to coalesce to form a larger spherical drop? (Stokes' law may be assumed to apply.) (L.)

9. The dimensions of *energy*, and also those of *moment of a force* are found to be 1 in *mass*, 2 in *length* and -2 in *time*. Explain and justify this statement.

(a) A sphere of radius a moving through a fluid of density ρ with *high* velocity V experiences a retarding force F given by $F = k . a^x . \rho^y . V^z$, where k is a non-dimensional coefficient. Use the method of dimensions to find the values of x, y and z.

(b) A sphere of radius 2 cm and mass 100 g, falling vertically through air of density $1 \cdot 2$ kg m^{-3}, at a place where the acceleration due to gravity is $9 \cdot 81$ m s^{-2}, attains a steady velocity of 30 m s^{-1}. Explain why a constant velocity is reached and use the data to find the value of k in this case. (O. &. C.)

10. Define *coefficient of viscosity* of a fluid.

When the flow is orderly the volume V of liquid which flows in time t through a tube of radius r and length l when a pressure difference p is maintained between its ends is given by the equation $\dfrac{V}{t} = \dfrac{\pi p r^4}{8 l \eta}$ where η is the coefficient of viscosity of the liquid. Describe an experiment based on this equation *either* (a) to determine the value of η for a liquid, *or* (b) to compare the values of η for two liquids, pointing out the precautions which must be taken in the experiment chosen to obtain an accurate

Water flows steadily through a horizontal tube which consists of two parts joined end to end; one part is 21 cm long and has a diameter of $0 \cdot 225$ cm and the other is $7 \cdot 0$ cm long and has a diameter of $0 \cdot 075$ cm. If the pressure difference between the ends of the tube is 14 cm of water find the pressure difference between the ends of each part. (L.)

11. Define *coefficient of viscosity*. Describe an experiment to compare the coefficients of viscosity of water and benzene at room temperature.

A small metal sphere is released from rest in a tall wide vessel of liquid. Discuss the forces acting on the sphere (a) at the moment of release, (b) soon after release, (c) after the terminal velocity has been attained.

Castor oil at 20°C has a coefficient of viscosity $2 \cdot 42$ N s m^{-2} and a density 940 kg m^{-3}. Calculate the terminal velocity of a steel ball of radius $2 \cdot 0$ mm falling under gravity in the oil, taking the density of steel as 7800 kg m^{-3}. (L.)

12. The viscous force acting on a small sphere of radius a moving slowly through a liquid of viscosity η with velocity v is given by the expression $6\pi\eta a v$. Sketch the general shape of the velocity-time graph for a particle falling from rest through a viscous fluid, and explain the form of the graph. List the observations you would make to determine the coefficient of viscosity of the fluid from the motion of the particle.

Some particles of sand are sprinkled on to the surface of the water in a beaker filled to a depth of 10 cm. Estimate the least time for which grains of diameter $0 \cdot 10$ mm remain in suspension in the water, stating any assumptions made.

[Viscosity of water $= 1 \cdot 1 \times 10^{-3}$ N s m^{-2}; density of sand $= 2200$ kg m^{-3}.] (C.)

Part Two
Heat

8 Introduction: Temperature, Heat, Energy

Temperature

We are interested in heat because it is the commonest form of energy, and because changes of temperature have great effects on our personal comfort, and on the properties of substances, such as water, which we use every day. *Temperature* is a scientific quantity which corresponds to primary sensations—hotness and coldness. These sensations are not reliable enough for scientific work, because they depend on contrast—the air in a thick-walled barn or church feels cool on a summer's day, but warm on a winter's day, although a thermometer may show that it has a lower temperature in the winter. A thermometer, such as the familiar mercury-in-glass instrument (Fig. 8.1), is a device whose readings depend on hotness or coldness, and which we choose to consider more reliable than our senses. We are justified in considering it more reliable because different thermometers of the same type agree with one another better than different people do.

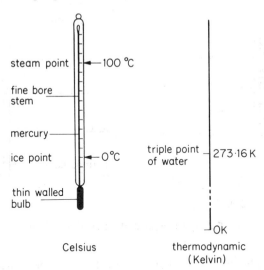

Fig. 8.1 Mercury-in-glass thermometer (left); °C and K scales

Types of Thermometers

The temperature of an object is not a fixed number but depends on the type of thermometer used and on the temperature scale adopted, discussed shortly.

In general, thermometers use some measurable property of a substance which is sensitive to temperature change. The *constant-volume gas thermometer*, for example, uses the pressure change with temperature of a gas at constant volume. The *resistance thermometer* uses the change of electrical resistance of a pure

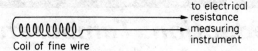

to electrical resistance measuring instrument

Coil of fine wire

Fig. 8.2 A resistance thermometer; the wire is usually of platinum

metal with temperature, Fig. 8.2. The *mercury-in-glass* thermometer depends on the change in volume of mercury with temperature relative to that of glass. A *thermoelectric thermometer* depends on the electromotive force change with temperature of two metals joined together.

Thermodynamic Temperature Scale

The *thermodynamic* temperature scale is the standard temperature scale adopted for scientific measurement. Thermodynamic temperature is denoted by the symbol T and is measured in *kelvin*, symbol K. The kelvin is the SI unit of temperature or of temperature change.

The thermodynamic temperature scale uses one fixed point, the *triple point of water*. This is the temperature at which saturated water-vapour, pure water and melting ice are all in equilibrium. The triple point temperature is *defined* as 273·16 K. Fig. 8.3 (i) shows an apparatus used for obtaining the triple point of water; distilled water (from which dissolved air has been driven out), water-vapour and ice are here in equilibrium. The kelvin is the fraction 1/273·16 of the thermodynamic temperature of the triple point of water.

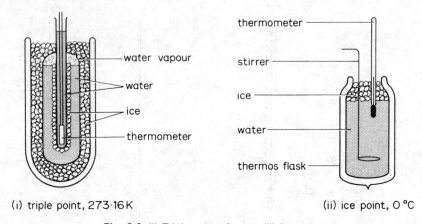

water vapour

water

ice

thermometer

thermometer

stirrer

ice

water

thermos flask

(i) triple point, 273·16 K

(ii) ice point, 0 °C

Fig. 8.3 (i) Triple point of water (ii) Ice point

On the thermodynamic scale, the ice point has a temperature 273·15 K. The slight difference from the triple point is due to the difference in pressure (4·6 mmHg at the triple point and 760 mmHg at the ice point) and to the removal of dissolved air from the distilled water used for the triple point.

Using the constant-volume gas thermometer, for example, the gas pressure p_{tr} is measured at the triple point of water, 273·16 K. If the pressure is p at an unknown temperature T on the thermodynamic scale, then, by definition,

$$T = \frac{p}{p_{tr}} \times 273 \cdot 16 \text{ K}.$$

With a platinum resistance thermometer, the resistance R is measured at the unknown temperature. The temperature T_{pt} on the thermodynamic scale is then given, if R_{tr} is the resistance at the triple point, by

$$T_{pt} = \frac{R}{R_{tr}} \times 273 \cdot 16 \text{ K}.$$

Celsius Temperature Scale

The *Celsius temperature*, symbol θ, is now defined by $\theta = T - 273 \cdot 15$, where T is the thermodynamic temperature (see *The International Practical Temperature Scale of 1968*, National Physical Laboratory, HMSO). The ice point is a Celsius temperature of 0°C and the steam point, the temperature of steam at 760 mmHg pressure, is 100°C.

It should be noted that if different types of thermometers are used to measure temperature, they only agree at the fixed points—273·16 K on the thermodynamic scale and 0°C and 100°C, for example, on the Celsius scale.

The temperature change or interval of one degree Celsius, 1°C, is exactly the same as the temperature interval 1 K on the thermodynamic scale. So '°C' may be replaced by 'K' in SI units, and '°C^{-1}' by 'K^{-1}'. Approximately, the temperature 0°C = 273 K and 100°C = 373 K. The absolute zero of temperature 0 K is approximately -273°C.

Heat and Energy

Heat and Temperature

If we run hot water into a lukewarm bath, we make it hotter; if we run in cold water, we make it cooler. The hot water, we say, gives heat to the cooler bath-water; but the cold water takes heat from the warmer bath-water. The quantity of heat which we can get from hot water depends on both the mass of water and on its temperature: a bucket-full at 80°C will warm the bath more than a cup-full at 100°C. Roughly speaking, temperature is analogous to electrical potential, and heat is analogous to quantity of electricity. We can detect temperature changes, and whenever the temperature of a body rises, that body has gained heat. The converse is not always true; when a body is melting or boiling, it is absorbing heat from the flame beneath it, but its temperature is not rising.

Thermal Equilibrium. The Zeroth Law

When two bodies are in thermal contact and there is no net flow of heat between them, they are said to be in *thermal equilibrium*. Experimentally, it is found that when two bodies A and B are each in thermal equilibrium with a third body C, then A and B are also in thermal equilibrium with each other. This is called the 'Zeroth Law of Thermodynamics'—the number 'zero' was used because this law logically precedes the 'first' and 'second' laws of thermodynamics and, in fact, is assumed in the two laws.

The Zeroth Law leads to the conclusion that temperature is a well-defined physical quantity. For example, suppose we wish to determine whether two bodies A and B are in thermal equilibrium. In practice, we do this by bringing each in turn into contact with a third body, a thermometer T say. Experimentally, then, we bring A and T into thermal equilibrium, and B and T into thermal equilibrium. If the temperature reading is the same in the two cases, we deduce, but only with the help of the Zeroth Law, that A and B are in thermal equilibrium. The Law thus enables temperature to be defined as that property of a body which decides whether or not it is in thermal equilibrium with another body.

Heat and Energy

The idea of heat as a form of energy was developed particularly by Benjamin Thompson (1753–1814); he was an American who, after adventures in Europe, became a Count of the Holy Roman Empire, and war minister of Bavaria. He is now generally known as Count Rumford. While supervising his arsenal, he noticed the great amount of heat which was liberated in the boring of cannon. The idea common at the time was that this heat was a fluid, pressed out of the chips of metal as they were bored out of the barrel. To measure the heat produced, Rumford used a blunt borer, and surrounded it and the end of the cannon with a wooden box, which was filled with water. From the mass of water, and the rate at which its temperature rose, he showed that the amount of heat liberated was in no way connected with the mass of metal bored away, and concluded that it depended only on the work done against friction. It followed that *heat was a form of energy.*

Rumford published the results of his experiments in 1798. No similar experiments were made until 1840, when Joule began his study of heat and other forms of energy. Joule measured the work done, and the heat produced, when water was churned, in an apparatus he designed for the experiment. He also measured the work done and heat produced when oil was churned, when air was compressed, when water was forced through fine tubes, and when cast iron bevel wheels were rotated one against the other. Always, within the limits of experimental error, he found that the heat liberated was proportional to the mechanical work done, and that the ratio of the two was the same in all types of experiment.

In other experiments, Joule measured the heat liberated by an electric current in flowing through a resistance; at the same time he measured the work done in driving the dynamo which generated the current. He obtained about the same ratio for work done to heat liberated as in his direct experiments. This work linked the ideas of heat, mechanical, and electrical energy. He also showed that the heat produced by a current is related to the chemical energy used up.

We can thus say (a) that an object gains *energy* when its temperature rises, (b) that *energy* (heat) passes from a warm to a cold object if they are placed in contact.

The metric unit of work or energy is the *joule*, J. Since experiment shows that heat is a form of energy, *the joule is now the scientific unit of heat*. 'Heat per second' is expressed in 'joules per second' or *watts*, W.

The Conservation of Energy

As a result of all his experiments, Joule developed the idea that energy in any one form could be converted into any other. There might be a loss of useful energy in the process—for example, some of the heat from the furnace of a steam-engine is lost up the chimney, and some more down the exhaust—but no energy is destroyed. The work done by the engine added to the heat lost as described and the heat developed as friction, it is equal to the heat provided by the fuel burnt. The idea underlying this statement is called the *Principle of the Conservation of Energy*. It implies that, if we start with a given amount of energy in any one form, we can convert it in turn into all other forms; we may not always be able to convert it completely, but if we keep an accurate balance-sheet we shall find that the total amount of energy, expressed in any one form—say heat or work—is always the same, and is equal to the original amount.

The principle of the conservation of energy is often expressed concisely in mathematical form by the equation

$$\Delta Q = \Delta U + \Delta W.$$

Here ΔQ is the *quantity of heat or energy* given to a system, ΔU is the consequent rise in *internal energy* of the system and ΔW is the *external work* done by the system. As we see later, the rise in internal energy, ΔU, of a system is shown by a temperature rise. If the system also expands, it does external work, ΔW, against the external forces. The principle of conservation of energy will be applied later to many cases of energy changes.

The conservation of energy applies to living organisms—plants and animals—as well as to inanimate systems. For example, we may put a man or a mouse into a box or a room, give him a treadmill to work, and feed him. His food is his fuel; if we burn a sample of it, we can measure its chemical energy, in heat units. And if we now add up the heat value of the work which the man does, and the heat which his body gives off, we find that their total is equal to the chemical energy of the food which the man eats. Because food is the source of man's energy, food values may be expressed in *kilojoules*. A man needs about 120 000 kilojoules per day.

All the energy by which we live comes from the sun. The sun's ultra-violet rays are absorbed in the green matter of plants, and make them grow; the animals eat the plants, and we eat them—we are all vegetarians at one remove. The

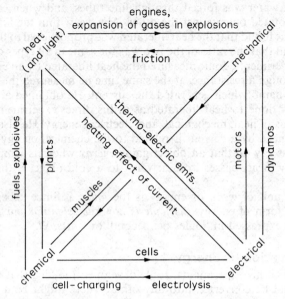

Fig. 8.4 Forms of energy, and their interconversions

plants and trees of an earlier age decayed, were buried, and turned into coal. Even water-power comes from the sun—we would have no lakes if the sun did not evaporate the water and provide the rainfall which fills the lakes. The relationship between the principal forms of energy are summarised in Fig. 8.4.

9 Heat Capacity. Latent Heat

Heat Capacity, Specific Heat Capacity

The *heat capacity* of a body, such as a lump of metal, is the quantity of heat required to raise its temperature by 1 degree. It is expressed in *joule per kelvin* ($J\ K^{-1}$) in SI units.

The *specific heat capacity* of a substance is the heat required to raise the temperature of 1 kg of it through 1 degree; it is the heat capacity per kg of the substance. Specific heat capacities are expressed in *joule per kilogram per kelvin* and the symbol is *c*. The specific heat capacity of water, c_w, is about 4200 $J\ kg^{-1}\ K^{-1}$, or 4·2 $kJ\ kg^{-1}\ K^{-1}$, where 1 kJ = 1 kilojoule = 1000 J. The megajoule, MJ, is a larger unit than the kilojoule used in industry. 1 MJ = 10^6 J = 1000 kJ.

From the definition of specific heat capacity, it follows that, for a particular object,

$$heat\ capacity,\ C = mass \times specific\ heat\ capacity.$$

The specific heat capacity of copper, for example, is about 400 $J\ kg^{-1}\ K^{-1}$. Hence the heat capacity of 5 kg of copper = $5 \times 400 = 2000\ J\ K^{-1} = 2\ kJ\ K^{-1}$. If the copper temperature rises by 10°C, then heat gained = $5 \times 400 \times 10$ = 20 000 J.

Generally, then, the heat Q gained (or lost) by an object is given by

$$Q = mc\theta,$$

where *m* is the mass of the object, *c* its specific heat capacity and θ its temperature change.

Temperature changes

From $Q = mc\ \theta$, the temperature change θ of an object of mass *m* which loses or gains a given quantity of heat Q is given by

$$\theta = \frac{Q}{mc},$$

where *c* is the specific heat capacity of the object. So for a given loss of heat Q to the surroundings, the temperature fall θ of a small mass of warm water in a room is greater than a large mass of water at the same temperature. We shall see later that the rate at which a hot solid or liquid cools depends on the nature and area of its surface, in addition to its temperature, mass and specific heat capacity.

When a thermoelectric thermometer is used, the thermometer junction is placed in contact with the object whose temperature is required. The junction has a small thermal capacity, *mc*, since its mass is small. The temperature of the junction thus quickly reaches the temperature of the object, which is an advantage of the thermometer.

Measurement of Specific Heat Capacity

Specific Heat Capacity of Solid by Electrical Method

The specific heat capacity of a metal can be found by an *electrical method* Fig. 9.1 shows a simple form of laboratory apparatus. M is a thick solid block of metal such as aluminium, with an electric heater element H completely inside a deep hole bored into the metal and a thermometer T inside another deep hole. Both H and T must make good thermal contact with the block. An insulating jacket J is placed round the metal.

In an experiment, suppose the voltmeter reads 12 V, the ammeter A reads 4·0 A and the block, mass 1·0 kg, rises by 16°C in 5 min or 300 s. Then (p. 637)

$$\text{heat supplied} = IVt = 12 \times 4 \times 300 = 14\,400 \text{ J}.$$

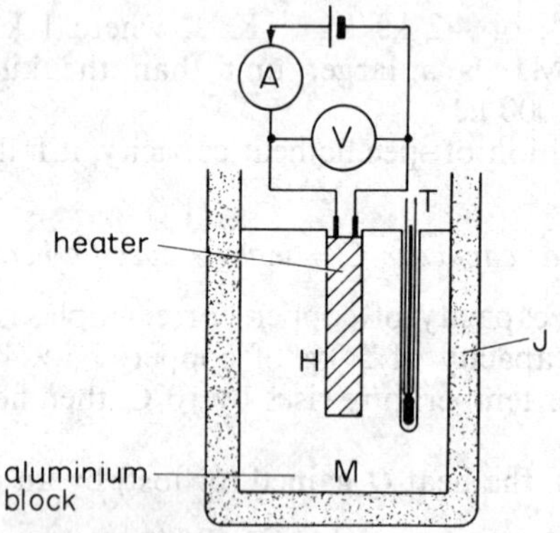

Fig. 9.1 Electrical method for specific heat capacity

If c is the specific heat capacity of the metal, then, assuming negligible heat losses,

$$Q = mc\theta = 1 \times c \times 16 = 14\,400$$

$$\therefore c = 900 \text{ J kg}^{-1} \text{ K}^{-1}.$$

If θ is the temperature rise corrected for heat losses (p. 186), then, generally,

$$IVt = mc\theta.$$

The electrical energy supplied, $\Delta Q = IVt$. From the principle of conservation of energy (p. 177), this is equal to the rise in internal energy ΔU of the metal ($mc\theta$) plus the heat losses h by cooling *plus* the external work ΔW done against external atmospheric pressure by the metal when it expands on warming. Since metals expand very slightly in volume on warming, ΔW can be neglected in this experiment. So $IVt = mc\theta + h$.

If θ is not corrected for heat losses, we see from $IVt = mc\theta$ that the result for c is too high. The cooling correction is discussed later.

Specific Heat Capacity by Mechanical Method

Figure 9.2 illustrates one simple form of apparatus for measuring the specific heat capacity of a metal by the transfer of *mechanical energy to heat energy*.

It consists of a small solid metal cylinder A, insulated from the rest of the apparatus by a nylon bush. The metal temperature can be read on a short-range thermometer B placed inside a hole bored axially in A. A flexible nylon cord C is wound several times round the cylinder. One end of C is attached to a rubber band D. The other end is attached to a heavy weight W which hangs from it.

By means of a handle H, the cylinder is rotated so that the rubber band goes

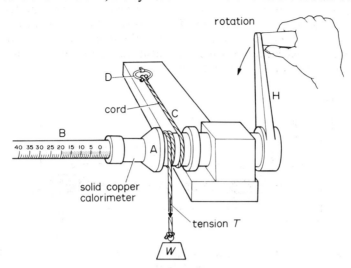

Fig. 9.2 Mechanical method for specific heat capacity of metal

slack. The tension in the cord C is then equal to the weight W and mechanical energy is expended against this force when the cylinder is rotated. After a suitable number of revolutions, the final temperature of the metal cylinder is noted, a cooling correction having been applied (p. 186).

Suppose $W = 50$ N, the mass of the metal cylinder is 0·200 kg, its diameter is 25 mm or 0·025 m, the temperature rise is 10·0 K and the number of revolutions is 200. Then, since the circumference of a cylinder $= \pi \times$ diameter,

$$\text{mechanical energy expended} = \text{force} \times \text{distance}$$
$$= 50 \times (\pi \times 0{\cdot}025) \times 200 \text{ J}.$$

From the principle of conservation of energy, we see that the mechanical energy ΔQ spent in turning the handle = the rise in internal energy ΔU of the metal *plus* the external work ΔW done against the external atmospheric pressure when the metal expands on warming. As explained on p. 208 the metal expansion is very small and so ΔW can be neglected in the equation for c. So $\Delta Q = \Delta U$.

Assuming negligible heat losses, $\Delta U =$ heat gained by metal $= mc\theta$, where c is its specific heat capacity.

$$\therefore \ 0{\cdot}2 \times c \times 10 = 50 \times \pi \times 0{\cdot}025 \times 200$$
$$\therefore \ c = 390 \text{ J kg}^{-1} \text{ K}^{-1} \text{ (approx.)}.$$

Specific Heat Capacity of a Liquid by Continuous Flow Method

In 1899, Callendar and Barnes devised an electrical method for the specific heat capacity of a liquid in which only steady temperatures are measured. They used platinum resistance thermometers, which are more accurate than mercury ones but take more time to read. In the measurement of steady temperatures, however, this is no drawback. As we shall see shortly *the heat capacity of the apparatus is not required*, which is a great advantage of the method.

Figure 9.3 shows Callendar and Barnes' apparatus used to measure the specific heat capacity of water. Water from the constant-head tank K flows through the glass tube U, and can be collected as it flows out. It is heated by the spiral resistance wire R, which carries a steady electric current I. Its temperature, as it enters and leaves, is measured by the Thermometers T_1 and T_2. (In a simplified laboratory experiment, these may be mercury thermometers.) Surrounding the apparatus is a glass jacket G, which is evacuated, so that heat cannot escape from the water by conduction or convection.

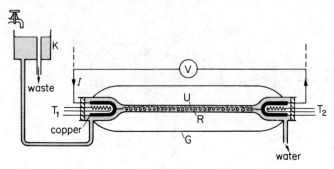

Fig. 9.3 Callendar and Barnes' apparatus (contracted several times in length relative to diameters)

When the apparatus is running, it settles down eventually to a steady state, in which the heat supplied by the current is all carried away by the water. *None is then taken in warming the apparatus, because every part of it is at a constant temperature.* The mass of water m, which flows out of the tube in t seconds, is then measured. If the water enters at a temperature θ_1 and leaves at θ_2, then if c_w is its mean specific heat capacity,

$$\text{heat gained by water} = Q = mc_w(\theta_2 - \theta_1) \text{ joules.}$$

The energy which liberates this heat is electrical. To find it, the current I, and the potential difference across the wire V, are measured with a potentiometer. If I and V are in amperes and volts respectively, then, in t seconds, energy supplied to the wire $= IVt$ joules.

If we ignore any heat losses and any external work, then, from the principle of conservation of energy,

$$mc_w(\theta_2 - \theta_1) = IVt$$

$$\therefore c_w = \frac{IVt}{m(\theta_2 - \theta_1)}.$$

Elimination of Heat Losses

To get the highest accuracy from this experiment, the small heat losses due to radiation, and conduction along the glass, must be allowed for. These are determined by the temperatures θ_1 and θ_2. For a given pair of values of θ_1 and θ_2, and constant-temperature surroundings (not shown), let the heat lost per second be h. Then, if m now represents the mass of liquid flowing *per second*,

$$\text{power supplied by heating coil} = mc_w(\theta_2 - \theta_1) + h, \quad . \qquad .$$

$$\therefore IV = mc_w(\theta_2 - \theta_1) + h \quad . \qquad . \qquad (1)$$

To allow for the loss h, the rate of flow of water is changed, to about half or twice its previous value. The current and voltage are then adjusted to bring θ_2 back to its original value. The inflow temperature θ_1, is fixed by the temperature of the water in the tank. If I', V', are the new values of I, V, and m' is the new mass of water flowing per second, then:

$$I'V' = m'c_w(\theta_2 - \theta_1) + h \qquad . \qquad . \qquad . \qquad . \qquad (2)$$

Note that h in equation (2) is the same as in equation (1) because the *mean* temperature of the liquid is the same for each experiment.

On subtracting equation (2) from equation (1), we find

$$(IV - I'V') = (m - m')c_w(\theta_2 - \theta_1),$$

$$\therefore c_w = \frac{(IV - I'V')}{(m - m')(\theta_2 - \theta_1)} \qquad . \qquad . \qquad . \qquad (3)$$

When the temperature rise, $\theta_2 - \theta_1$, is made small, for example, $\theta_1 = 20.0°C$, $\theta_2 = 22.0°C$, then c_w may be considered as the specific heat capacity at $21.0°C$, the mean temperature. If the inlet water temperature is now raised to say $\theta_1 = 40.0°C$ and θ_2 is then $42.0°C$, c_w is now the specific heat capacity at $41.0°C$. In this way it was found that *the specific heat capacity of water varies with temperature*. The continuous flow method can be used to find the variation in specific heat capacity of any liquid in the same way.

The table below shows the relative variation of the specific heat capacity of water, taking the value at 15°C as 1·0000 in magnitude.

SPECIFIC HEAT CAPACITY OF WATER

Temperature (°C)	5	15	25	40	70	100
c_w	1·0047	1·0000	0·9980	0·9973	1·0000	1·0057

Example

1. To measure the specific heat capacity of a liquid, a p.d. of 6·0 V was applied to the heating coil in a constant flow calorimeter. When the rate of flow of liquid was halved, a new p.d. V was required to produce the same inlet and outlet temperatures. Calculate V, assuming heat losses are negligible.

$$\text{Heat supplied per second} = IV = \frac{V^2}{R},$$

where R is the resistance of the heating coil. Since R is constant when the temperature is constant,

$$\text{heat supplied per second} \propto V^2$$

When the rate of flow is halved, *half* the heat per second is required to produce the same outlet temperature, assuming negligible heat losses. So, by proportion,

$$\frac{V^2}{6^2} = \frac{1}{2}$$

So

$$V = \sqrt{\frac{6^2}{2}} = \sqrt{18} = 4.2 \ V.$$

2. Water flows at the rate of 0·1500 kg min^{-1} through a tube and is heated by a heater dissipating 25·2 W. The inflow and outflow water temperatures are 15·2°C and 17·4°C respectively. When the rate of flow is increased to 0·2318 kg min^{-1} and the rate of heating to 37·8 W, the inflow and outflow temperatures are unaltered. Find (i) the specific heat capacity of water, (ii) the rate of loss of heat from the tube.

Suppose c_w is the specific heat of water in J kg^{-1} K^{-1} and h is the heat lost in J s^{-1}. Then, since 1 W = 1 J per second,

$$25·2 = \frac{0·1500}{60}c_w(17·4 - 15·2) + h \qquad . \qquad . \qquad . \qquad . \quad (1)$$

$$\text{and} \quad 37·8 = \frac{0·2318}{60}c_w(17·4 - 15·2) + h \qquad . \qquad . \qquad . \quad (2)$$

Subtracting (1) from (2),

$$\therefore 37·8 - 25·2 = \frac{0·2318 - 0·1500}{60}c_w(17·4 - 15·2)$$

$$\therefore c_w = \frac{12·6 \times 60}{0·0818 \times 2·2} = 4200 \text{ J kg}^{-1} \text{ K}^{-1}.$$

Substituting for c_w in (1),

$$\therefore h = 25·2 - \frac{0·15}{60} \times 4200 \times 2·2 = 2·1 \text{ J s}^{-1} = 2·1\text{ W}.$$

Method of Mixtures

A common way of measuring specific heat capacities of solids (or liquids) is the method of mixtures.

As an illustration of a specific heat capacity determination, suppose a metal of mass 0·2 kg at 100°C is dropped into 0·08 kg of water at 15°C contained in a calorimeter of mass 0·12 kg and specific heat capacity 400 J kg^{-1} K^{-1}. The final temperature reached is 35°C. Then, assuming negligible heat losses,

heat capacity of calorimeter = 0·12 × 400 = 48 J K^{-1}

heat capacity of water = 0·08 × 4200 = 336 J K^{-1}

∴ heat gained by water + cal. = (336 + 48) × (35 − 15) J

and heat lost by hot metal = 0·2 × c × (100 − 35) J

$$\therefore 0·2 \times c \times 65 = 384 \times 20$$

$$\therefore c = \frac{384 \times 20}{0·2 \times 65} = 590 \text{ J kg}^{-1} \text{ K}^{-1} \text{ (approx.)}.$$

Heat Losses

In a calorimetric experiment, some heat is always lost by leakage. Leakage of heat cannot be prevented, as leakage of electricity can, by insulation, because even the best insulator of heat still has appreciable conductivity (p. 262).

When convection is prevented, gases are the best thermal insulators. Hence calorimeters are often surrounded with a shield S and the heat loss due to conduction is made small by packing S with insulating material or by supporting the calorimeter on an insulating ring, or on threads. The loss by radiation is small at small excess temperatures over the surroundings. In some simple calorimetric experiments the final temperature of the mixture is reached quickly, so that the time for leakage is small. The total loss of heat is therefore negligible in laboratory experiments on the specific heats of metals, but not on the specific heat capacities of bad conductors, such as rubber, which give up their heat slowly. Here a 'cooling correction' must be added to the observed final temperature. When great accuracy is required, the loss of heat by leakage is always taken into account.

Newton's Law of Cooling

Newton was the first person to investigate the heat lost by a body in air. He found that *the rate of loss of heat is proportional to the excess temperature over the surroundings*. This result, called *Newton's law of cooling*, is approximately true in still air only for a temperature excess of about 20 K or 30 K; but it is true for all excess temperatures in conditions of forced convection of the air, i.e. in a draught. With natural convection Dulong and Petit found that the rate of loss of heat was proportional to $\theta^{5/4}$, where θ is the excess temperature, and this appears to be true for higher excess temperatures, such as from 50 K to 300 K.

To demonstrate Newton's law of cooling, we plot a temperature (θ)-time (t) cooling curve for hot water in a calorimeter placed in a draught (Fig. 9.4 (i)). If θ_R is the room temperature, then the excess temperature of the water is $(\theta - \theta_R)$. At various temperatures, such as θ in Fig. 9.4 (ii), we drew tangents such as APC to the curve. The slope of the tangent, in degrees per second, gives us the rate of fall of temperature, when the water is at the temperature θ:

$$\text{rate of fall} = \frac{\text{AB}}{\text{BC}} = \frac{\theta_1 - \theta_2}{t_2 - t_1}.$$

We then plot these rates against the excess temperature, $\theta - \theta_R$, as in Fig. 9.4 (iii), and find a straight line passing through the origin. Since the heat lost per second by the water and calorimeter is proportional to the rate of fall of the temperature, Newton's law is thus verified.

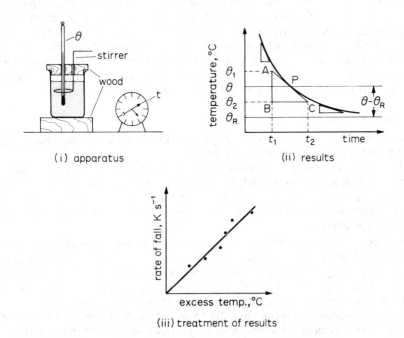

(i) apparatus

(ii) results

(iii) treatment of results

Fig. 9.4 Newton's law of cooling

Heat Loss and Temperature Fall

Besides the excess temperature, the rate of heat loss depends on the exposed area of the calorimeter, and on the nature of its surface: a dull surface loses heat a little faster than a shiny one, because it is a better radiator (p. 273). This can be shown by doing a cooling experiment twice, with equal masses of water, but once with the calorimeter polished, and once after it has been blackened in a candle-flame. In general, for any body with a uniform surface at a uniform temperature θ, we may write, if Newton's law is true,

$$\text{heat lost/second} = \frac{dQ}{dt} = kS(\theta - \theta_R) \qquad . \qquad . \qquad . \qquad (2)$$

where S is the area of the body's surface, θ_R is the temperature of its surroundings, k is a constant depending on the nature of the surface, and Q denotes the heat lost from the body.

When a body loses heat Q, its temperature θ falls; if m is its mass, and c its specific heat capacity, then its rate of fall of temperature, $d\theta/dt$, is given by

$$\frac{dQ}{dt} = -mc\frac{d\theta}{dt}.$$

Now the mass of a body is proportional to its volume. The rate of heat loss, however, is proportional to the surface area of the body. The rate of fall of temperature is therefore proportional to the ratio of surface to volume of the body. For bodies of similar shape, the ratio of surface to volume is inversely proportional to any linear dimension. If the bodies have surfaces of similar nature, therefore, the rate of fall of temperature is inversely proportional to the linear dimension; a small body cools faster than a large one. This is a fact of daily experience: a small coal which falls out of the fire can be picked up sooner than a large one; a tiny baby should be more thoroughly wrapped up than a grown man. In calorimetry by the method of mixtures, the fact that a small body cools faster than a large one means that the larger the specimen, the less serious is the heat loss in transferring it from its heating place to the calorimeter. It also means that the larger the scale of the whole apparatus, the less serious are the errors due to loss of heat from the calorimeter.

Cooling Correction

As we mentioned previously, a 'cooling correction' is needed in electrical heating experiments and the method of mixtures.

Newton's law can be used to make an approximation for the cooling correction where the heat gained by a solid, for example, is fairly rapid. Figure 9.5 shows the temperature rise from the initial temperature A to the observed maximum M when the solid gained heat, and the temperature fall from M when the solid cooled a few degrees.

We can see from the diagram that the average temperature excess θ_2 above the surroundings during cooling is about *twice* the average temperature excess θ_1 during heating. Hence, from Newton's law, the rate of cooling during the time t_1 of heating is about *half* that during cooling. So the cooling correction θ_c is *the temperature drop from the maximum M in half the time* t_1.

As an illustration, suppose that the temperature of the metal block in the electrical measurement of its specific heat capacity rose to an observed maximum of 21·5°C in 4·0 min. The cooling correction is then roughly the temperature

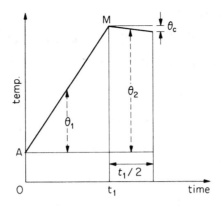

Fig. 9.5 Cooling correction (good conductor)

drop in 2·0 min. If this drop is 0·2 K, the final temperature corrected for cooling is 21·7°C.

The approximate cooling correction described is suitable only for good conductors. It can thus be used in determining the specific heat capacity of a metal by the electrical method. In similar experiments with poor conductors such as rubber or glass, however, it may take a long time for the solid to reach its final temperature. In such cases a more accurate way of making the cooling correction is needed.

Specific Latent Heat

Evaporation. Electrical Method for Specific Latent Heat

The *specific latent heat of evaporation* of a liquid is the heat required to convert unit mass of it, at its boiling point, into vapour at the same temperature. It is expressed in joule per kilogramme ($J\ kg^{-1}$), or, with high values, in $kJ\ kg^{-1}$.

A modern electrical method for the specific latent heat of evaporation of water is illustrated in Fig. 9.6. The liquid is heated in a vessel U by the heating coil R. As shown, the vapour escapes through H at the top of U into a surrounding jacket J and then passes down the tube T. Here the vapour is condensed by cold water flowing through the jacket K. When the apparatus has reached its steady state, the liquid is at its boiling-point, and the heat supplied by the coil is used in evaporating the liquid, and in offsetting the losses. The heat losses are considerably reduced by the use of the vapour jacket. The liquid emerging from the condenser is then collected for a measured time, and weighed.

If I and V are the current through the coil, and the potential difference across it, the electrical power supplied is IV. And if h is the heat lost from the vessel per second, and m the mass of liquid collected per second, then

$$IV = ml + h \qquad . \qquad . \qquad . \qquad . \qquad (1)$$

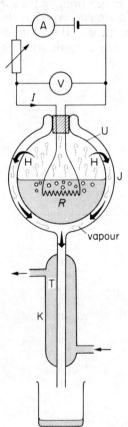

Fig. 9.6 Electrical method for specific latent heat of evaporation

The heat losses h are determined by the temperature of the vessel, which is fixed at the boiling-point of the liquid. Therefore they may be eliminated by a second experiment with a different rate of evaporation (cf. Callendar and Barnes, p. 182). If I', V' are the new current and potential difference, and if m' is the mass per second evaporated, then

$$I'V' = m'l + h$$

Hence by subtraction from equation (1)

$$l = \frac{(IV - I'V')}{(m - m')}.$$

It may be noted that a much higher power supply is needed for determining the specific latent heat of vaporisation of water than for alcohol, as water has a much higher value of l. The specific latent heat of evaporation of water is about $l = 2260$ kJ kg^{-1} or $2 \cdot 26 \times 10^6$ J kg^{-1}.

Method of Mixtures

To find the specific latent heat of evaporation of water by mixtures, we pass steam into a calorimeter with water (Fig. 9.7). On its way the steam passes

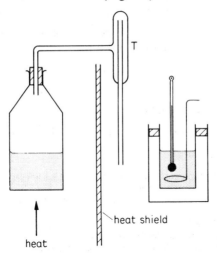

heat shield

heat

Fig. 9.7 Specific latent heat of evaporation of water by mixtures

through a vessel, T in the figure, which traps any water carried over by the steam and is called a steam-trap. The mass m of condensed steam is found by weighing. If θ_1 and θ_2 are the initial and final temperatures of the water, the specific latent heat l is given by:

$$\left. \begin{array}{l} \text{heat given by steam} \\ \text{condensing} \end{array} \right\} + \left\{ \begin{array}{l} \text{heat given by condensed} \\ \text{water cooling from} \\ 100°\text{C to } \theta_2 \end{array} \right\} = \left\{ \begin{array}{l} \text{heat taken by calori-} \\ \text{meter and water} \end{array} \right.$$

$$ml \qquad + \qquad mc_w(100 - \theta_2) \qquad = (m_1 c_w + C)(\theta_2 - \theta_1)$$

where m_1 is the mass of water in the calorimeter, c_w is the specific heat capacity of water and C is the heat capacity (mass × specific heat capacity) of the metal calorimeter.

Hence $$l = \frac{(m_1 c_w + C)(\theta_2 - \theta_1)}{m} - c_w(100 - \theta_2).$$

The result for l is only approximate as the steam-trap is not efficient.

Fusion

The specific latent heat of fusion of a solid is the heat required to convert unit mass of it, at its melting-point, into liquid at the same temperature. It is expressed in joules per kilogramme $(J\ kg^{-1})$. High values can be more conveniently expressed in $kJ\ kg^{-1}$.

Ice is one of the substances whose specific latent heat of fusion we are likely to have to measure. To do so, place warm water, at a temperature of θ_1 a few degrees above room temperature, inside a calorimeter. Then add small lumps of ice, dried by blotting paper, until the temperature reaches a value θ_2 as much below room temperature as θ_1 was above. In this case a 'cooling correction' is not necessary. Weigh the mixture, to find the mass m of ice which has been added. Then the specific latent heat l is given by:

$$\left\{\begin{array}{l}\text{heat given by calorimeter}\\ \text{and water in cooling}\end{array}\right\} = \left\{\begin{array}{l}\text{heat used in}\\ \text{melting ice}\end{array}\right\} + \left\{\begin{array}{l}\text{heat used in warming melted}\\ \text{ice from } 0°C \text{ to } \theta_2\end{array}\right\}$$

$$\therefore (m_1 c_w + C)(\theta_1 - \theta_2) = ml + mc_w(\theta_2 - 0),$$

where m_1 = mass of water and c_w = specific heat capacity, C = heat capacity of calorimeter, and θ_1 = initial temperature.

Hence $$l = \frac{(m_1 c_w + C)(\theta_1 - \theta_2)}{m} - c_w \theta_2.$$

A modern electrical method gives

$$l = 334\ kJ\ kg^{-1} \quad \text{or} \quad 3.34 \times 10^5\ J\ kg^{-1}.$$

Example

(a) Describe and give the theory of a mechanical method for finding the specific heat capacity of *either* water or copper. Explain how corrections for loss of heat might be made. Explain briefly how one could make use of the value so obtained to find the specific heat capacity of the other material.

(b) An electric kettle with a 2·0 kW heating element has a heat capacity of 400 J K^{-1}. 1·0 kg of water at 20°C is placed in the kettle. The kettle is switched on and it is found that 13 minutes later the mass of water in it is 0·5 kg. Ignoring heat losses calculate a value for the specific latent heat of water. (Specific heat capacity of water = 4.2×10^3 J kg^{-1}.) (L.)

Total heat supplied = $2000 \times 13 \times 60 = 1.56 \times 10^6$ J.

Heat used for kettle = $C\theta = 400 \times (100 - 20) = 32\,000 = 0.032 \times 10^6$ J.

Heat used to raise temperature of 1 kg of water from 20°C to 100°C

$$= mc\theta = 1 \times 4200 \times (100 - 20) = 0.336 \times 10^6\ J.$$

So total heat to change water at 100°C to steam at 100°C

$$= 1.56 \times 10^6 - (0.032 \times 10^6 + 0.336 \times 10^6)\ J$$
$$= 1.192 \times 10^6\ J.$$

Since mass of water changed to steam = $1.0 - 0.5 = 0.5$ kg, then

$$l = \frac{1.192 \times 10^6}{0.5} = 2.38 \times 10^6\ J\ kg^{-1}.$$

Exercises 9

1. A metal cylinder of mass 0·5 kg is heated electrically by a 12 W heater in a room at 15°C. The cylinder temperature rises uniformly to 25°C in 5 min and later becomes constant at 45°C.

(i) What is the rate of loss of heat of the cylinder to the surroundings at 45°C? Explain your answer. (ii) Assuming the rate of heat loss is proportional to the excess temperature over the surroundings calculate the rate of loss of heat of the cylinder at 20°C. (iii) Calculate the specific heat capacity of the metal, taking into account the loss of heat to the surroundings.

2. An electrical heater of 2 kW is used to heat 0·5 kg of water in a kettle of heat capacity 400 J K^{-1}. The initial water temperature is 20°C.

Neglecting heat losses, (i) how long will it take to heat the water to its boiling point, 100°C? (ii) starting from 20°C, what mass of water is boiled away in 5 min? (Assume for water, specific heat capacity = 4200 J kg^{-1} K^{-1} and specific latent heat of vaporisation = 2 × 10^6 J kg^{-1}.)

3. In an electrical constant flow experiment to determine the specific heat capacity of a liquid, heat is supplied to the liquid at a rate of 12 W. When the rate of flow is 0·060 kg min^{-1} the temperature rise along the flow is 2·0 K. Use these figures to calculate a value for the specific heat capacity of the liquid.

If the true value of the specific heat capacity is 5400 J kg^{-1} K^{-1}, estimate the percentage of heat lost in the apparatus. Explain briefly how, in practice, you would reduce or make allowance for this heat loss. (*L.*)

4. What is meant by the specific latent heat of vaporisation of a liquid? Explain how latent heat of vaporisation can be regarded as molecular potential energy.

Calculate the potential energy per molecule released when 18 g of steam condenses to water at 100°C. (Specific latent heat of vaporisation of water = 2·26 × 10^6 J kg^{-1}. Mass of 1 mole of water = 18 g. Number of molecules in a mole of molecules = 6·02 × 10^{23}.) (*L.*)

5. In a constant flow calorimeter, being used for measuring the specific heat capacity of a liquid, a p.d. of 4·0 V was applied to the heating coil. The rate of flow was now doubled and, by adjusting the applied p.d., the same inlet and outlet temperatures were obtained. Assuming heat losses to be negligible calculate the new value of the applied p.d. (*L.*)

6. Describe how you might measure, by an electrical method, the specific heat capacity of copper provided in the form of a cylinder 4 cm long and 1 cm in diameter.

Describe the procedure you would use to make an allowance for heat loss and explain how you would derive the specific heat capacity from your measurements.

When a metal cylinder of mass 2·0 × 10^{-2} kg and specific heat capacity 500 J kg^{-1} K^{-1} is heated by an electrical heater working at constant power, the initial rate of rise of temperature is 3·0 K min^{-1}. After a time the heater is switched off and the initial rate of fall of temperature is 0·3 K min^{-1}. What is the rate at which the cylinder gains heat energy immediately before the heater is switched off? (*N.*)

7. In an experiment to measure the specific heat capacity of water, a stream of water flows at a steady rate of 5·0 g s^{-1} over an electrical heater dissipating 135 W, and a temperature rise of 5·0 K is observed. On increasing the rate of flow to 10·0 g s^{-1}, the same temperature rise is produced with a dissipation of 240 W. Explain why the power in the second case is not twice that needed in the first case, and deduce a value for the specific heat capacity of water.

Discuss the advantages of using such a continuous flow method for measuring the specific heat capacity of a liquid. Describe how a temperature rise of the order of 5 K in the region between 0°C and 100°C may be measured to an accuracy of better than $\frac{1}{2}$%. (*O. & C.*)

8. (*a*) The conservation of energy principle may be written in equation form as $\Delta Q = \Delta U + \Delta W$. Express this relationship in words, making clear the meaning of each term.

(*b*) Outline a continuous flow method for measuring the specific heat capacity of water. Show how the conservation of energy principle can be used to calculate the specific heat capacity. Explain one important advantage of your method.

(c) In an experiment to determine the specific heat capacity of aluminium, a cylindrical 1 kg block of aluminium was heated electrically by a 17·3 W immersion heater inserted into a hole in the centre of the block. The block was suspended in a draught-free room at 20°C. The temperature of the block at first rose steadily (10 K in 10 min), then, more slowly, finally stabilising at 85°C.

(i) Explain, using the conservation of energy principle, why the temperature of the block stabilised, although the heater was still switched on.

(ii) Assuming the rate of heat loss from the block was proportional to the excess temperature of the block above that of the room, calculate (1) the rate of heat loss from the block at 25°C, and (2) the specific heat capacity of aluminium (corrected for heat loss).

(iii) When a similar experiment was performed under the same conditions with a cylindrical 1 kg block of iron, the final steady temperature was considerably higher than 85°C. Suggest the most likely reason for this. (L.)

9. Define *specific heat capacity*.

In an experiment to determine the specific heat capacity of a liquid, the liquid flows past an electric heating coil and in the steady state the inlet and outlet temperatures are 10·4°C and 13·5°C respectively. When the mass rate of flow of the liquid is $3·2 \times 10^{-3}$ kg s^{-1} the power supplied to the coils is 27·4 W. The flow rate is then changed to $2·2 \times 10^{-3}$ kg s^{-1} and, in order to maintain the same inlet and outlet temperatures the power supplied is adjusted to 19·3 W. Explain why two sets of data are obtained, and calculate the specific heat of the liquid.

Why are the temperatures made the same in each part of the experiment? What are the advantages of this method over the method of mixtures?

What is the rate of loss of heat in the above experiment? If in a further experiment the surrounding temperature is 11·95°C, the rate of loss of heat will be zero. Why is this so? (L.)

10. Give an account of an electrical method of finding the specific latent heat of vaporisation of a liquid boiling at about 60°C. Point out any causes of inaccuracy and explain how to reduce their effect.

Ice at 0°C is added to 200 g of water initally at 70°C in a vacuum flask. When 50 g of ice has been added and has all melted the temperature of the flask and contents is 40°C. When a further 80 g of ice has been added and has all melted the temperature of the whole becomes 10°C. Calculate the specific latent heat of fusion of ice, neglecting any heat lost to the surroundings.

In the above experiment the flask is well shaken before taking each temperature reading. Why is this necessary? (C.)

11. In the absence of bearing friction a winding engine would raise a cage weighing 1000 kg at 10 m s^{-1}, but this is reduced by friction to 9 m s^{-1}. How much oil, initially at 20°C, is required per second to keep the temperature of the bearings down to 70°C? (Specific heat capacity of oil = 2100 J kg^{-1} K^{-1}; $g = 9·81$ m s^{-1}.) (O. & C.)

12. Figure 9A represents a laboratory apparatus for determining the specific latent heat of vaporisation of water. Draw a diagram of a suitable electric circuit in which the heater

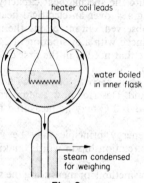

heater coil leads

water boiled in inner flask

steam condensed for weighing

Fig. 9A

coil could be incorporated, and describe briefly how the experiment is carried out and how the result is calculated from the observations made.

A pupil performing the experiment finds that, when the heat supply is 16 W, it takes 30 minutes for the temperature of the water to rise from 20°C to 100°C, and that the rate of vaporisation is very slow even at the latter temperature. Estimate an upper limit to the value of the heat capacity of the inner flask and its contents. Calculate the mass of water collected after 30 minutes of steady boiling when the power supply is 60 W. (Take specific latent heat of vaporisation of water to be $2 \cdot 26 \times 10^6$ J kg^{-1}.) (O.)

13. Give an account of a method of determining the specific latent heat of evaporation of water, pointing out the ways in which the method you describe achieves, or fails to achieve, high accuracy.

A 600 watt electric heater is used to raise the temperature of a certain mass of water from room temperature to 80°C. Alternatively, by passing steam from a boiler into the same initial mass of water at the same initial temperature the same temperature rise is obtained in the same time. If 16 g of water were being evaporated every minute in the boiler, find the specific latent heat of steam, assuming that there were no heat losses. (O. & C.)

14. Describe, giving the necessary theory, how the specific latent heat of a liquid may be determined by a method which involves a constant rate of evaporation. What are the advantages of this method over a method of mixtures?

In calorimetry experiments a cooling correction has to be applied for heat losses. Explain how this is done in the experiment you describe.

Figure 9B shows a cooling curve for a substance which, starting as a liquid, eventually solidifies. Explain the shape of the curve and use the following data to obtain a value for the specific latent heat of fusion of the substance: Room temperature = 20°C,

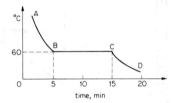

Fig. 9B

slope of tangent to curve when temperature is 70°C = 10 K min^{-1}, specific heat capacity of liquid = $2 \cdot 0 \times 10^3$ J kg^{-1} K^{-1}, mass of liquid = $1 \cdot 50 \times 10^{-2}$ kg. (You may assume that Newton's law of cooling holds.) (L.)

10 Gases: Gas Laws and Heat Capacities

In this chapter we shall be concerned with the relationship between the temperature, pressure and volume of a gas. Unlike the case of a solid or liquid this can be expressed in very simple laws, called the Gas Laws, and reduced to a simple equation, called the Equation of State. We shall also deal in this chapter with the heat capacities of gases.

The Gas Laws and the Equation of State

Pressure and Volume: Boyle's Law

In 1660 Robert Boyle—whose epitaph reads 'Father of Chemistry, and Nephew of the Earl of Cork'—published the results of his experiments on the 'natural spring of air'. He meant what we now call the relationship between the pressure of air and its volume.

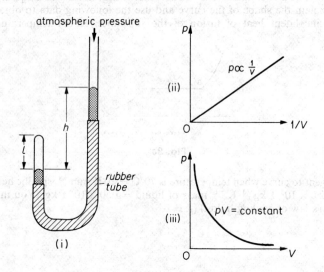

Fig. 10.1 Boyle's law apparatus

We can repeat Boyle's experiment with the apparatus shown in Fig. 10.1 (i), which contains dry air above mercury in a closed uniform tube. We set the open limb of the tube at various heights above and below the closed limb and measure the difference in level, h, of the mercury. When the mercury in the open limb is below that in the closed, we reckon h as negative. At each value of h we measure the corresponding length l of the air column in the closed limb. To find the pressure of the air we add the difference in level h to the height of the barometer, H; their sum gives the pressure p of the air in the closed limb:

$$p = g\rho(H + h)$$

where g is the acceleration of gravity and ρ is the density of mercury.

If S is the area of cross-section of the closed limb, the volume of the trapped air is

$$V = lS.$$

To interpret our measurements we may either plot $H + h$, which is a measure of p, against $1/l$, or make a table of the product $(H + h)l$. We find that the plot is a straight line, and therefore

$$(H + h) \propto \frac{1}{l} \qquad . \qquad . \qquad . \qquad . \qquad . \qquad (1)$$

Alternatively, we find

$$(H + h)l = \text{constant}, \qquad . \qquad . \qquad . \qquad . \qquad (2)$$

which means the same as (1).

Since g, ρ, and S are constants, the relationships (1) and (2) give

$$p \propto \frac{1}{V}$$

or
$$pV = \text{constant}.$$

We shall see shortly that the pressure of a gas depends on its temperature as well as its volume. To express the results of the above experiments, therefore, we say that *the pressure of a given mass of gas, at constant temperature, is inversely proportional to its volume*, or $pV = \text{constant}$. This is *Boyle's law*. See Fig. 10.1 (ii), (iii).

Example

A faulty barometer tube has some air at the top above the mercury. When the length of the air column is 250 mm, the reading of the mercury above the outside level is 750 mm. When the length of the air column is decreased to 200 mm, the reading of the mercury above the outside level becomes 746 mm. Calculate the atmospheric pressure.

Initial volume V_1 of air is proportional to the length 250 mm. Initial air pressure $p_1 = (A - 750)$ mmHg, where A is the atmospheric pressure. Also, new volume V_2 of air is proportional to length 200 mm, and new pressure of air, $p_2 = (A - 746)$ mmHg.

From Boyle's law, $\qquad p_1 V_1 = p_2 V_2$

So $\qquad\qquad 250 \times (A - 750) = 200 \times (A - 746)$

Thus $\qquad\qquad 5(A - 750) = 4(A - 746)$

So $\qquad\qquad 5A - 4A = A = 3750 - 2984 = 766$ mmHg

Mixture of Gases: Dalton's Law

Figure 10.2 shows an apparatus with which we can study the pressure of a mixture of gases. A is a bulb, of volume V_1, containing air at atmospheric pressure, p_1. C is another bulb, of volume V_2, containing carbon dioxide at a pressure p_2. The pressure p_2 is measured on the manometer M; in millimetres of mercury it is

$$p_2 = h + H,$$

where H is the height of the barometer. (In the same units, the air pressure, $p_1 = H$.)

When the bulbs are connected by opening the tap T, the gases mix, and reach

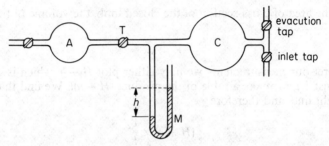

Fig. 10.2 Apparatus for demonstrating Dalton's law of partial pressure

the same pressure, p; this pressure is given by the new height of the manometer. Its value is found to be given by

$$p = p_1 \frac{V_1}{V_1 + V_2} + p_2 \frac{V_2}{V_1 + V_2}.$$

Now the quantity $p_1 V_1/(V_1 + V_2)$ is the pressure which the air originally in A would have, if it expanded to occupy A and C; for, if we denote this pressure by p', then $p'(V_1 + V_2) = p_1 V_1$. Similarly $p_2 V_2/(V_1 + V_2)$ is the pressure which the carbon dioxide originally in C would have, if it expanded to occupy A and C. Thus the total pressure of the mixture is the sum of the pressures which the individual gases exert, when they have expanded to fill the vessel containing the mixture.

The pressure of an individual gas in a mixture is called is *partial pressure*: it is the pressure which would be observed if that gas alone occupied the volume of the mixture, and had the same temperature as the mixture. The experiment described shows that *the pressure of a mixture of gases is the sum of the partial pressures of its constituents*. This statement was first made by Dalton, in 1801, and is called *Dalton's Law of Partial Pressures*.

Volume and Temperature: Charles's Law

Measurements of the change in volume of a gas with temperature, at constant pressure, were published by Charles in 1787 and by Gay-Lussac in 1802.

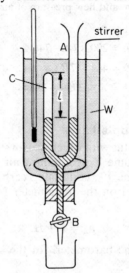

Fig. 10.3 Charles' law experiment

Figure 10.3 shows an apparatus which we may use to repeat their experiments. Dry air is trapped by mercury in the closed limb C of the tube AC; a scale engraved upon C enables us to measure the length of the air column, l. The tube is surrounded by a water-bath W, which we can heat by passing in steam. After making the temperature uniform by stirring, we level the mercury in the limbs A and C, by pouring mercury in at A, or running it off at B. The air in C is then always at atmospheric (constant) pressure. We measure the length l and plot it against the temperature, θ (Fig. 10.4).

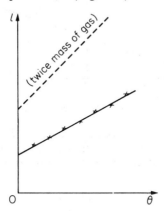

Fig. 10.4 Results of experiment

If S is the constant cross-section of the tube, the volume of the trapped air is

$$V = lS.$$

The cross-section S, and the distance between the divisions on which we read l, both increase with the temperature θ. But their increases are very small compared with the expansion of the gas. So we may say that the volume of the gas is proportional to the scale-reading of l. The graph then shows that the volume of the gas, at constant pressure, increases uniformly with its temperature as measured on the mercury-in-glass scale. A similar result is obtained with twice the mass of gas, as indicated in Fig. 10.4.

Expansivity of Gas (Volume Coefficient)

The rate at which the volume of a gas increases with temperature can be defined by a quantity called its *expansivity at constant pressure*, α_p, or *volume coefficient*:

$$\alpha_p = \frac{\text{volume at } \theta°C - \text{volume at } 0°C}{\text{volume at } 0°C} \times \frac{1}{\theta}.$$

Thus, if V is the volume at $\theta°C$, and V_0 the volume at $0°C$, then

$$\alpha_p = \frac{V - V_0}{V_0 \theta},$$

or
$$V = V_0(1 + \alpha_p \theta).$$

Charles, and Gay-Lussac, found that α_p had the same numerical value, $\frac{1}{273}$, for all gases. This observation is now called Charles's or Gay-Lussac's Law: *The volume of a given mass of any gas, at constant pressure, increases by $\frac{1}{273}$ of its value at $0°C$, for every degree Celsius rise in temperature.*

Absolute Temperature

Charles's Law shows that, if we plot the volume V of a given mass of any gas at constant pressure against its temperature θ, we shall get a straight line graph A as shown in Fig. 10.5. If we produce this line backwards, it will meet the temperature axis at $-273°C$. This temperature is called the *absolute zero*.

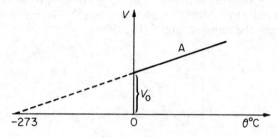

Fig. 10.5 Absolute zero

If a gas is cooled, it liquefies before it reaches $-273°C$, and Charles's Law no longer holds; but that fact does not affect the form of the relationship between the volume and temperature at higher temperatures. We may express this relationship by writing

$$V \propto (273 + \theta).$$

The quantity $(273 + \theta)$ is called the *absolute temperature* of the gas, and is denoted by T. The idea of absolute temperature was developed by Lord Kelvin, and absolute temperatures, T, are hence expressed in kelvin:

$$T = (273 + \theta)$$

From Charles's Law, we see that the volume of a given mass of gas at constant pressure is proportional to its absolute or kelvin temperature, since

$$V \propto (273 + \theta) \propto T.$$

Thus, if a given mass of gas has a volume V_1 at $\theta_1°C$, and is heated at constant pressure to $\theta_2°C$, its new volume is given by

$$\frac{V_1}{V_2} = \frac{273 + \theta_1}{273 + \theta_2} = \frac{T_1}{T_2}.$$

Pressure and Temperature

The effect of temperature on the pressure of a gas, at constant volume, can be investigated with the apparatus shown in Fig. 10.6 (i). The bulb B contains air, which can be brought to any temperature θ by heating the water in the surrounding bath W. When the temperature is steady, the mercury in the closed limb of the tube is brought to a fixed level D, so that the volume of the air is fixed. The difference in level, h, of the mercury in the open and closed limbs is then added to the height of the barometer, H, to give the pressure p of the gas in cm of mercury. If p, $(h + H)$, is plotted against the temperature, the plot is a straight line, Fig. 10.6 (ii).

The coefficient of pressure increase at constant volume, α_V, known as the *pressure coefficient*, is given by

$$\alpha_V = \frac{p - p_0}{p_0\theta}$$

where p_0 is the pressure at $0°C$. The coefficient α_V, which expresses the change

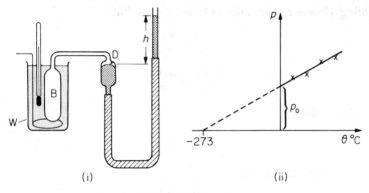

Fig. 10.6 Pressure and temperature

of pressure with temperature, at constant volume, has practically the same value for all gases: $\frac{1}{273}$ K^{-1}. It is thus numerically equal to the expansivity α_p. We may therefore say that, at constant volume, the pressure of a given mass of gas is proportional to its absolute or kelvin temperature T, since

$$p \propto (273 + \theta).$$

$$\therefore \frac{p_1}{p_2} = \frac{273 + \theta_1}{273 + \theta_2} = \frac{T_1}{T_2}.$$

The Equation of State

Figure 10.7 illustrates the argument by which we may find the general relationship between pressure, volume and temperature of a given mass of gas. This relationship is called the *equation of state*.

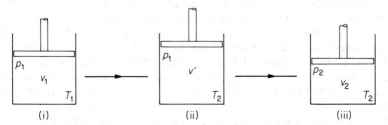

Fig. 10.7 Changing temperature and pressure of a gas

At (i) we have the gas occupying a volume V_1 at a pressure p_1, and an absolute temperature T_1. We wish to calculate its volume V_2 at an absolute temperature T_2 and pressure p_2, as at (iii). We proceed via (ii), raising the temperature of T_2 while keeping the pressure constant at p_1. If V is the volume of the gas at (ii) then, by Charles's law:

$$\frac{V'}{V_1} = \frac{T_2}{T_1} \qquad \cdot \quad \cdot \quad \cdot \quad \cdot \quad \cdot \quad (4)$$

We proceed now to (iii), by increasing the pressure to p_2, while keeping the temperature constant at T_2. By Boyle's law,

$$\frac{V_2}{V'} = \frac{p_1}{p_2} \qquad \cdot \quad \cdot \quad \cdot \quad \cdot \quad \cdot \quad (5)$$

Eliminating V' between equations (4) and (5), we find

$$\frac{V_2}{V_1} = \frac{T_2}{T_1} \cdot \frac{p_1}{p_2}$$

or

$$\frac{p_2 V_2}{T_2} = \frac{p_1 V_1}{T_1}.$$

In general, therefore,

$$\frac{pV}{T} = \mathbf{R} \qquad . \qquad . \qquad . \qquad . \qquad (6)$$

where $\mathbf{R}$ is a constant, so that

$$pV = \mathbf{R}T \qquad . \qquad . \qquad . \qquad . \qquad (7)$$

Ideal Gas Equation. The Gas Constant

Equation (7) is the *ideal gas equation* or *equation of state*. The magnitude of the constant in the equation depends on the nature of the gas—air, hydrogen, etc.—and on the amount—number of moles or mass—of the gas.

For *one mole* (p. 201) of a particular gas, the corresponding *molar gas constant* is given the symbol R. So if 1 mole occupies a volume V equal to V_m at a pressure p and absolute temperature T, we write

$$pV_m = RT \qquad . \qquad . \qquad . \qquad . \qquad (8)$$

n moles of the gas at the same pressure p and temperature T occupy a volume V where $V = nV_m$. So the gas constant here is nR. The equation of state for n moles is thus

$$pV = nRT \qquad . \qquad . \qquad . \qquad . \qquad (8A)$$

It is sometimes more useful for the engineer to consider a mass of gas, such as unit mass or 1 kg. Here it would be useful to recall that the 'relative molecular mass' of any substance X is the mass of a molecule of X compared with the mass of one-twelfth of the mass of the atom of the carbon-12 isotope (see p. 201). The 'molecular mass' M is the number of grams equal to this ratio, the relative molecular mass. For the gases hydrogen, helium, oxygen and carbon dioxide, the molecular mass M is respectively about 2, 4, 32 and 44 g, or, in SI units, 2×10^{-3}, 4×10^{-3}, 32×10^{-3} and 44×10^{-3} kg respectively.

We shall denote the gas constant per unit mass by the symbol r. Thus if M is the molecular mass in kg (SI unit), $r = R/M$. So the equation of state for unit mass of gas, 1 kg, may be written

$$pV = \frac{R}{M}T = rT \qquad . \qquad . \qquad . \qquad (9)$$

For a mass m kg of gas, the equation of state is

$$pV = m\frac{R}{M}T = mrT \qquad . \qquad . \qquad (9A)$$

Since $m/M = n$, the number of moles, equation (9A) is identical to (8A).

For a mole of gas, the *density* $\rho = M/V_m$, or $V_m = M/\rho$. From (8), it follows that the equation of state can be written

$$\frac{p}{\rho} = \frac{R}{M}T \qquad . \qquad . \qquad . \qquad . \qquad (10)$$

The same equation as (10) is obtained from (9), since with unit mass, $V = 1/\rho$.

Units of Gas Constant

The unit of

$$\frac{pV}{T} \text{ or } \frac{\text{pressure} \times \text{volume}}{\text{temperature}} = \frac{\text{N m}^{-2} \times \text{m}^3}{\text{K}} = \text{N m K}^{-1}$$

$$= \text{J K}^{-1},$$

since 1 newton $\times$ 1 metre $= 1$ joule. So

unit of molar gas constant $= \text{J mol}^{-1} \text{ K}^{-1}$

and   unit of gas constant per unit mass $= \text{J kg}^{-1} \text{ K}^{-1}$.

Avogadro's Hypothesis: Molar Gas Constant

Amedeo Avogadro, with one simple-looking idea, illuminated chemistry as Newton illuminated mechanics. In 1811 he suggested that chemically active gases, such as oxygen, existed not as single atoms, but as pairs: he proposed to distinguish between an atom, O, and a molecule, O_2. Ampere proposed the same distinction, independently, in 1814. Avogadro also put forward another idea, now called *Avogadro's hypothesis*: that equal volumes of all gases, at the same temperature and pressure, contained equal numbers of molecules. The number of molecules in 1 cm^3 of gas at s.t.p. is called Loschmidt's number; it is $2 \cdot 69 \times 10^{19}$.

The amount of a substance which contains as many elementary units as there are atoms in 0·012 kg (12 g) of carbon-12 is called a *mole*, symbol 'mol'. The number of molecules per mole is the same for all substances. It is called the *Avogadro* constant, symbol N_A, and is equal to $6 \cdot 02 \times 10^{23} \text{ mol}^{-1}$.

From Avogadro's hypothesis, it follows that the mole of all gases, at the same temperature and pressure, occupy equal volumes. Experiment confirms this; at s.t.p. 1 mole of any gas occupies 22·4 litres. Consequently, if V_m is the volume of 1 mole, then the ratio pV_m/T is the same for all gases. So the molar gas constant R, which is equal to this ratio, is the *same for all gases*.

At s.t.p. $V_m = 22 \cdot 4 \text{ litres} = 22 \cdot 4 \times 10^{-3}$

$$p = 760 \text{ mmHg} = 1 \cdot 013 \times 10^5 \text{ N m}^{-2} \text{ (p. 103)}$$

$$T = 273 \text{ K}.$$

$$\therefore R = \frac{pV_m}{T} = \frac{1 \cdot 013 \times 10^5 \times 22 \cdot 4 \times 10^{-3}}{273}$$

$$= 8 \cdot 31 \text{ J mol}^{-1} \text{ K}^{-1}.$$

Gas Constant per Unit Mass

To calculate the gas constant per unit mass, r, we can use the density of the gas at a given temperature and pressure. Very often, in dealing with gases, we specify the pressure not in newton per metre^2 (N m^{-2}) but simply in millimetres of mercury, mmHg. 1 mmHg pressure is called 1 *torr*. We do so because we are concerned only with relative values. A pressure of 760 mmHg, which is about the average pressure of the atmosphere, is sometimes called 'standard' pressure. A temperature of 0°C, or 273 K, is likewise called standard temperature. The conditions 273 K and 760 mmHg pressure are together called standard temperature and pressure (s.t.p.). A pressure of 760 mmHg is given, in newton per metre^2, by

$$p = g\rho H$$
$$= 9\cdot 8 \times 13\,600 \times 0\cdot 76 = 1\cdot 013 \times 10^5 \text{ N m}^{-2},$$

using ρ = mercury density = $13\,600$ kg m^{-3}; $g = 9\cdot 8$ m s^{-2}; $H = 0\cdot 76$ m.

At s.t.p. the density of hydrogen is about $0\cdot 09$ g/litre or $0\cdot 09$ kg m^{-3}. The gas constant r for *unit mass* of hydrogen, volume $(1/0\cdot 09)$ m^3, is then

$$r = \frac{pV}{T} = \frac{1\cdot 013 \times 10^5}{0\cdot 09 \times 273}$$

$$= 4\cdot 16 \times 10^3 \text{ J kg}^{-1} \text{ K}^{-1}.$$

The molar gas constant, which is the same for all gases, is $8\cdot 3$ J mol^{-1} K^{-1} approximately. Hence it follows that

$$\text{molecular mass of hydrogen} = \frac{R}{r} = \frac{8\cdot 3}{4\cdot 16 \times 10^3}$$

$$= 2 \times 10^{-3} \text{ kg (approx.).}$$

Applying the Equation of State

We can apply the equation $pV = nRT$, where n is the number of moles, to calculate the mass of a gas. Suppose oxygen gas, contained in a cylinder of volume V of 1×10^{-2} m^3 has a temperature T of 300 K and a pressure p_1 of $2\cdot 5 \times 10^5$ N m^{-2}. After some of the oxygen is used at constant temperature, the pressure falls to $1\cdot 3 \times 10^5$ N m^{-2}, p_2.

From $pV = nRT$, the number of moles in the cylinder initially is given by

$$n_1 = \frac{p_1 V}{RT},$$

and finally by

$$n_2 = \frac{p_2 V}{RT}.$$

Hence the number of moles used $= n_1 - n_2$

$$= \frac{(p_1 - p_2)V}{RT}$$

$$= \frac{(2\cdot 5 - 1\cdot 3) \times 10^5 \times 1 \times 10^{-2}}{8\cdot 3 \times 300} = 0\cdot 48,$$

using the value $R = 8\cdot 3$ J mol^{-1} K^{-1}.

But   molecular mass of oxygen = 32 g = 32×10^{-3} kg

$\therefore$ mass of oxygen used = $0\cdot 48 \times 32 \times 10^{-3}$ kg

$$= 0\cdot 015 \text{ kg (approx.).}$$

Example

A cylinder containing 19 kg of compressed air at a pressure $9\cdot 5$ times that of the atmosphere is kept in a store at 7°C. When it is moved to a workshop where the temperature is 27°C a safety valve on the cylinder operates, releasing some of the air. If the valve allows air to escape when its pressure exceeds 10 times that of the atmosphere, calculate the mass of air that escapes. (*L.*)

From $pV = nRT$, if A is the atmospheric pressure and V is the cylinder volume,

$$\text{initial number of moles of air, } n_1 = \frac{pV}{RT} = \frac{9\cdot 5A \times V}{R \times 280} \qquad . \qquad . \qquad . \qquad (1)$$

and   final number of moles, $n_2 = \dfrac{pV}{RT} = \dfrac{10A \times V}{R \times 300}$.   .   .   .   (2)

Suppose m is the mass of air left in the cylinder. Then, since the mass is proportional to the number of moles, it follows that $m/19$ kg $= n_2/n_1$. Dividing (2) by (1) and simplifying, then

$$\frac{m}{19 \text{ kg}} = \frac{10 \times 280}{9 \cdot 5 \times 300}.$$

So $$m = \frac{19 \times 10 \times 280}{9 \cdot 5 \times 300} = 18 \cdot 67 \text{ kg.}$$

So mass of air escaped $= 19 - 18 \cdot 67 = 0 \cdot 33$ kg.

Connected Gas Containers

The mass m of a gas can also be found from the relation

$$m = \frac{pV}{rT},$$

where $r = R/M =$ gas constant per unit mass (p. 200). In a closed system of connected gas containers, some gas may flow out of one container when its temperature rises, to other containers. The *total* mass of gas in all the containers remains constant, however, no matter what changes take place in individual containers. Hence, in a closed system, the sum of the values of pV/rT *for all the containers is constant.* This is used in the following example.

Example
State the laws of gases usually associated with the names of Boyle, Charles, Dalton and Graham. Two gas containers with volumes of 100 cm³ and 1000 cm³ respectively are connected by a tube of negligible volume, and contain air at a pressure of 1000 mm mercury. If the temperature of both vessels is originally 0°C, how much air will pass through the connecting tube when the temperature of the smaller is raised to 100°C? Give your answer in cm³ measured at 0°C and 760 mm mercury. (L.)

The *total* mass of air in the two containers remains constant, although some air is transferred from the hot to the cold container on heating.

Suppose r is the gas constant per unit mass of air.
Then, from $pV = mrT$,

$$\text{initially, total mass of air} = \frac{pV}{rT} = \frac{1000 \times 1100}{r \times 273}$$

$$\text{and finally, total mass of air} = \frac{p \times 100}{r \times 373} + \frac{p \times 1000}{r \times 273},$$

where p is the new pressure in both containers after heating.
Equating the two masses, which are unchanged by any transfer,

$$\therefore \frac{p \times 100}{r \times 373} + \frac{p \times 1000}{r \times 273} = \frac{1000 \times 1100}{r \times 273}.$$

Cancelling r and simplifying, then

$$p = \frac{11\,000 \times 373}{4003} = 1025 \text{ mmHg.}$$

As the question requires, we now have to convert the initial volume of 100 cm³ of air in the smaller container at 0°C and 1000 mmHg to 0°C and 760 mmHg, and the final volume of 100 cm³ at 100°C and 1025 mmHg to 0°C and 760 mmHg.

$$\text{Initial volume at 0°C and 760 mmHg} = 100 \times \frac{1000}{760} = 131 \cdot 6 \text{ cm}^3$$

$$\text{Final volume at } 0°C \text{ and } 760 \text{ mmHg } = 100 \times \frac{1025}{760} \times \frac{273}{373} = 98 \cdot 7 \text{ cm}^3$$

∴ volume of air at 0°C and 760 mmHg flowing out = 33 cm³ (approx.).

Work Done by Gas

Consider some gas, at a pressure p, in a cylinder fitted with a piston (Fig. 10.8).

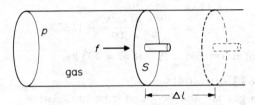

Fig. 10.8 Work done by gas in expansion

If the piston has an area S, the force on it is

$$f = pS.$$

If we allow the piston to move outwards a distance Δl, the gas will expand, and its pressure will fall. But by making the distance very short, we can make the fall in pressure so small that we may consider the pressure constant. The force f is then constant, and the work done is

$$\Delta W = f.\Delta l = pS.\Delta l.$$

The product $S.\Delta l$ is the increase in volume, ΔV, of the gas, so that

$$\Delta W = p.\Delta V \quad . \quad . \quad . \quad . \quad . \quad (1)$$

The product of pressure and volume, in general, therefore represents work. If the pressure p is in N m⁻², and the area S is in m², the force f is in newton. And if the movement Δl is in m, the work $f.\Delta l$ is in newton × metre or *joule* (J). The increase of volume, ΔV, is in m³. Thus the product of pressure in N m⁻², and volume in m³, represents work in joule.

From (1), ·   *work done = pressure × volume change.*

So if the volume of a gas at constant pressure of 10⁵ N m⁻² expands by 0·01 m³, then

$$\text{work done} = 10^5 \times 0 \cdot 01 = 1000 \text{ J.}$$

Latent Heat and Internal Energy

The volume of 1 g of steam at 100°C and 760 mmHg pressure is 1672 cm³. Therefore when 1 g of water turns into steam, it expands by 1671 cm³; in doing so, it does work against the atmospheric pressure. The heat equivalent of this work is that part of the latent heat which must be supplied to the water to make it overcome atmospheric pressure as it evaporates; it is called the 'external latent heat'. The rest of the specific latent heat—the internal part —is the equivalent of the work done in separating the molecules, against their mutual attractions. This amount of energy, then, is a measure of the increase in potential energy of the molecules of water in the gaseous state over that in the liquid state, at the same temperature.

The work done, W, in the expansion of 1 g from water to steam is the product of the atmospheric pressure p and the increase in volume ΔV:

$$W = p.\Delta V.$$

Normal atmospheric pressure corresponds to a barometer height H of 760 mmHg. Hence, as on p. 202,

$$p = H\rho g = 0{\cdot}76 \times 13\,600 \times 9{\cdot}8$$
$$= 1{\cdot}013 \times 10^5 \text{ N m}^{-2}$$

and $$W = p.\Delta V = 1{\cdot}013 \times 10^5 \times 1671 \times 10^{-6} \text{ J.}$$

The external specific latent heat is therefore

$$l_{ex} = 1{\cdot}013 \times 10^5 \times 1671 \times 10^{-6}$$
$$= 170 \text{ J g}^{-1} = 170 \text{ kJ kg}^{-1}.$$

This result shows that the external part of the specific latent heat is much less than the internal part. Since the total specific latent heat of vaporisation l is 2270 J g^{-1}, the internal part is

$$l_{in} = l - l_{ex} = 2270 - 170$$
$$= 2100 \text{ J g}^{-1} = 2100 \text{ kJ kg}^{-1}.$$

For 1 mole of water, 18 g, the number of molecules is about 6×10^{23}, the Avogadro constant. Hence, from above, the gain in potential energy of a molecule of water when changing from liquid at 100°C to vapour at 100°C is

$$\frac{18 \times 2100}{6 \times 10^{23}} = 6{\cdot}3 \times 10^{-20} \text{ J molecule}^{-1}.$$

Internal Energy of Gas

The *internal energy* of an ideal gas is the kinetic energy of thermal motion of its molecules. As we see later, the magnitude depends on the temperature of the gas and on the number of atoms in its molecule.

The thermal motion of the molecules is a random motion—it is often called the thermal 'agitation' of the molecules. We must appreciate that the energy of the gas which we call internal energy is quite independent of any motion of the gas in bulk. When a cylinder of oxygen is being carried by an express train, its kinetic energy as a whole is greater than when it is standing on the platform; but the random motion of the molecules relative to the cylinder is unchanged—and so is the temperature of the gas.

The same is true of a liquid; in a water-churning experiment to convert mechanical energy into heat, baffles must be used to prevent the water from acquiring any mass-motion—all the work done must be converted into random motion, if it is to appear as heat. Likewise, the internal energy of a solid is the kinetic energy of the vibration of its atoms about their mean positions: throwing a lump of metal through the air does not raise its temperature, but hitting it with a hammer does.

We shall use the symbol U for the internal energy. Although its absolute magnitude is not known, we are mainly concerned with *changes* in internal energy, denoted by ΔU. We can often calculate ΔU, as we soon show.

First Law of Thermodynamics

The First Law of Thermodynamics states that the *total energy in a closed system is constant*, that is, the energy is conserved in any transfer of energy from one form to another. This law is also known as the Principle of Conservation of Energy (p. 177).

When we warm a gas so that it expands, the heat ΔQ we give to it appears partly as an increase ΔU to its *internal energy*—and hence its temperature—and partly as the energy required for the *external work* done, ΔW. Thus from the First Law of Thermodynamics, we may write

$$\Delta Q = \Delta U + \Delta W \qquad . \qquad . \qquad . \qquad . \qquad (1)$$

If the expansion of the gas occurs reversibly, then no friction forces are present (p. 210). In this case, $\Delta W = p.\Delta V$. Thus, from (1),

$$\Delta Q = \Delta U + p.\Delta V \qquad . \qquad . \qquad . \qquad . \qquad (2)$$

Equation (2), then, is a mathematical statement of the First Law of Thermodynamics applied to the case of energy changes associated with a gas.

If the volume of a gas is kept *constant* when it is warmed, then no external work is done. So *all* the heat supplied goes in raising the internal energy of the gas. In this special case, then, $\Delta U = \Delta Q$, the heat supplied. If we have 1 mole of gas whose molar heat capacity at constant volume is C_V (discussed shortly), and ΔT is its temperature rise, then $\Delta Q = C_V.\Delta T$. At constant volume, then, the change in internal energy is given by

$$\Delta U = C_V.\Delta T \qquad . \qquad . \qquad . \qquad . \qquad (3)$$

It should be noted that this is always the change in internal energy of an ideal gas for a temperature change ΔT, no matter how the change has occurred. *The internal energy of an ideal gas is independent of its volume*—it depends only on its temperature.

A perfect, or ideal, gas is one which obeys Boyle's law exactly and whose internal energy is independent of its volume. No such gas exists, but at room temperature, and under moderate pressures, many gases approach the ideal closely enough for most purposes.

Molar and Specific Heat Capacities

Heat Capacities at Constant Volume and Constant Pressure

When we warm a gas, we may let it expand or not, as we please. If we do not let it expand—if we warm it in a closed vessel—then it does no external work, and all the heat we give it goes to increase its internal energy. *The heat required to warm one mole of gas through one degree, when its volume is kept constant, is called the molar heat capacity of the gas at constant volume.* It is denoted by C_V and is generally expressed in J mol^{-1} K^{-1}.

A similar definition applies to the *specific heat capacity* at constant volume, where the mass of gas warmed is 1 kg. The symbol used in this case is c_V and the unit is J kg^{-1} K^{-1}.

We can also warm a gas while keeping its pressure constant, and define the corresponding heat capacity. *The molar heat capacity of a gas at constant pressure is the heat required to warm one mole of it by one degree, when its pressure is kept constant.* It is denoted by C_p, and is expressed in the same units as C_V.

In the case of the specific heat capacity at constant pressure, which applies to a mass of 1 kg, the symbol used is c_p.

It can be seen that $c_V = C_V/M$ and $c_p = C_p/M$, where M is the numerical value of the mass of one mole expressed in kg.

Any number of heat capacities can be defined for a gas, according to the mass and the conditions imposed upon its pressure and volume. For unit mass, 1 kg, of a gas, the heat capacities at constant pressure c_p, and at constant volume c_V, are called the *principal specific heat capacities.*

Molar Heat Capacities: their Difference

Figure 10.9 shows how we can find a relationship between the molar heat capacities of a gas. We first consider 1 mole of the gas warmed through 1 K

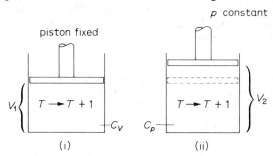

Fig. 10.9 Molar heat capacity at constant volume and pressure

at constant volume, (i). The heat required is C_V joules, and goes wholly to increase the internal energy U.

We next consider 1 mol warmed through 1 K at constant pressure, (ii). It expands from V_1 to V_2, and does an amount of external work given by

$$W = p(V_2 - V_1) \quad \text{(equation (1), p. 204.)}$$

The work W is in joule if p is in N m^{-2}, and the volume in m^3. Thus the amount of heat in joules required for this work is

$$W = p(V_2 - V_1).$$

Further, since the temperature rise of the gas is 1 K, and the internal energy of the gas is independent of volume, the rise in internal energy is C_V, the molar heat capacity at constant volume. Hence, from $\Delta Q = \Delta U + p.\Delta V$, the total amount of heat required to warm the gas at constant pressure is therefore

$$C_p = C_V + p(V_2 - V_2) \qquad (1)$$

We can simplify the last term of this expression by using the equation of state for one mole:

$$pV = RT,$$

where T is the absolute temperature of the gas, and R is the molar gas-constant. If T_1 is the absolute temperature before warming, then

$$pV_1 = RT_1 \qquad (2)$$

The absolute temperature after warming is $T_1 + 1$; therefore

$$pV_2 = R(T_1 + 1), \qquad (3)$$

and on subtracting (2) from (3) we find

$$p(V_2 - V_1) = R.$$

Equation (1) now gives $\qquad C_p = C_V + R$

or $\qquad C_p - C_V = R \qquad (4)$

Equation (4) was first derived by Robert Mayer in 1842.

A similar expression to (4) can be derived for the difference in the principal specific heat capacities of a gas, $c_p - c_V$. Thus if r is the gas constant per unit mass,

$$c_p - c_V = r \qquad (4A)$$

On p. 180, the specific heat capacity of a metal is measured at constant atmospheric pressure. So c_p was measured. The volume expansion of a metal at constant pressure is very small compared to that of a gas. So the external work done is very small. Hence it follows that there is not much difference between c_p and c_V for a metal.

Example

The density of a gas is 1·775 kg m^{-3} at 27°C and 10^5 N m^{-2} pressure and its specific heat capacity at constant pressure is 846 J kg^{-1} K^{-1}. Find the ratio of its specific heat capacity at constant pressure to that at constant volume.

$$r = \frac{pV}{T} = \frac{10^5 \times 1}{1\cdot775 \times 300} \text{ J kg}^{-1}\text{ K}^{-1}$$

since $V = 1$ m^3/1·775 for 1 kg, $T = 273 + 27 = 300$ K. Simplifying

$$\therefore r = 188 \text{ J kg}^{-1}\text{ K}^{-1}.$$

Now $\qquad c_p - c_V = r$

$$\therefore 846 - c_V = 188$$

$$\therefore c_V = 846 - 188 = 658 \text{ J kg}^{-1}\text{ K}^{-1}$$

$$\therefore \gamma = \frac{c_p}{c_V} = \frac{846}{658} = 1\cdot29.$$

Changes of Pressure, Volume and Temperature

In a car engine, for example, gases expand and are compressed, cool and are heated, in ways more complicated than those which we have already described. We shall now consider some of these ways.

Isothermal Changes

We have seen that the pressure p, and volume V of a mole of gas are related by the equation

$$pV = RT,$$

where T is the absolute temperature of the gas, and R is the molar gas constant. If the temperature is constant the curve of pressure against volume is a rectangular hyperbola,

$$pV = \text{constant},$$

representing Boyle's law. Such a curve is called an *isothermal* for the given mass of the given gas, at the temperature T. Figure 10.10 shows a family of isothermals, for 1 g of air at different temperatures. When a gas expands, or is compressed, at *constant temperature*, its pressure and volume vary along the appropriate isothermal, and the gas is said to undergo an *isothermal compression or expansion*.

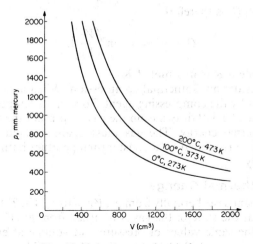

Fig. 10.10 Isothermals for lg of air

When a gas expands, it does work—for example, in driving a piston (Fig. 10.8, p. 204). The molecules of the gas bombard the piston, and if the piston moves they give up some of their kinetic energy to it; when a molecule bounces off a *moving* piston, it does so with a velocity less in magnitude than that with which it struck. The change in velocity is small, because the piston moves much more slowly than the molecule; but there are many molecules striking the piston at any instant, and their total loss of kinetic energy is equal to the work done in driving the piston forward. The work done by a gas in expanding, therefore, is done at the expense of its internal energy. The temperature of the gas will consequently fall during expansion, unless heat is supplied to it.

Conversely, if a gas is compressed, its temperature rises. The molecules now rebound from the forward-moving piston with a velocity greater than their incident velocity. The total increase in kinetic energy of all the molecules is equal to the work done in moving the piston. In an isothermal compression or expansion, the gas must be held in a thin-walled, highly conducting vessel, surrounded by a constant temperature bath. And the expansion must take place slowly, so that heat can pass into the gas to maintain its temperature at every instant during the expansion.

External Work done in Expansion

The heat taken in when a gas expands isothermally is the heat equivalent of the mechanical work done. If the volume of the gas increases by a small amount ΔV, at the pressure p, then the work done is (p. 204)

$$\Delta W = p\Delta V.$$

In an expansion from V_1 to V_2, therefore, the work done is

$$W = \int dW = \int_{V_1}^{V_2} p\, dV.$$

Assuming we have 1 mole of gas, $p = \dfrac{RT}{V}$,

so
$$W = \int_{V_1}^{V_2} p\, dV = \int_{V_1}^{V_2} RT\frac{dV}{V} = RT \ln\left(\frac{V_2}{V_1}\right).$$

The heat required, Q, is therefore

$$Q = W = RT \ln\left(\frac{V_2}{V_1}\right),$$

where W is in joule if R is in J mol^{-1} K^{-1}.

Now let us consider an isothermal compression. When a gas is compressed, work is done on it by the compressing agent. To keep its temperature constant, therefore, heat must be withdrawn from the gas, to prevent the work done from increasing its internal energy. The gas must again be held in a thin well-conducting vessel, surrounded by a constant-temperature bath; and it must be compressed slowly.

Reversible Isothermal Change

Suppose a gas expands isothermally from p_1, V_1, T to p_2, V_2, T. If the change can be reversed so that the state of the gas is returned from p_2, V_2, T to p_1, V_1, T through exactly the same values of pressure and volume at every stage, then the isothermal change is said to be *reversible*. A reversible isothermal change is an ideal one. It requires conditions such as a light frictionless piston, so that the pressure inside and outside the gas can always be equalised and no work is done against friction; very slow expansion, so that no eddies are produced in the gas to dissipate the energy; and a constant temperature reservoir with very thin good-conducting walls, as we have seen. In a reversible isothermal change of 1 mole of gas, $pV = $ constant $= RT$.

Reversible Adiabatic Change

Let us now consider a change of volume in which the conditions are at the opposite extreme from isothermal; no heat is allowed to enter or leave the gas.

An expansion of contraction in which *no heat* enters or leaves the gas is called an *adiabatic expansion or contraction*. In an adiabatic expansion, the external work is done wholly at the expense of the internal energy of the gas, and the gas therefore cools. In an adiabatic compression, all the work done on the gas by the compressing agent appears as an increase in its internal energy and therefore as a rise in its temperature. We have already discussed a reversible isothermal change. A *reversible adiabatic change* is an adiabatic change which can be exactly reversed in the sense explained on p. 210. As noted there, a reversible change is an ideal case.

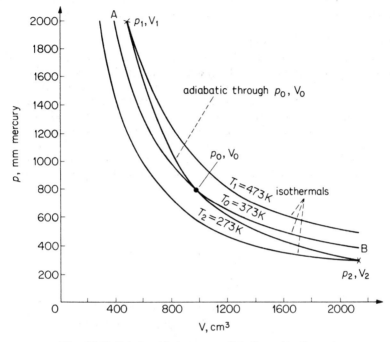

Fig. 10.11 Relationship between adiabatic and isothermals

The curve relating pressure and volume for a given mass of a given gas for adiabatic changes is called an 'adiabatic'. In Fig. 10.11, the curve shown is an adiabatic for 1 kg of air; it is steeper, at any point, than the isothermal through that point. The curve AB is the isothermal for the temperature $T_0 = 373$ K, which cuts the adiabatic at the point p_0, V_0. If the gas is adiabatically compressed from V_0 to V_1, its temperature rises to some value T_1. Its representative point p_1, V_1 now lies on the isothermal for T_1, since $p_1 V_1 = RT_1$. Similarly, if the gas is expanded adiabatically to V_2, it cools to T_2 and its representative point p_2, V_2 lies on the isothermal for T_2. Thus the adiabatic through any point— such as p_0, V_0—is steeper than the isothermal. We soon show (p. 213) that the equation for a reversible adiabatic p–V change of a given mass of gas is

$$pV^\gamma = \text{constant}, \qquad . \qquad . \qquad . \qquad . \qquad (1)$$

where $\gamma = C_p/C_V$ = the ratio of the molar heat capacities of the gas. Now the slope or gradient of the isothermal curve, $pV = \text{constant}$, at any point, is given, on differentiation, by

$$\frac{dp}{dV} = -\frac{p}{V}.$$

The gradient of the adiabatic curve at any point is given by differentiating (1). Then, after simplifying, we find

$$\frac{dp}{dV} = -\frac{\gamma p}{V}.$$

So the gradient of the adiabatic curve is γ times that of the isothermal curve at their point of intersection. Since γ has a value greater than 1 for any gas (p. 230), it follows that the adiabatic gradient at the point of intersection is greater than the isothermal gradient at the same point, as stated before.

Ideal and Real p–V Curves

The condition for an adiabatic change is that no heat must enter or leave the gas. The gas must therefore be held in a thick-walled, badly conducting vessel; and the change of volume must take place rapidly, to give as little time as possible for heat to escape. However, in a rapid compression, for example, eddies may be formed, so that some of the work done appears as kinetic energy of the gas in bulk, instead of as random kinetic energy of its molecules. All the work done then does not go to increase the internal energy of the gas, and the temperature rise is less than in a truly adiabatic compression. If the compression is made slowly, then more heat leaks out, since no vessel has perfectly insulating walls.

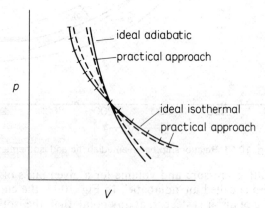

Fig. 10.12 Ideal and real p–V curves for a gas

Perfectly adiabatic changes are therefore impossible; and so, we have seen, are perfectly isothermal ones. Any practical expansion or compression of a gas must lie between isothermal and adiabatic. It may lie anywhere between them, but if it approximates to isothermal, the curve representing it will always be a little steeper than the ideal (Fig. 10.12); if it approximates to adiabatic, the curve representing it will never be quite as steep as the ideal.

Equation of Reversible Adiabatic

Before considering adiabatic changes in particular, let us first consider a change of volume and temperature which takes place in an arbitrary manner. For simplicity, we consider one mole of the gas, and we suppose that its volume expands from V to $V + \Delta V$, and that an amount of heat ΔQ is supplied to it. In general, the internal energy of the gas will increase by an amount ΔU. And the gas will do an amount of external work equal to $p\Delta V$, where p is its pressure. The heat supplied is equal to the increase in internal energy, plus the external work done:

$$\Delta Q = \Delta U + p\Delta V \qquad . \qquad . \qquad . \qquad (1)$$

The increase in internal energy represents a temperature rise, ΔT. We have seen already that the internal energy is independent of the volume, and is related to the temperature by the molar heat capacity at constant volume, C_V (p. 207). Therefore

$$\Delta U = C_V \Delta T.$$

Equation (1) becomes

$$\Delta Q = C_V \Delta T + p \Delta V \qquad . \qquad . \qquad . \qquad . \qquad (2)$$

Equation (2) is the fundamental equation for any change in state of one mole of an ideal gas.

For a reversible isothermal change, $\Delta T = 0$, and $\Delta Q = p \Delta V$.

For a reversible adiabatic change, $\Delta Q = 0$ and therefore

$$C_V \Delta T + p \Delta V = 0 \qquad . \qquad . \qquad . \qquad . \qquad (3)$$

To eliminate ΔT we use the general equation, relating pressure, volume and temperature:

$$pV = RT,$$

where R is the molar gas constant. Since both pressure and volume may change, when we differentiate this to find ΔT we must write

$$p \Delta V + V \Delta p = R \Delta T . \qquad . \qquad . \qquad . \qquad . \qquad (4)$$

Substituting in (4) for $R(= C_p - C_V)$ and for $\Delta T (= -p.\Delta V/C_V$ from (3)), and then simplifying, we obtain

$$C_V V \Delta p + C_p p \Delta V = 0$$

or

$$V \Delta p + \gamma p \Delta V = 0 \left(\text{where } \gamma = \frac{C_p}{C_V} \right).$$

Integrating,

$$\int \frac{dp}{p} + \gamma \int \frac{dV}{V} = 0$$

or

$$\ln p + \gamma \ln V = A,$$

where A is a constant.

Therefore,

$$pV^\gamma = \text{constant} \qquad . \qquad . \qquad . \qquad . \qquad (5)$$

This is the equation of a reversible adiabatic; the value of the constant can be found from the initial pressure and volume of the gas. Although we have considered one mole of gas, the same equation is true for any other given mass of gas, such as 1 kg (unit mass). Note that γ is the ratio of *either* the molar heat capacities *or* the principal specific heat capacities.

Equation for Temperature Change in an Adiabatic

If we wish to introduce the temperature, T, into equation (5), we use the gas equation for one mole of gas,

$$pV = RT.$$

Thus

$$p = \frac{RT}{V}$$

and

$$pV^\gamma = \frac{RT}{V}.V^\gamma = RTV^{\gamma-1}.$$

Thus equation (5) becomes

$$RTV^{\gamma-1} = \text{constant},$$

and since R is a constant for the gas, the equation for an adiabatic temperature change becomes

$$TV^{\gamma-1} = \text{constant}.$$

Although we have considered a mole of gas in deriving the adiabatic relation, the same relations between p, V and T are true for any given mass of gas such as 1 kg.

The applications of $pV^{\gamma} = \text{constant}$ and $TV^{\gamma-1} = \text{constant}$ are illustrated in the examples which follow.

Examples

1. An ideal gas at 17°C has a pressure of 760 mmHg, and is compressed (i) isothermally, (ii) adiabatically until its volume is halved, in each case reversibly. Calculate in each case the final pressure and temperature of the gas, assuming $c_p = 2100$, $c_V = 1500$ J kg^{-1} K^{-1}.

(i) Isothermally, $pV = \text{constant}$.

$$\therefore p \times \frac{V}{2} = 760 \times V$$

$$\therefore p = 1520 \text{ mmHg}.$$

The temperature is constant at 17°C.

(ii) Adiabatically, $pV^{\gamma} = \text{constant}$, and $\gamma = 2100/1500 = 1\cdot4$.

$$\therefore p \times \left(\frac{V}{2}\right)^{1\cdot4} = 760 \times V^{1\cdot4}$$

$$\therefore p = 760 \times 2^{1\cdot4} = 2010 \text{ mmHg}.$$

Since $TV^{\gamma-1} = \text{constant}$,

$$\therefore T \times \left(\frac{V}{2}\right)^{0\cdot4} = (273 + 17) \times V^{0\cdot4}$$

$$\therefore T = 290 \times 2^{0\cdot4} = 383 \text{ K}.$$

$$\therefore \text{ temperature} = 110°C.$$

2. Distinguish between *isothermal* and *adiabatic* changes. Show that for an ideal gas the curves relating pressure and volume for an adiabatic change have a greater slope than those for an isothermal change, at the same pressure.

A quantity of oxygen is compressed isothermally until its pressure is doubled. It is then allowed to expand adiabatically until its original volume is restored. Find the final pressure in terms of the initial pressure. (The ratio of the molar heat capacities of oxygen is to be taken as 1·40.) (L.)

Calculation. Let p_0, $V_0 = $ the original pressure and volume of the oxygen.
Since $pV = \text{constant}$ for an isothermal change,

$$\therefore \text{ new volume} = \frac{V_0}{2} \text{ when new pressure is } 2p_0.$$

Suppose the gas expands adiabatically to its volume V_0, when the pressure is p.

Then
$$p \times V_0^{1\cdot4} = 2p_0 \times \left(\frac{V_0}{2}\right)^{1\cdot4}$$

$$\therefore p = 2p_0 \times \left(\frac{1}{2}\right)^{1\cdot4} = 0\cdot8\ p_0.$$

Heat and Mechanical Work in Engines

If one metal surface is rubbed against another, practically the whole of the mechanical work done against friction is transformed into heat (see p. 181). Thus the conversion from mechanical energy to heat can be almost 100% efficient. As we now show, however, a practical machine or *engine* which converts heat into mechanical energy, the opposite process, can never be 100% efficient.

If the engine operates in a *cycle*, so that it returns to its original state after a series of operations, then it can be made to do work continuously.

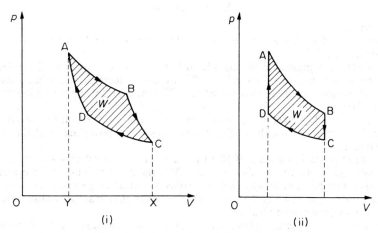

Fig. 10.13 Heat and work. Carnot cycle

Consider a gas taken through a cycle of reversible pressure (p)–volume (V) changes. Two examples of a cycle are shown in Fig. 10.13 (i) and (ii). Starting from the state represented by A, the gas expands reversibly from A to B. In Fig. 10.13 (i), it then expands reversibly from B to C. Here the mechanical work done *by* the gas, which is the integral of $p.dV$ for the two stages of the cycle, is represented by the *area* ABCXYA between the curves and the V-axis (see p. 204). To restore the gas to its initial state at A the gas must now be compressed reversibly along CD and DA. In this case the work done *on* the gas is represented by the area CDAYXC. The gas now gives up a quantity of heat Q_2, whereas it takes in a quantity of heat Q_1 along AB and BC while doing work.

Since the gas returns to its initial state at A, there is no change in the internal energy of the gas at the end of a cycle. From the First Law of Thermodynamics, then, the net work done in a cycle = heat gained by gas = $Q_1 - Q_2$. The *net* work done W is represented by the area ABCD in both Fig. 10.13 (i) and (ii).

The efficiency E of an engine is defined by

$$E = \frac{\text{work obtained}}{\text{energy supplied}} \times 100\%$$

$$= \frac{Q_1 - Q_2}{Q_1} \times 100\% = \left(1 - \frac{Q_2}{Q_1}\right) \times 100\%.$$

It follows that an engine can never be 100% efficient, that is, all the heat supplied can never be transferred or converted into mechanical energy during a complete cycle. This is one statement of the *Second Law of Thermodynamics*.

In 1824, the French scientist Carnot showed that the most efficient engine, that is, the engine which transfers the maximum amount of the heat supplied between two given temperatures to mechanical energy, was one working under *reversible* conditions (p. 210). Figure 10.13 (i) represents a *Carnot cycle* if AB is a reversible isothermal expansion and CD a reversible isothermal contraction, and BC, DA are respectively a reversible adiabatic expansion and contraction. Figure 10.13 (ii) represents two reversible adiabatic changes of a gas, and two reversible constant volume changes, known as the *Otto cycle*; this has maximum efficiency like the Carnot cycle because the changes are reversible.

Exercises 10

Gas Laws

1. A container of gas has a volume of 0.10 m^3 at a pressure of $2.0 \times 10^5 \text{ N m}^{-2}$ and a temperature of $27°C$.

(i) Find the new pressure if the gas is heated at constant volume to $87°C$. (ii) The gas pressure is now reduced to $1.0 \times 10^5 \text{ N m}^{-2}$ at constant temperature. What is the new volume of the gas? (iii) The gas is cooled to $-73°C$ at constant pressure. Find the new volume of the gas.

2. Using a vertical (y) axis to represent pressure, p, and a horizontal axis to represent volume, V, illustrate by rough sketches the changes which take place in (i), (ii) and (iii) in question **1**.

3. A cylinder of gas has a mass of 10.0 kg and a pressure of 8.0 atmospheres at $27°C$. When some gas is used in a cold room at $-3°C$, the gas remaining in the cylinder at this temperature has a pressure of 6.4 atmospheres. Calculate the mass of gas used.

4. State Boyle's law and Charles's law and show how they may be combined to give the equation of state of an ideal gas.

Two glass bulbs of equal volume are joined by a narrow tube and are filled with a gas at s.t.p. When one bulb is kept in melting ice and the other is placed in a hot bath, the new pressure is 877.6 mm mercury. Calculate the temperature of the bath. (*L.*)

5. The formula $pv = mrT$ is often used to describe the relationship between the pressure p, volume v, and temperature T of a mass m of a gas, r being a constant. Referring in particular to the experimental evidence how do you justify (*a*) the use of this formula, (*b*) the usual method of calculating T from the temperature t of the gas on the centigrade (Celsius) scale?

Two vessels of capacity 1.00 litre are connected by a tube of negligible volume. Together they contain 3.42×10^{-4} kg of helium at a pressure of 800 mm of mercury and temperature $27°C$. Calculate (i) a value for the constant r for helium, (ii) the pressure developed in the apparatus if one vessel is cooled to $0°C$ and the other heated to $100°C$, assuming that the capacity of each vessel is unchanged. (*N.*)

6. State *Boyle's law* and *Charles' law*, and show how they lead to the gas equation $PV = RT$. Describe an experiment you would perform to measure the thermal expansion coefficient of dry air.

What volume of liquid oxygen (density 1140 kg m^{-3}) may be made by liquefying completely the contents of a cylinder of gaseous oxygen containing 100 litres of oxygen at 120 atmospheres pressure and $20°C$? Assume that oxygen behaves as an ideal gas in this latter region of pressure and temperature.

[1 atmosphere $= 1.01 \times 10^5 \text{ N m}^{-2}$; molar gas constant $= 8.31 \text{ J mol}^{-1} \text{ K}^{-1}$; relative molecular mass of oxygen $= 32.0$.] (*O. & C.*)

7. A sealed bottle full of water is placed in a strong container full of air at standard atmospheric pressure, $1.0 \times 10^5 \text{ N m}^{-2}$, and at a temperature of $10°C$. The temperature in the container is raised to and maintained at $100°C$. Neglecting the expansion of the bottle and the container, what is the new pressure in the container?

If the bottle breaks what will the pressure be? (*L.*)

Heat Capacities of Gases

8. For hydrogen, the molar heat capacities at constant volume and constant pressure are respectively $20.5 \text{ J mol}^{-1} \text{ K}^{-1}$ and $28.8 \text{ J mol}^{-1} \text{ K}^{-1}$. What does this mean?

(i) Which heat capacity is related to internal energy assuming hydrogen is an ideal gas? Explain your answer. (ii) From the values given, calculate the molar gas constant. (iii) Is C_p always greater than C_V? Explain your answer.

9. From the values of C_V and C_p given in question **8**, calculate (i) the heat needed to raise the temperature of 8 g of hydrogen from 10°C to 15°C at constant pressure, (ii) the increase in internal energy of the gas, (iii) the external work done. (Molar mass of hydrogen = 2 g.)

10. A gas has a volume of 0·02 m^3 at a pressure of 2×10^5 Pa(N m^{-2}) and temperature of 27°C. It is heated at constant pressure until its volume increases to 0·03 m^3.

Calculate (i) the external work done, (ii) the new temperature of the gas, (iii) the increase in internal energy of the gas if its mass is 16 g, its molar heat capacity at constant volume is 0·8 J mol^{-1} K^{-1} and its molar mass is 32 g.

11. Why is the energy needed to raise the temperature of a given mass of gas by a certain amount greater if the pressure is kept constant than if the volume is kept constant? (*L.*)

12. What happens to the energy added to an ideal gas when it is heated (*a*) at constant volume, and (*b*) at constant pressure?

Show from this that a gas can have a number of values of specific heat capacity. Deduce an expression for the difference between the specific heat capacities of a gas at constant pressure and at constant volume.

If the ratio of the principal specific heat capacities of a certain gas is 1·40 and its density at s.t.p. is 0·090 kg m^{-3} calculate the values of the specific heat capacity at constant pressure and at constant volume. (Standard atmospheric pressure $= 1·01 \times 10^5$ N m^{-2}.) (*L.*)

13. Explain why the values of the specific heat capacities of a gas when measured at constant pressure and at constant volume respectively are different. Derive an expression for the difference, for an ideal gas, in terms of its relative molecular mass M and the molar gas constant R.

Given that the volume of a gas at s.t.p. is $2·24 \times 10^{-2}$ m^3 mol^{-1} and that standard pressure is $1·01 \times 10^5$ N m^{-2}, calculate a value for the molar gas constant R and use it to find the difference between the quantities of heat required to raise the temperature of 0·01 kg of oxygen from 0°C to 10°C when (*a*) the pressure, (*b*) the volume of the gas is kept constant. (Relative molecular mass of oxygen = 32.) (*O. & C.*)

14. Explain why the specific heat capacity of a gas is greater if it is allowed to expand while being heated than if the volume is kept constant. Discuss whether it is possible for the specific heat capacity of a gas to be zero.

When 1 g of water at 100°C is converted into steam at the same temperature 2264 J must be supplied. How much of this energy is used in forcing back the atmosphere? Explain what happens to the remainder of the energy. [1 g of water 100°C occupies 1 cm^3. 1 g of steam at 100°C and 760 mmHg occupies 1601 cm^3. Density of mercury = 13 600 kg m^{-3}.] (*C.*)

15. Write down the relation between the *heat input to a system*, the *work done by the system* and its increase in *internal energy*; define the quantities in italics.

How is the heat capacity of a system related to the heat input and the resulting change in temperature? Why, in considering the heat capacity of a gas, is it necessary to specify the conditions (e.g. constant volume or constant pressure) under which the temperature change takes place?

A rod, of uniform cross section A and length l, is made of a material of Young's modulus E and linear expansivity α. It is clamped securely at its ends and is heated through T°C without being allowed to expand along its length. Find expressions for the force exerted on the clamps and the strain energy stored in the rod after the heating. (*O. & C.*)

16. By considering the expansion of an ideal gas contained in a cylinder and enclosed by a piston, show that the work done in a small expansion is equal to the pressure times the volume change.

The increase in internal energy of an ideal gas of mass m when the temperature rises by T is $mc_V T$. In this expression c_V is the specific heat capacity of the gas at constant volume but the expression holds under all conditions. Explain this.

An ideal gas, at a temperature of 290 K and a pressure of $1·0 \times 10^5$ N m^{-2}, occupies a volume of $1·0 \times 10^{-3}$ m^3. Its density under these conditions is 0·30 kg m^{-3}. It expands at constant pressure to a volume of 1.5×10^{-3} m^3. Calculate the energy added.

It is now compressed isothermally to its original volume. Calculate (a) its final pressure and temperature, and (b) the difference between its final and initial internal energies. (Specific heat capacity at constant volume of this gas $= 7.1 \times 10^2$ J kg^{-1} K^{-1}.) (L.)

17. Explain why the molar heat capacity of a gas at constant pressure is different from that at constant volume.

The density of an ideal gas is 1.60 kg m^{-3} at 27°C and 1.00×10^5 newton metre^{-2} pressure and its specific heat capacity at constant volume is 312 J kg^{-1} K^{-1}. Find the ratio of the specific heat capacity at constant pressure to that at constant volume. Point out any significance to be attached to the result. (N.)

Isothermal and Adiabatic Changes

18. Explain why the temperature of a gas is liable to change when it expands.

Explain what energy changes take place in (a) *reversible isothermal*, (b) *reversible adiabatic* alterations in the volume of a gas. What conditions would be needed to produce such changes?

Gas in a cylinder, initially at a temperature of 17°C and a pressure of 1.01×10^5 N m^{-2}, is to be compressed to one-eighth of its volume. What would be the difference between the final pressures if the compression were done (a) isothermally, (b) adiabatically? What would be the final temperature in the latter case? (Ratio of the molar heat capacities $= 1.40$.) (L.)

19. A litre of air, initially at 20°C and at 760 mm of mercury pressure, is heated at constant pressure until its volume is doubled. Find (a) the final temperature, (b) the external work done by the air in expanding, (c) the quantity of heat supplied.

[Assume that the density of air at s.t.p. is 1.293 kg m^{-3} and that the specific heat capacity of air at constant volume is 714 J kg^{-1} K^{-1}.] (L.)

20. Distinguish between an *isothermal change* and an *adiabatic change*. In each instance state, for a reversible change of an ideal gas, the relation between pressure and volume.

A mass of air occupying initially a volume 2×10^{-3} m^3 at a pressure of 760 mm of mercury and a temperature of 20·0°C is expanded adiabatically and reversibly to twice its volume, and then compressed isothermally and reversibly to a volume of 3×10^{-3} m^3. Find the final temperature and pressure, assuming the ratio of the specific heat capacities of air to be 1.40. (L.)

21. Define the terms (a) *isothermal change*, and (b) *adiabatic change*, as applied to the expansion of a gas. Explain how these changes may be approximated to in practice. What would be the relationship between the *pressure* and *temperature* for each of them for an ideal gas?

Sketch, using the same axes, the *pressure-volume* curves for each of (a) and (b) for the expansion of a gas from a volume V_1 and pressure p_1 to a volume V_2. How, from these graphs, could you calculate the work done in each of the expansions?

Explain why the temperature falls during an adiabatic expansion and discuss whether or not the temperature fall would be the same for an ideal gas and a real gas. (L.)

22. (a) The first law of thermodynamics may be written $\delta Q = \delta U + \delta W$. Explain the meaning of this equation as applied to the heating of a gas. Use the equation to justify the fact that the molar heat capacity of a gas at constant pressure is greater than the molar heat capacity at constant volume.

(b) Explain the meaning of the terms isothermal change and adiabatic change. What is meant by a reversible change?

A mass of gas is expanded isothermally and then compressed adiabatically to its original volume. What further operation must be performed on the gas to restore it to its original state? Sketch a labelled p–V graph to represent the series of operations. What quantity is represented by the area enclosed?

(c) An ideal gas at an initial temperature of 15°C and pressure of 1.10×10^5 Pa is compressed isothermally to one quarter of its original volume. What will be its final pressure and temperature? What would have been the pressure and temperature if the compression had been adiabatic? (Ratio of principal specific heat capacities of the gas $= 1.40$.) (AEB, 1982)

11 Kinetic Theory of Gases

In the kinetic theory of gases, we explain the behaviour of gases by considering the motion of their molecules. Thus we suppose that the pressure of a gas is due to the molecules bombarding the walls of its container. Whenever a molecule bounces off a wall, its momentum at right-angles to the wall is reversed; the force which it exerts on the wall is equal to the rate of change of its momentum. The average force exerted by the gas on the whole of its container is the average rate at which the momentum of its molecules is changed by collision with the walls. Since pressure is force per unit area, to find the pressure of the gas we must find this force, and then divide it by the area of the walls.

The following assumptions are made to simplify the calculation:

(a) The attraction between the molecules is negligible.

(b) The volume of the molecules is negligible compared with the volume occupied by the gas.

(c) The molecules are like perfectly elastic spheres.

(d) The duration of a collision is negligible compared with the time between collisions.

Calculation of Pressure

Consider for convenience a cube of side l containing N molecules of gas each of mass m, Fig. 11.1. A typical molecule will have a velocity c at any instant and this will have components of u, v, w respectively in the direction of the three perpendicular axes Ox, Oy, Oz as shown. Thus $c^2 = u^2 + v^2 + w^2$.

Consider the force exerted on the face X of the cube due to the component u. Just before impact, the momentum of the molecule due to u is mu. After

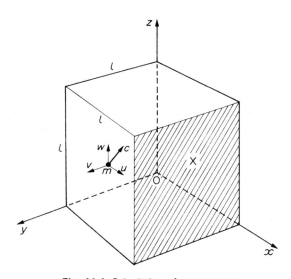

Fig. 11.1 Calculation of gas pressure

impact, the momentum is $-mu$, since the momentum reverses. Thus

$$\text{momentum change on impact} = mu - (-mu) = 2mu.$$

The time taken for the molecule to move across the cube to the opposite face and back to X is $2l/u$.

This is the time to make one impact. So the number of impacts per second n' on a given face $= 1 \div 2l/u = u/2l$. So at face X

$$\text{momentum change per second} = n' \times \text{one momentum change}$$

$$= \frac{u}{2l} \times 2mu = \frac{mu^2}{l}$$

$$\therefore \text{ force on X} = \frac{mu^2}{l}$$

$$\therefore \text{ pressure on X} = \frac{\text{force}}{\text{area}} = \frac{mu^2}{l \times l^2} = \frac{mu^2}{l^3}. \qquad . \qquad . \qquad (i)$$

We now take account of the N molecules in the cube. Each has a different velocity and hence a component of different magnitude in the direction Ox. If these are represented by $u_1, u_2, u_3, \ldots, u_N$, it follows from (i) that the total pressure on X, p, is given by

$$p = \frac{mu_1{}^2}{l^3} + \frac{mu_2{}^2}{l^3} + \frac{mu_3{}^2}{l^3} + \ldots + \frac{mu_N{}^2}{l^3}$$

$$= \frac{m}{l^3}(u_1{}^2 + u_2{}^2 + u_3{}^2 + \ldots + u_N{}^2) \qquad . \qquad . \qquad . \qquad (ii)$$

Let the symbol $\overline{u^2}$ represent the average or mean value of all the squares of the components in the Ox direction, that is,

$$\overline{u^2} = \frac{u_1{}^2 + u_2{}^2 + u_3{}^2 + \ldots + u_N{}^2}{N}.$$

Then $\qquad\qquad N\overline{u^2} = u_1{}^2 + u_2{}^2 + u_3{}^2 + \ldots + u_N{}^2$

Hence, from (ii),

$$p = \frac{Nm\overline{u^2}}{l^3} \qquad . \qquad . \qquad . \qquad . \qquad (iii)$$

Now with a large number of molecules of varying speed in random motion, the mean square of the component speed in any one of the three axes is the *same*.

$$\therefore \overline{u^2} = \overline{v^2} = \overline{w^2}.$$

But, for each molecule, $c^2 = u^2 + v^2 + w^2$, so that the mean square $\overline{c^2}$ is given by $\overline{c^2} = \overline{u^2} + \overline{v^2} + \overline{w^2}$.

$$\therefore \overline{u^2} = \tfrac{1}{3}\overline{c^2}.$$

Hence, from (iii),

$$p = \tfrac{1}{3}\frac{Nm\overline{c^2}}{l^3}.$$

The *number of molecules per unit volume, n,* $= N/l^3$. Thus we may write

$$p = \tfrac{1}{3}nm\overline{c^2} \qquad . \qquad . \qquad . \qquad . \qquad . \qquad (1)$$

If n is in molecules per metre^3, m in kilogram and c in metre per second, then the pressure p is in *newton per metre*^{2} (N m^{-2}).

In our calculation, we assumed that molecules of a gas do not collide with other molecules as they move to-and-fro across the cube. If, however, we assume that their collisions are perfectly elastic, both the kinetic energy and the momentum are conserved in them. The average momentum with which all the molecules strike the walls is then not changed by their collisions with one another; what one loses, another gains. The important effect of collisions between molecules is to distribute their individual speeds; on the average, the fast ones lose speed to the slow. We suppose, then, that different molecules have different speeds, and that the speeds of individual molecules vary with time, as they make collisions with one another; but we also suppose that the average speed of all the molecules is constant. These assumptions are justified by the fact that the kinetic theory leads to conclusions which agree with experiment.

Root-Mean-Square (R.M.S.) Speed

In equation (1) the factor nm is the product of the number of molecules per unit volume and the mass of one molecule. It is therefore the total mass of the gas per unit volume, its density ρ. Thus the equation gives

$$p = \tfrac{1}{3}\rho\overline{c^2} \qquad . \qquad . \qquad . \qquad . \qquad . \qquad (2)$$

or

$$\overline{c^2} = \frac{3p}{\rho} \qquad . \qquad . \qquad . \qquad . \qquad . \qquad (3)$$

If we substitute known values of p and ρ in equation (3), we can find $\overline{c^2}$. For hydrogen at s.t.p., $\rho = 0.09$ kg m^{-3}. The pressure in $N\ m^{-2}$ is $p = H\rho g$, where $g =$ acceleration of gravity $= 9.81$ m s^{-2}, $\rho =$ density of mercury $= 13\,600$ kg m^{-3}, $H =$ barometer height $= 760$ mm $= 0.76$ m.

$$\therefore \overline{c^2} = \frac{3p}{\rho} = \frac{3 \times 9.81 \times 13\,600 \times 0.76}{9 \times 10^{-2}}$$

$$= 3.37 \times 10^6 \text{ m}^2 \text{ s}^{-2}.$$

The square root of $\overline{c^2}$ is called the *root-mean-square speed*; it is of the same magnitude as the average speed, but not quite equal to it. See p. 222. Its value is

$$\sqrt{\overline{c^2}} = \sqrt{3.37} \times 10^3 = 1840 \text{ m s}^{-1} \text{ (approx.)}$$

$$= 1.84 \text{ km s}^{-1}.$$

From (3), note that $\sqrt{\overline{c^2}} = \sqrt{\dfrac{3p}{\rho}} = $ r.m.s. speed.

Molecular speeds were first calculated in this way by Joule in 1848; they turn out to have a magnitude which is high, but reasonable. The value is reasonable because it has the same order of magnitude as the speed of sound (1.30 km s^{-1} in hydrogen at 0°C). The speed of sound is the speed with which the molecules of a gas pass on a disturbance from one to another, and this we expect to be of the same magnitude as the speeds of their natural motion.

Root-mean-square Speed and Mean Speed

It should be carefully noted that the pressure p of the gas depends on the 'mean square' of the speed. This is because (a) the momentum change at a wall is proportional to u, as previously explained, and (b) the number of impacts per second on a given face is proportional to u. So the rate of change of momentum is proportional to $u \times u$ or to u^2. Further, the mean square speed is *not* equal to the square of the average speed. As an example, let us suppose that the speeds of six molecules are, 1, 2, 3, 4, 5, 6 units. Their mean speed $\bar{c}$ is given by

$$\bar{c} = \frac{1+2+3+4+5+6}{6} = \frac{21}{6} = 3 \cdot 5,$$

and its square is

$$(\bar{c})^2 = 3 \cdot 5^2 = 12 \cdot 25.$$

Their mean square speed, however, is

$$\overline{c^2} = \frac{1^2 + 2^2 + 3^2 + 4^2 + 5^2 + 6^2}{6} = \frac{91}{6} = 15 \cdot 2$$

So the root-mean-square speed, $\sqrt{\overline{c^2}} = \sqrt{15 \cdot 2} = 3 \cdot 9$, which is about 12% different from the mean speed $\bar{c}$ in this simple case.

Introduction of Temperature in Kinetic Theory

Consider a volume V of gas, containing N molecules. The number of molecules per unit volume $n = N/V$. So the pressure of the gas, by equation (1) is

$$p = \tfrac{1}{3}nm\overline{c^2} = \tfrac{1}{3}\frac{N}{V}m\overline{c^2}$$

$$\therefore pV = \tfrac{1}{3}Nm\overline{c^2} \qquad . \qquad . \qquad . \qquad . \qquad . \qquad (4)$$

But the ideal gas equation for 1 mole is

$$pV = RT.$$

We can therefore make the kinetic theory consistent with the observed behaviour of a gas, if we write

$$\tfrac{1}{3}Nm\overline{c^2} = RT \qquad . \qquad . \qquad . \qquad . \qquad (5)$$

Essentially, we are here assuming that the mean square speed of the molecules, $\overline{c^2}$, is proportional to the absolute (kelvin) temperature of the gas. This is a reasonable assumption, because we have learnt that heat is a form of energy; and the translational kinetic energy of a molecule, due to its random motion within its container, is proportional to the square of its speed. When we heat a gas, we expect to speed-up its molecules. So we write

$$pV = \tfrac{1}{3}Nm\overline{c^2} = RT \qquad . \qquad . \qquad . \qquad (6)$$

Variation of R.M.S. Speed

Since N molecules each have a mass m in the mole of gas considered, the molar mass M is Nm. Thus, from (5),

$$\tfrac{1}{3}M\overline{c^2} = RT$$

$$\therefore \text{ r.m.s. speed, } \sqrt{\overline{c^2}} = \sqrt{\frac{3RT}{M}}.$$

So

(1) the r.m.s. speed or velocity of the molecules of a *given gas* $\propto \sqrt{T}$, and

(2) the r.m.s. speed of the molecules of different gases at the *same temperature* $\propto 1/\sqrt{M}$, so that gases of higher molecular mass have smaller r.m.s. speeds.

To illustrate the numerical changes, hydrogen of relative molecular mass about 2 has a r.m.s speed at s.t.p. (273 K and 760 mmHg pressure) of roughly 1840 m s^{-1}. At 100°C or 373 K and 760 mmHg, the r.m.s. speed c_r is given by

$$\frac{c_r}{1840} = \sqrt{\frac{373}{273}}$$

or
$$c_r = 1840 \times \sqrt{\frac{373}{273}} = 1930 \text{ m s}^{-1} \text{ (approx.)}.$$

Oxygen has a relative molecular mass of about 32. From (2) above, it follows that at s.t.p. the r.m.s. speed of oxygen molecules is given by

$$\frac{c_r}{1840} = \sqrt{\frac{2}{32}}$$

or
$$c_r = 1840 \times \sqrt{\frac{2}{32}} = 460 \text{ m s}^{-1}.$$

Boltzmann Constant. Mean Energy of Molecule

The kinetic energy of a molecule moving at an instant with a speed c is $\frac{1}{2}mc^2$; the average kinetic energy of translation of the random motion of the molecule of a gas is therefore $\frac{1}{2}m\overline{c^2}$. To relate this to the temperature, we put equation (5) into the form

$$RT = \frac{1}{3}Nm\overline{c^2} = \frac{2}{3}N(\frac{1}{2}m\overline{c^2}),$$

so
$$\frac{1}{2}m\overline{c^2} = \frac{3}{2}\frac{R}{N}T \qquad . \qquad . \qquad . \qquad . \qquad . \qquad (7)$$

Thus, *the average kinetic energy of translation of a molecule is proportional to the absolute temperature of the gas.*

The ratio R/N in equation (7) is a universal constant, since $R = $ molar gas constant, 8·31 J mol^{-1} K^{-1} for all gases, and $N = N_A$, the Avogadro constant, 6·02 × 10^{23} mol^{-1} for all gases. Thus

$$\frac{R}{N_A} = k.$$

The constant k, the gas constant per molecule, is a universal constant called the *Boltzmann constant*. In terms of k equation (7) becomes

$$\frac{1}{2}m\overline{c^2} = \frac{3}{2}kT \qquad . \qquad . \qquad . \qquad . \qquad . \qquad (8)$$

The Boltzmann constant is usually given in joule per degree, since it relates energy to temperature: $k = \frac{1}{2}m\overline{c^2}/\frac{3}{2}T$. Its value is $k = 1·38 \times 10^{-23}$ J K^{-1}.

Diffusion: Graham's Law

When a gas passes through a porous plug, a cotton-wool wad, for example, it is said to 'diffuse'. Diffusion differs from the flow of a gas through a wide tube, in that it is not a motion of the gas in bulk, but is a result of the motion of its individual molecules.

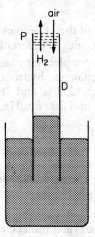

Fig. 11.2 Graham's apparatus for diffusion

Figure 11.2 shows an apparatus devised by Graham (1805–69) to compare the rates of diffusion of different gases. D is a glass tube, closed with a plug P of plaster of Paris. It is first filled with mercury, and inverted over mercury in a bowl. Hydrogen is then passed into it until the mercury levels are the same on each side; the hydrogen is then at atmospheric pressure. The volume of hydrogen, V_H, is proportional to the length of the tube above the mercury. The apparatus is now left; hydrogen diffuses out through P, and air diffuses in. Ultimately no hydrogen remains in the tube D. The tube is then adjusted until the level of mercury is again the same on each side, so that the air within it is at atmospheric pressure. The volume of air, V_A, is proportional to the new length of the tube above the mercury.

The volumes V_A and V_H are, respectively, the volumes of air and hydrogen which diffused through the plug in the same time. Therefore the rates of diffusion of the gases air and hydrogen are proportional to the volumes V_A and V_H:

$$\frac{\text{rate of diffusion of air}}{\text{rate of diffusion of hydrogen}} = \frac{V_A}{V_H}.$$

Graham found in his experiments that the volumes were inversely proportional to the square roots of the densities of the gases, ρ:

$$\frac{V_A}{V_H} = \sqrt{\frac{\rho_H}{\rho_A}}$$

thus

$$\frac{\text{rate of diffusion of air}}{\text{rate of diffusion of hydrogen}} = \sqrt{\frac{\rho_H}{\rho_A}},$$

In general:

$$\text{rate of diffusion} \propto \frac{1}{\sqrt{\rho}};$$

and in words: *the rate of diffusion of a gas is inversely proportional to the square root of its density.* This is *Graham's Law.*

Graham's law of diffusion is readily explained by the kinetic theory. At the same kelvin temperature T, the mean kinetic energies of the molecules of different gases are equal, since

$$\tfrac{1}{2}m\overline{c^2} = \tfrac{3}{2}kT$$

and k is a universal constant. Therefore, if the subscripts A and H denote air and hydrogen respectively,

$$\tfrac{1}{2}m_A\overline{c_A}^2 = \tfrac{1}{2}m_H\overline{c_H}^2,$$

so

$$\frac{\overline{c_A}^2}{\overline{c_H}^2} = \frac{m_H}{m_A}.$$

At a given temperature and pressure, the density of a gas, ρ, is proportional to the mass of its molecule, m, since equal volumes contain equal numbers of molecules.

Therefore

$$\frac{m_H}{m_A} = \frac{\rho_H}{\rho_A},$$

so

$$\frac{\overline{c_A}^2}{\overline{c_H}^2} = \frac{\rho_H}{\rho_A}$$

$$\therefore \quad \frac{\sqrt{\overline{c_A}^2}}{\sqrt{\overline{c_H}^2}} = \frac{\sqrt{\rho_H}}{\sqrt{\rho_A}} \qquad \qquad . \qquad . \qquad . \qquad . \qquad (1)$$

The average speed of the molecules of a gas is roughly equal to—and strictly proportional to—the square root of its mean square speed. Equation (1) therefore shows that the average molecular speeds are inversely proportional to the square roots of the densities of the gases. And so it explains why the rates of diffusion—which depend on the molecular speeds—are also inversely proportional to the square roots of the densities.

Example

Helium gas occupies a volume of 0.04 m³ at a pressure of 2×10^5 Pa (N m⁻²) and temperature 300 K.

Calculate (i) the mass of helium, (ii) the r.m.s. speed of its molecules, (ii) the r.m.s. speed at 432 K when the gas is heated at constant pressure to this temperature, (iv) the r.m.s. speed of hydrogen molecules at 432 K. (Relative molecular mass of helium and hydrogen = 4 and 2 respectively, molar gas constant = 8.3 J mol⁻¹ K⁻¹.)

(i) *Mass* For n mols, $pV = nRT$

So

$$n = \frac{pV}{RT} = \frac{2 \times 10^5 \times 0.04}{8.3 \times 300} = 3.2.$$

Hence   mass of helium = 3.2×4 g = 12.8 g.

(ii) *r.m.s. speed* Pressure $p = 2 \times 10^5$ Pa

$$\text{density } \rho = \frac{\text{mass}}{\text{volume}} = \frac{12.8 \times 10^{-3} \text{ kg}}{0.04 \text{ m}^3} = 0.32 \text{ kg m}^{-3}.$$

So   r.m.s. speed $= \sqrt{\dfrac{3p}{\rho}}$

$$= \sqrt{\frac{3 \times 2 \times 10^5}{0.32}} = 1369 \text{ m s}^{-1}.$$

(iii) *Temperature 432 K*   Since r.m.s. speed $\propto \sqrt{T}$, the new value c_r at 432 K is given by

$$\frac{c_r}{1369} = \sqrt{\frac{432}{300}} = \sqrt{1.44} = 1.2.$$

So $c_r = 1.2 \times 1369 = 1643$ m s⁻¹.

(iv) *Hydrogen* One mole of hydrogen has a mass of 2 g and one mole of helium has a mass of 4 g. So ratio of molar masses = 2:4 = 1:2.

But r.m.s. speed at a given temperature $\propto 1/\sqrt{M}$, where M is the molar mass. So at 432 K,

$$\text{r.m.s. speed of hydrogen molecules} = \sqrt{2} \times 1643$$

$$= 2324 \text{ m s}^{-1}.$$

Distribution of Molecular Speeds

So far we have used the 'root-mean-square' speed and the 'mean' speed of the large number of molecules in a given mass of gas. The actual distribution of the speeds among the numerous molecules can be investigated by an apparatus whose principle is illustrated in Fig. 11.3 (i) and from which all the air is evacuated.

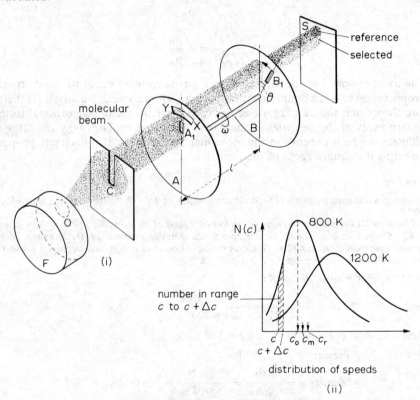

Fig. 11.3 Distribution of molecular speeds

A furnace F maintains a sample of molten metal at a constant high temperature *T*. Molecules from the vapour emerge from an opening O in the furnace and pass through a narrow collimator slit C which produces a parallel beam of molecules. This beam is incident on a wheel A, which rotates with the same angular speed as another wheel B on the same axle and is distant *l* from B. The wheel A has a narrow slit. A_1, and above it a wide slit YX through which all the vapour molecules may pass. The wheel B has a narrow slit B_1 in it which is displaced from the slit A_1 by an angle θ when both wheels are viewed end-on.

With B_1 originally displaced by an angle θ from A_1, both wheels are rotated at the same high angular velocity ω. Only molecules emerging from A_1, and which cross the distance l in the same time as the wheel B takes to turn through an angle θ, will pass through the slit B_1. These molecules have a velocity v given by

$$\frac{l}{v} = \frac{\theta}{\omega},$$

so

$$v = \frac{\omega l}{\theta}.$$

The molecules passing through A_1 and B_1 are thus a 'velocity selected' beam.

The slit YX, however, is wide enough to allow molecules of practically all speeds to pass through itself and through B_1.

These molecules form an 'unselected' or reference beam. The 'velocity selected' and reference beams are incident on a surface S cooled by liquid nitrogen and the ratio of the intensities of the two beams is a measure of the fraction of all the molecules which have velocities close to v in magnitude. Thus by rotating the wheels at different speeds, measurements can be made of the distribution of velocities among the molecules. Further, by varying the temperature T of the furnace, the molecular distribution can be found at different temperatures.

Maxwellian Distribution

The results are shown roughly in Fig. 11.3 (ii). The quantity $N(c)$ plotted on the vertical axis represents the number of molecules ΔN in a small range of speeds c to $c + \Delta c$, so that $\Delta N = N(c)$. Δc = area of strip shaded in Fig. 11.3 (ii). The distribution of velocities agrees with the Maxwellian distribution derived theoretically from advanced kinetic theory of gases. In Fig. 11.3 (ii), the value c_0 at the maximum of a curve is called the *most probable velocity*, because more molecules have velocities in the range c_0 to $c_0 + \Delta c$ than in any other similar range, Δc, of velocities. The value c_m is the mean velocity and c_r is the root-mean-square. For a Maxwellian distribution, calculation shows that

$$c_0 : c_m : c_r = 1 \cdot 00 : 1 \cdot 13 : 1 \cdot 23.$$

As already seen, the *mean-square velocity* is concerned in macroscopic gas properties such as 'pressure' and 'specific heat capacity'. This is because (a) the pressure of a gas is proportional to the momentum change per molecule and to the number of molecules per second arriving at the walls of the container, which together are proportional to the square of the individual velocities, (b) the specific heat capacity is proportional to the energy gained, which is proportional to $\frac{1}{2}M\overline{c^2}$ where M is the mass of gas. On the other hand, the *mean velocity* is concerned in macroscopic gas properties such as 'diffusion' through porous partitions, since this rate of diffusion is proportional to the mean velocity, and in 'viscosity', since there is a transfer of momentum from fast to slow moving gas layers in gas flow through pipes, for example.

Thermal Agitations and Internal Energy

The random motion of the molecules of a gas, whose kinetic energy depends upon the temperature, is often called the *thermal agitation* of the molecules. And the kinetic energy of the thermal agitation is called the *internal energy* of the gas, which we discussed earlier on p. 205.

The internal energy of a gas depends on the number of atoms in its molecule. A gas whose molecules consist of single atoms is said to be monatomic: for example, chemically inert gases and metallic vapours, Hg, Na, He, Ne, A. A gas with two atoms to the molecule is said to be diatomic: O_2, H_2, N_2, Cl_2, CO. And a gas with more than two atoms to the molecule is said to be polyatomic: H_2O, O_3, H_2S, CO_2, CH_4. The molecules of a monatomic gas we may regard as points, but those of a diatomic gas we must regard as 'dumb-bells', and those of a polyatomic gas as more complicated structures (Fig. 11.4). A molecule which extends appreciably in space—a diatomic or polyatomic molecule—has an appreciable moment of inertia. It may therefore have kinetic energy of rotation, as well as of translation. A monatomic molecule, however, must have a much smaller moment of inertia than a diatomic or polyatomic; its kinetic energy of rotation can therefore be neglected.

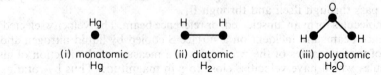

Fig. 11.4 Types of gas molecule

Figure 11.5 shows a monatomic molecule whose velocity c has been resolved into its components u, v, w along the x, y, z axes:

$$c^2 = u^2 + v^2 + w^2.$$

The x, y, z axes are called the molecules' degrees of freedom: they are three directions such that the motion of the molecule along any one is independent of its motion along the others.

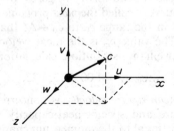

Fig. 11.5 Components of velocity

If we average the speed c, and the components u, v, w, over all the molecules in a gas, we have

$$\overline{c^2} = \overline{u^2} + \overline{v^2} + \overline{w^2}.$$

And since the molecules do not pile up in any corner of the vessel containing the gas, their average velocities in all directions must be the same. We may therefore write

$$\overline{u^2} = \overline{v^2} = \overline{w^2},$$

so

$$\overline{u^2} = \overline{v^2} = \overline{w^2} = \tfrac{1}{3}\overline{c^2}.$$

The average kinetic energy of a molecule of the gas is given by (p. 223)

$$\tfrac{1}{2}m\overline{c^2} = \tfrac{3}{2}kT.$$

Therefore the average kinetic nergy of a monatomic molecule, in each degree of freedom, is

$$\tfrac{1}{2}m\overline{u^2} = \tfrac{1}{2}m\overline{v^2} = \tfrac{1}{2}m\overline{w^2} = \tfrac{1}{2}kT.$$

Thus the molecule has kinetic energy $\tfrac{1}{2}kT$ per degree of freedom.

Rotational Energy

Let us now consider a diatomic or polyatomic gas. When two of its molecules collide, they will, in general, tend to rotate, as well as to deflect each other. In some collisions, energy will be transferred from the translations of the molecules to their rotations; in others, from the rotations to the translations. We may assume, then, that the internal energy of the gas is shared between the rotations and translations of its molecules.

To discuss the kinetic energy of rotation, we must first extend the idea of degrees of freedom to it. A diatomic molecule can have kinetic energy of rotation about any axis at right-angles to its own. Its motion about any such axis can be resolved into motions about two such axes at right-angles to each other, Fig. 11.6 (i). Motions about these axes are independent of each other, and a diatomic molecule therefore has two degrees of rotational freedom. A polyatomic molecule, unless it happens to consist of molecules all in a straight line, has no axis about which its moment of inertia is negligible. It can therefore have kinetic energy of rotation about three mutually perpendicular axes, Fig. 11.6 (ii). It has three degrees of rotational freedom.

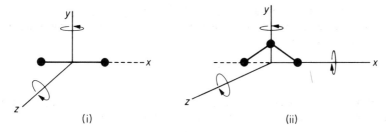

Fig. 11.6 Rotational energy of molecules

We have seen that the internal energy of a gas is shared between the translations and rotations of its molecules. Maxwell assumed that the average kinetic energy of a molecule, in each degree of freedom, *rotational as well as translational*, was $\tfrac{1}{2}kT$. This assumption is called the *principle of equipartition of energy*; experiment shows, as we shall find, that it is true at room temperature and above. At very low temperatures, when the gas is near liquefaction, it fails. At ordinary temperatures, then, we have:

average k.e. of monatomic molecule $= \dfrac{3}{2}kT$ (trans.);

average k.e. of diatomic molecule $= \dfrac{3}{2}kT$ (trans.) $+ \dfrac{2}{2}kT$ (rot.) $= \dfrac{5}{2}kT$;

average k.e. of polyatomic molecule $= \dfrac{3}{2}kT$ (trans.) $+ \dfrac{3}{2}kT$ (rot.) $= \dfrac{6}{2}kT$.

Internal Energy of a Gas

From the average kinetic energy of its molecules, we can find the internal energy of a mole of gas. Its internal energy, U, is the total kinetic energy of its molecules' random motions; so

$$U = N_A \times \text{average k.e. of molecule,}$$

since there are N_A molecules per mole where N_A is the Avogadro constant. For a monatomic gas, therefore,

$$U = \frac{3}{2}N_A kT \text{ (monatomic).}$$

The constant k is the gas constant per molecule; the product $N_A k$ is therefore the gas constant R for one mole. So the internal energy per mole is

$$U = \frac{3}{2}RT \text{ (monatomic)} \qquad . \qquad . \qquad . \qquad . \qquad (1)$$

Similarly, for 1 mole of a diatomic gas,

$$U = \frac{5}{2}N_A kT = \frac{5}{2}RT \qquad . \qquad . \qquad . \qquad . \qquad (2)$$

Similar relations to (1) and (2) also apply to the internal energy of unit mass (1 kg) of a gas. In this case $U = 3rT/2$ (monatomic) and $U = 5rT/2$ (diatomic), where r is the gas constant per unit mass.

Molar Heat Capacities

We have seen that the internal energy U of a gas, at a given temperature, depends on the number of atoms in its molecule. For 1 mole of a monatomic gas its value is, from equation (1),

$$U = \frac{3}{2}RT \qquad . \qquad . \qquad . \qquad . \qquad (i)$$

The molar heat capacity at constant volume is the heat required to increase the internal energy of 1 mole of the gas through 1 K. So, from (1),

$$C_V = \frac{3}{2}R.$$

The molar heat capacity of a monatomic gas at constant pressure is therefore (p. 208)

$$C_p = C_V + R = \frac{3}{2}R + R = \frac{5}{2}R.$$

Ratio of Molar Heat Capacities

Let us now divide C_p by C_V; their quotient is called the ratio of the molar heat capacities, and is denoted by γ (p. 211).

For a monatomic gas, its value is

$$\gamma = \frac{C_p}{C_V} = \frac{\frac{5}{2}R}{\frac{3}{2}R} = \frac{5}{3} = 1.667.$$

Similar relations apply to the principal specific heat capacities of a monatomic gas. In this case, we have, if r is the gas constant per kg,

$$c_V = \frac{3}{2}r, \; c_p = \frac{5}{2}r, \text{ and } \gamma = \frac{c_p}{c_V} = \frac{5}{3}.$$

For a diatomic molecule, from equation (2),

$$U = \frac{5}{2}RT \qquad \cdot \qquad \cdot \qquad \cdot \qquad \cdot \qquad \text{(ii)}$$

So
$$C_V = \frac{5}{2}R$$

and
$$C_p = C_V + R = \frac{7}{2}R.$$

Hence
$$\gamma = \frac{C_p}{C_V} = \frac{7}{5} = 1{\cdot}40.$$

Similarly, we find that the principal specific heats of a diatomic gas are given by

$$c_V = \frac{5}{2}r \text{ and } c_p = \frac{7}{2}r, \text{ so } \gamma = \frac{c_p}{c_V} = 1{\cdot}40.$$

In general, if the molecules of a gas has f degrees of freedom, the average kinetic energy of a molecule is $f \times \frac{1}{2}kT$ (p. 229). So

$$U = \frac{f}{2}RT$$

$$C_V = \frac{f}{2}R, \ C_p = C_V + R = \left(\frac{f}{2} + 1\right)R,$$

and
$$\gamma = \frac{C_p}{C_V} = \frac{\frac{f}{2} + 1}{\frac{f}{2}} = 1 + \frac{2}{f} \qquad \cdot \qquad \cdot \qquad \cdot \qquad \cdot \qquad \text{(iii)}$$

The ratio of the molar heat capacities of a gas thus gives us a measure of the number of atoms in its molecule, at least when the number is less than three. This ratio is fairly easy to measure, as we shall see later in this chapter. The poor agreement between the observed and theoretical values of γ for some of the polyatomic gases shows that, in its application to such gases, the theory is oversimple.

Mean Free Path

Although gas molecules move with high speeds such as several hundred metres per second, it still takes some time for two gases, initially separated, to diffuse into each other. This is due to numerous *collisions* which a molecule makes with other molecules surrounding it. Although the distance moved between successive collisions is not constant, we can consider the molecules of a gas under given conditions to have some average *mean free path*, λ.

Molecules have size. As an approximation, suppose a molecule in a gas is represented by a rigid sphere of diameter σ, and let us assume that a molecule moves a distance l while other molecules remain stationary. Then molecules such as A and B, situated at a distance σ from C, are those with which C *just* makes a collision, Fig. 11.7. Consequently C makes a collision with all those molecules whose centres lie inside a volume $\pi\sigma^2 l$.

Suppose n is the number of molecules per unit volume in the gas. Then the number of collisions = the number of molecules in a volume $\pi\sigma^2 l$.

$$\therefore \lambda = \frac{\text{distance moved}}{\text{number of collisions}} = \frac{l}{n\pi\sigma^2 l}$$

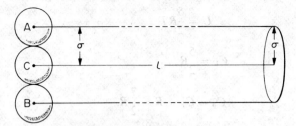

Fig. 11.7 Mean free path of molecule

$$\therefore \lambda = \frac{1}{n\pi\sigma^2} \qquad \qquad (1)$$

Since the molecules are not stationary as we have assumed but moving about in all directions, the chance of a collision by a molecule is greater. Taking this into account, the mean free path can be shown to be $\sqrt{2}$ less than that given in (1). Thus

$$\lambda = \frac{1}{\sqrt{2}n\pi\sigma^2} \qquad \qquad (2)$$

The relation in (2) shows that the mean free path λ decreases in inverse proportion to the number of molecules per unit volume, n, or the pressure of the gas.

Calculation of Mean Free Path

Suppose that, for hydrogen, $\sigma = 3 \times 10^{-10}$ m approximately. At a standard pressure of 760 mmHg, a mole of the gas (2 g) has 6×10^{23} molecules approximately (the Avogadro constant) and occupies a volume of about 22·4 litres or $22·4 \times 10^{-3}$ m³. Then, from (2),

$$\lambda = \frac{22·4 \times 10^{-3}}{\sqrt{2} \times 6 \times 10^{23} \times \pi \times 9 \times 10^{-20}}$$

$$= 9 \times 10^{-8} \text{ m (approx.)}.$$

Thus the mean free path of hydrogen molecules at standard atmospheric pressure is about 10^{-7} m. Since the mean speed, the average distance travelled per second, of the molecule is about 2×10^3 m s^{-1}, we see that a molecule makes about 2×10^{10} collisions per second.

At extremely low pressures the mean free path becomes comparable with the dimensions of the container and no meaning is then attached to the term 'mean free path' of the molecules.

Viscosity of Gas

The viscosity of a gas is due to a transfer of momentum by molecules (p. 168). On kinetic theory, the coefficient of viscosity η of a gas can be shown to be given by the formula

$$\eta = \frac{1}{3}mn\bar{c}\lambda \qquad \qquad (3)$$

where n is the number of molecules per unit volume, m is the mass of a molecule and $\bar{c}$ is the mean velocity and λ is the mean free path. But, from (2), $\lambda \propto 1/n$. It follows from (3) that η is *independent* of the pressure. This surprising result was shown to be true experimentally for normal pressures by Maxwell. He found that the damping of oscillations of a suspended horizontal disc was independent of the pressure of the gas when this was varied. The viscosity experiment is considered as one piece of sound evidence for the kinetic theory of gases, which leads to the relation in (3).

Exercises 11

1. Show that the relation $p = \frac{1}{3}\rho\overline{c^2}$ is dimensionally correct, where p is the pressure of a gas of density ρ and $\overline{c^2}$ is the mean square velocity of all its molecules.
   Write down (i) two assumptions made in deriving this relation from a simple kinetic theory, (ii) the meaning of (a) mean velocity and (b) mean square velocity.
2. Calculate the root-mean-square speed at 0°C of (i) hydrogen molecules and (ii) oxygen molecules, assuming 1 mole of a gas occupies a volume of 2×10^{-2} m³ at 0°C and 10^5 N m^{-2} pressure. (Relative molecular masses of hydrogen and oxygen = 2 and 32 respectively.)
3. Assuming helium molecules have a root-mean-square speed of 900 m s^{-1} at 27°C and 10^5 N m^{-2} pressure, calculate the root-mean-square speed at (i) 127°C and 10^5 N m^{-2} pressure, (ii) 27°C and 2×10^5 N m^{-2} pressure.
4. Using the kinetic theory, show that (i) the pressure of an ideal gas is doubled when its volume is halved at constant temperature, (ii) the pressure of an ideal gas decreases when it expands in a thermally insulated vessel.
5. Explain what is meant by the *root mean square velocity* of the molecules of a gas. Use the concepts of the elementary kinetic theory of gases to derive an expression for the root mean square velocity of the molecules in terms of the pressure and density of the gas.
   Assuming the density of nitrogen at s.t.p. to be 1·251 kg m^{-3}, find the root mean square velocity of nitrogen molecules at 127°C. (L.)
6. Describe the simple kinetic theory model of an ideal gas, stating four assumptions on which your model is based. (Derivation of equation *not* required.)
   This theory leads to the equation $p = \frac{1}{3}\rho\overline{c^2}$ where p is the pressure of the gas, ρ is its density and $\overline{c^2}$ is the mean square speed of the molecules. Explain how this equation is related to the equation of state, $pV = RT$, for an ideal gas?
   Describe how you would investigate experimentally the relationship between the volume of a mass of dry air maintained at atmospheric pressure and its temperature as the temperature is varied over the range 0°C to 100°C as measured on a mercury-in-glass thermometer. Why does the result of your experiment correspond closely to that which would be expected if an ideal gas could have been used? (L.)
7. Write down *four* assumptions about the properties and behaviour of molecules that are made in the kinetic theory in order to define an ideal gas. On the basis of this theory derive an expression for the pressure by an ideal gas.
   Use the kinetic theory to explain why hydrogen molecules diffuse out of a porous container into the atmosphere even when the pressure of the hydrogen is equal to the atmospheric pressure outside the container.
   Air at 273 K and 1·01 × 10^5 N m^{-2} pressure contains 2·70 × 10^{25} molecules per cubic metre. How many molecules per cubic metre will there be at a place where the temperature is 223 K and the pressure is 1·33 × 10^{-4} N m^{-2}? (L.)
8. Use the kinetic theory to derive an expression for the pressure exerted by an ideal gas, stating clearly the assumptions which you make.
   What direct evidence is available to justify the belief that the molecules of a gas are in a continual state of random motion? Use the kinetic theory to explain qualitatively: (a) Boyle's law, (b) why the pressure of a gas increases if the temperature is increased at constant volume, (c) why the temperature of a gas rises if it is compressed in a thermally insulated container. (L.)
9. The kinetic theory of gases leads to the equation $p = \frac{1}{3}\rho\overline{c^2}$, where p is the *pressure*,

ρ is the *density* and $\overline{c^2}$ is the *mean square molecular speed*. Explain the meaning of the terms in italics and list the simplifying assumptions necessary to derive this result. Discuss how this equation is related to Boyle's Law.

Air may be taken to consist of 80% nitrogen molecules and 20% oxygen molecules of relative molecular masses 28 and 32 respectively. Calculate (a) the ratio of the root mean square speed of nitrogen molecules to that of oxygen molecules in air, (b) the ratio of the partial pressures of nitrogen and oxygen molecules in air, and (c) the ratio of the root mean square speed of nitrogen molecules in air at 10°C to that at 100°C (*O. & C.*)

10. Calculate the pressure in mm of mercury exerted by hydrogen gas if the number of molecules per cm^3 is 6.80×10^{15} and the root mean square speed of the molecules is 1.90×10^3 m s^{-1}. Comment on the effect of a pressure of this magnitude (a) above the mercury in a barometer tube; (b) in a cathode ray tube. (Avogadro constant = 6.02×10^{23} mol^{-1}. Relative molecular mass of hydrogen = 2.02.) (*N.*)

11. Define the term *mean square speed* as applied to the molecules of a gas. Explain why the pressure exerted by a gas is proportional to the mean square speed of its molecules.

Figure 11A shows apparatus designed to measure speeds of molecules. Atoms of vapour of a heavy metal emerge from oven O into an evacuated space. The atoms pass through fixed slit S′ in a well-defined beam and enter radially through slit S″ in the curved surface of a cylindrical drum D. When the drum is stationary the atoms strike the inner surface of the drum, giving a well-defined trace T. The drum is then set into rotation about its axis C and maintained at a high constant speed until a second trace has been produced. State and explain the ways in which this trace differs from the first trace.

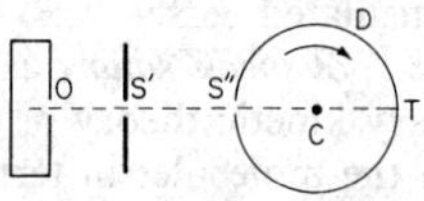

Fig. 11A

Oven O contains bismuth vapour (atomic weight 208) and is maintained at 1500°C. Calculate the root mean square speed of the atoms in the oven, assuming the vapour to be monatomic. (Take the gas constant R to be 8.314 J mol^{-1} K^{-1}.)

At what angular speed must the drum rotate if the traces for atoms having speeds of 400 and 800 m s^{-1} are to be separated by a distance of 10 mm on the drum surface? The drum diameter is 0.5 m. (*O. & C.*)

12. Use a simple treatment of the kinetic theory of gases, stating any assumptions you make, to derive an expression for the pressure exerted by a gas on the walls of its container. Thence deduce a value for the root mean square speed of thermal agitation of the molecules of helium in a vessel at 0°C. (Density of helium at s.t.p. = 0.1785 kg m^{-3}; 1 atmosphere = 1.013×10^5 N m^{-2}.)

If the total translational kinetic energy of all the molecules of helium in the vessel is 5×10^{-6} joule, what is the temperature in another vessel which contains twice the mass of helium and in which the total kinetic energy is 10^{-5} joule? (Assume that helium behaves as a perfect gas.) (*O. & C.*)

13. What do you understand by the term *ideal gas*? Describe a molecular model of an ideal gas and derive the expression $p = \frac{1}{3}\rho c^2$ for such a gas. What is the reasoning which leads to the assertion that the temperature of an ideal monatomic gas is proportional to the mean kinetic energy of its molecules?

The Doppler broadening of a spectral line is proportional to the r.m.s. speed of the atoms emitting light. Which source would have less Doppler broadening, a mercury lamp at 300 K or a krypton lamp at 77 K? (Take the mass numbers of Hg and Kr to be 200 and 84 respectively.)

What causes the behaviour of real gases to differ from that of an ideal gas? Explain qualitatively why the behaviour of all gases at very low pressures approximates to that of an ideal gas. (*O. & C.*)

14. Explain in terms of the kinetic theory what happens to the energy supplied to a gas when it is heated (a) at constant volume, (b) at constant pressure.

Deduce the total kinetic energy of the molecules in 1 g of an ideal gas at 0°C if its specific heat capacity at constant volume is 600 J kg^{-1} K^{-1}.

An iron rod 1 metre long is heated without being allowed to expand lengthwise. When the temperature has been raised by 500°C the rod exerts a force of $1·2 \times 10^4$ newtons on the walls preventing its expansion. How much work could be obtained if it were possible to maintain it at this temperature and allow it to expand gradually until free from stress? [Linear expansivity of iron = $1·0 \times 10^{-5}$ K^{-1}.] (C.)

15. The First Law of Thermodynamics is usually expressed in the form $\Delta Q = \Delta U + \Delta W$, where ΔQ represents heat supplied to the system, ΔU represents the increase in the internal energy of the system, and ΔW represents the external work done by the system.

Assuming the formula $pV = \frac{1}{3}mN_A\overline{c^2}$:

(a) evaluate ΔU and ΔW when one mole of an ideal monatomic gas has its temperature increased by T K (i) at constant volume, (ii) at constant pressure; (b) obtain values for the specific heat capacities c_V and c_p at constant volume and constant pressure respectively for argon, regarded as an ideal monatomic gas of relative molecular mass (molecular weight) 40; (c) find (i) the heat capacity at constant volume of 0·5 m^3 of argon at pressure 10^5 N m^{-2} when its density is 1·60 kg m^{-3}, (ii) the external work which must be done on this volume of gas in order to compress it isothermally to a volume of 0·05 m^3. (Take the value of the molar gas constant R to be 8·3 J mol^{-1} K^{-1}; $\log_e 10 = 2·303$.) (O.)

12 Changes of State. Vapours. Real Gases

Solid to Liquid: Fusion

The Solid State

Substances exist in the solid, liquid or gaseous state. In the *solid state*, a body has a regular, geometrical structure. Sometimes this structure gives the body a regular outward form, as in a crystal of alum; sometimes, as in a strand of wool, it does not. But X-rays can reveal to us the arrangement of the individual atoms or molecules in a solid; and whether the solid is wool or alum, we find that its atoms or molecules are arranged in a regular pattern. This pattern we call a *space-lattice*; its form may be simple, as in metals, or complicated, as in wool, proteins, and other chemically complex substances.

We consider that the atoms or molecules of a solid are vibrating about their mean positions in its space-lattice. And we consider that the kinetic energy of their vibrations increases with the temperature of the solid: its increase is the heat energy supplied to cause the rise in temperature. When the temperature reaches the melting-point, the solid liquefies. Lindemann has suggested that, at the melting-point, the atoms or molecules vibrate so violently that they collide with one another. The attractive forces between them can then no longer hold them in their pattern, the space-lattice collapses, and the solid melts. The work necessary to overcome the forces between the atoms or molecules of the solid, that is, to break-up the space-lattice, is the latent heat of melting or fusion.

The Liquid State

In the liquid state, a body has no form, but a fixed volume. It adapts itself to the shape of its vessel, but does not expand to fill it. We consider that its molecules still dart about at random, as in the gaseous state, and we consider that their average kinetic energy rises with the liquid's temperature. But we think that they are now close enough together to attract one another—by forces of an electrical nature.

Any molecule approaching the surface of the liquid experiences a resultant force opposing its escape (see p. 138, Surface Tension). Nevertheless, some molecules do escape, as is shown by the fact that the liquid evaporates: even in cold weather, a pool of water does not last for ever. The molecules which escape are the fastest, for they have the greatest kinetic energy, and therefore the greatest chance of overcoming the attraction of the others. Since the fastest escape, the slower, which remain, begin to predominate: the average kinetic energy of the molecules falls, and the liquid cools. The faster a liquid evaporates, the colder it feels on the hand—petrol feels colder than water, water feels colder than paraffin. To keep a liquid at constant temperature as it evaporates, heat must be supplied to it; the heat required is the latent heat of evaporation.

Melting and Freezing

When a solid changes to a liquid, we say it undergoes a *change of state* or *phase*. Pure crystalline solids melt and freeze sharply. If, for example, paradichlor-

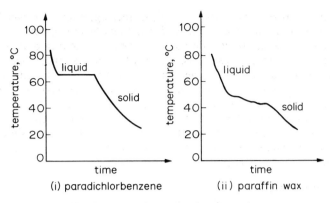

Fig. 12.1 Cooling curves, showing freezing

benzene is warmed in a test tube it melts, and then allowed to cool, its temperature falls as shown in Fig. 12.1 (i). A well-defined plateau in the cooling curve indicates the freezing (or melting) point. While the substance is freezing or solidifying, it is evolving its latent heat of fusion, which compensates for the heat lost by cooling, and its temperature does not fall. An impure substance such as paraffin wax, on the other hand, has no definite plateau on its cooling curve; it is a mixture of several waxes, which freeze out from the liquid at slightly different temperatures, Fig. 12.1 (ii).

Supercooling

If we try to find the melting-point of hypo from its cooling curve, we generally fail; the liquid goes on cooling down to room temperature. But if we drop a crystal of solid hypo into the liquid the temperature rises to the melting-point of hypo, and the hypo starts to freeze. While the hypo is freezing, its temperature stays constant at the melting-point; when all the hypo has frozen, its temperature starts to fall again, Fig. 12.2.

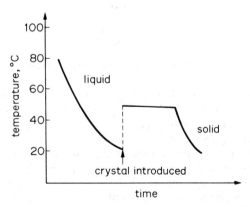

Fig. 12.2 Cooling curve of hypo

The cooling of a liquid below its freezing-point is called *supercooling*; the molecules of the liquid lose their kinetic energy as it cools, but do not take up the rigid geometric pattern of the solid. Shaking or stirring the liquid, or dropping grit or dust into it, may cause it to solidify; but dropping in a crystal of its own solid is more likely to make it solidify. As soon as the substance begins to solidify, it returns to its melting (or freezing) point. *The melting-point is the only temperature at which solid and liquid can be in equilibrium.*

No one has succeeded in warming a solid above its melting-point—or, if he has, he has failed to report his success. We may therefore suppose that to super-heat a solid is not possible; and we need not be surprised. For the melting-point of a solid is the temperature at which its atoms or molecules have enough kinetic energy to break up its crystal lattice: as soon as the molecules are moving fast enough, they burst from their pattern. On the other hand, when a liquid cools to its melting-point, there is no particular reason why its molecules should spon-taneously arrange themselves. They may readily do so, however, around a crystal in which their characteristic pattern is already set up.

Pressure and Melting

The melting-point of a solid is affected by increase of pressure. If we run a copper wire over a block of ice, and hang a heavy weight from it, as in Fig. 12.3, we find that the wire slowly works through the block. It does not cut its way through, for the melted ice freezes up behind it; the pressure of the wire makes the ice under it melt, and above the wire, where the pressure is released, the melted ice freezes again. The freezing again after melting by pressure is called *regelation*.

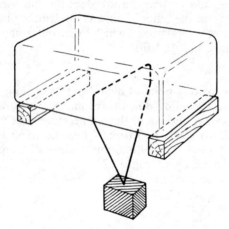

Fig. 12.3 Melting of ice under pressure

This experiment shows that increasing the pressure on ice makes it melt more readily; that is to say, *it lowers the melting-point* of the ice. We can understand this effect when we remember that ice shrinks when it melts (see p. 106); pressure encourages shrinking, and therefore melting.

The fall in the melting-point of ice with increase in pressure is small: 0·0072°C per atmosphere. It is interesting, because it explains the slipperiness of ice; skates for example, are hollow ground, so that the pressure on the line of contact is very high, and gives rise to a lubricating film of water. Ice which is much colder than 0°C is not slippery, because to bring its melting-point down to its actual temperature would require a greater pressure than can be realised. Most sub-stances swell on melting; an increase of pressure opposes the melting of such substances, and raises their melting-point.

Liquid to Gas: Evaporation

Evaporation differs from melting in that it takes place at all temperatures; as long as the weather is dry, a puddle will always clear away. In cold weather the puddle lasts longer than in warm, as the rate of evaporation falls rapidly with the temperature.

Solids as well as liquids evaporate. Tungsten evaporates from the filament of an electric lamp, and blackens its bulb; the blackening can be particularly well seen on the headlamp bulb of a bicycle dynamo set, if it has been frequently overrun through riding down-hill. The rate of evaporation of a solid is negligible at temperatures well below its melting-point, as we may see from the fact that bars of metal do not gradually disappear.

Saturated and Unsaturated Vapours

Figure 12.4 (i) shows an apparatus with which we can study vapours and their pressures. A is a glass tube, about a metre long, dipping in a mercury trough and backed by a scale S. Its upper end carries a bulb B, which is fitted with three taps T, of which T_1 and T_2 should be as close together as possible. Above T_1 is a funnel F. With T_1 closed but T_2 open, we evacuate the bulb and tube through T_3, with a rotary pump. If the apparatus is clean, the mercury in A rises to the barometer height H. Meanwhile we put some ether in the funnel F. When the apparatus is evacuated, we close T_3 and T_2. We now open and close T_1, so that a little ether flows into the space C. Lastly, we open T_2, so

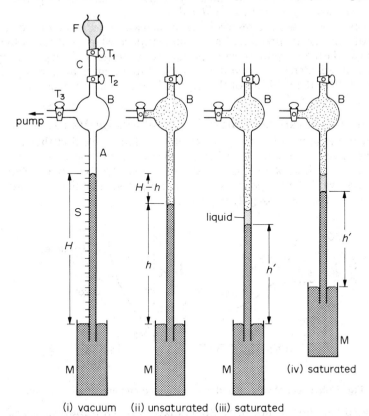

(i) vacuum (ii) unsaturated (iii) saturated

Fig. 12.4 Apparatus for studying vapours

that the ether evaporates into the bulb B. As it does so, the mercury in A falls, showing that the ether-vapour is exerting a pressure, Fig. 12.4 (ii). *If h is the new height of the mercury in A, then the pressure of the vapour in mm of mercury is equal to $H - h$.*

By closing T_2, opening and closing T_1, and then opening T_2 again, we can introduce more ether into the space B. At first, we find that, with each introduction, the pressure of the vapour, $H - h$, increases. But we reach a point at which the introduction of more ether does not increase the pressure, the height of the mercury column remains constant at h'. At this point we notice that liquid ether appears above the mercury in A, Fig. 12.4 (iii). We say that the vapour in B is now *saturated*; *a saturated vapour is one that is in contact with its own liquid.*

Before the liquid appeared in the above experiment, the pressure of the vapour could be increased by introducing more ether, and we say that the vapour in B was then *unsaturated*.

Behaviour of Saturated Vapour

To find out more about the saturated vapour, we may try to expand or compress it. We can try to compress it by raising the mercury reservoir M. But when we do, we find that the height h' does not change: the pressure of the vapour, $H - h'$, is therefore constant, Fig. 12.4(iv). The only change we notice is an increase in the volume of liquid above the mercury. We conclude, therefore, that *reducing the volume of a saturated vapour does not increase its pressure, but merely makes some of it condense to liquid.*

Similarly, if we lower the reservoir M to increase the volume of the vapour we do not decrease its pressure. Its pressure stays constant, but the volume of liquid above the mercury now decreases; liquid evaporates, and keeps the vapour saturated. If we increase the volume of the vapour until all the liquid has evaporated, then the pressure of the vapour begins to fall, because it becomes unsaturated (see Fig. 12.5 (i)).

Effect of Temperature: Validity of Gas-laws for Vapours

If we warm the apparatus of Fig. 12.4 with our hands, we find that the ether above the mercury evaporates further and the pressure of the vapour increases. Experiments which we shall describe later show that the pressure of a saturated vapour rises, with the temperature, at a rate much greater than that given by Charles's law. Its rise is roughly exponential.

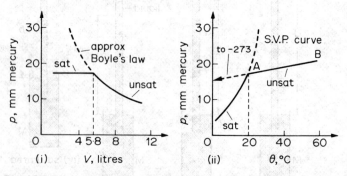

Fig. 12.5 Effect of volume and temperature on pressure of water vapour

Saturated vapours do not obey Boyle's law: *their pressure is independent of their volume.* Unsaturated vapours obey Boyle's law roughly, as they also obey roughly Charles's law, Fig. 12.5 (i). Vapours, saturated and unsaturated, are gases because they spread throughout their vessels; but we find it convenient to distinguish them by name from gases such as air, which obey Charles's and Boyle's laws closely. We shall elaborate this distinction later.

Figure 12.5 (ii) shows the effect of heating a saturated vapour. More and more of the liquid evaporates, and the pressure rises very rapidly. As soon as all the liquid has evaporated, however, the vapour becomes unsaturated, and its pressure rises more steadily along a straight line AB. Well away from the saturated state, the unsaturated vapour obeys the relation $p \propto T$ at constant volume, which is a gas law.

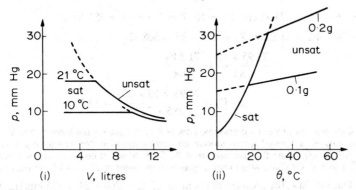

Fig. 12.6 Relationship between pressure, temperature, total mass and volume for water vapour and liquid

Figure 12.6 (i) shows isothermals (p-V curves at constant temperature) for a given mass of liquid and vapour at two temperatures, $\theta_1 = 10°C$, and $\theta_2 = 21°C$. The temperatures are chosen so that the saturated vapour pressure at θ_2 is double that at θ_1. The absolute temperatures are $T_1 = 273 + \theta_1 = 283$ K, and $T_2 = 273 + \theta_2 = 294$ K. Because the saturated vapour pressure rises so rapidly with temperature, the absolute temperature T_2 is not nearly double the absolute temperature T_1. Consequently the isothermals for the unsaturated vapour are fairly close together, as shown; and the change from saturated to unsaturated vapour occurs at a smaller volume at the higher temperature.

Figure 12.6 (ii) shows pressure-temperature curves for a vapour, initially in contact with different amounts of liquid, in equal total volumes. The more liquid present, the greater is the density of the vapour when it becomes unsaturated, and therefore the higher the pressure and temperature at which it does so.

Gas Laws Applied to Mixture of Vapour and Gas

The following examples illustrate how the gas laws are applied to the case of a vapour in a mixture with other gases. Note the following:

(i) *Vapour unsaturated.* Here the gas laws give a good approximation to changes in pressure, volume and temperature;

(ii) *Vapour saturated.* If the vapour is in a mixture with a gas such as air, we apply the gas laws to the *air* after using Dalton's law of partial pressures to allow for the saturated vapour. We do not apply the gas laws to the saturated vapour because its mass changes due to condensation or evaporation as conditions change, and the gas laws apply to constant mass of gas (p. 200).

Examples

1. Describe an experiment which demonstrates that the pressure of a vapour in equilibrium with its liquid depends on the temperature.

A narrow tube of uniform bore, closed at one end, has some air entrapped by a small quantity of water. If the pressure of the atmosphere is 760 mmHg, the equilibrium vapour pressure of water at 12°C and at 35°C is 10·5 mmHg and 42·0 mmHg respectively, and the length of the air column at 12°C is 10 cm, calculate its length at 35°C. (*L.*)

For the given mass of air,

$$\frac{p_1 V_1}{T_1} = \frac{p_2 V_2}{T_2}$$

$p_1 = 76 - 1·05 = 74·95$ cm, $V_1 = 10$, $T = 273 + 12 = 285$ K;

$p_2 = 76 - 4·2 = 71·8$ cm, $T_2 = 273 + 35 = 308$ K;

$$\therefore \frac{74·95 \times 10}{285} = \frac{71·8 V_2}{308}$$

$$\therefore V_2 = \frac{74·95 \times 10 \times 308}{285 \times 71·8} = 11·3.$$

2. State Dalton's law of partial pressures; how is it explained on the kinetic theory? A closed vessel contains air, saturated water-vapour, and an excess of water. The total pressure in the vessel is 760 mmHg when the temperature is 25°C; what will it be when the temperature has been raised to 100°C? (Saturation vapour pressure of water at 25°C is 24 mmHg.) (*C.*)

From Dalton's law, the pressure of the air at 25°C = 760 − 24 = 736 mmHg. Suppose the pressure is p mm at 100°C. Then, since pressure is proportional to absolute temperature for a fixed mass of air, we have

$$\frac{p}{736} = \frac{373}{298},$$

from which $p = 921$ mmHg.

Now the saturation vapour pressure of water at 100°C = 760 mmHg.

$$\therefore \text{ total pressure in vessel} = 921 + 760 = 1681 \text{ mmHg.}$$

Kinetic Theory of Saturation

Let us consider a vapour in contact with its liquid, in an otherwise empty vessel which is closed by a piston, Fig. 12.7. The molecules of the vapour, we suppose, are rushing randomly about, like the molecules of a gas, with kinetic energies whose average value is determined by the temperature of the vapour. They bombard the walls of the vessel, giving rise to the pressure of the vapour, and they also bombard the surface of the liquid.

The molecules of the liquid, we further suppose, are also rushing about with kinetic energies determined by the temperature of the liquid. The fastest of them escape from the surface of the liquid. At the surface, therefore, there are molecules leaving the liquid, and molecules arriving from the vapour. To complete our picture of the conditions at the surface, we suppose that the vapour molecules bombarding it are not reflected—as they are at the walls of the vessel—but are absorbed into the liquid. We may expect them to be, because we consider that molecules near the surface of a liquid are attracted towards the body of the liquid.

We shall assume that the liquid and vapour have the same temperature. Then the proportions of liquid and vapour will not change, if the temperature and the total volume are kept constant. Therefore, at the surface of the liquid, molecules must be arriving and departing at the same rate, and hence evaporation from the liquid is balanced by condensation from the vapour. This state of affairs is called a *dynamic equilibrium.* In terms of it, we can explain the behaviour of a saturated vapour.

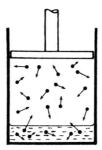

Fig. 12.7 Dynamic equilibrium

The rate at which molecules leave unit area of the liquid depends simply on their average kinetic energy, and therefore on the temperature. The rate at which molecules strike unit area of the liquid, from the vapour, likewise depends on the temperature; but it also depends on the concentration of the molecules in the vapour, that is to say, on the density of the vapour. The density and temperature of the vapour also determine its pressure; the rate of bombardment therefore depends on the pressure of the vapour.

Now let us suppose that we decrease the volume of the vessel in Fig. 12.7 by pushing in the piston. Then we momentarily increase the density of the vapour, and hence the number of its molecules striking the liquid surface per second. The rate of condensation thus becomes greater than the rate of evaporation, and the liquid grows at the expense of the vapour. As the vapour condenses its density falls, and so does the rate of condensation. The dynamic equilibrium is restored when the rate of condensation, and the density of the vapour, have returned to their original values. The pressure of the vapour will then also have returned to its original value. *Thus the pressure of a saturated vapour is independent of its volume.* The proportion of liquid to vapour, however, increases as the volume decreases.

Let us now suppose that we warm the vessel in Fig. 12.7, but keep the piston fixed. Then we increase the rate of evaporation from the liquid, and increase the proportion of vapour in the mixture. Since the volume is constant, the pressure of the vapour rises, and increases the rate at which molecules bombard the liquid. Thus the dynamic equilibrium is restored, at a higher pressure of vapour. The increase of pressure with temperature is rapid, because the rate of evaporation of the liquid increases rapidly—almost exponentially—with the temperature. A small rise in temperature causes a large increase in the proportion of molecules in the liquid moving fast enough to escape from it.

Boiling

A liquid boils when its saturated vapour pressure is equal to the atmospheric pressure. To see that this is true, we take a closed J-shaped tube, with water trapped in its closed limb, Fig. 12.8 (i). We heat the tube in a beaker of water, and watch the water in the J-tube. It remains trapped as at (i) until the water in the beaker is boiling. Then the water in the J-tube comes to the same level in each limb, showing that the pressure of the vapour in the closed limb is equal to the pressure of the air outside, Fig. 12.8 (ii).

The J-tube gives a simple means of measuring the boiling-point of a liquid which is inflammable, or which has a poisonous vapour, or of which only a small quantity can be had. A few drops of the liquid are imprisoned by mercury in the closed limb of the tube, all entrapped air having been shaken out, Fig. 12.8 (iii). The tube is then heated in a bath, and the temperature observed

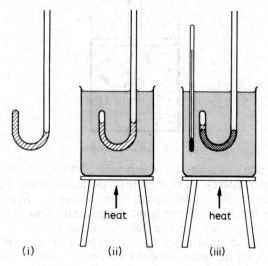

Fig. 12.8 Use of J-tube for boiling point

at which the mercury comes to the same level in both limbs. The bath is warmed a little further, and then a second observation made as the bath cools; the mean of the two observations is taken as the boiling-point of the liquid.

Boiling differs from evaporation in that a liquid evaporates from its surface alone, but it boils throughout its volume. If we ignore the small hydrostatic pressure of the liquid itself, we may say that the pressure throughout a vessel of liquid is the atmospheric pressure. Therefore, when the saturated vapour pressure is equal to the atmospheric pressure, a bubble of vapour can form anywhere in the liquid. Generally the bottom of the liquid is the hottest part of it, and bubbles form there and rise through the liquid to the surface. Just before the liquid boils, its bottom part may be at the boiling-point, and its upper part below. Bubbles of vapour then form at the bottom, rise to the colder liquid, and then collapse. The collapsing gives rise to the singing of a kettle about to boil.

Variation of Saturated Vapour Pressure with Temperature

We can now see how the relationship between the pressure of a saturated vapour and its temperature can be measured. We must apply various known air pressures to the liquid, heat the liquid, and measure the temperature of its vapour. Figure 12.9 shows a suitable apparatus, due to Regnault. The flask F contains the liquid, water in a laboratory experiment, and the flask R is an air reservoir. The pressure of the air in R is shown by the mercury manometer M; if its height is h, the pressure in mmHg is

$$p = H - h,$$

where H is the barometric height.

We first withdraw some air from R through the tap T, with a filter pump, until p is about 700 mm. We then close T and heat the water gently. The water vapour condenses in the condenser, and runs back to the flask. After a few minutes the water boils steadily. The temperature of the vapour, θ, and the pressure, p, become constant and we record their values. We next remove the flame from the flask F, and let the apparatus cool for a minute or two. Then we withdraw some more air from R, close T again, and repeat the observations.

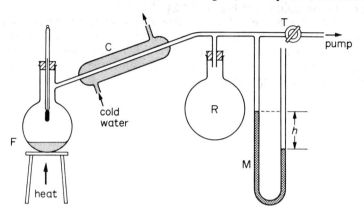

Fig. 12.9 Apparatus for variation of s.v.p. with temperature

If we wish to find the saturated vapour pressure when it is above atmospheric, that is to say, when the temperature is above the normal boiling-point of the liquid, air is pumped into the reservoir R—with a bicycle pump—instead of drawing it out. The manometer M then shows the excess pressure, and

$$p = H + h.$$

With simple glass apparatus we cannot go far in this direction.

Effect of Altitude on Boiling-Point

The pressure of the atmosphere decreases with increasing height above the earth's surface, because the thickness, and therefore the weight, of the belt of air above the observer decreases. The rate of fall in pressure is almost uniform over fairly small heights—approximately 85 mmHg per km. But at great altitudes the rate of fall diminishes. At the height of Everest, 9000 m, the atmospheric pressure is about 280 mmHg. On account of the fall in atmospheric pressure, the boiling-point of water falls with increasing height. Cooking-pots for use in high mountainous districts, such as the Andes, are therefore fitted with clamped lids. As the water boils, the steam accumulates in the pot, and its pressure rises above atmospheric. At about 760 mmHg a safety valve opens, so that the pressure does not rise above that value, and the cooking is done at 100°C.

The fall in the boiling-point with atmospheric pressure gives a simple way of determining one's height above sea-level. One observes the steam point with a thermometer. Knowing how the steam point falls with pressure, and how atmospheric pressure falls with increasing height, one can then find one's altitude.

Variation of Latent Heat with Temperature

When we speak of the latent heat of evaporation of a liquid, we usually mean the heat required to vaporise unit mass of it at its normal boiling-point, that is to say, under normal atmospheric pressure. But since evaporation takes place at all temperatures, the latent heat has a value for every temperature. Regnault measured the latent heat of steam over a range of temperatures, by boiling water at controlled pressures, as in measuring its saturated vapour pressure. His apparatus was in principle similar to Fig. 12.9. Modern measurements give, approximately,

$$l = 2520 - 2 \cdot 5\theta$$

where l is the specific latent heat in kJ kg^{-1} at $\theta°$C.

The relationship between l and θ shows that, at temperatures below the boiling-point at normal or standard atmospheric pressure, the latent heat of evaporation is less than at normal pressure. When the pressure above water in a closed flask is made very low by pumping air out of the flask, the water begins to boil at the room temperature.

Real Gases. Critical Phenomena

So far we have discussed the ideal or perfect gas. An ideal gas is one which obeys Boyle's law (pV = constant at constant temperature) and whose internal energy is independent of its volume. No such gas exists but at room temperature and moderate pressures, many gases approach the ideal closely for most purposes.

We shall now consider the behaviour of *real gases*; in doing so we shall come to appreciate better the relationship between liquid, vapour, and gas, and we shall see how gases such as air can be liquified.

Andrews' Experiments on Carbon Dioxide

In 1869 Andrews made experiments on carbon dioxide which have become classics. Figure 12.10 shows his apparatus. In the glass tube A he trapped carbon

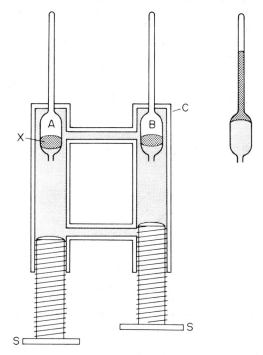

Fig. 12.10 Andrews' apparatus for isothermals of carbon dioxide at high pressures

dioxide above the pellet of mercury X. To do this, he started with the tube open at both ends and passed dry gas through it for a long time. Then he sealed the end of the capillary. He introduced the mercury pellet by warming the tube, and allowing it to cool with the open end dipping into mercury. Similarly, he trapped nitrogen in the tube B.

Andrews then fitted the tubes into the copper casing C, which contained water. By turning the screws S, he forced water into the lower parts of the tubes A and B, and drove the mercury upwards. The wide parts of the tubes were under the same pressure inside and out, and so were under no stress. The capillary extensions were strong enough to withstand hundreds of atmospheres. Andrews actually obtained 108 atmospheres.

When the screws S were turned far into the casing, the gases were forced into the capillaries, as shown on the right of the figure, and greatly compressed. From the known volumes of the wide parts of the tubes, and the calibrations of the capillaries, Andrews determined the volumes of the gases. He estimated the pressure from the compression of the nitrogen, assuming that it obeyed Boyle's law. Air was also used in place of nitrogen.

For work above and below room temperature, Andrews surrounded the capillary part of A with a water bath, which he kept at a constant temperature between about 10°C and 50°C.

p-V Curves of Carbon Dioxide. Critical Temperature

Andrews' results for the pressure (p)-volume (V) curves at various constant temperatures, called *isothermals*, are shown in Fig. 12.11.

Let us consider the one for 21·5°C, ABCD. Andrews noticed that, when the pressure reached the value corresponding to B, a meniscus appeared above the mercury in the capillary containing the carbon dioxide. He concluded that the liquid had begun to form. From B to C, he found no change in pressure as the screws were turned, but simply a decrease in the volume of the carbon dioxide. At the same time the meniscus moved upwards, suggesting that the proportion of liquid was increasing. At C the meniscus disappeared at the top of the tube, suggesting that the carbon dioxide had become wholly liquid. Beyond C the pressure rose very rapidly; this confirmed that the carbon dioxide was wholly liquid, since liquids are almost incompressible.

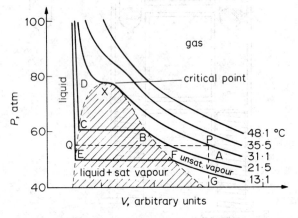

Fig. 12.11 Andrews' isothermals for carbon dioxide

Thus the part CBA of the isothermal for 21·5°C is a curve of volume against pressure for a liquid and vapour, showing saturation at B; it is like the isothermal for water given in Fig. 12.5 (i), p. 240. And the curve GFE is another such isothermal, for the lower temperature 13·1°C; the two curves are like the two in Fig. 12.6 (i), p. 241.

The isothermal for 31·1°C has no extended plateau; it merely shows a point of inflection at X. At that temperature, Andrews observed no meniscus; he concluded that it was the *critical temperature*. The isothermals for temperature above 31·3°C never become horizontal, and show no breaks such as B or F. At temperatures above the critical, no change from gas to liquid can be seen.

The isothermal for 48·1°C conforms fairly well to Boyle's law; even when the gas is highly compressed its behaviour is not far from ideal.

The point X in Fig. 12.11 is called the critical point. The pressure and volume (of unit mass) corresponding to it are called the *critical pressure* and *critical volume*; the reciprocal of the critical volume is the critical density.

Gases and Vapours. Continuity of State

We can now see the importance of Andrews' experiments. A gas *above its critical temperature* can not be liquefied by pressure. Early attempts to liquefy gases such as air, by compression without cooling, failed; and the gases were wrongly called 'permanent' gases. We still, for convenience, refer to a gas as a vapour when it is below its critical temperature, and as a gas when it is above it. The critical temperature of nitrogen is − 147°C (126 K). So nitrogen must be cooled below − 147°C to liquefy it by pressure. Air has a critical temperature of − 190°C (183 K) and oxygen a critical temperature of − 118°C (153 K).

Andrews' experiments also helped to illustrate the continuity between the liquid and gaseous states of a substance. Thus carbon dioxide is a gas above 31·1°C. As shown in Fig. 12.11, carbon dioxide can exist below 31·1°C as a liquid or as a vapour or as a mixture of liquid and vapour. One can also go directly from vapour to liquid along the line GP and then along PQ.

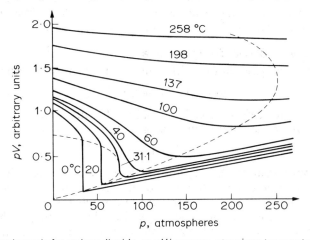

Fig. 12.12 Isothermals for carbon dioxide, as pV/p curves, at various temperatures. The small dotted loop passes through the ends of the vertical parts. The large dotted loop is the locus of the minima of pV

Figure 12.12 shows some of the isothermals for carbon dioxide obtained by Andrews over a wide temperature range, this time with pV plotted against p. Corrections were made for the departure of nitrogen from Boyle's law. The vertical parts of the isothermals below the critical temperature of 31·1°C correspond to the change from the gaseous to the liquid state or to saturated vapour. If the gas obeyed Boyle's law, pV = constant at constant temperature, the graph of pV against p would be a straight horizontal line. At the highest temperature shown, this is most nearly true.

Departures from Boyle's Law at High Pressure

In 1847 Regnault measured the volume of various gases at pressures of several atmospheres. He found that, to halve the volume of the gas, he did not have quite to double the pressure on it. The product pV, therefore, instead of being constant, decreased slightly with the pressure. He found one exception to this rule: hydrogen. By compressing the gases further, Regnault found the variation

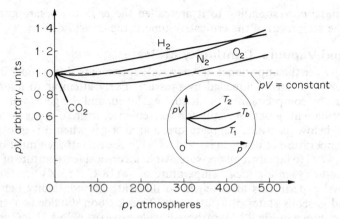

Fig. 12.13 Isothermals for various gases, at room temperature and high pressure

of pV with p at constant temperature, and obtained results which are represented by the early parts of the curves in Fig. 12.13. The complete curves in the figure show some of the results obtained by Amagat in 1892.

Real Gas Laws. Boyle Temperature

The results for real gases in Fig. 12.12 can be expressed by the general relation

$$pV = A + Bp + Cp^2 + ...,$$

where A, B, C are coefficients called 'virial coefficients' which are functions of temperature. At moderate pressures the gas obeys the relation

$$pV = A + Bp$$

to a good approximation. At very low pressures, $pV = A = RT$, that is, all real gases obey the same gas law. In these conditions the volume of the molecules themselves is negligible and the attractive forces between them are negligible, so the law, in fact, is that for an ideal gas. For this reason, gas thermometer measurements are extrapolated to zero pressure in accurate work (p. 298). From $pV = A + Bp + Cp^2 + ...$, it follows that the coefficient B is the *gradient* at $p = 0$ of the pV graph. Figure 12.13 shows that B is negative for some gases such as carbon dioxide or oxygen and positive for others such as hydrogen. From the gradient at $p = 0$ in Fig. 12.12, we see that the value of B becomes less negative as the temperature rises.

At one particular temperature T_b called the *Boyle temperature*, $pV =$ constant for moderate pressures, to a good approximation, Fig. 12.13, inset. For an ideal gas, the curve would be a straight line parallel to the axis of p, as shown in Fig. 12.13. Below the Boyle temperature, the pV against p curves have a minimum, as at T_1. Above the Boyle temperature, the pV values increase, as at T_2. Hydrogen has a Boyle temperature of $-167°C$; for many other gases the Boyle temperature is above room temperature.

At high pressures and high temperatures, then, the laws for real gases depart considerably from those for real gases. At very low pressures, however, real gases obey the ideal gas laws.

Internal Energy and Volume

In our simple account of the kinetic theory of gases, we assumed that the molecules of a gas do not attract one another. If they did, any molecule approaching the boundary of the gas would be pulled towards the body of it, as is a molecule of water approaching the surface (see p. 138). The attractions of the molecules would thus reduce the pressure of the gas.

Since the molecules of a substance are presumably the same whether it is liquid or gas, the molecules of a gas must attract one another. But except for brief instants when they collide, the molecules of a gas are much further apart than those of a liquid. In 1 cubic centimetre of gas at s.t.p. there are $2 \cdot 69 \times 10^{19}$ molecules, and in 1 cubic centimetre of water there are $3 \cdot 33 \times 10^{22}$; there are a thousand times as many molecules in the liquid, and so the molecules in the gas are ten times further apart. We may therefore expect that the mutual attraction of the molecules of a gas, for most purposes, can be neglected, as experiment, in fact, shows.

The experiment consists in allowing a gas to expand without doing external work; that is, to expand into a vacuum. Then, if the molecules attract one another, work is done against their attractions as they move further apart. But if the molecular attractions are negligible, the work done is also negligible. If any work is done against the molecular attractions, it will be done at the expense of the molecular kinetic energies. So as the molecules move apart the internal energy of the gas, and therefore its temperature, will fall.

The expansion of a gas into a vacuum is called a 'free expansion'. If a gas does not cool when it makes a free expansion, then the mutual attractions of its molecules are negligible. Joule and Kelvin showed that most gases, in expanding from high pressure to low, do lose a little of their internal energy. The loss represents work done against the molecular attractions, which are therefore not quite negligible.

If the internal energy of a gas is independent of its volume, it is determined only by the temperature of the gas. The simple expression for the pressure, $p = \frac{1}{3}\rho c^2$, then holds and the gas obeys Boyle's law. Such a gas is called an *ideal* gas. All gases, when far from liquefaction, behave for most practical purposes as though they were ideal.

Van der Waals' Equation

In deriving the ideal gas equation $pV = RT$ from the kinetic theory of gases, a number of assumptions were made. These are listed on p. 219. Van der Waals modified the ideal gas equation to take account that two of these assumptions may not be valid:

(1) *The volume of the molecules may not be negligible in relation to the volume V occupied by the gas.*
(2) *The attractive forces between the molecules may not be negligible.*

In the bulk of the gas, the resultant force of attraction between a particular molecule and those all round it is zero when averaged over a period. Molecules which strike the wall of the containing vessel, however, are retarded by an unbalanced force due to molecules behind them. The observed pressure p of a gas is thus *less* than the pressure in the ideal case, when the attractive forces due to molecules is zero.

Van der Waals derived an expression for this pressure 'defect'. He considered that it was proportional to the product of the number of molecules per second striking unit area of the wall and the number per unit volume behind them, since this is a measure of the force of attraction. For a given volume of gas, both these numbers are proportional to the *density* of the gas. Consequently the pressure defect, p_1 say, is proportional to $\rho \times \rho$ or ρ^2. For a fixed mass of gas, $\rho \propto 1/V$, where V is the volume. Thus $p_1 = a/V^2$, where a is a constant for the particular gas. Taking into account the attractive forces between the molecules, it follows that, if p is the observed pressure,

the gas pressure in the bulk of the gas $= p + a/V^2$.

The attraction of the walls on the molecules arriving there is to increase their velocity from v say to $v + \Delta v$. Immediately after rebounding from the walls, however, the force of attraction decreases the velocity to v again. Thus the attraction of the walls has no net effect on the momentum change due to collision. Likewise, the increase in momentum of the walls due to their attraction by the molecules arriving is lost after the molecules rebound.

Molecules have a particular diameter or volume because repulsive forces occur when they approach very closely and hence they cannot be compressed indefinitely. The volume of the space inside a container occupied by the molecules is thus not V but $(V - b)$, where b is a factor depending on the actual volume of the molecules. The magnitude of b is not the actual volume of the molecules, as if they were swept into one corner of the space, since they are in constant motion. b has been estimated to be about four times the actual volume.

Thus *van der Waals' equation* for real gases is:

$$\left(p + \frac{a}{V^2}\right)(V - b) = RT.$$

At high pressures, when the molecules are relatively numerous and close together, the volume factor b and pressure 'defect' a/V^2 both become important. Conversely, at low pressures, where the molecules are relatively few and far apart on the average, a gas behaves like an ideal gas and obeys the equation $pV = RT$.

Real Gas and van der Waals' Equation

As stated on p. 250, the relation of the product pV to p for a real gas can be expressed by

$$pV = A + Bp + Cp^2 + \ldots,$$

where $A, B, C, \ldots$ are coefficients decreasing in magnitude. The most important coefficients are A and B. So, at moderate pressures, $pV = A + Bp$. Further, when a real gas has a very low pressure, that is, p approaches zero, a real gas has the properties of an ideal gas. So $pV = A = RT$. At moderate pressures, then, we can write

$$pV = A + Bp = RT + Bp,$$

or $$p(V - B) = RT.$$

From this relation we see that B represents a volume. Comparing this with the van der Waals equation, in which the term $(V - b)$ occurs, then B is roughly equal to the volume of the gas molecules themselves. Suppose 1 mole of a particular gas has a value B of 3×10^{-5} m^3. Since the Avogadro constant is about 6×10^{23} mol^{-1}, then roughly

$$\text{volume of 1 molecule} = \frac{3 \times 10^{-5}}{6 \times 10^{23}} = 5 \times 10^{-29} \text{ m}^3.$$

The linear size of a molecule is of the order of the cube root of its volume. So an estimated size $= (5 \times 10^{-29})^{1/3} = 4 \times 10^{-10}$ m.

If required, we can expand van der Waals' equation to find a more exact relation for the coefficient B in terms of the constants a and b. Removing the brackets from the equation, then

$$pV - bp + \frac{a}{V} - \frac{ab}{V^2} = RT$$

or
$$pV = RT + bp - \frac{a}{V} + \frac{ab}{V^2}.$$

Using the approximation $pV = RT$, then $1/V = p/RT$. Substituting for $1/V$ and $1/V^2$,

$$pV = RT + \left(b - \frac{a}{RT}\right)p + \frac{ab}{R^2T^2}p^2.$$

Comparing this relation with $pV = A + Bp + Cp^2 + \ldots$, we see that

$$B = b - \frac{a}{RT} \quad \text{and} \quad C = \frac{ab}{R^2T^2}.$$

The Boyle temperature T_b is the temperature when a real gas obeys the ideal gas laws (p. 250). From $pV = A + Bp$, this occurs when $B = 0$ since $A = RT$. So, from above, the Boyle temperature is given by $(b - a/RT_b) = 0$, or $T_b = a/bR$. So T_b can be calculated when the constants a, b and R are known.

Isothermals of Real Gas

A graph of pressure p against volume V at constant temperature is called an *isothermal* or *isotherm*. Figure 12.14 (i) shows some isothermals for an ideal gas, which obeys the perfect gas law $pV = RT$. Figure 12.14 (ii) shows a number of isothermals for a gas which obeys van der Waals' equation, $(p + a/V^2)(V - b) = RT$.

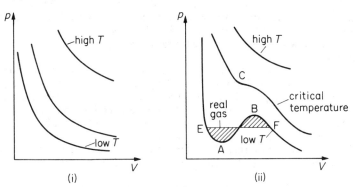

Fig. 12.14 Isothermals for ideal and van der Waals gases

At high temperatures the isothermals are similar. As the temperature is lowered, however, the isothermals in Fig. 12.14 (ii) change in shape. One curve has a point of inflexion at C, which corresponds to the *critical point* of a real gas. The isothermals thus approximate to those obtained by Andrews in his experiments on actual gases such as carbon dioxide, described on p. 248.

Below this temperature, however, isothermals such as EABF are obtained by using van der Waals' equation. These are unlike the isothermals obtained with real gases, because in the region AB the pressure increases with the volume, which is impossible. However, an actual isothermal in this region corresponds to a straight line EF, as shown. Here the liquid and vapour are in equilibrium (see p. 248) and the line EF is drawn to make the shaded areas above and below it equal. Thus van der Waals' equation roughly fits the isothermals of actual gases above the critical temperature but below the critical temperature it must be modified considerably. Many other gas equations have been suggested for real gases but quantitative agreement is generally poor.

Exercises 12

Vapours

1. In terms of the kinetic theory of matter explain (a) what is meant by *saturated vapour* and *saturation vapour pressure*, (b) how the saturation vapour pressure varies with temperature.

Describe an experiment to measure the saturation vapour pressure of water vapour at 300 K (27°C). Discuss one practical difficulty in using the apparatus you describe to measure the saturation vapour pressure of water vapour at 275 K (2°C).

State, with reasons, two advantages of using mercury in a barometer. (N.)

2. What is meant by a saturated vapour?

A closed vessel of constant volume contains a mixture of air and water vapour. The air and water vapour exert pressures of 5·00 kN m^{-2} and 1·50 kN m^{-2} respectively at 20°C. Using the data given below, determine, by a graphical method, the temperature at which the water vapour becomes saturated as the temperature is gradually reduced to 0°C. You may assume that an unsaturated vapour obeys the gas laws until saturation occurs.

S.V.P. of water vapour in kN m^{-2}	0·61	0·86	1·21	1·70	2·33
Temperature in °C	0	5	10	15	20

Calculate (a) the pressure of the air alone at 0°C, and
(b) the pressure of the air and water vapour at 0°C.

Describe an experiment you could perform in order to check the values of the saturated vapour pressures given in the table. (L.)

3. What is (a) a saturated vapour (b) an unsaturated vapour? Describe how the saturated vapour pressure of a substance varies with the temperature and volume of its container. How does its behaviour differ from that of a perfect gas?

Describe fully how you might measure the saturated vapour pressure of water between 20°C and 100°C.

Summarise the requirements that a physical quantity must satisfy in order that its temperature variation could form the basis of a practical thermometer over a particular temperature range. Compare the merits of a water vapour pressure thermometer and another thermometer of your own choice, for use between 0°C and 100°C. (O. & C.)

4. The saturation pressure of water vapour is 12 mmHg at 14°C and 24 mmHg at 25°C. Describe and explain the experiment you would perform to verify these data.

Explain qualitatively in terms of the kinetic theory of matter (a) what is meant by saturation vapour pressure, (b) the relative magnitudes of the saturation vapour pressures quoted.

Sketch a graph showing how the saturation pressure of water vapour varies between 0°C and 110°C. (N.)

5. Compare the properties of saturated and unsaturated vapours. By means of diagrams show how the pressure of (a) a gas, and (b) a vapour, vary with change (i) of volume at constant temperature, and (ii) of temperature at constant volume.

The saturation vapour pressure of ether vapour at 0°C is 185 mm of mercury and at 20°C it is 440 mm. The bulb of a constant volume gas thermometer contains dry air and sufficient ether for saturation. If the observed pressure in the bulb is 1000 mm at 20°C, what will it be at 0°C? (L.)

6. What is meant by a *saturated vapour*?

Draw a graph showing the variation of the saturated vapour pressure of water with temperature over the range 0°C to 100°C. Give a physical explanation of the boiling process and explain why a liquid boils when its saturated vapour pressure is equal to the surrounding pressure.

Describe an experiment you would perform to determine the saturated vapour pressure of water at temperatures in the range 80°C to 100°C.

A closed vessel contains a mixture of air and water vapour at 27°C at a total pressure of 1.070×10^5 N m^{-2}. The water vapour is just saturated at this temperature. Calculate the pressure exerted by the air alone in the vessel when the temperature is raised to 40°C. If the temperature is now lowered to 17°C what will be the total pressure in the vessel? You may neglect any expansion or contraction of the vessel. (s.v.p. of water at 17°C $= 1.9 \times 10^3$ N m^{-2}; s.v.p. of water at 27°C $= 3.7 \times 10^3$ N m^{-2}.) (L.)

7. Distinguish between a *saturated* and an *unsaturated* vapour. Describe an experiment to investigate the effect of pressure on the boiling-point of water and draw a sketch graph to show the general nature of the results to be expected.

A column of air was sealed into a horizontal uniform-bore capillary tube by a water index. When the atmospheric pressure was 762·5 torr (mmHg) and the temperature was 20°C, the air-column was 15·6 cm long; with the tube immersed in a water bath at 50°C, it was 19·1 cm long, the atmospheric pressure remaining the same. If the s.v.p. of water at 20°C is 17·5 torr, deduce its value at 50°C. (O. & C.)

8. State *Boyle's law* and *Dalton's law of partial pressures*.

The space above the mercury in a Boyle's law apparatus contains air together with alcohol vapour and a little liquid alcohol. Describe how the saturation vapour pressure of alcohol at room temperature may be determined with this apparatus.

A mixture of air and saturated alcohol vapour in the presence of liquid alcohol exerts a pressure of 128 mm of mercury at 20°C. When the mixture is heated at constant volume to the boiling-point of alcohol at standard pressure (i.e. 78°C), the vapour remaining saturated, the pressure becomes 860 mmHg. Find the saturation vapour pressure of alcohol at 20°C. (L.)

Critical Temperature. Real Gases

9. Oxygen can be liquefied by first allowing the gas to expand very rapidly (to reduce its temperature to below 154 K) and then applying sufficient external pressure to it.

(a) Explain why a sudden expansion can lead to cooling. (b) What is the significance of the temperature 154 K? (L.)

10. Define *critical temperature*.

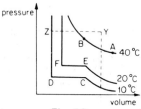

Fig. 12A

Figure 12A shows the form of the relationship between the pressure and volume for isothermal changes in volume of carbon dioxide at three different temperatures. Explain qualitatively, using kinetic theory, (a) the rise in pressure from A to B, (b) the constancy of the pressure from C to D, and (c) why the pressure at E and F is greater than the pressure at C and D. Explain the energy interchanges which would take place in going from A to B and from C to D.

If the carbon dioxide were in a glass container, so that its condition could be seen, what would be observed as if was taken (i) along the path CD, and (ii) along the path indicated from 10°C to Y and then to Z? (L.)

11. What are the conditions under which the equation $pV = RT$ gives a reasonable description of the relationship between the pressure p, the volume V and the temperature T of a real gas?

Sketch p–V isothermals for the gas-liquid states and indicate the region in which $pV = RT$ applies. Indicate the state of the substance in the various regions of the p–V

diagram. Mark and explain the significance of the critical isothermal.

Discuss a way in which the equation $pV = RT$ may be modified so that it can be applied more generally. Explain and justify on a molecular basis the additional terms introduced. Discuss the success of this modification. (L.)

12. What is meant by *critical temperature*?

A substance which is gaseous initially is at a low pressure and its volume is then reduced at constant temperature. Draw graphs showing the variation of volume with pressure if the substance is (a) above its critical temperature, (b) below its critical temperature. Explain how your graphs show the state of the substance.

Use the kinetic theory to explain the graph in case (b) and also to explain the effect of a rise in temperature of a few degrees on this graph, the temperature remaining below its critical value. (L.)

13. Describe experiments in which the relation between the pressure and volume of a gas has been investigated at constant temperature over a wide range of pressure. Sketch the form of the isothermal curves obtained.

Explain briefly how far van der Waals' equation accounts for the form of these isothermals. (L.)

14. Describe, with a diagram, the essential features of an experiment to study the departure of a real gas from ideal gas behaviour.

Give freehand, labelled sketches of the graphs you would expect to obtain on plotting (a) pressure P against volume V, (b) PV against P for such a gas at its critical temperature and at one temperature above and one below the critical temperature.

Explain van der Waals' attempt to produce an equation of state which would describe the behaviour of real gases.

Show that van der Waals' equation is consistent with the statement that all gases approach ideal gas behaviour at low pressures. (O. & C.)

15 (a) With pressure p as ordinate and volume V as abscissa, draw sketch graphs of $p–V$ isothermals for a real gas indicating (i) the region in which the gas approximately obeys Boyle's Law, (ii) the critical isothermal, (iii) the region in which the gas liquefies and gas and liquid are in equilibrium. Justify your answers in each case.

(b) At high pressure, the equation of state is modified from that of an ideal gas so that it becomes $pV = A + Bp$. Explain (i) how A depends on the temperature and (ii) why B is related to the size of the gas molecules.

In a particular experiment on 1 mole of oxygen, the value of B was found to be 20.2×10^{-6} m³. Estimate the volume of the oxygen molecule. The Avogadro constant $N_A = 6.03 \times 10^{23}$ mol⁻¹. (N.)

16. (a) Draw a fully labelled diagram of the apparatus used in Andrew's experiments to investigate the relationship between the pressure p and the volume V for a fixed mass of a real substance over a range of pressures to produce a change of state, and for several different temperatures. Explain carefully how the pressure was varied and how corresponding values of p and V were determined.

(b) Sketch a graph of the results obtained with this apparatus with pV as ordinate and p as abscissa for three temperatures; one above, one below and one equal to the critical temperature. Indicate on your graph the corresponding isothermals you would expect for an ideal gas.

(c) (i) Which assumptions of the kinetic theory of an ideal gas need to be modified for a real gas? (ii) Show how these modifications lead to an equation of state for a real gas different from that for an ideal gas. (N.)

17. Draw a sketch graph showing a series of isothermals relating the pressure and volume of a given mass of carbon dioxide for a wide range of these variables. On your graph indicate the state of the carbon dioxide represented by the various regions. Mark the approximate temperatures corresponding to your isothermals. What is the relevance of your curves to the general problem of liquefaction of gases? (AEB, 1982)

13 Transfer of Heat: Conduction and Radiation

Conduction

If we put a poker into the fire, and hold on to it, then heat reaches us along the metal. We say the heat is *conducted*; and we soon find that some substances—metals—are good conductors, and others—such as wood or glass—are not. Good conductors feel cold to the touch on a cold day, because they rapidly conduct away the body's heat.

Temperature Distribution along a Conductor

In order to study conduction in more detail consider Fig. 13.1 (i), which shows a metal bar AB whose ends have been soldered into the walls of two metal tanks, H, C. H contains boiling water, and C contains ice-water. Heat flows along the bar from A to B, and *when conditions are steady* the temperature θ of the bar is measured at points along its length. The measurements may be made with thermojunctions, not shown in the figure, which have been soldered to the rod. The curve in the upper part of the figure shows how the temperature falls along the bar, less and less steeply from the hot end to the cold.

The Fig. 13.1 (ii) shows how the temperature varies along the bar, if the bar is *well lagged* with a bad conductor, such as asbestos wool. It now falls uniformly. from the hot to the cold end.

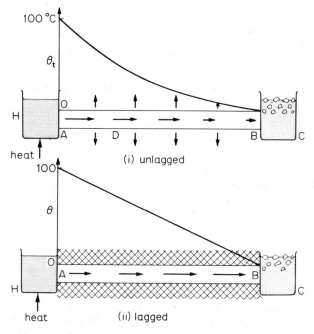

Fig. 13.1 Temperature fall along lagged and unlagged bars

The difference between the temperature distributions is due to the fact that, when the bar is unlagged, heat escapes from its sides, by convection in the surrounding air. So the heat flowing past D per second is less than that entering the bar at A by the amount which escapes from the surface AD. The arrows in the figure represent the heat escaping per second from the surface of the bar, and the heat flowing per second along its length. The heat flowing per second along the length decreases from the hot end to the cold. But when the bar is lagged, the heat escaping from its sides is negligible, and the flow per second is constant along the length of the bar.

We thus see that the temperature gradient along a bar is greatest where the heat flow through it is greatest. We also see that the temperature gradient is uniform only when there is a negligible loss of heat from the sides of the bar.

Thermal Conductivity

Consider a very large thick bar, of which AB in Fig. 13.2 (i) is a part, and along which heat is flowing steadily. We suppose that the loss of heat from the sides of the bar is made negligible by lagging. XY is a slice of the bar, of thickness

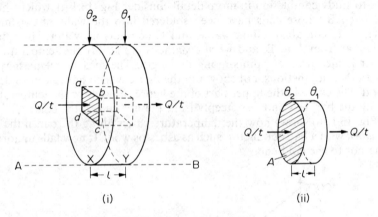

Fig. 13.2 Definition of thermal conductivity

l, whose faces are at temperatures θ_2 and θ_1. Then the *temperature gradient* over the slice is

$$\frac{\theta_2 - \theta_1}{l}.$$

We now consider an element *abcd* of the slice of unit cross-sectional area, and we denote by Q/t the heat flowing through it *per second*. The value of Q/t depends on the temperature gradient, and, since some substances are better conductors than others, it also depends on the material of the bar.

We therefore write

$$\frac{Q}{t} = k\frac{\theta_2 - \theta_1}{l}$$

where k is a factor depending on the material.

To a fair approximation the factor k is a constant for a given material; that is to say, it is independent of θ_2, θ_1, and l. It is called the *thermal conductivity* of the material concerned. From the above relation for Q/t, when the heat flow is *normal* to an area inside the material, k may be defined as *the heat flow per second per unit area per unit temperature gradient.*

This definition leads to a general equation for the flow of heat through any parallel-sided slab of the material, if no heat is lost from the sides of the slab. If the cross-sectional area of the slab is A in Fig. 13.2 (ii), its thickness is l, and the temperature of its faces are θ_1 and θ_2, then the heat flowing through it per second is

$$\frac{Q}{t} = \frac{kA(\theta_2 - \theta_1)}{l} \qquad . \qquad . \qquad . \qquad . \qquad (1)$$

A useful form of this equation is

$$\frac{Q}{At} = k\frac{\theta_2 - \theta_1}{l} \qquad . \qquad . \qquad . \qquad . \qquad (2)$$

or

heat flow per m^2 per second = conductivity × temperature gradient. . (2a)

Lagged and Unlagged Bars

In terms of the calculus, (2) may be re-written

$$\frac{1}{A}\frac{dQ}{dt} = -k\frac{d\theta}{dl} \qquad . \qquad . \qquad . \qquad . \qquad (3)$$

the temperature gradient being negative since θ diminishes as l increases.

If a bar is *lagged* perfectly, as in Fig. 13.1 (ii), then the heat per second, dQ/dt, flowing through every cross-section from the hot to the cold end is constant since no heat escapes through the sides. Hence, from (3), the temperature gradient, $d\theta/dl$, is constant along the bar. This is illustrated in Fig. 13.1 (ii); the temperature variation with distance along the bar is a straight line.

If the bar is *unlagged*, as in Fig. 13.1 (i), then heat is lost from the sides of the bar. In this case the heat per second, dQ/dt, flowing through each section decreases from the hot to the cold end. Hence, from (3), the temperature gradient, $d\theta/dl$, decreases with distance along the bar. This is shown by Fig. 13.1 (i); the gradient at a point of the curve decreases with distance from the hot end of the bar.

Units and Magnitude of Conductivity

Equation (2) enables us to find the unit of thermal conductivity. We have

$$k = \frac{Q/At \text{ (J m}^{-2}\text{ s}^{-1})}{(\theta_2 - \theta_1)/l \text{ (K m}^{-1})}.$$

Thus the unit of thermal conductivity = $J\,s^{-1}\,m^{-1}\,K^{-1}$, or since joule second^{-1} = watt (W), the unit of k is $W\,m^{-1}\,K^{-1}$.

The thermal conductivity of copper, cardboard, water and air are roughly 380, 0·2, 0·6 and 0·03 $W\,m^{-1}\,K^{-1}$ respectively. To a rough approximation we may say that the conductivities of metals are about 1000 times as great as those of other solids, and of liquids; and they are about 10 000 times as great as those of gases.

Example

Calculate the quantity of heat conducted through 2 m² of a brick wall 12 cm thick in 1 hour if the temperature on one side is 8°C and on the other side is 28°C. (Thermal conductivity of brick = 0·13 W m⁻¹ K⁻¹.)

$$\text{Temperature gradient} = \frac{28 - 8}{12 \times 10^{-2}} \text{ K m}^{-1} \text{ and } t = 3600 \text{ s.}$$

$$\therefore Q = kAt \times \text{temperature gradient}$$

$$= 0{\cdot}13 \times 2 \times 3600 \times \frac{28 - 8}{12 \times 10^{-2}} \text{ J}$$

$$= 156\,000 \text{ J.}$$

Thermal and Electrical Conductivity

We can make a useful analogy between thermal conductivity and electrical conductivity.

The electric current I flowing along a conductor $= Q/t$, where Q is the quantity of charge passing a given section in a time t. Also, $I = V/R$, where V is the potential difference between the ends of the conductor and R is its resistance (p. 621). Now $R = \rho l/A$, where ρ is the resistivity of the material, l is the length and A is its cross-sectional area (p. 631). So

$$I = \frac{Q}{t} = \frac{V}{\rho l/A}.$$

Thus

$$\frac{Q}{t} = \frac{1}{\rho} A \frac{V}{l}.$$

The quantity V/l is the *potential gradient* along the conductor. So

$$\frac{Q}{t} = \frac{1}{\rho} A \times \text{potential gradient} \qquad . \qquad . \qquad . \qquad (1)$$

For heat conduction,

$$\frac{Q}{t} = kA \times \text{temperature gradient.} \qquad . \qquad . \qquad (2)$$

Comparing (1) with (2), we see that $1/\rho$ is analogous to k. The inverse of resistivity is defined as the electrical *conductivity*, symbol σ. So thermal conductivity k is analogous to electrical conductivity σ.

Wiedemann and Franz discovered a law which states that, at a given temperature, the *ratio of the thermal to electrical conductivity is the same for all metals*. So a metal which is a good thermal conductor is also a good electrical conductor. This suggests that electrons are the carriers in both thermal and electrical conduction in metals. Thus on heating a metal bar the free electrons gain thermal energy and distribute this energy by collision with the fixed positive metal ions in the solid lattice.

Effect of Thin Layer of Bad Conductor

Figure 13.3 shows a lagged copper bar AB, whose ends are pressed against metal tanks at 0° and 100°C, but are separated from them by layers of dirt. The length of the bar is 10 cm or 0·1 m, and the dirt layers are 0·1 mm or $0{\cdot}1 \times 10^{-3}$ m thick. Assuming that the conductivity of dirt is 1/1000 that of copper, let us find the temperature of each end of the bar.

Suppose $\quad\quad\quad k =$ conductivity of copper,

$\quad\quad\quad\quad\quad\quad A =$ cross-section of copper,

$\quad\quad\quad\quad\quad\quad \theta_2, \theta_1 =$ temperature of hot and cold ends.

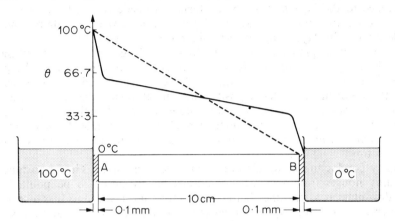

Fig. 13.3 Temperature gradients in good and bad conductors

Since the bar is lagged, the heat flow per second Q/t is constant from end to end. Therefore,

$$\frac{Q}{t} = \frac{k}{1000}A\frac{100 - \theta_2}{0\cdot1 \times 10^{-3}} = kA\frac{\theta_2 - \theta_1}{0\cdot1} = \frac{k}{1000}A\frac{\theta_1 - 0}{0\cdot1 \times 10^{-3}}.$$

Dividing through by kA, these equations give

$$\frac{100 - \theta_2}{0\cdot1} = \frac{\theta_2 - \theta_1}{0\cdot1} = \frac{\theta_1}{0\cdot1},$$

or $\quad\quad\quad\quad\quad 100 - \theta_2 = \theta_2 - \theta_1 = \theta_1,$

from which $\quad\quad\quad \theta_2 = 66\cdot7°C, \theta_1 = 33\cdot3°C.$

Thus the total temperature drop, 100°C, is divided equally over the two thin layers of dirt and the long copper bar. The heavy lines in the figure show the temperature distribution; the broken line shows what it would be if there were no dirt.

Good and Bad Conductors

This numerical example shows what a great effect a thin layer of a bad conductor may have on thermal conditions; 0·1 mm of dirt causes as great a temperature fall as 10 cm of copper. We can generalise this result with the help of equation (2a):

heat flow/m^2 s = conductivity × temperature gradient.

The equation shows that, if the heat flow is uniform, the temperature gradient is inversely proportional to the conductivity. So if the conductivity of dirt is 1/1000 that of copper, the temperature gradient in it is 1000 times that in copper; thus 1 mm of dirt sets up the same temperature fall as 1 m of copper. In general terms we express this result by saying that the dirt prevents a good thermal contact, or that it provides a bad one. The reader who has already studied electricity will see an obvious analogy here. We can say that a dirt layer has a high thermal resistance, and hence causes a great temperature drop.

Boiler plates are made of steel, not copper, although copper is about eight times as good a conductor of heat. The material of the plates makes no noticeable difference to the heat flow from the furnace outside the boiler to the water inside it, because there is always a layer of gas between the flame and the boiler plate. This layer may be very thin, but its conductivity is about 1/10 000 that of steel; if the plate is a centimetre thick, and the gas-film 1/1000 centimetre, then the temperature drop across the film is ten times that across the plate. Thus the rate at which heat flows into the boiler is determined mainly by the gas.

If the water in the boiler deposits scale on the plates, the rate of heat flow is further reduced. For scale is a bad conductor, and, though it may not be as bad a conductor as gas, it can build up a much thicker layer. Scale must therefore be prevented from forming, if possible; and if not, it must be removed from time to time.

Badly conducting materials are often called *insulators*. The importance of building houses from insulating materials hardly needs to be pointed out. Window-glass is a ten-times better conductor than brick, and it is also much thinner. A room with large windows therefore requires more heating in winter than one with small windows. Wood is as bad a conductor (or as good an insulator) as brick, but it also is thinner. Wooden houses therefore have double walls, with an air-space between them. Air is an excellent insulator, and the walls prevent convection. In polar climates, wooden huts must not be built with steel bolts going right through them; otherwise the inside ends of the bolts grow icicles from the moisture in the explorer's breath.

Examples

1. A cavity wall is made of bricks 0·1 m thick with an air space 0·1 m thick between them. (i) Assuming the thermal conductivity of brick is 20 times that of air, calculate the thickness of brick which conducts the same quantity of heat per second per unit area as 0·1 m of air, (ii) if the thermal conductivity of brick is 0·5 W m^{-1} K^{-1}, calculate the rate of heat conducted per unit area through the cavity wall when the outside surfaces of the brick walls are respectively 19°C and 4°C.

(i) Suppose θ_1, θ_2 are the respective temperatures at the ends of a brick of thickness l_B and thermal conductivity k_B, and at the ends of air of thickness l_A and thermal conductivity k_A. Then, with the usual notation,

$$\frac{Q}{t} = k_B A \frac{\theta_1 - \theta_2}{l_B} = k_A A \frac{\theta_1 - \theta_2}{l_A}.$$

So

$$\frac{k_B}{l_B} = \frac{k_A}{l_A}.$$

Then

$$l_B = l_A \times \frac{k_B}{k_A} = 0·1 \text{ m} \times 20 = 2 \text{ m.}$$

(ii) Since the two bricks and the air are thermally in series, we can replace the thickness of 0·1 m of air by 2 m of brick and add the three thicknesses of brick. So

total brick thickness = 0·1 + 2 + 0·1 = 2·2 m.

Then

$$\frac{1}{A} \cdot \frac{Q}{t} = k_B \frac{\theta_1 - \theta_2}{l_B} = 0·5 \frac{19 - 4}{2·2}$$

$$= 3·4 \text{ W m}^{-2}.$$

2. Define *thermal conductivity*. Describe and give the theory of a method of measuring the thermal conductivity of copper.

A sheet of rubber and a sheet of cardboard, each 2 mm thick, are pressed together and their outer faces are maintained respectively at 0°C and 25°C. If the thermal conductivities of rubber and cardboard are respectively 0·13 and 0·05 W m⁻¹ K⁻¹, find the quantity of heat which flows in 1 hour across a piece of the composite sheet of area 100 cm². (*L.*)

We must first find the temperature, θ°C, of the junction of the rubber and cardboard. The temperature gradient across the rubber $= (\theta - 0)/2 \times 10^{-3}$; the temperature gradient across the cardboard $= (25 - \theta)/2 \times 10^{-3}$.

$\therefore$ Q per second per m² across rubber $= 0\cdot13 \times (\theta - 0)/2 \times 10^{-3}$

and Q per second per m² across cardboard $= 0\cdot05 \times (25 - \theta)/2 \times 10^{-3}$.

But in the steady state the quantities of heat above are the same.

$$\therefore \frac{0\cdot13(\theta - 0)}{2 \times 10^{-3}} = \frac{0\cdot05(25 - \theta)}{2 \times 10^{-3}}$$

$$\therefore 13\theta = 125 - 5\theta$$

$$\therefore \theta = \frac{125}{18} = 7°C.$$

Now area $= 100$ cm² $= 100 \times 10^{-4}$ m². So, using the rubber alone, Q through area in 1 hour (3600 seconds)

$$= \frac{0\cdot13 \times 100 \times 10^{-4} \times 7 \times 3600}{2 \times 10^{-3}} = 16\,380 \text{ J.}$$

Measurement of High Conductivity: Metals

When the thermal conductivity of a metal is to be measured, two conditions must usually be satisfied: heat must flow through the specimen at a measurable rate, and the temperature gradient along the specimen must be measurably steep. These conditions determine the form of the apparatus used.

When the conductor is a *metal*, it is easy to get a fast enough heat flow. The problem is to build up a temperature gradient. It is solved by having as the specimen a bar long compared with its diameter. Figure 13.4 shows the apparatus, which is due to Searle. AB is the specimen, about 4 cm diameter and 20 cm long. In one form of apparatus it is heated by a coil at H, and cooled by circulating water at B. The whole apparatus is heavily lagged with felt. To measure the temperature gradient, thermometers are placed in the two mercury-

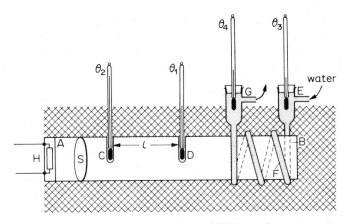

Fig. 13.4 Experiment for thermal conductivity of a metal

filled cups C, D; the cups are made of copper, and are soldered to the specimen at a known distance apart. Alternatively, thermometers are placed in holes bored in the bar, which are filled with mercury. In this way errors due to bad thermal contact are avoided.

The cooling water flows in at E, round the copper coil F which is soldered to the specimen, and out at G. The water leaving at G is warmer than that coming in at E, so that the temperature falls continuously along the bar: if the water came in at G and out at E, it would tend to reverse the temperature gradient at the end of the bar, and might upset it as far back as D or C.

The whole apparatus is left running, with a steady flow of water, until all the temperatures have become constant: the temperature θ_2 and θ_1, at C and D in the bar, and θ_4 and θ_3 of the water leaving and entering. The steady rate of flow of the cooling water is measured with a measuring cylinder and a stop-clock.

Calculation. If A is the cross-sectional area of the bar and k is conductivity, then the heat flow per second through a section such as S is

$$\frac{Q}{t} = kA\frac{\theta_2 - \theta_1}{l}.$$

This heat is carried away by the cooling water; if a mass m of specific heat capacity c_w flows through F in 1 second, the heat carried away is $mc_w(\theta_4 - \theta_3)$.

Therefore $$kA\frac{\theta_2 - \theta_1}{l} = mc_w(\theta_4 - \theta_3).$$

With this apparatus we can show that the conductivity k is a constant over small ranges of temperature. To do so we increase the flow of cooling water, and thus lower the outflow temperature θ_4. The gradient in the bar then steepens, and $(\theta_2 - \theta_1)$ increases. When the new steady state has been reached, the conductivity k is measured as before. Within the limits of experimental error, it is found to be unchanged.

Measurement of Low Conductivity: Non-metallic Solids

In measuring the conductivity of a bad conductor, the difficulty is to get an adequate heat flow. The specimen is therefore made in the form of a *thin* disc, D, about 10 cm in diameter and a few millimetres thick (Fig. 13.5 (i)). It is heated by a steam-chest C, whose bottom is thick enough to contain a hole for a thermometer.

The specimen rests on a thick brass slab B, also containing a thermometer. The whole apparatus is hung in mid air by three strings tied to B.

To ensure good thermal contact, the adjoining faces of C, D and B must be flat and clean; those of C and B should be polished. A trace of vaseline smeared over each face improves the contact.

When the temperatures have become steady, the heat passing from C through D escapes from B by radiation and convection. Its rate of escape from B is roughly proportional to the excess temperature of B over the room (Newton's law). Thus B takes up a steady temperature θ_1 such that its rate of loss of heat to the outside is just equal to its gain through D. The rate of loss of heat from the sides of D is negligible, because their surface area is small.

This apparatus is derived from one due to Lees, and simplified for elementary work. If we use glass or ebonite for the specimen, the temperature θ_1 is generally about 70°C; θ_2 is, of course, about 100°C. After these temperatures have become steady, and we have measured them, the problem is to find the rate of heat loss

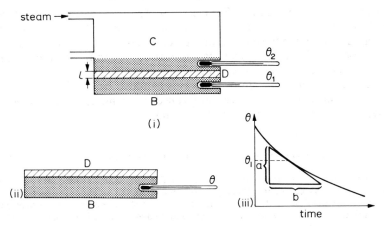

Fig. 13.5 Apparatus for thermal conductivity of a bad conductor

from B. To do this, we take away the specimunt D and heat B directly from C until its temperature has risen by about 10°C. We then remove C, and cover the top part of B again with the specimen D (Fig. 13.5 (ii)). At intervals of a minute—or less—we measure the temperature of B, and afterwards plot it against the time (Fig. 13.5 (iii)).

Calculation. (1) While the slab B is cooling it is losing heat by radiation and convection. It is doing so under the same conditions as in the first part of the experiment, because the felt prevents heat escaping from the top surface. Thus when the slab B passes through the temperature θ_1, it is losing heat at the same rate as in the first part of the experiment. The heat which it loses is now drawn from its own heat content, whereas before it was supplied from C via D; this is why the temperature of B is now falling, whereas before it was steady. The rate at which B loses heat at the temperature θ_1 is given by:

$$\text{heat lost/second} = Mc \times \text{temperature fall/second},$$

where M, c are respectively the mass and specific heat capacity of the slab.

(2) To find the rate of fall of temperature at θ_1, we draw the tangent to the cooling curve at that point. If, as shown in Fig. 13.5 (iii), its gradient at θ_1 would give a fall of a kelvin in b seconds, then the rate of temperature fall is a/b kelvin per second.

(3) We then have, if A is the cross-sectional area of the specimen, l its thickness, and k its conductivity,

$$kA\frac{\theta_2 - \theta_1}{l} = Mc\frac{a}{b}.$$

Thus k can be calculated.

Thermal Conduction in Solids

Metals. Metals are good thermal conductors and good electrical conductors. Wiedemann and Franz showed that, at a given temperature, *the ratio of thermal to electrical conductivity is the same for all metals* (p. 260).

Since electrons are the carriers in electrical conduction, it is considered that electrons transport thermal energy through metals. Thus on heating a metal bar the free electrons gain thermal energy and distribute this energy by collision with the fixed positive metal ions in the solid lattice.

Poor conductors. These have no free electrons. The transport of thermal energy through solids such as crystals is mainly due to waves. They are produced by lattice vibrations due to the thermal motion of the atoms. The waves are scattered by the atoms or by defects such as dislocations or impurity atoms and so distribute thermal energy to the solid.

The energy and momentum of the waves can also be considered carried by particles (p. 868). These particles are called *phonons*. Like the waves they represent, they travel with the speed of sound.

Example

Define *thermal conductivity* and explain how you would measure its value for a poorly conducting solid.

In order to minimise heat losses from a glass container, the walls of the container are made of two sheets of glass, each 2 mm thick, placed 3 mm apart, the intervening space being filled with a poorly conducting solid. Calculate the ratio of the rate of conduction of heat per unit area through this composite wall to that which would have occurred had a single sheet of the same glass been used under the same internal and external temperature conditions. (Assume that the thermal conductivity of glass and the poorly conducting solid = 0.63 and 0.049 W m^{-1} K^{-1} respectively.) (*L.*)

We can replace the 3 mm thick solid ($k = 0.049$) by a thermally equivalent *greater* thickness of x mm of glass ($k_g = 0.63$). The value of x is given (p. 262) by

$$x = \frac{0.63}{0.049} \times 3 \text{ mm} = 38\tfrac{4}{7} \text{ mm}.$$

So total equivalent glass thickness $= 2 + 2 + 38\tfrac{4}{7} = \dfrac{298}{7}$ mm.

If θ_2 and θ_1 are the outside temperatures of the respective glass sheets, then, with the usual notation,

$$\frac{1}{A}\frac{Q}{t} = k_g\frac{\theta_2 - \theta_1}{d} = k_g\frac{\theta_2 - \theta_1}{(298/7) \times 10^{-3}} \qquad . \qquad . \qquad . \qquad (1)$$

and for the single glass sheet of thickness 2 mm,

$$\frac{1}{A}\frac{Q}{t} = k_g\frac{\theta_2 - \theta_1}{d} = k_g\frac{\theta_2 - \theta_1}{2 \times 10^{-3}} \qquad . \qquad . \qquad . \qquad (2)$$

Dividing (1) by (2),

$$\text{ratio} = \frac{2 \times 7}{298} = 0.05 \text{ (approx.)}.$$

Radiation

All heat comes to us, directly or indirectly, from the sun. The heat which comes directly travels through 150 million km of space, mostly empty, and travels in straight lines, as does the light: the shade of a tree coincides with its shadow. Because heat and light travel with the same speed, they are both cut off at the same instant in an eclipse. Since light is propagated by waves of some kind we conclude that the heat from the sun is propagated by similar waves, and we say it is 'radiated'.

As we show later, more radiation is obtained from a dull black body than from a transparent or polished one. Black bodies are also better absorbers of radiation than polished or transparent ones, which either allow radiation to pass through themselves, or reflect it away from themselves. If we hold a piece of white card, with a patch of black drawing ink on it, in front of the fire, the black patch soon comes to feel warmer than its white surround.

Reflection and Refraction

If we focus the sun's light on our skin with a converging lens or a concave mirror, we feel heat at the focal spot. The heat from the sun has therefore been reflected or refracted in the same way as the light.

If we wish to show the reflection of heat unaccompanied by light, we may use two searchlight mirrors, set up as in Fig. 13.6. At the focus of one, F_1, we

Fig. 13.6 Reflection of radiant heat

put an iron ball heated to just below redness. At the focus of the other, F_2, we put the bulb of a thermometer, which has been blackened with soot to make it a good absorber (p. 274). The mercury rises in the stem of the thermometer. If we move either the bulb or the ball away from the focus, the mercury falls back; the bulb has therefore been receiving heat from the ball, by reflection at the two mirrors. We can show that the foci of the mirrors are the same for heat as for light if we replace the ball and thermometer by a lamp and screen. (In practice we do this first, to save time in finding the foci for the main experiment.)

To show the refraction of heat apart from the refraction of light is more difficult. It was first done by the astronomer Herschel in 1800. Herschel passed a beam of sunlight through a prism, as shown diagrammatically in Fig. 13.7, and explored the spectrum with a sensitive thermometer, whose bulb he had blackened. He found that in the visible part of the spectrum the mercury rose, showing that the light energy which it absorbed was converted into heat. But the mercury rose more when he carried the bulb into the darkened portion a little beyond the red of the visible spectrum. The sun's rays therefore carried energy which was not light.

Ultraviolet and Infrared

The radiant energy which Herschel found beyond the red is now called *infrared* radiation, because it is less refracted than the red. Radiant energy is also found

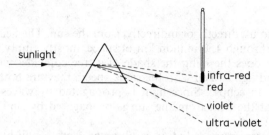

Fig. 13.7 Infrared and ultraviolet (diagrammatic)

beyond the violet and it is called *ultraviolet* radiation, because it is refracted more than the violet.

Ultraviolet radiation is absorbed by the human skin and causes sun-burn; more importantly, it stimulates the formation of vitamin D, which is necessary for the assimilation of calcium and the prevention of rickets. It is also absorbed by green plants; in them it enables water to combine with carbon dioxide to form carbohydrates. This process is called photosynthesis. Ultraviolet radiation causes the emission of electrons from metals, as in photoelectric cells; and it produces a latent image on a photographic emulsion. It is harmful to the eyes.

Ultraviolet radiation is strongly absorbed by glass—spectacle-wearers do not sunburn round the eyes—but enough of it gets through to affect a photographic film. It is transmitted with little absorption by quartz.

Infrared radiation is transmitted by rock-salt, but most of it is absorbed by glass. A little near the visible red passes easily through glass—if it did not, Herschel would not have discovered it. When infrared radiation falls on the skin, it gives the sensation of warmth. It is what we usually have in mind when we speak of heat radiation, and it is the main component of the radiation from a hot body; but it is in no essential way different from the other components, visible and ultraviolet radiation, as we shall now see.

Wavelengths of Radiation

In the section on Optics, we show how the wavelength of light can be measured with a diffraction grating—a series of fine close lines ruled on glass. The wavelength ranges from $4 \cdot 0 \times 10^{-7}$ m for the violet, to $7 \cdot 5 \times 10^{-7}$ m for the red. The first accurate measurements of wavelength were published in 1868 by Angstrom, and in his honour a distance of 10^{-10} m is called an *Angstrom unit* (A.U.). The wavelengths of infrared radiation can be measured with a grating made from fine wires stretched between two screws of close pitch. They range from 7500 A.U. to about 1 000 000 A.U. Often they are expressed in a longer unit than the Angstrom, such as the micron (μm) or the nanometre (nm).

$$1 \ \mu m = 10^{-6} \ m = 10^4 \ \text{A.U.}$$

$$1 \ nm = 10^{-9} \ m = 10 \ \text{A.U.}$$

We denote wavelength by the sumbol λ; its value for visible light ranges from 400 nm to 750 nm, or $0 \cdot 4 \ \mu$m to $0 \cdot 75 \ \mu$m, and for infrared radiation from 750 nm to about 10^5 nm, or $0 \cdot 75 \ \mu$m to about 100 μm.

We now consider that X-rays and radio waves also have the same nature as light, and so do the γ-rays from radioactive substances. For reasons which we cannot here discuss, we consider all these waves to be due to oscillating electric and magnetic fields. Figure 13.8 shows roughly the range of their wavelengths: it is a diagram of the *electromagnetic spectrum*.

Fig. 13A, B Infrared photography or thermal imaging. Photos A and B are of the same scene; a Boeing 747 aircraft being loaded at night. A is a normal (visible wavelengths) photo; it shows relatively little detail. B, however, is a clear photo. It uses only infrared wavelengths from the scene and is produced with infrared lenses and plates. In this photo, black is hot and white is cold. As can be seen, the aircraft has not been parked for very long as its tyres and engines are still warm. Note also the hot parts of the light beacon above the aircraft. (*Courtesy of Barr and Stroud Limited*)

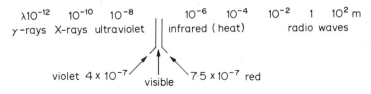

Fig. 13.8 The electromagnetic spectrum

Detection of Heat (Infrared) Radiation

A thermometer with a blackened bulb is a sluggish and insensitive detector of radiant heat. More satisfactory detectors, however, are electrical. One kind consists of a long thin strip of blackened platinum foil arranged in a compact zigzag on which the radiation falls (Fig. 13.9).

The foil is connected in a Wheatstone bridge, to measure its electrical resistance. When the strip is heated by the radiation, its resistance increases, and the increase is measured on the bridge. The instrument was devised by Langley in 1881; it is called a bolometer, *bole* being Greek for a ray.

The other commoner, type of radiation detector is called a *thermopile*. Its action depends on the electromotive force, which appears between the junctions of two different metals, when one junction is hot and the other cold. The modern thermopile consists of many junctions between the fine wires, as shown dia-

Fig. 13.9 Bolometer strip

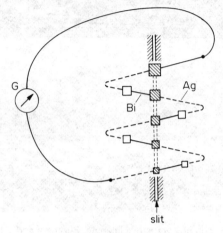

Fig. 13.10 Thermopile

grammatically in Fig. 13.10; the wires are of silver and bismuth, 0·1 mm or less in diameter. Their junctions are attached to thin discs of tin, about 0·2 mm thick, and about 1 mm square. One set of discs is blackened and mounted behind a slit, through which radiation can fall on them; the junctions attached to them become the hot junctions of the thermopile. The other, cold, junctions are shielded from the radiation to be measured; the discs attached to them help to keep them cool, by increasing their surface area.

When radiation falls on the blackened discs of a thermopile, it warms the junctions attached to them, and sets up an e.m.f. This e.m.f. can be measured with a potentiometer, or, for less accurate work, it can be used to deflect a galvanometer, G, connected directly to the ends of the thermopile (Fig. 13.10).

Reflection and Refraction: Inverse Square Law

With a thermopile and galvanometer, we can repeat Herschel's experiment more strikingly than with a thermometer. Using the simple apparatus of Fig.

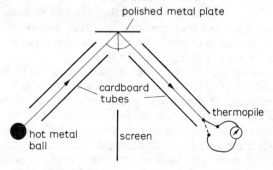

Fig. 13.11 Demonstration of reflection

13.11 we can show that, when infra-red is reflected, the angle of reflection equals the angle of incidence. We can also show the first law of reflection; that the incident and reflected rays are in the same plane as the normal to the reflector at the point of incidence.

If heat is radiant energy, its intensity should fall off as the inverse square of the distance from a point source. We can check that it does so by setting up an electric lamp, with a compact filament, in a dark room preferably with black walls. When we put a thermopile at different distances from the lamp, the deflection of the galvanometer is found to be inversely proportional to the square of the distance.

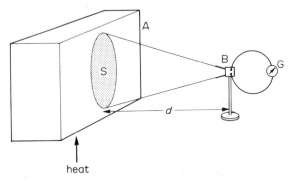

Fig. 13.12 Verification of inverse square law

If we wish to do this experiment with radiation that includes no visible light, we must modify it. Instead of the lamp, we use a large blackened tank of boiling water, A, and we fit the thermopile, B, with a conical mouthpiece, blackened on the inside, Fig. 13.12. The blackening prevents any radiation from reaching the pile by reflection at the walls of the mouthpiece. We now find that the deflection of the galvanometer, G, does not vary with the distance of the pile from the tank, *provided that the tank occupies the whole field of view of the cone* (Fig. 13.12). The area S of the tank from which radiation can reach the thermopile is then proportional to the square of the distance d. And since the deflection is unchanged when the distance is altered, the total radiation from each element of S must therefore fall off as the inverse square of the distance d.

The Infrared Spectrometer

Infrared spectra are important in the study of molecular structure. They are observed with an infrared spectrometer, whose principle is shown in Fig. 13.13. Since glass is opaque to the infrared, the radiation is focused by concave mirrors instead of lenses; the mirrors are plated with copper or gold on their front surfaces. The source of light is a Nernst filament, a metal filament coated with alkaline-earth oxides, and heated electrically. The radiation from such a filament is rich in infrared.

The slit S of the spectrometer is at the focus of one mirror which acts as a collimator. After passing through the rock-salt prism, A, the radiation is focused on to the thermopile P by the mirror M_2, which replaces the telescope of an optical spectrometer. Rotating the prism brings different wavelengths on to the slit of the thermopile; the position of the prism is calibrated in wavelengths with the help of a grating.

To a fair approximation, the deflection of the galvanometer is proportional to the radiant power carried in the narrow band of wavelengths which fall on

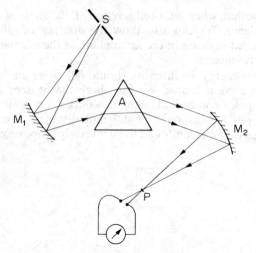

Fig. 13.13 Infrared spectrometer

the thermopile. If an absorbing body, such as a solution of an organic compound, is placed between the source and the slit, it weakens the radiation passing through the spectrometer, in the wavelengths which it absorbs. These wavelengths are therefore shown by a fall in the galvanometer deflection.

Reflection, Transmission, Absorption

Measurements have been made which give the amount of radiant energy approaching the earth from the sun. At the upper limit of our atmosphere, it is about 80 000 J m^{-2} min^{-1} or about 1340 W m^{-2}.

At the surface of the earth it is always less than this because of absorption in the atmosphere. Even on a cloudless day it is less, because the ozone in the upper atmosphere absorbs much of the ultraviolet.

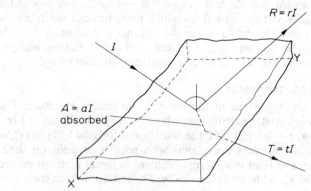

Fig. 13.14 Reflection, transmission and absorption

In Fig. 13.14, XY represents a body on which radiant energy is falling. The symbol I represents the radiation intensity in say watt per metre^2.

Some of the energy is reflected by the glass (R), some is absorbed (A), and some is transmitted (T). The total energy transmitted, absorbed and reflected per metre^2 per second is equal to the energy falling on the body over the same area and in the same time:

$$T + A + R = I.$$

If we denote by t, a, and r, the fractions of energy which are respectively transmitted, absorbed, and reflected by the body, then

$$tI + aI + rI = I$$

or
$$t + a + r = I \qquad . \qquad . \qquad . \qquad . \qquad . \qquad (6)$$

This equation expresses common knowledge. If a body is transparent ($t \to 1$), it is not opaque, and it is not a good reflector ($a \to 0$, $r \to 0$). But also, if the body is a good absorber of radiation ($a \to 1$), it is not transparent, and its surface is dull ($t \to 0$, $r \to 0$). And if it is a good reflector ($r \to 1$), it is neither transparent nor a good reflector ($t \to 0$, $a \to 0$). The term opaque, as commonly used, simply means not transparent; we see that it does not necessarily mean absorbent.

Equation (6), as we have written it above, is over-simplified. For a body may transmit some wavelengths (colours, if visible) and absorb or reflect others. If we now let I denote the intensity of radiation of a particular wavelength λ, then by repeating the argument we get

$$t_\lambda + a_\lambda + r_\lambda = 1 \qquad . \qquad . \qquad . \qquad . \qquad . \qquad (7)$$

where the coefficients t_λ, etc., all refer to the wavelength λ.

The truth of equation (7) is well shown by the metal gold, which reflects yellow light better than other colours. In thin films, gold is partly transparent, and the light which it transmits is green. Green is the colour complementary to yellow; gold removes the yellow from white light by reflection, and passes on the rest by transmission.

Radiation and Absorption

We have already pointed out that black surfaces are good absorbers and radiators of heat, and that polished surfaces are bad absorbers and radiators. This can be demonstrated by the apparatus in Fig. 13.15, called a *Leslie cube*. It is a cubical metal tank whose sides have a variety of finishes: dull black, dull white, highly polished. It contains boiling water, and therefore has a constant temperature. Facing it is a thermopile, P, which is fitted with the blackened conical mouth-piece described on p. 271.

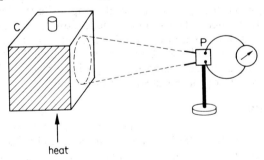

Fig. 13.15 Comparing radiators

Provided the face of the cube occupies the whole field of view of the cone, its distance from the thermopile does not matter (p. 271). The galvanometer deflection is greatest when the thermopile is facing the dull black surface of the cube, and least when it is facing the highly polished surface. The highly polished surface is therefore the worst radiator of all, and the dull black is the best.

Leslie's cube can also be used in an experiment to compare the absorbing properties of surfaces, Fig. 13.16. The cube, C, full of boiling water, is placed between two copper plates, A, B, of which A is blackened and B is polished. The temperature differences between A and B is measured by making each of them one element in a thermojunction: they are joined by a constantan wire, XY, and connected to a galvanometer, by copper wires, AE, DB. If A is hotter than B, the junction, X, is hotter than the junction, Y, and a current flows through the galvanometer in one direction. If B is hotter than A, the current is reversed.

The most suitable type of Leslie's cube is one which has two opposite faces similar—say grey—and the other two opposite faces very dissimilar—one black, one polished. At first the plates A, B are set opposite similar faces. The blackened plate, A, then becomes the hotter, showing that it is the better absorber.

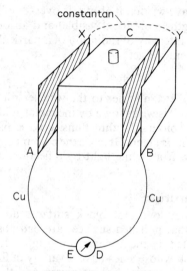

Fig. 13.16 Comparing absorbers

The cube is now turned so that the blackened plate, A, is opposite the polished face of the cube, while the polished plate, B, is opposite the blackened face of the cube. The galvanometer then shows no deflection; the plates thus reach the same temperature. It follows that the good radiating property of the blackened face of the cube, and the bad absorbing property of the polished plate, are just compensated by the good absorbing property of the blackened plate, and the bad radiating property of the polished face of the cube.

The Black Body

The experiments described before lead us to the idea of a *perfectly black body*; one which absorbs all the radiation that falls upon it, and reflects and transmits none. The experiments also lead us to suppose that such a body would be the best possible radiator.

A good black body can be made simply by punching a small hole in the lid of a closed empty tin. The hole looks almost black, although the shining tin is a good reflector. The hole looks black because the light which enters through it is reflected many times round the walls of the tin, before it passes through the hole again, Fig. 13.17. At each reflection, about 80 per cent of the light energy is reflected, and 20 per cent is absorbed. After two reflection, 64 per cent of the original light goes on to be reflected a third time; 36 per cent has been absorbed. After ten reflections, the fraction of the original energy which has been absorbed is 0.8^{10}, or 0.1. So the closed tin with a hole in it is a very good absorber of radiation.

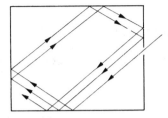

Fig. 13.17 Multiple reflections make a black body

Any space which is almost wholly enclosed approximates to a black body. And, since a good absorber is also a good radiator, an almost closed space is the best radiator we can find.

A form of black body used in radiation measurements is shown in Fig. 13.18. It consists of a porcelain sphere, S, with a small hole in it. The inside is blackened with soot to make it as good a radiator and as bad a reflector as possible. (The effect of multiple reflections is then to convert the body from nearly black to very nearly black indeed.) The sphere is surrounded by a high-temperature bath of, for example, molten salt (the melting-point of common salt is 800°C).

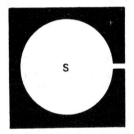

Fig. 13.18 A black body

Quality of Radiation

The deepest parts of a coal or wood fire are black bodies. Anyone who has looked into a fire knows that the deepest parts of it look brightest—they are radiating most power. In the hottest part, no detail of the coals or wood can be seen. So the radiation from an almost enclosed space is uniform; its character does not vary with the nature of the surface of the space. This is so because the radiation coming out from any area is made up partly of the radiation emitted by that area, and partly of the radiation from other areas, reflected at the area in question. And if the hole in the body is small, the radiations from every area inside it are well mixed by reflection before they can escape. So the intensity and quality of the radiation escaping thus does not depend on the particular surface from which it escapes.

When we speak of the *quality* of radiation we mean the relative intensities of the different wavelengths in it; the proportion of red to blue, for example. The quality of the radiation from a perfectly black body depends *only on its temperature*. When the body is made hotter, its radiation becomes not only more intense, but also more nearly white; the proportion of blue to red in it increases. Because its quality is determined only by its temperature, black body radiation is sometimes called 'temperature radiation'.

Properties of Temperature Radiation

The quality of the radiation from a black body was investigated by Lummer and Pringsheim in 1899. They used a black body represented by B in Fig. 13.19 and measured its temperature with a thermopile; they took it to 2000°C. To measure the intensities of the various wavelengths, Lummer and Pringsheim used an infrared spectrometer and a linear bolometer (p. 270) consisting of a single platinum strip.

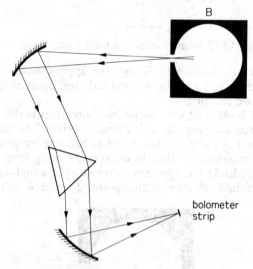

Fig. 13.19 Lummer and Pringsheim's apparatus for study of black body radiation (diagrammatic)

The results of the experiments are shown in Fig. 13.20 (i). Each curve gives the relative intensities of the different wavelengths, for a given temperature of the body. The curves show that, as the temperature rises, the intensity of every wavelength increases, but the intensities of the shorter wavelengths increase more rapidly. Thus the radiation becomes, as we have already observed, less red, that is to say, more nearly white. The curve for sunlight has its peak at about 5×10^{-7} m in the visible green; from the position of this peak we conclude that the surface temperature of the sun is about 6000 K. Stars which are hotter than the sun, such as Sirius and Vega, look blue, not white, because the peaks of their radiation curves lie further towards the visible blue than does the peak of sunlight.

The actual intensities of the radiations are shown on the right of the graph in Fig. 13.20. To speak of the intensity of a single wavelength is meaningless. The slit of the spectrometer always gathers a *band of wavelengths*—the narrower the slit the narrower the band—and we always speak of the intensity of a given band. We express it as follows ('s' represents 'second'):

$$\text{energy radiated m}^{-2}\text{ s}^{-1}, \text{ in band } \lambda \text{ to } \lambda + \Delta\lambda = E_\lambda\Delta\lambda. \qquad (8)$$

The quantity E_λ is called *emissive power* of a black body for the wavelength λ and at the given temperature; its definition follows from equation (8):

$$E_\lambda = \frac{\text{energy radiated m}^{-2}\,\text{s}^{-1},\ \text{in band } \lambda \text{ to } \lambda + \Delta\lambda}{\text{bandwidth, } \Delta\lambda}.$$

The expression 'energy per second' can be replaced by the word 'power', whose unit is the watt. Thus

$$E_\lambda = \frac{\text{power radiated m}^{-2} \text{ in band } \lambda \text{ to } \lambda + \Delta\lambda}{\Delta\lambda}.$$

In the figure E_λ is expressed in watt per m² per Angstrom unit. SI units may be 'watt per metre² per nanometre (10^{-9} m)'.

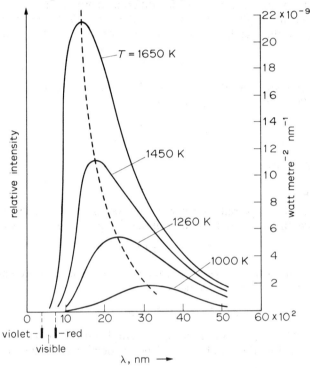

Fig. 13.20 Distribution of intensity in black body radiation

The quantity $E_\lambda\Delta\lambda$ in equation (8) is the *area* beneath the radiation curve between the wavelengths λ and $\lambda + \Delta\lambda$ (Fig. 13.21). Thus the energy radiated per metre² per second between those wavelengths is proportional to that area. Similarly, the *total radiation* emitted per metre² per second over all wavelengths is proportional to *the area under the whole curve*.

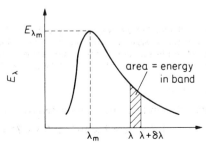

Fig. 13.21 Definition of E, λ_m and E_{λ_m}

Laws of Black Body Radiation

The curves of Fig. 13.20 can be explained only by Planck's quantum theory of radiation, which is outside our scope. Both theory and experiment lead to three generalisations, which together describe well the properties of black body radiation:

(i) If λ_m is the wavelength of the peak of the curve for T (in K), then

$$\lambda_m T = \text{constant} \qquad . \qquad . \qquad . \qquad . \qquad . \qquad (9)$$

The value of the constant is $2 \cdot 9 \times 10^{-3}$ m K. In Fig. 13.20 the dotted line is the locus of the peaks of the curves for different temperatures.

The relationship in (9) is sometimes called *Wien's displacement law*.

(ii) If E_{λ_m} is the height of the peak of the curve for the temperature T, then

$$E_{\lambda_m} \propto T^5 \qquad . \qquad . \qquad . \qquad . \qquad . \qquad (10)$$

(iii) The curve showing the variation of E_λ with λ at constant temperature T in Fig. 13.20 obeys the *Planck formula*

$$E_\lambda = \frac{c_1}{\lambda^5 (e^{c_2/\lambda T} - 1)}$$

where c_1 and c_2 are constants.

Stefan's Law

If E is the *total* energy radiated per metre^2 per second at a temperature T, which is represented by the *total area* under the particular $E_\lambda - \lambda$ curve, then

$$E = \sigma T^4,$$

where σ is a constant. This result is called *Stefan's law*, and the constant σ is called the *Stefan constant*. Its value is

$$\sigma = 5 \cdot 7 \times 10^{-8} \text{ W m}^{-2} \text{ K}^{-4}.$$

So in Fig. 13.20, which shows four $E_\lambda - \lambda$ graphs at different temperatures T, the total area below the graphs should be proportional to the corresponding value of T^4.

The energy per second or *power P* radiated by an area A of a black body radiator is thus given by

$$P = A\sigma T^4,$$

where P is in watts. This is illustrated in the examples which follow.

Example

1. The tungsten filament of an electric lamp has a length of $0 \cdot 5$ m and a diameter 6×10^{-5} m. The power rating of the lamp is 60 W. Assuming the radiation from the filament is equivalent to 80% that of a perfect black body radiator at the same temperature, estimate the steady temperature of the filament. (Stefan constant = $5 \cdot 7 \times 10^{-8}$ W m^{-2} K^{-4}.)

When the temperature is steady,

$$\text{power radiated from filament} = \text{power received} = 60 \text{ W}$$
$$\therefore \ 0 \cdot 8 \times 5 \cdot 7 \times 10^{-8} \times 2\pi \times 3 \times 10^{-5} \times 0 \cdot 5 \times T^4 = 60,$$

since surface area of cylindrical wire is $2\pi rh$ with the usual notation.

$$\therefore T = \left(\frac{60}{0.4 \times 5.7 \times 10^{-8} \times 2\pi \times 3 \times 10^{-5}} \right)^{1/4}$$

$$= 1933 \text{ K}.$$

2. The solar constant, which is the energy arriving per second at the earth from the sun, is about 1400 W m^{-2}. Estimate the surface temperature of the sun, given that the sun's radius $= 7 \times 10^5$ km, the distance of the sun from the earth $= 1.5 \times 10^8$ km and Stefan constant $= 5.7 \times 10^{-8}$ W m^{-2} K^{-4}.

Suppose T is the kelvin temperature of the sun's surface. Then

total energy per second radiated from sun's surface $= A\sigma T^4 = 4\pi r_s^2 \sigma T^4$,

since the sun's surface is $4\pi r_s^2$ if r_s is its radius.

This energy falls all round a sphere of radius r_0 where r_0 is the radius of the earth's circular orbit round the sun. Since the area of the sphere is $4\pi r_0^2$, the energy per second falling on unit area

$$= \frac{1}{4\pi r_0^2} \times 4\pi r_s^2 \sigma T^4 = \left(\frac{r_s}{r_0} \right)^2 \sigma T^4.$$

So

$$\left(\frac{r_s}{r_0} \right)^2 \sigma T^4 = 1400.$$

Hence

$$T^4 = \frac{1400}{5.7 \times 10^{-8}} \times \left(\frac{1.5 \times 10^8}{7 \times 10^5} \right)^2.$$

So

$$T = 5800 \text{ K (approx.)}.$$

Prévost's Theory of Exchanges

In 1792 Prévost applied the idea of dynamic equilibrium to radiation. He asserted that a body radiates heat at a rate which depends only on the nature of its surface and its temperature, and that it absorbs heat at a rate depending on the nature of its surface and the temperature of its surroundings. *When the temperature of a body is constant, the body is losing heat by radiation, and gaining it by absorption, at equal rates.*

Figure 13.21 shows a simple experiment to demonstrate Prévost's theory. A high vacuum, electric lamp is carefully put into a can of water (Fig. 13.22 (i)). The temperature of the lamp's filament is found by measuring its resistance. We find that, whatever the temperature of the water, the filament comes to that

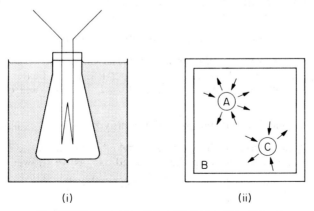

(i)   (ii)

Fig. 13.22 Illustrating Prévost's theory of exchanges

temperature, if we leave it long enough. When the water is cooler than the fila-ment, the filament cools down; when the water is hotter, the filament warms up.

In the abstract language of theoretical physics, Prévost's theory is easy enough to discuss. If a hot body A (Fig. 13.22 (ii)) is placed in an evacuated enclosure B, at a lower temperature than A, then A cools until it reaches the temperature of B. If a body C, cooler than B, is put in B, then C warms up to the temperature of B. We conclude that radiation from B falls on C, and therefore also on A, even though A is at a higher temperature. Thus A and C each come to equi-librium at the temperature of B when each is absorbing and emitting radiation at equal rates.

Now let us suppose that, after it has reached equilibrium with B, one of the bodies, say C, is transferred from B to a cooler evacuated enclosure D (Fig. 13.23 (i)). It loses heat and cools to the temperature of D. Therefore it is radiating heat. But if C is transferred from B to a warmer enclosure F, then C gains heat and warms up to the temperature of F (Fig. 13.23 (ii)). It seems unreasonable to suppose that C stops radiating when it is transferred to F. It is more reason-able to suppose that it goes on radiating but, while it is cooler than F, it absorbs more than it radiates.

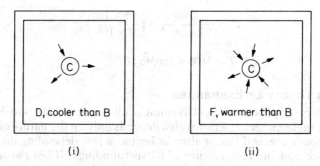

Fig. 13.23 Illustrating Prévost's theory

Hot Object in Enclosure

Consider a black body at a temperature of T_0 where T_0 is the temperature of the room or enclosure containing the body. Since the body is in temperature equilibrium, the energy per second it radiates must equal the energy per second it absorbs, from Prévost's theory of exchanges. If A is the surface area of the body, then, from Stefan's law,

$$\text{energy per second radiated} = \sigma A T_0{}^4.$$

So the *energy per second absorbed from the surroundings or enclosure* $= \sigma A T_0{}^4$.

Now suppose the black body X is heated electrically by a heater of power W watts and finally reaches a constant temperature T. In this case, from Prévost's theory,

energy per second from heater, W = net energy per second radiated by X.

The net energy per second radiated by X $= \sigma A T^4 - \sigma A T_0{}^4$, since $\sigma A T_0{}^4$ is the energy per second absorbed from the surroundings, as we showed before. So

$$W = \sigma A T^4 - \sigma A T_0{}^4 = \sigma A(T^4 - T_0{}^4).$$

Example

A metal sphere with a black surface and radius 30 mm, is cooled to $-73°C$ (200 K) and placed inside an enclosure at a temperature of $27°C$ (300 K). Calculate the initial rate of temperature rise of the sphere, assuming the sphere is a black body. (Assume density of metal = 8000 kg m^{-3} specific heat capacity of metal = 400 J kg^{-1} K^{-1}, and Stefan constant = 5.7×10^{-8} W m^{-2} K^{-4}.)

$$\text{Energy per second radiated by sphere} = \sigma A(T^4 - T_0{}^4)$$

where A is the surface area $4\pi r^2$ of the sphere of radius r, $T = 200$ K and $T_0 = 300$ K. Since the temperature of the surroundings is greater than that of the sphere, the energy per second, Q, *gained* from the surroundings is given by

$$Q = \sigma . 4\pi r^2 . (300^4 - 200^4).$$

The mass m of the sphere = volume × density = $\frac{4}{3}\pi r^3 \rho$, where ρ is the density. If c is the specific heat capacity of the metal, and θ is the initial rise per second of its temperature, then

$$Q = mc\theta = \frac{4}{3}\pi r^3 \rho c\theta = \sigma 4\pi r^2 (300^4 - 200^4).$$

Dividing by $4\pi r^2$, and then simplifying,

$$\theta = \frac{\sigma(300^4 - 200^4) \times 3}{r\rho c}$$

$$= \frac{5.7 \times 10^{-8} \times (300^4 - 200^4) \times 3}{30 \times 10^{-3} \times 8000 \times 400}$$

$$= 0.012 \text{ K s}^{-1} \text{ (approx.)}.$$

Emissivity

Consider a body B, in equilibrium with an enclosure A, at a temperature T (Fig. 13.24). If the body is perfectly black, it emits radiation characteristics of the temperature T. Suppose the total intensity of this radiation over all wavelengths is E watt/m^2. Since the body is in equilibrium with the enclosure, it is absorbing as much as its radiates. And since it absorbs all the radiation that falls upon it, the energy falling on it per m^2 per second must be equal to E. This conclusion need not surprise us, since the enclosure A is full of black body radiation characteristic of its temperature T.

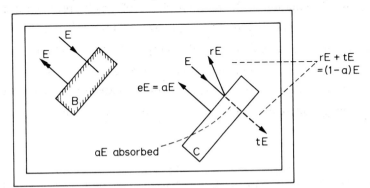

Fig.13.24 Radiation equilibrium in an enclosure

Now let us consider, in the same enclosure, a body of C which is not black. On each square metre of the body's surface, E watt of radiation fall (Fig. 13.24). Of this, suppose that the body absorbs a fraction, a, that is to say, it absorbs

aE watt per m^2. We may call a the total absorption factor of the body C, 'total' because it refers to the total radiation. The radiation which the body does not absorb, $(1 - a)E$, it reflects or transmits.

So: $\left.\begin{matrix} \text{power reflected or} \\ \text{transmitted/m}^2 \end{matrix}\right\} = E - aE.$

For equilibrium, the total power leaving the body per m^2 must be equal to the total power falling upon it, E in W/m^2. The power emitted by the body, which must be added to that reflected and transmitted, is therefore:

$$\text{total power transmitted/m}^2 = aE \qquad . \qquad . \qquad . \qquad (1)$$

The ratio of the total power radiated per m^2 by a given body, to that emitted by a black body at the same temperature, is called the *total emissivity* of the given body. Hence, by equation (1),

$$e = \frac{aE}{E} = a.$$

We have therefore shown that the total emissivity of a body is equal to its total absorption factor.

This was shown by experiment on page 274. If we combine it with Stefan's law, we find that the total energy E radiated per m^2 per second by a body of emissivity e at a temperature T K is

$$E = e\sigma T^4.$$

Spectral Emissivity: Kirchhoff's Law

Most bodies are coloured; they transmit or reflect some wavelengths better than others. We have already seen that they must absorb these wavelengths weakly; we now see that, because they absorb them weakly, they must also radiate them weakly. To show this, we have only to repeat the foregoing argument, but restricting it to a narrow band of wavelengths between λ and $\lambda + \Delta\lambda$. The energy falling per m^2 per second on the body, in this band, is $E_\lambda \Delta\lambda$ where E_λ is the emissive power of a black body in the neighbourhood of λ, at the temperature of the enclosure. If the body C absorbs a fraction a_λ of this, we call a_λ the spectral absorption factor of the body, for the wavelength λ. In equilibrium, the body emits as much radiation in the neighbourhood of λ as it absorbs; thus:

$$\text{energy radiated} = a_\lambda E_\lambda \Delta\lambda \text{ watts per m}^2.$$

We define the spectral emissivity of the body e_λ, by the equation

$$e_\lambda = \frac{\text{energy radiated by body in range } \lambda \text{ to } \lambda + \Delta\lambda}{\text{energy radiated in same range, by black body at same temperature}}$$

$$= \frac{\text{energy radiated by body in range } \lambda \text{ to } \lambda + \Delta\lambda}{E_\lambda \Delta\lambda} = \frac{a_\lambda E_\lambda \Delta\lambda}{E_\lambda \Delta\lambda}.$$

Thus $$e_\lambda = a_\lambda \qquad . \qquad . \qquad . \qquad . \qquad (1)$$

Equation (1) expresses a law due to Kirchhoff:

The spectral emissivity of a body for a given wavelength is equal to its spectral absorption factor for the same wavelength.

Kirchhoff's law is not easy to demonstrate by experiment. If a plate when cold shows a red pattern on a blue background, it glows blue on a red ground when heated in a furnace. Not all such plates do this, because the spectral emissivity of many coloured pigments vary with their temperature. However,

Fig. 13.25 shows two photographs of a decorated plate, one taken by reflected light at room temperature (left), the other by its own light when heated to a temperature of about 1100 K (right).

 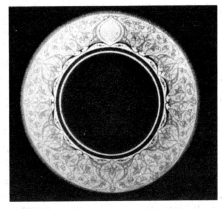

Fig. 13.25 Photographs showing how a decorated plate appears (left) by reflected light at room temperature and (right) by its own emitted light when incandescent at about 1100 K

Absorption by Gases

An experiment which shows that, if a body radiates a given wavelength strongly, it also absorbs that wavelength strongly, can be made with sodium vapour. A sodium vapour lamp runs at about 220°C; compared with the sun, or even an arc-lamp, it is cool. The experiment consists of passing sunlight or arc-light through a spectroscope, and observing its continuous spectrum. The sodium lamp is then placed in the path of the light, and a black line appears in the yellow. If the white light is now cut off, the line which looked black comes up brightly—it is the sodium yellow line.

The process of absorption by sodium vapour—or any other gas—is not, however, the same as the process of absorption by a solid. When a solid absorbs radiation, it turns it into heat—into the random kinetic energy of its molecules. It then re-radiates it in all wavelengths, but mostly in very long ones, because the solid is cool. When a vapour absorbs light of its characteristic wavelength, however, its atoms are excited; they then re-radiate the absorbed energy, in the same wavelength (589·3 nm for sodium). But they re-radiate it in all directions, and therefore less of it passes on in the original direction than before (Fig. 13.26). Thus the yellow component of the original beam is weakened, but the yellow light radiated sideways by the sodium is strengthened. The sideways strengthening is hard to detect, but it was shown by R. W. Wood in 1906. He used mercury vapour instead of sodium. See *Fraunhofer lines*, p. 377.

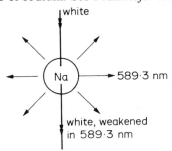

Fig. 13.26 Absorption by sodium vapour

Emission and Absorption in Non-equilibrium Conditions

It is sometimes said that since a good absorber is also a good emitter of radiation, the two effects should cancel each other out. For example, a blackened cloth exposed to radiation will absorb heat more quickly than a white cloth but it will also radiate more quickly. Why, then, should the temperature of the black cloth rise more rapidly? The answer is that we have been describing a *non-equilibrium* situation.

When two bodies with different emissive powers e_λ and absorptive powers a_λ, but otherwise identical, are in *equilibrium* with their surroundings, each emits exactly the amount of radiation which it receives from the surroundings. No difference in appearance can then be seen between the two bodies. We can illustrate this by considering a white china plate with a decorated pattern in radiative equilibrium inside a furnace. The plate appears uniformly bright—no pattern can be seen.

However, when the plate is at room temperature in an illuminated room, the pattern can be clearly seen against the white china background. There is now no radiative equilibrium. The sun, or lamp, which illuminates the plate has a high temperature. The plate, however, is at room temperature. It emits very little infra-red (invisible) radiation and we therefore see only light reflected from its surface. Thus any absorbing parts such as the decorated pattern are seen dark compared with the surrounding white china, whose reflection factor is high. In a room lit by the sun, the plate is only slightly heated by absorption and never reaches radiative equilibrium.

If the plate is taken out of a furnace and into a dark room, it is then not in radiative equilibrium. It now emits more radiation from the dark patches such as the pattern than the light patches, and since there is little radiation from the surroundings to absorb, the pattern looks brighter than the surrounding white china.

We can now account for the temperature changes of the two cloths mentioned at the beginning of the section. If black and white cloths are exposed to sunlight, they are not in radiative equilibrium with the sun. The black cloth absorbs more energy per second from the sun than the white cloth. The black cloth also emits more energy per second than the white cloth. But since the cloths are much cooler than the sun, each emits only a very small amount of radiation compared to what is received. The net effect, then, is that the black cloth absorbs heat at a faster rate, and so rises in temperature more quickly than the white cloth.

Example

Estimate the temperature T_e of the earth, assuming it is in radiative equilibrium with the sun. (Assume radius of sun, $r_s = 7 \times 10^8$ m, temperature of solar surface = 6000 K, distance of earth from sun = $1 \cdot 5 \times 10^{11}$ m.)

Power radiated from sun = $\sigma \times$ surface area $\times T^4$

$$= \sigma \times 4\pi r_s^2 \times T_s^4$$

Power received by earth = $\dfrac{\pi r_e^2}{4\pi R^2} \times$ power radiated by sun,

since πr_e^2 is the effective area of the earth on which the sun's radiation is incident *normally*, Fig. 13.27, and $4\pi R^2$ is the total area on which the sun's radiation falls at a distance R from the sun where the earth is situated.

Now power radiated by earth = $\sigma . 4\pi r_e^2 . T_e^4$.

Assuming radiative equilibrium,

power radiated by earth = power received by earth

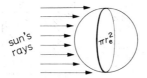

Fig. 13.27 Example

$$\therefore \sigma.4\pi r_e^2.T_e^4 = \sigma.4\pi r_s^2.T_s^4 \times \frac{\pi r_e^2}{4\pi R^2}.$$

Cancelling r_e^2 and simplifying, then

$$T_e^4 = T_s^4 \times \left(\frac{r_s^2}{4R^2}\right)$$

$$\therefore T_e = T_s \times \left(\frac{r_s}{2R}\right)^{1/2}$$

$$= 6000 \times \left(\frac{7 \times 10^8}{2 \times 1\cdot 5 \times 10^{11}}\right)^{1/2}$$

$$= 290 \text{ K}.$$

Note that the calculation is approximate, for example, the earth and the sun are not perfect black body radiators and the earth receives heat from its interior.

Exercises 13

Conduction
1. A closed metal vessel contains water (i) at 30°C and then (ii) at 75°C. The vessel has a surface area of 0·5 m² and a uniform thickness of 4 mm. If the outside temperature is 15°C, calculate the heat loss per minute by conduction in each case. (Thermal conductivity of metal = 400 W m⁻¹ K⁻¹.)
2. A uniform metal bar has one end kept at 100°C and the other at 0°C. Draw sketches showing how the temperature varies along the bar in the steady state (*a*) when its sides are well lagged, (*b*) when the sides are unlagged, (*c*) if the bar is not uniform but tapers or narrows from the hot to the cold end. Explain your answers in each case.
3. In measuring thermal conductivity of a metal, a long, thick, lagged bar is used. In measuring thermal conductivity of a bad conductor such as cardboard, a thin disc of large surface area is used which is not lagged.
  Explain the reasons for these practical arrangements.
4. A metal cylinder, containing water at 60°C, has a thickness of 4 mm and thermal conductivity 400 W m⁻¹ K⁻¹. It is lagged by felt of thickness 2 mm and thermal conductivity 0·002 W m⁻¹ K⁻¹. The room temperature is 10°C.
  Using the relation $Q/t = kAg$ to find the temperature gradient g for the metal and for the felt, show that (*a*) the temperature θ of the metal–felt interface is practically 60°C and (*b*) the rate of loss of heat by conduction is practically unaffected when the metal cylinder is replaced by a metal of smaller thermal conductivity 100 W m⁻¹ K⁻¹ and the same thickness.
5. (*a*) The diagram (Fig. 13A (i)) shows a section of a house wall one brick thick, the surfaces of the brick being at the temperatures shown. If the thermal conductivity of the brick is 0·6 W m⁻¹ K⁻¹, what is the rate of heat flow per unit area (W m⁻²) through the bricks if steady state conditions apply?
  Explain why, under these conditions, the temperature of the outer surface of the wall must be greater than the air temperature. At what rate must the outer surface be losing heat?
  (*b*) The diagram (Fig. 13A (ii)) shows a section of a cavity wall made up of brick, air and brick. The thermal conductivity of air is 0·02 W m⁻¹ K⁻¹. Explain why, when steady state conditions apply, the rate of heat flow across each layer is the same. Assuming

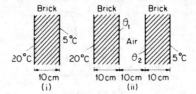

Fig. 13A

this to be the case, draw a sketch graph to show how the temperature changes between the brick surface at 20°C and that of 5°C.

Calculate, showing your working, the thickness of brick equivalent to 10 cm of air. Hence, or otherwise, calculate values for θ_1 and θ_2, the inner surfaces of the bricks, and compare the rate of heat loss through the cavity wall with that through the single brick described in (a).

Explain why, in practice, the introduction of a cavity does not produce the improvement suggested by the calculation unless, for example, the cavity is filled with plastic foam. (L.)

6. Define *thermal conducitivity*.

Figure 13B represents, in outline, the apparatus used in Searle's bar method for determining the thermal conductivity of copper. (a) Why is a thick bar used in this determination? (b) Why must it be well insulated except at its two ends? (c) Why does one wait for some time before taking readings? (d) Does it matter where the thermometers T_1 and T_2 are placed along the bar? Explain.

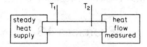

Fig. 13B

(e) One end of the insulated copper bar, which is of length 0·2 m and cross-sectional area $1\cdot2 \times 10^{-3}$ m², is maintained at a steady temperature by an electric heater which is supplying heat to the bar at the rate of 100 W. Thermometer T_1 is 0·06 m from the hot end and thermometer T_2 is 0·14 m from the hot end. At the cool end, water flows into a circulating coil at 15·3°C and leaves it at 16·7°C. Taking the thermal conductivity of copper to be 400 W m^{-1} K^{-1} and specific heat capacity of water 4200 J kg^{-1} K^{-1}, estimate the rate at which water is flowing through the circulating coil and also the reading of each of the thermometers T_1 and T_2. (O.)

7. Define *coefficient of thermal conductivity*. Describe a method of measuring this coefficient for a metal.

Assuming that the thermal insulation provided by a woollen glove is equivalent to a layer of quiescent air 3 mm thick, determine the heat loss per minute from a man's hand, surface area 200 cm² on a winter's day when the atmospheric air temperature is −3°C. The skin temperature is to be taken as 34°C and the thermal conductivity of air as 24×10^{-3} W m^{-1} K^{-1}. (L.)

8. The ends of a bar of uniform cross-section are maintained at steady different temperatures, both being above room temperature. Explain how the temperature varies along the bat if the bar is (a) ideally lagged, (b) unlagged.

A bar 0·20 m in length and of cross-sectional area $2\cdot5 \times 10^{-4}$ m² (2·5 cm²) is ideally lagged. One end is maintained at 373 K (100°C) while the other is maintained at 273 K (0°C) by immersion in melting ice. Calculate the rate at which the ice melts owing to the flow of heat along the bar.

Thermal conductivity of the material of the bar = $4\cdot0 \times 10^2$ W m^{-1} K^{-1}.

Specific latent heat of fusion of ice = $3\cdot4 \times 10^5$ J kg^{-1}. (N.)

9. Give a critical account of an experiment to determine the thermal conductivity of a material of low thermal conductivity such as cork. Why is it that most cellular materials, such as cotton wool, felt, etc., all have approximately the same thermal conductivity?

One face of a sheet of cork, 3 mm thick, is placed in contact with one face of a sheet of glass 5 mm thick, both sheets being 20 cm square. The outer faces of this square composite sheet are maintained at 100°C and 20°C, the glass being at the higher mean temperature. Find (a) the temperature of the glass-cork interface, and (b) the rate at which heat is conducted across the sheet, neglecting edge effects.

[Thermal conductivity of cork = $6\cdot3 \times 10^{-2}$ W m^{-1} K^{-1}, thermal conductivity of glass = $7\cdot2 \times 10^{-1}$ W m^{-1} K^{-1}.] (*O. & C.*)

10. Define *thermal conductivity* and state a unit in which it is expressed.

Explain why, in an experiment to determine the thermal conductivity of copper using a Searle's arrangement, it is necessary (a) that the bar should be thick, of uniform cross-section and have its sides well lagged, (b) that the temperatures used in the calculation should be the steady values finally registered by the thermometers.

Straight metal bars X and Y of circular section and equal in length are joined end to end. The thermal conductivity of the material of X is twice that of the material of Y, and the uniform diameter of X is twice that of Y. The exposed ends of X and Y are maintained at 100°C and 0°C, respectively and the sides of the bars are ideally lagged. Ignoring the distortion of the heat flow at the junction, sketch a graph to illustrate how the temperature varies between the ends of the composite bar when conditions are steady. Explain the features of the graph and calculate the steady temperature of the junction. (*N.*)

11. Outline an experiment to measure the thermal conductivity of a solid which is a poor conductor, showing how the result is calculated from the measurements.

Calculate the theoretical percentage change in heat loss by conduction achieved by replacing a single glass window by a double window consisting of two sheets of glass separated by 10 mm of air. In each case the glass is 2 mm thick. (The ratio of the thermal conductivities of glass and air is 3:1.)

Suggest why, in practice, the change would be less than that calculated. (*L.*)

12. Outline, with the necessary theory, a simple method for finding the thermal conductivity of a bad conductor.

Estimate the rate of heat loss from a room through a glass window of area 2 m^2 and thickness 3 mm when the temperature of the room is 20°C and that of the air outside is 5°C.

This estimate is too high because it is based on the assumption that the inner glass surface is at room temperature. A better approximation to actual conditions is to suppose that there is a thin uniform layer of still air in contact with the inner glass surface of the window. In practice, it is found that the rate of loss of heat through this window when the inside and outside temperatures have the above values is only 3 kW. Using the suggested approximation, find (a) the temperature difference between the faces of the glass, (b) the temperature difference across the air film, (c) the thickness of this film. (Take thermal conductivity of glass to be $1\cdot2$ W m^{-1} K^{-1} and that of air $2\cdot4 \times 10^{-2}$ W m^{-2} K^{-4}.) (*L.*)

Radiation

13. Explain why a body at 1000 K is 'red hot' whereas at 2000 K it is 'white hot'. (*C.*)

14. The silica cylinder of a radiant wall heater is $0\cdot6$ m long and has a radius of 5 mm. If it is rated at $1\cdot5$ kW estimate its temperature when operating. State *two* assumptions you have made in making your estimate. (The Stefan constant, $\sigma = 6 \times 10^{-8}$ W m^{-2} K^{-4}.) (*L.*)

15. What is a *black body radiator*? Draw a diagram of a simple laboratory form of black body radiator.

Sketch roughly the energy distribution among the wavelengths of (a) a black body radiator, (b) a non-black body radiator and state the main differences.

16. Explain the difference between the *total radiation E* of a given black body and the *intensity of radiation E$_\lambda$* and state a law which each obeys.

Write down the units of E and of E_λ.

17. What is the ratio of the energy per second radiated by the filament of a lamp at 2500 K to that radiated at 2000 K, assuming the filament is a black body radiator?

The filament of a particular electric lamp can be considered as a 90% black body radiator. Calculate the energy per second radiated when its temperature is 2000 K if its surface area is 10^{-6} m². (Stefan constant = 5.7×10^{-8} W m^{-2} K^{-4}.)

18. The sun is a black body of surface temperature about 6000 K. If the sun's radius is 7×10^8 m, calculate the energy per second radiated from its surface.

The earth is about 1.5×10^{11} m from the sun. Assuming all the radiation from the sun falls on a sphere of this radius, estimate the energy per second per metre² received by the earth. (Stefan constant = 5.7×10^{-8} W m^{-2} K^{-4}.)

19. Sketch graphs showing the distribution of energy in the spectrum of black body radiation at three temperatures, indicating which curve corresponds to the highest temperature. If such a set of graphs were obtained experimentally, how would you use information from them to attempt to illustrate Stefan's law? (*L*.)

20. Describe briefly how you would compare the thermal radiation emitted by a body at different temperatures.

How would you show that a matt black surface is a better radiator of heat than other surfaces? (*L*.)

21. Explain what is meant by the *Stefan constant*, defining any symbols used.

A sphere of radius 2·00 cm with a black surface is cooled and then suspended in a large evacuated enclosure the black walls of which are maintained at 27°C. If the rate of change of thermal energy of the sphere is 1.85 J s^{-1} when its temperature is $-73°$C, calculate a value for the *Stefan constant*. (*N*.)

22. Give an account of Stefan's law of radiation, explaining the character of the radiating body to which it applies and how such a body can be experimentally realised.

If each square cm of the sun's surface radiates energy at the rate of 6.3×10^3 J s^{-1} cm^{-2} and the Stefan constant is 5.7×10^{-8} W m^{-2} K^{-4}, calculate the temperature of the sun's surface in degrees centigrade, assuming Stefan's law applies to the radiation. (*L*.)

23. As the temperature of a black body rises what changes take place in (*a* the total energy radiated from it and (*b*) the energy distribution among the wavelengths radiated? Illustrate (*b*) by suitable graphs and explain how the information required in (*a*) could be obtained from these graphs.

Use your graphs to explain how the appearance of the body changes as its temperature rises and discuss whether or not it is possible for a black body to radiate white light.

The element of an electric fire, with an output of 1.0 kW, is a cylinder 25 cm long and 1·5 cm in diameter. Calculate its temperature when in use, if it behaves as a black body. (The Stefan constant = 5.7×10^{-8} W m^{-2} K^{-4}.) (*L*.)

24. State Prévost's theory of exchanges.

An enclosure contains a black body *A* which is in equilibrium with it. A second black body *B*, at a higher temperature than *A*, is then also placed in the enclosure. If the enclosure is maintained at constant temperature and all heat exchange is by radiation state and explain how the temperatures of *A* and *B* will change with time.

Describe an electrical instrument which can be used to detect heat radiation and explain its action.

A solid copper sphere, of diameter 10 mm, is cooled to a temperature of 150 K and is then placed in an enclosure maintained at 290 K. Assuming that all interchange of heat is by radiation, calculate the initial rate of rise of temperature of the sphere. The sphere may be treated as a black body.

(Density of copper = 8.93×10^3 kg m^{-3}.

Specific heat capacity of copper = 3.70×10^2 J kg^{-1} K^{-1}.

The Stefan constant = 5.70×10^{-8} W m^{-2} K^{-4}.) (*L*.)

25. What is meant by a perfectly black body? Outline an experiment to investigate the way in which the energy radiated per second by a black body is distributed throughout the wavelengths of the spectrum.

If such an experiment is carried out at various temperatures, sketch the graphs which would be obtained when the energy radiated per second at various wavelengths is plotted against the wavelength, for the different temperatures. From one such graph, how would you obtain the total energy radiated per second at that temperature? How does this total energy radiated per second depend on the temperature?

The element of a 1 kW electric fire has a surface area of 0·006 m². Estimate its working temperature. (The Stefan constant = 5.7×10^{-8} W m^{-2} K^{-4}.) (*L*.)

26. What is Prévost's Theory of Exchanges? Describe some phenomenon of theoretical or practical importance to which it applies.

A metal sphere of 1 cm diameter, whose surface acts as a black body, is placed at the focus of a concave mirror with aperture of diameter 60 cm directed towards the sun. If the solar radiation falling normally on the earth is at the rate of 0.14 watt cm^{-2}, Stefan's constant is taken as 6×10^{-8} W m^{-2} K^{-4} and the mean temperature of the surroundings is 27°C, calculate the maximum temperature which the sphere could theoretically attain, stating any assumptions you make. (*O. & C.*)

27. Explain what is meant by (*a*) a *black body*, (*b*) *black body radiation*.

State *Stefan's law* and draw a diagram to show how the energy is distributed against wavelength in the spectrum of a black body for two different temperatures. Indicate which temperature is the higher.

A roof measures 20 m × 50 m and is blackened. If the temperature of the sun's surface is 6000 K, Stefan's constant = 5.72×10^{-8} W m^{-2} K^{-4}, the radius of the sun is 7.8×10^8 m and the distance of the sun from the earth is 1.5×10^{11} m, calculate how much solar energy is incident on the roof per minute, assuming that half is lost in passing through the earth's atmosphere, the roof being normal to the sun's rays. (*O. & C.*)

28. How can the temperature of a furnace be determined from observations on the radiation emitted?

Calculate the apparent temperature of the sun from the following information:

Sun's radius: 7.04×10^5 km. Distance from earth: 14.72×10^7 km.

Solar constant: 1400 W m^{-2}, Stefan constant: 5.7×10^{-8} W m^{-2} K^{-4}. (*N.*)

14 Thermal Expansion of Solids and Liquids

Solids

Linear Expansion

Most solids increase in length when they are warmed. Suppose we measure the length l_1 of a metal rod at room temperature θ_1, and then measure the expansion e of the rod at a higher temperature θ_2. The increase in length, λ, of unit length of the material for one degree temperature rise is then given by

$$\lambda = \frac{\text{expansion}}{\text{original length} \times \text{temperature rise}} = \frac{e}{l_1(\theta_2 - \theta_1)}.$$

The quantity λ is called the *mean linear expansivity* of the metal, over the range θ_1 to θ_2. If this range is not too great—say less than 100 K—the quantity λ may, to a first approximation, be taken as constant.

The linear expansivity of a solid, like the pressure and volume expansivity of a gas, has the unit 'K^{-1}' in SI units; its dimensions are

$$[\lambda] = \frac{[\text{length}]}{[\text{length}] \times [\text{temp.}]} = [\text{temp.}]^{-1}.$$

From the definition of λ, we can estimate the new length of a rod, l_2, at a temperature θ_2 from the equation

$$l_2 = l_1\{1 + \lambda(\theta_2 - \theta_1)\}, \qquad . \qquad . \qquad . \qquad . \qquad (1)$$

where l_1 is the length of the rod at the temperature θ_1, and λ is the mean value over a range which includes θ_2 and θ_1.

For accurate work, however, the length of a solid at a temperature θ must be represented by an equation of the form

$$l = l_0(1 + a\theta + b\theta^2 + c\theta^3 + \ldots), \qquad . \qquad . \qquad . \qquad (2)$$

where l_0 is the length at 0°C, and a, b, c are constants. The constant a is of the same order of magnitude as the mean coefficient λ; the other constants are smaller.

When a solid is subjected to small changes of temperature about a mean value θ, its linear expansivity λ_θ in the neighbourhood of θ may be defined by the equation

$$\lambda_\theta = \frac{1}{l}\frac{dl}{d\theta},$$

where l is the length of the bar at the temperature θ. The following table shows how the linear expansivity of copper varies with temperature.

VALUES OF λ_θ COPPER

θ	-87	0	100	400	600	°C
λ_θ	14·1	16·1	16·9	19·3	20·9	$\times 10^{-6}$ K^{-1}

Accurate Measurement of Expansion

An instrument for accurately measuring the length of a bar, at a controlled temperature, is called a *comparator*, Fig. 14.1. It consists of two microscopes, M_1, M_2, rigidly attached to steel or concrete pillars P_1, P_2. Between the pillars are rails R_1, R_2, carrying water baths such as B. One of these baths contains the bar under test, X, which has scratches near its ends; the scratches are nominally a metre apart. Another water bath contains a substandard metre. The eyepieces of the microscopes are fitted with cross-webs carried on micrometer screws, m_1, m_2.

First the substandard metre is run under the microscopes, and the temperature of its bath is adjusted to that at which the bar was calibrated (usually about 18°C).

When the temperature of the bar is steady, the eyepiece webs are adjusted to intersect the scratches on its ends (Fig. 14.1 (ii)), and their micrometers are read. The distance between the cross-webs is then 1 metre.

The substandard is now removed, and the unknown bar put in its place; the temperature of the bar is brought to 0°C by filling its bath with ice-water. When the temperature of the bar is steady, the eyepiece webs are re-adjusted to intersect the scratches on its ends, and their micrometers are read. If the right-hand web

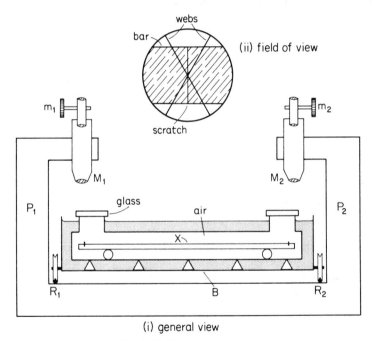

(ii) field of view

(i) general view

Fig. 14.1 The comparator

has been shifted x mm to the right, and the left-hand y mm, also to the right, then the length of the bar at 0°C is

$$l_0 = 1 \text{ metre} + (x - y) \text{ mm.}$$

The bath is now warmed to say 10°C, and the length of the bar again measured. In this way the length can be measured at small intervals of temperature, and the mean linear expansivity, or the coefficients a, b, c in equation (2), can be determined.

Metal-Glass Seals

In radio valves and many other pieces of physical apparatus, it is necessary to seal metal cones into glass tubes, with a vacuum-tight joint. The seal must be made at about 400°C, when the glass is soft; as it cools to room temperature, the glass will crack unless the glass and metal contract at the same rate. This condition requires that the metal and the glass have the same linear expansivity at every temperature between room temperature and the melting-point of glass. It is satisfied nearly enough by platinum and soda glass (mean $\lambda = 9$ and 8.5×10^{-6} K^{-1}, respectively), and by tungsten and some types of hard glass similar to pyrex (mean linear expansivity $= 3 - 4 \times 10^{-6}$ K^{-1}).

Modern seals through soft glass are not made with platinum, but with a wire of nickel-iron alloy, which has about the same linear coefficient as the glass. The wire has a thin coating of copper, which adheres to glass more firmly than the alloy. Also, being soft, the copper takes up small differences in expansion between the alloy and the glass.

In transmitting valves, and large vacuum plants, glass and metal tubes several centimetres in diameter must be joined end-to-end. The metal tubes are made of copper, chamfered to a fine taper at the end where the joint is to be made. The glass is sealed on to the edge of the chamfer; the copper there is thin enough to distort, with the difference in contraction, without cracking the glass.

Example

The metal of a pendulum clock has a linear expansivity of 2×10^{-5} K^{-1}. If the period is 2 s at 15°C, calculate the loss or gain in 10 hours when the temperature rises to 25°C.

Owing to the temperature rise, the length of the pendulum increases from l at 15°C (period T) to l_1 at 25°C (period T_1). Now the period $\propto \sqrt{\text{length}}$ (p. 72). So

$$\frac{T_1}{T} = \sqrt{\frac{l_1}{l}} \qquad \qquad . \qquad . \qquad . \qquad (1)$$

The new length $l_1 = l(1 + \alpha\theta) = l(1 + 10\alpha)$, since the temperature rise is 10 K. So

$$\frac{l_1}{l} = 1 + 10\alpha = 1 + \quad 10 \times 2 \times 10^{-5} = 1.0002.$$

From (1), $$\frac{T_1}{T} = \frac{T_1}{2} = \sqrt{1.0002} = 1.0001.$$

So $$T_1 = 2.0002 \text{ s.}$$

The clock therefore records 2 s when the time change is actually 2·0002 s, that is, the clock *loses* 0·0002 s in 2·0002 s or 2 s to a good approximation. So in 10 hours or 10×3600 s,

$$\text{loss} = \frac{10 \times 3600}{2} \times 0.0002 = 3.6 \text{ s.}$$

Liquids

Cubic Expansivity

The temperature of a liquid determines its volume, but its vessel determines its shape. The only expansivity which we can define for a liquid is therefore its cubic expansivity, γ. Most liquids, like most solids, do not expand uniformly, and γ is not constant over a wide range of temperature. Over a given range θ_1 to θ_2, the *mean coefficient* γ is defined as

$$\gamma = \frac{V_2 - V_1}{V_1(\theta_2 - \theta_1)},$$

where V_1 and V_2 are the volumes of a given mass of liquid at the temperatures θ_1 and θ_2.

MEAN EXPANSIVITIES OF WATER AND MERCURY
(Near room temperature)

	γ (K^{-1})		γ (K^{-1})
	$\times 10^{-4}$		$\times 10^{-4}$
Water: 5–10°C .	0·53	Mercury: 0–30°C .	1·81
10–20°C .	1·50	0–100°C .	1·82
20–40°C .	3·02	0–300°C .	1·87
40–60°C	4·58		
60–80°C	5·87		

Absolute and Apparent Expansion: Change of Density

If we try to find the cubic expansivity of a liquid by warming it in a vessel, the vessel also expands. The expansion which we observe is the difference between the increases in volume of the liquid and the vessel. This is true whether we start with the vessel full and catch the overflow, or observe the creep of the liquid up the vessel (Fig. 14.2). The expansion we observe we call the *apparent expansion*; it is always less than the true or absolute expansion of the liquid.

Most methods of measuring the expansion of a liquid, whether true or apparent, depend on the change in density of the liquid when it expands. We therefore consider this change, before describing the measurements in detail.

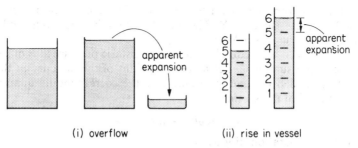

(i) overflow   (ii) rise in vessel

Fig. 14.2 Apparent expansion of a liquid

The mean true or absolute expansivity of a liquid, γ, is defined in the same way as the mean cubic expansivity of a solid:

$$\gamma = \frac{\text{increase in volume}}{\text{initial volume} \times \text{temperature rise}}.$$

Thus, if V_1 and V_2 are the volumes of unit mass of the liquid at θ_1 and θ_2, then

$$V_2 = V_1\{1 + \gamma(\theta_2 - \theta_1)\}.$$

The densities of the liquid at the two temperatures are

$$\rho_1 = \frac{1}{V_1}, \; \rho_2 = \frac{1}{V_2};$$

so that

$$\frac{1}{\rho_2} = \frac{1}{\rho_1}\{1 + \gamma(\theta_2 - \theta_1)\}$$

or

$$\rho_2 = \frac{\rho_1}{1 + \gamma(\theta_2 - \theta_1)}. \qquad \cdot \qquad \cdot \qquad \cdot \qquad (1)$$

Measurement of Absolute Expansivity

The first measurement of the absolute expansion of a liquid was made by Dulong and Petit in 1817. A simple form of their apparatus is shown in Fig. 14.3. It

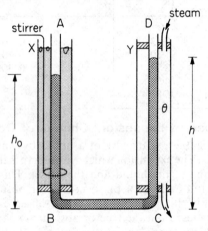

Fig. 14.3 Apparatus for absolute expansivity of mercury

consists of a glass tube ABCD, a foot or two high, containing mercury, and surrounded by glass jackets XY. The jacket X contains ice-water, and steam is passed through the jacket Y.

For the mercury to be in equilibrium, its hydrostatic pressure at B must equal its hydrostatic pressure at C. Let h_0 be the height of the mercury in the limb at 0°C and ρ_0 its density; and let h and ρ be the corresponding quantities at the temperature θ of the steam.

Then

$$g\rho_0 h_0 = g\rho h,$$

where g is the acceleration of gravity.

Hence

$$\frac{\rho}{\rho_0} = \frac{h_0}{h}.$$

But, by equation (1),

$$\frac{\rho}{\rho_0} = \frac{1}{1 + \gamma\theta}.$$

$$\therefore \frac{h_0}{h} = \frac{1}{1 + \gamma\theta}$$

$$\therefore \gamma = \frac{h - h_0}{h_0\theta}.$$

The height $h - h_0$ is measured with a cathetometer (a travelling telescope on a vertical column).

This simple apparatus is inaccurate because:

   (i) the expansion of CD throws BC out of the horizontal;
   (ii) the wide separation of A and D makes the measurement of $(h - h_0)$ inaccurate;
   (iii) surface tension causes a differences of pressure across each free surface of mercury; and these do not cancel one another, because the surface tensions are different at the temperatures of the hot and cold columns.

More accurate apparatus was devised by Regnault and others, following the principle of Dulong and Petit.

Correction of the Barometer

The hydrostatic pressure of a column of mercury, such as that in a barometer, depends on its density as well as its height. When we speak of a pressure of 760 mmHg, therefore, we must specify the temperature of the mercury; we choose 0°C. In practice barometers are generally warmer than that, and the readings must therefore be reduced to what they would be at 0°C. Also we must allow for the expansion of the scale with which the height is measured.

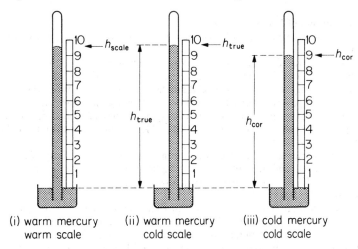

(i) warm mercury   (ii) warm mercury   (iii) cold mercury
    warm scale   cold scale   cold scale

Fig. 14.4 Reduction of barometer height to 0°C

The scale of a barometer may be calibrated at 0°C. At any higher temperature, θ, the height which it indicates, h_{scale} is less than the true height, h_{true}, of the mercury meniscus above the free surface in the reservoir (Fig. 14.4 (i) (ii)). The true height is given by

$$h_{true} = h_{scale}(1 + \lambda\theta) \qquad . \qquad . \qquad . \qquad . \qquad (1)$$

where λ is the linear expansivity of the metal scale. If ρ_θ is the density of mercury at θ, then the pressure of the atmosphere is

$$p = g\rho_\theta h_{true}.$$

The height of a mercury column at 0°C which would exert the same pressure is called the *corrected height, $h_{cor.}$.* (Fig. 14.4 (ii) (iii)). It is given by

$$p = g\rho_0 h_{cor.},$$

where ρ_0 is the density of mercury at 0°C.

Therefore

$$\rho_\theta h_{true} = \rho_0 h_{cor.}.$$

If γ is the coefficient of expansion of mercury, then

$$\rho_\theta = \frac{\rho_0}{1 + \gamma\theta}.$$

Therefore, by equation (1),

$$h_{cor.} = \frac{h_{scale}(1 + \lambda\theta)}{1 + \gamma\theta}.$$

$$= h_{scale}(1 + \lambda\theta)(1 + \gamma\theta)^{-1}.$$

$$= h_{scale}\{1 - (\gamma - \lambda)\theta\}$$

by binomial expansion, ignoring θ^2 and higher terms. The reader should notice that the correction depends on the difference between the cubic expansivity of the mercury and the linear expansivity of the scale; as in Dulong and Petit's experiment, there is no question of apparent expansion.

Exercises 14

Thermal Expansion of Solids
1. Define the *linear expansivity* of a solid, and describe a method by which it may be measured.

Show how the superficial (area) expansivity can be derived from this value.

A 'thermal tap' used in certain apparatus consists of a silica rod which fits tightly inside an aluminium tube whose internal diameter is 8 mm at 0°C. When the temperature is raised, the fit is no longer exact. Calculate what change in temperature is necessary to produce a channel whose cross-section is equal to that of a tube of 1 mm internal diameter. [Linear expansivity of silica = 8×10^{-6} K^{-1}. Linear expansivity of aluminium = 26×10^{-6} K^{-1}.] (*O. & C.*)

2. Describe an accurate method for determining the linear expansivity of a solid in the form of a rod. The pendulum of a clock is made of brass whose linear expansivity is 1.9×10^{-5} K^{-1}. If the clock keeps correct time at 15°C, how many seconds per day will it lost at 20°C? (*O. & C.*)

3. A steel wire 8 metres long and 4 mm in diameter is fixed to two rigid supports. Calculate the increase in tension when the temperature falls 10°C. [Linear expansivity of steel = 12×10^{-6} K^{-1}, Young modulus for steel = 2×10^{11} N m^{-2}.] (*O. & C.*)

4. Describe in detail how you would determine the linear expansivity of a metal rod or tube. Indicate the chief sources of error and discuss the accuracy you would expect to obtain.

A steel cylinder has an aluminium alloy piston and, at a temperature of 20°C when the internal diameter of the cylinder is exactly 10 cm, there is an all-round clearance of 0.05 mm between the piston and the cylinder wall. At what temperature will the fit be perfect? (The linear expansivity of steel and the aluminium alloy are 1.2×10^{-5} and 1.6×10^{-5} K^{-1} respectively.) (*O. & C.*)

Thermal Expansion of Liquids
5. Describe and explain how the absolute expansivity of a liquid may be determined without a previous knowledge of any other expansivity.

Aniline is a liquid which does not mix with water, and when a small quantity of it is poured into a beaker of water at 20°C it sinks to the bottom, the densities of the two liquids at 20°C being 1021 and 998 kg m^{-3} respectively. To what temperatures must the beaker and its contents be uniformly heated so that the aniline will form a globule which just floats in the water? (The mean absolute expansivity of aniline and water over the temperature range concerned are 0·00085 and 0·00045 K^{-1}, respectively.) (L.)

6. Define the *cubic expansivity* of a liquid. Find an expression for the variation of the density of a liquid with temperature in terms of its expansivity.

Describe, without experimental details, how the cubic expansivity of a liquid may be determined by the use of balanced columns.

A certain Fortin barometer has its pointers, body and scales made from brass. When it is at 0°C it records a barometric pressure of 760 mmHg. What will it read when its temperature is increased to 20°C if the pressure of the atmosphere remains unchanged? [Cubic expansivity of mercury = 1·8 × 10^{-4} K^{-1}; linear expansivity of brass = 2 × 10^{-5} K^{-1}.] (O. & C.)

7. Using the following data, determine the temperature at which wood will just sink in benzene.

Density of benzene at 0°C = 9·0 × 10^2 kg m^{-3}. Density of wood at 0°C = 8·8 × 10^2 kg m^{-3}. Cubical expansivity of benzene = 1·2 × 10^{-3} K^{-1}. Cubical expansivity of wood = 1·5 × 10^{-4} K^{-1}. (L.)

15 Thermometry

Realisation of Temperature Scale

In Chapter 8 we discussed the general idea of a temperature scale. To establish such a scale we need:

(i) some physical property of a substance—such as the volume of a gas or the electrical resistance of pure platinum—which increases continuously with increasing temperature, but is constant at constant temperature;

(ii) fixed temperatures—fixed points—which can be accurately reproduced in the laboratory.

The Fixed Points

On the thermodynamic scale, the triple point of water is chosen as the fixed point and is defined at 273·16 K (p. 174). The absolute zero is 0 K, the ice point is 273·15 K and the steam point is 373·15 K.

Suppose P is the chosen temperature-measuring quantity, such as a gas pressure at constant volume or the electrical resistance of a pure metal. If P_{tr} is the value of P at the triple point and P_T is the value at an unknown temperature T on the absolute thermodynamic scale, then, by definition,

$$T = \frac{P_T}{P_{tr}} \times 273\cdot16 \text{ K}.$$

The temperatures of melting ice and pure boiling water at 760 mmHg were chosen as the fixed points on the Celsius scale. Nowadays, the temperature θ on the Celsius scale is defined by $\theta = T - 273\cdot15$.

If P is the chosen temperature-measuring quantity, its values P_0 at the ice point and P_{100} at the steam point determine the fundamental interval of the centigrade scale, $P_{100} - P_0$. And the temperature θ_P on the centigrade scale which corresponds to a value P_θ is given by

$$\theta_P = \frac{P_\theta - P_0}{P_{100} - P_0} \times 100.$$

Gas Thermometry

In most accurate work, temperatures are measured by *gas thermometers*; for example, by the changes in pressure of a gas at constant volume.

At pressures of the order of one atmosphere, different gases give slightly different temperature scales, because none of them obeys the gas laws perfectly. But as the pressure is reduced, the gases approach closely to the ideal, and their temperature scales come together (see p. 250). By observing the departure of a gas from Boyle's law at moderate pressures it is possible to allow for its departure from the ideal. Temperatures measured with the gas in a constant volume thermometer can then be converted to the values which would be given by the same thermometer if the gas were ideal.

The Constant Volume Gas Thermometer

Figure 15.1 shows a constant volume hydrogen thermometer. B is a bulb of platinum–iridium, holding the gas. The volume is defined by the level of the index I in the glass tube A. The pressure is adjusted by raising or lowering the mercury reservoir R. A barometer CD is fitted directly into the pressure-measuring system; if H_1 is its height, and h the difference in level between the

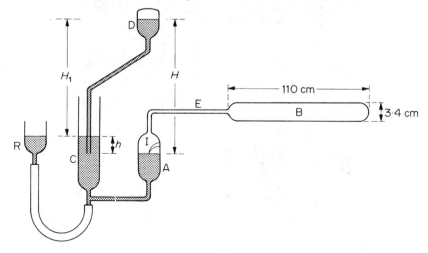

Fig. 15.1 Constant volume hydrogen thermometer (*not to scale*)

mercury surfaces in A and C, then the pressure H of the hydrogen, in mm mercury is

$$H = H_1 + h.$$

H is measured with a cathetometer (a travelling telescope with vernier).

The glass tubes A, C, D, all have the same diameter to prevent errors due to surface tension; and A and D are optically worked to prevent errors due to refraction (as in looking through common window-glass).

Observations made with a constant volume gas thermometer must be corrected for the following errors:

(i) the expansion of the bulb B;
(ii) the temperature of the gas in the tube E and A, which lies between the temperature of B and the temperature of the room;
(iii) the temperature of the mercury in the barometer and manometer.

The expansion of the bulb can be estimated from its coefficient of cubical expansion, by using the temperature shown by the gas thermometer. Since the expansion appears only as a small correction to the observed temperature, the uncorrected value of the temperature may be used in estimating it.

The tube E is called the 'dead-space' of the thermometer. Its diameter is made small, about 0·7 mm, so that it contains only a small fraction of the total mass of gas. Its volume is known, and the temperatures at various points in it are measured with mercury thermometers. The effect of the gas in it is then allowed for in a calculation similar to that used to calculate the pressure of a gas in two bulbs at different temperatures (p. 203). Mercury thermometers may be used to measure the temperatures because the error due to the dead-space is small; any error in allowing for it is of the second order of small quantities. For the same reason, mercury thermometers may be used to measure the temperature of the manometer and barometer.

Fig. 15A *Gas Thermometry.* The photograph shows the whole assembly of the National Physical Laboratory constant volume gas thermometer, used to measure a low temperature in the range about 2 K to 27 K. The working parts such as the gas bulb are immersed in liquid helium in the stainless steel Dewar vessel on the left. The pressure of the gas is measured by a sensitive pressure balance top right. The value in Pa is recorded on the circular dial above the Dewar vessel. (*Crown copyright. Courtesy of the National Physical Laboratory*)

A gas thermometer is a large awkward instrument, demanding much skill and time, and useless for measuring changing temperatures. In practice, gas thermometers are used only for calibrating electrical thermometers—resistance thermometers and thermocouples. The readings of these, when they are used to measure unknown temperatures, can then be converted into temperatures on the ideal gas scale. Helium gas is widely used in gas thermometers.

Examples

1. The pressure recorded by a constant volume gas thermometer at a kelvin temperature T is 4.80×10^4 N m^{-2}. Calculate T if the pressure at the triple point, 273·16 K, is 4.20×10^4 N m^{-2}.

$$T = \frac{p_T}{p_{tr}} \times 273.16 \text{ K}$$

$$= \frac{4.80 \times 10^4}{4.20 \times 10^4} \times 273.16$$

$$= 312 \text{ K}.$$

2. How is centigrade temperature defined (*a*) on the scale of a constant-pressure gas thermometer, (*b*) on the scale of a platinum resistance thermometer? A constant mass of gas maintained at constant pressure has a volume of 200·0 cm³ at the temperature of melting ice, 273·2 cm³ at the temperature of water boiling under standard pressure, and 525·1 cm³ at the normal boiling-point of sulphur. A platinum wire has resistances of 2·000, 2·778 and 5·280 Ω at the temperatures. Calculate the values of the boiling-point of sulphur given by the two sets of observations, and comment on the results. (*N.*)

On the gas thermometer scale, the boiling-point of sulphur is given by

$$\theta = \frac{V_\theta - V_0}{V_{100} - V_0} \times 100$$

$$= \frac{525 \cdot 1 - 200 \cdot 0}{273 \cdot 2 - 200 \cdot 0} \times 100$$

$$= 444 \cdot 1°C.$$

On the platinum resistance thermometer scale, the boiling-point is given by

$$\theta_p = \frac{R_\theta - R_0}{R_{100} - R_0} \times 100$$

$$= \frac{5 \cdot 280 - 2 \cdot 000}{2 \cdot 778 - 2 \cdot 000} \times 100$$

$$= 421 \cdot 6°C.$$

The temperatures recorded on the thermometers are therefore different. This is due to the fact that the variation of gas pressure with temperature at constant volume is different from the variation of the electrical resistance of platinum with temperature.

Electric Thermometers

Electrical thermometers have great advantages over other types. They are more accurate than any except gas thermometers, and are quicker in action and less cumbersome than those.

The measuring element of a *thermoelectric thermometer* is the welded junction of two fine wires. It is very small in size, and can therefore measure the temperature almost at a point. It causes very little disturbance wherever it is placed, because the wires leading from it are so thin that heat loss along them is usually negligible. It has a very small heat capacity, and can therefore follow a rapidly changing temperature. To measure such a temperature, however, the e.m.f. of the junction must be measured with a galvanometer, instead of a potentiometer, and some accuracy is then lost.

The measuring element of a *resistance thermometer* is a spiral of fine wire. It has a greater size and heat capacity than a thermojunction, and cannot therefore measure a local or rapidly changing temperature. But, over the range from about room temperature to a few hundred degrees Celsius, it is more accurate.

The platinum-resistance scale differs appreciably from the mercury-in-glass scale, for example, as the following table shows:

Mercury-in-glass	0	50	100	200	300	°C
Platinum-resistance	0	50·25	100	197	291	°C

Resistance Thermometers

Resistance thermometers are usually made of platinum. The wire is wound on two strips of mica, arranged crosswise as shown in Fig. 15.2 (i). The ends of the coil are attached to a pair of leads A, for connecting them to a Wheatstone bridge. A similar pair of leads B is near to the leads from the coil, and connected in the adjacent arm of the bridge (Fig. 15.2 (ii)). At the end near the coil, the pair of leads B is short-circuited. If the two pairs of leads are identical, their resistances are equal, whatever their temperature. Thus if $P = Q$ the dummy pair, B, just compensates for the pair A going to the coil; and the bridge measures the resistance of the coil alone.

The platinum resistance thermometer is used to measure temperature on the International Practical Temperature Scale (p. 306) between $-259 \cdot 34°C$ and

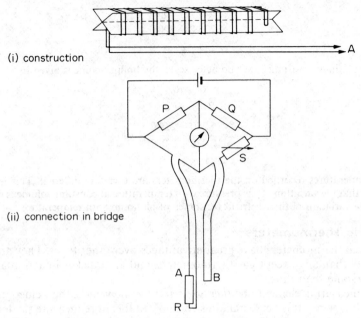

(i) construction

(ii) connection in bridge

Fig. 15.2 Platinum resistance thermometer ($P = Q$ and B compensates A so that $S = R$)

630·74°C. The platinum used in the coil must be strain-free and annealed pure platinum. Its purity and reliability are judged by the increase in its resistance from the ice point to the steam point. Thus if R_0 and R_{100} are the resistances of the coil at these points, then the coil is suitable if

$$\frac{R_{100}}{R_0} > 1\cdot 39250.$$

Various formulae and tables are provided to obtain the temperature over the wide temperature range.

Example

Explain how a centigrade temperature scale is defined, illustrating your answer by reference to a platinum resistance thermometer.

The resistance R_t of a platinum wire at temperature $t°C$, measured on the gas scale, is given by $R_t = R_0(1 + at + bt^2)$, where $a = 3\cdot 800 \times 10^{-3}$ and $b = -5\cdot 6 \times 10^{-7}$. What temperature will the platinum thermometer indicate when the temperature on the gas scale is 200°C? (*O. & C.*)

$$R_t = R_0(1 + at + bt^2)$$

$$\therefore R_{200} = R_0(1 + 200a + 200^2 b)$$

and

$$R_{100} = R_0(1 + 100a + 100^2 b).$$

$$\therefore \theta_p = \frac{R_{200} - R_0}{R_{100} - R_0} \times 100$$

$$= \frac{R_0(1 + 200a + 200^2 b) - R_0}{R_0(1 + 100a + 100^2 b) - R_0} \times 100$$

$$= \frac{200a + 200^2 b}{a + 100b} = \frac{200(a + 200b)}{a + 100b}$$

$$= 200\frac{(3\cdot 8 \times 10^{-3} - 11\cdot 2 \times 10^{-5})}{3\cdot 8 \times 10^{-3} - 5\cdot 6 \times 10^{-5}}$$

$$= \frac{200 \times 0.003688}{0.003744} = 197°C.$$

Thermocouples

Between $630.74°C$ and the freezing point of gold ($1064.43°C$), the international temperature scale is expressed in terms of the electromotive force of a thermo-couple. The wires of the thermocouple are platinum, and platinum–rhodium alloy (90% Pt.: 10% Rh.). Since the e.m.f. is to be measured on a potentiometer, care must be taken that thermal e.m.f.'s are not set up at the junctions of the thermocouple wires and the copper leads to the potentiometer. To do this three junctions are made, as shown in Fig. 15.3 (i). The junctions of the copper leads to the thermocouple wires are both placed in melting ice. The electromotive force E of the whole system is then equal to the e.m.f. of two platinum/platinum–rhodium junctions, one in ice and the other at the unknown temperature (Fig. 15.3 (ii)).

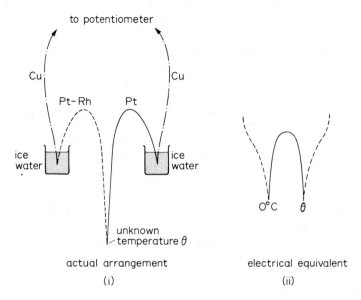

Fig. 15.3 Use of thermocouples

The temperature θ measured by this thermocouple is defined by the relation

$$E = a + b\theta + c\theta^2,$$

where a, b and c are constants. The values of the constants are determined by measurements of E at the gold point ($1064.43°C$), the silver point ($960.8°C$), and the temperature of freezing antimony (about $630.74°C$). See p. 306.

Other Thermocouples

Because of their convenience, thermocouples are used to measure temperatures outside their range on the international scale, when the highest accuracy is not required. The arrangement of three junctions and potentiometer may be used, but for less accurate work the potentiometer may be replaced by a galvanometer G, in the simpler arrangement of Fig. 15.4 (i). The galvanometer scale may be calibrated to read directly in temperatures, the known melting-points of metals like tin and lead being used as subsidiary fixed points. For rough work, particu-

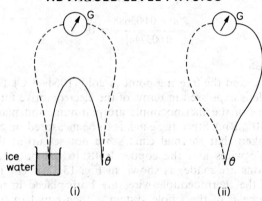

ice
water

(i)   (ii)

Fig. 15.4 Simple thermojunction thermometer

larly at high temperatures, the cold junction may be omitted (Fig. 15.4 (ii)). An uncertainty of a few degrees in a thousand is often of no importance.

High Temperature Measurement

High temperatures are usually measured by observing the radiation from the hot body, and the name *Pyrometry* is given to this measurement. Before describing pyrometers, however, we may mention some other, rough, methods which are sometimes used. One method is to insert in the furnace a number of ceramic cones, of slightly different compositions; their melting-points increase from one to the next by about 20°C. The temperature of the furnace lies between the melting-points of adjacent cones, one of which softens and collapses, and the other of which does not.

The temperature of steel, when it is below red heat, can be judged by its colour, which depends on the thickness of the oxide film upon it. Temperatures below red heat can also be estimated by the use of paints, which change colour at known temperatures.

Radiation pyrometers can be used only above red-heat (about 600°C). They fall into two classes:

(i) *total radiation pyrometers*, which respond to the total radiation from the hot body, heat and light;

(ii) *optical pyrometers*, which respond only to the visible light.

Total Radiation Pyrometers

Figure 15.5 illustrates the principle of a Féry total radiation pyrometer. The blackened tube A is open at the end B; at the other end C it carries an eyepiece E. D is a thermocouple attached to a small blackened disc of copper, which faces the end C of the tube and is shielded from direct radiation. M is a gold-plated mirror, pierced at the centre to allow light to reach the eyepiece, and moveable by a rack and pinion P.

In use, the eyepiece is first focused upon the disc D. The mirror M is then adjusted until the furnace S is also focused upon D. Since a body which is black or nearly so shows no detail, focusing it upon D by simply looking at the image would be almost impossible. To make the focusing easier, two small plane mirrors m′, m are fitted in front of D. They are inclined with their normals at about 5° to the axis of the tube, and are pierced with semi-circular holes to allow radiation from M to reach the disc. The diameter of the resulting circular

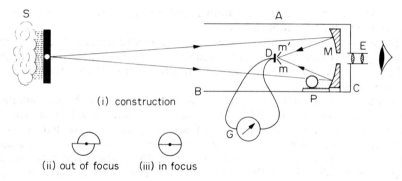

Fig. 15.5 Total radiation pyrometer

hole is less than that of the disc. When the source of heat is not focused on the disc, the two mirrors appear as at (ii) in the figure; when the focusing is correct, they appear as at (iii). The source must be of such a size that its image completely fills the hole.

The radiation from the source warms the junction and sets up an electromotive force. A galvanometer G connected to the junction is then deflected, and can be calibrated to read directly the temperature of the source.

The calibration gives the correct temperature if the source is a black body. If the source is not black, its total radiation is equal to that of a black body at some lower temperature. The pyrometer therefore reads too low and a correction is applied.

Optical Pyrometers

Figure 15·6 illustrates the principle of the commonest type of optical pyrometer, called a *disappearing filament pyrometer*. It consists essentially of a low power telescope, OE, and a tungsten filament lamp L. The eyepiece E is focused upon

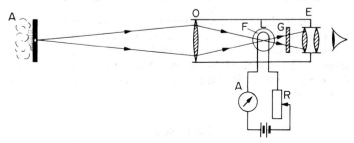

Fig. 15.6 Optical radiation pyrometer (*not to scale*)

the filament F. The hot body A whose temperature is to be found is then focused by the lens O so that its image lies in the plane of F. The light from both the filament and the hot body passes through a filter of red glass G before reaching the eye. If the body is brighter than the filament, the filament appears dark on a bright ground. If the filament is brighter than the body, it appears bright on a dark ground. The temperature of the filament is adjusted, by adjusting the current through it, until it merges as nearly as possible into its background. It is then as bright as the body. The rheostat R which adjusts the current is mounted on the body of the pyrometer, and so is the ammeter A which measures the current. The ammeter is calibrated directly in degrees Celsius. A pyrometer of this type can be adjusted to within about 5°C at 1000°C; more elaborate types can be adjusted more closely.

The range of an optical pyrometer can be extended by introducing a filter of green glass between the objective O and the lamp L; this reduces the brightness of the red light. A second scale on the ammeter is provided for use when the filter is inserted.

The scale of a radiation pyrometer is calibrated by assuming that the radiation is black body radiation (p. 278). If—as usual—the body is not black, then it will be radiating less intensely than a black body at the same temperature. Conversely, a black body which radiates with the same intensity as the actual body will be cooler than the actual body. Thus the temperature indicated by the pyrometer will be lower than the true temperature of the actual, not black, body and a correction must be applied to the pyrometer reading.

International Practical Temperature Scale

An accepted practical scale of temperature, called the International Practical Temperature Scale, was adopted in 1968 (IPTS-68). In this scale a number of fixed points are given values which agree as closely as possible with thermodynamic temperatures. The temperature measuring instruments between these fixed points are specified. When they are used, certain formulae and tables must be applied to calculate the temperature on the international scale from the measurements taken.

The following table shows the temperature range and the corresponding standard instrument required.

Temperature Range	Standard Instrument
13·81 K (−259·34°C) to 630·74°C	Platinum resistance thermometer
630·74°C to 1064·43°C	Thermocouple
Above 1064·43°C	Radiation pyrometer

−259·34°C is the triple point of equilibrium of hydrogen; 630·74°C is the freezing point of antimony; 1064·43°C is the freezing point of gold.

The interested reader is recommended for details to *The International Temperature Scale of 1968*, published by HMSO, London, and to *The Calibration of Thermometers* by C. R. Barber, National Physical Laboratory, published by HMSO, London.

Exercises 15

1.

	Steam point 100°C	Ice point 0°C	Room temperature
Resistance of resistance thermometer	75·000 Ω	63·000 Ω	64·992 Ω
Pressure recorded by constant volume gas thermometer	$1·10 \times 10^5$ N m^{-2}	$8·00 \times 10^4$ N m^{-2}	$8·51 \times 10^4$ N m^{-2}

Using the above data, which refers to the observations of a particular room temperature using two types of thermometer, calculate the room temperature on the scale of the resistance thermometer and on the scale of the constant volume gas thermometer. Why do these values differ slightly? (*L.*)

2. (*a*) Explain how a temperature scale is defined.

(*b*) Discuss the relative merits of (i) a mercury-in-glass thermometer, (ii) a platinum resistance thermometer, (iii) a thermocouple, for measuring the temperature of an oven which is maintained at about 300°C. (*N.*)

3. Tabulate various physical properties used for measuring temperature. Indicate the temperature range for which each is suitable.

Discuss the fact that the numerical value of a temperature expressed on the scale of the platinum resistance thermometer is not the same as its value on the gas scale except at the fixed points.

If the resistance of a platinum thermometer is 1·500 ohms at 0°C, 2·060 ohms at 100°C and 1·788 ohms at 50°C on the gas scale, what is the difference between the numerical values of the latter temperature on the two scales? (*N.*)

4. Describe the structure of a simple constant volume gas thermometer. Discuss how it would be used to establish a scale of temperature.

Explain why the same temperature measured on two different scales need not have the same value.

Discuss the circumstances in which (*a*) a gas thermometer and (*b*) a thermocouple might be used. Why is it generally not sensible to use a thermoelectric e.m.f. as the physical property used to *define* a scale of temperature? (*L.*)

5. How is a scale of temperature defined? What is meant by a temperature of 15°C?

On what evidence do you accept the statement that there is an absolute zero of temperature at about −273°C?

In a special type of thermometer a fixed mass of gas has a volume of 100·0 cm³ and a pressure of 81·6 cm of mercury at the ice point, and volume 124·0 units with pressure 90·0 units at the steam point. What is the temperature when its volume is 120·0 units and pressure 85·0 units, and what value does the scale of this thermometer give for absolute zero? Explain the principle of your calculation. (*O. & C.*)

6. On the kinetic theory model of a gas what is the interpretation of temperature? How does this theory explain (*a*) the rise in pressure when the temperature of a gas is increased at constant volume and (*b*) the rise in temperature which occurs when a gas is compressed quickly?

How may *a scale of temperature* be defined?

Draw a clearly labelled diagram of a simple constant volume gas thermometer which could be used to calibrate a thermocouple on the constant volume gas scale of temperature. Describe the procedure you would adopt to measure a temperature on the constant volume gas scale.

Describe a simple form of thermocouple suitable for use at about room temperature. (*L.*)

7. Explain what is meant by a change in temperature of 1 °C on the scale of a platinum resistance thermometer.

Draw and label a diagram of a platinum resistance thermometer together with a circuit in which it is used.

Give *two* advantages of this thermometer and explain why, in its normal form, it is unsuited for measurement of varying temperatures.

The resistance R_t of platinum varies with the temperature t°C as measured by a constant volume gas thermometer according to the equation

$$R_t = R_0(1 + 8000\alpha t - \alpha t^2)$$

where α is a constant. Calculate the temperature on the platinum scale corresponding to 400°C on this gas scale. (*N.*)

8. Give the essential steps involved in setting up a scale of temperature. Explain why scales based on different properties do not necessarily agree at all temperatures. At what temperature or temperatures do these different scales agree?

The volume of some air at constant pressure, and also the length of an iron rod, are measured at 0°C and again at 100°C with the following results:

	0°C	100°C
Volume of air (cm³)	28·5	38·9
Length of rod (cm)	100·00	100·20

Calculate (a) the absolute zero of this air thermometer scale, and (b) the length of the iron rod at this temperature if its expansion is uniform according to the air scale. (C.)

9. Describe briefly how temperature is measured on each of the following types of thermometer: (a) resistance thermometer, (b) thermocouple, and (c) optical pyrometer. Details of structure and circuitry are *not* required. State, with reasons, the use for which each of the three above types of thermometer is particularly suitable.

A liquid-in-glass thermometer uses liquid of which the volume varies with temperature according to the relationship $V_\theta = V_0(1 + a\theta + b\theta^2)$ where V_θ and V_0 are the volumes at $\theta°C$ and $0°C$ on the gas scale respectively and a and b are constants. If $a = b \times 10^3$, what temperature will be indicated on the liquid-in-glass scale when that on the gas thermometer is $60°C$? (L.)

10. Give a brief account of the principles underlying the establishment of a scale of temperature and explain precisely what is meant by the statements that the temperature of a certain body is (a) $t°C$ on the constant volume air scale, (b) $t_P°C$ on the platinum resistance scale, and (c) $t_T°C$ on the Cu–Fe thermocouple scale. Why are these three temperatures usually different?

Describe an optical pyrometer and explain how it is used to measure the temperature of a furnace. (N.)

Part Three
Geometrical Optics

16 Introduction. Reflection at Plane Surfaces and Curved Mirrors

Introduction

Light Rays and Beams

Light is a form of energy. We know this is the case because plants and vegetables grow when they absorb sunlight. Further, electrons are emitted by certain metals when light is incident on them, showing that there was some energy in the light; this phenomenon is the basis of the *photoelectric cell* (p. 847). Substances like wood or brick which allow no light to pass through them are called 'opaque' substances; unless an opaque object is perfectly black, some of the light falling on it is reflected (p. 314). A 'transparent' substance, like glass, is one which allows some of the light energy incident on it to pass through, the remainder of the energy being absorbed and (or) reflected.

A ray of light is the direction along which the light energy travels; and though rays are represented in diagrams by straight lines, in practice a ray has a finite width. A *beam* of light is a collection of rays. A searchlight emits a *parallel beam* of light, and the rays from a point on a very distant object like the sun are substantially parallel, Fig. 16.1 (i). A lamp emits a *divergent beam* of light; while a source of light behind a lens, as in a projection lantern, can provide a *convergent beam*, Fig. 16.1 (ii), (iii).

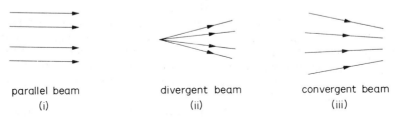

parallel beam	divergent beam	convergent beam
(i)	(ii)	(iii)

Fig. 16.1 Beams of light

Direction of Image seen by Eye

When a fish is observed in water, rays of light coming from a point such as O on it pass from water into air, Fig. 16.2 (i). At the boundary of the water and air, the rays OA, OC proceed along new directions AB, CD respectively and enter the eye. Similarly, a ray OC from an object O observed in a mirror is reflected along a new direction CD and enters the eye, Fig. 16.2 (ii).

These phenomena are studied more fully later, but the reader should take careful note that the eye sees an object *in the direction in which the rays enter the eye*. In Fig. 16.2 (i), for example, the object O is seen in the water at I, which lies on BA and DC produced slightly to the right of O; in Fig. 16.2 (ii), O is seen behind the mirror at I, which lies on DC produced. In either case, all rays from O which enter the eye appear to come from I, which is called the image of O.

Reversibility of Light

If a ray of light is directed along DC towards a mirror, experiment shows that the ray is reflected along the path CO, Fig. 16.2 (ii). If the ray is incident along OC, it is reflected along CD, as shown. Thus if a light ray is reversed it always travels along its original path, and this is known as *the principle of the reversibility of light*. In Fig. 16.2 (i), a ray BA in air is refracted into the water along the path AO, since it follows the reverse path to OAB. We shall have occasion to use the principle of the reversibility of light later in the book.

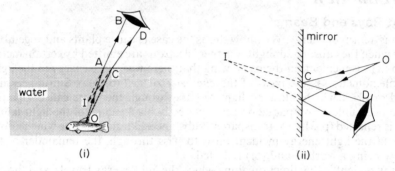

(i)   (ii)

Fig. 16.2 Images observed by eye

Luminous Flux, Lumen, Candela, Lux

We conclude this introductory section with a brief account of some of the main points in *photometry*, the science of light measurement. Here we are concerned with the *luminous energy* emitted by a source of light, which stimulates the sensation of vision; and not with any other radiations it may emit, such as infra-red rays, for example, which are invisible.

A source of light such as a lamp emits a continuous stream of luminous energy. We give the name *luminous flux*, symbol Φ, to the 'luminous energy emitted per second'. The unit of luminous flux is the *lumen*, symbol lm. A lumen is a unit of energy per second or power, so it must be related to the watt. Experiment shows that about 621 lumens of green light of wavelength 5.540×10^{-10} m is equal to 1 watt.

A light source as a lamp radiates luminous flux in all directions round it. If we consider a small lamp S and a particular direction SA, an amount of luminous flux Φ is radiated in a small cone of 'solid angle' ω drawn round SA with S at the apex, Fig. 16.3 (i). The *luminous intensity*, I, of the lamp is defined as the ratio Φ/ω or the 'luminous flux per unit solid angle'. Since a solid angle is measured in steradian, sr, the unit of I is 'lm sr^{-1}'.

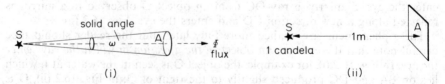

(i)   (ii)

Fig. 16.3 (i) Luminous intensity (ii) Lux

A practical unit of luminous intensity is the *candela*, symbol cd. It is defined as the luminous intensity of $1/600\,000$ metre^2 ($1/60$ cm^2) of the surface of a black body at the temperature of freezing platinum under $101\,325$ newton per metre^2 pressure. A standard is maintained at the National Physical Laboratory and here the luminous intensity of manufacturers' lamps are measured in terms of the standard. From previous, $I = \Phi/\omega$, so $\Phi = I\omega$. Thus 1 lm = 1 cd sr.

We now consider the surface on which the luminous flux falls. The *illuminance* (or *illumination*), E, of a surface is defined as the 'luminous flux per unit area'. If we imagine concentric spheres of different radii r drawn round a small lamp S as centre, the total flux from S will fall on areas equal to $4\pi r^2$. So we can see that the illuminance varies *inversely as the square* of the distance from S. The unit of illuminance is the *lux*, lx. This is the illuminance of a surface A 1 m away from a lamp S of 1 cd when the light falls normally on A, Fig. 16.3 (ii).

The *luminance*, L, of a surface is the 'luminous flux per unit area' coming from that surface. The illuminance of white chalk on a blackboard is the same as the surrounding surface. The luminance of the chalk, however, is very much higher than that of the board since the reflection factor of the chalk is much greater than that of the board.

The following table summarises some of the units discussed.

	Symbol	Unit
Luminous flux	Φ	lumen (lm)
Luminous intensity	I	candela (cd)
Illuminance	E	lux (lx)
Luminance	L	cd m^{-2}

Reflection at Plane Surfaces

Highly polished metal surfaces reflect about 80 to 90% of the light incident on them; *mirrors* in everyday use are therefore usually made by depositing silver on the back of glass. In special cases the front of the glass is coated with the metal; for example, the largest reflector in the world is a curved mirror nearly 5 metres across, the front of which is coated with aluminium (p. 388). Glass by itself will also reflect light, but the percentage reflected is small compared with the case of a silvered surface; it is about 5% for an air-glass surface.

Laws of Reflection

If a ray of light, AO, is incident on a plane mirror XY at O, the angle AON made with the *normal* ON to the mirror is called the 'angle of incidence', *i*, Fig. 16.4. The angle BON made by the reflected ray OB with the normal is called the 'angle of reflection', *r*; and experiments show that:

(1) *The reflected ray, the incident ray, and the normal to the mirror at the point of incidence all lie in the same plane.*
(2) *The angle of incidence = the angle of reflection.*
These are called the two *laws of reflection.*

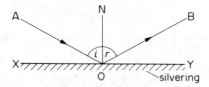

Fig. 16.4 Plane mirror

Regular and Diffuse Reflection

In the case of a plane mirror or glass surface, it follows from the laws of reflection that a ray incident at a given angle on the surface is reflected in a definite direction. Thus a parallel beam of light incident on a plane mirror in the direction AO is reflected as a parallel beam in the direction OB; this is known as a case of *regular reflection*, Fig. 16.5 (i). On the other hand, if a parallel beam of light is incident on a sheet of paper in a direction AO, the light is reflected in all different directions from the paper: this is an example of *diffuse reflection*, Fig. 16.5 (ii). Objects in everyday life, such as flowers, books, people, are seen by light diffusely reflected from them. The explanation of the diffusion of light is that the surface of paper, for example, is not perfectly smooth like a mirrored surface; the 'roughness' in a paper surface can be seen with a microscope. At each point on the paper the laws of reflection are obeyed, but the angle of incidence varies, unlike the case of a mirror.

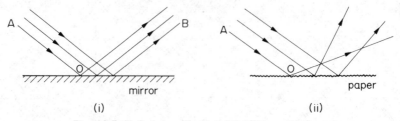

(i)   (ii)
Fig. 16.5 (i) Regular reflection (ii) Diffuse reflection

Deviation of Light by Plane Mirror

Besides other purposes, plane mirrors are used in the sextant, in simple periscopes, and in signalling at sea. These instruments use a plane mirror to change or deviate light from one direction to another.

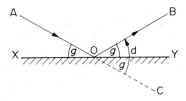

Fig. 16.6 Deviation of light by plane mirror

Consider a ray AO incident at O on a plane mirror XY, Fig. 16.6. The angle AOX made by AO with XY is known as the *glancing angle, g,* with the mirror; and since the angle of reflection is equal to the angle of incidence, the glancing angle BOY made by the reflected ray OB with the mirror is also equal to *g.*

The light has been deviated from a direction AO to a direction OB. Since angle COY = angle XOA = *g,* it follows that

$$\text{angle of deviation, } d = 2g \qquad . \qquad . \qquad . \qquad . \qquad (1)$$

so that, in general, *the angle of deviation of a ray by a plane surface is twice the glancing angle.*

Deviation of Reflected Ray by Rotated Mirror

Consider a ray AO incident at O on a plane mirror M_1, α being the glancing angle with M_1, Fig. 16·7. If OB is the reflected ray, then, as shown above, the angle of deviation COB = $2g = 2\alpha$.

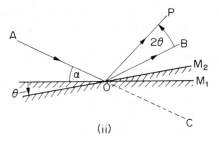

(ii)

Fig. 16.7 Rotation of reflected ray

Suppose the mirror is rotated through an angle θ to a position M_2, the direction of the incident ray AO being *constant.* The ray is now reflected from M_2, in a direction OP, and the glancing angle with M_2 is $(\alpha + \theta)$. Hence the new angle of deviation COP = $2g = 2(\alpha + \theta)$. The reflected ray has thus been rotated through an angle BOP when the mirror rotated through an angle θ; and since

$$\angle\, \text{BOP} = \angle\, \text{COP} - \angle\, \text{COB},$$

then $\angle\, \text{BOP} = 2(\alpha + \theta) - 2\alpha = 2\theta.$

Thus, *if the direction of an incident ray is constant, the angle of rotation of the reflected ray is twice the angle of rotation of the mirror.* If the mirror rotates through 4°, the direction of the incident ray being kept unaltered, the reflected ray turns through 8°.

Optical Lever in Mirror Galvanometer

In a number of instruments a beam of light is used as a 'pointer'; this has a negligible weight and so is sensitive to deflections of the moving system. In a mirror galvanometer, used for measuring very small electric currents, a small mirror M_1 is rigidly attached to a system which rotates when a current flows in it, and a beam of light from a fixed lamp L shines on the mirror, Fig. 16.8. If the light is incident normally on the mirror at A, the beam is reflected directly back, and a spot of light is obtained at O on a graduated scale S placed just above L. Suppose that the moving system, to which the mirror is attached, undergoes a rotation θ. The mirror is then rotated through this angle to a position M_2, and the spot of light is deflected through a distance x, say to a position P on the scale.

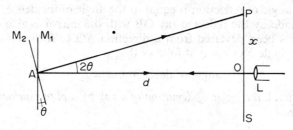

Fig. 16.8 Optical lever principle

Since the direction OA of the incident light is constant, the rotation of the reflected ray is twice the angle of rotation of the mirror (p. 315). Thus angle $OAP = 2\theta$. Now $\tan 2\theta = x/d$, where d is the distance OA. Thus 2θ can be calculated from a knowledge of x and d, and hence θ is obtained. If 2θ is small, then $\tan 2\theta$ is approximately equal to 2θ in radians, and in this case θ is equal to $x/2d$ radians.

In conjunction with a mirror, a beam of light used as a 'pointer' is known as an 'optical lever'. Besides a negligible weight, it has the advantage of magnifying by two the rotation of the system to which the mirror is attached, as the angle of rotation of the reflected light is twice the angle of rotation of the mirror. An optical lever can be used for measuring small increases of length due to the expansion or contraction of a solid.

Images in Plane Mirrors

So far we have discussed the deviation of light by a plane mirror. We now consider the *images* in plane mirrors.

Suppose that a *point object* A is placed in front of a mirror M, Fig. 16.9. A ray AO from A, incident on M, is reflected along OB so that angle AON = angle BON, where ON is the normal at O to the mirror. A ray AD incident normally on the mirror at D is reflected back along DA. Thus the rays reflected from M appear to come from a point I *behind* the mirror, where I is the point of intersection of BO and AD produced. As we prove shortly, any ray from A, such as AP, is also reflected as if it comes from I, and so an observer at E sees the image of A at I.

Since angle AON = alternate angle DAO, and angle BON = corresponding angle DIO, it follows that angle DAO = angle DIO. The two triangles ODA

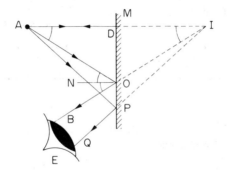

Fig. 16.9 Image in plane mirror

and ODI, in fact, are congruent, and therefore AD = ID. For a given position of the object, A and D are fixed points. Consequently, since AD = ID, the point I is a fixed point. So *any* ray reflected from the mirror must pass through I, as was stated above.

We have just shown that *the object and image in a plane mirror are at equal perpendicular distances from the mirror.* It should also be noted that AO = OI in Fig. 16.9. So the object and image are at equal distances from any point on the mirror.

Image of Finite-sized Object. Lateral Inversion

If a right-handed tennis player observes his (or her) stance in a plane mirror, he (or she) appears left-handed. Again, the words on a piece of blotting-paper become legible when the paper is viewed in a mirror. This phenomenon can be explained by considering an E-shaped object placed in front of a mirror M, Fig. 16·10. The image of a point *a* on the object is at *a'* at an equal distance behind the mirror, and the image of a point *b* on the left of *a* is at *b'*, which is on the *right* of *a'*. The left-hand side of the image thus corresponds to the right-hand side of the object, and vice-versa, and the object is said to be *laterally inverted* to an observer.

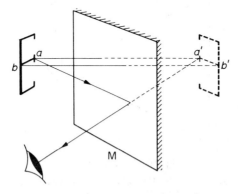

Fig. 16.10 Laterally inverted image

Virtual and Real Images

As was shown on p. 316, an object O in front of a mirror has an image I behind the mirror. The rays reflected from the mirror do not actually pass through I, but only *appear* to do so, and the image cannot be received on a screen because

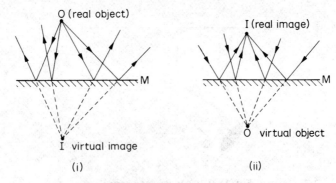

Fig. 16.11 Virtual and real image in plane mirror

the image is behind the mirror, Fig. 16.11 (i). This type of image is therefore called an unreal or *virtual* image.

It must not be thought, however, that only virtual images are obtained with a plane mirror. If a *convergent* beam is incident on a plane mirror M, the reflected rays pass through a point I *in front of* M, Fig. 16.11 (ii). If the incident beam converges to the point O, then O is called a 'virtual' object; I is called a *real* image because it can be received on a screen. Figure 16.11 (i) and (ii) should now be compared. In the former, a real object (divergent beam) gives rise to a virtual image; in the latter, a virtual object (convergent beam) gives rise to a real image. In each case the image and object are at equal distances from the mirror.

Example

A man 2 m tall, whose eye level is 1·84 m above the ground, looks at his image in a vertical mirror. What is the minimum vertical length of the mirror if the man is to be able to see the whole of himself? Indicate its position accurately in a diagram.

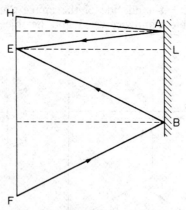

Fig. 16.12 Example

Suppose the man is represented by HF, where H is his head and F is his feet; suppose that E represents his eyes, Fig. 16.12. Since the man sees his head H, a ray HA from H to the top A of the mirror is reflected to E. Thus A lies on the perpendicular bisector of HE, and hence $AL = \frac{1}{2} HE = 0·08$ m, where L is the point on the mirror at the same level as E. Since the man sees his feet F, a ray FB from F to the bottom B of the mirror is also reflected to E. Thus the perpendicular bisector of EF passes through B, and hence $BL = \frac{1}{2} FE = \frac{1}{2} \times 1·84$ m $= 0·92$ m.

$$\therefore \text{ length of mirror} = AL + LB = 0·08 \text{ m} + 0·92 \text{ m} = 1 \text{ m.}$$

Reflection at Curved Mirrors

Curved mirrors are widely used as driving mirrors in cars. Make-up and dentists mirrors are curved mirrors. The largest telescope in the world uses an enormous curved mirror to collect light from distant stars.

Convex and Concave Mirrors. Definitions

In Optics we are mainly concerned with curved mirrors which are parts of *spherical* surfaces. In Fig. 16.13 (i), the mirror APB is part of a sphere whose centre C is in front of the reflecting surface; in Fig. 16.13 (ii), the mirror KPL is part of a sphere whose centre C is behind its reflecting surface. To a person in front of it APB curves inwards and is known as a *concave* mirror, while KPL bulges outwards and is known as a *convex* mirror.

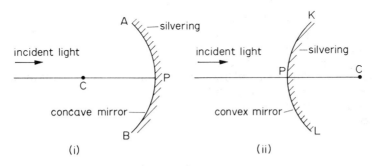

Fig. **16.13** Concave (converging) and convex (diverging) mirrors

The mid-point, P, of the mirror is called its *pole*; C, the centre of the sphere of which the mirror is part, is known as the *centre of curvature*; and AB is called the *aperture* of the mirror. The line PC is known as the *principal axis*, and plays an important part in the drawing images in the mirrors; lines parallel to PC are called *secondary axes*.

Narrow and Wide Beams. Parabolic Mirror

When a very narrow beam of rays, parallel to the principal axis and close to it, is incident on a concave mirror, experiment shows that all the reflected rays converge to a point F on the principal axis, which is therefore known as the *principal focus* of the mirror, Fig. 16.14 (i). On this account a concave mirror is better described as a 'converging' mirror. An image of the sun, whose rays on the earth are parallel, can hence be received on a screen at F, and thus a concave mirror has a *real* focus.

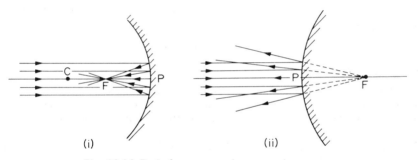

Fig. **16.14** Foci of concave and convex mirrors

If a narrow beam of parallel rays is incident on a convex mirror, experiment shows that the reflected rays form a divergent beam which appear to come from a point F *behind* the mirror, Fig. 16·14 (ii). A convex mirror has thus a *virtual* focus, and the image of the sun cannot be received on a screen using this type of mirror. To express its action on a parallel beam of light, a convex mirror is often called a 'diverging' mirror.

When a *wide* beam of light, parallel to the principal axis, is incident on a concave spherical mirror, experiment shows that reflected rays do not pass through a single point, as was the case with a narrow beam. So if a small lamp is placed at the focus F of a concave spherical mirror, it follows from the principle of the reversibility of light that those rays from the lamp which strike the mirror at points well away from the pole P will be reflected in different directions and not as a parallel beam. In this case the reflected beam diminishes in intensity as its distance from the mirror increases. So a concave spherical mirror is useless as a searchlight mirror. For this reason a mirror whose section is the shape of a parabola (the path of a ball thrown forward into the air) is used in searchlights. A parabolic mirror has the property of reflecting the wide beam of light from a lamp at its focus F as a perfectly parallel beam, in which case the intensity of the reflected beam is practically undiminished as the distance from the mirror increases, Fig. 16.15. For the same reason, motor headlamp reflectors are parabolic in shape.

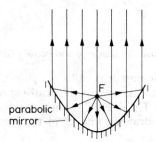

Fig. 16.15 Parabolic mirror

Focal Length (*f*) and Radius of Curvature (*r*)

From now onwards we shall be concerned with curved spherical mirrors of small aperture, so that a parallel incident beam will pass through the focus after reflection. The diagrams which follow are exaggerated for purposes of clarity.

The distance PC from the pole to the centre of curvature is known as the *radius of curvature* (*r*) of a mirror; the distance PF from the pole to the focus is known as the *focal length* (*f*) of the mirror. For both concave and convex mirrors, geometry shows that

$$r = 2f, \qquad \text{or } f = r/2.$$

Images in Concave Mirrors

Concave mirrors produce images of different sizes; sometimes they are inverted and real, and on other occasions they are upright and virtual. As we shall see, the nature of the image formed depends on the distance of the object from the mirror.

Consider an object of finite size OH placed at O perpendicular to the principal axis of the mirror, Fig. 16.16 (i). The image, R, of the top point H can be located

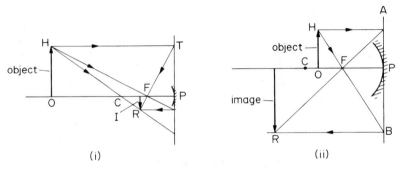

Fig. 16.16 Images in concave mirrors

by the intersection of two reflected rays coming initially from H, and the rays usually chosen are two of the following: (1) The ray HT parallel to the principal axis, which is reflected to pass through the focus, F, (2) the ray HC passing through the centre of curvature, C, which is reflected back along its own path because it is a normal to the mirror, (3) the ray HF passing through the focus, F, which is reflected parallel to the principal axis. Since the mirror has a small aperture, and we are considering a narrow beam of light, the mirror must be represented in accurate image drawings by a *straight* line. Thus PT in Fig. 16.16 (i) represents a perfect mirror.

When the object is a very long distance away (at infinity), the image is small and is formed inverted at the focus (p. 319). As the object approaches the centre of curvature, C, the image remains real and inverted, and is formed in front of the object, Fig. 16.16 (i). When the object is between C and F, the image is real, inverted, and larger than the object; it is now further from the mirror than the object, Fig. 16.16 (ii).

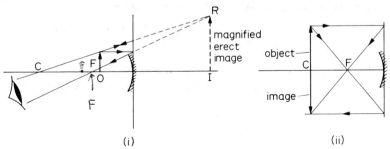

Fig. 16.17 Images in concave mirrors

As the object approaches the focus, the image recedes further from the mirror, and when the object is at the focus, the image is at infinity. When the object is nearer to the mirror than the focus the image IR becomes *upright* and *virtual*, as shown in Fig. 16.17 (i). In this case the image is *magnified*. The concave mirror can thus be used as a make-up mirror or a shaving mirror.

A special case occurs when the object is at the centre of curvature, C. The image is then real, inverted, and the same size of the object, and it is also situated at C, Fig. 16.17 (ii). This provides a simple way of measuring the radius of curvature of a concave mirror.

Images in Convex Mirrors

Experiment shows that the image of an object in a convex mirror is upright virtual, and diminished in size, no matter where the object is situated. Suppose

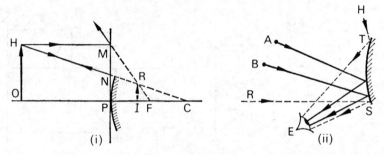

Fig. 16.18 Images in convex mirrors

an object OH is placed in front of a convex mirror, Fig. 16.18 (i). A ray HM parallel to the principal axis is reflected as if it appeared to come from the virtual focus, F, and a ray HN incident towards the centre of curvature, C, is reflected back along its path. The two reflected rays intersect *behind* the mirror at R, and IR is a virtual and upright image.

Objects well outside the principal axis of a convex mirror, such as A, B in Fig. 16.18 (ii), can be seen by an observer at E, whose *field of view* is that between HT and RS, where T, S are the edges of the mirror. Thus in addition to providing an upright image the convex mirror has a wide field of view, and is hence used as a driving mirror.

Sign Rule and Formulae

In using optical formulae, we apply a *sign rule or convention*. We shall adopt the following rule for curved mirrors and lenses:

A real object or image distance is a positive distance.

A virtual object or image distance is a negative distance.

In brief, 'real is positive, virtual is negative'. The focal length of a concave mirror is thus a positive distance; the focal length of a convex mirror is a negative distance.

Using this sign rule, the following formula is obtained, where u is the object distance, v is the image distance and f is the focal length of the curved mirror or lens:

$$\frac{1}{v} + \frac{1}{u} = \frac{1}{f} \qquad . \qquad . \qquad . \qquad . \qquad . \qquad (1)$$

The transverse magnification, m, produced by a mirror is defined by

$$m = \frac{height\ of\ image}{height\ of\ object}.$$

From simple ray diagrams, we find that *numerically*

$$m = \frac{v}{u} \qquad . \qquad . \qquad . \qquad . \qquad . \qquad . \qquad (2)$$

Some Applications of Mirror Formulae

The following examples will assist the reader to understand how to apply the formulae $\frac{1}{v} + \frac{1}{u} = \frac{1}{f}$ and $m = \frac{v}{u}$ correctly. They will also assist the reader when the section on Lenses is reached.

1. An object is placed 10 cm in front of a concave mirror of focal length 15 cm. Find the image position and the magnification.

Since the mirror is concave, $f = +15$ cm. The object is real, and hence $u = +10$ cm.

Substituting in
$$\frac{1}{v} + \frac{1}{u} = \frac{1}{f},$$

$$\frac{1}{v} + \frac{1}{(+10)} = \frac{1}{(+15)}$$

$$\therefore \frac{1}{v} = \frac{1}{15} - \frac{1}{10} = -\frac{1}{30}$$

$$\therefore v = -30.$$

Since v is negative in sign the image is *virtual*, and it is 30 cm from the mirror. See Fig. 16.17 (i). The magnification, $m = \frac{v}{u} = \frac{30}{10} = 3$, so that the image is three times as high as the object.

2. The image of an object in a convex mirror is 4 cm from the mirror. If the mirror has a radius of curvature of 24 cm, find the object position and the magnification.

The image in a convex mirror is always virtual (p. 322). Hence $v = -4$ cm. The focal length of the mirror $= \frac{1}{2}r = 12$ cm; and since the mirror is convex, $f = -12$ cm.

Substituting in
$$\frac{1}{v} + \frac{1}{u} = \frac{1}{f}$$

$$\frac{1}{(-4)} + \frac{1}{u} = \frac{1}{(-12)}$$

$$\therefore \frac{1}{u} = -\frac{1}{12} + \frac{1}{4} = \frac{1}{6}$$

$$\therefore u = 6.$$

Since u is positive in sign the object is real, and it is 6 cm from the mirror. The magnification, $m = \frac{v}{u} = \frac{4}{6} = \frac{2}{3}$, and hence the image is two-thirds as high as the object. See Fig. 16.18 (i).

3. An erect (upright) image, three times the size of the object, is obtained with a concave mirror of radius of curvature 36 cm. What is the position of the object?

If x cm is the numerical value of the distance of the object from the mirror, the image distance must be $3x$ cm, since the magnification $m = \dfrac{\text{image distance}}{\text{object distance}} = 3$. Now an *erect* image is obtained with a concave mirror only when the image is *virtual* (p. 321).

$$\therefore \text{ image distance, } v = -3x$$

$$\text{Also, object distance, } u = +x$$

$$\text{and focal length, } f = \tfrac{1}{2}r = +18 \text{ cm.}$$

Substituting in $\dfrac{1}{v} + \dfrac{1}{u} = \dfrac{1}{f}$,

$$\frac{1}{(-3x)} + \frac{1}{(+x)} = \frac{1}{(+18)}$$

$$\therefore -\frac{1}{3x} + \frac{1}{x} = \frac{1}{18}$$

$$\therefore \frac{2}{3x} = \frac{1}{18}$$

$$\therefore x = 12.$$

Thus the object is 12 cm from the mirror.

Virtual Object and Convex Mirror

We have already seen that a convex mirror produces a virtual image of an object in front of it, which is a real object. A convex mirror may sometimes produce a real image of a *virtual* object.

As an illustration, consider an incident beam of light bounded by AB, DE, converging to a point O *behind* the mirror, Fig. 16.19. O is regarded as a virtual

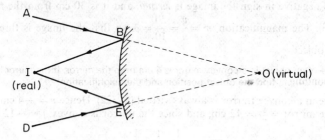

Fig. 16.19 Real image in convex mirror

object, and if its distance from the mirror is 10 cm, then the object distance $u = -10$. Suppose the convex mirror has a focal length of 15 cm, i.e., $f = -15$.

Since

$$\frac{1}{v} + \frac{1}{u} = \frac{1}{f},$$

$$\frac{1}{v} + \frac{1}{(-10)} = \frac{1}{(-15)}$$

$$\therefore \frac{1}{v} = -\frac{1}{15} + \frac{1}{10} = +\frac{1}{30}$$

$$\therefore v = +30.$$

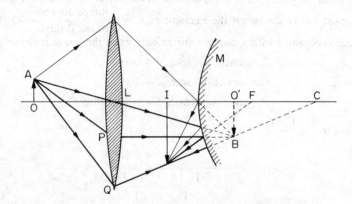

Fig. 16.20 Ray formation of real image

The point image, I, is thus 30 cm from the mirror, and is *real*. The beam reflected from the mirror is hence a convergent beam, Fig. 16.19; a similar case with a plane mirror is shown in Fig. 16.11 (ii). Figure 16.20 shows how the real image I of an object O of finite size is formed. A lens L provides a convergent beam on to the mirror M and produces a virtual object O′. Note that (*a*) the ray from P which passes through B is parallel to the axis and is therefore reflected from the mirror as if it came from F, (*b*) the ray from Q passes through B and C, the centre of curvature, and is therefore reflected from the mirror along its incident path.

Exercises 16

1. Show that for a person of given height standing upright the minimum length of a vertical plane mirror in which he can see his feet and the top of his head at the same time is independent of the distance between his eyes and the top of his head. (*L.*)

2. State the laws of reflection of light. Two plane mirrors are parallel and face each other. They are *a* cm apart and a small luminous object is placed *b* cm from one of them. Find the distance from the object of an image produced by four reflections. Deduce the corresponding distance for an image produced by 2*n* reflections. (*L.*)

3. Two plane mirrors are inclined to each other at a fixed angle. If a ray travelling in a plane perpendicular to both mirrors is reflected first from one and then from the other, show that the angle through which it is deflected does not depend on the angle at which it strikes the first mirror.

Describe and explain the action of a rear reflector on a bicycle. (*L.*)

Curved Mirrors

4. An object is placed (i) 10 cm, (ii) 4 cm from a concave mirror of radius curvature 12 cm. Calculate the image position in each case, and the respective magnifications.

5. An object is placed 15 cm from a convex mirror of focal length 10 cm. Calculate the image distance and the magnification produced.

6. A mirror forms an erect image 30 cm from the object and twice its height. Where must the mirror be situated? What is its radius of curvature? Assuming the object to be real, determine whether the mirror is convex or concave. (*L.*)

7. What do you understand by *linear magnification*? Prove that linear magnification produced by a concave mirror is equal to the ratio of the image distance to the object distance.

A coin 2·54 cm in diameter held 254 cm from the eye just covers the full moon. What is the diameter of the image of the moon formed by a concave mirror of radius of curvature 1·27 m? Describe carefully what happens to this image if the aperture is reduced. (*C.*)

8. What are the advantages of a concave mirror over a lens for use in an astronomical telescope?

A driving mirror consists of a cylindrical mirror of radius 10 cm and length (over the curved surface) of 10 cm. If the eye of the driver be assumed at a great distance from the mirror, find the angle of view. (*O. & C.*)

9. Describe an experiment to determine the radius of curvature of a convex mirror by an optical method. Illustrate your answer with a ray diagram and explain how the result is derived from the observations.

A small convex mirror is placed 60 cm from the pole and on the axis of a large concave mirror, radius of curvature 200 cm. The position of the convex mirror is such that a real image of a distant object is formed in the plane of a hole drilled through the concave mirror at its pole. Calculate (*a*) the radius of curvature of the convex mirror, (*b*) the height of the real image if the distant object subtends an angle of 0·50° at the pole of the concave mirror. Draw a ray diagram to illustrate the action of the convex mirror in producing the image of a non-axial point of the object and suggest a practical application of this arrangement of mirrors. (*N.*)

17 Refraction at Plane Surfaces

Laws of Refraction

When a ray of light AO is incident at O on the plane surface of a glass medium, observation shows that some of the light is reflected from the surface along OC in accordance with the laws of reflection, while the rest of the light travels along a new direction, OB, in the glass, Fig. 17.1. On account of the change in direction the light is said to be 'refracted' on entering the glass; and the *angle of refraction*, *r*, is the angle made by the refracted ray OB with the normal at O.

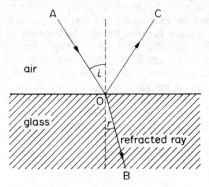

Fig. 17.1 Refraction at plane surface

SNELL, a Dutch professor, discovered in 1620 that the sines of the angles bear a constant ratio to each other. The *laws of refraction* are:

1. *The incident and refracted rays, and the normal at the point of incidence, all lie in the same plane.*

2. *For two given media, $\dfrac{\sin i}{\sin r}$ is a constant, where i is the angle of incidence and r is the angle of refraction* (Snell's law).

Refractive Index

The constant ratio $\sin i / \sin r$ is known as the *refractive index* for the two given media; and as the magnitude of the constant depends on the colour of the light, it is usually specified as that obtained for a particular yellow light (see p. 378). If the medium containing the incident ray is denoted by 1, and that containing the refracted ray by 2, the refractive index can be denoted by $_1n_2$.

Scientists have drawn up tables of refractive indices when the incident ray is travelling *in vacuo* and is then refracted into the medium concerned, for example, glass or water. The values thus obtained are known as the *absolute* refractive indices of the media; and as a vacuum is always the first medium, the subscripts for the absolute refractive index, symbol *n*, can be dropped. The magnitude of *n* for glass is about 1·5, *n* for water is about 1·33, and *n* for air at normal pressure is about 1·00028. As the magnitude of the refractive index of a medium is only very slightly altered when the incident light is in air instead of a vacuum, experiments to determine the absolute refractive index *n* are usually performed with the light incident from air into the medium. Thus we can take $_{air}n_{glass}$ as equal to $_{vacuum}n_{glass}$ for most practical purposes.

Light is refracted because it has different velocities in different media. The

Wave Theory of Light, discussed later, shows that the refractive index $_1n_2$ for two given media 1 and 2 is given by

$$_1n_2 = \frac{velocity\ of\ light\ in\ medium\ 1}{velocity\ of\ light\ in\ medium\ 2} \qquad \cdot \qquad \cdot \qquad \cdot \qquad (1)$$

This is a *definition* of refractive index which can be used instead of the ratio sin *i*/sin *r*. An alternative definition of the absolute refractive index, *n*, of a medium is then

$$n = \frac{velocity\ of\ light\ in\ a\ vacuum,\ c}{velocity\ of\ light\ in\ medium,\ v} \qquad \cdot \qquad \cdot \qquad \cdot \qquad (2)$$

In practice the velocity of light in air can replace the velocity *in vacuo* in this definition.

Relations between Refractive Indices

(1) Consider a ray of light, AO, refracted from *glass to air* along the direction OB; observation then shows that the refracted ray OB is bent away from the normal, Fig. 17.2. The refractive index from glass to air, $_gn_a$, is given by sin *x*/sin *y*,

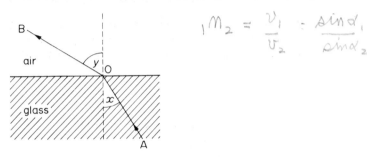

$$_1m_2 = \frac{v_1}{v_2} = \frac{\sin d_1}{\sin d_2}$$

Fig. 17.2 Refraction from glass to air

by definition, where *x* is the angle of incidence in the glass and *y* is the angle of refraction in the air.

From the principle of the reversibility of light (p. 312), it follows that a ray travelling along BO in air is refracted along OA in the glass. The refractive index from air to glass, $_an_g$, is then given by sin *y*/sin *x*, by definition. But $_gn_a = $ sin *x*/sin *y*, from the previous paragraph.

$$\therefore \ _gn_a = \frac{1}{_an_g} \qquad \cdot \qquad \cdot \qquad \cdot \qquad \cdot \qquad (3)$$

If $_an_g$ is 1·5, then $_gn_a = 1/1\cdot5 = 0\cdot67$. Similarly, if the refractive index from air to water is 4/3, the refractive index from water to air is 3/4.

(2) Consider a ray AO incident in air on a plane glass boundary, then refracted from the glass into a water medium, and finally emerging along a direction CD into air. *If the boundaries of the media are parallel, experiment shows that the emergent ray CD is parallel to the incident ray AO*, although there is a relative displacement, Fig. 17.3. Thus the angles made with the normals by AO, CD are equal, and we shall denote them by i_a.

Suppose i_g, i_w are the angles made with the normals by the respective rays in the glass and water media. Then, by definition, $_gn_w = $ sin i_g/sin i_w.

But
$$\frac{\sin i_g}{\sin i_w} = \frac{\sin i_g}{\sin i_a} \times \frac{\sin i_a}{\sin i_w},$$

and $\qquad \dfrac{\sin i_g}{\sin i_a} = {}_gn_a$, and $\dfrac{\sin i_a}{\sin i_w} = {}_an_w$

$$\therefore \ {}_gn_w = {}_gn_a \times {}_an_w \qquad . \qquad . \qquad . \qquad . \qquad . \qquad (i)$$

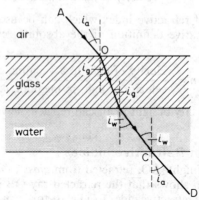

Fig. 17.3 Refraction at a parallel plane surfaces

We can derive this relation more simply from the definition of refractive index n in terms of the velocity of light, using $n = c/v$ (p. 444). Assuming the velocity of light in air is practically the same as the velocity c, then

$$_gn_a \times {}_an_w = \frac{v_g}{c} \times \frac{c}{v_w} = \frac{v_g}{v_w} = {}_gn_w.$$

Further, as $_gn_a = \dfrac{1}{{}_an_g}$, we can write from above

$$_gn_w = \frac{{}_an_w}{{}_an_g}.$$

Since $_an_w = 1 \cdot 33$ and $_an_g = 1 \cdot 5$, it follows that $_gn_w = \dfrac{1 \cdot 33}{1 \cdot 5} = 0 \cdot 89.$

From (i) above, it follows that in general

$$_1n_3 = {}_1n_2 \times {}_2n_3 \qquad . \qquad . \qquad . \qquad . \qquad (4)$$

The order of the suffixes enables this formula to be easily memorised.

General Relation between n and Sin i

From Fig. 17.3, $\sin i_a / \sin i_g = {}_an_g$

$$\therefore \sin i_a = {}_an_g \sin i_g \qquad . \qquad . \qquad . \qquad . \qquad (i)$$

Also, $\sin i_w / \sin i_a = {}_wn_a = 1/{}_an_w$

$$\therefore \sin i_a = {}_an_w \sin i_w \qquad . \qquad . \qquad . \qquad . \qquad (ii)$$

Hence, from (i) and (ii),

$$\sin i_a = {}_an_g \sin i_g = {}_an_w \sin i_w.$$

If the equations are re-written in terms of the absolute refractive indices of air (n_a), glass (n_g), and water (n_w), we have

$$n_a \sin i_a = n_g \sin i_g = n_w \sin i_w$$

since $n_a = 1$. This relation shows that when a ray is refracted from one medium to another, *the boundaries being parallel*,

$$n \sin i = constant \qquad . \qquad . \qquad . \qquad . \qquad (5)$$

where n is the absolute refractive index of a medium and i is the angle made by the ray with the normal in that medium.

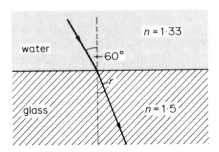

Fig. 17.4 Refraction from water to glass

This relation applies also to the case of light passing directly from one medium to another. As an illustration of its use, suppose a ray is incident on a water-glass boundary at an angle of 60°, Fig. 17.4. Then, applying '$n \sin i$ is a constant', we have

$$1 \cdot 33 \sin 60° = 1 \cdot 5 \sin r \qquad . \qquad . \qquad . \qquad . \qquad (iii)$$

where r is the angle of refraction in the glass, and 1·33, 1·5 are the respective values of n_w and n_g. Thus $\sin r = 1 \cdot 33 \sin 60°/1 \cdot 5 = 0 \cdot 7679$, from which $r = 50 \cdot 1°$.
Alternatively,

$$\frac{\sin 60°}{\sin r} = {}_w n_g = {}_w n_a \times {}_a n_g = \frac{1}{{}_a n_w} \times {}_a n_g = \frac{1 \cdot 5}{1 \cdot 33}.$$

So $\sin r = 1 \cdot 33 \sin 60°/1 \cdot 5$. From Tables, $r = 50 \cdot 1°$.

Multiple Images in Mirrors

If a candle or other object is held in front of a plane mirror, a series of faint or 'ghost' images are observed in addition to one bright image. Suppose O is an object placed in front of a mirror with silvering on the back surface M, as shown in Fig. 17.5. A ray OA from O is then reflected from the front (glass) surface along AD and gives rise to a faint image I_1, while the remainder of the

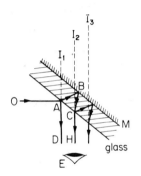

Fig. 17.5 Multiple images

light energy is refracted at A along AB. Reflection then takes place at the silvered surface, and after refraction into the air along CH a bright image is observed at I_2. A small percentage of the light is reflected at C, however, and re-enters the glass again, thus forming a faint image at I_3. Other faint images are formed in the same way. Thus a series of *multiple images* is obtained, the brightest being I_2. The images lie on the normal from O to the mirror, the distances depending on the thickness of the glass and its refractive index and the angle of incidence.

Apparent Depth

Swimmers in particular are aware that the bottom of a pool of water appears nearer the surface than is actually the case; the phenomenon is due to the refraction of light.

Consider an object O at a distance below the surface of a medium such as water or glass, which has a refractive index n, Fig. 17.6. A ray OM from O perpendicular to the surface passes straight through into the air along MS. A

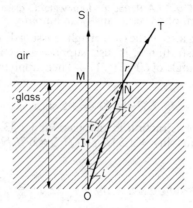

Fig. 17.6 Apparent depth (inclination of ON to OM exaggerated)

ray ON very close to OM is refracted at N into the air away from the normal, in a direction NT; and an observer viewing O directly overhead sees it in the position I, which is the point of intersection of SM and TN produced. Though we have only considered two rays in the air, a *cone* of rays, with SM as the axis, actually enters the observer's eye.

Suppose the angle of incidence in the glass is i, and the angle of refraction in the air is r. Then, since 'n sin i is a constant' (p. 329), we have

$$n \sin i = 1 \times \sin r \qquad . \qquad . \qquad . \qquad . \qquad \text{(i)}$$

where n is the refractive index of glass; the refractive index of air is 1. Since i = angle NOM, and r = MIN, sin i = MN/ON and sin r = MN/IN. From (i), it follows that

$$n\frac{\text{MN}}{\text{ON}} = \frac{\text{MN}}{\text{IN}}$$

$$\therefore n = \frac{\text{ON}}{\text{IN}} \qquad . \qquad . \qquad . \qquad . \qquad \text{(ii)}$$

Since we are dealing with the case of an observer directly above O, the rays

ON, IN are *very* close to the normal OM. Hence to a very good approximation, ON = OM and IN = IM. From (ii),

$$\therefore n = \frac{ON}{IN} = \frac{OM}{IM}.$$

Since the real depth of the object O = OM, and its apparent depth = IM,

$$\therefore n = \frac{\text{real depth}}{\text{apparent depth}} \qquad . \qquad . \qquad . \qquad . \qquad (6)$$

If the real depth, OM = t, the apparent depth = t/n, from (6). The *displacement*, OI, of the object, which we shall denote by d, is thus given by $t - t/n$, i.e.,

$$d = t\left(1 - \frac{1}{n}\right) \qquad . \qquad . \qquad . \qquad . \qquad (7)$$

If an object is 6 cm below water of refractive index $n = 1\frac{1}{3}$, it appears to be displaced upward to an observer in air by an amount, $d = 6\left(1 - \frac{1}{1\frac{1}{3}}\right) = 1\frac{1}{2}$ cm.

General case. If, in Fig. 17.6, MN is the boundary between media and refractive indices n_1 (in place of glass) and n_2 (in place of air), then we must replace n in (6) by $_2n_1$, the relative refractive index between the two media. Hence in this case

$$\frac{OM}{IM} = {_2n_1} = \frac{n_1}{n_2} \qquad . \qquad . \qquad . \qquad . \qquad (8)$$

since $_2n_1 = {_2n_a} \times {_an_1} = {_an_1}/{_an_2}$, where a represents a vacuum (p. 328). In (8), note that n_1 refers to the medium in which the object is situated and n_2 to the medium into which the rays are refracted.

Measurement of Refractive Index by Apparent Depth Method

The formula for the refractive index of a medium in terms of the real and apparent depths can be used to measure refractive index. A *travelling microscope*, S (a microscope which can travel in a vertical direction and which has a fixed graduated scale T beside it) is focused on lycopodium particles at O on a sheet of white paper, and the reading on T is noted, Fig. 17.7. Suppose it is c cm. If the refractive index of glass is required, a glass block A is placed on the paper, and the microscope is raised until the particles are refocused at I.

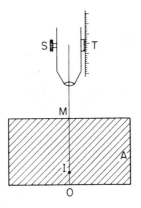

Fig. 17.7 Refractive index by apparent depth

Suppose the reading on T is b cm. Some lycopodium particles are then sprinkled at M on the top of the glass block, and the microscope is raised until they are focused, when the reading on T is noted. Suppose it is a cm.

Then   real depth of O $= OM = (a - c)$ cm

and   apparent depth $= IM = (a - b)$ cm

$$\therefore n = \frac{\text{real depth}}{\text{apparent depth}} = \frac{a - c}{a - b}.$$

The high accuracy of this method for n lies mainly in the fact that the objective of the microscope collects only those rays near to its axis, so that the object O, and its apparent position I, are seen by rays very close to the normal OM. The experiment thus fulfils the theoretical conditions assumed in the proof of the formula $n = $ real depth/apparent depth, p. 331.

The refractive index of water can also be obtained by an apparent depth method. The block A is replaced by a dish, and the microscope is focused first on an object on the bottom of the dish and then on lycopodium powder sprinkled on the surface of water poured into the dish. The apparent position of the bottom of the dish is also noted, and the refractive index of the water n_w is calculated from the relation

$$n_w = \frac{\text{real depth of water}}{\text{apparent depth of water}}.$$

Object below Parallel-sided Glass Block

Consider an object O placed some distance in air below a parallel-sided glass block of thickness t, Fig. 17.8. The ray OMS normal to the surface emerges along MS, while the ray OO_1 close to the normal is refracted along O_1N in the glass and emerges in air along NT in a direction parallel to OO_1 (see p. 327). An observer (not shown) above the glass thus sees the object at I, the point of intersection of TN and SM.

Suppose the normal at O_1 intersects IN at I_1. Then, since O_1I_1 is parallel to OI and IT is parallel to OO_1, OII_1O_1 is a parallelogram. Thus $OI = O_1I_1$. But OI is the displacement of the object O. Hence O_1I_1 is equal to the displacement. Since the apparent position of an object at O_1 is at I_1 (compare Fig. 17.6), we

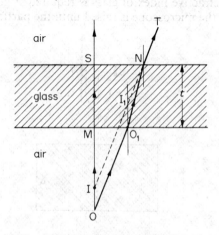

Fig. 17.8 Object below glass block

conclude that *the displacement OI of O is independent of the position of O below the glass,* and is given by $OI = t\left(1 - \dfrac{1}{n}\right)$, as shown on p. 331.

If there are parallel layers of different material resting on top of each other, for example, a layer of oil on a layer of water which in turn rests on a block of glass, the apparent position of an object at the bottom of the glass can be found by *adding the separate displacements due to each layer.* An example will illustrate the method.

Example

Find an expression for the distance through which an object appears to be displaced towards the eye when a plate of glass of thickness t and refractive index n is interposed.

A tank contains a slab of glass 8 cm thick and of refractive index 1·6. Above this is a depth of 4·5 cm of a liquid of refractive index 1·5 and upon this floats 6 cm of water ($n = 4/3$). To an observer looking down from above, what is the apparent position of a mark on the bottom of the tank? (*O. & C.*)

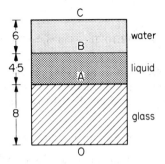

Fig. 17.9 Example

Suppose O is the mark at the bottom of the tank, Fig. 17.9. Then since the boundaries of the media are parallel, the total displacement of O is the sum of the displacements due to each of the media.

For glass, displacement, $d = t\left(1 - \dfrac{1}{n}\right) = 8\left(1 - \dfrac{1}{1\cdot6}\right) = 3$ cm.

For liquid, $d = t\left(1 - \dfrac{1}{n}\right) = 4\cdot5\left(1 - \dfrac{1}{1\cdot5}\right) = 1\cdot5$ cm

For water, $d = 6\left(1 - \dfrac{1}{4/3}\right) = 1\cdot5$ cm

$\therefore$ total displacement $= 3 + 1\cdot5 + 1\cdot5 = 6$ cm

$\therefore$ apparent position of O is 6 cm from bottom.

Total Internal Reflection. Critical Angle

If a ray AO in glass is incident at a small angle α on a glass–air plane boundary, observation shows that part of the incident light is reflected along OE in the glass, while the remainder of the light is refracted away from the normal at an angle β into the air. The reflected ray OE is weak, but the refracted ray OL is bright, Fig. 17.10 (i). This means that most of the incident light energy is transmitted, and a little is reflected.

When the angle of incidence, α, in the glass is increased, the angle of emergence, β, is increased at the same time; and at some angle of incidence c in the glass the refracted ray OL travels along the glass–air boundary, making the angle of refraction 90°, Fig. 17.10 (ii). The reflected ray OE is still weak in intensity, but as the angle of incidence in the glass is increased slightly the reflected ray suddenly becomes bright, and no refracted ray is then observed, Fig. 17.10 (iii). Since *all* the incident light energy is now reflected, *total reflection* is said to take place in the glass at O.

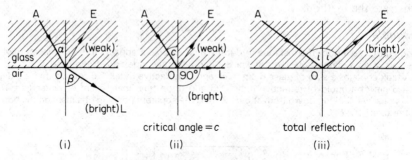

Fig. 17.10 Total internal reflection

When the angle of refraction in air is 90°, a critical stage is reached at the point of incidence O. The angle of incidence *in the glass* is accordingly known as the *critical angle* for glass and air, Fig. 17.10 (ii). Since '$n \sin i$ is a constant' (p. 329), we have

$$n \sin c = 1 \times \sin 90°,$$

where n is the refractive index of the glass. As $\sin 90° = 1$, then

$$n \sin c = 1,$$

or,
$$\sin c = \frac{1}{n} \qquad . \qquad . \qquad . \qquad . \qquad (8)$$

Crown glass has a refractive index of about 1·51 for yellow light, and thus the critical angle for glass to air is given by $\sin c = 1/1·51 = 0·667$. Consequently $c = 41·5°$. Thus if the incident angle in the glass is greater than c, for example 45°, total reflection occurs, Fig. 17.10 (iii). The critical angle between two media for blue light is less than for red light, since the refractive index for blue light is greater than that for red light (see p. 378).

The phenomenon of total reflection may occur when light in glass ($n_g = 1·51$, say) is incident on a boundary with water ($n_w = 1·33$). Applying '$n \sin i$ is a constant' to the critical case, Fig. 17.11, we have

$$n_g \sin c = n_w \sin 90°,$$

where c is the critical angle. As $\sin 90° = 1$

$$n_g \sin c = n_w$$

$$\therefore \sin c = \frac{n_w}{n_g} = \frac{1·33}{1·51} = 0·889$$

Thus if the angle of incidence in the glass exceeds 63°, total internal reflection occurs.

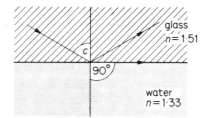

Fig. 17.11 Critical angle for water and glass

It should be carefully noted that the phenomenon of total internal reflection can occur only when light travels from one medium to another which has a *smaller* refractive index, i.e., which is optically less dense. The phenomenon cannot occur when light travels from one medium to another optically denser, for example from air to glass, or from water to glass, as a refracted ray is then always obtained.

Total reflecting prisms are usually made of crown glass of refractive index n about 1·5. The prism angles are 45°, 45° and 90°. Used in prism binoculars (p. 391) for example, light passes normally into the glass prism and is incident at 45° *in the glass* on one face, from the geometry of the prism. Now the critical angle c is given by $\sin c = 1/n = 1/1·5$, from which $c = 42°$ approximately. So the angle of incidence 45° is greater than the critical angle. Hence the light is totally reflected and so it is deviated through 90° and passes out of the prism. Submarine periscopes contain total reflecting prisms. They are used in place of plane mirrors for deviating light as the mirrors produce multiple images (p. 329). See also Fig. 17A.

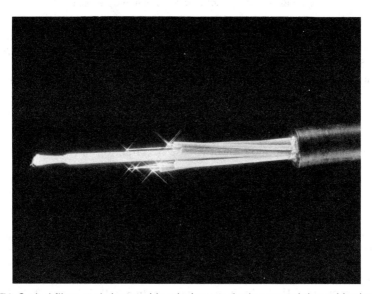

Fig. 17A *Optical fibres* are being used in telephone and other transmitting cables by British Telecom in a new network. The fibres are hair thin strands of specially coated glass. They can transmit a laser or other light beam from one end to the other as a result of repeated total internal reflections at the glass boundary, even if the fibre is bent or twisted. Each fibre can carry as many as 2000 telephone conversations, with less signal loss than in conventional telephone cables. (*By courtesy of The Post Office*)

Measurement of Refractive Index of a Liquid by Air-cell Method

The phenomenon of total internal reflection is used in many methods of measuring refractive index. Figure 17.12 (i) illustrates how the refractive index of a *liquid* can be determined. Two thin plane-parallel glass plates, such as microscope slides, are cemented together so as to contain a thin film of air of constant thickness between them, thus forming an air-cell, X. The liquid whose refractive index is required is placed in a glass vessel having thin plane-parallel sides, and X is placed in the liquid. A bright source of light, S, provides rays which are incident on one side of X in a constant direction SO, and the light through X is observed by a person on the other side at E.

When the light is incident normally on the sides of X, the light passes straight through X to E. When X is rotated slightly about a vertical axis, light is still observed; but as X is rotated farther, the light is suddenly cut off from E, and hence no light now passes through X, Fig. 17.12 (i).

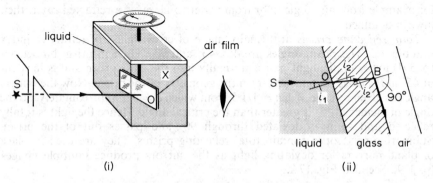

Fig. 17.12 Refractive index of liquid by air-cell

Theory. Figure 17.12 (ii) shows the behaviour of the light when this happens. The ray SO is refracted along OB in the glass, but at B *total internal reflection begins*. Suppose i_1 is the angle of incidence in the liquid, i_2 is the angle of incidence in the glass, and n, n_g are the corresponding refractive indices. Since the boundaries of the media are parallel we can apply the relation '$n \sin i$ is a constant'. Hence

$$n \sin i_1 = n_g \sin i_2 = 1 \times \sin 90° \qquad \qquad \text{(i)}$$

the last product corresponding to the case of refraction in the air-film.

$$\therefore n \sin i_1 = 1 \times \sin 90° = 1 \times 1 = 1$$

$$\therefore n = \frac{1}{\sin i_1} \qquad \qquad \qquad \text{(ii)}$$

It should now be carefully noted that i_1 is the angle of incidence in the *liquid* medium. This is the angle of rotation when X is turned from its position normal to SO to the position when the light is cut off. In practice, it is better to rotate X in opposite directions and determine the angle θ between the *two* positions for the extinction of the light. The angle i_1 is then half the angle θ, and hence $n = 1/\sin\dfrac{\theta}{2}$.

From equations (i) and (ii), it will be noted that i_1 is the critical angle between the liquid and air, and i_2 is the critical angle between the glass and air. We cannot measure i_2, however, as we can i_1, and hence the method provides the refractice index of the *liquid*.

The source of light, S, in the experiment should be a monochromatic source, that is, it should provide light of one colour, for example, yellow light. The extinction of the light is then sharp. If white light is used, the colours in its spectrum are cut off at slightly different angles of incidence, since refractive index depends upon the colour of the light (p. 378). The extinction of the light is then gradual and ill-defined.

Atmospheric Refraction. Refraction of Electromagnetic Waves

The optical density of air depends on its density. On a hot day the warm air near the ground expands and becomes less dense than the cooler air higher up. So light from the sky is refracted continually away from the normal as it enters layers of air near the ground and at one stage a critical angle is reached. The light is now totally reflected in an upward direction. When it enters the eye of an observer, a virtual image of part of the sky is seen. This gives the impression of a pool of water on the ground. Atmospheric refraction also produces the mirage seen in a hot desert.

Radio waves are electromagnetic waves like light waves (p. 269). Sent skywards from England, for example, radio waves are refracted away from the normal when, high above the earth, they enter a 'belt' or layers of electrons and ions. At one stage total reflection occurs and the waves now travel downwards to Australia, for example, on the other side of the globe where they can be detected. The main layers of electrical particles are about 110 km and 270 km above the earth and are called respectively the *Heaviside* and *Appleton* layers after the discoverers.

Exercises 17

1. A ray of light is incident at 60° in air on an air–glass plane surface. Find the angle of refraction in the glass (n for glass $= 1.5$).
2. A ray of light is incident in water at an angle of 30° on a water–air plane surface. Find the angle of refraction in the air (n for water $= 4/3$).
3. A ray of light is incident in water at an angle of (i) 30°, (ii) 70° on a water–glass plane surface. Calculate the angle of refraction in the glass in each case ($_an_g = 1.5$, $_an_w = 1.33$).
4. What is the apparent position of an object below a rectangular block of glass 6 cm thick if a layer of water 4 cm thick is on top of the glass (refractive index of glass and water $= 1\frac{1}{2}$ and $1\frac{1}{3}$ respectively)?   ✓ 23 oct
5. Calculate the critical angle for (i) an air–glass surface, (ii) an air–water surface, (iii) a water–glass surface; draw diagrams in each case illustrating the total reflection of a ray incident on the surface ($_an_g = 1.5$, $_an_w = 1.33$).
6. Explain what happens in general when a ray of light strikes the surface separating transparent media such as water and glass. Explain the circumstances in which total reflection occurs and show how the critical angle is related to the refractive index.

Describe a method for determining the refractive index of a medium by means of critical reflection. (*L.*)
7. Explain carefully why the apparent depth of the water in a tank changes with the position of the observer.

A microscope is focused on a scratch on the bottom of a beaker. Turpentine is poured into the beaker to a depth of 4 cm, and it is found necessary to raise the microscope through a vertical distance of 1.28 cm to bring the scratch again into focus. Find the refractive index of the turpentine. (*C.*)
8. (*a*) State the conditions under which total reflection occurs. Show that the phenomenon will occur in the case of light entering normally one face of an isosceles right-angle prism

of glass, but not in the case when light enters similarly a similar hollow prism full of water. (b) A concave mirror of small aperture and focal length 8 cm lies on a bench and a pin is moved vertically above it. At what point will image and object coincide if the mirror is filled with water of refractive index 4/3? (N.)

9. Explain the meaning of critical angle and total internal reflection. Describe fully (a) one natural phenomenon due to total internal reflection, (b) one practical application of it. Light from a luminous point on the lower face of a rectangular glass slab, 2·0 cm thick, strikes the upper face and the totally reflected rays outline a circle of 3·2 cm radius on the lower face. What is the refractive index of the glass? (N.)

10. Explain the meaning of *critical angle*, and describe how you would measure the critical angle for a water–air boundary.

ABCD is the plan of a glass cube. A horizontal beam of light enters the face AB at grazing incidence. Show that the angle θ which any rays emerging from BC would make with the normal to BC is given by sin θ = cot a, where a is the critical angle. What is the greatest value that the refractive index of glass may have if any of the light is to emerge from BC? (N.)

11. (a) For light travelling in a medium of refractive index n_1 and incident on the boundary with a medium of refractive index n_2, explain what is meant by total internal reflection and state the circumstances in which it occurs.

(b) A cube of glass of refractive index 1·500 is placed on a horizontal surface separated from the lower face of the cube by a film of liquid, as shown in Fig. 17A. A ray of light from outside and in a vertical plane parallel to one face of the cube strikes another vertical face of the cube at an angle of incidence $i = 48°\ 27'$ and, after refraction, is totally reflected at the critical angle at the glass–liquid interface. Calculate (i) the critical angle at the glass–liquid interface and (ii) the angle of emergence of the ray from the cube. (N.)

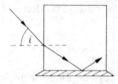

Fig. 17A

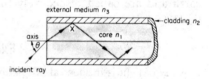

Fig. 17B

12. What do you understand by *angle of refraction* and *refractive index*?

A ray of light crosses the interface between two transparent media of refractive indices n_A and n_B. Give a formula relating the directions of the ray on the two sides of the interface. Show the angles you use in your formula on a diagram. Hence, deduce the conditions necessary for total internal reflection to take place at an interface.

In the simple 'light pipe' shown in Fig. 17B, a ray of light may be transmitted (with little loss) along the core by repeated internal reflection. The diagram shows a cross-section through the diameter of the 'pipe' with a ray incident in that plane. The core, cladding and external medium have refractive indices n_1, n_2 and n_3 respectively. Show that total internal reflection takes place at X provided that the angle θ is smaller than a value θ_m given by the expression sin $\theta_m = \sqrt{(n_1{}^2 - n_2{}^2)}/n_3$. Explain why the pipe does not work for rays for which $\theta > \theta_m$. (Reminder: $\sin^2\theta + \cos^2\theta = 1$.) (C.)

18 Refraction Through Prisms

In Optics, a *prism* is a transparent object usually made of glass which has two plane surfaces, XDEY, XDFZ, inclined to each other, Fig. 18.1. Prisms are used in many optical instruments, for example, prism binoculars. They are also used for separating the colours of the light emitted by glowing objects, which would then give an accurate knowledge of their chemical composition. A prism of glass enables the refractive index of the glass to be measured very accurately.

The angle between the inclined plane surfaces XDFZ, XDEY is known as the *angle of the prism*, or the *refracting angle*, the line of intersection XD of the planes is known as the *refracting edge*, and any plane in the prism perpendicular to XD, such as PQR, is known as a *principal section* of the prism. A ray of light *ab*, incident on the prism at *b* in a direction perpendicular to XD, is refracted towards the normal along *bc* when it enters the prism, and is refracted away from the normal along *cd* when it emerges into the air. From the law of refraction (p. 326), the rays *ab*, *bc*, *cd* all lie in the same plane, which is PQR in this case. If the incident ray is directed towards the refracting angle, as in Fig. 18.1, the light is always deviated by the prism towards its base.

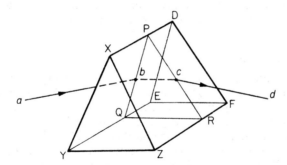

Fig. 18.1 Prism

Refraction through a Prism

Consider a ray HM incident in air on a prism of refracting angle A, and suppose the ray lies in the principal section PQR, Fig. 18.2. Then, if i_1, r_1 and i_2, r_2 are the angles of incidence and refraction at M, N as shown, and n is the prism refractive index,

$$\sin i_1 = n \sin r_1 \qquad . \qquad . \qquad . \qquad . \qquad \text{(i)}$$

$$\sin i_2 = n \sin r_2 \qquad . \qquad . \qquad . \qquad . \qquad \text{(ii)}$$

Further, as MS and NS are normals to PM and PN respectively, angle MPN + angle MSN = 180°, considering the quadrilateral PMSN. But angle NST + angle MSN = 180°.

$$\therefore \text{ angle NST} = \text{angle MPN} = A$$
$$\therefore A = r_1 + r_2 \qquad . \qquad . \qquad . \qquad . \qquad \text{(iii)}$$

as angle NST is the exterior angle of triangle MSN.

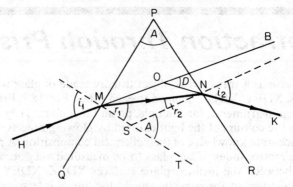

Fig. 18.2 Refraction through prism

In the following sections, we shall see that the *angle of deviation*, D, of the light, caused by the prism, is used considerably. The angle of deviation at M = angle OMN = $i_1 - r_1$; the angle of deviation at N = angle MNO = $i_2 - r_2$. Since the deviations at M, N are in the same direction, the total deviation, D (angle BOK), is given by

$$D = (i_1 - r_1) + (i_2 - r_2) \qquad . \qquad . \qquad . \qquad . \qquad \text{(iv)}$$

Equations (i)–(iv) are the general relations which hold for refraction through a prism. In deriving them, it should be noted that the geometrical form of the prism base plays no part.

Minimum Deviation

The angle of deviation, D, of the incident ray HM is the angle BOK in Fig. 18.2. The variation of D with the angle of incidence, i, can be obtained experimentally by placing the prism on paper on a drawing board and using a ray AO from a ray-box (or two pins) as the incident ray, Fig. 18.3 (i). When the direction AO is kept constant and the drawing board is turned so that the ray is always incident at O on the prism, the angle of incidence i is varied; the corresponding emergent rays CE, HK, LM, NP can be traced on the paper.

Experiment shows that as the angle of incidence i is increased from zero, the deviation D begins to decrease continuously to some minimum value D_{min}, and then increases to a maximum as i is increased further to 90°. A minimum deviation, corresponding to the emergent ray NP, is thus obtained. A graph of D plotted

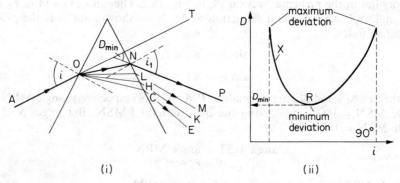

(i)   (ii)

Fig. 18.3 Minimum deviation

against i has the appearance of the curve X, which has a minimum value at R, Fig. 18.3 (ii).

Experiment and theory show that *the minimum deviation, D_{min}, of the light occurs when the ray passes symmetrically through the prism.* Suppose the ray is AONP in Fig. 18.3 (i). Then the incident angle, i, is equal to the angle of emergence, i_1, into the air at N for this special case.

A proof of symmetrical passage of ray at minimum deviation. Experiment shows that minimum deviation is obtained at *one* particular angle of incidence. On this assumption it is possible to prove that the angle of incidence is equal to the angle of emergence in this case. In Fig. 18.4, suppose the deviation of the

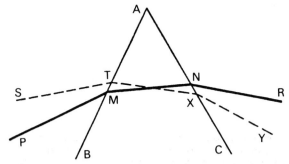

Fig. 18.4 Minimum deviation proof

ray PM is not a minimum, so angle PMB is not equal to angle RNC. It then follows that a ray YX, incident on AC at an angle CXY equal to angle PMB, will emerge along TS, where angle BTS = angle CNR; and from the principle of the reversibility of light, a ray incident along ST on the prism emerges along XY. We therefore have *two* angles of incidence which give the same deviation through the prism.

Since there is only *one* angle of incidence which gives minimum deviation, we now see that this is obtained when the angle of emergence is exactly equal to the angle of incidence. So in the minimum deviation case the ray passes *symmetrically* through the prism.

Relation between A, D_{min} and n

A very convenient formula for refractive index, n, can be obtained in the minimum deviation case. The ray PQRS then passes symmetrically through the prism, and the angles made with the normal in the air and in the glass at Q,

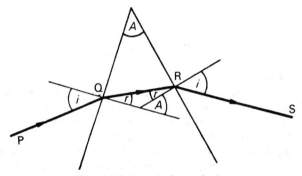

Fig. 18.5 Formula for n of prism

R respectively are equal, Fig. 18.5. Suppose the angles are denoted by i, r, as shown. Then, as explained on p. 340,

$$i - r + i - r = D_{min} \qquad \qquad \text{(i)}$$

and
$$r + r = A \qquad \qquad \text{(ii)}$$

From (ii),
$$r = \frac{A}{2}$$

Substituting for r in (i),
$$2i = A + D_{min}$$

$$\therefore i = \frac{A + D_{min}}{2}$$

$$\therefore n = \frac{\sin i}{\sin r} = \frac{\sin \dfrac{A + D_{min}}{2}}{\sin \dfrac{A}{2}} \qquad \qquad \text{(1)}$$

Example

Describe a good method of measuring the refractive index of a substance such as glass and give the theory of the method. A glass prism of angle 72° and index of refraction 1·66 is immersed in a liquid of refractive index 1·33. What is the angle of minimum deviation for a parallel beam of light passing through the prism? (L.)

$$n = \frac{\sin\left(\dfrac{A + D_{min}}{2}\right)}{\sin \dfrac{A}{2}}$$

where n is the *relative refractive index* of glass with respect to the liquid.

But
$$n = \frac{1 \cdot 66}{1 \cdot 33}$$

$$\therefore \frac{1 \cdot 66}{1 \cdot 33} = \frac{\sin\left(\dfrac{72° + D_{min}}{2}\right)}{\sin \dfrac{72°}{2}} = \frac{\sin\left(\dfrac{72° + D_{min}}{2}\right)}{\sin 36°}$$

$$\therefore \sin\left(\frac{72° + D_{min}}{2}\right) = \frac{1 \cdot 66}{1 \cdot 33}\sin 36° = 0 \cdot 7335$$

$$\therefore \frac{72° + D_{min}}{2} = 47° \; 11'$$

$$\therefore D_{min} = 22° \; 22'.$$

The Spectrometer

The spectrometer is an optical instrument which is mainly used to study the light from different sources. As we shall see soon, it can be used to measure accurately the refractive index of glass in the form of a prism.

The instrument consists essentially of a *collimator*, C, a *telescope*, T, and a *table*, on which a prism PMN can be placed. The lenses in C, T are achromatic lenses (p. 378). The collimator is fixed, but the table and the telescope can be rotated round a circular scale graduated in half-degrees (not shown) which has a common vertical axis with the table, Fig. 18.6. A vernier is also provided for this scale. The *source* of *light*, S, used in the experiment is placed in front of a narrow slit at one end of the collimator, so that the prism is illuminated by light from S.

Before the spectrometer can be used, however, three adjustments must be made: (1) The collimator C must be adjusted so that parallel light emerges from it; (2) the telescope T must be adjusted so that parallel rays entering it are brought to a focus at cross-wires near its eyepiece; (3) the refracting edge of

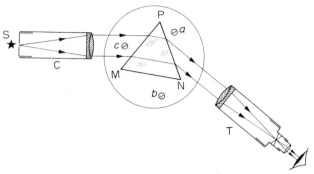

Fig. 18.6 Spectrometer

the prism must be parallel to the axis of rotation of the telescope, i.e., the table must be 'levelled'.

Adjustments of Spectrometer

The telescope adjustment is made by first moving its eyepiece until the cross-wires are distinctly seen, and then sighting the telescope on to a *distant* object through an open window. The length of the telescope is now altered by a screw arrangement until the object is clearly seen at the same place as the cross-wires, so that parallel rays now entering the telescope are brought to a focus at the cross-wires.

The collimator adjustment. With the prism removed from the table, the telescope is now turned to face the collimator, C, and the slit in C is illuminated by a sodium flame which provides yellow light. The edges of the slit are usually blurred, showing that the light emerging from the lens of C is not a parallel beam. The position of the slit is now adjusted by moving the tube in C, to which the slit is attached, until the edges of the latter are sharp.

'Levelling' the table. If the rectangular slit is not in the centre of the field of view when the prism is placed on the table, the refracting edge of the prism is not parallel to the axis of rotation of the telescope. The table must then be adjusted, or 'levelled', by means of the screws *a, b, c* beneath it. One method of procedure consists of placing the prism on the table with one face MN approximately perpendicular to the line joining two screws *a, b*, as shown in Fig. 18.6. The table is turned until MN is illuminated by the light from C, and the telescope T is then moved to receive the light reflected from MN. The screw *b* is then adjusted until the slit appears in the centre of the field of view. With C and T fixed, the table is now rotated until the slit is seen by reflection at the face NP of the prism, and the screw *c* is then adjusted until the slit is again in the middle of the field of view. The screw *c* moves MN in its own plane, and hence the movement of *c* will not upset the adjustment of MN in the perpendicular plane.

Measurement of the Angle, A, of a Prism

The angle A of a prism can be measured very accurately by a spectrometer. The refracting edge, P, of the prism is turned so as to face the collimator lens, which then illuminates the two surfaces containing the refracting angle A with parallel light, Fig. 18.7 (i). An image of the collimator slit is hence observed with the telescope in positions T_1, T_2, corresponding to reflection of light at the respective faces of the prism. It is shown below that the angle of rotation of the telescope from T_1 to T_2 is equal to $2A$, and hence the angle of the prism, A, can be obtained.

Proof. Suppose the incident ray MN makes a glancing angle α with one face of the prism, and a parallel ray at K makes a glancing angle β with the other face, Fig. 18.7 (ii). The reflected ray NQ then makes a glancing angle α with the prism surface, and hence the deviation of MN is 2α (see p. 315). Similarly, the deviation by reflection at K is 2β. Thus the reflected rays QN, LK are inclined at an angle equal to $2\alpha + 2\beta$, corresponding to the angle of rotation of the

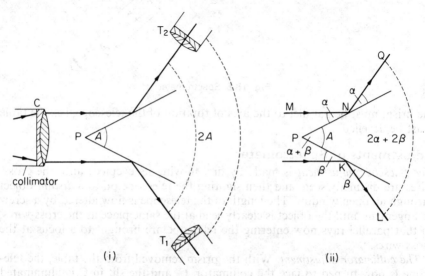

Fig. 18.7 Measurement of angle of prism

telescope from T_1 to T_2. But the angle, A, of the prism $= \alpha + \beta$, as can be seen by drawing a line through P parallel to MN, and using alternate angles. Hence the rotation of the telescope $= 2\alpha + 2\beta = 2A$.

Measurement of the Minimum Deviation, D_{min}

In order to measure the minimum deviation, D_{min}, caused by refraction through the prism, the latter is placed with its refracting angle A pointing *away* from the collimator, as shown in Fig. 18.8 (i). The telescope is then turned until an image of the slit is obtained on the cross-wires, corresponding to the position T_1. The table is now slowly rotated so that the angle of incidence on the left side of the prism decreases, and the image of the slit is kept on the cross-wires by moving the telescope at the same time. The image of the slit, and the telescope, then slowly approach the fixed line XY. But at one position, corresponding to T_2, the image of the slit begins to move *away* from XY. If the table is now turned in the opposite direction the image of the slit again moves back when the telescope reaches the position T_2. The angle between the emergent ray CH and the XY is hence the smallest angle of deviation caused by the prism, and is thus equal to D_{min}.

The minimum deviation is obtained by finding the angle between the positions of the telescope (i) at T_2, (ii) at T; the prism is removed in the latter case so as to view the slit directly. Alternatively, the experiment to find the minimum deviation is repeated with the refracting angle pointing the opposite way, the prism being represented by dotted lines in this case, Fig. 18.8 (ii). If the position of the telescope for minimum deviation is now O, it can be seen that the angle

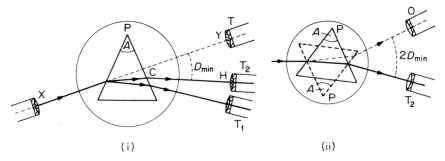

Fig. 18.8 Measurement of minimum deviation

between the position O and the other minimum deviation position T_2 is $2D_{min}$. The value of D_{min} is thus easily calculated.

The Refractive Index of a Prism Material

The refractive index, n, of the material of the prism can be easily calculated once A and D_{min} have been determined, since, from p. 342.

$$n = \sin \frac{A + D_{min}}{2} \Big/ \sin \frac{A}{2}.$$

In an experiment of this nature, the angle, A, of a glass prism was found to be 59° 52′, and the minimum deviation, D_{min}, was 40° 30′. Thus

$$n = \sin \frac{59° \ 52′ + 40° \ 30′}{2} \Big/ \sin \frac{59° \ 52′}{2}$$

$$= \sin 50° \ 11′ / \sin 29° \ 56′$$

$$= 1 \cdot 539.$$

The spectrometer prism method of measuring refractive index is capable of providing an accuracy of one part in a thousand. The refractive index of a liquid can also be found by this method, using a hollow glass prism made from thin parallel-sided glass strips.

Since n for glass is less for red light than for blue light, it follows from the formula for n that D_{min} has a smaller value for red than for blue light.

Grazing Incidence for a Prism

We shall now leave any further considerations of minimum deviation, and shall consider briefly other special cases of refraction through a prism.

Maximum deviation, D_{max}, occurs when the angle of incidence on the face of the prism is 90°. This is shown in the graph of D against i in Fig. 18.3 (ii). In this case the ray has 'grazing incidence' on the face of the prism, as shown in

Fig. 18.9 (i), and it emerges making an angle i to the normal at the second face. From the principle of the reversibility of light, it follows that a ray making an angle of incidence i on the face of the prism emerges making an angle of 90° to the normal, Fig. 18.9 (ii). So maximum deviation occurs for two angles of incidence on the face of a prism, 90° and i. Also, from Fig. 18.9 (i) or (ii),

$$D_{max} = d_1 + d_2 = (90° - c) + (i - r),$$

since the angle in the glass is the critical angle c when the angle in the air is 90°.

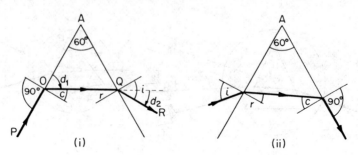

Fig. 18.9 Maximum deviation

Example

Calculate the angle of emergence i, and the deviation, when light is incident at 90° on the face of a 60° prism of refractive index 1·5.

In Fig. 18.9 (i), the incident ray PO is refracted at the critical angle c along OQ in the glass. From $\sin c = 1/n = 1/1·5 = 0·6667$, then $c = 41·8°$.

Since $A = 60° = c + r$, where r is the angle of incidence at Q, then $r = 60° - 41·8° = 18·2°$. From $\sin i/\sin r = n$, the angle of emergence i is given by

$$\frac{\sin i}{\sin 18·2°} = 1·5$$

or $\sin i = 1·5 \times \sin 18·2° = 0·4685.$

So $i = 27·9°.$

Deviation, $D_{max} = d_1 + d_2 = 90° - c + i - r$
$$= 90° - 41·8° + 27·9° - 18·2°$$
$$= 57·9°.$$

Grazing Incidence and Grazing Emergence

If a ray BM is at grazing incidence on the face of a prism, and the angle A of the prism is increased, a calculation shows that the refracted ray MN in the glass will make a bigger and bigger angle of incidence on the other face PR, Fig. 18.10. This is left as an exercise for the reader.

At a certain value of A, MN will make the critical angle, c, with the normal at N, and the emergent ray NR will then graze the surface PR, as shown in Fig. 18.10. As A is increased further, the rays on the glass strike PR at angles of incidence greater than c, and hence no emergent rays are obtained. Thus Fig. 18.10 illustrates the largest angle of a prism for which emergent rays are obtained, and this is known as the *limiting angle* of the prism. It can be seen from the geometry of Fig. 18.10 that $A = c + c$ in this special case, and hence *the limiting angle of a prism is twice the critical angle*. For crown glass of $n = 1·51$ the critical angle c is 41° 30′, and hence transmission of light through a prism of crown glass is impossible if the angle of the prism exceeds 83°.

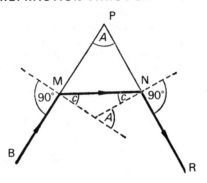

Fig. 18.10 Maximum deviation of prism

Total Reflecting Prisms

When a plane mirror silvered on the back is used as a reflector, multiple images are obtained (p. 329). This disadvantage is overcome by using right-angled isosceles prisms as reflectors of light in optical instruments such as prism binoculars (see p. 391).

Consider a ray OQ incident normally on the face AC of such a prism, Fig. 18.11 (i). The ray is undeviated, and is therefore incident at P in the glass at an

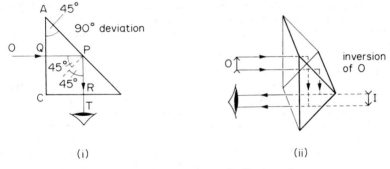

(i)   (ii)

Fig. 18.11 Inversion by total reflecting prisms

angle of 45° to the normal at P. If the prism is made of crown glass its critical angle is 41° 30′. Hence the incident angle, 45°, in the glass is greater than the critical angle, so the light is *totally* reflected in the glass at P. A bright beam of light thus emerges from the prism along RT, and since the angle of reflection at P is equal to the incident angle, RT is perpendicular to OQ. The prism thus deviates the light through 90°. If the prism is positioned as shown in Fig. 18.11 (ii), an inverted bright image I of the object O is seen by total reflection at the two surfaces of the prism.

There is no loss of brightness when total internal reflection occurs at a surface, whereas the loss may be as much as 10 per cent or more in reflection at a silver surface.

Exercises 18

1. A ray of light is refracted through a prism of angle 70°. If the angle of refraction in the glass at the first face is 28°, what is the angle of incidence in the glass at the second face?

2. (i) The angle of a glass prism is 60°, and the minimum deviation of light through the prism is 39°. Calculate the refractive index of the glass. (ii) The refractive index of a glass prism is 1·66, and the angle of the prism is 60°. Find the minimum deviation.

3. A narrow beam of light is incident normally on one face of an equilateral prism (refractive index 1·45) and finally emerges from the prism. The prism is now surrounded by water (refractive index 1·33). What is the angle between the directions of the emergent beam in the two cases? (L.)

4. By means of a labelled diagram show the paths of rays from a monochromatic source to the eye through a correctly adjusted prism spectrometer.

Obtain an expression relating the deviation of the beam by the prism to the refracting angle and the angles of incidence and emergence.

A certain prism is found to produce a minimum deviation of 51° 0′, while it produces a deviation of 62° 48′ for two values of the angle of incidence, namely 40° 6′ and 82° 42′ respectively. Determine the refracting angle of the prism, the angle of incidence at minimum deviation and the refractive index of the material of the prism. (L.)

5. The deviation of a ray of light in passing through a triangular prism depends on the angle of incidence with the first face. Experiment shows that the deviation has a minimum value. Explain why this occurs when the light passes symmetrically through the prism.

Derive an expression for the refractive index of the material of the prism in terms of the refracting angle and the angle of minimum deviation.

Explain how you would use the spectrometer to determine the refractive index of the material of a prism. State the initial adjustments which are required to be made to the spectrometer before you can make your measurements. (Details of experimental procedure are not required.) (L.)

6. A ray of light passing symmetrically through a glass prism of refracting angle A is deviated through an angle D. Derive an expression for the refractive index of the glass.

A prism of refracting angle about 60° is mounted on a spectrometer table and all the preliminary adjustments are made to the instrument. Describe and explain how you would then proceed to measure the angles A and D.

PQR represents a right-angled isosceles prism of glass of refractive index 1·50. A ray of light enters the prism through the hypotenuse QR at an angle of incidence i, and is reflected at the critical angle from PQ to PR. Calculate and draw a diagram showing the path of the ray through the prism. (Only rays in the plane of PQR need be considered.) (N.)

7. Give a labelled diagram showing the essential optical parts of a prism spectrometer. Describe the method of adjusting a spectrometer and using it to measure the angle of a prism.

A is the vertex of a triangular glass prism, the angle at A being 30°. A ray of light OP is incident at P on one of the faces enclosing the angle A, in a direction such that the angle $OPA = 40°$. Show that, if the refractive index of the glass is 1·50, the ray cannot emerge from the second face. (L.)

8. Show that the refractive index, n, of the material of a glass prism is given by $n = \sin(A + D_m)/2 \div \sin(A/2)$, where A is the refracting angle of the prism and D_m is the angle of minimum deviation for light passing through the prism.

Rays of red and blue light are used with a given prism. Explain which will have the larger value of D_m. If both rays pass through at minimum deviation will they be parallel anywhere? Explain your answer with the aid of a diagram.

Describe how, using pins, paper and a drawing board, you would measure (a) the refracting angle of the prism, and (b) the angle of minimum deviation for light passing through the prism. (L.)

9. Explain how you would adjust the telescope of a spectrometer before making measurements.

Draw and label a diagram of the optical parts of a prism spectrometer after the adjustments have been completed. Indicate the position of the crosswires and show the paths through the instrument of two rays from a monochromatic source when the setting for minimum deviation has been obtained.

The refracting angle of a prism is 62·0° and the refractive index of the glass for yellow light is 1·65°. What is the smallest possible angle of incidence of a ray of this yellow light which is transmitted without total internal reflection? Explain what happens if white light is used instead, and the angle of incidence is varied in the neighbourhood of this minimum. (N.)

19 Lenses and Defects. Spectra

Converging and Diverging Lenses

A *lens* is an object, usually made of glass, bounded by one or two spherical surfaces. Figure 19.1 (i) illustrates three types of *converging* lenses, which are thicker in the middle than at the edges. Figure 19.1 (ii) shows three types of *diverging* lenses, which are thinner in the middle than at the edges.

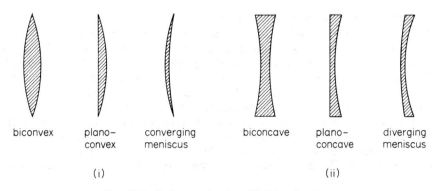

| biconvex | plano–convex | converging meniscus | biconcave | plano–concave | diverging meniscus |

(i)   (ii)

Fig. 19.1 (i) Converging lens (ii) Diverging lens

The *principal axis* of a lens is the line joining the centres of curvature of the two surfaces, and passes through the middle of the lens. Experiments with a ray-box show that a thin converging lens brings an incident parallel beam of rays to a *principal focus*, F, on the other side of the lens when the beam is narrow and incident close to the principal axis, Fig. 19.2 (i). On account of the convergent beam contained with it, the lens is better described as a 'converging' lens. If a similar parallel beam is incident on the other (right) side of the lens, it converges to a focus F′, which is at the same distance from the lens as F when the lens is thin. To distinguish F from F′ the latter is called the 'first principal focus'; F is known as the 'second principal focus'.

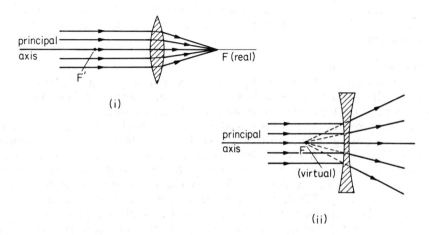

Fig. 19.2 Focus of (i) converging, and (ii) diverging lenses

When a narrow parallel beam, close to the principal axis, is incident on a thin diverging lens, experiment shows that a beam is obtained which appears to diverge from a point F on the same side as the incident beam, Fig. 19.2 (ii). F is known as the principal 'focus' of the diverging lens.

Signs of Focal Length, f

From Fig. 19.2 (i), it can be seen that a converging lens has a real focus; the focal length, f, of a *converging* lens is thus *positive* in sign. Since the focus of a diverging lens is virtual, the focal length of such a lens is negative in sign, Fig. 19.2 (ii). The reader should memorise the sign of f for a converging and diverging lens respectively, as this is always required in lens formulae.

Explanation of Effects of Lenses

A thin lens may be regarded as made up of a very large number of *small-angle prisms* placed together, as shown in the exaggerated sketches of Fig. 19.3. If the spherical surfaces of the various truncated prisms are imagined to be produced, the angles of the prisms can be seen to increase from zero at the middle to a small value at the edge of the lens. The truncated prisms farther away from the middle of the lens deviates an incident ray more than those prisms near the middle, and this is how the converging lens brings the parallel rays to a focus F in Fig. 19.3 (i). It will be noted that a ray AC incident on the middle, C, of the lens emerges parallel to AC, since the middle acts like a rectangular piece of glass (p. 328). This fact is used in drawing images in lenses, discussed shortly.

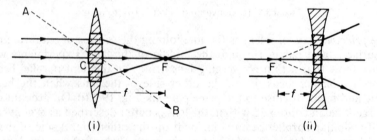

Fig. 19.3 Action of converging and diverging lenses

Since the *diverging* lens is made up of truncated prisms pointing the opposite way to the converging lens, the deviation of the light is in the opposite direction, Fig. 19.3 (ii). So a divergent beam is obtained when parallel rays are refracted by this lens.

Images in Lenses

Converging lens. (i) When an object is a very long way from this lens, i.e., at infinity, the rays arriving at the lens from the object are parallel. Thus the image is formed at the focus of the lens, and is real and inverted.

(ii) Suppose an object OP is placed at O perpendicular to the principal axis of a thin converging lens, so that it is farther from the lens than its principal focus, Fig. 19.4 (i). A ray PC incident on the middle, C, of the lens is very slightly displaced by its refraction through the lens, as the opposite surfaces near C are parallel (see Fig. 19.3 (i), which is an exaggerated sketch of the passage of the ray). We therefore consider that PC passes *straight through* the lens, and this is true for any ray incident on the middle of a thin lens.

A ray PL parallel to the principal axis is refracted so that it passes through

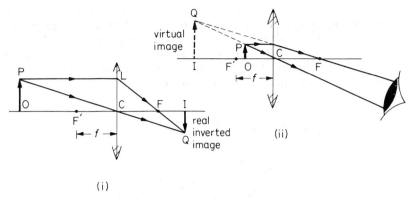

Fig. 19.4 Images in converging lenses

the focus F. Thus the image, Q, of the top point P of the object is formed below the principal axis, and hence the whole image IQ is real and inverted. In making accurate drawings the lens should be represented by a straight line, as illustrated in Fig. 19.4, as we are only concerned with thin lenses and a narrow beam incident close to the principal axis.

(iii) The image formed by a converging lens is always real and inverted until the object is placed nearer the lens than its focal length, Fig. 19.4 (ii). In this case the rays from the top point P *diverge* after refraction through the lens, and hence the image Q is *virtual*. The whole image, IQ, is erect (the same way up as the object) and magnified, besides being virtual, and hence the converging lens can be used as a simple 'magnifying glass' (see p. 392).

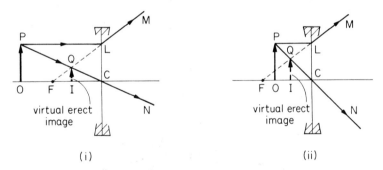

Fig. 19.5 Images in diverging lenses

Diverging lens. In the case of a converging lens, the image is sometimes real and sometimes virtual. In a diverging lens, the image is always virtual; in addition, the image is always erect and diminished. Fig. 19.5 (i), (ii) illustrate the formation of two images. A ray PL appears to diverge from the focus F after refraction through the lens, a ray PC passes straight through the middle of the lens and emerges along CN, and hence the emergent beam from P appears to diverge from Q on the same side of the lens as the object. The image IQ is thus *virtual*.

The rays entering the eye from a point on an object viewed through a lens can easily be traced. Suppose L is a converging lens, and IQ is the image of the object OP, drawn as already explained, Fig. 19.6. If the eye E observes the top point P of the object through the lens, the cone of rays entering E are those bounded by the image Q or P and the pupil of the eye. If these rays are produced back to meet the lens L, and the points of incidence are joined to P, the rays entering E are shown shaded in the beam. The method can be applied to trace the beam of light entering the eye from any other point on the object; the important thing to remember is to *work back from the eye*.

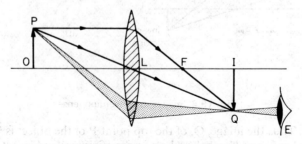

Fig. 19.6 Rays entering the eye

Deviation by Small-angle Prism

As we saw on p. 350, a thin lens can be considered to be made up of a large number of *small-angle prisms*.

Consider a ray PM of monochromatic light incident almost normally on the face TM of a prism of small angle A, so that the angle of incidence, i_1, is small, Fig. 19.7. Then $\sin i_1/\sin r_1 = n$, where r_1 is the angle of refraction in the prism, and n is the refractive index for the colour of the light. As r_1 is less than i_1, r_1 also is a small angle. Now the sine of a small angle is practically equal to the angle measured in radians. Thus $i_1/r_1 = n$, or

$$i_1 = nr_1 \qquad . \qquad . \qquad . \qquad . \qquad . \qquad (i)$$

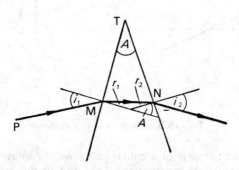

Fig. 19.7 Deviation by small-angle prism

From the geometry of Fig. 19.7, the angle of incidence r_2 on the face TN of the prism is given by $r_2 = A - r_1$; and since A and r_1 are both small, it follows that r_2 is a small angle. The angle of emergence i_2 is thus also small, and since $\sin i_2/\sin r_2 = n$, we may state that $i_2/r_2 = n$, or

$$i_2 = nr_2 \qquad . \qquad . \qquad . \qquad . \qquad . \qquad (ii)$$

The deviation, d, of the ray on passing through the prism is given by $d = (i_1 - r_1) + (i_2 - r_2)$. Substituting for i_1 and i_2 from (i) and (ii),

$$\therefore d = nr_1 - r_1 + nr_2 - r_2 = n(r_1 + r_2) - (r_1 + r_2)$$
$$\therefore d = (n-1)(r_1 + r_2).$$

But
$$r_1 + r_2 = A$$
$$\therefore \mathbf{d = (n-1)A} \qquad . \qquad . \qquad . \qquad . \qquad (1)$$

This is the magnitude of the deviation produced by a *small*-angle prism for *small* angles of incidence. If A is expressed in radians, then d is in radians; if A is expressed in degrees, then d is in degrees. If $A = 6°$ and $n = 1\cdot6$, the deviation d of the light for small angles of incidence is given by $d = (1\cdot6 - 1)\ 6° = 3\cdot6°$. From (1), it should be noted that the deviation is *independent* of the magnitude of the small angle of incidence on the prism. Thus rays with different small angles of incidence are all deviated by the *same* amount. We shall now apply this result to a lens.

Relations between Image and Object Distances for Thin Lens

We can now derive a relation between the object and image distances when a lens is used. We shall limit ourselves to the case of a *thin* lens, i.e., one whose thickness is small compared with its other dimensions, and consider narrow beams of light incident on its central portion.

Suppose O is a point object on the principal axis, at a distance u from the lens greater than the focal length f, Fig. 19.8 (i). A ray OP incident on the lens at a small height h above the axis is refracted at P and Q by the lens surfaces, and then passes through the image I, distance v from the lens.

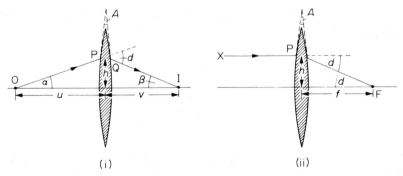

Fig. 19.8 Lens formula

We can see that the small deviation d of the ray OA may be considered due to a *prism of small* angle A, enclosed by the tangent planes to the lens surfaces at P and Q. Now from Fig. 19.8 (i), $d = \alpha + \beta$, where α and β are the angles shown. But $\alpha = h/u$ and $\beta = h/v$ in radians, since $\tan \alpha = \alpha$ in radians when the angle is small and $\tan \beta = \beta$ for a similar reason.

$$\therefore d = \frac{h}{v} + \frac{h}{u} \qquad . \qquad . \qquad . \qquad . \qquad (1)$$

Now consider a ray XP parallel to the principal axis and incident on the lens at the small height h, Fig. 19.8 (ii). After refraction, this ray passes through F, the principal focus, distance f from the lens. As before, the deviation of the ray can be considered due to a prism of small angle A, and the angle of deviation $= h/f$ in this case. Now we showed before that the deviation by a small angle prism is *independent* of the angle of incidence when this is small, which is the case here. Hence from (1),

$$\frac{h}{v} + \frac{h}{u} = \frac{h}{f}.$$

Dividing by h,

$$\therefore \frac{1}{v} + \frac{1}{u} = \frac{1}{f}.$$

This lens equation applies to both converging and diverging lenses, and to all cases of real and virtual objects and images, provided we use the sign rule stated on p. 322, namely, *real* object and image distances are given a $+$ sign and *virtual* object and image distances are given a $-$ sign.

The sign rule also applies to focal lengths. A converging lens has a real focus. So $f = +10$ cm for a converging lens of focal length 10 cm. A diverging lens has a virtual focus. So $f = -20$ cm for a diverging lens of focal length 20 cm.

Some Application of the Lens Equation

The following examples illustrate how to apply correctly the lens equation $1/v + 1/u = 1/f$. The cases of virtual object should be carefully noted.

Examples

1. *Converging lens. Real object*
An object is placed 12 cm from a converging lens of focal length 18 cm. Find the position of the image.

Since the lens is converging, $f = +18$ cm. The object is real, and therefore $u = +12$ cm.
Substituting in $\dfrac{1}{v} + \dfrac{1}{u} = \dfrac{1}{f}$,

$$\therefore \frac{1}{v} + \frac{1}{(+12)} = \frac{1}{(+18)}$$

$$\therefore \frac{1}{v} = \frac{1}{18} - \frac{1}{12} = -\frac{1}{36}$$

$$\therefore v = -36.$$

Since v is negative in sign the image is *virtual*, and it is 36 cm from the lens. See Fig. 19.4 (ii).

2. *Converging lens. Virtual object*
A beam of light, converging to a point 10 cm behind a converging lens, is incident on the lens. Find the position of the point image if the lens has a focal length of 40 cm.

If the incident beam converges to the point O, then O is a *virtual object*, Fig. 19.9.

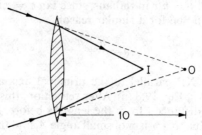

Fig. 19.9 Virtual object

See p. 324.
Thus $u = -10$ cm. Also, $f = +40$ cm, since the lens is converging. Substituting in $\dfrac{1}{v} + \dfrac{1}{u} = \dfrac{1}{f}$,

$$\frac{1}{v} + \frac{1}{(-10)} = \frac{1}{(+40)}$$

$$\therefore \frac{1}{v} = \frac{1}{40} + \frac{1}{10} = \frac{5}{40}$$

$$\therefore v = \frac{40}{5} = 8.$$

Since v is positive in sign the image is *real*, and it is 8 cm from the lens. The image is I in Fig. 19.9.

3. *Diverging lens. Real object*
An object is placed 6 cm in front of a diverging lens of focal length 12 cm. Find the image position.

Since the lens is diverging, $f = -12$ cm. The object is real, and hence $u = +6$ cm. Substituting in $\dfrac{1}{v} + \dfrac{1}{u} = \dfrac{1}{f}$,

$$\therefore \frac{1}{v} + \frac{1}{(+6)} = \frac{1}{(-12)}$$

$$\therefore \frac{1}{v} = -\frac{1}{12} - \frac{1}{6} = -\frac{3}{12}$$

$$\therefore v = -\frac{12}{3} = -4.$$

Since v is negative in sign the image is virtual, and it is 4 cm from the lens. See Fig. 19.5 (i).

4. *Diverging lens. Virtual object*
A converging beam of lights is incident on a diverging lens of focal length 15 cm. If the beam converges to a point 3 cm behind the lens, find the position of the point image.

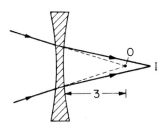

Fig. 19.10 Virtual object

If the beam converges to the point O, then O is a virtual object, as in example 2, Fig. 19.10. Thus $u = -3$ cm. Since the lens is diverging, $f = -15$ cm. Substituting in $\dfrac{1}{v} + \dfrac{1}{u} = \dfrac{1}{f}$,

$$\therefore \frac{1}{v} + \frac{1}{(-3)} = \frac{1}{(-15)}$$

$$\therefore \frac{1}{v} = -\frac{1}{15} + \frac{1}{3} = \frac{4}{15}$$

$$\therefore v = \frac{15}{4} = 3\tfrac{3}{4}.$$

Since v is positive in sign the point image, I, is *real*, and it is $3\tfrac{3}{4}$ cm from the lens, Fig. 19.10.

Lateral Magnification

The lateral or transverse or linear magnification, m, of an object produced by a lens is defined by

$$m = \frac{height\ of\ image}{height\ of\ object} \qquad . \qquad . \qquad . \qquad . \qquad (1)$$

Thus $m = IQ/OP$ in Fig. 19.4 or Fig. 19.5 on p. 351. Since triangles QIC, POC are similar in either of the diagrams,

$$\frac{IQ}{OP} = \frac{CI}{CO} = \frac{v}{u},$$

where v, u are the respective image and object distances from the lens.

$$\therefore \mathbf{m} = \frac{\mathbf{v}}{\mathbf{u}} \qquad . \qquad . \qquad . \qquad . \qquad . \qquad (2)$$

Equation (2) provides a simple formula for the magnitude of the magnification; there is no need to consider the signs of v and u in this case.

Other formulae for magnification. Since $1/v + 1/u = 1/f$, we have, by multiplying throughout by v,

$$1 + \frac{v}{u} = \frac{v}{f}$$

$$\therefore 1 + m = \frac{v}{f}$$

$$\therefore m = \frac{v}{f} - 1 \qquad . \qquad . \qquad . \qquad . \qquad (3)$$

Thus if a real image is formed 25 cm from a converging lens of focal length 10 cm, the magnification, m, $= \dfrac{+25}{+10} - 1 = 1 \cdot 5.$

Object at Distance 2f from Converging Lens

When an object is placed at a distance of $2f$ from a converging lens, drawing shows that the real image obtained is the same size as the image and is also formed at a distance $2f$ from the lens, Fig. 19.11. This result can be accurately checked by using the lens equation $1/v + 1/u = 1/f$. Substituting $u = +2f$, and noting that the focal length, f, of a converging lens is positive, we have

$$\frac{1}{v} + \frac{1}{2f} = \frac{1}{f}$$

$$\therefore \frac{1}{v} = \frac{1}{f} - \frac{1}{2f} = \frac{1}{2f}$$

$$\therefore v = 2f = \text{image distance.}$$

$$\therefore \text{ lateral magnification, } m = \frac{v}{u} = \frac{2f}{2f} = 1,$$

showing that the image is the same size as the object.

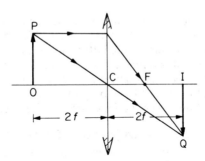

Fig. 19.11 Object and image of same size

Least possible Distance between Object and Real Image with Converging Lens

It is not always possible to obtain a real image on a screen, although the object and the screen may both be at a greater distance from a converging lens than its focal length. The theory below shows that the distance between an object and a screen must be equal to, or greater than, *four times the focal length* if a real image is required.

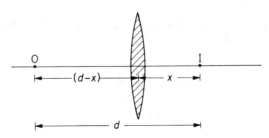

Fig. 19.12 Minimum distance between object and image

Theory. Suppose I is the real image of a point object O in a converging lens. If the image distance $= x$, and the distance OI $= d$, the object distance $= (d - x)$, Fig. 19.12. Thus $v = +x$, and $u = +(d - x)$. Substituting in the lens equation $1/v + 1/u = 1/f$, in which f is positive, we have

$$\frac{1}{x} + \frac{1}{d - x} = \frac{1}{f}$$

$$\therefore \frac{d}{x(d - x)} = \frac{1}{f}$$

$$\therefore x^2 - dx + df = 0 . \qquad . \qquad . \qquad . \qquad . \qquad \text{(i)}$$

For a real image, the roots of this quadratic equation for x must be real roots. Applying to (i) the condition $b^2 - 4ac > 0$ for the general quadratic $ax^2 + bx + c = 0$, then

$$d^2 - 4df > 0$$
$$\therefore d^2 > 4df$$
$$\therefore d > 4f.$$

Thus the distance OI between the object and screen must be greater than $4f$, otherwise no image can be formed on the screen. Hence $4f$ is the minimum distance between object and screen; the latter case is illustrated by Fig. 19.9, in which $u = 2f$ and $v = 2f$. If it is difficult to obtain a real image on a screen when a converging lens is used, possible causes may be (i) the object is nearer to the lens than its focal length, Fig. 19.4 (ii), or (ii) the distance between the screen and object is less than four times the focal length of the lens.

Displacement of Lens when Object and Screen are Fixed

Suppose that an object O, in front of a converging lens A, gives rise to an image on a screen at I, Fig. 19.13. Since the image distance AI (v) is greater than the object distance AO (u), the image is larger than the object. If the object and the screen are kept fixed at O, I respectively, another clear image can be obtained on the screen by moving the lens from A to a position B. This time the image is smaller than the object, as the new image distance BI is less than the new object distance OB.

Since O and I can be interchanged in position with respect to the lens, it follows that $OB = IA$ and $IB = OA$. If the *displacement*, AB, of the lens $= d$, and the constant distance $OI = l$, then $OA + BI = l - d$. But, from above, $OA = IB$. Hence $OA = (l - d)/2$. Further, $AI = AB + BI = OA + AB = (l - d)/2 + d = (l + d)/2$.

But $u = OA$, and $v = AI$ for the lens in the position A. Substituting for OA and AI in $1/v + 1/u = 1/f$,

$$\frac{1}{(l + d)/2} + \frac{1}{(l - d)/2} = \frac{1}{f}$$
$$\therefore \frac{2}{l+d} + \frac{2}{l-d} = \frac{1}{f}$$
$$\therefore \frac{4l}{l^2 - d^2} = \frac{1}{f}$$
$$\therefore f = \frac{l^2 - d^2}{4l} \qquad \qquad (1)$$

Thus if the displacement d of the lens, and the distance l between the object and the screen, are measured, the focal length f of the lens can be found from equation (1). This provides a very useful method of measuring the focal length of a lens whose surfaces are inaccessible (for example, when the lens is in a tube), when measurements of v and u cannot be made (see p. 366).

Magnification. When the lens is in the position A, the lateral magnification m_1 of the object $= v/u = AI/OA$, Fig. 19.11. So the image is bigger than the object and

$$\frac{h_1}{h} = \frac{AI}{AO} \qquad \qquad (i)$$

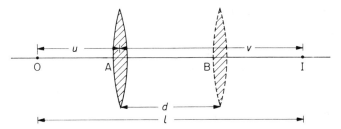

Fig. 19.13 Displacement of lens

where h_2 is the length of the image. But, from our previous discussion, AI = OB

When the lens is in the position B, the image is smaller than the object. The lateral magnification, m_2 = BI/OB.

$$\therefore \frac{h_2}{h} = \frac{BI}{OB} \qquad . \qquad . \qquad . \qquad . \qquad \text{(ii)}$$

where h_2 is the length of the image. but, from our previous discussion, AI = OB and OA = BI. From (i) and (ii) it follows that, by inverting (i),

$$\frac{h}{h_1} = \frac{h_2}{h}$$

$$\therefore h^2 = h_1 h_2$$

$$\therefore h = \sqrt{h_1 h_2}. \qquad . \qquad . \qquad . \qquad . \qquad \text{(2)}$$

The length, h, of an object can hence be found by measuring the lengths h_1, h_2 of the images for the two positions of the lens. This method of measuring h is most useful when the object is inaccessible, for example, when the width of a slit inside a tube is required.

Examples

1. Give an account of a method of measuring the focal length of a diverging lens, preferably without the aid of an auxiliary converging lens. A luminous object and a screen are placed on an optical bench and a converging lens is placed between them to throw a sharp image of the object on the screen; the linear magnification of the image is found to be 2·5. The lens is now moved 30 cm nearer the screen and a sharp image again formed. Calculate the focal length of the lens. (*N.*)

If O, I are the object and screen positions, respectively, and L_1, L_2 are the two positions of the lens, then $OL_1 = IL_2$, Fig. 19.14. See previous discussion. Suppose $OL_1 = x = L_2I$.

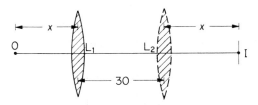

Fig. 19.14 Example

For the lens in the position L_1, $u = OL_1 = x$, and $v = L_1I = 30 + x$.

But magnification, $m = \dfrac{v}{u} = 2\cdot 5$

$$\frac{30 + x}{x} = 2 \cdot 5$$

$$\therefore x = 20 \text{ cm}$$

$$\therefore u = OL_1 = 20 \text{ cm}$$

$$v = L_1I = 30 + x = 50 \text{ cm}.$$

Substituting in $\dfrac{1}{v} + \dfrac{1}{u} = \dfrac{1}{f}$,

$$\therefore \frac{1}{20} + \frac{1}{50} = \frac{1}{f}$$

from which $f = 14 \cdot 3$ cm.

2. An object O is placed 15 cm from a converging lens A of focal length 10 cm and an image I is formed on a screen S on the other side of the lens, Fig. 19.15 (i). A diverging lens B is now placed half-way between A and I, and the screen is moved back 10 cm so that a clear image is now formed at I_1.

Calculate the focal length of the lens B and the magnification of the final image at I_1.

Using an object of finite size at O, draw a ray diagram showing how the final image I_1 is formed.

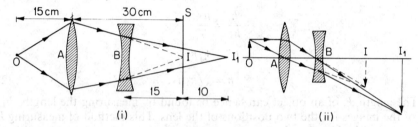

Fig. 19.15 Example

(*a*) *Take one lens at a time.* For lens A, $u = OA = +15$ cm, $f = +10$ cm. So, from $1/v + 1/u = 1/f$, we have

$$\frac{1}{v} + \frac{1}{+15} = \frac{1}{+10}$$

Solving, $\qquad\qquad v = +30 \text{ cm} = AI$

For lens B, I is a *virtual* object and I_1 is a real image. Since B is half-way between A and I, $BI = u = -\frac{1}{2} \times 30 = -15$ cm. Also, $BI_1 = v = +(15 + 10) = +25$ cm. So, from $1/v + 1/u = 1/f$, we have,

$$\frac{1}{+25} + \frac{1}{-15} = \frac{1}{f}$$

Solving, $\qquad\qquad f = -37 \cdot 5$ cm.

(*b*) Magnification of $I_1 = m_A \times m_B$, where m_A and m_B are the respective magnifications due to A and B.

For Lens A, $\qquad\qquad m_A = \dfrac{v}{u} = \dfrac{30}{15} = 2$

For Lens B, $\qquad\qquad m_B = \dfrac{v}{u} = \dfrac{25}{15} = 1\dfrac{2}{3}$

So total magnification $= 2 \times 1\frac{2}{3} = 3\frac{1}{3}$.

Figure 19.15 (ii) shows the ray diagram. The rays converging to the top point of I are intercepted by the diverging lens B and form the top point of the final real image I_1.

Focal Length of Lens. Small-angle Prism Method

The focal length f of a lens depends on the radii of curvature, r_1, r_2, of its surfaces and on the refractive index n of the glass used.

The relation for f can be found by using the deviation formula due to a small-angle prism. Consider a ray PQ parallel to the principal axis at a height h above it, Fig. 19.16 (i). This ray is refracted to the principal focus, and thus undergoes a small deviation through an angle d given by

$$d = \frac{h}{f} \qquad . \qquad . \qquad . \qquad . \qquad . \qquad \text{(i)}$$

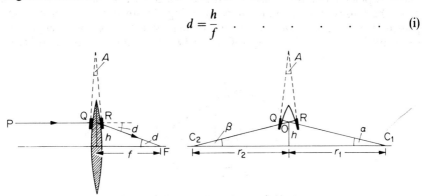

Fig. 19.16 Focal length relation

This is the deviation through a prism of small angle A formed by the tangents at Q, R to the lens surfaces, as shown. Now for a small angle of incidence, which is the case for a thin lens and a ray close to the principal axis, $d = (n-1)A$. See p. 353.

From (i),
$$\frac{h}{f} = (n-1)A$$

$$\therefore \frac{1}{f} = (n-1)\frac{A}{h} \qquad . \qquad . \qquad . \qquad \text{(ii)}$$

The normals at Q, R pass respectively through the centres of curvatures C_1, C_2 of the lens surfaces. From the geometry, angle $ROC_1 = A = \alpha + \beta$, where α, β are the angles with the principal axis at C_1, C_2 respectively, Fig. 19.16 (ii). But $\alpha = h/r_1$, $\beta = h/r_2$.

$$\therefore A = \alpha + \beta = \frac{h}{r_1} + \frac{h}{r_2} \qquad . \qquad . \qquad . \qquad \text{(iii)}$$

$$\therefore \frac{A}{h} = \frac{1}{r_1} + \frac{1}{r_2}.$$

Substituting in (ii),
$$\frac{1}{f} = (n-1)\left(\frac{1}{r_1} + \frac{1}{r_2}\right)$$

Generally, n is the *relative refractive index* of the lens material to the medium outside, that is, if the lens is made of glass of $n_2 = 1\cdot5$, and it is placed in water of $n_1 = 1\cdot33$, then the relative refractive index $n = n_2/n_1 = 1\cdot5/1\cdot33 = 1\cdot13$.

In practice, however, lenses are usually situated in air; in which case $n_1 = 1$. If the glass has a refractive index, n_2, equal to n, the relative refractive index, $n_2/n_1 = n/1 = n$. Generally, then

$$\frac{1}{f} = (n-1)\left(\frac{1}{r_1}+\frac{1}{r_2}\right) \qquad . \qquad . \qquad . \qquad . \qquad (1)$$

Combined Focal Length of Two Thin Lenses in Contact

In order to diminish the colouring of the image due to dispersion when an object is viewed through a single lens, the lenses of telescopes and microscopes are made by placing two thin lenses together (see p. 378). The combined focal length, F, of the lenses can be found by considering a point object O placed on the principal axis of two *thin lenses in contact*, which have focal lengths f_1, f_2 respectively, Fig. 19.17. A ray OC from O passes through the middle, C, of

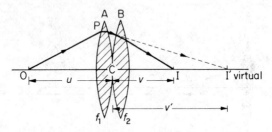

Fig. 19.17 Focal length of combined lenses

both lenses undeviated. A ray OP from O is refracted through the first lens A to intersect OC at I', which is therefore the image of O in A. If $OC = u$, $CI' = v'$,

$$\therefore \frac{1}{v'}+\frac{1}{u}=\frac{1}{f_1} \qquad . \qquad . \qquad . \qquad . \qquad (i)$$

The beam of light incident on the second lens B converges to I', which is therefore a *virtual* object for this lens. The image is formed at I at a distance CI, or v, from the lens. Thus since the object distance CI' is virtual, $u = -v'$ for refraction in this case. For lens B, therefore, we have

$$\frac{1}{v}+\frac{1}{(-v')}=\frac{1}{f_2},$$

or

$$\frac{1}{v}-\frac{1}{v'}=\frac{1}{f_2} \qquad . \qquad . \qquad . \qquad . \qquad (ii)$$

Adding (i) and (ii) to eliminate v',

$$\therefore \frac{1}{v}+\frac{1}{u}=\frac{1}{f_1}+\frac{1}{f_2}.$$

Since I is the image of O by refraction through both lenses,

$$\frac{1}{v}+\frac{1}{u}=\frac{1}{F},$$

where F is the *focal length of the combined lenses*. Hence

$$\frac{1}{F}=\frac{1}{f_1}+\frac{1}{f_2} \qquad . \qquad . \qquad . \qquad . \qquad (1)$$

This formula for F applies to any two thin lenses in contact, such as two diverging lenses, or a converging and diverging lens. When the formula is used, the signs of the focal lengths must be inserted. As an illustration, suppose that a thin converging lens of 8 cm focal length is placed in contact with a diverging lens of 12 cm focal length. Then $f_1 = +8$ cm, and $f_2 = -12$ cm. The combined focal length, F, is thus given by

$$\frac{1}{F} = \frac{1}{(+8)} + \frac{1}{(-12)} = \frac{1}{8} - \frac{1}{12} = +\frac{1}{24}$$

$$\therefore F = +24 \text{ cm.}$$

The positive sign shows that the combination acts like a converging lens.

Refractive Index of a Small Quantity of Liquid

The refractive index of a small amount of liquid can be found by placing a drop on a plane mirror and placing a converging lens on top, as shown in the exaggerated sketch of Fig. 19.18. An object O is then moved along the principal axis until the inverted image I seen looking down into the mirror is coincident with O in position. In this case the rays which pass through the lens and liquid are incident *normally* on the mirror, and the distance from O to the lens is now the focal length, F, of the *lens and liquid* combination (see p. 365). If the experiment is repeated with the glass lens alone on the mirror, the focal length f_1 of the lens can be measured. But $1/F = 1/f_1 + 1/f_2$, where f_1 is the focal length of the liquid lens. Thus, knowing f_1 and F, f_2 can be calculated.

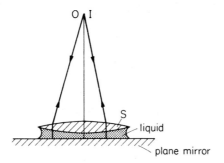

Fig. 19.18 Refractive index of liquid

From Fig. 19.18, it can be seen that the liquid lens is a plano-concave type; its lower surface corresponds to the plane surface of the mirror, while the upper surface corresponds to the surface S of the converging lens. If the latter has a radius of curvature r, then, from equation (1) on p. 362,

$$\frac{1}{f_2} = (n-1)\left(\frac{1}{r} + \frac{1}{\infty}\right).$$

$$\therefore \frac{1}{f_2} = (n-1)\frac{1}{r}$$

$$\therefore n-1 = \frac{r}{f_2}$$

$$\therefore n = 1 + \frac{r}{f_2} \qquad . \qquad . \qquad . \qquad . \qquad \text{(i)}$$

The radius of curvature r of the surface S of the lens can be measured. Since

f_2 has already been found, the refractive index n of the liquid can be calculated from (i). This method of measuring n is useful when only a small quantity of the liquid is available.

Example

Describe two methods for the determination of the focal length of a diverging lens. A thin equiconvex lens is planed on a horizontal plane mirror, and a pin held 20 cm vertically above the lens coincides in position with its own image. The space between the under surface of the lens and the mirror is filled with water (refractive index 1·33) and then, to coincide with its image as before, the pin has to be raised until its distance from the lens is 27·5 cm. Find the radius of curvature of the surfaces of the lens. (*N.*)

The focal length, f_1, of the lens = 20 cm, and the focal length, F, of the water and glass lens combination = 27·5 cm. See Fig. 19.18. The focal length, f, of the water lens is given by

$$\frac{1}{F} = \frac{1}{f} + \frac{1}{f_1}$$

$$\therefore \frac{1}{(+27\cdot5)} = \frac{1}{f} + \frac{1}{(+20)}.$$

Solving,

$$\therefore \frac{1}{f} = \frac{1}{27\cdot5} - \frac{1}{20} = -\frac{3}{220}$$

the minus showing that the water lens is a diverging lens.

But

$$\frac{1}{f} = (n-1)\frac{1}{r},$$

when $n = 1\cdot33$, and r = radius of the curved face of the lens.

$$\therefore -\frac{3}{220} = (1\cdot33 - 1)\frac{1}{r}$$

$$\therefore r = -24\cdot2 \text{ cm.}$$

The glass lens is equiconvex, and hence the radii of its surfaces are the same.

Measuring Focal Lengths of Lenses

Converging Lens

(1) *Plane mirror method.* In this method a plane mirror M is placed on a table, and the lens L is placed on the mirror, Fig. 19.19. A pin O is then moved along the axis of the lens until its image I is observed to coincide with O when they are both viewed from above, the method of no parallax being used. The distance from the pin O to the lens is then the focal length, f, of the lens, which can thus be measured.

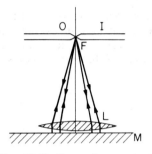

Fig. 19.19 Plane mirror method

The explanation of the method is as follows. In general, rays from O pass through the lens, are reflected from the mirror M, and then pass through the lens again to form an image at some place. When O and the image coincide in position, the rays incident on M from a point *on the principal axis* must have returned along their incident path after reflection from the mirror. This is only possible if the rays are incident *normally* on M. Consequently the rays entering the lens after reflection are all parallel, and hence the point to which they converge must be the focus, F, Fig. 19.19. It will thus be noted that the mirror provides a simple method of obtaining parallel rays incident on the lens.

(2) *Lens formula method.* In this method five or six values of u and v are obtained by using an illuminated object and a screen, or by using two pins and the method of no parallax. The focal length, f, can then be calculated from the equation $1/v + 1/u = 1/f$, and the average of the values obtained. Alternatively, the values of $1/u$ can be plotted against $1/v$, and a straight line drawn through the points. When $1/u = 0$, $1/v = OA = 1/f$, from the lens equation; thus $f = 1/OA$, and hence can be calculated, Fig. 19.20. Since $1/u = 1/f$ when $1/v = 0$, then, from the lens equation, $OB = 1/f$. Thus f may also be evaluated from $1/OB$.

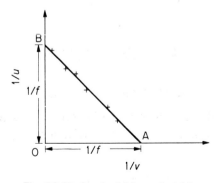

Fig. 19.20 Graph of $1/u$ against $1/v$

(3) *Displacement method.* In this method, an illuminated object O is placed in front of the lens, A, and an image I is obtained on a screen. Keeping the object and screen fixed, the lens is then moved to a position B so that a clear image is again obtained on the screen, Fig. 19.21. From our discussion on p. 358, it follows that a magnified sharp image is obtained at I when the lens is in position A, and a diminished sharp image when the lens is in the position B. If the displacement of the lens is d, and the distance between the object and the screen is l, the focal length, f, is given by $f = \dfrac{l^2 - d^2}{4l}$, from p. 358. Thus f can be calculated. The experiment can be repeated by altering the distance between the object and the screen, and the average value of f is then calculated. It should be noted that the screen must be at a distance from the object of at least four times the focal length of the lens, otherwise an image is unobtainable on the screen (p. 358).

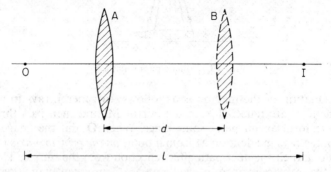

Fig. 19.21 Displacement method for focal length

Since no measurements need be made to the surfaces of the lens (the 'displacement' is simply the distance moved by the holder of the lens), this method can be used for finding the focal length of (i) a thick lens, (ii) an inaccessible lens, such as that fixed inside an eyepiece or telescope tube. Neither of the two methods previously discussed could be used for such a lens.

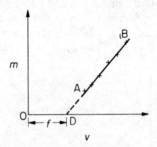

Fig. 19.22 Graph of m against v

Lateral Magnification Method of Measuring Focal Length

On p. 356, we showed that the lateral magnification, m, produced by a lens is given by

$$m = \frac{v}{f} - 1 \qquad \qquad \text{(i)}$$

where f is the focal length of the lens and v the distance of the image. If an illuminated glass scale is set up as an object in front of a lens, and the image is received on a screen, the magnification, m, can be measured directly. From (i) a straight line graph BA is obtained when m is plotted against the corresponding image distance v, Fig. 19.22. Further, from (i), $v/f - 1 = 0$ when $m = 0$; thus $v = f$ in this case. Hence, by producing BA to cut the axis of v in D, it follows that $OD = f$; the focal length of the lens can thus be found from the graph.

Diverging Lens

(1) *Converging lens method.* By itself, a diverging lens always forms a virtual image of a real object. A real image may be obtained, however, if a *virtual object* is used, and a converging lens can be used to provide such an object, as shown in Fig. 19.23. An object S is placed at a distance from M greater than its focal length, so that a beam converging to a point O is obtained. O is thus a virtual object for the diverging lens L placed as shown in Fig. 19.23, and a real image I can now be obtained. I is farther away from L than O, since the concave lens makes the incident beam on it diverge more.

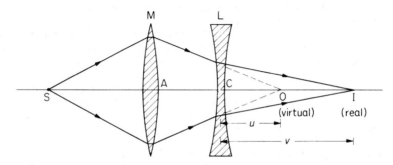

Fig. 19.23 Focal length of diverging lens

The image distance, v, from the diverging lens is CI and can be measured; v is +ve in sign as I is real. The object distance, u, from this lens = CO = AO − AC, and AC can be measured. The length AO is obtained by removing the lens L, leaving the converging lens, and noting the position of the real image now formed at O by the lens M. Thus u (= CO) can be found; it is a −ve distance, since O is a virtual object for the diverging lens. Substituting for u and v in $1/v + 1/u = 1/f$, the focal length of the diverging lens can be calculated.

(2) *Concave mirror method.* In this method a real object is placed in front of a diverging lens, and the position of the virtual image is located with the aid of a concave mirror. An object O is placed in front of the lens, L, and a concave mirror M is placed behind the lens so that a divergent beam is incident on it, Fig. 19.24. With L and M in the same position, the object O is moved until an image is obtained coincident with it in position, i.e., beside O. The distances CO, CM are then measured.

As the object and image are coincident at O, the rays must be incident *normally* on the mirror M. The rays BA, ED thus pass through the centre of curvature of M, and this is also the position of the virtual image I. The image distance, v, from the lens = IC = IM − CM = r − CM, where r is the radius of curvature; CM can be measured, while r can be determined by means of a separate experiment, as mentioned on p. 321. The object distance, u, from the

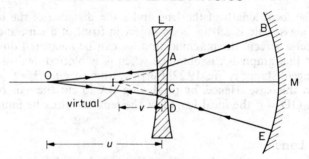

Fig. 19.24 Focal length of diverging lens

lens = OC, and by substituting for u and v in the formula $1/v + 1/u = 1/f$, the focal length f can be calculated. Of course, v is negative as I is a virtual image for the lens.

Exercises 19A

1. An object is placed (i) 12 cm, (ii) 4 cm from a converging lens of focal length 6 cm. Calculate the image position and the magnification in each case, and draw sketches illustrating the formation of the image.

2. What do you know about the image obtained with a diverging lens? The image of a real object in a diverging lens of focal length 10 cm is formed 4 cm from the lens. Find the object distance and the magnification. Draw a sketch to illustrate the formation of the image.

3. The image obtained with a converging lens is erect and three times the length of the object. The focal length of the lens if 20 cm. Calculate the object and image distances.

4. A beam of light converges to a point 9 cm behind (i) a converging lens of focal length 12 cm, (ii) a diverging lens of focal length 15 cm. Find the image position in each case, and draw sketches illustrating them.

5. Draw a ray diagram to show how a converging lens produces an image of finite size of the moon clearly focused on a screen. If the moon subtends an angle of $9 \cdot 1 \times 10^{-3}$ radian at the centre of the lens, which has a focal length of 20 cm, calculate the diameter of this image. With the screen removed, a second converging lens of focal length $5 \cdot 0$ cm is placed coaxial with the first and 24 cm from it on the side remote from the moon. Find the position, nature and size of the final image. (*N.*)

6. What do you understand by a *virtual object* in optics? Describe a direct method, not involving any calculation, of finding (*a*) the focal length of a double concave lens, and (*b*) the radius of curvature of one of its faces. You may use other lenses and mirrors if you wish.

A converging lens of 20 cm focal length is arranged coaxially with a diverging lens of focal length $8 \cdot 0$ cm. A point object lies on the same side as the converging lens and very far away on the axis. What is the smallest possible distance between the lenses if the combination is to form a real image of the object?

If the lenses are placed $6 \cdot 0$ cm apart, what is the position and nature of the final image of this distant object? Draw a diagram showing the passage of a wide beam of light through the system in this case. (*C.*)

7. Describe in detail how you would determine the focal length of a diverging lens with the help of (*a*) a converging lens, (*b*) a concave mirror.

A converging lens of 6 cm focal length is mounted at a distance of 10 cm from a screen placed at right angles to the axis of the lens. A diverging lens of 12 cm focal length is then placed coaxially between the converging lens and the screen so that an image of an object 24 cm from the converging lens is focused on the screen. What is the distance between the two lenses? Before commencing the calculation state the sign convention you will employ. (*N.*)

8. Define *focal length, conjugate foci, real image.* Obtain an expression for the transverse magnification produced by a thin converging lens.

Light from an object passes through a thin converging lens, focal length 20 cm, placed 24 cm from the object and then through a thin diverging lens, focal length 50 cm, forming a real image 62·5 cm from the diverging lens. Find (*a*) the position of the image due to the first lens, (*b*) the distance between the lenses, (*c*) the magnification of the final image. (*L.*)

9. Describe how you would determine the focal length of a diverging lens if you were provided with a converging lens (*a*) of shorter focal length, (*b*) of longer focal length.

An illuminated object is placed at right angles to the axis of a converging lens, of focal length 15 cm, and 22·5 cm from it. On the other side of the converging lens, and coaxial with it, is placed a diverging lens of focal length 30 cm. Find the position of the final image (*a*) when the lenses are 15 cm apart and a plane mirror is placed perpendicular to the axis 40 cm beyond the diverging lens, (*b*) when the mirror is removed and the lenses are 35 cm apart. (*N.*)

10. Why is a sign convention used in geometrical optics?

A thin equiconvex lens of glass of refractive index 1·50 whose surfaces have a radius curvature of 24·0 cm is placed on a horizontal plane mirror. When the space between the lens and mirror is filled with a liquid, a pin held 40·0 cm vertically above the lens is found to coincide with its own image. Calculate the refractive index of the liquid. (*N.*)

11. A converging equiconvex lens of glass of refractive index 1·5 is laid on a horizontal plane mirror. A pin coincides with its inverted image when it is 1·0 m above the lens. When some liquid is placed between the lens and the mirror the pin has to be raised by 0·55 m for the coincidence to occur again. What is the refractive index of the liquid?

(The focal length of two lenses in contact is given by $1/F = 1/f_1 + 1/f_2$.) (*L.*)

12. Find the relation between the focal lengths of two thin lenses in contact and the focal length of the combination.

The curved face of a planoconvex lens ($n = 1·5$) is placed in contact with a plane mirror. An object at 20 cm distance coincides with the image produced by the lens and reflection by the mirror. A film of liquid is now placed between the lens and the mirror and the coincident object and image are at 100 cm distance. What is the index of refraction of the liquid? (*L.*)

Defects of Vision

We can now give a brief account of the optical principles of some defects of vision and their correction.

Far and Near Points of Eye. Accommodation

We shall assume that the reader is familiar with the main features of the eye. The cornea, which is the surface in front of the eye, appears to refract the light rays entering the eye much more than the eye-lens itself. The image formed by the eye-lens L must appear on the retina R, the light-sensitive screen at the back of the eye, in order to be clearly seen, Fig. 19.25. The ciliary muscles, which are attached to the lens surfaces, can alter the radii of curvature of the surfaces and hence the focal length of the lens. This enables the eye to focus objects at different distances on the retina, a property of the eye known as its power of *accommodation*.

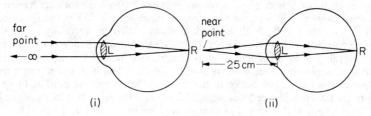

(i)   (ii)

Fig. 19.25 Normal eye

The most distant point it can focus, called the *far point*, is at infinity for a normal eye, Fig. 19.25 (i). In this case the lens L is slim and parallel rays entering the eye are focused on the retina as shown. On the other hand, an object is seen in great detail when it is placed as near as possible to the eye while remaining in focus. This distance D from the eye is known as its *least distance of distinct vision* and is about 25 cm for a normal eye. The point at this distance is called the *near point* of the eye, Fig. 19.25 (ii).

Short Sight

If the focal length of the eye is too short, owing to the eye-ball being too long,

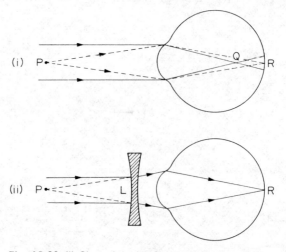

Fig. 19.26 (i) Short sight (ii) Correction for far point

parallel rays will be brought to a focus at a point Q in front of the retina, Fig. 19.26 (i). In this case the far point of the eye is not at infinity, but at a point P nearer to the eye. The defect of vision is known as *short sight*.

A suitable *diverging* lens, L, is required to correct for short sight, Fig. 19.26 (ii). Parallel rays refracted through L are now made divergent, and if they appear to come from the far point P of the eye, the rays are brought to a focus on the retina R. From Fig. 19.26 (ii), it can be seen that the focal length of the required lens is equal to PL, which is practically equal to the distance of the far point from the eye.

Long Sight

If a person's far point is normal, i.e., at infinity, but his near point is farther from the eye than the normal least distance of distinct vision, 25 cm, the person is said to be 'long-sighted'. In Fig. 19.27 (i), X is the near point of a person suffering from long sight, due to a short eye-ball, for example. Rays from X are brought to a focus on the retina R; whereas rays from the normal near point A, 25 cm from the eye, are brought to a focus at B behind the retina.

A suitable *converging* lens, L, is required to correct for this defect of vision, Fig. 19.27 (ii). Rays from A then appear to come from X after refraction through L, and an image is thus now formed on the retina. It can be seen that X is the virtual image of A in the lens L. Thus if XL = 50 cm, and AL = 25 cm, the focal length of L is given from the lens formula by

$$\frac{1}{(-50)} + \frac{1}{(+25)} = \frac{1}{f}$$

$$\therefore \frac{1}{50} = \frac{1}{f}$$

$$\therefore f = 50 \text{ cm.}$$

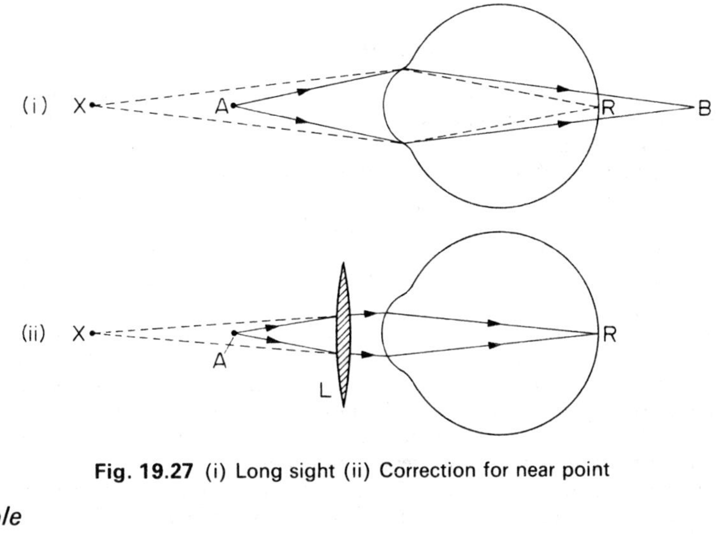

Fig. **19.27** (i) Long sight (ii) Correction for near point

Example

Draw diagrams to illustrate *long sight* and *short sight*. Draw also diagrams showing the corrections of these defects by suitable lenses. A person can focus objects only when they lie between 50 cm and 300 cm from his eyes. What spectacles should he use (*a*) to increase his maximum distance of distinct vision to infinity, (*b*) to reduce his least distance of distinct vision to 25 cm? Find this range of distinct vision using each pair. (*N.*)

(a) To increase the maximum distance of distinct vision from 300 cm to infinity, the person requires a *diverging* lens. See Fig. 19.26 (ii). Assuming the lens is close to his eye, the focal length PL = 300 cm as P is 300 cm from the eye. One limit of the range of distinct vision is now infinity. The other limit is the object (*u*) corresponding to an image distance (*v*) of 50 cm from the lens, as the person can see distinctly objects 50 cm from his eyes. In this case, then, $v = -50$ cm, $f = -300$ cm. Substituting in the lens equation, we have

$$\frac{1}{(-50)} + \frac{1}{u} = \frac{1}{(-300)}$$

from which $\qquad\qquad u = 60$ cm.

The range of distinct vision is thus from 60 cm to infinity.

(b) To reduce his least distance of distinct vision from 50 cm to 25 cm, the person requires a *converging* lens. See Fig. 19.27 (ii). In this case, assuming the lens is close to the eye, $u = +25$ cm, $v = -50$ cm, as the image must be formed 50 cm from the eye on the same side as the object, making the image virtual. The focal length of the lens is thus given by

$$\frac{1}{(-50)} + \frac{1}{(+25)} = \frac{1}{f}$$

from which $\qquad\qquad f = +50$ cm.

Objects placed at the focus of this lens appear to come from infinity. The maximum distance of distinct vision, *u*, is given by substituting $v = -300$ cm and $f = +50$ cm in the lens formula.

Thus $\qquad\qquad$ $$\frac{1}{(-300)} + \frac{1}{u} = \frac{1}{(+50)}$$

from which $\qquad\qquad u = \frac{300}{7} = 42 \cdot 9$ cm.

The range of distinct vision is thus from 25 to 42·9 cm.

Exercises 19B

1. Explain how the eye is focused for viewing objects at different distances. Describe and explain the defects of vision known as *long sight* and *short sight*, and their correction by the use of spectacles.

Explain the advantages we gain by the use of two eyes instead of one.

A certain person can see clearly objects at distances between 20 cm and 200 cm from his eye. What spectacles are required to enable him to see distant objects clearly, and what will be his least distance of distinct vision when he is wearing them? (*L.*)

2. Give an account of the common optical defects of the human eye and explain how their effects may be corrected.

An elderly person cannot see clearly, without the use of spectacles, objects nearer than 200 cm. What spectacles will he need to reduce this distance to 25 cm? If his eyes can focus rays which are converging to points not less than 150 cm behind them, calculate his range of distinct vision when using the spectacles. (*N.*)

3. Describe the optical functions of the cornea and lens in the human eye and explain how the corresponding purposes are served in a camera.

In order to correct his near point to 25 cm a man is given spectacles with converging lenses of 50 cm focal length, and to correct his far point to infinity he is given diverging lenses of 200 cm focal length. Ignoring the separation of lens and eye determine the distances of his near and far points when not wearing spectacles and suggest reasons for his defects of vision. (*N.*)

Dispersion. Spectra. Chromatic Aberration

Spectrum of White Light

In 1666, NEWTON made a great scientific discovery. He found that sunlight, or white light, was made up of different colours, consisting of red, orange, yellow, green, blue, indigo, violet. Newton first made a small hole in a shutter in a darkened room, and received a white circular patch of sunlight on a screen in the path of the light. He then placed a glass prism between the hole A and the screen S, and on moving the screen round as shown in Fig. 19.28 (i), he observed a series of overlapping coloured patches in place of the white patch. The total

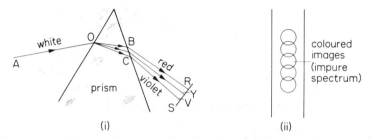

(i) (ii)

Fig. 19.28 Impure spectrum (diagrammatic)

length of the coloured images was several times their width, Fig. 19.28 (ii). By separating one colour from the rest, Newton demonstrated that the colours themselves could not be changed by refraction through a prism, and he concluded that the colours were not *introduced* by the prism, but were components of the white light. The *spectrum* (colours) of white light consists of red, orange, yellow, green, blue, indigo, and violet, and the separation of the colours by the prism is known as *dispersion*.

The red rays are the least deviated by the prism, and the violet rays are the most deviated, as shown in the exaggerated sketch of Fig. 19.28 (i) for one typical ray in the incident beam. Since the angle of incidence at O in the air is the same for the red and violet rays, and the angle of refraction made by the red ray OB in the glass is greater than that made by the violet ray OC, it follows from $\sin i/\sin r$ that the refractive index of the prism material for red light is less than for violet light. Similarly, the refractive index for yellow light lies between the refractive index values for red and violet light (see also p. 378).

Production of Pure Spectrum

Newton's spectrum of sunlight is an *impure spectrum* because the different coloured images overlap, Fig. 19.28 (ii). A *pure spectrum* is one in which the different coloured images contain light of one colour only, i.e., they are monochromatic images. In order to obtain a pure spectrum (i) the white light must be admitted through a very narrow opening, so as to assist in the reduction of the overlapping of the images, (ii) the beams of coloured rays emerging from the prism must be parallel, so that each beam can be brought to a separate focus.

The spectrometer can be used to provide a pure spectrum. The collimator slit is made very narrow, and the collimator C and the telescope T are both adjusted for parallel light, Fig. 19.29. A bright source of white light S is placed near the slit, and the prism P is usually set in the minimum deviation position for yellow light, although this is not essential. The rays refracted through P are now separated into a number of different coloured parallel beams of light, each travelling in slightly different directions, and the telescope brings each coloured beam to a separate focus. A pure spectrum can now be seen through T, consisting of a series of monochromatic images of the slit.

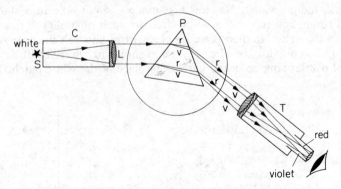

Fig. 19.29 Pure spectrum of white light

If only one lens, L, is available, the prism P *must* be placed in the minimum deviation position for yellow light in order to obtain a fairly pure spectrum, Fig. 19.30. The prism is then also approximately in the minimum deviation position for the various colours in the incident convergent beam, and hence the rays of one colour are approximately deviated by the same amount by the prism, thus forming an image of the slit S at rougly the same place.

Infra-red and ultraviolet rays. In 1800 HERSCHEL discovered the existence of *infra-red rays*, invisible rays beyond the red end of the spectrum. Fundamentally, they are of the same nature as rays in the visible spectrum but having longer wavelengths than the red, and produce a sensation of heat (see p. 268). Their existence may be demonstrated in the laboratory by means of an arc light in place of S in Fig. 19.30, a rocksalt lens at L and a rocksalt prism at P. A suitable

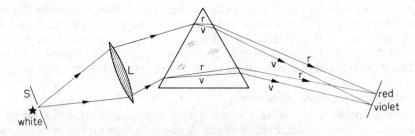

Fig. 19.30 Fairly pure spectrum

phototransistor, connected to an amplifier and galvanometer, is very sensitive to infra-red light. When this detector is moved into the dark beyond the red end of the spectrum, a deflection is obtained in the galvanometer. Since they are not scattered by fine particles as much as the rays in the visible spectrum, infra-red rays can penetrate fog and mist. Clear pictures have been taken in mist by using infra-red filters and photographic plates.

About 1801 RITTER discovered the existence of invisible rays beyond the violet end of the visible spectra. *Ultraviolet rays*, as they are known, affect photographic plates and cause certain minerals to fluoresce. They can also eject electrons from metal plates (see *Photoelectric effect*, p. 843). Ultraviolet rays can be detected in the laboratory by using an arc light in the place of S in Fig. 19.30, a quartz lens at L, and a quartz prism at P. A sensitive detector is a photoelectric cell connected to a galvanometer and battery. When the cell is moved beyond the violet into the dark part of the spectrum a deflection is observed in the galvanometer.

The Origin of Spectra

The study of the wavelengths of the radiation from a hot body comes under the general heading of *Spectra*. The number of spectra of elements and compounds which have been recorded runs easily into millions. As discussed more fully in atomic physics (p. 855), the origin of spectra is linked with energy changes of the atom.

It is now considered that an atom consists of a nucleus of positive charge surrounded by electrons in various orbits, and that a particular electron in an orbit has a definite amount of energy. In certain circumstances the electron may move from this orbit to another, where it has a smaller amount of energy. When this occurs radiation is emitted, and the energy in the radiation is equal to the difference in energy of the atom between its initial and final states. The displacement of an electron from one orbit to another occurs when a substance is raised to a high temperature, in which case the atoms present collide with each other very violently. Light of a definite wavelength will then be emitted, and will be characteristic of the electron energy changes in the atom. There is usually more than one wavelength in the light from a hot body (iron has more than 400 different wavelengths in its spectrum), and each wavelength corresponds to a change in energy of the atom. A study of spectra should therefore reveal much important information concerning the structure and properties of atoms (p. 855).

Every element has a unique spectrum characteristic of its atoms. Consequently a study of the spectrum of a substance enables its composition to be readily determined. *Spectroscopy* is the name given to the exact analysis of mixtures or compounds by a study of their spectra. The science has developed to such an extent that the presence in a substance of less than a millionth of a milligram of sodium can be detected.

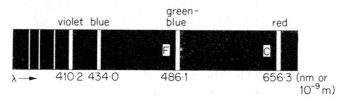

Fig. 19.31 Visible line spectra of hydrogen

Types of emission spectra. There are three different types of spectra, which are easily recognised. They are known as (*a*) *line spectra*, (*b*) *band spectra*, (*c*) *continuous spectra*.

(*a*) *Line spectra*. When the light emitted by the atoms of a glowing substance (such as vaporised sodium or helium gas) is examined by a prism and spectrometer, lines of various wavelengths are obtained. These lines, it should be noted, are images of the narrow slit of the spectrometer on which the light is incident. The spectra of hydrogen, Fig. 19.31, and helium are line spectra. Generally, line spectra are obtained from *atoms*. See also Fig. 19A, p. 376.

(*b*) *Band spectra*. Band spectra are obtained from *molecules*, and consist of

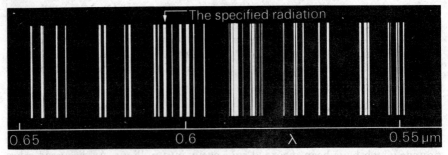

Fig. 19A *Line Spectrum.* Part of the spectrum of the gas krypton-86 in the range 0·65 μm (650 nm) to 0·55 μm (550 nm). The wavelength in a vacuum of the specified radiation shown is used in the definition of the metre (p. 474). (*Crown copyright. Courtesy of National Physical Laboratory*)

a series of bands each sharp at one end but 'fading' at the other end, Fig. 19.32. The term 'fluting' is often used to describe the way in which the bands are spaced. Careful examination reveals that the bands are made up of numerous fine lines very close to each other. Two examples of band spectra are those usually obtained from nitrogen and oxygen.

Fig. 19.32 Diagrammatic representation of band spectra

(*c*) *Continuous spectra.* The spectrum of the sun is an example of a continuous spectrum, and, in general, these spectra are obtained from *solids and liquids*. In these states of matter the atoms and molecules are close together, and the energy changes in a particular atom are influenced by neighbouring atoms to such an extent that radiations of all different wavelengths are emitted. In a gas the atoms are comparatively far apart, and each atom is uninfluenced by any other. The gas therefore emits radiations of wavelengths which result from energy changes in the atom due solely to the high temperature of the gas, and a line spectrum is obtained. When the temperature of a gas is decreased and pressure applied so that the liquid state is approached, the line spectrum of the gas is observed to broaden out consideraly.

Absorption Spectra. Kirchhoff's Law

The spectra just discussed are classified as *emission spectra*. There is another class of spectra known as *absorption spectra*, which we shall now briefly consider.

If light from a source having a continuous spectrum is examined after it has passed through a sodium flame, the spectrum is found to be crossed by a dark line; this dark line is in the position corresponding to the bright line emission spectrum obtained with the sodium flame alone. The continuous spectrum with the dark line is naturally characteristic of the absorbing substance, in this case sodium, and it is known as an *absorption spectrum*. An absorption spectrum is obtained when red glass is placed in front of sunlight, as it allows only a narrow band of red rays to be transmitted.

KIRCHHOFF's investigations on absorption spectra in 1855 led him to formulate a simple law concerning the emission and absorption of light by a substance. This states: *A substance which emits light of a certain wavelength at a given tem-*

perature can also absorb light of the same wavelength at that temperature. In other words, a good emitter of a certain wavelength is also a good absorber of that wavelength. From Kirchhoff's law it follows that if the radiation from a hot source emitting a continuous spectrum is passed through a vapour, the absorption spectrum obtained is deficient in those wavelengths which the vapour would emit if it were raised to the same high temperature. Thus if a sodium flame is observed through a spectrometer in a darkened room, a bright yellow line is seen; if a strong white arc light, richer in yellow light than the sodium flame, is placed behind the flame, a dark line is observed in the place of the yellow lines. The sodium absorbs yellow light from the white light, and re-radiates it in all directions. Consequently there is less yellow light in front of the sodium flame than if it were removed, and a dark line is thus observed.

Fraunhofer Lines

In 1814 FRAUNHOFER noticed that the sun's spectrum was crossed by many hundreds of dark lines. These *Fraunhofer lines,* as they are called, were mapped out by him on a chart of wavelengths, and the more prominent were labelled by the letters of the alphabet. Thus the dark line in the blue part of the spectrum was known as the F line, the dark line in the yellow part as the D line, and the dark line in the red part as the C line.

The Fraunhofer lines indicate the presence in the sun's atmosphere of certain elements in a vaporised form. The vapours are cooler than the central hot portion of the sun, and they absorb their own characteristic wavelengths from the sun's continuous spectrum. Now every element has a characteristic spectrum of wavelengths. Accordingly, it became possible to identify the elements round the sun from a study of the wavelengths of the Fraunhofer (dark) lines in the sun's spectrum, and it was then found that hydrogen and helium were present. This was how helium was first discovered. The D line is the yellow sodium emission line.

The incandescent gases round the sun can be seen as flames enormously high during a total eclipse of the sun, when the central portion of the sun is cut off from the observer. If the spectrum of the sun is observed just before an eclipse takes place, a continuous spectrum with Fraunhofer lines is obtained, as already stated. At the instant when the eclipse becomes total, however, bright emission lines are seen in exactly the same position as those previously occupied by the Fraunhofer lines, and they correspond to the emission spectra of the vapours alone. This is an illustration of Kirchhoff's law, p. 282. The F and C lines are in the emission spectrum of hydrogen.

Dispersive Power

Red and blue are two extreme colours visible in the spectrum of white light. The *dispersive power,* ω, of a particular material such as glass for particular wavelengths in these two colours is defined by the ratio:

$$\omega = \frac{n_b - n_r}{n - 1},$$

where n_b and n_r are the refractive indices for blue and red respectively and n is the refractive index of the glass for yellow light. Various types of glass, some with greater refractive index and dispersive power than others, are used in the optical industry. The table which follows shows values of refractive index of a particular type of crown glass and flint glass.

	n_b	n_r	n
Crown glass, X	1·521	1·510	1·517
Flint glass, Y	1·665	1·645	1·655

The refractive index for each colour is measured to a high degree of accuracy by making a large angle prism of the material, and then using the spectrometer method on p. 344 to measure the angle A and the minimum deviation D_{min}.

'Blue', 'red' and 'yellow' are bands of colours. Each covers a narrow range of wavelengths. So the refractive index of glass will be slightly different for the wavelengths in the blue, for example. Consequently, for accuracy, a *particular* wavelength or line is conventionally chosen in each of the blue, red and yellow bands. These monochromatic lines are called F, C and D lines respectively, following Fraunhofer (see p. 377).

For a certain crown glass, $n_F = 1·521$, $n_C = 1·510$ and $n_D = 1·517$. Hence, for the F(blue) and C(red) lines,

$$\text{dispersive power, } \omega_1 = \frac{n_F - n_C}{n_D - 1} = 0·021.$$

For a certain flint glass, $n_F = 1·665$, $n_C = 1·645$ and $n_D = 1·655$. So

$$\text{dispersive power, } \omega_2 = \frac{n_F - n_C}{n_D - 1} = 0·031.$$

So the flint glass has about 1·5 times the dispersive power of the crown glass.

Achromatic lenses

When white light from an object is refracted by a lens, a coloured image is formed. This is because the glass refracts different colours such as red, r, and blue, b, to a different focus (Fig. 19.33). The coloured images are formed at slightly different places and this is called the *chromatic aberration* (colour defect) of a single lens.

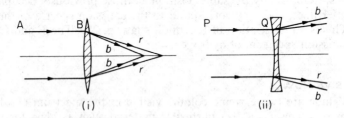

Fig. 19.33 Dispersion produced by converging and diverging lens

A converging lens deviates an incident ray such as AB towards its principal axis, Fig. 19.33 (i). A diverging lens, however, deviates a ray PQ away from its principal axis, Fig. 19.33 (ii). The dispersion between two colours produced by a converging lens can thus be neutralised by placing a suitable diverging lens beside it. Two such lenses which together eliminate the chromatic aberration of a single lens are called an *achromatic* combination of lenses. Figure 19.34 illustrates an achromatic lens combination, known as an *achromatic doublet*. The biconvex lens is made of crown glass, while the diverging lens is made of

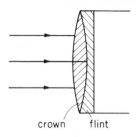

Fig. 19.34 Achromatic doublet in telescopic objective

flint glass and is a plano-concave lens. So that the lenses can be cemented together with Canada balsam, the radius of curvature of the curved surface of the plano-concave lens is made numerically the same as that of one surface of the converging lens. The achromatic combination acts as a converging lens when used as the objective lens in a high-quality telescope or microscope.

It should be noted that chromatic aberration would occur if the diverging and converging lenses were made of the *same* material, as the two lenses together would then constitute a single thick lens of one material.

Achromatic lenses were first made about 1729. It can be shown that: *Two lenses form an achromatic doublet for two colours, if the ratio of their focal lengths is numerically equal to the ratio of the corresponding dispersive powers of their materials.* Hence, since f_1, f_2 are of opposite signs because one lens must be converging and the other diverging,

$$\frac{f_1}{f_2} = -\frac{\omega_1}{\omega_2}$$

where ω_1 and ω_2 are the respective dispersive powers.

Condition for achromatic lenses. Since $\dfrac{1}{f} = (n-1)\left(\dfrac{1}{r_1} + \dfrac{1}{r_2}\right)$ with the usual notation, with blue and then red light we have

$$\frac{1}{f_b} = (n_b - 1)\left(\frac{1}{r_1} + \frac{1}{r_2}\right) \text{ and } \frac{1}{f_r} = (n_r - 1)\left(\frac{1}{r_1} + \frac{1}{r_2}\right).$$

So
$$\frac{1}{f_b} - \frac{1}{f_r} = \Delta\left(\frac{1}{f}\right) = (n_b - n_r)\left(\frac{1}{r_1} + \frac{1}{r_2}\right) \qquad (i)$$

$$\Delta\left(\frac{1}{f}\right) = \frac{1}{f_b} - \frac{1}{f_r}$$

where $\Delta\left(\dfrac{1}{f}\right)$ is the small change in $\dfrac{1}{f}$ when blue, and then red, rays are incident on the lens.

But
$$\frac{1}{f} = (n-1)\left(\frac{1}{r_1} + \frac{1}{r_2}\right) \qquad . \qquad . \qquad . \qquad (ii)$$

where f is the focal length of the lens when yellow light is incident on the lens, and n is the refractive index for yellow light. Dividing (i) by (ii) and simplifying, we obtain

$$\Delta\left(\frac{1}{f}\right) = \frac{n_b - n_r}{n-1} \cdot \frac{1}{f} = \frac{\omega}{f} \qquad . \qquad . \qquad . \qquad (iii)$$

since the dispersive power, ω, of a material for the two colours is defined by

$$\omega = \frac{n_b - n_r}{n - 1}.$$

Combined lenses. Suppose f_1, f_2 are the respective focal lengths of two thin lenses in contact, ω_1, ω_2 are the corresponding dispersive powers of their materials, and F is the combined focal length. If the combination is achromatic for blue and red light, the focal length F_b for blue light is the same as the focal length F_r for red light, that is, $F_b = F_r$.

$$\therefore \frac{1}{F_b} - \frac{1}{F_r} = \Delta\left(\frac{1}{F}\right) = 0 \qquad . \qquad . \qquad . \qquad \text{(iv)}$$

But

$$\frac{1}{F} = \frac{1}{f_1} + \frac{1}{f_2}$$

So

$$\Delta\left(\frac{1}{F}\right) = \Delta\left(\frac{1}{f_1}\right) + \Delta\left(\frac{1}{f_2}\right).$$

From (iv),

$$0 = \Delta\left(\frac{1}{f_1}\right) + \Delta\left(\frac{1}{f_2}\right).$$

Now

$$\Delta\left(\frac{1}{f_1}\right) = \frac{\omega_1}{f_2} \text{ and } \Delta\left(\frac{1}{f_2}\right) = \frac{\omega_2}{f_2} \text{ from (iii)}$$

$$\therefore 0 = \frac{\omega_1}{f_1} + \frac{\omega_2}{f_2} \text{ from above}$$

$$\therefore \frac{f_1}{f_2} = -\frac{\omega_1}{\omega_2} \qquad . \qquad . \qquad . \qquad . \qquad \text{(v)}$$

Thus the ratio of the focal lengths is equal to the ratio of the dispersive powers of the corresponding lens materials. Since ω_1, ω_2 are positive numbers, it follows from (v) that f_1 and f_2 must have opposite signs. So a diverging lens must be combined with a converging lens to form an achromatic combination (see Fig. 19.34).

Exercises 19C

1. Explain the difference in appearance between a *line spectrum* and a *continuous spectrum*. State the kind of laboratory light source which you would use to produce each type of spectrum.

Why are a number of dark lines seen in the spectrum of light from the sun?

A spectrometer, using a prism, is to be set up to produce a pure spectrum. State the essential adjustments of the spectrometer necessary to achieve this result. (Experimental details are not required.) (*L.*)

2. Describe the optical system of a simple prism spectrometer. Illustrate your answer with a diagram showing the paths through the spectrometer of the pencils of rays which form the red and blue ends of the spectrum of a source of white light. (Assume in your diagram that the lenses are achromatic.)

The prism of a spectrometer has a refracting angle of 60° and is made of glass whose refractive indices for red and violet are respectively 1·514 and 1·530. A white source is used and the instrument is set to give minimum deviation for red light. Determine (*a*) the angle of incidence of the light on the prism, (*b*) the angle of emergence of the violet light, (*c*) the angular width of the spectrum. (*N.*)

3. Distinguish between emission spectra and absorption spectra. Describe the spectrum of the light emitted by (i) the sun, (ii) a car headlamp fitted with yellow glass, (iii) a sodium vapour street lamp.

What are the approximate wavelength limits of the visible spectrum? How would you demonstrate the existence of radiations whose wavelengths lie just outside these limits? (*O. & C.*)

4. Describe the processes which lead to the formation of numerous dark lines (Fraunhofer lines) in the solar spectrum. Explain why the positions of these lines in the spectrum differ very slightly when the light is received from opposite ends of an equatorial diameter of the sun. (*N.*)

5. A parallel beam of white light is incident on a converging glass lens which has a focal length of 20·0 cm for yellow light. A white screen is placed 20·0 cm from the lens on the other side of the beam. With the aid of a diagram, explain the change in the appearance of the coloured image seen on the screen when the screen is moved (*a*) towards the lens and (*b*) away from the lens.

6. What is *chromatic aberration* of a lens? Explain why chromatic aberration is obtained. Draw a sketch of an *achromatic doublet* and explain, with the aid of diagrams, why it reduces chromatic aberration.

7. Is the focal length of a converging lens greater or smaller for blue light than for red?

A white object is placed farther from a converging lens than its average focal length. Taking first red, and then blue, rays from the object, draw a ray diagram showing how red and blue coloured real images are formed by the lens. Why is this a disadvantage?

20 Optical Instruments

When a telescope or a microscope is used to view an object, the appearance of the final image is determined by the cone of rays entering the eye. A discussion of optical instruments and their behaviour must therefore be preceded by a consideration of the image formed by the eye. As we now need to use them again, we list some of the points about the eye mentioned in previous pages.

Firstly, the image formed by the eye lens L must appear on the retina R at the back of the eye if the object is to be clearly seen, Fig. 20.1. Secondly, the normal eye can focus an object at infinity (the 'far point' of the normal eye). Thirdly, the eye can see an object in greatest detail when it is placed at a certain distance D from the eye, known as the *least distance of distinct vision*, which is about 25 cm, for a normal eye (p. 370). The point at a distance D from the eye is known as its 'near point'.

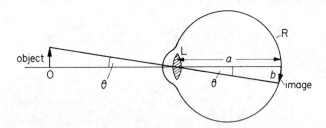

Fig. 20.1 Length of image on retina, and visual angle

Visual Angle

Consider an object O placed some distance from the eye, and suppose θ is the angle in radians subtended by it at the eye, Fig. 20.1. Since the opposite angles at L are equal, it follows that the length b of the image on the retina is given by $b = a\theta$, where a is the distance from R to L. But a is a constant; hence $b \propto \theta$. We thus arrive at the important conclusion that *the length of the image formed by the eye is proportional to the* **angle** *subtended at the eye by the object*. This angle is known as the *visual angle*; the greater the visual angle, the greater is the apparent size of the object.

Figure 20.2 (i) illustrates the case of an object moved from A to B, and viewed by the eye in both positions. At B the angle β subtended at the eye is greater than the visual angle α subtended at A. Hence the object appears larger at B than at A, although its physical size is the same. Figure 20.2 (ii) illustrates the case of two objects, at P, Q respectively, which subtend the same visual angle θ at the eye. The objects thus appear to be of equal size, although the object at P is physically bigger than that at Q. It should be remembered that an object is not clearly seen if it is brought closer to the eye than the near point.

Angular Magnification of Telescopes

Telescopes and microscopes are instruments designed to increase the visual angle, so that the object viewed can be made to appear much larger with their aid. Before they are used the object may subtend a small angle α at the eye; when they are used the final images may subtend an increased angle α' at the

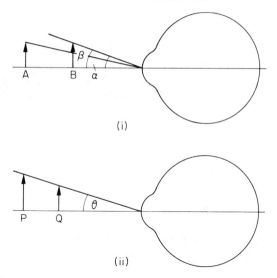

(i)

(ii)

Fig. 20.2 Relation between visual angle and length of image

eye. The *angular magnification*, M, of the instrument is defined as the ratio

$$M = \frac{\alpha'}{\alpha} \qquad \qquad (1)$$

This is also popularly known as the *magnifying power* of the instrument. It should be carefully noted that we are concerned with visual angles in the theory of optical instruments, and not with the physical sizes of the object and the image obtained.

Telescopes are instruments used for viewing distant objects. They are used extensively at astronomical observatories. The first telescope is reputed to have been made about 1608, and in 1609 Galileo made a telescope through which he observed the satellites of Jupiter and the rings of Saturn. The telescope paved the way for great astronomical discoveries, particularly in the hands of KEPLER. Newton also designed telescopes. He was the first person to suggest the use of curved mirrors for telescopes free from chromatic aberration (see p. 388).

If α is the angle subtended at the unaided eye by a *distant* object, and α' is the angle subtended at the eye by its image when a telescope is used, the angular magnification M (also called the 'magnifying power') of the instrument is given by

$$M = \frac{\alpha'}{\alpha}.$$

Astronomical Telescope in Normal Adjustment

An astronomical telescope made from lenses consists of an *objective* of long focal length and an *eyepiece* of short focal length, for a reason given on p. 384. Both lenses are converging. *The telescope is in normal adjustment when the final image is formed at infinity.* The eye is then relaxed or unaccommodated when viewing the image, so the image is not seen distinctly. The unaided eye is also relaxed when the distance object viewed can be considered to be at infinity.

The objective lens O collects parallel rays from the distant object. So it forms an image I at its focus F_0. Figure 20.3 shows three of the many non-axial rays *a* from the *top* point of the object, which pass through the top point T of the

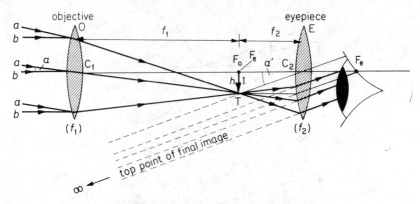

Fig. 20.3 Telescope in *normal* adjustment

image. The three rays b from the foot of the object would pass through the foot of I (not shown). As the final image is at infinity, I must be at the focus F_e of the eyepiece. So F_e and F_0 are at the same place.

To draw the final image, take one lens at a time. (*a*) For O, draw a central ray a straight through C_1 to T, the top of the objective image below F_0. Then draw the other two rays a to pass through T, as shown. (*b*) For E, draw a line from T to pass straight through C_2 and another line from T parallel to the principal axis to pass through F_e. The lines emerging from E are parallel; so the final image is at infinity. (*c*) Now continue the rays passing through T from O so that they meet the lens E; then draw each refracted ray *parallel to* TC_2 because they must pass through the top of the image of T when extended back. Note carefully that the two lines first drawn from T to E to find the image position are construction lines and *not* actual light rays, and so should not have arrows on them.

To find the angular magnification M of the telescope, assume that the eye is close to the eyepiece. Since the telescope length is very small compared with the distance of the object from either lens, we can take the angle α subtended at the unaided eye by the object as that subtended at the *objective* lens, as shown. Since I is distance f_1 from C_1, where f_1 is the focal length of O, we see that $\alpha = h/f_1$, where h is the length of I. Also, the angle α' subtended at the eye when the telescope is used is given by h/f_2, where f_2 is the focal length of the eyepiece E. So

$$M = \frac{\alpha'}{\alpha} = \frac{h/f_2}{h/f_1}$$

$$\therefore M = \frac{f_1}{f_2} \qquad . \qquad . \qquad . \qquad . \qquad . \qquad (2)$$

Thus the angular magnification is equal to the ratio of the focal length of the objective (f_1) to that of the eyepiece (f_2). For high angular magnification, it follows from (2) that the objective should have a long focal length and the eyepiece a short focal length. Note that the separation of the lenses is $f_1 + f_2$.

The Eye-ring, and Relation to Angular Magnification

When an object is viewed by an optical instrument, only those rays from the object which are bounded by the perimeter of the objective lens enter the instrument. The lens thus acts as a *stop* to the light from the object. Similarly, the only rays from the image causing the sensation of vision are those which enter the pupil of the eye. The pupil thus acts as a natural stop to the light from the image. With a given objective, the best position of the eye is one where it collects as much light as possible from that passing through the objective.

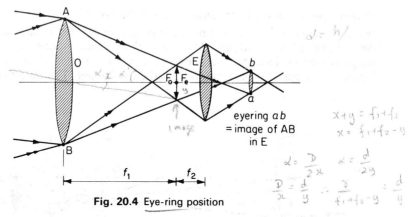

Fig. 20.4 Eye-ring position

Figure 20.4 shows the rays from the field of view which are refracted at the *boundary* of the objective O to form an image at F_0 or F_e with the telescope in normal adjustment. These rays are again refracted at the boundary of the eyepiece E to form a small image ab. From the ray diagram, we see that a is the image of A on the objective and b is an image of B on the objective. *So ab is the image of the objective AB in the eyepiece.*

The small circular image ab is called the *eye–ring*. It is the best position for the eye. Here the eye can collect the maximum amount of light entering the objective from outside so that it has a *wide field of view*. If the eye were placed closer to the eyepiece than the eye-ring the observer would have a smaller field of view.

If the telescope is in normal adjustment, the distance u of the objective from the eyepiece E, focal length f_2, is $(f_1 + f_2)$. From the lens equation, the eye-ring distance v from E is given by

$$\frac{1}{v} + \frac{1}{+(f_1 + f_2)} = \frac{1}{+(f_2)}$$

from which
$$v = \frac{f_2}{f_1}(f_1 + f_2).$$

Now the objective diameter:eye-ring diameter $= AB:ab = u:v$

$$= (f_1 + f_2) \cdot \frac{f_2}{f_1}(f_1 + f_2)$$

$$= f_1/f_2.$$

But the angular magnification of the telescope $= f_1/f_2$ (p. 384). Thus the angular magnification, M, is also given by

$$M = \frac{\text{diameter of objective}}{\text{diameter of eye-ring}} \qquad . \qquad . \qquad . \qquad . \qquad (3)$$

the telescope being in normal adjustment.

The relation in (3) provides a simple way of measuring M for a telescope.

The telescope objective is placed in front of a clear window. A screen is then moved behind the eyepiece until a sharp circular eye-ring is obtained. By measuring the diameter of the eye-ring and of the objective, M can be found from (3).

Telescope with Final Image at Near Point

When a telescope is used, the final image can be formed at the near point of the eye instead of at infinity. The eye is then 'accommodated', and now the image is seen in detail but the telescope is *not* in normal adjustment (p. 383). Figure 20.5 illustrates the formation of the final image. The objective forms an

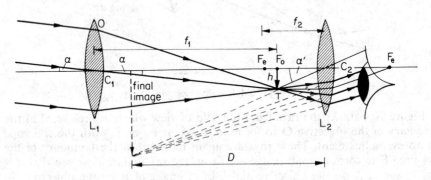

Fig. 20.5 Final image at near point

image of the distant object at its focus F_o, and the eyepiece is moved so that the image is nearer to it than its focus F_e, thus acting as a magnifying glass.

To draw the final image, take one lens at a time. (a) First draw the image F_oT formed at F_o by the objective O, using the central ray C_1T. (b) Using construction lines (see p. 384), draw the final virtual image of F_oT formed by E. (c) Finally, draw the three rays through T from O to meet E and extend each of them back to meet the top point of the final image, as shown.

The angle α subtended at the unaided eye is practically that subtended at the objective L_1. Thus $\alpha = h/f_1$, where h is the length of the image in the objective and f_1 its focal length. The angle α' subtended at the eye by the final image $= h/u$, if the eye is close to the eyepiece, where $u =$ the distance of the image at F_o from the eyepiece.

Thus angular magnification, $M = \dfrac{\alpha'}{\alpha} = \dfrac{h/u}{h/f_1}$

$$\therefore M = \frac{f_1}{u} \qquad\qquad\qquad \text{(i)}$$

As the final image is formed at a numerical distance D from the eyepiece L_2, we have $v = -D$ when $f = +f_2$. Thus, from $1/v + 1/u = 1/f$,

$$\frac{1}{-D} + \frac{1}{u} = \frac{1}{+f_2},$$

from which $\qquad\qquad u = \dfrac{f_2 D}{f_2 + D}.$

Substituting in (i) for u,

$$\therefore M = \frac{f_1}{f_2}\left(\frac{f_2 + D}{D}\right)$$

$$\therefore M = \frac{f_1}{f_2}\left(1 + \frac{f_2}{D}\right) \qquad . \quad . \quad . \quad . \quad (4)$$

The angular magnification when the telescope is in normal adjustment (i.e., final image at infinity) is f_1/f_2 (p. 384). Hence, from (4), the angular magnification is increased in the ratio $\left(1 + \frac{f_2}{D}\right):1$ when the final image is formed at the near point.

Examples

1. An astronomical telescope in normal adjustment has an objective of focal length 50 cm and an eye-lens of focal length 5 cm. What is its angular magnification? If it is assumed that the eye is placed very close to the eye lens and that the pupil of the eye has a diameter of 3 mm, what will be the diameter of the objective if all the light passing through the objective is to emerge as a beam which fills the pupil of the eye? Assume that the telescope is pointed directly at a particular star.

Since the telescope is in *normal* adjustment, the final image is formed at infinity. The angular magnification of the telescope is then $\dfrac{50 \text{ cm}}{5 \text{ cm}}$, or 10.

If all the light emerging from the eyepiece fills the pupil of the eye, the pupil is at the eye-ring. See p. 385. The eye-ring is the image of the objective in the eyepiece. Since the distance, u, from the objective to the eyepiece $= 50 + 5 = 55$ cm, the eye-ring distance, v, is given by

$$\frac{1}{v} + \frac{1}{(+55)} = \frac{1}{(+5)}$$

from which $\qquad\qquad v = 5 \cdot 5 \text{ cm.}$

This is the position of the pupil of the eye. The magnification is given by

$$\frac{\text{eye-ring diameter}}{\text{objective diameter}} = \frac{v}{u} = \frac{5 \cdot 5}{55} = \frac{1}{10}$$

$$\therefore \frac{3 \text{ mm}}{\text{objective diameter}} = \frac{1}{10}$$

$$\therefore \text{objective diameter} = 3 \text{ cm.}$$

2. A simple astronomical telescope has an objective of 250 cm and an eyepiece of 8 cm. The eyepiece is adjusted so that a real image of the sun is formed on a screen 24 cm from the eyepiece.

(i) Calculate the distance between the objective and eyepiece. (ii) If the sun's image on the screen has a diameter of 10 cm, calculate the angle subtended by the sun at the objective.

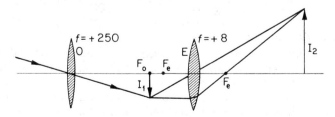

Fig. 20.6 Example

Figure 20.6 shows a ray from the sun passing through the top point of I_1, the image formed by the objective O, and the line passing through the eyepiece E from the top of I_1 to the real image I_2 formed by E.

(i) Applying the lens equation $1/v + 1/u = 1/f$ to E, then the object distance u of I_1 from E is given by

$$\frac{1}{+24} + \frac{1}{u} = \frac{1}{+8}.$$

Solving, $u = 12$ cm.

So   distance between lenses $= 250 + 12 = 262$ cm.

(ii) Since the image diameter is 10 cm, then, from the magnification formula $m = v/u$ applied to lens E,

$$\frac{\text{diameter of } I_1}{\text{diameter of } I_2} = \frac{12 \text{ cm}}{24 \text{ cm}} = \frac{1}{2}.$$

So   diameter of $I_1 = \frac{1}{2} \times 10$ cm $= 5$ cm.

Then   angle α subtended by sun at O $= \dfrac{5 \text{ cm}}{250 \text{ cm}} = 0\cdot02$ rad.

Resolving Power

If two distant objects are close together, it may not be possible to see their images apart through a telescope even though the lenses are perfect. This is due to the phenomenon of diffraction and is explained on page 485. Here we can state that the smallest angle θ subtended at a telescope by two distant objects which can just be seen separated is given approximately by

$$\theta = \frac{1\cdot22\lambda}{D},$$

where λ is the mean wavelength of the light from the distant objects and D is the diameter of the *objective* lens.

θ is called the *resolving power* of the telescope. The smaller the value of θ, the *greater* is the resolving power because two distant objects which are closer together can be seen separated through the telescope. Note that the formula for θ only depends on the diameter of the objective and not on its focal length, and that it does not concern the eyepiece. As we have seen, the focal lengths of the objective and eyepiece affect the angular magnification of a telescope but high angular magnification does not produce high resolving power. Higher resolving power is obtained by using an objective lens of greater diameter.

So if the objective of a telescope has a diameter of 200 mm, and the mean wavelength of the light from distant stars is 6×10^{-7} m, then

$$\theta = \frac{1\cdot22 \times 6 \times 10^{-7}}{0\cdot200} = 4 \times 10^{-6} \text{ rad (approx.)}.$$

Reflector Telescope

The astronomical telescope so far discussed has a lens objective and is therefore a *refractor* telescope. A *reflector* telescope, with a large curved mirror as its objective, was first suggested by Newton.

The construction of the Hale telescope at Mount Palomar, the largest telescope in the world, is one of the most fascinating stories of scientific skill and invention. The major feature of the telescope is a *parabolic mirror*, 5 metres

Fig. 20A *Radcliffe Refractor Telescope.* The larger telescope has an objective of 60 cm diameter and focal length about 7 m. The resolving power is of the order 10^{-6} rad. This telescope acts as a camera whereas the smaller telescope shown in use, which has a 50 cm objective, acts as a guide telescope.

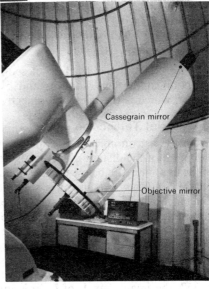

Cassegrain mirror

Objective mirror

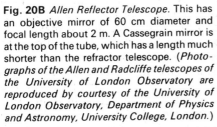

Fig. 20B *Allen Reflector Telescope.* This has an objective mirror of 60 cm diameter and focal length about 2 m. A Cassegrain mirror is at the top of the tube, which has a length much shorter than the refractor telescope. (*Photographs of the Allen and Radcliffe telescopes of the University of London Observatory are reproduced by courtesy of the University of London Observatory, Department of Physics and Astronomy, University College, London.*)

across, which is made of pyrex, a low expansion glass. The glass itself took more than six years to grind and polish, and the front of the mirror is coated with aluminium, instead of being covered with silver, as it lasts much longer. The huge size of the mirror enables enough light from very distant stars and planets to be collected and brought to a focus for them to be photographed. Special cameras are incorporated in the instrument. This method has the advantage that plates can be exposed for hours, if necessary, to the object to be studied, enabling records to be made. It is used to obtain useful information about the building-up and breaking-down of the elements in space, to investigate astronomical theories of the universe, and to photograph planets such as Mars. The Hale telescope was built on the top of Mount Palomar, California, where the air is particularly free of mist and other hindrances to night vision.

Besides the main parabolic mirror O, which is the telescope objective, seven other mirrors are used in the 5 metre telescope. Some are plane, Fig. 20.7 (i), while others are convex, Fig. 20.7 (ii), and they are used to bring the light to a more convenient focus, where the image can be photographed, or magnified several hundred times by an eyepiece E for observation. The various methods of focusing the image were suggested respectively by *Newton, Cassegrain*, and *Coudé*, the last thing being a combination of the former two methods, Fig. 20.7 (iii).

The reflecting telescope is free from the chromatic aberration of lenses (p. 378), which colours images in the refractor telescope. The image is also

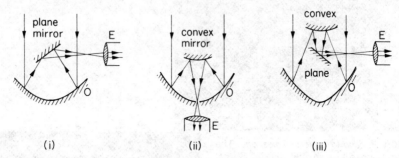

Fig. 20.7 (i) Newton reflector (ii) Cassegrain reflector (iii) Coudé reflector

brighter than in a refractor telescope, where some loss of light occurs by reflection at the lens' surfaces and by absorption. The large diameter of the mirror, which is the telescope objective, also produces high resolving power.

Terrestrial Telescope

From the astronomical telescope diagram in Fig. 20.5, it can be seen that the top point of the distant object is above the axis of the lenses, but the top point of the final image is below the axis. Thus the image in an astronomical telescope is *inverted*. This instrument is suitable for astronomy because it makes little difference if the image of a star, for example, is inverted, but it is useless for viewing objects on the earth or sea, in which case an erect image is required.

A *terrestrial telescope* provides an erect image. In addition to the objective and eyepiece of the astronomical telescope, it has a converging lens L of focal length f between them, Fig. 20.8. L is placed at a distance $2f$ in front of the inverted real image I_1 formed by the objective, in which case, as shown on p. 357, the image I in L of I_1 (i) is inverted, real, and the same size as I_1, (ii) is also at a distance $2f$ from L.

Thus the image I is now the same way up as the distant object. If I is at the focus of the eyepiece, the final image is formed at infinity and is also erect.

The lens L is often known as the 'erecting' lens of the telescope, as its only function is that of inverting the image I_1. Since the image I produced by L is the same size as I_1, the presence of L does not affect the magnitude of the angular magnification of the telescope, which is thus f_1/f_2 (p. 384). The erecting lens, however, reduces the intensity of the light emerging through the eyepiece, as light is reflected at the lens surfaces. Yet another disadvantage is the increased length of the telescope when L is used; the distance from the objective to the eyepiece is now $(f_1 + f_2 + 4f)$, Fig. 20.8, compared with $(f_1 + f_2)$ in the astronomical telescope.

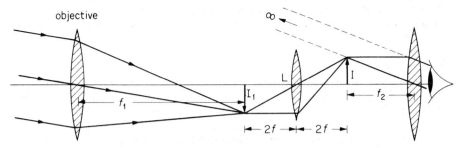

Fig. 20.8 Terrestrial telescope

Prism Binoculars

Prism binoculars are widely used as field glasses, and consist of short astronomical telescopes containing two right-angled isosceles prisms between the objective eyepiece, Fig. 20.9. These lenses are both converging, and they would produce an inverted image of the distant object if they alone were used. The purpose of the two prisms is to invert the image and obtain a final *upright* image with *no lateral inversion*.

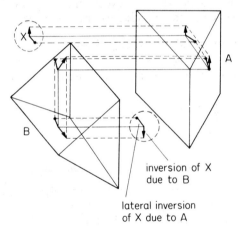

Fig. 20.9 Prism binoculars

One prism A, is placed with its refracting edge vertical, while the other, B, is placed with its refracting edge horizontal. As shown in Fig. 20.5, the image formed by the objective alone is inverted. Prism A, however, turns it round in a horizontal direction, and prism B inverts it in a vertical direction, both prisms acting as reflectors of light (see p. 347). The image produced after reflection at B is now the same way up, and the same way round, as the original object. Since the eyepiece is a converging lens acting as a magnifying glass, it produces a final image the same way up as the image in front of it, and hence the final image is the same way up as the distant object.

Figure 20.9 illustrates the path of some rays through the optical system. Since the optical path of a ray is about 3 times the distance d between the objective and the eyepiece, the system is equivalent optically to an astronomical telescope of length $3d$. The focal lengths of the objective and eyepiece in the prism binoculars thus provide the same angular magnification as an astronomical telescope 3 times as long. The compactness of the prism binocular is one of its advantages; another is the wide field of view obtained, as it is an astronomical telescope.

Microscopes

At the beginning of the seventeenth century single lenses were developed as powerful magnifying glasses, or *microscopes*, and many important discoveries in human and animal biology were made with their aid. Shortly afterwards two or more converging lenses were combined to form powerful microscopes, and with their use HOOKE, in 1648, discovered the existence of 'cells' in animal and vegetable tissue.

A microscope is an instrument used for viewing *near* objects. When it is in normal use, therefore, the image formed by the microscope is usually at the least distance of distinct vision, D, from the eye, i.e., at the near point of the eye. With the unaided eye (i.e., without the instrument), the object is seen clearest when it is placed at the near point. Consequently the angular magnification of a microscope in *normal* use is given by

$$M = \frac{\alpha'}{\alpha},$$

where α' is the angle subtended at the eye by the image at the near point, and α is the angle subtended at the unaided eye by the object at the near point.

Simple Microscope or Magnifying Glass

Suppose that an object of length h is viewed at the near point, A, by the unaided eye, Fig. 20.10. The visual angle, α, is then h/D in radian measure. Now

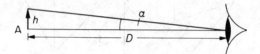

Fig. 20.10 Visual angle with unaided eye

suppose that a converging lens L is used as a magnifying glass to view the same object. An erect, magnified image is obtained when the object O is nearer to L than its focal length (p. 351), and the observer moves the lens until the image at I is situated at his near point. If the observer's eye is close to the lens at C, the distance IC is then equal to D, the least distance of distinct vision, Fig. 20.11. Thus the new visual angle α' is given by h'/D, where h' is the length of the virtual image, and it can be seen that α' is greater than α by comparing Fig. 20.10 with Fig. 20.11.

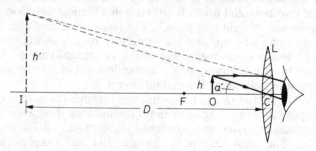

Fig. 20.11 Simple microscope, or magnifying glass

The angular magnification, M, of this simple microscope can be found in terms of D and the focal length f of the lens. From definition, $M = \alpha'/\alpha$.

But
$$\alpha' = \frac{h'}{D}, \quad \alpha = \frac{h}{D}$$

$$\therefore M = \frac{h'}{D}\bigg/\frac{h}{D} = h'/h \qquad . \qquad . \qquad . \qquad . \qquad (1)$$

Now h'/h is the 'linear magnification' produced by the lens, and is given by $h'/h = v/u$, where v is the image distance CI and u is the object distance CO (see p. 356). Since $1/v + 1/u = 1/f$, with the usual notation, we have

$$1 + \frac{v}{u} = \frac{v}{f},$$

by multiplying throughout by v. Since the image is virtual, $v = \text{CI} = -D$, where D is the *numerical* value of the least distance of distinct vision,

$$\therefore \frac{v}{u} = \frac{v}{f} - 1 = -\frac{D}{f} - 1$$

$$\therefore \frac{h'}{h} = -\frac{D}{f} - 1.$$

$$\therefore M = -\left(\frac{D}{f} + 1\right) \qquad . \qquad . \qquad . \qquad . \qquad (2)$$

from (1) above. So numerically, $\quad M = \left(\frac{D}{f} + 1\right)$

If the magnifying glass has a focal length of 5 cm, $f = +5$ as it is converging; also, if the least distance of distinct vision is 25 cm, $D = 25$ numerically. Substituting in (2),

$$M = -\left(\frac{25}{5} + 1\right) = -6.$$

Thus the angular magnification is 6. The position of the object O is given by substituting $v = -25$ and $f = +5$ in the lens equation $1/v + 1/u = 1/f$, from which the object distance u is found to be $+4.2$ cm.

From the formula for M in (2), it follows that a lens of *short* focal length is required for high angular magnification.

When an object OA is viewed through a converging lens acting as a *magnifying glass*, various coloured virtual images, corresponding to I_R, I_V for red and violet rays for example, are formed, Fig. 20.12. The top point of each image lies on

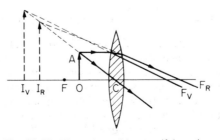

Fig. 20.12 Dispersion with magnifying glass

the line CA. Hence each image subtends the same angle at the eye close to the lens, so that the colours received by the eye will practically overlap. Thus the virtual image seen in a magnifying glass is almost free of chromatic aberration. A little colour is observed at the edges as a result of spherical aberration. A *real* image formed by a lens has chromatic aberration, as explained on p. 378.

Magnifying Glass with Image at Infinity

We have just considered the normal use of the simple microscope, where the image formed is at the near point of the eye and the eye is accommodated (p. 392). When the image is formed at infinity, however, which is not a normal use of the microscope, the eye is undergoing the least strain and is then un-accommodated (p. 370). In this case the object must be placed at the focus, F, of the lens, Fig. 20.13.

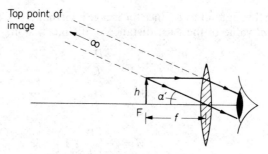

Fig. 20.13 Final image at infinity

Suppose that the focal length of the lens is *f*. The visual angle α' now subtended at the eye is then h/f if the eye is close to the lens, and hence the angular magnification, M, is given by

$$M = \frac{\alpha'}{\alpha} = \frac{h/f}{h/D},$$

as $\alpha = h/D$, Fig. 20.10.

$$\therefore M = \frac{D}{f} \qquad \qquad . \quad . \quad . \quad . \quad . \quad (3)$$

When $f = +5$ cm and $D = 25$ cm, $M = 5$. The angular magnification was 6 when the image was formed at the near point (p. 393). It can easily be verified that the angular magnification varies between 5 and 6 when the image is formed between infinity and the near point. The maximum angular magnification is thus obtained when the image is at the near point.

Compound Microscope

From the formula $M = -\left(\dfrac{D}{f} + 1\right)$, M is greater numerically the smaller the focal length of the lens. As it is impracticable to decrease f beyond a certain limit, owing to the mechanical difficulties of grinding a lens of short focal length (great curvature), *two* separated lenses are used to obtain a high angular magnification, and constitute a *compound* microscope. The lens nearer to the object is called the *objective*; the lens through which the final image is viewed is called the *eyepiece*. The objective and the eyepiece are both converging, and both have small focal lengths for a reason explained later (p. 396).

When the microscope is used, the object O is placed at a slightly *greater* distance from the objective than its focal length. In Fig. 20.14, F_o is the focus of this lens. An inverted real image is then formed at I_1 in the microscope tube, and the eyepiece is adjusted so that a large virtual image is formed by it at I_2. Thus I_1 is *nearer* to the eyepiece than the focus F_e of this lens. It can now be seen that the eyepiece functions as a simple magnifying glass, used for viewing the image formed at I_1 by the objective.

To draw the final image I_2, we first draw construction lines from the top of I_1 to the eyepiece as shown in Fig. 20.14. The actual rays, shown by heavy lines in Fig. 20.14, can then be drawn as we explained for the telescope on p. 386, to which the reader should refer.

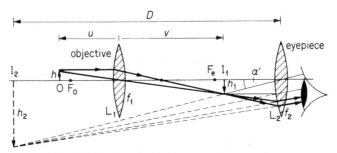

Fig. 20.14 Compound microscope in *normal* use

Figure 20.14 illustrates only the basic principle of a compound microscope. The single lens objective shown would produce a real image of the object which is coloured (see *chromatic aberration*, p. 378). The single lens eyepiece would produce a virtual image fairly free of colour (p. 394). In practice, both the objective and eyepiece of microscopes (and telescopes) are made of several lenses which together reduce chromatic aberration as well as spherical aberration (see p. 379).

The best position for the eye is at *the image of the objective in the eyepiece* or *eye-ring*. All the rays from the object pass through this image. See p. 385. Suppose the objective is 16 cm from L_2, which has a focal length of 2 cm. The image distance, v, in L_2 is given by $\dfrac{1}{v} + \dfrac{1}{(+16)} = \dfrac{1}{(+2)}$, from which $v = 2{\cdot}3$ cm. Thus the eye-ring is a short distance from the eyepiece, and in practice the eye should be farther from the eyepiece than in Fig. 20.14. This is arranged in commercial microscopes by having a circular opening fixed at the eye-ring distance from the eyepiece, so that the observer's eye has automatically the best position when it is placed close to the opening.

Angular Magnification with Microscope in Normal Use

When the microscope is in normal use the image at I_2 is formed at the least distance of distinct vision, D, from the eye (p. 392). Suppose that the eye is close to the eyepiece, as shown in Fig. 20.14. The visual angle α' subtended by the image at I_2 is then given by $\alpha' = h_2/D$, where h_2 is the height of the image. With the unaided eye, the object subtends a visual angle given by $\alpha = h/D$, where h is the height of the object, see Fig. 20.10.

$$\therefore \text{ angular magnification, } M = \frac{\alpha'}{\alpha}$$

$$= \frac{h_2/D}{h/D} = \frac{h_2}{h}.$$

Now $\dfrac{h_2}{h}$ can be written as $\dfrac{h_2}{h_1} \times \dfrac{h_1}{h}$, where h_1 is the length of the intermediate image formed at I_1.

$$\therefore M = \frac{h_2}{h_1} . \frac{h_1}{h} \qquad \text{.} \qquad \text{.} \qquad \text{.} \qquad \text{.} \qquad \text{(i)}$$

The ratio h_2/h_1 is the linear magnification of the 'object' at I_1 produced by the *eyepiece*, and we have shown on p. 356 that the linear magnification is also given by $v/f_2 - 1$, where v is the image distance from the lens and f_2 is the focal length. Since $v = -D$ where D is the numerical value of the least distance of distinct vision, it follows that

$$\frac{h_2}{h_1} = \frac{D}{f_2} - 1 = -\left(\frac{D}{f_2} + 1\right) \qquad \text{.} \qquad \text{.} \qquad \text{.} \qquad \text{(ii)}$$

Also, the ratio h_1/h is the linear magnification of the object at O produced by the *objective* lens. Thus if the distance of the image I_1 from this lens is denoted by v, we have

$$\frac{h_1}{h} = \frac{v}{f_1} - 1 \qquad \text{.} \qquad \text{.} \qquad \text{.} \qquad \text{.} \qquad \text{(iii)}$$

$$\therefore M = \frac{h_2}{h_1} . \frac{h_1}{h} = -\left(\frac{D}{f_2} + 1\right)\left(\frac{v}{f_1} - 1\right) \qquad \text{.} \qquad \text{.} \qquad \text{(4)}$$

It can be seen that if f_1 and f_2 are small, M is large. Thus the angular magnification is high if the focal lengths of the objective and the eyepiece are small.

Example

What do you understand by (a) the apparent size of an object, and (b) the magnifying power of a microscope? A model of a compound microscope is made up of two converging lenses of 3 and 9 cm focal length at a fixed separation of 24 cm. Where must the object be placed so that the final image may be at infinity? What will be the magnifying power if the microscope as thus arranged is used by a person whose nearest distance of distinct vision is 25 cm? State what is the best position for the observer's eye and explain why. (L.)

(i) Suppose the objective A is 3 cm focal length, and the eyepiece B is 9 cm focal length, Fig. 20.15. If the final image is at infinity, the image I_1 in the objective must be 9 cm from B, the focal length of the eyepiece. See p. 394, thus the image distance LI_1, from the objective A $= 24 - 9 = 15$ cm. The object distance OL is thus given by

$$\frac{1}{(+15)} + \frac{1}{u} = \frac{1}{(+3)},$$

from which $\qquad\qquad u = \text{OL} = 3\tfrac{3}{4}$ cm.

(ii) The angle α' subtended at the observer's eye is given by $\alpha' = h_1/9$, where h_1 is the height of the image at I_1, Fig. 20.15. Without the lenses, the object subtends an angle α at the eye given by $\alpha = h/25$, where h is the height of the object, since the least distance of distinct vision is 25 cm.

$$\therefore \text{ magnifying power } M = \frac{\alpha'}{\alpha} = \frac{h_1/9}{h/25} = \frac{25}{9} \times \frac{h_1}{h}.$$

But $\qquad\qquad\qquad \dfrac{h_1}{h} = \dfrac{LI_1}{LO} = \dfrac{15}{3\tfrac{3}{4}} = 4$

$$\therefore M = \frac{25}{9} \times 4 = 11 \cdot 1$$

The best position of the eye is at the eye-ring, which is the image of the objective A in the eyepiece B (p. 385).

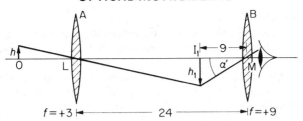

Fig. 20.15 Example

Spherical Aberration

If a single lens was used for the objective or eyepiece of a telescope or microscope, the image produced would have *chromatic aberration* or defect due to colour. Achromatic lenses, which reduce the colour effect considerably, are therefore used in optical instruments (p. 378).

We have now to consider another defect of an image due to a single lens, known as *spherical aberration*.

The lens formula $1/v + 1/u = 1/f$ has been obtained by considering a narrow beam of rays incident on the central portion of a lens. In this case the angles of incidence and refraction at the surfaces of the lens are small, and sin i and sin r can then be replaced respectively by i and r in radians, as shown on p. 353. This leads to the lens formula and a unique focus, F. If a *wide* parallel beam of light is incident on the lens, however, experiment shows that the rays are not all brought to the same focus, Fig. 20.16. It therefore follows that the image of an object is distorted if a wide beam of light falls on the lens, and this is known as *spherical aberration*. The aberration may be reduced by surrounding the lens with an opaque disc having a hole in the middle, so that light is incident only on the middle of the lens, but this method reduces the brightness of the image since it reduces the amount of light energy passing through the lens.

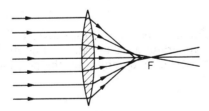

Fig. 20.16 Spherical aberration

As rays converge to a single focus for small angles of incidence, spherical aberration can be diminished if the angles of incidence on the lens' surfaces are diminished. In general, then, the *deviation* of the light by a lens should be shared as equally as possible by its surfaces, as each angle of incidence would then be as small as possible. A practical method of reducing spherical aberration is to use *two* lenses, when four surfaces are obtained, and to share the deviation equally between the lenses. The lenses are usually plano-convex.

In the compound microscope, the slide or other object is placed close to the objective (p. 395). A large angle is then subtended by the object at the lens and so the angle of incidence of rays is large. Correction for spherical aberration is hence more important for the objective of this instrument than chromatic aberration; the reverse is the case for the objective of a refractor telescope. A compound microscope of good quality has several lenses which help to correct the aberrations.

Ramsden's Eyepiece

The eyepiece of a telescope should be designed to reduce chromatic and spherical aberration; in practice this can most conveniently be done by using *two* lenses as the eyepiece.

Ramsden's eyepiece consists of two plano-convex lenses of equal focal length f, the distance between them being $2f/3$, Fig. 20.17. The achromatic condition requires that the distance between the two lenses should be f, the average of the focal lengths. If the field lens F were at the focus of the eye lens E, however, E would magnify any dust on F, and vision would then be obscured. F is placed at a distance $f/4$ from the focus of the objective of the telescope, where the real image I is formed, in which case the image in F is formed at a distance from E equal to f, its focal length, and parallel rays emerge from E.

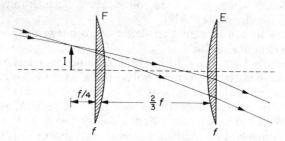

Fig. 20.17 Ramsden's eyepiece

The chromatic aberration of Ramsden's eyepiece is small and so is the spherical aberration. The advantage of the eyepiece, however, lies in the fact that crosswires can be used with it; they are placed outside the combination at the place where the real image I is formed.

Projector

The projector or projection lantern is used for showing slides on a screen. The *essential features* of the apparatus are illustrated in Fig. 20.18. S is a slide whose image is formed on the screen A by adjusting the position of an achromatic objective lens L.

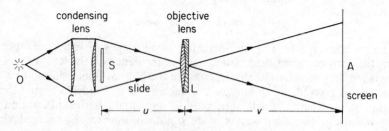

Fig. 20.18 Projector

The illumination of the slide must be as high as possible, otherwise the image of it on the screen is difficult to see clearly. For this purpose a very bright point source of light, O, is placed near a *condensing lens* C, and the slide S is placed immediately in front of C. The condensing lens consists of a plano-convex lens arrangement which concentrates the light energy from O in the direction of S, and it has a short focal length. The lens L and the source O are arranged to be conjugate foci for the lens C (i.e., the image of O is formed at L), in which case (i) all the light passes through L, and (ii) an image of O is not formed on the screen. Fig. 20.18 illustrates the path of the beam of light from O which illuminates the screen.

The linear magnification, m, of the slide is given by $m = \dfrac{v}{u}$, where v, u are the respective screen and slide distances from L. Now $\dfrac{v}{u} = \dfrac{v}{f} - 1$ (see p. 356). Thus the required high magnification is obtained by using an objective whose focal length is small compared with v.

Example

A projector is required to project slides 7·5 cm square onto a screen 4·2 m square. The distance between the slide and the screen is to be 10 m. What focal length of projection lens would you consider most suitable?

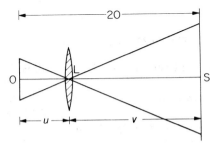

Fig. 20.19 Example

Suppose O is the slide, L is the projection lens, and S is the screen, Fig. 20.19. The linear magnification, m, due to the lens is given by

$$m = \frac{420 \text{ cm}}{7 \cdot 5 \text{ cm}} = 56$$

$$\therefore \text{LS} : \text{LO} = 56 : 1$$

$$\therefore \text{LS} = v = \frac{56}{57} \times 10 \text{ m}$$

and

$$\text{LO} = u = -\frac{1}{57} \times 10 \text{ m}.$$

Applying the lens equation,

$$\therefore \frac{1}{560/57} + \frac{1}{10/57} = \frac{1}{f}$$

from which

$$f = 0 \cdot 17 \text{ m (approx.)}.$$

Photographic Camera; f-number

The photographic camera consists essentially of a *lens system* L, a *light-sensitive film* F at the back, a *focusing* device for adjusting the distance of the lens from F, and an *exposure* arrangement which provides the correct exposure for a given lens aperture, Fig. 20.20. The lens system may contain an achromatic doublet and separated lenses which together reduce considerably chromatic and spherical aberration (p. 398). An *aperture* or *stop* of diameter d is provided

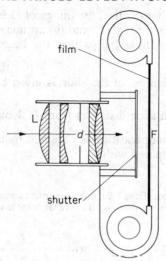

Fig. 20.20 Photographic camera

so that the light is incident centrally on the lens, thus diminishing distortion (p. 397).

The amount of luminous flux falling on the image in a camera is proportional to the area of the lens aperture, or to d^2, where d is the diameter of the aperture. The area of the image formed is proportional to f^2, where f is the focal length of the lens, since the length of the image formed is proportional to the focal length, as illustrated by Fig. 20.21. It therefore follows that the luminous flux

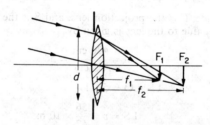

Fig. 20.21 Brightness of image

per unit area of the image, or *brightness B*, of the image, is proportional to d^2/f^2. The time of exposure, t, for activating the chemicals on the given negative is inversely proportional to B. Hence

$$t \propto \frac{f^2}{d^2} \qquad . \qquad . \qquad . \qquad . \qquad . \qquad \text{(i)}$$

The *relative aperture* of a lens is defined as the ratio d/f, where d is the diameter of the aperture and f is the focal length of the lens. The aperture is usually expressed by its *f-number*. If the aperture is *f*-4, this means that the diameter d of the aperture is $f/4$, where f is the focal length of the lens. An aperture of *f*-8 means a diameter d equal to $f/8$, which is a smaller aperture than $f/4$.

Since the time t of exposure is proportional to f^2/d^2, from (i) it follows that the exposure required with an aperture *f*-8 ($d = f/8$) is 16 times that required with an aperture f-2 ($d = f/2$). The f-numbers on a camera are 2, 2·8, 3·5, 4, 4·8, for example. On squaring the values of f/d for each number, we obtain 4, 8, 12, 16, 20, or 1, 2, 3, 4, 5, which are the relative exposure times.

Depth of Field

An object will not be seen by the eye until its image on the retina covers at least the area of a single cone, which transmits along the optic nerve light energy just sufficient to produce the sensation of vision. As a basis of calculation in photography, a circle of finite diameter about 0·25 mm viewed 250 mm away will just be seen by the eye as a fairly sharp point, and this is known as the *circle of least confusion*. It corresponds to an angle of about 1/1000th radian subtended by an object at the eye.

On account of the lack of resolution of the eye, a camera can take clear pictures of objects at different distances. Consider a point object O in front of a camera lens A which produces a point image I on a film, Fig. 20.22. If XY

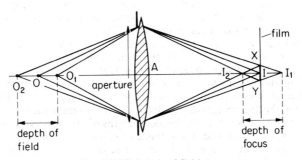

Fig. 20.22 Depth of field

represents the diameter of the circle of least confusion round I, the eye will see all points in the circle as reasonably sharp points. Now rays from the lens aperture to the edge of XY meet at I_1 beyond I, and also at I_2 in front of I. The point images I_1, I_2 correspond to point objects O_1, O_2 on either side of O, as shown. Consequently the images of all objects between O_1, O_2 are seen clearly on the film.

The distance O_1O_2 is therefore known as the *depth of field*. The distance I_1I_2 is known as the *depth of focus*. The depth of field depends on the lens aperture. If the aperture is made smaller, and the diameter XY of the circle of least confusion is unaltered, it can be seen from Fig. 20.22 that the depth of field increases. If the aperture is made larger, the depth of field decreases.

Exercises 20

Telescopes
1. A simple astronomical telescope in normal adjustment has an objective of focal length 100 cm and an eyepiece of focal length 5 cm.
(i) Where is the final image formed? (ii) Calculate the angular magnification. (iii) How would you increase the *resolving power* of the telescope?
2. Draw a ray diagram showing how the image of a distant star is formed at the least distance of distinct vision of an observer using a simple astronomical telescope. In your sketch show the principal focus of the two lenses.
The same telescope is now required to produce the image of the star on a photographic plate beyond the eyepiece. What adjustment is required? Draw a diagram to explain your answer.

11·25 cm

3. What is the *eye-ring* of a telescope? Draw a ray diagram showing how the eye-ring is formed in a simple astronomical telescope and explain why this telescope has a wide field of view.

Calculate the distance of the eye-ring from the eyepiece of a simple astronomical telescope in normal adjustment whose objective and eyepiece have focal lengths of 80 cm and 10 cm respectively.

4. Draw a sketch of a *reflector telescope* and show with a ray diagram how an observer sees the final image of a distant star.

State (i) the advantages of a reflector telescope over a refractor telescope, (ii) how the resolving power of the reflector telescope can be increased, (iii) the purpose of a radio reflector telescope.

5. Explain the term *angular magnification* as related to an optical instrument. Describe, with the aid of a ray diagram, the structure and action of an astronomical telescope. Derive an expression for its angular magnification when used so that the final image is at infinity. With such an instrument what is the best position for the observer's eye? Why is this the best position?

Even if the lenses in such an instrument are perfect it may not be possible to produce clear separate images of two points which are close together. Explain why this is so. Keeping the focal lengths of the lenses the same, what could be changed in order to make the separation of the images more possible? (*L.*)

0·0091 rad

6. A refracting telescope has an objective of focal length 1·0 m and an eyepiece of focal length 2·0 cm. A real image of the sun, 10 cm in diameter, is formed on a screen 24 cm from the eyepiece. What angle does the sun subtend at the objective? (*L.*)

7. Draw a diagram showing the passage of rays through a simple astronomical refracting telescope when it is used to view a distant extended object such as the moon, and is adjusted so that the final image is at infinity. Using the diagram, show how the magnifying power of the telescope is related to the focal lengths of the objective and eyepiece lenses.

The objective of a telescope has a diameter of 100 mm. Estimate the approximate angular separation of two stars which can just be resolved by the telescope.

What are the advantages of using a reflecting (rather than a refracting) objective in an astronomical telescope? (*O. & C.*)

8. Explain the essential features of the astronomical telescope. Define and deduce an expression for the magnifying power of this instrument.

A telescope is made of an object glass of focal length 20 cm and an eyepiece of 5 cm, both converging lenses. Find the magnifying power in accordance with your definition in the following cases: (*a*) when the eye is focused to receive parallel rays, and (*b*) when the eye sees the image situated at the nearest distance of distinct vision which may be taken as 25 cm. (*L.*)

4
4·8

9. An astronomical telescope may be constructed using as objective either (*a*) a converging lens, or (*b*) a concave mirror. Draw diagrams to illustrate the optical system of both types of telescope. Include in each diagram at least three rays reaching the instrument from an off-axial direction.

Define the magnifying power of a telescope. A telescope consists of two thin converging lenses of focal lengths 0·3 m and 0·03 m separated by 0·33 m. It is focused on the moon, which subtends an angle of 0·5° at the objective. Starting from first principles, find the angle subtended at the observer's eye by the image of the moon formed by the instrument.

Explain why one would expect this image to be coloured. Suggest how this defect might be rectified. (*O. & C.*)

5°

10. What is the *eye-ring* of a telescope?

For an astronomical telescope in normal adjustment deduce expressions for the size and position of the eye-ring in terms of the diameter of the object glass and the focal lengths of the object glass and eye-lens.

Discuss the importance of (i) the magnitude of the diameter of the object glass, (ii) the structure of the object glass, (iii) the position of the eye. (*L.*)

11. Show, by means of a ray diagram, how an image of a distant extended object is formed by an astronomical refracting telescope in normal adjustment (i.e. with the final image at infinity).

A telescope objective has focal length 96 cm and diameter 12 cm. Calculate the focal length and minimum diameter of a simple eyepiece lens for use with the telescope, if the magnifying power required is × 24, and all the light transmitted by the objective from a distant point on the telescope axis is to fall on the eyepiece. Derive any formulae you use. (*O. & C.*)

12. An astronomical telescope consisting of an objective focal length 60 cm and an eyepiece of focal length 3 cm is focused on the moon so that the final image is formed at the minimum distance of distinct vision (25 cm) from the eyepiece. Assuming that the diameter of the moon subtends an angle of $\frac{1}{2}°$ at the objective, calculate (*a*) the angular magnification, (*b*) the actual size of the image seen.

22·4

4·95

How, with the same lenses, could an image of the moon, 10 cm in diameter, be formed on a photographic plate? (*C.*)

13. Explain, with the aid of a ray diagram, how a simple astronomical telescope employing two converging lenses may form an apparently enlarged image of a distant extended object. State with reasons where the eye should be placed to observe the image.

A telescope constructed from two converging lenses, one of focal length 250 cm, the other of focal length 2 cm, is used to observe a planet which subtends an angle of 5×10^{-5} radian. Explain how these lenses would be placed for normal adjustment and calculate the angle subtended at the eye of the observer by the final image. $6·25 \times 10^{-3}$

How would you expect the performance of this telescope for observing a star to compare with one using a concave mirror as objective instead of a lens, assuming that the mirror had the same diameter and focal length as the lens. (*O. & C.*)

Microscopes

14. A converging lens of focal length 5 cm is used as a magnifying glass. If the near point of the observer is 25 cm from the eye and the lens is held close to the eye, calculate (i) the distance of the object from the lens, (ii) the angular magnification.

4 2 cm

6

What is the angular magnification when the final image is formed at infinity?

5

15. Explain what is meant by the magnifying power of a magnifying glass.

Derive expressions for the magnifying power of a magnifying glass when the image is (*a*) 25 cm from the eye and (*b*) at infinity. In each case draw the appropriate ray diagram. (*N.*)

16. Draw a labelled ray diagram to illustrate the action of a compound microscope.

State and explain how you would arrange simple converging lenses, one of focal length 2 cm and one of focal length 5 cm, to act as a compound microscope with magnifying power (angular magnification) × 42, the final image being 25 cm from the eye lens.

Assume that the focal lengths quoted, and your calculations, relate to the image formed when the object is illuminated by monochromatic red light. Without further calculation, state and explain the changes in the position and size of the image formed by the objective and in the apparent size of the final image (i.e. the angle it subtends at the centre of the eyelens) which would occur on changing the illumination of the object through the spectral range from red to violet, the setting of the microscope remaining unchanged. (*O. & C.*)

17. A point object is placed on the axis of, and 3·6 cm from, a thin converging lens of focal length 3·0 cm. A second thin converging lens of focal length 16·0 cm is placed coaxial with the first and 26·0 cm from it on the side remote from the object. Find the position of the final image produced by the two lenses.

16

Why is this not a suitable arrangement for a compound microscope used by an observer with normal eyesight?

For such an observer wishing to use the two lenses as a compound microscope with the eye close to the second lens decide, by means of a suitable calculation, where the second lens must be placed relative to the first. (*N.*)

18. Draw the path of two rays, from a point on an object, passing through the optical system of a compound microscope to the final image as seen by the eye.

If the final image formed coincides with the object, and is at the least distance of distinct vision (25 cm) when the object is 4 cm from the objective, calculate the focal lengths of the objective and eye lenses, assuming that the magnifying power of the microscope is 14. (*L.*)

19. Give a detailed description of the optical system of the compound microscope,

8·7cm
46·7

explaining the problems which arise in the design of an objective lens for a microscope.

A compound microscope has lenses of focal length 1 cm and 3 cm. An object is placed 1·2 cm from the object lens; if a virtual image is formed 25 cm from the eye, calculate the separation of the lenses and the magnification of the instrument. (*O. & C.*)

Other instruments

20. A projection lantern contains a condensing lens and a projection lens. Show clearly in a ray diagram the function of these lenses.

A lantern has a projection lens of focal length 25 cm and is required to be able to function when the distance from lantern to screen may vary from 6 m to 12 m. What range of movement for the lens must be provided in the focusing arrangement? What is the approximate value of the ratio of the magnifications at the two extreme distances?

21. Briefly describe an optical instrument which includes a reflecting prism. What is the function of the prism and what is the principle governing its action?

What are the advantages in using a prism rather than a silvered mirror in the apparatus you describe?

Parallel rays of light fall normally on the face *AC* of a total reflection prism *ABC* of refractive index 1·5 which has angle *A* exactly 45°, angle *B* approximately 90° and angle *C* approximately 45°. After total internal reflections in the prism two beams of parallel light emerge from the hypotenuse face, the angle between them being 60°. Calculate the value of angle *B*. You may assume that for small angles sin *i*/sin *r* equals *i*/*r*.

How would you discover whether the angle is more or less than 90°? (*O. & C.*)

22. Under certain conditions a suitable setting for a camera is: exposure time 1/125 second, aperture $f/5.6$. If the aperture is changed to $f/16$ what would be the new exposure time in order to achieve the same film image density? What other effect would this change in *f*-number produce? (*L.*)

23. (*a*) Define (i) linear magnification and (ii) magnifying power. Why is the latter appropriate in considering optical instruments such as telescopes?

(*b*) Draw a ray diagram to illustrate the action of an astronomical telescope consisting of two convex (converging) lenses, the instrument being in normal adjustment. On your diagram indicate the positions of the principal foci of the lenses.

(*c*) When a single convex lens is used as a magnifying glass, and the eye is placed close to the lens, the angles subtended by object and image are approximately the same. This being the case, explain why the magnifying glass produces a magnified image.

When white light is refracted on passing through a lens it undergoes dispersion and each colour produces a separate image. Why, then, is a series of coloured images not observed when the eye is placed close to a magnifying glass?

(*d*) A camera is set at $f 5.6$, 1/120 s. If the aperture is changed to $f 16$, to what value should the exposure time be set to achieve the same exposure? What other effect would the change of aperture have? (*AEB*, 1982)

Part Four
Waves, Wave Optics, Sound

21 Oscillations and Waves

In this chapter we shall study the properties of oscillations and waves in general. Topics in waves which concern particular branches of the subject are discussed elsewhere in this book. We begin with a summary of the results relating to simple harmonic motion already derived (p. 65).

S.H.M.

Simple harmonic motion (s.h.m.) occurs when the force acting on an object or system is directly proportional to its displacement x from a fixed point and is always directed towards this point. If the object executes s.h.m., then the variation of the displacement x with time t can be written as

$$x = a \sin \omega t \qquad . \qquad . \qquad . \qquad . \qquad . \qquad (1)$$

Here a is the greatest displacement from the mean or equilibrium position; a is the *amplitude*, Fig. 21.1. The constant $\omega = 2\pi f$ where f is the *frequency* of vibration or number of cycles per second. The period T of the motion, or time to undergo one complete cycle, is equal to $1/f$, so that $\omega = 2\pi/T$.

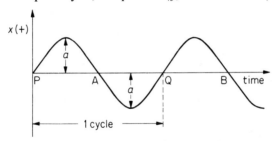

Fig. 21.1 Sine curve

The small oscillation of a pendulum bob or vibrating layer of air is a *mechanical oscillation*, so that x is a displacement from a mean fixed position. Later, *electrical oscillations* are considered; x may then represent the instantaneous charge on the plates of a capacitor when the charge alternates about a mean value of zero. In an *electromagnetic wave*, x may represent the component of the electric or magnetic field vectors at a particular place.

Energy in S.H.M.

On p. 67, it was shown that the sum of the potential and kinetic energies of a body moving with s.h.m. is *constant* and equal to the total energy in the vibration. Further, it was shown that the time averages of the potential energy (p.e.) and kinetic energy (k.e.) are equal; each is half the total energy. In any mechanical oscillation, *there is a continuous interchange or exchange of energy from p.e. to k.e. and back again.*

For vibrations to occur, therefore, an agency is required which can possess and store p.e. and another which can possess and store k.e. This was the case for a mass oscillating on the end of a spring, as we saw on p. 67. The mass stores k.e. and the spring stores p.e.; and interchange occurs continuously from one to the other as the spring is compressed and released alternately. In the oscillations of a simple pendulum, the mass stores k.e. as it swings downwards from the end of an oscillation, and this is changed to p.e. as the height of the bob increases above its mean position.

Note that some agency is needed to accomplish the transfer of energy. In the case of the mass and spring, the force in the spring causes the transfer. In the case of the pendulum, the component of the weight along the arc of the circle causes the change from p.e. to k.e.

Electrical Oscillations

So far we have dealt with mechanical oscillations and energy. The energy in electrical oscillations takes a different form. There are still two types of energy. One is the energy stored in the electric field, and the other that stored in the magnetic field. To obtain electrical oscillations, an inductor (a coil) is used to produce the magnetic field and a capacitor to produce the electric field (see also p. 834).

Suppose the capacitor is charged and there is no current at this moment, Fig. 21.2 (i). A p.d. then exists across the capacitor and an electric field is present between the plates. At this instant all the energy is stored in the electric field, and since the current is zero there is no magnetic energy. Because of the p.d. a current will begin to flow and magnetic energy will begin to be stored in the inductor. Thus there will be a change from electric to magnetic energy. The p.d. is the agency which causes the transfer of energy.

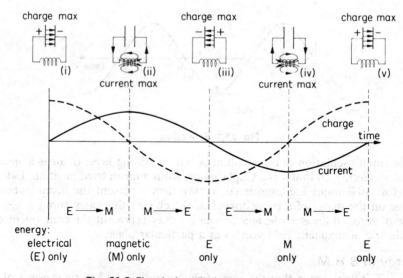

Fig. 21.2 Electrical oscillations—energy exchanges

One quarter of a cycle later the capacitor will be fully discharged and the current will be at its greatest, so that the energy is now entirely stored in the magnetic field, Fig. 21.2 (ii). The current continues to flow for a further quarter-cycle until the capacitor is fully charged in the opposite direction, when the energy is again completely stored in the electric field, Fig. 21.2 (iii). The current then reverses and the processes occur in reverse order, Fig. 21.2 (iv), after which the original state is restored and a complete oscillation has taken place, Fig. 21.2 (v). The whole process then repeats, giving continuous oscillations.

Phase of Vibrations

Consider an oscillation given by $x_1 = a \sin \omega t$. Suppose a second oscillation has the same amplitude, a, and angular frequency, ω, but reaches the end of its oscillation a fraction, β, of the period T later than the first one. The second oscillation thus *lags behind* the first by a time βT, and so its displacement x_2 is given by

$$x_2 = a \sin \omega(t - \beta T)$$
$$= a \sin (\omega t - \varphi), \qquad . \quad . \quad . \quad . \quad (2)$$

where $\varphi = \omega \beta T = 2\pi \beta T/T = 2\pi \beta$. If the second oscillation *leads* the first by a time βT, the displacement is given by

$$x_2 = a \sin (\omega t + \varphi) \qquad . \quad . \quad . \quad . \quad (3)$$

φ is known as the *phase angle* of the oscillation. It represents the *phase difference* between the oscillations $x_1 = a \sin \omega t$ and $x_2 = a \sin (\omega t - \varphi)$. Graphs of displacement against time are in Fig. 21.3. Curve 1 represents $x_1 = a \sin \omega t$. Curve 2 represents $x_2 = a \sin (\omega t + \pi/2)$, so that its phase lead

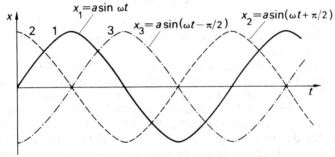

Fig. 21.3 Phase difference

is $\pi/2$; this is a lead of one quarter of a period. Curve 3 represents $x_3 = a \sin (\omega t - \pi/2)$ so that its phase lag is $\pi/2$; this is a lag of one quarter of a period on curve 1. If the phase difference is 2π, the oscillations are effectively in phase.

Note that if the phase difference is π, the displacement of one oscillation reaches a positive maximum value at the same instant as the other oscillation reaches a *negative* maximum value. The two oscillations are thus sometimes said to be 'antiphase'.

Damped Vibrations

In practice, the amplitude of vibration in simple harmonic motion does not remain constant but becomes progressively smaller. Such a vibration is said to be *damped*. The diminution of amplitude is due to loss of energy; for example, the amplitude of the bob of a simple pendulum diminishes slowly owing to the viscosity (friction) of the air. This is shown by curve 1 in Fig. 21.4.

The general behaviour of mechanical systems subject to various amounts of damping may be conveniently investigated using a coil of a ballistic galvanometer (p. 749). If a resistor is connected to the terminals of a ballistic galvanometer when the coil is swinging, the induced e.m.f. due to the motion of the coil in the magnetic field of the galvanometer magnet causes a current to flow through the resistor. This current, by Lenz's Law (p. 743), opposes the motion of the coil and so causes damping. The smaller the value of the resistor, the greater is the degree of damping. The galvanometer coil is set swinging by discharging a capacitor through it. The time period, and the time taken for

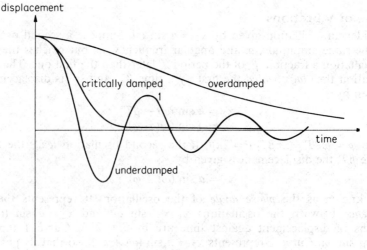

Fig. 21.4 Damped motion

the amplitude to be reduced to a certain fraction of its original value, are then measured. The experiment can then be repeated using different values of resistor connected to the terminals.

It is found that as the damping is increased the time period increases and the oscillations die away more quickly. As the damping is increased further there is a value of resistance which is just sufficient to prevent the coil from vibrating past its rest position. This degree of damping, called the *critical damping, reduces the motion to rest in the shortest possible time.* If the resistance is lowered further, to increase the damping, no vibrations occur but the coil takes a longer time to settle down to its rest position. Graphs showing the displacement against time for 'underdamped', 'critically damped', and 'overdamped' motion are shown in Fig. 21.4.

When it is required to use a galvanometer as a current-measuring instrument, rather than ballistically to measure charge, it is generally critically damped. The return to zero is then as rapid as possible.

These results, obtained for the vibrations of a damped galvanometer coil, are quite general. All vibrating systems have a certain critical damping, which brings the motion to rest in the shortest possible time.

Forced Oscillations. Resonance

In order to keep a system, which has a degree of damping, in continuous oscillatory motion, some external periodic force must be used. The frequency of this force is called the *forcing frequency*. In order to see how systems respond to a forcing oscillation, we may use an electrical circuit comprising a coil L, capacitor C and resistor R, shown in Fig. 21.5 (see also p. 784).

The applied oscillating voltage is displayed on the Y_2 plates of a double-beam oscilloscope (p. 816). The voltage across the resistor R is displayed on the Y_1 plates. Since the current I through the resistor is given by $I = V/R$, the voltage across R is a measure of the current through the circuit. The frequency of the oscillator is now set to a low value and the amplitude of the Y_1 display is recorded. The frequency is then increased slightly and the amplitude again measured. By taking many such readings, a graph can be drawn of the current through the circuit as the frequency is varied. A typical result is shown in Fig. 21.6 (i).

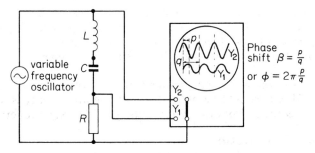

Fig. 21.5 Demonstration of oscillations

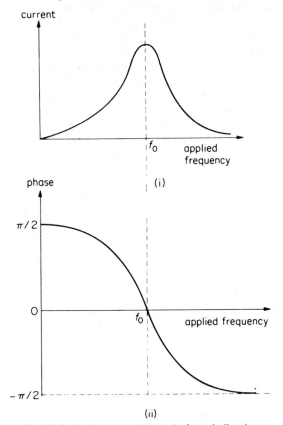

Fig. 21.6 Amplitude and phase in forced vibrations

The phase difference, φ, between the Y_1 and Y_2 displays can be found by measuring the horizontal shift p between the traces, and the length q occupied by one complete waveform. φ is given by $(p/q) \times 2\pi$. A graph of the variation of phase difference between current and applied voltage can then be drawn. Fig. 21.6 (ii) shows a typical curve.

The following observations may be made:

1. The current is greatest at a certain frequency f_0. This is the frequency of undamped oscillations of the system, when it is allowed to oscillate on its own. f_0 is called the *natural frequency* of the system. When the forcing frequency is equal to the natural frequency, *resonance* is said to occur. The largest current is then produced.

2. At resonance, the current and voltage are in phase. Well below resonance, the current leads the voltage by $\pi/2$; at very high frequencies the current lags by $\pi/2$. The behaviour of other resonant systems is similar.

3. The forced oscillations always have the same frequency as the forcing oscillations.

Examples of resonance occur in sound and in optics. These are discussed later (see p. 540). It should be noted that considerable energy is absorbed at the resonant frequency from the system supplying the external periodic force.

Waves and Wave-motion

A *wave* allows energy to be transferred from one point to another some distance away without any particles of the medium travelling between the two points. For example, if a small weight is suspended by a string, energy to move the weight may be obtained by repeatedly shaking the other end of the string up and down through a small distance. Waves, which carry energy, then travel along the string from the top to the bottom. Likewise, water waves may spread along the surface from one point A to another point B, where an object floating on the water will be disturbed by the wave. No particles of water at A actually travel to B in the process. The energy in the electromagnetic spectrum, comprising X-rays and light waves, for example, may be considered to be carried by electromagnetic waves from the radiating body to the absorber. Again, sound waves carry energy from the source to the ear by disturbance of the air (p. 413).

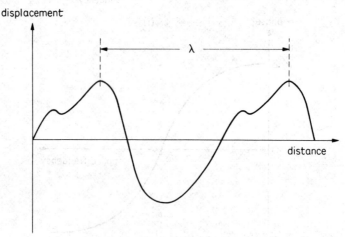

Fig. 21.7 Wave and wavelength

If the source or origin of the wave oscillates with a frequency f, then each point in the medium concerned oscillates with the same frequency. A snapshot of the wave profile or waveform may appear as in Fig. 21.7 at a particular instant. The source repeats its motion f times per second, so a repeating *waveform* is observed spreading out from it. The distance between corresponding points in successive waveforms, such as two successive crests or two successive troughs, is called the *wavelength*, λ. Each time the source vibrates once, the waveform moves forward a distance λ. Thus in one second, when f vibrations occur, the wave moves forward a distance $f\lambda$. Hence the velocity v of the waves, which is the distance the profile moves in one second, is given by:

$$v = f\lambda.$$

displacement

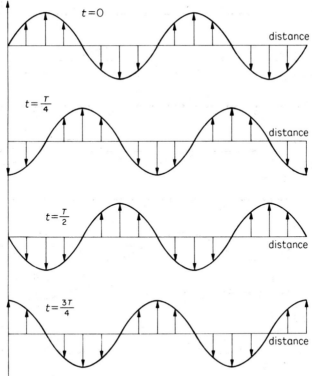

Fig. 21.8 Progressive transverse wave

This equation is true for all wave motion, whatever its origin, that is, it applies to sound waves, electromagnetic waves and mechanical waves.

Transverse Waves

A wave which is propagated by vibrations *perpendicular* to the direction of travel of the wave is called a *transverse* wave. Examples of transverse waves are waves on plucked strings and on water. Electromagnetic waves, which include light waves, are transverse waves.

The propagation of a transverse wave is illustrated in Fig. 21.8. Each particle vibrates perpendicular to the direction of propagation with the same amplitude and frequency, and the wave is shown successively at $t = 0$, $T/4$, $T/2$, $3T/4$, in Fig. 21.8, where T is the period.

Longitudinal Waves

In contrast to a transverse wave, a *longitudinal* wave is one in which the vibrations occur in the *same* direction as the direction of travel of the wave. Fig. 21.9 illustrates the propagation of a longitudinal wave. The row of dots shows the actual positions of the particles whereas the *graph* shows the *displacement* of the particles from their equilibrium positions. The positions at time $t = 0$, $t = T/4$, $t = T/2$ and $t = 3T/4$ are shown. The diagram for $t = T$ is, of course, the same as $t = 0$. With displacements to R (right) and to L (left), see graph axis, note that:

(i) The displacements of the particles cause regions of high density (*compressions* C) and of low density (*rarefactions* R) to be formed along the wave.

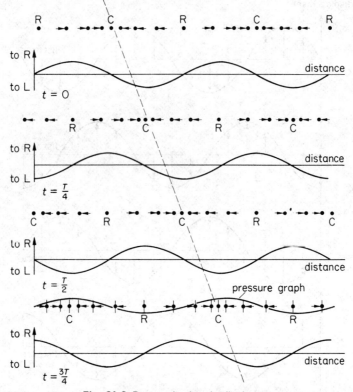

Fig. 21.9 Progressive longitudinal wave

(ii) These regions move along with the speed of the wave, as shown by the broken diagonal line.

(iii) Each particle vibrates about its mean position with the same amplitude and frequency.

(iv) The regions of greatest compression are one-quarter wavelength (90° phase) ahead of the greatest displacement in the direction of the wave. Compare the *pressure graph* at $t = 3T/4$ with the displacement graph. This result is useful in sound waves.

The most common example of a longitudinal wave is a sound wave. This is propagated by alternate compressions and rarefactions of the air.

Progressive Waves

Both the transverse and longitudinal waves described above are *progressive*. This means that the wave profile moves along with the speed of the wave. If a snapshot is taken of a progressive wave, it repeats at equal distances. The repeat distance is the *wavelength* λ. If one point is taken, and the profile is observed as it passes this point, then the profile is seen to repeat at equal intervals of time. The repeat time is the *period, T*.

The vibrations of the particles in a progressive wave are of the same amplitude and frequency. But *the phase of the vibrations changes for different points along the wave*. This can be seen by considering Figs. 21.8 and 21.9. The phase difference may be demonstrated by the following experiment, in which sound waves of the order of 1000 to 2000 Hz may be used.

An audio-frequency (af) oscillator is connected to the loudspeaker L and to

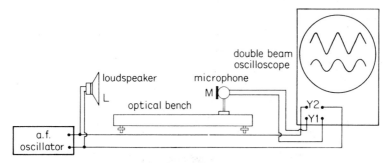

Fig. 21.10 Demonstration of phase in progressive wave

the Y_2 plates of a double-beam oscilloscope, Fig. 21.10. A microphone M, mounted on an optical bench, is connected to the Y_1 plates. When M is moved away from or towards L, the two traces on the screen are as shown in Fig. 21.11 (i) at one position. This occurs when the distance LM is equal to a whole number of wavelengths, so that the signal received by M is in phase with that sent out by L. When M is now moved further away from L through distance $\lambda/4$, where λ is the wavelength, the appearance on the screen changes to that shown in Fig. 21.11 (ii). The resultant phase change is $\pi/2$, so that the signal now arrives a quarter of a period later. When M is moved a distance $\lambda/2$ from its 'in-phase' position, the signal arrives half a period later, a phase change of π, Fig. 21.11 (iii).

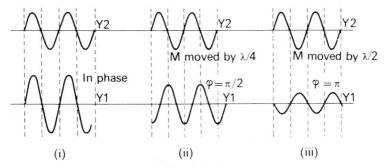

Fig. 21.11 Phase difference and wavelength

Velocity of Sound in Free Air

The velocity of sound in free air can be found from this experiment. Firstly, a position of the microphone M is obtained when the two signals on the screen are in phase, as in Fig. 21.10. The reading of the position of M on the optical bench is then taken. M is now moved slowly until the phase of the two signals on the screen is seen to change through $\pi/2$ to π and then to be in phase again. The shift of M is then measured. It is equal to λ, the wavelength. From several measurements the average value of λ is found, and the velocity of sound is calculated from $v = f\lambda$, where f is the frequency obtained from the oscillator dial. Another method for finding the velocity of sound in free air is given on p. 427.

Progressive Wave Equation

An equation can be formed to represent generally the displacement y of a vibrating particle in a medium which a *wave* passes. Suppose the wave moves

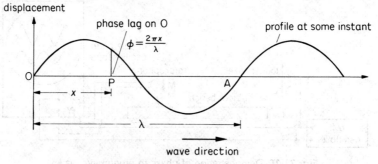

Fig. 21.12 Progressive wave equation

from left to right and that a particle at the origin O then vibrates according to the equation $y = a \sin \omega t$, where t is the time and $\omega = 2\pi f$ (p. 409).

At a particle P at a distance x from O to the right, the phase of the vibration will be different from that at O, Fig. 21.12. A distance λ from O corresponds to a phase difference of 2π (p. 411). Thus the phase difference φ at P is given by $(x/\lambda) \times 2\pi$ or $2\pi x/\lambda$. Hence the displacement of any particle at a distance x from the origin is given by

$$y = a \sin (\omega t - \varphi)$$

or
$$y = a \sin \left(\omega t - \frac{2\pi x}{\lambda} \right) \qquad \qquad (4)$$

Since $\omega = 2\pi f = 2\pi v/\lambda$, where v is the velocity of the wave, this equation may be written:

$$y = a \sin \left(\frac{2\pi v t}{\lambda} - \frac{2\pi x}{\lambda} \right)$$

or
$$y = a \sin \frac{2\pi}{\lambda} (vt - x) \qquad \qquad (5)$$

Also, since $\omega = 2\pi/T$, equation (4) may be written:

$$y = a \sin 2\pi \left(\frac{t}{T} - \frac{x}{\lambda} \right) \qquad \qquad (6)$$

Equations (5) or (6) represent a *plane-progressive wave*. The negative sign in the bracket indicates that, since the wave moves from left to right, the vibrations at points such as P to the right of O will lag on that at O. A wave travelling in the *opposite direction*, from right to left, arrives at P before O. Thus the vibration at P leads that at O. Consequently a wave travelling in the opposite direction is given by

$$y = a \sin 2\pi \left(\frac{t}{T} + \frac{x}{\lambda} \right), \qquad \qquad (7)$$

that is, the sign in the bracket is now a plus sign.

As an illustration of calculating the constants of a wave, suppose a wave is represented by

$$y = a \sin \left(2000\pi t - \frac{\pi x}{17} \right),$$

where t is in seconds, y in cm. Then, comparing it with equation (5),

$$y = a \sin \frac{2\pi}{\lambda}(vt - x),$$

we have
$$\frac{2\pi v}{\lambda} = 2000\pi \text{ and } \frac{2\pi}{\lambda} = \frac{\pi}{17}.$$

$$\therefore \lambda = 2 \times 17 = 34 \text{ cm}$$

and
$$v = 1000\lambda = 1000 \times 34 = 34000 \text{ cm s}^{-1}$$

$$\therefore \text{ frequency, } f, = \frac{v}{\lambda} = \frac{34000}{34} = 1000 \text{ Hz}$$

$$\therefore \text{ period, } T, = \frac{1}{f} = \frac{1}{1000} \text{ s.}$$

If two layers of the wave are 180 cm apart, they are separated by 180/34 wavelengths, or by $5\frac{10}{34}\lambda$. Their *phase difference* for a separation λ is 2π; and hence, for a separation $10\lambda/34$, omitting 5λ from consideration, we have:

$$\text{phase difference} = \frac{10}{34} \times 2\pi = \frac{10\pi}{17} \text{ radians.}$$

Since y represents the displacement of a particle as the wave travels, the *velocity* v of the particle at any instant is given by dy/dt. From equation (6),

$$v = \frac{dy}{dt} = \frac{2\pi a}{T} \cos 2\pi \left(\frac{t}{T} - \frac{x}{\lambda}\right).$$

So the graph of v against x is 90° out of phase with the graph of y against x. Some microphones are 'velocity' types, in the sense that the audio current produced is proportional to the velocity of the particles of air. Other microphones may be 'pressure' types—the audio current is here proportional to the pressure changes in the air.

Principle of Superposition
When two waves travel through a medium, their combined effect at any point can be found by the *Principle of Superposition*. This states that *the resultant displacement at any point is the sum of the separate displacements due to the two waves.*

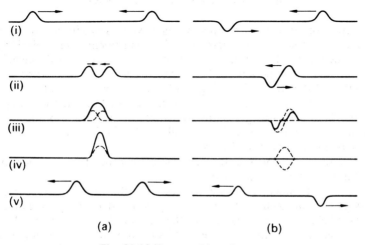

(a)   (b)

Fig. 21.13 Superposition of waves

The principle can be illustrated by means of a long stretched spring ('Slinky'). If wave pulses are produced at each end simultaneously, the two waves pass through the wire. Fig. 21.13(a) shows the stages which occur as the two pulses pass each other. In Fig. 21.13(a) (i), they are some distance apart and are approaching each other, and in Fig. 21.13(a) (ii) they are about to meet. In Fig. 21.13(a) (iii), the two pulses, each shown by broken lines, are partly overlapping. The resultant is the sum of the two curves. In Fig. 21.13(a) (iv), the two pulses exactly overlap and the greatest resultant is obtained. The last diagram shows the pulses receding from one another. The diagrams in Fig. 21.13(b) show the same sequence of events (i)–(v) but the pulses are equal and opposite. The Principle of Superposition is widely used in discussion of wave phenomena such as interference, as we shall see (p. 456).

Stationary Waves

We have already discussed progressive waves and their properties. Fig. 21.14 shows an apparatus which produces a different kind of wave (see also p. 531). If the weights on the scale-plan are suitably adjusted, a number of *stationary vibrating loops* are seen on the string when one end is set vibrating. This time the wave-like profile on the string does *not* move along the medium, which is the string, and the wave is therefore called a *stationary* (or *standing*) *wave*. In Fig. 21.14 a wave travelling along the string to one end is reflected here. So the stationary wave is due to the superposition of *two waves* of equal frequency and amplitude travelling in *opposite* directions along the string. This is discussed more fully on p. 531.

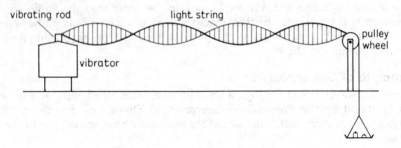

Fig. 21.14 Demonstration of stationary wave

The motion of the string when a stationary wave is produced can be studied by using a Xenon stroboscope (strobe). This instrument gives a flashing light whose frequency can be varied. The apparatus is set up in a darkened room and illuminated with the strobe. When the frequency of the strobe is nearly equal to that of the string, the string can be seen moving up and down slowly. Its observed frequency is equal to the difference between the frequency of the strobe and that of the string. Progressive stages in the motion of the string can now be seen and studied, and these are illustrated in Fig. 21.15.

The following points should be noted:

1. There are points such as B where the displacement is permanently zero. These points are called *nodes* of the stationary wave.

2. At points between successive nodes the vibrations *are in phase*. This property of the stationary wave is in sharp contrast to the progressive wave, where the phase of points near each other are all different (see p. 414). Thus when one point of a stationary wave is at its maximum displacement, *all* points are then at their maximum displacement. When a point (other than a node)

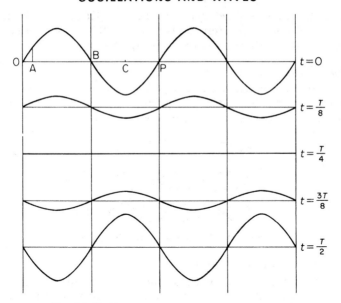

Fig. 21.15 Changes in motion of stationary wave

has zero displacement, *all* points then have zero displacement.

3. Each point along the wave has a *different amplitude* of vibration from neighbouring points. Again this is different from the case of a progressive wave, where every point vibrates with the same amplitude. Points, e.g. C, which have the greatest amplitude are called *antinodes*.

4. The wavelength is equal to the distance OP, Fig. 21.15. Thus the wavelength λ is twice the *distance between successive nodes or successive antinodes*. The distance between successive nodes or antinodes is $\lambda/2$; the distance between a node and a neighbouring antinode is $\lambda/4$.

Stationary Longitudinal Waves

In Sound, stationary longitudinal waves can be set up in a pipe closed at one end (closed pipe). We shall study this in more detail in a later chapter. Here we may note that all the possible frequencies obtained from the pipe are subject to the condition that the closed end must be a displacement node of the stationary wave formed, since the air cannot move here, and the open end must be a displacement antinode as the air is most free to move here. Figure 21.16 (i) shows the stationary wave formed for the lowest possible frequency f_0 and other possible or allowed frequencies $3f_0$ and $5f_0$.

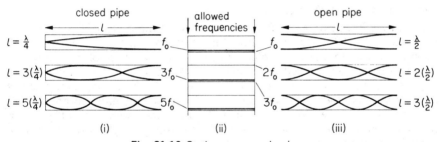

Fig. 21.16 Stationary waves in pipes

Fig. 21.16 (ii) shows the possible frequencies, f_0, $2f_0$, $3f_0$,..., for the case of the pipe open at both ends. Here the two ends of the pipes must be antinodes and so the possible stationary waves are those shown.

Stationary Transverse Waves

Fig. 21.17 shows the possible frequencies for stationary transverse waves produced by plucking in the middle a string fixed at both ends. Here the ends must always be displacement nodes and the middle an antinode.

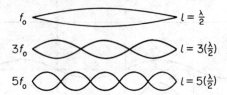

Fig. 21.17 Stationary waves in strings

Pressure in Stationary Wave

Consider the instant corresponding to curve 1 of the displacement graph of a stationary wave, Fig. 21.18 (i). At the node a, the particles on either side produce a compression (increase of pressure), from the direction of their displacement. At the same instant the pressure at the antinode b is normal and that at the node c is a rarefaction (decrease in pressure). Fig. 21.18 (ii) shows the pressure variation along the stationary wave—the displacement nodes are the pressure antinodes. So the closed end of the pipe in Fig. 21.16 (i) is a pressure antinode.

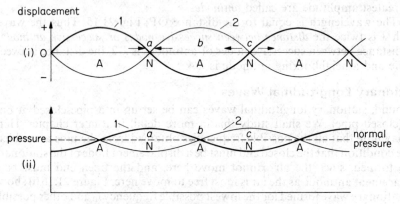

Fig. 21.18 Pressure variation due to stationary wave

Stationary Light Waves

Light waves have extremely short wavelengths of the order of 5×10^{-7} m or 0·0005 mm. In 1890 Wiener succeeded in detecting stationary light waves. He deposited a very thin photographic film, about one-twentieth of the wavelength of light, on glass and placed it in a position XY inclined at the extremely small angle of about 4′ to a plane mirror CD. Fig. 21.19 is an exaggerated sketch for clarity. When the mirror was illuminated normally by monochromatic light and the film was developed, bright and dark bands were seen. These were respectively antinodes A and nodes N of the stationary light waves formed by reflection at the mirror (see Fig. 21.20).

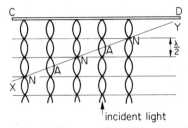

Fig. 21.19 Stationary light waves

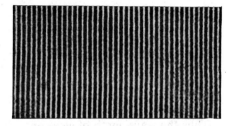

Fig. 21.20 Stationary light waves due to the mercury line of wavelength 546 nm

Stationary Waves in Aerials

Stationary waves, due to oscillating electrons, are produced in *aerials* tuned to incoming radio waves, or aerials transmitting radio waves. Figure 21.21 illustrates the stationary wave obtained on a vertical metal rod acting as a 'quarter-wave' aerial. Here the electrons cannot move at the top of the rod, so this is a current (I) node. The current antinode is near the other end of the rod. The current node corresponds to a voltage (V) antinode, as shown (compare 'displacement' and 'pressure' for the case of the closed pipe on p. 534).

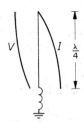

Fig. 21.21 Stationary waves in aerials

Stationary Waves in Electron Orbits

Moving electrons have wave properties (see p. 867). If we consider a circular orbit of the simplest atom, the hydrogen atom, there must be a complete number of such waves in the orbit for a stable atom; otherwise some of the waves or energy would be radiated as the electron rushed round the orbit and the atom would then lose its energy. So stationary waves are formed in the orbit. Fig. 21.22.

Thus if the radius is r and there are n waves of wavelength λ, we must have

$$2\pi r = n\lambda.$$

Now Bohr suggested that the angular momentum about the centre $= nh/2\pi$ (p. 853), where n is an integer and h is the Planck constant.

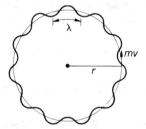

Fig. 21.22 Stationary waves in electron orbit

$$\therefore mv \times r = \frac{nh}{2\pi}.$$

From above,
$$r = \frac{n\lambda}{2\pi} = \frac{nh}{2\pi mv}$$

$$\therefore \lambda = \frac{h}{mv}.$$

Thus the wavelength of electrons with momentum mv is h/mv, as de Broglie first proposed (p. 868).

Stationary Wave Equation

In deriving the wave equation of a progressive wave, we used the fact that the phase changes from point to point (p. 416). In the case of a stationary wave, we may find the equation of motion by considering the *amplitude* of vibration at each point because the amplitude varies while the phase remains constant.

As we have seen, if ω is a constant, the vibration of each particle may be represented by the equation.

$$y = Y \sin \omega t, \qquad . \qquad . \qquad . \qquad . \qquad (8)$$

where Y is the amplitude of the vibration at the point considered. Y varies along the wave with the distance x from some origin. If we suppose the origin to be at an antinode, then the origin will have the greatest amplitude, A, say. Now the wave repeats at every distance λ, and it can be seen that the amplitudes at different points vary sinusoidally with their particular distance x. An equation representing the changing amplitude Y along the wave is thus:

$$Y = A \cos \frac{2\pi x}{\lambda} = A \cos kx, \qquad . \qquad . \qquad . \qquad (9)$$

where $k = 2\pi/\lambda$. When $x = 0$, $Y = A$; when $x = \lambda$, $Y = A$. When $x = \lambda/2$, $Y = -A$. This equation hence correctly describes the variation in amplitude along the wave, Fig. 21.15. Hence the equation of motion of a stationary wave is, with (8),

$$y = A \cos kx . \sin \omega t \qquad . \qquad . \qquad . \qquad . \qquad (10)$$

From (10), $y = 0$ at all times when $\cos kx = 0$. Thus $kx = \pi/2, 3\pi/2, 5\pi/2, \ldots$, in this case. This gives values of x corresponding to $\lambda/4, 3\lambda/4, 5\lambda/4, \ldots$ These points are *nodes* since the displacement at a node is always zero (p. 418). Thus equation (10) gives the correct distance, $\lambda/2$, between nodes.

A stationary wave can be considered as produced by the superposition of *two progressive waves, of the same amplitude and frequency, travelling in opposite directions*, as we now show.

Mathematical proof of stationary wave properties. The properties of the stationary wave just deduced can be obtained by a mathematical treatment.

Suppose $y_1 = a \sin 2\pi \left(\dfrac{t}{T} - \dfrac{x}{\lambda} \right)$ is a plane-progressive wave travelling in one

direction along the x-axis (p. 416). Then $y_2 = a \sin 2\pi \left(\dfrac{t}{T} + \dfrac{x}{\lambda} \right)$ represents a

wave of the same amplitude and frequency travelling in the opposite direction. The resultant displacement, y, is hence given by

$$y = y_1 + y_2 = a \left[\sin 2\pi \left(\frac{t}{T} - \frac{x}{\lambda} \right) + \sin 2\pi \left(\frac{t}{T} + \frac{x}{\lambda} \right) \right]$$

from which
$$y = 2a \sin \frac{2\pi t}{T} . \cos \frac{2\pi x}{\lambda} \qquad . \qquad . \qquad . \qquad (i)$$

using the transformation of the sum of two sine functions to a product.

$$\therefore y = Y \sin \frac{2\pi t}{T} \qquad . \qquad . \qquad . \qquad (ii)$$

where
$$Y = 2a \cos \frac{2\pi x}{\lambda} \qquad . \qquad . \qquad . \qquad (iii)$$

From (ii), Y is the magnitude of the *amplitude* of vibration of the various layers; and from (iii) it also follows that the amplitude is a maximum and equal to $2a$ at $x = 0$, $x = \lambda/2$, $x = \lambda$, and so on. These points are thus antinodes, and consecutive antinodes are hence separated by a distance $\lambda/2$. The amplitude Y is zero when $x = \lambda/4$, $x = 3\lambda/4$, $x = 5\lambda/4$, and so on. These points are thus nodes, and they are hence midway between consecutive antinodes.

Wave Properties. Reflection

Any wave motion can be *reflected*. The reflection of light waves, for example, is discussed on p. 314.

Like light waves, sound waves are reflected from a plane surface so that the angle of incidence is equal to the angle of reflection. This can be demonstrated by placing a tube T_1 in front of a plane surface AB and blowing a whistle gently at S, Fig. 21.23. Another tube T_2, directed towards N, is placed on the other side of the normal NQ, and moved until a sensitive microphone, connected to a cathode-ray oscilloscope, is considerably affected at R, showing that the reflected wave is in the direction NR. It will then be found that angle RNQ = angle SNQ.

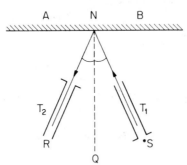

Fig. 21.23 Reflection of sound

It can also be shown that sound waves come to a focus when they are incident on a curved concave mirror. A surface shaped like a parabola reflects sound waves to long distances if the source of sound is placed at its focus (see also p. 320). The famous whispering gallery of St. Paul's is a circular-shaped chamber whose walls repeatedly reflected sound waves round the gallery, so that a person talking quietly at one end can be heard distinctly at the other end.

Electromagnetic waves of about 21 cm wavelength from outer space are now detected by radio-telescopes. The waves are reflected by a large parabolic 'dish' to a sensitive receiver (see p. 486). A demonstration of the reflection of 3 cm electromagnetic waves is shown on p. 429.

Refraction

Waves can also be *refracted*, that is, their direction changes when they enter a new medium. This is due to the change in velocity of the waves on entering a different medium. Refraction of light is discussed on p. 326.

Sound waves can be refracted as well as reflected. TYNDALL placed a watch in front of a balloon filled with carbon dioxide, which is heavier than air, and found that the sound was heard at a definite place on the other side of the balloon. The sound waves thus converged to a focus on the other side of the balloon, which therefore has the same effect on sound waves as a converging lens has on light waves (see p. 349). If the balloon is filled with hydrogen, which is lighter than air, the sound waves diverge on passing through the balloon. The latter thus acts similarly to a diverging lens when light waves are incident on it (see p. 349).

The refraction of sound explains why sounds are easier to hear at night than during day-time. In the day-time, the upper layers of air are colder than the layers near the earth. Now sound travels faster the higher the temperature (see 434), and sound waves are hence refracted in a direction away from the earth. The intensity of the sound waves thus diminishes. At night-time, however, the layers of air near the earth are colder than those higher up, and hence sound waves are now refracted towards the earth, with a consequent increase in intensity.

For a similar reason, a distant observer O hears a sound from a source S more easily when the wind is blowing towards him than away from him, Fig. 21.24. When the wind is blowing towards O, the bottom of the sound wavefront

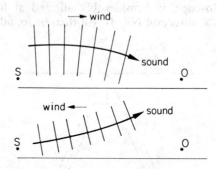

Fig. 21.24 Refraction of sound

is moving slower than the upper part, and hence the wavefronts turn towards the observer, who therefore hears the sound easily. When the wind is blowing in the opposite direction the reverse is the case, and the wavefronts turn upwards away from the ground and O. The sound intensity thus diminishes. This phenomenon shows how wavefronts may change direction due to variation in wind velocity.

Radio (electromagnetic) waves are refracted in the ionosphere high above the Earth when they are transmitted from one side of the globe to the other. A demonstration of the refraction of microwaves, 3 cm electromagnetic waves, is shown on p. 429.

Diffraction

Waves can also be 'diffracted'. *Diffraction* is the name given to the spreading of waves when they pass through apertures or around obstacles.

The general phenomenon of diffraction may be illustrated by using water waves in a ripple tank, with which we assume the reader is familiar. Fig. 21.25 (i) shows the effect of widening the aperture and Fig. 21.25 (ii) the effect of shortening the wavelength and keeping the same width of opening. In certain circumstances in diffraction, reinforcement of the waves, or complete cancellation occurs in particular directions from the aperture, as shown in Fig. 21.25 (i) and (ii). These patterns are called 'diffraction bands' (p. 481).

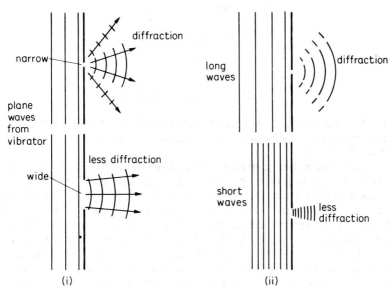

Fig. 21.25 Diffraction of waves

Generally, the smaller the width of the aperture in relation to the wavelength, the greater is the spreading or diffraction of the waves. This explains why we cannot see round corners. The wavelength of *light waves* is about 6×10^{-7} m (p. 462). This is so short that no appreciable diffraction is obtained around obstacles of normal size. With very small obstacles or narrow apertures, however, diffraction of light may be appreciable (see p. 481). *Electromagnetic waves* can be diffracted, as shown on p. 429.

Sound waves are diffracted round wide openings such as doorways because their wavelength is comparable with the width of the opening. For example, the wavelength for a frequency of, say, 680 Hz is about 0·5 m and the width of a door may be about 0·8 m. Generally, the diffraction increases with longer wavelength. For this reason, the low notes of a band marching away out of sight round a corner are heard for a longer time than the high notes. Similarly, the low notes of an orchestra playing in a hall can be heard through a doorway better than the high notes by a listener outside the hall.

Interference

When two or more waves of the same frequency overlap, the phenomenon of *interference* occurs. Interference is easily demonstrated in a ripple tank. Two

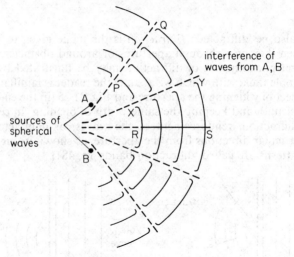

Fig. 21.26 Interference of waves

sources, A and B, of the same frequency are used. These produce circular waves which spread out and overlap, and the pattern seen on the water surface is shown in Fig. 21.26.

The interference pattern can be explained from the Principle of Superposition (p. 417). If the oscillations of A and B are in phase, crests from A will arrive at the same time as crests from B at any point on the line RS. Hence by the Principle of Superposition there will be reinforcement or a large wave along RS. Along XY, however, crests from A will arrive before corresponding crests from B. In fact, every point on XY is half a wavelength, λ, nearer to A than to B, so that crests from A arrive at the same time as troughs from B. Thus, by the Principle of Superposition, the resultant is *zero*. Generally, reinforcement (constructive interference) occurs at a point C when the path difference AC − BC

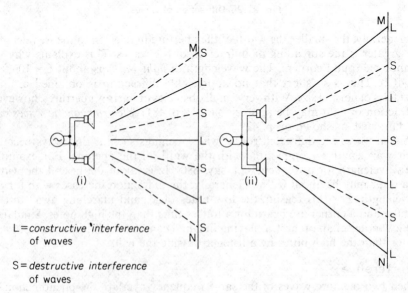

L = *constructive interference* of waves

S = *destructive interference* of waves

Fig. 21.27 Interference of sound waves

= 0 or λ or 2λ, and cancellation (destructive interference) when AC − BC = $\lambda/2$ or $3\lambda/2$ or $5\lambda/2$.

Interference of *light waves* is discussed in detail on p. 456. An experiment to demonstrate the interference of *electromagnetic waves* (microwaves) is given on p. 429. The interference of *sound waves* can be demonstrated by connecting two loudspeakers in parallel to an audio-frequency oscillator, Fig. 21.27 (i). As the ear or microphone is moved along the line MN, alternate loud (L) and soft (S) sounds are heard according to whether the receiver of sound is on a line of reinforcement (constructive interference) or cancellation (destructive inter-ference) of waves. Fig. 21.27 (i) indicates the positions of loud and soft sounds if the two speakers oscillate in phase. If the connections to *one* of the speakers is reversed, so that they oscillate out of phase, then the pattern is altered as shown in Fig. 21.27 (ii). The reader should try to account for this difference.

Example

Two small loudspeakers A, B, 1·00 m apart, are connected to the same oscillator so that both emit sound waves of frequency 1700 Hz in phase, Fig. 21.28. A sensitive detector, moving parallel to the line AB along PQ 2·40 m away, detects a maximum wave at P on the perpendicular bisector MP of AB and another maximum wave when it first reaches a point Q directly opposite to B.
  Calculate the speed c of the sound waves in air from these measurements.

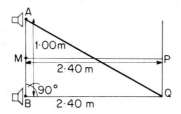

Fig. 21.28 Example

There is constructive interference of the sound waves at P and Q. Since Q is the first maximum after P, where AP = BP, it follows that

$$AQ - BQ = \lambda, \qquad . \qquad . \qquad . \qquad . \qquad (1)$$

where λ is the wavelength of the sound waves. Now BQ = 2·40 m, AB = 1·00 m and angle ABQ = 90°. So

$$AQ = \sqrt{BQ^2 + AB^2} = \sqrt{2·40^2 + 1·00^2} = 2·60 \text{ m}.$$

From (1), $\lambda = 2·60 - 2·40 = 0·20$ m.

So   wave speed $c = f\lambda = 1700 \times 0·20 = 340$ m s^{-1}.

Measurement of Velocity of Sound

Fig. 21.29 shows one method of measuring the velocity of sound in free air by an interference method.

Sound waves of constant frequency, such as 1500 Hz, travel from a loud-speaker L towards a vertical board M. Here the waves are reflected and interfere with the incident waves. As explained on p. 418, the two waves travelling in opposite directions produce a *stationary wave* between the board M and L.

A small microphone, positioned in front of the board, is connected to the Y-plates of an oscilloscope. As the microphone is moved back from M towards L, the amplitude of the waveform seen on the screen increases to a maximum at one position A, as shown. This is an antinode of the stationary wave. When

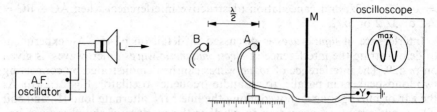

Fig. 21.29 Velocity of sound in free air—interference method

the microphone is moved on, the amplitude diminishes to a minimum (a node) and then increases to a maximum again at a position B, the next antinode. The distance between successive antinodes is $\lambda/2$ (p. 419). Thus by measuring the average distance d between successive maxima, the wavelength λ can be found. Knowing the frequency f of the note from the loudspeaker, the velocity v of the sound wave can be calculated from $v = f\lambda$.

Velocity of Sound by Lissajous' Figures

The velocity of sound can be measured by a different method using an oscilloscope. In this case, the loudspeaker L is connected to the X-plates of the oscilloscope and the microphone A to the Y-plates. The board M in Fig. 21.29 is completely removed.

When A faces L, the oscilloscope beam is affected (a) horizontally by a voltage V_L due to the sound waves from L and (b) vertically by a voltage V_A due to sound waves arriving at A. These two sets of waves together produce a resultant waveform on the screen, whose geometrical form is called a *Lissajous figure*.

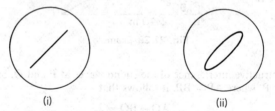

(i)   (ii)

Fig. 21.30 Lissajous' figures

If V_L and V_A are exactly in phase, a straight inclined line is seen on the screen. Fig. 21.30 (i). If V_L and V_A are out of phase, an ellipse is seen, Fig. 21.30 (ii). By moving the microphone, the average distance between successive positions when a sloping straight line appears on the screen can be measured. This distance is λ, the wavelength. The velocity of sound V is then calculated from $V = f\lambda$, where f is the known frequency of the sound. This method is more accurate than the method described earlier, as the Lissajous' straight line position of the microphone can be found with far greater accuracy than the maxima (or minima) positions required in Fig. 21.29.

Wave Properties of Electromagnetic Waves

Electromagnetic waves, like all waves, can undergo reflection, refraction, interference and diffraction. In laboratory demonstrations, *microwaves* of about 3 cm wavelength may be used. These are radiated from a horn waveguide T and are received by a similar waveguide R or by a smaller *probe* X. The detected wave then produces a deflection in a connected meter. Some experiments which can be performed in a school laboratory are illustrated in Fig. 21.31 (i)–(v).

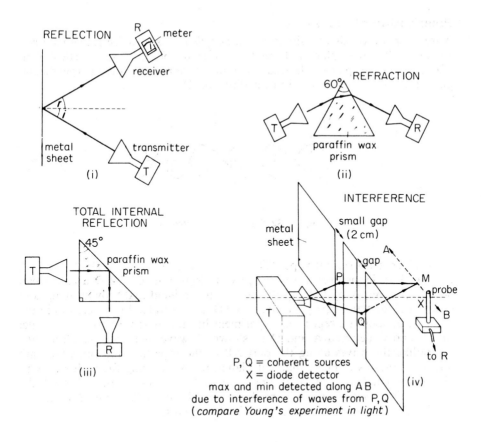

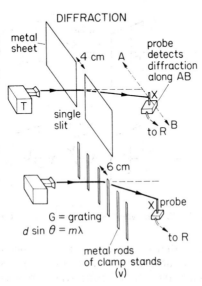

Fig. 21.31 Experiments with microwaves: (i) reflection; (ii) refraction; (iii) total internal reflection; (iv) interference; (v) diffraction

Polarization of Waves

A transverse wave due to vibrations in *one plane* is said to be *plane-polarized*. Figure 21.32 shows a plane-polarized wave due to vibrations in the vertical plane yOx and another plane-polarized wave due to vibrations in the perpendicular plane zOx. Both waves travel in the direction Ox.

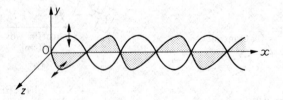

Fig. 21.32 Plane-polarized waves

Consider a horizontal rope AD attached to a fixed point D at one end, Fig. 21.33 (i). Transverse waves due to vibrations in many different planes can be set up along A by holding the end A in the hand and moving it up and down in all directions perpendicular to AD, as illustrated by the arrows in the plane X. Suppose we repeat the experiment but this time we have two parallel slits B and C between A and D as shown. A wave then emerges along BC, but unlike the waves along the part AB of the rope (not shown), which are due to vibrations in many different planes, the wave along BC is due only to vibrations parallel to the slit B. This plane-polarized wave passes through the parallel slit C. But when C is turned so that it is *perpendicular* to B, as shown in Fig. 21.33 (ii), no wave is now obtained beyond C.

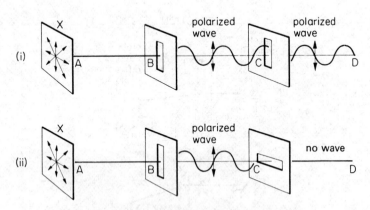

Fig. 21.33 Transverse waves and polarization

It is important to note that *no polarization can be obtained with longitudinal waves*. Fig. 21.33 illustrates how transverse waves can be distinguished by experiment from longitudinal waves. If the rope AD is replaced by a thick elastic cord, and longitudinal waves are produced along AD, then turning the slit C round from the position shown in Fig. 21.33 (i) to that shown in Fig. 21.32 (ii) makes no difference to the wave—it travels through B and C undisturbed.

Since sound waves are longitudinal waves, no polarization of sound waves can be produced. As we shall see in a later chapter, however, because light waves are transverse waves they can be polarized. The phenomena of interference and diffraction occur both with sound and light waves, but only the phenomenon of polarization can distinguish between waves which may be longitudinal or transverse.

Fig. 21.34 illustrates an experiment on polarization carried out with *electromagnetic waves*. Here a grille of parallel metal rods is rotated between (*a*) a source T of 3 cm electromagnetic waves or microwaves and a (*b*) detector, a probe, with a meter connected to it. When the rods are horizontal the meter reading is high, Fig. 21.34 (i). So a wave travels past the grille. When the grille is turned round so that the rods are vertical, there is no deflection in the meter, Fig. 21.34 (ii). Thus the wave does not travel past the grille. This experiment is analogous to that illustrated in Fig. 21.33 for mechanical waves travelling along a rope. It shows that the electromagnetic waves produced by T are plane-polarized and so they are transverse waves.

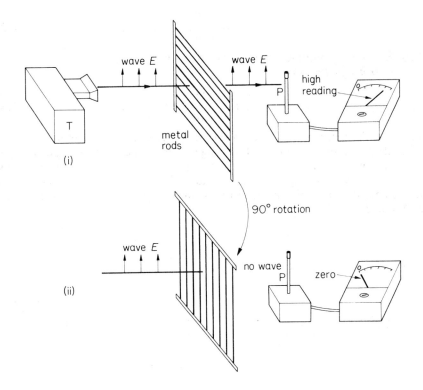

Fig. 21.34 Plane-polarized electromagnetic waves (microwaves)

We can also see that the electromagnetic waves from T are plane-polarized by placing T in front of the probe P *without* using the grille, so that the meter joined to P indicates a high reading. If T is now rotated about its axis through 90°, the meter reading falls to zero. The probe P detects vibrations in one plane, so that when T is rotated through 90° the waves are not detected as they are plane-polarised.

Velocity of Waves

We now list, for convenience, the velocity v of waves of various types, some of which are considered more fully in other sections of the book:

1. *Transverse wave on string*

$$v = \sqrt{\frac{T}{m}}, \qquad \qquad \text{(11)}$$

where T is the tension and m is the mass per unit length.

2. *Sound waves in gas*

$$v = \sqrt{\frac{\gamma p}{\rho}}, \qquad \qquad \text{(12)}$$

where p is the pressure, ρ is the density and γ is the ratio of the molar heat capacities of the gas (p. 434).

3. *Longitudinal waves in solid*

$$v = \sqrt{\frac{E}{\rho}}, \qquad \qquad \text{(13)}$$

where E is Young modulus and ρ is the density.

4. *Electromagnetic waves*

$$v = \sqrt{\frac{1}{\mu\varepsilon}}, \qquad \qquad \text{(14)}$$

where μ is the permeability and ε is the permittivity of the medium.

Fig. 21.35 (i) shows roughly the range of frequencies in the spectrum of *mechanical waves*—waves due to vibrations in solids, liquids and gases including sound waves. Sound frequencies in air range from about 20 to 20 000 Hz for those detected by the human ear but very much higher particle frequencies can be obtained in solids.

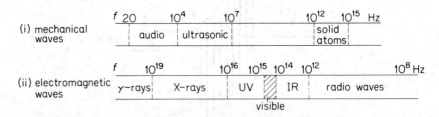

Fig. **21.35** Frequency spectrum

Fig. 21.35 (ii) shows roughly the range of frequencies in the spectrum of *electromagnetic waves*. The various waves are discussed later. The frequency range detected by the eye is about 4×10^{14} to 7×10^{14} Hz, a range factor of about 2. The human ear, however, has a range factor of about 1000.

Velocity of Sound in a Medium

When a sound wave travels in a medium, such as a gas, a liquid, or a solid, the particles in the medium are subjected to varying stresses, with resulting strains (p. 413). The velocity of a sound wave is thus partly governed by the *modulus of elasticity*, E, of the medium, which is defined by the relation

$$E = \frac{\text{stress}}{\text{strain}} = \frac{\text{force per unit area}}{\text{change in length (or volume)/original length (or volume)}} \quad \text{(i)}$$

The velocity, v, also depends on the density, ρ, of the medium, and it can be shown that

$$v = \sqrt{\frac{E}{\rho}} \qquad . \qquad . \qquad . \qquad . \qquad (1)$$

When E is in newton per metre^2 (N m^{-2}) and ρ in kg m^{-3}, then v is in metre per second (m s^{-1}). The relation (1) was first derived by Newton.

For a *solid*, E is Young modulus of elasticity. The magnitude of E for steel is about 2×10^{11} N m^{-2}, and the density ρ of steel is 7800 kg m^{-3}. Thus the velocity of sound in steel is given by

$$v = \sqrt{\frac{E}{\rho}} = \sqrt{\frac{2 \times 10^{11}}{7800}} = 5060 \text{ m s}^{-1}.$$

For a *liquid*, E is the bulk modulus of elasticity. Water has a bulk modulus of $2{\cdot}04 \times 10^9$ N m^{-2}, and a density of 1000 kg m^{-3}. The calculated velocity of sound in water is thus given by

$$v = \sqrt{\frac{2{\cdot}04 \times 10^9}{1000}} = 1430 \text{ m s}^{-1}.$$

The proof of the velocity formula requires advanced mathematics, and is beyond the scope of this book. It can partly be verified by the method of dimensions, however. Thus since density, ρ, = mass/volume, the dimensions of ρ are given by ML^{-3}. The dimensions of force (mass × acceleration) are MLT^{-2}, the dimensions of area are L^2; and the denominator in (i) has zero dimensions since it is the ratio of two similar quantities. Thus the dimensions of modulus of elasticity, E, are given by

$$\frac{\text{ML}}{\text{T}^2\text{L}^2} \text{ or } \text{ML}^{-1}\text{T}^{-2}.$$

Suppose the velocity, v, = $kE^x\rho^y$, where k is a constant. The dimensions of v are LT^{-1}

$$\therefore \text{ LT}^{-1} \equiv (\text{ML}^{-1}\text{T}^{-2})^x \times (\text{ML}^{-3})^y$$

using the dimensions of E and ρ obtained above. Equating the respective indices of M, L, T on both sides, then

$$x + y = 0 \quad . \qquad . \qquad . \qquad . \qquad . \quad \text{(ii)}$$
$$-x - 3y = 1 \quad . \qquad . \qquad . \qquad . \qquad . \quad \text{(iii)}$$
$$-2x = -1 \quad . \qquad . \qquad . \qquad . \qquad . \quad \text{(iv)}$$

From (iv), $x = 1/2$; from (ii), $y = -1/2$. Thus, as $v = kE^x \rho^y$,

$$v = kE^{\frac{1}{2}}\rho^{-\frac{1}{2}}$$
$$\therefore v = k\sqrt{\frac{E}{\rho}}.$$

It is not possible to find the magnitude of k by the method of dimensions, but a rigid proof of the formula by calculus shows that $v = \sqrt{E/\rho}$, so $k = 1$.

Velocity of Sound in a Gas. Laplace's Correction

The velocity of sound in a *gas* is also given by $v = \sqrt{E/\rho}$ where E is the *bulk modulus* of the gas and ρ is its density. Now it is shown on p. 123 that $E = p$, the pressure of the gas, if the stresses and strains in the gas take place isothermally. The formula for the velocity then becomes $v = \sqrt{p/\rho}$ and as the density, ρ, of air is 1·29 kg m^{-3} at s.t.p. and $p = 0\cdot76 \times 13\,600 \times 9\cdot8$ N m^{-2},

$$v = \sqrt{\frac{0\cdot76 \times 13\,600 \times 9\cdot8}{1\cdot29}} = 280 \text{ m s}^{-1} \text{ (approx.)}.$$

This calculation for v was first performed by Newton, who saw that the above theoretical value was well below the experimental value of about 330 m s^{-1}. The discrepancy remained unexplained for more than a century, when LAPLACE suggested in 1816 that E should be the *adiabatic* bulk modulus of a gas, not its isothermal bulk modulus as Newton had assumed. We shall see shortly that adiabatic conditions are maintained in a gas because of the relative slowness of sound wave oscillations. On p. 123, we showed that the adiabatic bulk modulus of a gas is γp where γ is the ratio of the molar heat capacities (i.e., $\gamma = C_p/C_V$). The formula for the velocity of sound in a gas thus becomes

$$v = \sqrt{\frac{\gamma p}{\rho}} \qquad \qquad (2)$$

The magnitude of γ for air is 1·40, and *Laplace's correction*, as it is known, then amends the value of the velocity in air at 0°C to

$$v = \sqrt{\frac{1\cdot40 \times 0\cdot76 \times 13\,600 \times 9\cdot8}{1\cdot29}} = 331 \text{ m s}^{-1}.$$

This is in good agreement with the experimental value.

Effect of Pressure and Temperature on Velocity of Sound in a Gas

Suppose that a mole of gas has a mass M and a volume V. The density is then M/V and so the velocity of sound, v, is

$$v = \sqrt{\frac{\gamma p}{\rho}} = \sqrt{\frac{\gamma p V}{M}}.$$

But $pV = RT$, where R is the molar gas constant and T is the absolute temperature. Hence

$$v = \sqrt{\frac{\gamma R T}{M}} \qquad \qquad (i)$$

Since γ, M and R are constants for a given gas, it follows that *the velocity of sound in a gas is independent of the pressure* if the temperature remains constant. This has been verified by experiments which showed that the velocity of sound at the top of a mountain is about the same as at the bottom. It also follows from (i) that *the velocity of sound is proportional to the square root of its absolute temperature*. Thus if the velocity in air at 16°C is 338 m s^{-1} by experiment, the velocity, v, at 0°C is calculated from

$$\frac{v}{338} = \sqrt{\frac{273}{289}},$$

from which $$v = 338 \sqrt{\frac{273}{289}} = 328\cdot5 \text{ m s}^{-1}.$$

Adiabatic and Isothermal Sound Waves in a Gas

Consider a progressive sound wave travelling in a gas. At an instant, suppose ABC represents a wavelength λ, with a compression at A and a rarefaction at B. Fig. 21.36. Under compression the gas temperature at A is raised; at the same instant the gas temperature at B is lowered. Thus heat tends to flow from A to B. If the temperatures can be equalised in the time taken by the wave to reverse the pressure conditions or temperatures at A and B, shown by the broken line in Fig. 21.36, then the pressure–volume changes in the gas will take place isothermally. If not, the changes will be adiabatic.

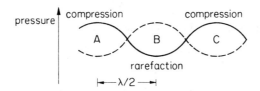

Fig. 21.36 Adiabatic and isothermal sound waves

Consider, therefore, a time equal to half a period, $T/2$. Since $T = \lambda/v$, where v is the constant velocity of the wave, then $T/2 \propto \lambda$, the wavelength. From the formula for conduction of heat (p. 259), assuming a linear temperature gradient for simplification,

heat flow from A to B $\propto$ temperature gradient × time

$$\propto \frac{\text{temperature difference}}{\lambda/2} \times \lambda$$

$$\propto \text{temperature difference.}$$

Hence the *temperature change* of the mass of gas in the region AB

$$= \frac{\text{heat flowing}}{\text{mass} \times \text{sp. ht. capacity}}$$

$$\propto \frac{\text{temperature difference}}{\lambda} \propto \frac{1}{\lambda},$$

since the heat flowing is proportional to the temperature difference and the mass of gas between A and B is proportional to the wavelength.

At audio-frequencies, the relatively long wavelength produces a small temperature change. No equalisation of temperature then occurs between B and C by the time the conditions reverse. Thus the wave travels under *adiabatic* conditions. This is why Laplace's formula, $v = \sqrt{\gamma p/\rho}$, holds for the velocity of sound waves (p. 434). As the wavelength decreases, however, the temperature change increases. Thus at ultrasonic frequencies, for example, the wave travels under conditions which are more isothermal than adiabatic and the velocity is then $v = \sqrt{p/\rho}$.

A more detailed analysis shows that isothermal conditions would be obtained if the frequency is extremely high, which is not the case for normal sound waves (see *Gases, Liquids and Solids* by D. Tabor (Penguin)).

Examples

1. How does the velocity of sound in a medium depend upon the elasticity and density? Illustrate your answer by reference to the case of air and of a long metal rod. The velocity of sound in air being 330.0 m s^{-1} at $0°C$, find the change in velocity per $°C$ rise of temperature. (*L.*)

First part. The velocity of sound, v, is given by $v = \sqrt{E/\rho}$, where E is modulus of elasticity of the medium and ρ is its density. In the case of air, a gas, E represents the bulk modulus of the air under adiabatic conditions and $E = \gamma p$ (see p. 434). Thus $v = \sqrt{\gamma p / \rho}$ for air.

For a long metal rod, E is Young modulus for the metal, assuming the sound travels along the length of the rod.

Second part. The velocity of sound in a gas is proportional to the square root of its absolute temperature, and hence

$$\frac{v}{v_0} = \sqrt{\frac{274}{273}},$$

where v is the velocity at $1°C$ and v_0 is the velocity at $0°C$.

$$\therefore v = v_0 \sqrt{\frac{274}{273}} = 330 \times \sqrt{\frac{274}{273}} = 330.6 \text{ m s}^{-1}$$

$$\therefore \text{ change in velocity} = 0.6 \text{ m s}^{-1}.$$

2. State briefly how you would show by experiment that the characteristics of the transmission of sound are such that (*a*) a finite time is necessary for transmission, (*b*) a material medium is necessary for propagation, (*c*) the disturbance may be reflected and refracted. The wavelength of the note emitted by a tuning-fork, frequency 512 Hz, in air at 17°C is 66.5 cm. If the density of air at s.t.p. is 1.293 kg m^{-3}, calculate the ratio of the molar heat capacities of air. Assume that the density of mercury is 13 600 kg m^{-3}. (*N.*)

Since $v = f\lambda$, the velocity of sound at $17°C$ is given by

$$v = 512 \times 0.665 \text{ m s}^{-1} \qquad \qquad \text{(i)}$$

Now

$$\frac{v_0}{v} = \sqrt{\frac{273}{290}},$$

where v_0 is the velocity at $0°C$, since the velocity is proportional to the square root of the absolute temperature.

$$\therefore v_0 = \sqrt{\frac{273}{290}} \times v = \sqrt{\frac{273}{290}} \times 512 \times 0.665 \qquad \text{(ii)}$$

But

$$v_0 = \sqrt{\frac{\gamma p}{\rho}},$$

where $p = 0.76$ m of mercury $= 0.76 \times 13\,600 \times 9.8$ N m^{-2}, and $\rho = 1.293$ kg m^{-3}.

$$\therefore \gamma = \frac{v_0{}^2 \times \rho}{p}$$

$$= \frac{273 \times 512^2 \times 0.665^2 \times 1.293}{290 \times 0.76 \times 13\,600 \times 9.8}$$

$$= 1.39.$$

1. A small piece of cork in a ripple tank oscillates up and down as ripples pass it. If the ripples travel at 0.20 m s^{-1}, have a wavelength of 15 mm and an amplitude of 5.0 mm, what is the maximum velocity of the cork? (*L.*)

2. At certain definite engine speeds parts of a car, such as a door panel, may vibrate strongly. Explain *briefly* the physical phenomenon of which this is an example and give *three* further examples of it. (*L.*)

3. If the velocity of sound in air is 340 metres per second, calculate (i) the wavelength when the frequency is 256 Hz, (ii) the frequency when the wavelength is 0.85 m.

   Prove the velocity formula used in your calculations.

4. State and explain the differences between progressive and stationary waves.

   A progressive and a stationary simple harmonic wave each have the same frequency of 250 Hz and the same velocity of 30 m s^{-1}. Calculate (i) the phase difference between two vibrating points on the progressive wave which are 10 cm apart, (ii) the equation of motion of the progressive wave if its amplitude is 0.03 metre, (iii) the distance between nodes in the stationary wave.

5. Explain carefully what is meant by the term *resonance* and give *two* specific examples of its occurrence.

   There are several ways in which a metre rule, clamped at one end, may vibrate. With each of these ways is associated a resonant frequency. Draw sketches to illustrate *two* resonant modes for the rule and explain how the rule could be induced to vibrate in these modes. (*L.*)

6. Two waves of equal frequency and amplitude travel in opposite directions in a medium. (*a*) Why is the resultant wave called 'stationary'? (*b*) State *two* differences between a stationary and a progressive wave. (*c*) Explain briefly, using the principle of superposition, why nodes and antinodes of displacement are obtained in a stationary wave.

   If the amplitudes of the two waves are 3 units and 1 unit respectively, show by the principle of superposition that the ratio of the amplitudes of the stationary wave at an antinode and node respectively is $2:1$.

7. Two small loudspeakers A and B are 0.50 m apart. Fig. 21A (i). Both emit sound waves of the same frequency f. A detector moves along a line CD 1.20 m from AB and parallel to it. A maximum wave is detected at C where OC is the perpendicular bisector of AB and again at D directly opposite A. Calculate f. (Velocity of sound $= 340$ m s^{-1}.)

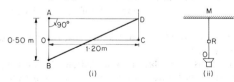

Fig. 21A

8. A small source O of electromagnetic waves is placed some distance from a plane metal reflector M. Fig. 21A (ii). A receiver R, moving between O and M along the line normal to the reflector, detects successive maximum and minimum readings on the meter joined to it.

   (*a*) Explain why these readings are obtained. (*b*) Calculate the frequency of the source O if the average distance between successive minima is 1.5 cm and the speed of electromagnetic waves in air $= 3.0 \times 10^8$ m s^{-1}.

9. Explain what is meant by the statement that 'sound is propagated in air as longitudinal progressive waves', and outline the experimental evidence in favour of this statement. Compare the mode of propagation of sound in air with that of (*a*) waves travelling along a long metal rod, produced by tapping one end, (*b*) water waves.

   Two loudspeakers face each other at a separation of about 100 m and are connected to the same oscillator, which gives a signal of frequency 110 Hz. Describe and explain the variation of sound intensity along the line joining the speakers. A man walks along the line with a uniform speed of 2.0 m s^{-1}. What does he hear? (Speed of sound $= 330$ m s^{-1}.) (*O.*)

10. Explain the terms *damped oscillation, forced oscillation* and *resonance*. Give one example of each.

Describe an experiment to illustrate the behaviour of a simple pendulum (or pendulums) undergoing forced oscillation. Indicate qualitatively the results you would expect to observe.

What factors determine (*a*) the period of free oscillations of a mechanical system, and (*b*) the amplitude of a system undergoing forced oscillation? (*O. & C.*)

11. What is the principle of superposition as applied to wave motion?

Discuss as fully as you can the result of superposing two waves of equal amplitude (*a*) of the same frequency travelling in opposite directions, (*b*) of slightly different frequencies travelling in the same direction. Describe how you would demonstrate the validity of your conclusions in *one* of these cases.

Two plane sound waves of the same frequency travelling in opposite directions have different amplitudes. When an observer moves along the direction of travel of one of the waves, the amplitude of the sound he hears fluctuates by a factor of two. Explain this and find the ratio of the amplitudes of the two travelling waves. (*O. & C.*)

12. Write an equation that represents a progressive sinusoidal wave motion. With the aid of suitable diagrams, explain the meanings of the quantities appearing in your equation.

On axes immediately above each other, sketch two similar sinusoidal waves each of amplitude A and with phase differences such that, when superposed, the waves would produce (*a*) maximum constructive interference, (*b*) maximum destructive interference. Give the value (magnitude and unit) of the phase difference in case (*b*).

If two such waves have exactly one-third of the phase difference relevant to case (*b*) and are superposed, find (by means of a phasor diagram or otherwise)

(i) the amplitude of the resultant wave in terms of A,

(ii) the ratio of the power carried by the resultant wave to the total power carried by the two component waves considered separately. Comment on your result in relation to the principle of conservation of energy. (*C.*)

13. (*a*) A small source S emits electromagnetic waves of wavelength about 3 cm which can be detected by an aerial A (a straight wire) connected to a meter measuring the intensity of the radiation. Initially the distance SA = d. When the distance between the source and the aerial is increased to $2d$, the meter reading falls to one-quarter of its original value. What conclusion can be drawn from this?

What would you expect the meter reading to be if the aerial were moved until SA = $3d$?

If the source is rotated through 90° about the line SA, the meter reading falls to zero. Explain briefly the reason for this.

(*b*) A metal reflecting screen is now placed some distance beyond A with its plane perpendicular to the line SA. It is found that as the screen is moved slowly away from A, alternate maximum and minimum readings are shown on the meter. Explain briefly the reason for this. If the screen is displaced a distance of 8·7 cm between a first and a seventh minimum, calculate the wavelength and the frequency of the wave. (Speed of electromagnetic waves in air = $3·0 \times 10^8$ m s^{-1}).

(*c*) The source of the electromagnetic waves is assumed to be *monochromatic*. Explain what monochromatic means. What would you have observed in (*b*) if the source had emitted simultaneously waves of wavelength 3 cm and 6 cm, of the same intensity? (*L.*)

14. A plane-progressive wave is represented by the equation

$$y = 0·1 \sin (200\pi t - 20\pi x/17),$$

where y is the displacement in millimetres, t is in seconds and x is the distance from a fixed origin O in metres (m).

Find (i) the frequency of the wave, (ii) its wavelength, (iii) its speed, (iv) the phase difference in radians between a point 0·25 m from O and a point 1·10 m from O, (v) the equation of a wave with double the amplitude and double the frequency but travelling exactly in the opposite direction.

15. The equation $y = a \sin(\omega t - kx)$ represents a plane wave travelling in a medium along the x-direction, y being the displacement at the point x at time t.

Deduce whether the wave is travelling in the positive x-direction or in the negative x-direction.

If $a = 1\cdot0 \times 10^{-7}$ m, $\omega = 6\cdot6 \times 10^3$ s^{-1} and $k = 20$ m^{-1}, calculate (a) the speed of the wave, (b) the maximum speed of a particle of the medium due to the wave. (N.)

Sound Waves

16. A source of sound of frequency 550 Hz emits waves of wavelength 600 mm in air at 20°C. What is the velocity of sound in air at this temperature? What would be the wavelength of the sound from the source in air at 0°C? (L.)

17. If a detonator is exploded on a railway line an observer standing on the rail 2·0 km away hears two reports. Why is this so? What is the time interval between these reports?

(The Young modulus for steel $= 2\cdot0 \times 10^{11}$ N m^{-2}. Density of steel $= 8\cdot0 \times 10^3$ kg m^{-3}. Density of air $= 1\cdot4$ kg m^{-3}. Ratio of the molar heat capacities of air $= 1\cdot40$. Atmospheric pressure $= 10^5$ N m^{-2}.) (L.)

18. Describe how a *sound wave* passes through air, using graphs which illustrate and compare the variation of (i) the *displacement* of the air particles, (ii) the *pressure changes*, while the wave travels.

Using the same axes as the displacement graph, draw a sketch of the graph showing the variation of *velocity* of the air particles.

19. Describe a determination (other than resonance) of the velocity of sound in air. How is the velocity dependent upon atmospheric conditions? Give Newton's expression for the velocity of sound in a gas, and Laplace's correction. Hence calculate the velocity of sound in air at 27°C. (Density of air at s.t.p. $= 1\cdot29$ kg m^{-3}; ratio of molar heat capacities $= 1\cdot4$.) (L.)

20. Describe the factors on which the velocity of sound in a gas depends. A man standing at one end of a closed corridor 57 m long blew a short blast on a whistle. He found that the time from the blast to the sixth echo was two seconds. If the temperature was 17°C, what was the velocity of sound at 0°C? (C.)

21. Write down an expression for the speed of sound in an ideal gas. Give a consistent set of units for the quantities involved.

Discuss the effect of changes of pressure and temperature on the speed of sound in air.

Describe an experimental method for finding a *reliable* value for the speed of sound in free air. (N.)

22. Describe an experiment to measure the velocity of sound in the open air. What factors may affect the value obtained and in what way may they do so?

It is noticed that a sharp tap made in front of a flight of stone steps gives rise to a ringing sound. Explain this and, assuming that each step is 0·25 m deep, estimate the frequency of the sound. (The velocity of sound may be taken to be 340 m s^{-1}.) (L.)

22 Wave Theory of Light. Velocity of Light

Historical

It has already been mentioned that light is a form of energy which stimulates our sense of vision. One of the early theories of light, about 400 B.C., suggested that particles were emitted from the eye when an object was seen. It was realised, however, that something is *entering* the eye when a sense of vision is caused, and about 1660 Newton proposed that particles, or corpuscles, were emitted from a luminous object. The *corpuscular theory of light* was adopted by many scientists of the day owing to the authority of Newton, but HUYGENS, an eminent Dutch scientist, proposed about 1680 that light energy travelled from one place to another by means of a wave-motion. If the *wave theory of light* was correct, light should bend round a corner, just as sound travels round a corner. The experimental evidence for the wave theory in Huygens' time was very small, and the theory was dropped for more than a century. In 1801, however, THOMAS YOUNG obtained evidence that light could produce wave effects (p. 460), and he was among the first to see clearly the close analogy between sound and light waves. As the principles of the subject became understood other experiments were carried out which showed that light could spread round corners, and Huygens' wave theory of light was revived. Newton's corpuscular theory was rejected since it was incompatible with experimental observations (see p. 445). The wave theory of light has played, and is still playing, an important part in the development of the subject.

In 1905 the eminent mathematical physicist EINSTEIN suggested that the energy in light could be carried from place to place by 'particles' whose energy depended on the wavelength of the light. This was a return to a corpuscular theory, though it was completely different from that of Newton, as we see later. Experiments carried out showed that Einstein's theory was true, and the particles of light energy are known as 'photons' (p. 845). It is now considered that *either* the wave theory *or* the particle theory of light can be used in a problem on light, depending on the circumstances of the problem. In this section we shall consider Huygens' wave theory, which led to many notable advances in the subject.

Wavefronts

Consider a point source of light, S, in air, and suppose that a disturbance, or wave, originates at S as a result of vibrations occurring inside the atoms of the source, and travels outwards. After a time t the wave has travelled a distance ct, where c is the velocity of light in air, and the light energy has thus reached the surface of a sphere of centre S and radius ct, Fig. 22.1. The surface of the sphere is called the *wavefront* of the light at this instant, and every point on it is vibrating 'in step' or *in phase* with every other point. As time goes on the wave travels farther and new wavefronts are obtained which are the surfaces of spheres of centre S.

At points a long way from S, such as C or D, the wavefronts are portions of a sphere of very large radius, and the wavefronts are then substantially *plane*. Light from the sun reaches the earth in plane wavefronts because the sun is so far away; plane wavefronts also emerge from a converging lens when a point source of light is placed at its focus.

The significance of the wavefront, then, is that it shows how the light energy travels from one place in a medium to another. A *ray* is the name given to the direction along which the energy travels, and consequently a ray of light passing through a point is perpendicular to the wavefront at that point. The rays diverge near S, but they are approximately parallel a long way from S, as the curved wavefronts are then approximately plane, Fig. 22.1.

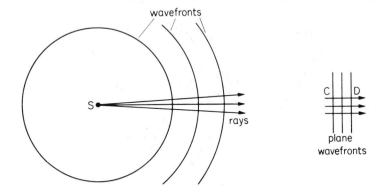

Fig. 22.1 Wavefronts and rays

Huygens' Construction for the New Wavefront

Suppose that the wavefront from a centre of disturbance S had reached the surface AB in a medium at some instant, Fig. 22.2. To obtain the position of the new wavefront after a further time *t*, Huygens postulated that *every point, A,..., C,..., E,..., B, on AB becomes a new or 'secondary' centre of disturbance.* The wavelet from A then reaches the surface M of a sphere of radius *vt* and centre A, where *v* is the velocity of light in the medium; the wavelet from C reaches the surface D of a sphere of radius *vt* and centre C; and so on for every point on AB. According to Huygens, *the new wavefront is the surface MN which touches all the wavelets from the secondary sources*; and in the case considered, it is the surface of a sphere of centre S.

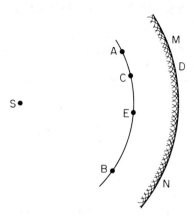

Fig. 22.2 Huygens' construction

In this simple example of obtaining the new wavefront, the light travels in the same medium. Huygens' construction, however, is especially valuable for deducing the new wavefront when the light travels from one medium to another, as we shall soon show.

Reflection at Plane Surface

Suppose that a beam of parallel rays between HA and LC is incident on a plane mirror, and imagine a plane wavefront AB which is normal to the rays, reaching the mirror surface, Fig. 22.3. At this instant the point A acts as a centre of

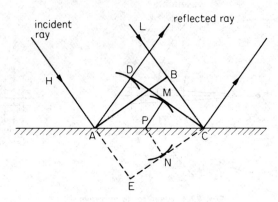

Fig. 22.3 Reflection at plane surface

disturbance. Suppose we require the new wavefront at a time corresponding to the instant when the disturbance at B reaches C. The wavelet from A reaches the surface of a sphere of radius AD at this instant; and as other points between AC on the mirror, such as P, are reached by the disturbances originating on AB, wavelets of smaller radius than AD are obtained at the instant we are considering. The new wavefront is the surface CMD which touches all the wavelets.

In the absence of the mirror, the plane wavefront AB would reach the position EC in the time considered. Thus AD = AE = BC, and PN = PM, where PN is perpendicular to EC. The triangles PMC, PNC are hence congruent, as PC is common, angles PMC, PNC are each 90°, and PN = PM. Thus angle PCM = angle PCN. But triangles ACD, AEC are congruent. Consequently angle ACD = angle ACE = angle PCN = angle PCM, since EC is a plane. Hence CMD is a *plane* surface.

Law of reflection. We can now deduce the law of reflection concerning the angles of incidence and reflection. From the above, it can be seen that the triangles ABC, AEC are congruent, and that triangles ADC, AEC are congruent. The triangles ABC, ADC are hence congruent, and therefore angle BAC = angle DCA. Now these are the angles made by the wavefront AB, CD respectively with the mirror surface AC. Since the incident and reflected rays, for example HA, AD, are normal to the wavefronts, these rays also make equal angles with AC. So it follows that the angles of incidence and reflection are equal.

Point Object

Consider now a point object O in front of a plane mirror M, Fig. 22.4. A spherical wave spreads out from O, and at some time the wavefront reaches ABC. In the absence of the mirror the wavefront would reach a position DEF in a time *t* thereafter, but every point between D and F on the mirror acts

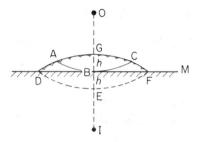

Fig. 22.4 Point object

as a secondary centre of disturbance and wavelets are reflected back into the air. At the end of the time t, a surface DGF is drawn to touch all the wavelets, as shown. DGF is part of a spherical surface which advances into the air, and it appears to have come from a point I as a centre below the mirror, which is therefore a virtual image.

The sphere of which DGF is part has a chord DF. Suppose the distance from B, the midpoint of the chord, to G is h. The sphere of which DEF is part has the same chord DF, and the distance from B to E is also h. It follows, from the theorem of product of intersection of chords of a circle, that DB.BF = $h(2r - h) = h(2R - h)$, where r is the radius OE and R is the radius IG. Thus $R = r$, or IG = OE, and hence IB = OB. The image and object are thus equidistant from the mirror.

Refraction at Plane Surface

Consider a beam of parallel rays between LO and PD incident on the plane surface of a water medium from air in the direction shown, and suppose that a plane wavefront has reached the position OA at a certain instant, Fig. 22.5.

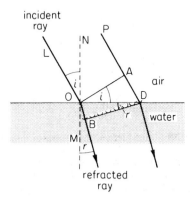

Fig. 22.5 Refraction at plane surface

Each point between O, A becomes a new centre of disturbance as the wavefront advances to the surface of the water, and the wavefront changes in direction when the disturbance enters the liquid.

Suppose that t is the time taken by the light to travel from A to D. The disturbance from O travels a distance OB, or vt, in water in a time t, where v is the velocity of light in water. At the end of the time t, the wavefronts in the water from the other secondary centres between O, D reach the surfaces of spheres to each of which DB is a tangent. Thus DB is the new wavefront in the water, and the ray OB which is normal to the wavefront is consequently the refracted ray.

Since c is the velocity of light in air, $AD = ct$. Now

$$\frac{\sin i}{\sin r} = \frac{\sin LON}{\sin BOM} = \frac{\sin AOD}{\sin ODB}$$

$$\therefore \frac{\sin i}{\sin r} = \frac{AD/OD}{OB/OD} = \frac{AD}{OB} = \frac{ct}{vt} = \frac{c}{v} \qquad . \qquad . \qquad . \qquad \text{(i)}$$

But c, v are constants for the given media.

$$\therefore \frac{\sin i}{\sin r} \text{ is a constant,}$$

which is Snell's law of refraction (p. 326).

It can now be seen from (i) that the refractive index, n, of a medium is given by $n = \dfrac{c}{v}$, where c is the velocity of light in air and v is the velocity of light in the medium.

Newton's Corpuscular Theory of Light

Prior to the wave theory of light, Newton had proposed a corpuscular or particle theory of light. According to Newton, particles are emitted by a source of light, and they travel in a straight line until the boundary of a new medium is encountered.

In the case of *reflection at a plane surface*, Newton stated that at some very small distance from the surface M, represented by AB, the particles were acted upon by a repulsive force, which gradually diminished the component of the velocity v in the direction of the normal and then reversed it, Fig. 22.6. The

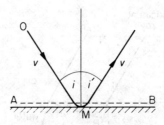

Fig. 22.6 Newton's corpuscular theory of reflection

horizontal component of the velocity remained unaltered, and hence the velocity of the particles of light as they moved away from M is again v. Since the horizontal components of the incident and reflected velocities are the same, it follows that

$$v \sin i = v \sin i' \qquad . \qquad . \qquad . \qquad . \qquad \text{(i)}$$

where i' is the angle of reflection.

$$\therefore \sin i = \sin i', \text{ or } i = i'.$$

Thus the corpuscular theory explains the law of reflection at a plane surface.

To explain *refraction at a plane surface* when light travels from air to a denser medium such as water, Newton stated that a force of attraction acted on the particles as they approached beyond a line DE very close to the boundary N,

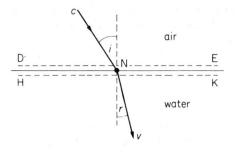

Fig. 22.7 Newton's corpuscular theory of refraction

Fig. 22.7. The vertical component of the velocity of the particles was thus increased on entering the water, the horizontal component of the velocity remaining unaltered, and beyond a line HK close to the boundary the vertical component remained constant at its increased value. The resultant velocity, v, of the particles in the water is thus *greater* than its velocity, c, in air.

Suppose i, r are the angles of incidence and refraction respectively. Then, as the horizontal components of the velocity are unaltered,

$$c \sin i = v \sin r$$

$$\therefore \frac{\sin i}{\sin r} = \frac{v}{c}$$

$$\therefore n = \frac{v}{c} = \text{the refractive index.}$$

Since n is greater than 1, the velocity of light (v) in water is greater than the velocity (c) in air, as was stated above. This is according to Newton's corpuscular theory. On the wave theory, however, $n = c/v$ (see p. 444); and hence the velocity of light (v) in water is *less* than the velocity (c) in air according to the wave theory. The corpuscular theory and wave theory are thus in conflict. In an experiment carried out about 150 years later, Foucault obtained a value for v which showed that the corpuscular theory of Newton could not be true (see p. 451).

Dispersion

The dispersion of colours produced by a medium such as glass is due to the difference in speeds of the various colours in the medium. Thus suppose a plane wavefront AC of white light is incident in air on a plane glass surface, Fig. 22.8.

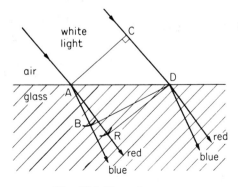

Fig. 22.8 Dispersion

In the time the light takes to travel in air from C to D, the red light from the centre of disturbance A reaches a position shown by the wavelet at R. The blue light from A reaches another position shown by the wavelet at B, since the speed of blue light in glass is less than that of red light, so that AB is less than AR. On drawing the new wavefronts DB, DR, it can be seen that the blue wavefront BD is refracted more in the glass than the red wavefront DR. The refracted blue ray is AB and the refracted red ray is AR, and hence dispersion occurs.

Refraction through Prism at Minimum Deviation

Consider a plane wavefront HB incident on the face HB of a prism of angle A, Fig. 22.9. If the emerging wavefront is EC, then the light travels a distance HXE in air in the same time as the light travels a distance BC in glass.

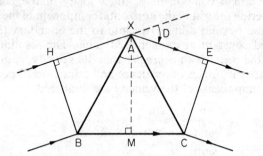

Fig. 22.9 Refraction at minimum deviation

$$\therefore \frac{HX + XE}{c} = \frac{BC}{v},$$

where c is the velocity of light in air and v is the velocity in glass

$$\therefore HX + XE = \frac{c}{v} BC = n\ BC \quad . \qquad . \qquad . \qquad . \qquad (i)$$

At minimum deviation, the wavefront passes symmetrically through the prism (p. 341).

$$\therefore HX = XE.$$

From (i), $\qquad\qquad\qquad \therefore 2HX = n\ BC$

$$\therefore n = \frac{2\ HX}{BC} \quad . \qquad . \qquad . \qquad . \qquad . \qquad (ii)$$

But $HX = XB \cos BXH = XB \cos \left[\dfrac{180° - (A + D_{min})}{2} \right]$

$$= XB \sin \left(\frac{A + D_{min}}{2} \right),$$

and $\qquad\qquad\qquad BC = 2\ BM = 2\ XB \sin \dfrac{A}{2}.$

From (ii), $\qquad\qquad\qquad \therefore n = \sin \dfrac{\left(\dfrac{A + D_{min}}{2} \right)}{\sin \dfrac{A}{2}}.$

Power of a Lens

We have now to consider the effect of lenses on the *curvature* of wavefronts. The curvature of a spherical wavefront is defined as $1/r$, where r is the radius of the wavefront surface.

When a plane wavefront is incident on a converging lens L, a spherical wavefront, S, of radius f emerges from L, where f is the focal length of the lens, Fig. 22.10 (i). Parallel rays, which are normal to the plane wavefront, are thus

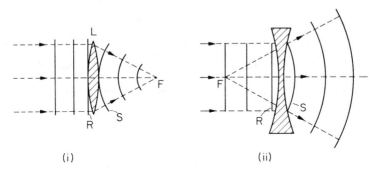

(i)   (ii)

Fig. 22.10 (i) Converging lens (ii) Diverging lens

refracted towards F, the focus of the lens. Now the curvature of a plane wavefront is zero, and the curvature of the spherical wavefront S is $1/f$. Thus the converging lens adds a curvature of $1/f$ to a wavefront incident on it. $1/f$ is defined as the *converging power* of the lens:

$$\text{Power } P = \frac{1}{f} \qquad . \qquad . \qquad . \qquad . \qquad (1)$$

Fig. 22.10 (ii) illustrates the effect of a *diverging* lens on a plane wavefront R. The front S emerging from the lens has a curvature opposite to S in Fig. 22.10 (i), and it appears to be diverging from a point F behind the diverging lens, which is its focus. The curvature of the emerging wavefront is thus $1/f$, where f is the focal length of the lens, and the powers of the converging and diverging lens are opposite in sign.

The power of a converging lens is positive, since its focal length is positive, while the power of a diverging lens is negative. The unit of power is the *dioptre*, D, which is the power of a lens of 1 metre focal length. A lens of $+8$ dioptres, or $+8D$, is therefore a converging lens of focal length $1/8$ m or $12 \cdot 5$ cm, and a lens of $-4D$ is a diverging lens of $1/4$ m or 25 cm focal length.

The Lens Equation

Suppose that an object O is placed a distance u from a converging lens, Fig. 22.11. The spherical wavefront A from O which reaches the lens has a radius

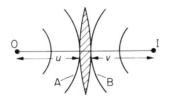

Fig. 22.11 Effect of lens on wavefront

of curvature u, and hence a curvature $1/u$. Since the converging lens adds a curvature of $1/f$ to the wavefront as we proved, the spherical wavefront B emerging from the lens into the air has a curvature $\left(\dfrac{1}{u}+\dfrac{1}{f}\right)$. But the curvature is also given by $\dfrac{1}{v}$, where v is the image distance IB from the lens.

$$\therefore \frac{1}{v} = \frac{1}{u} + \frac{1}{f}$$

It can be seen that the curvature of A is of an opposite sign to that of B; and taking this into account, the lens equation $\dfrac{1}{v}+\dfrac{1}{u}=\dfrac{1}{f}$ is obtained. A similar method can be used for a diverging lens, which is left as an exercise for the student.

Velocity of Light

For many centuries the velocity of light was thought to be infinitely large; from about the end of the seventeenth century, however, evidence began to be obtained which showed that the speed of light, though enormous, was a finite quantity. Galileo, in 1600, attempted to measure the velocity of light by covering and uncovering a lantern at night, and timing how long the light took to reach an observer a few miles away. Owing to the enormous speed of light, however, the time was too small to measure, and the experiment was a failure. The first successful attempt to measure the velocity of light was made by RÖMER, a Danish astronomer, in 1676.

Römer's Astronomical Method

Römer was engaged in recording the eclipses of one of Jupiter's satellites or moons, which has a period of 1·77 days round Jupiter. The period of the satellite is thus very small compared with the period of the earth round the sun (one year), and the eclipses of the satellite occur very frequently while the earth moves only a very small distance in its orbit. Thus the eclipses may be regarded as *signals sent out from Jupiter* at comparatively short intervals, and observed on the earth; almost like a bright lamp covered at regular intervals at night and viewed by a distant observer.

Fig. 22.12 shows the position of the earth E_1 and Jupiter J_1 when they are closest together in line with the sun S. At this position, Römer noted the date and time of the eclipse of one of Jupiter's satellites. About $6\frac{1}{2}$ months later, when the earth E_2 and Jupiter J_2 were again in line with the sun S but on opposite sides, Römer found that the expected eclipse of the satellite was $16\frac{1}{2}$ minutes late. He deduced that this was the time taken by the light to travel across the diameter of the earth's orbit. The diameter is about 3×10^{11} m. So with these figures,

$$\text{speed of light, } c = \frac{3 \times 10^{11} \text{ m}}{16\cdot 5 \times 60 \text{ s}} = 3 \times 10^8 \text{ m s}^{-1} \text{ (approx).}$$

About $6\frac{1}{2}$ months later, when the earth is at E_3 and Jupiter is at J_3 in line with the sun S, the expected eclipse was about $16\frac{1}{2}$ minutes earlier than in the

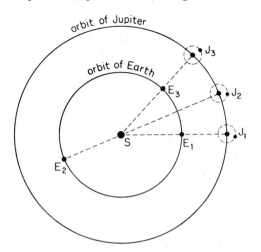

Fig. 22.12 Römer's method

positions E_2 and J_2, owing to the time taken for light to travel across the diameter of the earth's orbit.

Foucault's Rotating Mirror Method

In 1862, Foucault devised a fairly accurate method of measuring the velocity of light. In his method a plane mirror M_1 is rotated at a high constant angular velocity about a vertical axis at A, Fig. 22.13. A lens L is placed so that light from a bright source at O_1 is reflected at M_1 and comes to a focus at a point P on a concave mirror C. The centre of curvature of C is at A, and consequently the light is reflected back from C along its original path, giving rise to an image coincident with O_1. In order to see the image, a plate of glass G is placed at 45° to the axis of the lens, from which the light is reflected to form an image at B_1.

Suppose the plane mirror M_1 begins to rotate. The light reflected by it is then incident on C for a fraction of a revolution, and if the speed of rotation is 2 rev per second, an intermittent image is seen. As the speed of M_1 is increased to about 10 rev per second the image is seen continuously as a result of the rapid impressions on the retina. As the speed is increased further, the light reflected from the mirror flashes across from M_1 to C, and returns to M_1 to

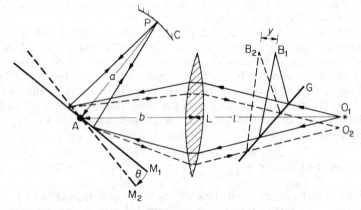

Fig. 22.13 Foucault's rotating mirror method

find it displaced by a very small angle θ to a new position M_2. An image is now observed at B_2, and by measuring the displacement, $B_1 B_2$ or y of the image, Foucault was able to calculate the velocity of light, c. The formula for c, proved shortly, is

$$c = \frac{8\pi m a^2 l}{(a + b)y},$$

where y is the displacement of the image, m is the number of revolutions per second of the mirror, l is the distance from L to O_1, b is the distance from L to the plane mirror at A, and a is the radius of curvature of the mirror C.

As m, a, l, b are known, and the displacement $y = O_1 O_2 = B_1 B_2$ and can be measured, the velocity of light c can be found.

To measure the velocity of light *in water*, Foucault placed a long pipe with water between the plane mirror and C. He found that, with the number of revolutions per second of the mirror the same as when air was used, the displacement y of the image B was *greater*. Since the velocity of light $= 8\pi m a^2 l/(a + b)y$, it follows that the velocity of light in water is *less* than in air. Newton's 'corpuscular theory' of light predicted that light should travel faster in water than in air (p. 445), whereas the 'wave theory' of light predicted that light should travel slower in water than in air. The direct observation of the velocity of light in water by Foucault's method showed that the corpuscular theory of Newton could not be true.

Theory of Foucault's Method

Consider the point P on the curved mirror from which the light is always reflected back to the plane mirror, Fig. 22.14. When the plane mirror is at M_1, the image

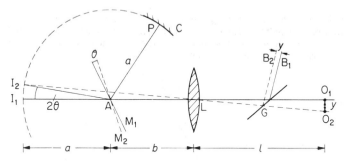

Fig. 22.14 Explanation of Foucault's method

of P in it is at I_1, where $AI_1 = AP = a$, the radius of curvature of C (see p. 321). The rays incident on the lens L from the plane mirror appear to come from I_1. When the mirror is at M_2 the image of P in it is at I_2, where $AI_2 = AP = a$, and the rays incident on L from the mirror now appear to come from I_2. Now the mirror has rotated through an angle θ from M_1 to M_2, and the direction PA of the light incident on it is constant. The angle between the reflected rays is thus 2θ (see p. 315), and hence $I_2AI_1 = 2\theta$.

$$\therefore I_2I_1 = a \times 2\theta = 2a\theta \qquad . \qquad . \qquad . \qquad . \qquad \text{(i)}$$

The images O_1, O_2 formed by the *lens*, L, are the images of I_1, I_2 in it, as the light incident on L from the mirror appear to come from I_1, I_2. So if $y = O_1O_2$, $AL = b$, and $LO_1 = l$,

$$\therefore \frac{I_2I_1}{y} = \frac{(a+b)}{l}$$

$$\therefore I_2I_1 = \frac{(a+b)y}{l} \qquad . \qquad . \qquad . \qquad . \qquad \text{(ii)}$$

From (i) and (ii), it follows that

$$2a\theta = \frac{(a+b)y}{l}$$

$$\therefore \theta = \frac{(a+b)y}{2al} \qquad . \qquad . \qquad . \qquad . \qquad \text{(iii)}$$

The angle θ can also be expressed in terms of the velocity of light, c, and the number of revolutions per second, m, of the plane mirror. The angular velocity of the mirror is $2\pi m$ radian per second, and hence the time taken to rotate through an angle θ radian is $\theta/2\pi m$ second. But this is the time taken by the light to travel from the mirror to C and back, which is $2a/c$ seconds.

$$\therefore \frac{\theta}{2\pi m} = \frac{2a}{c}$$

$$\therefore \theta = \frac{4\pi ma}{c} \qquad . \qquad . \qquad . \qquad . \qquad (iv)$$

From (iii) and (iv), we have

$$\frac{(a+b)y}{2al} = \frac{4\pi ma}{c}$$

$$\therefore c = \frac{8\pi ma^2 l}{(a+b)y} \qquad . \qquad . \qquad . \qquad . \qquad (1)$$

Example

A beam of light is reflected by a rotating mirror on to a fixed mirror, which sends it back to the rotating mirror from which it is again reflected, and then makes an angle of 18° with its original direction. The distance between the two mirrors is 10^4 m, and the rotating mirror is making 375 revolutions per second. Calculate the velocity of light. (*L.*)

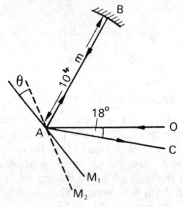

Fig. 22.15 Example

Suppose OA is the original direction of the light, incident at A on the mirror in the position M_1, B is the fixed mirror, and AC is the direction of the light reflected from the rotating mirror when it reaches the position M_2, Fig. 22.15.

The angle θ between M_1, M_2 is $\frac{1}{2} \times 18°$, since the angle of rotation of a mirror is half the angle of deviation of the reflected ray when the incident ray (BA in this case) is kept constant. Thus $\theta = 9°$.

$$\text{Time taken by mirror to rotate } 360° = \frac{1}{375} \text{ s.}$$

$$\therefore \text{ time taken to rotate } 9° = \frac{9}{360} \times \frac{1}{375} \text{ s.}$$

But this is also the time taken by the light to travel from A to B and back, which is given by $2 \times 10^4/c$, where c is the velocity of light in m s^{-1}.

$$\therefore \frac{2 \times 10^4}{c} = \frac{9}{360} \times \frac{1}{375}$$

$$\therefore c = \frac{2 \times 10^4 \times 360 \times 375}{9} = 3 \times 10^8 \text{ m s}^{-1}.$$

Michelson's Method for the Velocity of Light

The velocity of light, c, is a quantity which appears in many fundamental formulae in advanced Physics, especially in connection with the theories concerning particles in atoms and calculations on atomic (nuclear) energy. EINSTEIN has shown, for example, that the energy W released from an atom is given by $W = mc^2$ joules, where m is the decrease in mass of the atom in kilograms and c the velocity of light in metres per second. A knowledge of the magnitude of c is thus important. A. A. MICHELSON, an American physicist, spent many years of his life in measuring the velocity of light, and the method he devised was considered as one of the most accurate.

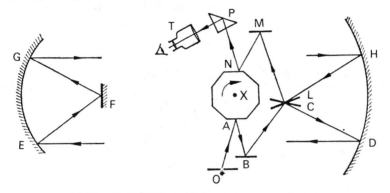

Fig. 22.16 Michelson's rotating prism method

The essential features of Michelson's apparatus are shown in Fig. 22.16. X is an equiangular octagonal steel prism which can be rotated at constant speed about a vertical axis through its centre. The faces of the prism are highly polished, and the light passing through a slit from a very bright source O is reflected at the surface A towards a plane mirror B. From B the light is reflected to a plane mirror L, which is placed so that the image of O formed by this plane mirror is at the focus of a large concave mirror HD. The light then travels as a parallel beam to another concave mirror GE a long distance away, and it is reflected to a plane mirror F at the focus of GE. The light is then reflected by the mirror, travels back to H, and is there reflected to a plane mirror C placed just below L and inclined to it as shown. From C the light is reflected to a plane mirror M, and is then incident on the face N of the octagonal prism opposite to A. The final image thus obtained is viewed through T with the aid of a totally reflecting prism P.

The image is seen by light reflected from the top surface of the octagonal prism X. When the latter is rotated the image disappears at first, as the light reflected from A when the prism is just in the position shown in Fig. 22.16 arrives at the opposite face to find this surface in some position inclined to that shown. When the speed of rotation is increased and suitably adjusted, however, the image reappears and is seen in the same position as when the prism X is at rest. *The light reflected from A now arrives at the opposite surface in the time taken for the prism to rotate through 45°, or ⅛th of a revolution*, as in this case the surface on the left of N, for example, will occupy the position of N when the light arrives at the upper surface of X.

Suppose d is the total distance in metres travelled by the light in its journey from A to the opposite face; the time taken is then d/c, where c is the velocity of light. But this is the time taken by X to make ⅛th of a revolution, which is $1/8m$ seconds if the number of revolutions per second is m.

$$\therefore \frac{1}{8m} = \frac{d}{c}$$

$$\therefore c = 8md \text{ metre per second.}$$

Thus c can be calculated from a knowledge of m and d.

Michelson performed the experiment in 1926, and again in 1931, when the light path was enclosed in an evacuated tube 1·6 km long. Multiple reflections were obtained to increase the effective path of the light. A prism with 32 faces was also used, and Michelson's result for the velocity of light *in vacuo* was $2·99\,774 \times 10^8$ m s^{-1}. Michelson died in 1931 while he was engaged with Pease and Pearson in another accurate measurement of the velocity of light.

Exercises 22

Wave Theory

1. Using Huygens' principle of secondary wavelets explain, making use of a diagram, how a refracted wavefront is formed when a beam of light, travelling in glass, crosses the glass-air boundary. Show how the sines of the angles of incidence and refraction are related to the speeds of light in air and glass. (*L.*)

2. A parallel beam of monochromatic radiation travelling through glass is incident on the plane boundary between the glass and air. Using Huygens' principle draw diagrams (one in each case) showing successive positions of the wave fronts when the angle of incidence is (*a*) 0°, (*b*) 30°, (*c*) 60°. Indicate clearly and explain the constructions used. (The refractive index of glass for the radiation used is 1·5.) (*N.*)

3. A plane wavefront of monochromatic light is incident normally on one face of a glass prism, of refracting angle 30°, and is transmitted. Using Huygens' construction trace the course of the wavefront. Explain your diagram and find the angle through which the wavefront is deviated. (Refractive index of glass = 1·5.) (*N.*)

4. State *Snell's law of refraction* and define *refractive index*.

Show how refraction of light at a plane interface can be explained on the basis of the wave theory of light.

Light travelling through a pool of water in a parallel beam is incident on the horizontal surface. Its speed in water is $2·2 \times 10^8$ m s^{-1}. Calculate the maximum angle which the beam can make with the vertical if light is to escape into the air where its speed is $3·0 \times 10^8$ m s^{-1}.

At this angle in water, how will the path of the beam be affected if a thick layer of oil, of refractive index 1·5, is floated on to the surface of the water? (*O. & C.*)

5. Discuss briefly the arguments by which the speed of light in glass may be expressed in terms of its speed in air and the refractive index of the glass (*a*) from the point of view of the wave theory of light, (*b*) from the point of view of Newton's corpuscular theory of light.

Describe an experimental method of determining the speed of light in air.

Fig. 22A represents a plane wavefront AB striking a plane surface in air. The refractive

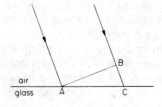

Fig. 22A

index of the glass is 1·5, and the speed of light in air is 3×10^8 m s^{-1}. The distance BC is 3 cm. Taking the time from the instant shown in the diagram, and considering only refraction, (*a*) construct accurately the wavefront at time 10^{-10} s; (*b*) draw the position of the wavefront at time 2×10^{-10} s; (*c*) draw the position of the wavefront at time 5×10^{-11} s. (*O.*)

6. How did Huygens explain the reflection of light on the wave theory? Using Huygens' conceptions, show that a series of light waves diverging from a point source will, after reflection at a plane mirror, appear to be diverging from a second point, and calculate its position. (*C.*)

7. Explain what is meant by Huygens' principle.

Use the principle to show that a plane wave incident obliquely on a plane mirror is reflected (a) as a plane wave, (b) so that the angle of incidence is equal to the angle of reflection. (N.)

8. What is Huygens' principle?

Draw and explain diagrams which show the positions of a light wavefront at successive equal time intervals when (a) parallel light is reflected from a plane mirror, the angle of incidence being about 60°, (b) monochromatic light originating from a small source in water is transmitted through the surface of the water into the air.

Describe an experiment, and add the necessary theoretical explanation, to show that in air the wavelength of blue light is less than that of red light. (N.)

9. Using Huygens' concept of secondary wavelets show that a plane wave of monochromatic light incident obliquely on a plane surface separating air from glass may be refracted and proceed as a plane wave. Establish the physical significance of the refractive index of the glass.

In what circumstances does dispersion of light occur? How is it accounted for by the wave theory?

If the wavelength of yellow light in air is $6 \cdot 0 \times 10^{-7}$ m, what is its wavelength in glass of refractive index $1 \cdot 5$? (N.)

Velocity of Light

10. Write down two advantages and two disadvantages of (a) Foucault's rotating mirror method and (b) Michelson's rotating prism method of determining the speed of light.

11. Draw a diagram of Foucault's method of measuring the velocity of light. How has the velocity of light in water been shown to be less than in air? The radius of curvature of the curved mirror is 20 metres and the plane mirror is rotated at 20 revs per second. Calculate the angle in degrees between a ray incident on the plane mirror and then reflected from it after the light has travelled to the curved mirror and back to the plane mirror (velocity of light = 3×10^8 m s^{-1}).

12. (a) In an experiment to determine the speed of light in air, light from a point source is reflected from one face of a sixteen-sided mirror M, travels a distance d to a stationary mirror from which it returns and, after a second reflection from M, forms an image of the source on a screen. When M is rotated at certain speeds, the image is still seen in the same position. Explain how this can occur and show that, if the lowest speed of rotation for which the image remains in the same position is n (in revolutions per second), the speed of light, c, is given by $c = 32 \ nd$.

(b) Using the above arrangement, an image is seen on the screen when the speed of rotation is 900 revolutions per second. The speed of rotation is gradually increased until at 1200 revolutions per second the image is again seen. If c = $3 \cdot 00 \times 10^8$ m s^{-1}, calculate a value for d consistent with these figures. What is the lowest speed of rotation for which an image will be seen on the screen? (N.)

13. Describe an experiment to determine the speed of light in a vacuum or in air. Show how the result is calculated from the measurements made, estimate the errors to be expected in the measurements, and deduce the maximum possible error in the result.

What result would be obtained in a determination of the speed of light in water? How have the results of such experiments influenced views as to the nature of light? (O. & C.)

14. A beam of light after reflection at a plane mirror, rotating 2000 times per minute, passes to a distant reflector. It returns to the rotating mirror from which it is reflected to make an angle of 1° with its original direction. Assuming that the velocity of light is 300 000 km s^{-1}, calculate the distance between the mirrors. (L.)

15. Describe a method of measuring the speed of light. Explain precisely what observations are made and how the speed is calculated from the experimental data.

A horizontal beam of light is reflected by a vertical plane mirror A, travels a distance of 250 metres, is then reflected back along the same path and is finally reflected again by the mirror A. When A is rotated with constant angular velocity about a vertical axis in its plane, the emergent beam is deviated through an angle of 18 minutes. Calculate the number of revolutions per second made by the mirror.

If an atom may be considered to radiate light of wavelength 500 nm for a time of 10^{-10} second, how many cycles does the emitted wave train contain? (O. & C.)

23 Interference of Light Waves

The beautiful colours seen in thin films of oil in the road, or in soap bubbles, are due to a phenomenon in light called *interference*. Newton discovered that circular coloured rings were obtained when white light illuminated a converging lens of large radius of curvature placed on a sheet of plane glass (p. 466), which is another example of interference. Interference of light has many applications in industry.

Coherent Sources

As we see later (p. 854), light waves from a sodium lamp, for example, are due to energy changes in the sodium atoms. The emitted waves occur in bursts lasting about 10^{-8} second. The light waves produced by the different atoms are out of phase with each other, as they are emitted randomly and rapidly. We call such sources of light waves as these atoms *incoherent sources* on account of the continual change of phase.

Two sodium lamps X and Y both emit light waves of the same colour or wavelength. But owing to the random emission of light waves from their atoms, their resultant light waves are constantly out of phase. So X and Y are incoherent sources. *Coherent* sources are those which emit light waves of the same wavelength which are always in phase with each other or have a constant phase difference. As we now show, two coherent sources can together produce the phenomenon of interference.

Interference of Light Waves. Constructive Interference

Suppose two sources of light, A, B, have exactly the same frequency and amplitude of vibration, and that their vibrations are always in phase with each other, Fig. 23.1. The two sources A and B are therefore *coherent* sources. *Their*

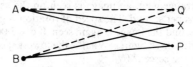

Fig. 23.1 Interference of waves

combined effect at a point is obtained by adding algebraically the displacements at the point due to the sources individually; this is known as the *Principle of Superposition*. Thus their resultant effect at X, for example, is the algebraic sum of the vibrations at X due to the source A alone and the vibrations at X due to the source B alone. If X is equidistant from A and B, the vibrations at X due to the two sources are *always* in phase as (i) the distance AX travelled by the wave originating at A is equal to the distance BX travelled by the wave originating at B, (ii) the sources A, B are assumed to have the same frequency and to be always in phase with each other.

Fig. 23.2 (i), (ii) illustrate the vibrations at X due to A, B, which have the same amplitude. The resultant vibration at X is obtained by adding the two curves, and has an amplitude double that of either curve and a frequency the same as either, Fig. 23.2 (iii). Now the energy of a vibrating source is proportional

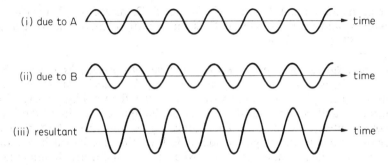

(i) due to A · · · time

(ii) due to B · · · time

(iii) resultant · · · time

Fig. 23.2 Vibrations at X— constructive interference

to the square of its amplitude (p. 521). Consequently the light energy at X is four times that due to A or B alone. A bright band of light is thus obtained at X. As A and B are coherent sources, the bright band is *permanent*. We say it is due to *constructive interference* of the light waves from A and B at X.

If Q is a point such that BQ is greater than AQ by a whole number of wavelengths (Fig. 23.1), the vibration at Q due to A is in phase with the vibration there due to B (see p. 415). A permanent bright band is then obtained at Q. Generally, a permanent bright band is obtained at any point Y if the *path difference*, BY − AY, is given by

$$BY - AY = m\lambda,$$

where λ is the wavelength of the sources A, B, and m is an integer.

We now see that permanent interference between two sources of light can only take place if they are *coherent* sources, i.e. they must have the same wavelength and be always in phase with each other or have a constant phase difference. This implies that the two sources of light must have the same colour. As we see later, two coherent sources of light can be produced by using a single primary source of light.

Destructive Interference

Consider now a point P in Fig. 23.1 whose distance from B is half a wavelength longer than its distance from A, i.e., AP − BP = $\lambda/2$. The vibration at P due to B will then be 180° out of phase with the vibration there due to A (see p. 415), Fig. 23.3 (i), (ii). The resultant effect at P is thus zero, as the displacements at any instant are equal and opposite to each other, Fig. 23.3 (iii). No light is therefore seen at P. The permanent dark band here is said to be due to *destructive interference* of the waves from A and B.

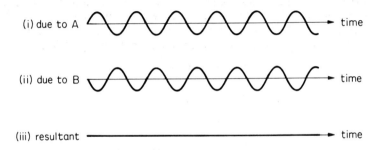

(i) due to A · · · time

(ii) due to B · · · time

(iii) resultant · · · time

Fig. 23.3 Vibrations at P—destructive interference

If the path difference, AP − BP, were $3\lambda/2$ or $5\lambda/2$, instead of $\lambda/2$, a permanent dark band would again be seen at P as the vibrations there due to A and B would be 180° out of phase.

Summarising:

If the path-difference is zero or a whole number of wavelengths, a bright band is obtained; if it is an odd number of half-wavelengths, a dark band is obtained.

From the principle of the conservation of energy, the total light energy from the sources A and B above must be equal to the light energy in all the bright bands of the interference pattern. The light energy missing from the dark bands is therefore found in the bright bands. It follows that the bright bands on a screen appear *brighter* than the screen when this is uniformly illuminated by A and B without the formation of the interference pattern.

Optical Path. Reflection of Waves.

The phase of a wave arriving at a point is affected by the medium through which it travels. For example, part of its path may be in air and part in glass. Since the velocity of light is less in glass than in air, there are more waves in a given length in glass than in an equal length in air.

Suppose light travels a distance t in a medium of refractive index n. Then if λ is the wavelength in the medium, the phase difference Δ due to this path (p. 416) is

$$\Delta = \frac{2\pi t}{\lambda} \qquad . \qquad . \qquad . \qquad . \qquad . \qquad (1)$$

If the wave travels from a vacuum (or air) to this medium, its frequency does not alter but its wavelength and velocity become smaller. Suppose λ_0 is the wavelength and c is the velocity in a vacuum. Then if v is the velocity in the medium,

$$\text{frequency} = \frac{c}{\lambda_0} = \frac{v}{\lambda}$$

$$\therefore \lambda = \frac{v}{c}\lambda_0 \qquad . \qquad . \qquad . \qquad . \qquad . \qquad (2)$$

Substituting for λ from (2) in (1),

$$\therefore \Delta = \frac{2\pi ct}{v\lambda_0} = \frac{2\pi nt}{\lambda_0} \qquad . \qquad . \qquad . \qquad (3)$$

since $n = c/v$.

From (1) and (3), we see that a light path of geometric length t in a medium of refractive index n produces the same phase change as a light path of length nt in a vacuum. We call 'nt', the product of the refractive index and path length, the *optical path* in the medium. In interference phenomena, we must always calculate the optical paths travelled by the coherent light rays. With the notation

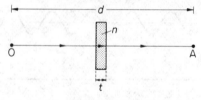

Fig. 23.4 Optical path

on p. 457, constructive interference occurs if their optical path difference is $m\lambda$.

As an illustration of optical path, suppose light travels from O to A, a distance d, in air, Fig. 23.4. The optical path $= n_0 d = d$, since the refractive index n_0 of air is practically 1. Now suppose a thin slab of glass of thickness t and refractive index n is placed between O and A so that the light passes through a length t in the glass. The optical path between O and A is now

$$(d - t) + nt = d + (n - 1)t,$$

since the light travels a distance $(d - t)$ in air and a distance t in glass.

Reflection of waves. Light waves may also undergo phase change by reflection at some point in their path. If the waves are reflected at a *denser* medium, for example, at an air–glass interface (boundary) after travelling in air, the reflected waves have a phase change of π or 180° compared to the incident waves, Fig. 23.5 (i). This phase change also occurs with matter waves such as sound waves,

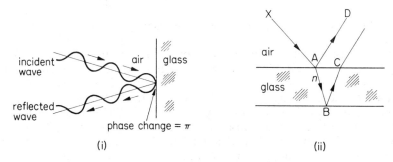

Fig. 23.5 Phase difference and reflection

as shown in Fig. 27.2. From (1), it follows that the equivalent path change t is given by

$$\Delta = \pi = \frac{2\pi t}{\lambda}$$

$$\therefore t = \frac{\lambda}{2}.$$

To take into account reflection at a *denser* medium, then, we must add (or subtract) $\lambda/2$ to the optical path.

Fig. 23.5 (ii) shows an incident ray of light XA partly refracted at A from air to glass and then reflected at B, the glass-air interface. The optical path from A to C is $n(AB + BC)$; there is no phase change by reflection at B since this occurs at an interface with the *less* dense medium air. By contrast, a phase change equivalent to a path of $\lambda/2$ occurs when XA is reflected at A along AD, since this is reflection at a denser medium, glass.

Young's Two-Slit Experiment

From our previous discussion, it can be understood that two conditions are essential to obtain an interference phenomenon: (i) Two coherent sources of light must be produced, (ii) the coherent sources must be very close to each other as the wavelength of light is very small, otherwise the bright and dark pattern produced some distance away would be too fine to see and no interference pattern is obtained.

One of the first demonstrations of the interference of light waves was given

by YOUNG in 1801. He placed a source, S, of monochromatic light in front
of a narrow slit C, and arranged two very narrow slits A, B, close to each other,
in front of C. Much to his delight, Young observed bright and dark bands on
either side of O on a screen T, where O is on the perpendicular bisector of
AB, Fig. 23.6.

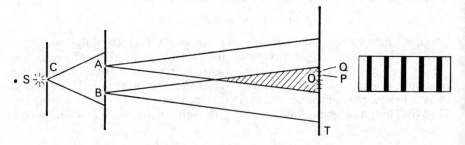

Fig. 23.6 Young's experiment—photographed fringes are shown on right

Young's observations can be explained by considering the light from S illuminating the two slits A, B. Since the light diverging from A has exactly the same
frequency as, and is always in phase with, the light diverging from B, A and
B act as two close coherent sources. Interference thus takes place in the shaded
region, where the light beams overlap, Fig. 23.6. As AO = OB, a bright band
is obtained at O. At a point close to O, such that BP − AP = $\lambda/2$, where λ is
the wavelength of the light from S, a dark band is obtained. At a point Q such
that BQ − AQ = λ, a bright band is obtained; and so on for either side of O.
Young demonstrated that the bands or *fringes* were due to interference by
covering A or B, when the fringes disappeared. Young's two-slit experiment is
an example of interference by *division of wavefront*. Here the wavefront from
C is 'divided' at A and B.

Separation of Fringes

Suppose P is the position of the mth bright fringe, so that BP − AP = $m\lambda$, Fig.
23.7. Let OP = x_m = distance from P to O, the centre of the fringe system, where
MO is the perpendicular bisector of AB. If a length PN equal to PA is described
on PB, then BN = BP − AP = $m\lambda$. Now in practice AB is very small, and PM
is very much larger than AB. Thus AN meets PM practically at right angles.
It then follows that

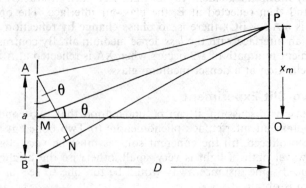

Fig. 23.7 Theory of Young's fringes (*exaggerated*)

angle PMO = angle BAN = θ say.

From triangle BAN, $\sin \theta = \dfrac{BN}{AB} = \dfrac{m\lambda}{a}$,

where $a = AB$ = the distance between the slits or their separation. From triangle PMO,

$$\tan \theta = \frac{PO}{MO} = \frac{x_m}{D},$$

where $D = MO$ = the distance from the screen to the slits. Since θ is very small, about 0·01 radian or 0·5° for 20 fringes when D is 1 metre, $\tan \theta = \sin \theta$.

$$\therefore \frac{x_m}{D} = \frac{m\lambda}{a}$$

$$\therefore x_m = \frac{mD\lambda}{a}.$$

If Q is the neighbouring or $(m-1)$th bright fringe, it follows that

$$OQ = x_{m-1} = \frac{(m-1)D\lambda}{a}$$

$\therefore$ separation y between successive fringes $= x_m - x_{m-1} = \dfrac{\lambda D}{a}$.   . (i)

$$\therefore \lambda = \frac{ay}{D}$$.   .   .   .   .   . (ii)

Measurement of Wavelength by Young's Interference Fringes

A laboratory experiment to measure wavelength by Young's interference fringes is shown in Fig. 23.8. Light from a small filament lamp is focused by a lens

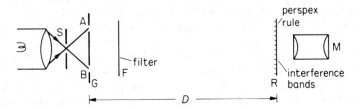

Fig. 23.8 Laboratory experiment on Young's interference fringes

on to a narrow slit S, such as that in the collimator of a spectrometer. Two narrow slits A, B, about 0·5 millimetre apart, are placed a short distance in front of S, and the light coming from A, B is viewed in a low-powered microscope or eyepiece M about one metre away. Some coloured interference fringes are then observed by M. A red and then a blue filter, F, placed in front of the slits, produces red and then blue fringes. Observation shows that the separation of the red fringes is more than that of the blue fringes. Now $\lambda = ay/D$, from (ii), where y is the separation of the fringes. It follows that the wavelength of red light is *longer* than that of blue light.

An approximate value of the wavelength of red or blue light can be found by placing a Perspex rule R in front of the eyepiece and moving it until the graduations are clearly seen, Fig. 23.8. The average distance, y, between the fringes is then measured on R. The distance a between the slits can be found by magnifying the distance by a converging lens, or by using a travelling microscope. The distance D from the slits to the Perspex rule, where the fringes are formed, is measured with a metre rule. The wavelength λ can then be calculated from $\lambda = ay/D$; it is of the order 6×10^{-7} m. Further details of the experiment can be obtained from *Advanced Level Practical Physics* by Nelkon and Ogborn (Heinemann).

Measurements can also be made using a spectrometer, with the collimator and telescope adjusted for parallel light (p. 343). The narrow collimator slit is illuminated by sodium light, for example, and the double slits placed on the table. Young's fringes can be seen through the telescope after alignment. From the theory on p. 461, the angular separation of the fringes is λ/a. Thus by measuring the average angular separation of a number of fringes with the telescope, λ can be calculated from $\lambda = a\theta$, if a is known or measured.

The wavelengths of the extreme colours of the visible spectrum vary with the observer. This may be 4×10^{-7} m for violet and 7×10^{-7} m for red; an 'average' value for visible light is 5.5×10^{-7} m, which is a wavelength in the green.

Appearance of Young's Interference Fringes

The experiment just outlined can also be used to demonstrate the following points:

1. If the source slit S is moved nearer the double slits the separation of the fringes is unaffected but their intensity increases. This can be seen from the formula y (separation) $= \lambda D/a$, since D and a are constant.

2. If the distance apart a of the slits is diminished, keeping S fixed, the separation of the fringes increases. This follows from $y = \lambda D/a$.

3. If the source slit S is widened the fringes gradually disappear. The slit S is then equivalent to a large number of narrow slits, each producing its own fringe system at different places. The bright and dark fringes of different systems therefore overlap, giving rise to uniform illumination. It can be shown that, to produce interference fringes which are recognisable, the slit width of S must be less than $\lambda D'/a$, where D' is the distance of S from the two slits A, B.

4. If one of the slits, A or B, is covered up, the fringes disappear.

5. If white light is used the central fringe is white, and the fringes either side are coloured. Blue is the colour nearer to the central fringe and red is farther away. The path difference to a point O on the perpendicular bisector of the two slits A, B is zero for all colours, and consequently each colour produces a bright fringe here. As they overlap, a white fringe is formed. Farther away from O, in a direction parallel to the slits, the shortest visible wavelengths, blue, produce a bright fringe first.

Example

In a Young's slits experiment, the separation of four bright fringes is 2·5 mm when the wavelength used is 6.2×10^{-7} m. The distance from the slits to the screen is 0·80 m. Calculate the separation of the two slits.

From previous, $\lambda = \dfrac{ay}{D}$, where a is the slit separation

$$\therefore a = \frac{\lambda D}{y} = \frac{6.2 \times 10^{-7} \times 0.8}{2.5 \times 10^{-3}/4}$$

$$= 8 \times 10^{-4} \text{ m} = 0.8 \text{ mm.}$$

Interference in Thin Wedge Films

A very thin wedge of an air film can be formed by placing a thin piece of foil or paper between two microscope slides at one end Y, with the slides in contact at the other end X, Fig. 23.9. The wedge has then a very small angle θ, as shown. When the air-film is illuminated by monochromatic light from an extended source S, straight bright and dark fringes are observed which are parallel to the line of intersection X of the two slides.

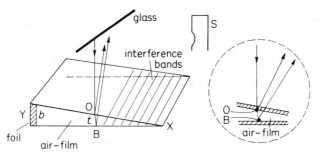

Fig. 23.9 Air-wedge fringes

The light reflected down towards the wedge is partially reflected upwards from the lower surface O of the top slide. The remainder of the light passes through the slide and some is reflected upward from the top surface B of the lower slide. The two trains of waves are coherent, since both have originated from the same centre of disturbance at O, and so they produce an interference pattern if brought together by the eye or in an eyepiece.

The path difference is $2t$, where t is the small thickness of the air-film at O. At X, where the path difference is apparently zero, we would expect a bright fringe. But a *dark* fringe is observed at X. This is due to a phase change of 180°, equivalent to an extra path difference of $\lambda/2$, which occurs when a wave is reflected at a denser medium. See p. 459. The optical path difference between the two coherent beams is thus actually $2t + \lambda/2$. Hence, if the beams are brought together to interfere, a bright fringe is obtained when $2t + \lambda/2 = m\lambda$, or $2t = (m - \frac{1}{2})\lambda$. A dark fringe is obtained at a thickness t given by $2t = m\lambda$.

The bands or fringes are located at the air-wedge film, and the eye or microscope must be focused here to see them. The appearance of a fringe is the contour of all points in the wedge air film where the optical path difference is the same. If the wedge surfaces make perfect optical contact at one edge, the fringes are straight lines parallel to the line of intersection of the surfaces. If the glass surfaces are uneven, and the contact at one edge is not regular, the fringes are not perfectly straight. A particular fringe still shows the locus of all points in the air-wedge which have the same optical path difference in the air-film.

In *transmitted light*, the appearance of the fringes is complementary to those seen by reflected light, from the law of conservation of energy. The bright fringes thus correspond in position to the dark fringes seen by reflected light, and the fringe where the surfaces touch is now bright instead of dark.

The wedge air film is an example of interference by *division of amplitude*. Here part of the wave is transmitted at O and the remainder is reflected at O, so that the amplitude of the wave, which is a measure of its energy (p. 521), is 'divided' into two parts. This method of producing interference is basically different from producing interference by division of wavefront (p. 460).

Fig. 23.10 Interference bands in air-wedges. The angle of the air-wedge on the right is about $2\frac{1}{2}$ times less than that on the left, so that the separation of the bands is correspondingly greater

Thickness of Thin Foil. Expansion of Crystal

If there is a bright fringe at Y at the edge of the foil, Fig. 23.9, the thickness b of the foil is given by $2b = (m + \frac{1}{2})\lambda$, where m is the number of bright fringes between X and Y. If there is a dark band at Y, then $2b = m\lambda$. Thus by counting m, the thickness b can be found. The small angle θ of the wedge is given by b/a, where a is the distance XY, and by measuring a with a travelling microscope focused on the air-film, θ can be found.

The angle θ of the wedge can also be found from the separation s of the bright bands. In Fig. 23.11, B_1 and B_2 are consecutive bright bands. So the

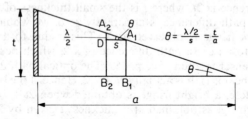

Fig. 23.11 Air-wedge theory

extra path difference in the air film from one band to the other is λ. The extra path difference is $2t_2 - 2t_1$, or $2A_2D$, where A_1D is the perpendicular from A_1 to A_2B_2. So $A_2D = \lambda/2$. Hence if s is the separation, B_1B_2, of the bands, which is equal to A_1D, then

$$\tan \theta = \frac{A_2D}{A_1D} = \frac{\lambda/2}{s} = \frac{\lambda}{2s}.$$

Since θ is very small, $\tan \theta = \theta$ in radians. So

$$\theta = \frac{\lambda}{2s}.$$

As an illustration, suppose an air wedge is illuminated normally by monochromatic light of wavelength $5\cdot8 \times 10^{-7}$ m (580 nm) and the separation of the bright bands is $0\cdot29$ mm or $2\cdot9 \times 10^{-4}$ m. Then, from $\theta = \lambda/2s$, we have

$$\theta = \frac{5{\cdot}8 \times 10^{-7}}{2 \times 2{\cdot}9 \times 10^{-4}} = 10^{-3} \text{ rad.}$$

If a *liquid wedge* is formed between the plates, the optical path difference becomes $2nt$, where the air thickness is t, n being the refractive index of the liquid. An optical path difference of λ now occurs for a change in t which is n times *less* than in the case of the air-wedge. The spacing of the bright and dark fringes is thus n times closer than for air. So measurement of the relative spacing enables n to be found.

The coefficient of expansion of a crystal can be found by forming an air-wedge of small angle between a fixed horizontal glass plate and the upper surface of the crystal, and illuminating the wedge by monochromatic light. When the crystal is heated a number of bright fringes, m say, cross the field of view in a microscope focused on the air-wedge. The increase in length of the crystal in an upward direction is $m\lambda/2$, since a change of λ represents a change in the thickness of the film is $\lambda/2$, and the coefficient of expansion can then be calculated.

Examples

1. A wedge air film is formed by placing aluminium foil between two glass slides at a distance of 75 mm from the line of contact of the slides. When the air wedge is illuminated normally by light of wavelength $5{\cdot}60 \times 10^{-7}$ m, interference fringes are produced parallel to the line of contact which have a separation of $1{\cdot}20$ mm. Calculate the angle of the wedge and the thickness of the foil.

If s is the separation of the bands, then

$$\text{angle of wedge, } \theta = \frac{\lambda/2}{s} = \frac{\lambda}{2s}$$

$$= \frac{5{\cdot}6 \times 10^{-7}}{2 \times 1{\cdot}2 \times 10^{-3}} = 2{\cdot}3 \times 10^{-4} \text{ rad.}$$

If t is the thickness of the foil, then

$$\frac{t}{75 \times 10^{-3}} = \theta = 2{\cdot}3 \times 10^{-4}.$$

So $\quad t = 75 \times 10^{-3} \times 2{\cdot}3 \times 10^{-4} = 1{\cdot}7 \times 10^{-5} \text{ m.}$

2. Two optically flat glass plates, in contact along one edge, make a very small angle with each other. They are illuminated by red light of wavelength 750 nm and blue light of wavelength 450 nm. Looking down on the wedge the first place where it appears purple is $5{\cdot}0$ mm from the line of contact.

If red and blue light together produce purple light, find the angle between the plates.

The first place where the wedge appears purple corresponds to a thickness t of the air wedge where both red and blue light first form a coincident bright interference band. The colours then mix and produce purple light.

At the thickness t, the path difference for bright interference band of wavelength λ is given by $2t = (m - \frac{1}{2})\lambda$, from p. 463. So for $\lambda = 750$ nm and $\lambda = 450$ nm,

$$2t = (m - \tfrac{1}{2})\ 750 \text{ nm} = (m + \tfrac{1}{2})\ 450 \text{ nm,}$$

since if m is the whole number for the 750 nm wavelength, then $(m + 1)$ is the whole number for the overlapping 450 nm (shorter) wavelength.

$$\therefore\ (m - \tfrac{1}{2})\ 750 = (m + \tfrac{1}{2})\ 450.$$

Solving, $\qquad\qquad\qquad m = 2.$

So $\quad 2t = (2 - \frac{1}{2}) \times 750 \text{ nm} = 1125 \text{ nm, and } t = 562{\cdot}5 \text{ nm} = 5{\cdot}6 \times 10^{-7} \text{ m (approx).}$

Hence $\quad$ angle of wedge $\theta = \dfrac{t}{5 \times 10^{-3}} = \dfrac{5{\cdot}6 \times 10^{-7}}{5 \times 10^{-3}} = 1{\cdot}1 \times 10^{-4} \text{ rad (approx).}$

Newton's Rings

Newton discovered an example of interference which is known as 'Newton's rings'. In this case a lens L is placed on a sheet of plane glass H, L having a lower surface of very large radius of curvature, Fig. 23.12. By means of a

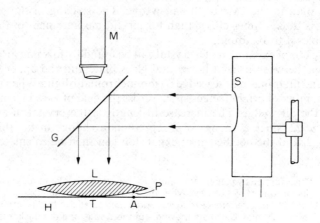

Fig. 23.12 Newton's rings

sheet of glass G monochromatic light from a sodium lamp S, for example, is reflected downwards towards L; and when the light reflected upwards is observed through a microscope M focused on H, a series of bright and dark rings is seen. The circles have increasing radius, and are concentric with the point of contact T of L with H. See Fig. 23.13.

Consider the air-film PA between A on the plate and P on the lower lens surface. Some of the incident light is reflected from P to the microscope, while the remainder of the light passes straight through to A, where it is also reflected to the microscope and brought to the same focus. The two rays of light have thus a net path difference of $2t$, where $t = $ PA. The same path difference is obtained at all points round T which are distant TA from T; and hence if $2t = $

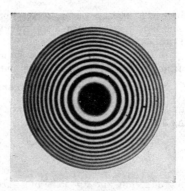

Fig. 23.13 Newton's rings, formed by interference of yellow light between converging lens and flat glass plate

$m\lambda$, where m is an integer and λ is the wavelength, we might expect a bright *ring* with centre T. Similarly, if $2t = (m + \frac{1}{2})\lambda$, we might expect a dark ring.

When a ray is reflected from an optically *denser* medium, however, a phase change of 180° occurs in the wave, which is equivalent to an extra path difference of $\lambda/2$ (see also p. 459). The truth of this statement can be seen by the presence of the dark spot at the centre, T, of the rings. At this point there is no geometrical path difference between the rays reflected from the lower surface of the lens and H, so that they should be in phase when they are brought to a focus and should form a bright spot. The dark spot means, therefore, that one of the rays suffers a phase change of 180°. Taking the phase change into account, it follows that

$$2t = m\lambda \text{ for a } dark \text{ ring} \qquad . \qquad . \qquad . \qquad . \qquad (1)$$

and

$$2t = (m + \tfrac{1}{2})\lambda \text{ for a } bright \text{ ring} \qquad . \qquad . \qquad . \qquad (2)$$

where m is an integer. Young verified the phase change by placing oil of sassafras between a crown and a flint glass lens. This liquid had a refractive index greater than that of crown glass and less than that of flint glass, so that light was reflected at an optically denser medium at each lens. A bright spot was then observed in the middle of the Newton's rings, showing that no net phase change had now occurred.

The grinding of a lens surface can be tested by observing the appearance of the Newton's rings formed between it and a flat glass plate when mono-chromatic light is used. If the rings are not perfectly circular as in Fig. 23.13, the grinding is imperfect (see Fig. 23.14). As in the case of the wedge air film, Newton's rings is an example of interference by division of amplitude (p. 463).

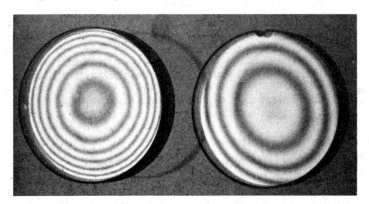

Fig. 23.14 Newton's rings. Newton's rings formed by transmitted light. The central spot is bright since there is no phase change on transmission (compare the dark central spot in Newton's rings formed by reflected light). If the lens surface is imperfect the rings are distorted, as shown on the right

Measurement of Wavelength by Newton's Rings

The radius r of a ring can be expressed in terms of the thickness, t, of the corresponding layer of air by simple geometry. Suppose TO is produced to D to meet the completed circular section of the lower surface PO of the lens of radius a, PO being perpendicular to the diameter TD through T, Fig. 23.15. Then, from the well-known theorem concerning the segments of chords in a circle, TO. OD = QO. OP. But AT $= r =$ PO, QO = OP $= r$, AP $= t =$ TO, and OD $= 2a -$ OT $= 2a - t$.

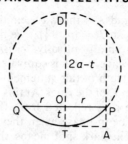

Fig. 23.15 Theory of radius of Newton's rings

$$\therefore\ t(2a - t) = r \times r = r^2$$

$$\therefore\ 2at - t^2 = r^2.$$

But t^2 is very small compared with $2at$, as a is large.

$$\therefore\ 2at = r^2$$

$$\therefore\ 2t = \frac{r^2}{a} \qquad\qquad . \quad . \quad . \quad . \quad . \quad (i)$$

But $\qquad\qquad 2t = (m + \tfrac{1}{2})\lambda$ for a bright ring.

$$\therefore\ \frac{r^2}{a} = (m + \tfrac{1}{2})\lambda \qquad\qquad . \quad . \quad . \quad . \quad (3)$$

The first bright ring obviously corresponds to the case of $m = 0$ in equation (3); the second bright ring corresponds to the case of $m = 1$. Thus the radius of the 15th bright ring is given from (3) by $r^2/a = 14\tfrac{1}{2}\lambda$, from which $\lambda = 2r^2/29a$. Knowing r and a, therefore, the wavelength λ can be calculated. Experiment shows that the rings become narrower when blue or violet light is used in place of red light, which proves, from equation (3), that the wavelength of violet light is shorter than the wavelength of red light. Similarly it can be proved that the wavelength of yellow light is shorter than that of red light and longer than the wavelength of violet light.

Visibility of Newton's Rings

When white light is used in Newton's rings experiment the rings are coloured, generally with violet at the inner and red at the outer edge. This can be seen from the formula $r^2 = (m + \tfrac{1}{2})\lambda a$, since $r^2 \propto \lambda$.

When Newton's rings are formed by sodium light, close examination shows that the clarity, or visibility, of the rings gradually diminishes as one moves outwards from the central spot, after which the visibility improves again. The variation in clarity is due to the fact that sodium light is not monochromatic but consists of *two wavelengths*, λ_2, λ_1, close to one another. These are (i) $\lambda_2 = 5890 \times 10^{-10}$ m (D_2), (ii) $\lambda_1 = 5896 \times 10^{-10}$ m (D_1). Each wavelength produces its own pattern of rings, and the ring patterns gradually separate as m, the number of the ring, increases. When $m\lambda_1 = (m + \tfrac{1}{2})\lambda_2$, the bright rings of one wavelength fall in the dark spaces of the other and the visibility is a minimum. In this case

$$5896m = 5890\,(m + \tfrac{1}{2}).$$

$$\therefore\quad m = \frac{5890}{12} = 490 \text{ (approx.).}$$

At a further number of ring m_1, when $m_1\lambda_1 = (m_1 + 1)\lambda_2$, the bright (and dark) rings of the two ring patterns coincide again, and the clarity, or visibility, of the interference pattern is restored. In this case

$$5896m_1 = 5890(m_1 + 1),$$

from which $m_1 = 980$ (approx.). Thus at about the 500th ring there is a minimum visibility, and at about the 1000th ring the visibility is a maximum.

It may be noted here that the fringes in films of varying thickness, such as Newton's rings and the air-wedge fringes, p. 463, appear to be formed in the film itself, and the eye must be focused on the film to see them. We say that the fringes are 'localized' at the film. With a thin film of uniform thickness, however, fringes are formed by parallel rays which enter the eye, and these fringes are therefore localized at infinity.

Fresnel's Biprism Experiment

Fresnel used a biprism R which had a very large angle of nearly 180°, and placed a narrow slit S, illuminated by monochromatic light, in front of it so that the refracting edge was parallel to the slit, Fig. 23.16. The light emerging after

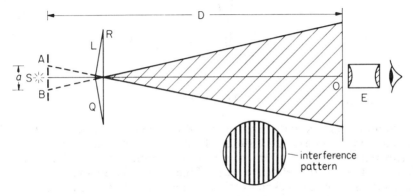

Fig. 23.16 Fresnel's biprism experiment (*not to scale*)

refraction from the two halves, L, Q, of the prism can be considered to come from two sources, A, B, which are the virtual images of the slit S in L, Q respectively. Thus A, B are coherent sources; further, as R has a very large obtuse angle, A and B are close together. Thus an interference pattern is observed in the region of O where the emergent light from the two sources overlap, as shown by the shaded portion of Fig. 23.16, and bright and dark fringes can be seen through an eyepiece E at O directed towards R. By using crosswires, and moving the eyepiece by a screw arrangement, the distance y between successive bright fringes can be measured. Now it was shown on p. 461 that $\lambda = ay/D$, where a is the distance between A, B and D is the distance of the source slit from the eyepiece. The distance D is measured with a metre rule. The distance a can be found by moving a converging lens between the fixed biprism and eyepiece until a magnified image of the two slits A, B is seen clearly, and the magnified distance b between them is measured. The magnification m is (image distance ÷ object distance) for the lens, and a can be calculated from $a = b/m$. Knowing a, y, D, the wavelength λ can be determined.

If A is the large angle, nearly 180°, of the biprism, each of the small base angles is $(180° - A)/2$, or $90° - A/2$. The small deviation d in radians of light from the slit S is $(n - 1)\,\theta$, where θ is the magnitude of the base angle in radians (p. 353), and hence the distance A, B between the virtual images of the slit = $2td = 2t\,(n - 1)\,\theta$, where t is the distance from S to the biprism. Fresnel's biprism experiment is another example of interference by division of wavefront.

'Blooming' of Lenses

Whenever lenses are used, a small percentage of the incident light is reflected from each surface. In compound lens systems, as in telescopes and microscopes, this produces a background of unfocused light, which results in a reduction in the clarity of the final image. There is also a reduction in the intensity of the image, since less light is transmitted through the lenses.

The amount of reflected light can be considerably reduced by evaporating a thin coating of a fluoride salt such as magnesium fluoride on to the surfaces, Fig. 23.17. Some of the light, of average wavelength λ, is then reflected from

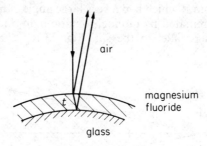

Fig. 23.17 Blooming of lens

the air-fluoride surface and the remainder penetrates the coating and is partially reflected from the fluoride-glass surface. Destructive interference occurs between the two reflected beams when there is a phase difference of 180°, or a path difference of $\lambda/2$, as the refractive index of the fluoride is less than that of glass. Thus if t is the required thickness of the coating and n' its refractive index, $2n't = \lambda/2$. Hence $t = \lambda/4n' = 6 \times 10^{-7}/(4 \times 1\cdot38)$, assuming λ is 6×10^{-7} m and n' is $1\cdot38$; thus $t = 1\cdot1 \times 10^{-7}$ m.

For best results n' should have a value equal to about $\sqrt{n}$, where n is the refractive index of the glass lens. The intensities of the two reflected beams are then equal, and hence complete interference occurs between them. No light is then reflected back from the lens. In practice, since complete interference is not possible simultaneously for every wavelength of white light, an average wavelength for λ, such as green-yellow, is chosen. The lens thus appears purple, a mixture of red and blue, since these colours in white light are reflected. 'Bloomed' lenses produce a marked improvement in the clarity of the final image in optical instruments.

Lloyd's Mirror

In 1834 LLOYD obtained interference fringes on a screen by using a plane mirror M, and illuminating it with light nearly at grazing incidence, coming from a slit O parallel to the mirror, Fig. 23.18. A point such as A on the screen is illuminated (i) by waves from O travelling along OA and (ii) by waves from O travelling along OM and then reflected along MA, which appear to come from the virtual image I of O in the mirror. Since O and I are close coherent sources interference fringes are obtained on the screen.

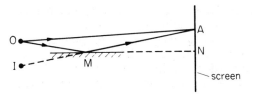

Fig. 23.18 Lloyd's mirror experiment

Experiment showed that the fringe at N, which corresponds to the point of intersection of the mirror and the screen, was *dark*; since ON = IN, this fringe might have been expected, before the experiment was carried out, to be bright. Lloyd concluded that a phase change of 180°, equivalent to half a wavelength, occurred by reflection at the mirror surface, which is a denser surface than air (see p. 459). Lloyd's mirror experiment is an example of interference by division of wavefront (p. 460).

Colours in Thin Films

The colours in thin films of oil or glass are due to interference from an extended source such as the sky or a cloud. Fig. 23.19 illustrates interference between

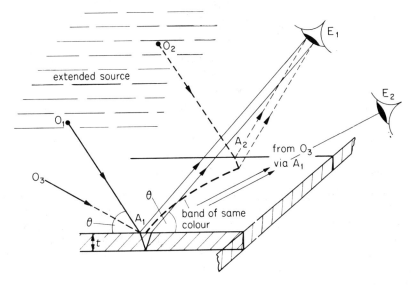

Fig. 23.19 Colours in thin films

rays from points O_1, O_2 respectively on the extended source. Each ray is reflected and refracted at A_1, A_2, on the film, and enter the eye at E_1. Although O_1, O_2 are non-coherent, the eye will see the same colour of a particular wavelength λ if $2nt \cos r = (m - \frac{1}{2})\lambda$, the condition for constructive interference, or the complementary colour if $2nt \cos r = m\lambda$, the condition for destructive interference. These conditions are proved shortly.

The separation of the two rays from A_1 or from A_2 must be less than the diameter of the eye-pupil for interference to occur, and this is the case only for thin films. The angle of refraction r is determined by the angle of incidence, or reflection, at the film. The particular colour seen thus depends on the position of the eye. At E_2, for example, a different colour will be seen from another point O_3 on the extended source. The variation of θ and hence r is small when the eye observes a particular area of the film. So a fringe of a particular colour is the contour of paths of *equal inclination* to the film such as A_1A_2. Since the angle θ is constant round the perpendicular line from the eye to the film, the fringe or band of a particular colour is *circular*. So if $2nt \cos r = m\lambda$ for a blue colour, the eye sees a circular band of the complementary colour such as green-yellow.

Interference in Thin Films

We can now prove the relation for constructive interference or bright bands and for destructive interference or dark bands which we used to account for the colours in thin films.

Consider a ray AO of monochromatic light incident on a thin parallel-sided film of thickness t and refractive index n. Fig. 23.20 is exaggerated for clarity.

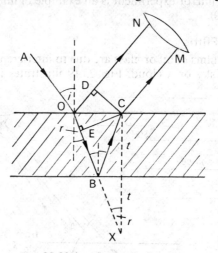

Fig. 23.20 Interference in thin films

Some of the light is reflected at O along ON, while the remainder is refracted into the film, where reflection occurs at B. The ray BC then emerges into the air along CM, which is parallel to ON. The incident ray AO thus divides at O into two beams of different amplitude which are coherent and if ON, CM are combined by a lens, or by the eye-lens, a bright or dark fringe is observed according to the path difference of the rays.

The *optical* path of a length y in a medium of refractive index n is ny (p. 458). The optical path difference between the two rays ON and OBCM is thus n (OB + BC) − OD, where CD is perpendicular to ON, Fig. 23.17. If CE is the perpendicular from C to OB, then OD/OE = $\sin i/\sin r = n$, so that $nOE = OD$.

$$\therefore \text{ optical path difference} = n(EB + BC) = n(EB + BX) = n.EX.$$

$$= 2nt \cos r,$$

where r is the angle of refraction in the film. With a phase change of $180°$ by reflection at a denser medium, a bright fringe is therefore obtained when $2nt \cos r + \lambda/2 = m\lambda$,

or
$$2nt \cos r = (m - \tfrac{1}{2})\lambda \qquad . \qquad . \qquad . \qquad . \qquad \text{(i)}$$

For a dark fringe,
$$2nt \cos r = m\lambda \qquad . \qquad . \qquad . \qquad . \qquad \text{(ii)}$$

These are the relations for bright and dark fringes for light reflected by the film. As explained for the case of the illuminated air-wedge (p. 463), in transmitted light bright fringes are seen corresponding to the positions of the dark fringes and dark fringes to the positions of the bright fringes.

Vertical Soap Film Colours

An interesting experiment on thin films, due to C. V. Boys, can be performed by illuminating a vertical soap film with monochromatic light. At first the film appears uniformly coloured. As the soap drains to the bottom, however, a wedge-shaped film of liquid forms in the ring, the top of the film being thinner than the bottom. The thickness of the wedge is constant in a horizontal direction, and thus horizontal bright and dark fringes are observed across the film. When the upper part of the film becomes extremely thin a black fringe is observed at the top (compare the dark central spot in Newton's rings experiment), and the film breaks shortly afterwards.

With white light, a succession of broad coloured fringes is first observed in the soap film. Each fringe contains colours of the spectrum, red to violet. The fringes widen as the film drains, and just before it breaks a black fringe is obtained at the top.

For normal incidence of white light, a particular wavelength λ is seen where the optical path difference due to the film $= (m - \frac{1}{2})\lambda$ and m is an integer. Thus a red colour of wavelength $7 \cdot 0 \times 10^{-7}$ m is seen where the optical path difference is $3 \cdot 5 \times 10^{-7}$ m, corresponding to $m = 1$. No other colour is seen at this part of the thin film. Suppose, however, that another part of the film is much thicker and the optical path difference here is $21 \times 3 \cdot 5 \times 10^{-7}$ m. Then a red colour of wavelength $7 \cdot 0 \times 10^{-7}$ m, $m = 11$, an orange colour of wavelength about $6 \cdot 4 \times 10^{-7}$ m, $m = 12$, a yellow wavelength about $5 \cdot 9 \times 10^{-7}$ m, $m = 13$, and other colours of shorter wavelengths corresponding to higher integral values of m, are seen at the same part of the film. These colours all overlap and produce a white colour. If the film is thicker still, it can be seen that numerous wavelengths throughout the visible spectrum are obtained and the film then appears uniformly white.

Michelson Interferometer

Michelson designed an apparatus which used long path lengths to produce interference fringes. The principle of the interferometer is shown in Fig. 23.21. Light from a monochromatic source S is incident at 45° on a thick glass plate A which is partially silvered on the back at P. Here the light splits up into

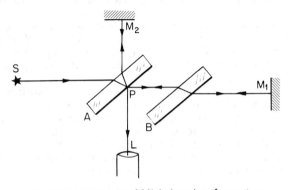

Fig. 23.21 Principle of Michelson interferometer

two beams. One is reflected to a plane mirror M_2 normal to its path which reflects the beam straight back. The other beam is transmitted through P and is incident normally on a plane mirror M_1 which also reflects the light beam back to P.

The light reflected from M_2 passes through the plate A along the direction PL. The light reflected from M_1 is reflected at P in the same direction PL. Since the light from M_2 has passed through two thicknesses of glass A in its journey from P to M_2 and back, a glass plate B of the same thickness as A and parallel to it is used to compensate for the extra path difference. The light from P to M_1 and back to P then has a similar path length as the beam from M_2.

The light beams along PL are coherent beams since both originate from P. An observer using a lens L or a low power telescope thus sees bright and dark interference fringes, which are circular (see p. 472). A change in path difference of one wavelength λ of the two beams reaching L produces a shift of one fringe across the field of view. This occurs when the mirror M_2 is moved slowly back a distance d from A equal to $\lambda/2$. By counting the number of fringes which move across the field of view, and measuring the distance M_2 has moved by means of a micrometer screw, the wavelength λ of the light source can be measured. The Michelson interferometer is very sensitive. A fringe shift of only 0·01 can be measured with it. In this way the wavelengths in the green, blue and red lines in the spectrum of cadmium were measured to one part in 100 million.

The standard metre was originally the distance between two scratches on a particular metal bar. The centres of the scratches were not precise and Michelson suggested that the length of the standard metre could be measured very accurately in terms of a particular wavelength. In a series of experiments, he determined the number of wavelengths in 1 metre of a particular green line in the cadmium spectrum. Today, from measurements with improved types of interferometer, the metre is defined as the length equal to 1 650 763·73 waves in a vacuum of the radiation corresponding to a particular wavelength in the spectrum of the krypton-86 atom. We therefore have a very accurate reproducible standard metre, even if the original standard metre bar is damaged or lost.

The Ether. Principle of Michelson-Morley Experiment

Michelson designed his sensitive interferometer in 1881. At that time many scientists believed that, since light was a wave-motion, it needed a medium to transmit it, just as sound waves required air or other matter for transmission. They called the medium *ether*. If the ether existed and was stationary in space, it could provide a frame of reference to measure the absolute velocity of moving objects such as the earth or the planets in their orbits.

The velocity of the earth in its orbit round the sun is about 30 km s^{-1}. If the earth moved through a stationary ether, at one time in its orbit an observer using a light beam in a laboratory would detect an ether wind directly opposite to the velocity of the light. Some time later in its orbit, the observer would detect the ether wind directly assisting the velocity of the light beam.

Michelson and Morley carried out a crucial experiment to test whether the ether existed. The principle used was similar to the case of two swimmers X and Y moving with the same speed through equal distances in a stream of water, Fig. 23.22. X travels a distance d upstream against the stream velocity v, which takes a time $d/(c-v)$, where c is the actual velocity of the swimmer in still water, Fig. 23.22 (i). When X returns downstream to the same place his velocity is now assisted by the stream velocity and the time taken is $d/(c+v)$. The total time t_1 to travel a distance $2d$ is

$$t_1 = \frac{d}{c-v} + \frac{d}{c+v} = \frac{2dc}{c^2 - v^2} \cdot \qquad . \qquad . \qquad . \qquad (1)$$

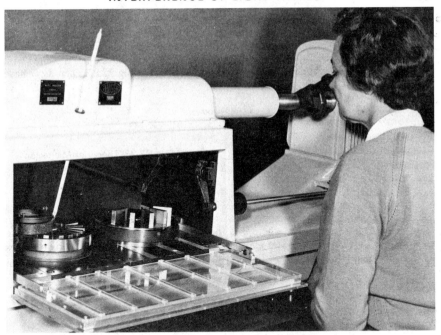

Fig. 23A *Interferometer measurement of length*. The photograph shows an interferometer used to measure the accuracy of block (slip) guages, widely used for precision measurements in the engineering industry. As shown, the gauges are placed on turntables. Optical flats are used with them to form interference bands; a circular one is shown ready for use above the guages on the left. Any error in length is deduced from an accurately-known wavelength in the visible spectrum of cadmium, which is used to form the bands. Correction is needed for temperature changes, observed by the long mercury thermometer shown. (*Crown copyright. Courtesy of National Physical Laboratory*)

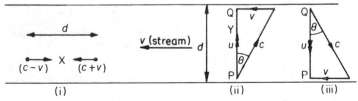

Fig. 23.22 Principle of Michelson–Morley experiment

Y, however swims *at right angles* to the stream through a distance $2d$. To do this, the velocity c must be directed at a suitable angle θ to the stream, Fig. 23.22 (ii). The actual speed u from P to Q is the resultant of c and v, and is therefore given by $u = \sqrt{c^2 - v^2}$. To return from Q to P the velocity c must again be directed at an angle θ to the stream as shown in Fig. (iii) and the resultant velocity is $\sqrt{c^2 - v^2}$ again. So the total time t_2 to travel a distance $2d$ is given by

$$t_2 = \frac{2d}{\sqrt{c^2 - v^2}} \qquad \qquad . \qquad . \qquad . \qquad . \qquad . \qquad . \qquad (2)$$

Dividing (2) by (1), we obtain

$$\frac{t_2}{t_1} = \frac{\sqrt{c^2 - v^2}}{c} = \sqrt{\frac{c^2 - v^2}{c^2}} = \sqrt{1 - \frac{v^2}{c^2}} . \qquad . \qquad . \qquad (3)$$

If v is less than c, then v^2/c^2 is less than 1 and from (3) we see that t_2 is less than t_1. So the swimmer X, moving at right angles to the stream, takes a shorter time to travel the same distance $2d$ than the swimmer Y, who moves with and then against the stream.

Michelson-Morley Experiment and Result

Using his interferometer, Michelson and Morley carried out with two light beams an experiment to test the ether theory which was based on the principle just discussed.

Suppose the light beam transmitted through P to M_1 and back to P travelled with velocity c in a direction opposite to the ether wind of velocity v Fig. 23.23. Then, as we have

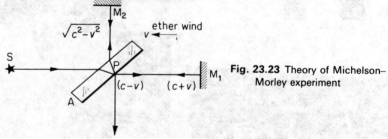

Fig. 23.23 Theory of Michelson–Morley experiment

shown, the time to travel $2d$, where d is the optical path from P to M_1, is given by $t_1 = 2dc/(c^2 - v^2)$. The light beam reflected at P to M_2 and back travels at right angles to the ether wind. So the shorter time taken, $t_2, = 2d/\sqrt{c^2 - v^2}$, as shown. Now from equation (3),

$$t_2 = t_1 \sqrt{1 - \frac{v^2}{c^2}}.$$

So time difference, $t_1 - t_2 = t_1 - t_1 \sqrt{1 - \frac{v^2}{c^2}} = t_1 - t_1 \left(1 - \frac{v^2}{c^2}\right)^{1/2}$

$$= t_1 - t_1 \left(1 - \frac{v^2}{2c^2}\right)$$

by binomial expansion, since c^2 is very much greater than v^2. So

$$t_1 - t_2 = t_1 \frac{v^2}{2c^2}.$$

From equation (1), $t_1 = 2dc/(c^2 - v^2) = 2dc/c^2$, since v^2 is very small compared with c^2, so $t_1 = 2d/c$. Hence the small difference Δt in the time is given by

$$t_1 - t_2 = \Delta t = \frac{dv^2}{c^3}. \qquad \qquad . \qquad . \qquad . \qquad . \qquad (1)$$

If the ether velocity v is zero, there would be no difference in the time between the two light beams arriving back at P, that is, $t_1 = t_2$. If there is an ether wind velocity v, the time difference Δt represents a shift in fringes. The distance travelled by light in a time Δt is $c.\Delta t$ and if λ is the wavelength of the light, the number of waves ΔN in this distance is $c.\Delta t/\lambda$. So, from (1),

$$\Delta N = \frac{c.\Delta t}{\lambda} = \frac{dv^2}{\lambda c^2}.$$

Michelson used a distance d from P to M_1 or M_2 of about 10 m. Suppose the wavelength λ of the light waves was 5×10^{-7} m. Then, with $c = 3 \times 10^8$ m s^{-1} and an ether velocity of 30 km s^{-1} or 3×10^4 m s^{-1}, the earth orbital velocity, the number of fringes displaced is

$$\Delta N = \frac{10 \times (3 \times 10^4)^2}{5 \times 10^{-7} \times (3 \times 10^8)^2} = 0.2.$$

The Michelson interferometer is so sensitive that a fringe shift of only 0.01, which is 20 times less than the estimated fringe shift of 0.2, could be detected.

The interferometer was placed on a large stone and floated on mercury in a trough. It was then slowly rotated through 90°, so that the light paths in the two beams reaching the observer were interchanged and the fringe shift was then 2×0.2 or 0.4. Numerous observations of the fringe pattern, however, taken at different times of the earth's orbital path during the seasons, showed that *no shift occurred*.

The ether did not therefore exist. Explanations were proposed for the negative result of the Michelson-Morley experiment based on relative motion but none was satisfactory. The ether theory was therefore abandoned. Einstein used the negative result to state that the velocity of light was a constant value irrespective of the motion of the observer and used this as a fundamental principle in his theory of special relativity, proposed in 1905. Among other results of the theory, he declared that mass and energy are related by the equation $E = mc^2$, where E is the energy produced by a mass change m. This was verified in experiments on radioactivity (p. 896).

Exercises 23

1. In Young's two-slit experiment using red light, state the effect of the following procedure on the appearance of the fringes:
  (a) The separation of the slits is decreased.
  (b) The screen is moved closer to the slits.
  (c) The source slit is moved closer to the two slits.
  (d) Blue light is used in place of red light.
  (e) One of the two slits is covered up.
  (f) The source slit is made wider.

2. In a Young's two-slit experiment using light of wavelength 6.0×10^{-7} m, the slits were 0.40 mm apart and the distance of the slits to the screen was 1.20 m. Find the separation of the fringes.
  What is the angle in radians subtended by a central pair of bright fringes at the slits?

3. In an air-wedge experiment using white light, the following are observed:
  (a) A dark band is obtained where the two slides touch.
  (b) The bands nearest to the dark band are coloured blue.
  (c) The bands are straight and parallel.
Explain each of these effects.

4. (a) A wedge-shaped film of air is formed between two, thin, parallel-sided, glass plates by means of a straight piece of wire. The two plates are in contact along one edge of the film and the wire is parallel to this edge.
  (i) Draw and label a diagram of the experimental arrangement you would use to observe and make measurements on interference fringes produced with light incident normally on the film. (ii) Explain the function of each part of the apparatus.
  (b) In such an experiment using light of wavelength 589 nm, the distance between the seventh and one hundred and sixty-seventh dark fringes was 26.3 mm and the distance between the junction of the glass plates and the wire was 35.6 mm. Calculate the angle of the wedge and the diameter of the wire. (*N.*)

5. (a) Describe, with the aid of a diagram, how you would produce and view interference fringes using a monochromatic light source, a double slit and any other essential apparatus. Describe the appearance of the fringes.
  (b) State the measurements you would make in order to determine the wavelength of the light and indicate the instrument you would use for each measurement, justifying your choice in each case.
  (c) Derive an expression for the wavelength in terms of the relevant measurements.
  (d) Describe the effect on the appearance of the fringes of reducing the slit separation and discuss how this would affect the accuracy of your measurements in (b) above. (*N.*)

6. Two plane glass plates which are in contact at one edge are separated by a piece of metal foil 12.50 cm from that edge. Interference fringes parallel to the line of contact are observed in reflected light of wavelength 5.46×10^{-7} m and are found to be 1.50 mm apart. Find the thickness of the foil. (*L.*)

7. Explain why, for visible interference effects, it is normally necessary for the light to

come from a single source and to follow different optical paths. Include a statement of the conditions required for complete destructive interference.

Explain *either* the colours seen in thin oil films *or* the colours seen in soap bubbles.

A wedge-shaped film of air between two glass plates gives equally spaced dark fringes, using reflected sodium light, which are 0·22 mm apart. When monochromatic light of another wavelength is used the fringes are 0·24 mm apart. Explain why the two fringe spacings are different. (The incident light falls normally on the air film in both cases.) Discuss what you would expect to see if the air film were illuminated by both sources simultaneously. Calculate the wavelength of the second source of light.

[Wavelength of sodium light = 589 nm (589 × 10⁻⁹ m).] (*L.*)

8. Describe how to set up apparatus to observe and make measurements on the interference fringes produced by Young's slits. Explain how (i) the wavelengths of two monochromatic light sources could be compared, (ii) the separation of the slits could be deduced using a source of known wavelength. Establish any formula required.

State, giving reasons, what you would expect to observe (*a*) if a white light source were substituted for a monochromatic source, (*b*) if the source slit were then displaced slightly at right angles to its length in the plane parallel to the plane of the Young's slits. (*L.*)

9. Describe, with the aid of a labelled diagram, how the wavelength of monochromatic light may be found using Young's slits. Give the theory of the experiment.

State, and give physical reasons for the features which are common to this method and to *either* the method based on Lloyd's mirror *or* that based on Fresnel's biprism.

In an experiment using Young's slits the distance between the centre of the interference pattern and the tenth bright fringe on either side is 3·44 cm and the distance between the slits and the screen is 2·00 m. If the wavelength of the light used is 5·89 × 10⁻⁷ m determine the slit separation. (*N.*)

10. Explain what is meant by the term *path-difference* with reference to the interference of two wave-motions.

Why is it not possible to see interference where the light beams from the headlamps of a car overlap?

Interference fringes were produced by the Young's slits method, the wavelength of the light being 6 × 10⁻⁷ m. When a film of material 3·6 × 10⁻³ cm thick was placed over *one* of the slits, the fringe pattern was displaced by a distance equal to 30 times that between two adjacent fringes. Calculate the refractive index of the material. To which side are the fringes displaced?

(When a layer of transparent material whose refractive index is *n* and whose thickness is *d* is placed in the path of a beam of light, it introduces a path difference equal to (*n* − 1)*d*.) (*O. & C.*)

11. Explain the formation of interference fringes by an air wedge and describe how the necessary apparatus may be arranged to demonstrate them.

Fringes are formed when light is reflected between the flat top of a crystal resting on a fixed base and sloping glass plate. The lower end of the plate rests on the crystal and the upper end on a fixed knife-edge. When the temperature of the crystal is raised the fringe separation changes from 0·96 mm to 1·00 mm. If the length of the glass plate from knife-edge to crystal is 5·00 cm, and the light of wavelength 6·00 × 10⁻⁷ m is incident normally on the wedge, calculate the expansion of the crystal. (*L.*)

12. What do you understand by (i) *interference*, (ii) *coherence* between two separate wave trains, (iii) *coherence* along one wave train?

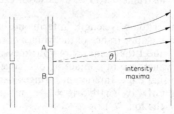

Fig. 23A

In a 'Young's slits' experiment, the centres of the double slits are 0·25 mm apart and the wavelength of the light used is $6·0 \times 10^{-4}$ mm. Calculate the angle θ subtended at the slits by adjacent maxima of the fringe pattern (see Fig. 23A). Describe and explain what happens to these fringes if (a) slit A is covered with a thin sheet of transparent material of high refractive index, (b) the light emerging from slit A is reduced in intensity to half that emerging from the other slit B, (c) A and B are each covered with a thin film of Polaroid and one of these films is slowly rotated, (d) the distance between slit A and slit B is slowly increased. (C).

13. What are the necessary conditions for interference of light to be observable? Describe with the aid of a labelled diagram how optical interference may be demonstrated using Young's slits. Indicate suitable values for all the distances shown.

How are the colours observed in thin films explained in terms of the wave nature of light? Why does a small oil patch on the road often show approximately circular coloured rings? (L.)

Newton's Rings
14. Draw a labelled diagram of the apparatus you would use to view Newton's rings. Explain why darkness is produced at certain points and state the conditions required for this.

Also explain why
(a) the interference fringes are circular;
(b) the centre of the system is normally black;
(c) the radii of the rings are proportional to the square roots of the natural numbers.

State, with reasons, what you would expect to see if the spherical lens were replaced by a cylindrical one.

Newton's rings were produced using a plano-convex lens, made of glass of refractive index 1·48 resting on a flat glass plate. The diameter of the 10th dark ring from the centre was measured and found to be 3·36 mm. The diameter of the 30th dark ring was found to be 5·82 mm. The wavelength of the light used was 589 nm (589×10^{-9} m). What was the focal length of the lens? (L.)

15. A glass converging lens rests in contact with a horizontal plane sheet of glass. Describe how you would produce and view Newton's rings, using reflected sodium light.

Explain how the rings are formed and derive a formula for their diameters. Describe the measurements you would make in order to determine the radii of curvature of the faces of the lens, assuming that the wavelength of sodium light is known. Show how the result is derived from the observations.

How would the ring pattern change if (a) the lens were raised vertically one quarter of a wavelength, (b) the space between the lens and the plate were filled with water? (N.)

16. In the interference of light what is meant by the requirement of coherency? How is this usually achieved in practice?

A thin spherical lens of long focal length is placed on a flat piece of glass. How, using this arrangement, would you demonstrate interference by reflection?

Explain how these fringes are formed and describe their appearance. What would be the effect if the spherical lens were replaced by a cylindrical one?

Such a system using a spherical lens is illuminated with light of wavelength 600 nm. When the lens is carefully raised from the plate 50 extra fringes appear at and move away from the centre of the fringe system. By what distance was the lens raised? (L.)

17. Explain how Newton's rings are formed, and describe how you would demonstrate them experimentally. How is it possible to predict the appearance of the centre of the ring pattern when (a) the surfaces are touching, and (b) the surfaces are not touching?

In a Newton's rings experiment one surface was fixed and the other movable along the axis of the system. As the latter surface was moved the rings appeared to contract and the centre of the pattern, initially at its darkest, became alternately bright and dark, passing through 26 bright phases and finishing at its darkest again. If the wavelength of the light was $5·461 \times 10^{-7}$ m, how far was the surface moved and did it approach, or recede from the fixed surface? Suggest one possible application of this experiment. (O. & C.)

24 Diffraction of Light Waves

In 1665 GRIMALDI observed that the shadow of a very thin wire in a beam of light was much broader than he expected. The experiment was repeated by Newton, but the true significance was only recognised more than a century later, after Huygens' wave theory of light had been revived. The experiment was one of a number which showed that light could bend round corners in certain circumstances.

We have seen how interference patterns, for example, bright and dark fringes, can be obtained with the aid of two sources of light close to each other. These sources must be coherent sources, i.e., they must have the same amplitude and frequency, and always be in phase with each other. Consider two points on the *same wavefront*, for example the two points A, B, on a plane wavefront arriving at a narrow slit in a screen, Fig. 24.1. A and B can be considered as

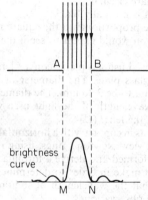

Fig. 24.1 Diffraction of light

Fig. 24.2 Diffraction rings in the shadow of a small circular disc. The bright spot is at the centre of the geometrical shadow

secondary sources of light, an aspect introduced by Huygens in his wave theory of light (p. 441); and as they are on the same wavefront, A and B have identical amplitudes and frequencies and are in phase with each other. Consequently A, B, are coherent sources, and we can expect to find an interference pattern on a screen in front of the slit, provided its width is small compared with the wavelength of light. For a short distance beyond the edges M, N, of the projection of AB, i.e., in the geometrical shadow, observation shows that there are some alternate bright and dark fringes. See Fig. 24.3, 24.4.

Fig. 24.3 (below) The variation in intensity of the bright diffraction bands due to a single small rectangular aperture is shown roughly in the diagram. The central bright band, in which most of the light is concentrated, has maximum intensity in the direction of P, the middle of the band. The intensity of this band diminishes to a minimum at Q or Q_1, which are at an angle of diffraction to the direction of P

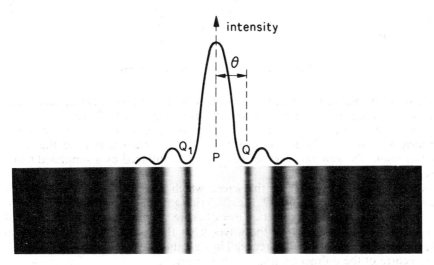

Fig. 24.4 Diffraction bands formed by a single small rectangular aperture of width *a*

Thus light can travel round corners. The phenomenon is due to *diffraction*, and it has enabled scientists to measure accurately the wavelength of light as we see later.

If a source of white light is observed through the eyelashes, a series of coloured images can be seen. These images are due to interference between sources on the same wavefront, and the phenomenon is thus an example of diffraction. Another example of diffraction was unwittingly deduced by POISSON at a time when the wave theory was new. Poisson considered mathematically the combined effect of the wavefronts round a circular disc illuminated by a distant small source of light, and he came to the conclusion that the light should be visible beyond the disc in the middle of the geometrical shadow. Poisson thought this was impossible; but experiment confirmed his deduction, and he became a supporter of the wave theory of light. See Fig. 24.2.

Diffraction at Single Slit

We now consider in more detail the image produced by diffraction at an opening or slit. The image of a distant star produced by a telescope objective is due to diffraction at a circular opening. On this account a study of the image has application in Astronomy, as we see later.

Suppose parallel wavefronts from a distant object are diffracted at a rectangular slit AB of width *a*, Fig. 24.5 (i). This is called *Fraunhofer diffraction*. (The diffraction of circular wavefronts from a near object is called *Fresnel diffraction*, which we shall not study.) If the light passing through the slit is received on a screen S a long way from AB, we can consider that parallel wavefronts have travelled to S to form the image of the slit.

Consider a wavefront reaching the opening AB. All points on it between A

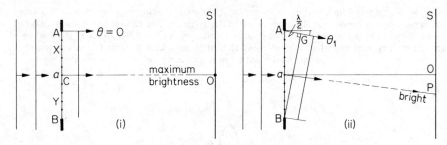

Fig. 24.5 Rectangular slit diffraction image

and B are in phase, that is, they are coherent. These points act as secondary centres, sending out waves beyond the slit. Their combined effect at any distant point can be found by summing the numerous waves arriving there, from the Principle of Superposition. The mathematical treatment is beyond the scope of this book. The general effect, however, can be obtained by a simplified treatment.

Consider first a point O on the screen which lies on the normal to the slit passing through its midpoint C, Fig. 24.5 (i). O is the centre of the diffraction image. It corresponds to a direction $\theta = 0$, where θ is measured from the normal to AB. Now in this direction the waves from the secondary sources such as A, X, Y, B have no path difference. Thus all the waves arrive *in phase* at O. The centre of the diffraction image is hence bright.

Diffraction Pattern of Image

Along the screen from O, the brightness of the diffraction image of the slit diminishes. Consider, for example, a direction $\theta = \theta_1$ corresponding to a point P on the image of the slit, Fig. 24.5 (ii). The diffracted waves travelling in this direction are then parallel to BG, as shown. If $AG = \lambda/2$, where λ is the wavelength of the incident light, the waves from the ends A, B of the slit then have a path difference of $\lambda/2$. This is equivalent to a phase difference of 180°. Hence the resultant amplitude on adding these two waves is zero. However, it can be seen that waves from all other points between A and B have a path difference less than $\lambda/2$. Thus although the waves have a resultant amplitude at P, the brightness of the image at P is less than at the centre O.

Consider now a point Q farther than P from O, Fig. 24.6. This corresponds to a direction $\theta = \theta_2$ greater than θ_1. The diffracted waves now travel in a

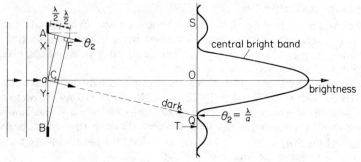

Fig. 24.6 Variation of brightness of diffraction image

direction parallel to AF. Suppose the path difference AF of the waves from A and B is λ.

To find the resultant amplitude of the waves arriving at Q, divide the wavefront AB into two halves. The top point A of the upper half CA, and the top point C of the lower half BC, send out waves to Q which have a path difference $\lambda/2$, from above. Thus the resultant amplitude at Q is zero. All other pairs of corresponding points in the two halves of AB, for example, X and Y where CX = BY, also have a path difference $\lambda/2$. Hence it follows that the brightness at Q is zero. This point, then, is one *edge* of the central bright fringe on the screen. The other edge is R on the opposite side of O, where OR = OQ.

It should be noticed from Fig. 24.6 that the path difference AF = $\lambda/2 + \lambda/2$ = λ. We shall use this path difference shortly.

Secondary Fringes

Secondary bright and dark fringes are also obtained on the screen beyond Q, as shown in the photograph in Fig. 24.4. Consider, for example, a point T which lies in a direction θ_3 where the path difference of waves starting from A and B is $3\lambda/2$, Fig. 24.6. We can imagine the wavefront AB divided into *three* equal parts. Now the waves from the extreme ends of the upper two parts have a path difference λ. Thus, as we explained for the dark fringes at Q, these two parts of the wavefront produce darkness at T. The third part produces a fringe of light at T much less bright than the central fringe, which was due to the whole wavefront between A and B. Calculation shows that the intensity at T is less than 5 per cent of the intensity at O, the middle of the central bright fringe. Thus most of the light incident on AB is diffracted into the central bright fringe.

Rectilinear and Non-Rectilinear Propagation

The *angular width* of the central bright fringe is 2θ, where θ is the angle between the direction CO of maximum intensity and the direction CQ of minimum intensity or the edge of the bright fringe, Fig. 24.6. This angle θ is equal to θ_2 and so is given by

$$\sin \theta = \frac{AF}{AB} = \frac{\lambda}{a},$$

where a = width of slit AB. The angle θ is the angular half-width of the central fringe.

When the slit is widened and a becomes large compared with λ, then $\sin \theta$ is very small and hence θ is very small. In this case the directions of the minimum and maximum intensities of the central fringe are very close to each other. Practically the whole of the light is thus confined to a direction immediately in front of the incident direction, that is, no spreading occurs. This explains the *rectilinear propagation of light*. When the slit width a is very small and equal to 2λ, for example, then $\sin \theta = \lambda/a = 1/2$, or $\theta = 30°$. The light waves now spread round through $30°$ on either side of the slit, that is, the diffraction is appreciable.

These results are true for any wave phenomenon. In the case of an electromagnetic wave of 3 cm wavelength, a slit of these dimensions produces sideways spreading. Sound waves of a particular frequency 256 Hz have a wavelength of about 1·3 m. Consequently, sound waves spread round apertures such as a doorway or an open window, which have comparable dimensions to their wavelengths.

Diffraction in Telescope Objective

When a parallel beam of light from a distant object such as a star S_1 enters a telescope objective L, the lens collects light through a circular opening and forms a diffraction pattern of the star round its principal focus, F. This is illustrated in the exaggerated diagram of Fig. 24.7.

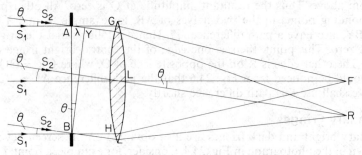

Fig. 24.7 Diffraction in telescope objective

Consider an incident plane wavefront AB from the star S_1, and suppose for a moment that the aperture is rectangular. The diffracted rays such as AG, BH normal to the wavefront are incident on the lens in a direction parallel to the principal axis LF. The optical paths AGF, BHF are equal. This is true for all other diffracted rays from points between A, B which are parallel to LF, since the optical paths to an image produced by a lens are equal. The central part F of the star pattern is therefore bright.

Now consider those diffracted rays from all points between AB which enter the lens at an angle θ to the principal axis. This corresponds to a diffracted plane wavefront BY at an angle θ to AB. As explained previously, if $AY = \lambda$, then R is the dark edge of the central maximum of the diffraction pattern of the star S_1.

The angle θ corresponding to the edge R is given by

$$\sin \theta = \frac{\lambda}{D},$$

where D is the diameter of the lens aperture. This is the case where the aperture can be divided into a number of rectangular slits. For a *circular* opening such as a lens, or the concave mirror of the Palomar telescope, the formula becomes $\sin \theta = 1 \cdot 22\lambda/D$. As λ is small compared to D, then θ is small and so we may write $\theta = 1 \cdot 22\lambda/D$, where θ is in radians.

Resolving Power

Suppose now that another distant star S_2 is at an angular distance θ from S_1, Fig. 24.8. The maximum intensity of the central pattern of S_2 then falls on the minimum or edge of the central pattern of the star S_1, corresponding to R in Fig. 24.8 (i). Experience shows that the two stars can then just be distinguished or *resolved*. Lord Rayleigh stated a criterion for the resolution of two objects, which is generally accepted: *Two objects are just resolved when the maximum intensity of the central pattern of one object falls on the first minimum or dark edge of the other.* Fig. 24.8 (i) shows the two stars just resolved. The resultant intensity in the middle dips to about 0·8 of the maximum, and the eye is apparently sensitive to the change here. Fig. 24.8 (ii) shows two stars S_1, S_2 unresolved, and Fig. 24.8 (iii) the same stars completely resolved. See also Fig. 24.9.

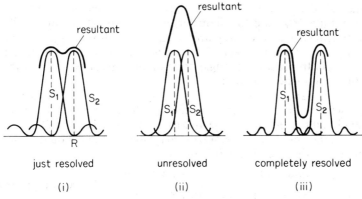

Fig. 24.8 Resolving power—Rayleigh criterion

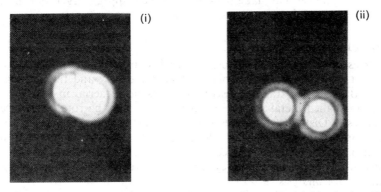

Fig. 24.9 Resolving (i) Two sources just resolved according to Rayleigh's criterion (ii) Two sources completely resolved

The angular distance θ between two distant stars just resolved is thus given by $\sin \theta = \theta = 1 \cdot 22\ \lambda/D$, where D is the diameter of the objective. This is an expression for the *limit of resolution*, or *resolving power*, of a telescope. The limit of resolution or resolving power increases when θ is *smaller*, as two stars closer together can then be resolved. Consequently telescope objectives of large diameter D give high resolving power.

The Yerkes Observatory has a large telescope objective diameter of about 1 metre. The angular distance θ between two stars which can just be resolved is thus given by

$$\theta = \frac{1 \cdot 22\lambda}{D} = \frac{1 \cdot 22 \times 6 \times 10^{-7}}{1} = 7 \cdot 3 \times 10^{-7} \text{ radians,}$$

assuming 6×10^{-7} m for the wavelength of light. The Mount Palomar telescope has a parabolic mirror objective of aperture 5 metres. The resolving power is thus five times as great as the Yerkes Observatory telescope. In addition to the advantage of high resolving power, a large aperture collects more light from the distant source and thus provides an image of higher intensity.

Magnifying Power of Telescope and Resolving Power

If the width of the emergent beam from a telescope is greater than the diameter of the eye-pupil, rays from the outer edge of the objective do not enter the eye.

Hence the full diameter D of the objective is not used. If the width of the emergent beam is less than the diameter of the eye-pupil, the eye itself, which has a constant aperture, may not be able to resolve the distant objects. Theoretically, the angular resolving power of the eye is $1.22 \lambda/a$, where a is the diameter of the eye-pupil. In practice an angle of 1 minute is resolved by the eye, which is more than the theoretical value.

Now the angular magnification, or magnifying power, of a telescope is the ratio α'/α, where α' is the angle subtended at the eye by the final image and α is the angle subtended at the objective (p. 383). To make the fullest use of the diameter D of the objective, therefore, the magnifying power should be increased to the angular ratio given by, if D is in metre,

$$\frac{\text{resolving power of eye}}{\text{resolving power of objective}} = \frac{\pi/(180 \times 60)}{1.22 \times 6 \times 10^{-7}/D} = 400\ D\ \text{(approx.)}.$$

In this case the telescope is said to be in 'normal adjustment'. Any further increase in magnifying power will make the distant objects appear larger, but there will be no increase in definition or resolving power.

Radio Telescope

Radio telescopes are used to investigate and map the sources of radio waves arriving at the earth from the solar system, the galaxy and the extragalactic nebulae. Basically the radio telescope consists of an aerial in the form of a paraboloid-shaped metal surface or 'dish', which may be rotated to face any part of the sky. Distant radio waves, like distant light waves, are reflected towards a focus from all parts of the paraboloid surface (p. 320). The converging waves are then passed to a sensitive receiver and after detection the signals may be recorded on a paper chart or they may be recorded on tape for feeding into a computer for analysis.

The Jodrell Bank radio telescope has a dish about 75 m in diameter. The hydrogen line emitted from interstellar space has a wavelength of about 21 cm or 0.21 m. Thus the angular resolution

$$= \frac{1.22\lambda}{D} = \frac{1.22 \times 0.21}{75} = 0.0034\ \text{rad} = 0.2°.$$

Larger telescopes can provide greater resolving power but the technical problems and cost make this unpractical. A technique using interferometry principles, in which the dishes need not be large, provides much greater resolving power. This has been developed particularly at the Mullard Radio Astronomy Observatory, Cambridge, England.

Radio Interferometers

The principle of the interferometer type of radio telescope is illustrated in Fig. 24.10.

Two dishes A and B, separated by a distance of say 5 km, are connected to the same receiver R, Fig. 24.10 (i). Radiation from a moving source will reach both aerials or dishes. If the path difference is a whole number of wavelengths or zero, the two signals are in phase and a maximum resultant signal is obtained (constructive interference); if the signals are in antiphase, the resultant signal is a minimum or zero (destructive interference). Thus as a source S moves across the sky, the resultant signal varies in intensity. The principles concerned are similar to a Young's two-slit experiment, except that in this case we are dealing

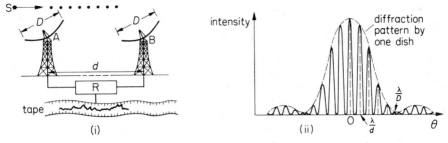

Fig. 24.10 Radio telescope—interferometer type

with two 'point receivers' of radiation instead of 'point emitters' of light. The mathematics of the interference is the same in both cases.

The variation in intensity with angle is shown roughly in Fig. 24.10 (ii). It is similar to the variation in intensity of the interference fringes in a Young's two-slit experiment. Successive maxima or peaks are thus obtained at angular separations equal to λ/d, where d is the separation of the two dishes. So if $\lambda = 3$ cm $= 0.03$ m and $d = 5$ km $= 5000$ m,

$$\text{angular separation} = \frac{\lambda}{d} = \frac{0.03}{5000} = 6 \times 10^{-6} \text{ rad} = 0.0003° \text{ (approx.).}$$

Further, the angular resolution θ, the angle from the central maximum to the first minimum (p. 484), is given by

$$\theta = \frac{\lambda}{2d} = 3 \times 10^{-6} \text{ rad.}$$

If only one dish were used, with a diameter $D = 13$ m for example, the single slit intensity pattern (p. 485) shows that the angular resolution is now

$$\theta = \frac{1.22\lambda}{D} = \frac{1.22 \times 0.03}{13} = 3 \times 10^{-3} \text{ rad (approx.),}$$

which is 1000 times less resolution than that obtained with the two dishes. Thus the angular diameter of a source, or of two sources, may be found more accurately by the interferometer method. Further, the method enables the direction of a moving source to be tracked more accurately—the fringe of the central maximum is much narrower with two dishes than with one and this helps to locate the source better.

The interferometer principle has been successfully applied to surveying regions of the sky and to mapping radio galaxies. The Mullard Radio Astronomy Observatory has a telescope consisting of eight parabolic reflectors or dishes, each about 13 m in diameter and mounted on rails about $1\frac{1}{4}$ km long, acting together as a 'grating interferometer'. The effective baseline of the telescope is 5 km. The dishes can be moved along the rails to different sets of positions to map a region of the sky. The signals received from the different aerials are recorded on tape and then synthesised by means of a computer. In this way a telescope of very large aperture (5 km) can be simulated. This is called *aperture synthesis*. With this telescope extremely accurate maps have been made of radio sources in outer space. The Nobel Prize was awarded in 1974 to Sir Martin Ryle, the director of the Mullard Radio Astronomy Observatory, and to Professor A. Hewish of Cambridge for their contributions to radio astronomy, especially aperture synthesis.

Increasing Number of Slits

On p. 482 we saw that the image of a single narrow rectangular slit is a bright central or principal maximum diffraction fringe, together with subsidiary maxima diffraction fringes which are much less bright. Suppose that parallel light is incident on two more parallel close slits, and the light passing through the slits is received by a telescope focused at infinity. Since each slit produces a similar diffraction effect in the same direction, the observed diffraction pattern will have an intensity variation identical to that of a single slit. This time, however, the pattern is crossed by a number of interference fringes, which are due to interference between slits (see *Young's experiment*, p. 460). The envelope of the intensity variation of the interference fringes follow the diffraction pattern variation due to a single slit. In general, if I_s is the intensity at a point due to interference between slits and I_d that due to diffraction of a *single* slit, then the resultant intensity I is given by $I = I_d \times I_s$. Hence if $I_d = 0$ at any point, then $I = 0$ at this point irrespective of the value of I_s.

As more parallel equidistant slits are introduced, the intensity and sharpness of the principal maxima increase and those of the subsidiary maxima decrease. The effect is illustrated roughly in Fig. 24.11. With several hundred lines per millimetre, only a few sharp principal maxima are seen in directions discussed shortly. Their angular separation depends only on the distance between successive slits. The slit width affects the intensity of the higher order principal maxima; the narrower the slit, the greater is the diffraction of light into the higher orders.

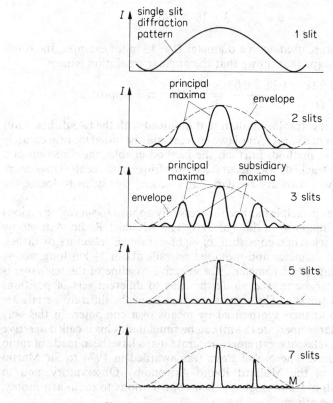

Fig. 24.11 Principal maxima with increasing slits

As shown, the single slit diffraction pattern modulates the maxima. In the direction corresponding to M (7 slits), the single slit intensity $I_d = 0$. Hence, as stated above, the resultant intensity $I = 0$. This means that no light is diffracted in this direction from any of the slits. See also Fig. 24.12.

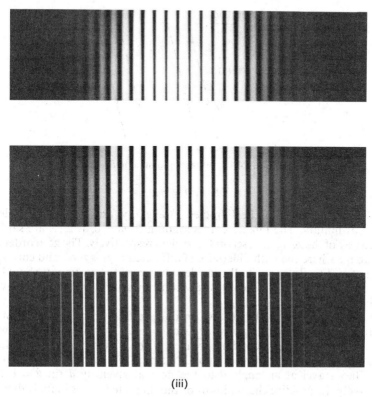

(iii)

Fig. 24.12 Diffraction by gratings (i) 3 slits; (ii) 5 slits; (iii) 20 slits. Diffraction patterns by gratings with different numbers of slits, all of the same width and separation. As the number of slits increases, the bright lines become sharper

Principal Maxima of Diffraction Grating

A *diffraction grating* is a large number of close parallel equidistant slits, ruled on glass or metal; it provides a very valuable means of studying spectra. If the width of a slit or clear space is a and the thickness of a ruled opaque line is b, the spacing d of the slits is $(a + b)$. Thus with a grating of 600 lines per millimetre, the spacing $d = 1/600$ millimetre $= 17 \times 10^{-7}$ m, or a few wavelengths of visible light.

The angular positions of the principal maxima produced by a diffraction grating can easily be found. Suppose X, Y are corresponding points in consecutive slits, where $XY = d$, and the grating is illuminated normally by monochromatic light of wavelength λ, Fig. 24.13. In a direction θ, the diffracted rays XL, YM have a path difference XA of $d \sin \theta$. The diffracted rays from all other corresponding points in the two slits have a path difference of $d \sin \theta$ in the same direction. Other pairs of slits throughout the grating can be treated in the same way. Hence bright or principal maxima are obtained when

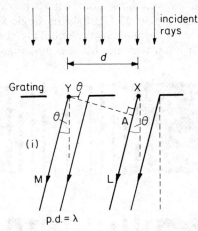

Fig. 24.13 Diffraction grating images

$$d \sin \theta = m\lambda, \qquad . \qquad . \qquad . \qquad . \qquad . \qquad (i)$$

where m is an integer, if all the diffracted parallel rays are collected by a telescope focused at infinity. The images corresponding to $m = 0, 1, 2, \ldots$ are said to be respectively of the zero, first, second ... orders respectively. The zero order image is the image where the path difference of diffracted rays is zero, and corresponds to that seen directly opposite the incident beam on the grating. It should again be noted that all points in the slits are secondary centres on the same wavefront and therefore coherent sources.

Fig. 24.14 shows how the grating produces diffracted waves which interfere constructively in directions given by $d \sin \theta = m\lambda$. Only six slits are shown and only six corresponding secondary sources in the slits, separated by a distance d. The diffracted wavefronts correspond to a path difference of λ and 2λ respectively. They travel at an angle θ to the normal given by $d \sin \theta = \lambda$ and 2λ respectively. In practice the build-up of the diffracted wavefronts is due to the effect of many slits, for example, 300 or more per millimetre, and to the numerous secondary sources within each slit.

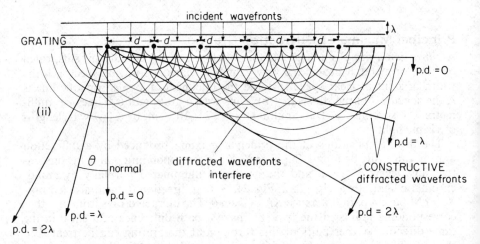

Fig. 24.14 Grating action—diffraction and interference of wavefronts

Diffraction Images

The *first order* diffraction image is obtained when $m = 1$. Thus

$$d \sin \theta = \lambda,$$

or

$$\sin \theta = \frac{\lambda}{d}.$$

If the grating has 600 lines per millimetre (600 mm^{-1}), the spacing of the slits, d, is $1/600$ mm $= 1/(600 \times 10^3)$ m. Suppose yellow light, of wavelength $\lambda = 5.89 \times 10^{-7}$ m, is used to illuminate the grating. Then

$$\sin \theta = \frac{\lambda}{d} = 5.89 \times 10^{-7} \times 600 \times 10^3 = 0.3534$$

$$\therefore \theta = 20.7°.$$

The *second order* diffraction image is obtained when $m = 2$. In this case $d \sin \theta = 2\lambda$.

$$\sin \theta = \frac{2\lambda}{d} = 2 \times 5.89 \times 10^{-7} \times 600 \times 10^3 = 0.7068$$

$$\therefore \theta = 45.0°.$$

If $m = 3$, $\sin \theta = 3\lambda/d = 1.060$. Since the sine of an angle cannot be greater than 1, it is impossible to obtain a third order image with the diffraction grating for this wavelength.

With a grating of 1200 lines per mm the diffraction images of sodium light would be given by $\sin \theta = m\lambda/d = m \times 5.89 \times 10^{-7} \times 12 \times 10^5 = 0.7068\,m$. Thus only $m = 1$ is possible here. As all the diffracted light is now concentrated in one image, instead of being distributed over several images, the first order image is very bright, which is an advantage.

Missing Orders in Principal Maxima

In a diffraction grating, the order m of the principal maxima obtained is given by $\sin \theta = m\lambda/d$. As we saw previously, however, the intensity of the principal maxima will be zero if the intensity due to a *single slit* of the grating is zero, as this modulates the over-all intensity variation (see Fig. 24.11). Now the intensity of the single-slit pattern is zero where $\sin \theta = \lambda/a$, if a is the width of the slit. So the *missing order m* in the principal maxima obtained by a diffraction grating is given by

$$\sin \theta = \frac{m\lambda}{d} = \frac{\lambda}{a}.$$

Thus

$$m = \frac{d}{a}.$$

If the width a of the slit is one-quarter of the separation d of the grating lines, then $m = 4$ from above. The 4th order will be missing from the principal maxima of the grating. With a particular grating, missing orders occur only if d/a is a whole number.

Measurement of Wavelength by Diffraction Grating

The wavelength of monochromatic light can be measured by a diffraction grating in conjunction with a spectrometer.

The collimator C and telescope T of the instrument are first adjusted for parallel light (p. 343), and the grating P is then placed on the table so that its plane is perpendicular to two screws, Q, R, Fig. 24.15 (i). To level the table

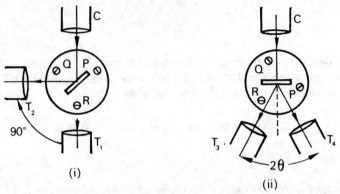

Fig. 24.15 Measurement of wavelength by diffraction grating

so that the plane of P is parallel to the axis of rotation of the telescope, the latter is first placed in the position T_1 directly opposite the illuminated slit of the collimator, and then rotated exactly through 90° to a position T_2. The table is now turned until the slit is seen in T_2 by reflection at P, and one of the screws Q, R turned until the slit image is in the middle of the field of view. The plane of P is now parallel to the axis of rotation of the telescope. The table is then turned through 45° so that the plane of the grating is exactly perpendicular to the light from C, and the telescope is turned to a position T_3 to receive the first diffraction image, Fig. 24.15 (ii). If the lines of the grating are not parallel to the axis of rotation of the telescope, the image will not be in the middle of the field of view. The third screw is then adjusted until the image is central.

The readings of the first diffraction image are observed on both sides of the normal. The angular difference is 2θ, and the wavelength is calculated from $\lambda = d \sin \theta$, where d is the spacing of the slits, obtained from the number of lines per centimetre of the grating. If a second order image is obtained for a diffraction angle θ_1, then $\lambda = \frac{1}{2} d \sin \theta_1$.

Diffraction gratings with suitable values of d have been used to measure a wide range of wavelengths. For very short wavelengths such as X-rays, a grating of similar linear dimensions is provided by a crystal (see p. 862).

Resolving Power of Grating

The resolving power of a grating is a measure of how effectively it can separate or *resolve* two wavelengths in a given order of their spectrum. Using the Rayleigh criterion for resolving power (p. 484), it can be shown that the resolving power depends on the *total lines or width Nd* of the grating, where N is the number of lines and d is their spacing.

So a grating 5 cm wide with 3000 lines per cm has twice the resolving power of the same grating if it was masked to be only 2·5 cm wide. In both gratings the diffraction images of two close wavelengths would be formed at the same angles for a given order, since $d \sin \theta = m\lambda$ and d is the same. But the images with the 2·5 cm grating are not as sharp as with the 5 cm grating and the two close wavelengths may not be seen separated or resolved.

Note that $Nd = Nm\lambda/\sin \theta$, so the resolving power increases with the order m of the image.

Example

A parallel beam of sodium light is incident normally on a diffraction grating. The angle between the two first order spectra on either side of the normal is 27° 42'. Assuming that the wavelength of the light is 5.893×10^{-7} m, find the number of rulings per mm on the grating.

The first order spectrum occurs at an angle $\theta = \frac{1}{2} \times 27° 42' = 13° 51'$.

$$\therefore d = \frac{\lambda}{\sin \theta} = \frac{5.893 \times 10^{-7}}{\sin 13° 51'} \text{ m}$$

$$\therefore \text{ number of rulings per metre} = \frac{1}{d} = \frac{\sin 13° 51'}{5.893 \times 10^{-7}} = 406\,000$$

$$= 406 \text{ per millimetre.}$$

Spectra in Grating

If white light is incident normally on a diffraction grating several coloured spectra are observed on either side of the normal, Fig. 24.16 (i). The first order

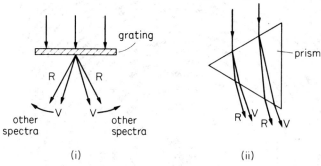

Fig. 24.16 Spectra in grating and prism

diffraction images are given by $d \sin \theta = \lambda$, and as violet has a shorter wavelength than red, θ is less for violet than for red. Consequently the spectrum colours on either side of the incident white light are violet to red. In the case of a spectrum produced by dispersion in a glass prism, the colours range from red, the least deviated, to violet, Fig. 24.16 (ii). Second and higher order spectra are obtained with a diffraction grating on opposite sides of the normal, whereas only one spectrum is obtained with a glass prism. The angular spacing of the colours is also different in the grating and the prism.

If $d \sin \theta = m_1\lambda_1 = m_2\lambda_2$, where m_1, m_2 are integers, then a wavelength λ_1 in the m_1 order spectrum overlaps the wavelength λ_2 in the m_2 order. The extreme violet in the visible spectrum has a wavelength about 3.8×10^{-7} m. The violet direction in the second order spectrum would thus correspond to $d \sin \theta = 2\lambda$ $= 7.6 \times 10^{-7}$ m, and this would not overlap the extreme colour, red, in the first order spectrum, which has a wavelength about 7.0×10^{-7} m. In the second order spectrum, a wavelength λ_2 would be overlapped by a wavelength λ_3 in the third order if $2\lambda_2 = 3\lambda_3$. If $\lambda_2 = 6.9 \times 10^{-7}$ m (red), then $\lambda_3 = 2\lambda_2/3 = 4.6 \times 10^{-7}$ m (blue). Thus overlapping of colours occurs in spectra of higher orders than the first.

Holography

When an object is photographed by a camera, only the *intensity* of the light from its different points is recorded on the photographic film to form the image. Now the intensity is a measure of the mean square value of the amplitude of the original light wave from the object. Consequently the *phase* of the wave arriving at the film from the different points on the object is lost.

In *holography*, however, both the phase and the amplitude of the light waves are recorded on the film. The resulting photograph is called a *hologram*. As we see later, it is a speckled pattern of fine dots. Dr. D. Gabor, who laid the foundations of holography in 1948, gave this name from the Greek 'holos' meaning 'the whole', because it contains the whole information about the light wave, that is, its phase as well as its amplitude.

Gabor's method of producing a hologram and of reconstructing the original wave were difficult to put in practice in 1948 because sources which remain coherent only over very short path differences were then available. In 1962, however, the laser was invented. This gave a powerful source of light which remained coherent over long paths (see Fig. 24.17) and the subject of holography then developed rapidly.

Fig. 24.17 Newton's rings. Obtained simply by reflecting a helium–neon laser beam from the front and back surfaces of a low power lens and so demonstrating the coherency of a laser beam over long paths

Making a Hologram

Fig. 24.18 shows the *basic principle* of making a hologram. By means of a half-silvered mirror M, part of the coherent light from a laser is reflected towards the object O and the remainder passes straight through as shown. The photographic plate or film P is thus illuminated by (*a*) light waves scattered or diffracted from O, called the 'object beam', and (*b*) by a direct beam called the 'reference beam'.

The numerous points which make up the image on P are then formed by interference between the overlapping coherent waves of the object beam and the reference beam. As shown inset diagrammatically, the hologram consists

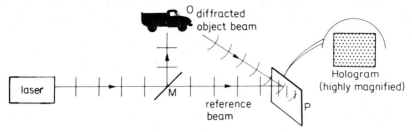

Fig. 24.18 Making a hologram

of a very large number of closely-spaced points, invisible to the naked eye but seen under a powerful microscope. The 'structure' of the hologram is like a diffraction grating; it has opaque and transparent regions very closely spaced.

In commercial broadcasting, audio-frequency currents are arranged to modulate the carrier radio wave from the transmitter. By analogy, in making a hologram the reference beam may be considered analogous to a 'carrier wave' which is 'modulated' at the photographic film by interference with the object beam. Thus, to simplify matters, suppose the object and reference beams arriving at the plate are plane waves represented by $y = O \sin (\omega t + \Delta)$ and $y = R \sin \omega t$ respectively, where Δ is their phase difference due to the path difference. The resultant amplitude at the plate is given, from vector addition, by $(O^2 + R^2 + 2OR \cos \Delta)^{1/2}$. The third term $2OR \cos \Delta$ contains (i) the phase angle Δ and (ii) the amplitude O of the object wave modulated by the amplitude R of the reference beam. After the plate is exposed and developed, the object wave, accompanied by the others due to interference, is 'recorded' on the emulsion. It may be noted that if the object moves very slightly (of the order of a fraction of a wavelength) while the photograph is taken, the diffracted waves from the object no longer produces a hologram. Rigid mounting of the object is therefore essential.

Reconstructing the Hologram Image

Figure 24.19 shows how the object wavefront or image is 'reconstructed'. The optical arrangement is simply reversed and the hologram H is illuminated by coherent light from the laser, which was the original reference beam. Now as a general principle, if an image produced by a grating is recorded on photographic film, then the image formed by illuminating the resulting grating on the negative will be identical to the original grating. In a similar way, we may consider the object as a '3-D grating' and its 3-D image is formed when the hologram is illuminated as in Fig. 24.19.

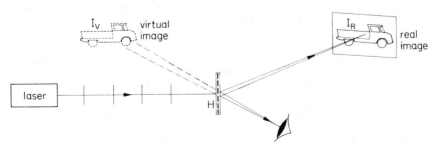

Fig. 24.19 Reconstructing hologram image

The points on the hologram H thus act as a diffraction grating. The waves diffracted through H carry the phase and amplitude of the waves originally diffracted from the object O when the hologram was made. The object wavefronts have thus been 'reconstructed'. One of the diffracted beams forms a real image I_R, as shown. Another diffracted beam forms a virtual image I_V. This image can be seen on looking through the hologram H. The hologram thus acts like a 'window' through which the image can be seen.

One of the most remarkable features of the hologram is that by moving the head while looking through it, one can see more of the object originally hidden from view. Thus a three-dimensional (3D) view is recorded on a two-dimensional photographic film. See Fig. 24.20. This is due to the fact that all parts of the object originally photographed have sent diffracted waves to the photographic film. Further, the 'grain' of a particular hologram is repeated regularly throughout the whole of the film. Thus if the hologram is cut into small pieces, although the quality is poor the whole of the image can be seen through one piece.

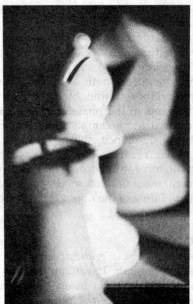

Fig. 24.20 Hologram of chess pieces (i) Focused on the castle (ii) Focused on the knight and taken from a different angle and showing the 3D nature of the image

Exercises 24

1. A diffraction grating has 400 lines per mm and is illuminated normally by monochromatic light of wavelength 600 nm (6×10^{-7} m).

Calculate (a) the grating spacing, (b) the angle to the normal at which the first order maximum is seen, (c) the number of diffraction maxima obtained.

2. A diffraction grating is illuminated normally by monochromatic light of wavelength λ. Show in a diagram using waves (i) where diffraction occurs, (ii) where interference occurs, (iii) the directions in which the first and second order diffraction maxima are seen, together with the corresponding path difference.

3. A plane diffraction grating is illuminated by a source which emits two spectral lines of wavelengths 420 nm (420×10^{-9} m) and 600 nm (600×10^{-9} m). Show that the 3rd order line of one of these wavelengths is diffracted through a greater angle than the 4th order of the other wavelength. (L.)

4. A spectral line of known wavelength 5.792×10^{-7} m emitted from a mercury vapour lamp is used to determine the spacing between the lines ruled on a plane diffraction grating. When the light is incident normally on the grating the third order spectrum, measured using a spectrometer, occurs at an angle of $60°$ $19'$ to the normal. Calculate the grating spacing.

Why is the value obtained using the third order spectrum likely to be more accurate than if the first order were used? (*L.*)

5. When a plane wavefront of monochromatic light meets a diffraction grating, the diffraction pattern produced is caused by the interference of beams diffracted through the various grating slits. Explain, with the aid of a diagram, exactly where and under what conditions diffraction occurs.

Explain why only beams diffracted in certain directions interfere constructively. (*L.*)

6. Light of wavelength λ falls normally on a transmission diffraction grating of spacing d. Show, with the aid of a suitable diagram, how light from the various slits reinforces at certain values of the diffraction angle, θ, and that then $n\lambda = d \sin \theta$, where n is an integer.

A spectrometer is used with the grating to determine the wavelength of monochromatic light. You may assume that all the initial adjustments for the spectrometer and for positioning the grating have been made, and that the grating spacing, d, is known. List the readings you would take in order to use the apparatus to find the best value of the wavelength.

A stationary ultrasonic (high frequency sound) plane wave of frequency 5.0×10^6 Hz is set up in a transparent tank of liquid. This produces a stationary pattern of density variation. Monochromatic light of wavelength 390 nm (in the liquid) passes through the tank at right angles to the ultrasonic wave. A diffraction pattern similar to that of a grating is produced in the light, the deviation of the sixth order ray being $1.0°$ (0.017 rad). Explain why a diffraction pattern is produced and calculate the speed of ultrasound in the liquid. (*L.*)

7. Give a labelled sketch and a brief description of the essential features of a spectrometer incorporating a plane diffraction grating. What part is played by (*a*) diffraction, and (*b*) interference, in the operation of the diffraction grating?

A source emitting light of two wavelengths is viewed through a grating spectrometer set at normal incidence. When the telescope is set at an angle of $20°$ to the incident direction, the second order maximum for one wavelength is seen superposed on the third order maximum for the other wavelength. The shorter wavelength is 400 nm. Calculate the longer wavelength and the number of lines per metre in the grating. At what other angles, if any, can superposition of two orders be seen using this source? (*O. & C.*)

8. A pure spectrum is one in which there is no overlapping of light of different wavelengths. Describe how you would set up a diffraction grating to display on a screen as close an approximation as possible to a pure spectrum. Explain the purpose of each optical component which you would use.

A grating spectrometer is used at normal incidence to observe the light from a sodium flame. A strong yellow line is seen in the first order when the telescope axis is at an angle of $16°$ $26'$ to the normal to the grating. What is the highest order in which the line can be seen?

The grating has 4800 lines per cm; calculate the wavelength of the yellow radiation.

What would you expect to observe in the spectrometer set to observe the first-order spectrum if a small but very bright source of white light is placed close to the sodium flame so that the flame is between it and the spectrometer? (*O. & C.*)

9. Explain what is meant by diffraction, and discuss the parts played by diffraction and by interference in the action of a diffraction grating. Describe briefly the diffraction patterns observed when a parallel beam of monochromatic light illuminates (*a*) a fairly narrow slit, (*b*) a very narrow slit, (*c*) a straight edge.

A diffraction grating is set up on a spectrometer table so that parallel light is incident on it normally. For a light of wavelength 5.5×10^{-7} m, first order reinforcement is observed in directions making angles θ of $18°$ with the normal on each side of the normal. (i) Find the value of the grating spacing d. (ii) Calculate the wavelength of monochromatic light which would give a first order spectral line at $\theta = 25°$. (iii) What would be the values of θ for second order spectral lines for each of these two wavelengths? (iv) Is a third order spectrum possible for each? Explain. (O.)

10. What is *Huygens' theory*? Indicate with the aid of diagrams how the theory accounts for (a) the reflection of waves from a plane surface, and (b) the diffraction of waves by a grating.

Describe how you would demonstrate the diffraction of waves at a single slit. How does the width of the slit affect the results?

A grating has 500 lines per mm and is illuminated normally with monochromatic light of wavelength 589 nm. How many diffraction maxima may be observed? Calculate their angular positions. (L.)

11. What are the advantages and disadvantages of a diffraction grating as compared with a prism for the study of spectra?

A rectangular piece of glass 2 cm × 3 cm has 18 000 evenly spaced lines ruled across its whole surface, parallel to the shorter side, to form a diffraction grating. Parallel rays of light of wavelength 5×10^{-7} m fall normally on the grating. What is the highest order of spectrum in the transmitted light?

What is the minimum diameter of a camera lens which can accept all the light of this wavelength in this order which leaves the grating on one side of the normal? (O. & C.)

12. (a) What is meant by (i) *diffraction*, (ii) *superposition* of waves? Describe *one* phenomenon to illustrate each.

(b) The floats of two men fishing in a lake from boats are 22·5 metres apart. A disturbance at a point in line with the floats sends out a train of waves along the surface of the water, so that the floats bob up and down 20 times per minute. A man in a third boat observes that when the float of one of his colleagues is on the crest of a wave that of the other is in a trough, and that there is then one crest between them. What is the velocity of the waves? (O. & C.)

13. Describe and give the theory of an experiment to compare the wavelengths of yellow light from a sodium and red light from a cadmium discharge lamp, using a diffraction grating. Derive the required formula from first principles.

White light is reflected normally from a soap film of refractive index 1·33 and then directed upon the slit of a spectrometer employing a diffraction grating at normal incidence. In the first-order spectrum a dark band is observed with minimum intensity at an angle of 18° 0′ to the normal. If the grating has 500 lines per mm, determine the thickness of the soap film assuming this to be the minimum value consistent with the observations. (L.)

14. Describe how you would determine the wavelength of monochromatic light using a diffraction grating and a spectrometer. Give the theory of the method.

A filter which transmits only light between $6·3 \times 10^{-7}$ and $6·0 \times 10^{-7}$ m is placed between a source of white light and the slit of a spectrometer; the grating has 5000 lines to the centimetre; and the telescope has an objective of focal length 15 cm with an eyepiece of focal length 3 cm. Find the width in millimetres of the first-order spectrum formed in the focal plane of the objective. Find also the angular width of this spectrum seen through the eyepiece. (O.)

15. (a) Parallel monochromatic light is incident normally on a thin slit of width d and focused on to a screen. Derive the relationship between the wavelength λ and the angle of diffraction θ for the first minimum of intensity on the screen.

(b) If the light has a wavelength of 540 nm and is focused by a converging lens of focal length 0·50 m placed immediately in front of the slit which has a width of 0·10 mm, calculate the distance from the centre of the intensity distribution to the first minimum.

(c) Explain how diffraction effects, similar to those referred to above, limit the sharpness of the image produced by a telescope. (N).

25 Polarization of Light Waves

We have shown that light can behave as a wave-motion of some kind, i.e., that it is a travelling vibration. For a long time after the wave-theory was revived it was thought that the vibrations of light occurred in the same direction as the light wave travelled, analogous to sound waves. Thus light waves were thought to be longitudinal waves (p. 413). Observations and experiments, however, to be described shortly, showed that the vibrations of light occur in planes *perpendicular* to the direction along which the light wave travels, and thus light waves are *transverse* waves.

Polarized Light

Polaroid is an artificial crystalline material which can be made in thin sheets. As we shall see shortly, it has the property of allowing light vibrations only of a particular polarization to pass through.

Suppose two Polaroids, P and Q, are placed one behind the other in front of a window, and the light passing through P and Q is observed, Fig. 25.1.

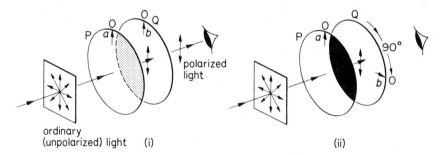

Fig. 25.1 Plane-polarization of light by Polaroid

When the Polaroids have their axes *a* and *b* parallel, the light passing through Q appears slightly darker, Fig. 25.1 (i). If Q is now rotated slowly about the line of vision with its plane parallel to P, the light passing through Q becomes darker and darker and disappears at one stage. In this case the axes *a* and *b* are perpendicular, Fig. 25.1 (ii). When Q is rotated further the light reappears, and becomes brightest when the axes *a*, *b* are again parallel.

This simple experiment leads to the conclusion that light waves are *transverse* waves; otherwise the light emerging from Q could never be extinguished by simply rotating the Polaroid. The experiment, in fact, is analogous to that illustrated in Fig. 21.32, where transverse waves were set up along a rope and plane-polarized waves were obtained by means of a slit. Polaroid, because of its internal molecular structure, transmits only those vibrations of light in a particular plane. Consequently (*a*) plane-polarized light is obtained beyond the crystal P, and (*b*) no light emerges beyond Q when its axis is perpendicular to that of P.

Vibrations in Unpolarized and Polarized Light

Fig. 25.2 (i) is an attempt to represent diagrammatically the vibrations of ordinary or unpolarized light at a point A when a ray travels in a direction AB. X is a plane perpendicular to AB, and ordinary (unpolarized) light may be imagined

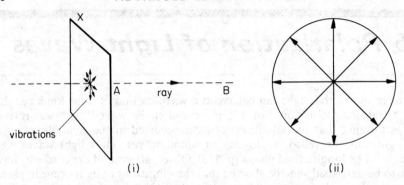

Fig. 25.2 (i) Vibrations occur in every plane perpendicular to ray (ii) Vibrations in ordinary light

as due to vibrations which occur in every one of the millions of planes which pass through AB and are perpendicular to X. As represented in Fig. 25.2 (ii), the amplitudes of the vibrations are all equal.

Consider the vibrations in ordinary light when it is incident on the Polaroid P in Fig. 25.1 (i). Each vibration can be resolved into two components, one in a direction parallel to *a*, the direction of easy transmission of light through P, and the other in a direction *m* perpendicular to *a*. Fig. 25.3. Polaroid absorbs the light due to vibrations parallel to *m*, known as the *ordinary rays*, allowing light due to the other vibrations, known as the *extraordinary rays*, to pass through. So plane-polarized light is produced, as illustrated in Fig. 25.1.

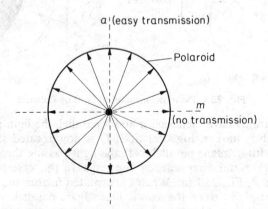

Fig. 25.3 Plane-polarized waves by selective absorption

Polaroid thus absorbs light due to vibrations in a particular direction and allows light due to vibrations in a perpendicular direction to pass through. This 'selective absorption' is also shown by certain natural crystals such as tourmaline.

Theory and experiment show that the vibrations of light are *electromagnetic* in origin. This is discussed later at the end of the chapter.

Polarized Light by Reflection

In 1808 MALUS discovered that polarized light is obtained when ordinary light is reflected by a plane sheet of glass (p. 503). The most suitable angle of incidence is about 57°, Fig. 25.4. If the reflected light is viewed through a Polaroid which is slowly rotated about the line of vision, the light is practically extinguished

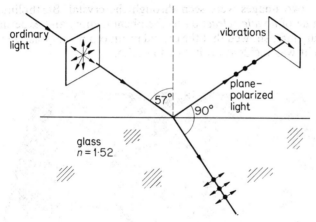

Fig. 25.4 Plane-polarization by reflection

at one position of the Polaroid. This proves that the light reflected by the glass is practically plane-polarized. Light reflected from the surface of a table becomes darker when viewed through a rotated Polaroid, showing it is partially plane-polarized.

The production of the polarized light by the glass is explained as follows. Each of the vibrations of the incident (ordinary) light can be resolved into a component parallel to the glass surface and a component perpendicular to the surface. The light due to the components parallel to the glass is largely reflected, but the remainder of the light, due mainly to the components perpendicular to the glass, is *refracted* into the glass. Thus the light reflected by the glass is partially plane-polarized.

Brewster's Law. Polarization by Pile of Plates

The particular angle of incidence i on a transparent medium when the reflected light is almost completely plane-polarized is called the *polarizing angle*. BREWSTER found that, in this case, $\tan i = n$, where n is the refractive index of the medium (*Brewster's law*), so that, with crown glass of $n = 1.52$, $i = 57°$ (approx.). Since n varies with the colour of the light, white light can not be completely plane-polarized by reflection. As $\sin i / \sin r = n$, where r is the angle of refraction, it follows from Brewster's law that $\cos i = \sin r$, or $i + r = 90°$. Thus the reflected and refracted beams are at $90°$ to each other at the polarizing angle. See Fig. 25.4.

The refracted beam contains light mainly due to vibrations perpendicular to that reflected and is therefore partially plane-polarized. Since refraction and reflection occur at both sides of a glass plate, the transmitted beam contains a fair percentage of plane-polarized light. A *pile of plates* increases the percentage, and thus provides a simple method of producing plane-polarized light. They are mounted inclined in a tube so that the ordinary (unpolarized) light is incident at the polarizing angle, and the transmitted light is then practically plane-polarized.

Polarization by Double Refraction

We have already considered two methods of producing polarized light. The first observation of polarized light, however, was made by BARTHOLINUS in 1669, who placed a crystal of Iceland spar on some words on a sheet of paper. To his surprise, two images were seen through the crystal. Bartholinus therefore gave the name of *double refraction* to the phenomenon, and experiments more than a century later showed that the crystal produced plane-polarized light when ordinary light was incident on it. See Fig. 25.5.

Fig. 25.5 Double refraction. A ring with a spot in the centre, photographed through a crystal of Iceland spar. The two plane-polarized beams of light emerging from the crystal form two rings and two spots

Iceland spar is a crystalline form of calcite (calcium carbonate) which cleaves in the form of a 'rhomboid' when it is lightly tapped; this is a solid whose opposite faces are parallelograms. When a beam of unpolarized light is incident on one face of the crystal, its internal molecular structure produces two beams of polarised light, E, O, whose vibrations are perpendicular to each other, Fig. 25.6. If the incident direction AB is parallel to a plane known as the 'principal

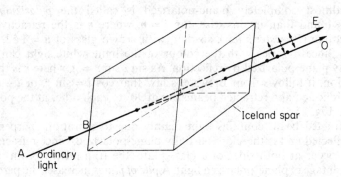

Fig. 25.6 Plane-polarized light by double refraction

section' of the crystal, one beam O emerges parallel to AB, while the other beam E emerges in a different direction. As the crystal is rotated about the line of vision the beam E revolves round O. On account of this abnormal behaviour the rays in E are called 'extraordinary' rays; the rays in O are known as 'ordinary' rays (p. 503). Thus two images of a word on a paper, for example, are seen when an Iceland spar crystal is placed on top of it; one image is due to the ordinary rays, while the other is due to the extraordinary rays.

With the aid of an Iceland spar crystal Malus discovered the polarization of light by reflection (p. 500). While on a visit to Paris he looked through the crystal at the light of the sun reflected from the windows of the Palace of Luxemburg, and observed that only one image was obtained for a particular position of the crystal when it was rotated slowly. The light reflected from the windows could not therefore be ordinary (unpolarized) light, and Malus found that it was plane-polarized.

Nicol Prism

We have seen that a Polaroid produces polarized light, and that the Polaroid can be used to detect light (p. 499). NICOL designed a form of Iceland spar crystal which was widely used for producing and detecting polarized light, and it is known as a *Nicol prism*. A crystal whose faces contain angles of 72° and 108° is broken into two halves along the diagonal AB, and the halves are cemented together by a layer of Canada balsam, Fig. 25.7. The refractive index

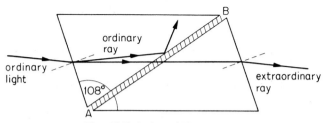

Fig. 25.7 Action of Nicol prism

of the crystal for the ordinary rays is 1·66, and is 1·49 for the extraordinary rays; the refractive index of the Canada balsam is about 1·55 for both rays, since Canada balsam does not polarize light. A critical angle thus exists between the crystal and Canada balsam for the ordinary rays, but not for the extraordinary rays. Hence total reflection of the former rays takes place at the Canada balsam if the angle of incidence is large enough, as it is with the Nicol prism. The emergent light is then due to the extraordinary rays, and is polarized.

The prism is used like a Polaroid to detect plane-polarized light, namely, the prism is held in front of the beam of light and is rotated. If the beam is plane-polarized the light seen through the Nicol prism varies in intensity, and is extinguished at one position of the prism.

Differences between Light and Sound Waves

We are now in a position to distinguish fully between light and sound waves. The physical difference, of course, is that light waves are due to varying electric and magnetic fields, while sound waves are due to vibrating layers or particles of the medium concerned. Light can travel through a vacuum, but sound cannot travel through a vacuum. Another very important difference is that the vibrations of the particles in sound waves are in the same direction as that along which the sound travels, whereas the vibrations in light waves are perpendicular to the direction along which the light travels. Sound waves are therefore *longitudinal* waves, whereas light waves are *transverse* waves. As we have seen, sound waves can be reflected and refracted, and can give rise to interference phenomena; but no polarization phenomena can be obtained with sound waves since they are longitudinal waves, unlike the case of light waves.

Polarization and Electric Vector

As we have just stated, light is a transverse electromagnetic wave and thus contains electric and magnetic fields. Experiment shows that only the electric field is concerned in the blackening of a photographic film when exposed to light. Hence we usually define *the direction of light vibrations to be that of the electric vector E*. The 'plane of polarization' is then defined as the plane containing the light ray and *E*.

Figure 25.8 (i) shows *plane-polarized* light due to vertical vibrations or a vertical vector *E*. Figure 25.8 (ii) shows plane-polarized light due to horizontal vibrations; in this case a dot is used to indicate the vibrations or vector *E* perpendicular to the paper. Figure 25.8 (iii) shows *ordinary* or *unpolarized* light.

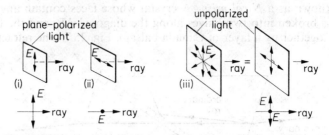

Fig. 25.8 Electric vectors in plane-polarized and unpolarized light

The associated field vectors *E* in ordinary light act in all directions in a plane perpendicular to the ray and vary in phase. Since each vector can be resolved into components in perpendicular directions, the total effect is equivalent to two perpendicular vectors equal in magnitude. Thus ordinary or unpolarized light can be represented in a diagram by an arrow and a dot, as shown in Fig. 25.8 (iii).

Polaroid Transmission and Light Intensity

Suppose a Polaroid A produces polarized light whose electric vector vibrates in a particular direction, say AY in Fig. 25.9 (i). If another Polaroid B is placed

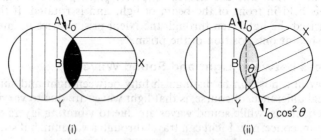

Fig. 25.9 Intensity variation by rotating Polaroid

in front of A so that its 'easy' direction of transmission BX is perpendicular to AY, the overlapping areas appear black because the electric vector has no component in a perpendicular direction and so no light is transmitted.

Suppose B is turned so that BX makes an angle θ with AY, Fig. 25.9 (ii). If E_0 is the initial amplitude of the electric vector, the amplitude of the vector transmitted by B is the component *E* given by $E = E_0 \cos \theta$. The intensity is proportional to the square of the amplitude (p. 521). Since $E^2 = E_0^2 \cos^2 \theta$, the intensity *I* of the light transmitted by B is given by $I = I_0 \cos^2 \theta$, where I_0 is the light intensity incident on B from A, Fig. 25.9 (ii). It may be noted that the intensity of the unpolarized light *incident* on A is $2I_0$, since, from Fig. 25.9, one of the perpendicular vectors is absorbed by A.

Applications of Polarized Light

Polaroids are used in many practical applications of polarized light. For example, they are used in sunglasses to reduce the intensity of incident sunlight and to eliminate reflected light or glare.

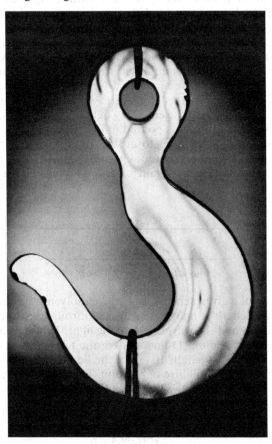

Fig. 25A Photoelasticity. The pattern in a perspex hook between polaroids when the hook is under stress. Similar patterns in transparent models help the engineer to investigate the stresses in mechanical structures subjected to loads or forces. (*Courtesy of Kodak Limited*)

Photoelasticity, or photoelastic stress analysis, utilises polarized light. Under mechanical stress, certain isotropic substances such as glass and celluloid become doubly refracting. Ordinary white light does not pass through crossed Polaroids (p. 499). However, if a celluloid model with a cut-out design is placed between the crossed Polaroids and then subjected to compression by a vice, coloured bright and dark fringes can now be seen or projected on a screen. The fringes spread from places where the stress is most concentrated. Now the pattern of the fringe varies with the stress. So a study of the pattern using a model provides the engineer with valuable information on design. This is particularly useful with complicated shapes, where mathematical computation is difficult.

In *films*, it is possible to give the illusion of three-dimensions or 3-D by projecting two overlapping pictures, with slightly different views, on to the same screen. Each picture has been taken by light polarized respectively in perpendicular directions. The viewer is provided with special spectacles, with perpendicular Polaroids in the respective frames. The two pictures received simultaneously by the two eyes provide a 3-D view of the scene.

Saccharimetry is the measurement of the concentration of sugars such as cane sugar in solution. Due to the molecular structure of the sugar, these solutions rotate the plane of polarization of plane-polarized light as the light passes through. Solids such as quartz produce the same effect, which is called *optical activity*. The rotation of the plane of polarization when the incident light is viewed may be right-handed (clockwise) or left-handed (anticlockwise).

There are various types of saccharimeter. Here we are only concerned with the basic principle of their action. A saccharimeter usually consists of a nicol prism P, which produces plane-polarized light from an incident source S of monochromatic light and is called the *polarizer*; a *tube* T containing the solution; and a nicol A through which the emerging light is observed, called the *analyser*. Fig. 25.10. Before the solution is poured into T, the analyser A is rotated until the plane-polarized light emerging from P is completely extinguished. T is then filled with the sugar solution. On looking through A, light can now be seen. A is then rotated until the light is again just extinguished and the angle of rotation θ is measured.

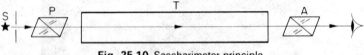

Fig. 25.10 Saccharimeter principle

For a solution of a given substance in a given solvent, the amount of rotation θ depends on the length of light path travelled through the liquid, the temperature of the solution and the wavelength of the light. Tables provide the rotation in degrees when the sodium D-line is used, the temperature is 20°C, and the light travels a column of length 10 cm of the liquid which contains 1 gram of the active substance per millilitre of solution. This is called the *specific rotation* or *rotary power*. Knowing its value, the concentration of a sugar solution can be found by a polarimeter experiment.

Exercises 25

1. With the aid of suitable diagrams, explain the difference between transverse waves which are polarised and those which are unpolarised. State one useful application of plane polarised waves, specifying the type of wave involved (e.g. electromagnetic). (*L.*)

2. What is meant by *plane of polarization*? Explain why the phenomenon of polarization is met with in dealing with light waves, but not with sound waves.

Describe and explain the action of (*a*) a Nicol prism, (*b*) a sheet of Polaroid.

How can a pair of Polaroid sheets and a source of natural light be used to produce a beam of light the intensity of which may be varied in a calculable manner? (*L.*)

3. Explain what is meant by the statement that a beam of light is *plane-polarised*. Describe *one* experiment in each instance to demonstrate (*a*) polarisation by reflexion, (*b*) polarisation by double refraction, (*c*) polarisation by scattering.

The refractive index of diamond for sodium light is 2·417. Find the angle of incidence for which the light reflected from diamond is completely plane polarized. (*L.*)

4. Explain the terms *linearly polarised light* and *unpolarised light*.

A parallel beam of light is incident on the surface of a transparent medium of refractive index n at an angle of incidence i such that the reflected and refracted beams are at right angles to each other. Assuming the laws of reflection and refraction, find the relation between n and i.

At this value of i, the Brewster angle, the reflected beam is found to be linearly polarised. Describe the apparatus and procedure you would use to find the refractive index of a material, using the Brewster angle. Has the method any advantage over other methods for measurement of n? (O. & C.)

5. Give an account of the action of (a) a single glass plate. (b) a Nicol prism, in producing plane-polarized light. State *one* disadvantage of *each* method.

Mention *two* practical uses of polarizing devices. (N.)

6. Explain what is meant by *polarization* and *interference*.

Describe (a) an experiment to demonstrate polarisation and (b) an application in which polarized light is involved. Explain why electromagnetic waves can be polarised but sound waves cannot.

A series of interference fringes is formed on a screen placed 1·6 m away from a double slit illuminated by a narrow line source. The slits are 1·0 mm apart and are illuminated by light of wavelength 640 nm (640×10^{-9} m). Calculate the spacing of the fringes. Explain why the fringes disappear if the line source is made too wide. (L.)

7. Give one piece of experimental evidence in each case, with a brief account of how that evidence is obtained, that light behaves (a) as a wave, (b) as an electromagnetic wave, (c) as a stream of particles.

When an unpolarised beam of light travelling in air is incident on a glass plate at an angle of incidence 30° the reflected light is found to be *partially linearly polarised*. Explain the meaning of the term in italics and describe how you would verify this statement experimentally. (O. & C.)

8. A beam of plane-polarized light falls normally on a sheet of Polaroid, which is at first set so that the intensity of the transmitted light, as estimated by a photographer's light-meter, is a maximum. (The meter is suitably shielded from all other illumination.) Describe and explain the way in which you would expect the light-meter readings to vary as the Polaroid is rotated in stages through 180° about an axis at right angles to its plane.

How would you show experimentally (a) that calcite is doubly refracting, (b) that the two refracted beams are plane-polarized, in planes at right angles to one another, and (c) that in general the two beams travel through the crystal with different velocities? (O.)

9. Explain what is meant by *plane-polarised* electromagnetic radiation. How many plane-polarised (a) light and (b) radio waves be produced and detected?

Two polarizing sheets initially have their polarization directions parallel. Through what angle must one sheet be turned so that the intensity of transmitted light is reduced to a third of the original transmitted intensity.

A vertical dipole is connected to a radio frequency generator. An identical dipole, similarly connected to the same generator, is arranged horizontally a quarter of a wavelength in front of the first so that both dipoles are normal to the horizontal line joining their midpoints. Describe as fully as you can the nature of the radiation at a distant point on this line. (C.)

10. Explain as fully as you can the nature of linearly polarised light.

When unpolarised light falls on the surface of a block of glass the reflected light is partially polarised. If the angle of incidence is $\tan^{-1} n$, where n is the refractive index of the glass, the reflected light is completely linearly polarized. Describe the apparatus you would use and the experiments you would perform in order to verify these statements for a sample of glass of known refractive index.

Why would it be necessary, if very accurate results were required, to use monochromatic light to verify the second statement?

Show that, when the condition for completely polarised light is satisfied, the reflected and refracted beams are at right angles to one another. (O. & C.)

Sound

26 Characteristics of Sound Waves

Characteristics of Notes

Notes may be similar to or different from each other in three respects: (i) *pitch*, (ii) *loudness*, (iii) *quality*. These three quantities define or 'characterise' a note.

Pitch and Frequency

Pitch is analogous to colour in light, which depends on the wavelength or frequency of the light wave (p. 462). Similarly, the pitch of a note depends only on the frequency of the sound vibrations. A high frequency gives rise to a high-pitched note; a low frequency produces a low-pitched note. Thus the high-pitched whistle of a boy may have a frequency of several thousand Hz, whereas a low-pitched hum due to a.c. mains frequency when first switched on may be 100 Hz. The range of sound frequencies is about 15 to 20 000 Hz depending on the observer.

Musical Intervals

If a note of frequency 300 Hz, and then a note of 600 Hz, are sounded by a siren, the pitch of the higher note is recognised to be an upper octave of the lower note. A note of frequency 1000 Hz is recognised to be an upper octave of a note of frequency 500 Hz. So the *musical interval* between two notes is an upper octave if the ratio of their frequencies is 2:1. It can be shown that the musical interval between two notes depends on the *ratio* of their frequencies, and not on the actual frequencies.

Ultrasonics and their Production

There are sound waves of higher frequency than 20 000 Hz, which are inaudible to a human being. These are known as *ultrasonics*; and since velocity = wavelength × frequency, ultrasonics have short wavelengths compared with sound waves in the audio-frequency range.

In 1881 CURIE discovered that a thin plate of quartz increased or decreased in length if an electrical battery was connected to its opposite faces. By correctly cutting the plate, the expansion or contraction could be made to occur along the axis of the faces to which the battery was applied. When an alternating voltage of ultrasonic frequency was connected to the faces of such a quartz crystal the faces vibrated at the same frequency, and ultrasonic sound waves were produced.

Another method of producing ultrasonics is to place an iron or nickel rod inside a solenoid carrying an alternating current of ultrasonic frequency. Since the length of a magnetic specimen increases slightly when it is magnetised, ultrasonic sound waves are produced by the vibrations of the rod.

In recent years ultrasonics have been utilised for a variety of industrial purposes. They are used on board coasting vessels for depth sounding, the time taken by the wave to reach the bottom of the sea from the surface and back being determined. Ultrasonics are also used to kill bacteria in liquids, and they are used extensively to locate faults and cracks in metal castings, following a method similar to that of radar. Ultrasonic waves are sent into the metal under investigation, and the beam reflected from the fault is picked up on a cathode-ray tube screen together with the reflection from the other end of the metal. The position of the fault can then easily be located.

Beats

If two notes of nearly equal frequency are sounded together, a periodic rise and fall in intensity can be heard. This is known as the phenomenon of *beats*. The frequency of the beats is the number of intense sounds heard per second.

Consider a layer of air some distance away from two pure notes of nearly equal frequency, say 48 and 56 Hz respectively, which are sounding. The variation of the displacement, y_1, of the layer due to one fork alone is shown in Fig. 26.1 (i); the variation of the displacement y_2, of the layer due to the second

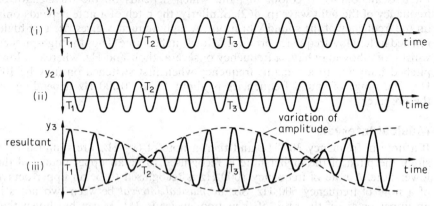

Fig. 26.1 Beats (*not to scale*)

fork alone is shown in Fig. 26.1 (ii). According to the Principle of Superposition (p. 417), the variation of the resultant displacement, y, of the layer is the algebraic sum of the two curves, which varies in amplitude in the way shown in Fig. 26.1 (iii). To understand the variation of y, suppose that the displacements y_1, y_2 are in phase at some instant T_1, Fig. 26.1. Since the frequency of the curve in Fig. 26.1 (i) is 48 cycles per sec the variation y_1 undergoes 3 complete cycles in $\frac{1}{16}$th second; in the same time, the variation y_2 undergoes $3\frac{1}{2}$ cycles, since its frequency is 56 cycles per second. Thus y_1 and y_2 are 180° out of phase with each other at this instant, and their resultant y is then a minimum at some instant T_2. Thus T_1T_2 represents $\frac{1}{16}$th of a second in Fig. 26.1 (iii). In $\frac{1}{8}$th of a second from T_1, y_1 has undergone 6 complete cycles and y_2 has undergone 7 complete cycles. The two waves are hence in phase again at T_3 where T_1T_3 represents $\frac{1}{8}$th of a second, and their resultant at their instant is again a maximum, Fig. 26.1 (iii). In this way it can be seen that a loud sound is heard after every $\frac{1}{8}$ second, and thus the beat frequency is 8 cycles per second. This is the difference between the frequencies, 48, 56, of the two notes. We show soon that *the beat frequency is always equal to the difference of the two nearly equal frequencies*. From Fig. 26.1 we see that beats are a phenomenon of interference.

Beat Frequency Formula

Suppose two sounding tuning-forks have frequencies f_1, f_2 cycles per second which are close to each other. At some instant the displacement of a particular layer of air near the ear due to each fork will be a maximum to the right. The resultant displacement is then a maximum, and a loud sound or beat is heard. After this, the vibrations of air due to each fork go out of phase, and t seconds later the displacement due to each fork is again a maximum to the right, so that a loud sound or beat is heard again. One fork has then made exactly one cycle more than the other. But the number of cycles made by each fork in t seconds is $f_1 t$ and $f_2 t$ respectively. Assuming f_1 is greater than f_2,

$$\therefore f_1 t - f_2 t = 1$$

$$\therefore f_1 - f_2 = \frac{1}{t}.$$

Now 1 beat has been made in t seconds, so that $1/t$ is the number of beats per second or beat frequency.

$$\therefore f_1 - f_2 = beat\ frequency.$$

Uses of Beats

The phenomenon of beats can be used to measure the unknown frequency, f_1, of a note. For this purpose a note of known frequency f_2 is used to provide beats with the unknown note, and the frequency f of the beats is obtained by counting the number made in a given time. Since f is the difference between f_2 and f_1, it follows that $f_1 = f_2 - f$, or $f_1 = f_2 + f$. Thus suppose $f_2 = 1000$ Hz, and the number of beats per second made with a tuning-fork of unknown frequency f_1 is 4. Then $f_1 = 1004$ or 996 Hz.

To decide which value of f_1 is correct, the end of the tuning-fork prong is loaded with a small piece of plasticine, which diminishes the frequency a little, and the two notes are sounded again. If the beat frequency is *increased*, a little thought indicates that the frequency of the note must have been originally 996 Hz. If the beats are decreased, the frequency of the note must have been originally 1004 Hz. The tuning-fork must not be overloaded, as the frequency may decrease, if it was 1004 Hz, to a frequency such as 995 Hz, in which case the significance of the beats can be wrongly interpreted.

Beats are also used to 'tune' an instrument to a given note. As the instrument note approaches the given note, beats are heard. The instrument may be regarded as 'tuned' when the beats occur at a very slow rate.

Doppler Effect

The whistle of a train or a jet aeroplane appears to increase in pitch as it approaches a stationary observer; as the moving object passes the observer, the pitch changes and becomes lower. The apparent alteration in frequency was first predicted by DOPPLER in 1845. He stated that a change of frequency of the wave-motion should be observed when a source of sound or light was moving, and this is known as the *Doppler effect*.

The Doppler effect occurs whenever a source of sound or light moves *relative* to an observer. In light, the effect was observed when measurements were taken of the wavelength of the colour of a moving star; they showed a marked variation. In sound, the Doppler effect can be demonstrated by placing a whistle in the end of a long piece of rubber tubing, and whirling the tube in a horizontal circle above the head while blowing the whistle. The open end of the tube acts as a moving source of sound, and an observer hears a rise and fall in pitch as the end approaches and recedes from him.

A complete calculation of the apparent frequency in particular cases is given

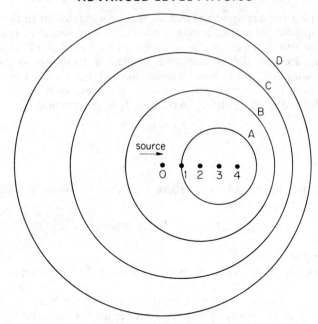

Fig. 26.2 Doppler effect

shortly, but Fig. 26.2 illustrates how the change of wavelengths, and hence frequency, occurs when a source of sound is moving towards a stationary observer. At a certain instant the position of the moving source is at 4. At four successive seconds *before* this instant the source had been at the positions 3, 2, 1, 0 respectively. If v is the velocity of sound, the wavefront from the source when in the position 3 reaches the surface A of a sphere of radius v and centre 3 when the source is just at 4. In the same way, the wavefront from the source when it was in the position 2 reaches the surface B of a sphere of radius $2v$ and centre 2. The wavefront C corresponds to the source when it was in the position 1, and the wavefront D to the source when it was in the position O. Thus if the observer is on the right of the source S, he receives wavefronts which are relatively more crowded together than if S were stationary; the frequency of S thus appears to increase. When the observer is on the left of S, in which case the source is moving away from him, the wavefronts are farther apart than if S were stationary, and hence the observer receives correspondingly fewer waves per second. The apparent frequency is thus lowered.

Calculation of Apparent Frequency

Suppose V is the velocity of sound in air, u_s is the velocity of the source of sound S, u_0 is the velocity of an observer O, and f is the true frequency of the source.

(i) *Source moving towards stationary observer.* If the source S were stationary, the f waves sent out in one second towards the observer O would occupy a distance V, and the wavelength would be V/f, Fig. 26.3 (i). If S moves with a velocity u_s towards O, however, the f waves sent out occupy a distance $(V - u_s)$, because S has moved a distance u_s towards O in 1 s, Fig. 26.3 (ii). Thus the wavelength λ' of the waves reaching O is now $(V - u_s)/f$.

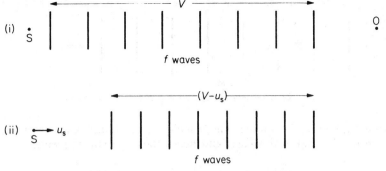

Fig. 26.3 Source moving towards stationary observer

But   velocity of sound waves $= V$.

$$\therefore \text{ apparent frequency, } f' = \frac{\text{velocity of sound relative to O}}{\text{wavelength of waves reaching O}}$$

$$= \frac{V}{\lambda'} = \frac{V}{(V - u_s)/f}$$

$$\therefore f' = \frac{V}{V - u_s} f \qquad . \qquad . \qquad . \qquad . \qquad (1)$$

Since $(V - u_s)$ is less than V, f' is greater than f; the apparent frequency thus appears to increase when a source is moving towards an observer.

(ii) *Source moving away from stationary observer.* In this case the f waves sent out towards O in 1 s occupy a distance $(V + u_s)$, Fig. 26.4. The wavelength λ' of the waves reaching O is thus $(V + u_s)/f$, and hence the apparent frequency f' is given by

$$f' = \frac{V}{\lambda'} = \frac{V}{(V + u_s)/f}$$

$$\therefore f' = \frac{V}{V + u_s} \cdot f \qquad . \qquad . \qquad . \qquad . \qquad (2)$$

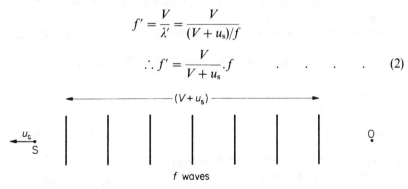

Fig. 26.4 Source moving away from stationary observer

Since $(V + u_s)$ is greater than V, f' is less than f, and hence the apparent frequency decreases when a source moves away from an observer.

(iii) *Source stationary, and observer moving towards it.* Since the source is stationary, the f waves sent out by S towards the moving observer O occupies a distance V, Fig. 26.5. The wavelength of the waves reaching O is hence V/f, and thus unlike the cases already considered, the wavelength is unaltered. The velocity of the sound waves relative to O is not V, however, as O is

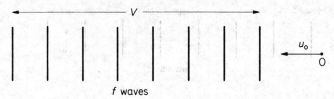

f waves

Fig. 26.5 Observer moving towards stationary source

moving relative to the source. The velocity of the sound waves relative to O is given by $(V + u_0)$ in this case, and hence the apparent frequency f' is given by

$$f' = \frac{\text{velocity of sound relative to O}}{\text{wavelength of waves reaching O}}$$

$$= \frac{V + u_0}{V/f}$$

$$\therefore f' = \frac{V + u_0}{V} \cdot f \qquad . \qquad . \qquad . \qquad . \qquad . \qquad (3)$$

Since $(V + u_0)$ is greater than V, f' is greater than f; thus the apparent frequency is increased.

(iv) *Source stationary, and observer moving away from it*, Fig. 26.6. As in the case just considered, the wavelength of the waves reaching O is unaltered, and is given by V/f.

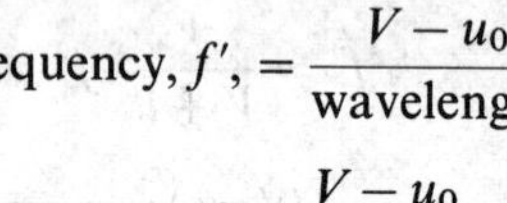

Fig. 26.6 Observer moving away from stationary source

The velocity of the sound waves relative to $O = V - u_0$, and hence

$$\text{apparent frequency}, f', = \frac{V - u_0}{\text{wavelength}} = \frac{V - u_0}{V/f}$$

$$\therefore f' = \frac{V - u_0}{V} \cdot f \qquad . \qquad . \qquad . \qquad . \qquad (4)$$

Since $(V - u_0)$ is less than V, the apparent frequency f' appears to be decreased.

Source and Observer Both Moving

If the source and the observer are both moving, the apparent frequency f' can be found from the formula

$$f' = \frac{V'}{\lambda'}$$

where V' is the velocity of the sound waves relative to the observer, and λ' is the wavelength of the waves reaching the observer. This formula can also be used to find the apparent frequency in any of the cases considered before.

Suppose that the observer has a velocity u_0, the source a velocity u_s, and that both are moving in the *same* direction. Then

$$V' = V - u_0$$

and
$$\lambda' = (V - u_s)/f$$

as was deduced in case (i), p. 514.

$$\therefore f' = \frac{V'}{\lambda'} = \frac{V - u_0}{(V - u_s)/f} = \frac{V - u_0}{V - u_s} \cdot f \qquad (i)$$

If the observer is moving towards the source, $V' = V + u_0$, and the apparent frequency f' is given by

$$f' = \frac{V + u_0}{V - u_s} \cdot f \qquad (ii)$$

From (i), it follows that $f' = f$ when $u_0 = u_s$, in which case there is no relative velocity between the source and the observer. It should also be noted that the motion of the observer affects only V', the velocity of the waves reaching the observer, while the motion of the source affects only λ', the wavelength of the waves reaching the observer.

The effect of the wind can also be taken into account in the Doppler effect. Suppose the velocity of the wind is u_w, in the direction of the line SO joining the source S to the observer O. Since the air has then a velocity u_w relative to the ground, and the velocity of the sound waves relative to the air is V, the velocity of the waves relative to ground is $(V + u_w)$ if the wind is blowing in the same direction as SO. All our previous expressions for f' can now be adjusted by replacing the velocity V in it by $(V + u_w)$. If the wind is blowing in the opposite direction to SO, the velocity V must be replaced by $(V - u_w)$.

Example

A car, sounding a horn producing a note of 500 Hz, approaches and then passes a stationary observer O at a steady speed of 20 m s^{-1}. Calculate the change in pitch of the note heard by O (velocity of sound = 340 m s^{-1}).

Towards O. Velocity of sound relative to O, $V' = 340$ m s^{-1}

Wavelength of waves reaching O, $\lambda' = (340 - 20)/500$ m

$$\therefore \text{ apparent frequency to O}, f' = \frac{V'}{\lambda'}$$

$$= \frac{340}{320} \cdot 500 = 531 \text{ Hz} \qquad (1)$$

Away from O. With the above notation, $V' = 340$ m s^{-1}

and $\lambda' = (340 + 20)/500$ m

$$\therefore \text{ apparent frequency to O}, f'' = \frac{340}{(340 + 20)} \cdot 500$$

$$= 472 \text{ Hz} \qquad (2)$$

From (1) and (2), change in pitch $= \dfrac{f''}{f'} = 0 \cdot 9$ (approx.).

Reflection of Waves

Consider a source of sound A approaching a fixed reflector R such as a wall or bridge, for example. The reflected waves then appear to travel from R to A as if they came from the 'mirror image' A' of A in R, Fig. 26.7.

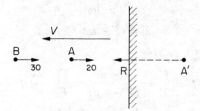

Fig. 26.7 Doppler's principle and reflection of waves

As an illustration, suppose a car A approaches R with a velocity of 20 m s^{-1} when sounding a note of 1000 Hz from its horn, and that another car B behind A is travelling towards A with a velocity of 30 m s^{-1}, Fig. 26.7.

In B, the driver hears a note from R which has an apparent frequency $f' = V'/\lambda'$, where V' is the velocity of sound relative to B and λ' is the wavelength of the waves reaching B. If the velocity of sound V is 340 m s^{-1}, then

$$V' = 340 + 30 = 370 \text{ m s}^{-1}$$

and
$$\lambda' = (340 - 20)/1000 = 320/1000 \text{ m}$$

$$\therefore f' = \frac{V'}{\lambda'} = \frac{370}{320} \cdot 1000 = 1156 \text{ Hz.}$$

In B, the driver also hears a note directly from A. In this case,

$$V' = 340 + 30 = 370 \text{ m s}^{-1}$$

and
$$\lambda' = (340 + 20)/1000 = 360/1000 \text{ m}$$

$$\therefore f'' = \frac{V'}{\lambda'} = \frac{370}{360} \cdot 1000 = 1028 \text{ Hz.}$$

Doppler Principle in Light

The speed of distant stars and planets has been estimated from measurements of the wavelengths of the spectrum lines which they emit. Suppose a star or planet is moving with a velocity v away from the earth and emits light of wavelength λ. If the frequency of the vibrations is f cycles per second, then f waves are emitted in one second, where $c = f\lambda$ and c is the velocity of light *in vacuo*. Owing to the velocity v, the f waves occupy a distance $(c + v)$. Thus the *apparent wavelength* λ' to an observer on the earth in line with the star's motion is

$$\lambda' = \frac{c+v}{f} = \frac{c+v}{c} \cdot \lambda = \left(1 + \frac{v}{c}\right)\lambda$$

$$\therefore \lambda' - \lambda = \text{'shift' in wavelength} = \frac{v}{c}\lambda, \qquad . \qquad . \qquad . \qquad \text{(i)}$$

and hence
$$\frac{\lambda' - \lambda}{\lambda} = \text{fractional change in wavelength} = \frac{v}{c}. \qquad . \qquad \text{(ii)}$$

From (i), it follows that λ' is greater than λ when the star or planet is moving away from the earth, that is, there is a 'shift' or displacement *towards the red*. The position of a particular wavelength in the spectrum of the star is compared with that obtained in the laboratory, and the difference in the wavelengths, $\lambda' - \lambda$, is measured. From (i), knowing λ and c, the velocity v can be calculated.

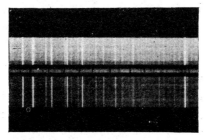

Fig. 26.8 Doppler shift. The central band of dark lines is the absorption spectrum of the star Eta Cephei. The bright lines in the wider bands above and below are the same lines in the emission spectrum of iron as obtained in the laboratory. Because the star is moving away from the earth, on account of the Doppler effect each dark line has a wavelength slightly greater than if the star were stationary

If the star is moving *towards* the earth with a velocity u, the apparent wavelength λ'' is given by

$$\lambda'' = \frac{c-u}{f} = \frac{c-u}{c}.\lambda = \left(1 - \frac{u}{c}\right)\lambda.$$

$$\therefore \lambda - \lambda'' = \frac{u}{c}\lambda. \qquad . \qquad . \qquad . \qquad . \qquad . \qquad . \qquad . \qquad (iii)$$

Since λ'' is less than λ, there is a displacement towards the blue in this case*.

In measuring the speed of a star, a photograph of its spectrum is taken. The spectral lines are then compared with the same lines obtained by photographing in the laboratory an arc or spark spectrum of an element present in the star. If the lines are displaced towards the red, the star is receding from the earth; if displaced towards the violet, the star is approaching the earth. By this method the velocities of the stars have been found to be between about 10 km s^{-1} and 300 km s^{-1}.

The Doppler effect has also been used to measure the speed of rotation of the sun. Photographs are taken of the east and west edges of the sun; each contains absorption lines due to elements such as iron vaporised in the sun, and also some absorption lines due to oxygen in the earth's atmosphere. When the two photographs are put together so that the oxygen lines coincide, the iron lines in the two photographs are displaced relative to each other. In one case the edge of the sun approaches the earth, and in the other the opposite edge recedes from the earth. Measurements show a rotational speed of about 2 km s^{-1}.

Measurement of Plasma Temperature

In very hot gases or plasma, used in thermonuclear fusion experiments, the temperature is of the order of millions of degrees Celsius. At these high temperatures molecules of the glowing gas are moving away and towards the observer with very high speeds and, owing to the Doppler effect, the wavelength λ of a particular spectral line is apparently changed. One edge of the line now corresponds to an apparently increased wavelength λ_1 due to molecules moving directly away from the observer, and the other edge to an apparent decreased

* Equations (i)–(iii) apply for velocities much less than c, otherwise relativistic corrections are required. See *Introduction to Relativity*, Rosser (Butterworth).

wavelength λ_2 due to molecules moving directly towards the observer. The line is thus observed to be *broadened*.

From our previous discussion, if v is the velocity of the molecules,

$$\lambda_1 = \frac{c+v}{c} . \lambda$$

and

$$\lambda_2 = \frac{c-v}{c} . \lambda$$

$$\therefore \text{ breadth of line, } \lambda_1 - \lambda_2 = \frac{2v}{c} . \lambda \qquad . \qquad . \qquad . \qquad \text{(i)}$$

The breadth of the line can be measured by a diffraction grating, and as λ and c are known, the velocity v can be calculated. By the kinetic theory of gases, the velocity v of the molecules is roughly the root-mean-square velocity, or $\sqrt{3RT/M}$, where T is the absolute temperature, R is the molar gas constant and M is the mass of one mole.

Examples

1. Obtain the formula for the Doppler effect when the source is moving with respect to a stationary observer. Give examples of the effect in sound and light. A whistle giving out 500 Hz moves away from a stationary observer in a direction towards and perpendicular to a flat wall with a velocity of 1·5 m s⁻¹. How many beats per second will be heard by the observer? (Take the velocity of sound as 336 m s⁻¹ and assume there is no wind.) (*C.*)

The observer hears a note of apparent frequency f' from the whistle directly, and a note of apparent frequency f'' from the sound waves reflected from the wall.

Now

$$f' = \frac{V'}{\lambda'}$$

where V' is the velocity of sound in air relative to the observer and λ' is the wavelength of the waves reaching the observer. Since $V' = 336$ m s⁻¹ and $\lambda' = \dfrac{336 + 1\cdot5}{500}$ m

$$\therefore f' = \frac{336 \times 500}{337\cdot5} = 497\cdot8 \text{ Hz.}$$

The note of apparent frequency f'' is due to sound waves moving towards the observer with a velocity of 1·5 m s⁻¹.

$$\therefore f'' = \frac{V'}{\lambda'} = \frac{336}{(336 - 1\cdot5)/500}$$

$$= \frac{336 \times 500}{334\cdot5} = 502\cdot2 \text{ Hz}$$

$$\therefore \text{ beats per second} = f'' - f' = 502\cdot2 - 497\cdot8 = 4\cdot4$$

2. Two observers A and B are provided with sources of sound of frequency 500. A remains stationary and B moves away from him at a velocity of 1·8 m s⁻¹. How many beats per second are observed by A and by B, the velocity of sound being 330 m s⁻¹? Explain the principles involved in the solution of this problem. (*L.*)

Beats observed by A. A hears a note of frequency 500 due to its own source of sound. He also hears a note of apparent frequency f' due to the moving source B. With the usual notation,

$$f' = \frac{V'}{\lambda'} = \frac{330}{(330 + 1\cdot8)/500}$$

since the velocity of sound, V', relative to A is 330 m s^{-1} and the wavelength λ' of the waves reaching him is $(330 + 1\cdot8)/500$ m.

$$\therefore f' = \frac{330 \times 500}{331\cdot8} = 497\cdot3$$

$\therefore$ beats observed by A $= 500 - 497\cdot29 = 2\cdot71$ Hz.

Beats observed by B. The apparent frequency f' of the sound from A is given by

$$f' = \frac{V'}{\lambda'}.$$

In this case $V' =$ velocity of sound relative to B $= 330 - 1\cdot8 = 328\cdot2$ m s^{-1} and the wavelength λ' of the waves reaching B is unaltered. Since $\lambda' = 330/500$ m, it follows that

$$f' = \frac{328\cdot2}{330/500} = \frac{328\cdot2 \times 500}{330} = 497\cdot27$$

$\therefore$ beats heard by B $= 500 - 497\cdot27 = 2\cdot73$ Hz.

Intensity and Amplitude

The *intensity* of a sound at a place is defined as the energy per second flowing through one square metre held normally at that place to the direction along which the sound travels. At a distance r from a *small* source of sound, the energy passes through the surface, area $4\pi r^2$, of a surrounding sphere. So in this case, intensity $\propto 1/r^2$.

Suppose the displacement y of a vibrating layer of air is given by $y = a \sin \omega t$, where $\omega = 2\pi/T$ and a is the amplitude of vibration, see equation (1), p. 415. The velocity, v, of the layer is given by

$$v = \frac{dy}{dt} = \omega a \cos \omega t,$$

and hence the kinetic energy, W, is given by

$$W = \tfrac{1}{2}mv^2 = \tfrac{1}{2}m\omega^2 a^2 \cos^2 \omega t \qquad . \qquad . \qquad . \qquad \text{(i)}$$

where m is the mass of the layer. The layer also has potential energy as it vibrates. Its total energy, W_0, which is constant, is therefore equal to the maximum value of the kinetic energy. From (i), it follows that

$$W_0 = \tfrac{1}{2}m\omega^2 a^2 \qquad . \qquad . \qquad . \qquad \text{(ii)}$$

In 1 second, the air is disturbed by the wave over a distance V m, where V is the velocity of sound in m s^{-1}; and if the area of cross-section of the air is 1 m^2, the volume of air disturbed is V m^3. The mass of air disturbed per second is thus $V\rho$ kg, where ρ is the density of air in kg m^{-3}, and hence, from (ii),

$$W = \tfrac{1}{2}V\rho\omega^2 a^2 \qquad . \qquad . \qquad . \qquad . \qquad \text{(iii)}$$

It therefore follows that *the intensity of a sound due to a wave of given frequency is proportional to the square of its amplitude of vibration.*

At the beginning of this section we showed that the intensity due to a small source of sound at a distance r from it was proportional to $1/r^2$. But we have just proved that the intensity is also proportional to a^2, where a is the amplitude of vibration at this distance from the source. So $a^2 \propto 1/r^2$, or

$$a \propto \frac{1}{r}.$$

Since the amplitude is inversely-proportional to the distance r, it follows that the amplitude of vibration at a point 5 m from a small loudspeaker is *twice* that at a distance of 10 m.

It can be seen from (ii) that the greater the mass m of air in vibration, the greater is the intensity of the sound obtained. For this reason the sound set up by the vibration of the diaphragm of a telephone earpiece cannot be heard except with the ear close to the earpiece. On the other hand, the cone of a loudspeaker has a large surface area, and thus disturbs a large mass of air when it vibrates, giving rise to a sound of much larger intensity than the vibrating diaphragm of the telephone earpiece. It is difficult to hear a vibrating tuning-fork a small distance away from it because its prongs set such a small mass of air vibrating. If the fork is placed with its end on a table, however, a much louder sound is obtained, which is due to the large mass of air vibrating by contact with the table.

Intensity unit. Loudness

Observation shows that the change in intensity of two sound sources depends on the *ratio* of their intensities, I_2/I_1 say. The unit of intensity change is the *bel*, symbol B, and by definition,

$$\text{number of bels} = \log_{10}\left(\frac{I_2}{I_1}\right).$$

In practice, the bel is too large a unit and the *decibel*, symbol dB, is used. Since $1\ \text{B} = 10\ \text{dB}$, it follows that

$$\text{number of dB} = 10\log_{10}\left(\frac{I_2}{I_1}\right).$$

The standard 'zero' of intensity level is defined as 10^{-12}W m^{-2}. This is the intensity of the lowest audible sound and is called the *threshold value of hearing*. Suppose we call this intensity value I_0 and another sound has an intensity I 50 dB above it. Then, from above,

$$10\log_{10}\left(\frac{I}{I_0}\right) = 50.$$

So
$$I = 10^5 \times I_0 = 10^5 \times 10^{-12}$$
$$= 10^{-7}\ \text{W m}^{-2}.$$

Loudness, unlike intensity, is subjective because it depends on the observer. It is measured in units called *phons*. The loudness of a sound varies with frequency and intensity. So scientists adopt a standard source of frequency 1000 Hz and intensity $10^{-12}\ \text{W m}^{-2}$ with which the loudness of other sounds are compared. The source H whose loudness is required is placed near the standard source, which is then altered until the loudness is the same as H. The intensity or power level of the standard source is then measured, and if this is n decibels above the threshold value ($10^{-12}\ \text{W m}^{-2}$) the loudness is said to be n *phons*. The 'threshold of feeling', when sound produces a painful sensation to the ear, corresponds to a loudness of about 120 phons.

Quality or Timbre

If the same note is sounded on the violin and then on the piano, an untrained listener can tell which instrument is being used, without seeing it. We say that the *quality* or *timbre* of the note is different in each case.

The waveform of a note is never simple harmonic in practice; the nearest approach is that obtained by sounding a tuning-fork, which thus produces what may be called a 'pure' note, Fig. 26.9 (i). If the same note is played on a violin and piano respectively, the waveforms produced might be represented by Fig. 26.9 (ii), (iii), which have the same frequency and amplitude as the waveform in Fig. 26.9 (i). Now curves of the shape of Fig. 26.9 (ii), (iii) can be analysed

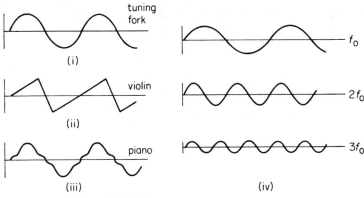

Fig. 26.9 Waveforms of notes

mathematically into the sum of a number of *simple harmonic* curves, whose frequencies are multiples of f_0, the frequency of the original waveform; the amplitudes of these curves diminish as the frequency increases. Fig. 26.9 (iv), for example, might be an analysis of a curve similar to Fig. 26.9 (iii), corresponding to a note on a piano. The ear is able to detect simple harmonic waves and therefore it registers the presence of notes of frequencies $2f_0$ and $3f_0$, in addition to f_0, when the note is sounded on the piano. The amplitude of the curve corresponding to f_0 is greatest, Fig. 26.9 (iv), and the note of frequency f_0 is heard predominantly because the intensity is proportional to the square of the amplitude (p. 521). In the background, however, are the notes of frequencies $2f_0$, $3f_0$, which are called the *overtones*. The frequency f_0 is called the *fundamental*.

As the waveform of the same note is different when it is obtained from different instruments, it follows that the analysis of each will differ; for example, the waveform of a note of frequency f_0 from a violin may contain overtones of frequencies $2f_0$, $4f_0$, $6f_0$. The musical 'background' to the fundamental note is therefore different when it is sounded on different instruments, and hence *the overtones present in a note determine its quality or timbre*.

A *harmonic* is the name given to a note whose frequency is a simple multiple of the fundamental frequency f_0. The latter is thus termed the 'first harmonic'; a note of frequency $2f_0$ is called the 'second harmonic', and so on. Certain harmonics of a note may be absent from its overtones; for example, the only possible notes obtained from an organ-pipe closed at one end are f_0, $3f_0$, $5f_0$, $7f_0$, and so on (p. 536).

Sound Reception and Reproduction

In this section we shall discuss briefly the principles of some instruments used in sound reception and reproduction. Details of the instruments concerned must be obtained from specialist works on them.

Microphone

Microphones convert sound energy to electrical energy. Fig. 26.10 shows diagrammatically a *ribbon microphone*, which is widely used. This is a sensitive

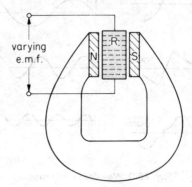

Fig. 26.10 Ribbon microphone

microphone, with a uniform response over practically the whole of the audio-frequency range from 40 to 15 000 Hz. It has a corrugated aluminium ribbon R clamped between two pole-pieces N, S of a powerful magnet. When sound waves are incident on R, the ribbon vibrates perpendicular to the magnetic field. A varying induced e.m.f. of the same frequency is therefore obtained, as shown, and this is passed to an amplifier.

The ribbon microphone is a 'velocity' type, since the movement of R depends on the velocity changes of the air particles which carry the sound wave. In contrast, the *carbon microphone*, used in the hand set of commercial telephones, is a 'pressure' type of microphone because the varying audio-frequency current produced depends on the pressure changes in the sound wave.

Loudspeaker

The *moving-coil loudspeaker* is used to reproduce sound energy from the electrical energy obtained with a microphone. It has a coil C or speech coil, wound on a cylindrical former, which is positioned symmetrically in the radial field of a pot magnet M, Fig. 26.11. A thin cardboard cone D is rigidly attached to the former and loosely connected to a large baffle-board B which surrounds it.

When C carries audio-frequency current, it vibrates at the same frequency in the direction of its axis. This is due to the force on a current-carrying conductor in a magnetic field, Fig. 26.12, whose direction is given by Fleming's left-hand rule. Since the surface area of the cone is large, the large mass of air in contact with it is disturbed and hence a loud sound is produced.

When the cone moves forward, a compression of air occurs in front of the cone and simultaneously a rarefaction behind it. The wave generated behind the cone is then 180° out of phase with that in front. If this wave reaches the front of the cone quickly, it will interfere appreciably with the wave there. Hence the intensity of the wave is diminished. This effect will be more noticeable at

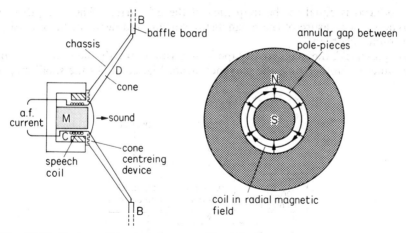

Fig. 26.11 Moving coil loudspeaker   **Fig. 26.12** Speech coil in magnetic field

low frequency, or long wavelength, as the wave behind then has time to reach the front before the next vibration occurs. Generally, then, the sound would lack low note or bass intensity. The large baffle reduces this effect appreciably. It makes the path from the rear to the front so much longer that interference is negligible.

Tape Recording and Reproduction

We now consider the principles of recording sound on tape and on film, and its reproduction. Details are beyond the scope of this book and must be obtained from manuals on the subject.

Tape recorder. Fig. 26.13 illustrates the principle of tape recording in which flexible tape is used. It is coated with a fine uniform layer of a special form of ferric oxide which can be magnetised. The backing is a smooth plastic-base tape. When recording, the tape moves at a constant speed past the narrow gap between the poles of a ring of soft iron which has a coil round it. The coil carries the audio-frequency (a.f.) current due to the sound recorded. On one half of a cycle, that part of the moving tape then in the gap is unmagnetised. On the other half of the same cycle, the next piece of tape in the gap is magnetised in the opposite direction, as shown. The rate at which pairs of such magnets

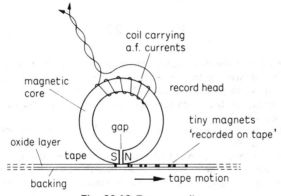

Fig. 26.13 Tape recording

is produced is equal to the frequency of the a.f. current. The strength of the magnets is a measure of the magnetising current and hence of the intensity of the sound recorded.

In 'playback', the magnetised tape is now run at exactly the same speed past the same or another ring, the playback head, Fig. 26.14. As the small magnets

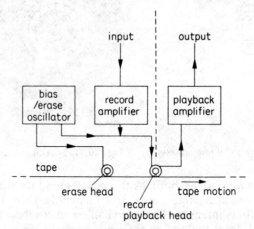

Fig. 26.14 Sound reproduction

pass the gap between the poles, the flux in the iron changes. An induced e.m.f. is thus obtained of the same frequency and strength as that due to the original tape recording. This is amplified and passed to the loudspeaker, which reproduces the sound.

To obtain high-quality sound reproduction from tape, the output from a special high frequency *bias oscillator* is applied to the recording head in addition to the recording signal. This ensures that the magnets formed on the tape have strengths which are proportional to the recording signal. If the bias oscillator was not used, severe distortion due to non-linearity would occur. The bias oscillator can also be used to *erase* the recording on the tape. This is done by applying its output to a coil in a special *erase head*. The erase head is similar to the record head but has a larger gap, so that the tape is in the magnetic field for a longer period. The bias signal takes the magnetic material on the tape through many thousands of hysteresis loops. These become progressively smaller until the magnetism disappears (p. 767).

Exercises 26

Sound Waves
1. Describe the nature of the disturbance set up in air by a vibrating tuning-fork and show how the disturbance can be represented by a sine curve. Indicate on the curve the points of (a) maximum particle velocity, (b) maximum pressure.

What characteristics of the vibration determine the pitch, intensity, and quality respectively of the note? (*N.*)

2. Define *frequency* and explain the term *harmonics*. How do harmonics determine the *quality* of a musical note?

It is much easier to hear the sound of a vibrating tuning fork if it is (a) placed in contact with a bench, or (b) held over a *certain* length of air in a tube. Explain why this is so in both these cases and give two further examples of the phenomenon occurring in (b).

Describe how you would measure the wavelength in the air in a tube of the note emitted by the fork. How would the value obtained be affected by changes in (i) the temperature of the air and (ii) the pressure of the air? (*L*.)

3. Each of the following is an expression for the speed of a wave

$$v = \sqrt{\frac{T}{m}}; \qquad v = \sqrt{\frac{\gamma p}{\rho}}; \qquad v = \sqrt{\frac{E}{\rho}}.$$

State a situation to which each expression applies, giving in each case the meanings of the symbols employed. Show that the right hand side of each expression has the dimensions of velocity.

Describe an experiment, giving the theory, by which the speed of sound in air may be determined. Explain how you would ensure that your result was as accurate as possible.

A column of air is set into vibration and the note emitted gives 10 beats per second when a tuning fork of frequency 440 Hz is sounded, the temperature being 20°C. The frequency of the beats decreases when the tuning fork is loaded with a small piece of plasticine. At what temperature will the unloaded fork and the air column be in unison? (Assume that the wavelength of the note emitted by the air column remains constant and that the frequency of the fork is independent of temperature.) (*L*.)

4. Distinguish between *longitudinal* and *transverse* wave motions, giving examples of each type. Find a relationship between the frequency, wavelength and velocity of propagation of a wave motion.

Describe experiments to investigate quantitatively for sound waves the phenomena of (*a*) reflection, (*b*) refraction, (*c*) interference. (*C*.)

5. Explain why sounds are heard very clearly at great distances from the source (*a*) on still mornings after a clear night, and (*b*) when the wind is blowing from the source to the observer.

6. Continuous sound waves of a single frequency are emitted from two small loudspeakers A and B, fed by the same signal generator and located as shown in Fig. 26A (i), which

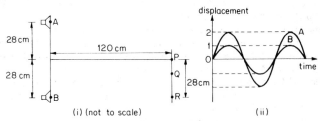

(i) (not to scale)   (ii)

Fig. 26A

is not to scale. A small sensitive microphone placed at P is connected (via a pre-amplifier) to a cathode ray oscilloscope. The graphs in Fig. 26A (ii) show the traces that appear on the screen of the c.r.o. due to A alone, then B alone, and represent the variation of the displacement at P produced by each wave separately with time.

(*a*) Calculate the relative intensities of the waves emitted by A and B.

The width of the trace on the c.r.o. is 12·0 cm and the time base 'speed' is 50 μs cm^{-1}. Calculate the frequency of the waves.

(*b*) The two waves interfere at P. Using the superposition principle, construct on the graph (Fig. 26A (ii)) the resultant displacement-time curve, using the same axes as for the original waves. What is the intensity of the resultant wave at P compared with that caused by B alone?

(*c*) Using axes on a separate sheet, construct the resultant displacement for a point Q where the waves from A and B arrive 180° out of phase with each other. (Assume that the amplitudes of the waves arriving at Q are the same as those arriving at P.) What is the intensity of the resultant wave compared with that caused by B alone in this case?

(*d*) Explain why, although the displacement of the resultant wave produced at P or Q varies with time, the sound intensity does not.

(e) A maximum of sound intensity occurs when the microphone is at R. From the dimensions given in the diagram determine the largest possible value for the wavelength of the sound waves.

(f) The experimental arrangement shown in Fig. 26A (i) could be used to measure the speed of sound, c, in air. Explain briefly how you would use the apparatus to measure c. Explain how by altering the frequency of the sound a more reliable value of c might be obtained than by using only one frequency. (L.)

7. A thin, vertical rod is partially immersed in a large deep pool of water. It moves vertically with simple harmonic motion of small amplitude. Describe the waves produced on the water surrounding the rod. State and explain how (i) the wavelength and (ii) the amplitude of the waves depend on the distance from the rod of the point at which they are measured.

Describe and briefly explain what happens when a second rod, similar to the first and vibrating with the same frequency and amplitude, and in phase with it, is placed in the water at a distance d from the first rod.

Discuss the difficulties encountered in attempting to demonstrate similar behaviour for two sources of visible light and describe an experiment you would perform to achieve this. (O. & C.)

8. Describe a method for the accurate measurement of the velocity of sound in *free* air. Indicate the factors which influence the velocity and how they are allowed for or eliminated in the experiment you describe.

At a point 20 m from a small source of sound the intensity is 0·5 microwatt cm^{-2}. Find a value for the rate of emission of sound energy from the source, and state the assumptions you make in your calculation. (N.)

9. Explain the origin of the beats heard when two tuning-forks of slightly different frequency are sounded together. Deduce the relation between the frequency of the beats and the difference in frequency of the forks. How would you determine which fork had the higher frequency?

A simple pendulum set up to swing in front of the 'seconds' pendulum ($T = 2$ s) of a clock is seen to gain so that the two swing in phase at intervals of 21 s. What is the time of swing of the simple pendulum? (L.)

10. What is meant by (a) the *amplitude*, (b) the *frequency* of a wave in air? What are the corresponding characteristics of the musical sound associated with the wave? How would you account for the difference in quality between two notes of the same pitch produced by two different instruments, e.g., by a violin and by an organ pipe?

What are 'beats'? Given a set of standard forks of frequencies 256, 264, 272, 280, and 288, and a tuning-fork whose frequency is known to be between 256 and 288, how would you determine its frequency accurately?

11. Describe a method for determining the frequency of a tuning-fork using an audio-frequency oscillator.

Two forks A and B vibrate in unison but when two slits are fixed to the prongs of A, so that they are in line when the prongs are at rest, 9 beats in 10 s are heard when the forks are sounded together. A is then made to vibrate in front of a stroboscopic disc, on which are marked 50 equally spaced radial lines. The disc is viewed through the slits and the lines appear at rest when the disc rotates at 25 rev s^{-1}. What is the frequency of B?

Describe and explain what would be seen if the speed of rotation of the disc were slightly decreased.

Doppler's Principle

12. An observer travels with constant velocity of 30 m s^{-1} towards a distant source of sound, which has a frequency of 1000 Hz. Calculate the apparent frequency of the sound heard by the observer. What frequency is heard after passing the source of sound? (Assume velocity of sound = 330 m s^{-1}.)

13. A car travelling at 20 m s^{-1} sounds its horn which has a frequency of 600 Hz. What frequency is heard by a stationary distant observer as the car approaches? What frequency is heard after the car has passed? (Velocity of sound = 340 m s^{-1}.)

14. The wavelength of a particular line in the emission spectrum of a distant star is measured as 600·80 nm. The true wavelength is 600·00 nm. (a) Is the star moving away from or towards the observer? (b) Calculate the speed of the star. (Velocity of light = $3·0 \times 10^8$ m s^{-1}.)

15. An observer, travelling with a constant velocity of 20 m s^{-1}, passes close to a stationary source of sound and notices that there is a change of frequency of 50 Hz as he passes the source. What is the frequency of the source?

(Speed of sound in air = 340 m s^{-1}.) (*L.*)

16. Deduce expressions for the frequency heard by an observer (a) when he is stationary and a source of sound is moving towards him and (b) when he is moving towards a stationary source of sound. Explain your reasoning carefully in each case.

Give an example of change of frequency due to motion of source or observer from some other branch of physics, explaining either, a use which is made of it, or a deduction from it.

A car travelling at 10 m s^{-1} sounds its horn, which has a frequency of 500 Hz, and this is heard in another car which is travelling behind the first car, in the same direction, with a velocity of 20 m s^{-1}. The sound can also be heard in the second car by reflection from a bridge head. What frequencies will the driver of the second car hear? (Speed of sound in air = 340 m s^{-1}.) (*L.*)

17. An object, vibrating vertically with a frequency of 10 Hz, is moving in a horizontal straight line with a velocity of 2·0 cm s^{-1}. It is producing waves, which travel with a speed of 12 cm s^{-1}, on a water surface. Draw a diagram showing the instantaneous positions of the waves emitted during the previous half second.

Calculate the frequency of the waves in the direction of motion of the object.

A boy sitting on a swing which is moving to an angle of 30° from the vertical is blowing a whistle which has a frequency of 1·0 kHz. The whistle is 2·0 m from the point of support of the swing. A girl stands in front of the swing. Calculate the maximum and minimum frequencies she will hear. (Speed of sound = 330 m s^{-1}, g = 9·8 m s^{-2}.) (*L.*)

18. (a) State the conditions necessary for 'beats' to be heard and derive an expression for their frequency.

(b) A fixed source generates sound waves which travel with a speed of 330 m s^{-1}. They are found by a distant stationary observer to have a frequency of 500 Hz. What is the wavelength of the waves? From first principles find (i) the wavelength of the waves in the direction of the observer, and (ii) the frequency of the sound heard if (1) the source is moving towards the stationary observer with a speed of 30 m s^{-1}, (2) the observer is moving towards the stationary source with a speed of 30 m s^{-1}, (3) both source and observer move with a speed of 30 m s^{-1} and approach one another. (*N.*)

19. Explain how *beats* are produced by two notes sounding together and obtain an expression for the number of beats heard per second.

A whistle of frequency 1000 Hz is sounded on a car travelling towards a cliff with a velocity of 18 m s^{-1}, normal to the cliff. Find the apparent frequency of the echo as heard by the car driver. Derive any relations used. (Assume velocity of sound in air to be 330 m s^{-1}.) (*L.*)

20. Explain in each case the change in the apparent frequency of a note brought about by the motion of (i) the source, (ii) the observer, relative to the transmitting medium.

Derive expressions for the ratio of the apparent to the real frequency in the cases where (a) the source, (b) the observer, is at rest, while the other is moving along the line joining them.

The locomotive of a train approaching a tunnel in a cliff face at 95 km h^{-1} is sounding a whistle of frequency 1000 Hz. What will be the apparent frequency of the echo from the cliff face heard by the driver? What would be the apparent frequency of the echo if the train were emerging from the tunnel at the same speed? (Take the velocity of sound in air as 330 m s^{-1}.) (*L.*)

27 Waves in Pipes, Strings, Rods

Introduction

The music from an organ, a violin, or a xylophone is due to vibrations in the air set up by oscillations in these instruments. In the organ, air is blown into a pipe, which sounds its characteristic note as the air inside it vibrates; in the violin, the strings are bowed so that they oscillate; and in a xylophone a row of metallic rods are struck in the middle with a hammer, which sets them into vibration.

Before considering each of the above cases in more detail, it would be best to consider the feature common to all of them. A violin string is fixed at both ends, A, B, and waves travel along m, n to each end of the string when it is bowed and are there reflected, Fig. 27.1 (i).

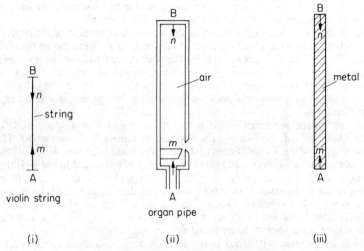

Fig. 27.1 Reflection of waves in instruments

The vibrations of the particles of the string are hence due to *two waves of the same frequency and amplitude travelling in opposite directions*. A similar effect is obtained with an organ pipe closed at one end B, Fig. 27.1 (ii). If air is blown into the pipe at A, a wave travels along the direction m and is reflected at B in the opposite direction n. The vibrations of the air in the pipe are thus due to two waves travelling in opposite directions. If a metal rod is fixed at its middle in a vice and stroked at one end A, a wave travels along the rod in the direction m and is reflected at the other end B in the direction n, Fig. 27.1 (iii). The vibrations of the rod, which produce a high-pitched note, are thus due to two waves travelling in opposite directions.

Stationary Waves and Wavelength

In Chapter 21 on Waves, we showed that a *stationary wave* is formed when two waves of equal amplitude and frequency travel in opposite directions in

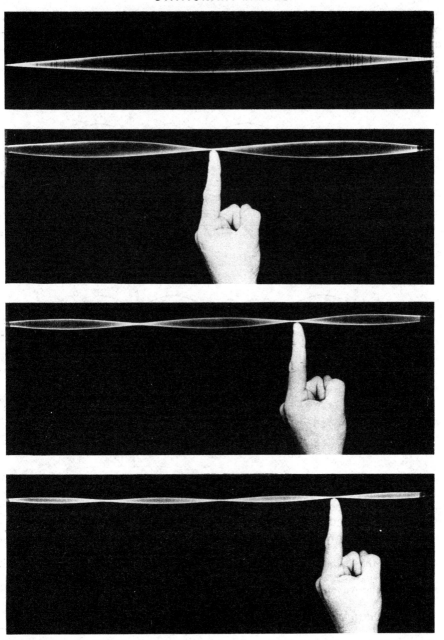

Fig. 27A (a)–(d) The photographs illustrate how 'stopping' with a light touch of a finger at different points of a vibrating cord produces successive harmonics. At the top, the mode of vibration corresponds to the fundamental frequency f_0 and the others to $2f_0$, $3f_0$, $4f_0$. As shown, a displacement node is produced at the stop in each case. (*Courtesy of Prof. C. A. Taylor, Cardiff University*)

a medium. Fig. 27.2 shows a plane-progressive sound wave *a* in air incident on a smooth wall W, together with the reflected wave *b*. When the displacements due to the two waves are added together at the times shown, where *T* is the period of *a* or *b*, the resultant wave S has points N of permanent zero displacement called *nodes*. Other points A, half way between the nodes N, have a maximum amplitude of vibration; they are called *antinodes*.

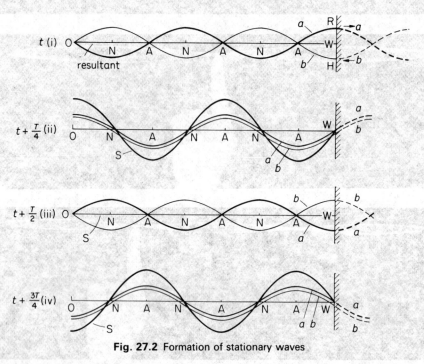

Fig. 27.2 Formation of stationary waves

Fig. 27.2 shows displacement nodes and antinodes. As we saw on p. 413, the pressure nodes occur at displacement antinodes; the pressure antinodes occur at displacement nodes.

The importance of the nodes and antinodes in a stationary wave lies in their simple connection with the wavelength. We have shown in Chapter 21 that

$$\text{the distance between consecutive nodes, } NN = \frac{\lambda}{2} \qquad . \qquad . \qquad \text{(i)}$$

where λ is the wavelength of the progressive wave;

$$\text{the distance between consecutive antinodes, } AA = \frac{\lambda}{2}, \qquad . \qquad . \qquad \text{(ii)}$$

and

$$\text{the distance from a node to the next antinode, } NA = \frac{\lambda}{4} \qquad . \qquad . \qquad \text{(iii)}$$

Figure 27A on page 531 shows various stationary waves on a vibrating cord and the nodes and antinodes in the waves.

Example

Distinguish between progressive and stationary wave motion. Describe and illustrate with an example how stationary wave motion is produced. Plane sound waves of frequency 100 Hz fall normally on a smooth wall. At what distances from the wall will the air particles have (*a*) maximum, (*b*) minimum amplitude of vibration? Give reasons for your answer. (The velocity of sound in air may be taken as 340 m s^{-1}.) (*L*.)

A stationary wave is set up between the source and the wall, due to the production of a reflected wave. The wall is a displacement node, since the air in contact with it cannot move; and other nodes are at equal distances, d, from the wall. Now if the wave-length is λ,

$$d = \frac{\lambda}{2}$$

Since

$$\lambda = \frac{v}{f} = \frac{340}{100} = 3 \cdot 4 \text{ m}$$

$$\therefore d = \frac{3 \cdot 4}{2} = 1 \cdot 7 \text{ m}$$

Thus minimum amplitude of vibration is obtained 1·7, 3·4, 5·1 m ... from the wall.

The antinodes are midway between the nodes. Thus maximum amplitude of vibration is obtained 0·85, 2·55, 4·25 m, ... from the wall.

Waves in Pipes

Closed Pipe

A *closed* or *stopped organ pipe* consists essentially of a metal pipe closed at one end Q, and a blast of air is blown into it at the other end P, Fig. 27.3 (i). A wave thus travels up the pipe to Q, and is reflected at this end down

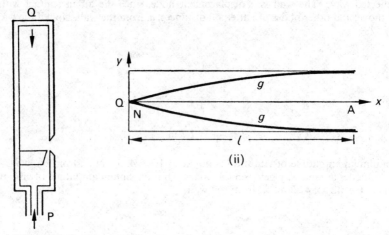

Fig. 27.3 (i) Closed (stopped) pipes (ii) Fundamental of closed (stopped) pipe

the pipe, so that a *stationary wave* is obtained. The end Q of the closed pipe must be a node N, since the layer in contact with Q must be permanently at rest, and the open end A, where the air is free to vibrate, must be an antinode A. The simplest stationary wave in the air in the pipe is hence represented by *g* in Fig. 27.3 (ii), where the pipe is positioned horizontally to show the relative displacement, *y*, of the layers at different distances, *x*, from the closed end Q; the axis of the stationary wave is Q*x*.

It can now be seen that the length *l* of the pipe is equal to the distance between a node N and a consecutive antinode A of the stationary wave. But NA = $\lambda/4$, where λ is the wavelength (p. 532).

$$\therefore \frac{\lambda}{4} = l \text{ or } \lambda = 4l$$

But the frequency, *f*, of the note is given by $f = v/\lambda$, where *v* is the velocity of sound in air.

$$\therefore f = \frac{v}{4l}$$

This is the frequency of the lowest note obtainable from the pipe, and it is known as its *fundamental*. We shall denote the fundamental frequency by f_0, so that

$$f_0 = \frac{v}{4l} \cdot \qquad . \qquad . \qquad . \qquad . \qquad . \qquad (1)$$

Fig. 27.3 (ii) shows the stationary wave of *displacement* of air molecules along the closed pipe. The *pressure* variation is also a stationary wave. But in contrast to Fig. 27.3 (ii), the pressure node is at the open end since the air pressure here is constant and equal to the external atmospheric pressure; and the pressure antinode is at the closed end since here the layers of air are compressed.

Overtones of Closed Pipe

If a stronger blast of air is blown into the pipes, notes of higher frequency can be obtained which are simple multiples of the fundamental frequency f_0. Two possible cases of stationary waves are shown in Fig. 27.4. In each, the closed

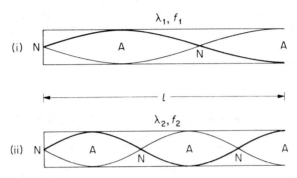

Fig. 27.4 Overtones in closed pipe

end of the pipe is a node, and the open end is an antinode. In Fig. 27.4 (i), however, the length l of the pipe is related to the wavelength λ_1 of the wave by

$$l = \frac{3}{4}\lambda_1$$

$$\therefore \lambda_1 = \frac{4l}{3}$$

The frequency f_1 of the note is thus given by

$$f_1 = \frac{v}{\lambda_1} = \frac{3v}{4l} \qquad \qquad . \qquad . \qquad . \qquad . \qquad \text{(i)}$$

But $$f_0 = \frac{v}{4l}$$

$$\therefore f_1 = 3f_0 \qquad \qquad . \qquad . \qquad . \qquad . \qquad . \qquad \text{(ii)}$$

As we previously explained, the stationary *pressure* wave in the air has pressure nodes at the displacement antinodes A in Fig. 27.4 (i) and pressure antinodes at the displacement nodes N.

In Fig. 27.4 (ii), when a note of frequency f_2 is obtained, the length l of the pipe is related to the wavelength λ_2 by

$$l = \frac{5\lambda_2}{4} \text{ or } \lambda_2 = \frac{4l}{5}$$

$$\therefore f_2 = \frac{v}{\lambda_2} = \frac{5v}{4l} \qquad \qquad . \qquad . \qquad . \qquad . \qquad \text{(iii)}$$

$$\therefore f_2 = 5f_0 \qquad \qquad . \qquad . \qquad . \qquad . \qquad . \qquad \text{(iv)}$$

By drawing other sketches of stationary waves, with the closed end as a node and the open end as an antinode, it can be shown that higher frequencies can be obtained which have frequencies of $7f_0$, $9f_0$, and so on. They are produced by blowing harder at the open end of the pipe. The frequencies obtainable at a closed pipe are hence f_0, $3f_0$, $5f_0$, and so on, i.e., the closed pipe gives only odd harmonics, and hence the frequencies $3f_0$, $5f_0$, etc. are possible *overtones*.

Open Pipe

An 'open' pipe is one which is open at both ends. When air is blown into it at P, a wave m travels to the open end Q, where it is reflected in the direction n on encountering the free air, Fig. 27.5 (i). A stationary wave is hence set up

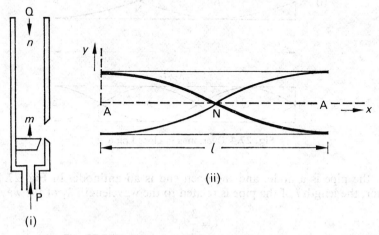

Fig. 27.5 (i) Open pipe (ii) Fundamental of open pipe

in the air in the pipe, and as the two ends of the pipe are open, they must both be *antinodes*. The simplest type of wave is hence that shown in Fig. 27.5 (ii), the x-axis of the wave being drawn along the middle of the pipe, which is horizontal. A node N is midway between the two antinodes.

The length l of the pipe is the distance between consecutive antinodes. But the distance between consecutive antinodes $= \lambda/2$, where λ is the wavelength (p. 532).

$$\therefore \frac{\lambda}{2} = l \text{ or } \lambda = 2l$$

Thus the frequency f_0 of the note obtained from the pipe is given by

$$f_0 = \frac{v}{\lambda} = \frac{v}{2l} \qquad \qquad . \qquad . \qquad . \qquad . \qquad (2)$$

This is the frequency of the fundamental note of the pipe.

Overtones of Open Pipe

Notes of higher frequencies than f_0 can be obtained from the pipe by blowing harder. The stationary wave in the pipe has always an antinode A at each end, and Fig. 27.6 (i) represents the case of a note of a frequency f_1.

The length l of the pipe is equal to the wavelength λ_1 of the wave in this case.

Thus
$$f_1 = \frac{v}{\lambda_1} = \frac{v}{l}$$

But
$$f_0 = \frac{v}{2l}, \text{ from (2) above.}$$

$$\therefore f_1 = 2f_0 \qquad . \qquad . \qquad . \qquad . \qquad \text{(i)}$$

In Fig. 27.6 (ii), the length $l = 3\lambda_2/2$, where λ_2 is the wavelength in the pipe, so $\lambda_2 = 2l/3$. The frequency f_2 is thus given by

$$f_2 = \frac{v}{\lambda_2} = \frac{3v}{2l}$$

$$\therefore f_2 = 3f_0 \qquad . \qquad . \qquad . \qquad . \qquad \text{(ii)}$$

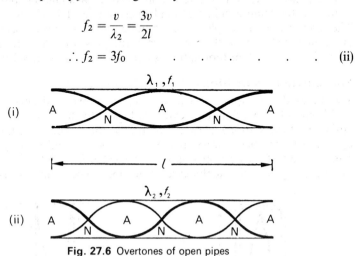

Fig. 27.6 Overtones of open pipes

The frequencies of the overtones in the open pipe are thus $2f_0$, $3f_0$, $4f_0$, and so on, i.e., all harmonics are obtainable. The frequencies of the overtones in the closed pipe are $3f_0$, $5f_0$, $7f_0$, and so on, and hence the *quality* of the same note obtained from a closed and an open pipe is different (see p. 523).

Detection of Nodes and Antinodes, and Pressure Variation, in Pipes

The *nodes and antinodes* in a sounding pipe have been detected by suspending inside it a very thin piece of paper with lycopodium or fine sand particles on it, Fig. 27.7 (i). The particles are considerably agitated at the antinodes, but they are motionless at the nodes.

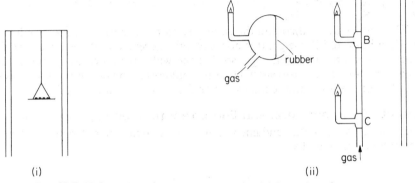

Fig. 27.7 (i) Detection of nodes and antinodes (ii) Detection of pressure

The *pressure variation* in a sounding pipe has been examined by means of a sensitive flame, designed by Lord Rayleigh. The length of the flame can be made sensitive to the pressure of the gas supplied, so that if the pressure changes the length of flame is considerably affected. Several of the flames can be arranged at different parts of the pipe, with a thin rubber or mica diaphragm in the pipe, such as at B, C, Fig. 27.7 (ii). At a place of maximum pressure variation, which is a node (p. 535), the length of flame alters accordingly. At a place of constant (normal) pressure, which is an antinode, the length of flame remains constant.

The pressure variation at different parts of a sounding pipe can also be examined by using a suitable small microphone at B, C, instead of a flame. The microphone is coupled to a cathode-ray tube and a wave of maximum amplitude is shown on the screen when the pressure variation is a maximum. At a place of constant (normal) pressure, no wave is observed on the screen.

End-correction of Pipes

The air at the open end of a pipe is free to move, and hence the vibrations at this end of a sounding pipe extend a little into the air outside the pipe. The antinode of the stationary wave due to any note is thus a distance c from the open end in practice, known as the *end-correction*, and hence the wavelength λ in the case of a closed pipe is given by $\lambda/4 = l + c$, where l is the length of the pipe, Fig. 27.8 (i). In the case of an open pipe sounding its fundamental

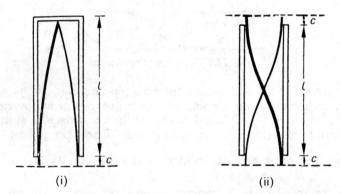

Fig. 27.8 (i) Closed pipe (ii) Open pipe

note, the wavelength λ is given by $\lambda/2 = l + c + c$, since *two* end-corrections are required, assuming the end-corrections are equal, Fig. 27.8 (ii). Thus $\lambda = 2(l + 2c)$. See also p. 536.

The mathematical theory of the end-correction was developed independently by Helmholtz and Rayleigh. It is now generally accepted that $c = 0.58r$, or $0.6r$, where r is the radius of the pipe, so that the wider the pipe, the greater is the end-correction. It was also shown that the end-correction depends on the wavelength λ of the note, and tends to vanish for very short wavelengths.

Effect of Temperature, and End-correction, on Pitch of Pipes

The frequency, f_0, of the fundamental note of a closed pipe of length l and end-correction c is given by

$$f_0 = \frac{v}{\lambda} = \frac{v}{4(l + c)} \qquad \qquad \text{(i)}$$

with the usual notation, since $\lambda = 4(l + c)$. See above. Now the velocity of sound, v, in air at $\theta°C$ is related to its velocity v_0 at $0°C$ by

$$\frac{v}{v_0} = \sqrt{\frac{273 + \theta}{273}} = \sqrt{1 + \frac{\theta}{273}} \qquad . \qquad . \qquad . \qquad \text{(ii)}$$

since the velocity is proportional to the square root of the kelvin temperature. Substituting for v in (i),

$$\therefore f_0 = \frac{v_0}{4(l + c)} \sqrt{1 + \frac{\theta}{273}} \qquad . \qquad . \qquad . \qquad \text{(iii)}$$

From (iii), it follows that, with a given pipe, *the frequency of the fundamental increases as the temperature increases.* Also, for a given temperature and length of pipe, the frequency decreases as c increases. Now $c = 0.6r$ where r is the radius of the pipe. Thus *the frequency of the note from a pipe of given length is lower the wider the pipe*, the temperature being constant. The same results hold for an open pipe.

Resonance

If a diving springboard is bent and then allowed to vibrate freely, it oscillates with a frequency which is called its *natural frequency*. When a diver on the edge of the board begins to jump up and down repeatedly, the board is forced to vibrate at the frequency of the jumps; and at first, when the amplitude is small, the board is said to be undergoing *forced vibrations*. As the diver jumps up and down to gain increasing height for his dive, the frequency of the periodic downward force reaches a stage where it is practically the same as the natural frequency of the board. The amplitude of the board then becomes very large, and the periodic force is said to have set the board in *resonance* (see also p. 411).

A mechanical system which is free to move, like a wooden bridge or the air in pipes, has a natural frequency of vibration, f_0, which depends on its dimensions. When a periodic force of a frequency different from f_0 is applied to the system, the latter vibrates with a small amplitude and undergoes forced vibrations. When the periodic force has a frequency equal to the natural frequency f_0 of the system, the amplitude of vibration becomes a maximum, and the system is then set into resonance. Fig. 27.9 is a typical curve showing the variation of amplitude

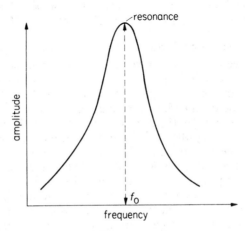

Fig. 27.9 Resonance curve

with frequency. Some time ago it was reported in the newspapers that a soprano who was broadcasting had broken a glass tumbler on the table of a listener when she had reached a high note. This is an example of resonance. The glass had a natural frequency equal to that of the note sung, and was thus set into a large amplitude of vibration sufficient to break it.

The phenomenon of resonance occurs in branches of Physics other than Sound and Mechanics. When an electrical circuit containing a coil and capacitor is 'tuned' to receive the radio waves from a distant transmitter, the frequency of the radio waves is equal to the natural frequency of the circuit and resonance is therefore obtained. A large current then flows in the electrical circuit (p. 785).

Sharpness of Resonance

As the resonance condition is approached, the effect of the frictional or *damping* forces on the amplitude increases. Damping prevents the amplitude from becoming infinitely large at resonance. The lighter the damping, the sharper is the resonance, that is, the amplitude diminishes considerably at a frequency slightly different from the resonant frequency, Fig. 27.10. A heavily-damped system has a fairly flat resonance curve. Tuning is therefore more difficult in a system which has light damping.

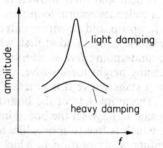

Fig. 27.10 Sharpness of resonance

The effect of damping can be illustrated by attaching a simple pendulum carrying a pith bob, and one of the same length carrying a lead bob of equal size, to a horizontal string. The pendula are set into vibration by a third pendulum of equal length attached to the same string, and it is then seen that the amplitude of the lead bob is much greater than that of the pith bob. The damping of the pith bob due to air resistance is much greater than for the lead bob.

Resonance in a Tube or Pipe

If a person blows gently down a pipe closed at one end, the air inside vibrates freely, and a note is obtained from the pipe which is its fundamental (p. 534). A stationary wave then exists in the pipe, with a node N at the closed end and an antinode A at the open end, as explained on p. 534.

If the prongs of a tuning-fork are held over the top of the pipe, the air inside it is set into vibration by the periodic force exerted on it by the prongs. In general, however, the vibrations are feeble, as they are *forced* vibrations, and the intensity of the sound heard is correspondingly small. But when a tuning-fork of the same frequency as the fundamental frequency of the pipe is held over the latter, the air inside is set in *resonance* by periodic force, and the amplitude of the vibrations is large. A loud note, which has the same frequency as the fork, is then heard coming from the pipe, and a stationary wave is set up with the top of the pipe acting as an antinode and the fixed end as a node,

Fig. 27.11. If a sounding tuning-fork is held over a pipe open at both ends, resonance occurs when the stationary wave in the pipe has antinodes at the two open ends, as shown by Fig. 27.5; the frequency of the fork is then equal to the frequency of the fundamental of the open pipe. A similar case to the closed pipe, but using electrical oscillations, was discussed on p. 421.

Fig. 27.11 Resonance in closed pipe

Resonance Tube Experiment. Measurement of Velocity of Sound and 'End-Correction' of Tube

If a sounding tuning-fork is held over the open end of a tube T filled with water, resonance is obtained at some position as the level of water is gradually lowered, Fig. 27.12 (i). The stationary wave set up is then as shown. If c is the end-correction of the tube (p. 538), and l is the length from the water level to the top of the tube, then

$$l + c = \frac{\lambda}{4} \qquad . \qquad . \qquad . \qquad . \qquad . \qquad . \qquad \text{(i)}$$

But
$$\lambda = \frac{v}{f},$$

where f is the frequency of the fork and v is the velocity of sound in air.

$$\therefore l + c = \frac{v}{4f} \qquad . \qquad . \qquad . \qquad . \qquad . \qquad \text{(ii)}$$

If different tuning-forks of known frequency f are taken, and the corresponding values of l obtained when resonance occurs, it follows from equation (ii) that a graph of $1/f$ against l is a straight line, Fig. 27.12 (ii). Now from equation (ii),

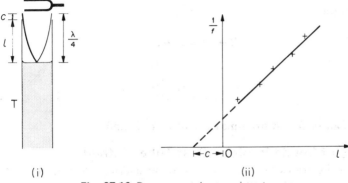

Fig. 27.12 Resonance tube experiment

the gradient of the line is $4/v$; thus v can be determined. Also, the negative intercept of the line on the axis of l is c, from equation (ii); hence the end-correction can be found.

If only one fork is available, and the tube is sufficiently long, another method for v and c can be adopted. In this case the level of the water is lowered further from the position in Fig. 27.12 (i), until resonance is again obtained at a level L_1, Fig. 27.13. Since the stationary wave set up is that shown and the new length to the top from L_1 is l_1, it follows that

$$l_1 + c = \frac{3\lambda}{4} \qquad \qquad \text{(iii)}$$

But
$$l + c = \frac{\lambda}{4}, \text{ from (ii).}$$

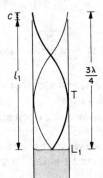

Fig. 27.13 Resonance at new water level

Subtracting,
$$l_1 - l = \frac{\lambda}{2}$$

$$\therefore \lambda = 2(l_1 - l)$$

$$\therefore v = f\lambda = 2f(l_1 - l) \qquad \qquad \text{(3)}$$

In this method for v, therefore, the end-correction c is eliminated. The magnitude of c can be found from equations (ii) and (iii). Thus, from (ii),

$$3l + 3c = \frac{3\lambda}{4}$$

But, from (iii),
$$l_1 + c = \frac{3\lambda}{4}$$

$$\therefore 3l + 3c = l_1 + c$$

$$\therefore 2c = l_1 - 3l$$

$$\therefore c = \frac{l_1 - 3l}{2} \qquad \qquad \text{(4)}$$

Hence c can be found from measurements of l_1 and l.

Velocity of Sound in Air by Dust Tube Method

Fig. 27.14 illustrates another method for measuring the velocity of sound in air by means of stationary waves.

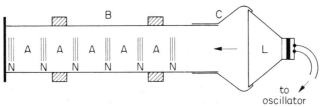

Fig. 27.14 Velocity of sound in air

A measuring cylinder B is placed on its side and is arranged to lie horizontally on supports such as plasticene. The inside of the cylinder is coated lightly with lycopodium powder or cork dust along its length. A paper cone C, attached to a loudspeaker L, is fitted over the open end of B. By connecting a suitable oscillator to L, sound waves are produced which travel to the closed end of B and are reflected, so that stationary waves are formed in the air.

The frequency of the oscillator is varied. At a frequency of the order of a kilohertz or more depending on the length of the measuring cylinder, the dust suddenly settles into regularly spaced heaps at positions along the cylinder. These are nodes, N, of the stationary wave (zero displacement positions). Midway between the nodes are antinodes, A, where the air has maximum amplitude of vibration and so little dust settles there.

Measurement of the average distance NN between successive nodes $= \lambda/2$, where λ is the wavelength. So λ can be found. The velocity in air $v = f\lambda$, where f is known, and hence v can be calculated. This is an approximate method for v as (a) the sound waves are damped by the sides of the tube and so this is not the velocity in free air (see p. 415), (b) the distance NN cannot be measured to a high degree of accuracy. The method outlined here is a dust tube method due originally to Kundt (p. 552). The dust and the tube must be dry.

Examples

1. Describe the natural modes of vibration of the air in an organ pipe closed at one end, and explain what is meant by the term 'end-correction'. A cylindrical pipe of length 28 cm closed at one end is found to be at resonance when a tuning fork of frequency 864 Hz is sounded near the open end. Determine the mode of vibration of the air in the pipe, and deduce the value of the end-correction. [Take the velocity of sound in air as 340 m s^{-1}.] (L.)

Let $\lambda =$ the wavelength of the sound in the pipe.

Then
$$\lambda = \frac{v}{f} = \frac{34\,000}{864} = 39 \cdot 35 \text{ cm}$$

If the pipe is resonating to its fundamental frequency f_0, the stationary wave in the pipe is that shown in Fig. 27.12 (i) and the wavelength λ_0, is given by $\lambda_0/4 = 28$ cm. Thus $\lambda_0 = 112$ cm. Since $\lambda = 39 \cdot 35$ cm, the pipe cannot be sounding its resonant frequency. The first overtone of the pipe is $3f_0$, which corresponds to a wavelength λ_1 given by $3\lambda/4 = 28$ (see Fig. 27.4).

$$\therefore \lambda_1 = \frac{112}{3} = 37\tfrac{1}{3} \text{ cm}$$

Consequently, allowing for the effect of an end correction, the pipe is sounding its first overtone.

Let $c =$ the end-correction in cm.

Then
$$28 + c = \frac{3\lambda_1}{4}$$

But, accurately, $$\lambda_1 = \frac{v}{f} = \frac{34\,000}{864} = 39\cdot35$$

$$\therefore 28 + c = \tfrac{3}{4} \times 39\cdot35$$

$$\therefore c = 1\cdot5 \text{ cm.}$$

2. Explain, with diagrams, the possible states of vibration of a column of air in (*a*) an open pipe, (*b*) a closed pipe. An open pipe 30 cm long and a closed pipe 23 cm long, both of the same diameter, are each sounding its first overtone, and these are in unison. What is the end-correction of these pipes? (*L*.)

Suppose v is the velocity of sound in air, and f is the frequency of the note. The wave-length, λ, is thus v/f.

When the open pipe is sounding its first overtone, the length of the pipe plus end-corrections $= \lambda$.

$$\therefore \frac{v}{f} = 30 + 2c \quad . \qquad . \qquad . \qquad . \qquad . \qquad \text{(i)}$$

since there are two end-corrections.

When the closed pipe is sounding its first overtone,

$$\frac{3\lambda}{4} = 23 + c$$

$$\therefore \frac{3v}{4f} = 23 + c \quad . \qquad . \qquad . \qquad . \qquad . \qquad \text{(ii)}$$

From (i) and (ii), it follows that

$$23 + c = \tfrac{3}{4}\,(30 + 2c)$$

$$\therefore 92 + 4c = 90 + 6c$$

$$\therefore c = 1 \text{ cm.}$$

Waves in Strings

If a horizontal rope is fixed at one end, and the other end is moved up and down, a wave travels along the rope. The particles of the rope are then vibrating vertically, and since the wave travels horizontally, this is an example of a *transverse* wave (see p. 412). The waves propagated along the surface of the water when a stone is dropped into it are also transverse waves, as the particles of the water are moving up and down while the wave travels horizontally. A transverse wave is also obtained when a stretched string, such as a violin string, is plucked. Before we can study waves in strings, we require to know the velocity of transverse waves travelling along them.

Velocity of Transverse Waves Along a Stretched String

Suppose that a transverse wave is travelling along a thin string of length l and mass s under a constant tension T. If we assume that the string has no 'stiffness', i.e., that the string is perfectly flexible, the velocity v of the transverse wave along it depends only on the values of T, s, l. The velocity is given by

$$v = \sqrt{\frac{T}{s/l}},$$

$$\text{or } v = \sqrt{\frac{T}{m}} \qquad . \qquad . \qquad . \qquad . \qquad . \qquad (5)$$

where m is the 'mass per unit length' of the string.

When T is in *newton* and m in *kilogram per metre*, then v is in *metre per second*.

The formula for v may be partly deduced by the method of dimensions, in which all the quantities concerned are reduced to the fundamental units of mass, M, length, L, and time, T. Suppose that

$$v = kT^x s^y l^z \qquad . \qquad . \qquad . \qquad . \qquad . \qquad (i)$$

where k, x, y, z, are numbers. The dimensions of velocity v are LT^{-1}, the dimensions of tension T, a force, are MLT^{-2}, the dimension of s is M, and the dimension of l is L. As the dimensions on both sides of (i) must be equal, it follows that

$$LT^{-1} = (MLT^{-2})^x (M^y)(L^z)$$

Equating the indices of M, L, T on both sides, we have

for M, $x + y = 0$

for L, $x + z = 1$

for T, $2x = 1$

$$\therefore x = \tfrac{1}{2}, \ z = \tfrac{1}{2}, \ y = -\tfrac{1}{2}$$

Thus, from (i)

$$v = kT^{\frac{1}{2}} s^{-\frac{1}{2}} l^{\frac{1}{2}}$$

$$\therefore v = k\sqrt{\frac{Tl}{s}} = k\sqrt{\frac{T}{s/l}}$$

A rigid mathematical treatment shows that the constant $k = 1$, so $v = \sqrt{\dfrac{T}{s/l}}$.

Since s/l is the 'mass per unit length' of the string, it follows that

$$v = \sqrt{\frac{T}{m}},$$

where m is the mass per unit length.

Modes of Vibration of Stretched String

If a wire is stretched between two points N, N and is plucked in the middle, a transverse wave travels along the wire and is reflected at the fixed end. A *stationary wave* is thus set up in the wire, and the simplest mode of vibration is one in which the fixed ends of the wire are nodes, N, and the middle is an antinode, A, Fig. 27.15. Since the distance between consecutive nodes is $\lambda/2$,

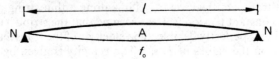

Fig. 27.15 Fundamental of stretched string

where λ is the wavelength of the transverse wave in the wire, it follows that

$$l = \frac{\lambda}{2},$$

where l is the length of the wire. Thus $\lambda = 2l$. The frequency f of the vibration is hence given by

$$f = \frac{v}{\lambda} = \frac{v}{2l},$$

where v is the velocity of the transverse wave. But $v = \sqrt{T/m}$, from previous.

$$\therefore f = \frac{1}{2l}\sqrt{\frac{T}{m}}$$

This is the frequency of the *fundamental* note obtained from the string; and if we denote the frequency by the usual symbol f_0, we have

$$f_0 = \frac{1}{2l}\sqrt{\frac{T}{m}} \qquad \qquad . \qquad . \qquad . \qquad . \qquad (6)$$

Overtones of Stretched String

The first overtone f_1 of a string plucked in the middle corresponds to a stationary wave shown in Fig. 27.16, which has nodes at the fixed ends and an antinode in the middle. If λ_1 is the wavelength, it can be seen that

$$l = \frac{3}{2}\lambda_1,$$

or $\lambda_1 = \dfrac{2l}{3}.$

The frequency f_1 is thus given by

$$f_1 = \frac{v}{\lambda_1} = \frac{3v}{2l} = \frac{3}{2l}\sqrt{\frac{T}{m}} \qquad \qquad . \qquad . \qquad (i)$$

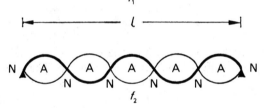

Fig. 27.16 Overtones of stretched string plucked in middle

But the fundamental frequency, f_0, $= \dfrac{1}{2l}\sqrt{\dfrac{T}{m}}$, from equation (6).

$$\therefore f_1 = 3f_0$$

The second overtone f_2 of the string when plucked in the middle corresponds to a stationary wave shown in Fig. 27.16. In this case $l = \dfrac{5}{2}\lambda_2$, where λ_2 is the wavelength.

$$\therefore \lambda_2 = \frac{2l}{5}$$

$$\therefore f_2 = \frac{v}{\lambda_2} = \frac{5v}{2l}$$

where f_2 is the frequency. But $v = \sqrt{T/m}$.

$$\therefore f_2 = \frac{5}{2l}\sqrt{\frac{T}{m}} = 5f_0$$

The overtones are thus $3f_0$, $5f_0$, and so on.

Other notes than those considered above can be obtained by touching or 'stopping' the string lightly at its midpoint, for example, so that the latter becomes a node in addition to those at the fixed ends. If the string is plucked one-quarter of the way along it from a fixed end, the simplest stationary wave set up is

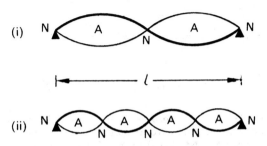

Fig. 27.17 Even harmonics in stretched string

that illustrated in Fig. 27.17 (i) (see also page 531). Thus the wavelength $\lambda = l$, and hence the frequency f is given by

$$f = \frac{v}{\lambda} = \frac{v}{l} = \frac{1}{l}\sqrt{\frac{T}{m}}$$

$$\therefore f = 2f_0, \text{ since } f_0 = \frac{1}{2l}\sqrt{\frac{T}{m}}.$$

If the string is plucked one-eighth of the way from a fixed end, a stationary wave similar to that in Fig. 27.17 (ii) may be set up. The wavelength, $\lambda' = l/2$, and hence the frequency $f' = \frac{v}{\lambda'} = \frac{2v}{l}$.

$$\therefore f' = \frac{2}{l}\sqrt{\frac{T}{m}} = 4f_0$$

Verification of the Laws of Vibration of a Fixed String. The Sonometer

As we have already shown (p. 546), the frequency of the fundamental of a stretched string is given by $f = \frac{1}{2l}\sqrt{\frac{T}{m}}$, writing f for f_0. It thus, follows that:

(1) $f \propto \dfrac{1}{l}$ for a given tension (T) and string $(m$ constant$)$.

(2) $f \propto \sqrt{T}$ for a given length (l) and string $(m$ constant$)$.

(3) $f \propto \dfrac{1}{\sqrt{m}}$ for a given length (l) and tension (T).

These are known as the 'laws of vibration of a fixed string', first completely given by MERSENNE in 1636. The *sonometer*, or *monochord*, was designed to verify them.

The sonometer consists of a hollow wooden box Q, with a thin horizontal wire attached to A at one end, Fig. 27.18. The wire passes over a grooved wheel

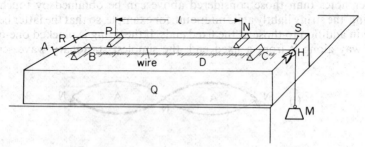

Fig. 27.18 Sonometer verification of $f \propto 1/l$ and $\sqrt{T}$

H, and is kept taut by a mass M hanging down at the other end. Wooden bridges, B, C, can be placed beneath the wire so that a definite length of wire is obtained, and the length of wire can be varied by moving one of the bridges. The length of wire between B, C can be read from a fixed horizontal scale D, graduated in millimetres, on the box below the wire.

(1) *To verify $f \propto 1/l$ for a given tension (T) and mass per unit length (m)*, the mass M is kept constant so that the tension, T, in the wire AH is constant.

The length, l, of the wire between B, C is varied by moving C until the note obtained by plucking BC in the middle is the same as that produced by a sounding tuning-fork of known frequency f. If the observer lacks a musical ear, the 'tuning' can be recognised by listening for beats when the wire and the tuning-fork are both sounding, as in this case the frequencies of the two notes are nearly equal (p. 512). Alternatively, a small piece of paper in the form of an inverted V can be placed on the middle of the wire, and the end of the sounding tuning-fork then placed on the sonometer box. The vibrations of the fork are transmitted through the box to the wire, which vibrates in resonance with the fork if its length is 'tuned' to the note. The paper will then vibrate considerably and may be thrown off the wire.

Different tuning-forks of known frequency f are used, and the lengths, l. of the wire are observed when they are tuned to the corresponding note. A graph of f against $1/l$ is then plotted, and is found to be a straight line within the limits of experimental error. Thus $f \propto 1/l$ for a given tension and mass per unit length of wire.

(2) *To verify* $f \propto \sqrt{T}$ *for a given length and mass per unit length*, the length BC between the bridges is kept fixed, so that the length of wire is constant, and the mass M is varied to alter the tension. To obtain a measure of the frequency f of the note produced when the wire between B, C is plucked in the middle, a second wire, fixed to R, S on the sonometer, is utilised. This usually has a weight (not shown) attached to one end to keep the tension constant, Fig. 27.18. The wire RS has bridges P, N beneath it, and N is moved until the note from the wire between P, N is the same as the note from the wire between B, C. Now the tension in PN is constant as the wire is fixed to R and S. Thus, since frequency, $f \propto 1/l$ for a given tension and wire, the frequency of the note from BC is proportional to $1/l$, where l is the length of PN.

By varying the mass M, the tension T in BC is varied. A graph of $1/l$ against $\sqrt{T}$ is found to be a straight line passing through the origin. So $f \propto \sqrt{T}$ for a given length of wire.

(3) *To verify* $f \propto 1/\sqrt{m}$ *for a given length of tension*, wires of different material are connected to B, C, and the same mass M and the same length BC are taken. The frequency, f, of the note obtained from BC is again found by using the second wire RS in the way already described. The mass per unit length, m, is the mass per metre length of wire, and is given by $\pi r^2 \rho$ kg m^{-1}, where r is the radius of the wire in m and ρ is its density in kg m^{-3}, as $(\pi r^2 \times 1)$ m^3 is the volume of 1 m of the wire.

When $1/l$ is plotted against $1/\sqrt{m}$, the graph is found to be a straight line passing through the origin. So $f \propto 1/\sqrt{m}$ for a given length and tension.

Measurement of the Frequency of A.C. Mains

The frequency of the alternating current (a.c.) mains can be determined with the aid of a sonometer wire. The alternating current is passed into the wire MP, and the poles N, S of a powerful magnet are placed on either side of the wire so that the magnetic field due to it is perpendicular to the wire, Fig. 27.19. As a result of the magnetic effect of the current, a force acts on the wire which is perpendicular to the directions of both the magnetic field and the current, and hence the wire is subjected to a transverse force. If the current is an alternating one of 50 Hz, the magnitude of the force varies at the rate of 50 Hz. By adjusting the tension in the sonometer wire by varying weights in a scale-pan A, a position can be reached when the wire is seen to be vibrating through a large amplitude; in this case the wire is *resonating* to the applied force, Fig. 27.19.

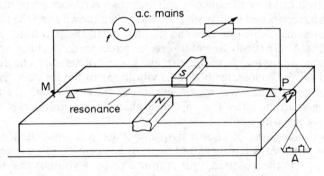

Fig. 27.19 Mains frequency by vibrating wire

The length l of wire between the bridges is now measured, and the tension T and the mass per unit length, m, are also found. The frequency f of the alternating current is then calculated from

$$f = \frac{1}{2l}\sqrt{\frac{T}{m}}.$$

Velocity of Longitudinal Waves in Wires

If a sonometer wire is stroked along its length by a rosined cloth, a high-pitched note is obtained. This note is due to *longitudinal* vibrations in the wire, and must be clearly distinguished from the note produced when the wire is plucked, which sets up *transverse* vibrations of the wire and a corresponding transverse wave. As we saw on p. 433, the velocity v of a longitudinal wave in a medium is

$$v = \sqrt{\frac{E}{\rho}},$$

where E is Young modulus for the wire and ρ is its density. The wavelength, λ, of the longitudinal wave is $2l$, where l is the length of the wire, since a stationary longitudinal wave is set up. Thus the frequency f of the note is given by

$$f = \frac{v}{\lambda} = \frac{1}{2l}\sqrt{\frac{E}{\rho}}.$$

The frequency of the note may be obtained approximately with the aid of an audio oscillator, and thus the velocity of sound in the wire, or its Young modulus, can be roughly calculated.

Examples

1. Explain the meaning of the term *resonance*, giving in illustration two methods of obtaining resonance between the stretched string of a sonometer and a tuning-fork of fixed frequency. A sonometer wire of length 76 cm is maintained under a tension of value 40 N and an alternating current is passed through the wire. A horse-shoe magnet is placed with its poles above and below the wire at its midpoint, and the resulting forces set the wire in resonant vibration. If the density of the material of the wire is 8800 kg m⁻³ and the diameter of the wire is 1 mm, what is the frequency of the alternating current? (*L.*)

The wire is set into resonant vibration when the frequency of the alternating current is equal to its natural frequency, f.

Now $$f = \frac{1}{2l}\sqrt{\frac{T}{m}} \qquad . \qquad . \qquad . \qquad . \qquad . \qquad \text{(i)}$$

where $l = 0.76$ m, $T = 40$ N, and m = mass per metre in kg m^{-1}.

Also, mass of 1 metre = volume × density

$$= \pi r^2 \times 1 \times 8800 \text{ kg,}$$

where radius r of wire $= \frac{1}{2}$ mm $= 0.5 \times 10^{-3}$ m

From (i), $\therefore f = \dfrac{1}{2 \times 0.76}\sqrt{\dfrac{40}{\pi \times 0.5^2 \times 10^{-6} \times 1 \times 8800}}$

$$= 50 \text{ Hz.}$$

2. A piano string has a length of 2·0 m and a density of 8000 kg m^{-3}. When the tension in the string produces a strain of 1%, the fundamental note obtained from the string in transverse vibration is 170 Hz. Calculate the Young modulus value for the material of the string.

If E is the Young modulus, A is the cross-section area of the string, l is the length of the string and e is the extension due to a force (tension) T, then, from page 117,

$$T = EA\frac{e}{l} = EA \times \frac{1}{100}$$

since the strain $e/l = 1\% = 1/100$. So

$$\text{frequency,} f = \frac{1}{2l}\sqrt{\frac{T}{m}} = \frac{1}{2l}\sqrt{\frac{EA}{100 \, A\rho}}$$

since m = mass per unit length $= A \times 1 \times \rho = A\rho$. So cancelling A,

$$f = \frac{1}{2l}\sqrt{\frac{E}{100\rho}}.$$

Squaring, $\qquad\qquad E = 4 f^2 l^2 \times 100\rho$

$$= 4 \times 170^2 \times 2^2 \times 100 \times 8000$$

$$= 3.7 \times 10^{11} \text{ N m}^{-2}.$$

Waves in Rods

Sound waves travel through liquids and solids, as well as through gases, and in the nineteenth century an experiment to measure the velocity of sound in iron was carried out by tapping one end of a very long iron tube. The speed of sound in iron is much greater than in air, and the sound through the iron thus arrived at the other end of the pipe before the sound transmitted through the air. From a knowledge of the interval between the sounds, the length of the pipe, and the velocity of sound in air, the velocity of sound in iron was determined. More accurate methods were soon forthcoming for the velocity of sound in substances such as iron, wood, and glass, and they depend mainly on the formation of stationary waves in rods of these materials.

Consider a rod AA fixed by a vice B at its mid-point N, Fig. 27.20. If the

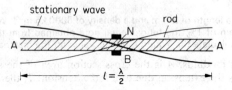

Fig. 27.20 Stationary wave in rod

rod is stroked along its length by a rosined cloth, a stationary longitudinal wave is set up in the rod on account of reflection at its ends, and a high-pitched note is obtained. Since the mid-point of the rod is fixed, this is a node, N, of the stationary wave; and since the ends of the rod are free, these are antinodes, A. Thus the length l of the rod is equal to half the wavelength, $\lambda/2$, of the wave in the rod, and hence $\lambda = 2l$. Thus the velocity of the sound in the rod, $v = f\lambda = f \times 2l$, where f is the frequency of the note from the rod.

Kundt's Dust Tube

About 1868, KUNDT devised a simple method of showing the stationary waves in air or any other gas. He used a closed tube T containing the gas, and sprinkled some dry lycopodium powder, or cork dust, along the entire length, Fig. 27.21.

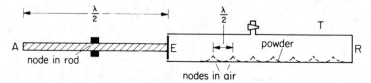

Fig. 27.21 Dust tube

A rod AE, clamped at its mid-point, is placed with one end projecting into T, and a disc E is attached at this end so that it just clears the sides of the tube, Fig. 27.21. When the rod is stroked at A by a rosined cloth in the direction EA, the rod vibrates longitudinally and a high-pitched note can be heard. The end E acts as a vibrating source of the same frequency, and a sound wave thus travels through the air in T and is reflected at the fixed end R. If the rod is moved so that the position of E alters, a position can be found when the stationary wave in the air in T causes the lycopodium powder to become violently agitated. The powder then settles into definite small heaps at the nodes, which are the positions of permanent rest of the stationary wave, and the distance between consecutive nodes can best be found by measuring the distance between several of them and dividing by the appropriate number.

Determination of Velocity of Sound in a Rod

Kundt's tube can be used to determine the velocity of sound, v_r, in the rod. Suppose the length of the rod is l; then $\lambda/2 = l$, or $\lambda = 2l$, where λ is the wavelength of the sound wave *in the rod* (p. 531). Thus the frequency of the high-pitched note obtained from the rod is given by

$$f = \frac{v_r}{\lambda} = \frac{v_r}{2l} \qquad \cdot \qquad \cdot \qquad \cdot \qquad \cdot \qquad \text{(i)}$$

If l_1 is the distance between consecutive nodes of the stationary wave in the air, we have $\lambda_1/2 = l_1$, where λ_1 is the wavelength of the sound wave *in the air*. Thus $\lambda_1 = 2l_1$, and hence the frequency of the wave, which is also f, is given by

$$f = \frac{v_a}{\lambda} = \frac{v_a}{2l_1}, \qquad \cdot \qquad \cdot \qquad \cdot \qquad \cdot \qquad \text{(ii)}$$

where v_a is the velocity of sound in air. From (i) and (ii) it follows that

$$\frac{v_r}{2l} = \frac{v_a}{2l_1}$$

$$\therefore v_r = \frac{l}{l_1} v_a \qquad \cdot \qquad \cdot \qquad \cdot \qquad \cdot \qquad \cdot \qquad \text{(7)}$$

Thus knowing v_a, l, l_1, the velocity of sound in the rod, v_r, can be calculated. By using glass, brass, copper, steel and other substances in the form of a rod, the velocity of sound in these media have been determined. Kundt also used liquids in the tube T instead of air, and employed fine iron filings instead of lycopodium powder to detect the nodes in the liquid. In this way he determined the velocity of sound in liquids.

Determination of Young Modulus of a Rod

On p. 433, it was shown that the velocity of sound, v, in a medium is always given by

$$v = \sqrt{\frac{E}{\rho}},$$

where E is the appropriate modulus of elasticity of the medium and ρ is its density. In the case of a rod undergoing longitudinal vibrations, as in Kundt's tube experiment, E is the Young modulus (see p. 490). Thus if v_r is the velocity of sound in the rod,

$$v_r = \sqrt{\frac{E}{\rho}},$$

and $\qquad\qquad\qquad\qquad \therefore E = v_r^2 \rho \qquad \cdot \qquad \cdot \qquad \cdot \qquad \cdot \qquad \cdot \qquad \text{(8)}$

Since v_r is obtained by the method explained above, and ρ can be obtained from tables, it follows that E can be calculated.

Determination of Velocity of Sound in a Gas

If the air in Kundt's tube T is replaced by some other gas, and the rod stroked, the average distance l' between the piles of dust in T is the distance between consecutive nodes of the stationary wave in the gas. The wavelength, λ_g, in the gas is thus $2l'$, and the frequency f is given by

$$f = \frac{v_g}{\lambda_g} = \frac{v_g}{2l'},$$

where v_g is the velocity of sound in the gas. But the wavelength, λ, of the wave in the rod $= 2l$, where l is the length of the rod (p. 532); hence f is also given by

$$f = \frac{v_r}{\lambda} = \frac{v_r}{2l}$$

$$\therefore \frac{v_g}{2l'} = \frac{v_r}{2l}$$

$$\therefore v_g = \frac{l'}{l} v_r \qquad . \qquad . \qquad . \qquad . \qquad . \qquad (9)$$

Knowing l', l, and v_r, the latter obtained from a previous experiment (p. 552), the velocity of sound in a gas, v_g, can be calculated. The velocity of sound in a gas can also be found by the more direct method described below.

Determination of Ratio of Molar Heat Capacities of a Gas, and its Molecular Structure

The velocity of sound in a gas, v_g, is given by

$$v_g = \sqrt{\frac{\gamma p}{\rho}},$$

where γ is the ratio (C_p/C_V) of the two molar heat capacities of the gas, p is its pressure, and ρ is its density. See p. 434. Thus

$$\gamma = \frac{v_g^2 \rho}{p} \qquad . \qquad . \qquad . \qquad . \qquad . \qquad (10)$$

Now it has already been shown that v_g can be found; and knowing ρ and p, γ can be calculated. The determination of γ is one of the most important applications of Kundt's tube, as kinetic theory shows that $\gamma = 1 \cdot 66$ for a monatomic gas and $1 \cdot 40$ for a diatomic gas. Thus Kundt's tube provides valuable information about the molecular structure of a gas. When RAMSEY isolated the hitherto unobtainable argon from the air, Lord Rayleigh in 1895 suggested a Kundt's tube experiment for finding the ratio γ for the gas. It was then discovered that γ was about $1 \cdot 65$, showing that argon was a monatomic gas. The dissociation of the molecules of a gas at high temperatures has been investigated by containing it in Kundt's tube surrounded by a furnace, and measuring the magnitude of γ when the temperature was changed.

Comparison of Velocities of Sound in Gases by Kundt's Tube

The ratio of the velocities of sound in two gases can be found from a Kundt's tube experiment. The two gases, air and carbon dioxide for example, are contained in tubes A, B respectively, into which the ends of a metal rod R project, Fig. 27.22. The middle of the rod is clamped. By stroking the rod, and adjusting the positions of the movable discs Y, X in turn, lycopodium powder in each tube can be made to settle into heaps at the various nodes. The average distances, d_a, d_b, between successive nodes in A, B respectively are then measured.

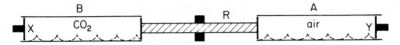

Fig. 27.22 Comparison of velocities of sound in gases

The frequency f of the sound wave in A, B is the same, being the frequency of the note obtained from R. Since $f = v/\lambda$, it follows that

$$\frac{v_g}{\lambda_g} = \frac{v_a}{\lambda_a}, \qquad \cdot \quad \cdot \quad \cdot \quad \cdot \quad \cdot \qquad \text{(i)}$$

where v_g, v_a are the velocities of sound in carbon dioxide and air respectively, and λ_g, λ_a are the corresponding wavelengths.

Now

$$\frac{\lambda_g}{\lambda_a} = \frac{d_b}{d_a} \qquad \cdot \quad \cdot \quad \cdot \quad \cdot \quad \cdot \qquad \text{(ii)}$$

since the distance between successive nodes is half a wavelength. From (i),

$$\frac{v_g}{v_a} = \frac{\lambda_g}{\lambda_a}$$

$$\therefore \frac{v_g}{v_a} = \frac{d_b}{d_a} \qquad \cdot \quad \cdot \quad \cdot \quad \cdot \quad \cdot \qquad \text{(11)}$$

The two velocities can thus be compared as d_b, d_a are known; and if the velocity of sound, v_a, in air is known, the velocity in carbon dioxide can be calculated.

Examples

1. Describe and explain the way in which a Kundt tube may be used to determine the ratio of the molar heat capacities of a gas. A Kundt tube is excited by a brass rod 150 cm long and the distance between successive nodes in the tube is 13·6 cm; what is the ratio of the velocity of sound in brass to that in air? (*L.*)

Since both ends of the rod are successive antinodes, the wavelength λ_1 in the rod $= 2 \times 150 = 300$ cm. The wavelength λ_2 in the air $= 2 \times 13\cdot6 = 27\cdot2$ cm.

The frequency f of the note in the rod and the air is the same.

$$\therefore f = \frac{v_1}{\lambda_1} = \frac{v_2}{\lambda_2}$$

where v_1, v_2 are the velocities of sound in the rod and in the air.

$$\therefore \frac{v_1}{v_2} = \frac{\lambda_1}{\lambda_2} = \frac{300}{27\cdot2} = 11\cdot0.$$

2. Describe the dust tube experiment. How may it be used to compare the velocities of sound in different gases? The fundamental frequency of longitudinal vibration of a rod clamped at its centre is 1500 Hz. If the mass of the rod is 96·0 g, find the increase in its total length produced by a tension due to a load of mass 10 kg ($g = 9\cdot8$ m s^{-2}). (*L.*)

The wavelength of the wave in the rod $= 2l$, where l metre is its length, since the ends are antinodes. The velocity, v, of the wave is given by

$$v = f\lambda = 1500 \times 2l = 3000 \, l \qquad \cdot \quad \cdot \quad \cdot \quad \cdot \qquad \text{(i)}$$

Since the vibrations of the rod are longitudinal,

$$v = \sqrt{\frac{E}{\rho}}.$$

E is Young modulus in N m^{-2} and ρ is the density of the rod in kg m^{-3}

$$\therefore v = \sqrt{\frac{E}{0 \cdot 096/V}} = \sqrt{\frac{EV}{0 \cdot 096}} \qquad . \qquad . \qquad . \qquad . \qquad \text{(ii)}$$

where V is the volume of the rod in metre^3.
From (i) and (ii),

$$\sqrt{\frac{EV}{0 \cdot 096}} = 3000 \, l$$

$$\therefore \frac{EV}{l^2} = 0 \cdot 096 \times 3000^2$$

$$\therefore \frac{EA}{l} = 0 \cdot 096 \times 3000^2 \qquad . \qquad . \qquad . \qquad . \qquad . \qquad \text{(iii)}$$

since $V = Al$, where A is the area of cross-section of the rod. Now if x is the increase in length produced by $10 \times 9 \cdot 8$ newtons, it follows from the definition of E (p. 117) that

$$\text{force} = \frac{EAx}{l} = 10 \times 9 \cdot 8 \text{ newtons}$$

$$\therefore \frac{EA}{l} = \frac{10 \times 9 \cdot 8}{x}$$

From (iii), $0 \cdot 096 \times 3000^2 = \dfrac{10 \times 9 \cdot 8}{x}$

$$\therefore x = \frac{10 \times 9 \cdot 8}{0 \cdot 096 \times 3000^2} = 1 \cdot 1 \times 10^{-4} \text{ metre.}$$

Exercises 27

Pipes

1. Write down in terms of wavelength, λ, the distance between (i) consecutive nodes, (ii) a node and an adjacent antinode, (iii) consecutive antinodes. Find the frequency of the fundamental of a closed pipe 15 cm long if the velocity of sound in air is 340 m s^{-1}.

2. Discuss what is meant by the statement that *sound is a wave motion*. Use the example of the passage of a sound wave through air to explain the terms wavelength (λ), frequency (f), and velocity (v) of a wave. Show that $v = f\lambda$.

Explain the increase in loudness (or 'resonance') which occurs when a sounding tuning-fork is held near the open end of an organ pipe when the length of the pipe has certain values, the other end of the pipe being closed. Find the shortest length of such a pipe which resonates with a 440 Hz tuning-fork, neglecting end corrections. (Velocity of sound in air = 350 m s^{-1}.) (*O. & C.*)

3. Explain the conditions necessary for the creation of stationary waves in air.

Describe how (*a*) the displacement, (*b*) the pressure vary at different points along a stationary wave in air and describe how these effects might be demonstrated experimentally.

A tube is closed at one end and closed at the other by a vibrating diaphragm which may be assumed to be a displacement node. It is found that when the frequency of the diaphragm is 2000 Hz a stationary wave pattern is set up in the tube and the distance between adjacent nodes is then 8·0 cm. When the frequency is gradually reduced the stationary wave pattern disappears but another stationary wave pattern reappears at a frequency of 1600 Hz. Calculate (i) the speed of sound in air, (ii) the distance between adjacent nodes at a frequency of 1600 Hz, (iii) the length of the tube between the diaphragm and the closed end, (iv) the next lower frequency at which a stationary wave pattern will be obtained. (*N.*)

4. What are the chief characteristics of a progressive wave motion? Give your reasons for believing that sound is propagated through the atmosphere as a longitudinal wave motion, and find an expression relating the velocity, the frequency, and the wavelength.

Neglecting end effects, find the lengths of (a) a closed organ pipe, and (b) an open organ pipe, each of which emits a fundamental note of frequency 256 Hz. (Take the speed of sound in air to be 330 m s^{-1}.) (O.)

5. (a) Explain in terms of the properties of a gas, but without attempting mathematical treatment, how the vibration of a sound source, such as a loudspeaker diaphragm, can be transmitted through the air around it.

Explain, also, the reflection which occurs when the vibration reaches a fixed barrier, such as a wall.

(b) Plane, simple harmonic, progressive sound waves of wavelength 1·2 m and speed 348 m s^{-1}, are incident normally on a plane surface which is a perfect reflector of sound. What statements can be made about the amplitude of vibration and about air pressure changes at points distant (i) 30 cm, (ii) 60 cm, (iii) 90 cm, (iv) 10 cm from the reflector? Justify your answers. (O. & C.)

6. Describe the motion of the air in a tube closed at one end and vibrating in its fundamental mode. An observer (a) holds a vibrating tuning-fork over the open end of a tube which resounds to it, (b) blows lightly across the mouth of the tube. Describe and explain the difference in the quality of the notes that he hears.

A uniform tube, 60·0 cm long, stands vertically with its lower end dipping into water. When the length above water is 14·8 cm, and again when it is 48·0 cm, the tube resounds to a vibrating tuning-fork of frequency 512 Hz. Find the lowest frequency to which the tube will resound when it is open at both ends. (L.)

7. Discuss the factors which determine the pitch of the note given by a 'closed' pipe. Explain why the fundamental frequency and the quality of the note from a 'closed' pipe differ from those of the note given under similar conditions by a pipe of the same length which is open at both ends. (N.)

8. What do you understand by (a) forced vibrations, (b) free vibrations, and (c) resonance? Illustrate your answer by giving three distinct examples, one for each of (a), (b) and (c).

Explain how a stationary sound wave may be set up in a gas column and how you would demonstrate the presence of nodes and antinodes. State what measurements would be required in order to deduce the speed of sound in air from your demonstration, and show how you would calculate your result.

The speed of sound, c, in an ideal gas is given by the formula $c = \sqrt{\gamma p/\rho}$, where p is the pressure, ρ is the density of the gas and γ is a constant. By considering this formula explain the effect of a change in (i) temperature, and (ii) pressure, on speed of sound. (L.)

9. Distinguish between the formation of an echo and the formation of a stationary sound wave by reflection, explaining the general circumstances in which each is produced.

Describe an experiment in which the velocity of sound in air may be determined by observations on stationary waves.

An organ pipe is sounded with a tuning-fork of frequency 256 Hz. When the air in the pipe is at a temperature of 15°C, 23 beats are heard in 10 seconds; when the tuning-fork is loaded with a small piece of wax, the beat frequency is found to decrease. What change of temperature of the air in the pipe is necessary to bring the pipe and the unloaded fork into unison? (C.)

10. What is meant by (a) a *stationary wave motion* and (b) a *node*?

Describe how the phenomenon of resonance may be demonstrated using a loudspeaker, a source of alternating voltage of variable frequency and a suitable tube open at one end and closed at the other. Explain how resonance occurs in the arrangement you describe, draw a diagram showing the position of the nodes in the tube in a typical case of resonance and state clearly the meaning of the diagram. How would you demonstrate the position of the nodes experimentally? (O. & C.)

11. Describe and give the theory of one experiment in each instance by which the velocity of sound may be determined, (a) in free air, (b) in the air in a resonance tube.

What effect, if any, do the following factors have on the velocity of sound in free air; frequency of the vibrations; temperature of the air; atmospheric pressure; humidity?

State the relationship between this velocity and temperature. (L.)

Strings. Rods

12. Describe the differences between stationary waves and progressive waves. Outline an experimental arrangement to illustrate the formation of a stationary wave in a string.

Waves of wavelength λ, from a source S, reach a common point P by two different routes. At P the waves are found to have a phase difference $3\pi/4$ rad. Show graphically what this means. What is the minimum path difference between the two routes?

A string fixed at both ends is vibrating in the lowest mode of vibration for which a point a quarter of its length from one end is a point of maximum vibration. The note emitted has a frequency of 100 Hz. What will be the frequency emitted when it vibrates in the next mode such that this point is again a point of maximum vibration? (*L.*)

13. Explain what is meant by the *wavelength*, the *frequency*, and the *speed* of a sinusoidal travelling wave and derive a relation between them.

What is meant by a stationary wave? A stationary sinusoidal wave of period T is set up on a stretched string so that there are nodes only at the two ends of the string and at its midpoint. The displacement of each point of the string has its maximum value at $t = 0$. Show on a single sketch the shape taken by the string at times $t = 0$, $T/8$, $T/4$, $3T/8$ and $T/2$.

A piano string 1·5 m long is made of steel of density $7\cdot7 \times 10^3$ kg m^{-3} and Young's modulus 2×10^{11} N m^{-2}. It is maintained at a tension which produces an elastic strain of 1% in the string. What is the fundamental frequency of transverse vibration of the string? (*O. & C.*)

14. Describe the motion of the particles of a string under constant tension and fixed at both ends when the string executes transverse vibrations of (*a*) its fundamental frequency, (*b*) the first overtone (second harmonic). Illustrate your answer with suitable diagrams.

A horizontal sonometer wire of fixed length 0·50 m and mass $4\cdot5 \times 10^{-3}$ kg is under a fixed tension of $1\cdot2 \times 10^2$ N. The poles of a horse-shoe magnet are arranged to produce a horizontal transverse magnetic field at the midpoint of the wire, and an alternating sinusoidal current passes through the wire. State and explain what happens when the frequency of the current is progressively increased from 100 to 200 Hz. Support your explanation by performing a suitable calculation. Indicate how you would use such an apparatus to measure the fixed frequency of an alternating current. (*N.*)

15. Explain the meaning of the following terms in relation to wave motion: displacement, amplitude, wavelength, frequency, phase.

What is the nature of a wave motion in which (*a*) the amplitude is the same at all points, but the phase varies with position, (*b*) the phase is the same at all points, but the amplitude varies with position?

The velocity of transverse waves along a string depends only on the tension F, the radius r and the density ρ of the material. Use the method of dimensions to determine the form of the dependence, and describe briefly how you would attempt to verify your result experimentally. (*O. & C.*)

16. What is meant by (*a*) a forced vibration, (*b*) resonance? Give an example of each from (i) mechanics, (ii) sound.

Using the same axes sketch graphs showing how the amplitude of a forced vibration depends upon the frequency of the applied force when the damping of the system is (*a*) light, (*b*) heavy. Point out any special features of the graphs.

A sonometer wire is stretched by hanging a metal cylinder of density 8000 kg m^{-3} at the end of the wire. A fundamental note of frequency 256 Hz is sounded when the wire is plucked.

Calculate the frequency of vibration of the same length of wire when a vessel of water is placed so that the cylinder is totally immersed. (*N.*)

17. Distinguish between a *progressive* wave and a *stationary* wave. Explain in detail how you would use a sonometer to establish the relation between the fundamental frequency of a stretched wire and (*a*) its length, (*b*) its tension. You may assume a set of standard tuning-forks and set of weights in steps of half a kilogram to be available.

A pianoforte wire having a diameter of 0·90 mm is replaced by another wire of the same material but with diameter 0·93 mm. If the tension of the wire is the same as before, what is the percentage change in the frequency of the fundamental note? What percentage change in the tension would be necessary to restore the original frequency? (*L.*)

18. Describe an experiment to determine the velocity of sound in a gas, e.g. nitrogen. How would you expect the velocity to be affected by (*a*) temperature, (*b*) pressure and (*c*) humidity? Give reasons for your answers.

What information about the nature of a gas can be obtained from a measurement of the velocity of sound in that gas, the pressure and density being known? (*L.*)

19. Describe experiments to illustrate the differences between (*a*) *transverse* waves, (*b*) *longitudinal* waves, (*c*) *progressive* waves and (*d*) *stationary* waves? To which classes belong (i) the vibrations of a violin string, (ii) the sound waves emitted by the violin into the surrounding air?

A wire whose mass per unit length is 10^{-3} kg m^{-1} is stretched by a load of 4 kg over the two bridges of a sonometer 1 m apart. If it is struck at its middle point, what will be (*a*) the wavelength of its subsequent fundamental vibrations, (*b*) the fundamental frequency of the note emitted? If the wire were struck at a point near one bridge what further frequencies might be heard? (Do not derive standard formulae.) (Assume $g = 10$ m s^{-2}.) (*O. & C.*)

20. A uniform wire vibrates transversely in its fundamental mode. On what factors, other than the length does the frequency of vibration depend, and what is the form of the dependence for each factor?

Describe the experiment you would perform to verify the form of dependence for *one* factor.

A wire of diameter 0·040 cm and made of steel of density 8000 kg m^{-3} is under constant tension of 80 N. A fixed length of 50 cm is set in transverse vibration. How would you cause the vibration of frequency about 840 Hz to predominate in intensity? (*N.*)

21. (*a*) The velocity of sound in air being known, describe how Young modulus for brass may be found using Kundt's tube. (*b*) Discuss how the frequency of a note heard by an observer is affected by movement of (i) the source, (ii) the observer along the line joining source and observer. (*L.*)

22. Give an expression for the velocity of a transverse wave along a thin flexible string and show that it is dimensionally correct. Explain how reflexion may give rise to transverse *standing waves* on a stretched string and use the expression for the velocity to drive the frequency of the fundamental mode of vibration.

A steel wire of length 40·0 cm and diameter 0·0250 cm vibrates transversely in unison with a tube, open at each end and of effective length 60·0 cm, when each is sounding its fundamental note. The air temperature is 27°C. Find the tension in the wire. (Assume that the velocity of sound in air at 0°C is 331 m s^{-1} and the density of steel is 7800 kg m^{-3}.) (*L.*)

Part Five
Electricity and Atomic Physics

28 Electrostatics

General Phenomena

If a rod of ebonite is rubbed with fur, or a fountain-pen with a coat-sleeve, it gains the power to attract light bodies, such as pieces of paper or tin-foil or a suspended pith-ball. The discovery that a body could be made attractive by rubbing is attributed to Thales (640–548 B.C.). He seems to have been led to it through the Greeks' practice of spinning silk with an amber spindle; the rubbing of the spindle in its bearings caused the silk to adhere to it. The Greek word for amber is *elektron*, and a body made attractive by rubbing is said to be 'electrified' or *charged*. This branch of Electricity, the earliest discovered, is called *Electrostatics*.

Conductors and Insulators

Little progress was made in the study of electrification until the sixteenth century A.D. Then Gilbert (1540–1603), who was physician-in-ordinary to Queen Elizabeth, found that other substances besides amber could be electrified: for example, glass when rubbed with silk. He failed to electrify metals, however, and concluded that to do so was impossible.

More than 100 years later—in 1734—he was shown to be wrong. Du Fay found that a metal could be charged by rubbing with fur or silk, but only if it were held in a handle of glass or amber; it could not be charged if it were held directly in the hand. His experiments followed the discovery, by Gray in 1729, that electric charges could be transmitted through the human body, water, and metals. These are examples of *conductors*; glass and amber are examples of *insulators*.

Positive and Negative Charges

In the course of his experiments du Fay also discovered that there were two kinds of electrification or charges. He showed that electrified glass and amber tended to oppose one another's attractiveness. To illustrate how he did so, we may use ebonite instead of amber, which has the same electrical properties. We suspend a pith-ball, and attract it with an electrified ebonite rod E, Fig. 28.1

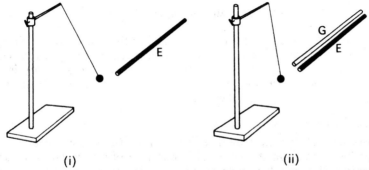

(i)   (ii)

Fig. 28.1 Demonstrating that an electrified glass or acetate rod tends to oppose effect of electrified ebonite or polythene rod (i); we then bring an electrified glass rod G towards the ebonite rod, and the pith-ball falls away, Fig. 28.1 (ii).

Benjamin Franklin, a pioneer of electrostatics, gave the name of 'positive electricity' to the charge on a glass rod rubbed with silk, and 'negative electricity' to that on an ebonite rod rubbed with fur. Rubbed by a duster, a cellulose acetate rod obtains a positive charge and a polythene rod obtains a negative charge.

Electrons and Electrostatics

Towards the end of the nineteenth century Sir J. J. Thomson discovered the existence of the *electron* (p. 804). This is a particle of very low mass—it is about 1/1840th of the mass of the hydrogen atom—and experiments show that it carries a tiny quantity of *negative* charge. Later experiments showed that electrons are present in all atoms.

The detailed structure of atoms is complicated, but, generally, electrons exist round a very tiny core or nucleus carrying positive charge. Normally, atoms are electrically neutral, that is, there is no surplus of charge on them. Consequently the total negative charge on the electrons is equal to the positive charge on the nucleus. In insulators, all the electrons appear to be firmly 'bound' to the nucleus under the attraction of the unlike charges. In metals, however, some of the electrons appear to be relatively 'free'. These electrons play an important part in electrical phenomena concerning metals.

The theory of electrons (negatively charged particles) gives a simple explanation of electrification by friction. If the silk on which a glass rod has been rubbed is brought near to a charged and suspended ebonite rod it repels it; the silk must therefore have a negative charge. We know that the glass has a positive charge. We therefore suppose that when the two were rubbed together electrons from the surface atoms were transferred from the glass to the silk. Likewise we suppose that when fur and ebonite are rubbed together, electrons go from the fur to the ebonite. Similar explanations hold for rubbed acetate and polythene.

Attraction of Charged Body for Uncharged Bodies

To explain the attraction of a charged body for an uncharged one, we shall suppose that the uncharged body is a conductor—a metal. If it is brought near to a charged polythene rod, say, then the negative charge on the rod repels the free electrons in the metal to its far end (Fig. 28.2). A positive charge is thus left

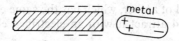

Fig. 28.2 Attraction by charged body

on the near end of the metal; since this is nearer than the negative charge on the far end, it is attracted more strongly than the negative charge is repelled. On the whole, therefore, the metal is attracted. If the uncharged body is not a conductor, the mechanism by which it is attracted is more complicated; we shall postpone its description to a later chapter.

Electrostatics Today

The discovery of the electron has led, during the present century, to a considerable increase in the practical importance of electrostatics. In devices such as radio valves and cathode-ray tubes, for example, electrons are moving under the influence of electrostatic forces. The problems of preventing sparks and the breakdown of insulators are essentially electrostatic. These problems occur in high voltage electrical engineering. Later, we shall also describe a modern electrostatic generator, of the type used to provide a million volts or more for X-ray work and nuclear bombardment. Such generators work on principles of electrostatics discovered over a hundred years ago.

Gold-leaf Electroscope

One of the earliest instruments used for testing positive and negative charges consisted of a metal rod A to which gold leaves L were attached (Fig. 28.3).

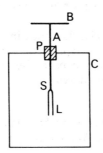

Fig. 28.3 A gold-leaf electroscope

The rod was fitted with a circular disc or cap B, and was insulated with a plug P from a metal case C which screened L from outside influences other than those brought near to B.

When B is touched by a polythene rod rubbed with a duster, some negative charge on the rod passes to the cap and L; and since like charges repel, the leaves open or diverge, Fig. 28.4 (i). If an unknown charge X is now brought near to B, an increased divergence implies that X is negative, Fig. 28.4 (ii). A positive charge is tested in a similar way; the electroscope is first given a positive charge and an increased divergence indicates a positive charge.

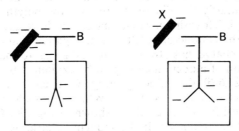

Fig. 28.4 Testing charge with electroscope

Induction

We shall now show that it is possible to obtain charges, called *induced charges*, without any contact with another charge. An experiment on *electrostatic induction* as the phenomenon is called, is shown in Fig. 28.5 (i).

Two insulated metal spheres A, B are arranged so that they touch one another and a negatively charged polythene rod C is brought near to A. The spheres are now separated, and then the rod is taken away. Tests with a charged pith-ball now show that A has a positive charge and B a negative charge, Fig. 28.5 (ii). If the spheres are placed together so that they touch, it is found that they now have no effect on a pith-ball held near. Their charges must therefore have

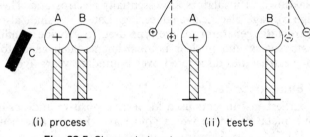

(i) process   (ii) tests

Fig. 28.5 Charges induced on a conductor

neutralised each other completely, thus showing that the induced positive and negative charges are equal. This is explained by the movement of electrons from A to B when the rod C is brought near, Fig. 28.5 (i). B has then a negative charge and A an equal positive charge.

Charging by Induction

Figure 28.6 shows how a conductor can be given a permanent charge by induction, without dividing it in two. We first bring a charged polythene rod C, say,

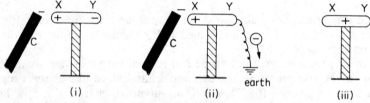

Fig. 28.6 Charging permanently by induction

near to the conductor XY, (i); next we connect the conductor to earth by touching it momentarily (ii). Finally we remove the polythene. We then find that the conductor is left with a positive charge (iii). If we use a charged acetate rod, we find that the conductor is left with a negative charge; the charge left, called the induced charge, has always the *opposite* sign to the inducing charge.

This phenomenon of induction can again be explained by the movement of electrons. In Fig. 28.6 (i), the inducing charge C repels electrons to Y, leaving an equal positive charge at X as shown. When we touch the conductor XY, electrons are repelled from it to earth, as shown in Fig. 28.6 (ii), and a positive charge is left on the conductor. If the inducing charge is positive, then the electrons are attracted up from the earth to the conductor, which then becomes negatively charged.

Induction and the Electroscope

It is always observed that the leaves of an electroscope diverge when a charged body is brought near its cap, without touching it. We can now easily understand what happens. If, for example, we bring a negatively charged rod near the cap, it induces a positive charge on the cap, and a negative one on the leaves: the leaves then repel each other, Fig. 28.7 (i).

We can use induction to give a permanent charge to the cap and leaves of an electroscope, by momentarily earthing the cap while hold the inducing charge near it. The electrons are repelled to earth as shown, Fig. 28.7 (ii). When the

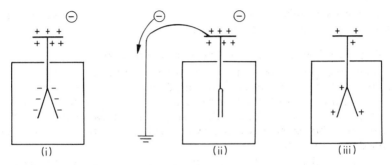

Fig. 28.7 Charging electroscope by induction

negative charge is finally taken away, the positive charge spreads and the leaves open.

The Electrophorus

A device which provides an almost unlimited supply of charge, by induction, was invented by Volta about 1800; it is called an *electrophorus*.

It consists of a polythene or perspex base, E in Fig. 28.8, and a metal disc

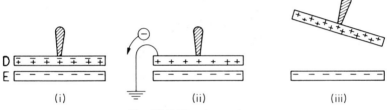

Fig. 28.8 The electrophorus

D on an insulating handle. The polythene is charged negatively by rubbing it very vigorously with a duster. The disc is then laid upon it, and acquires induced charges, positive underneath and negative on top, (i). Very little negative charge escapes from the polythene to the disc, because the natural unevenness of their surfaces prevents them touching at more than a few points; charge escapes from these points only, because the polythene is a non-conductor. After it has been placed on the polythene, the disc is earthed with the finger, and the negative charge on its upper surface flows away, (ii). The disc can then be removed, and carries with it the positive charge which was on its underside, (iii).

An electrophorus produces sufficient charge to given an audible—and some-times visible—spark. The disc can be discharged and charged again repeatedly, until the charge on the polythene has disappeared by leakage. Apparently, therefore, it is in principle an inexhaustible source of energy. However, work is done in raising the disc from the polythene, against the attraction of their opposite charges, and this work must be done each time the disc is charged. The electrophorus is therefore not a source of energy, but a device for converting it from a mechanical into an electrical form.

The action of the electrophorus illustrates the advantages of charging by induction. First, the supply of charge is almost inexhaustible, because the original charge is not carried away. Second, a great charge—nearly equal to the charge on the whole of the polythene—can be concentrated on to the conducting disc.

As we have seen, only a very small charge could be transferred by contact, because the polythene is not a conductor.

The Action of Points, Van de Graaff Generator

Sometimes in experiments with an electroscope connected to other apparatus by a wire, the leaves of the electroscope gradually collapse, as though its charge were leaking away. This behaviour can often be traced to a sharp point on the wire—if the point is blunted, the leakage stops. Charge leaks away from a sharp point through the air, being carried by molecules away from the point. This is explained later (p. 579).

Points are used to collect the charges produced in *electrostatic generators*. These are machines for continuously separating charges by induction, and thus building up very great charges and potential differences. Figure 28.9 is a

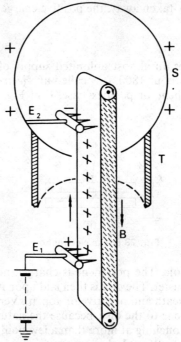

Fig. 28.9 Principle of Van de Graaff generator

simplified diagram of one such machine, due to Van de Graaff. A hollow metal sphere S is supported on an insulating tube T, many metres high. A silk belt B runs over the pulleys shown, of which the lower is driven by an electric motor. Near the bottom and top of its run, the belt passes close to the electrodes E, which are sharply pointed combs, pointing towards the belt. The electrode E_1 is made about 10 000 volts positive with respect to the earth by a battery.

As shown later, the high electric field at the points ionises the air there; and so positive charges are repelled on to the belt, which carries it up into the sphere. There it induces a negative charge on the points of electrode E_2 and a positive charge on the sphere to which the blunt end of E_2 is connected. The high electric field at the points ionises the air there, and negative charges, repelled to the belt, discharges the belt before it passes over the pulley. In this way the sphere gradually charges up positively, until its potential is about a million volts relative to the earth.

Large machines of this type are used with high-voltage X-ray tubes, and for atom-splitting experiments. They have more elaborate electrode systems, stand about 15 m high, and have 4 m spheres. They can produce potential differences up to 5 000 000 volts and currents of about 50 microamperes. The electrical energy which they deliver comes from the work done by the motor in drawing the positively charged belt towards the positively charged sphere, which repels it.

In all types of high-voltage equipment sharp corners and edges must be avoided, except where points are deliberately used as electrodes. Otherwise, corona discharges may break out from the sharp places. All such places are therefore enlarged by metal globes called stress-distributors. See also p. 579.

Fig. 28.10 Van de Graaff electrostatic generator at Aldermaston, England. The dome is the high-voltage terminal. The insulated rings are equipotentials, and provide a uniform potential gradient down the column. Beams of protons or deutrons, produced in the dome, are accelerated down the column to bombard different materials at the bottom, thereby producing nuclear reactions which can be studied

Ice-pail Experiment

A famous experiment on electrostatic induction was made by Faraday in 1843. In it he used the ice-pail from which it takes its name; but it was a modest pail, 27 cm high—not a bucket. He stood the pail on an insulator, and connected it to a gold-leaf electroscope, as in Fig. 28.11 (i). He next held a metal ball on the end of a long silk thread, and charged it positively by a spark from an electrophorus. Then he lowered the ball into the pail, without letting it touch the sides or bottom, Fig. 28.11 (ii). A positive charge was induced on the outside of the pail and the leaves, and made the leaves diverge. Once the ball was well inside the pail, Faraday found that the divergence of the leaves did not change

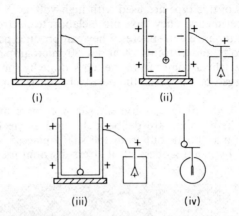

(i)   (ii)

(iii)   (iv)

Fig. 28.11 Faraday's ice-pail experiment

when he moved the ball about—nearer to or farther from the walls or the bottom. This showed that the amount of the induced positive charge did not depend on the position of the ball, once it was well inside the pail.

Faraday then allowed the ball to touch the pail, and noticed that the leaves of the electroscope still did not move (Fig. 28.11 (iii)). When the ball touched the pail, therefore, no charge was given to, or taken from, the outside of the pail. Faraday next lifted the ball out of the pail, and tested it for charge with another electroscope. He found that the ball had no charge whatever, Fig. 28.11 (iv). The induced negative charge on the inside of the pail must therefore have been equal in magnitude to the original positive charge on the ball.

Faraday's experiment does not give these simple results unless the pail—or whatever is used in place of it—very nearly surrounds the charged ball, Fig. 28.12 (i). If, for example, the ball is allowed to touch the pail before it is well inside, as in Fig. 28.12 (iii), then it does not lose all its charge.

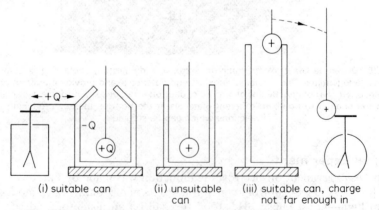

(i) suitable can   (ii) unsuitable   (iii) suitable can, charge
                               can   not far enough in

Fig. 28.12 Experimental conditions in Faraday's ice-pail experiment

Conclusions

The conclusions to be drawn from the experiment therefore apply, strictly, to a *hollow closed conductor*. They are:

(*i*) When a charged body is enclosed in a hollow conductor it induces on the inside of that conductor a charge equal but opposite to its own; and on the outside a charge equal and similar to its own Fig. 28.11 (i).

(*ii*) The *total* charge inside a hollow conductor is always zero: either there are equal and opposite charges on the inside walls and within the volume (before the ball touches), or there is no charge at all (after the ball has touched).

Comparison and Collection of Charges

Faraday's ice-pail experiment gives us a method of comparing quantities of electric charges. The experiment shows that if a charged body is lowered well inside a tall narrow can then it gives to the outside of the can a charge equal to its own. If the can is connected to the cap of an electroscope, the divergence of the leaves is a measure of the charge on the body. Thus we can compare the magnitudes of charges, without removing them from the bodies which carry them: we merely lower those bodies, in turn, into a tall insulated can, connected to an electroscope.

Sometimes we may wish to discharge a conductor completely, without letting its charge run to earth. We can do this by letting the conductor touch the bottom of a tall can on an insulating stand. The whole of the body's charge is then transferred to the outside of the can.

Charges Produced by Separation; Lines of Force

The ice-pail experiment suggests that a positive electric charge, for example, is always accompanied by an equal negative charge. Farday repeated his experiment with a nest of hollow conductors, insulated from one another, and showed that equal and opposite charges were induced on the inner and outer walls of each (Fig. 28.13).

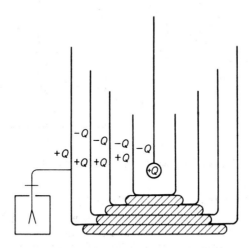

Fig. 28.13 Extension of ice-pail experiment

Faraday also showed that equal and opposite charges are produced when a body is electrified by rubbing. He fitted an ebonite rod with a fur cap, which he rotated by a silk thread or string wrapped round it (Fig. 28.14 (i)); he then compared the charges produced with an ice-pail and electroscope, Fig. 28.14 (i), (ii), (iii), (iv).

In describing the conclusions from this last experiment, we now say, as indeed we have done already, that electrons flow from the fur to the ebonite, carrying

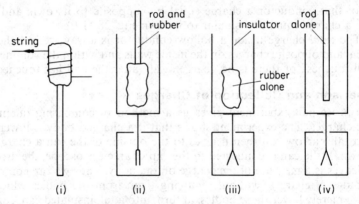

Fig. 28.14 Stages in showing that equal and opposite charges are produced by friction

to it a negative charge, and leaving on the fur a positive charge. It appears, therefore, that free charges are always produced by separating equal amounts of the opposite kinds of electricity.

The idea that charges always occur in equal opposite pairs affects our drawing of lines of force diagrams. Lines of force radiate outwards from a positive charge, and inwards to a negative one; from any positive charge, therefore, we draw lines of force ending on an equal negative charge. Fig. 28.15 gives illustrations of this procedure.

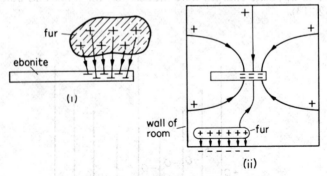

Fig. 28.15 Charging by friction—lines of force

Distribution of Charge; Surface Density

By using a can connected to an electroscope we can find how electricity is distributed over a charged conductor of any form—pear-shaped, for example. We take a number of small leaves of tin-foil, all of the same area, but differently shaped to fit closely over the different parts of the conductor, and mounted on polythene handles, Fig. 28.16 (i). These are called *proof-planes* We charge the body from an electrophorus, press a proof-plane against the part which it fits, and then lower the proof-plane into a can connected to an electroscope, Fig. 28.16 (ii). After noting the divergence of the leaves we discharge the can and electroscope by touching one of them, and repeat the observation with a proof-plane fitting a different part of the body. Since the proof-planes have equal areas, each of them carries away a charge proportional to the charge per unit area of the body, over the region which it touched.

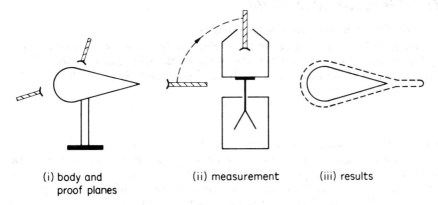

(i) body and   (ii) measurement   (iii) results
proof planes

Fig. 28.16 Investigating charge distribution

The charge per unit area over a region of the body is called the *surface-density* of the charge in that region. We find that the surface-density increases with the curvature of the body, as shown in Fig. 28.16 (iii); the distance of the dotted line from the outline of the body is roughly proportional to the surface-density of charge.

The Electrostatic Field

Law of Force between Two Charges

The magnitude of the force between two electrically charged bodies was studied by Coulomb in 1875. He showed that, if the bodies were small compared with the distance between them, then the force F was inversely proportional to the square of the distance r, i.e.

$$F \propto \frac{1}{r^2}. \qquad \qquad \qquad (1)$$

This result is known as the *inverse square law*, or Coulomb's law.

It is not possible to verify the law accurately by direct measurement of the force between two charged bodies. In 1936 Plimton and Lawton showed, by an indirect method, that the power in the law cannot differ from 2 by more than $\pm 2 \times 10^{-9}$. We have no reason to suppose, therefore, that the inverse square law is other than exactly true.

Quantity of Charge in Law of Force

The SI unit of charge is the *coulomb* (C.). The *ampere* (A), the unit of current, is defined later (p. 713). The coulomb is defined as that quantity of charge which passes a section of a conductor in one second when the current flowing is one ampère.

By measuring the force F between two charges when their respective magnitudes Q and Q' are varied, it is found that F is proportional to the product QQ'. Thus

$$F \propto QQ' . \qquad \qquad \qquad (2)$$

Combining (1) and (2), we have

$$F \propto \frac{QQ'}{r^2}$$

$$\therefore F = k\frac{QQ'}{r^2}, \qquad \qquad \qquad (3)$$

where k is a constant. For reasons explained later, k is written as $1/4\pi\varepsilon_0$, where ε_0 is a constant called the *permittivity* of free space if we suppose the charges are situated in a vacuum. Thus

$$F = \frac{1}{4\pi\varepsilon_0}\frac{QQ'}{r^2} \qquad \qquad \qquad (4)$$

In this expression, F is measured in newton (N), Q in coulomb (C) and r in metre (m). Now, from (4),

$$\varepsilon_0 = \frac{QQ'}{4\pi F r^2}.$$

Hence the units of ε_0 are coulomb^2 newton^{-1} metre^{-2} ($C^2\,N^{-1}\,m^{-2}$). Another unit of ε_0, more widely used, is *farad metre*$^{-1}$ ($F\,m^{-1}$). See p. 604.

We shall see later that ε_0 has the numerical value of $8\cdot854 \times 10^{-12}$, and $1/4\pi\varepsilon_0$ then has the value 9×10^9 approximately.

Permittivity

So far we have considered charges in a vacuum. If charges are situated in other media such as water, then the force between the charges is reduced. Equation (4) is true only in a vacuum. In general, we write

$$F = \frac{1}{4\pi\varepsilon}\frac{QQ'}{r^2} \qquad . \qquad . \qquad . \qquad . \qquad . \qquad (1)$$

where ε is the *permittivity* of the medium. The permittivity of air at normal pressure is only about 1·005 times that, ε_0, of a vacuum. For most purposes, therefore we may assume that value of ε_0 for the permittivity of air. The permittivity of water is about eighty times that of a vacuum. Thus the force between charges situated in water is eighty times less than if they were situated the same distance apart in a vacuum.

Example

(a) Calculate the value of two equal charges if they repel one another with a force of 0·1 N when situated 50 cm apart in a vacuum.

(b) What would be the size of the charges if they were situated in an insulating liquid whose permittivity was ten times that of a vacuum?

(a) From (4),
$$F = \frac{1}{4\pi\varepsilon_0}\frac{QQ'}{r^2}.$$

Since $Q = Q'$ here,

$$0·1 = \frac{9 \times 10^9\ Q^2}{(0·5)^2}$$

or
$$Q^2 = \frac{0·1 \times (0·5)^2}{9 \times 10^9}$$

$$Q = 1·7 \times 10^{-6}\ \text{C} = 1·7\ \mu\text{C (microcoulomb)}.$$

(b) The permittivity of the liquid $\varepsilon = 10\varepsilon_0$. So the force between charges is *reduced* 10 times in the liquid. To maintain the force of 0·1 N, it follows that the product of the charges Q in the liquid must be 10 times that in a vacuum. So

$$Q \times Q = Q^2 = 10 \times (1·7 \times 10^{-6})^2.$$

Therefore
$$Q = 3·1 \times 1·7 \times 10^{-6}$$
$$= 5·3 \times 10^{-6}\ \text{C} = 5·3\ \mu\text{C}.$$

Electric Intensity or Field-strength. Lines of Force

An 'electric field' can be defined as a region where an electric force is experienced. As in magnetism, electric fields can be mapped out by electrostatic lines of force, which may be defined as a line such that the tangent to it is in the direction of the force on a small positive charge at that point. Arrows on the lines of force show the direction of the force on a positive charge; the force on a negative charge is in the opposite direction. Figure 28.17 shows the lines of force, also called *electric flux*, in some electrostatic fields.

The force exerted on a charged body in an electric field depends on the charge of the body and on the *intensity* or *strength* of the field. If we wish to explore the variation in intensity of an electric field, then we must place a test charge Q' at the point concerned which is small enough not to upset the field by its introduction. The intensity E of an electrostatic field at any point is defined as *the force per unit charge* which it exerts at that point. Its direction is that of the force exerted on a positive charge.

From this definition,

$$E = \frac{F}{Q'}$$

$$F = EQ' \qquad . \qquad . \qquad . \qquad . \qquad . \qquad (1)$$

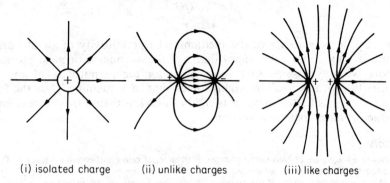

(i) isolated charge   (ii) unlike charges   (iii) like charges

Fig. 28.17 Lines of electrostatic force

Since F is measured in newtons and Q' in coulombs, it follows that *intensity E has units of newton per coulomb* ($N \ C^{-1}$). We shall see later that a more practical unit of E is *volt metre*$^{-1}$ ($V \ m^{-1}$) (see p. 584).

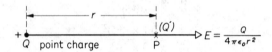

Fig. 28.18 Electric field intensity due to point charge

We can easily find an expression for the strength E of the electric field due to a point charge Q situated in a vacuum (Fig. 28.18). We start from equation (4) p. 574, for the force between two such charges:

$$F = \frac{1}{4\pi\varepsilon_0} \frac{QQ'}{r^2}$$

If the test charge Q' is situated at the point P in Fig. 28.18, the electric field strength at that point is given by (1).

$$\therefore E = \frac{F}{Q'} = \frac{Q}{4\pi\varepsilon_0 r^2} \qquad \cdot \qquad \cdot \qquad \cdot \qquad (2)$$

The direction of the field is radially outward if the charge Q is positive (Fig. 28.17 (i)); it is radially inward if the charge Q is negative. If the charge were surrounded by a material of permittivity ε then,

$$E = \frac{Q}{4\pi\varepsilon r^2} \qquad \cdot \qquad \cdot \qquad \cdot \qquad (3)$$

Flux from a Point Charge

We have already shown how electric fields can be described by lines of force. From Fig. 28.17 (i) it can be seen that the density of the lines increases near the charge where the field intensity is high. The intensity E at a point can thus be represented by *the number of lines per unit area* through a surface perpendicular to the lines of force at the point considered. The *flux* through an area perpendicular to the lines of force is the name given to the product of $E \times area$, where E is the intensity at that place. This is illustrated in Fig. 28.19 (i).

Consider a sphere of radius r drawn in space concentric with a point charge,

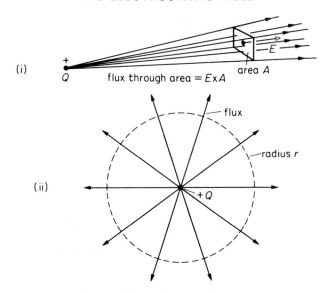

Fig. 28.19 Flux from a point charge

Fig. 28.19 (ii). The value of E at this place is given by (3), p. 576. The total flux through the sphere is,

$$E \times \text{area} = E \times 4\pi r^2$$

$$= \frac{Q}{4\pi \varepsilon r^2} \times 4\pi r^2 = \frac{Q}{\varepsilon}$$

$$= \frac{\text{charge inside sphere}}{\text{permittivity}} \qquad . \qquad . \qquad . \qquad (1)$$

This demonstrates the important fact that the total flux crossing any sphere drawn outside and concentrically around a point charge is a constant. It does not depend on the distance from the charged sphere. It should be noted that this result is only true if the inverse square law is true.

To see this, suppose some other force law were valid, i.e. $E = Q/4\pi \varepsilon r^n$. Then the total flux through the area

$$= \frac{Q}{4\pi \varepsilon r^n} \times 4\pi r^2 = \frac{Q}{\varepsilon} r^{(2-n)}$$

This is only independent of r if $n = 2$.

Field due to Charged Sphere and Plane Conductor

Equation (1) can be shown to be generally true. Thus the total flux passing through any *closed* surface whatever its shape, is always equal to Q/ε, where Q is the total charge enclosed by the surface. This relation, called *Gauss's Theorem*, can be used to find the value of E in other common cases.

(1) *Outside a charged sphere*

The flux across a spherical surface of radius r, concentric with a small sphere carrying a charge Q (Fig. 28.20), is given by,

$$\text{Flux} = \frac{Q}{\varepsilon}$$

$$\therefore E \times 4\pi r^2 = \frac{Q}{\varepsilon}$$

$$\therefore E = \frac{Q}{4\pi \varepsilon r^2}$$

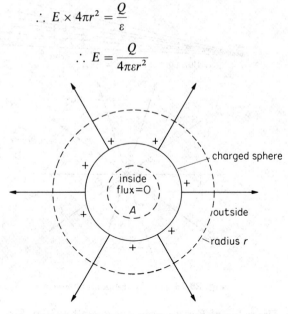

Fig. 28.20 Electric field of a charged sphere

This is the same answer as that for a point charge. This means that *outside* a charged sphere, the field behaves as if all the charge on the sphere were concentrated at the centre.

(2) *Inside a charged empty sphere*

Suppose a spherical surface A is drawn *inside* a charge sphere, as shown in Fig. 28.20. Inside this sphere there are no charges and so Q in equation (1), p. 577, is zero. This result is independent of the radius drawn, provided that it is less than that of the charged sphere. Hence from (1), *E must be zero everywhere inside a charged sphere.*

(3) *Outside a charged plane conductor*

Now consider a charged *plane* conductor S, with a surface charge density of σ coulomb metre^{-2}. Figure 28.21 shows a plane surface P, drawn outside S, which is parallel to S and has an area A metre^2. Applying equation (1),

$$\therefore E \times \text{area} = \frac{\text{Charge inside surface}}{\varepsilon}$$

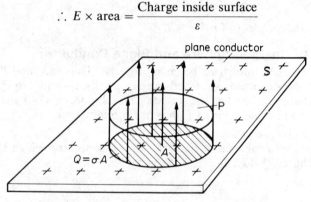

Fig. 28.21 Field of a charged plane conductor

Now by symmetry, the intensity in the field must be perpendicular to the surface. Further, the charges which produce this field are those in the projection of the area P on the surface S, i.e. those within the shaded area A in Fig. 28.21. The total charge here is thus σA coulomb.

$$\therefore E.A = \frac{\sigma A}{\varepsilon}$$

$$\therefore E = \frac{\sigma}{\varepsilon}$$

Field Round Points

On p. 573 we saw that the surface-density of charge (charge per unit area) round a point of a conductor is very great. Consequently, the strength of the electric field near the point is very great. The intense electric field breaks down the insulation of the air, and sends a stream of charged molecules away from the point. The mechanism of the breakdown, which is called a 'corona discharge', is complicated, and we shall not discuss it here; some of the processes in it are similar to those in conduction through a gas at low pressure, which we shall describe in Chapter 38. Corona breakdown starts when the electric field strength is about 3 million volt metre^{-1}. The corresponding surface-density is about $2\cdot 7 \times 10^{-5}$ coulomb metre^{-2}.

Example

An electron of charge $1\cdot6 \times 10^{-19}$ C is situated in a uniform electric field of intensity 120 000 V m^{-1}. Find the force on it, its acceleration, and the time it takes to travel 20 mm from rest (electron mass, $m = 9\cdot1 \times 10^{-31}$ kg).

Force on electron $F = eE$.

Now $E = 120\,000$ V m^{-1}.

$$\therefore F = 1\cdot6 \times 10^{-19} \times 1\cdot2 \times 10^5$$

$$= 1\cdot92 \times 10^{-14} \text{ N}$$

Acceleration,
$$a = \frac{F}{m} = \frac{1\cdot92 \times 10^{-14}}{9\cdot1 \times 10^{-31}}$$

$$= 2\cdot12 \times 10^{16} \text{ m s}^{-2}$$

Time for 20 mm or 0·02 m travel is given by

$$s = \tfrac{1}{2}at^2$$

$$\therefore t = \sqrt{\frac{2s}{a}} = \sqrt{\frac{2 \times 0\cdot02}{2\cdot12 \times 10^{16}}}$$

$$= 1\cdot37 \times 10^{-9} \text{ s}.$$

The extreme shortness of this time is due to the fact that the ratio of charge-to-mass for an electron is very great:

$$\frac{e}{m} = \frac{1\cdot6 \times 10^{-19}}{9\cdot1 \times 10^{-31}} = 1\cdot8 \times 10^{11} \text{ C kg}^{-1}.$$

In an electric field, the charge e determines the force on an electron, while the mass m determines its inertia. Because of the large ratio e/m, the electron moves almost instantaneously, and requires very little energy to displace it. Also it can respond to changes in an electric field which take place even millions of times per second. Thus it is the large value of e/m for electrons which makes electronic tubes, for example, useful in electrical communication and remote control.

Electric Potential

Potential in Fields

When an object is held at a height above the earth it is said to have potential energy. A heavy body tends to move under the force of attraction of the earth from a point of great height to one of less, and we say that points in the earth's gravitational field have potential values depending on their height.

Electric potential is analogous to gravitational potential, but this time we think of points in an electric field. Thus in the field round a positive charge, for example, a positive charge moves from points near the charge to points further away. Points round the charge are said to have an 'electric potential'.

Potential Difference

In mechanics we are always concerned with differences of height; if a point A on a hill is *h* metre higher than a point B, and our weight is *w* newton, then we do *wh* joule of work in climbing from B to A, Fig. 28.22 (i). Similarly in electricity we are often concerned with differences of potential; and we define these also in terms of work.

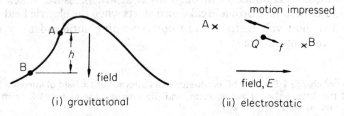

(i) gravitational   (ii) electrostatic

Fig. 28.22 Work done, in gravitational and electrostatic fields

Let us consider two points A and B in an electrostatic field, and let us suppose that the force on a positive charge *Q* has a component *f* in the direction AB, Fig. 28.22 (ii). Then if we move a positively charged body from B to A, we do work against this component of the field *E*. *We define the potential difference between A and B as the work done in moving a unit positive charge from B to A.* We denote it by the symbol V_{AB}.

The work done will be measured in joules (J). The unit of potential difference is called the volt and may be defined as follows: *The potential difference between two points A and B is one volt if the work done in taking one coulomb of positive charge from B to A is one joule.*

From this definition, if a charge of *Q* coulomb is moved through a p.d. of *V* volt, then the work done *W*, in joule, is given by

$$W = QV \qquad . \qquad . \qquad . \qquad . \qquad . \qquad (1)$$

Potential and Energy

Let us consider two points A and B in an electrostatic field, A being at a higher potential than B. The potential difference between A and B we denote as usual by V_{AB}. If we take a positive charge *Q* from B to A, we do work on it of amount QV_{AB}: the charge gains this amount of potential energy. If we now let the charge go back from A to B, it loses that potential energy: work is done on it by the electrostatic force, in the same way as work is done on a falling stone by gravity. This work may become kinetic energy, if the charge moves freely, or external work if the charge is attached to some machine, or a mixture of the two.

The work which we must do in first taking the charge from B to A does not depend on the path along which we carry it, just as the work done in climbing a hill does not depend on the route we take. If this were not true, we could devise a perpetual motion machine, in which we did less work in carrying a charge from B to A via X than it did for us in returning from A to B via Y, Fig. 28.23.

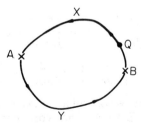

Fig. 28.23 A closed path in an electrostatic field

The fact that the potential differences between two points is a constant, independent of the path chosen between the points, is the most important property of potential in general; we shall see why later on. This property can be conveniently expressed by saying that the work done in carrying a charge round a closed path in an electrostatic field, such as BXAYB in Fig. 28.23 is zero.

The Electron-Volt

The kinetic energy gained by an electron which has been accelerated through a potential difference of 1 volt is called an *electron-volt* (eV). Since the energy gained in moving a charge Q through a p.d. $V = QV$,

$\therefore$ 1 eV = electronic charge $\times$ 1 = $1 \cdot 6 \times 10^{-19} \times 1$ joule = $1 \cdot 6 \times 10^{-19}$ J.

The electron-volt is a useful unit of energy in atomic physics. For example, the work necessary to extract a conduction electron from tungsten is $4 \cdot 52$ eV. This quantity determines the magnitude of the thermionic emission from the metal at a given temperature (p. 810); it is analogous to the latent heat of evaporation of a liquid.

Potential Difference Formula

To obtain a formula for potential difference, let us calculate the potential difference between two points in the field of a single point positive charge, Q in Fig. 28.24. For simplicity we will assume that the points, A and B, lie on a

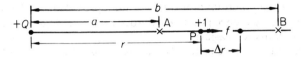

Fig. 28.24 Calculation of potential

line of force at distances a and b respectively from the charge. When a unit positive charge is at a distance r from the charge Q in free space the force on it is

$$f = \frac{Q \times 1}{4\pi\varepsilon_0 r^2}$$

The work done in taking the charge from B to A, against the force f, is equal to the work which the force f would do if the charge were allowed to go from A to B. Over the short distance Δr, the work done by the force f is

$$\Delta W = f\,\Delta r.$$

Over the whole distance AB, therefore, the work done by the force on the unit charge is

$$\int_A^B \Delta W = \int_{r=a}^{r=b} f \; dr = \int_a^b \frac{Q}{4\pi\varepsilon_0 r^2} dr$$

$$= -\left[\frac{Q}{4\pi\varepsilon_0 r}\right]_a^b = \frac{Q}{4\pi\varepsilon_0 a} - \frac{Q}{4\pi\varepsilon_0 b}$$

This, then, is the value of the work which an external agent must do to carry a unit positive charge from B to A. The work per coulomb is the potential difference V_{AB} between A and B.

$$\therefore \; V_{AB} = \frac{Q}{4\pi\varepsilon_0}\left(\frac{1}{a} - \frac{1}{b}\right) \qquad . \qquad . \qquad . \qquad . \qquad (1)$$

V_{AB} will be in volt if Q is in coulomb, a and b are in metres and ε_0 is taken as $8\cdot85 \times 10^{-12}$ F m^{-1} or $1/4\pi\varepsilon_0$ as 9×10^9 m F^{-1} approximately (see p. 574).

Example

Two positive point charges, of 12 and 8 microcoulomb respectively, are 10 cm apart. Find the work done in bringing them 4 cm closer. (Assume $1/4\pi\varepsilon_0 = 9 \times 10^9$ m F^{-1}.)

Suppose the 12 μC charge is fixed in position. Since 6 cm $= 0\cdot06$ m and 10 cm $= 0\cdot1$ m, then the potential difference between points 6 and 10 cm from it is given by (1).

$$\therefore \; V = \frac{12 \times 10^{-6}}{4\pi\varepsilon_0}\left(\frac{1}{0\cdot06} - \frac{1}{0\cdot1}\right)$$

$$= 12 \times 10^{-6} \times 9 \times 10^9 (16\tfrac{2}{3} - 10)$$

$$= 720\,000 \text{ V.}$$

(Note the very high potential difference due to quite small charges.)

The work done in moving the 8 μC charge from 10 cm to 6 cm away from the other is given by, using $W = QV$,

$$W = 8 \times 10^{-6} \times V$$

$$= 8 \times 10^{-6} \times 720\,000 = 5\cdot8 \text{ J.}$$

Zero Potential

Instead of speaking continually of potential differences between pairs of points, we may speak of the potential at a single point—provided we always refer it to some other, agreed, reference point. This procedure is analogous to referring the heights of mountains to sea-level.

For practical purposes we generally choose as our reference point the electric potential of the surface of the earth. Although the earth is large it is all at the same potential, because it is a good conductor of electricity; if one point on it were at a higher potential than another, electrons would flow from the lower to the higher potential. As a result, the higher potential would fall, and the lower would rise; the flow of electricity would cease only when the potentials became equalised.

In general it is difficult to calculate the potential of a point relative to the earth. This is because the electric field due to a charged body near a conducting surface is complicated, as shown by the lines of force diagram in Fig. 28.25. In theoretical calculations, therefore, we often find it convenient to consider charges so far from the earth that the effect of the earth on their field is negligible; we call these 'isolated' charges.

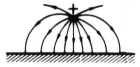

Fig. 28.25 Electric field of positive charge near earth

Thus we define the potential at a point A as V volt if V joule of work is done in bringing one coulomb of positive charge from infinity to A.

Potential Formula

Equation (1), p. 582, gives the potential difference between two points in the field of an isolated point charge Q. If we let the point B retreat to infinity, then $b \gg a$, and the equation gives for the potential at A:

$$V_A = \frac{Q}{4\pi\varepsilon_0 a} \qquad . \qquad . \qquad . \qquad . \qquad . \qquad (1)$$

The derivation of this equation shows us what we mean by the word 'infinity'; the distance b is infinite if $1/b$ is negligible compared with $1/a$. If a is 1 cm, and b is 1 m, we make an error of only 1 per cent in ignoring it; if b is 100 m, then for all practical purposes the point B is at infinity. In atomic physics, where the distances concerned have the order of 10^{-10} m, a fraction of a millimetre is considered infinity.

In the neighbourhood of an isolated negative charge, the potential is negative, because Q in equation (1) is negative. The potential is also negative in the neighbourhood of a negative charge near the earth: the earth is at zero potential, and a positive charge will tend to move from it towards the negative charge. A negative potential is analogous to the depth of a mine below sea-level. Fig. 28.26 shows the potential variation near a positive charge C before and after a conductor AB is brought near. Fig. 28.27 shows the potential variation when AB is earthed.

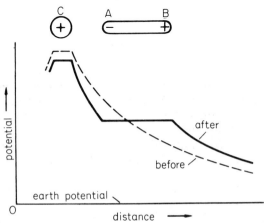

Fig. 28.26 Potential distribution near a positive charge before and after bringing up an uncharged conductor

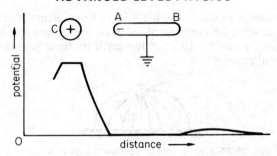

Fig. 28.27 Potential distribution near a positive charge in the presence of an earthed conductor

Potential Gradient and Intensity

We shall now see how potential difference is related to intensity or field-strength. Suppose A, B are two neighbouring points on a line of force, so close together that the electric field-intensity between them is constant and equal to E (Fig. 28.28). If V is the potential at A, $V + \Delta V$ is that at B, and the respective distances of A, B from the origin are x and $x + \Delta x$, then

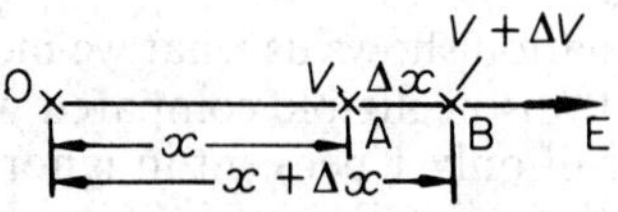

Fig. 28.28 Field strength and potential gradient

$$V_{AB} = \text{potential difference between A, B}$$
$$= V_A - V_B = V - (V + \Delta V) = -\Delta V.$$

The work done in taking a unit charge from B to A

$$= \text{force} \times \text{distance} = E \times \Delta x = V_{AB} = -\Delta V.$$

Hence
$$E = -\frac{\Delta V}{\Delta x},$$

or, in the limit,

$$E = -\frac{dV}{dx} \qquad . \qquad . \qquad . \qquad . \qquad (1)$$

The quantity dV/dx is the rate at which the potential rises with distance, and is called the potential gradient. Equation (1) shows that the strength of the electric field is equal to the negative of the potential gradient, and strong and weak fields in relation to potential are illustrated in Fig. 28.29.

In Fig. 28.30 the electric intensity $= V/h$, the potential gradient, and this is uniform in magnitude in the middle of the plates. At the edge of the plates the field becomes non-uniform.

We can now see why E is usually given in units of 'volt per metre' (V m^{-1}). From (1), $E = -(dV/dx)$. Since V is measured in volts and x in metres, then E will be in volt per metre (V m^{-1}). From the original definition of E, summarised by equation (1) on p. 575, the units of E were newton coulomb^{-1}. To show that these are equivalent, from (1),

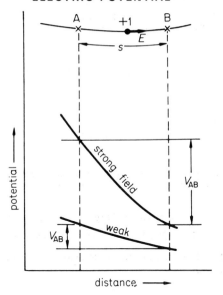

Fig. 28.29 Relationship between potential and field strength

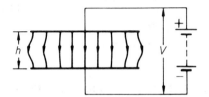

Fig. 28.30 Electric field between parallel plates

$$1 \text{ volt} = 1 \text{ joule coulomb}^{-1}$$

$$= 1 \text{ newton metre coulomb}^{-1}$$

Since $\qquad\qquad 1 \text{ joule} = 1 \text{ newton} \times 1 \text{ metre}$

$$\therefore 1 \text{ volt metre}^{-1} = 1 \text{ newton coulomb}^{-1}$$

Examples

1. An electron is liberated from the lower of two large parallel metal plates separated by a distance $h = 20$ mm. The upper plate has a potential of 2400 V relative to the lower. How long does the electron take to reach it? (Assume charge-mass ratio, e/m, for electron $= 1 \cdot 8 \times 10^{11}$ C kg^{-1}.)

Between large parallel plates, close together, the electric field is uniform except near the edges of the plates, as shown in Fig. 28.30. Except near the edges, therefore, the potential gradient between the plates is uniform; its magnitude is V/h, where $h = 0 \cdot 02$ m, so

$$\text{electric intensity } E = \text{potential gradient}$$

$$= 2400/0 \cdot 02 \text{ V m}^{-1}$$

$$= 1 \cdot 2 \times 10^5 \text{ V m}^{-1}.$$

Force on electron of charge e is given by $F = Ee$.

$$\text{acceleration, } a = \frac{F}{m} = \frac{Ee}{m}$$

$$= 1.2 \times 10^5 \times 1.8 \times 10^{11}$$

$$= 2.16 \times 10^{16} \text{ m s}^{-2}.$$

Then, from $s = \frac{1}{2}at^2$,

$$t = \sqrt{\frac{2s}{a}} = \sqrt{\frac{2 \times 20 \times 10^{-3}}{2.16 \times 10^{16}}}$$

$$= 1.4 \times 10^{-9} \text{ s.}$$

2. An electron is liberated from a hot filament, and attracted by an anode, of potential 1200 volts positive with respect to the filament. What is the speed of the electron when it strikes the anode?

$$e = \text{electronic charge} = 1.6 \times 10^{-19} \text{ C.}$$

$$V = 1200 \text{ V, } m = \text{mass of electron} = 9.1 \times 10^{-31} \text{ kg.}$$

The energy which the electron gains from the field $= QV = eV$.
    Kinetic energy gained

$$= \frac{1}{2}mv^2 = eV,$$

where v is the speed gained from rest.

$$\therefore v = \sqrt{\frac{2eV}{m}} = \sqrt{\frac{2 \times 1.6 \times 10^{-19} \times 1200}{9.1 \times 10^{-31}}}$$

$$= 2.1 \times 10^7 \text{ m s}^{-1}.$$

Equipotentials

We have already said that the earth must have the same potential all over, because it is a conductor. In a conductor there can be no differences of potential, because these would set up a potential gradient or electric field and electrons would then redistribute themselves throughout the conductor, under the influence of the field, until they had destroyed the field. This is true whether the conductor has a net charge, positive or negative, or whether it is uncharged.

Any surface or volume over which the potential is constant is called an *equipotential*. The volume or surface may be that of a material body, or simply a surface or volume in space. For example, as we see later, the space inside a hollow charged conductor is an equipotential volume. Equipotential surfaces can be drawn throughout any space in which there is an electric field, as we shall now explain.

Let us consider the field of an isolated point-charge Q. At a distance a from the charge, the potential is $Q/4\pi\varepsilon_0 a$; a sphere of radius a and centre at Q is therefore an equipotential surface, of potential $Q/4\pi\varepsilon_0 a$. In fact, all spheres centred on the charge are equipotential surfaces, whose potentials are inversely proportional to their radii, Fig. 28.31. An equipotential surface has the property that, along any direction lying in the surface, there is no electric field; for there is no potential gradient. *Equipotential surfaces are therefore always at right angles to lines of force*, as shown in Fig. 28.31. This also shows numerical values proportional to their potentials. Since conductors are always equipotentials, if any conductors appear in an electric-field diagram the lines of force must always be drawn to meet them at right angles.

Fig. 28.31 Equipotentials and lines of force around a point charge

Potential due to a System of Charges

When we consider the electric field due to more charges than one, we see the advantages of the idea of potential over the idea of field-strength. If we wish to find the field-strength E at the point P in Fig. 28.32, due to the two charges Q_1 and Q_2, we have first to find the force exerted by each on a unit charge at P, and then to compound these forces by the parallelogram method. See Fig. 28.32. On the other hand, if we wish to find the potential at P, we merely calculate the potential due to each charge, and *add the potentials algebraically.*

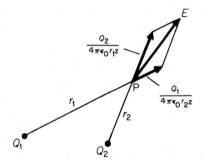

Fig. 28.32 Finding resultant field of two point charges

Quantities which can be added algebraically are called 'scalars'; they may have signs—positive or negative, like a bank balance—but they have no direction: they do not point north, east, south, or west. Quantities which have direction, like forces, are called 'vectors'; they have to be added either by resolution into components, or by the parallelogram method. Either way is slow and clumsy, compared with the addition of scalars.

When we have plotted the equipotentials, they turn out to be more useful than lines of force. A line of force diagram appeals to the imagination, and helps us to see what would happen to a charge in the field. But it tells us little about the strength of the field—at the best, if it is more carefully drawn than most, we can only say that the field is strongest where the lines are closest. But equipotentials can be labelled with the values of potential they represent; and from their spacing we can find the actual value of the potential gradient, and hence the field-strength. The only difficulty in interpreting equipotential diagrams lies in visualising the direction of the force on a charge; this is always at right angles to the equipotential curves.

Field inside Hollow Conductor. Potential Difference and Gold-leaf Electroscope

If a hollow conductor contains no charged bodies, then, whatever charge there may be on its outside, there is none on its inside. Inside it, therefore, there is no electric field; the space within the conductor is an equipotential volume. If the conductor has an open end, like a can, then most of the space inside it is equipotential, but near its mouth there is a weak field (Fig. 28.33).

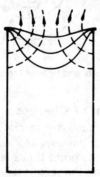

Fig. 28.33 Equipotentials and lines of force near mouth of an open charged can

The behaviour of the *gold-leaf electroscope* illustrates this point. If we stand the case on an insulator, and connect the cap to it with a wire, then, no matter what charge we give to the cap, the leaves do not diverge (Fig. 28.34). Any charge we give to the cap spreads over the case of the electroscope, but none appears on the leaves, and there is no force acting to diverge them. When, as usual, the cap is insulated and the case earthed, charging the cap sets up a potential difference between it and the case. Charges appear on the leaves, and the field between them and the case makes them diverge (p. 565). If the case is insulated from earth, as well as from the cap, the leaves diverge less; the charge on them and the cap raises the potential of the case and reduces the potential difference between it and the leaves. The field acting on the leaves is thus made weaker, and the force on the leaves less. We can sum up these observations by saying that *the electroscope indicates the potential difference between its leaves and its case.*

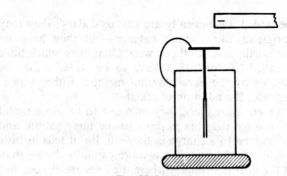

Fig. 28.34 Electroscope cap joined to case

Potential of Pear-shaped Conductor

On p. 573 we saw that the surface-density of the charge on a pear-shaped conductor was greatest where the curvature was greatest. The potential of the conductor at various points can be examined by means of the gold-leaf electroscope, the case being earthed. One end of a wire is connected to the cap; some of the wire is then wrapped round an insulating rod, and the free end of the wire is placed on the conductor. As the free end is moved over the conductor, it is observed that the divergence of the leaf remains constant. This result was explained on pp. 586, 587.

Electrostatic Shielding

The fact that there is no electric field inside a closed conductor, when it contains no charged bodies, was demonstrated by Faraday in a spectacular manner. He made for himself a large wire cage, supported it on insulators, and sat inside it with his electroscopes. He then had the cage charged by an induction machine —a forerunner of the type we described on p. 568—until painful sparks could be drawn from its outside. Inside the cage Faraday sat in safety and comfort, and there was no deflection to be seen on even his most sensitive electroscope.

If we wish to protect any persons or instruments from intense electric fields, therefore, we enclose them in hollow conductors; these are called 'Faraday cages', and are widely used in high-voltage measurements in industry.

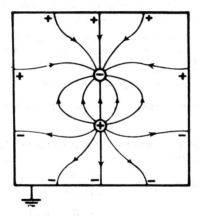

Fig. 28.35 Lines of force round charges

We may also wish to prevent charges in one place from setting up an electric field beyond their immediate neighbourhood. To do this we surround the charges with a Faraday cage, and connect the cage to earth (Fig. 28.35). The charge induced on the outside of the cage then runs to earth, and there is no external field. (When a cage is used to shield something *inside* it, it does not have to be earthed.)

Comparison of Static and Current Phenomena

Broadly speaking, we may say that in electrostatic phenomena we meet small quantities of charge, but great differences of potential. On the other hand in the phenomena of current electricity discussed later, the potential differences are small but the amounts of charge transported by the current are great. Sparks and shocks are common in electrostatics, because they require great potential differences; but they are rarely dangerous, because the total amount of energy available is usually small. On the other hand, shocks and sparks in current electricity are rare, but, when the potential difference is great enough to cause them, they are likely to be dangerous.

These quantitative differences make problems of insulation much more difficult in electrostatic apparatus than in apparatus for use with currents. The high
potentials met in electrostatics make leakage currents relatively great, and the
small charges therefore tend to disappear rapidly. Any wood, for example, ranks
as an insulator for current electricity, but a conductor in electrostatics. In
electrostatic experiments we sometimes wish to connect a charged body to earth;
all we have then to do is to touch it.

Examples

1. Explain why (a) the magnitude of the electric field intensity E varies along the surface of
a pear-shaped charged conductor, (b) the direction of E is always normal to the surface at the
point concerned.

(a) The magnitude of E at a given point on the surface is proportional to the charge
density σ round the point, since $E = \sigma/\varepsilon_0$ in air. Now σ varies with the curvature of
the surface. So E varies along the surface of the pear-shaped conductor and has the
greatest value over the pointed area.

(b) A charged conductor, irrespective of its shape, has the same potential V at all
points. So the rate of change of its potential along its surface is zero. As the intensity
is numerically equal to the potential gradient, the electric intensity *in the direction of
the surface* is zero. So the intensity E at a point in the surface has no component along
the surface. Therefore E is normal to the surface at the point.

2. Two horizontal parallel plates, 10·0 mm apart, have a p.d. of 1000 V between them the
upper plate being at a +ve potential. A negatively-charged oil drop, mass $4 \cdot 8 \times 10^{-15}$ kg, is
situated between the plates. Calculate the number of electrons on the drop if it is stationary
in the air, neglecting the density of the air.
(Assume $g = 10$ m s^{-2}, e = electron charge = $-1 \cdot 6 \times 10^{-19}$ C.)

Since the drop is stationary, upward force on drop = weight. So

$$EQ = mg$$

where E is the electric intensity between the plates, Q is the charge on the drop and m
is the mass of the drop. Now

$$E = \text{potential gradient} = \frac{1000 \text{ V}}{10 \times 10^{-3} \text{ m}} = 10^5 \text{ V m}^{-1}.$$

So

$$Q = \frac{mg}{E} = \frac{4 \cdot 8 \times 10^{-15} \times 10}{10^5}$$

$$= 4 \cdot 8 \times 10^{-19}.$$

Hence number of electrons $= \dfrac{Q}{e} = \dfrac{4 \cdot 8 \times 10^{-19}}{1 \cdot 6 \times 10^{-19}} = 3.$

3. A small positively-charged particle X, moving with a velocity of 10^7 m s^{-1}, approaches
head-on a fixed particle Y having a positive charge of 2×10^{-17} C. If the mass of X is 7×10^{-27}
kg and its charge is $3 \cdot 2 \times 10^{-19}$ C, calculate the closest distance of approach of X to Y.
(Assume $4\pi\varepsilon_0 = 1 \cdot 1 \times 10^{-10}$ F m^{-1}.)

When the particle X reaches its closest distance r to Y, all its initial kinetic energy is
changed to electrical potential energy at r in the field of Y.

Initial k.e. $= \frac{1}{2}mv^2 = \frac{1}{2} \times 7 \times 10^{-27} \times (10^7)^2 = 3 \cdot 5 \times 10^{-13}$ J.

Potential energy at r from Y = potential at r × charge on X

$$= \frac{Q}{4\pi\varepsilon_0 r} \times \text{charge on X}$$

$$= \frac{2 \times 10^{-17}}{1 \cdot 1 \times 10^{-10} r} \times 3 \cdot 2 \times 10^{-19}$$

$$= \frac{5 \cdot 8 \times 10^{-26}}{r}.$$

So $$\frac{5 \cdot 8 \times 10^{-26}}{r} = 3 \cdot 5 \times 10^{-13}.$$

Hence $$r = \frac{5 \cdot 8 \times 10^{-26}}{3 \cdot 5 \times 10^{-13}} = 1 \cdot 7 \times 10^{-13} \text{ m}.$$

Exercises 28

(*Where necessary, assume* $\varepsilon_0 = 8 \cdot 85 \times 10^{-12} \ F \ m^{-1}$)

1. What is the *potential gradient* between two parallel plane conductors when their separation is 20 mm and a p.d. of 400 V is applied to them? Calculate the force on an oil drop between the plates if the drop carries a charge of 8×10^{-19} C.

2. Using the same graphical axes in each case, draw sketches showing the variation of potential (i) inside and outside an isolated hollow spherical conductor A which has a positive charge, (ii) between A and an insulated sphere B brought near to A, (iii) between A and B if B is now earthed.

3. A charged oil drop remains stationary when situated between two parallel horizontal metal plates 25 mm apart and a p.d. of 1000 V is applied to the plates. Find the charge on the drop if it has a mass of 5×10^{-15} kg. (Assume $g = 10$ N kg^{-1}.)

Draw a sketch of the electric field between the plates and state if the field is everywhere uniform.

4. How do (*a*) the magnitude of the gravitational field, and (*b*) the magnitude of the electrostatic field, vary with distance from a point mass and a point charge respectively?

Sketch a graph illustrating the variation of electrostatic field strength E with distance r from the *centre* of a uniformly solid metal sphere of radius r_0 which is positively charged. Explain the shape of your graph (i) for $r > r_0$, and (ii) for $r < r_0$. (*L*.)

5. Define (*a*) electric intensity, (*b*) difference of potential. How are these quantities related?

A charged oil-drop of radius $1 \cdot 3 \times 10^{-6}$ m is prevented from falling under gravity by the vertical field between two horizontal plates charged to a difference of potential of 8340 V. The distance between the plates is 16 mm, and the density of oil is 920 kg m^{-3}. Calculate the magnitude of the charge on the drop ($g = 9 \cdot 81$ m s^{-2}). (*O. & C.*)

6. Show how (i) the surface density, (ii) the intensity of electric field, (iii) the potential, varies over the surface of an elongated conductor charged with electricity. Describe experiments you would perform to support your answer in cases (i) and (iii).

Describe and explain the action of points on a charged conductor; and give two practical applications of the effect. (*L*.)

7. Describe, with the aid of a labelled diagram, a Van de Graaff generator, explaining the physical principles of its action.

The high voltage terminal of such a generator consists of a spherical conducting shell of radius 0·50 m. Estimate the maximum potential to which it can be raised in air for which electrical breakdown occurs when the electric intensity exceeds 3×10^6 V m^{-1}.

State two ways in which this maximum potential could be increased. (*N*.)

8. Define *potential at a point* in an electric field.

Sketch a graph illustrating the variation of potential along a radius from the centre of a charged isolated conducting sphere to infinity.

Assuming the expression for the potential of a charged isolated conducting sphere in air, determine the change in the potential of such a sphere caused by surrounding it with an earthed concentric thin conducting sphere having three times its radius. (*N*.)

9. Two plane parallel conducting plates 15·0 mm apart are held horizontal, one above the other, in air. The upper plate is maintained at a positive potential of 1500 V while the lower plate is earthed. Calculate the number of electrons which must be attached to a small oil drop of mass $4 \cdot 90 \times 10^{-15}$ kg, if it remains stationary in the air between the plates. (Assume that the density of air is negligible in comparison with that of oil.)

If the potential of the upper plate is suddenly changed to -1500 V what is the initial acceleration of the charged drop? Indicate, giving reasons, how the acceleration will change. (*L.*)

10. Describe carefully Faraday's ice-pail experiments and discuss the deductions to be drawn from them. How would you investigate experimentally the charge distribution over the surface of a conductor? (*C.*)

11. What is an *electric field*? With reference to such a field define *electric potential*.

Two plane parallel conducting plates are held horizontal, one above the other, in a vacuum. Electrons having a speed of $6 \cdot 0 \times 10^6$ m s^{-1} and moving normally to the plates enter the region between them through a hole in the lower plate which is earthed. What potential must be applied to the other plate so that the electrons just fail to reach it? What is the subsequent motion of these electrons? Assume that the electrons do not interact with one another.

(Ratio of charge to mass of electron is $1 \cdot 8 \times 10^{11}$ C kg^{-1}.) (*N.*)

12. An isolated conducting spherical shell of radius $0 \cdot 10$ m, in vacuo, carries a positive charge of $1 \cdot 0 \times 10^{-7}$ C. Calculate (*a*) the electric field intensity, (*b*) the potential, at a point on the surface of the conductor. Sketch a graph to show how one of these quantities varies with distance along a radius from the centre to a point well outside the spherical shell. Point out the main features of the graph. (*N.*)

13. Electric field strength at a point may be defined as 'force per unit charge' on a charge placed at the point. Show how this leads to a general expression relating E_x, the x-component of the field strength, to the potential gradient in the x-direction. Explain the significance of the sign in your expression.

A potential difference V is maintained between two large horizontal metal plates separated by a small distance d. Between them is a small oil drop of mass m carrying a charge q. Neglecting the upthrust of the air, derive the condition for the drop to remain at rest. Explain why it is reasonable to neglect the upthrust.

The value of q found in one such experiment was $1 \cdot 15 \times 10^{-18}$ C. The mass m was determined with an error of $+5 \%$, and the distance d with an error of -2%. The other quantities involved were measured with negligible error. Find the consequent absolute error in the calculated value of q.

When values of q are determined from many such experiments, they show a certain irregularity. Describe this regularity and explain what may be deduced from it. (*C.*)

14. Define the *electric potential V* and the *electric field strength E* at a point in an electrostatic field. How are they related? Write down an expression for the electric field strength at a point close to a charged conducting surface, in terms of the surface density of charge.

Corona discharge into the air from a charged conductor takes place when the potential gradient at its surface exceeds 3×10^6 V m^{-1}; a potential gradient of this magnitude also breaks down the insulation afforded by a solid dielectric. Calculate the greatest charge that can be placed on a conducting sphere of radius 20 cm supported in the atmosphere on a long insulating pillar; also calculate the corresponding potential of the sphere. Discuss whether this potential could be achieved if the pillar of insulating dielectric was only 50 cm long. (Take ε_0 to be $8 \cdot 85 \times 10^{-12}$ F m^{-1}.) (*O.*)

29 Capacitors

A capacitor is a device for storing charge. The earliest capacitor was invented—almost accidentally—by van Musschenbroek of Leyden, in about 1746, and became known as a Leyden jar. One form of it is shown in Fig. 29.1 (i); J is a

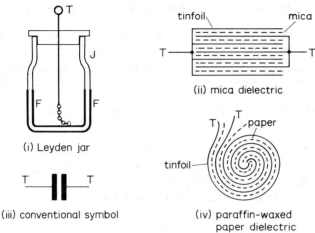

(i) Leyden jar

(ii) mica dielectric

(iii) conventional symbol

(iv) paraffin-waxed paper dielectric

Fig. 29.1 Types of capacitor

glass jar, FF are tin-foil coatings over the lower parts of its walls, and T is a knob connected to the inner coating. Modern forms of capacitor are shown at (ii) and (iv) in the figure. Essentially, all capacitors consist of two metal plates separated by an insulator. The insulator is called the dielectric; in some capacitors it is polystyrene, oil or air. Fig. 29.1 (iii) shows the conventional symbol for such a capacitor; T, T are terminals joined to the plates.

Charging and Discharging Capacitor

Figure 29.2 (i) shows a circuit which may be used to study the action of a capacitor. C is a large capacitor such as 500 microfarad (see later), R is a large resistor such as 100 kilohms (10^5 Ω), A is a current meter reading 100–0–100 microamperes (100 μA), K is a two-way key, and D is a 6V d.c. supply.

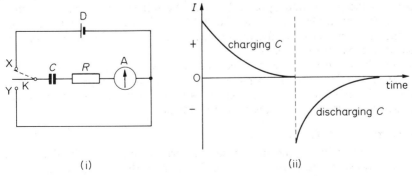

(i)

(ii)

Fig. 29.2 Charging and discharging capacitor

When the battery is connected to C by contact at X, the current I in the meter A is observed to be initially about 60 μA and then, as shown in Fig. 29.2 (ii), it slowly decreases to zero. Thus current flows to C for a short time when the battery is connected to it, even though the capacitor plates are separated by an insulator.

We can disconnect the battery from C by opening X. If contact with Y is now made, so that in effect the plates of C are joined together through R and A, the current in the meter is observed to be about 60 μA initially in the opposite direction to before and then slowly decreases to zero. Figure 29.2 (ii). This flow of current shows that C *stored charge when it was connected to the battery* originally.

Generally, a capacitor is *charged* when a battery or p.d. is connected to it. When the plates of the capacitor are joined together, the capacitor becomes *discharged*. Large values of C and R in the circuit of Fig. 29.2 (i) help to slow the current flow so that we can see the charging and discharging which occurs, as explained more fully on p. 611.

We can also show that a charged capacitor has stored electricity by connecting its terminals by a piece of wire. A spark passes just as the wire makes contact.

Charging and Discharging Processes

When we connect a capacitor to a battery, electrons flow from the negative terminal of the battery on to the plate A of the capacitor connected to it (Fig. 29.3); and, at the same rate, electrons flow from the other plate B of the capacitor

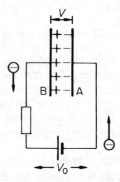

Fig. 29.3 A capacitor charging (resistance is shown because some is always present, even if only that of the connecting wires)

towards the positive terminal of the battery. Positive and negative charges thus appear on the plates, and oppose the flow of electrons which causes them. As the charges accumulate, the potential difference between the plates increases, and the charging current falls to zero when the potential difference becomes equal to the battery voltage V_0.

When the battery is disconnected and the plates are joined together by a wire, electrons flow back from plate A to plate B until the positive charge on B is completely neutralised. A current thus flows for a time in the wire, and at the end of the time the charges on the plates become zero.

Capacitors in A.C. Circuits

Capacitors are widely used in alternating current and radio circuits, because they can transmit alternating currents. To see how they do so, let us consider the circuit of Fig. 29.4, in which the capacitor may be connected across either

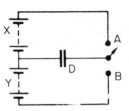

Fig. 29.4 Reversals of voltage applied to capacitor

of the batteries X, Y. When the key is closed at A, current flows from the battery X, and charges the plate D of the capacitor positively. If the key is now closed at B instead, current flows from the battery Y; the plate D loses its positive charge and becomes negatively charged. Thus if the key is rocked rapidly between A and B, current surges backwards and forwards along the wires connected to the capacitor. An alternating voltage, as we shall see in Chapter 37, is one which reverses many times a second; when such a voltage is applied to a capacitor, therefore, an alternating current flows in the connecting wires.

Variation of Capacitor Charge with P.D.

Figure 29.5 shows a circuit which may be used to investigate how the charge Q stored on a capacitor C varies with the p.d. V applied. The d.c. supply D can be altered in steps from a value such as 10 V to 25 V, C is a capacitor

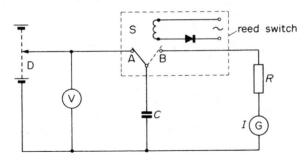

Fig. 29.5 Variation of Q with V—vibrating reed switch

consisting of two large square metal plates separated by small pieces of polythene at the corners, G is a sensitive current meter such as a light beam galvanometer, and R is a protective high resistor in series with G.

The capacitor can be charged and discharged rapidly by means of a *vibrating reed switch* S. The vibrator charges C by contact with A and discharges C through G by contact with B. When the vibrator frequency f is made suitably high, such as several hundred hertz, a steady current I flows in G. Its magnitude is given by

$$I = \text{charge per second} = fQ,$$

where Q is the charge on the capacitor each time it is charged, since f is the number of times per second it is charged. Thus, for a given value of f, the charge Q is proportional to the current I in G.

When V is varied and values of I are observed, results show that $I \propto V$. Thus experiment shows that, for a given capacitor, $Q \propto V$.

Capacitance Definition and Units

Since $Q \propto V$, then Q/V is a constant for the capacitor. The ratio of the charge on either plate to the potential difference between the plates is called the *capacitance*, C, of the capacitor:

$$C = \frac{Q}{V}. \quad \quad \quad \quad (1)$$

Thus
$$Q = CV, \quad \quad \quad \quad (2)$$

and
$$V = \frac{Q}{C}. \quad \quad \quad \quad (3)$$

When Q is in coulomb (C) and V in volt (V), then capacitance C is in farad (F). One farad (1F) is the capacitance of an extremely large capacitor. In practical circuits, such as in radio receivers, the capacitance of capacitors used are therefore expressed in *microfarad* (μF). One microfarad is one millionth part of a farad, that is $1 \ \mu F = 10^{-6}$ F. It is also quite usual to express small capacitors, such as those used on record players, in *picofarad* (pF). A picofarad is one millionth part of a microfarad, that is $1 \ pF = 10^{-6} \ \mu F = 10^{-12}$ F.

Comparison of Capacitances

The vibrating reed circuit shown in Fig. 29.5 can be used to compare large capacitances (of the order of microfarads) or to compare small capacitances. With large capacitances, a meter with a suitable range of current of the order of milliamperes may be required. With smaller capacitances, a meter such as the light beam galvanometer in Fig. 33.14 (i) may be more suitable, as the current flowing is then much smaller. In both cases suitable values for the applied p.d. V and the frequency f must be chosen.

Suppose two large, or two small, capacitances, C_1 and C_2, are to be compared. Using C_1 first in the vibrating reed circuit, the current flowing is I_1 say. When C_1 is replaced by C_2, suppose the new current is I_2.

Now $Q_1 = C_1 V$ and $Q_2 = C_2 V$; hence $Q_1/Q_2 = C_1/C_2$. But from p. 595, $Q \propto I$. Thus $Q_1/Q_2 = I_1/I_2$.

$$\therefore \frac{C_1}{C_2} = \frac{I_1}{I_2}$$

Thus the ratio C_1/C_2 can be found from the current readings I_1 and I_2.

Ballistic galvanometer method

Large capacitances, of the order of microfarads, can also be compared with the aid of a *ballistic galvanometer*. In this instrument, as explained on p. 749, the first 'throw' or deflection is proportional to the quantity of charge (Q) passing through it.

The circuit required is shown in Fig. 29.6. The capacitor of capacitance C_1 is charged by a battery of e.m.f. V, and then discharged through the ballistic galvanometer G. The corresponding first deflection θ_1 is observed. The capacitor is now replaced by another of capacitance C_2, charged again by the battery, and the new deflection θ_2 is observed when the capacitor is discharged. Now

$$\frac{Q_1}{Q_2} = \frac{\theta_1}{\theta_2}.$$

$$\therefore \frac{C_1 V}{C_2 V} = \frac{C_1}{C_2} = \frac{\theta_1}{\theta_2}.$$

If C_2 is a standard capacitor, whose value is known, then the capacitance of C_1 can be found.

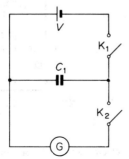

Fig. 29.6 Comparison of capacitances—ballistic galvanometer

Electrometer and D. C. Amplifier

An *electrometer* or *d.c. amplifier* is an electronic device which acts as a voltmeter of extremely high input resistance. The instrument is thus a 'perfect' voltmeter (see p. 630). It contains a special form of valve or transistor and amplifying circuit, with a zero adjustment, but we shall not be concerned with the circuit action. Used as an electrometer, the instrument measures charge. Used as a d.c. amplifier, the instrument measures very small currents.

Figure 29.7 shows in a block diagram the electrometer/d.c. amplifier. The

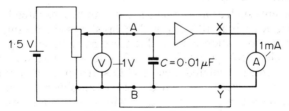

Fig. 29.7 Calibration of electrometer

input terminals are A, B and the output terminals are X, Y; the triangle is the symbol for an amplifier. As shown, with an input capacitor of 0·01 μF, it is arranged to read charge Q. The instrument first requires calibration. The potential divider across the 1·5 V battery is varied to give a reading of 1 V on the voltmeter, so that the input is 1 V. Most school laboratory electrometers can then be adjusted to give a full scale deflection of 1 mA on a 0–1 mA meter joined to X, Y, so that the output is 1 mA for 1 V input.

To measure small currents I, the capacitor is replaced by a high input resistor, such as 100 megohms (10^8 Ω) to 100 000 megohms (10^{11} Ω). The calibration is again arranged so that an input of 1 V produces an output of 1 mA on the milliammeter. In using the instrument as an electrometer or d.c. amplifier, the zero adjuster is needed to ensure zero output current for zero input p.d.

Measurement of Charge

To measure charge, an input capacitor C_i, such as 0·01 μF or 0·1 μF, is used, Fig. 29.8 (i). The charge Q may be on a capacitor C as shown, or on a charged

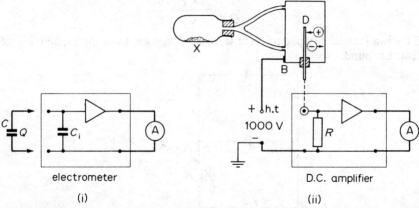

Fig. 29.8 Measurement of (i) charge (electrometer), (ii) ionisation current (d.c. amplifier)

metal sphere such as the terminal of a model Van de Graaff generator. The charge is transferred to the input and the steady reading on the milliammeter is then taken. Suppose this is 0·30 mA. Then the p.d. across the input is 0·30 V. If C_i is 0·1 μF, for example, the charge $Q = CV = 0\cdot1 \times 10^{-6} \times 0\cdot3 = 3 \times 10^{-8}$ C.

It should be noted that C_i must be *much* greater than C if practically the whole of the charge on C is required to be transferred to C_i by contact. With $C_i = 0\cdot01$ μF and $C = 0\cdot001$ μF, for example, so that C_i is 10 times as large as C, calculation shows that about 91% of the charge on C is transferred to C_i by contact. The electrometer thus reads 9% less than the charge on C.

Measurement of Current

Very small currents such as *ionisation currents* can be measured by using a high input resistance R such as 10^9 Ω. Fig. 29.8 (ii) shows the arrangement needed for measuring the ionisation current due to some radioactive gas Radon-220, which was expelled from a plastic bottle X into an ionisation chamber. This chamber has a central insulated metal rod D surrounded by a metal casing B, and is joined to one input terminal of the d.c. amplifier. In Fig. 29.8 (ii), the separated electrodes D and B are in series with R across a high-tension supply such as 1000 V.

The radioactive gas ionises the air in the ionisation chamber. The $+$ve ions drift towards D and the $-$ve ions drift towards B. The current I flowing through R produces a p.d. V across R. If the milliammeter reading initially is 0·40 mA, then $V = 0\cdot40$ V (see p. 597). Thus the ionisation current is given by

$$I = \frac{V}{R} = \frac{0\cdot4}{10^9} = 4 \times 10^{-10} \text{ A}$$

Factors determining Capacitance

We can now find out by experiment what factors influence capacitance.

Distance between plates. Figure 29.9 (i) shows two parallel metal plates X and

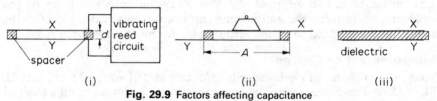

Fig. 29.9 Factors affecting capacitance

Y separated by a distance d equal to the thickness of the polythene spacers shown in Fig. 29.5. The observed current I in G is proportional to Q, the charge to $2d$, $3d$ and $4d$.

A measure of the capacitance C can be found using the vibrating reed circuit shown in Fig. 29.5. The observed current I in G is proportional to Q the charge on the capacitor, and Q is proportional to C for a given applied p.d. V. So $C \propto I$. Experiment shows that, allowing for error, $C \propto 1/d$ where d is the *separation* between the plates.

Area between plates. By placing a weight on the top plate X and moving sideways, Fig. 29.9 (ii), the area A of overlap, or common area between the plates, can be varied while d is kept constant. Alternatively, pairs of plates of different area can be used which have the same separation d. By using the vibrating reed circuit, experiment shows that $C \propto A$.

Dielectric. Let us now replace the air between the plates by completely filling the space with a 'dielectric' such as polystyrene or polythene sheets or glass, Fig. 29.9 (iii). In this case the area A and distance d remain constant. The vibrating reed experiment then shows that the capacitance has increased appreciably when the dielectric is used.

Some Practical Capacitors

As we have just seen, the simplest capacitor consists of two flat parallel plates with an insulating medium between them. Practical capacitors have a variety of forms but basically they are all forms of parallel-plate capacitors.

A capacitor in which the effective area of the plates can be adjusted is called a *variable capacitor*. In the type shown in Fig. 29.10, the plates are semicircular

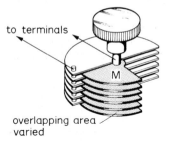

Fig. 29.10 Variable air capacitor

although other shapes may be used, and one set can be swung into or out of the other. The capacitance is proportional to the area of overlap of the plates. The plates are made of brass or aluminium, and the dielectric may be air or oil or mica. The *variable air capacitor* is used in radio receivers for tuning to the different wavelengths of commercial broadcasting stations.

Figure 29.1 (ii), p. 593, shows a *multiple capacitor* with a mica dielectric. The capacitance is n times the capacitance between two successive plates where n is the number of dielectrics between all the plates. The whole arrangement is sealed into a plastic case.

Figure 29.1 (iv), p. 593, shows a *paper* capacitor—it has a dielectric of paper impregnated with paraffin wax or oil. Unlike the mica capacitor, the papers can be rolled and sealed into a cylinder of relatively small volume. To increase the stability and reduce the power losses, the paper is now replaced by a thin layer of *polystyrene*.

Electrolytic capacitors are widely used. Basically, they are made by passing a direct current between two sheets of aluminium foil, with a suitable electrolyte or liquid conductor between them, Fig. 29.11. A very thin film of aluminium oxide is then formed on the anode plate, which is on the *positive* side of the d.c. supply (see p. 682). This film is an insulator. It forms the dielectric between the two plates, the electrolyte being a good conductor, Fig. 29.11 (i). Since the

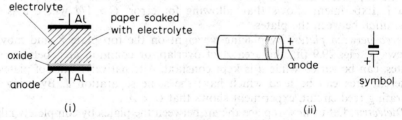

Fig. **29.11** Electrolytic capacitor

dielectric thickness d is so very small, and $C \propto 1/d$, the capacitance value can be very high. Several thousand microfarads may easily be obtained in a capacitor of small volume. To maintain the oxide film, the anode terminal is marked in red or by a + sign, Fig. 29.11 (ii). This terminal must be connected to the positive side of the circuit in which the capacitor is used, otherwise the oxide film will break down. It is represented by the unblacked rectangle in the symbol for the electrolytic capacitor shown in Fig. 29.11 (iii).

Parallel Plate Capacitor

We now proceed to derive a formula for the capacitance of a parallel-plate capacitor.

Suppose two parallel plates of a capacitor each have a charge numerically equal to Q, Fig. 29.12. The surface density σ is then Q/A where A is the area of either plate, and the intensity between the plates, E, is given, from p. 579, by

$$E = \frac{\sigma}{\varepsilon} = \frac{Q}{\varepsilon A}$$

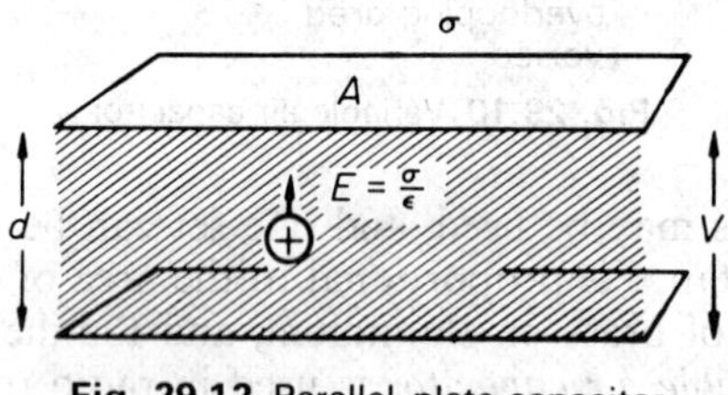

Fig. **29.12** Parallel-plate capacitor

Now E is numerically equal to the potential gradient V/d, p. 584.

$$\therefore \frac{V}{d} = \frac{Q}{\varepsilon A}$$

$$\therefore \frac{Q}{V} = \frac{\varepsilon A}{d}$$

$$\therefore C = \frac{\varepsilon A}{d} \qquad . \qquad . \qquad . \qquad . \qquad (1)$$

It should be noted that this formula for C is approximate, as the field becomes non-uniform at the edges. See Fig. 28.30, p. 585.

Thus a capacitor with parallel plates, having a vacuum (or air, if we assume the permittivity of air is the same as a vacuum) between them, has a capacitance given by

$$C = \frac{\varepsilon_0 A}{d}$$

where C = capacitance in farad (F), A = area of overlap of plates in metre^2, d = distance between plates in metre and $\varepsilon_0 = 8\cdot854 \times 10^{-12}$ farad metre^{-1}.

Capacitance of Isolated Sphere

Suppose a sphere of radius r metre situated in air is given a charge of Q coulomb. We assume, as on p. 577, that the charge on a sphere gives rise to potentials *on and outside* the sphere as if all the charge were concentrated at the *centre*. From p. 586, the surface of the sphere thus has a potential relative to that 'at infinity' (or, in practice, to that of the earth) given by:

$$V = \frac{Q}{4\pi\varepsilon_0 r}$$

$$\therefore \frac{Q}{V} = 4\pi\varepsilon_0 r$$

$$\therefore \text{ Capacitance, } C = 4\pi\varepsilon_0 r. \qquad . \qquad . \qquad . \qquad . \qquad (2)$$

The other 'plate' of the capacitor is the earth.

Suppose $r = 10$ cm $= 0\cdot1$ m. Then

$$C = 4\pi\varepsilon_0 r = 4\pi \times 8\cdot85 \times 10^{-12} \times 0\cdot1 \text{ F}$$

$$= 11 \times 10^{-12} \text{ F (approx.)} = 11 \text{ pF.}$$

Concentric Spheres

Faraday used two concentric spheres to investigate the dielectric constant (p. 602) of liquids. Suppose a, b are the respective radii of the inner and outer spheres, Fig. 29.13. Let $+Q$ be the charge given to the inner sphere and let the outer sphere be earthed, with air between them.

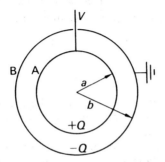

Fig. 29.13 Concentric spherical capacitor

The induced charge on the outer sphere is $-Q$ (see p. 571). The potential V_a of the inner sphere = potential due to $+Q$ plus potential due to $-Q =$

$+\dfrac{Q}{4\pi\varepsilon_0 a} - \dfrac{Q}{4\pi\varepsilon_0 b}$, since the potential due to the charge $-Q$ is $-Q/4\pi\varepsilon_0 b$ everywhere inside the larger sphere (see p. 586).

But $V_1 = 0$, as the outer sphere is earthed.

$\therefore$ potential difference, $V, = V_1 - V_1 = \dfrac{1}{4\pi\varepsilon_0}\left(\dfrac{Q}{a} - \dfrac{Q}{b}\right)$

$$\therefore V = \dfrac{Q}{4\pi\varepsilon_0}\left(\dfrac{b-a}{ab}\right).$$

$$\therefore \dfrac{Q}{V} = \dfrac{4\pi\varepsilon_0 ab}{b-a},$$

$$C = \dfrac{4\pi\varepsilon_0 ab}{b-a} \qquad \qquad . \qquad . \qquad . \qquad . \qquad (3)$$

As an example, suppose $b = 10$ cm $= 0.1$ m and $a = 9$ cm $= 0.09$ m.

$$\therefore C = \dfrac{4\pi\varepsilon_0 ab}{b-a}$$

$$= \dfrac{4\pi \times 8.85 \times 10^{-12} \times 0.1 \times 0.09}{(0.1 - 0.09)} \text{ F}$$

$$= 100 \text{ pF (approx.)}.$$

Note that the inclusion of a nearby second plate to the capacitor increases the capacitance. For an *isolated* sphere of radius 10 cm, the capacitance was 11 pF (p. 601).

Relative Permittivity (Dielectric Constant) and Dielectric Strength

The ratio of the capacitance with and without the dielectric between the plates is called the *relative permittivity* (or *dielectric constant*) of the material used. The expression 'without a dielectric' strictly means 'with the plates in a vacuum'; but the effect of air on the capacitance of a capacitor is so small that for most purposes it may be neglected. The relative permittivity of a substance is denoted by the letter ε_r; thus

$$\varepsilon_r = \dfrac{\text{capacitance of given capacitor, with space between plates filled with dielectric}}{\text{capacitance of same capacitor with plates } in \; vacuo}$$

If we take the case of a parallel plate capacitor as an example, then

$$\varepsilon_r = \dfrac{\varepsilon A/d}{\varepsilon_0 A/d} = \dfrac{\varepsilon}{\varepsilon_0}$$

Thus the relative permittivity is the ratio of the permittivity of the substance to that of free space. Note that relative permittivity is a pure number and has no dimensions, unlike ε and ε_0.

The following table gives the value of relative permittivity, and also of *dielectric strength*, for various substances. The strength of a dielectric is the potential gradient at which its insulation breaks down, and a spark passes through it. A solid dielectric is ruined by such a breakdown, but a liquid or gaseous one heals up as soon as the applied potential difference is reduced.

Water is not suitable as a dielectric in practice, because it is a good insulator only when it is very pure, and to remove all matter dissolved in it is almost impossible.

PROPERTIES OF DIELECTRICS

Substance	Relative permittivity	Dielectric strength, kilovolts per mm
Glass	5–10	30–150
Mica	6	80–200
Ebonite	2·8	30–110
Ice*	94	—
Paraffin wax	2	15–50
Paraffined paper . . .	2	40–60
Methyl alcohol* . . .	32	—
Water*	81	—
Air (*normal pressure*) . .	1·0005	—

* Polar molecules (see p. 604).

Action of Dielectric

The explanation of dielectric action which we shall now give is similar in principle to Faraday's, but expressed in modern terms—there was no knowledge of electrons in his day.

We regard a molecule as a collection of atomic nuclei, positively charged, and surrounded by a cloud of negative electrons. When a dielectric is in a charged capacitor, its molecules are in an electric field; the nuclei are urged in the direction of the field, and the electrons in the opposite direction, Fig. 29.14 (i). Thus each

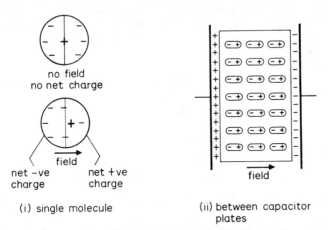

no field
no net charge

net −ve charge   net +ve charge

field

(i) single molecule

(ii) between capacitor plates

field

Fig. 29.14 Polarization of dielectric

molecule is distorted, or polarized: one end has an excess of positive charge, the other an excess of negative. At the surfaces of the dielectric, therefore, charges appear, as shown in Fig. 29.14 (ii). These charges are of opposite sign to the charges on the plates, and so reduce the potential difference between the plates.

If the capacitor is connected to a battery, then its potential difference is constant; but the surface charges on the dielectric still increase its capacitance. They do so because they offset the charges on the plates, and so enable greater charges to accumulate there before the potential difference rises to the battery voltage.

Some molecules, we believe, are permanently polarized: they are called *polar molecules*. For example, the water molecule consists of an oxygen atom, O, with two hydrogen atoms H, making roughly a right-angled structure, Fig. 29.15 (i).

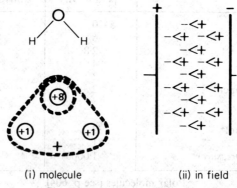

(i) molecule   (ii) in field

Fig. 29.15 Water as a dielectric

Oxygen has a nuclear charge of $+8e$, where e is the electronic charge, and has eight electrons. Hydrogen has a nuclear charge of $+e$, and one electron. In the water molecule, the two electrons from the hydrogen atom move in paths which surround the oxygen nucleus. Thus they are partly added to the oxygen atom, and partly withdrawn from the hydrogen atoms. On the average, therefore, the apex of the triangle is negatively charged, and its base is positively charged. In an electric field, water molecules tend to turn as shown in Fig. 29.15 (ii). The effect of this, in a capacitor, is to increase the capacitance in the way already described. The increase is, in fact, much greater than that obtained with a dielectric which is polarized merely by the action of the field.

ε_0 and its Measurement

We can now see how the unit of ε_0 may be stated in a more convenient manner and how its magnitude may be measured.

Unit. From $C = \dfrac{\varepsilon_0 A}{d}$, we have $\varepsilon_0 = \dfrac{Cd}{A}$.

Thus the unit of $\varepsilon_0 = \dfrac{\text{farad} \times \text{metre}}{\text{metre}^2}$

$= \text{farad metre}^{-1}$ (see also p. 574).

Measurement. In order to find the magnitude of ε_0, the circuit in Fig. 29.16 is used.

C is a parallel plate capacitor, which may be made of sheets of glass or perspex coated with aluminium foil. The two conducting surfaces are placed facing inwards, so that only air is present between these plates. The area A of the plates in metre^2, and the separation d in metres, are measured. P is a high tension supply capable of delivering about 200 V, and G is a calibrated sensitive galvanometer, such as a 'Scalamp' type. S is a *vibrating switch* unit, energised by a low a.c. voltage from the mains. When operating, the vibrating bar X touches D and then B, and the motion is repeated at the mains frequency, fifty times a second. When the switch is in contact with D, the capacitor is charged from the supply P to a potential difference of V volt, measured on the voltmeter. When the contact

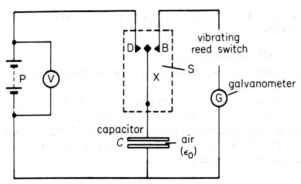

Fig. 29.16 Measurement of ε_0

moves over to B, the capacitor discharges through the galvanometer. The galvanometer thus receives fifty pulses of charge per second. This gives an average steady reading on the galvanometer, corresponding to a mean current I.

Now from previous, $C = \varepsilon_0 A/d$. Thus on charging, the charge stored, Q, is given by

$$Q = CV = \frac{\varepsilon_0 VA}{d}.$$

The capacitor is discharged fifty times per second. Since the current is the charge flowing per second.

$$\therefore I = \frac{\varepsilon_0 VA.50}{d} \text{ ampere}$$

$$\therefore \varepsilon_0 = \frac{Id}{50VA} \text{ farad metre}^{-1}.$$

The following results were obtained in one experiment:

$$A = 0.0317 \text{ m}^2, \ d = 1.0 \text{ cm} = 0.010 \text{ m}, \ V = 150 \text{ V}, \ I = 0.21 \times 10^{-6} \text{ A}$$

$$\therefore \varepsilon_0 = \frac{Id}{50VA}$$

$$= \frac{0.21 \times 10^{-6} \times 0.01}{50 \times 150 \times 0.0317}$$

$$= 8.8 \times 10^{-12} \text{ farad metre}^{-1}.$$

As very small currents are concerned, care must be taken to make the apparatus of high quality insulating material, otherwise leakage currents will lead to serious error.

This method can also be used to find the permittivity of various materials. Thus if the experiment is repeated with a material of permittivity ε completely filling the space between the plates, then $\varepsilon = I'd/50VA$, where I' is the new current.

If only the relative permittivity, ε_r, is required, there is no need to know the p.d. supplied or the dimensions of the capacitor. In this case,

$$\varepsilon_r = \frac{\varepsilon}{\varepsilon_0} = \frac{I'd/50VA}{Id/50VA} = \frac{I'}{I}$$

Thus ε_r is the ratio of the respective currents in G with and without the dielectric between the plates.

Measurement of Capacitance

If a standard capacitor is available, an unknown capacitance can be measured as described on p. 596.

If no standard capacitor is available, the method of the last section can be employed. The unknown capacitor replaces C in the circuit of Fig. 29.16. The current I in the galvanometer, and the p.d. supplied, V, are then measured. Now the charge $Q = CV$. This is discharged fifty times per second. Since the current is the charge flowing per second,

$$\therefore I = 50CV$$

$$\therefore C = \frac{I}{50V}.$$

With I in ampere and V in volt, then C is in farad.

Arrangements of Capacitors

In radio circuits, capacitors often appear in arrangements whose resultant capacitances must be known. To derive expressions for these, we need the equation defining capacitance in its three possible forms:

$$C = \frac{Q}{V}, \quad V = \frac{Q}{C}, \quad Q = CV.$$

In Parallel. Figure 29.17 shows three capacitors, having all their left-hand

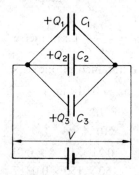

Fig. 29.17 Capacitors in parallel

plates connected together, and all their right-hand plates likewise. They are said to be connected in parallel. If a cell is not connected across them, they all have the same potential difference V. (For, if they had not, current would flow from one to another until they had.) The charges on the individual capacitors are respectively

$$\left.\begin{array}{l} Q_1 = C_1 V \\ Q_2 = C_2 V \\ Q_3 = C_3 V \end{array}\right\} \qquad . \qquad . \qquad . \qquad . \qquad . \qquad (1)$$

The total charge on the system of capacitors is

$$Q = Q_1 + Q_2 + Q_3 = (C_1 + C_2 + C_3)V.$$

And the system is therefore equivalent to a single capacitor, of capacitance

$$C = \frac{Q}{V} = C_1 + C_2 + C_3.$$

Thus when capacitors are connected in parallel, their resultant capacitance is the sum of their individual capacitances. It is greater than the greatest individual one.

In Series. Figure 29.18 shows three capacitors having the right-hand plate

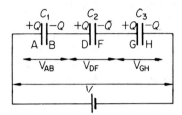

Fig. 29.18 Capacitors in series

of one connected to the left-hand plate of the next, and so on—connected in series. When a cell is connected across the ends of the system, a charge Q is transferred from the plate H to the plate A, a charge $-Q$ being left on H. This charge induces a charge $+Q$ on plate G; similarly, charges appear on all the other capacitor plates, as shown in the figure. (The induced and inducing charges are equal because the capacitor plates are very large and very close together; in effect, either may be said to enclose the other.) The potential differences across the individual capacitors are, therefore, given by

$$\left. \begin{aligned} V_{AB} &= \frac{Q}{C_1} \\[2mm] V_{DF} &= \frac{Q}{C_2} \\[2mm] V_{GH} &= \frac{Q}{C_3} \end{aligned} \right\} \qquad . \quad . \quad . \quad . \quad (2)$$

The sum of these is equal to the applied potential difference V because the work done in taking a unit charge from H to A is the sum of the work done in taking it from H to G, from F to D, and from B to A. Therefore

$$V = V_{AB} + V_{DF} + V_{GH}$$

$$= Q\left(\frac{1}{C_1} + \frac{1}{C_2} + \frac{1}{C_3}\right) \qquad . \quad . \quad . \quad (3)$$

The resultant capacitance of the system is the ratio of the charge stored to the applied potential difference, V. The charge stored is equal to Q, because, if the battery is removed, and the plates HA joined by a wire, a charge Q will pass through that wire, and the whole system will be discharged. The resultant capacitance is therefore given by

$$C = \frac{Q}{V}, \quad \text{or} \quad \frac{1}{C} = \frac{V}{Q},$$

so, by equation (3),

$$\frac{1}{C} = \frac{1}{C_1} + \frac{1}{C_2} + \frac{1}{C_3} \qquad . \qquad . \qquad . \qquad . \qquad (4)$$

Thus, to find the resultant capacitance of capacitors in series, we must add the reciprocals of their individual capacitances. The resultant is less than the smallest individual.

Comparison of Series and Parallel Arrangements. Let us compare Figs. 29.17 and 29.18. In Fig. 29.18, where the capacitors are in series, all the capacitors carry the same charge, which is equal to the charge carried by the system as a whole, Q. The potential difference applied to the system, however, is divided amongst the capacitors, in inverse proportion to their capacitances (equations (2)). In Fig. 29.17, where the capacitors are in parallel, they all have the same potential difference; but the charge stored is divided amongst them, in direct proportion to the capacitances (equations (1)).

Example

Find the charges on the capacitors in Fig. 29.19, and the potential differences across them.

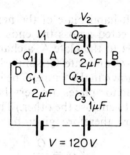

Fig. 29.19 Example

Capacitance between A and B,

$$C' = C_2 + C_3 = 3\ \mu F.$$

Overall capacitance B to D, since C_1 and C' are in series, is

$$C = \frac{C_1 C'}{C_1 + C'} = \frac{2 \times 3}{2 + 3} = 1\cdot 2\ \mu F$$

Charge stored in this capacitance C

$$= Q_1 = Q_2 + Q_3 = CV = 1\cdot 2 \times 10^{-6} \times 120$$
$$= 144 \times 10^{-6}\ C,$$

$$\therefore V_1 = \frac{Q_1}{C_1} = \frac{144 \times 10^{-6}}{2 \times 10^{-6}} = 72\ V.$$

So
$$V_2 = V - V_1 = 120 - 72 = 48\ V,$$
$$Q_2 = C_2 V_2 = 2 \times 10^{-6} \times 48 = 96 \times 10^{-6}\ C,$$
$$Q_3 = C_3 V_2 = 10^{-6} \times 48 = 48 \times 10^{-6}\ C.$$

Energy of a Charged Capacitor

A charged capacitor is a store of electrical energy, as we may see from the vigorous spark it can give on discharge. This can also be shown by charging a large electrolytic capacitor C, such as $10\,000\ \mu F$, to a p.d. of 6 V, and then discharging it through a small or toy electric motor A, Fig. 29.20 (i). A small mass M such as 10 g, suspended from a thread tied round the motor wheel, now rises as the motor functions. Some of the stored energy in the capacitor is thus transferred to gravitational potential energy of the mass; the remainder is transferred to kinetic energy and heat in the motor.

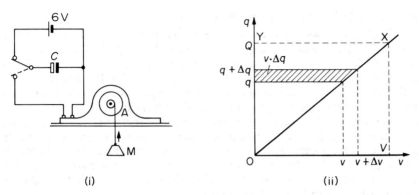

Fig. 29.20 Energy in charged capacitor

To find the energy stored in the capacitor, we note that since q (charge) is proportional to v (p.d. across the capacitor) at any instant, the graph OX showing how q varies with v is a straight line, Fig. 29.20 (ii). We may therefore consider that the final charge Q on the capacitor moved from one plate to the other through an *average* p.d. equal to $\frac{1}{2}(0 + V)$, since there is zero p.d. across the plates at the start and a p.d. V at the end. So

work done, W = energy stored = charge × p.d. = $Q \times \frac{1}{2}V$.

So $$W = \tfrac{1}{2}QV.$$

From $Q = CV$, other expressions for the energy stored are

$$W = \tfrac{1}{2}CV^2 = \frac{Q^2}{2C}.$$

Alternative Proof of Energy Formulae

We can also calculate the energy stored in a charged capacitor by a calculus method.

At any instant of the charging process, suppose the charge on the plates is q and the p.d. across the plates is then v. If an additional charge Δq now flows from the negative to the positive plate, the p.d. increases to $v + \Delta v$. Since Δv is very small compared to v, we may say that the charge Δq has moved through a p.d. equal to v. So

work done in displacing the charge $\Delta q = v.\Delta q$.

Therefore total work done = energy stored = $\displaystyle\int_{0}^{Q} v.dq$

where the limits are $q = Q$, final charge, and $q = 0$, as shown. To integrate, we substitute $v = q/C$. Then

$$\text{energy stored } W = \int_{0}^{Q} \frac{q.dq}{C} = \frac{1}{C}\left[\frac{q^2}{2}\right]_{0}^{Q} = \frac{Q^2}{2C}.$$

Using $Q = CV$, other expressions for W are

$$W = \tfrac{1}{2}CV^2 \quad \text{or} \quad W = \tfrac{1}{2}QV.$$

If C is measured in farad, Q in coulomb and V in volt, then the formulae will give the energy W in joules.

Energy and Q-V Graph. Heat Produced

Figure 29.20 (ii) shows the variation of the charge q on the capacitor and its corresponding p.d. v while the capacitor is charged to a final value q. The small shaded area shown $= v.\Delta q$ and so the area represents the small amount of work done or energy stored during a change from q to $q + \Delta q$. If therefore follows that the total energy stored by the capacitor is represented by the area of the triangle OXY. This area $= \tfrac{1}{2}QV$, as previously obtained.

If a high resistor R is included in the charging circuit, the rate of charging is slowed. See p. 612. When the charging current ceases to flow, however, the final charge Q on the capacitor is the same as if negligible resistance was present in the circuit, since the whole of the applied p.d. V is the p.d. across the capacitor when the current in the resistor is zero. Thus the energy stored in the capacitor is $\tfrac{1}{2}QV$ whether the resistor is large or small.

It is important to note that the energy in the capacitor comes from the battery. This supplies an amount of energy equal to QV during the charging process. Half of the energy, $\tfrac{1}{2}QV$, goes to the capacitor. The other half is transferred to *heat* in the circuit resistance. If this is a high resistance, the charging current is slow and the capacitor gains its final charge after a long time. If it is a low resistance, the charging current is higher and the capacitor gains its final charge in a quicker time. In both cases, however, the total amount of heat produced is the same, $\tfrac{1}{2}QV$.

Loss of Energy

Consider a capacitor C_1 of 2 μF charged to a p.d. of 50 V, and a capacitor C_2 of 3 μF charged to a p.d. of 100 V, Fig. 29.21 (i). Then

$$\text{charge } Q_1 \text{ on } C_1 = C_1 V_1 = 2 \times 10^{-6} \times 50 = 10^{-4} \text{ C}$$

and

$$\text{charge } Q_2 \text{ on } C_2 = C_2 V_2 = 3 \times 10^{-6} \times 100 = 3 \times 10^{-4} \text{ C}$$

$$\therefore \text{ total charge} = 4 \times 10^{-4} \text{ C} \qquad . \qquad . \qquad . \qquad . \qquad (1)$$

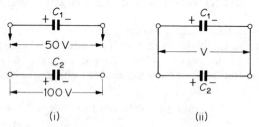

(i)   (ii)

Fig. 29.21 Loss of energy in joined capacitors

Suppose the capacitors are now joined with plates of like charges connected together, Fig. 29.21 (ii). Then some charge will flow from C_1 to C_2 until the p.d. across each capacitor becomes *equal* to some value V. Further, since charge is conserved, the total charge on C_1 and C_2 after connection = the total charge before connection. Now after connection,

$$\text{total charge} = C_1 V + C_2 V = (C_1 + C_2)V = 5 \times 10^{-6} \text{ V} \qquad . \qquad (2)$$

Hence, from (1),

$$5 \times 10^{-6} \, V = 4 \times 10^{-4}$$

$$\therefore \; V = 80 \text{ V}$$

∴ total energy of C_1 and C_2 after connection

$$= \tfrac{1}{2}(C_1 + C_2)V^2$$

$$= \tfrac{1}{2} \times 5 \times 10^{-6} \times 80^2 = 0{\cdot}016 \text{ J} \; . \qquad . \qquad . \qquad (3)$$

The total energy of C_1 and C_2 *before* connection

$$= \tfrac{1}{2}C_1 V_1{}^2 + \tfrac{1}{2}C_2 V_2{}^2$$

$$= \tfrac{1}{2} \times 2 \times 10^{-6} \times 50^2 + \tfrac{1}{2} \times 3 \times 10^{-6} \times 100^2$$

$$= 0{\cdot}0025 + 0{\cdot}015 = 0{\cdot}0175 \text{ J} \qquad . \qquad . \qquad . \qquad (4)$$

Comparing (4) with (3), we can see that a *loss of energy* occurs when the capacitors are connected. This loss of energy is converted to *heat* in the connecting wires.

Discharge in *C-R* Circuit

We now consider in more detail the discharge of a capacitor C through a resistor R. Suppose the capacitor is initially charged to a p.d. V_0 so that its charge is then $Q = CV_0$. At a time t after the discharge through R has begun, the current I flowing $= V/R$ where V is then the p.d. across C, Fig. 29.22 (i). Now

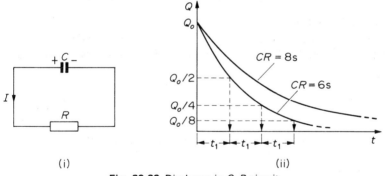

Fig. 29.22 Discharge in *C–R* circuit

$$V = \frac{Q}{C} \quad \text{and} \quad I = -\frac{dQ}{dt} \quad \text{(the minus shows } Q \text{ decreases with increasing } t\text{)}$$

Hence, from $I = V/R$, we have

$$-\frac{dQ}{dt} = \frac{1}{CR}Q$$

Integrating,

$$\therefore \int_{Q_0}^{Q} \frac{dQ}{Q} = -\frac{1}{CR}\int_{0}^{t} dt$$

$$\therefore \ln\left(\frac{Q}{Q_0}\right) = -\frac{t}{CR}$$

$$\therefore Q = Q_0 e^{-t/CR} \qquad\qquad (1)$$

Hence Q decreases exponentially with time t, Fig. 29.22 (ii). Since the p.d. V across C is proportional to Q, it follows that $V = V_0 e^{-t/CR}$. Further, since the current I in the circuit is proportional to V, then $I = I_0 e^{-t/CR}$, where I_0 is the initial current value, V_0/R.

From (1), Q decreases from Q to half its value, $Q_0/2$, in a time t_1 given by

$$e^{-t/CR} = \tfrac{1}{2} = 2^{-1}$$

$$\therefore t_1 = CR \ln 2$$

Similarly, Q decreases from $Q_0/2$ to half this value, $Q_0/4$, in a time $t_1 = CR\ln 2$. This is the same time from Q_0 to $Q_0/2$. Thus the time for a charge to diminish to half its initial value, no matter what the initial value may be, is always the same. See Fig. 29.22 (ii). This is true for fractions other than one-half. It is typical of an exponential variation or 'decay' which also occurs in radioactivity (p. 881).

Time Constant

The *time constant T* of the discharge circuit is defined as CR second, where C is in farad and R is in ohm. Thus if $C = 4\ \mu F$ and $R = 2\ M\Omega$, then $T = (4 \times 10^{-6}) \times (2 \times 10^6) = 8$ seconds. Now, from (1), if $t = CR$, then

$$Q = Q_0 e^{-1} = \frac{1}{e} Q_0$$

Thus the time constant may be defined as the time for the charge to decay to $1/e$ times its initial value ($e = 2\cdot72$ approximately, so that $1/e = 0\cdot37$ approx.). If the time constant is high, then the charge will diminish slowly; if the time constant is small, the charge will diminish rapidly.

Charging C through R

Consider now the charging of a capacitor C through a resistance R in series, and suppose the applied battery has an e.m.f. E and a negligible internal resistance, Fig. 29.23 (i).

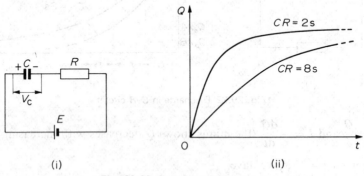

(i)   (ii)

Fig. 29.23 Charging in C–R circuit

At the instant of making the circuit, there is no charge on C and hence no p.d. across it. So the p.d. across $R = E$, the applied circuit p.d. Thus the initial current flowing, $I_0 = E/R$. Suppose I is the current flowing after a time t. Then, if V_C is the p.d. now across C,

$$I = \frac{E - V_C}{R}.$$

Now $I = dQ/dt$ and $V_C = Q/C$. Substituting in the above equation and simplifying,

$$\therefore\ CR\frac{dQ}{dt} = CE - Q = Q_0 - Q,$$

where $Q_0 = CE =$ final charge on C, when no further current flows.

Integrating,

$$\therefore\ \frac{1}{CR}\int_0^t dt = \int_0^Q \frac{dQ}{Q_0 - Q}$$

$$\therefore\ \frac{t}{CR} = -\ln\left(\frac{Q_0 - Q}{Q_0}\right)$$

$$\therefore\ Q = Q_0(1 - e^{-t/CR}) \qquad . \qquad . \qquad . \qquad (2)$$

As in the case of the discharge circuit, the *time constant* T is defined as CR seconds with C in farad and R in ohm. If T is high, it takes a long time for C to reach its final charge, that is, C charges slowly. If T is small, C charges rapidly. See Fig. 29.23 (ii). The voltage V_C follows the same variation as Q, since $V_C \propto Q$.

Rectangular Pulse Voltage and *C-R* Circuit

We can apply our results to find how the voltages across a capacitor C and resistor R vary when a *rectangular pulse voltage*, shown in Fig. 29.24 (i), is applied to a C–R series circuit. This type of circuit is used in analogue computers.

On one half of a cycle, the p.d. is constant along AB at a value E say. We can therefore consider that this is similar to the case of *charging* a C–R circuit by a battery of e.m.f. E. The p.d. V_C across the capacitor hence *rises* along exponential curve, Fig. 29.24 (ii). During the same time, the p.d. across R, V_R, falls as shown in Fig. 29.24 (iii), since $V_R = E - V_C$; that is, the curves for V_R and V_C together add up to the straight line graph AB in Fig. 29.24 (i).

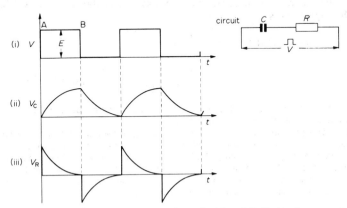

Fig. 29.24 Rectangular pulse voltage and C-R circuit

Examples

1. A capacitor of capacitance C is fully charged by a 200 V battery. It is then discharged through a small coil of resistance wire embedded in a thermally insulated block of specific heat capacity 2.5×10^2 J kg^{-1} K^{-1} and of mass 0·1 kg. If the temperature of the block rises by 0·4 K, what is the value of C? (*L.*)

$$\text{Energy in capacitor} = \tfrac{1}{2}CV^2 = \tfrac{1}{2} \times C \times 200^2 = 20\,000\ C.$$

$$\text{Energy through coil} = mc\theta = 0.1 \times 2.5 \times 10^2 \times 0.4 = 10\ J$$

So
$$20\,000\ C = 10$$

$$C = \frac{10}{20\,000} = \frac{1}{2000}\ F$$

$$= 500\ \mu F.$$

2. In a vibrating reed experiment, two parallel plates have an area 0·12 m² and are separated 2 mm by a dielectric. The battery of 150 V charges and discharges the capacitor at a frequency of 50 Hz, and a current of 20 μA is produced. Calculate the relative permittivity of the dielectric if the permittivity of free space is 8.9×10^{-12} F m^{-1}.
  What is the new capacitance if the dielectric is half withdrawn from the plates?

Suppose C is the capacitance between the plates. Then, with the usual notation,

$$\text{current } I = 50CV$$

So
$$C = \frac{I}{50V} = \frac{20 \times 10^{-6}}{50 \times 150} = \frac{4 \times 10^{-8}}{15} \qquad . \quad . \quad . \quad (i)$$

But
$$C = \frac{\varepsilon_r \varepsilon_0 A}{d} = \frac{\varepsilon_r \times 8.9 \times 10^{-12} \times 0.12}{2 \times 10^{-3}} \qquad . \quad . \quad (ii)$$

So, from (i) and (ii),

$$\varepsilon_r = \frac{4 \times 10^{-8} \times 2 \times 10^{-3}}{15 \times 8.9 \times 10^{-12} \times 0.12} = 5.$$

If the dielectric is half withdrawn, the common area of each of the two capacitors formed is now $0.5A$. One capacitor, with air dielectric, has a capacitance given by $0.5\varepsilon_0 A/d$. The other, with dielectric of $\varepsilon_r = 5$, has a capacitance given by $2.5\varepsilon_0 A/d$. So, adding the capacitances,

$$\text{total capacitance, } C = \frac{3\varepsilon_0 A}{d}$$

$$= \frac{3 \times 8.9 \times 10^{-12} \times 0.12}{2 \times 10^{-3}} = 1.6 \times 10^{-9}\ F.$$

3. Define *electrical capacitance*. Describe experiments to demonstrate the factors which determine its value for a parallel plate capacitance.
  The plates of a parallel plate air capacitor consisting of two circular plates, each of 10 cm radius, placed 2 mm apart, are connected to the terminals of an electrostatic voltmeter. The system is charged to give a reading of 100 on the voltmeter scale. The space between the plates is then filled with oil of dielectric constant 4·7 and the voltmeter reading falls to 25. Calculate the capacitance of the voltmeter. You may assume that the voltage recorded by the voltmeter is proportional to the scale reading. (*N.*)

Suppose V is the initial p.d. across the air capacitor and voltmeter, and let C_1 be the voltmeter capacitance.

Then
$$\text{total charge} = CV + C_1 V = (C + C_1)V \qquad . \quad . \quad (i)$$

When the plates are filled with oil the capacitance increases to 4·7C, and the p.d. fall to V_1. But the total charge remains constant.

$$\therefore 4{\cdot}7CV_1 + C_1V_1 = (C + C_1)V, \quad \text{from (i),}$$

$$\therefore (4{\cdot}7C + C_1)V_1 = (C + C_1)V,$$

$$\therefore \frac{4{\cdot}7C + C_1}{C + C_1} = \frac{V}{V_1} = \frac{100}{25} = 4,$$

$$\therefore 0{\cdot}7C = 3C_1,$$

$$\therefore C_1 = \frac{0{\cdot}7C}{3} = \frac{7}{30}C.$$

Now $C = \varepsilon_0 A/d$, where A is in metre^2 and d is in metre.

$$\therefore C = \frac{8{\cdot}85 \times 10^{-12} \times \pi \times (10 \times 10^{-2})^2}{2 \times 10^{-3}} \text{ F}$$

$$= 1{\cdot}4 \times 10^{-10} \text{ F (approx.)}.$$

$$\therefore C_1 = \frac{7}{30} \times 1{\cdot}4 \times 10^{-10} \text{ F} = 3{\cdot}3 \times 10^{-11} \text{ F}.$$

Exercises 29

1. A capacitor charged from a 50 V d.c. supply is discharged across a charge-measuring instrument and found to have carried a charge of 10 μC. What was the capacitance of the capacitor and how much energy was stored in it? (*L.*)

2. A 300 V battery is connected across capacitors of 3 μF and 6 μF (*a*) in parallel, and then (*b*) in series. Calculate the charge and energy stored in each capacitor in (*a*) and (*b*).

3. A parallel-plate capacitor with air as the dielectric has a capacitance of 6×10^{-4} μF and is charged by a 100 V battery. Calculate (*a*) the charge, (*b*) the energy stored in the capacitor, (*c*) the energy supplied by the battery. What accounts for the difference in the answers for (*b*) and (*c*)?

The battery connections are now removed, leaving the capacitor charged, and a dielectric of relative permittivity 3 is then carefully placed between the plates. What is the new energy stored in the capacitor?

4.

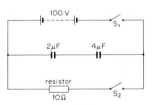

Fig. 29A

If S_2 is left open and S_1 is closed, calculate the quantity of charge on each capacitor, Fig. 29A,

If S_1 is now opened and S_2 is closed, how much charge will flow through the 10 Ω resistor?

If the entire process were repeated with the 10 Ω resistor replaced by one of much larger resistance what effect would this have on the flow of charge? (*L.*)

5. (*a*) Define *capacitance*. Describe briefly the structure of (i) a variable air capacitor, (ii) an electrolytic capacitor, and (iii) a simple paper capacitor. (*b*) The circuit in Fig. 29B (i)

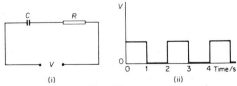

Fig. 29B

shows a capacitor C and a resistor R in series. The applied voltage V varies with time as shown in Fig. 29B (ii). The product CR is of the order 1 s. Sketch graphs showing the way the voltages across C and R vary with time.

If the product CR were made considerably smaller than 1 s what would be the effect on the graphs?

(c) A capacitor of capacitance 4 μF is charged to a potential of 100 V and another of capacitance 6 μF is charged to a potential of 200 V. These capacitors are now joined, with plates of like charge connected together. Calculate (i) the potential across each after joining, (ii) the total electrical energy stored before joining, and (iii) the total electrical energy stored after joining. Explain why the energies calculated in (ii) and (iii) are different. (L.)

6. Explain what is meant by *dielectric constant* (*relative permittivity*). State two physical properties desirable in a material to be used as the dielectric in a capacitor.

A sheet of paper 40 mm wide and 1.5×10^{-2} mm thick between metal foil of the same width is used to make a 2·0 μF capacitor. If the dielectric constant (relative permittivity) of the paper is 2·5, what length of paper is required? ($\varepsilon_0 = 8.85 \times 10^{-12}$ F m^{-1}.) (N.)

7.

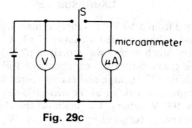

Fig. 29c

In the circuit shown in Fig. 29C, S is a vibrating reed switch and the capacitor consists of two flat metal plates parallel to each other and separated by a small air-gap. When the number of vibrations per second of S is n and the potential difference between the battery terminals is V, a steady current I is registered on the microammeter.

(a) Explain this and show that $I = nCV$, where C is the capacitance of the parallel plate arrangement. (b) Describe how you would use the apparatus to determine how the capacitance C depends on (i) the area of overlap of the plates, (ii) their separation, and show how you would use your results to demonstrate the relationships graphically. (c) Explain how you could use the measurements made in (b) to obtain a value for the permittivity of air. (d) In the above arrangement, the microammeter records a current I when S is vibrating. A slab of dielectric having the same thickness as the air-gap is slid between the plates so that one-third of the volume is filled with dielectric. The current is now observed to be $2I$. Ignoring the edge effects, calculate the relative permittivity of the dielectric. (N.)

8. A student is provided with two square plates of glass of uniform thickness, each about 1·0 m^2 in area, and some thin aluminium foil, and is asked to construct a parallel plate air capacitor. Explain briefly how this should be done and list any additional items of equipment needed.

How should the plates be arranged to obtain the highest possible value of capacitance? Estimate the highest capacitance value that could be achieved with this apparatus in practice. (Permittivity of air $= 9 \times 10^{-12}$ F m^{-1}.)

In order to further increase the capacitance the student is given a sheet of P.V.C. of thickness 0·2 mm with which to separate the glass plates. If the relative permittivity of P.V.C. is 4 calculate by what factor the capacitance of the student's capacitor would be increased. Describe, with the aid of a labelled circuit diagram, how you would compare experimentally the capacitances in these two cases. (L.)

9.

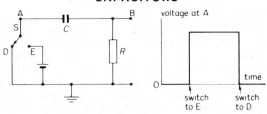

Fig. 29D

In the Fig. 29D circuit, C is a capacitor and R is a high resistor. By operating the switch S the voltage at A is made to vary with time as shown in the diagram. Sketch the voltage-time graph you would expect to obtain at B and explain its form. (*L*.)

10. Derive an expression for the energy stored in a capacitor C when there is a potential difference V between the plates. If C is in microfarad and V is in volt, express the result in joule.

Show that when a battery is used to charge a capacitor through a resistor, the heat dissipated in the circuit is equal to the energy stored in the capacitor.

Describe the structure of a 1 microfarad capacitor and describe an experiment to compare the capacitance of two capacitors of this type. (*N*.)

11. Explain the meaning of the terms *capacitor* and *dielectric*. Give a brief qualitative explanation, based on the behaviour of molecules, for the change in capacitance of a capacitor when the space between its plates is filled with a dielectric.

Give *one* practical example of the use of a capacitor, explaining how the properties of the capacitor are utilized in the example you choose.

Two capacitors, of capacitance 2 μF and 3 μF, are connected as shown (Fig. 29E)

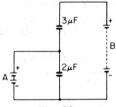

Fig. 29E

to batteries A and B which have e.m.f. 6 V and 10 V respectively. What is the energy stored in each of the capacitors? Calculate also the stored energy in each capacitor (i) when the terminals of battery A are reversed, and (ii) when the battery B is disconnected and the capacitors are connected in parallel with A. (*O. & C.*)

12. A plane sheet of metal is fixed in an insulating stand and is given an electric charge. A similar but uncharged sheet, also in an insulating stand, is now placed with its plane close to and parallel with it. Explain why the potential difference between the first plate and earth changes.

The second sheet is subsequently, and on separate occasions, (*a*) earthed, and (*b*) connected to the first sheet. Explain the effect which each of these operations would have on the potential difference between the first plate and the earth.

Explain the effect on the capacitance of the capacitor formed by the two plates produced by increasing the distance between the plates and describe how you would demonstrate the effect.

Two capacitors, of capacitance 4·0 μF and 12·0 μF respectively, are connected in series and the combination connected momentarily across a 200 V battery. The charged capacitors are now isolated and connected in parallel, similar charged plates being connected together. What would be the resulting potential difference across the combination? (*L*.)

13. Define electric field strength and potential at a point in an electric field.

Explain what is meant by the *relative permittivity* of a material. How may its value be determined experimentally?

A capacitor of capacitance 9·0 μF is charged from a scource of e.m.f. 200 V. The capacitor is now disconnected from the source and connected in parallel with a second capacitor of capacitance $3 \cdot 0$ μF. The second capacitor is now removed and discharged. What charge remains on the $9 \cdot 0$ μF capacitor? How many times would the process have to be performed in order to reduce the charge on the $9 \cdot 0$ μF capacitor to below 50% of its initial value? What would the p.d. between the plates of the capacitor now be? (*L.*)

14. Define the capacitance of a parallel plate capacitor. Write down an expression for this capacitance and explain why your expression is only approximately correct.

A potential difference of 600 V is established between the top cap and the case of a calibrated electroscope by means of a battery which is then removed, leaving the electroscope isolated. When a parallel plate capacitor with air dielectric is connected across the electroscope, one plate to the top cap and the other plate to the case, the p.d. across the electroscope is found to drop to 400 V. If the capacitance of the parallel plate capacitor is $1 \cdot 0 \times 10^{-11}$ F, calculate

(*a*) the capacitance of the electroscope;

(*b*) the change in electrical energy which results from the sharing of the charge. Explain why the total energy is different after sharing.

If the space between the parallel plates of the capacitor were then filled with material of relative permittivity 2, what would then be the potential of the electroscope? (*N.*)

15. Two horizontal parallel plates, each of area 500 cm^2, are mounted 2 mm apart in a vacuum. The lower plate is earthed and the upper one is given a positive charge of $0 \cdot 05$ μC. Neglecting edge effects, find the electric field strength between the plates and state in what direction the field acts.

Deduce values for (*a*) the potential of the upper plate, (*b*) the capacitance between the two plates and (*c*) the electrical energy stored in the system.

If the separation of the plates is doubled, keeping the lower plate earthed and the charge on the upper plate fixed, what is the effect on the field between the plates, the potential of the upper plate, the capacitance and the electrical stored energy?

Discuss how the change in energy can be accounted for. (*O. & C.*)

16. Define *potential, capacitance.*

Obtain from first principles a formula for the capacitance of a parallel-plate capacitor.

The plates of such a capacitor are each $0 \cdot 4$ m square, and separated by 10^{-3} m, the space between being filled with a medium of relative permittivity 5. A vibrating contact, with frequency 50 second^{-1}, repeatedly connects the capacitor across a 120-volt battery and then discharges it through a galvanometer whose resistance is of the order of 50 ohm. Calculate the current recorded, and explain why this is independent of the actual value of the galvanometer resistance. (Take the permittivity of vacuum to be $8 \cdot 85 \times 10^{-12}$ F m^{-1}.) (*O.*)

17. (*a*) Define (i) electric field strength, (ii) electric potential at a point. Show that the electric field strength at a point is equal to the negative potential gradient at that point.

(*b*) Why is the capacitance of a single isolated metal plate less than the capacitance of an arrangement consisting of the same plate with a similar earthed metal plate placed close to and parallel with it?

(*c*) State the three factors which affect the capacitance of a parallel plate capacitor and describe how the effect of each of these factors may be investigated experimentally. Compare the energy stored in a 100 μF capacitor, charged to a p.d. of 400 V, with that stored in a 12 volt 40 ampere-hour car battery. (*AEB, 1982*)

30 Current Electricity

Ohm's and Joule's Laws: Resistance and Power

Discovery and Electric Current

By the middle of the eighteenth century, electrostatics was a well-established branch of physics. Machines had been invented which could produce by friction great amounts of charge, giving sparks and electric shocks. The momentary current (as we would now call it) carried by the spark or the body was called a 'discharge'.

In 1786 Galvani, while dissecting a frog, noticed that its leg-muscle twitched when one of his assistants produced an electric spark in another part of the room.

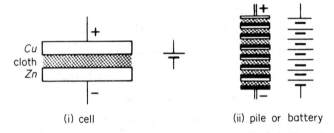

Fig. 30.1 Voltaic cell and pile, with conventional symbols

He also found that, when a frog's leg-muscle was hung by a copper hook from an iron stand, the muscle twitched whenever it swung so as to touch the stand. Galvani supposed that the electricity which caused the twitching was generated within the muscle, but his fellow-Italian Volta believed that it arose from the contact of the two different metals. Volta turned out to be right, and in 1799 he discovered how to obtain from two metals a continuous supply of electricity: he placed a piece of cloth soaked in brine between copper and zinc plates, Fig. 30.1 (i). The arrangement is called a *voltaic cell*, and the metal plates its 'poles'; the copper is known as the positive pole, the zinc as the negative. Volta increased the power by building a pile of cells, with the zinc of one cell resting on the copper of the other, Fig. 30.1 (ii). From this pile he obtained sparks and shocks similar to those given by electrostatic machines.

Shortly after, it was found that water was decomposed into hydrogen and oxygen when connected to a voltaic pile. This was the earliest discovery of the chemical effect of an electric current. The heating effect was also soon found, but the magnetic effect, the most important effect, was discovered some twenty years later.

Ohm's Experiment

The properties of an electric circuit, as distinct from the effects of a current, were first studied by Ohm in 1826. He set out to find how the length of wire in a circuit affected the current through it—in modern language, he investigated electrical resistance. In his first experiment he used voltaic piles as sources of current, but he found that the current which they gave fluctuated considerably, and he later replaced them by thermocouples (p. 650). The voltaic pile or battery and the thermocouple are 'electrical generators'. As we see later, a battery converts chemical energy to electrical energy and a thermocouple converts heat energy to electrical energy. These, and other, electrical generators produce a *potential difference* (p.d.), V, at their terminals. When a length of wire is joined to the terminals an electric current, I, flows along the wire whose magnitude depends on the magnitude of V.

Using a constant p.d. from a thermocouple made of copper (Cu) and bismuth (Bi) wires, Ohm passed currents through various lengths of brass wire, 0·37 mm in diameter, and observed the current in a galvanometer G, Fig. 30.2 (i). He

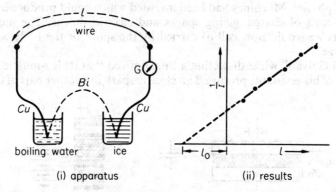

Fig. 30.2 Ohm's experiment

found that the current I in his experiments was almost inversely proportional to the length of wire, l, in the circuit. He plotted the reciprocal of the current (in arbitrary units) against the length l, and got a straight line, as shown in Fig. 30.2 (ii). Thus

$$I \propto \frac{1}{l_0 + l},$$

where l_0 is the intercept of the line on the axis of length. Ohm explained this result by supposing, naturally, that the thermocouples and galvanometer, as well as the wire, offered resistance to the current. He interpreted the constant l_0 as the length of wire equal in resistance to the galvanometer and thermocouples.

Mechanism of Metallic Conduction

The conduction of electricity in metals is due to free electrons. Free electrons have thermal energy which depends on the metal temperature, and they wander randomly through the metal from atom to atom. When a battery is connected across the ends of the metal, an electric field is set up. The electrons are now accelerated by the field, so they gain velocity and energy. When they 'collide' with an atom vibrating about its fixed mean position (called a 'lattice site'), they give up some of their energy to it. The amplitude of the vibrations is then increased and the temperature of the metal rises. The electrons are then again accelerated by the field and again give up some energy. Although their movement is erratic, on the average the electrons drift in the direction of the field with a mean speed we calculate shortly. This drift constitutes an 'electric current'. It will be noted that heat is generated by the collision of electrons whichever way they flow. Thus the heating effect of a current—called *Joule heating* (p. 636)—is irreversible, that is, it still occurs when the current in a wire is reversed.

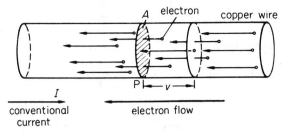

conventional
current

electron flow

Fig. 30.3 Theory of metallic conduction

A simple calculation enables the average drift speed to be estimated. Figure 30.3 shows a portion of a copper wire of cross-sectional area A through which a current I is flowing. We suppose that there are n electrons per unit volume, and that each electron carries a charge e. Now in one second all those electrons within a distance v to the right of the plane at P, that is, in a volume Av, will flow through this plane, as shown. This volume contains nAv electrons and hence a charge $nAve$. Thus a charge of $nAve$ per second passes P, and so the current I is given by

$$I = nAve \qquad . \qquad . \qquad . \qquad . \qquad . \qquad (1)$$

To find the order of magnitude of v, suppose $I = 10$ A, $A = 1$ mm$^2 = 10^{-6}$ m^2, $e = 1 \cdot 6 \times 10^{-19}$ C, and $n = 10^{28}$ electrons m^{-3}. Then, from (1),

$$v = \frac{I}{nAe} = \frac{10}{10^{28} \times 10^{-6} \times 1 \cdot 6 \times 10^{-19}}$$

$$= \frac{1}{160} \text{ m s}^{-1} \text{ (approx.)}$$

This is a surprisingly slow drift compared with the average thermal speeds, which are of the order of several hundred metres per second (p. 221).

Resistance

The *resistance* R of a conductor is *defined* as the ratio V/I, where V is the p.d. across the conductor and I is the current flowing in it. Thus if the same p.d. V is applied to two conductors A and B, and a smaller current I flows in A, then the resistance of A is greater than that of B. We write, then,

$$\frac{V}{I} = R \qquad . \qquad . \qquad . \qquad . \qquad . \qquad (2)$$

The unit of potential difference, V is the *volt*, symbol V; the unit of current, I, is the *ampere*, symbol A; the unit of resistance, R, is the *ohm*, symbol Ω. The ohm is thus the resistance of a conductor through which a current of one ampere flows when a potential difference (p.d.) of one volt is maintained across it. Figure 30.4 shows some symbols which may be used for different types of resistors, and for ammeters, voltmeters and galvanometers (sensitive current-measuring meters).

From the above equation, it also follows that

$$V = IR, \quad \text{and} \quad I = \frac{V}{R} \qquad . \qquad . \qquad . \qquad . \qquad (3)$$

Smaller units of current are the milliampere (one-thousandth of an ampere),

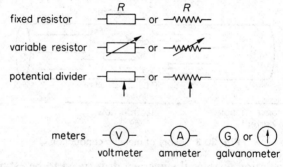

Fig. 30.4 Symbols for resistors and meters

symbol mA, and the micro-ampere (one-millionth of an ampere), symbol μA. Smaller units of p.d. are the millivolt (1/1000 V) and the microvolt ($1/10^6$ V). A small unit of resistance is the microhm ($1/10^6$ or 10^{-6} Ω); larger units are the kilohm (1000 Ω) and the megohm (10^6 ohms).

   Conductance is defined as the ratio I/V, and is therefore the inverse of resistance, or $1/R$ in numerical value. The unit of conductance is the *siemens*, symbol S.

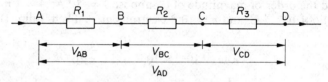

Fig. 30.5 Resistances in series

Series Resistors

The resistors of an electric circuit may be arranged in series, so that the charges carrying the current flow through each in turn (Fig. 30.5); or they may be arranged in parallel, so that the flow of charge divides between them as in Fig. 30.6, p. 623.

   Fig. 30.5 shows three passive resistors in series, carrying a current I. If V_{AD} is the potential difference across the whole system, the electrical energy supplied to the system per second is IV_{AD} (p. 638). This is equal to the electrical energy dissipated per second in all the resistors; therefore

$$IV_{AD} = IV_{AB} + IV_{BC} + IV_{CD},$$

whence $$V_{AD} = V_{AB} + V_{BC} + V_{CD} \qquad . \qquad . \qquad . \qquad (1)$$

The individual potential differences are given, from previous, by

$$V_{AB} = IR_1, \ V_{BC} = IR_2, \ V_{CD} = IR_3 \qquad . \qquad . \qquad (2)$$

Hence, by equation (1),

$$V_{AD} = IR_1 + IR_2 + IR_3$$
$$= I(R_1 + R_2 + R_3) \qquad . \qquad . \qquad (3)$$

And the effective resistance of the system is

$$R = \frac{V_{AD}}{I} = R_1 + R_2 + R_3 \qquad . \qquad . \qquad (4)$$

The physical facts are:

(i) *Current same through all resistors.*

(ii) *Total potential difference = sum of individual potential differences* (equation (1)).

(iii) *Individual potential differences directly proportional to individual resistances* (equation (2)).

(iv) *Total resistance greater than greatest individual resistance* (equation (4)).

(v) *Total resistance = sum of individual resistances.*

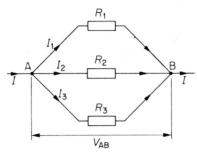

Fig. 30.6 Resistances in parallel

Resistors in Parallel

Figure 30.6 shows three passive resistors connected in parallel, between the points A, B. A passive device is one which produces no energy. A current I enters the system at A and leaves at B, setting up a potential difference V_{AB} between those points. The current branches into I_1, I_2, I_3, through the three elements, and

$$I = I_1 + I_2 + I_3 \qquad . \qquad . \qquad . \qquad . \qquad . \qquad (5)$$

Now
$$I_1 = \frac{V_{AB}}{R_1}, \quad I_2 = \frac{V_{AB}}{R_2}, \quad I_3 = \frac{V_{AB}}{R_3}.$$

$$\therefore I = V_{AB}\left(\frac{1}{R_1} + \frac{1}{R_2} + \frac{1}{R_3}\right).$$

$$\therefore \frac{I}{V_{AB}} = \frac{1}{R} = \frac{1}{R_1} + \frac{1}{R_2} + \frac{1}{R_3} \qquad . \qquad . \qquad . \qquad (6)$$

where R is the effective resistance (V_{AB}/I) of the system.

The physical facts about resistors in parallel may be summarised as follows:

(i) *Potential difference same across each resistor.*

(ii) *Total current = sum of individual current* (equation (5)).

(iii) *Individual currents inversely proportional to individual resistances.*

(iv) *Effective resistance less than least individual resistance* (equation (6)).

Resistance Boxes

In many electrical measurements variable known resistances are required; they are called resistance boxes. As shown in Fig. 30.7 (i) ten coils, each of resistance 1 ohm, for example, are connected in series. A rotary switch with eleven contacts enables any number of these coils to be connected between the terminals AA'. A resistance box contains several sets of coils and switches, the first giving resistances 0–10 ohms in steps of 1 ohm, the next 0–100 ohms in steps of 10 ohms, and so on. These are called decade boxes.

The switches used in a decade box are of very high quality; their contact resistances are negligible compared with the resistances of the coils which they select. In older boxes no switches were used. Instead, the resistances were varied by means of plugs. As shown in Fig. 30.7 (ii), the resistance coils are joined across gaps in a thick brass bar, and the gaps are formed into tapered sockets to receive short-circuiting plugs P. The resistance between the terminals A and B in Fig. 30.7 (ii) is the sum of the unplugged resistances between them—3 ohms in this example.

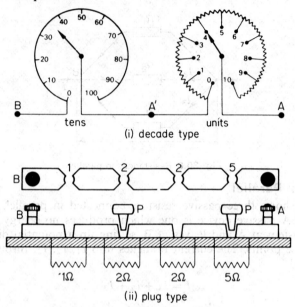

(i) decade type

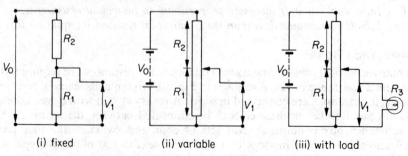

(ii) plug type

Fig. 30.7 Resistance boxes

The coils of a resistance box are wound in a particular way, which we shall describe and explain later (p. 753). They are not intended to carry large currents, and must not be allowed to dissipate a power of more than one watt. The greatest safe current for a 1-ohm coil is 1 A, and for a 10-ohm coil about 0·3 A. If the one-watt limit is exceeded, the insulation will be damaged, or the wire burnt out.

The Potential Divider

Two resistance boxes in series are often used in the laboratory to provide a known fraction of a given potential difference—for example, of one which is

(i) fixed   (ii) variable   (iii) with load

Fig. 30.8 Potential divider

too large to measure easily. Fig. 30.8 (i) shows the arrangement, which is called a resistance 'potential divider'. The current flowing, I, is given by

$$I = \frac{V_0}{R_1 + R_2},$$

$$\therefore V_1 = IR_1 = \frac{R_1}{R_1 + R_2} V_0 \quad . \quad . \quad . \quad . \quad (7)$$

A resistor with a sliding contact can similarly be used, as shown in Fig. 30.8 (ii), to provide a continuously variable potential difference, from zero to the full supply value V_0. This is a convenient way of controlling the voltage applied to a load such as a lamp, Fig. 30.8 (iii). The resistance of the load, R_3, however, acts in parallel with the resistance R_1; equation (7) is therefore no longer true, and the voltage V_1 must be measured with a voltmeter. It can be calculated, as in the following example, if R_3 is known; but if the load is a lamp its resistance varies greatly with the current through it, because its temperature varies.

Example

A load of 2000 Ω is connected, via a potential divider of resistance 4000 Ω, to a 10 V supply, Fig. 30.9. What is the potential difference across the load when the slider is (a) one-quarter, (b) half-way up the divider?

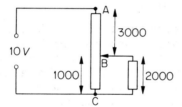

Fig. 30.9 A loaded potential divider

(a) Since

$$\frac{1}{R} = \frac{1}{2000} + \frac{1}{1000},$$

$$R_{BC} = \frac{2000 \times 1000}{2000 + 1000} = \frac{2000}{3}\ \Omega.$$

$$\therefore R_{AC} = R_{AB} + R_{BC} = 3000 + \frac{2000}{3}\ \Omega,$$

$$\therefore V_{BC} = \frac{R_{BC}}{R_{AC}} V_{AC}$$

$$= \frac{2000/3}{11\,000/3} \times 10 = \frac{2}{11} \times 10$$

$$= 1{\cdot}8\ \text{V}.$$

If the load were removed, V_{BC} would be 2·5 V.

(b) It is left for the reader to show similarly that $V_{BC} = 3{\cdot}3$ V. Without the load it would be 5 V.

Conversion of a Milliammeter into a Voltmeter

We will now see how to use a milliammeter as a voltmeter. Let us suppose that we have a moving-coil instrument which requires 5 milliamperes for full-scale deflection (f.s.d.). And let us suppose that the resistance of its coil, r is 20 ohms,

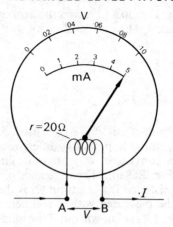

Fig. 30.10 P.d. across moving-coil ammeter

Fig. 30.10. Then, when it is fully deflected, the potential difference across it is

$$V = rI$$
$$= 20 \times 5 \times 10^{-3} = 100 \times 10^{-3} \text{ V}$$
$$= 0\cdot1 \text{ V}.$$

If the coil resistance is constant, the current through it is proportional to the potential difference across it; and since the deflection of the pointer is proportional to the current it is therefore also proportional to the potential difference. Thus the instrument can be used as a voltmeter, giving full-scale deflection for a potential difference of 0·1 volt, or 100 millivolts. Its scale could be engraved as shown at the top of Fig. 30.10.

The potential differences to be measured in the laboratory are usually greater than 100 millivolts, however. To measure such a potential difference we insert a resistor R in series with the coil, as shown in Fig. 30.11. If we wish to measure

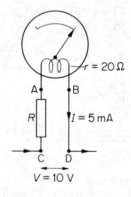

single − range

Fig. 30.11 Single-range voltmeter

up to 10 volts we must choose the resistance R so that, when 10 volts are applied between the terminals CD, then a current of 5 milliamperes flows through the moving coil. Now

$$V = (R + r)I,$$

$$\therefore \ 10 = (R + 20) \times 5 \times 10^{-3}$$

or
$$R + 20 = \frac{10}{5 \times 10^{-3}} = 2 \times 10^3 = 2000 \ \Omega.$$

$$\therefore \ R = 2000 - 20$$

$$= 1980 \ \Omega. \qquad . \qquad . \qquad . \qquad . \qquad . \qquad . \qquad (8)$$

The resistance R is called a *multiplier*. Many voltmeters contain a series of multipliers of different resistances, which can be chosen by a switch or plug-and-socket arrangement, Fig. 30.12.

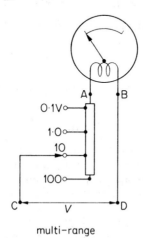

multi-range

Fig. 30.12 Multi-range voltmeter

Conversion of a Milliammeter into an Ammeter

Moving-coil instruments give full-scale deflection for currents smaller than those generally encountered in the laboratory. If we wish to measure a current of the order of an ampere or more we connect a low resistance S, called a *shunt*, across the terminals of a moving-coil meter, Fig. 30.13. The shunt diverts most of the current to be measured, I_C, away from the coil—hence its name. Let us

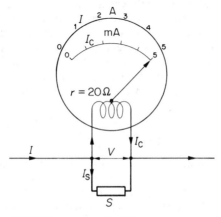

Fig. 30.13 Conversion of milliammeter to ammeter

suppose that, as before, the coil of the meter has a resistance r of 20 ohms and is fully deflected by a current, I_C, of 5 milliamperes. And let us suppose that we wish to shunt it so that it gives f.s.d. for 5 amperes to be measured. Then the current through the shunt is

$$I_S = I - I_C$$
$$= 5 - 0.005$$
$$= 4.995 \text{ A.}$$

The potential difference across the shunt is the same as that across the coil, which is

$$V = rI_C = 20 \times 0.005 = 0.1 \text{ V.}$$

The resistance of the shunt must therefore be

$$S = \frac{V}{I_S} = \frac{0.1}{4.995} = 0.020\,02 \ \Omega \qquad . \qquad . \qquad (9)$$

The ratio of the current measured to the current through the coil is

$$\frac{I}{I_C} = \frac{5}{5 \times 10^{-3}} = 1000.$$

This ratio is the same whatever the current I, because it depends only on the resistances S and r; the reader may easily show that its value is $(S + r)/S$. The deflection of the coil is therefore proportional to the measured current, as indicated in the figure, and the shunt is said to have a 'power' of 1000 when used with this instrument.

The resistance of shunts and multipliers are always given with four-figure accuracy. The moving-coil instrument itself has an error of the order of 1%; a similar error in the shunt or multiplier would therefore double the error in the instrument as a whole. On the other hand, there is nothing to be gained by making the error in the shunt less than about 0·1%, because at that value it is swamped by the error of the moving system.

Multimeters

A *multimeter* is an instrument which is adapted for measuring both current and voltage. It has a shunt R as shown, and a series of voltage multipliers R', Fig. 30.14. The shunt is connected permanently across the coil, and the resistances in R' are adjusted to give the desired full-scale voltages with the shunt in position. A switch or plug enables the various full-scale values of current or voltage to be chosen, but the user does the mental arithmetic. The instrument shown in the figure is reading 1·7 volts; if it were on the 10-volt range, it would be reading 6·4.

The terminals of a meter, multimeter or otherwise, are usually marked $+$ and $-$; the pointer is deflected to the right when current passes through the meter from $+$ to $-$.

The multimeters are generally arranged to measure resistance as well as current and voltage. An extra position on the switch, marked 'R' or 'ohms', puts a dry cell C and a variable resistor R'' in series with the moving coil, Fig. 30.15. Before the instrument is used to measure a resistance, its terminals TT are short-circuited, and R'' is adjusted until the pointer is fully deflected. As shown in the figure, it is then opposite the zero on the ohms scale. The short-circuit is next removed, and the unknown resistance R_x is connected across the terminals. The current falls, and the pointer moves to the left, indicating on the

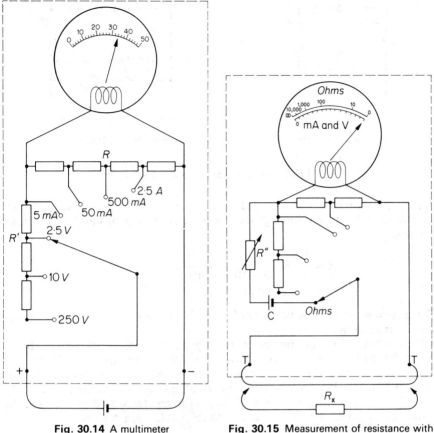

Fig. 30.14 A multimeter   **Fig. 30.15** Measurement of resistance with multimeter

ohms scale the value of R_x. The ohms scale is calibrated by the makers with known resistances.

Use of Voltmeter and Ammeter

A moving-coil voltmeter is a current-operated instrument. It can be used to measure potential differences if we assume that the current which it draws is always proportional to the potential difference applied to it as the current varies. Since its action depends on Ohm's law, a moving-coil voltmeter cannot be used in any experiment to demonstrate that law (see p. 633).

We use moving-coil voltmeters as they are more sensitive and more accurate than other forms of voltmeters. The current which they take does, however, sometimes complicate their use. To see how it may do so, let us suppose that we wish to measure a resistance R of about 100 ohms. As shown in Fig. 30.16 (i), we connect it in series with a cell, a milliammeter, and a variable resistance; across it we place the voltmeter. We adjust the current until the voltmeter reads, say, $V_1 = 1$ volt; let us suppose that the milliammeter then reads $I = 12$ mA. The value of the resistance then appears to be

$$R = \frac{V_1}{I} = \frac{1}{12 \times 10^{-3}} = \frac{10^3}{12}$$
$$= 83 \ \Omega \ \text{(approx.)}.$$

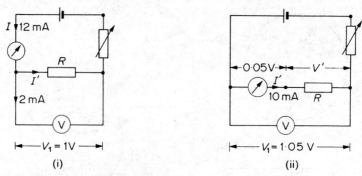

Fig. 30.16 Use of (i) ammeter and (ii) voltmeter

But the milliammeter reading *includes the current drawn by the voltmeter.* If that is 2 mA, then the current through the resistor, I', is only 10 mA and its resistance is actually

$$R = \frac{V_1}{I'} = \frac{1}{10 \times 10^{-3}} = \frac{1}{10^{-2}}$$

$$= 100 \ \Omega.$$

The current drawn by the voltmeter has made the resistance appear 17% lower than its true value.

In an attempt to avoid this error, we might connect the voltmeter as shown in Fig. 30.16 (ii): across both the resistor and the milliammeter. But its reading would then include the potential difference across the milliammeter. Let us suppose that this is 0·05 V when the current through the milliammeter is 10 mA. Then the potential difference V' across the resistor would be 1 V, and the voltmeter would read 1·05 V. The resistance would appear to be

$$R = \frac{1·05}{10 \times 10^{-3}} = \frac{1·05}{10^{-2}}$$

$$= 105 \ \Omega.$$

Thus the voltage drop across the milliammeter would make the resistance appear 5% higher than its true value.

Errors of this kind are negligible only when the voltmeter current is much less than the current through the resistor, or when the voltage drop across the ammeter is much less than the potential difference across the resistor. If we were measuring a resistance of about 1 Ω, for example, the current I' in Fig. 30.16 (i) would be 1 A, and I would be 1·002 A. The error in measuring R would then be only 0·2%—less than the intrinsic error of the meter. But the circuit of Fig. 30.16 (i) would give the same error as before. It could do so because, as we saw when considering shunts, the shunt across the milliammeter would have been chosen to make the voltage drop still 0·05 V. Thus V_1 would still be 1·05 V when V' was 1 V, and the error would be 5% as before.

In low-resistance circuits, therefore, the voltmeter should be connected as in Fig. 30.16 (i), so that its reading does not include the voltage drop across the ammeter. But in high-resistance circuits the voltmeter should be connected as in Fig. 30.16 (ii) so that the ammeter does not carry its current.

A potentiometer method of measuring V in Fig. 30.16 (i), discussed in the next chapter, would increase the accuracy of measuring R. This method is equivalent to using a voltmeter with an infinitely-high resistance, so that no current is drawn.

Resistivity

Ohm showed, by using wires of different length and diameter, that the resistance of a wire, R, is proportional to its length, l, and inversely proportional to its cross-sectional area A. The truth of this can easily be demonstrated today by experiments with a Wheatstone bridge (see p. 666) and suitable lengths of wire. We have, then, for a given wire,

$$R \propto \frac{l}{A};$$

we may therefore write

$$R = \rho \frac{l}{A}, \qquad . \qquad . \qquad . \qquad . \qquad (1)$$

where ρ is a constant for the material of the wire. It is called the *resistivity* of that material.

To define it in words, we imagine a rectangular prism of the material of unit length and unit cross-section. Then $l = 1$, $A = 1$, and so $R = \rho$. Thus the resistivity of a substance is the resistance between the faces of a rectangular prism of the substance, which has unit length and unit cross-sectional area. The SI unit of resistivity is *ohm metre* (Ω m), because, from equation (1),

$$\rho = R \frac{A}{l}, \qquad . \qquad . \qquad . \qquad . \qquad (2)$$

which has the units

$$\text{ohm} \times \frac{\text{metre}^2}{\text{metre}} = \text{ohm} \times \text{metre}.$$

RESISTIVITIES

Substance	Resistivity ρ, Ω m (at 20°C)	Temperature coefficient α, K^{-1}
Aluminium	$2 \cdot 82 \times 10^{-8}$	$0 \cdot 0039$
Brass	$c.\ 8 \times 10^{-8}$	$c.\ 0 \cdot 0015$
Constantan[1]	$c.\ 49 \times 10^{-8}$	$0 \cdot 00001$
Copper	$1 \cdot 72 \times 10^{-8}$	$0 \cdot 0043$
Iron	$c.\ 9 \cdot 8 \times 10^{-8}$	$0 \cdot 0056$
Manganin[2] . . .	$c.\ 44 \times 10^{-8}$	$c.\ 0 \cdot 00001$
Mercury	$95 \cdot 77 \times 10^{-8}$	$0 \cdot 00091$
Nichrome[3]	$c.\ 100 \times 10^{-8}$	$0 \cdot 0004$
Silver	$1 \cdot 62 \times 10^{-8}$	$0 \cdot 0039$
Tungsten[4]	$5 \cdot 5 \times 10^{-8}$	$0 \cdot 0058$
Carbon (graphite) . . .	33 to 185×10^{-8}	$-0 \cdot 0006$ to $-0 \cdot 0012$

[1] Also called Eureka; 60% Cu, 40% Ni.
[2] 84% Cu, 12% Mn, 4% Ni; used for resistance boxes and shunts.
[3] Ni–Cu–Cr; used for electric fires—does not oxidize at 1000°C.
[4] Used for lamp filaments—melts at 3380°C.

In equation (1), R is in ohm when l is in metre, A is in metre^2 and ρ is in ohm metre.

The resistivity of a metal is increased by even small amounts of impurity; and alloys, such as Constantan, may have resistivities far greater than any of their constituents.

Conductivity. Relation between J and E

Conductivity, symbol σ, is the inverse of resistivity or $1/\rho$; its SI unit is thus $\Omega^{-1} \text{ m}^{-1}$.

When a p.d. V is applied to a conductor of resistance R, the current I flowing $= V/R$. If the conductor has a uniform area of cross-section A and length l, then $R = \rho l/A = l/\sigma A$, where σ is the conductivity. Hence

$$I = \frac{V}{R} = \frac{\sigma A V}{l}$$

$$\therefore \frac{I}{A} = \sigma \frac{V}{l}$$

The *current density* J along the conductor is defined as the current per unit area of cross-section or I/A. The *electric intensity* E along the wire = potential gradient numerically $= V/l$ (p. 584). Hence, from above,

$$J = \sigma E \qquad . \qquad . \qquad . \qquad . \qquad (1)$$

It may be noted that the relation in (1) is analogous to the relationship between the quantity of heat per second per unit cross-sectional area through a thermal conductor and the temperature gradient, stated on p. 259. It then follows that the thermal conductivity k is analogous to σ, the electrical conductivity.

Resistivity and Conductivity in Electron Theory

Suppose a p.d. V is applied to a metal of length l and uniform cross-section area A. Then a current I flows which is given, from p. 621, by

$$I = nAve \qquad . \qquad . \qquad . \qquad . \qquad (1)$$

On the simple theory given on p. 621, electrons accelerate from zero velocity to a velocity v_1 say between collisions with the fixed atoms in the metal. Thus if a is the acceleration and t is the mean time between collisions, then $v_1 = at$. The drift velocity v in (1) is the *mean* or average velocity of the electrons and is hence given by

$$v = \frac{v_1}{2} = \frac{at}{2} \qquad . \qquad . \qquad . \qquad . \qquad (2)$$

The force F on the electron is Ee, where E is the electric intensity of the field in the metal set up by the p.d. V between the ends of the metal. Now $E =$ potential gradient along metal $= V/l$ (p. 584). If m is the mass of an electron,

$$\therefore a = \frac{F}{m} = \frac{Ee}{m} = \frac{Ve}{lm}$$

Substituting for a in (2), and using the result for v in (1), we obtain

$$I = \frac{Vne^2tA}{2ml}$$

Resistance, R, is defined as the ratio V/I (p. 621).

$$\therefore \text{ resistance } R = \frac{V}{I} = \frac{2ml}{ne^2tA} \qquad . \qquad . \qquad . \qquad (3)$$

When the temperature of a pure metal rises, the ions vibrate with greater amplitude. So the time t between electron collisions is reduced. From (3), it follows that the resistance R then increases, as experiment shows (see p. 668). The resistivity ρ of the metal is given by $R = \rho l/A$. Hence, from (3),

$$\rho = \frac{2m}{ne^2t} \qquad . \qquad . \qquad . \qquad . \qquad (4)$$

The *conductivity*, σ, of the metal $= 1/\rho$, and thus

$$\sigma = \frac{ne^2t}{2m} \qquad . \qquad . \qquad . \qquad . \qquad (5)$$

It should be noted that an accurate treatment of the movement of electrons in metals requires the use of the quantum theory, which is outside the scope of this book.

Mean Free Path of Electrons

Equation (4) or (5) enables us to estimate the time t between *collisions*. Using approximate values for copper at room temperature, we have

$$\rho = 1{\cdot}7 \times 10^{-8} \ \Omega \text{ m}, \ n = 8{\cdot}5 \times 10^{28} \text{ m}^{-3}, \ e = 1{\cdot}6 \times 10^{-19} \text{ C}, \ m = 9{\cdot}1 \times 10^{-31} \text{ kg}$$

From (4),

$$t = \frac{2m}{\rho ne^2} = \frac{2 \times 9{\cdot}1 \times 10^{-31}}{1{\cdot}7 \times 10^{-8} \times 8{\cdot}5 \times 10^{28} \times (1{\cdot}6 \times 10^{-19})^2}$$

$$= 5 \times 10^{-14} \text{ s}$$

The *mean free path* λ_e of the electrons is the average distance travelled between collisions. From quantum theory, the average thermal velocity of a free electron is of the order 10^6 m s^{-1}. Hence

$$\lambda_e = 5 \times 10^{-14} \times 10^6 = 5 \times 10^{-8} \text{ m}$$

The interatomic distance for copper is about 2×10^{-10} m. Hence the mean free path of the electrons is several hundred times the interatomic distance, that is, relatively very long compared with the distance between two atoms.

Ohm's Law

Ohm investigated how the current I in a given metal varied with the p.d. V across it and came to a conclusion about their relationship, stated later, called *Ohm's law*. Let us consider an experiment which can easily be done with modern apparatus. As shown in Fig. 30.17 (i), we connect in series the following apparatus:

    (i) one or more accumulators, S;
    (ii) a milliammeter A reading to 15 milliamperes;
    (iii) a wire-wound resistor Q of the order of 50 ohms;
    (iv) a suitable variable resistance or *rheostat* P of the same order of resistance.

Across the resistor Q we connect a device to measure the potential difference V across it, such as a potentiometer (p. 661) whose calibration does not depend on Ohm's law, otherwise the experiment would not be valid. The milliammeter

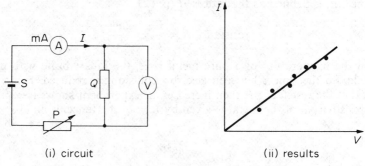

(i) circuit   (ii) results

Fig. 30.17 Demonstration of Ohm's law

calibration likewise must not depend on Ohm's law. By adjusting the resistor we vary the current I through the circuit, and at each value of I we measure V. On plotting V against I we get a straight line through the origin, as in Fig. 30.17 (ii); this shows that the potential difference across the resistor Q is proportional to the current through it:

$$V \propto I$$

This relation was found by Ohm to hold for many conductors. So their resistance R, which is the ratio V/I, is a constant independent of V or I. This is known as *Ohm's law*. Taking into account that resistance depends on temperature and other physical conditions such as mechanical strain, Ohm's law for these type of conductors can be stated as follows:

Under constant physical conditions, the resistance V/I is a constant independent of V or I.

Ohmic and Non-ohmic Conductors

Ohm's law is obeyed by the most important class of conductors, metals. These are called *ohmic conductors*. In this type of conductor the current I is reversed in direction when the p.d. V is reversed but the magnitude of I is unchanged. The characteristic or I–V graph is thus a straight line *passing through the origin*, as shown in Fig. 30.18 (i). As we see later, an electrolyte such as copper sulphate solution with copper electrodes obeys Ohm's law, Fig. 30.18 (ii).

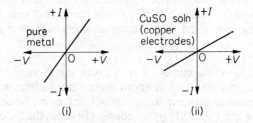

(i)   (ii)

Fig. 30.18 Characteristics of ohmic conductors

Non-ohmic conductors are those which do not obey Ohm's law ($V \propto I$). Many useful components in the electrical industry must be non-ohmic; for example, a non-ohmic component is essential in a radio receiver circuit. A non-ohmic characteristic or $I - V$ graph may have a curve instead of a straight line; or it may not pass through the origin as in the ohmic characteristic; or it may conduct poorly or not at all when the p.d. is reversed ($- V$). Figure 30.19 illustrates the

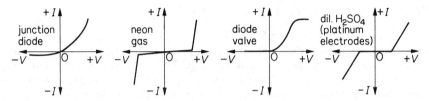

Fig. 30.19 Characteristics of some non-ohmic conductors

non-ohmic characteristics of a junction (semiconductor) diode, neon gas, a diode valve, and the electrolyte dilute sulphuric acid with platinum electrodes where, unlike Fig. 30.18 (ii), an e.m.f. is produced at the electrodes by the chemicals liberated there (p. 683).

Example

Some electrically-conducting solids have a resistance which rises when the temperature is increased whilst for others the resistance decreases when the temperature is increased. Give an example of each type and explain the difference in behaviour.

Obtain, from first principles, the formulae for the resistance of a combination of resistors of resistance R_1, R_2, ..., R_n when they are connected (a) in series, (b) in parallel.

A cell C, having an e.m.f. 2·2 V and negligible internal resistance, is connected to the combination of resistors shown below. What is the effective value of the resistance connected across the terminals of the cell? What are the values of the currents i_1, i_2 and i_3?

If the two 5 Ω resistors in the above circuit are in the form of two straight, parallel wires, 2 m long and 10 mm apart, what, approximately, is the magnitude of the force between them? (O. & C.)

The resistance of copper *increases* when its temperature is increased. See p. 668.
The resistance of silicon *decreases* when its temperature is increased. See p. 824.
Calculation (i) Resistance along DEF = 10 + 5 = 15 Ω.
Since DEF is in parallel with the 5 Ω resistor between D,G (G and F are connected together), the combined resistance R is given by

$$\frac{1}{R} = \frac{1}{15} + \frac{1}{5} = \frac{4}{15}, \text{ so } R = 15/4 = 3.75 \ \Omega.$$

Thus total resistance between terminals of C = 10 + 3·75 = 13·75 Ω . . (1)

So
$$i_1 = \frac{E}{R} = \frac{2 \cdot 2}{13 \cdot 75} = 0 \cdot 16 \text{ A} \qquad \qquad (2)$$

Also, since DEF (15 Ω) is in parallel with DG (5 Ω),

$$i_2 = \frac{15}{5 + 15} \times 0 \cdot 16 \text{ A} = 0 \cdot 12 \text{ A}$$

and $i_3 = 0 \cdot 16 - 0 \cdot 12 = 0 \cdot 04$ A.

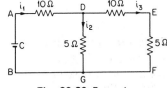

Fig. 30.20 Example

(ii) Force between two parallel straight long conductors carrying currents I_1, I_2 respectively is given approximately by

$$F = \frac{\mu_0 I_1 I_2 l}{2\pi r}$$

where l is the length of one conductor, r is the distance apart and $\mu_0 = 4\pi \times 10^{-7}$ numerically (see p. 713). So

$$F = \frac{4\pi \times 10^{-7} \times 0 \cdot 12 \times 0 \cdot 04 \times 2}{2\pi \times 10 \times 10^{-3}}$$

$$= 1 \cdot 92 \times 10^{-7} \text{ N}.$$

Heat and Power

Electrical Heating. Joule's Laws

In 1841 Joule studied the heating effect of an electric current by passing it through a coil of wire in a jar of water, Fig. 30.21. He used various currents, measured by an early form of galvanometer G, and various lengths of wire, but always the same mass of water. The rise in temperature of the water, in a given time, was then proportional to the heat developed by the current in that time Joule found that the heat produced in a given time, with a given wire, was proportional to I^2, where I is the current flowing. If H is the heat produced per second, then

$$H \propto I^2 \qquad . \qquad . \qquad . \qquad . \qquad . \qquad (1)$$

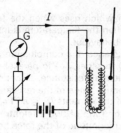 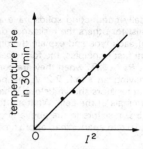

Fig. 30.21 Joule's experiment on heating effect of current

Joule also made experiments on the heat produced by a given current in different wires. He used wires of different lengths, but of the same diameter, and of the same material; he found that the rate at which heat was produced, by a given current, was proportional to the length of the wire. That is to say, he found that the rate of heat production was proportional to what Ohm had already called the *resistance* of the wire:

$$H \propto R \qquad . \qquad . \qquad . \qquad . \qquad (2)$$

Relationships (2) and (3) together give

$$H \propto I^2 R \qquad . \qquad . \qquad . \qquad . \qquad (3)$$

Mechanism of the Heating Effect

Heat is a form of energy. The heat produced per second by a current in a wire is therefore a measure of the energy which it liberates in one second, as it flows through the wire.

The heat is produced, we suppose, by the free electrons as they move through the metal. On their way they collide frequently with atoms; at each collision they lose some of their kinetic energy, and give it to the atoms which they strike. Thus, as the current flows through the wire, it increases the kinetic energy of vibration of the metal atoms: it generates heat in the wire. The electrical resistance of the metal is due, we say, to its atoms obstructing the drift of the electrons past them: it is analogous to mechanical friction. As the current flows through the wire, the energy lost per second by the electrons is the electrical power supplied by the battery which maintains the current. That power comes, as we shall see later, from the chemical energy liberated by these actions within the battery.

Potential Difference and Energy

On p. 580 we defined the potential difference V_{AB} between two points, A and B, as the work done by an external agent in taking a unit positive charge from B to A, Fig. 30.22 (i). This definition applies equally well to points in an electrostatic field and to points on a conductor carrying a current.

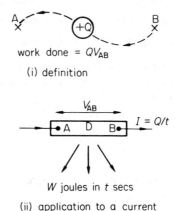

work done $= QV_{AB}$

(i) definition

V_{AB}

$I = Q/t$

W joules in t secs

(ii) application to a current

Fig. 30.22 Potential difference and energy

In Fig. 30.22 (ii), D represents any electrical device or circuit element: a lamp, motor, or battery on charge, for example. A current of I ampere flows through it from the terminal A to the terminal B; if it flows for t second, the charge Q which it carries from A to B is, since a current is the quantity of electricity per second flowing,

$$Q = It \text{ coulomb.} \qquad . \qquad . \qquad . \qquad . \qquad (1)$$

Let us suppose that the device D liberates a total amount of energy W joules in the time t; this total may be made up of heat, light, sound, mechanical work, chemical transformation, and any other forms of energy. Then W is the amount of electrical energy given up by the charge Q in passing through the device D from A to B.

$$\therefore W = QV_{AB}, \qquad . \qquad . \qquad . \qquad . \qquad . \qquad (2)$$

where V_{AB} is the potential difference between A and B in volts.

The work, in all its forms, which the current I does in t seconds, as it flows through the device, is therefore

$$W = IV_{AB}t, \qquad . \qquad . \qquad . \qquad . \qquad . \qquad (3)$$

by equations (1) and (2).

Electrical Power

The energy liberated per second in the device is defined as its electrical *power*. The electrical power, P, supplied is given, from above, by (Fig. 30.23 (i)).

$$P = \frac{W}{t} = \frac{IV_{AB}t}{t},$$

or

$$P = IV_{AB} \qquad . \qquad . \qquad . \qquad . \qquad . \qquad (1)$$

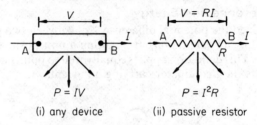

Fig. 30.23 Power equations

When an electric current flows through a wire or 'passive' resistor, all the power which it conveys to the wire appears as heat. If I is the current, R is the resistance, then $V_{AB} = IR$, Fig. 30.23 (ii)

$$\therefore P = I^2 R \qquad \qquad \qquad (2)$$

Also,

$$P = \frac{V_{AB}^{2}}{R} \qquad \qquad \qquad (3)$$

The power, P, is in *watts* (W) when I is in ampere, R is in ohm, and V_{AB} is in volt. 1 kilowatt (kW) = 1000 watts.

The formulae for power, $P = I^2 R$ or V^2/R, is true only when all the electrical power supplied is dissipated as heat. As we shall see, the formulae do not hold when part of the electrical energy supplied is converted into mechanical work, as in a motor, or into chemical energy, as in an accumulator being charged. A device which converts all the electrical energy supplied to it into heat is called a 'passive' resistor; it may be a wire, or a strip of carbon, or a liquid which conducts electricity but is not decomposed by it. Since the joule (J) is the unit of heat, it follows that, for a resistor, the heat H in it in joules is given by

$$H = IVt,$$

or by

$$H = I^2 Rt \qquad \qquad \qquad (4)$$

or by

$$H = \frac{V^2 t}{R}.$$

The units of I, V, R are ampere (A), volt (V), ohm (Ω) respectively.

High-tension Transmission

When electricity has to be transmitted from a source, such as a power station, to a distant load, such as a factory, the two must be connected by cables. These cables have resistance, which is in effect added to the internal resistance of the generator; power is wasted in them as heat. If r is the total resistance of the cables, and I the supply current, the power wasted is $I^2 r$. The power delivered to the factory is IV, where V is the potential difference at the factory. Economy requires the waste power, $I^2 r$, to be small; but it also requires the cables to be thin, and therefore cheap to buy and erect. The thinner the cables, however, the higher their resistance r. Thus the most economical way to transmit the power is to make the current, I, as small as possible; this means making the potential difference V as high as possible. When large amounts of power are to be transmitted, therefore, very high voltages are used: 400 000 volts on the main lines of the British grid, 23 000 volts on subsidiary lines. These voltages are much too high to be brought into a house, or even a factory. They are stepped down by transformers, in a way which we shall describe later; stepping-down in that way is possible only with alternating current, which is one of the main reasons why alternating current is so widely used.

Summary of Formulae Related to Power

In any device whatever (see Fig. 30.23 (i)):

Electrical power consumed = power developed in other forms,

$$P = IV,$$

watts = amperes × volts.

In a passive resistor (see Fig. 30.23 (ii)):

(i) $$V = IR; \ I = \frac{V}{R}; \ R = \frac{V}{I},$$

(ii) Power consumed = heat developed per second, in watts.

$$P = I^2R = IV = \frac{V^2}{R}.$$

(iii) Heat developed in time t:

Electrical energy consumed = heat developed in joules

$$I^2Rt = IVt = \frac{V^2}{R}t.$$

Board of Trade (commercial) unit = kilowatt hour (kWh) = kilowatt × hour
$$= 3 \cdot 6 \times 10^6 \text{ joule.}$$

Example

An electric heating element to dissipate 480 watts on 240 V mains is to be made from Nichrome ribbon 1 mm wide and thickness 0·05 mm. Calculate the length of ribbon required if the resistivity of Nichrome is $1 \cdot 1 \times 10^{-6}$ ohm metre.

Power, $$P = \frac{V^2}{R},$$

$$\therefore R = \frac{V^2}{P} = \frac{240^2}{480} = 120 \ \Omega.$$

The area A of cross-section of the ribbon = $1 \times 0 \cdot 05$ mm^2 = $0 \cdot 05 \times 10^{-6}$ m^2.

From $$R = \frac{\rho l}{A},$$

$$\therefore l = \frac{R.A}{\rho} = \frac{120 \times 0 \cdot 05 \times 10^{-6}}{1 \cdot 1 \times 10^{-6}} = 5 \cdot 45 \text{ m.}$$

Electromotive Force

E.M.F. and Internal Resistance

An electrical generator provides energy and power. This is considered later. Here we consider the current and potential difference, p.d., in circuits connected to a generator such as a battery.

If a high resistance voltmeter is connected across the terminals of a dry battery B, the meter may read about 1·5 V, Fig. 30.24 (i). Since practically no current flows from the battery in this case we say it is on 'open circuit'. The p.d. across the terminals of a battery (or any other generator) on open circuit is called its *electromotive force* or *e.m.f.*, symbol E. We define e.m.f. in terms of energy later (p. 645).

When a resistor is connected to the battery, the current flows through the resistor and through the *internal resistance*, r, of the battery to complete the circuit flow.

The e.m.f. of a battery depends on the nature of the chemicals used and not on its size. A tiny battery has the same e.m.f. as a large battery made of the same chemicals. The internal resistance of the tiny battery, however, is much less than the large battery. Provided only a small current is taken from a battery, its e.m.f. and internal resistance are fairly constant.

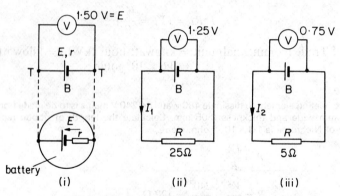

Fig. 30.24 E.m.f. and internal resistance

Any electrical generator, then, has two important properties, an e.m.f. E and an internal resistance r. As shown in Fig. 30.24 (i), E and r may be represented separately in a diagram, though in practice they are inseparable between the terminals T,T.

Circuit Principles. Terminal p.d.

In Fig. 30.24 (ii), a resistor of 25 Ω is connected to the battery B so that a current I_1 flows in the circuit. The voltmeter reading across the battery terminals, or terminal p.d., may then be 1·25 V, although the e.m.f. is 1·5 V. When the resistor is replaced by one of 5 Ω, Fig. 30.24 (iii), a larger current I_2 flows and the voltmeter reading or terminal p.d. is now 0·75 V.

To understand why the terminal p.d. varies when a current flows from a battery, it is important to realise that the voltmeter is connected across the *external* or outside resistance in Fig. 30.24 (ii). So 1·25 V is the p.d. across the 25 Ω resistor. Now the *e.m.f.*, 1·5 V maintains the current in the *whole* circuit, that is, through the external *and* internal resistance r. So we deduce that the p.d. across the internal resistance $r = 1·5 - 1·25 = 0·25$ V.

Similarly, in Fig. 30.24 (iii) 0·75 V is the p.d. across the external resistance 5 Ω. So in this case the p.d. across the internal resistance $r = 1·5 - 0·75 = 0·75$ V. A common error is to think that the voltmeter across the terminals reads the e.m.f. This is not the case here as there is a p.d. across the internal resistance when a current flows, which cannot be observed directly using a voltmeter.

In Fig. 30.24 (ii), the p.d. across the 25 Ω external resistor R is 1·25 V and the p.d. across the internal resistance r is 0·25 V. Since the same current flows in R and r it follows that $R = 5r$, or $r = R/5 = 5$ Ω.

Similarly, the p.d. across the external resistor R of 5 Ω in Fig. 30.24 (iii) is 0·75 V and that across the internal resistance r is 0·75 V. So $r = R = 5$ Ω, as previously calculated.

You should now see that as the external resistance R increases, the terminal p.d. increases. When R is an infinitely-high value, the terminal p.d. is equal to the e.m.f. value E.

Circuit Formulae

In Fig. 30.24 (ii), a battery of e.m.f. E and internal resistance r is joined to an external resistor R, and a current I flows in the circuit. The p.d. across $R = IR$ and the p.d. across $r = Ir$. So

$$E = IR + Ir \qquad . \qquad . \qquad . \qquad . \qquad (1)$$

or

$$I = \frac{E}{R + r} \qquad . \qquad . \qquad . \qquad . \qquad (2)$$

Note carefully that when the e.m.f. value E is used to find the current I, the resistance $(R + r)$ of the *whole* circuit is required.

On the other hand, the terminal p.d., $V = $ p.d. across external resistor R. So

$$\text{terminal p.d. } V = IR = \frac{ER}{R + r}.$$

Further, from (1),

$$V = E - Ir \qquad . \qquad . \qquad . \qquad . \qquad (3)$$

This is a useful formula for the terminal p.d. when the e.m.f. E and internal resistance r are known. For example, suppose a current of 0·5 A flows from a battery of e.m.f. E of 3 V and internal resistance 4 Ω. Then

$$\text{terminal p.d. } V = E - Ir = 3 - (0·5 \times 4) = 1 \text{ V.}$$

The internal resistance r can be found from the e.m.f. E, the terminal p.d. V and the current I. From (1),

$$Ir = E - IR = E - V$$

So

$$r = \frac{E - V}{I} \qquad . \qquad . \qquad . \qquad . \qquad (4)$$

If I is needed, we may use $I = V/R$.

Example on Circuit

A battery of e.m.f. 1·50 V has a terminal p.d. of 1·25 V when a resistor of 25 Ω is joined to it. Calculate the current flowing, the internal resistance r and the terminal p.d. when a resistor of 10 Ω replaces the 25 Ω resistor.

We have

$$I = \frac{V}{R} = \frac{1·25}{25} = 0·05 \text{ A.}$$

Also, $$r = \frac{\text{p.d.}}{\text{current}} = \frac{E - V}{I}$$

$$= \frac{1 \cdot 50 - 1 \cdot 25}{0 \cdot 05} = \frac{0 \cdot 25}{0 \cdot 05} = 5 \ \Omega.$$

When the external resistor is 10 Ω, the current I flowing is

$$I = \frac{E}{R + r} = \frac{1 \cdot 50}{10 + 5} = 0 \cdot 1 \ A.$$

So   terminal p.d. $V = IR = 0 \cdot 1 \times 5 = 0 \cdot 5$ V.

Terminal p.d. with Current in Opposition to E.M.F.

So far we have considered the terminal p.d. when the battery e.m.f. maintains the current. Suppose, however, that a current is passed through a battery in *opposition* to its e.m.f., a case which occurs in re-charging an accumulator, for example.

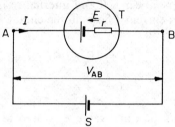

Fig. 30.25 Terminal p.d.

Figure 30.25 shows a supply S sending a current I through a battery T in opposition to its e.m.f. E. The terminal p.d. V_{AB} must be greater than E in this case. Since the net p.d., $V_{AB} - E$, across the terminals must maintain the current I in r, then the net p.d. $= Ir$. So

$$V_{AB} - E = Ir.$$

Hence $$V_{AB} = E + Ir.$$

In contrast, when the battery e.m.f. itself maintains a current, so that the current is in the same direction as the e.m.f., then the terminal p.d. $V = E - Ir$, as we have already seen.

Example on Terminal p.d.

Figure 30.26 shows a circuit with two batteries in opposition to each other. One has an e.m.f.

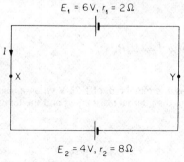

Fig. 30.26 Example

E_1 of 6 V and internal resistance r_1 of 2 Ω and the other an e.m.f. E_2 of 4 V and internal resistance r_2 of 8 Ω. Calculate the p.d. V_{XY} across XY.

Net e.m.f. in circuit $= E_1 - E_2 = 6 - 4 = 2$ V.

So   current, $I = \dfrac{E_1 - E_2}{r_1 + r_2} = \dfrac{6 - 4}{2 + 8} = 0\cdot2$ A

The e.m.f. E_1 is in the same direction as the current. So

terminal p.d., $V_{XY} = E_1 - Ir_1 = 6 - (0\cdot2 \times 2) = 5\cdot6$ V.

If we consider the battery of e.m.f. E_2, we see that I flows in *opposition* to E_2. In this case,

terminal p.d., $V_{XY} = E_2 + Ir_2 = 4 + (0\cdot2 \times 8) = 5\cdot6$ V.

This result agrees with the terminal p.d. value obtained by using E_1.

Cells in Series and Parallel

When cells or batteries are in series and assist each other, then the total e.m.f.

$$E = E_1 + E_2 + E_3 + \ldots, \qquad . \quad . \quad . \quad (1)$$

and the total internal resistance

$$r = r_1 + r_2 + r_3 + \ldots, \qquad . \quad . \quad . \quad (2)$$

where E_1, E_2 are the individual e.m.f.s and r_1, r_2 are the corresponding internal resistances. If one cell, e.m.f. E_2 say, is turned round 'in opposition' to the others, then $E = E_1 - E_2 + E_3 + \ldots$; but the total internal resistance remains unaltered.

When *similar* cells are in parallel, the total e.m.f. $= E$, the e.m.f. of any one of them. The internal resistance r is here given by

$$\frac{1}{r} = \frac{1}{r_1} + \frac{1}{r_1} + \ldots, \qquad . \quad . \quad . \quad . \quad (3)$$

where r_1 is the internal resistance of each cell. If different cells are in parallel, there is no simple formula for the total e.m.f. and the total internal resistance, and any calculations involving circuits with such cells are dealt with by applying Kirchhoff's laws (p. 648).

Examples on Circuits

1. Two similar cells A and B are connected in series with a coil of resistance 9·8 Ω. A voltmeter of very high resistance connected to the terminals of A reads 0·96 V and when connected to the terminals of B it reads 1·00 V. Find the internal resistance of each cell. (Take the e.m.f. of a cell as 1·08 V.) (*L.*)

The p.d. across both cells $= 0\cdot96 + 1\cdot00 = 1\cdot96$ V

$= $ p.d. across 9·8 Ω.

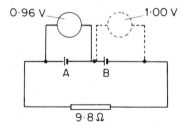

Fig. 30.27 Example

$$\therefore \text{ current flowing, } I, = \frac{V}{R} = \frac{1 \cdot 96}{9 \cdot 8} = 0 \cdot 2 \text{ A.}$$

Now terminal p.d. across each cell $= E - Ir$.

$$\therefore \text{ for cell A, } 0 \cdot 96 = 1 \cdot 08 - 0 \cdot 2r, \text{ or } r = 0 \cdot 6 \ \Omega.$$

$$\text{for cell B, } 1 \cdot 00 = 1 \cdot 08 - 0 \cdot 2r, \text{ or } r = 0 \cdot 4 \ \Omega.$$

2. What is meant by the *electromotive force* of a cell?

A voltmeter is connected in parallel with a variable resistance, R, which is in series with an ammeter and a cell. For one value of R the meters read 0·3 A and 0·9 V. For another value of R the readings are 0·25 A and 1·0 V. Find the values of R, the e.m.f. of the cell, and the internal resistance of the cell. What assumptions are made about the resistance of the meters in the calculation?

If in this experiment the ammeter had a resistance of 10 Ω and the voltmeter a resistance of 100 Ω and R was 2 Ω, what would the meters read? (*L.*)

The voltmeter reads the p.d. across the cell if the resistances of the meters are neglected. Thus, with the usual notation,

$$E - Ir = 0 \cdot 9, \text{ or } E - 0 \cdot 3r = 0 \cdot 9 \qquad \qquad . \qquad . \qquad . \qquad \text{(i)}$$

and
$$E - 0 \cdot 25r = 1 \cdot 0 \qquad \qquad . \qquad . \qquad . \qquad . \qquad \text{(ii)}$$

Subtracting (i) from (ii),

$$0 \cdot 05r = 0 \cdot 1, \text{ i.e. } r = 2 \ \Omega.$$

Also, from (i),

$$E = 0 \cdot 3r + 0 \cdot 9 = 0 \cdot 6 + 0 \cdot 9 = 1 \cdot 5 \text{ V.}$$

Further,

$$R_1 = \frac{V}{I} = \frac{0 \cdot 9}{0 \cdot 3} = 3 \ \Omega,$$

and

$$R_2 = \frac{1 \cdot 0}{0 \cdot 25} = 4 \ \Omega.$$

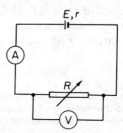

Fig. 30.28 Example

If the voltmeter has 100 Ω resistance and is in parallel with the 2 Ω resistance, the combined resistance R is given by

$$\frac{1}{R} = \frac{1}{2} + \frac{1}{100} = \frac{51}{100}, \text{ or } R = \frac{100}{51} \ \Omega.$$

$$\therefore \text{ current, } I = \frac{E}{\text{Total resistance}}$$

$$= \frac{1 \cdot 5}{\dfrac{100}{51} + 10 + 2} = 0 \cdot 11 \text{ A.}$$

$$\text{Also, voltmeter reading} = IR = 0 \cdot 11 \times \frac{100}{51} = 0 \cdot 21 \text{ V.}$$

Electromotive Force and Energy

We can get a rigorous definition of electromotive force from energy principles.

Consider a circuit, such as Fig. 30.24 (ii), in which a battery of e.m.f. E and internal resistance r is connected to an external resistor R so that a current I flows in the circuit. When a charge Q has passed right round the circuit including the battery, the energy delivered by the battery comes from the chemical energy used inside the battery. The current, in passing round the circuit, will have converted this energy into heat. Some of the heat will have been dissipated in the external resistance R, some in the internal resistance r.

We can define the e.m.f. E of a battery or any other generator as the *total energy per coulomb it delivers round a circuit joined to it*. So if a device has an electromotive force E, then, in passing a charge Q round a circuit joined to it, it liberates an amount of electrical energy equal to QE. If a charge Q is passed through the source against its e.m.f., then the work done against the e.m.f. is QE. The above definition of e.m.f. does not depend on any assumptions about the nature of its source.

If a device of e.m.f. E passes a steady current I for a time t, then the charge that it circulates is

$$Q = It.$$

Thus:

$$\text{electrical energy liberated, } W, = QE = IEt, \qquad . \qquad . \qquad (1)$$

and

$$\text{electrical power generated, } P = \frac{W}{t} = EI \qquad . \qquad . \qquad . \qquad (2)$$

We can now define e.m.f. in terms of power and current, and therefore in a way suitable for dealing with circuit problems. From equation (2)

$$P = EI,$$

or

$$E = \frac{P}{I}.$$

Thus *the e.m.f. of a device is the ratio of the electrical power which it generates to the current which it delivers*. If current is forced through a device in opposition to its e.m.f., then equation (2) gives the power consumed in overcoming the e.m.f.

Electromotive force resembles potential difference in that both can be defined as the ratio of power to current. The unit of e.m.f. is therefore 1 watt per ampere, or 1 volt; and *the e.m.f. of a source, in volt, is numerically equal to the power which it generates when it delivers a current of 1 ampere*.

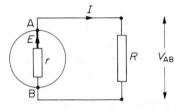

Fig. 30.29 A complete circuit

We can apply the definition of E in terms of power to the circuit in Fig. 30.29. The total power supplied by the source is EI. The power delivered to the external resistor is called the *output power*. Its value $= IV_{AB} = I \times IR = I^2R$. The power delivered to the internal resistance $r = I^2r$. So

$$EI = I^2R + I^2r \qquad . \qquad . \qquad . \qquad . \qquad (1)$$

Dividing by I,
$$E = IR + Ir \qquad . \qquad . \qquad . \qquad . \qquad (2)$$

and
$$I = \frac{E}{R+r} \qquad . \qquad . \qquad . \qquad . \qquad (3)$$

The results in (2) and (3) were obtained earlier.

Output Power and Efficiency

As we have just seen, the power delivered to the external resistor R, often called the *load* of a battery or other generator, is given by $P_{out} = IV_{AB}$ in Fig. 30.29. The power supplied by the source or generator $P_{gen} = IE$. The difference between the power generated and the output is the power wasted as heat in the source: I^2r. The ratio of the power output to the power generated is the efficiency, η, of the circuit as a whole:

$$\eta = \frac{P_{out}}{P_{gen}} \qquad . \qquad . \qquad . \qquad . \qquad (1)$$

So
$$\eta = \frac{P_{out}}{P_{gen}} = \frac{IV_{AB}}{IE} = \frac{V_{AB}}{E}.$$

Now $V_{AB} = IR = ER/(R+r)$. So

$$\eta = \frac{R}{R+r} \qquad . \qquad . \qquad . \qquad . \qquad (2)$$

This shows that the efficiency tends to unity (or 100 per cent) as the load resistance R tends to infinity. For high efficiency the load resistance must be several times the internal resistance of the source. When the load resistance is equal to the internal resistance, the efficiency is 50 per cent. (See Fig. 30.30 (i).)

Power Variation. Maximum Power

Now let us consider how the power output varies with the load resistance. Since the power output $= IV_{AB} = I \times IR$, then

$$P_{out} = I^2R.$$

Also
$$I = \frac{E}{R+r},$$

so
$$P_{out} = \frac{E^2R}{(R+r)^2}.$$

If we take fixed values of E and r, and plot P_{out} as a function of R, we find that it passes through a maximum when $R = r$, Fig. 30.30 (i). We can get the same result in a more general way by differentiating P_{out} with respect to R, and equating the differential coefficient to zero. Physically, this result means that the power output is very small when R is either very large or very small, compared with r. When R is very large, the terminal potential difference, V_{AB}, approaches a constant value equal to the e.m.f. E (Fig. 30.30 (ii)); as R is increased the current falls, and the power IV_{AB} falls with it. When R is very small, the current approaches the constant value E/r, but the potential difference (which

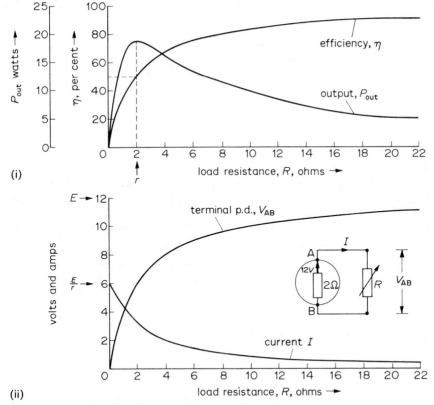

(i)

(ii)

Fig. 30.30 Effects of varying load resistance in circuit

is equal to IR) falls steadily with R; the power output therefore falls likewise. Consequently the power output is greater for a moderate value of R; the mathematics shows that this value is actually $R = r$.

To prove $R = r$, differentiate the expression for P_{out} with respect to R. Then

$$E^2 \frac{(R + r)^2 - R.2(R + r)}{(R + r)^4} = 0, \text{ for a maximum.}$$

From the numerator, $r^2 - R^2 = 0$, or $R = r$.

Examples of Loads in Electrical Circuits

The loading on a dynamo or battery is generally adjusted for high efficiency, because that means greatest economy. Also, if a large dynamo were used with a load not much greater than its internal resistance, the current would be so large that the heat generated in the internal resistance would ruin the machine. With batteries and dynamos, therefore, the load resistance is made many times greater than the internal resistance.

Loading for greatest *power output* is common in communication engineering. For example, the last transistor in a receiver delivers electrical power to the loudspeaker, which the speaker converts into mechanical power as sound waves (p. 524). Because it converts electrical energy into mechanical energy, and not heat, the loudspeaker is not a passive resistor, and the simple equations above do not apply to it. Nevertheless, circuit conditions can be specified which enable the transistor to deliver the greatest power to the speaker; these are similar to the condition of equal load and internal resistances, and are usually satisfied in practice.

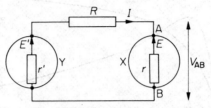

Fig. 30.31 Accumulator charging

Load not a Passive Resistor

As an example of a load which is not a passive resistor, we shall take an accumulator being charged. The charging is done by connecting the accumulator X in opposition to a source of greater e.m.f., Y in Fig. 30.31, via a controlling resistor R. If E, E' and r, r' are the e.m.f. and internal resistances of X and Y respectively, then the current I is given by the equation:

$$\left.\begin{array}{c}\text{power generated}\\ \text{in Y}\end{array}\right\} = \left\{\begin{array}{c}\text{power converted to}\\ \text{chemical energy}\\ \text{in X}\end{array}\right\} + \left\{\begin{array}{c}\text{power dissipated}\\ \text{as heat in all}\\ \text{resistances}\end{array}\right.$$

$$E'I \quad = \quad EI \quad + \quad I^2R + I^2r' + I^2r \quad (1)$$

Thus $$(E' - E)I = I^2(R + r' + r),$$

whence $$I = \frac{E' - E}{R + r' + r} \qquad . \qquad . \qquad . \qquad (2)$$

The potential difference across the accumulator, V_{AB}, is given by

$$\left.\begin{array}{c}\text{power delivered}\\ \text{to X}\end{array}\right\} = \left\{\begin{array}{c}\text{power converted to}\\ \text{chemical energy}\end{array}\right\} + \left\{\begin{array}{c}\text{power dissipated}\\ \text{as heat}\end{array}\right.$$

$$IV_{AB} \quad = \quad IE \quad + \quad I^2r$$

Hence $$V_{AB} = E + Ir \qquad . \qquad . \qquad . \qquad . \qquad (3)$$

Equation (3) shows that, when current is driven through a generator in opposition to its e.m.f., then the potential difference across the generator is equal to the *sum* of its e.m.f. and the voltage drop across its internal resistance. This result follows at once from energy considerations, as we have just seen.

Kirchhoff's Laws

A 'network' is usually a complicated system of electrical conductors. Kirchhoff (1824–87) extended Ohm's law to networks, and gave two laws, which together enabled the current in any part of the network to be calculated.

The *first law* refers to any point in the network, such as A in Fig. 30.32 (i); it states that the total current flowing into the point is equal to the total current flowing out of it:

$$I_1 = I_2 + I_3$$

The law follows from the fact that electric charges do not accumulate at the points of a network. It is often put in the form that *the algebraic sum of the currents at a junction of a circuit is zero*, or

$$\Sigma I = 0$$

where a current, I, is reckoned positive if it flows towards the point, and negative if it flows away from it. Thus at A in Fig. 30.32 (i),

$$I_1 - I_2 - I_3 = 0.$$

Kirchhoff's first law gives a set of equations which contribute towards the solving of the network; in practice, however, we can shorten the work by putting the first law straight into the diagram, as shown in Fig. 30.32 (ii) for example, since

$$\text{current along AC} = I_1 - I_g.$$

Kirchhoff's *second law* is a generalisation for e.m.f. and p.d. in the complete circuit. It refers to any closed loop, such as AYCA in Fig. 30.32 (ii); and it states that, round such a loop, the algebraic sum of the voltage drops is equal to the algebraic sum of the e.m.f.s:

$$\Sigma RI = \Sigma E.$$

Thus, clockwise round the loop,

$$R_{AC}(I_1 - I_g) - R_g I_g = E_2.$$

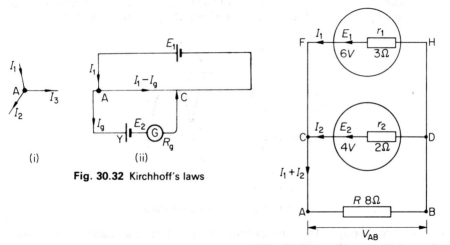

(i) (ii)

Fig. 30.32 Kirchhoff's laws

Fig. 30.33 Load across cells in parallel

Example

Figure 30.33 shows a network in which the currents I_1, I_2, can be found from Kirchhoff's laws

From the first law the current in the 8 Ω wire is $(I_1 + I_2)$, assuming I_1, I_2 are the currents through the cells.

Taking closed circuits formed by each cell with the 8 Ω wire, we have, from the second law,

$$E_1 = 6 = 3I_1 + 8(I_1 + I_2) = 11I_1 + 8I_2$$

and

$$E_2 = 4 = 2I_2 + 8(I_1 + I_2) = 8I_1 + 10I_2.$$

Solving the two equations, we find

$$I_1 = \frac{14}{23} = 0.61 \text{ A}, \ I_2 = -\frac{2}{23} = -0.09 \text{ A}.$$

The minus sign indicates that the current I_2 flows in the sense opposite to that shown in the diagram; i.e. it flows against the e.m.f. of the generator E_2.

The Thermoelectric Effect

Seebeck Effect

The heating effect of the current converts electrical energy into heat, but we have not so far described any mechanism which converts heat into electrical energy. This was discovered by Seebeck in 1822.

In his experiments he connected a plate of bismuth between copper wires leading to a galvanometer, as shown in Fig. 30.34 (i). He found that if one of the bismuth-copper junctions was heated, while the other was kept cool, then a current flowed through the galvanometer. The direction of the current was from the copper to the bismuth at the cold junction. We can easily repeat Seebeck's experiment, using copper and iron wires and a galvanometer capable of indicating a few microamperes (p. 700), Fig. 30.34 (ii).

Thermocouples

Seebeck went on to show that a current flowed, without a battery, in any circuit containing two different metals, with their two junctions at different temperatures. Currents obtained in this way are called thermoelectric currents, and a pair of metals, with their junctions at different temperatures, are said to form a thermocouple. The following is a list of metals, such that any two of them

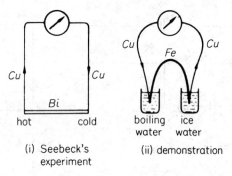

(i) Seebeck's
experiment

(ii) demonstration

Fig. 30.34 The thermoelectric effect

form a thermocouple, then the current will flow from the higher to the lower in the list, across the cold junction:

Antimony, Iron, Zinc, Lead, Copper, Platinum, Bismuth.

Thermoelectric currents often appear when they are not wanted; they may arise from small differences in purity of two samples of the same metal, and from small differences of temperature—due, perhaps, to the warmth of the hand. They can cause a great deal of trouble in circuits used for precise measurements, or for detecting other small currents, not of thermal origin. As sources of electrical energy, thermoelectric currents are neither convenient nor economical, but they have been used—e.g. in solar batteries. Their only wide application is in the measurement of temperature, and of other quantities, such as radiant energy, which can be measured by a temperature rise.

Variation of Thermoelectric E.M.F. with Temperature

On p. 662 we shall see how thermoelectric e.m.f.s are measured. When the cold junction of a given thermocouple is kept constant at 0°C, and the hot junction temperature θ°C is varied, the e.m.f. E is found to vary as $E = a\theta + b\theta^2$, where a, b are constants. This is a parabola-shaped curve (Fig. 30.35). The temperature

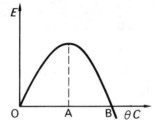

Fig. 30.35 Thermoelectric e.m.f. variation with temperature

A corresponding to the maximum e.m.f. is known as the *neutral temperature*; it is about 250°C for a copper-iron thermocouple. Beyond the temperature B, known as the *inversion temperature*, the e.m.f. reverses. Thermoelectric thermo-meters, which utilise thermocouples, are used only as far as the neutral tem-perature, as the same e.m.f. is obtained at two different temperatures, from Fig. 30.35. See also p. 303.

Peltier and Thomson Effects

When a current flows along the junction A of two metals in series, heat is evolved or absorbed at A depending on the current direction. This is known as the *Peltier effect*. It has no connexion with the usual heating or Joule effect of a current, discussed on p. 636. The Joule effect is irreversible, that is, heat is obtained in both directions of the current. In the Peltier effect, however, the effect is reversed when the current is reversed; that is, a cooling is produced at the junction of two metals in one direction, and an evolution of heat in the other direction.

Sir William Thomson, later Lord Kelvin, also found that heat was evolved or absorbed when a current flows along a metal whose ends are kept at different temperatures. The *Thomson effect*, like the Peltier effect, is also reversible.

Exercises 30

Circuit Calculations

1. A battery of e.m.f. 4 V and internal resistance 2 Ω is joined to a resistor of 8 Ω. Calculate the terminal p.d.

What additional resistance in series with the 8 Ω resistor would produce a terminal p.d. of 3·6 V?

2. A battery of e.m.f. 24 V and internal resistance r is connected to a circuit having two parallel resistors of 3 Ω and 6 Ω in series with an 8 Ω resistor, Fig. 30A (i). The current flowing in the 3 Ω resistor is then 0·8 A. Calculate (i) the current in the 6 Ω resistor, (ii) r, (iii) the terminal p.d. of the battery.

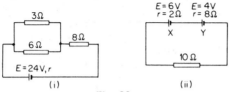

Fig. 30A

3. A battery X of e.m.f. 6 V and internal resistance 2 Ω is in series with a battery Y of e.m.f. 4 V and internal resistance 8 Ω so that the two e.m.f.s act in the same direction, Fig. 30A (ii). A 10 Ω resistor is connected to the batteries. Calculate the terminal p.d. of each battery.

If Y is reversed so that the e.m.f.s now oppose each other, what is the new terminal p.d. of X and Y?

4. Two resistors of 1200 Ω and 800 Ω are connected in series with a battery of e.m.f. 24 V and negligible internal resistance, Fig. 30B (i). What is the p.d. across each resistor?

A voltmeter V of resistance 600 Ω is now connected firstly across the 1200 Ω resistor as shown, and then across the 800 Ω resistor. Find the p.d. recorded by the voltmeter in each case.

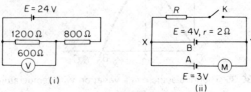

Fig. 30B

5. In Fig. 30B (ii), A has an e.m.f. of 3 V and negligible internal resistance and B has an e.m.f. of 4 V and internal resistance 2·0 Ω. With the switch K open, what current flows in the meter M?

When K is now closed, no current flows in M. Calculate the value of R.

(*Hint.* When no current flows in M, p.d. across XY = p.d. across A.)

6. A moving coil meter has a resistance of 25 Ω and indicates full scale deflection when a current of 4·0 mA flows through it. How could this meter be converted to a milliammeter having a full-scale deflection for a current of 50 mA? (*L.*)

7. Explain, with the aid of diagrams, how the moving coil meter in question 6 could be converted (*a*) to a voltmeter with a 0–3 V range, (*b*) to an ammeter with a 0–1 A range.

8.

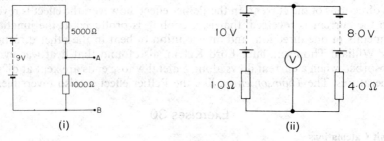

Fig. 30c

(i) What is the final potential difference between *A* and *B* in the circuit in Fig. 30c (i): (*a*) in the circuit as shown, (*b*) if an additional 500 Ω resistor were connected from *A* to *B*, (*c*) if the 500 Ω were replaced by a 2 μF capacitor? For what purpose would the circuit in (*a*) be useful?

(ii) In the circuit of Fig. 30c (ii), the batteries have negligible internal resistance and the voltmeter V has a very high resistance. What would be the reading of the voltmeter? (*L.*)

9. (*a*) Define *the volt* and *the ohm*. Water in a barrel may be released at a variable rate by an adjustable tap at the bottom. If this system is compared to an electric circuit what in the system would be analogous to (i) the potential difference, (ii) the charge flowing, (iii) the current, and (iv) the resistance, in the circuit?

Give *two* examples, illustrated by appropriate graphs, of conductors or components which do not obey Ohm's law.

(*b*)

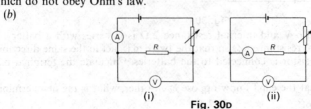

Fig. 30D

The circuits in Fig. 30D (i) and (ii) may be used to measure the resistance of the resistor, R. If both meters are of the moving coil type, explain why in each case the value for R obtained would not be correct. What alternative method would you use to obtain a better value for R? (No circuit details are required.)

(c) In Fig. 30E, what is the potential difference between the points B and D? What

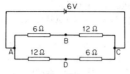

Fig. 30E

resistor could you add to the 12 Ω resistor in branch ADC in order to make the potential difference between B and D zero? (L.)

10. State Ohm's law. Why is it illogical to attempt to verify it using a moving coil voltmeter? Deduce a formula for the effective resistance of two resistors in parallel.

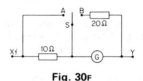

Fig. 30F

The galvanometer G in Fig. 30F has a resistance of 20 Ω. The terminals XY are connected to a thermopile generating an e.m.f. of 0·50 V with an internal resistance of 20 Ω. Calculate the current flowing in the galvanometer (i) when the switch S is in position A, (ii) when the switch S is in position B. Calculate the potential difference across the terminals of the thermopile (iii) when the switch S is in the position A, (iv) when the switch is in the position B.

This circuit is sometimes labelled 'divide by two' when used in a multi-range meter. Explain the particular advantage of this form of shunt. (C.)

11. State Ohm's law and describe an experiment to verify it.

A resistance of 1000 Ω and one of 2000 Ω are placed in series with a 100 V mains supply. What will be the reading on a voltmeter of internal resistance 2000 Ω when placed across (a) the 1000 Ω resistance, (b) the 2000 Ω resistance? (L.)

12. Twelve cells each of e.m.f. 2 V and of internal resistance $\frac{1}{2}$ Ω are arranged in a battery of n rows and an external resistance of $\frac{3}{8}$ Ω is connected to the poles of the battery. Determine the current flowing through the resistance in terms of n.

Obtain numerical values of the current for the possible values which n may take and draw a graph of current against n by drawing a smooth curve through the points. Give the value of the current corresponding to the maximum of the curve and find the internal resistance of the battery when the maximum current is produced. (L.)

13. Describe with full experimental details an experiment to test the validity of Ohm's law for a metallic conductor.

An accumulator of e.m.f. 2 V and of negligible internal resistance is joined in series with a resistance of 500 Ω and an unknown resistance $X\Omega$. The readings of a voltmeter successively across the 500 Ω resistance and X are 2/7 and 8/7 V respectively. Comment on this and calculate the value of X and the resistance of the voltmeter. (N.)

14. State with reasons the essential requirement for the resistance of (a) an ammeter, (b) a voltmeter.

A voltmeter having a resistance of 1800 Ω is used to measure the potential difference across a 200 Ω resistance which is connected to the terminals of a d.c. power supply having an e.m.f. of 50 V and an internal resistance of 20 Ω. Determine the percentage change in the potential difference across the 200 Ω resistor as a result of connecting the voltmeter across it. (N.)

15. State Ohm's law, and describe the experiments you would make in order to verify it. The positive poles A and C of two cells are connected by a uniform wire of resistance

4 ohms and their negative poles B and D by a uniform wire of resistance 6 Ω. The middle point of BD is connected to earth. The e.m.f.s of the cells AB and CD are 2 V and 1 V respectively, their resistances 1 Ω and 2 Ω respectively. Find the potential at the middle point of AC. (*O. & C.*)

Power. Heating effect. Thermoelectric e.m.f.

16. The maximum power dissipated in a 10 000 Ω resistor is 1 W. What is the maximum current?

17. Two heating coils A and B, connected in parallel in a circuit, produce powers of 12 W and 24 W respectively. What is the ratio of their resistances, R_A/R_B, when used?

18. A heating coil of power rating 10 W is required when the p.d. across it is 20 V. Calculate the length of nichrome wire needed to make the coil if the cross-sectional area of the wire used is 1×10^{-7} m^2 and the resistivity of nichrome is 1×10^{-6} Ω m.

What length of wire would be needed if its diameter was half that previously used?

19. Two resistors, of resistance R_1 and R_2, where R_1 is considerably greater than R_2, are to be connected to a battery of negligible internal resistance. Which of them would you expect to become hotter if they are connected (*a*) in series, and (*b*) in parallel? In each case justify your answer. (*L.*)

20. A thin film resistor in a solid-state circuit has a thickness of 1 μm and is made of nichrome of resistivity 10^{-6} Ω m. Calculate the resistance available between opposite edges of a 1 mm^2 area of film (*a*) if it is square shaped, (*b*) if it is rectangular, 20 times as long as it is wide. (*C.*)

21. State the laws of the development of heat when an electric current flows through a wire of uniform material.

An electrical heating coil is connected in series with a resistance of X Ω across the 240 V mains, the coil being immersed in a kilogram of water at 20°C. The temperature of the water rises to boiling-point in 10 minutes. When a second heating experiment is made with the resistance X short-circuited, the time required to develop the same quantity of heat is reduced to 6 minutes. Calculate the value of X. (Heat losses may be neglected.) (*L.*)

22. Deduce an expression for the heat developed in a wire by the passage of an electric current.

The temperature of 0·3 kg of paraffin oil in a vacuum flask rises 1·0°C per minute with an immersion heater of 12·3 watts input. On repeating with 0·4 kg of oil the temperature rises by 1·20°C per minute for an input of 19·2 watts. Find the specific heat capacity of the oil and the thermal capacity (assumed constant) of the flask. (*L.*)

23. (*a*) Explain what is meant by (i) the electrical resistance of a conductor, and (ii) the resistivity of the material of a conductor.

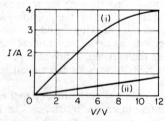

Fig. 30G

(*b*) The graphs in Fig. 30G show how the current varies with applied potential difference across (i) a 12 V, 36 W filament lamp, and (ii) a metre length of nichrome wire of cross-section 0·08 mm^2. Using the graphs, find the ratio of the values of the electrical resistance of the filament lamp to the nichrome wire (1) when the potential difference across them is 12 V, and (2) when the potential difference across them is 0·5 V.

How does the resistance of the filament lamp change as the current increases? Suggest a physical explanation for this change.

(*c*) The resistivity of copper is about $1·8 \times 10^{-8}$ Ω m at 20°C. Show, using the information in (*b*) above, that the resistivity of nichrome is approximately 60 times this value. Explain why, in a domestic circuit containing a fire element and connecting cable, only the element becomes appreciably hot. (*L.*)

24. Indicate, by means of graphs, the relation between the current and voltage (*a*) for a uniform manganin wire; (*b*) for a water voltameter; (*c*) for a diode valve. How do you account for the differences between the three curves?

An electric hot plate has two coils of manganin wire, each 20 metres in length and 0.23 mm^2 cross-sectional area. Show that it will be possible to arrange for three different rates of heating, and calculate the wattage in each case when the heater is supplied from 200 V mains. The resistivity of manganin is 4.6×10^{-7} Ω m. (*O. & C.*)

25. Describe an experiment for determining the variation of the resistance of a coil of wire with temperature.

An electric fire dissipates 1 kW when connected to a 250 V supply. Calculate to the nearest whole number the percentage change that must be made in the resistance of the heating element in order that it may dissipate 1 kW on a 200 V supply. What percentage change in the length of the heating element will produce this change of resistance if the consequent increase in the temperature of the wire causes its resistivity to increase by a factor 1.05? The cross-sectional area may be assumed constant. (*N.*)

26. Describe an instrument which measures the strength of an electric current by making use of its heating effect. State the advantages of this method.

A surge suppressor is made of a material whose conducting properties are such that the current passing through is directly proportional to the fourth power of the applied voltage. If the suppressor dissipates energy at a rate of 6.0 W when the potential difference across it is 240 V, estimate the power dissipated when the potential difference rises to 1200 V. (*C.*)

27. Describe an experimental method of producing a thermoelectric e.m.f.

How may a thermojunction be used to measure temperatures?

Why is a copper–iron junction not used to measure temperatures above 250°C, although a copper–constantan junction is often so employed? (*L.*)

28. Give a short account of the thermoelectric effect first discovered by Seebeck.

If you were given ice, boiling water, a thermocouple, a variable resistor and sensitive galvanometer (with a linear scale) but no thermometer, describe how you would determine the temperature inside a domestic refrigerator. How would you test the assumption you are making? No other apparatus is available, but you may assume body temperature is 36.9°C.

Describe how you would measure, in the laboratory, the resistance of a pair of headphones, which is damaged if the current through it exceeds 100 milliamp. (*C.*)

31 Measurements by Potentiometer and Wheatstone Bridge

In this chapter we discuss the accurate measurement of potential difference and resistance and their applications.

The Potentiometer

Pointer instruments are useless for very accurate measurements: the best of them have an intrinsic error of about 1% of full scale. Where greater accuracy than this is required, elaborate measuring circuits are used.

One of the most versatile of these circuits is the *potentiometer*. It consists of a uniform wire, AB in Fig. 31.1 (i), about a metre long. An accumulator X maintains a steady current I in AB. Since the wire is uniform, its resistance per centimetre, R, is constant; the potential difference across 1 cm of the wire, RI, is therefore also constant. The potential difference between the end A of the

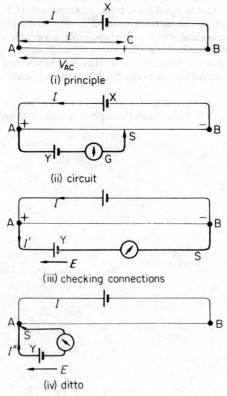

(i) principle

(ii) circuit

(iii) checking connections

(iv) ditto

Fig. 31.1 The potentiometer

wire, and any point C upon it, is thus proportional to the length of wire l between A and C:

$$V_{AC} \propto l. \qquad . \qquad . \qquad . \qquad . \qquad . \qquad (1)$$

Comparison of E.M.F.s

To illustrate the use of the potentiometer, suppose we take a cell, Y in Fig. 31.1 (ii), and join its positive terminal to the point A (to which the positive terminal of X is also joined). We connect the negative terminal of Y through a sensitive galvanometer to a slider S, which we can press on to any point in the wire.

Let us suppose that the cell Y has an e.m.f. E, which is less than the potential difference V_{AB} across the whole of the wire. Then if we press the slider on B, a current I' will flow through Y in opposition to its e.m.f., Fig. 31.1 (iii). This current will deflect the galvanometer G—let us say to the *right*. If we now press the slider on A, the cell Y will be connected straight across the galvanometer, and will deliver a current I'' in the direction of its e.m.f., Fig. 31.1 (iv). The galvanometer will therefore show a deflection to the *left*. If the deflections at A and B are not opposite, then either the e.m.f. of Y is greater than the potential difference across the whole wire, or we have connected the circuit wrongly. The commonest mistake in connecting up is not joining both positive poles to A.

Now let us suppose that we place the slider on to the wire at a point a few centimetres from A, then at a point a few centimetres farther on, and so forth.

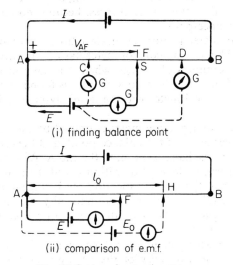

(i) finding balance point

(ii) comparison of e.m.f.

Fig. 31.2 Use of potentiometer

(We do not run the slider continuously along the wire, because the scraping would destroy the uniformity.)

When the slider is at a point C near A (Fig. 31.2 (i)) the potential difference V_{AC} is less than the e.m.f. E of Y. So current flows through G in the direction of E, and G may deflect to the left. When the slider is at D near B, V_{AD} is greater than E, current flows through G in opposition to E, and G deflects to the right.

By trial and error (but no scraping of the slider) we can find a point F such that, when the slider is pressed upon it, the galvanometer shows no deflection. *The potential difference* V_{AF} *is then equal to the e.m.f.* E; no current flows through the galvanometer because E and V_{AF} act in opposite directions in the galvanometer circuit, Fig. 31.2 (i). Because no current flows, the resistance of the galvanometer, and the internal resistance of the cell, cause no voltage drop. So the full e.m.f. E therefore appears between the points A and S, and is balanced by V_{AF}, that is,

$$E = V_{AF}.$$

If we now take another cell of e.m.f. E_0, and balance it in the same way, at a point H (Fig. 31.2 (ii)), then

$$E_0 = V_{AH}.$$

Therefore

$$\frac{E}{E_0} = \frac{V_{AF}}{V_{AH}}.$$

But, from previous, the potential differences V_{AF}, V_{AH} are proportional to the lengths l, l_0 from A to F, and from A to H, respectively. Therefore

$$\frac{E}{E_0} = \frac{l}{l_0}. \qquad . \qquad . \qquad . \qquad . \qquad . \qquad (2)$$

So the ratio of the e.m.f.s is proportional to the ratio of the balancing lengths and can therefore be calculated.

Accuracy

The following points should be noted:

(1) When the potentiometer is used to compare the e.m.f.s of cells, no errors are introduced by the internal resistances, because no current flows through the cells at the balance-points.

(2) The potentiometer is more accurate than the moving-coil voltmeter for measuring e.m.f. The moving-coil voltmeter has a resistance and this lowers the p.d. between the terminals of the cell when it is connected. In contrast, since no current flows from the cell when a balance is found, the potentiometer may be considered to be a voltmeter with an infinitely-high resistance, which is the ideal voltmeter.

(3) The accuracy of a potentiometer is limited by the non-uniformity of the slide-wire, the uncertainty of the balance-point, and the error in measuring the length l of wire from the balance-point to the end A. With even crude apparatus, the balance-point can be located to within about 0·5 mm; if the length l is 50 cm, or 500 mm, then the error in locating the balance-point is 1 : 1000. If the wire has been carefully treated, its non-uniformity may introduce an error of about the same magnitude. The overall error is then about ten times less than that of a pointer instrument.

(4) The *precision* with which the balance-point of a potentiometer can be found depends on the *sensitivity* of the galvanometer; with a very sensitive galvanometer a very small current can be detected.

A moving-coil galvanometer must be protected by a series resistance R of several thousand ohms, which is shorted out when the balance is nearly reached, Fig. 31.3. A series resistance is preferable to a shunt, because it reduces the current drawn from the cell under test when the potentiometer is unbalanced. Looking for the balance-point then causes less change in the chemical condition of the cell, and therefore in its e.m.f.

It is important to realise that the accuracy of a potentiometer does not depend on the accuracy of the galvanometer, but only on its sensitivity. The galvanometer is used not to measure a current but merely to show one when the potentiometer is off balance. It is said to be used as a null-indicator, and the potentiometer method of measurement is called a null method.

(5) The current through the potentiometer wire must be steady—it must not

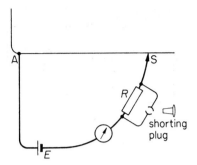

Fig. 31.3 Use of protective resistance with galvanometer

change appreciably between the finding of one balance-point and the next. The accumulator which provides it should therefore be neither freshly charged nor nearly run-down; when an accumulator is in either of those conditions its e.m.f. falls with time.

Errors in potentiometer measurements may be caused by non-uniformity of the wire, and by the resistance of its connection to the terminal at A. This resistance is added to the resistance of the length l of the wire between A and the balance-point, and if it is appreciable it makes equation (2) invalid.

Uses of the Potentiometer. E.M.F. and Internal Resistance

All the uses of the potentiometer depend on the fact that it can measure potential difference accurately, and without drawing current from the circuit under test.

(a) If one of the cells in Fig. 31.2 (ii) has a known e.m.f., say E_0, then the e.m.f. of the other, E, is given by equation:

$$\frac{E}{E_0} = \frac{l}{l_0}. \qquad . \qquad . \qquad . \qquad . \qquad . \qquad (1)$$

A cell of known e.m.f. is called a *standard cell*. The e.m.f.s of standard cells are determined absolutely—that is to say, without reference to the e.m.f.s of any other cell—by methods which depend, in principle, on the definition of e.m.f. (power/current, p. 645). Standard cells are described on p. 684, along with the precautions which must be taken in their use.

Equation (1) is true only if the current I through the potentiometer wire has remained constant. The easiest way to check that it has done so is to balance the standard cell against the wire before and after balancing the unknown cell. If the lengths to the balance-point are equal—within the limits of experimental error—then the current I may be taken as constant. A check of this kind should be made in each of the experiments to be described.

(b) The *internal resistance of a cell*, r, can be found with a potentiometer by balancing first its e.m.f., E, when the cell is on open circuit. Suppose the balance length is l. A known resistance R is then connected to the cell, as shown in Fig. 31.4. The terminal p.d. V is now balanced by a smaller length l' than l since a current flows from the cell. Now

$$V = IR = \frac{E}{R+r}R.$$

So

$$\frac{V}{E} = \frac{R}{R+r}. \qquad . \qquad . \qquad . \qquad . \qquad (2)$$

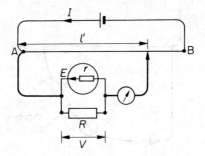

Fig. 31.4 Measurement of internal resistance

But
$$\frac{V}{E} = \frac{l'}{l'}, \qquad \cdots \cdots \cdots \quad (3)$$

where l and l' are the lengths of potentiometer wire required to balance E and V. From equations (2) and (3), $l/l' = (R + r)/R$. So r can be found from

$$r = \left(\frac{l}{l'} - 1\right)R.$$

Also, since $r\left(\dfrac{1}{R}\right) = \dfrac{l}{l'} - 1$, we can vary R and measure l' for each value of R. A graph of l/l' against $1/R$ is a straight line whose *gradient* is equal to r.

Measurement of Current

A current can be measured on a potentiometer by measuring the potential difference V which it sets up across a standard known resistance R in Fig. 31.5 (i), and then using $I = V/R$. A low resistance R is chosen so that it does not disturb the circuit in which it is placed.

Figure 31.5 (ii) shows in detail the standard resistance used. It consists of broad strip of alloy, such as manganin, whose resistance varies very little with temperature (p. 631). The current is led in and out at the terminals i, i. The terminals v, v are connected to fine wires soldered to points PP on the strip; they are called the potential terminals. The marked value R of the resistance is the value between the points PP.

Figure 31.5 (i) shows how an ammeter M can be calibrated by a potentiometer. The rheostat S is adjusted until the required ammeter reading is obtained and the p.d. V between the terminals v, v of R is balanced on the potentiometer wire. Suppose this gives a balance length l. The e.m.f. E_0 of a standard cell is now balanced on the wire (see Fig. 31.6 (ii)). If this balance length is l_0, then

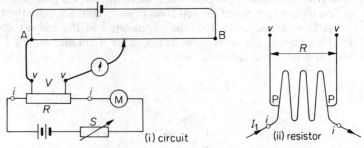

Fig. 31.5 Calibration of ammeter with potentiometer

$$\frac{V}{E_0} = \frac{l}{l_0}.$$

So V can be found since E_0, l_0 and l are known and the true current I is then calculated from $I = V/R$.

The resistance of the wires connecting the potential terminals to the points PP, and to the potentiometer circuit, do not affect the result, because at the balance-point the current through them is zero.

Calibration of Voltmeter

Figure 31.6 (i) shows how a potentiometer can be used to calibrate a voltmeter. A standard cell is first used to find the p.d. per cm or volt per cm of the wire (Fig. 31.6 (ii)): if its e.m.f. E_0 is balanced by a length l_0, then

$$\text{volt per cm} = \frac{E_0}{l_0}. \qquad . \qquad . \qquad . \qquad . \qquad (1)$$

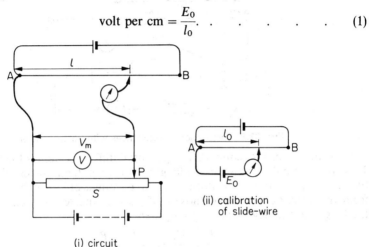

(i) circuit

(ii) calibration of slide-wire

Fig. 31.6 Calibration of voltmeter with potentiometer

Different voltages V_m are now applied to the voltmeter by the adjustable potential divider or rheostat S, Fig. 31.6 (i), which has a high resistance. If l is the length of potentiometer wire which balances a p.d. V_m then

$$V_m = l \times (\text{volt/cm of wire})$$

$$= l\frac{E_0}{l_0}. \qquad . \qquad . \qquad . \qquad . \qquad (2)$$

The value of V_m is the true value of the p.d. across the voltmeter terminals. If the voltmeter reading is V_{obs}, then the correction to be added to it is $V_m - V_{obs}$. This is plotted against V_{obs}, as in Fig. 31.7, and this provides a correction curve for the voltmeter readings when the meter is used.

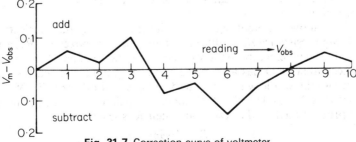

Fig. 31.7 Correction curve of voltmeter

Comparison of Resistances

A potentiometer can be used to compare resistances, by comparing the potential differences across them when they are carrying the same current I_1, Fig. 31.8. This method is particularly useful for very *low resistances*, because, as we have just seen, the resistances of the connecting wires do not affect the result of the experiment. It can, however, be used for higher resistances if desired. With low resistances the ammeter A' and rheostat P are necessary to adjust the current to a value which will neither exhaust the accumulator Y, nor overheat the resistors, and a series resistor (not shown) is required with X in the potentiometer circuit.

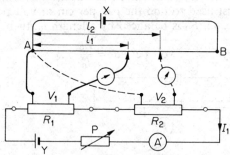

Fig. 31.8 Comparison of resistances with potentiometer

No standard cell is required. The potential difference across the first resistor, $V_1 = R_1 I_1$, is balanced against a length l_1 of the potentiometer wire, as shown by the full lines in the figure. Both potential terminals of R_1 are then disconnected from the potentiometer, and those of R_2 are connected in their place. If l_2 is the length to the new balance-point, then

$$\frac{l_1}{l_2} = \frac{V_1}{V_2} = \frac{R_1 I_1}{R_2 I_1} = \frac{R_1}{R_2}.$$

This result is true only if the current I_1 is constant, as well as the potentiometer current. The accumulator Y, as well as X, must therefore be in good condition. To check the constancy of the current I_1, the ammeter A' is not accurate enough. The reliability of the experiment as a whole can be checked by balancing the potential V_1 a second time, after V_2. If the new value of l_1 differs from the original then at least one of the accumulators is running down and must be replaced.

Measurement of Thermoelectric E.M.F.

The e.m.f.s of thermojunctions (p. 650) are small—of the order of a millivolt. If we attempted to measure such an e.m.f. on a simple potentiometer we should find the balance-point very near one end of the wire, so that the end-error would be serious.

Figure 31.9 shows a potentiometer circuit for measuring thermoelectric e.m.f. A suitable high resistance R, produced by two resistance boxes R_1, R_2, is needed in series with the wire. Suppose the wire has a resistance of $3\cdot0$ Ω and we assume that the accumulator D has an e.m.f. 2 V and negligible internal resistance. If the p.d. across the whole wire is required to be, say, 4 mV or $0\cdot004$ V, then the p.d. across R is $2 - 0\cdot004 = 1\cdot996$ V. Since the p.d. across resistors in a series circuit is proportional to the resistance, then R is given by

$$\frac{R}{3} = \frac{1\cdot996}{0\cdot004}$$

Thus $R = 1497 \, \Omega$ on calculation.

Calibration experiment. First, we find the 'p.d. per ohm' in the circuit. For this purpose a standard cell, of e.m.f. $E_s = 1 \cdot 018$ V say, is placed across one resistance R_1 with a galvanometer G in one lead, Fig. 31.9. The total resistance from the boxes R_1 and R_2 must be kept constant at 1497 Ω so that the potentiometer current is constant. Initially, then, $R_1 = 750 \, \Omega$ and $R_2 = 747 \, \Omega$, for example, and by taking out a resistor from one box and replacing an equal resistor in

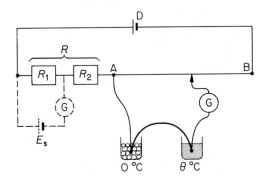

Fig. 31.9 Measurement of thermoelectric e.m.f.

the other box, a balance in G can soon be obtained. Suppose $R_1 = 779 \, \Omega$ in this case. Then the p.d. across the 100 cm length of potentiometer wire, $3 \cdot 0 \, \Omega$, is

$$\frac{3}{779} \times 1 \cdot 018 = 0 \cdot 003 \, 92 \text{ V} = 3 \cdot 92 \times 10^{-3} \text{ V.}$$

Thermoelectric e.m.f. After removing the standard cell, we now proceed to measure the thermoelectric e.m.f. E of a thermocouple at various temperatures $t°$C of the hot junction, the other junction being kept constant at $0°$C, Fig. 31.9. Suppose the balance length on the wire at a particular temperature is 62·4 cm. Then, from above,

$$E = \frac{62 \cdot 4}{100} \times 3 \cdot 92 \times 10^{-3} = 2 \cdot 45 \times 10^{-3} \text{ V.}$$

Use of the standard cell overcomes the error in assuming that the accumulator e.m.f. is 2 V. If the thermoelectric e.m.f. is not required accurately, then we can assume that the p.d. across 100 cm length of potentiometer wire is 4 mV when the series resistance R is 1497 Ω, as calculated above, and not use the calibration part of the experiment.

Thermoelectric E.M.F. and Temperature

Figure 31.10 shows the results of measuring the e.m.f. E when the cold junction is at $0°$C and the hot junction is at various temperatures θ in °C. The curves approximate to parabolas:

$$E = a\theta + b\theta^2 \qquad . \qquad . \qquad . \qquad . \qquad (1)$$

Since the same value of E is obtained at *two* different temperatures θ, the thermopile is never used for measuring temperature greater than the value corresponding to its maximum e.m.f.

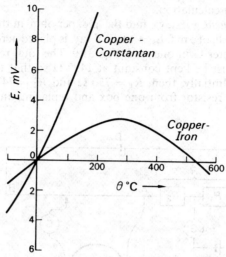

Fig. 31.10 E.m.f.s of thermocouples (reckoned positive when into copper at the cold junction)

THERMOELECTRIC E.M.F.s
(E in microvolt when θ is in °C
and cold junction at 0°C)

Junction	a	b	Range for a and b, °C	Limits of use, °C
Cu/Fe . . .	14	−0·02	0–100	*See* 1
Cu/Constantan[2] .	41	0·04	−50 to +300	−200 to +300
Pt/Pt–Rh[3] . .	6·4	0·006	0–200	0–1700
Chromel[4]/Alumel[5] .	41	0·001	0–900	0–1300

[1] Simple demonstrations. [2] See p. 631.
[3] 10% Rh; used only for accurate work or very high temperatures.
[4] 90% Ni, 10% Cr.
[5] 94% Ni, 3% Mn, 2% Al, 1% Si.

Example

In the circuit shown, the e.m.f. E_s of a standard cell is 1·02 V and this is balanced by the p.d. across a resistance of 2040 Ω in series with a potentiometer wire AB. If AB is 1·00 m long and has a resistance of 4 Ω, calculate the length AC on it which balances the e.m.f. 1·2 mV of the thermocouple XY.

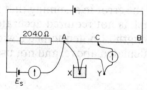

Fig. 31.11 Example

Since 1·02 V is the p.d. across 2040 Ω, and the 4 Ω wire AB is in series with 2040 Ω, then

$$\text{p.d. across AB} = \frac{4}{2040} \times 1·02 \text{ V} = \frac{4}{2000} \text{ V} = 2 \text{ mV}$$

So thermocouple e.m.f., 1·2 mV, is balanced by a length AC on AB (1 m) given by

$$\frac{AC}{AB} = \frac{1·2 \text{ mV}}{2 \text{ mV}} = \frac{3}{5}$$

$$\therefore AC = \tfrac{3}{5} \times AB = \tfrac{3}{5} \times 1·00 \text{ m} = 0·60 \text{ m}$$

Wheatstone Bridge: Measurement of Resistance

Wheatstone Bridge Circuit

About 1843 Wheatstone designed a circuit called a 'bridge circuit' which gave an accurate method for measuring resistance. We shall deal later with the practical aspects. In Fig. 31.12, X is the unknown resistance, and P, Q, R are

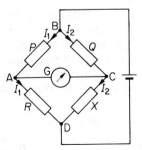

Fig. 31.12 Wheatstone bridge

resistance boxes. One of these—usually R—is adjusted until the galvanometer G between A and C shows no deflection, a so-called 'balance' condition. In this case the current I_g in G is zero. Then, as we shall show,

$$\frac{P}{Q} = \frac{R}{X},$$

so

$$X = \frac{Q}{P}R.$$

Wheatstone Bridge Proof

At balance, since no current flows through the galvanometer, the points A and C must be at the same potential, Fig. 31.12. Therefore

$$V_{AB} = V_{CB} \quad \text{and} \quad V_{AD} = V_{CD}$$

So

$$\frac{V_{AB}}{V_{AD}} = \frac{V_{CB}}{V_{CD}}. \qquad \qquad \text{(i)}$$

Also, since $I_g = 0$, P and R carry the same current, I_1, and X and Q carry the same current, I_2. Therefore

$$\frac{V_{AB}}{V_{AD}} = \frac{I_1 P}{I_1 R} = \frac{P}{R}$$

and

$$\frac{V_{CB}}{V_{CD}} = \frac{I_2 Q}{I_2 X} = \frac{Q}{X}. \qquad \qquad \text{(ii)}$$

From equations (i) and (ii),

$$\frac{P}{R} = \frac{Q}{X}.$$

So

$$X = \frac{Q}{P}R.$$

Exactly the same relationship between the four resistances are obtained if the galvanometer and cell positions are interchanged. Further analysis of the circuit shows that the bridge is most sensitive when the galvanometer is connected between the junction of the highest resistances and the junction of the lowest resistances.

The Slide-wire (Metre) Bridge

Figure 31.13 shows a simple form of Wheatstone bridge; it is sometimes called a slide-wire or metre bridge, since the wire AB is often a metre long. The wire is uniform, as in a potentiometer, and can be explored by a slider S.

The unknown resistance X and a known resistance R are connected as shown in the figure; heavy brass or copper strip is used for the connections AD, FH, KB, whose resistances are generally negligible. When the slider is at a point C in

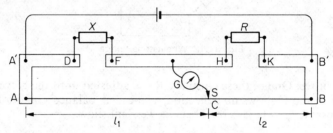

Fig. 31.13 Slide-wire (metre) bridge

the wire it divides the wire into two parts, of resistances R_{AC} and R_{CB}; these, with X and R, form a Wheatstone bridge. (The galvanometer and battery are interchanged relative to the circuits we have given earlier; that enables the slider S to be used as the galvanometer key. We have already seen that the interchange does not affect the condition for balance in G.) The connections are checked by placing S first on A, then on B. The balance-point is found by trial and error—not by scraping S along AB. At balance,

$$\frac{X}{R} = \frac{R_{AC}}{R_{CB}}.$$

Since the wire is uniform, the resistances R_{AC} and R_{CB} are proportional to the lengths of wire, l_1 and l_2. Therefore

$$\frac{X}{R} = \frac{l_1}{l_2}. \qquad . \qquad . \qquad . \qquad . \qquad (1)$$

The resistance R should be chosen so that the balance-point C comes fairly near to the centre of the wire—within, say, its middle third. If either l_1 or l_2 is small, the resistance of its end connection AA′ or BB′ in Fig. 31.13 is not negligible in comparison with its own resistance; equation (1) then does not hold. Some idea of the accuracy of a particular measurement can be got by interchanging R and X, and balancing again. If the new ratio agrees with the old within about 1%, then their average may be taken as the value of X.

Since the galvanometer G is a sensitive current-reading meter, a high protective resistor is required in series with it until a *near* balance is found on the wire. At this stage the high resistor is shunted or removed and the final balance-point found.

The lowest resistance which a bridge of this type can measure with reasonable accuracy is about 1 ohm. Resistances lower than about 1 ohm cannot be measured accurately on a Wheatstone bridge, because of the resistances of the wires connecting them to the X terminals, and of the contacts between those wires and the terminals to which they are, at each end, attached. This is the reason why the potentiometer method is more satisfactory for comparing and measuring low resistances.

Lorenz Absolute Method for Resistance

Lorenz devised a method of measuring resistance in which no electrical quantities are needed, that is, this is an *absolute method*. It is therefore adopted for measuring resistance in national physical laboratories. In contrast, when measuring resistance by V/I, one relies upon the accuracy of the voltmeter and ammeter used. In measuring resistance by a Wheatstone bridge method, one relies upon the accuracy of the standard resistance provided.

Figure 31.14 shows the principle of the method. A long coil A with n turns per metre is placed in series with the resistance R so that each carries a current I. A circular metal disc D is placed with its plane perpendicular to the magnetic field B inside the coil. By means of brushes, connections are made to R from the axle or centre of the disc and the circumference or edge. A galvanometer G is included in one lead.

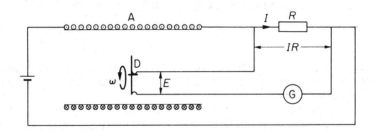

Fig. 31.14 Absolute method for measuring resistance

As explained on p. 734, the disc acts as a generator (dynamo) when it is rotated steadily in the field B. By varying the angular velocity ω, the induced e.m.f. E between the centre and circumference is used to balance the constant p.d. IR across R.

The e.m.f. $E = \omega r^2 B/2$, where r is the radius of the disc (p. 735). The field value $B = 4\pi n I \times 10^{-7}$ (p. 709). Since there is a balance,

$$IR = \frac{Br^2\omega}{2} = \frac{4\pi n I r^2 \omega \times 10^{-7}}{2}$$

$$\therefore R = 2\pi n r^2 \omega \times 10^{-7}$$

Thus by measuring ω in rad s^{-1}, r in metre and n, the resistance R can be calculated in ohm.

Temperature Coefficient of Resistance

We have seen that the resistance of a given wire increases with its temperature. If we put a coil of fine copper wire into a water bath, and use a Wheatstone bridge to measure its resistance at various moderate temperatures θ, we find that the resistance, R, increases with the temperature, Fig. 31.15. We may

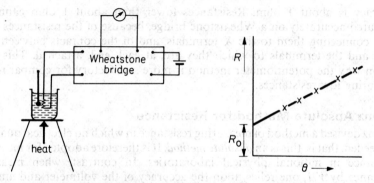

Fig. 31.15 Measurement of temperature coefficient

therefore define a *temperature coefficient of resistance*, α, such that

$$R_t = R_0(1 + \alpha\theta), \qquad . \qquad . \qquad . \qquad . \qquad (1)$$

where R_0 is the resistance at $0°C$. In words, starting with the resistance at $0°C$,

$$\alpha = \frac{\text{increase of resistance per K rise of temperature}}{\text{resistance at } 0°C}$$

If R_1 and R_2 are the resistances at $\theta_1°C$ and $\theta_2°C$, then, from (1),

$$\frac{R_1}{R_2} = \frac{1 + \alpha\theta_1}{1 + \alpha\theta_2}. \qquad . \qquad . \qquad . \qquad . \qquad (2)$$

Values of α for pure metals are of the order of $0·004 \text{ K}^{-1}$. They are much less for alloys than for pure metals, a fact which makes alloys useful materials for resistance boxes and shunts.

Equation (1) represents the change of resistance with temperature fairly well, but not as accurately as it can be measured. More accurate equations are given on p. 302 in the Heat section of this book, where resistance thermometers are discussed.

Thermistors

A *thermistor* is a heat-sensitive resistor usually made from semiconductors. One type of thermistor has a high positive temperature coefficient of resistance. Thus when it is placed in series with a battery and a current meter and warmed, the current is observed to decrease owing to the rise in resistance. Another type of thermistor has a negative temperature coefficient of resistance, that is, its resistance rises when its temperature is decreased, and falls when its temperature is increased. Thus when it is placed in series with a battery and a current meter and warmed, the current is observed to increase owing to the decrease in resistance.

Thermistors with a high negative temperature coefficient are used for resistance thermometers in very low temperature measurement of the order of 10 K, for example. The higher resistance at low temperature enables more accurate measurement to be made.

Thermistors with negative temperature coefficient may be used to safeguard against current surges in circuits where this could be harmful, for example, in a circuit where the heaters of radio valves are in series. A thermistor, T, is included in the circuit, as shown, Fig. 31.16. When the supply voltage is switched on, the thermistor has a high resistance at first because it is cold. It thus limits

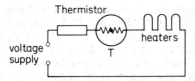

Fig. 31.16 Use of thermistor

Fig. 31.16 Use of thermistor

the current to a moderate value. As it warms up, the thermistor resistance drops appreciably and an increased current then flows through the heaters. Thermistors are also used in transistor receiver circuits to compensate for excessive rise in collector current.

Example

How would you compare the resistances of two wires A and B, using (a) a Wheatstone bridge method and (b) a potentiometer? For each case draw a circuit diagram and indicate the method of calculating the result.

In an experiment carried out at 0°C, A was 1·20 m of Nichrome wire of resistivity 100×10^{-8} Ω m and diameter 1·20 mm, and B a German silver wire 0·80 mm diameter and resistivity 28×10^{-8} Ω m. The ratio of the resistances A/B was 1·20. What was the length of the wire B?

If the temperature coefficient of Nichrome is $0\cdot000\,40$ K^{-1} and of German silver is $0\cdot000\,30$ K^{-1}, what would the ratio of resistances become if the temperature were raised by 100 K? (L.)

With usual notation,

for A,
$$R_1 = \frac{\rho_1 l_1}{A_1},$$

and for B,
$$R_2 = \frac{\rho_2 l_2}{A_2}$$

$$\therefore \frac{R_1}{R_2} = \frac{\rho_1}{\rho_2} \cdot \frac{l_1}{l_2} \cdot \frac{A_2}{A_1} = \frac{\rho_1}{\rho_2} \cdot \frac{l_1}{l_2} \cdot \frac{d_2{}^2}{d_1{}^2},$$

where d_2, d_1 are the respective diameters of B and A.

$$\therefore 1\cdot20 = \frac{100}{28} \times \frac{1\cdot20}{l_2} \times \frac{0\cdot8^2}{1\cdot20^2}.$$

$$\therefore l_2 = \frac{100 \times 1\cdot20 \times 0\cdot64}{1\cdot20 \times 28 \times 1\cdot44} = 1\cdot59 \text{ m.} \qquad . \qquad . \qquad . \qquad \text{(i)}$$

When the temperature is raised by 100 K, the resistance increases according to the relation $R_t = R_0(1 + \alpha\theta)$. Thus

new Nichrome resistance, $R_A = R_1(1 + \alpha\,.\,100) = R_1 \times 1\cdot04$,

and new German silver resistance, $R_B = R_2(1 + \alpha'\,.\,100) = R_2 \times 1\cdot03$.

$$\therefore \frac{R_A}{R_B} = \frac{R_1}{R_2} \times \frac{1\cdot04}{1\cdot03} = 1\cdot20 \times \frac{1\cdot04}{1\cdot03} = 1\cdot21 \qquad . \qquad . \qquad . \qquad \text{(ii)}$$

Exercises 31

1. The e.m.f. of a battery A is balanced by a length of 75·0 cm on a potentiometer wire. The e.m.f. of a standard cell, 1·02 V, is balanced by a length of 50·0 cm. What is the e.m.f. of A?

Calculate the new balance length if A has an internal resistance of 2 Ω and a resistor of 8 Ω is joined to its terminals.

2. A 1·0 Ω resistor is in series with an ammeter M in a circuit. The p.d. across the resistor is balanced by a length of 60·0 cm on a potentiometer wire. A standard cell of e.m.f. 1·02 V is balanced by a length of 50·0 cm. If M reads 1·10 A, what is the error in the reading?

3. The driver cell of a potentiometer has an e.m.f. of 2 V and negligible internal resistance. The potentiometer wire has a resistance of 3 Ω. Calculate the resistance needed in series with the wire if a p.d. of 5 mV is required across the whole wire.

The wire is 100 cm long and a balance length of 60 cm is obtained for a thermocouple e.m.f. *E*. What is the value of *E*?

4. In a potentiometer experiment, a balance length cannot be found. Write down *two* possible reasons, with explanations.

5. Explain the reasons for the following procedures in potentiometer experiments:

(*a*) The positive pole of a battery whose e.m.f. is required is connected to the same terminal of the potentiometer wire as the positive pole of the driver cell.

(*b*) The protective resistor is removed before a final balance point is determined.

(*c*) A rheostat is sometimes included in the potentiometer circuit with the driver cell but its resistance must not be too high.

(*d*) A standard cell is needed in an experiment to calibrate an ammeter but not in an experiment to measure the internal resistance of a cell.

(*e*) In comparing the resistances of two resistors A and B, the resistors are placed in series in a circuit.

6. (*a*) Figure 31A (i), in which AB is a uniform resistance wire, is a simple potentiometer circuit. Explain why a point X may be found on the wire which gives zero galvanometer deflection.

When the circuit was first set up it was impossible to find a balance point. State and explain two possible causes of this.

How would you use the circuit to compare the e.m.f.s of two cells with minimum error? Why is this circuit not suitable for the comparison of an e.m.f. of a few millivolts with an e.m.f. of about a volt?

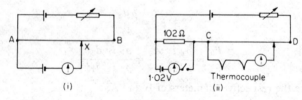

Fig. 31A

(*b*) The second circuit, Fig 31A (ii), may be used to measure the e.m.f. of a thermocouple provided that the resistance of CD is known. Describe how you would use it. If the resistance of CD were 2·00 Ω, its length were 1·00 m and the balance length were 79 cm, what would be the e.m.f. of the thermocouple? (*L.*)

7. A simple potentiometer circuit is set up as in Fig. 31B, using a uniform wire AB, 1·0 m long, which has a resistance of 2·0 Ω. The resistance of the 4-V battery is negligible. If the variable resistor *R* were given a value of 2·4 Ω, what would be the length AC for zero galvanometer deflection?

If *R* were made 1·0 Ω and the 1·5 V cell and galvanometer were replaced by a voltmeter of resistance 20 Ω, what would be the reading of the voltmeter if the contact *C* were placed at the mid-point of AB? (*L.*)

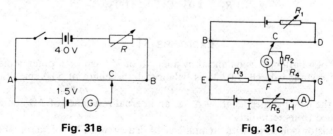

Fig. 31B   Fig. 31c

8. (*a*) Figure 31c shows a potentiometer circuit arranged to compare the values of two low resistance resistors.

(i) Which resistors are to be compared? (ii) What are the functions of the remaining three resistors? (iii) During the experiment what connection changes would need to be made? (iv) When the circuit was initially set up the galvanometer was found to be deflected in the same direction wherever along the wire BD the sliding contact C was placed. Suggest *two* possible reasons for this. (v) For the purpose of this experiment explain whether or not it is necessary to calibrate the potentiometer with a standard cell.

(*b*) A potentiometer may be regarded as equivalent to a voltmeter. Illustrate this by drawing the *basic* potentiometer circuit, marking the points which correspond to the positive and negative terminals of the equivalent voltmeter.

Describe how this basic circuit may be developed in order to measure the internal resistance of a cell. How may the observations be displayed in the form of a straight line graph and how could the internal resistance be found from this graph? (*L*.)

9. (*a*) Describe, with a circuit diagram, a potentiometer circuit arranged (i) to compare the e.m.f. of a cell with that of a standard cell, and (ii) to measure accurately a steady direct current of approximately 1 A. What factors determine the accuracy of the current measurement?

(*b*) A 12 V, 24 W tungsten filament bulb is supplied with current from *n* cells connected in series. Each cell has an e.m.f. of 1·5 V and internal resistance 0·25 Ω. What is the value of *n* in order that the bulb runs at its rated power?

An additional resistance *R* is introduced into the circuit so that the potential difference across the bulb is 6 V. Why is the power dissipated in the bulb not 6 W? Is it greater or less than 6 W? (*O. & C*.)

10. (*a*) Explain the principle of the slide-wire potentiometer.

(*b*) Discuss the relative merits of using (i) a potentiometer, (ii) a moving-coil voltmeter for determining the e.m.f. of a cell of e.m.f. approximately 1·5 V.

(*c*) A potentiometer wire is 1.00 m long and has a resistance of about 4 Ω. Explain how you could obtain a potential difference of about 4 mV across the wire using a 2 V accumulator of negligible internal resistance and any other essential apparatus. Support your explanation with a suitable calculation.

(*d*) A thermocouple generates an e.m.f. which is proportional to the temperature difference between its junctions and, for a temperature difference of 100 K, the e.m.f. is approximately 4 mV. Draw a labelled circuit diagram showing how you could use the potentiometer adapted in (*c*) to obtain temperature readings in the range 0°C to 100°C so that each centimetre of wire corresponds to 1 K and describe how you would use the apparatus to determine the room temperature. You may assume that apparatus is available for maintaining fixed temperatures of 0°C and 100°C.

(*e*) Give *two* reasons (other than faulty apparatus and poor electrical contact) why it may not be possible to obtain a balance point in a potentiometer experiment and indicate in each case how the fault can be remedied. (*N*.)

11. The circuit in Fig. 31D is being used to measure the e.m.f. of a thermocouple T. AB is a uniform wire of length 1·00 m and resistance 2·00 Ω. With K_1 closed and K_2 open, the balance length is 90·0 cm. With K_2 closed and K_1 open, the balance length is 45 cm. What is the e.m.f. of the thermocouple?

What is the value of *R* if the resistance of the driver cell is negligible? (*L*.)

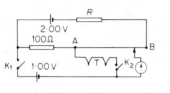

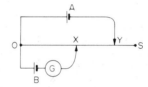

Fig. 31D	**Fig. 31**E

12. State *Ohm's law*. Discuss two examples of non-ohmic conductors.

Cells A and B and a galvanometer G are connected to a slide wire OS by two sliding contacts X and Y as shown in Fig. 31E. The slide wire is 1·0 m long and has a resistance of 12 Ω. With OY 75 cm, the galvanometer shows no deflection when OX is 50 cm. If Y is moved to touch the end of the wire at S, the value of OX which gives no deflection is 62·5 cm. The e.m.f. of cell B is 1·0 V.

Calculate (a) the p.d. across OY when Y is 75 cm from O (with the galvanometer balanced), (b) the p.d. across OS when Y touches S (with the galvanometer balanced), (c) the internal resistance of cell A, (d) the e.m.f. of cell A. (C.)

13. The circuit diagram in Fig. 31F represents the slide wire potentiometer used for the comparison of the e.m.f.s of the cells B and C.

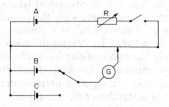

Fig. 31F

(a) What is the main advantage of using a potentiometer for this purpose? (b) What is the main quality required of the driver cell A? (c) What is the purpose of the rheostat R? (d) If, in practice, a balance point could not be found for cell B suggest two possible reasons. (e) Outline the experimental procedure, which you would adopt for comparing the e.m.f.s.

Draw a circuit diagram to show how a potentiometer may be adapted to measure an e.m.f. of a few millivolts. Explain how you would standardise this potentiometer using a standard cell. Indicate the approximate values of the components used, if the potentiometer wire has a resistance of 5 Ω (L.)

14. Describe and explain how a potentiometer is used to test the accuracy of the 1 V reading of a voltmeter.

A potentiometer consists of a fixed resistance of 2030 Ω in series with a slide wire of resistance 4 Ω metre^{-1}. When a constant current flows in the potentiometer circuit a balance is obtained when (a) a Weston cell of e.m.f. 1·018 V is connected across the fixed resistance and 150 cm of the slide wire and also when (b) a thermocouple is connected across 125 cm of the slide wire only. Find the current in the potentiometer circuit and the e.m.f. of the thermocouple.

Find the value of the additional resistance which must be present in the above potentiometer circuit in order that the constant current shall flow through it, given that the driver cell is a lead accumulator of e.m.f. 2 V and of negligible resistance and the length of the slide wire is 2 metres. (L.)

Wheatstone Bridge. Resistance

15. A copper coil has a resistance of 20·0 Ω at 0°C and a resistance of 28·0 Ω at 100°C. What is the temperature coefficient of resistance of copper?

Used in a circuit, the p.d. across the coil is 12 V and the power produced in it is 6 W. What is the temperature of the coil?

16. A tungsten coil has a resistance of 12·0 Ω at 15°C. If the temperature coefficient of resistance of tungsten is 0·004 K^{-1}, calculate the coil resistance at 80°C.

17. A heating coil is to be made, from nichrome wire, which will operate on a 12 V supply and will have a power of 36 W when immersed in water at 373 K. The wire available has an area of cross-section of 0·10 mm^2. What length of wire will be required? (Resistivity of nichrome at 273 K = 1·08 × 10^{-6} Ω m. Temperature coefficient of resistivity of nichrome = 8·0 × 10^{-5} K^{-1}.) (L.)

18. Describe how you would measure the temperature coefficient of resistance of a metal.

Give a short account of the platinum resistance thermometer.

A steady potential difference of 12 V is maintained across a wire which has a resistance of 3·0 Ω at 0°C; the temperature coefficient of resistance of the material is 4 × 10^{-3} K^{-1}. Compare the rates of production of heat in the wire at 0°C and at 100°C.

The wire is embedded in a body of constant heat capacity 600 J K^{-1}. Neglecting heat losses, and taking the thermal conductivity of the body to be large, find the time taken to increase the temperature of the body from 0°C to 100°C. (*O.*)

19. What do you understand by *temperature coefficient of resistance*?

Describe fully how you would use two equal resistors, one calibrated variable resistor, and other apparatus, to measure the temperature coefficient of resistance of copper, by means of a Wheatstone bridge circuit. Derive from first principles the equation which is satisfied when your bridge is 'balanced'.

To a good approximation, the resistivity of copper near room temperature is proportional to its absolute temperature. Calculate the temperature coefficient of resistance of copper, and explain your calculation. (Take 0°C as 273 K.) (*C.*)

20. Derive the balance condition for a Wheatstone bridge. Describe a practical form of Wheatstone bridge and explain how you would use it to determine the resistance of a resistor of nominal value 20 Ω.

An electric fire element consists of 4·64 m of nichrome wire of diameter 0·500 mm, the resistivity of nichrome at 15°C being 112 × 10^{-8} Ω m. When connected to a 240 V supply the fire dissipates 2·00 kW and the temperature of the element is 1015°C. Determine a value for the mean temperature coefficient of resistance of nichrome between 15°C and 1015°C. (*L.*)

21. Describe the Wheatstone bridge circuit and deduce the condition for 'balance'. State clearly the fundamental electrical principles on which you base your argument. Upon what factors do (*a*) the sensitivity of the bridge, (*b*) the accuracy of the measurement made with it, depend?

Using such a circuit, a coil of wire was found to have a resistance of 5 Ω in melting ice. When the coil was heated to 100°C, a 100 Ω resistor had to be connected in parallel with the coil in order to keep the bridge balanced at the same point. Calculate the temperature coefficient of resistance of the coil. (*C.*)

22. Define *resistivity* and *temperature coefficient of resistance*.

Explain, with the help of a clear circuit diagram, how you would use a bridge method to determine the resistance of an electric lamp filament at room temperature.

A carbon lamp filament was found to have a resistance 375 Ω at the laboratory temperature of 20°C. The lamp was then connected in series with an ammeter and a d.c. supply, and a voltmeter of resistance 1050 Ω was connected in parallel with the lamp. The ammeter and voltmeter indicated 0·76 A and 100 V respectively. The temperature of the carbon filament was estimated to be 1200°C. Estimate the mean value of the temperature coefficient of resistance of carbon between 20°C and 1200°C, and comment on your result. (*L.*)

23. (*a*) Define the volt and use your definition to derive an expression for the power dissipated (in watt) in a resistor of resistance *R* (in ohm) when a current *I* (in amp) flows through it.

(*b*) Explain how you would use a metre bridge to determine the resistivity of a metal in the form of a wire.

(*c*) Discuss the process of conduction in a metal. Derive an expression for the current flowing in a wire in terms of the number of free electrons per unit volume *n*, the area of cross section of the wire *A*, the electronic charge *e*, and the average drift velocity of the electrons *v*. (*AEB,* 1982)

32 The Chemical Effect of the Current

Electrolysis

The chemical effect of the electric current was first studied quantitatively by Faraday, who introduced most of the technical terms which are now used in describing it. A conducting solution is called an *electrolyte* and the chemical changes which occur when a current passes through it are called *electrolysis* (*lysis* = decomposition). Solutions in water of acids, bases, and salts are electrolytes. The plates or wires which dip into the electrolyte to connect it to the circuit are called electrodes; the one by which the current enters the solution is called the *anode*, and the one by which it leaves is called the *cathode* (Fig. 32.1 (i)).

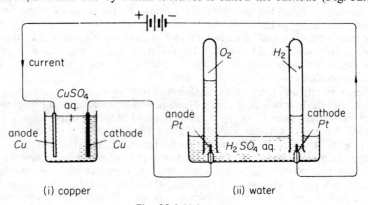

Fig. 32.1 Voltameters

The whole arrangement is called a *voltameter*; if the electrolyte is a solution of a copper or silver salt, the voltameter is called a copper or silver voltameter respectively. If the electrolyte is acidulated water, then the voltameter is called a water voltameter, because when a current passes through it, the water, not the acid, is decomposed, Fig. 32.2 (ii). We shall see why later.

Faraday's Laws of Electrolysis

When a current is passed through copper sulphate solution with copper electrodes, copper is deposited on the cathode and lost from the anode. Faraday showed that the mass dissolved off the anode by a given current in a given time is equal to the mass deposited on the cathode. He also showed that the mass is proportional to the product of the current, and the time for which it flows: that is to say, to the quantity of charge which passes through the voltameter. When he studied the electrolysis of water, he found that the masses of hydrogen and oxygen, though not equal, were each proportional to the quantity of charge that flowed. He therefore put forward his first law of electrolysis: *the mass of any substance liberated in electrolysis is proportional to the quantity of electric charge that liberated it.*

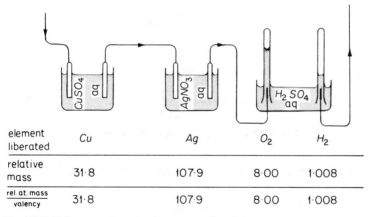

element liberated	Cu	Ag	O_2	H_2
relative mass	31·8	107·9	8·00	1·008
rel. at. mass / valency	31·8	107·9	8·00	1·008

Fig. 32.2 Voltameters in series (same quantity of charge passes through each)

Faraday's second law of electrolysis concerns the masses of different substances liberated by the same quantity of charge. An experiment to illustrate it is in Fig. 32.2. The experiment shows that *the masses of different substances, liberated in electrolysis by the same quantity of electric charge, are proportional to the ratio of the relative atomic mass to the valency.* This is Faraday's second law; it implies that the same quantity of charge is required to liberate one mole divided by the valency of any substance. Recent measurements give this quantity as 96 500 coulombs approximately; it is called the *Faraday constant*, symbol *F*. It is the charge required to liberate one mole of a monovalent element such as silver, or to liberate half a mole of a divalent element such as oxygen or copper II.

Mass Liberated in Electrolysis

The mass of a substance liberated by one coulomb. It is expressed in kilogram per coulomb (kg C^{-1}) in SI units. If z is the mass per coulomb, the mass m in kilogram liberated by a current I in ampere in a time t in seconds is

$$m = zQ = zIt \qquad . \qquad . \qquad . \qquad . \qquad (1)$$

For hydrogen, the ratio relative atomic mass/valency is 1·008. So 1·008 g or $1·008 \times 10^{-3}$ kg of hydrogen is liberated by 96 500 coulombs. Thus the mass of hydrogen liberated per coulomb, z_H, is

$$z_H = \frac{1·008 \times 10^{-3}}{96\,500} = 1·05 \times 10^{-8} \text{ kg } C^{-1}.$$

Similarly, since the ratio relative atomic mass/valency for copper II is 31·8, the mass of copper deposited per coulomb is given by

$$z_{Cu} = \frac{31·8 \times 10^{-3}}{96\,500} = 3·29 \times 10^{-7} \text{ kg } C^{-1}.$$

Measurement of Current by Electrolysis

In the absence of a standard ammeter, the chemical effect can be used to find the error in a particular reading on an ammeter A, Fig. 32.3. A copper voltameter is connected in series with the ammeter, and a steady current I passed for a known time t. The current is kept constant by adjusting the rheostat R to keep the deflection constant. The cathode is weighed before and after the experiment.

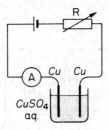

Fig. 32.3 Current measurement by electrolysis

If the increase in mass is m and the mass of copper liberated per coulomb is z. Then,

$$m = zIt,$$

$$I = \frac{m}{zt} \qquad\qquad (2)$$

The error in the ammeter is then the difference in the reading on A and the current calculated from (2).

Great care must be taken in this experiment over the cleanliness of the electrodes. They must be cleaned with emery paper at the start; and, at the finish, the cathode must be rinsed with water and dried with alcohol, or over a gentle spirit flame: strong heating will oxidise the copper deposit.

Example

A copper refining cell consists of two parallel copper plate electrodes, 6 cm apart and 1 metre square, immersed in a copper sulphate solution of resistivity 1.2×10^{-2} ohm metre. Calculate the potential difference which must be established between the plates to provide a constant current to deposit 0·48 kg of copper on the cathode in one hour (mass of copper liberated per coulomb, $z = 3.29 \times 10^{-7}$ kg C^{-1}).

From $m = zIt$,

$$I = \frac{m}{zt} = \frac{0.48}{3.29 \times 10^{-7} \times 3600} \text{ A} \qquad\qquad (i)$$

The resistance of the cell, $R, = \dfrac{\rho l}{A}$

$$= \frac{1.2 \times 10^{-2} \times 6 \times 10^{-2}}{1^2}.$$

Hence, from (i), the p.d. $V = IR = \dfrac{0.48 \times 1.2 \times 10^{-2} \times 6 \times 10^{-2}}{3.29 \times 10^{-7} \times 3600 \times 1^2}$

$$= 0.3 \text{ V (approx.).}$$

The Mechanism of Conduction: Ions

The theory of electrolytic conduction is generally attributed to Arrhenius (1859–1927), although Faraday had stated some of its essentials in 1834. Faraday suggested that the current through an electrolyte was carried by charged particles, which he called ions (Greek *ion* = go). A solution of silver nitrate, he supposed, contained silver ions and 'nitrate' ions. The silver ions were silver atoms with a positive charge; they were positive because silver was deposited at the cathode, or negative electrode, Fig. 32.4. The nitrate ions were groups

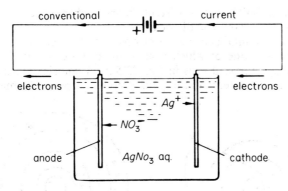

Fig. 32.4 Ions in electrolysis

of atoms—NO_3 groups—with a negative charge; they travelled towards the anode, or positive electrode, and, when silver electrodes are used, formed silver nitrate.

Nowadays, we consider that a silver ion is a silver atom which has lost an electron; this electron transfers itself to the $NO_3{}^-$ group when the silver nitrate molecule is formed, and gives the nitrate ion its negative charge. We denote nitrate and silver ions, respectively, by the symbols $NO_3{}^-$ and AG^+. When the ions appear at the electrodes of a voltameter they are discharged. The current in the external circuit brings electrons to the cathode, and takes them away from the anode, Fig. 32.4. At the anode silver atoms lose electrons and go into solution as positive ions. *In effect*, the negative charges carried across the cell by the NO_3^- ions flow away through the external circuit. At the cathode, each silver ion gains an electron, and becomes a silver atom, which is deposited upon the electrode.

Ionisation

The splitting up of a compound into ions in solution is called ionisation, or ionic dissociation. Faraday does not seem to have paid much attention to how it took place, and the theory of it was given by Arrhenius in 1887. For a reason which we will consider later, Arrhenius suggested that an electrolyte ionised as soon as it was dissolved: that its ions were not produced by the current through it, but were present as such in the solution, before ever the current was passed.

We now consider that salts of strong bases and acids, such as silver nitrate, copper sulphate, sodium chloride, ionise completely as soon as they are dissolved in water. That is to say, a solution contains no molecules of these salts, but only their ions. Such salts are called strong electrolytes; so are the acids and bases from which they are formed, for these also ionise completely when dissolved in water.

Other salts, such as sodium carbonate, do not appear to ionise completely on solution in water. They are the salts of weak acids, and are called weak electrolytes. The weak acids themselves are also incompletely ionised in water.

Formation of Ions: Mechanism of Ionisation

In the Heat section of this book we described the structure of the solid state (p. 236). In solid crystalline salts such as sodium chloride the structure is made up of sodium and chlorine ions: not of atoms, nor of NaCl molecules, but of Na^+ and Cl^- ions. In other words, we think today that ions exist in solid crystalline salts, as well as in their solutions. We do so for the reason that the idea enables us to build up a consistent theory of chemical combination, of the solid state, and of electrolytic dissociation.

A sodium atom contains eleven electrons, ten of which move in orbits close to the nucleus, and one of which ranges much more widely; for our present purposes we may represent it as in Fig. 32.5 (i). A chlorine atom has ten inner electrons and seven outer ones; for our present purposes we may lump these

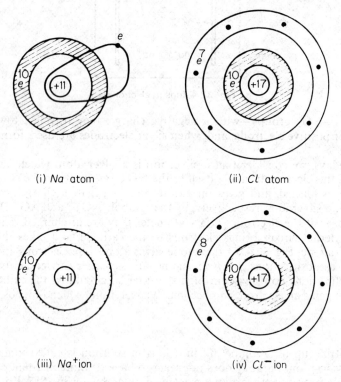

(i) Na atom   (ii) Cl atom

(iii) Na⁺ion   (iv) Cl⁻ ion

Fig. 32.5 Sodium and chlorine, atoms and ions

into two groups, as in Fig. 32.5 (ii). The outer electron of the sodium atom is weakly attracted to its nucleus, but the outer electrons of the chlorine atom are strongly attracted (because the ten inner electrons are a more effective shield round the $+11$ nucleus of sodium than round the $+17$ nucleus of chlorine). Therefore, when a sodium and a chlorine atom approach one another, the outer electron of the sodium atom is attracted more strongly by the chlorine nucleus than by the sodium nucleus. It leaves the sodium atom, and joins the outer electrons of the chlorine; the sodium atom becomes a positively charged sodium ion, Fig. 32.5 (iii), and the chlorine atom a negatively charged chlorine ion, Fig. 32.5 (iv). Between these two ions there now appears a strong electrostatic attraction, which holds them together as a molecule of NaCl. In the solid state, the ions are arranged alternately positive and negative; the forces between them bind the whole into a rigid crystal.

When such a crystal is dropped into water, it dissolves and ionises. We can readily understand this when we remember that water has a very high dielectric constant: 81 (p. 574). It therefore reduces the forces between the ions 81 times, and the crystal falls apart into ions. In the same way we explain the ionisation of other salts, and bases and acids. The idea that these dissociate because they are held together by electrostatic forces, which the solvent weakens, is supported by the fact that they ionise in some other solvents as well as water. These solvents also are liquids which have a high dielectric constant, such as methyl and ethyl alcohols (32 and 26 respectively). In these liquids, however, it seems that strong electrolytes behave as weak ones do in water: only a fraction of the dissolved molecules, not all of them, dissociate. In the electronic theory of atomic structure, the chemical behaviour of an element is determined by the number of its outer electrons. If it can readily lose one or two it is metallic, and forms positive ions; if it can readily gain one or two, it is acidic, and forms negative ions. Ions are not chemically active in the way that atoms are. Sodium atoms, in the form of a lump of the metal, react violently with water; but the hydroxide which they form ionises ($NaOH \rightarrow Na^+ + OH^-$), and the sodium ions drift peaceably about in the solution—which is still mainly water.

Pure water is a feeble conductor of electricity, and we consider that it is but feebly ionised into H^+ and OH^- ions. These, we believe, are continually joining up to form water molecules, and then dissociating again in a dynamical equilibrium.

$$H_2O \rightleftharpoons H^+ + OH^-.$$

If, as we shall find in the electrolysis of water, H^+ and OH^- ions are removed from water, then more molecules dissociate, to restore the equilibrium.

The concentrations of H^+ and OH^- in water are so small that they do not contribute appreciably to the conduction of electricity when an electrolyte is dissolved in the water; but, as we shall see, they sometimes take part in reactions at the electrodes.

Explanation of Faraday's Laws

The theory of dissociation neatly explains Faraday's laws and some other phenomena of electrolysis. If an $AgNO_3$ molecule splits up into Ag^+ and NO_3^- ions, then each NO_3^- ion that reaches the anode dissolves one silver atom off it. At the same time, one silver atom is deposited on the cathode. Thus the gain in mass of the cathode is equal to the loss in mass of the anode. Also the total mass of silver nitrate in solution is unchanged; experiment shows that this is true. The mass of silver deposited is proportional to the number of ions reaching the cathode; if all the ions carry the same charge—a reasonable assumption— then the number deposited is proportional to the quantity of charge which deposits them. This is Faraday's first law.

To see how the ionic theory explains Faraday's second law, let us again consider a number of voltameters in series, Fig. 32.6. When a current flows through them all, the same quantity of charge passes through each in a given time. Experiment shows that

$$\frac{\text{mass of silver deposited}}{\text{mass of hydrogen liberated}} = 107 \cdot 0.$$

From experiments on chemical combination, we know that

$$\frac{\text{mass of silver atom}}{\text{mass of hydrogen atom}} = \frac{107 \cdot 9}{1 \cdot 008} = 107 \cdot 0.$$

Therefore we may say that, each time a silver ion is discharged and deposited as an atom, a hydrogen ion is also discharged and becomes an atom. The hydrogen atoms thus formed join up in pairs, and escape as molecules of hydrogen gas. The theory fits the facts on the simple assumption that the hydrogen and silver atoms carry equal charges: we now say that each is an atom which has lost one electron.

But when we consider the copper voltameter in Fig. 32.6, we find a complication. For

$$\frac{\text{mass of copper deposited}}{\text{mass of hydrogen liberated}} = \frac{31\cdot8}{1\cdot008},$$

whereas

$$\frac{\text{mass of copper atom}}{\text{mass of hydrogen atom}} = \frac{63\cdot6}{1\cdot008}.$$

To explain this result we must suppose that only one copper atom is deposited for every two hydrogen atoms liberated. In terms of the ionic theory, therefore,

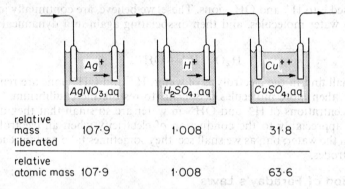

	Ag^+	H^+	Cu^{++}
	$AgNO_3$, aq	H_2SO_4, aq	$CuSO_4$, aq
relative mass liberated	107·9	1·008	31·8
relative atomic mass	107·9	1·008	63·6

Fig. 32.6 Illustrating Faraday's laws

only one copper ion is discharged for every two hydrogen ions. It follows that a copper ion must have twice as great a charge as a hydrogen ion: it must be an atom which has lost two electrons.

This conclusion fits in with our knowledge of the chemistry of copper. One atom of copper can replace two of hydrogen, as it does, for example, in the formation of copper sulphate, $CuSO_4$, from sulphuric acid, H_2SO_4. We therefore suppose that the sulphate ion also is double charged: $SO_4{}^{2-}$. When sulphuric acid is formed, two hydrogen atoms each lose an electron, and the SO_4 group gains two. When copper sulphate is formed, each copper atom gives up two electrons to an SO_4 group. And when copper sulphate ionises, each molecule splits into two double charged ions:

$$CuSO_4 \rightarrow Cu^2 + SO_4{}^{2-} \qquad . \qquad . \qquad . \qquad . \qquad (3)$$

In general, if we express the charge on an ion in units of the electronic charge, we find that it is equal to the valency of the atom from which the ion was formed. That is to say, it is equal to the number of hydrogen atoms which the atom can combine with or replace.

This assertion, which is illustrated in Fig. 32.7, explains Faraday's second law (mass deposited $\propto$ relative atomic mass/valency). For if a current I passes through a voltameter for a time t, the total charge carried through it is It. And if q is the charge on an ion, the number of ions liberated is It/q. If M is the mass of an ion, the mass liberated is $M(It/q)$, and is therefore proportional to M/q. But M is virtually equal to the relative atomic mass, since the mass of an electron is negative. And q, we have just seen, is equal in electronic units,

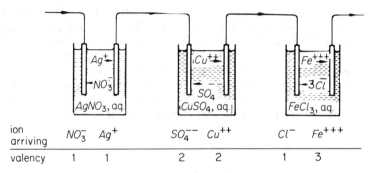

Fig. 32.7 Movement of ions in electrolysis

to the valency. Therefore the mass liberated is proportional to the ratio of relative atomic mass to valency. See also p. 793.

Calculation of Faraday and Avogadro Constants. Specific Charge

We can now calculate the Faraday constant F, the charge required to liberate 1 mole of a monovalent element, and the Avogadro constant N_A, the number of particles per mole.

The mass of copper per coulomb deposited in electrolysis is about $3 \cdot 3 \times 10^{-7}$ kg C^{-1} (p. 675). So the charge carried by 1 kg of copper $= (10^7/3 \cdot 3)$ C. But 1 mole of copper, a divalent element, has a mass of $63 \cdot 6$ g $= 63 \cdot 6 \times 10^{-3}$ kg.

$$\therefore \text{ charge carried by 1 mole } = \frac{10^7}{3 \cdot 3} \times 63 \cdot 6 \times 10^{-3}$$

$$= 192\,720 \text{ C (approx.)}$$

$$\therefore \text{ Faraday constant, } F = \frac{192\,720}{2} = 96\,360 \text{ C.}$$

As copper is divalent, the charge carried by this copper is $2e$, where e is the electronic charge, $1 \cdot 6 \times 10^{-19}$ C (p. 680).

$$\therefore \text{ number of atoms in 1 mole } = \frac{192\,720}{2 \times 1 \cdot 6 \times 10^{-19}}$$

$$= 6 \cdot 02 \times 10^{23}$$

$$= \text{Avogadro constant, } N_A.$$

It should be noted that about $96\,360$ C is the charge on 1 mole of electrons, that is, $N_A e = F$.

The *specific charge* of an ion is defined as the *charge per unit mass* or the ratio q/M where q is the charge on the ion and M is its mass. The mass per coulomb of the copper ion Cu^{2+} is about $3 \cdot 3 \times 10^{-7}$ kg C^{-1}; so the specific charge of the ion is 1 $C/(3 \cdot 3 \times 10^{-7})$ kg $= 3 \cdot 0 \times 10^6$ C kg (approx.). If the mass per coulomb for the hydrogen ion or proton, H^+, is $1 \cdot 04 \times 10^{-8}$ kg C^{-1}, the specific charge of the ion $= 1/(1 \cdot 04 \times 10^{-8}) = 9 \cdot 6 \times 10^7$ C kg^{-1}.

The Electrolytic Capacitor

An electrolytic capacitor is one in which the dielectric is formed by electrolysis— by a secondary reaction at an insoluble electrode. It is made from two coaxial aluminium tubes, A and K in Fig. 32.8, with a solution or paste of ammonium borate between them. A current is passed through from A (anode) to K (cathode) and a secondary reaction at A liberates oxygen. The oxygen does not come off as a gas, however, but combines with the aluminium to form a layer of aluminium oxide over the electrode A. This layer is about 1/40 m thick, and is an insulator. When the layer has been formed, the whole system can be used as a capacitor, one of whose electrodes is the cylinder A, and the other the surface of the liquid or paste adjacent to A. Because the dielectric layer is so thin, the capacitance is much greater than that of a paper capacitor of the same size.

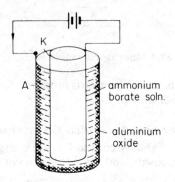

Fig. 32.8 Forming an electrolytic capacitor

In the use of an electrolytic capacitor, some precautions must be taken. The voltage applied to it must not exceed a value determined by the thickness of the dielectric, and marked on the capacitor; otherwise the layer of aluminium oxide will break down (see 'dielectric strength', p. 602). And the voltage must always be applied in the same sense as when the layer was being formed. If the plate A is made negative with respect to K, the oxide layer is rapidly dissolved away. Consequently an alternative voltage must never be applied to an electrolyte capacitor. This condition limits the usefulness of these capacitors.

Electrolytic capacitors are not very reliable, because the oxide layer is apt to break down with age. They are used in domestic radio receivers but in high-grade apparatus they are avoided.

Application of Ohm's Law to Electrolytes

Figure 32.9 (i) shows how the current I through an electrolyte, and the potential difference V across it, may be measured. If the electrodes are soluble—copper in copper sulphate, for example—experiment shows that *the current is proportional to the potential difference* (Fig. 32.9 (ii)).

The best results in this experiment are obtained with very small currents. If the current is large, the solution becomes non-uniform: it becomes stronger near the anode, where copper is dissolved by the attack of the SO_4^{2-} ions, and weaker near the anode, where copper is deposited, and SO_4^{2-} ions drift away.

The best way to do the experiment is to use an alternating current; the electrolysis reverses with the current fifty or more times per second, and no changes of concentration build up. Rectifier-type meters (p. 812) are most suitable for measuring the current and potential difference in this case.

When the measurements are properly made they show, as we have said, that the current is proportional to the potential difference: for example, copper sulphate solution, with copper electrodes, obeys Ohm's law. The voltameter behaves as a passive resistor; all the electrical energy delivered to it by the current appears as heat (pp. 636–9); no electrical energy is converted into mechanical

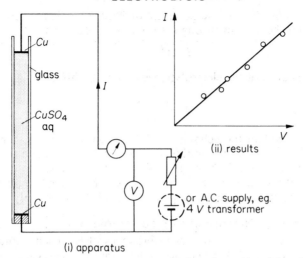

(i) apparatus

Fig. 32.9 Current/voltage characteristics of electrolyte with soluble electrodes

or chemical work. In particular, therefore, no electrical energy is used to break up the molecules of copper sulphate into ions. This is the argument which led Arrhenius to suggest that the electrolyte dissociates into ions as soon as it is dissolved; dissociation is a result of solution, not of electrolysis.

Water Voltameter. Back E.M.F.

If we set out to find the current/voltage relationship for a water voltameter with platinum electrodes, using a d.c. supply as in Fig. 32.10 (i), we find that the voltameter does not obey Ohm's law. If we apply to it a voltage less than 1·7 volts, the current flows only for a short time and then stops. It appears, therefore, that the voltameter is setting up a *back-e.m.f.* which prevents a steady current from flowing, unless the supply voltage is greater than 1·7 V. The existence of the back-e.m.f. can be demonstrated by disconnecting the battery and then shorting the anode and cathode through a suitable meter.

When the p.d. across the voltameter is increased, we find that the current I increases linearly with the potential difference V, when V is greater than 1·7

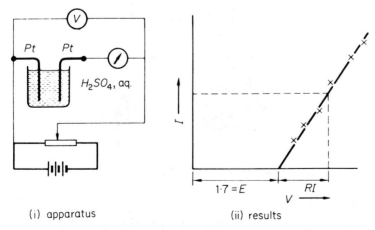

(i) apparatus   (ii) results

Fig. 32.10 Current/voltage characteristics of water voltameter

volts, Fig. 32.10 (ii). If V is the potential difference applied to it and E is the back-e.m.f. then the current through it obeys the relationship

$$I \propto (V - E).$$

If R is the resistance of the electrolyte, we can write this as

$$I = \frac{V - E}{R}.$$

We can now explain the behaviour of the water voltameter. While current was being sent through the voltameter, the cathode became covered with hydrogen, and the anode with oxygen. We may therefore suppose that oxygen and hydrogen at the platinum electrodes set up a 'back-e.m.f.', E, in the circuit; that is to say, it sets up an e.m.f. opposing that of the d.c. supply.

Electrical Energy Consumed in Decomposition

The concept of a back-e.m.f. occurs in branches of Electricity other than the chemical effect of current. We may notice here that the behaviour of the volta-meter is somewhat like that of an electric motor. When the armature of a motor rotates, a back-e.m.f. is induced in it, and the current through it is given by an equation similar to $I = (V - E)/R$. The back-e.m.f. in the motor, we shall see later, represents the electrical power converted into mechanical work. So here the back-e.m.f. E represents this electrical power converted into chemical work—used in breaking up the water molecules. The potential difference across the voltameter is

$$V = IR + E$$

from the equation for I; the power equation is therefore

$$IV = I^2R + EI.$$

The left-hand term is the electrical power input; the first term on the right is the heat produced per second in the electrolyte; and the second term is the work done per second in decomposing the water.

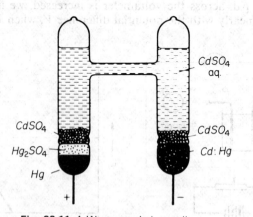

Fig. 32.11 A Weston cadmium cell

Standard Cells

A standard cell is one whose e.m.f. varies very little with time, and with tem-perature, so that it can be used as a standard of potential difference in potentiometer experiments. The commonest type if the Weston cadmium cell, Fig. 32.11. It is housed in an H-shaped glass tube because its electrodes are liquid or semi-liquid. The negative electrode is an amalgam of cadmium in mercury; the solution is of cadmium sulphate; the depolariser is a paste of mercurous sulphate; and the positive electrode is mercury. In some cells, crystals of cadmium sulphate are placed on top of the electrodes to keep the solution saturated. The e.m.f. of one of these, in volts at a temperature $\theta°C$, is

$$E = 1 \cdot 018\ 30 - 0 \cdot 000\ 040\ 66\ (\theta - 20)$$
$$+ 0 \cdot 000\ 000\ 95\ (\theta - 20)^2 + 0 \cdot 000\ 000\ 01\ (\theta - 20)^3.$$

The e.m.f. of the type without crystals is about 1·0186 volts between 0°C and 40°C.

A standard cell without crystals of cadmium sulphate is called an unsaturated cell; one with crystals is called a saturated cell, because the crystals keep the solution saturated. Saturated cells give an accurately reproducible e.m.f., because the concentration of the solution is sharply defined at any given temperature. Unsaturated cells do not agree among one another so well, because the solution may vary a little from one to the other.

The depolariser of a standard cell is effective only for very small currents, and the e.m.f. of the cell change appreciably if more than about 10 microamperes are drawn from it. *A standard cell must not, in any circumstances, be used as a source of current.* In the early stages of balancing a standard cell against a potentiometer wire, a protective resistance of about 100 000 ohms should be connected in series with the cell.

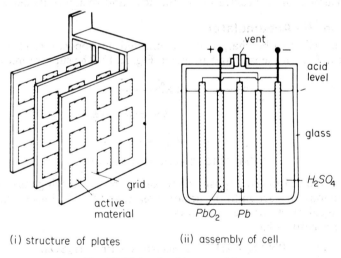

(i) structure of plates   (ii) assembly of cell

Fig. 32.12 Lead-acid accumulator

The Lead Accumulator

The cell just described is called a *primary cell.* A dry (Leclanché) cell, with which we assume the reader is familiar, is also a primary cell. When primary cells are run-down, their active materials must be renewed; the cells cannot be recharged by passing a current through them from another source. A *secondary cell,* however, is one which can be recharged in this way.

The commonest secondary cell is the lead-acid accumulator. Its active materials are spongy lead, Pb (for the negative plate), lead oxide, PbO_2 (for the positive plate), and sulphuric acid. The active materials of the plates are supported in grids of hard lead-antimony alloy, Fig. 32.12 (i). These are assembled in interchanging groups, closely spaced to give a low internal resistance, and often held apart by strips of wood or celluloid, Fig. 32.12 (ii).

When the cell is discharging—giving a current—hydrogen ions drift to the positive plate, and SO_4^{2-} ions to the negative. As they give up their charges they attack the plates, and reduce the active materials of each to lead sulphate.

At the negative plate the reaction is

$$Pb + SO_4^{2-} - 2 \text{ electrons} \rightarrow PbSO_4. \qquad . \qquad . \qquad (1)$$

The chemical action at the positive plate is generally given as

$$\text{(i) } PbO_2 + 2H^+ + 2 \text{ electrons} \rightarrow PbO + H_2O;$$
$$\text{(ii) } PbO + H_2SO_4 \rightarrow PbSO_4 + H_2O;$$

so altogether

$$PbO_2 + H_2SO_4 + 2H^+ + 2 \text{ electrons} \rightarrow PbSO_4 + 2H_2O. \qquad . \qquad (2)$$

However, H_2SO_4 molecules do not exist in the solution—they are dissociated into $2H^+$ and SO_4^- ions. We may therefore write equation (2) as

$$PbO_2 + SO_4^{2-} + 4H^+ + 2 \text{ electrons} \rightarrow PbSO_4 + 2H_2O. \qquad . \qquad (3)$$

The lead sulphate produced in these reactions is a soft form, which is chemically more active than the hard, insoluble lead sulphate familiar in the general chemistry of lead. In the discharging reactions water is formed and sulphuric acid consumed; the concentration of the acid, and therefore its density, fall.

Charging the Accumulator

When the cell is to be charged it is connected, in opposition, to a supply of greater e.m.f., via a rheostat and ammeter, Fig. 32.13. The supply forces a current

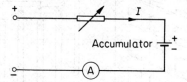

Fig. 32.13 Charging an accumulator

I through the cell in the opposite direction to the discharging current, so that hydrogen ions are carried to the negative plate, and SO_4^{2-} ions to the positive. The chemical reactions are as follows.

At the negative plate:

$$PbSO_4 + 2H^+ + 2 \text{ electrons} \rightarrow Pb + H_2SO_4. \qquad . \qquad (1)$$

At the positive plate:

$$\text{(i) } PbSO_4 + SO_4^{2-} - 2 \text{ electrons} \rightarrow PbO_2 + 2SO_3;$$
$$\text{(ii) } 2SO_3 + 2H_2O \rightarrow 2H_2SO_4;$$

altogether:

$$PbSO_4 + 2H_2O + SO_4^{2-} - 2 \text{ electrons} \rightarrow PbO_2 + 2H_2SO_4. \qquad . \qquad (2)$$

The active materials are converted back to lead and lead dioxide, water is consumed, and sulphuric acid is formed. The acid therefore becomes more concentrated during charge, and its density rises.

Properties and Care of the Lead Accumulator

The e.m.f. of a freshly charged lead accumulator is about 2·2 volts, and the relative density of the acid about 1·25. When the cell is being discharged its e.m.f. falls rapidly to about 2 volts, and then becomes steady (Fig. 32.14); but towards the end of the discharge the e.m.f. begins to fall again. When the terminal voltage load has dropped below about 1·9 volts, or the relative density of the acid below about 1·15, the cell should be recharged. If the cell is discharged too far, or left in a discharged condition, hard lead sulphate forms on its plates, and it becomes useless.

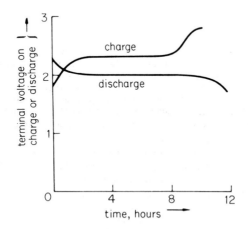

Fig. 32.14 Voltage-time curves of lead accumulator

The internal resistance of a lead accumulator, like that of any other cell, depends on the area and spacing of its plates. It is much lower than that of any primary cell, however, being usually of the order of 1/10 to 1/100 ohm. The amount of electricity which an accumulator can store is called its *capacity*. It is a vague quantity, but a particular accumulator may give, for example, 4 amperes for 20 hours before needing a recharge. The capacity of this accumulator would be 80 ampere-hours. (One ampere-hour = 3600 coulombs.) If the accumulator were discharged faster—at 8 amperes, say—then it would probably need recharging after rather less than 10 hours; and if it were discharged more slowly—say at 2 amperes—it might hold out for more than 40 hours. The capacity of an accumulator therefore depends on its rate of discharge; it is usually specified at the '10-hour' or '20-hour' rate. Discharging an accumulator faster than at about the 10-hour rate causes the active material to fall out of the plates.

Accumulators are usually charged at about the '8-hour' rate—say 5 amperes for the cell discussed above. The charging is continued until gas is bubbling freely off the plates. When the plates are gassing, the chemical reactions (1) and (2) have been completed, and the current through the cell is simply decomposing the water in it. Before the charge is started the vent-plugs in the cell-case must be removed to let the gases out; the gases are hydrogen and oxygen and naked lights near are dangerous. The water lost at the end of each charge must be made up by pouring in distilled water until the acid rises to the level marked on the case. If the relative density of the acid is then less than 1·25, the charging must be continued. Near the end of the charging the back-e.m.f. of the cell rises sharply to about 2·6 volts, Fig. 32.14. It never gives a forward e.m.f. as great as this: as soon as it is put on discharge, its e.m.f. falls to about 2·2 volts.

The Nickel-Iron Accumulator

The nickel-iron (NIFE) accumulator has active materials of nickel hydroxide (positive), iron (negative) and caustic potash solution. Its e.m.f. varies from 1·3 volt to 1·0 on discharge; it has a higher internal resistance than a lead accumulator of similar size; and it is less efficient. Its advantages are that it is more rugged, both mechanically and electrically. Very rapid charging and discharging do not harm it, nor do overdischarging and overcharging. Vibration does not make the active materials fall out of the plates, as it does with a lead cell. Nickel-iron accumulators are therefore used in electric trucks and at sea.

Examples

1. State Faraday's laws of electrolysis and show that the ionic dissociation theory offers an explanation of them. Acidulated water is electrolysed between platinum electrodes. Sketch a graph showing the relation between the strength of the current and the reading of a voltmeter connected to the electrodes. Comment on the nature of the graph.

Give a circuit diagram showing how you would charge a series battery of 12 lead accumulators, each of e.m.f. 2 V and internal resistance $1/24$ Ω, from 240 V d.c. mains, if the charging current is not to exceed 3 A. What percentage of the energy taken from the mains would be wasted? (*L*.)

First part (see text). When the water is electrolysed, no current flows until the p.d. is greater than about 1·7 V, when the back-e.m.f. of the liberated product is overcome. After this, a straight-line graph is obtained between V and I.

Second part. A series resistance R is required, given by

$$I = 3 = \frac{240 - 12 \times 2}{R + \dfrac{12}{24}}.$$

$$\therefore 3R + 1\cdot5 = 216.$$

$$\therefore R = \frac{214\cdot5}{3} = 71\cdot5 \ \Omega.$$

Energy taken from mains $= EIt = 240 \times 3t = 720t$, where t is the time.

$$\text{Energy wasted} = (I^2R + I^2r)t = (3^2 \times 71\cdot5 + 3^2 \times 0\cdot5)t$$

$$= 648t.$$

$$\therefore \text{percentage wasted} = \frac{648t}{720t} \times 100\% = 90\%.$$

2. From energy considerations, estimate the e.m.f. of a cell with a negative electrode of zinc and a copper sulphate depolariser, using the data that 1 g of zinc dissolved in copper sulphate solution liberates 3330 J and that in electrolysis 1 coulomb liberates $3\cdot4 \times 10^{-7}$ kg of zinc.

The e.m.f. E of the cell is the total energy per coulomb produced by the cell.
Now energy W liberated for 1 g of zinc $= 3330$ J.
To liberate 1 g or 10^{-3} kg of zinc, charge flowing is given by

$$Q = \frac{10^{-3}}{3\cdot4 \times 10^{-7}} = \frac{1}{3\cdot4 \times 10^{-4}} \ \text{C},$$

$$\therefore \text{energy liberated per coulomb, } E = \frac{W}{Q} = 3330 \times 3\cdot4 \times 10^{-4}$$

$$= 1\cdot1 \ \text{V}.$$

Exercises 32

1. Assuming the Faraday constant F is 96 500 C mol^{-1}, calculate (i) the charge needed to deposit 1·6 g of oxygen in the electrolysis of water, (ii) the time required if a steady current of 2·5 A is used, (iii) the mass of hydrogen deposited at the end of this time. (Relative molecular mass of hydrogen and oxygen = 2 and 32 respectively.)

2. Assuming the Faraday is 96 500 C mol^{-1} and that the relative atomic masses of copper and silver are 63 and 108 respectively, calculate (a) the number of atoms of copper, Cu^{2+}, and of silver which are liberated respectively by the Faraday, (b) the masses of these two elements liberated respectively by 0·5 A in 10 min. (Electronic charge, $e = -1·6 \times 10^{-19}$ C.)

3. (a) What do you understand by *the Faraday constant, F*? If 1 mole of electrons contains $6·02 \times 10^{23}$ electrons, calculate a value for F.

(b) In a copper plating system an electrolysis current of 3·0 A is used. How many atoms of Cu^{2+} are deposited in 1·5 h? (Electronic charge, $e = -1·6 \times 10^{-19}$ C.) (L.)

4. Explain the process by which an electrolyte conducts electricity. What do you understand by *the Faraday constant*? (L.)

5. Explain how the ionic theory of electrolysis shows that the quantity of charge passed through a voltameter must be an integral multiple of a basic quantity of charge. (L.)

6. What is meant by the *specific charge* of an ion? Describe an experiment to measure this quantity for the Cu^{2+} ion. Describe and explain what happens (a) when copper sulphate solution, (b) when dilute sulphuric acid, are electrolysed between platinum electrodes.

Using the information given below, calculate the volume at s.t.p. of hydrogen formed when a current of 0·50 A passes for 2·0 h in experiment (b) described above. Explain your calculation carefully. (The Avogadro constant = $6·0 \times 10^{23}$ mol^{-1}; charge on proton = $1·6 \times 10^{-19}$ C; volume of 1 mole of gas at s.t.p. = $2·24 \times 10^{-2}$ m^3.) (C.)

7. State the laws of electrolysis and give a concise account of an elementary theory of electrolysis which is consistent with the laws.

If an electric current passes through a copper voltameter and a water voltameter in series, calculate the volume of hydrogen which will be liberated in the latter, at 25°C and 780 mmHg pressure, whilst 5×10^{-5} kg of copper is deposited in the former. (Take mass of hydrogen deposited per coulomb as $1·04 \times 10^{-8}$ kg C^{-1}, of copper as $3·3 \times 10^{-7}$ kg C^{-1}, density of hydrogen as 9×10^{-2} kg m^{-3} at s.t.p.) (L.)

8. Explain what happens when an e.m.f. is applied to platinum electrodes immersed in dilute sulphuric acid. What is the relation between the e.m.f. and the current in such a cell?

If the mass of hydrogen deposited per coulomb is $1·04 \times 10^{-8}$ kg C^{-1}, and if 1 g of hydrogen on burning to form water liberates 147 000 J, calculate the back-e.m.f. produced in a water voltameter when it is connected to a 2 V accumulator. (C.)

9. Explain the general nature of the chemical changes that take place in a lead accumulator during charging and discharging.

A battery of accumulators, of e.m.f. 50 V and internal resistance 2 Ω, is charged on a 100-V direct-current mains. What series resistance will be required to give a charging current of 2 A? If the price of electrical energy is 1p per kilowatt-hour, what will it cost to charge the battery for 8 hours, and what percentage of the energy supplied will be wasted in the form of heat? (C.)

33 Magnetic Field and Force on Conductor

Natural magnets were known some thousands of years ago, and in the eleventh century A.D. the Chinese invented the magnetic compass. This consisted of a magnet, floating on a buoyant support in a dish of water. The respective ends of the magnet, where iron filings are attracted most, are called the north and south poles.

In the thirteenth century the properties of magnets were studied by Peter Peregrinus. He showed that

like poles repel and *unlike poles attract*.

His work was forgotten, however, and his results were rediscovered in the sixteenth century by Dr. Gilbert, who is famous for his researches in magnetism and electrostatics.

Ferromagnetism

About 1823 Sturgeon placed an iron core into a coil carrying a current, and found that the magnetic effect of the current was increased enormously. On switching off the current the iron lost nearly all its magnetism. Iron, which can be magnetised strongly, is called a *ferromagnetic* material. Steel, made by adding a small percentage of carbon to iron, is also ferromagnetic. It retains its magnetism, however, after removal from a current-carrying coil, and is more difficult to magnetise than iron.

Nickel and cobalt are the only other ferromagnetic elements in addition to iron, and are widely used for modern magnetic apparatus. A modern alloy for permanent magnets, called *alnico*, has the composition 54 per cent iron, 18 per cent nickel, 12 per cent cobalt, 6 per cent copper, 10 per cent aluminium. It retains its magnetism extremely well, and, by analogy with steel, is therefore said to be magnetically very hard. Alloys which are very easily magnetised, but do not retain their magnetism, are said to be magnetically soft. An example is *mumetal*, which contains 76 per cent nickel, 17 per cent iron, 5 per cent copper, 2 per cent chromium.

Magnetic Fields

The region round a magnet, where a magnetic force is experienced, is called a *magnetic field*. The appearance of a magnetic field is quickly obtained by iron filings, and accurately plotted with a small compass, as the reader knows. The *direction* of a magnetic field is taken as the direction of the force on a north pole if placed in the field.

Figure 33.1 shows a few typical fields. The field round a bar-magnet is 'non-uniform', that is, its strength and direction vary from place to place, Fig. 33.1 (i). The earth's field locally, however, is uniform, Fig. 33.1 (ii). A bar of soft iron placed north–south becomes magnetised by induction by the earth's field, and the lines of force become concentrated in the soft iron, Fig. 33.1 (iii). The tangent to a line of force at a point gives the direction of the magnetic field at that point.

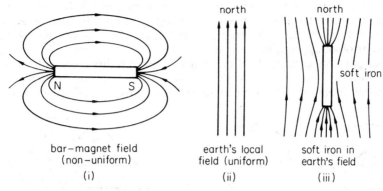

Fig. 33.1 Magnetic fields

Oersted's Discovery

The magnetic effect of the electric current was discovered by Oersted in 1820. Like many others, Oersted suspected a relationship between electricity and magnetism, and was deliberately looking for it. In the course of his experiments, he happened to lead a wire carrying a current over, but parallel to, a compass-needle, as shown in Fig. 33.2 (i); the needle was deflected. Oersted then found that if the wire was led under the needle, it was deflected in the opposite sense, Fig. 33.2 (ii).

From these observations he concluded that the magnetic field was *circular* round the wire. We can see this by plotting the lines of force of a long vertical wire, as shown in Fig. 33.3. To get a clear result a strong current is needed, and we must work close to the wire, so that the effect of the earth's field is negligible. It is then seen that the lines of force are circles, concentric with the wire.

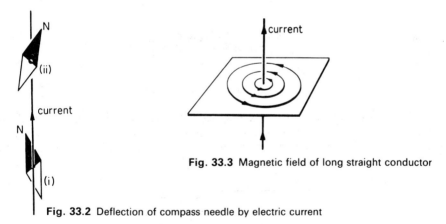

Fig. 33.3 Magnetic field of long straight conductor

Fig. 33.2 Deflection of compass needle by electric current

Directions of Current and Field; Corkscrew Rule

The relationship between the direction of the lines of force and of the current is expressed in Maxwell's corkscrew rule: if we imagine ourselves driving a corkscrew in the direction of the current, then the direction of rotation of the corkscrew is the direction of the lines of force. Figure 33.4 illustrates this rule, the small, heavy circle representing the wire, and the large light one a line of force.

At (i) the current is flowing into the paper; its direction is indicated by a cross, which stands for the tail of an arrow moving away from the reader. At (ii) the current is flowing out of the paper; the dot in the centre of the wire stands for the point of an approaching arrow.

If we plot the magnetic field of a circular coil carrying a current, we get the result shown in Fig. 33.5. Near the circumference of the coil, the lines of force are closed loops, which are not circular, but whose directions are still given

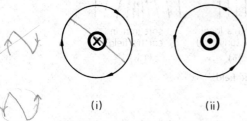

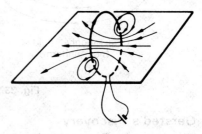

(i)   (ii)

Fig. 33.4 Illustrating corkscrew rule   **Fig. 33.5** Magnetic field of narrow coil

by the corkscrew rule, as in Fig. 33.5. Near the centre of the coil, the lines are almost straight and parallel. Their direction here is again given by the corkscrew rule, but the current and the lines of force are interchanged, that is, if we turn the screw in the direction of the current, then its point travels in the direction of the lines.

The Solenoid

The same is true of the magnetic field of a long cylindrical coil, shown in Fig. 33.6. Such a coil is called a *solenoid*; it has a field similar to that of a bar-magnet, whose poles are indicated in the figure. If an iron or steel core were put into the coil, it would become magnetised with the polarity shown.

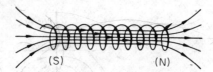

Fig. 33.6 Magnetic field of solenoid

If the terminals of a battery are joined by a wire which is simply doubled back on itself, as in Fig. 33.7, there is no magnetic field at all; each element of the outward run, such as AB, in effect cancels the field of the corresponding

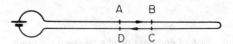

Fig. 33.7 A doubled-back current has no magnetic field

element of the inward run, CD. But as soon as the wire is opened out into a loop, its magnetic field appears, Fig. 33.8. Within the loop, the field is strong, because all the elements of the loop give magnetic fields in the same sense, as we can see by applying the corkscrew rule to each side of the square ABCD. Outside the loop, for example at the point P, corresponding elements of the loop give opposing fields (for example, DA opposes BC); but these elements

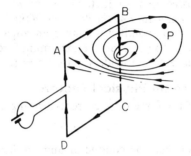

Fig. 33.8 An open loop of current has magnetic field

are at different distances from P (DA is farther away than BC). Thus there is a resultant field at P, but it is weak compared with the field inside the loop. A magnetic field can thus be set up either by wires carrying a current, or by the use of permanent magnets.

Force on Conductor. Fleming's Rule

When a conductor carrying a current is placed in a magnetic field due to some source other than itself, it experiences a mechanical force. To demonstrate this,

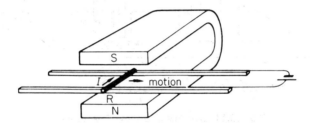

Fig. 33.9 Force on current in magnetic field

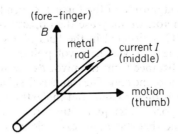

Fig. 33.10 Left-hand rule

a short brass rod R is connected across a pair of brass rails, as shown in Fig. 33.9. A horseshoe magnet is placed so that the rod lies in the field between its poles. When we pass a current I through the rod, from an accumulator, the rod rolls along the rails. The relative directions of the current, the applied field, and the motion are shown in Fig. 33.10; they are the same as those of the middle finger, the forefinger, and the thumb of the *left* hand when held all at right angles to one another. If we place the magnet so that its field lies in the *same* direction as the current, then the rod experiences no force.

Experiments like this were first made by Ampère in 1820. As a result of them, he concluded that the force on a conductor is always *at right angles to the plane which contains both the conductor and the direction of the field in which it is placed.* He also showed that, if the conductor makes an angle α with the field, the force on it is proportional to sin α. So the maximum force is exerted when the conductor is *perpendicular* to the field, when sin $\alpha = 1$.

Dependence of Force on Physical Factors

Since the magnitude of the force on a current-carrying conductor is given by

$$F \propto \sin \alpha, \qquad . \qquad . \qquad . \qquad . \qquad (1)$$

where α is the angle between the conductor and the field, it follows that F is zero when the conductor is parallel to the field direction. This defines the direction of the magnetic field. To find which way it points, we can apply Fleming's rule to the case when the conductor is placed at right angles to the field. The direction of the field then corresponds to the direction of the forefinger.

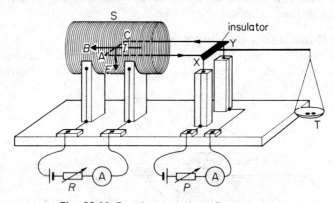

Fig. 33.11 Experiment to show F varies with I

Variation of *F* with *I*

To investigate how the magnitude of the force F depends on the current I and the length l of the conductor, we may use the apparatus of Fig. 33.11.

Here the conductor AC is situated in the field of a solenoid S. The current flows into, and out of, the wire via the pivot points Y and X. The scale pan T is placed at the same distance from the pivot as the straight wire AC, which is perpendicular to the axis of the coil. The frame is first balanced with no current flowing in AC. A current is then passed, and the extra weight needed to restore the frame to a horizontal position is equal to the force on the wire AC. By varying the current in AC with the rheostat P, for example, by doubling or halving the circuit resistance, it may be shown that:

$$F \propto I \qquad . \qquad . \qquad . \qquad . \qquad . \qquad (2)$$

If different frames are used so that the length, l, of AC is changed, it can be shown that, with constant current and field,

$$F \propto l \qquad . \qquad . \qquad . \qquad . \qquad . \qquad (3)$$

Effect of *B*

The magnetic field due to the solenoid will depend on the current flowing in it. If this current is varied by adjusting the rheostat R, it can be shown that the larger the current *in the solenoid*, S, the larger is the force F. It is reasonable to suppose that a larger current in S produces a stronger magnetic field. Thus the force F increases if the magnetic field strength is increased. The magnetic field is represented by a vector quantity which is given the symbol B. This is called the *flux density* in the field. We assume that:

$$F \propto B \qquad . \qquad . \qquad . \qquad . \qquad . \qquad (4)$$

Magnitude of F

From the results expressed in equations (1) to (4), we obtain

$$F \propto BIl \sin \alpha,$$

or

$$F = kBIl \sin \alpha \qquad . \qquad . \qquad . \qquad . \qquad . \qquad (5)$$

where k is a constant.

In the SI system of units, the unit of B is the tesla (T). $1\ T = 1$ weber metre^{-2} (Wb m^{-2}). One tesla may be defined as the flux density of a uniform field when the force on a conductor 1 metre long, placed perpendicular to the field and carrying a current of 1 ampere, is 1 newton. Substituting $F = 1$, $B = 1$, $l = 1$ and $\sin \alpha = \sin 90° = 1$ in (5), then $k = 1$. Thus in Fig. 33.12 (i), with the above units,

$$F = BIl \sin \alpha \qquad . \qquad . \qquad . \qquad . \qquad . \qquad (6)$$

When the whole length of the conductor is *perpendicular* to the field B, Fig. 33.12 (ii), then, since $\alpha = 90°$ in this case,

$$F = BIl \qquad . \qquad . \qquad . \qquad . \qquad . \qquad . \qquad (7)$$

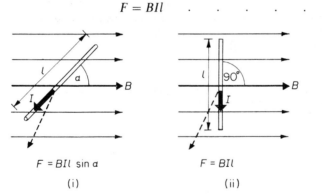

$$F = BIl \sin \alpha$$

(i)

$$F = BIl$$

(ii)

Fig. 33.12 Magnitude of F which acts towards reader

It may be noted that the apparatus of Fig. 33.11 can be used to determine the flux density B of the field in the solenoid. In this case, $\alpha = 90°$ and $\sin \alpha = 1$. So measurement of F, I and l enables B to be found from (7).

It may help the reader if we now summarize the main points about B:

1. When a current-carrying conductor XY is turned in a uniform magnetic field of flux density B until no force acts on it, then XY points in the direction of B.

2. When a straight conductor of length l carrying a current I is placed perpendicular to a uniform field and a force F acts on the conductor, then the magnitude B of the flux density is given by

$$B = \frac{F}{Il}.$$

Interaction of Magnetic Fields

The force on a conductor in a magnetic field can be accounted for by the interaction between magnetic fields.

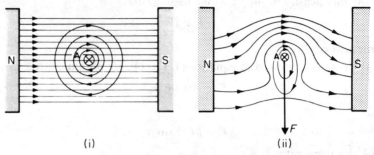

(i)   (ii)

Fig. 33.13 Interaction of magnetic fields

Figure 33.13 (i) shows a section A of a vertical conductor carrying a downward current. The field pattern consists of circles round A as centre (p. 691). When the conductor is in the uniform horizontal field B due to the poles N, S, the magnetic flux (lines) due to B, which consists of straight parallel lines, passes on either side of A. The two fields interact. As shown, the resultant field has a *greater* flux density above A in Fig. 33.13 (ii) and a *smaller* flux density below A. The conductor moves from the region of greater flux density to smaller flux density. So A moves downwards as shown. As the reader should verify, the direction of the force F on the conductor is given by Fleming's left hand rule.

If a current-carrying conductor is placed in the *same* direction as a uniform magnetic field, the flux-density on both sides of the conductor is the same, as the reader should verify. The conductor is now not affected by the field, that is, no force acts on it in this case.

Example

A wire carrying a current of 10 A and 2 metres in length is placed in a field of flux density 0·15 T. What is the force on the wire if it is placed (*a*) at right angles to the field, (*b*) at 45° to the field, (*c*) along the field.

From (5) $F = BIl \sin \alpha$

(*a*) $F = 0·15 \times 10 \times 2 \times \sin 90°$

                           $= 3$ N.

(*b*) $F = 0·15 \times 10 \times 2 \times \sin 45°$

                           $= 2·12$ N.

(*c*) $F = 0$, since $\sin 0° = 0$.

Torque on Rectangular Coil in Uniform Field

A rectangular coil of insulated copper wire is used in the moving-coil meter, which we discuss shortly. Industrial measurements of current and p.d. are made mainly with moving-coil meters.

Consider a rectangular coil situated with its plane *parallel* to a uniform magnetic field of flux density B. Suppose a current I is passed into the coil, Fig. 33.14 (i). Viewed from above, the coil appears as shown in Fig. 33.14 (ii).

The side PS of length l is perpendicular to B. So the force on it is given by $F = BIl$. If the coil has N turns, the length of the conductor is increased N times and so the force on the side PS, F, $= BIlN$.

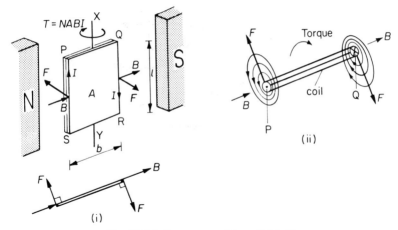

Fig. 33.14 Torque on coil in radial field

The force on the opposite side QR is also given by $F = BIlN$, but its direction is *opposite* to that on PS. There are no forces on the sides PQ and SR although they carry currents because PQ and SR are parallel to the field B.

The two forces F on the sides PS and QR tend to turn the coil about an axis XY passing through the middle of the coil. The two forces together are called a *couple* and their moment (turning-effect) or *torque* T is given, by definition, by

$$T = F \times p,$$

where p is the *perpendicular* distance between the two forces. Now from Fig. 33.14 (i), $p = b$, the width PQ or SR of the coil. So

$$T = F \times p = BIlN \times b.$$

But $l \times b =$ area A of the coil. So

$$torque\ T = BANI \quad . \qquad . \qquad . \qquad . \qquad . \qquad (1)$$

The unit of torque (force × distance) is newton metre, symbol N m. In using $T = BANI$, B must be in units of T (tesla), A in m^2 and I in A.

If there were no opposition to the torque, the coil PQRS would turn round and settle with its plane normal to B, that is, facing the poles N, S in Fig. 33.14 (i). As we see later, springs can control the amount of rotation of the coil.

Figure 33.14 (ii) is a plan view PQ of the rectangular coil with its plane in the same direction as the uniform magnetic field of the magnet N, S. As we explained previously, the magnetic field of the current in the straight sides PS, QR of the coil interacts with the field of the magnet. Figure 33.14 (ii) shows roughly the appearance of the resultant field round the vertical sides of the conductors whose tops are P and Q respectively. The current is downward in Q and upward towards the reader in P. The forces F act from the dense to the less dense flux and together they produce a torque on the coil.

Torque on Coil at Angle to Uniform Field

Suppose now that the plane of the coil is at an angle θ to the field B when it carries a current I. Fig. 33.15 (i) shows the forces F_1 on its vertical sides PS and QR; these two forces set up a torque which rotate the coil. The forces F_2 on its horizontal sides merely compress the coil and are resisted by its rigidity.

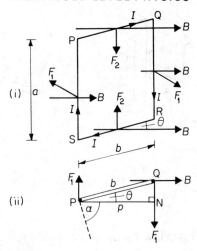

Fig. 33.15 Torque on coil at angle to uniform field

The forces F_1 on the sides PS and QR are still given by $F_1 = BIlN$ because PS and QR are perpendicular to B. But now the forces F_1 are not separated by a perpendicular distance b, the coil breadth. The perpendicular distance p is less than b and is given by

$$p = b \cos \theta.$$

So this time

$$\text{torque } T = F_1 \times p = BIlN \times b \cos \theta.$$

So

$$T = BANI \cos \theta. \qquad \qquad \qquad (2)$$

When the plane of the coil is *parallel* to B, then $\theta = 0°$ and $\cos \theta = 1$. So the torque $T = BANI$ as we have already shown. If the plane of the coil is *perpendicular* to B, then $\theta = 90°$ and $\cos \theta = 0$. So the torque $T = 0$ in this case.

If α is the angle between B and the *normal* to the plane of the coil, then $\theta = 90° - \alpha$. From (2), the torque T is given by

$$T = BANI \sin \alpha . \qquad \qquad \qquad (3)$$

Magnetic Moment

The torque on the coil thus depends on the magnitude of the flux density B, the current I in it and the area A. We can write equation (3) as

$$T = mB \sin \alpha \qquad \qquad \qquad (1)$$

where $m = NIA$. m is a property of the coil and the current in it and is called the *magnetic moment* of the coil. In general, the magnetic moment of a coil is defined as the torque exerted on it when it is placed with its plane parallel to a field of 1 T. In the case of $B = 1$, $\sin \alpha = \sin 90° = 1$ and hence $T = m$. It should be noted that the magnetic moment of a coil can be calculated from

$$m = NIA \qquad \qquad \qquad (2)$$

whatever the shape of the coil. From this expression the unit of m is ampere metre^2 (A m^2).

Magnetism is due to circulating and spinning electrons inside atoms (see p. 763). The moving charges are equivalent to electric currents. Consequently, like a current-carrying coil, permanent magnets also have a torque acting on them when they are placed with their axis at an angle to a magnetic field. Like the coil, they turn and settle in equilibrium with their axis along the field direction. Thus the magnetic compass needle will point magnetic north–south in the direction of the Earth's magnetic field.

Example

A vertical rectangular coil of sides 5 cm by 2 cm has 10 turns and carries a current of 2 A. Calculate the torque on the coil when it is placed in a uniform horizontal magnetic field of 0·1 T with its plane (*a*) parallel to the field, (*b*) perpendicular to the field, (*c*) 60° to the field.

The area A of the coil $= 5 \times 10^{-2}$ m $\times 2 \times 10^{-2}$ m $= 10^{-3}$ m^2.

So (*a*) $\qquad$ torque $T = NABI = 10 \times 10^{-3} \times 0 \cdot 1 \times 2$

$$= 2 \times 10^{-3} \text{ N m.}$$

(*b*) Here $\qquad T = 0.$

(*c*) $\qquad T = NABI \cos 60° \text{ or } NABI \sin 30°$

$$= 2 \times 10^{-3} \times 0 \cdot 5 = 10^{-3} \text{ N m.}$$

The Moving-coil Meter

All current measurements except the most accurate are made today with a moving-coil meter. In this instrument a rectangular coil of fine insulated copper wire is suspended in a strong magnetic field, Fig. 33.16 (i). The field is set up between soft iron pole-pieces, NS, attached to a powerful permanent magnet.

The pole-pieces are curved to form parts of a cylinder coaxial with the suspension of the coil. And between them lies a cylindrical core of soft iron, C; it is supported on a brass pin, T in Fig. 33.16 (ii), which is placed so that it does not foul the coil. As the diagram shows, the magnetic field B is *radial* to the core and pole-pieces, over the region in which the coil can swing. In this case the deflected coil always comes to rest with its plane *parallel* to the field in which it is then situated, as shown in Fig. 33.16 (ii).

The moving-coil milliammeter or ammeter have hair-springs and jewelled bearings. The coil is wound on a rigid but light aluminium frame, which also carries the pivots. The pivots are insulated from the former if it is aluminium, and the current is led in and out through the springs. The framework, which carries the springs and jewels, is made from brass or aluminium—if it were steel

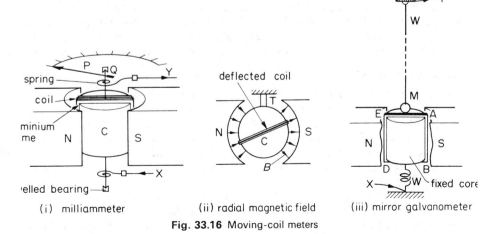

(i) milliammeter $\qquad$ (ii) radial magnetic field $\qquad$ (iii) mirror galvanometer

Fig. 33.16 Moving-coil meters

it would affect the magnetic field. An aluminium pointer, P, shows the deflection of the coil; it is balanced by a counterweight, Q, Fig. 33.16 (ii).

In the more sensitive instruments, the coil is suspended on a phosphor-bronze wire, WW, which is kept taut, Fig. 33.16 (iii). The current is led into and out of the coil EABD through the suspension, at X and Y, and the deflection of the coil is shown by a beam of light, reflected by a mirror M to a scale in front of the instrument.

Theory of Moving-coil Instrument

The rectangular coil is situated in the radial field B. When a current is passed into it, the coil rotates through an angle θ which depends on the strength of the springs. No matter where the coil comes to rest, the field B in which it is situated always lies along the *plane* of the coil because the field is radial. As we have previously seen, the forces F on the two sides of the coil are then separated by a distance equal to the breadth of the coil and the torque T is always given by $NABI$. So the torque $T \propto I$, since N, A, B are constant.

In equilibrium, the deflecting torque T on the coil is equal to the opposing torque due to the elastic forces in the spring. The opposing torque $= c\theta$, where c is a constant of the springs which depends on its elasticity under twisting forces and on its dimensions. So

$$NABI = c\theta$$

and
$$I = \frac{c}{NAB}\theta. \qquad . \qquad . \qquad . \qquad . \qquad (1)$$

Equation (1) shows that the deflection θ is proportional to the current I. So the scale showing current values is a uniform one, that is, equal divisions along the calibrated scale represent equal steps in current. This is an important advantage of the moving coil meter. It can be accurately calibrated and its subdivisions read accurately.

If the radial field were not present, for example, if the soft iron cylinder were removed, the torque would then be $NABI \cos \theta$ (p. 698) and I would be proportional to $\theta/\cos \theta$. The scale would then be *non-uniform* and difficult to calibrate or to read accurately.

The pointer type of instrument (Fig. 33.16 (i)) usually has a scale calibrated directly in milliamperes or microamperes. Full-scale reading on such an instrument corresponds to a deflection θ of 90° to 120°; it may represent a current of 50 microamperes to 15 milliamperes, according to the strength of the hair springs, the geometry of the coil, and the strength of the magnetic field. The less sensitive models are more accurate, because their pivots and springs are more robust, and therefore are less affected by dust, vibration, and hard use.

Sensitivity of Meter

The *sensitivity* of a current meter is the *deflection per unit current*, or θ/I. Small currents must be measured by a meter which gives an appreciable deflection. From $NABI = c\theta$, we have $\theta/I = NAB/c$. So greater sensitivity is obtained with a stronger field B, a low value of c, that is, weak springs, and a greater value of N and A. The size and number of turns of a coil would increase the resistance of the meter, which is not desirable. The elastic constant c of the springs can be varied, however.

When a galvanometer is of the suspended-coil type (Fig. 33.16 (iii)), its sensitivity is generally expressed in terms of the displacement of the spot of light reflected from the mirror on to the scale. A Scalamp or Edspot, a form of light beam galvanometer, may give a deflection of 25 mm per microampere.

All forms of moving-coil galvanometer have one disadvantage: they are easily damaged by overload. A current much greater than that which the instrument is intended to measure will burn out its hair-springs or suspension.

Sensitivity of Voltmeter

The sensitivity of a voltmeter is the deflection per unit p.d., or θ/V, where θ is the deflection produced by a p.d. V.

If the resistance of a moving coil meter is R, the p.d. V across its terminals when a current I flows through it is given by $V = IR$. From our expression for I given previously,

$$V = \frac{cR}{NAB}\theta.$$

So   voltage sensitivity $= \dfrac{\theta}{V} = \dfrac{NAB}{cR}.$

Example

A moving coil meter X has a coil of 20 turns and a resistance 10 Ω. Another moving coil meter Y has a coil of 10 turns and a resistance of 4 Ω. If the area of each coil, the strength of the springs and the field B are the same in each meter, which has (a) the greater current sensitivity, (b) the greater voltage sensitivity?

(a) The current sensitivity is given by

$$\frac{\theta}{I} = \frac{NAB}{c}.$$

Since the sensitivity $\propto N$, with A, c and B constant, then X (20 turns) has a greater sensitivity than Y (10 turns).

(b) The voltage sensitivity $= NAB/cR$. So with A, B, c constant,

$$\text{sensitivity} \propto \frac{N}{R}.$$

Now $N/R = 20/10 = 2$ numerically for X, and $N/R = 10/4 = 2.5$ for Y. So Y has the greater voltage sensitivity.

As we showed on p. 625, a moving-coil milliammeter can be converted to a voltmeter by adding a suitable high resistance in *series* with the meter, and to an ammeter by adding a suitable low resistance in *parallel* with the meter to act as a shunt. See pp. 627–630.

The Wattmeter

The wattmeter is an instrument for measuring electrical power. In construction and appearance it resembles a moving-coil voltmeter or ammeter, but it has no permanent magnet. Instead it has two fixed coils, FF in Fig. 33.17, these set up the magnetic field in which the suspended coil, M, moves. When the instrument is in use, the coils FF are connected in series with the device X whose power consumption is to be measured. The magnetic field B, set up by FF, is then proportional to the current I drawn by X:

$$B \propto I.$$

The moving-coil M is connected across the device X. In series with M is a high resistance R, similar to the multiplier of a voltmeter; M is, indeed, often called the volt-coil. The current I' through the volt-coil is small compared with

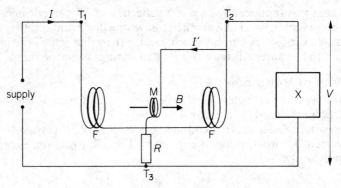

Fig. 33.17 Principle of wattmeter

the main current I, and is proportional to the potential difference V across the device X:

$$I' \propto V.$$

The torque acting on the moving-coil is proportional to the current through it, and to the magnetic field in which it is placed:

$$T \propto BI'.$$

Consequently $T \propto IV.$

That is to say, the torque on the coil is proportional to the product of the current through the device X, and the voltage across it. The torque is therefore proportional to the power consumed by X, and the power can be measured by the deflection of the coil.

The diagram shows that, because the volt-coil draws current, the current through the fixed coils is a little greater than the current through X. As a rule, the error arising from this is negligible; if not, it can be allowed for as when a voltmeter and ammeter are used separately.

Force on Charges moving in Magnetic Fields

As we explained earlier, an electric current in a wire can be regarded as a drift of electrons in the wire, superimposed on their random thermal motions. If the electrons in the wire drift with average velocity v, and the wire lies at right angles to the field, then the force on *each* electron, as we soon show, is given by

$$F = Bev \qquad . \qquad . \qquad . \qquad . \qquad (1)$$

Generally, the force F on a charge Q moving at right angles to a field of flux density B is given by

$$F = BQv \qquad . \qquad . \qquad . \qquad . \qquad (2)$$

If B is in tesla (T), e or Q is in coulomb (C) and v in metre second^{-1} (m s^{-1}), then F will be in newton (N) (Fig. 33.18 (i)).

The proof of equation (1) can be obtained as follows. Suppose a current I flows in a straight conductor of length l when it is perpendicular to a uniform field of flux density B. From p. 621, $I = nvAe$, where n is the number of electrons per unit volume, v is the drift velocity of the electrons, A is the area of cross-section of the conductor and e is the electron charge. Then the force F' on the conductor is given by

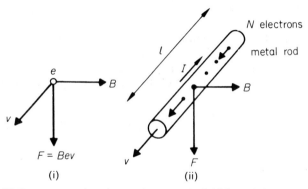

Fig. 33.18 Force on moving electron in magnetic field (v at right angles to page)

$$F' = BIl = BnevAl = Bev \times nAl.$$

Now Al is the volume of the wire. So nAl is the number N of electrons in the conductor.

So   force on one electron, $F = \dfrac{F'}{N} = Bev.$

Generally, a charge Q moving *perpendicular* to a magnetic field B with a velocity v has a force on it given by

$$F = BQv.$$

If the velocity v and the field B are inclined to each other at an angle θ,

$$F = BQv \sin \theta.$$

It should be carefully noted that the force F acts *perpendicular* to v and to B. This means that F is a *deflecting force*, that is, it changes the direction of motion of the moving charge when the charge enters the field B but does not alter the magnitude of v.

Further, since F is perpendicular to the direction of motion or displacement of the charge, *no work* is done by F as the charge moves in the field. So no energy is gained by a charge when it enters a magnetic field and forces act on it.

The *direction* of F is given by Fleming's left hand rule, The middle finger points in the direction of the conventional current or direction of motion of a *positive* charge. If a *negative* charge moves from X to Y, the middle finger points in the opposite direction, Y to X, since this is the equivalent positive charge movement.

An electron moving across a magnetic field experiences a force whether it is in a wire or not—for example, it may be one of a beam of electrons in a vacuum tube. Because of this force, a magnetic field can be used to focus or deflect an electron beam, instead of an electrostatic field as on p. 814. Magnetic deflection and focusing are common in cathode ray tubes used for television. In nuclear energy machines, protons may be deflected by a magnetic field. A proton is a hydrogen nucleus carrying a positive charge and its mass is much greater than an electron (p. 889).

Hall Effect

In 1879, Hall found that an e.m.f. is set up *transversely* or *across* a current-carrying conductor when a perpendicular magnetic field is applied. This is called the *Hall effect*.

To explain the Hall effect, consider a slab of metal carrying a current, Fig. 33.19. The flow of electrons is in the opposite direction to the conventional current. If the metal is placed in a magnetic field B at right angles to the face AGDC of the slab and directed out of the plane of the paper, a force Bev then acts on each electron in the direction from CD to AG. Thus electrons accumulate along the side AG of the metal, which will make AG negatively charged and lower its potential with respect to CD. Thus a potential difference or e.m.f. opposes the electron flow. The flow ceases when the e.m.f. reaches a particular value V_H called the *Hall voltage* as shown in Fig. 33.19, which may be measured by using a high impedance voltmeter.

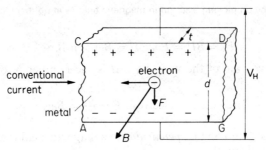

Fig. 33.19 Hall voltage

Magnitude of Hall Voltage

Suppose V_H is the magnitude of the Hall voltage and d is the width of the slab. Then the electric field intensity E set up across the slab is numerically equal to the potential gradient and hence $E = V_H/d$. Hence the force on each electron $= Ee = V_He/d$.

This force, which is directed upwards from AG to CD, is equal to the force produced by the magnetic field when the electrons are in equilibrium.

$$\therefore Ee = Bev$$

$$\therefore \frac{V_He}{d} = Bev$$

$$\therefore V_H = Bvd \qquad . \qquad . \qquad . \qquad . \qquad . \qquad (1)$$

From p. 621, the drift velocity of the electrons is given by

$$I = nevA, \qquad . \qquad . \qquad . \qquad . \qquad . \qquad (2)$$

where n is the number of electrons per unit volume and A is the area of cross-section of the conductor. In this case $A = td$ where t is the thickness. Hence, from (2),

$$v = \frac{I}{netd}$$

Substituting in (1),

$$\therefore V_H = \frac{BI}{net} \qquad . \qquad . \qquad . \qquad . \qquad . \qquad (3)$$

We now take some typical values for copper to see the order of magnitude of V_H. Suppose $B = 1$ T, a field obtained by using a large laboratory electromagnet. For copper, $n \simeq 10^{29}$ *electrons* per metre^3, and the charge on the electron is 1.6×10^{-19} coulomb. Suppose the specimen carries a current of 10 A and that its thickness is about 1 mm or 10^{-3} m. Then

$$V_H = \frac{1 \times 10}{10^{29} \times 1.6 \times 10^{-19} \times 10^{-3}} = 0.6 \ \mu\text{V (approx.)}.$$

This e.m.f. is very small and would be difficult to measure. The importance of the Hall effect becomes apparent when semiconductors are used, as we now see.

Hall Effect in Semiconductors

In semiconductors, the charge carriers which produce a current when they move may be positively or negatively charged (see p. 824). The Hall effect helps us to find the sign of the charge carried. In Fig. 33.19, p. 704, suppose that electrons were not responsible for carrying the current, and that the current was due to the movement of positive charges in the *same* direction as the conventional current. The magnetic force on these charges would also be downwards, in the same direction as if the current were carried by electrons. This is because the sign *and* the direction of movement of the charge carriers have both been reversed. Thus AG would now become positively charged, and the polarity of the Hall voltage would be reversed.

Experimental investigation of the polarity of the Hall voltage hence tells us whether the current is predominantly due to the drift of positive charges or to the drift of negative charges. In this way it was shown that the current in a metal such as copper is due to movement of negative charges, but that in impure semiconductors such as germanium or silicon, the current may be predominantly due to movement of either negative or positive charges (p. 824).

The magnitude of the Hall voltage V_H in metals was shown as above to be very small. In semiconductors it is much larger because the number n of charge carriers per metre^3 is much *less* than in a metal and $V_H = BI/net$. Suppose that n is about 10^{25} per metre^3 in a semiconductor, and $B = 1$ T, $t = 10^{-3}$ m, $e = 1.6 \times 10^{-19}$ C, as above. Then

$$V_H = \frac{1 \times 10}{10^{25} \times 1.6 \times 10^{-19} \times 10^{-3}} = 6 \times 10^{-3} \text{ V (approx.)} = 6 \text{ mV}.$$

The Hall voltage is thus much more measurable in semiconductors than in metals.

Use of Hall Effect

Apart from its use in semiconductor investigations, a *Hall probe* may be used to measure the flux density B of a magnetic field. A simple Hall probe is shown in Fig. 33.20. Here a wafer of semiconductor has two contacts on opposite sides which are connected to a high impedance voltmeter, V. A current, generally less than one ampere, is passed through the semiconductor and is measured on the ammeter, A. The 'araldite' encapsulation prevents the wires from being detached from the wafer. Now, from (3) on p. 704,

$$V_H = \frac{BI}{net}$$

$$\therefore B = \frac{V_H net}{I}$$

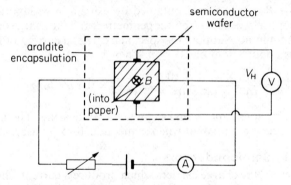

Fig. 33.20 Measurement of *B* by Hall voltage

Now *net* is a constant for the given semiconductor, which can be determined previously. Thus from the measurement of V_H and *I*, *B* can be found.

Exercises 33

1. A vertical straight conductor X of length 0·5 m is situated in a uniform horizontal magnetic field of 0·1 T.
(i) Calculate the force on X when a current of 4 A is passed into it. Draw a sketch showing the directions of the current, field and force. (ii) Through what angle must X be turned in a vertical plane so that the force on X is halved?

2. A straight horizontal rod X, of mass 50 g and length 0·5 m, is placed in a uniform horizontal magnetic field of 0·2 T perpendicular to X. Calculate the current in X if the force acting on it just balances its weight. Draw a sketch showing the directions of the current, field and force. ($g = 10$ N kg^{-1}.)

3. A narrow vertical rectangular coil is suspended from the middle of its upper side with its plane parallel to a uniform horizontal magnetic field of 0·02 T. The coil has 10 turns, and the lengths of its vertical and horizontal sides are 0·1 m and 0·05 m respectively. Calculate the torque on the coil when a current of 5 A is passed into it. Draw a sketch showing the directions of the current, field and torque.
What would be the new value of the torque if the plane of the vertical coil was initially at 60° to the magnetic field and a current of 5 A was passed into the coil?

4. A horizontal rod PQ, of mass 10 g and length 0·10 m, is placed on a smooth plane inclined at 60° to the horizontal, as shown in Fig. 33A.
A uniform vertical magnetic field of value *B* is applied in the region of PQ. Calculate *B* if the rod remains stationary on the plane when a current of 1·73 A flows in the rod.
What is the direction of the current in the rod?

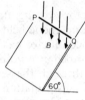

Fig. 33A

5. An electron beam, moving with a velocity of 10^6 m s^{-1}, moves through a uniform magnetic field of 0·1 T which is perpendicular to the direction of the beam. Calculate the force on an electron if the electron charge is $-1·6 \times 10^{-19}$ C. Draw a sketch showing the directions of the beam, field and force.

6. A current of 0·5 A is passed through a rectangular section of a semiconductor 4 mm thick which has majority carriers of negative charges or free electrons. When a magnetic field of 0·2 T is applied perpendicular to the section, a Hall voltage of 6·0 mV is produced between the opposite edges.

Draw a diagram showing the directions of the field, charge carriers and Hall voltage, and calculate the number of charge carriers per unit volume.

7. Figure 33B represents a cylindrical aluminium bar A resting on two horizontal aluminium rails which can be connected to a battery to drive a current through A. A magnetic field, of flux density 0·10 T, acts perpendicularly to the paper and into it. In which direction will A move if the current flows?

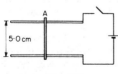

Fig. 33B

Calculate the angle to the horizontal to which the rails must be tilted to keep A stationary if its mass is 5·0 g, the current in it is 4·0 A and the direction of the field remains unchanged. (Acceleration of free fall, $g = 10 \text{ m s}^{-2}$.) (*L.*)

8. Describe an experiment to show that a force is exerted on a conductor carrying a current when it is placed in a magnetic field. Give a diagram showing the directions of the current, the field, and the force.

A rectangular coil of 50 turns hangs vertically in a uniform magnetic field of magnitude 10^{-2} T, so that the plane of the coil is parallel to the field. The mean height of the coil is 5 cm and its mean width 2 cm. Calculate the strength of the current that must pass through the coil in order to deflect it 30° if the torsional constant of the suspension is 10^{-9} newton metre per degree. Give a labelled diagram of a moving-coil galvanometer. (*L.*)

9. Describe with the aid of diagrams the structure and mode of action of a moving coil galvanometer having a linear scale and suitable for measuring small currents. If the coil is rectangular, derive an expression for the deflecting couple acting upon it when a current flows in it, and hence obtain an expression for the current sensitivity (defined as the deflection per unit current).

If the coil of a moving galvanometer having 10 turns and of resistance 4 Ω is removed and replaced by a second coil having 100 turns and of resistance 160 Ω calculate

(*a*) the factor by which the current sensitivity changes and

(*b*) the factor by which the voltage sensitivity changes.

Assume that all other features remain unaltered. (*N.*)

10. Define the coulomb. Deduce an expression for the current I in a wire in terms of the number of free electrons per unit volume, n, the area of cross-section of the wire, A, the charge on the electron, e, and its drift velocity, v.

A copper wire has $1·0 \times 10^{29}$ free electrons per cubic metre, a cross-sectional area of $2·0 \text{ mm}^2$ and carries a current of 5·0 A. Calculate the force acting on each electron if the wire is now placed in a magnetic field of flux density 0·15 T which is perpendicular to the wire. Draw a diagram showing the directions of the electron velocity, the magnetic field and this force on an electron.

Explain, without experimental detail, how this effect could be used to determine whether a slab of semiconducting material was *n*-type or *p*-type. (Charge on electron $= -1·6 \times 10^{-19}$ C.) (*L.*)

11. A moving coil galvanometer consists of a rectangular coil of N turns each of area A suspended in a radial magnetic field of flux density B. Derive an expression for the torque on the coil when a current I passes through it. (You may assume the expression for the force on a current-carrying conductor in a magnetic field.)

If the coil is suspended by a torsion wire for which the couple per unit twist is C, show that the instrument will have a linear scale.

How may the current sensitivity of the instrument be made as large as possible? What practical considerations limit the current sensitivity?

Two galvanometers, which are otherwise identical, are fitted with different coils. One has a coil of 50 turns and resistance 10 Ω while the other has 500 turns and a resistance of 600 Ω. What is the ratio of the deflections when each is connected in turn to a cell of e.m.f. 2·5 V and internal resistance 50 Ω? (L.)

12. Write down a formula for the magnitude of the force on a straight current-carrying wire in a magnetic field, explaining clearly the meaning of each symbol in your formula.

Derive an expression for the couple on a rectangular coil of n turns and dimensions $a \times b$ carrying a current I when placed in a uniform magnetic field of flux density B at right angles to the sides of the coil of length a and at an angle θ to the sides of length b. Describe briefly how you would demonstrate experimentally that the couple on a plane coil in a uniform field depends only on its area and not on its shape.

A circular coil of 50 turns and area $1·25 \times 10^{-3}$ m^2 is pivoted about a vertical diameter in a uniform horizontal magnetic field and carries a current of 2 A. When the coil is held with its plane in a north–south direction, it experiences a couple of 0·04 N m. When its plane is east–west, the corresponding couple is 0·03 N m. Calculate the magnetic flux density. (Ignore the earth's magnetic field.) (O. & C.)

13. A strip of metal 1·2 cm wide and $1·5 \times 10^{-3}$ cm thick carries a current of 0·50 A along its length. If it is assumed that the metal contains 5×10^{22} free electrons per cm^3, calculate the mean drift velocity of these electrons ($e = 1·6 \times 10^{-19}$ C).

The metal foil is placed normal to a magnetic field of flux density B. Explain why, in these circumstances, you might expect a p.d. to be developed across the foil. By equating the magnetic and electric forces acting on an electron when the p.d. has been established, derive an expression for the p.d. in terms of B, the current I, the electron charge e, the number of electrons per unit volume N and the thickness of the foil t. Illustrate your answer with a clear diagram. (N.)

14. Describe a moving-coil type of galvanometer and deduce a relation between its deflection and the steady current passing through it.

A galvanometer, with a scale divided into 150 equal divisions, has a current sensitivity of 10 divisions per milliampere and a voltage sensitivity of 2 divisions per millivolt. How can the instrument be adapted to serve (a) as an ammeter reading to 6 A, (b) as a voltmeter in which each division represents 1 V? (L.)

15. Explain the origin of the Hall effect. Include a diagram showing clearly the directions of the Hall voltage and other relevant vector quantities for a specimen in which electron conduction predominates.

A slice of indium antimonide is 2·5 mm thick and carries a current of 150 mA. A magnetic field of flux density 0·5 T, correctly applied, produces a maximum Hall voltage of 8·75 mV between the edges of the slice. Calculate the number of free charge carriers per unit volume, assuming they each have a charge of $-1·6 \times 10^{-19}$ C. Explain your calculation clearly.

What can you conclude from the observation that the Hall voltage in different conductors can be positive, negative or zero? (C.)

34 Magnetic Fields of Current-Carrying Conductors

We now consider the magnetic fields of three typical conductors. At first we shall state the values of the flux density B due to each and prove them later.

Solenoid

Solenoids, or relatively long coils of wire, are widely used in industry. For example, solenoids are used in telephone earpieces to carry the speech current and in magnetic relays used in telecommunications.

The magnetic field inside an infinitely-long solenoid is constant in magnitude. A form of coil which gives a very nearly uniform field is shown in Fig. 34.1 (i). It is a solenoid of N turns and length L metre wound on a circular support instead of a straight one, and is called a *toroid*. If its average diameter D is

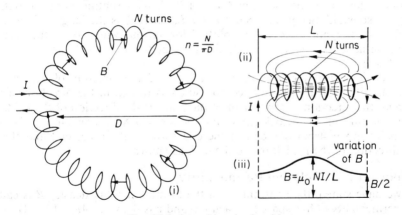

Fig. 34.1 A toroid and solenoid

several times its core diameter d, then the turns of wire are almost equally spaced around its inside and outside circumferences; their number per metre is therefore

$$n = \frac{N}{L} = \frac{N}{\pi D} \qquad . \qquad . \qquad . \qquad . \qquad . \quad (1)$$

The magnetic field within a toroid is very nearly uniform, because the coil has no ends. The coil is equivalent to an infinitely long solenoid. If I is the current, the flux density B at all points within it is given by

$$B = \mu_0 n I \qquad . \qquad . \qquad . \qquad . \qquad . \quad (2)$$

μ_0 is a constant known as the *permeability of free space* which has the value $4\pi \times 10^{-7}$ H m^{-1} (H is a unit called a 'henry' and is discussed later). The constant μ_0 is necessary to make the units correct, that is, B is then in teslas (T) when I is in amperes (A) and l is in metres (m).

In practice, solenoids cannot be made infinitely long. But if the length L of a solenoid is about ten times its diameter, the field near its middle is fairly uniform, and has the value given by equation (2). Figure 34.1 (ii) shows a solenoid of length L and N turns, so that $n = N/L$. The flux density in the *middle* of the coil is given approximately by

$$B = \mu_0 nI = \mu_0 \frac{NI}{L} \quad . \qquad . \qquad . \qquad . \qquad . \qquad (3)$$

If a long solenoid is imagined cut at any point R near the middle, the two solenoids on each side have the same field B at their respective centres since each has the same number of turns per unit length as the long solenoid. So each solenoid contributes equally to the field at R. Hence each solenoid provides a field $B/2$ at their end R. We therefore see that the field at the *end* of any long solenoid is *half* that at the centre and is given by

$$B = \frac{1}{2}\mu_0 \frac{NI}{L} \quad . \qquad . \qquad . \qquad . \qquad . \qquad (4)$$

Fig. 34.1 (iii) shows roughly the variation of B along the solenoid.

A 'Slinky' is a long coil which can be stretched to provide a solenoid with a varying number of turns per metre, n. A small search coil with many thousands of turns, placed coaxially inside the middle of the solenoid, can be connected to an oscilloscope to provide a measure of B when alternating current is passed into the solenoid. See p. 711. Since $n \propto 1/L$, where L is the length of the coil, a graph of B against $1/L$ can be plotted for various values of L, the current being the same each time. A straight line through the origin is obtained, showing that $B \propto n$.

Maxwell's corkscrew rule can be used to find the direction of the field in Fig. 34.1 (i), (ii). If a right-handed corkscrew is turned in the direction of the current, the direction of movement of the point is the direction of the field. So when the current flows clockwise round the turns at X as shown in Fig. 34.1 (i), the point moves from one end of the coil towards the other end and this is the direction of the field B inside the coil, as shown.

Effect on B of Relative Permeability

As we have stated, the constant μ_0 in the formula for flux density B is called the permeability of free space (or vacuum) and has the value $4\pi \times 10^{-7}$ H m^{-1}. The permeability of air at normal pressure is only very slightly different from that of a vacuum. So we can consider the permeability of air to be practically $4\pi \times 10^{-7}$ H m^{-1}.

If the solenoid is wound round soft iron, so that this material is now the core of the solenoid, the permeability is increased considerably. The name 'relative permeability', symbol μ_r, is given to the number of times the permeability has increased relative to that of free space or air. So if $\mu_r = 1000$, the value of B in the solenoid is 1000 times as great as with an air core. Generally, the permeability μ of an iron core would be given by

$$\mu = \mu_r \mu_0.$$

Note that μ_r is a number and has no units, unlike μ_0 and μ. See also p. 764.

Long Straight Conductor

We now consider the magnetic field of a long straight current-carrying conductor. A submarine cable carrying messages is an example of such a conductor.

All round a straight current-carrying wire, the field pattern consists of circles concentric with the wire. Figure 34.2 (i) shows the field round one section of

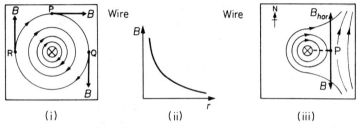

Fig. 34.2 Field due to long straight conductor

the conductor. Maxwell's corkscrew rule gives the field direction: If a right-handed corkscrew is turned so that the point moves along the current direction, the field direction is the same as the direction of turning.

The direction of B is along the tangent to a circle at the point concerned. So at P due north of the wire, B points east for a downward current. At a point due east, B points south and at a point due west, B points north.

At a point distance r from an infinitely-long wire, the value of B is given by

$$B = \frac{\mu_0 I}{2\pi r}.$$

So for a given current, $B \propto 1/r$, Fig. 34.2 (ii).

The earth's horizontal magnetic field B_{hor} is about 4×10^{-5} T and acts due north. When this cancels exactly the magnetic field of the current, a *neutral point* is obtained in the combined field of the earth and the current. Since the field due to the current must be due south, the neutral point P in Fig. 34.2 (iii) is due *east* of the wire. Suppose the current is 5 A. The distance r of the neutral point from the wire is then given by

$$\frac{\mu_0 I}{2\pi r} = B_{hor} = 4 \times 10^{-5}.$$

So

$$r = \frac{\mu_0 I}{2\pi \times 4 \times 10^{-5}} = \frac{4\pi \times 10^{-7} \times 5}{2\pi \times 4 \times 10^{-5}}$$

$$= 0.025 \text{ m} = 25 \text{ mm}.$$

Variation of B with Distance—A.C. Method

An apparatus suitable for finding the variation of B with distance r from a long straight wire CD is shown in Fig 34.3. Alternating current (a.c.) of the order of

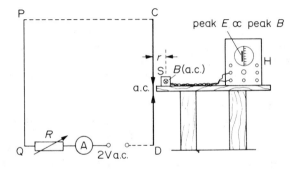

Fig. 34.3 Investigation of B due to long straight conductor

10 A, from a low voltage mains transformer, is passed through CD by using another long wire PQ at least one metre away, a rheostat R and an a.c. ammeter A. A small search coil S, with thousands of turns of wire, such as the coil from an output transformer, is placed near CD. It is positioned with its axis at a small distance r from CD and so that the flux from CD enters its face normally. S is joined by long twin flex to the Y-plates of an oscilloscope H and the greatest sensitivity, such as 5 mV/cm, is used.

When the a.c. supply is switched on, the varying flux through S produces an induced alternating e.m.f. E. The peak value of E can be determined by switching off the time-base and measuring the length of the line trace, Fig. 34.3. See p. 815. Now the peak value of the magnetic flux density B is proportional to the peak value of E, as shown in the case of the simple dynamo on p. 738. Thus the length of the trace gives a measure of the peak value of B.

The distance r of the coil CD is then increased and the corresponding length of the trace is measured. The length of the trace plotted against $1/r$ gives a straight line graph passing through the origin. Hence $B \propto 1/r$. A similar method can be used for investigating the field B for the case of a narrow circular coil or for a solenoid (p. 710).

Forces between Currents

In 1821, Ampere discovered by experiment that current-carrying conductors exert a force on each other. For example, when the currents in two long neighbouring straight conductors X and Y are in the same direction, there is a force of *attraction* between them. If the currents flow in opposite directions, there is a repulsive force between them. Each conductor has a force on it due to the magnetic field of the other.

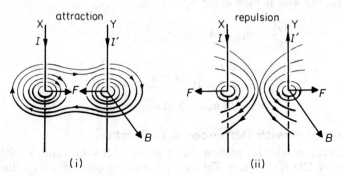

Fig. 34.4 Forces between currents

Figure 34.4 (i) shows the resultant magnetic flux round two long straight vertical conductors X,Y in a horizontal plane when the currents are both downwards. The lines tend to pull the conductors towards each other. In Fig. 34.4 (ii), the currents are in opposite directions. Here the lines tend to push the conductors apart.

Fleming's left hand rule confirms the direction of the forces. At Y, the flux-density B due to the conductor X is perpendicular to Y (the flux due to X alone consists of circles with X as centre and at Y the tangent to the circular line is perpendicular to Y). So, from Fleming's rule, the force F on Y in Fig. 34.4 (i) is towards X. From the law of action and reaction, the force F on X is towards Y and equal to that on Y. So the conductors attract each other.

Magnitude of force. The ampere

If two long straight conductors lie parallel and close together at a distance r apart, and carry currents I, I' respectively, then the current I is in a magnetic field of flux density B equal to $\mu_0 I'/2\pi r$ due to the other current I' (p. 711). The force per metre length, F, is hence given by

$$F = BIl = BI \times 1 = \frac{\mu_0 I'}{2\pi r} \times I \times 1$$

$$\therefore F = \frac{\mu_0 II'}{2\pi r} \qquad \cdot \qquad \cdot \qquad \cdot \qquad \cdot \qquad \cdot \qquad (1)$$

Nowadays the ampere is *defined* in terms of the force between conductors. It is *that current, which flowing in each of two infinitely-long parallel straight wires of negligible cross-sectional area separated by a distance of 1 metre* in vacuo, *produces a force* between the wires *of* 2×10^{-7} *newton metre*$^{-1}$.

Taking $I = I' = 1$ A, $r = 1$ metre, $F = 2 \times 10^{-7}$ newton metre^{-1}, then, from (1),

$$2 \times 10^{-7} = \frac{\mu_0 \times 1 \times 1}{2\pi \times 1}$$

$$\therefore \mu_0 = 4\pi \times 10^{-7} \text{ henry metre}^{-1},$$

which is the value used in formulae with μ_0.

It may be noted that the electrostatic force of repulsion between the negative charges of the moving electrons in the two wires is completely neutralised by the attractive force on them by the positive charges on the stationary metal ions. Thus the force between the two wires is only the *electromagnetic* force, due to the moving electrons (currents).

Example

A long straight conductor X carrying a current of 2 A is placed parallel to a short conductor Y of length 0·05 m carrying a current of 3 A. The two conductors are 0·10 m apart. Calculate (i) the flux density due to X at Y, (ii) the approximate force on Y.

(i) Due to X, $\quad B = \dfrac{\mu_0 I}{2\pi r} = \dfrac{4\pi \times 10^{-7} \times 2}{2\pi \times 0\cdot10}$

$$= 4 \times 10^{-6} \text{ T.}$$

(ii) On Y, length $l = 0\cdot05$ m,

$$\text{force } F = BIl = 4 \times 10^{-6} \times 3 \times 0\cdot05$$

$$= 6 \times 10^{-7} \text{ N.}$$

Absolute Determination of Current

A laboratory form of an *ampere balance*, which measures current by measuring the force between current-carrying conductors, is shown in Fig. 34.5.

With no current flowing, the zero screw is adjusted until the plane of ALCD is horizontal. The current I to be measured is then switched on so that it flows through ALCD and EHGM in series and HG repels CL. The mass m necessary to restore balance is then measured, and mg is the force between the conductors since the respective distances of CL and the scale pan from the pivot are equal. The equal lengths l of the straight wires CL and HG, and their separation r, are all measured.

From equation (1) above,

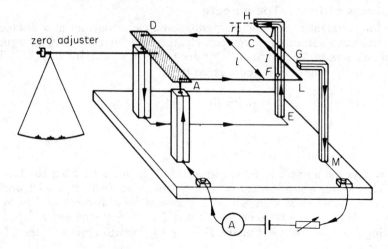

Fig. 34.5 Laboratory form of ampere balance

Fig. 34D *Ampere balance.* Current balance at the National Physical Laboratory. One large coil (bottom left) has been lowered so that the small suspended coil above it, at the end of the beam on the left, can be seen. To measure current, the large coils (left and right) and the two suspended coils above them are all connected in series, in such a way that one suspended coil is repelled upwards and the other is attracted downwards when the current flows. Equilibrium is restored by adding masses on one of the scale pans. These are placed on or lifted off the scale pan by rods controlled by the knobs outside the case. (*Crown copyright, Courtesy of National Physical Laboratory*)

$$\text{force per metre} = \frac{4\pi \times 10^{-7} I^2}{2\pi r}$$

$$\therefore mg = \frac{4\pi \times 10^{-7} I^2 l}{2\pi r}$$

$$\therefore I = \sqrt{\frac{mgr}{2 \times 10^{-7} l}}.$$

In this expression, I will be in ampere if m is in kilogram, $g = 9\cdot8$ m s^{-2} and l and r are measured in metre.

Narrow Circular Coil

The third of our typical conductors is the narrow circular coil.

Figure 34.6 (i) shows the magnetic field pattern round a narrow vertical circular coil C carrying a current I, in the horizontal (perpendicular) plane passing through the middle of the coil. In the middle M of the coil, the field is uniform for a short distance either side. Here the field value B is given by

$$B = \frac{\mu_0 I}{2r}$$

where r is the radius in metres.

Figure 34.6 (ii) shows how B varies as we move from the centre of the coil

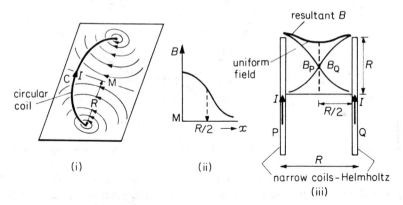

Fig. 34.6 Fields due to narrow circular coils

along a line perpendicular to the plane of the coil. The field value decreases continuously. Helmholtz, an eminent scientist of the 19th century, showed that two narrow circular coils of the same radius and carrying the same current could provide a *uniform* magnetic field between them. For this purpose they are placed facing each other at a distance apart equal to their radius R. As shown in Fig. 34.6 (iii), the resultant magnetic field B round a point half-way between the coils P and Q is fairly uniform for some distance on either side of the point. The flux density B of the uniform field is given approximately by

$$B = 0\cdot72\frac{\mu_0 NI}{R},$$

where N is the number of turns in each coil, I is the current in amperes and R is the radius in metres.

Helmholtz coils were used by Sir J. J. Thomson to obtain a uniform magnetic field of known value in a famous experiment to find the charge–mass ratio of an electron (see p. 805).

Earth's Magnetism

It was Dr. Gilbert who first showed that a magnetized needle, when freely suspended about its centre of gravity, dipped downwards towards the north at about 70° to the horizontal in England. He also found that this *angle of dip* increased with latitude, as shown in Fig. 34.7, and concluded that the earth itself was, or contained, a magnet. The points where the angle of dip is 90° are called the earth's magnetic poles; they are fairly near to the geographic poles, but their positions are continuously, though slowly, changing. Gilbert's simple

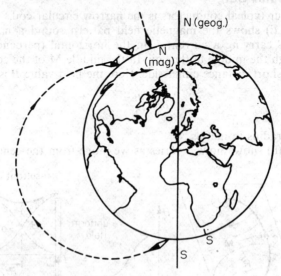

Fig. 34.7 Illustrating the angle of dip

idea of the earth as a magnet has had to be rejected. The earth's crust does not contain enough magnetic material to make a magnet of the required strength; the earth's core is, we believe, molten—and molten iron is non-magnetic. The origin of the earth's magnetism is, in fact, one of the great theoretical problems of the present day.

Horizontal and Vertical Components. Variation and Dip

Since a freely suspended magnetic needle dips downwards at some angle δ to the horizontal, the earth's *resultant magnetic field*, B_R, acts at an angle δ to the horizontal. The 'angle of dip', or *inclination*, can thus be defined as the angle between the resultant earth's field and the horizontal. The earth's field has a *vertical component*, B_V, given by

$$B_V = B_R \sin \delta \qquad\qquad (1)$$

and a *horizontal component*, B_H, given by

$$B_H = B_R \cos \delta \qquad\qquad (2)$$

Also,
$$\frac{B_V}{B_H} = \tan \delta \qquad\qquad (3)$$

To specify the earth's magnetic field at any point, we must state its strength and direction. To specify its direction we must give the direction of the magnetic meridian, and the angle of dip δ, Fig. 34.8. In most parts of the world the magnetic meridian does not lie along the geographic meridian (the vertical plane running geographically north–south). The angle between the magnetic and geographic meridians, ε, is called the magnetic variation, or sometimes the declination, at the place concerned; it is shown on the margins of maps. The horizontal and vertical components of the earth's field, and the angle of dip, can be measured by a large coil or 'earth inductor' (see p. 750). The earth's total of resultant flux density B_R in Britain is now about 5×10^{-5} T and the horizontal component B_H is about 2×10^{-5} T.

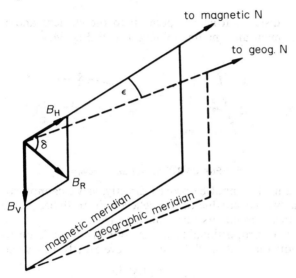

Fig. 34.8 Magnetic and geographic meridians. Dip

Magnitudes of B for Current-carrying Conductors

We conclude this chapter with proofs of the values of B used earlier for a narrow circular coil, a straight conductor and a solenoid.

Law of Biot and Savart

To calculate B for any shape of conductor, Biot and Savart gave a law which can now be stated as follows: The flux density ΔB at a point P due to a small element Δl of a conductor carrying a current is given by

$$\Delta B \propto \frac{I \Delta l \sin \alpha}{r^2}, \qquad \qquad (1)$$

where r is the distance from the point P to the element and α is the angle between the element and the line joining it to P, Fig. 34.9.

Fig. 34.9 Biot and Savart law

The formula in (1) cannot be proved directly, as we cannot experiment with an infinitesimally small conductor. We believe in its truth because the deductions for large practical conductors turn out to be true.

The constant of proportionality in equation (1) depends on the medium in which the conductor is situated. In air (or, more exactly, in a vacuum), we write

$$\Delta B = \frac{\mu_0}{4\pi} \frac{I \Delta l \sin \alpha}{r^2}. \qquad \qquad (2)$$

The value of μ_0, from p. 713, is

$$\mu_0 = 4\pi \times 10^{-7},$$

and its unit is 'henry per metre' (H m^{-1}) as will be shown later.

B for Narrow Coil

The formula for the value of B at the centre of a narrow circular coil can be immediately deduced from (2). Here the radius r is constant for all the elements Δl, and the angle α is constant and equal to $90°$, Fig. 34.10. If the coil has N turns, the length of wire in it is $2\pi r N$, and the field at its centre is therefore given, if the current is I, by

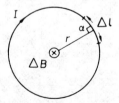

Fig. 34.10 Field of circular coil

$$B = \int dB = \frac{\mu_0}{4\pi} \int_0^{2\pi rN} \frac{Idl \sin 90°}{r^2}$$

$$= \frac{\mu_0 I}{4\pi r^2} \int_0^{2\pi rN} dl = \frac{\mu_0 I}{4\pi r^2} 2\pi rN$$

$$= \frac{\mu_0 NI}{2r} \qquad . \qquad . \qquad . \qquad . \qquad . \qquad . \qquad (1)$$

From (1), $B \propto I$ when r and N are constant, $B \propto 1/r$ when I and N are constant, and $B \propto N$ when I and r are constant.

B along Axis of a Narrow Circular Coil

We will now find the magnetic field at a point anywhere on the axis of a narrow circular coil (P in Fig. 34.11). We consider an element Δl of the coil, at right angles to the plane of the paper. This sets up a field ΔB at P, in the plane of the paper, and at right angles to the radius vector r. If β is the angle between

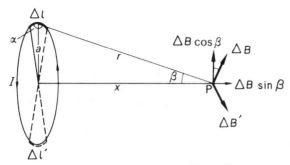

Fig. 34.11 Field on axis of flat coil

r and the axis of the coil, then the field ΔB has components $\Delta B \sin \beta$ along the axis, and $\Delta B \cos \beta$ at right angles to the axis. If we now consider the element $\Delta l'$ diametrically opposite to Δl, we see that it sets up a field $\Delta B'$ equal in magnitude to ΔB. This also has a component, $\Delta B' \cos \beta$, at right angles to the axis; but this component acts in the opposite direction to $\Delta B \cos \beta$ and therefore cancels it. By considering elements such as Δl and $\Delta l'$ all round the circumference of the coil, we see that the field at P can have no component at right angles to the axis. Its value along the axis is

$$B = \int dB \sin \beta.$$

From Fig. 34.11, we see that the length of the radius vector r is the same for all points on the circumference of the coil, and that the angle α is also constant, being 90°. Thus, if the coil has a single turn, and carries a current I,

$$\Delta B = \frac{\mu_0 I \Delta l \sin \alpha}{4\pi r^2} = \frac{\mu_0 I}{4\pi r^2} \Delta l.$$

And, if the coil has a radius a, then

$$B = \int dB \sin \beta = \int_0^{2\pi a} \frac{\mu_0 I}{4\pi r^2} dl \sin \beta$$

$$= \frac{\mu_0 I a \sin \beta}{2r^2} \qquad . \qquad . \qquad . \qquad . \qquad . \qquad . \qquad (i)$$

When the coil has more than one turn, the distance r varies slightly from one turn to the next. But if the width of the coil is small compared with all its other dimensions, we may neglect it, and write,

$$B = \frac{\mu_0 N I a \sin \beta}{2r^2} \qquad \cdot \qquad \cdot \qquad \cdot \qquad \cdot \qquad \cdot \qquad \text{(ii)}$$

where N is the number of turns.

Equation (ii) can be put into a variety of forms, by using the facts that

$$\sin \beta = \frac{a}{r},$$

and

$$r^2 = x^2 + a^2,$$

where x is the distance from P to the centre of the coil. Thus

$$B = \frac{\mu_0 N I a^2}{2r^3} = \frac{\mu_0 N I a^2}{2(x^2 + a^2)^{3/2}} \qquad \cdot \qquad \cdot \qquad \cdot \qquad \text{(1)}$$

Helmholtz Coils

The field along the axis of a single coil varies with the distance x from the coil. In order to obtain a *uniform* field, Helmholtz used two coaxial parallel coils of equal radius R, separated by a distance R. In this case, when the same current flows around each coil in the same direction, the resultant field B is uniform for some distance on either side of the point on their axis midway between the coils. See p. 807.

The magnitude of the resultant field B at the midpoint can be found from our previous formula for a single coil. We now have $a = R$ and $x = R/2$. Thus, for the two coils,

$$B = 2 \times \frac{\mu_0 N I R^2}{2(R^2/4 + R^2)^{3/2}} = \left(\frac{4}{5}\right)^{3/2} \times \frac{\mu_0 N I}{R}$$

$$= 0.72 \frac{\mu_0 N I}{R} \text{ (approx.).}$$

B on Axis of a Long Solenoid

We may regard a solenoid as a long succession of narrow coils; if it has n turns per metre, then in an element Δx of it there are $n\Delta x$ coils, Fig. 34.12. At a point

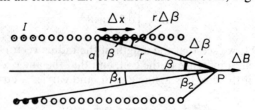

Fig. 34.12 Field on axis of solenoid

P on the axis of the solenoid, the field due to these is, by equation (ii),

$$\Delta B = \frac{\mu_0 I a \sin \beta}{2r^2} n\Delta x,$$

in the notation which we have used for the flat coil. If the element Δx subtends an angle $\Delta \beta$ at P, then, from the figure,

$$r\Delta\beta = \Delta x \sin \beta;$$

so

$$\Delta x = \frac{r\Delta\beta}{\sin \beta}.$$

Also,

$$a = r \sin \beta.$$

Thus

$$\Delta B = \frac{\mu_0 I r \sin^2 \beta}{2r^2} n\frac{r\Delta\beta}{\sin \beta}.$$

$$= \frac{\mu_0 n I}{2} \sin \beta\Delta\beta.$$

If the radii of the coil, at its ends, subtend the angles β_1 and β_2 at P, then the field at P is

$$H = \int_{\beta_1}^{\beta_2} \frac{\mu_0 n I}{2} \sin \beta d\beta$$

$$= \frac{\mu_0 n I}{2} \left[-\cos \beta \right]_{\beta_1}^{\beta_2}$$

$$= \frac{\mu_0 n I}{2}(\cos \beta_1 - \cos \beta_2) \qquad . \qquad . \qquad . \qquad . \qquad (1)$$

Fig. 34.13 A very long solenoid

If the point P inside a very long solenoid—so long that we may regard it as infinite—then $\beta_1 = 0$ and $\beta_2 = \pi$, as shown in Fig. 34.13. Then, by equation (1):

$$B = \frac{\mu_0 n I}{2} \left[-\cos \beta \right]_0^{\pi}.$$

so

$$B = \mu_0 n I \qquad . \qquad . \qquad . \qquad . \qquad . \qquad (2)$$

The quantity nI is often called the 'ampere-turns per metre'.

B due to Long Straight Wire

In Fig. 34.14, AC represents part of a long straight wire. P is taken as a point so near it that, from P, the wire looks infinitely long—it subtends very nearly 180°. An element XY of this wire, of length Δl, makes an angle α with the radius

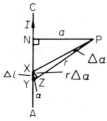

Fig. 34.14 Field of a long straight wire

vector, r, from P. It therefore contributes to the magnetic field at P an amount

$$\Delta B = \frac{\mu_0 I \Delta l \sin \alpha}{4\pi r^2} \qquad . \qquad . \qquad . \qquad . \qquad \text{(i)}$$

when the wire carries a current I. If α is the perpendicular distance, PN, from P to the wire, then

$$PN = PX \sin \alpha \quad \text{or} \quad a = r \sin \alpha,$$

so
$$r = \frac{a}{\sin \alpha} \qquad . \qquad . \qquad . \qquad . \qquad \text{(ii)}$$

Also, if we draw XZ perpendicular to PY, we have

$$XZ = XY \sin \alpha = \Delta l \sin \alpha.$$

If Δl subtends an angle $\Delta \alpha$ at P, then

$$XZ = r \Delta \alpha = \Delta l \sin \alpha.$$

From (i) $\qquad\qquad \therefore \Delta B = \dfrac{\mu_0 I \Delta l \sin \alpha}{4\pi r^2} = \dfrac{\mu_0 I r \Delta \alpha}{4\pi r^2} = \dfrac{\mu_0 I \Delta \alpha}{4\pi r}$

From (ii), $\qquad\qquad \therefore \Delta B = \dfrac{\mu_0 I \sin \alpha \Delta \alpha}{4\pi a}.$

When the point Y is at the bottom end A of the wire, $\alpha = 0$; and when Y is at the top C of the wire, $\alpha = \pi$. Therefore the total magnetic field at P is

$$B = \frac{\mu_0}{4\pi} \int_0^\pi \frac{I \sin \alpha \Delta \alpha}{a} = \frac{\mu_0 I}{4\pi a}\left[-\cos \alpha \right]_0^\pi$$

$$\therefore B = \frac{\mu_0 I}{2\pi a} \qquad . \qquad . \qquad . \qquad . \qquad \text{(1)}$$

Equation (1) shows that the magnetic field of a long straight wire, at a point near it, is inversely proportional to the distance of the point from the wire. The result was discovered experimentally by Biot and Savart, and led to their general formula in (i) which we used to derive equation (1).

Ampère's Theorem

In the calculation of magnetic flux density B, we have used so far only the Biot and Savart law. Another law useful for calculating B is *Ampère's theorem*.

Ampère showed that if a *continuous closed line or loop* is drawn round one or more current-carrying conductors, and B is the flux density in the direction of an element dl of the loop, then for free space

$$\oint \frac{B}{\mu_0}.dl = I$$

where the symbol $\oint$ represents the integral taken completely round the closed loop and I is the total current enclosed by the loop. So we can write

$$\oint B.dl = \mu_0 I \qquad . \qquad . \qquad . \qquad \text{(1)}$$

The proof of (1) is outside the scope of this book.

We now apply the theorem to two special cases of current-carrying conductors.

1. *Straight wire*
Figure 34.15 shows a circular loop L of radius r, drawn concentrically round

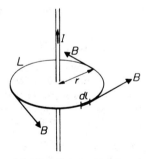

Fig. 34.15 *B* due to straight wire

a straight wire carrying a current I. The flux lines are circles and so, at every part of a closed line, B is directed along the tangent to the circle at that part. Further, by symmetry, B has the same value everywhere along the line.

So
$$\oint B.dl = B\oint dl = B.2\pi r$$

since B is constant. Hence, from (1),
$$B.2\pi r = \mu_0 I$$

and so
$$B = \frac{\mu_0 I}{2\pi r}.$$

This agrees with the result derived earlier.

2. *Toroid (Solenoid)*
Consider the closed loop M indicated by the broken line in Fig. 34.16. Again

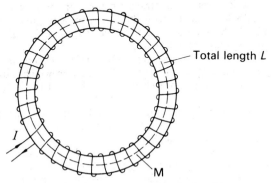

Total length L

Fig. 34.16 *B* due to toroid or solenoid

B is everywhere the same at M and is directed along the loop at every point.

So
$$\oint B.dl = B\oint dl = BL.$$

where L is the total length of the loop M. Hence, from (1),

$$BL = \mu_0 NI.$$

So
$$B = \frac{\mu_0 NI}{L} = \mu_0 nI,$$

where N is the total number of turns, and n is the number of turns per metre. This agrees with the result previously obtained on p. 721.

Exercises 34

1. A vertical conductor X carries a downward current of 5 A. (a) Draw the pattern of the magnetic flux in a horizontal plane round X. (b) What is the flux density due to the current alone at a point P 10 cm due east of X. (c) If the earth's horizontal magnetic flux density has a value 4×10^{-5} T, calculate the resultant flux density at P.

Is the resultant flux density at a point 10 cm due north of X greater *or* less than at P? Explain your answer.

2. A horizontal wire, of length 5 cm and carrying a current of 2 A, is placed in the middle of a long solenoid at right angles to its axis. The solenoid has 1000 turns per metre and carries a steady current I. Calculate I if the force on the wire is vertically downwards and equal to 10^{-4} N.

3. Two vertical parallel conductors X and Y are 0·12 m apart and carry currents of 2 A and 4 A respectively in a downward direction. Figure 34A (i). (a) Draw the resultant flux pattern between X and Y. (b) Ignoring the earth's magnetic field, find the distance from X of a point where the magnetic fields due to X and Y neutralise each other. (c) Calculate the force per metre on X and on Y, and show their directions in a sketch.

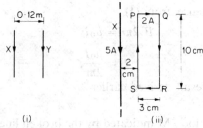

(i)   (ii)

Fig. 34A

4. In Fig. 34A (ii), X is a very long straight conductor carrying a current of 5 A. A metal rectangle PQRS is suspended with PS 2 cm from X as shown. The dimensions of PQRS are 10 cm by 3 cm, and a current of 2 A flows in the coil. Calculate the resultant force on PQRS in magnitude and direction.

5. Two very long thin straight parallel wires each carrying a current in the same direction are separated by a distance d. With the aid of a diagram which indicates the current directions, account for the force on each wire and show on the diagram the direction of one of the forces.

Write down an expression for the magnitude of the force per unit length of wire and hence define the ampere. Why is the electrostatic force between charges ignored in the definition? $(\mu_0 = 4\pi \times 10^{-7}$ H m^{-1}.) (N.)

6. Define the *ampere*. Write down expressions for (i) the magnitude of the flux density B at a distance of d from a very long straight conductor carrying a current I, and (ii) the mechanical force acting on a straight conductor of length l carrying a current I at right angles to a uniform magnetic field of flux density B.

Show how these two expressions may be used to deduce a formula for the force per unit length between two long straight parallel conductors *in vacuo* carrying currents I_1 and I_2 separated by a distance d.

A horizontal straight wire 5 cm long weighing 1·2 g m^{-1} is placed perpendicular to a uniform horizontal magnetic field of flux density 0·6 T. If the resistance of the wire is 3·8 Ω m^{-1}, calculate the p.d. that has to be applied between the ends of the wire to make it just self-supporting. Draw a diagram showing the direction of the field and the direction in which the current would have to flow in the wire ($g = 9.8$ m s^{-2}). (*C.*)

7. State the law of force acting on a conductor carrying an electric current in a magnetic field. Indicate the direction of the force and show how its magnitude depends on the angle between the conductor and the direction of the field.

Sketch the magnetic field due solely to two long parallel conductors carrying respectively currents of 12 and 8 A in the same direction. If the wires are 10 cm apart, find where a third parallel wire also carrying a current must be placed so that the force experienced by it shall be zero. (*L.*)

8. Define the *ampere*.

Two long vertical wires, set in a plane at right angles to the magnetic meridian, carry equal currents flowing in opposite directions. Draw a diagram showing the pattern, in a horizontal plane, of the magnetic flux due to the currents alone—that is, for the moment ignoring the earth's magnetic field.

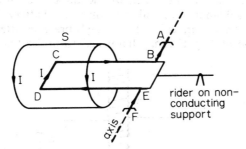

Fig. 34B

Next, taking into account the earth's magnetic field, discuss the various situations that can give rise to neutral points in the plane of the diagram.

Figure 34B shows a simple form of current balance. The 'long' solenoid S, which has 2000 turns per metre, is in series with the horizontal rectangular copper loop ABCDEF, where BC = 10 cm and CD = 3 cm. The loop, which is freely pivoted on the axis AF, goes well inside the solenoid, and CD is perpendicular to the axis of the solenoid. When the current is switched on, a rider of mass 0.2 g placed 5 cm from the axis is needed to restore equilibrium. Calculate the value of the current, *I*. (*O.*)

9. (*a*) A long straight wire of radius *a* carries a steady current. Sketch a diagram showing the lines of magnetic flux density (*B*) near the wire and the relative directions of the current and *B*. Describe, with the aid of a sketch graph, how *B* varies along a line from the surface of the wire at right-angles to the wire.

(*b*) Two such identical wires R and S lie parallel in a horizontal plane, their axes being 0.10 m apart. A current of 10 A flows in R in the opposite direction to a current of 30 A in S. Neglecting the effect of the earth's magnetic flux density calculate the magnitude and state the direction of the magnetic flux density at a point P in the plane of the wires if P is (i) midway between R and S, (ii) 0.05 m from R and 0.15 m from S. The permeability of free space, $\mu_0 = 4\pi \times 10^{-7}$ H m^{-1}. (*N.*)

10. Define the *ampere*.

Draw a labelled diagram of an instrument suitable for measuring a current absolutely in terms of the ampere, and describe the principle of it.

A very long straight wire PQ of negligible diameter carries a steady current I_1. A square coil ABCD of side *l* with *n* turns of wire also of negligible diameter is set up with sides AB and DC parallel to and coplanar with PQ; the side AB is nearest to PQ and is at a distance *d* from it. Derive an expression for the resultant force on the coil when a steady current I_2 flows in it, and indicate on a diagram the direction of this force when the current flows in the same direction in PQ and AB.

Calculate the magnitude of the force when $I_1 = 5$ A, $I_2 = 3$ A, $d = 3$ cm, $n = 48$ and $l = 5$ cm. (*O. & C.*)

35 Electromagnetic Induction

Faraday's Discovery

After Ampere and others had investigated the magnetic effect of a current Faraday attempted to find its converse: he tried to produce a current by means of a magnetic field. He began work on the problem in 1825 but did not succeed until 1831.

The apparatus with which he worked is represented in Fig. 35.1; it consists of two coils of insulated wire, A, B, wound on a wooden core. One coil was connected to a galvanometer, and the other to a battery. No current flowed through the galvanometer, as in all Faraday's previous attempts. But when he disconnected the battery Faraday happened to notice that the galvanometer

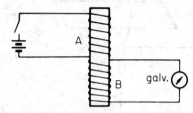

Fig. 35.1 Faraday's experiment on induction

needle gave a kick. And when he connected the battery back again, he noticed a kick in the opposite direction. However often he disconnected and reconnected the battery, he got the same results. The 'kicks' could hardly be all accidental—they must indicate momentary currents. Faraday had been looking for a steady current—that was why it took him six years to find it.

Conditions for Generation of Induced Current

The results of Faraday's experiments showed that a current flowed in coil B of Fig. 35.1 only while the magnetic field due to coil A was changing—the field building up as the current in A was switched on, decaying as the current in A was switched off. And the current which flowed in B while the field was decaying was in the opposite direction to the current which flowed while the field was building up. Faraday called the current in B an *induced current*. He found that it could be made much greater by winding the two coils on an iron core, instead of a wooden one.

Once he had realised that an induced current was produced only by a *change* in the magnetic field inducing it. Faraday was able to find induced currents wherever he had previously looked for them. In place of the coil A he used a magnet, and showed that as long as the coil and the magnet were at rest, there was no induced current, Fig. 35.2 (i). But when he moved either the coil or the magnet an induced current flowed as long as the motion continued, Fig. 35.2 (ii). If the current flowed one way when the north pole of the magnet was approaching the end X of the coil, it flowed the other way when the north pole was retreating from X, or the south pole approached X.

Since a flow of current implies the presence of an e.m.f., Faraday's experiments showed that an e.m.f. could be induced in a coil by moving it relatively to a

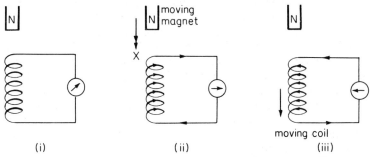

Fig. 35.2 Induced current by moving magnet or moving coil

magnetic field, Fig. 35.2 (iii). In discussing induction it is more fundamental to deal with the e.m.f. than the current, because the current depends on both the e.m.f. and the resistance.

Summarising, *relative motion* is needed between a magnet and a coil to produce induced currents. The induced current increases when the relative velocity increases and when a soft iron core is used inside the coil.

Direction of E.M.F.; Lenz's Law

Before considering the magnitude of an induced e.m.f., let us investigate its direction. To do so we must first see which way the galvanometer deflects when a current passes through it in a known direction: we can find this out with a battery and a megohm resistor, Fig. 35.3 (i). We then take a coil whose direction of winding we know, and connect this to the galvanometer. In turn we plunge each pole of a magnet into and out of the coil; and we get the results shown in Fig. 35.3 (ii), (iii), (iv).

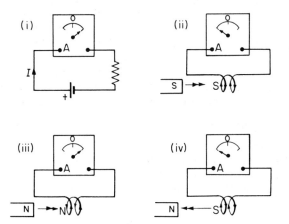

Fig. 35.3 Direction of induced currents

These results were generalised most elegantly into a rule by Lenz in 1835. He said that *the induced current flows always in such a direction as to oppose the change causing it.* For example, in Fig. 35.3 (ii), the clockwise current flowing in the coil makes this end an S pole. So it repels the approaching S pole. In Fig. 35.3 (iii), the induced anticlockwise current makes the end of the coil an N pole. So the approaching N pole is repelled. In Fig. 35.3 (iv), the induced clockwise current in the coil now attracts the N pole moving away from it.

Lenz's law is a beautiful example of the conservation of energy. The induced current sets up a force on the magnet, which the mover of the magnet must overcome; the work done in overcoming this force provides the electrical energy of the current. (This energy is dissipated as heat in the coil.)

If the induced current flowed in the opposite direction to that which it actually takes, then it would speed up the motion of the magnet and so the current would continuously increase the kinetic energy of the magnet. So both mechanical and electrical energy would be produced, without any agent having to do work. The system would be a perpetual motion machine and this is impossible. So the induced current always flows in a direction to oppose the motion and the electrical energy comes from the mechanical energy required to overcome the opposition to the motion.

The direction of the induced e.m.f., *E*, is specified by that of the current, as in Fig. 35.4 (i). If we wished to reword Lenz's law, substituting e.m.f. for current, we would have to speak of the e.m.f.s *tending* to oppose the change … etc., because there can be no opposing force unless the circuit is closed and a current can flow.

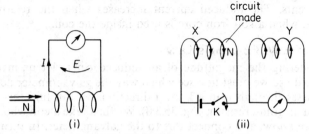

Fig. 35.4 Direction of induced e.m.f.

In Fig. 35.4 (ii), a coil X connected to a battery is placed near a coil Y. When the circuit in X is made by pressing the switch K, the current in the face of the coil near Y flows anticlockwise when viewed from Y. This is similar to bringing a N-pole suddenly near Y. So the induced current in Y is anticlockwise, as shown. If the current in X is switched off, this is similar to removing a N-pole suddenly from Y. So the current in Y is now clockwise, that is, in the opposite direction to before. The induced e.m.f. in Y, which follows the direction of the current, hence reverses when the current in X is switched on and off.

Magnitude of E.M.F.

Accurate experiments on induction are difficult to do with simple apparatus; but rough-and-ready experiments will show on what factors the magnitude depends. We require coils of the same diameter but different numbers of turns, coils of the same number of turns but different diameters, and two similar

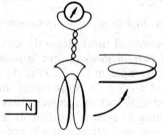

Fig. 35.5 E.m.f. induced by turning coil

magnets, which we can use singly or together. If we use a high-resistance galvano-meter, the current will not vary much with the resistance of the coil in which the e.m.f. is induced, and we can take the deflection as a measure of the e.m.f. There is no need to plunge the magnet into and out of the coil: we can get just as great a deflection by simply turning the coil through a right angle, so that its plane changes from parallel to perpendicular to the magnet, or vice versa, Fig. 35.5. We find that the induced e.m.f. increases with:

(i) the speed with which we turn the coil;
(ii) the area of the coil;
(iii) the strength of the magnetic field (two magnets give a greater e.m.f. than one);
(iv) the number of turns in the coil.

To generalise these results and to build up useful formulae, we use the idea of *magnetic flux*, or field lines, passing through a coil. Figure 35.6 shows a coil, of area A, whose normal makes an angle θ with a uniform field of flux density B. (The component of the field at right angles to the plane of the coil is $B \cos \theta$, and we say that the magnetic flux Φ through the coil is

$$\Phi = AB \cos \theta \qquad . \qquad . \qquad . \qquad . \qquad . \qquad (1)$$

(We get the same result if we multiply the flux density B by the area projected at right angles to the field, $A \cos \theta$.) If either the strength of the field is changed, or the coil is turned so as to change the angle θ, then the flux through the coil changes.

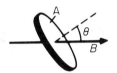

Fig. 35.6 Magnetic flux

Results (i) to (iii) above, therefore, show that the e.m.f. induced in a coil increases with the *rate of change of the magnetic flux* through it. More accurate experiments show that the induced e.m.f. is actually proportional to the rate of change of flux through the coil. This result is sometimes called *Faraday's*, or *Neumann's, law*: The induced e.m.f. is proportional to the *rate of change of magnetic flux linking the coil or circuit.*

The unit of magnetic flux Φ is the *weber* (Wb). Hence the unit of B, the flux density *or* flux per unit area, is the *weber per metre²* (Wb m⁻²) or *tesla* (T).

Flux Linkage

If a coil has more than one turn, then the flux through the whole coil is the sum of the fluxes through the individual turns. We call this the *flux linkage* through the whole coil. If the magnetic field is uniform, the flux through one turn is given, from (1), by $AB \cos \theta$. If the coil has N turns, the total flux linkage Φ is given by

$$\Phi = NAB \cos \theta \qquad . \qquad . \qquad . \qquad . \qquad . \qquad (2)$$

From Faraday's or Neumann's law, the e.m.f. induced in a coil is proportional to the rate of change of the flux linkage, Φ. Hence

$$E \propto \frac{\mathrm{d}\Phi}{\mathrm{d}t},$$

or
$$E = -k\frac{d\Phi}{dt}, \qquad \cdot \quad \cdot \quad \cdot \quad \cdot \quad (3)$$

where k is a positive constant. The minus sign expresses Lenz's law. It means that the induced e.m.f. is in such a direction that, if the circuit is closed, the induced current *opposes* the change of flux. Note that an induced e.m.f. exists across the terminals of a coil when the flux linkage changes, even though the coil is on 'open circuit'. A current, of course, does not flow in this case.

On p. 733, it is shown that $E = -kd\Phi/dt$ is consistent with the expression $F = BIl$ for the force on a conductor only if $k = 1$. We may therefore say that

$$E = -\frac{d\Phi}{dt}, \qquad \cdot \quad \cdot \quad \cdot \quad \cdot \quad (4)$$

where Φ is the flux linkage in webers, t is in seconds, and E is in volts.

From (4), it follows that one weber is the flux linking a circuit if the induced e.m.f. is one volt when the flux is reduced uniformly to zero in one second.

Example

(a) A narrow coil of 10 turns and area 4×10^{-2} m^2 is placed in a uniform magnetic field of flux density B of 10^{-2} T so that the flux links the turns normally. Calculate the average induced e.m.f. in the coil if it is removed completely from the field in 0·5 s.

(b) If the same coil is *rotated* about an axis through its middle so that it turns through 60° in 0·2 s in the field B, calculate the average induced e.m.f.

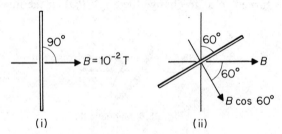

(i)   (ii)

Fig. 35.7 Example

(a) Flux linking coil initially $= NAB = 10 \times 4 \times 10^{-2} \times 10^{-2}$
$$= 4 \times 10^{-3} \text{ Wb (Fig. 35.7 (i))}$$

So   average induced e.m.f. $= \dfrac{\text{flux change}}{\text{time}} = \dfrac{4 \times 10^{-3}}{0\cdot5}$

$$= 8 \times 10^{-3} \text{ V.}$$

(b) When the coil is initially perpendicular to B, flux linking coil $= NAB$, Fig. 35.7 (ii). When the coil is turned through 60°, the flux density normal to the coil is now $B \cos 60°$. So

flux change through coil $= NAB - NAB \cos 60°$
$$= 4 \times 10^{-3} - 4 \times 10^{-3} \times 0\cdot5.$$

So   average induced e.m.f. $= \dfrac{\text{flux change}}{\text{time}} = \dfrac{2 \times 10^{-3}}{0\cdot2}$

$$= 10^{-2} \text{ V.}$$

E.M.F. Induced in Moving Rod

Generators at power stations produce high induced voltages by rotating long *straight conductors*. Figure 35.8 (i) shows a simple apparatus for demonstrating that an e.m.f. may be induced in a straight rod or wire, when it is moved across a magnetic field. The apparatus consists of a rod AC resting on rails XY, and lying between the poles NS of a permanent magnet. The rails are connected to a galvanometer G.

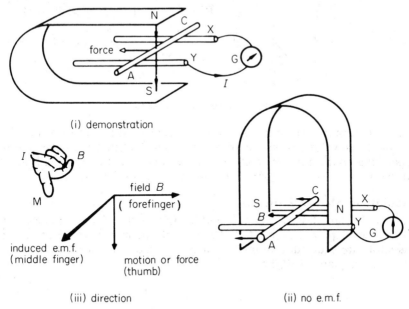

(i) demonstration

field *B*
(forefinger)

induced e.m.f.
(middle finger)

motion or force
(thumb)

(iii) direction

(ii) no e.m.f.

Fig. 35.8 E.m.f. induced in moving rod

If we move the rod to the left, so that it cuts across the field *B* of the magnet, a current *I* flows as shown. If we move the rod to the right, the current reverses. We notice that the current flows only while the rod is moving, and we conclude that the motion of the rod AC induces an e.m.f. in it.

By turning the magnet into a vertical position (Fig. 35.8 (ii)) we can show that no e.m.f. is induced in the rod when it moves parallel to the field *B*. We conclude that an e.m.f. is induced in the rod only when it *cuts across* the field. And, whatever the direction of the field, no e.m.f. is induced when we slide the rod parallel to its own length. The induced e.m.f. is greatest when we move the rod at right angles, both to its own length and to the magnetic field. These results may be summarised in *Fleming's right-hand rule*:

If we extend the thumb and first two fingers of the right hand, so that they are all at right angles to one another, then the directions of field, motion, and induced e.m.f. are related as in Fig. 35.8 (iii).

To show E.M.F. ∝ Rate of Change

The variation of the magnitude of the e.m.f. in a rod with the speed of 'cutting' magnetic flux can be demonstrated with the apparatus in Fig. 35.9 (i).

Here AC is a copper rod, which can be rotated by a wheel W round one pole N of a long magnet. Brush contacts at X and Y connect the rod to a galvanometer G and a series resistance R. When we turn the wheel, the rod AC cuts across the field *B* of the magnet, and an e.m.f. is induced in it. If we turn the wheel steadily, the galvanometer gives a steady deflection, showing that a steady current is flowing round the circuit.

To find how the current and e.m.f. depends on the speed of the rod, we keep

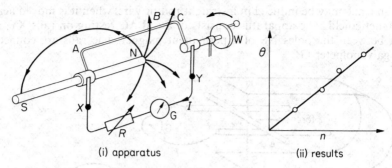

(i) apparatus   (ii) results

Fig. 35.9 Induced e.m.f. experiment

the circuit resistance constant, and vary the rate at which we turn the wheel. We time the revolutions with a stop-watch, and find that the deflection θ is proportional to the number of revolutions per second, n, Fig. 35.9 (ii). It follows that the induced e.m.f. is proportional to the speed of the rod.

Calculation of E.M.F. in Rod

Consider the circuit shown in Fig. 35.10. PQ is a straight wire touching the two connected parallel wires QR, PS and free to move over them. All the conductors are situated in a uniform magnetic field of flux density B, perpendicular to the plane of PQRS.

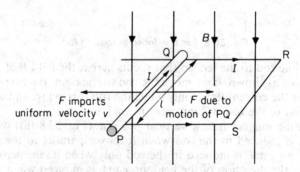

Fig. 35.10 Calculation of induced e.m.f.

E by energy principles. Suppose the rod PQ is pulled with a uniform velocity v by an external force F. There will then be a change of flux linkage in the area PQRS and so an e.m.f. will be induced in the circuit. This produces a current I which flows round the circuit. A *force* will now act on the wire PQ due to the current flowing and to the presence of the magnetic field (p. 693). By Lenz's law, the direction of this force will oppose the movement of PQ. If the current flowing is I, and the length of PQ is l, the force on PQ is BIl. This is equal to the external force F, since PQ is not accelerating.

Because energy is conserved, *the rate of working by the external force is equal to the rate at which energy is supplied to the electrical circuit.* Now in one second, PQ moves a distance v. Hence

work done per second = force × distance moved per second.

$$= BIlv.$$

If the induced e.m.f. is E, the electrical energy used in one second, or power, $= EI$.

$$\therefore EI = BIlv$$

$$E = Blv. \qquad . \qquad . \qquad . \qquad . \qquad . \qquad (1)$$

E by rate of cutting flux. This result has been derived without using the relation $E = -d\Phi/dt$. To see if the same result as (1) can be obtained, consider the flux changes. In one second the area of PQRS changes by vl. Hence the change in flux linkage per second, $d\Phi/dt = B \times$ area change per second $= Blv$. Hence, numerically,

$$\therefore E = Blv.$$

This means that the relation $E = -d\Phi/dt$ may be used to find the induced e.m.f. in a straight wire.

Induced E.M.F. and Force on Moving Electrons

We have seen that an electron moving across a magnetic field experiences a mechanical force (p. 703). As we now show, this explains the origin of the e.m.f. induced in a wire moving through a magnetic field.

When we move the wire downwards across the field B as in Fig. 35.11, each

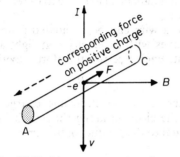

Fig. 35.11 Forces on a moving electron

electron moves downwards across the field. A downward movement of electron charge $-e$ is equivalent to an *upward* movement of positive charge or conventional current I, as shown. Applying Fleming's left hand rule (in which the force F is at right angles to the velocity v of the wire and to B), we see that F drives the electrons along the wire from A to C. So if the wire is not connected in a closed circuit, electrons will pile up at C. Thus the end C will gain a negative charge and A will be left with an equal positive charge. After a time the charge at C will oppose further electron movement along the wire and so the drift stops.

The charges between A and C produce an *electromotive force E*, Fig. 35.12. As in a battery, A is the 'positive pole' of the wire generator and C is the

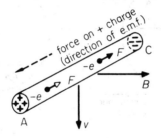

Fig. 35.12 Induced e.m.f. arising from force on moving electrons

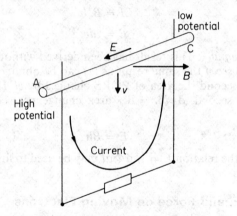

Fig. 35.13 E.m.f. and potential difference

'negative pole'. So when an external resistor is joined to A and C, the conventional current flows in it as shown in Fig. 35.13. We see now that the induced e.m.f. in a wire moving across a magnetic field is due to the electromagnetic force on the electrons inside the wire.

The e.m.f. induced in a wire can be calculated from the force on a moving electron. If the wire moves with a velocity v at right angles to a field B, then so do the electrons in it. Each electron therefore experiences a force

$$F = Bev,$$

shown on p. 703, where e is the electronic charge. The work which this force does in carrying the electron along the length l of the wire is Fl. But it is also, by definition, equal to the product of the e.m.f. E and the charge e (p. 645). Therefore

$$Ee = Fl = Bevl,$$

So $E = Blv.$

Homopolar or Disc Generator

Another type of generator, which gives a very steady e.m.f., is illustrated in Fig. 35.14 (i). It consists of a copper disc which rotates between the poles of a magnet; connections are made to its axle and circumference. We assume for simplification that the magnetic field B is uniform over the radius XY.

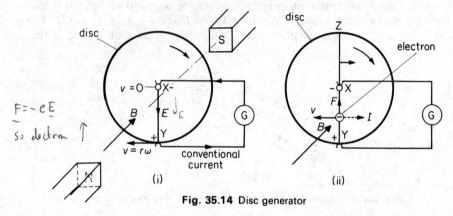

Fig. 35.14 Disc generator

E by flux cutting. The radius XY continuously cuts the magnetic flux between the poles of the magnet. For this straight conductor, the velocity at the end X is zero and that at the other end Y is $r\omega$, where ω is the angular velocity of the disc. So

$$\text{average velocity of XY, } v = \frac{1}{2}(0 + r\omega) = \frac{r\omega}{2}.$$

Now the induced e.m.f. E in a straight conductor of length l and moving with velocity v normal to a field B is given by $E = Blv$. Since in this case $l = r$ and $v = \frac{1}{2}r\omega$, then

$$E = B \times r \times \tfrac{1}{2}r\omega = \tfrac{1}{2}Br^2\omega. \qquad . \qquad . \qquad . \qquad (1)$$

Since $\omega = 2\pi f$, where f is the number of revolutions per second of the disc, we can say that

$$E = B.\pi r^2.f \qquad . \qquad . \qquad . \qquad . \qquad . \qquad . \qquad (2)$$

The direction of E is given by Fleming's *right* hand rule. Applying the rule, we find that E acts from X to Y so that Y is at the *higher* potential, as shown.

We can understand the origin of the e.m.f. by considering an electron between X and Y, Fig. 35.14 (ii). When the disc rotates, the electron moves to the left as shown. The equivalent conventional current I is then to the right. Applying Fleming's left hand rule, we find that the force F on the electron drives it to X. So X obtains a negative charge and Y a positive charge. The radius is thus a generator with Y as its positive pole.

E by energy conversion. When the disc is rotated with a steady angular velocity,

mechanical work done per second = electrical energy per second obtained.

If I is the current flowing to an external resistance joined to it, then the mechanical force F on the rotating radius $= BIl = BIr$. The rotating radius moves with an average velocity $r\omega/2$. So

$$\text{mechanical work per second or power} = F \times \text{velocity} = BIr \times r\omega/2$$

$$= BIr^2\omega/2.$$

But electrical energy produced per second or power $= EI$,

where E is the induced e.m.f. in the generator. So

$$EI = BIr^2\omega/2.$$

Hence

$$E = Br^2\omega/2 = B.\pi r^2.f$$

as in (2).

As the disc rotates clockwise, Fig. 35.14 (ii), the radius XY moves to the left at the same time as the radius XZ moves to the right. If the magnetic field covered the whole disc, the induced e.m.f. in the two radii would be in *opposite* directions. So the resultant e.m.f. between YZ would be *zero*. $B.\pi r^2.f$, the e.m.f. between the centre and rim of the disc, is the maximum e.m.f. which can be obtained from the dynamo.

If the disc had a radius r_1 and an axle at the centre of radius r_2, the area swept out by a rotating radius of the metal disc $= \pi r_1^2 - \pi r_2^2 = \pi(r_1^2 - r_2^2)$. In this case the induced e.m.f. would be $E = B.\pi(r_1^2 - r_2^2).f$.

Generators of this kind are called *homopolar* because the e.m.f. induced in the moving conductor is always in the same direction. They are sometimes used for electroplating, where only a small voltage is required, but they are not useful for most purposes, because they give too small an e.m.f. The e.m.f. of a commutator dynamo can be made large by having many turns in the coil; but the e.m.f. of a homopolar dynamo is limited to that induced in one radius of the disc.

Example

A circular metal disc is placed with its plane perpendicular to a uniform magnetic field of flux density B. The disc has a radius of 0·20 m and is rotated at 5 rev s^{-1} about an axis through its centre perpendicular to its plane. The e.m.f. between the centre and the rim of the disc is balanced by the p.d. across a 10 Ω resistor when carrying a current of 1·0 mA. Calculate B.

The induced e.m.f. $E = B \cdot \pi r^2 \cdot f = B \times 0 \cdot 2^2 \times 5 = 0 \cdot 2\,\pi B$

and p.d. across 10 Ω, $V = IR = 1 \times 10^{-3} \times 10 = 10^{-2}$ V.

$$\therefore\ 0 \cdot 2\,\pi B = 10^{-2}$$

$$\therefore\ B = 1 \cdot 6 \times 10^{-2} \text{ T.}$$

The Dynamo and Generator

Faraday's discovery of electromagnetic induction was the beginning of electrical engineering. Nearly all the electric current used today is generated by induction, in machines which contain coils moving continuously in a magnetic field.

Figure 35.15 illustrates the principle of such a machine, which is called a *dynamo*, or *generator*. A coil DEFG, shown for simplicity as having only one turn, rotates on a shaft, which is not shown, between the poles NS of a horseshoe magnet. The ends of the coil are connected to flat brass rings R, which are supported on the shaft by discs of insulating material, also not shown. Contact with the rings is made by small blocks of carbon B, supported on springs, and shown connected to a lamp L.

As the coil rotates, the flux linking it changes, and a current is induced in it which flows, via the carbon blocks B, to the lamp L. The magnitude (which we

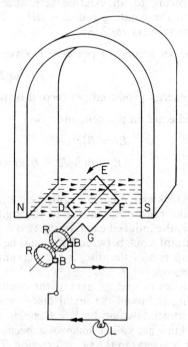

Fig. 35.15 A simple dynamo

study shortly) and the direction of the current are not constant. Thus when the coil is in the position shown, the limb ED is moving downwards through the lines of force, and GF is moving upwards. Half a revolution later, ED and GF will have interchanged their positions, and ED will be moving upwards. Consequently, applying Fleming's right-hand rule (p. 731), the current round the coil must *reverse* as ED changes from downward to upward motion. The actual direction of the current at the instant shown on the diagram is indicated by the double arrows, using Fleming's rule. By applying this rule, it can be seen that *the current reverses* every time the plane of the coil passes the vertical position.

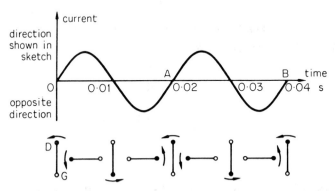

Fig. 35.16 Current generation by dynamo of Fig. 35.15 plotted against time and coil position

We shall see shortly that the magnitude of the e.m.f. and current varies with time as shown in Fig. 35.16; this diagram also shows the corresponding position of DG. This type of current is called an *alternating current* (a.c.). A complete alternation, such as from A to B in the figure, is called a 'cycle'; and the number of cycles which the current goes through in one second is called its 'frequency'. The frequency of the current represented in the figure is that of most domestic supplies in Britain—50 Hz (cycles per second). Thus from A to B, which is one cycle, the time taken $(0·04–0·02) = 0·02$ s $= 1/50$ s. So the frequency is 50 Hz.

E.M.F. in Dynamo

We can now calculate the e.m.f. in the rotating coil. If the coil has an area A, and its normal makes an angle θ with the magnetic field B, as in Fig. 35.17 (i), then the flux through the coil

$$= AB \cos \theta \text{ (see p. 729).}$$

The flux linkages with the coil, if it has N turns, are

$$\Phi = NAB \cos \theta$$

Figure 35.17 (ii) shows how the flux linkage Φ varies with the angle θ starting from $\theta = 0$, when the coil is vertical (V). Since $\theta = \omega t$, then $\theta \propto t$. So the horizontal axis can also represent the time t, as indicated. When $\theta = 90°$ the coil is horizontal (H) and no flux links the coil. As the coil rotates further the flux linking the face same reverses and so Φ becomes negative as shown.

If the coil turns with a steady angular velocity ω or $d\theta/dt$, then the e.m.f. induced in volts in the coil is given by $E = d\Phi/dt = -gradient$ of the $\Phi - t$ graph in Fig. 35.17 (ii). Figure 35.17 (iii) shows the negative gradient variation found from Fig. 35.17 (ii). This is the variation of E with time t.

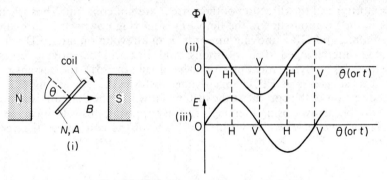

Fig. 35.17 Coil inclined to magnetic field

We can calculate E exactly as follows.

$$E = -\frac{d\Phi}{dt}$$

$$= -NAB\frac{d}{dt}(\cos\,\theta)$$

$$= NAB\,\sin\,\theta\frac{d\theta}{dt} \quad . \quad . \quad . \quad . \quad (1)$$

$d\theta/dt$ is ω, the angular velocity and $\omega = 2\pi f$ where f is the number of revolutions per second. Also, in a time t, $\theta = \omega t = 2\pi ft$. So, from (1), we can write

$$E = \omega NAB\,\sin\,\omega t \quad . \quad . \quad . \quad (2)$$

or $$E = 2\pi f NAB\,\sin\,2\pi ft. \quad . \quad . \quad (3)$$

Thus the e.m.f. varies sinusoidally with time, like the pressure in a soundwave, the frequency being f cycles per second.

The maximum (peak) value or amplitude of E occurs when $\sin\,2\pi ft$ reaches the value 1. If the maximum value is denoted by E_0, it follows that

$$E_0 = 2\pi f NAB,$$

and $$E = E_0\,\sin\,2\pi ft \quad . \quad . \quad . \quad (4)$$

The e.m.f. E sends an alternating current of a similar sine equation, through a resistor connected across the coil (see p. 733).

Dynamo E.M.F. from Energy Principles

The e.m.f. in a simple dynamo can also be found from energy principles.

At a time t, suppose the normal to the plane of the coil makes an angle θ with the field B, Fig. 35.17 (i). The torque (moment of couple) acting on the coil is then given by $NABI\,\sin\,\theta$ (p. 698), where I is the current flowing if the ends of the coil are connected to an external resistor.

The work done by a torque in rotation through an angle = torque × angle of rotation (p. 82). So if the coil is rotated through a small angle $\Delta\theta$ in a time Δt,

mechanical work done per second by torque = torque × $\Delta\theta/\Delta t$

$$= NABI\,\sin\,\theta \times \omega,$$

since $\omega = \Delta\theta/\Delta t$. But if E is the induced e.m.f. in the coil, the electrical energy per second generated in the coil = EI. So

$$EI = NABI \sin \theta \times \omega$$

or $$E = \omega NAB \sin \theta = \omega NAB \sin \omega t.$$

This result for E agrees with the calculation using $E = -\mathrm{d}\Phi/\mathrm{d}t$.

Alternators

Generators of alternating current are often called *alternators*. In all but the smallest, the magnetic field of an alternator is provided by an electromagnet called a field-magnet or *field*, as shown in Fig. 35.18; it has a core of cast steel, and is fed with direct current from a separate d.c. generator. The rotating coil, called the *armature*, is wound on an iron core, which is shaped so that it can turn within the pole-pieces of the field-magnet; with the field magnet, the armature core forms a system which is almost wholly iron, and can be strongly magnetised by a small current through the field winding. The field in which the armature turns is much stronger than if the coil had no iron core, and the e.m.f. is proportionately greater. In the small alternators used for bicycle lighting the armature is stationary, and the field is provided by permanent magnets, which rotate around it. In this way rubbing contacts, for leading the current into and out of the armature, are avoided.

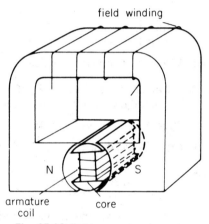

Fig. 35.18 Field magnet and armature

When no current is being drawn from a generator, the power required to turn its armature is merely that needed to overcome friction, since no electrical energy is produced. But when a current is drawn, the power required increases, to provide the electrical power. The current, flowing through the armature winding, causes the magnetic field to set up a couple which opposes the rotation of the armature, and so demands the extra power.

The Transformer

A transformer is a device for stepping up—or down—an alternating voltage. It has primary and secondary windings but no make-and-break, Fig. 35.19. It has an iron core, which is made from E-shaped laminations, interleaved so that the magnetic flux does not pass through air at all; in this way the greatest flux is obtained with a given current.

When an alternating e.m.f. E_p is connected to the primary winding, it sends an alternating curent through it. This sets up an alternating flux in the core of magnitude BA, where B is the flux density and A is the cross-sectional area.

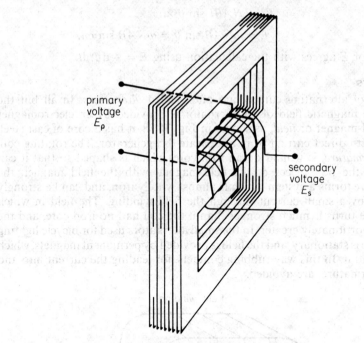

Fig. 35.19 Transformer with soft iron core

This induces an alternating e.m.f. in the secondary E_s. If N_p, N_s are the number of turns in the primary and secondary coils, their linkages with the flux Φ are:

$$\Phi_p = N_p AB \qquad \Phi_s = N_s AB.$$

The magnitude of the e.m.f. induced in the secondary is, from the formula on p. 730:

$$E_s = \frac{d\Phi_s}{dt} = N_s A \frac{dB}{dt}$$

The changing flux also induces a back-e.m.f. in the primary, whose magnitude is

$$E_p = \frac{d\Phi_p}{dt} = N_p A \frac{dB}{dt}$$

The voltage applied to the primary, from the source of current, is used simply in overcoming the back-e.m.f. E_p, if we neglect the resistance of the wire. Therefore it is equal in magnitude to E_p. (This is analogous to saying, in mechanics, that action and reaction are equal and opposite.) Consequently we have

$$\frac{\textit{e.m.f. induced in secondary}}{\textit{voltage applied to primary}} = \frac{E_s}{E_p} = \frac{N_s}{N_p} \quad . \qquad . \quad (1)$$

So the transformer steps voltage up or down according to its *'turns-ratio'*:

$$\frac{\text{secondary voltage}}{\text{primary voltage}} = \frac{\text{secondary turns}}{\text{primary turns}}.$$

The relation in (1) is only true when the secondary is on open circuit. When a load is connected to the secondary winding, a current flows in it. Thus the power drawn from the secondary is drawn, in turn, from the supply to which the primary is connected. So now a greater primary current is flowing than before the secondary was loaded.

Transformers are used to step up the voltage generated at a power station from 23 000 to 400 000 volts for high-tension transmission (p. 638). After transmission they are used to step it down again to a value safer to distribution (240 volts in houses). Inside a house a transformer may be used to step the voltage down from 240 V to 4 V, for ringing bells. Transformers with several secondaries are used in television receivers, where several different voltages are required.

D.C. Generators

Figure 35.20 (i) is a diagram of a direct-current generator or dynamo. Its essential difference from an alternator is that the armature winding is connected to a *commutator* instead of slip-rings.

The commutator consists of two half-rings of copper C, D, insulated from one another, and turning with the coil. Brushes BB, with carbon tips, press against the commutator and are connected to the external circuit. The commutator is oriented so that it *reverses* the connections from the coil to the circuit at the instant when the e.m.f. reverses in the coil.

Figure 35.20 (ii) shows several positions of the coil and commutator, and the

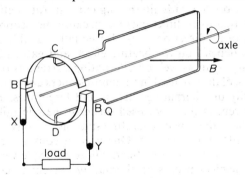

(i) principle

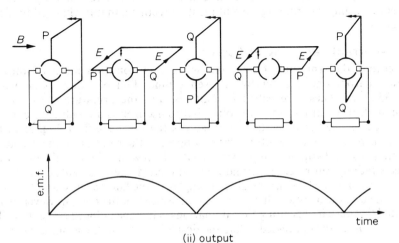

(ii) output

Fig. 35.20 D.c. generator

e.m.f. observed at the terminals XY. This e.m.f. pulsates in magnitude, but it acts always in the same sense round the circuit connected to XY. It is a pulsating direct e.m.f. The average value in this case can be shown to be $2/\pi$ of the maximum e.m.f. E_0, given in equation (3), p. 738.

In practice, as in an alternator, the armature coil is wound with insulated wire on a soft iron core, and the field-magnet is energised by a current. This current is provided by the dynamo itself. The steel of the field-magnet has always a small residual magnetism, so that as soon as the armature is turned an e.m.f. is induced in it. This then sends a current through the field winding, which increases the field and the e.m.f.; the e.m.f. rapidly builds up to its working value.

Most consumers of direct current wish it to be steady, not pulsating as in Fig. 35.20. A reasonably steady e.m.f. is given by an armature with many coils, inclined to one another, and a commutator with a correspondingly large number of segments. The coils are connected to the commutator in such a way that their e.m.f.s add round the external circuit.

Applications of Alternating and Direct Currents

Direct currents are less easy to generate than alternating currents, and alternating e.m.f.s are more convenient to step up and to step down, and to distribute over a wide area. The national grid system, which supplies electricity to the whole country, is therefore fed with alternating current. Alternating current is just as suitable for heating as is direct current, because the heating effect of a current is independent of its direction. It is also equally suitable for lighting, because filament lamps depend on the heating effect, and gas-discharge lamps—neon, sodium, mercury—run as well on alternating current as on direct.

Small motors, of the size used in vacuum-cleaners and common machine-tools, run satisfactorily on alternating current, but large ones, as a general rule, do not. Direct current is therefore used on most electric railway systems. These systems either have their own generating stations, or convert alternating current from the grid into direct current. One way of converting alternating current into direct current is to use a rectifier, whose principle we shall describe later.

For electrochemical processes alternating current is useless. The chemical effect of a current reverses with its direction, and if, therefore, we tried to deposit a metal by alternating current, we would merely cause a small amount of the metal to be alternately deposited and dissolved. For electroplating, and for battery charging, alternating current must be rectified, that is, changed to direct current.

Eddy-currents and Power Losses

The core of the armature of a dynamo is built up from thin sheets of soft iron insulated from one another by an even thinner film of oxide, as shown in Fig. 35.21 (i). These are called *laminations*, and the armature is said to be laminated. If the armature were solid, then, since iron is a conductor, currents would be induced in it by its motion across the magnetic field, Fig. 35.21 (ii). These currents would absorb power by opposing the rotation of the armature, and they would dissipate that power as heat, which would damage the insulation of the winding. But when the armature is laminated, these currents cannot flow, because the induced e.m.f. acts at right angles to the laminations, and therefore to the insulation between them. The magnetisation of the core, however, is not affected, because it acts along the laminations. Thus the induced currents, called *eddy-currents*, are suppressed, while the desired e.m.f.—in the armature coil—is not.

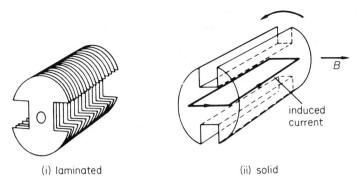

(i) laminated   (ii) solid

Fig. 35.21 Armature cores

Eddy-currents, by Lenz's law, always tend to oppose the motion of a solid conductor in a magnetic field. The opposition can be shown in many ways. One of the most impressive is to make a chopper with a thick copper blade, and to try to slash it between the poles of a stronger electromagnet; then to hold it delicately and allow it to drop between them. The resistance to the motion in the former case can be felt.

If a rectangular metal bob of a pendulum is set swinging between the poles of a strong magnet, it soon comes to rest. The eddy-currents circulating inside the metal oppose the motion. But if many deep slots are cut into the metal, as in a comb, the pendulum now keeps oscillating for a much longer time before coming to rest. The eddy-currents are considerably reduced in this case as they cannot flow across the many air gaps formed by the slots.

Damping of Moving-Coil Meters

Sometimes eddy-currents can be made use of—for example, in damping a galvanometer. When a current is passed through the coil of a galvanometer, a couple acts on the coil which sets it swinging. If the swings are opposed only by the viscosity of the air, they decay very slowly and are said to be naturally damped, Fig. 35.22. The pointer then takes a long time to come to its final steady deflection θ.

To bring the pointer more rapidly to rest, the damping must be increased. One way of increasing the damping is to wind the coil on a *metal* frame or former made of aluminium. Then, as the coil swings, the field of the permanent magnet induces eddy-currents in the former, and these, by Lenz's law, oppose the motion. They therefore slow down the turning of the coil towards its eventual position, and *also stop its swings about that position*. So in the end the deflected coil comes to rest sooner than if it were not damped. See Fig. 33.16.

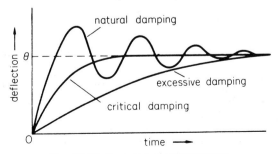

Fig. 35.22 Damping of galvanometer

Galvanometer coils which are wound on insulating formers can be damped by short-circuiting a few of their turns, or by joining the galvanometer terminals with connecting wire so that the whole coil is short-circuited. The meter can then be carried safely from one place to another without excessive swinging of the coil, which might otherwise damage the instrument.

If a shunt is connected to the meter, the eddy-currents circulate round the coil and shunt, independently of the current to be measured. The smaller the shunt, the greater the eddy-currents and the damping; if the coil is overdamped, as shown in Fig. 35.22, it may take almost as long to come to rest as when it is undamped. The damping which is just sufficient to prevent overshoot is called 'critical' damping.

Electric Motors

If a simple direct-current dynamo, of the kind described on p. 741, is connected to a battery it will run as a *motor*, Fig. 35.23. Current flows round the armature coil, and the magnetic field exerts a couple on this, as in a moving-coil galvanometer. The commutator reverses the current just as the limbs of the coil are changing from upward to downward movement and vice versa. Thus the couple on the armature is always in the same sense, and the shaft turns continuously. The reader should verify these statements with the help of Fig. 35.23.

The armature of a motor is laminated, in the same way and for the same reason, as the armature of a dynamo.

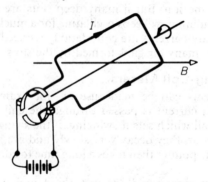

Fig. 35.23 Principle of d.c. motor

Back-e.m.f. in Motor

When the armature of a motor rotates, an e.m.f. is induced in its windings. By Lenz's law this e.m.f. opposes the current which is making the coil turn. It is therefore called a back-e.m.f. If its magnitude is E, and V is the potential difference applied to the armature by the supply, then the armature current is

$$I_a = \frac{V - E}{R_a} \qquad . \qquad . \qquad . \qquad . \qquad (1)$$

Here R_a is the resistance of the armature, which is generally small—of the order of 1 ohm.

The back-e.m.f. E is proportional to the strength of the magnetic field, and to the speed of rotation of the armature. When the motor is first switched on, the back-e.m.f. is zero: it rises as the motor speeds up. In a large motor the starting current would be ruinously great. To limit it, a variable resistance is therefore inserted in series with the armature, and this is gradually reduced to zero as the motor gains speed.

When a motor is running, the back-e.m.f. in its armature E is not much less than the supply voltage V. For example, a motor running off the mains ($V = 240$ V say) might develop a back-e.m.f. $E = 230$ V. If the armature had a resistance of 1 ohm, the armature current would then be 10 A (equation (1)). When the motor was switched on, the armature current would be 240 A if no starting resistor were used.

Example

A motor has an armature resistance of $4 \cdot 0\ \Omega$. On a 240 V supply and a light load, the motor speed is 200 rev min^{-1} and the armature current is 5 A. Calculate the motor speed at a full load when the armature current is 40 A.

Generally,
$$I = \frac{240 - e_1}{4}$$

where e_1 is the back-e.m.f. So at $I = 5$ A,

$$240 - e_1 = 5 \times 4 = 20$$

and
$$e_1 = 240 - 20 = 220 \text{ V}.$$

Suppose e_2 is the new back-e.m.f. when $I = 40$ A. Then, from the equation for I,

$$240 - e_2 = 40 \times 4 = 160.$$

So
$$e_2 = 240 - 160 = 80 \text{ V}.$$

Now the speed of the motor is proportional to the back-e.m.f.

So
$$\text{full load speed} = \frac{80}{220} \times 200 \text{ rev min}^{-1}$$

$$= 73 \text{ rev min}^{-1} \text{ (approx.)}.$$

Back-e.m.f. and Power

The back-e.m.f. in the armature of a motor represents the mechanical power which it develops. To see that this is so, we use an argument similar to that which we used in finding an expression for the e.m.f. induced in a conductor. We consider a rod AC, able to slide along rails, in a plane at right angles to a magnetic field B, Fig. 35.24. But we now suppose that a current I is maintained in the rod by a battery, which sets up a potential difference V between the rails. The magnetic field then exerts a force F on the rod, given by

$$F = BIl.$$

The force F makes the rod move; if its velocity is v, the mechanical power developed by the force F is

$$P_m = Fv = BIlv \qquad . \qquad . \qquad . \qquad . \qquad . \qquad (1)$$

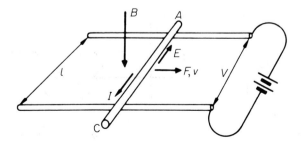

Fig. 35.24 Back-e.m.f. and mechanical power

As the rod moves, a back-e.m.f. is induced in it, whose magnitude E is given by the expression for the e.m.f. in a moving rod (p. 733):

$$E = Blv \qquad . \qquad . \qquad . \qquad . \qquad . \qquad (2)$$

Equations (1) and (2) together give

$$P_m = EI \qquad . \qquad . \qquad . \qquad . \qquad . \qquad (3)$$

Thus the mechanical power developed is equal to the product of the back-e.m.f. and the current.

We can now complete the analysis of the action represented in Fig. 35.24. If R is the resistance of the rails and rod, the heat developed in them is I^2R. The power supplied by the battery is IV, and the battery is the only source of power in the whole system. Therefore

$$IV = I^2R + P_m \qquad . \qquad . \qquad . \qquad . \qquad (4)$$

the power supplied by the battery goes partly into heat, and partly into useful mechanical power. Also, by equation (3),

$$IV = I^2R + EI, \qquad . \qquad . \qquad . \qquad . \qquad (5)$$

So
$$V = IR + E$$

or
$$I = \frac{V - E}{R}.$$

This is equation (1), p. 744, which we previously obtained simply from Ohm's law.

Series- and Shunt-Wound Motors

The field winding of a motor may be connected in series or in parallel with the armature. If it is connected in series, it carries the armature current, which is large, Fig. 35.25. The field winding therefore has few turns of thick wire, to keep down its resistance and so waste little power in it as heat. The few turns are enough to magnetise the iron, because the current is large.

Series motors are used where great torque is required in starting—for example, in cranes. They develop a great starting torque because the armature current flows through the field coil. At the start the armature back-e.m.f. is small, and the current is great—as great as the starting resistance will allow. The field-magnet is therefore very strongly magnetised. The torque on the armature is

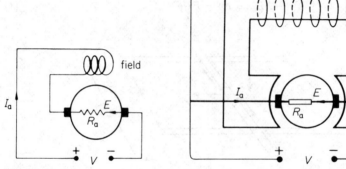

Fig. 35.25 Series-wound motor   **Fig. 35.26** Current and voltages in shunt-wound motor

proportional to the field and to the armature current; since both are great at the start, the torque is very great.

If the field coil is connected in parallel with the armature, as in Fig. 35.26, the motor is said to be 'shunt-wound'. The field winding has many turns of fine wire to keep down the current which it consumes.

Shunt-wound motors are used for driving machine-tools, and in other jobs where a steady speed is required. A shunt motor keeps a nearly steady speed for the following reason. If the load is increased, the speed falls a little; the back-e.m.f. then falls in proportion to the speed, and the current rises, enabling the motor to develop more power to overcome the increased load. A series motor does not keep such a steady speed as a shunt motor

Charge and Flux Linkage

Flux and Charge Relation

We have already seen that an electromotive force is induced in a circuit when the magnetic flux linked with it changes. If the circuit is closed, a current flows, and electric charge, Q, is carried round the circuit. As we shall now show, there is a simple relationship between the charge and the change of flux.

Consider a closed circuit of total resistance R ohm, which has a total flux linkage Φ with a magnetic field, Fig. 35.27. If the flux linkages start to change,

$$\text{induced e.m.f., } E = -\frac{d\Phi}{dt}.$$

$$\therefore \text{ current, } I = \frac{E}{R} = -\frac{1}{R}\frac{d\Phi}{dt} \qquad . \qquad . \qquad . \qquad (1)$$

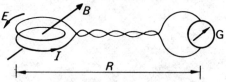

Fig. 35.27 Coil with changing flux

In general, the flux linkage will not change at a steady rate, and the current will not be constant. But, throughout its change, charge is being carried round the circuit. In a time t from zero, the charge carried round the circuit is

$$Q = \int_0^t I\,dt.$$

From (1),
$$\therefore Q = -\frac{1}{R}\int_0^t \frac{d\Phi}{dt}.dt$$

$$= -\frac{1}{R}\int_{\Phi_0}^{\Phi_t} d\Phi,$$

where Φ_0 is the number of linkages at $t = 0$, and Φ_t is the number of linkages at time t. Thus

$$Q = -\frac{\Phi_t - \Phi_0}{R} = \frac{\Phi_0 - \Phi_t}{R}.$$

The quantity $\Phi_0 - \Phi_t$ is the positive if the linkages Φ have decreased, and negative if they have increased. But as a rule we are interested only in the magnitude of the charge, and we may write

$$Q = \frac{change\ of\ flux\ linkage}{R} \qquad . \qquad . \qquad . \qquad (2)$$

Equation (2) shows that the charge circulated is proportional to the change of flux-linkages, and is *independent of the time*.

Ballistic Galvanometer

It can be seen from the last section that the charge which flows round a given circuit is directly proportional to the change of flux linkage. If the charge flowing is measured by a *ballistic galvanometer* G, as shown in Fig. 35.27, then we have a measure of the change in the flux linkage, Φ.

Ballistics is the study of the motion of an object, such as a bullet, which is set off by a blow, and then allowed to move freely. By freely, we mean without friction. A ballistic galvanometer is one used to measure an electrical blow, or impulse, for example, the charge Q which circulates when a capacitor is discharged through it.

A galvanometer which is intended to be used ballistically has (i) a heavier coil than one which is not, and (ii) has as little damping as possible—an insulating former, no short-circuited turns, no shunt. The mass of its coil makes it swing slowly; in the example above, for instance, the capacitor has discharged, and the charge has finished circulating, while the galvanometer coil is just beginning to turn. The galvanometer coil continues to turn, however; and as it does so it twists the controlling spring. The coil stops turning when its kinetic energy, which it gained from the forces set up by the current, has been converted into potential energy of the spring. The coil then swings back, as the spring untwists, and it continues to swing back and forth for some time. Eventually it comes to rest, but only because of the damping due to the viscosity of the air, and to the spring. Theory shows that, if the damping is negligible, *the first deflection of the galvanometer is proportional to the quantity of electricity, Q, that passed through its coil, as it began to move.* This first deflection, θ, is often called the 'throw' of the galvanometer; we have, then,

$$Q = k\theta, \qquad . \qquad . \qquad . \qquad . \qquad . \qquad (1)$$

where k is a constant of the galvanometer.

Equation (1) is true only if all the energy given to the coil is spent in twisting the suspension. If an appreciable amount of energy is used to overcome damping —i.e. dissipated as heat by eddy currents—then the galvanometer is not ballistic, and θ is not proportional to Q.

To calibrate the ballistic galvanometer, a capacitor of known capacitance, for example, 2 μF, is charged by a battery of known e.m.f. such as 50 volt, and then discharged through the instrument. See p. 596. Suppose the deflection is 200 divisions. The charge $Q = CV = 100$ microcoulomb, and thus the galvanometer sensitivity is 2 divisions per microcoulomb.

Measurement of Flux Density

Figure 35.28 illustrates the principle of measuring the flux density B in the field between the poles of a powerful magnet such as a loudspeaker magnet. A small coil, called a *search coil*, with a known area and number of turns, is connected to a ballistic galvanometer G. It is positioned at right angles to the field to be measured, as shown, so that the flux enters the coil face normally.

The coil is then pulled completely out of the field by moving it smartly down-

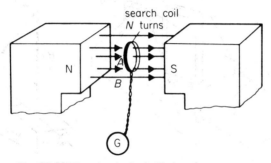

Fig. 35.28 Flux density by ballistic galvanometer

wards, for example, and the throw θ produced in the galvanometer is observed. The charge Q which passes round the circuit is proportional to θ, from above.

Suppose B is the field-strength in tesla (T), A is the area of the coil in m² and N is the number of turns. Then

$$\text{change of flux-linkages} = NAB$$

$$\therefore \text{ quantity, } Q, \text{ through galvanometer} = \frac{NAB}{R}$$

where R is the *total* resistance of the galvanometer and search coil. But

$$Q = c\theta,$$

where c is the quantity per unit deflection of the ballistic galvanometer.

$$\therefore \frac{NAB}{R} = c\theta$$

$$\therefore B = \frac{Rc\theta}{NA} \qquad . \qquad . \qquad . \qquad . \qquad (1)$$

The constant c is found by discharging a capacitor through the galvanometer (see p. 596). By knowing c, θ, R, N and A, the flux density B can be calculated from (1).

In this way, by using a suitable so-called 'search' coil connected to a ballistic galvanometer, the flux-density at various points between the poles of a large horse-shoe magnet can be compared.

The Earth Inductor

As another example of the use of a search coil and ballistic galvanometer, we describe a method that has been used for measuring the angle of dip of the earth's magnetic field (p. 716).

The earth's field is so nearly uniform that the search coil may be large, usually about 30 cm square. The field, however, is also so weak that even a large coil must have many turns—of the order of 100. The coil, called an *earth inductor*, is pivoted in a wooden frame, and this is fitted with stops so that the coil can be turned rapidly through 180°, Fig. 35.29.

To find the angle of dip, we connect the coil to a ballistic galvanometer G,

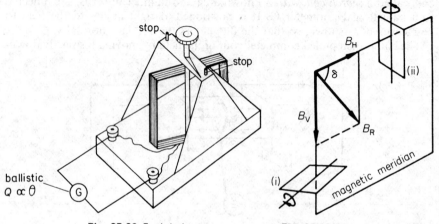

Fig. 35.29 Earth inductor   Fig. 35.30 Measurement of dip

and set it with its plane *horizontal*, as shown at (i) in Fig. 35.30. The flux linking the coil of N turns is then

$$\Phi = NAB_V,$$

where A is the area of the coil, and B_V is the vertical component of the earth's field. If we were to turn the coil through $90°$, the flux would fall to zero; and if we were to turn it through a further $90°$, the flux linkage would become NAB_V once more, but it would thread the coil in the opposite direction. So when we turn the coil through $180°$, the change the flux linkage is $2NAB_V$; at the same time we observe the throw, θ, of the galvanometer.

If R is the total resistance of galvanometer and search coil, the circulated charge is, by equation (2) on p. 748,

$$Q = \frac{\Phi}{R} = \frac{2NAB_V}{R} = c\theta,$$

where c is the constant of the galvanometer. So

$$\frac{2NAB_V}{R} = c\theta \qquad . \qquad . \qquad . \qquad . \qquad . \qquad (1)$$

We now set the frame of the earth inductor so that the axis of the coil is *vertical* and so that, when the coil is held by one of the stops, its plane lies East–West. See Fig. 35.30 (ii). The flux threading the coil is now NAB_H, where B_H is the horizontal component of the earth's field. Therefore, when we turn the coil through $180°$, the throw θ' of the galvanometer is given by

$$\frac{2NAB_H}{R} = c\theta' \qquad . \qquad . \qquad . \qquad . \qquad . \qquad (2)$$

Now the angle of dip, δ, is given by

$$\tan \delta = \frac{B_V}{B_H}.$$

Therefore, from equations (1) and (2),

$$\tan \delta = \frac{\theta}{\theta'}.$$

The horizontal and vertical components of the earth's total flux density can also be found from (1) and (2) respectively.

Example

A long solenoid carries a current which produces a flux-density B as its centre. A narrow coil Y of 10 turns and mean area 4.0×10^{-5} m² is placed in the middle of the solenoid so that the flux links its turns normally and the ends of Y are connected. If a charge of 1.6×10^{-6} C circulates through Y when the current in the solenoid is reversed, and the resistance of Y is 0.2 Ω, calculate B.

When current reverses,

$$\text{flux change } \Phi = NAB - (-NAB) = 2NAB$$

Since

$$\frac{\Phi}{R} = Q$$

$$\therefore \quad \frac{2 \times 10 \times 4 \times 10^{-5} B}{0.2} = 1.6 \times 10^{-6}$$

$$\therefore B = \frac{1 \cdot 6 \times 10^{-6} \times 0 \cdot 2}{2 \times 10 \times 4 \times 10^{-5}}$$

$$= 4 \times 10^{-4} \text{ T.}$$

Self-induction

The phenomenon called *self-induction* was discovered by the American, Joseph Henry, in 1832. He was led to it by a theoretical argument, starting from the phenomena of induced e.m.f., which he had discovered at about the same time as Faraday.

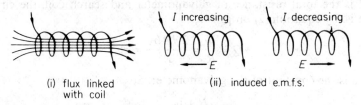

(i) flux linked with coil   (ii) induced e.m.f.s.

Fig. 35.31 Self-induction

When a current flows through a coil, it sets up a magnetic field. And that field threads the coil which produces it, Fig. 35.31 (i). If the current I through the coil is changed—by means of a variable resistance, for example—the flux linked with the turns of the coil changes. An e.m.f. is therefore induced in the coil. By Lenz's law the direction of the induced e.m.f. will be such as to oppose the change of current; the e.m.f. will be against the current if it is increasing, but in the same direction if it is decreasing, Fig. 35.31 (ii).

Back-e.m.f.

When an e.m.f. is induced in a circuit by a change in the current through that circuit, the e.m.f. induced is called a back-e.m.f. Self-induction opposes the growth of current in a coil, and so the current may increase gradually to its final value.

This effect can be demonstrated by the circuit shown in Fig. 35.32 (i). Two parallel arrangements are connected to a battery B and a key K. One consists of an iron-cored coil L with many turns in series with a small lamp A_1. The other

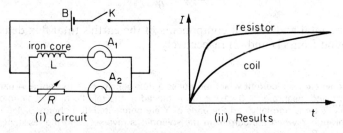

(i) Circuit   (ii) Results

Fig. 35.32 Self-induction experiment

has a variable resistor R in series with a similar lamp A_2. Initially R is adjusted so that the two lamps have the same brightness in their respective circuits with steady current flowing. With the circuit open as shown in Fig. 35.32 (i), K is closed, so that B is now connected. The lamp A_2 with R is seen to become bright almost immediately but the lamp A_1 with L increases slowly to full brightness. The induced or back-e.m.f. in the coil L opposes the growth of current so the glow in the lamp filament in A_1 increases slowly. The resistor R, however, has a negligible back-e.m.f. So its lamp A_2 glows fully bright as soon as K is closed. See Fig. 35.32 (ii).

Just as self-induction opposes the rise of an electric current when it is switched on, so also it opposes the decay of the current when it is switched off. When the circuit is broken, the current starts to fall very rapidly, and a correspondingly great e.m.f. is induced, which tends to maintain the current. This e.m.f. is often great enough to break down the insulation of the air between the switch contacts, and produce a spark. To do so, the e.m.f. must be about 350 volts or more, because air will not break down—not over any gaps, narrow or wide—when the voltage is less than that value. The e.m.f. at break may be much greater than the e.m.f. of the supply which maintained the current: a spark can easily be obtained, for example, by breaking a circuit consisting of an iron-cored coil and an accumulator.

Non-inductive Coils

In some circuits containing coils, self-induction is a nuisance. To minimise their

Fig. 35.33 Non-inductive winding

self-inductance, the coils of resistance boxes are wound so as to set up extremely small magnetic fields; as shown in Fig. 35.33, the wire is doubled-back on itself before being coiled up. Every part of the coil is then travelled by the same current travelling in opposite directions, and so its magnetic field is negligible. Such a coil is said to be *non-inductive*.

Self-inductance

To discuss the effects of self-induction more fully, we must define the property of a coil called its *self-inductance*. By definition,

$$self\text{-}inductance = \frac{back\text{-}e.m.f.\ induced\ in\ coil\ by\ a\ changing\ current}{rate\ of\ change\ of\ current\ through\ coil}$$

Self-inductance is denoted by the symbol L. Numerically, we may therefore write its definition as

$$L = \frac{E_{back}}{dI/dt}$$

or
$$E_{back} = L\frac{dI}{dt} \qquad . \qquad . \qquad . \qquad . \qquad . \qquad (1)$$

Equation (1) is the simplest form in which to remember the definiton.

The unit of self-inductance is the henry (H). It is defined by making each term in equation (1) equal to unity; thus *a coil has a self-inductance of 1 henry if the* back-e.m.f. in it is 1 volt, when the current through it is changing at the rate of 1 ampere per second. Equation (1) then becomes:

$$E_{back} \text{ (volts)} = L \text{ (henrys)} \times \frac{dI}{dt} \text{ (ampere/second)}.$$

The iron-cored coils used for smoothing the rectified supply current to a television receiver are usually very large and have an inductance of about 30 H.

L for Coil

Since the induced e.m.f. $E = d\Phi/dt = LdI/dt$, numerically, it follows by integration from a limit to zero that

$$\Phi = LI.$$

Thus $L = \Phi/I$. Hence the self-inductance may be defined as the *flux linkage per unit current*. When Φ is in weber and I in ampere, then L is in henry. Thus if a current of 2 A produces a flux linkage of 4 Wb in a coil, the inductance $L = 4 \text{ Wb}/2 \text{ A} = 2 \text{ H}$.

Earlier we saw that when a long coil of N turns and length l carries a current I, the flux density B inside the coil with an air core is given by $B = \mu_0 NI/l$, where μ_0 is the permeability of air, $4\pi \times 10^{-7}$ H m^{-1} (p. 709). With an iron core of *relative permeability* μ_r, the flux density is given by $B = \mu_r\mu_0 NI/l$ (p. 710). In this case

$$\text{flux linkage } \Phi = NAB = \frac{\mu_r\mu_0 N^2 AI}{l}$$

$$\therefore L = \frac{\Phi}{I} = \frac{\mu_r\mu_0 N^2 A}{l} \qquad \qquad (1)$$

This formula may be used to find the approximate value of the inductance of a coil. L is in henry when A is in metre^2, l in metre and μ_0 is $4\pi \times 10^{-7}$ henry metre^{-1}.

We can now show that the unit of μ_0 is H m^{-1}. Since $B = \mu_0 NI/l$ for an air-core coil, and the unit of B is weber metre^{-2}, it follows that the unit of μ_0 is

$$\frac{\text{weber meter}^{-2} \times \text{metre}}{\text{ampere}} \quad \text{or} \quad \text{Wb A}^{-1}\text{ m}^{-1} \qquad . \qquad (2)$$

Now the unit of inductance L is the henry (H), which can be defined from the relation $\Phi = LI$. Thus, since $L = \Phi/I$

$$1 \text{ H} = 1 \text{ Wb A}^{-1}.$$

It follows from (2) that the unit of μ_0 is H m^{-1}.

Energy Stored; E.M.F. at Break

When the current in a coil is interrupted by breaking the circuit, a spark passes across the gap and energy is liberated in the form of heat and light. This energy has been stored in the *magnetic field of the coil*, just as the energy of a charged capacitor is stored in the electrostatic field between its plates (p. 408). When the current in the coil is first switched on, the back-e.m.f. opposes the rise of current; the current flows against the back-e.m.f. and therefore does work against it (p. 645). When the current becomes steady, there is no back-e.m.f. and no more work done against it. The total work done in bringing the current to its final value is stored in the magnetic field of the coil. It is liberated when the current collapses; for then the induced e.m.f. tends to maintain the current, and to do external work of some kind.

To calculate the energy stored in a coil, suppose that the current through it is rising at a rate dI/dt ampere per second. Then, if L is its self-inductance in henrys, the back-e.m.f. across it is given numerically by

$$E = L\frac{dI}{dt}.$$

If the value of the current is I at that instant, then the rate at which work is being done against the back-e.m.f. is

$$P = EI = LI\frac{dI}{dt}.$$

The total work done in bringing the current from zero to a steady value I_0 is therefore

$$W = \int P dt = \int_0^{I_0} LI\frac{dI}{dt}dt = \int_0^{I_0} LI.dI$$

$$= \tfrac{1}{2}LI_0{}^2$$

This is the energy stored in the magnetic field of the coil.

To calculate the e.m.f. induced at break is, in general, a complicated business. But we can easily do it for one important practical circuit. To prevent sparking

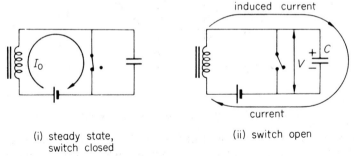

(i) steady state,   (ii) switch open
switch closed

Fig. 35.34 Prevention of sparking by capacitor

at the contacts of a switch in an inductive circuit, such as a relay used in tele-communications, a capacitor is often connected across the switch, Fig. 35.34 (i). When the circuit is broken, the collapsing flux through the coil tends to maintain the current; but now the current can continue to flow for a brief time because it can flow by charging the capacitor, Fig. 35.34 (ii). Consequently the current does not decay as rapidly as it would without the capacitor, and the back-e.m.f. never rises as high. If the capacitance of the capacitor is great enough, the potential difference across it (and therefore across the switch) never rises high enough to cause a spark.

To find the value to which the potential difference does rise, we assume that all the energy originally stored in the magnetic field of the coil is now stored in the electrostatic field of the capacitor.

If C is the capacitance of the capacitor in farad, and V_0 the final value of potential difference across it in volt, then the energy stored in it is $\tfrac{1}{2}CV_0{}^2$ joule (p. 609). Equating this to the original value of the energy stored in the coil, we have

$$\tfrac{1}{2}CV_0{}^2 = \tfrac{1}{2}LI_0{}^2$$

Let us suppose that a current of 1 ampere is to be broken, without sparking, in a circuit of self-inductance 1 henry and to prevent sparking, the potential difference across the capacitor must not rise above 350 volt. The least capacitance that must be connected across the switch is therefore given by

$$\tfrac{1}{2}C \times 350^2 = \tfrac{1}{2} \times 1 \times 1^2.$$

So
$$C = \frac{1}{350^2} = 8 \times 10^{-6}\ \text{F} = 8\ \mu\text{F}.$$

A capacitor of capacitance $8\ \mu\text{F}$, and able to withstand 350 volts, would therefore be required.

Current in *L* and *R* Series Circuit

Consider a coil of inductance $L = 2$ H and resistance $R = 5\ \Omega$ connected to a 10 V battery of negligible internal resistance, with a switch S in the circuit, Fig. 35.35 (i).

When the switch is closed so that current flows, part of the 10 V is needed to maintain the current in R and the rest of the p.d. is needed to maintain the growth of the current against the back e.m.f. E_b due to the inductance L.

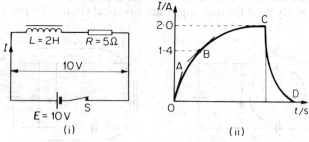

Fig. 35.35 Current variation in *L, R* series circuit

1. At the instant the switch is closed (time $t = 0$), there is no current in the circuit. So there is no p.d. across R, from $V = IR$. Hence the whole of the $10\ V = E_b$, the back e.m.f. So

$$E_b = 10 = L\frac{dI}{dt} = 2\frac{dI}{dt}$$

Hence
$$\frac{dI}{dt} = \frac{10}{2} = 5\ \text{A s}^{-1}$$

In Fig. 35.35 (ii), OA represents the rate of change of current with time at $t = 0$ and the line has a gradient of $5\ \text{A s}^{-1}$.

2. Suppose the current I rises to a value 1·4 A, which is represented by B in Fig. 35.35 (ii). The p.d. across R is then given by

$$V = IR = 1\cdot 4 \times 5 = 7\ \text{V}$$

So
$$\text{back e.m.f. } E_b = 10 - 7 = 3\ \text{V}$$

Hence
$$L\frac{dI}{dt} = 3 \quad \text{and} \quad \frac{dI}{dt} = \frac{3}{2} = 1\cdot 5\ \text{A s}^{-1}$$

So the current rise, shown by the gradient at B, decreases as time goes on.

3. When the current is finally established and is constant, there is no flux change in the coil and hence no back e.m.f. In this case the whole of the 10 V maintains the current in R. So if I_0 is the final or steady current, $V = 10 = IR$. Hence

$$I_0 = \frac{V}{R} = \frac{10}{5} = 2 \text{ A}$$

This is the current value corresponding to C in Fig. 35.35 (ii). The graph shows how I increases to its final value after a time t. It is an exponential graph (see below).

Generally, we see that $E = V + L \, dI/dt = IR + L \, dI/dt$. When we solve this differential equation (see *Scholarship Physics* by the author), the result is

$$I = I_0(1 - e^{-Rt/L})$$

When the circuit is broken, the flux in the coil decreases rapidly and the current falls quickly along CD in Fig. 35.35 (ii). A spark may then be obtained across the switch as previously explained.

Mutual Induction

We have already seen that an e.m.f. may be induced in one circuit by a changing current in another (Fig. 35.1, p. 726). The phenomenon is often called *mutual induction*, and the pair of circuits which show it are said to have mutual inductance. The *mutual inductance*, M, between two circuits is defined by the equation:

$$\left.\begin{array}{l}\text{e.m.f induced in B, by} \\ \text{changing current in A}\end{array}\right\} = M \times \left\{\begin{array}{l}\text{rate of change of} \\ \text{current in A.}\end{array}\right.$$

Exercises 35

1. A bar magnet M, with its S pole at the bottom, is dropped vertically through a horizontal flat coil C, Fig. 35A (i). Draw a sketch showing the direction of the induced current (*a*) just before M passes through C, (*b*) just after M has passed completely through C.

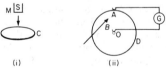

(i)   (ii)

Fig. 35A

2. Figure 35A (ii) shows a vertical copper disc D rotating clockwise in a uniform horizontal magnetic field B directed normally towards D. A galvanometer G is connected to contacts at O and A. The radius OA can be considered as a straight conductor moving in the field. Copy the diagram and show in your sketch the direction of the induced current flowing through G. Has A or O the higher potential?

3. A horizontal rod PQ of length 1·5 m is perpendicular to a uniform horizontal field B of 0·1 T, Fig. 35B (i). Calculate the induced e.m.f., if any, in PQ when the rod is moved

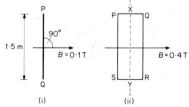

(i)   (ii)

Fig. 35B

through the field with a uniform velocity of 4 m s^{-1} (a) in the direction of B, (b) perpendicular to B and upwards. Which end of PQ has the higher potential?

4. Figure 35B (ii) shows a vertical rectangular coil PQRS with its plane parallel to a uniform horizontal magnetic field B of 0·4 T. The coil has 5 turns, PS is 10 cm long and SR is 5 cm long. Calculate the average induced e.m.f. in the coil, if any, (a) when it is moved sideways in the direction of B with a velocity of 2 m s^{-1}, (b) when it is rotated through 90° about the vertical axis XY in 0·1 s.

If the resistance of the coil PQRS is 10 Ω and its terminals are connected, what charge circulates in the coil in case (b)?

5. In Fig. 35B (ii) the coil PQRS is rotated about the axis XY at 50 rev s^{-1}. Calculate the maximum e.m.f. induced in the coil.

What is the instantaneous e.m.f. in the coil when its plane is (i) parallel to the direction of B, (ii) 60° to B, (iii) 90° to B?

6. Explain, using the case of a N pole of a magnet approaching a coil, why Lenz's law is a consequence of the law of the conservation of energy.

How is Lenz's law applied to explain why an induced e.m.f. is obtained in a straight conductor cutting flux in a magnetic field?

7. What difference is seen when a 12 V battery is connected (a) to a 12 V − 24 W filament lamp, (b) to a circuit containing the same lamp in series with a high inductance coil of low resistance? Explain your answer.

8. Figure 35C (i) shows a horizontal metal rod XY moving with a uniform velocity v perpendicularly to a uniform magnetic field B acting into the paper. By considering the force on electron, charge −e, explain why an induced e.m.f. is obtained along XY.

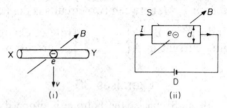

Fig. 35c

9. Figure 35C (ii) shows an n-type semiconductor S, with a battery D connected so that a current I flows through it. When a strong magnetic field B is applied perpendicularly to the plane of S, explain why an e.m.f. (Hall voltage) E is obtained.

Show that (a) $E = Bvd$, where v is the drift velocity of the n-charges, (b) $E = BI/net$, where n is the number of n-charges per metre^3 and t is the thickness of S.

10. A circular metal disc of area $3·0 \times 10^{-3}$ m^2 is rotated at 50 rev/s about an axle through its centre perpendicular to its plane. The disc is in a uniform magnetic field of flux density $5·0 \times 10^{-3}$ T in the direction of the axle. Between which points on the disc is the maximum e.m.f. induced? What is the value of this e.m.f.? How does this e.m.f. vary with time?

The disc is now replaced in the same position by a flat coil of wire and this is rotated at constant angular velocity about a diameter. Sketch a graph showing how the e.m.f. induced between the ends of the wire varies with time. (L.)

11. State the *laws of electromagnetic induction*. Show how Lenz's law is consistent with the principle of conservation of energy.

Draw four arrows, labelled A, B, C and D, showing the directions of the currents induced

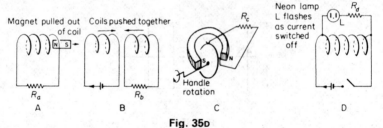

Fig. 35D

in the resistors in the experiments illustrated. Explain how the e.m.f. arises in cases C and D, Fig. 35D.

A copper disc of area A rotates at frequency f at the centre of a long solenoid of turns per unit length n and carrying a current I. The plane of the disc is normal to the flux. The rotation rate is adjusted so that the e.m.f. generated between the centre of the copper disc and its rim is 1% of the potential difference across the ends of the solenoid. Deduce an expression for the e.m.f. generated between the centre of the copper disc and its rim. Hence find the resistance of the solenoid in terms of μ_0, A, f and n. (C.)

12. (a) Show, by considering the force on an electron, that a potential difference will be established between the ends of a metal rod which is moving in a direction at right angles to a magnetic field. Draw a diagram in which the direction of motion of the rod is shown, the direction of the magnetic field is stated and the polarity of the ends of the conductor is shown.

(b) How would you show that an induced current in a conductor moving in a magnetic field is such a direction as to oppose the motion of the conductor? Explain why this follows from conservation of energy.

(c) The primary of a transformer is connected to a constant voltage a.c. supply and the secondary is on open circuit. Discuss the factors which determine the current flowing in the primary winding. (L.)

13. (a) A wire of length l is horizontal and oriented North–South. It moves East with velocity v through the earth's magnetic field which has a downward vertical component of flux density B. Write down an expression for the potential difference between the two ends of the wire. Which end of the wire is at the more positive potential?

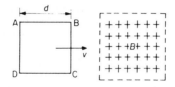

Fig. 35E

(b) A horizontal square frame ABCD, of side d, moves with velocity v parallel to sides AB, DC from a field-free region into a region of uniform magnetic field of flux density B, Fig. 35E. The boundaries of the field are parallel to the sides BC, AD of the frame and the field is directed vertically downward. Write down expressions for the electromotive force induced in the frame (i) when side BC has entered the field but side AD has not, (ii) when the frame is entirely within the field region, (iii) when side BC has left the field but side AD has not.

For each position derive an expression for the magnitude and direction of the current in the frame and the resultant force acting on the frame due to the current. The total resistance of the wire frame is R, and its self-inductance may be neglected. (O. &. C.)

14. A flat search coil containing 50 turns each of area $2 \cdot 0 \times 10^{-4}$ m^2 is connected to a galvanometer; the total resistance of the circuit is 100 Ω. The coil is placed so that its plane is normal to a magnetic field of flux density $0 \cdot 25$ T.

(a) What is the change in magnetic flux linking the circuit when the coil is moved to a region of negligible magnetic field? (b) What charge passes through the galvanometer? (C.)

15. (a) A long vertical straight wire carries a steady electric current supplied by a battery of cells. Sketch the magnetic flux pattern in a horizontal plane around the wire (i) when there is no other magnetic flux, (ii) when there is also a uniform horizontal magnetic flux.

In case (ii) indicate the direction of any force experienced by the wire. If the wire were allowed to move in the direction of this force would the current in the wire change? Explain your answer.

(b) An expression for the magnetic flux density at a distance r from a straight wire carrying a current I is $\mu_0 I / 2\pi r$, where $\mu_0 = 4\pi \times 10^{-7}$ H m^{-1}. In the situation (ii) above it is found that when the horizontal flux density is 2.0×10^{-5} T there is no resultant flux density at a point 15 cm from the wire. What is the value of the current in the wire? What would be the new value of the uniform horizontal flux density if the point of no resultant flux density were to be removed to 5·0 cm from the wire?

(c) The rectangular coil in Fig. 35F is in a vertical plane and is connected to the Y plates of an oscilloscope as shown. The coil is moved in a uniform horizontal magnetic flux (i) by taking it sideways in a vertical plane across the flux, without rotation, and (ii) by rotating it about the vertical axis AB at 10 rev s^{-1}. In each case describe and explain the appearance of the oscilloscope trace if a time base frequency of 5 Hz were applied to the X plates.

What would happen to the trace if the frequency of rotation of the coil were doubled to 20 rev s^{-1}? (L.)

Fig. 35F

16. (a) What is meant by the statement that a solenoid has an inductance of 2 H?

A 2·0 H solenoid is connected in series with a resistor, so that the total resistance is 0·50 Ω, to a 2·0 V d.c. supply. Sketch the graph of current against time when the current is switched on. What is (i) the final current, (ii) the initial rate of change of current with time, (iii) the rate of change of current with time when the current is 2·0 A?

Explain why an e.m.f. greatly in excess of 2·0 V will be produced when the current is switched off.

(b) A long air-cored solenoid has 1000 turns of wire per metre and a cross-sectional area of 8·0 cm². A secondary coil, of 2000 turns, is wound around its centre, and connected to a ballistic galvanometer, the total resistance of coil and galvanometer being 60 Ω. The sensitivity of the galvanometer is 2·0 divisions per microcoulomb. If a current of 4·0 A in the primary solenoid were switched off, what would be the deflection of the galvanometer? (Permeability of free space = $4\pi \times 10^{-7}$ H m^{-1}.) (L.)

17. Explain what is meant by the mutual inductance of two coils. If you were provided with a calibrated cathode ray oscilloscope (or a high resistance millivoltmeter) and a means of producing a steadily increasing current, how would you measure the mutual inductance of two coils? (Assume that normal laboratory equipment is also available.)

Five turns of wire are wound closely about the centre of a long solenoid of radius 20 mm. If there are 500 turns per metre in the solenoid, calculate the mututal inductance of the two coils. Show your reasoning. (The permeability of free space is $4\pi \times 10^{-7}$ H m^{-1}.) (L.)

18. A straight solenoid of length L and circular cross-section of radius r is uniformly wound with a single layer of N turns and carries a current I. The solenoid is so long that end-effects may be neglected. Write down expressions for the flux density B inside the solenoid and the total flux through the solenoid.

(a) The current in the solenoid is increased at a uniform rate from zero to I in a time t. Find an expression for the back-e.m.f. induced in the solenoid during the change.
(b) Find the electrical work done against this back-e.m.f. in establishing the current.
(c) The work found in (b) is the energy stored in the magnetic field of the solenoid. Show that this energy is $B^2/2\mu_0$ per unit volume. (d) What becomes of this energy when the circuit is broken? (O. & C.)

19. Define *electromotive force* and state the *laws of electromagnetic induction*. Using the definition and the laws, derive an expression for the e.m.f. induced in a conductor moving in a magnetic field.

When a wheel with metal spokes 1·2 m long is rotated in a magnetic field of flux density 5×10^{-5} T normal to the plane of the wheel, an e.m.f. of 10^{-2} V is induced between the rim and the axle. Find the rate of rotation of the wheel. (L.)

20. Explain the effect of an iron-cored solenoid upon the rate of growth of current in a d.c. circuit. (*L.*)

21. State the laws relating to the electromotive force induced in a conductor which is moving in a magnetic field.

Describe the mode of action of a simple dynamo.

Find in volts the e.m.f. induced in a straight conductor of length 20 cm, on the armature of a dynamo and 10 cm from the axis when the conductor is moving in a uniform radial field of 0·5 T and the armature is rotating at 1000 r.p.m. (*L.*)

22. State Lenz's law of electromagnetic induction and describe, with explanation, an experiment which illustrates its truth.

Describe the structure of a transformer suitable for supplying 12 V from 240-V mains and explain its action. Indicate the energy losses which occur in the transformer and explain how they are reduced to a minimum.

When the primary of a transformer is connected to the a.c. mains the current in it (*a*) is very small if the secondary circuit is open, but (*b*) increases when the secondary circuit is closed. Explain these facts. (*L.*)

23. Describe and account for two constructional differences between a moving-coil galvanometer used to measure current and the ballistic form of the instrument.

An electromagnet has plane-parallel pole faces. Give details of an experiment, using a search coil and ballistic galvanometer of known sensitivity, to determine the variation in the magnitude of the magnetic flux density along a line parallel to the pole faces and mid-way between them. Indicate in qualitative terms the variation you would expect to get.

A coil of 100 turns each of area $2·0 \times 10^{-3}$ m^2 has a resistance of 12 Ω. It lies in a horizontal plane in a vertical magnetic flux density of $3·0 \times 10^{-3}$ Wb m^{-2}. What charge circulates through the coil if its ends are short-circuited and the coil is rotated through 180° about a diametral axis? (*N.*)

24. State the laws of electromagnetic induction and describe experiments you would perform to illustrate the factors which determine the magnitude of the induced current set up in a closed circuit.

A simple electric motor has an armature of 0·1 Ω resistance. When the motor is running on a 50 V supply the current is found to be 5 A. Explain this and show what bearing it has on the method of starting large motors. (*L.*)

25. State what is meant by (*a*) self induction, and (*b*) mutual induction. Describe one experiment in each case to illustrate these effects.

In the circuit shown (Fig. 35G) A and B have equal ohmic resistance but A is of negligible self inductance whilst B has a high self inductance.

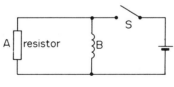

Fig. 35G

Describe and explain how the currents through A and B change with time (i) when the switch S is closed, and (ii) when it is opened. Illustrate your answers graphically.

Describe briefly *two* applications of self inductors. (*L.*)

26. Describe the phenomena of self induction and mutual induction.

Describe the construction and explain the action of a simple form of a.c. transformer.

In an a.c. transformer in which the primary and secondary windings are perfectly coupled and in which a negligible primary current flows when there is no load in the secondary, a current of 5 A (r.m.s.) was observed to flow in the primary under an applied voltage of 100 V (r.m.s.) when the secondary was connected to resistors only. If the primary contains 100 turns and the secondary 25 000 turns, calculate (*a*) the voltage, (*b*) the current in the secondary, stating any simplifying assumptions you make. (*O. & C.*)

27. A choke of large self inductance and small resistance, a battery and a switch are connected in series. Sketch and explain a graph illustrating how the current varies with time after the switch is closed. If the self inductance and resistance of the coil are 10 H and 5 Ω respectively and the battery has an e.m.f. of 20 V and negligible resistance, what are the greatest values after the switch is closed of (a) the current, (b) the rate of change of current? (N.)

28. State Lenz's law and describe how you would demonstrate it using a solenoid with two separate superimposed windings with clearly visible turns, a cell with marked polarity, and a centre-zero galvanometer. Illustrate your answer with diagrams.

A metal aircraft with a wing span of 40 m flies with a ground speed of 1000 km h^{-1} in a direction due east at constant altitude in a region of the northern hemisphere where the horizontal component of the earth's magnetic field is $1\cdot6 \times 10^{-5}$ T and the angle of dip is $71\cdot6°$. Find the potential difference in volts that exists between the wing tips and state, with reasons, which tip is at the higher potential. (N.)

29. Define *magnetic flux* (Φ) and *magnetic flux density* (B). What do you understand by the term flux linkage?

A 'long' solenoid of effective cross-sectional area 8×10^{-4} m^2 and length $0\cdot5$ m has 2000 turns wound uniformly in several layers on a hollow plastic tube. Closely wound over the middle region of this solenoid is a short secondary coil of 600 turns. Calculate: (a) the flux density within the solenoid when it carries a steady current of 5 A; (b) the flux linked with the solenoid when it carries a steady current of 5 A; (c) the self inductance of the solenoid: (d) the mutual inductance between the solenoid and the secondary coil; (e) a lower limit to the value of the self inductance of the secondary coil; (f) the magnitude of the e.m.f. induced in the secondary coil at the instant when primary current is increasing at the rate of 12 A s^{-1}. ($\mu_0 = 4\pi \times 10^{-7}$ H m^{-1}.) (O.)

30. A circuit contains an iron-cored inductor, a switch and a d.c. source arranged in series. The switch is closed and, after an interval, reopened. Explain why a spark jumps across the switch contacts.

In order to prevent sparking a capacitor is placed in parallel with the switch. The energy stored in the inductor at the instant when the circuit is broken is $2\cdot00$ J and to prevent sparking the voltage across the contacts must not exceed 400 V. Assuming there are no energy losses due to resistance in the circuit, calculate the minimum capacitance required. (AEB, 1982).

36 Magnetic Properties of Materials

The magnetic properties of materials require investigation to decide whether they are suitable for permanent magnets such as loudspeaker magnets, for temporary magnets such as electromagnets, or for cores of electromagnetic induction apparatus such as transformers.

Flux Density in Magnetic Material

Consider a toroid of length L, wound with N turns each carrying a current I round a ring of magnetic material, Fig. 36.1.

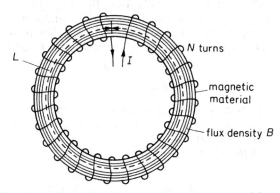

Fig. 36.1 Flux density (induction) in magnetic material

The *total flux density* B in the material is partly due to the currents flowing in the wire and partly due to the magnetisation of the material. We thus write

$$B = B_0 + B_M \qquad . \qquad . \qquad . \qquad . \qquad (1)$$

where B_0 is the flux density due directly to the current in the wire, and B_M is the flux density due to the magnetisation of the material.

We now assume that the flux density B_M is produced by many small circulating currents inside the magnetic material, due to the circulating and spinning electrons in the atoms. Figure 36.2 shows that the effect of many small adjacent current loops may be thought of as one current loop. In the same way, the internal circulating and spinning currents can be replaced by a single current I_M flowing in the coil wound round the core. This theoretical surface or magnetisation current is additional to the real or actual current I flowing in the coil.

By itself, the real current I produces a flux-density B_0. If the toroid has n turns per unit length ($n = N/L$), then, from p. 709,

$$B_0 = \mu_0 nI.$$

The surface current I_M may be imagined to flow in n turns per metre of the solenoid, as the real current does. Since a current I_M in the coil would produce a flux density B_M equal to that due to the magnetisation in the material, the flux density B_M is given by

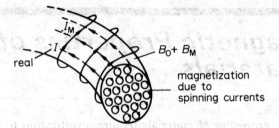

Fig. 36.2 Magnetising surface current

$$B_M = \mu_0 n I_M.$$

$$\therefore \text{ total } B = B_0 + B_M = \mu_0 n(I + I_M) \quad . \quad (2)$$

Intensity of Magnetisation

It is possible to write B_M, the flux density due to the magnetisation of the material, in a different way. By definition, the *magnetic moment* of each turn due to this imaginary surface current $= A \times I_M$, where A is the area of each turn. The magnetic moment of the whole toroid is then $nLAI_M$, since nL is the total number of turns. Hence the *magnetic moment per unit volume*

$$= \frac{nLAI_M}{\text{volume}} = \frac{nLAI_M}{LA} = nI_M.$$

The 'magnetic moment per unit volume' is called the *intensity of magnetisation*, M, of the magnetic core in the toroid. Thus $B_M = \mu_0 n I_M = \mu_0 M$. Further, nI is the magnetic field intensity, H, in the toroid due to the current I.

Thus the *total flux-density B* in the core when the coil carries a current I is given by

$$B = B_0 + B_M = \mu_0 n I + \mu_0 n I_M$$

$$\therefore B = \mu_0(H + M) \quad . \quad . \quad . \quad . \quad (3)$$

Equation (3) relates the total B to the magnetising field intensity H of the current and to the density of magnetisation M of the material produced. Note that H and M have the same units from (3). *Thus the unit of M is ampere per metre* $(A\ m^{-1})$.

Relative Permeability

The *permeability* of a material is defined by the relation $\mu = B/H$. Here B is the flux-density inside a magnetic material and H is the magnetising field intensity nI due to a current flowing with n turns per metre wound round the material. If the magnetic material is removed leaving a vacuum inside the coil, the flux density inside the coil is now reduced to a value B_0 say. So the permeability μ_0 of the vacuum $= B_0/H$.

The ratio μ/μ_0 is called the *relative permeability*, μ_r, of the material. The permeability μ of the material (B/H) has dimensions; its unit is henry per metre $(H\ m^{-1})$. The relative permeability μ_r, however, is a number and has no dimensions. Since

$$\mu = \mu_r \mu_0,$$

we see that if μ_r for a particular magnetic material is 3000, then its permeability μ is given by

$$\mu = 3000 \times 4\pi \times 10^{-7} \text{ H m}^{-1} = 3{\cdot}8 \times 10^{-3} \text{ H m}^{-1}.$$

In general, $B = \mu_0(H + M)$. Since $B = \mu H$,

$$\therefore \mu H = \mu_0(H + M)$$

or

$$\frac{\mu}{\mu_0} = 1 + \frac{M}{H}.$$

The ratio M/H is called the *susceptibility*, χ, of the material. Hence, from above,

$$\mu_r = 1 + \chi. \qquad \qquad \qquad \qquad (4)$$

Materials are often subjected to varying magnetic fields when they are used in electromagnetic machinery. As a magnetic material, originally unmagnetised, is subjected to an increasing field, the intensity of magnetisation M increases until it reaches a maximum value (Fig. 36.3 (i)). The material is then 'saturated'; that is, its magnetic 'domains' are completely aligned with the field H. B, however, continues to increase with H, since $B = \mu_0(H + M)$. Figure 36.3 (ii) also shows how the relative permeability μ_r and the susceptibility χ varies with H. It increases

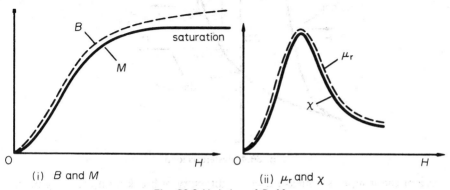

(i) *B* and *M* (ii) μ_r and χ

Fig. 36.3 Variation of B, M, μ_r, χ

at first, and passes through a maximum value. As the material approaches saturation the domains cannot yield much further, and the susceptibility falls to a low value.

Ferromagnetic materials are those which have a high relative permeability and susceptibility. This is generally very much greater than 1, for example, 3000. Some materials are capable of retaining their magnetisation and forming strong permanent magnets. Others form temporary magnets.

Variation of *B* with *H*—Hysteresis. Remanence. Coercive Force

As we saw in electromagnetic induction, many electrical appliances such as transformers have coils wound round magnetic materials similar to soft iron. The coils carry alternating current. Consequently the material goes through many *magnetic cycles* per second. For this reason we need to study how the magnetic flux density B in the material varies with the magnetising field H in a magnetic cycle.

Figure 36.4 shows the variation of B with applied field H when the specimen is taken through a complete cycle B of H. After the specimen has become saturated, and the field is reduced to zero, the iron is still quite strongly magnetised, setting up a flux-density B_r. This flux-density is called the *remanence*; it is due to the tendency of groups of molecules, or domains, to stay put once they have been aligned (see p. 770).

When the field is reversed, the residual magnetism is opposed. Each increase of magnetising field now causes a decrease of flux-density, as the domains are twisted farther out of alignment. Eventually, the flux-density is reduced to zero, when the opposing field H has the value H_c. This value of H is called the *coercive*

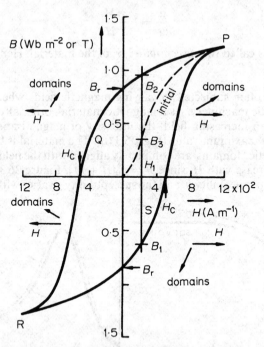

Fig. 36.4 Hysteresis loop

force of the iron; it is a measure of the difficulty of breaking up the alignment of the domains.

We now see that, when once the iron has been magnetised, its magnetisation curve never passes through the origin again. Instead, it forms the closed loop PQRS, which is called a *hysteresis loop*. Hysteresis, which comes from a Greek word meaning 'delayed', can be defined as *the lagging of B behind the magnetising field, H*, when the specimen is taken through a magnetic cycle.

Properties of Magnetic Materials

Figure 36.5 shows the hysteresis loops of iron and steel. Steel is more suitable for permanent magnets, because its high coercivity means that it is not easily demagnetised by shaking. The fact that the remanence of iron is a little greater than that of steel is completely outweighed by its much smaller coercivity, which makes it very easy to demagnetise. On the other hand, iron is much more suitable for electromagnets, which have to be switched on and off, as in relays. Iron is also more suitable for the cores of transformers and the armatures of machines. Both of these go through complete magnetising cycles continuously: transformer cores because they are magnetised by alternating current, armatures because they are turning round and round in a constant field. In each cycle the iron passes through two parts of its hysteresis loop (near Q and S in Fig. 36.4), where the magnetising field is having to demagnetise the iron. There the field is doing work against the internal friction of the domains. This work, like all work that is done against friction, is dissipated as heat. The energy dissipated

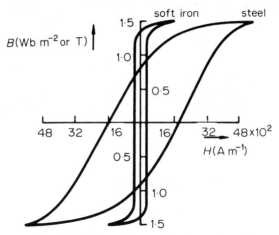

Fig. 36.5 Hysteresis curves for iron and steel

in this way, per cycle, is less for iron than for steel, because iron is easier to demagnetise. It is called the *hysteresis loss* and is proportional to the *area of the B-H loop*.

In a large transformer the hysteresis loss, together with the heat developed by the current in the resistance of the windings, liberates so much heat that the transformer must be artificially cooled. The cooling is done by circulating oil, which itself is cooled by the atmosphere: it passes through pipes which can be seen outside the transformer, running from top to bottom.

Subsidiary Hysteresis Loops

When a piece of iron is magnetised, first one way and then the other, it goes round a hysteresis loop even if it is not magnetised to saturation at any point, Fig. 36.6. The subsidiary loops, *ab* for example, may represent the magnetisation of a transformer core by an alternating current in the primary winding:

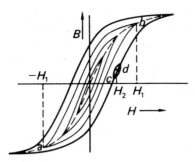

Fig. 36.6 Subsidiary hysteresis loops

the amplitude H_1 of the magnetising field is proportional to the amplitude of the current. A transformer core is designed so that it is never saturated under working conditions. For, if it were saturated, the flux through it would not follow the changes in primary current; and the e.m.f. induced in the secondary would be less than it should.

The energy dissipated as heat in going round a subsidiary hysteresis loop is proportional to the area of the loop, just as in going round the main one.

Another kind of subsidiary loop is shown at *cd* in the figure. The iron goes round such a loop when the field is varied above and below the value H_2. This happens in a transformer when the primary carries a fluctuating direct current.

Demagnetisation

The only satisfactory way to demagnetise a piece of iron or steel is to carry it round a series of hysteresis loops, shrinking gradually to the origin. If the iron is the core of a toroid, we can do this by connecting the winding to an a.c. supply via a potential divider, as in Fig. 36.7, and reducing the current to zero. Since the iron goes through fifty loops per second, we do not have to reduce the current very slowly.

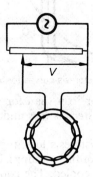

Fig. 36.7 Demagnetisation

To demagnetise a loose piece of iron or steel, such as a watch, we merely put it into, and take it out of, a coil of many turns connected to an a.c. supply. As we draw the watch out, it moves into an ever-weakening field, and is demagnetised.

Ferromagnetism. Magnetic Domain Theory

As we have already mentioned, atoms contain circulating and spinning electrons. Each electron possesses a resultant magnetic moment on account of its orbital motion and its spin. In a ferromagnetic material strong 'interactions' are present between the moments, the nature of which requires quantum theory to understand it and is outside the scope of this work. These cause neighbouring moments to align, even in the absence of an applied field, with the result that tiny regions of very strong magnetism are obtained inside the unmagnetised material called *magnetic domains*.

A crystal of ferromagnetic material is shown in Fig. 36.8 (i). If all the domains were aligned completely, the material would behave like one enormous domain and the energy in the magnetic field outside is then considerable, as represented by the flux shown. Now all physical systems settle in equilibrium when their energy is a minimum. Figure 36.8 (ii) is therefore more stable than Fig. 36.8 (i) because the external magnetic field energy is less. Thus the domains grow in the material, as shown in Fig. 36.8 (iii) and (iv). The region between two domains, where the magnetisation changes direction, is called a *domain wall* and also contains energy. When the formation of a new domain wall requires more energy than is gained by the reduction in the external magnetic field, no more domains are formed. Thus there is a limit to the number of domains formed. This occurs when the volume of the domains is of the order 10^{-4} cm^3 or less.

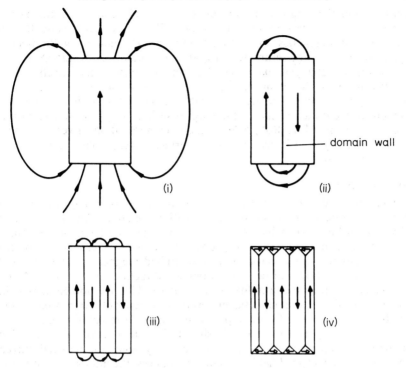

Fig. 36.8 Magnetic domains

Domains and Magnetisation

Some of the phenomena in magnetisation of ferromagnetic material can now be explained. In an unmagnetised specimen, the domains point in different directions. The net magnetisation is then zero. If a small magnetising field H is applied, there is some small rotation of the magnetisation within the domains, which produces an overall component of magnetisation in the direction of H. This occurs in the region AB of the magnetisation-field (M-H) curve shown in Fig. 36.9.

If the field H is removed, the domain magnetisation returns to its original

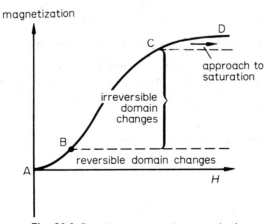

Fig. 36.9 Domain movement in magnetisation

direction. Thus the magnetisation returns to zero. The changes in the part AB of the curve are hence reversible. If the field H is increased beyond B in the region BC, the magnetisation becomes greater. On removal of the field the magnetisation does not return to zero, and so remanence occurs. Along BC, then, irreversible changes take place; the domains grow in the direction of the field, by movement of domain walls, at the expense of those whose magnetisation is in the opposite direction. At very high applied fields H there is complete alignment of the domains and so the magnetisation M approaches 'saturation' along CD. (See also Fig. 36.4.) A ferromagnetic material has a very high value of susceptibility, χ, and hence of relative permeability, μ_r. The value of μ_r can be several thousands.

Diamagnetism

If a magnetic field is produced in the neighbourhood of a magnetic material, a changing flux occurs in the current loops within the atoms. An e.m.f. or electric field will then be set up which causes the electrons to alter their motions, so that an extra or induced current is produced. By Lenz's law, this current gives rise to a magnetic field which *opposes* the applied magnetic field H. Thus the induced magnetisation will be in the opposite direction to H, that is, M/H is negative. Hence the susceptibility χ is negative. This phenomenon is called *diamagnetism.* For a diamagnetic material, χ is generally very small, about -0.000015 for bismuth, for example. The relative permeability, μ_r, which is given by $\mu_r = 1 + \chi$, is thus generally slightly less than 1. All substances have a diamagnetic contribution to their susceptibility, since the induced currents always oppose the applied field. In many substances, the diamagnetism is completely masked by another magnetic phenomenon (p. 771).

If a rod of diamagnetic material is placed in a non-uniform magnetic field,

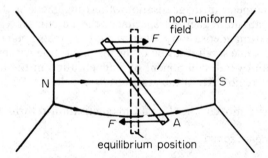

Fig. 36.10 Rod of diamagnetic material in strong field

it will settle at right angles to the field. Figure 36.10 shows the specimen slightly displaced from this position. The magnetisation will oppose the applied field so that the end A will now effectively be a weak S pole. It will then experience a force as shown by the arrow, so that a restoring couple turns the specimen back to its position at right angles to the field.

It should be noted that diamagnetism is a natural 'reaction' to an applied magnetic field and that it is independent of temperature.

Paramagnetism

In contrast to bismuth, a rod of a material such as platinum will settle along the same direction as the applied magnetic field. Further, the induced magnetism will be in the same direction as the field, Fig. 36.11. Platinum is an example of a *paramagnetic* material. The susceptibility, χ, of a paramagnetic substance is very small and positive, $+0.0001$ for example, so that its relative permeability μ_r is very slightly greater than 1 from $\mu_r = 1 + \chi$.

Fig. 36.11 Rod of paramagnetic material in strong field

In a diamagnetic atom, the resultant magnetic moment is zero. In a paramagnetic atom, however, there is a resultant magnetic moment. Generally, the thermal motions of the atoms will cause these magnetic moments to be oriented purely at random and there will be no resultant magnetisation. If, however, a field is applied, each atomic moment will try to set in the direction of the field but the thermal motions will prevent complete alignment. In this case there will be overall weak magnetisation in the direction of the applied field. This accounts for the phenomenon of paramagnetisation.

It is clear that paramagnetism is temperature dependent. At low temperatures, the thermal motions will be less successful at preventing the alignment of the atomic moments and so the susceptibility will be larger. At higher temperatures thermal motion will make alignment difficult. At very high temperatures, the material may become diamagnetic, for the diamagnetic contribution to χ is not affected by temperature whilst the paramagnetic contribution falls.

Above a critical temperature called the *Curie point*, ferromagnetics become paramagnetics.

Exercises 36

1. Give an account of the domain theory of magnetisation.

How does the theory explain the processes of magnetisation and demagnetisation?

What are the desirable magnetic properties for the material of (*a*) the core of an electromagnet, and (*b*) a permanent magnet?

What is meant by magnetic hysteresis? Sketch a typical hysteresis curve and explain what can be deduced from this about the magnetic properties of the material. (*L.*)

2. Define *intensity of magnetisation, flux density*.

Indicate graphically the relationship between the flux density and the strength of the magnetising field for a specimen of (*a*) soft iron, (*b*) steel, taken through a complete cycle of magnetisation.

A soft iron ring has a mean diameter of 0.20 m and an area of cross-section of 5.0×10^{-4} m². It is uniformly wound with 2000 turns carrying a current of 2.0 A and the magnetic flux in the iron is 8.0×10^{-3} Wb. What is the relative permeability of the iron? (*L.*)

3. Define *intensity of magnetisation, susceptibility* and *permeability*. Two substances, A and B, have relative permeabilities slightly greater and slightly less than unity respectively. What does this signify about their magnetic properties? To what group of magnetic substances do A and B each belong?

A soft iron ring of cross-sectional diameter 8 cm and mean circumference 200 cm has 400 turns of wire wound uniformly on it. Calculate the current necessary to produce magnetic flux to the value of 5×10^{-4} Wb if the relative permeability of the iron in the condition stated is 1800. Why is it not possible to say from this information what the flux would be if the current were reduced to $1/10$ of its calculated value? (*L.*)

4. With the aid of a sketch graph, explain the meanings of the following terms used in connection with magnetic materials: (i) remanence, (ii) coercivity, (iii) permeability, (iv) hysteresis. (*AEB*, 1982)

37 A.C. Circuits.

Measurement of A.C.

If an alternating current (a.c.) is passed through a moving-coil meter, the pointer does not move. The coil is urged clockwise and anticlockwise at the frequency of the current—50 times per second if it is drawn from the British grid—and does not move at all. In a sensitive instrument the pointer may be seen to vibrate with a small amplitude.

The relation between current I and pointer deflection θ in a moving-coil meter is $I \propto \theta$. This is unsuitable for measuring alternating current as the deflection reverses on the negative half of the cycle. Instruments for measuring alternating currents must be so made that the pointer deflects the same way when the current flows through the instrument in either direction. As we shall see, a suitable law of deflection is $\theta \propto I^2$, a square-law deflection.

Moving-iron Instrument

A fairly common type of such instrument, called the *moving-iron meter*, is shown in Fig. 37.1. It consists of two iron rods XY, PQ, surrounded by a coil

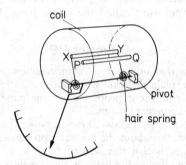

Fig. 37.1 Principle of moving-iron repulsion meter

which carries the current. The coil is fixed to the framework of the meter, and so is one of the rods PQ. The other rod is attached to an axle, which also carries the pointer; its motion is controlled by hair-springs.

When a current flows through the coil, it magnetises the rods, so their adjacent ends have the same polarity. The polarity of each pair of ends reverses with the current, but whichever direction the current has, the iron rods repel each other. The force on the pivoted rod is therefore always in the same direction, and the pointer is deflected through an angle which is proportional to the average force. To a fair approximation, the magnetisation of the rods at any instant is proportional to the current at that instant; the force between the rods is therefore roughly proportional to the *square* of the current. The deflection of the pointer is therefore roughly proportional to the average value of the square of the current.

Hot-wire Instrument

Another type of 'square law' instrument is the hot-wire ammeter, Fig. 37.2. In it the current flows through a fine resistance-wire XY, which it heats. The wire

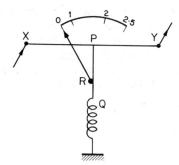

Fig. 37.2 Hot wire meter

warms up to such a temperature that it loses heat—mainly by convection—at a rate equal to the average rate at which heat is developed in the wire. The rise in temperature of the wire makes it expand and sag; the sag is taken up by a second fine wire PQ, which is held taut by a spring. The wire PQ passes round a pulley R attached to the pointer of the instrument, which rotates the pointer. The deflection of the pointer is roughly proportional to the average rate at which heat is developed in the wire XY; it is therefore roughly proportional to the average value of the square of the alternating current, and the scale is a square-law (non-uniform) one as shown.

Root-Mean-Square Value of A.C.

On p. 738 we saw that an alternating current I varied sinusoidally; that is, it could be represented by the equation $I = I_m \sin \omega t$, where I_m was the peak (maximum) value of the current. In commercial practice, alternating currents are always measured and expressed in terms of their *root-mean-square (r.m.s.)* value.

Consider two resistors of equal resistance R, one carrying an alternating current and the other a direct current. Suppose both are dissipating the same power, P, as heat. The root-mean-square (r.m.s.) value of the alternating current, I_r, is defined as equal to the direct current, I_d. Thus:

the root-mean-square value of an alternating current is defined as that value of steady current which would dissipate heat at the same rate in a given resistance.

Since the power dissipated by the direct current is

$$P = I_d^2 R,$$

our definition means that, in the a.c. circuit,

$$P = I_r^2 R \qquad . \qquad . \qquad . \qquad . \qquad . \qquad (1)$$

Whatever the wave-form of the alternating current, if I is its value at any instant, the power which it delivers to the resistance R at that instant is $I^2 R$. Consequently, the average power P is given by

$$P = \text{average value of } (I^2 R)$$

$$= R \times \text{average value of } (I^2),$$

since R is a constant. Therefore, by equation (1),

$$I_r^2 R = R \times \text{average value of } (I^2)$$

or
$$I_r^2 = \text{average value of } (I^2) \qquad . \qquad . \qquad . \qquad . \qquad (2)$$

The average value of (I^2) is called the *mean-square* current. Figure 37.3 (i) shows a sinusoidal (sine variation) current I from the a.c. mains and the way its I^2 values vary. The values are *positive* on the negative half cycle. Since this graph of I^2 is a symmetrical one, the mean or average value of I^2 is $I_m^2/2$, where I_m is the *maximum* or *peak* value of the current. So in this case, the root-mean-square (r.m.s.) value of the current is given by

$$I_r = \frac{I_m}{\sqrt{2}} = 0.71\ I_m \qquad \qquad \qquad \qquad (3)$$

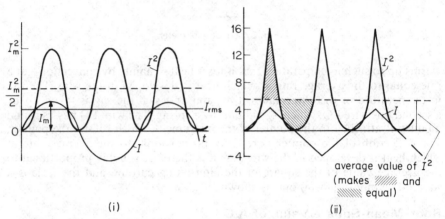

(i)   (ii)

Fig. 37.3 Mean-square values

If the r.m.s. value I_r is known, the peak value of the current I_m is calculated from

$$I_m = \sqrt{2}I_r.$$

In Britain, the a.c. mains supply is 240 V (r.m.s.). So the peak or maximum value of the voltage is

$$V_m = \sqrt{2}V_r = 1.41 \times 240 = 338 \text{ V}.$$

This means that an electrical appliance is unsuitable for use on the a.c. mains if it cannot withstand a voltage of about 338 V.

Figure 37.3 (ii) shows an alternating current I which is not sinusoidal and the way its I^2 values vary with time. Unlike the sine-wave current in Fig. 37.3 (i), the mean-square value is *not* half-way between the zero and the peak or maximum value. The mean-square value corresponds to the value which, for a cycle, makes the areas equal on both sides, as shown in Fig. 37.3 (ii). In this case it can be seen that the mean-square value is *less* than $I_m^2/2$, where I_m is the maximum value of the current. So the r.m.s. is less than $I_m/\sqrt{2}$.

Figure 37.4 (i) shows one form of a *square wave* alternating current. Unlike sinusoidal a.c., this has a constant positive current I_m for half a cycle OA and a constant negative current I_m for the other half of the cycle AB. The square of the current is positive on both halves of the cycle and equal to I_m^2 throughout. So the root-mean-square value is I_m. The power delivered to a pure resistance R would therefore be $I_m^2 R$.

A.C. Meters

We can see that for measuring alternating current, we require a meter whose

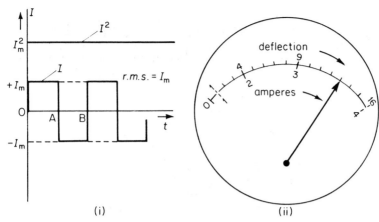

Fig. 37.4 (i) Square wave (ii) Scale of a.c. ammeter

deflection measures not the current through it but the average value of the square of the current. As we have already seen, moving-iron and hot-wire meters have just this property (p. 772).

For convenience, such meters are scaled to read amperes, not (amperes)^2, as in Fig. 37.4 (ii). The scale reading is then proportional to the square-root of the deflection, and indicates directly the root-mean-square value of the current, I_r. An a.c. meter of the moving-iron or hot-wire type can be calibrated by using direct current. This follows at once from the definition of the r.m.s. value of current as the value of direct current which produces the same heat per second in a resistor.

Moving-coil meters with semiconductor diode rectifiers are used in multi-meters for measuring alternating current and voltage, as described later (p. 812).

A.C. through a Capacitor

In many radio circuits, resistors, capacitors, and coils are present. An alternating current can flow through a resistor, but it is not obvious at first that it can flow through a capacitor. This can be demonstrated, however, by wiring a capacitor of 1 μF or more in series with a mains filament lamp of low rating such as 25 W. The lamp lights up, showing that a current is flowing through it. Direct current cannot flow through a capacitor because an insulating medium is between the plates. So with a mixture of a.c. and d.c., only the a.c. flows through a capacitor.

The current flows because the capacitor plates are being continually charged, discharged, and charged the other way round by the alternating voltage of the mains, Fig. 37.5 (i). The current thus flows round the circuit, and can be measured by an a.c. milliammeter inserted in any of the connecting wires.

Figure 37.5 (ii) shows how the alternating voltage V varies with time, t. Since the charge Q on the capacitor plates is given at any instant by $Q = CV$, and C is constant, the graph of Q is in phase with that of V as shown.

The current I is the rate of change of Q with time, that is, $I = dQ/dt$. So the value of I at any instant is the corresponding *gradient* of the $Q - t$ graph. At O, the gradient is a maximum, so I is then a maximum. From O to A, the gradient of the $Q - t$ graph decreases to zero. So I decreases to zero at N. From A to B, the gradient of the $Q - t$ curve is *negative* and so I is negative from N to R. In this way we see that the $I - t$ graph is PNRST. So I and V are 90° out of phase, *with I leading V by 90°*.

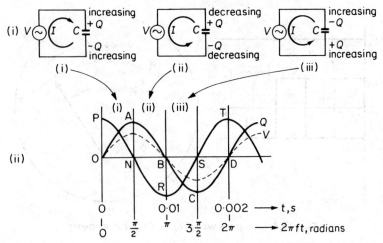

Fig. 37.5 Flow of a.c. through capacitor, frequency 50 Hz

If V is made bigger, we see that the gradient at O is bigger. So I_m increases when V_m increases. Also, if the frequency f of V is doubled, for example, so that there are now two cycles between O and B, the gradient at O becomes greater. So I_m increases when both f and V_m increase.

Calculation for *I*

To find the exact variation of I with time t, suppose the amplitude or peak of the voltage V applied to the capacitor C is V_m and its frequency is f. Then, assuming a sinusoidal voltage variation, the instantaneous voltage at any time t is

$$V = V_m \sin 2\pi ft.$$

If C is the capacitance of the capacitor, then the charge Q on its plates is

$$Q = CV,$$

so $$Q = CV_m \sin 2\pi ft.$$

The current, I, flowing at any instant, is equal to the rate at which charge is accumulating on the capacitor plates. Thus

$$I = \frac{dQ}{dt} = \frac{d}{dt}(CV_m \sin 2\pi ft)$$

$$= 2\pi fCV_m \cos 2\pi ft \quad . \quad . \quad . \quad . \quad (1)$$

Equation (1) shows that the peak or maximum value I_m of the current is $2\pi fCV_m$; so I_m is proportional to the frequency, the capacitance, and the voltage amplitude. These results are easy to explain. The greater the voltage, or the capacitance, the greater the charge on the plates, and therefore the greater the current required to charge or discharge the capacitor. And the higher the frequency, the more rapidly is the capacitor charged and discharged, and therefore again the greater is the current.

A more puzzling feature of equation (1) is the factor giving the time variation of the current, cos $2\pi ft$. It shows that *the current varies a quarter-cycle or 90° ($\pi/2$) out of phase with the voltage*. Figure 37.5 shows this variation, and also helps to explain it physically. When the voltage is a maximum, so is the charge on the capacitor. It is therefore not charging and the current is zero. When the voltage starts to fall, the capacitor starts to discharge; the rate of discharging, or current, reaches its maximum when the capacitor is completely discharged and the voltage across it is zero. Since the current I passes its maximum a quarter-cycle ahead of the voltage V, we see that I leads V by $90°$ ($\pi/2$).

Reactance of C

The *reactance* of a capacitor is its opposition in ohms to the passage of alternating current. We do not use the term 'resistance' in this case because this is the opposition to direct current.

The reactance, symbol X_C, is defined by

$$X_C = \frac{V_m}{I_m}$$

where V_m and I_m are the peak or maximum of the a.c. voltage and current. Since the ratio $V_m/I_m = V_r/I_r$, where V_r and I_r are the r.m.s. voltage and current respectively, we can also define reactance X_C by the ratio V_r/I_r. We shall omit the suffix r when using r.m.s. values and so

$$X_C = \frac{V}{I}.$$

Here X_C is in ohms when V is in volts (r.m.s.) and I is in amperes (r.m.s.).

We have already seen (p. 776) that the amplitude or peak value of the current through a capacitor is given by

$$I_m = 2\pi f C V_m.$$

The reactance of the capacitor is therefore

$$X_C = \frac{V_m}{I_m} = \frac{1}{2\pi f C} = \frac{1}{2\pi f C}.$$

X_C is in ohms when f is in Hz (cycles per second), and C in farads.

For convenience we often write $\omega = 2\pi f$. The quantity ω is called the angular frequency of the current and voltage. It is expressed in radians per second. Then an alternating voltage, for example, may be written as

$$V = V_m \sin \omega t.$$

The reactance of a capacitor is written as

$$X_C = \frac{1}{\omega C}$$

Calculations with X_C

As an illustration, suppose a capacitor C of $0\cdot1$ μF is used on the mains frequency of 50 Hz. Then the reactance is

$$X_C = \frac{1}{2\pi f C} = \frac{1}{2 \times 3\cdot14 \times 50 \times 0\cdot1 \times 10^{-6}}$$

$$= \frac{10^6}{2 \times 3\cdot14 \times 50 \times 0\cdot1} = 32\,000 \ \Omega \ \text{(approx.)}.$$

From the formula for reactance we note that $X_C \propto 1/C$ for a given frequency. So if a 1 μF capacitor is used on the 50 Hz mains, its reactance is 10 times

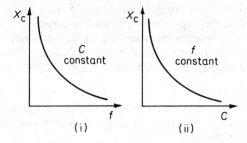

Fig. 37.6 Reactance of capacitor

less than that of 0·1 μF, which is 32 000 Ω/10 or 3200 Ω. Figure 37.6 (i) shows how X_C varies with C.

Also, since $X_C \propto 1/f$ for a given capacitor, at $f = 1000$ Hz a capacitor of 1 μF has 20 times *less* reactance than at $f = 50$ Hz. So $X_C = 3200 \ \Omega/20 = 160 \ \Omega$. See Fig. 37.6 (ii).

Since $X_C = V/I$, where V and I are both r.m.s. (or peak) values, we can see that

$$I = \frac{V}{X_C} \quad \text{and} \quad V = IX_C.$$

These are similar formulae to d.c. circuit formulae. The difference with d.c. circuits is that we must always consider the phase difference between V and I, and V and I are 90° out of phase as we have previously shown.

Example

A capacitor C of 1 μF is used in a radio circuit where the frequency is 1000 Hz and the current flowing is 2 mA (r.m.s.). Calculate the voltage across C.

What current flows when an a.c. voltage of 20 V r.m.s., $f = 50$ Hz is connected to this capacitor?

(i) Reactance, $X_C = \dfrac{1}{2\pi f C} = \dfrac{1}{2\pi \times 1000 \times 1/10^6} = 159 \ \Omega$ (approx.).

$$\therefore V = IX_C = \frac{2}{1000} \times 159 = 0\cdot32 \text{ V (approx.)}.$$

(ii) When 20 V r.m.s., $f = 50$ Hz, is connected to C, the reactance of C changes. Since $X_C \propto 1/f$,

$$X_C \text{ at } f = 50 \text{ Hz is 20 times } X_C \text{ at } f = 1000 \text{ Hz.}$$

So $$X_C = 20 \times 159 \ \Omega = 3180 \ \Omega$$

$$\therefore I = \frac{V}{X_C} = \frac{20}{3180} = 6\cdot3 \times 10^{-3} \text{ A r.m.s.}$$

A.C. through an Inductor

Since a coil is made from conducting wire, we have no difficulty in seeing that an alternating current can flow through it. However, if the coil has appreciable self-inductance, the current is less than would flow through a non-inductive coil of the same resistance. We have already seen how self-inductance opposes changes of current; it must therefore oppose an alternating current, which is continuously changing.

Let us suppose that the resistance of the coil is negligible, a condition which can be satisfied in practice. We can simplify the theory by considering first the current, and then finding the potential difference across the coil. Let us therefore denote the current by

$$I = I_m \sin 2\pi ft \qquad . \qquad . \qquad . \qquad . \qquad (1)$$

where I_m is its amplitude, Fig. 37.7. If L is the inductance of the coil, the changing current sets up a back-e.m.f. in the coil, of magnitude

$$E = L\frac{dI}{dt}.$$

To maintain the current, the applied supply voltage must be equal to the back-e.m.f. The voltage applied to the coil must therefore be given by

$$V = L\frac{dI}{dt}.$$

Since L is constant, it follows that V is proportional to dI/dt.

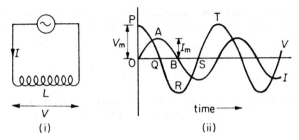

Fig. 37.7 Flow of a.c. through a coil

Figure 37.7 (ii) shows how the current I varies with time t. The values of dI/dt are the *gradients* of the $I - t$ graph at the time concerned. At O the gradient is a maximum; so the maximum voltage of V, or V_m, occurs at O and is represented by OP as shown. From O to A, the gradient of the $I - t$ graph decreases to zero. So the voltage V decreases from P to Q. From A to B the gradient of the $I - t$ graph is *negative* (downward slope). So the voltage decreases along QR. We now see that V *leads* I by $90°$ $(\pi/2)$.

We can find the value of V_m from the equation $V = LdI/dt$. From (1),

$$I = I_m \sin 2\pi ft.$$

So, by differentiation with respect to t,

$$V = L\frac{dI}{dt}$$

Thus

$$V = L\frac{d}{dt}(I_m \sin 2\pi ft) = 2\pi fLI_m \cos 2\pi ft.$$

So

$$V_m = \text{maximum (peak) voltage} = 2\pi fLI_m.$$

Hence the reactance of the inductor is

$$X_L = \frac{V_m}{I_m} = 2\pi fL.$$

X_L is in ohms when f is in Hz, and L is in henrys (H). An *iron-cored* coil has a high inductance L such as 20 H. Used on the mains frequency f of 50 Hz, its reactance $X_L = 2\pi fL = 2 \times 3{\cdot}14 \times 50 \times 20 = 6280\,\Omega$. Since this type of inductor provides a high reactance, it is sometimes called a 'choke'.

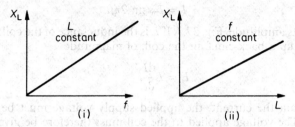

Fig. 37.8 Variation of reactance X_L

Since $X_L = 2\pi fL$, it follows that $X_L \propto f$ for a given inductance, Fig. 37.8 (i), and that $X_L \propto L$ for a given frequency, Fig. 37.8 (ii).

Example

An inductor of 2 H and negligible resistance is connected to a 12 V mains supply, $f = 50$ Hz. Find the current flowing. What current flows when the inductance is changed to 6 H?

$$\text{Reactance, } X_L = 2\pi fL = 2\pi \times 50 \times 2 = 628\,\Omega$$

$$\therefore I = \frac{V}{X_L} = \frac{12}{628}\text{A} = 19 \text{ mA (approx.)}.$$

When the inductance is increased to 6 H, its reactance X_L is increased 3 times since $X_L \propto L$ for a given frequency. So the current is reduced to 1/3rd of its value. So now $I = 6$ mA (approx.).

Phasor Diagrams

In the Mechanics section of this book, it is shown that a quantity which varies sinusoidally with time may be represented as the projection of a rotating vector

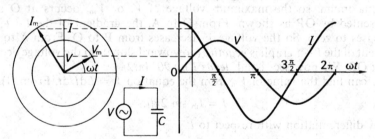

Fig. 37.9 Phasor diagram for capacitor

(p. 62). These quantities are called *phasors*, as the phase angle must also be represented. Alternating currents and voltages may therefore be represented as phasors. Figure 37.9 shows, on the left, the phasors representing the current through a capacitor, and the voltage across it. Since the current leads the voltage by $\pi/2$, the current vector I is displaced by 90° ahead of the voltage vector V.

Figure 37.10 shows the phasor diagram for a pure inductor. In drawing it, the voltage has been taken as $V = V_m \sin \omega t$, and the current drawn lagging $\pi/2$ behind it. This enables the diagram to be readily compared with that for a capacitor. To show that it is essentially the same as Fig. 37.7 (ii), we have only to shift the origin by $\pi/2$ to the right, from 0 to 0′.

When an alternating voltage is connected to a pure resistance R, the current

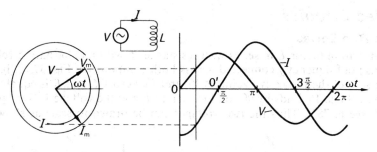

Fig. 37.10 Phasor diagram for pure inductance

I at any instant $= V/R$, where V is the voltage at that instant. So I is zero when V is zero and I is a maximum when V is a maximum. Hence I and V are in phase, Fig. 37.11 (i). Since the phase angle is zero, we draw the vector or phasor I in the same direction as V, where the phasors represent either peak values or r.m.s. values.

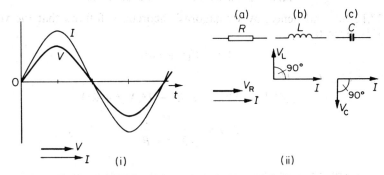

Fig. 37.11 Phasor diagrams for R, L, C

Figure 37.11 (ii) summarises the phasor diagrams for (a) a pure resistance R, (b) a pure inductance L and (c) a pure capacitance C. They should be memorised by the reader for use in a.c. circuits.

Series Circuits

L and R in Series

Consider an inductor L in series with resistance R, with an alternating voltage V (r.m.s.) of frequency f connected across both components, Fig. 37.12 (i).

The sum of the respective voltages V_L and V_R across L and R is equal to V. But the voltage V_L leads by 90° on the current I, and the voltage V_R is in phase with I (see p. 781). Thus the two voltages can be drawn to scale as shown in

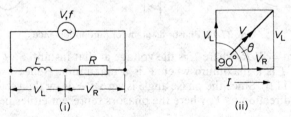

Fig. 37.12 Inductance and resistance in series

Fig. 37.12 (ii), and hence, by Pythagoras' theorem, it follows that the vector sum V is given by

$$V^2 = V_L{}^2 + V_R{}^2.$$

But $V_L = IX_L$, $V_R = IR$.

$$\therefore V^2 = I^2 X_L{}^2 + I^2 R^2 = I^2 (X_L{}^2 + R^2),$$

$$\therefore I = \frac{V}{\sqrt{X_L{}^2 + R^2}} \quad . \qquad . \qquad . \qquad . \qquad (i)$$

Also, from Fig. 37.11 (ii), the current I lags on the applied voltage V by an angle θ given by

$$\tan \theta = \frac{V_L}{V_R} = \frac{IX_L}{IR} = \frac{X_L}{R} \quad . \qquad . \qquad . \qquad . \qquad (ii)$$

From (i), it follows that the 'opposition' Z to the flow of alternating current is given in ohms by

$$Z = \frac{V}{I} = \sqrt{X_L{}^2 + R^2} \qquad . \qquad . \qquad . \qquad . \qquad (iii)$$

This 'opposition', Z, is known as the *impedance* of the circuit.

Example

An iron-cored coil of 2 H and 50 Ω resistance is placed in series with a resistor of 450 Ω, and a 100 V, 50 Hz, a.c. supply is connected across the arrangement. Find (a) the current flowing in the coil, (b) its phase angle relative to the voltage supply, (c) the voltage across the coil.

(a) The reactance $X_L = 2\pi f L = 2\pi \times 50 \times 2 = 628$ Ω.
Total resistance $R = 50 + 450 = 500$ Ω.

$$\therefore \text{ circuit impedance } Z = \sqrt{X_L{}^2 + R^2} = \sqrt{628^2 + 500^2} = 803 \text{ Ω}$$

$$\therefore I = \frac{V}{Z} = \frac{100}{803} \text{ A} = 12 \cdot 5 \text{ mA (approx.).}$$

(b) $$\tan \theta = \frac{X_L}{R} = \frac{628}{500} = 1 \cdot 256.$$

So $$\theta = 51 \cdot 5°.$$

(c) For the coil, $X_L = 628 \ \Omega$ and $R = 50 \ \Omega$.

So coil impedance $Z = \sqrt{X_L{}^2 + R^2} = \sqrt{628^2 + 50^2} = 630 \ \Omega$.

Thus   voltage across coil $V = IZ = 12 \cdot 5 \times 10^{-3} \times 630$

$$= 7 \cdot 9 \ \text{V (approx.)}.$$

C and R in Series

A similar analysis enables the impedance to be found of a capacitance C and resistance R in series, Fig. 37.13 (i). In this case the voltage V_C across the capacitor

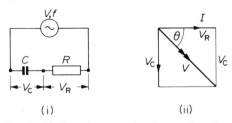

(i)   (ii)

Fig. 37.13 Capacitance and resistance in series

lags by 90° on the current I (see p. 781), and the voltage V_R across the resistance is in phase with the current I. As the vector sum is V, the applied voltage, it follows by Pythagoras' theorem that

$$V^2 = V_C{}^2 + V_R{}^2 = I^2 X_C{}^2 + I^2 R^2 = I^2(X_C{}^2 + R^2),$$

$$\therefore I = \frac{V}{\sqrt{X_C{}^2 + R^2}} \quad . \qquad . \qquad . \qquad . \qquad . \quad \text{(i)}$$

Also, from Fig. 37.12 (ii), the current I leads on V by an angle θ given by

$$\tan \theta = \frac{V_C}{V_R} = \frac{IX_C}{IR} = \frac{X_C}{R} \quad . \qquad . \qquad . \qquad . \quad \text{(ii)}$$

It follows from (i) that the impedance Z of the C–R series circuit is

$$Z = \frac{V}{I} = \sqrt{X_C{}^2 + R^2}.$$

It should be noted that although the impedance formula for a C–R series circuit is of the same mathematical form as that for a L–R series circuit, the current in the former case leads on the applied voltage but the current in the latter case lags on the applied voltage.

L, C, R in Series

The most general series circuit is the case of L, C, R in series, Fig. 37.14 (i). The phasor diagram has V_L leading by 90° on V_R V_C lagging by 90° on V_R with the current I in phase with V_R, Fig. 37.14 (ii). If V_L is greater than V_C, their resultant is $(V_L - V_C)$ in the direction of V_L, as shown. Thus, from Pythagoras' theorem for triangle ODB, the applied voltage V is given by

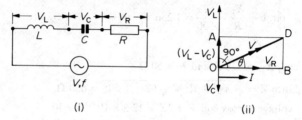

(i)   (ii)

Fig. 37.14 *L, C, R* in series

$$V^2 = (V_L - V_C)^2 + V_R{}^2.$$

But $V_L = IX_L$, $V_C = IX_C$, $V_R = IR$.

$$\therefore \ V^2 = (IX_L - IX_C)^2 + I^2 R^2 = I^2 [(X_L - X_C)^2 + R^2],$$

$$\therefore \ I = \frac{V}{\sqrt{(X_L - X_C)^2 + R^2}} \qquad . \qquad . \qquad . \qquad \text{(i)}$$

Also, I lags on V by an angle θ given by

$$\tan \theta = \frac{DB}{OB} = \frac{V_L - V_C}{V_R} = \frac{IX_L - IX_C}{IR} = \frac{X_L - X_C}{R} \qquad . \qquad . \qquad \text{(ii)}$$

Resonance in the *L, C, R* Series Circuit

From (i), it follows that the impedance Z of the circuit is given by

$$Z = \sqrt{(X_L - X_C)^2 + R^2}.$$

The impedance varies as the frequency, f, of the applied voltage varies, because X_L and X_C both vary with frequency. Since $X_L = 2\pi f L$, then $X_L \propto f$, and thus the variation of X_L with frequency is a straight line passing through the origin, Fig. 37.15 (i). Also, since $X_C = 1/2\pi f C$, then $X_C \propto 1/f$, and thus the variation of X_C with frequency is a curve approaching the two axes, Fig. 37.15 (i). The resistance R is independent of frequency, and is thus represented by a

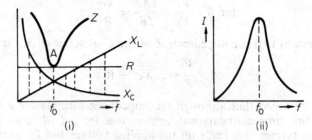

(i)   (ii)

Fig. 37.15 Resonance curves

line parallel to the frequency axis. The difference $(X_L - X_C)$ is represented by the dotted lines shown in Fig. 37.15 (i), and it can be seen that $(X_L - X_C)$ decreases to zero for a particular frequency f_0, and thereafter increases again. Thus, from $Z = \sqrt{(X_L - X_C)^2 + R^2}$, the impedance diminishes and then increases as the frequency f is varied. The variation of Z with f is shown in Fig. 37.15 (i), and since the current $I = V/Z$, the current varies as shown in Fig. 37.15 (ii). Thus the current has a maximum value at the frequency f_0, and this is known as the *resonant frequency* of the circuit.

The magnitude of f_0 is given by $X_L - X_C = 0$, or $X_L = X_C$.

$$\therefore 2\pi f_0 L = \frac{1}{2\pi f_0 C} \quad \text{or} \quad 4\pi^2 L C f_0 = 1.$$

$$\therefore f_0 = \frac{1}{2\pi\sqrt{LC}}.$$

At frequencies above and below the resonant frequency, the current is less than the maximum current, see Fig. 37.15 (ii), and the phenomenon is thus basically the same as the forced and resonant vibrations obtained in Sound or Mechanics (p. 410).

The series resonance circuit is used for tuning a radio receiver. In this case the incoming waves of frequency f say from a distant transmitting station induces a varying voltage in the aerial, which in turn induces a voltage V of the same frequency in a coil and capacitor circuit in the receiver, Fig. 37.16. When

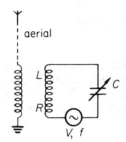

Fig. 37.16 Tuning a receiver

the capacitance C is varied the resonant frequency is changed; and at one setting of C the resonant frequency becomes f, the frequency of the incoming waves. The maximum current is then obtained, and the station is now heard very loudly.

Parallel Circuits

We now consider briefly the principles of a.c. parallel circuits. In d.c. parallel circuits, the currents in the individual branches are added arithmetically to find their total. In a.c. circuits, however, we add the currents by vector methods, taking into account the phase angle between them.

L, R in Parallel

In the parallel circuit in Fig. 37.17 (i), the supply current I is the vector sum of

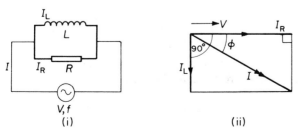

Fig. 37.17 *L, R* in parallel

I_L and I_R, Fig. 37.17 (ii) shows the vector addition. I_L is 90° out of phase with I_R, since I_R is in phase with V, and I_L lags 90° behind V. So

$$I^2 = I_R{}^2 + I_L{}^2 = \left(\frac{V}{R}\right)^2 + \left(\frac{V}{X_L}\right)^2.$$

Then
$$I = V\sqrt{\frac{1}{R^2} + \frac{1}{X_L{}^2}}.$$

Also, from Fig. 37.17 (ii), I lags behind the applied voltage V by an angle φ given by

$$\tan \varphi = \frac{I_L}{I_R} = \frac{R}{X_L}.$$

Similar analysis shows that when an a.c. voltage V r.m.s. is applied to a parallel C, R circuit, the supply I is given by

$$I = V\sqrt{\frac{1}{R^2} + \frac{1}{X_C^2}} \quad \text{and} \quad \tan \varphi = \frac{R}{X_C},$$

where I now leads V by the angle φ.

L, C in Parallel

A parallel arrangement of coil (L, R) and capacitor (C) is widely used in transistor oscillators and in radio-frequency amplifier circuits. To simplify matters, let us assume that the resistance of the coil is negligible compared with its reactance. We then have effectively an inductor L in parallel with a capacitor C, Fig. 37.18 (i).

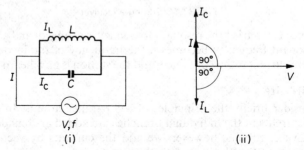

Fig. 37.18 L, C in parallel

Figure 37.18 (ii) shows the two currents in the components. I_L lags by 90° on V but I_C leads by 90° on V. If I_C is greater than I_L at the particular frequency f, then

$$I = I_C - I_L = \frac{V}{X_C} - \frac{V}{X_L}.$$

Since I leads by 90° on V in this case, we say that the circuit is 'net capacitive'. If, however, I_L is greater than I_C, then

$$I = I_L - I_C = \frac{V}{X_L} - \frac{V}{X_C}.$$

Since I lags by 90° on V in this case, the circuit is 'net inductive'.

Suppose V, L and C are kept constant and the frequency f of the supply is varied from a low to a high value. The magnitude and phase of I then varies according to the relative magnitudes of $X_L(2\pi fL)$ and $X_C(1/2\pi fC)$, as shown in Fig. 37.19 (i). A special case occurs when $X_L = X_C$. Then $I_L = I_C$ and so $I = 0$,

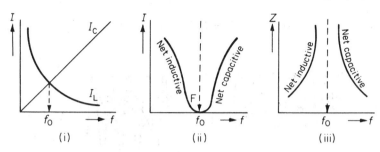

Fig. 37.19 L, C in parallel—variation of Z and I

Fig. 37.19 (ii). At this frequency f_0, we have $X_L = X_C$, so

$$2\pi f_0 L = \frac{1}{2\pi f_0 C}, \quad \text{or} \quad f_0 = \frac{1}{2\pi\sqrt{LC}}.$$

Figure 37.19 (ii) shows how the current I varies with the frequency f. Figure 37.19 (iii) shows how the *impedance* Z of the parallel L, C circuit varies with frequency f. Since $Z = V/I$, and I is zero at the frequency f_0, it follows that Z is infinitely high at f_0. The parallel inductor-capacitor circuit has therefore a *resonant frequency* (f_0) so far as its impedance Z is concerned.

In practice, when the resistance R of the coil is taken into account, a similar variation of Z with frequency f is obtained. The maximum value of Z is now finite and theory shows that $Z = L/CR$. The resonant frequency f_0 is practically still given by $f_0 = 1/2\pi\sqrt{LC}$. In contrast, the *series* L, C, R circuit gives a maximum *current* I at resonance as the circuit impedance is then a minimum (p. 784).

Power in A.C. Circuits

Resistance R. The power absorbed is usually $P = IV$. In the case of a resistance, $V = IR$, and $P = I^2 R$. The variation of power is shown in Fig. 37.20 (i), where $I_0 = I_m$ = the peak (maximum) value of the current. On p. 773 we explained the reason for choosing the root-mean-square value of alternating current. So the average power absorbed in R is given by

$$P = I^2 R,$$

where I is the r.m.s. value.

Inductance L. In the case of a pure inductor, the voltage V across it leads by $90°$ on the current I. Thus if $I = I_m \sin \omega t$, then $V = V_m \sin(90° + \omega t) = V_m \cos \omega t$. Hence, at any instant,

$$\text{power absorbed} = IV = I_m V_m \sin \omega t . \cos \omega t = \tfrac{1}{2} I_m V_m \sin 2\omega t.$$

The variation of power, P, with time t is shown in Fig. 37.20 (ii); it is a sine curve with an average of zero. *Hence no power is absorbed in a pure inductance.* This is explained by the fact that on the first quarter of the current cycle, power is absorbed ($+$) in the magnetic field of the coil (see p. 755). On the next quarter-cycle the power is returned ($-$) to the generator, and so on.

Capacitance. With a pure capacitance, the voltage V across it lags by 90°
In the current I (p. 781). Thus if $I = I_m \sin \omega t$,

$$V = V_m \sin (\omega t - 90°) = - V_m \cos \omega t.$$

Hence, numerically,

$$\text{power at an instant, } P = IV = I_m V_m \sin \omega t \cos \omega t = \frac{I_m V_m}{2} \sin 2\omega t.$$

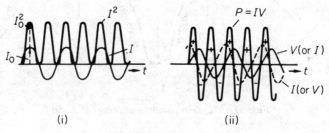

| (i) | (ii) |

Fig. 37.20 Power in a.c. circuits

Thus, as in the case of the inductance, *the power absorbed in a cycle is zero*,
Fig. 37.20 (ii). This is explained by the fact that on the first quarter of the cycle,
energy is stored in the electrostatic field of the capacitor. On the next quarter
the capacitor discharges, and the energy is returned to the generator.

Formulae for A.C. Power

It can now be seen that, if I is the r.m.s. value of the current in amps in a circuit
containing a resistance R ohms, the power absorbed is $I^2 R$ watts. Care should
be taken to exclude the inductances and capacitances in the circuit, as no power
is absorbed in them. So if a current of 2 A r.m.s. flows in a circuit containing
a coil of 2 H and resistance 10 Ω in series with a capacitor of 1 μF, the power
absorbed in the circuit $= I^2 R = 2^2 \times 10 = 40$ W.

If the voltage V across a circuit leads by an angle θ on the current I, the
voltage can be resolved into a component $V \cos \theta$ in phase with the current,
and a voltage $V \sin \theta$ perpendicular to the current, Fig. 37.21. The former

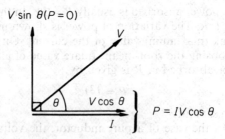

Fig. 37.21 Power absorbed

component, $V \cos \theta$, represents that part of the voltage across the total *resistance*
in the circuit, and hence the power absorbed is

$$P = IV \cos \theta.$$

The component $V \sin \theta$ is that part of the applied voltage across the total
inductance and capacitance. Since the power absorbed here is zero, it is sometimes called the 'wattless component' of the voltage.

Examples

1. A circuit consists of a capacitor of 2 μF and a resistor of 1000 Ω. An alternating e.m.f. of 12 V (r.m.s.) and frequency 50 Hz is applied. Find (1) the current flowing, (2) the voltage across the capacitor, (3) the phase angle between the applied e.m.f. and current, (4) the average power supplied.

The reactance X_C of the capacitor is given by

$$X_C = \frac{1}{2\pi f C} = \frac{1}{2\pi \times 50 \times 2/10^6} = 1590 \ \Omega \ (approx.)$$

$\therefore$ total impedance $Z = \sqrt{R^2 + X_C^2} = \sqrt{1000^2 + 1600^2} = 1880 \ \Omega$ (approx.).

(1) $\therefore$ current, $I = \dfrac{V}{Z} = \dfrac{12}{1880} = 6\cdot4 \times 10^{-3}$ A.

(2) voltage across C, $V_C = IX_C = \dfrac{12}{1880} \times 1590 = 10\cdot2$ V (approx.).

(3) The phase angle θ is given by

$$\tan\theta = \frac{X_C}{R} = \frac{1590}{1000} = 1\cdot59$$

$$\therefore \theta = 58° \ (approx.).$$

(4) Power supplied $= I^2 R = \left(\dfrac{12}{1880}\right)^2 \times 1000 = 0.04$ W (approx.)

2. A capacitor of capacitance C, a coil of inductance L and resistance $\bar{R}$, and a lamp are placed in series with an alternating voltage V. Its frequency f is varied from a low to a high value while the magnitude of V is kept constant. Describe and explain how the brightness of the lamp varies.

If $V = 0\cdot01$ V (r.m.s.) and $C = 0\cdot4$ μF, $L = 0\cdot4$ H, $R = 10$ Ω, calculate (i) the resonant frequency, (ii) the maximum current, (iii) the voltage across C at resonance, neglecting the lamp resistance.

When f is varied, the impedance Z of the circuit decreases to a minimum value (resonance) and then increases. Z is a minimum when $X_L = X_C$, so that $Z = R$ at resonance. Since the *current* flowing in the circuit increases to a maximum and then decreases, the brightness of the lamp increases to a maximum at resonance and then decreases.

(i)   Resonant frequency $f_0 = \dfrac{1}{2\pi\sqrt{LC}} = \dfrac{1}{2\pi\sqrt{0\cdot4 \times 0\cdot4 \times 10^{-6}}}$

$$= \frac{10^3}{2\pi \times 0\cdot4} = 400 \text{ Hz (approx.)}$$

(ii)   Maximum current $I = \dfrac{V}{R} = \dfrac{0\cdot01}{10} = 0\cdot001$ A (r.m.s.)

(iii)'   Voltage across $C = IX_C = 0\cdot001 \times \dfrac{1}{2\pi \times 400 \times 0\cdot4 \times 10^{-6}}$

$$= \frac{0\cdot001 \times 10^6}{2\pi \times 400 \times 0\cdot4} = 1 \text{ V.}$$

Exercises 37

1. Figure 37A represents alternating currents of different wave shapes, each of peak (amplitude) value 3·0 A. (i) is a sinusoidal a.c., (ii) is a square wave and (iii) is a rectangular wave. Calculate the r.m.s. value of the current in each case.

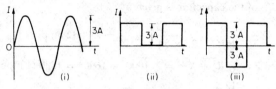

Fig. 37A

2. An alternating voltage of 10 V r.m.s. and frequency 50 Hz is applied to (i) a resistor of 5 Ω, (ii) an inductor of 2 H, and (iii) a capacitor of 1 μF. Determine the r.m.s. current flowing in each case and draw a phasor diagram of the current and voltage for each.

3. An alternating current of 0·2 A r.m.s. and frequency $100/2\pi$ Hz flows in a circuit consisting of a series arrangement of a resistor R of 20 Ω, an inductor L of 0·15 H and a capacitor C of 500 μF. Calculate the a.c. voltage (i) across each component, (ii) across R and L together, (iii) across L and C together, (iv) the total voltage across L, C, R.

What power is dissipated in each component?

4. A coil of inductance L and negligible resistance is in series with a resistance R. A supply voltage of 40 V (r.m.s.) is connected to them. If the voltage across L is equal to that across R, calculate (a) the voltage across each component, (b) the frequency f of the supply, (c) the power absorbed in the circuit, if $L = 0·1$ H and $R = 40$ Ω.

5. An inductor L of negligible resistance is connected in *parallel* with a capacitor C and an a.c. voltage V of constant r.m.s. value is connected across the arrangement. With the aid of a phasor diagram, explain why the current I drawn from the a.c. supply is zero at a particular frequency.

Draw a sketch showing the variation of I with frequency f when f is varied from a very low value to a very high value.

6. If a sinusoidal current, of peak value 5 A, is passed through an a.c. ammeter the reading will be $5/\sqrt{2}$ A. Explain this.

What reading would you expect if a square-wave current, switching rapidly between $+0·5$ and $-0·5$ A, were passed through the instrument? (*L.*)

7. (a) The *impedance* of a circuit containing a capacitor C and a resistor R connected to an alternating voltage supply is given by $\sqrt{R^2 + X^2}$, where X is the *reactance* of the capacitor. Define the two terms in italics.

The current in the circuit leads the voltage by a phase angle θ, where $\tan \theta = X/R$. Explain, using a vector diagram, why this is so.

An inductor is put in series with the capacitor and resistor and a source of alternating voltage of constant value but variable frequency. Sketch a graph to show how the current will vary as the frequency changes from zero to a high value.

(b) An alternating voltage of 10 V r.m.s. and 5·0 kHz is applied to a resistor, of resistance 4·0 Ω, in series with a capacitor of capacitance 10 μF. Calculate the r.m.s. potential differences across the resistor and the capacitor. Explain why the sum of these potential differences is not equal to 10 V. (Assume $\pi^2 = 10$.) (*L.*)

8. A coil having inductance and resistance is connected to an oscillator giving a fixed sinusoidal output voltage of 5·00 V r.m.s. With the oscillator set at a frequency of 50 Hz, the r.m.s. current in the coil is 1·00 A and at a frequency of 100 Hz, the r.m.s. current is 0·625 A.

(a) Explain why the current through the coil changes when the frequency of the supply is changed. (b) Determine the inductance of the coil. (c) Calculate the ratio of the powers dissipated in the coil in the two cases. (*N.*)

9. An inductor and a capacitor are connected one at a time to a variable-frequency power source. State how, and explain in non-mathematical terms why, the current through the inductor and the capacitor varies as the frequency is varied.

A circuit is set up containing an inductor, a capacitor, a lamp and a variable-frequency source with the components arranged in series. Explain why, as the frequency of the supply is varied, the lamp is found to increase in brightness, reach a maximum and then become less bright. Explain why the inductor is heated by the passage of the current while the capacitor remains cool. (*L.*)

10. (*a*) Explain why a moving coil ammeter cannot be used to measure an alternating current even if the frequency is low. Draw a diagram of a bridge rectifier circuit which could be used with such an ammeter and explain its action. (You are not required to explain the mode of operation of an individual diode in the bridge.)

(*b*) An alternating voltage is connected to a resistor and an inductor in series. By using a vector diagram, and explaining the significance of each vector, derive an expression for the impedance of the circuit.

A 50 V, 50 Hz a.c. supply is connected to a resistor, of resistance 40 Ω, in series with a solenoid whose inductance is 0·20 H. The p.d. between the ends of the resistor is found to be 20 V. What is the resistance of the wire of the solenoid? (Assume $\pi^2 = 10$.) (*L.*)

11. (*a*) Define the *impedance* of a coil carrying an alternating current. Distinguish between the *impedance* and *resistance* of a coil and explain how they are related. Describe and explain how you would use a length of insulated wire to make a resistor having an appreciable resistance but negligible inductance.

(*b*) Outline how you would determine the impedance of a coil at a frequency of 50 Hz using a resistor of known resistance, a 50 Hz a.c. supply and a suitable measuring instrument. Show how to calculate the impedance from your measurements.

(*c*) A coil of inductance L and resistance R is connected in series with a capacitor of capacitance C and a variable frequency sinusoidal oscillator of negligible impedance. Sketch qualitatively how the current in the circuit varies with the applied frequency and account for the shape of the curve. Sketch on the same axes the curve you would expect for a considerably larger value of R, the values of L and C remaining unchanged, taking care to indicate which curve refers to the larger value of R. (*N.*)

12. (*a*) A sinusoidal alternating potential difference of which the peak value is 20 V is connected across a resistor of resistance 10 Ω. What is the mean power dissipated in the resistor?

(*b*) A sinusoidal alternating potential difference is to be rectified using the circuit in Fig. 37B (i), which consists of a diode D, a capacitor C and a resistor R. Sketch the variation with time of the potential difference between A and B which you would expect and explain why it has that form.

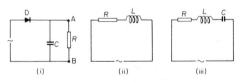

(i)   (ii)   (iii)

Fig. 37B

(*c*) A sinusoidal alternating difference of a constant amplitude is applied across the resistor R and inductor L, Fig. 37B (ii). Explain why the amplitude of the current through the circuit decreases as the frequency of the alternating potential difference is increased.

(*d*) A sinusoidal alternating potential difference of constant frequency and amplitude is applied to the circuit in Fig. 37B (iii). Describe and explain how the amplitude of the current through the circuit changes as the capacitance C is increased slowly from a very small value to a very large value. (*O. & C.*)

13. Describe and explain the mode of action of an ammeter suitable for the measurement of alternating current.

A constant a.c. supply is connected to a series circuit consisting of a resistance of 300 Ω in series with a capacitance of 6·67 μF, the frequency of the supply being $3000/2\pi$ Hz. It is desired to reduce the current in the circuit to half its value. Show how this could be done by placing either (*a*) an additional resistance, or (*b*) an inductance, in series. Calculate in each case the magnitude of the extra component. (*L.*)

14. Explain what is meant by the *peak value* and *root mean square value* of an alternating current. Establish the relation between these quantities for a sinusoidal waveform.

What is the r.m.s. value of the alternating current which must pass through a resistor immersed in oil in a calorimeter so that the initial rate of rise of temperature of the coil is three times that produced when a direct current of 2 A passes through the resistor under the same conditions? (*N*.)

15. Define the *impedance* of an a.c. circuit.

A 2.5 μF capacitor is connected in series with a non-inductive resistor of 300 Ω across a source of p.d. of r.m.s. value 50 V alternating at $1000/2\pi$ Hz. Calculate (*a*) the r.m.s. values of the current in the circuit and the p.d. across the capacitor, (*b*) the mean rate at which energy is supplied by the source. (*N*.)

16. A source of a.c. voltage is connected by wires of negligible resistance across a capacitor. Explain, without the use of mathematical expressions, why

(*a*) a current flows;

(*b*) the current is not in phase with the voltage;

(*c*) the size of the current depends upon the frequency of the supply voltage;

(*d*) the power output of the source is zero.

If a resistor, of resistance R, is connected in series with a capacitor, of capacitance C, to an a.c. voltage of frequency f, derive an expression for the phase difference between the voltage and the current.

If the supply voltage were 10 V, the frequency 1.0 kHz and the capacitance 2.0 μF, what value of R in the circuit would allow a current of 0.10 A to flow? (*L*.)

17. Explain what is meant in an alternating current circuit by (*a*) reactance, (*b*) impedance, (*c*) resonance. Describe an experiment by which (*c*) may be demonstrated.

A coil of self-inductance of 0.200 H and resistance 50.0 Ω is to be supplied with a current of 1.00 A from a 240 V, 50 Hz, supply and it is desired to make the current in phase with the potential difference of the source. Find the values of the components that must be put in series with the coil. Illustrate the conditions in the circuit with a phasor (vector) diagram. (*L*.)

18. Explain what is meant by the *frequency, amplitude, phase* and *root mean square value* of a sinusoidal alternating current.

(*a*) A perfect capacitor and a coil of negligible resistance are connected in parallel to a variable frequency sinusoidal alternating voltage supply. Explain why, as the frequency is varied from a very small to a very high value, there is a frequency at which no current is drawn from the supply. Explain also how the current would vary with frequency if the capacitor and coil are re-connected in series instead of in parallel. (*O. & C.*)

19. Explain what is meant by the *reactance* of an inductor or capacitor.

An alternating potential difference is applied (i) to in inductor, (ii) to a capacitor. Describe and explain the phase lag or lead between the current and the applied potential difference in each case.

Calculate the reactance of an inductor L of inductance 100 mH and of a capacitor C of capacitance 2 μF, both at a frequency of 50 Hz. At what frequency f_0 are their reactances equal in magnitude?

The inductor L and capacitor C are connected in parallel and an alternating potential difference of constant amplitude and variable frequency f is applied, Fig. 37c. (*a*) What

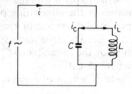

Fig. 37c

is the phase relationship between i_L and i_C? (*b*) What are the relative magnitudes of i_L and i_C when $f = f_0$ and what is then the value of i? (*c*) What are the relative magnitudes of i, i_L and i_C when f is very much greater than f_0? Explain briefly your conclusions in each case. (*O. & C.*)

38 Electrons: Motion in Fields. Electron Tubes

Basic Unit of Charge

As we discussed earlier, matter has a particle nature—it consists of extremely small but separate particles which are atoms or molecules. Helmholtz, an eminent German scientist, stated about a century ago that charges were built up of basic units, that is, electricity is 'granular' and not continuous. He was led to this conclusion from studying the laws of electrolysis.

Particle Nature of Electricity

In electrolysis, we assume that the carriers of current through an acid or salt solution are ions, which may be positively and negatively charged (p. 677). From Faraday's laws of electrolysis, the charge carried by each ion is proportional to its valency (p. 681). We can find the charge on a monovalent ion using the following argument.

The Avogadro constant, about $6 \cdot 02 \times 10^{23}$, is the number of molecules in one mole. In electrolysis, 96 500 coulombs (the Faraday constant) is the quantity of electricity required to deposit one mole of a monovalent element (see p. 681). When the element is monatomic, the number of ions of one kind carrying this charge is equal to the number of molecules. Thus the charge on each ion is given by $96 500/6 \cdot 02 \times 10^{23}$ or $1 \cdot 6 \times 10^{-19}$C. If $1 \cdot 6 \times 10^{-19}$C is denoted by the symbol e, the charge on any ion is then e, $2e$, or $3e$, etc., depending on its valency. Thus e is a basic unit of charge.

All charges, whether produced in electrostatics, current electricity or any other method, are multiples of the basic unit e. Evidence that this is the case was obtained by Millikan, who, in 1909, designed an experiment to measure the unit e.

Theory of Millikan's Experiment

Millikan first measured the terminal velocity of an oil-drop through air. He then charged the oil-drop and applied an electric field to oppose gravity. The drop now moved with a different terminal velocity, which was again measured.

Suppose the radius of the oil-drop is a, the densities of oil and air are ρ and σ respectively, and the viscosity of air is η. When the drop, without a charge, falls steadily under gravity with a terminal velocity v_1, the upward viscous force $= 6\pi\eta a v_1$, from Stokes' law (p. 167) $= k v_1$, where k is $6\pi\eta a$. This is equal to $m'g$, the weight of the drop less the upthrust due to the air. So

$$m'g = k v_1 \qquad . \qquad . \qquad . \qquad . \qquad . \qquad \text{(i)}$$

Suppose the drop now gains a charge q and an electric field of intensity E is applied to oppose gravity and slow the falling drop. The drop then has a smaller terminal velocity v_2. Since the force on the drop due to E is Eq, then, if the mass of the drop has remained constant,

$$m'g - Eq = k v_2 \qquad . \qquad . \qquad . \qquad . \qquad . \qquad \text{(ii)}$$

$$\therefore Eq = m'g - kv_2 = k(v_1 - v_2)$$

So
$$q = \frac{k}{E}(v_1 - v_2) \qquad . \qquad . \qquad . \qquad . \qquad . \qquad \text{(iii)}$$

The weight of the oil-drop = volume × density × $g = \frac{4}{3}\pi a^3 \rho g$; the upthrust on the drop due to the air = weight of air displaced = $\frac{4}{3}\pi a^3 \sigma g$. So

$$m'g = \text{weight} - \text{upthrust}$$

$$= \tfrac{4}{3}\pi a^3 (\rho - \sigma)g = 6\pi\eta av_1, \text{ from (i)}$$

$$\therefore a = \left[\frac{9\eta v_1}{2(\rho - \sigma)g}\right]^{\frac{1}{2}}$$

Since $k = 6\pi\eta a$, it follows from (iii) that

$$q = \frac{6\pi\eta}{E}\left[\frac{9\eta v_1}{2(\rho - \sigma)g}\right]^{\frac{1}{2}}(v_1 - v_2) \qquad . \qquad . \qquad . \qquad . \qquad \text{(iv)}$$

Experiment

In his experiments Millikan used two horizontal plates A, B about 20 cm in diameter and 1·5 cm apart, with a small hole H in the centre of the upper plate, Fig. 38.1. He used a fine spray to 'atomise' the oil and create tiny drops above

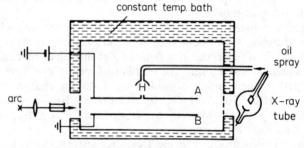

constant temp. bath

Fig. 38.1 Principle of Millikan's experiment

H, and occasionally one would find its way through H, and would be observed in a low-power microscope by reflected light when the chamber was brightly illuminated. The drop was seen as a pin-point of light, and its downward velocity was measured by timing its fall through a known distance by means of a scale in the eyepiece. The field was applied by connecting a battery of several thousand volts across the plates A, B, and its intensity E was known, since $E = V/d$, where V is the p.d. between the plates and d is their distance apart. Millikan found that the friction between the drops when they were formed by the spray created electric charge, but to give a drop an increased charge an X-ray tube was operated near the chamber to ionise the air. In elementary laboratory experiments, increased charges may be obtained by using a radioactive source to ionise the air.

In his experiment, Millikan found that, when an oil-drop gained charges, the *velocity change, $v_1 - v_2$,* varied by integral multiples of a basic unit. Now the charge q is proportional to $(v_1 - v_2)$, from (iii), for a particular drop. Consequently *the charge is a multiple of some basic unit.* So a particular charge consists of 'grains' or multiples of this basic unit.

From equation (iv), it follows that when v_1, v_2, E, ρ, σ and η are all known, the charge q on the drop can be calculated. Millikan found, working with hundreds of drops, that the charge q was always a simple multiple of a basic unit, which was about $1 \cdot 6 \times 10^{-19}$ C. Earlier, Sir J. J. Thomson had discovered the existence of the electron, a particle inside atoms which carried a negative charge $-e$ (p. 805). Millikan concluded that the charge e was $1 \cdot 6 \times 10^{-19}$ C.

Example

Calculate the radius of a drop of oil, density 900 kg m^{-3}, which falls with a terminal velocity of $2 \cdot 9 \times 10^{-4}$ m s^{-1} through air of viscosity $1 \cdot 8 \times 10^{-5}$ N s m^{-2}. Ignore the density of the air.

If the charge on the drop is $-3e$, what p.d. must be applied between two plates 5 mm apart ($e = 1 \cdot 6 \times 10^{-19}$ C.)

When the drop falls with a terminal velocity, force due to viscous drag = weight of sphere. With the usual notation, if ρ is the oil density, we have

$$6\pi\eta av = \text{volume} \times \text{density} \times g = \tfrac{4}{3}\pi a^3 \rho g$$

$$\therefore a = \sqrt{\frac{9\eta v}{2\rho g}} = \sqrt{\frac{9 \times 1 \cdot 8 \times 10^{-5} \times 2 \cdot 9 \times 10^{-4}}{2 \times 900 \times 9 \cdot 8}}$$

$$= 1 \cdot 6 \times 10^{-6} \text{ m} \quad . \qquad . \qquad . \qquad . \qquad . \qquad . \qquad . \quad (1)$$

since $g = 9 \cdot 8$ m s^{-2}.

Suppose the upper plate is V volt higher than the lower plate when the drop is stationary, so that the electric field intensity E between the plates is V/d. Then upward force on drop $= E \times 3e = $ weight of drop.

$$\therefore E \times 3e = \tfrac{4}{3}\pi a^3 \rho g$$

$$\therefore E = \frac{4\pi a^3 \rho g}{9e} = \frac{V}{d}$$

$$\therefore V = \frac{4\pi a^3 \rho g d}{9e}$$

$$= \frac{4\pi \times (1 \cdot 6 \times 10^{-6})^3 \times 900 \times 5 \times 10^{-3} \times 9 \cdot 8}{9 \times 1 \cdot 6 \times 10^{-19}}$$

$$= 1576 \text{ V.}$$

Cathode Rays (Electrons) and Properties

Atomic physics can be said to have begun with the study of the conduction of electricity through gases. The passage of electricity through a gas, called a 'discharge', was familiar to Faraday, but the steady conduction—as distinct from sparks—takes place when the pressure of the gas is less than about 50 mmHg; in a neon lamp it is about 10 mmHg.

The Gaseous Discharge at Various Pressures

Figure 38.2 (i) represents a glass tube, about 0·5 metre long, connected to a vacuum pump P and a pressure gauge G. It contains an anode A and a cathode K, connected respectively to the positive and negative terminals of the secondary of an induction coil. As the air is pumped out, nothing happens until the pressure has fallen to about 100 mmHg. Then thin streamers of luminous gas appear between the electrodes, Fig. 38.2 (ii).

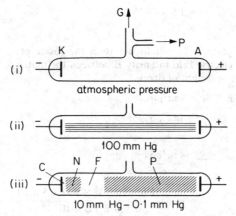

Fig. 38.2 Stages in development of gaseous discharge

At about 10 mmHg the discharge becomes a steady glow, spread through the tube, Fig. 38.2 (iii). It is broken up by two darker regions, of which the one, C, nearest the cathode K is narrow and hard to see. The dark region C is called the cathode dark space, or sometimes, after its discoverer, the Crookes' dark space. Beyond the cathode dark space is a bright region N called the negative glow, and beyond that the Faraday dark space F—also called after its discoverer. Beyond the Faraday dark space stretches a luminous column P, called the positive column, which fills the rest of the discharge tube. Sometimes the positive column breaks up into alternating bright and dark segments, called striations, shown in Fig. 38.2. In all the photographs of Fig. 38.3 the cathode is on the left. The cathode dark space can hardly be seen—it lies just around the cathode— but the negative glow and Faraday dark space are clear.

The positive column is the most striking part of the discharge, but the cathode dark space is electrically the most important. In it the electrons from the cathode are being violently accelerated by the electric field, and gaining energy with which to ionise the gas atoms. In the positive column some atoms are being ionised by collisions with electrons; others are being excited, in a way which we cannot describe here, and made to emit their characteristic spectra.

When the pressure of the gas in the discharge tube is reduced still further, the dark spaces swell, and the positive column shrinks. At about 1 mmHg the

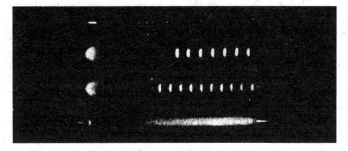

Fig. 38.3 The lowest photograph shows the positive column. As the pressure decreases, the positive column breaks up into striations and shrinks towards the anode on the right. The top photograph shows the dark space completely filling the tube, when the pressure is about 0·01 mm Hg

cathode dark space becomes distinct, and at 0·1 mmHg it is several centimetres long. Eventually, as the pressure falls, the cathode dark space stretches from the anode to the cathode, and the negative glow and positive column vanish. This happens at about 0·01 mmHg in a tube about a half-metre long.

When the cathode dark space occupies the whole discharge tube, the walls of the glass tube fluoresce. The electrons flying across the space and hitting the glass are called *cathode rays*. Where they strike the anode they produce X-rays (p. 861).

The Mechanism of Conduction

How are the ions and electrons in a gaseous discharge produced? A luminous discharge requires a voltage, V, of at least a hundred volts across the gas at pressures of about 1 mmHg. At much higher or lower pressures, it may require thousands of volts. But with a voltage of about ten, although there is no glow, a very weak current, I, can be detected—of the order of 10^{-15} A, Fig. 38.4 (1). This we attribute to electrons emitted from the cathode by the photoelectric effect (p. 843); a trace of ultraviolet light in the laboratory would account for the emission.

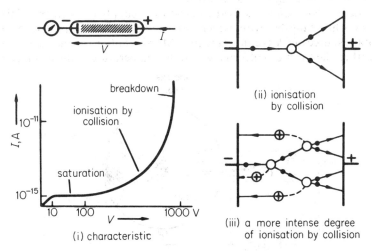

Fig. 38.4 Discharge through gas at low pressure. In (i) currents and voltages are order of magnitude only: in (ii) and (iii), the numerous non-ionising collisions are not shown

When the voltage is increased the electrons are accelerated by the electric field to a higher speed, and strike the gas atoms more violently on their way to the anode. When the voltage is high enough the electrons strike the atoms with sufficient kinetic energy to knock electrons out of them, Fig. 38.4 (ii). This process is called *ionization by collision*; the atoms become ions, and move towards the cathode; the extra electrons join the original ones in their flight to the anode. At higher voltages the knocked-out electrons are accelerated enough to produce more ions and electrons on the way, Fig. 38.4 (iii). Eventually a point is reached at which the current grows uncontrollably—the gas is said to break down. In practice, the current is limited by a resistor, in series with the discharge tube; in a commercial neon lamp this resistor, of resistance about 5000 Ω, is hidden in the base.

The current through a gas, like that through an electrolyte, is carried by carriers of both signs—positive and negative. At the anode, the negative electrons enter the wires of the outside circuit, and eventually come round to the cathode. There they meet positive ions, which they now enter, and so re-form neutral gas atoms. Positive ions arriving at the cathode knock off some of the atoms, which diffuse into the body of the discharge, and there, eventually, they are ionised again. Thus a limited amount of gas can carry a current indefinitely.

Once a gas has broken down, current can continue to pass through it even in the dark, that is to say, when there is no ultraviolet light to make the cathode emit electrons. The electrons from the cathode are now simply knocked out of it by the violent bombardment of the positive ions.

Ultraviolet light is not, as a rule, necessary even for starting a gaseous discharge. The somewhat mysterious cosmic rays, which reach the earth from outer space, are able to ionise a gas; they may therefore enable a discharge to start. Once it has started, the emission of electrons by bombardment of the cathode keeps it going.

Modern Production of Cathode Rays

The discharge tube method is not a convenient one for producing and studying cathode rays or electrons. Firstly, a gas is needed at the appropriate low pressure; secondly, a very high p.d. is needed across the tube; thirdly, X-rays are produced (p. 861) which may be dangerous.

Nowadays a *hot cathode* is used to produce a supply of electrons. This may consist of a fine tungsten wire, which is heated to a high temperature when a low voltage source of 6 V is connected to it. Metals contain free electrons, moving about rather like the molecules in a gas. If the temperature of the metal is raised,

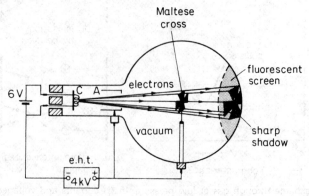

Fig. 38.5 Electrons travel in straight lines

the thermal velocities of the electrons will be increased. The chance of electrons escaping from the attraction of the positive ions, fixed in the lattice, will then also be raised. Thus by heating a metal such as tungsten to a high temperature, electrons can be 'boiled off'. This is called *thermionic emission* (see also p. 810).

Figure 38.5 shows a tungsten filament C inside an evacuated tube. When heated by a low voltage supply, electrons are produced, and they are accelerated by a positive voltage of several thousand volts applied between C and a metal cylinder A. The electrons travel unimpeded across the tube past A, and produce a glow when they collide with a fluorescent screen and give up their energy.

Properties of Cathode Rays

Fast-moving electrons emitted from C produce a sharp shadow of a Maltese cross on the fluorescent screen, as shown in Fig. 38.5. Thus the cathode rays travel in straight lines. They also produce heat when incident on a metal—a fine piece of platinum glows, for example.

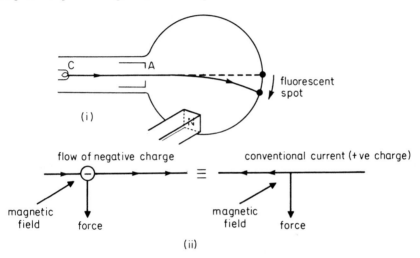

Fig. 38.6 Deflection shows electrons are negatively charged

When a magnet is brought near to the electron beam, the glow on the fluorescent screen moves, Fig. 38.6 (i). If Fleming's left hand rule is applied to

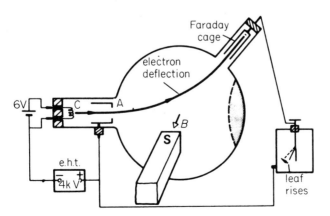

Fig. 38.7 Perrin tube. Direct method for testing electron charge

the motion to deduce the conventional current (+ ve charge) direction, the middle finger points in a direction *opposite* to the electron flow, Fig. 38.6 (ii). So electrons appear to be particles which carry a *negative* charge.

This is confirmed by collecting electrons inside a *Perrin tube*, shown in Fig. 38.7. The electrons are deflected by the magnet S until they pass into a metal cylinder called a 'Faraday cage' (see p. 589). The cylinder is connected to the plate of an electroscope, which has been negatively charged using an ebonite rod and fur. As soon as the electrons are deflected into the cage the leaf rises further, showing that an extra *negative* charge has been collected by the cage. This supports the idea that cathode rays are electrons.

Electron Motion in Electric and Magnetic Fields

Deflection in an Electric Field

Suppose a horizontal beam of electrons, moving with velocity v, passes between two parallel plates, Fig. 38.8. If the p.d. between the plates is V and their distance apart is d, the field intensity $E = V/d$. Hence the force on an electron of charge e moving between the plates $= Ee = eV/d$ and is directed towards the positive plate.

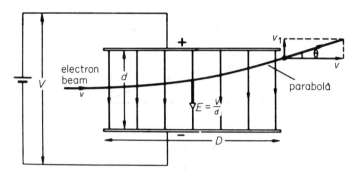

Fig. 38.8 Deflection in electric field

Since the electricity intensity E is vertical, no horizontal force acts on the electron entering the plates. Thus the horizontal velocity, v, of the beam is unaffected. This is similar to the motion of a projectile projected horizontally under gravity. The vertical acceleration due to gravity does not affect the horizontal motion.

In a vertical direction the displacement, $y = \frac{1}{2}at^2$, where $a = $ acceleration $= $ force/mass $= Ee/m_e$ if m_e is the mass of an electron and t is the time.

$$\therefore y = \frac{1}{2}\frac{Ee}{m_e}t^2 \qquad . \qquad . \qquad . \qquad . \qquad . \qquad \text{(i)}$$

In a horizontal direction, the displacement, $x = vt$. $\qquad$. $\qquad$ (ii)

Eliminating t between (i) and (ii), we obtain

$$y = \frac{1}{2}\left(\frac{Ee}{m_e}\right)\frac{x^2}{v^2} = \frac{Ee}{2m_ev^2}x^2.$$

The path is therefore a *parabola*.

When the electron just passes the plates, $x = D$. The value of y is then $y = EeD^2/2m_ev^2$. The beam then moves in a straight line, as shown in Fig. 38.8. The time for which the electron is between the plates is D/v. Thus the component of the velocity v_1, gained in the direction of the field during this time, is given by

$$v_1 = \text{acceleration} \times \text{time} = \frac{Ee}{m_e} \times \frac{D}{v}.$$

Hence the angle θ at which the beam emerges from the field is given by:

$$\tan \theta = \frac{v_1}{v} = \frac{EeD}{m_ev}\cdot\frac{1}{v} = \frac{EeD}{m_ev^2}.$$

As the reader can verify, we can also write $\tan \theta = y/\frac{1}{2}D$, where y is the vertical displacement produced.

The energy of the electron is increased by an amount of $\frac{1}{2}mv_1^2$ as it passes through the plates, since the energy due to the horizontal motion is unaltered.

Example

A beam of electrons, moving with a velocity of $1 \times 0 \times 10^7$ m s^{-1}, enters midway between two horizontal parallel plates P, Q in a direction parallel to the plates, Fig. 38.9. P and Q are 5 cm long, 2 cm apart and have a p.d. V applied between them. Calculate V if the beam is deflected so that it just grazes the edge of the low plate Q. (Assume $e/m_e = 1\cdot8 \times 10^{11}$ C kg^{-1}.)

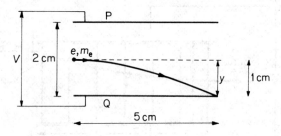

Fig. **38.9** Example

Electric intensity between plates, $E = \dfrac{V}{d} = \dfrac{V}{2 \times 10^{-2}}$.

Downward acceleration on electron, $a = \dfrac{\text{force}}{\text{mass}} = \dfrac{Ee}{m_e}$.

So vertical distance, $y = \frac{1}{2}at^2 = \frac{1}{2}\dfrac{Ee}{m_e}t^2$.

But $y = 1$ cm $= 10^{-2}$ m and $t = \dfrac{5 \times 10^{-2}}{1 \times 10^7} = 5 \times 10^{-9}$ s, since the horizontal velocity is not affected by the vertical electric field and remains constant.

So $\qquad\qquad 10^{-2} = \dfrac{1}{2} \times \dfrac{V}{2 \times 10^{-2}} \times 1\cdot8 \times 10^{11} \times (5 \times 10^{-9})^2$.

Simplifying, $\qquad\qquad V = 89$ V (approx.).

Deflection in a Magnetic Field

Consider an electron beam, moving with a speed v, which enters a uniform magnetic field of magnitude B acting perpendicular to the direction of motion, Fig. 38.10. The force F on an electron is then Bev. The direction of the force is perpendicular to both B and v. Consequently, unlike the electric force, the

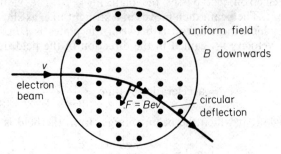

Fig. **38.10** Circular motion in uniform magnetic field

magnetic force cannot change the *energy* of the electron. It *deflects* the electron but does not change its speed or kinetic energy.

The force Bev is always normal to the path of the beam. If the field is uniform the force is constant in magnitude and the beam then travels in a *circle* of radius r. Since Bev is the centripetal force (towards the centre),

$$Bev = \frac{m_e v^2}{r}$$

$$\therefore r = \frac{m_e v}{Be} = \frac{\text{momentum}}{Be}.$$

If the velocity v of the electron decreases continuously due to collisions, for example, its momentum decreases. So, from the relation for r above, the radius of its path decreases and the electron will thus tend to spiral instead of moving in a circular path of constant radius.

Example

Protons, with a charge-mass-ratio of 1.0×10^8 C kg^{-1}, are rotated in a circular orbit of radius r when they enter a uniform magnetic field of 0.5 T. Show that the number of revolutions per second, f, is independent of r and calculate f.

Suppose m_p is the proton mass and e is the charge. Then, with the usual notation,

$$Bev = \frac{m_p v^2}{r}$$

So
$$v = \frac{Ber}{m_p} = r\omega,$$

where ω is the angular velocity. Since r cancels,

$$\omega = \frac{Be}{m_p} = 2\pi f.$$

Then
$$f = \frac{Be}{2\pi m_p}.$$

This result for f shows that it is independent of r. Also,

$$f = \frac{0.5 \times 1 \times 10^8}{2 \times 3.14} = 8 \times 10^6 \text{ rev s}^{-1}.$$

Fig. 38A Parabolic path of an electron beam after entering a uniform perpendicular *electric field*. The upper plate is +ve in potential relative to the lower plate. The screen, which is coated with a luminous paint, has squares marked on it for measuring the electron deflection in a charge/mass ratio experiment. Helmholtz coils, the large circular coils in front and behind the tube, are used to apply a uniform *magnetic field* to the electron beam. (*Courtesy of Teltron Limited*)

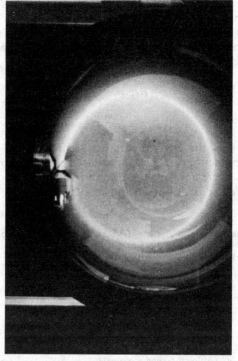

Fig. 38B Circular deflection of electron beam in a uniform magnetic field due to Helmholtz coils (not visible). Faster electrons produce the outer diffuse beam.

Unlike the electric deflection tube (Fig. 38A), this tube has a small amount of helium gas inside it. The gas molecules are ionised after collision with high-speed electrons and emit light. The electron track is thus made visible as a fine straight beam before the field is applied.

(*Courtesy of Teltron Limited*)

Thomson's Experiment for *e/m*

In 1897, Sir J. J. Thomson devised an experiment for measuring the ratio *charge/mass* or e/m_e for an electron, called its *specific charge*.

Thomson's apparatus is shown simplified in Fig. 38.11. C and A are the cathode and anode respectively, and narrow slits are cut in opposite plates at A so that the cathode rays passing through are limited to a narrow beam. The rays then strike the glass at O, producing a glow there. The rays can be deflected electrostatically by means of connecting a large battery to the horizontal plates P, Q,

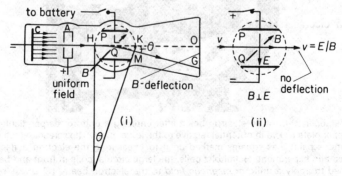

Fig. 38.11 Thomson's determination of e/m_e for electron (*not to scale*)

or magnetically by means of a current passing through two Helmholtz coils on either side of the tube near P and Q (see Fig. 38A, p. 803).

The magnetic field B is perpendicular to the paper, and if it is uniform a constant force acts on the cathode rays (electrons) normal to its motion. With B alone, the particles thus begin to move along the arc HM of a circle of radius r, Fig. 38.11 (i). When they leave the field, the particles move in a straight line MG (see Fig. 38.10) and strike the glass at G.

With the usual notation, see p. 803,

$$\text{force } F = Bev = \frac{m_e v^2}{r},$$

where e is the charge on an electron and m_e is its mass.

$$\therefore \frac{e}{m_e} = \frac{v}{rB} \qquad \qquad \text{(i)}$$

To find the radius r, from Fig. 38.11 (i), $\tan \theta = OG/OL = HK/r$.

$$\therefore r = \frac{HK.OL}{OG}.$$

L is about the middle of the magnetic field coils.

The electric field E between P and Q was now varied until the beam was returned to O. There is now no deflection, Fig. 38.11 (ii). Since the electric field direction is vertical and the magnetic field B is horizontal, E is perpendicular to B; this is sometimes called 'crossed' fields. Hence the upward force $Ee =$ the downward force Bev. So

$$Ee = Bev$$

$$\therefore v = \frac{E}{B}.$$

Thomson found that v was considerably less than the velocity of light, 3×10^8 m s^{-1}, so that cathode rays were certainly not electromagnetic waves.

On substituting for v and r in (i), the ratio charge/mass (e/m_e) for an electron was obtained. Modern determinations show that

$$\frac{e}{m_e} = 1\cdot76 \times 10^{11} \text{ C kg}^{-1}$$

or $$\frac{m_e}{e} = \frac{1}{1\cdot76} \times 10^{-11} \text{ kg per coulomb (kg C}^{-1}) \qquad . \qquad . \qquad \text{(ii)}$$

Now from electrolysis the mass-charge ratio for a hydrogen ion is $1\cdot04 \times 10^{-8}$ kg C^{-1}.

$$\therefore \frac{m_H}{e} = 1\cdot04 \times 10^{-8} \text{ kg C}^{-1},$$

assuming the hydrogen ion carries a charge e numerically equal to that on an electron, m_H being the mass of the hydrogen ion. Hence, with (ii),

$$\frac{m_e}{m_H} = \frac{1}{1\cdot76 \times 1\cdot04 \times 10^3} = \frac{1}{1830}.$$

Thus the mass of an electron is nearly two–thousandths that of the hydrogen atom. Until Sir J. J. Thomson's experiment, it was believed that the hydrogen atom was the lightest particle in existence.

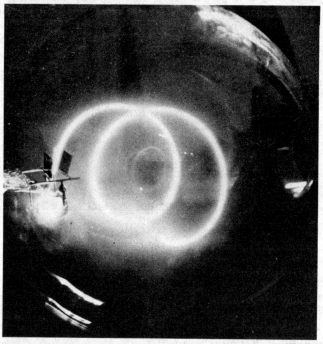

Fig. 38C Spiral electron path, produced when an electron beam enters a *magnetic field* at an angle to the field. The component of the velocity normal to the field produces the circular motion; the component parallel to the field produces the translational motion. Initially the beam passes through the two plates shown, which can be used to apply an electric field. (*Courtesy of Leybold-Heraeus GMB and Co*)

Specific Charge of Electron by Magnetic Deflection

The specific charge of an electron (e/m_e) can be measured in the school laboratory by a TELTRON tube designed for this purpose, Fig. 38.12.

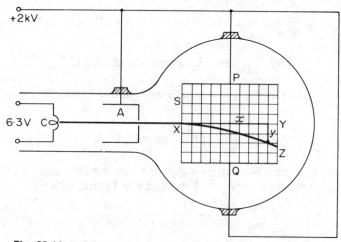

Fig. 38.12 Deflection of electron beam (Helmholtz coils not shown)

The evacuated tube has (i) a hot cathode C which produces an electron beam, (ii) an accelerating anode A, (iii) a vertical screen S coated with luminous paint, between two plates P, Q not used in this experiment. The screen S has horizontal and vertical lines equally spaced and is slightly inclined to the electron beam incident on it. The beam then produces a luminous glow and its path can be seen.

Two Helmholtz coils H, H connected in series are on opposite sides of the tube (see p. 803). Their purpose is to produce a uniform magnetic field B perpendicular to the beam over the region of S, so that the beam is deflected in a circular path. The circuit for the coils is shown in Fig. 38.13. The value of

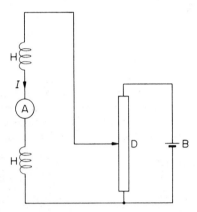

Fig. 38.13 Circuit for Helmholtz coils

B is directly porportional to the current I in the coils, which is measured by the ammeter A, and this is varied by means of a potential divider D with a battery B connected to it. Calculation shows that the magnitude of B is given by, approximately,

$$B = 0.72 \frac{\mu_0 N I}{R}$$

where N is the number of turns in each coil, I is the current in amperes and R is the radius in metres of the coils (p. 715).

Experiment. In an experiment, the cathode C is heated by a 6·3 V supply. The anode A has a high potential V relative to C of say 2 kV or 2000 V. The plates P, Q are joined together and kept at the same potential as A; this eliminates electric fields beyond A, which would alter the speed of the electron beam. After passing A, the beam produces a luminous horizontal line XY on S.

A current is now passed into the Helmholtz coils, H, H. The electron beam was then deflected along a circular path XZ seen on S, Fig. 38.12. By altering the current in the coils, the magnetic field B can be varied to produce a path XZ of suitable radius of curvature r. The horizontal and vertical distances x, y from X of a convenient point on XZ are then read from the graduations on S. As shown on p. 468, r is given by

$$r = \frac{x^2 + y^2}{2y}$$

Theory. If v is the velocity of the electrons at the anode A, then, assuming zero velocity at C, their gain in kinetic energy $= \frac{1}{2}m_e v^2$. So if V is the anode potential,

$$\frac{1}{2}m_e v^2 = eV \qquad . \qquad . \qquad . \qquad . \qquad . \qquad (1)$$

The radius r of the circular path XZ of the beam is given by

$$Bev = \frac{m_e v^2}{r} \qquad . \quad . \quad . \quad . \quad . \quad (2)$$

From (1)

$$v^2 = 2V\left(\frac{e}{m_e}\right)$$

From (2)

$$v = Br\left(\frac{e}{m_e}\right)$$

So

$$B^2 r^2 \left(\frac{e}{m_e}\right)^2 = v^2 = 2V\left(\frac{e}{m_e}\right)$$

Cancelling e/m_e,

$$\frac{e}{m_e} = \frac{2V}{B^2 r^2} \qquad . \quad . \quad . \quad . \quad (3)$$

Knowing V, r and B, then e/m_e can be calculated.

The main errors in the experiment are: (i) the difficulty of measuring the radius r with accuracy, (ii) B may not be uniform over the whole region of S, (iii) error in the voltmeter measuring V. An error in r or in B would produce double the percentage error in e/m_e since r^2 and B^2 occur in (3) above.

Helical Path of Electrons

FIG. 38.13A Helical electron path

Consider an electron P of charge e, mass m_e entering a uniform magnetic field B at a small angle θ with velocity v, Fig. 38.13A. The component $v \cos \theta$ parallel to B produces a *translational* (linear) motion, since no electromagnetic force acts in this case. The component $v \sin \theta$ normal to B, however, produces a *circular* motion. So the electron path is a helix (spiral). See Fig. 38c, page 806.

The pitch XY of the helix, the distance between neighbouring turns, is given by $XY = v \cos \theta . T$, where T is the period of rotation of the electron. For circular motion (see p. 803),

$$Bev \sin \theta = \frac{m_e(v \sin \theta)^2}{r}$$

so

$$\frac{r}{v \sin \theta} = \frac{m_e}{Be}$$

Hence

$$T = \frac{2\pi r}{v \sin \theta} = \frac{2\pi m_e}{Be}$$

So

$$XY = v \cos \theta . T = \frac{2\pi m_e v \cos \theta}{Be}$$

If $v = 10^6 \text{ m s}^{-1}$, $e/m_e = 1.8 \times 10^{11} \text{ C kg}^{-1}$, $\theta = 10°$, $B = 2 \times 10^{-4} \text{ T}$,

then

$$XY = \frac{2\pi \times 10^6 \times \cos 10°}{1.8 \times 10^{11} \times 2 \times 10^{-4}} = 0.17 \text{ m}$$

Examples

1. Describe and give the theory of a method to determine e the electronic charge. Why is it considered that all electric charges are multiples of e?

An electron having 450 eV of energy moves at right angles to a uniform magnetic field of flux density $1 \cdot 50 \times 10^{-3}$ T. Show that the path of the electron is a circle and find its radius. Assume that the specific charge of the electron is $1 \cdot 76 \times 10^{11}$ C kg^{-1}. (L.)

With the usual notation, the velocity v of the electron is given by

$$\tfrac{1}{2}m_e v^2 = eV, \text{ where } V \text{ is } 450 \text{ V.}$$

$$\therefore v = \sqrt{\frac{2eV}{m_e}} \qquad . \qquad . \qquad . \qquad . \qquad . \qquad (1)$$

The path of the electron is a circle because the force Bev is constant and always normal to the electron path. Its radius r is given by

$$Bev = \frac{m_e v^2}{r}$$

$$\therefore r = \frac{m_e}{e} \cdot \frac{v}{B} = \frac{m_e}{e} \cdot \frac{1}{B}\sqrt{\frac{2eV}{m_e}}, \text{ from (1)}$$

$$\therefore r = \frac{1}{B}\sqrt{2\frac{m_e V}{e}}$$

Now $e/m_e = 1 \cdot 76 \times 10^{11}$ C kg^{-1}, $V = 450$ V, $B = 1 \cdot 5 \times 10^{-3}$ T

$$\therefore r = \frac{1}{1 \cdot 5 \times 10^{-3}}\sqrt{\frac{2 \times 450}{1 \cdot 76 \times 10^{11}}} \text{ m}$$

$$= 4 \cdot 8 \times 10^{-2} \text{ m.}$$

2. A charged oil drop of mass 6×10^{-15} kg falls vertically in air with a steady velocity between two long parallel vertical plates 5 mm apart. When a p.d. of 3000 V is applied between the plates, the drop now falls with a steady velocity at an angle of 58° to the vertical. Calculate the charge Q on the drop. (Assume $g = 10$ m s^{-2}.)

The drop falls with steady velocity due to the viscosity of the air. Neglecting the upthrust, if v_1 is the vertical velocity

$$\text{downward force on drop} = \text{upward viscous force}$$

or $$mg = kv_1$$

Similarly, when the electric field of intensity E is applied,

$$\text{horizontal force on drop} = EQ = kv_2,$$

where v_2 is the horizontal velocity.

By vector addition, if θ is the angle to the vertical,

$$\tan \theta = \frac{v_2}{v_1} = \frac{EQ}{mg}$$

Now $$E = \frac{V}{d} = \frac{3000}{5 \times 10^{-3}} = 6 \times 10^5 \ V \ m^{-1}$$

$$\therefore \tan 58° = \frac{EQ}{mg} = \frac{6 \times 10^5 \ Q}{6 \times 10^{-15} \times 10} = 10^{19} \ Q$$

$$\therefore Q = \frac{\tan 58°}{10^{19}} = 1 \cdot 6 \times 10^{-19} \text{ C}$$

Radio Valves. Cathode Ray Oscilloscope

Rectification of A.C. Diode Valve

Alternating current is easier to distribute than direct current, because alternating voltage can be transformed easily up or down. For electrolysis, battery-charging, and the operation of radio-receivers and transmitters, however, direct current is essential. It can be obtained from an alternating current supply by means of a *rectifier*, which is a device that will only pass current in one direction.

A thermionic type of rectifier is normally a *diode valve*. It contains a metal filament, F, surrounded by a metal anode A, shown diagrammatically in Fig. 38.14 (i). These are contained inside a glass bulb from which all the air has been removed, thus leaving a vacuum. The filament is heated by a current drawn from a low voltage supply B, and emits electrons. The anode potential can be varied, as shown.

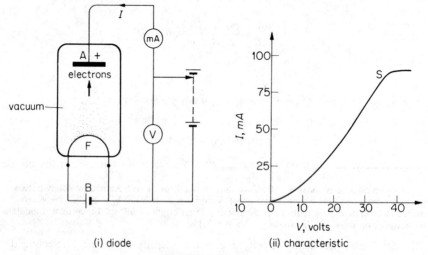

(i) diode   (ii) characteristic

Fig. 38.14 Diode valve

Since electrons are negative charges, such a device passes current when its anode is made positive with respect to its filament, but not when the anode is made negative. Figure 38.14 (ii) shows the curve of anode current I against anode potential V for a small diode; it is called the diode's characteristic curve, or simply its *characteristic*. The current increases with the positive anode potential as far as the point S. Beyond this point the current does not increase, because the anode is collecting all the electrons emitted by the filament; the current is said to be saturated.

At first sight we might expect that any positive anode potential, however small, would draw the full saturation current from the filament to the anode. But it does not, because the charges on the electrons make them repel one another. Thus the cloud of electrons between the anode and filament forms a negative 'space charge' and repels the electrons leaving the filament, turning some of them back. The electrons round the filament are like the molecules in a cloud of vapour above a liquid; they are continually escaping from it and returning to it. The positive anode draws some electrons away from the electron cloud or space charge so that more electrons escape from the filament than return. The higher the anode potential, the fewer electrons return to the filament; as the anode potential rises, the current increases, to its saturation limit.

Rectifier Circuit

When a diode is used as a rectifier it is connected in a circuit such as Fig. 38.15 (i). The low-voltage secondary of the transformer simply provides the heating current from the filament. The current to be rectified is drawn from the high-voltage secondary. One end of this secondary is connected to the load, which could be, as shown, as accumulator on charge; the other end of the secondary is connected to the anode of the diode, and the other end of the load to one of

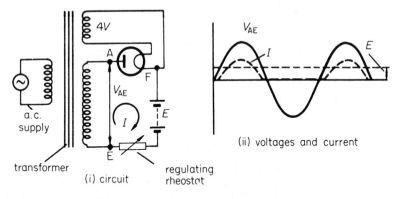

(ii) voltages and current

transformer   (i) circuit   regulating rheostat

Fig. 38.15 Rectifier circuit with diode

the filament connexions. When the transformer secondary voltage V_{AE} is greater than the e.m.f. E of the accumulator, the anode is positive with respect to the filament; the electrons from the hot filament are then drawn to the anode, and a current flows through the load, Fig. 38.15 (ii). On half-cycles when the anode is negative, the electrons are repelled, and no current flows. Because it only allows current to flow through it in one direction, a thermionic diode is often called a valve.

Some rectifying valves contain a little mercury vapour. When electrons flow through them, they ionise the mercury atoms, as explained on p. 798. The ions and electrons thus produced make the valve a very good conductor, and reduce the voltage drop across it; they therefore allow more of the voltage from the transformer to appear across the load.

The current from a rectifier flows in pulses, whenever the anode is positive with respect to the filament. Sometimes a smoother current is required, as, for example, in a television receiver, where the pulses would cause a humming sound in the loudspeaker. The current can be smoothed by connecting an inductance coil of about 30 henrys in series with the load. The inductance prevents rapid fluctuations in current. So also does a capacitor of about 2000 microfarads connected across the load. Generally the two are used together to give a very smooth output, as explained later on p. 827.

Other Rectifiers

There are other rectifiers which are not thermionic. One such type consists of an oxidized copper disc, Cu_2O/Cu, pressed against a disc of lead, Pb, Fig. 38.16 (i). These conduct well when the lead is made positive, but very badly when it is made negative; they are called metal rectifiers and are used in small battery chargers to change the mains a.c. to d.c. voltage so that current flows one way through the battery (p. 827).

A semiconductor (silicon) diode is another type of rectifier (p. 826). In Fig. 38.16 (ii), four such rectifiers are arranged in a bridge circuit to convert a

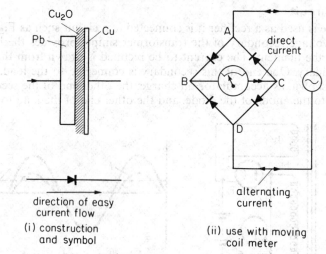

(i) construction
and symbol

direction of easy
current flow

(ii) use with moving
coil meter

alternating
current

direct
current

Fig. 38.16 Metal rectifier and use

moving-coil meter into an alternating current meter. The current flows only one way in the *diagonal arm* of the bridge; ABCD on one half of the a.c. cycle and DBCA on the other half of the same cycle. The meter records the average value of the current.

Cathode-Ray Oscilloscope (C.R.O.)

We now discuss an important type of electron tube, widely used, called a *cathode ray oscilloscope.*

An oscilloscope is an instrument for plotting one varying physical quantity—potential difference, sound-pressure, heart-beat—against another—current, displacement, time. A cathode-ray oscillograph, of the kind we are about to describe, plots alternating potential difference against time. It is called a cathode-ray oscilloscope because it traces the required wave-form with a beam of electrons, and beams of electrons were originally called cathode rays.

A cathode-ray oscilloscope is essentially an electrostatic instrument. It consists of a highly evacuated glass tube, T in Fig. 38.17, one end of which opens out to form a screen S which in internally coated with zinc sulphide or other fluorescent material. A hot filament F, at the other end of the tube, emits electrons. These are then attracted by the cylinders A_1 and A_2, which have increasing positive potentials with respect to the filament. Many of the electrons, however, shoot through the cylinders and strike the screen; where they do so, the zinc sulphide fluoresces in a green spot. On their way to the screen, the electrons pass through two pairs of metal plates, XX and YY, called the deflecting

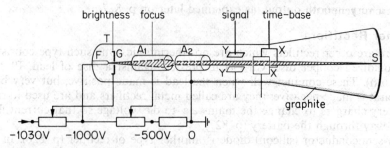

Fig. 38.17 A cathode-ray oscilloscope tube

plates. The inner walls of the tube are coated with graphite, which is connected to the final anode A_2. This makes the space between A_2 and S an equipotential volume so that the speed of the electrons is maintained to S from A_2.

The *brightness* of the light on the screen is controlled by an electrode G in front of the filament F. If G is made more negative in potential relative to F, the increased repulsive force reduces the number of electrons per second passing G.

In practice, the screen S, the tube and A_2 are all earthed to avoid danger due to high voltages. Further, touching the outside of S with earthed fingers does not then alter the electrostatic field inside the tube and affect the deflection of the beam. Thus, as indicated, the filament F is at a high negative potential relative to earth and is therefore dangerous to touch. The filament, electrode G and accelerating electrodes are often called the 'gun assembly' of a tube. Figure 38.17 illustrates a potential divider arrangement for the simple cathode-ray oscilloscope shown. The anode A_2, tube and screen are earthed; A_1 has a varying voltage for the focus control as explained soon; and the brightness control G has a varying negative potential relative to the potential of F, which is about -1000 V.

Deflection; Time-base

If a battery were connected between the Y-plates, so as to make the upper one positive, the electrons in the beam would be attracted towards that plate, and the beam would be deflected upwards. In the same way, the beam can be deflected horizontally by a potential difference applied between the X-plates.

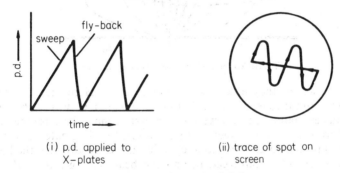

(i) p.d. applied to X–plates

(ii) trace of spot on screen

Fig. 38.18 Action of a C.R.O.

When the oscilloscope is in use, the alternating potential difference to be examined is applied between the Y-plates. If that were all, then the spot would be simply drawn out into a vertical line. To trace the wave-form of the alternating potential difference, the X-plates are used to provide a *time-axis*. A special type of circuit generates a potential difference which rises steadily to a certain value, as shown in Fig. 38.18 (i), and then falls rapidly to zero; it can be made to go through these changes tens, hundreds, thousands or millions of times per second. This potential difference is applied between the X-plates, so that the spot is swept steadily to the right, and then flies swiftly back and starts out again. This horizontal motion provides what is called the *time-base* of the oscillograph. On it is superimposed the vertical motion produced by the Y-plates. Then as shown in Fig. 38.18 (ii), the wave-form of the potential difference to be examined can be displayed on the screen.

In a *double beam oscilloscope*, the two Y-plates are joined to terminals labelled Y_1 and Y_2 respectively. An earthed plate between the plates splits the beam into two halves. One half can be deflected by an input voltage connected to Y_1 and the other half can be deflected by another input voltage connected to Y_2. With a common time-base applied to the X-plates, two traces can be obtained on the screen and two different waveforms, from Y_1 and Y_2 respectively, can thus be compared.

Focusing

To give a clear trace on the screen, the electron beam must be focused to a sharp spot. This is the function of the cylinders A_1 and A_2, called the first and second anodes. Fig. 38.19 shows the equipotentials of the field between them, when their difference of potential is 500 volts. Electrons entering the field from the filament experience forces from low potential to high at right angles to the equipotentials. They have, however, considerable momentum, because they have been accelerated by a potential difference of about 500 volts, and are travelling fast. Consequently the field merely deflects them, and, because of its cylindrical symmetry, it converges the beam towards the point P. Before they can reach

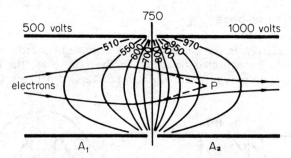

Fig. 38.19 Focusing in an oscilloscope tube

this point, however, they enter the second cylinder. Here the potential rises from the axis, and the electrons are deflected outwards. However, they are now travelling faster than when they were in the first cylinder, because the potential is everywhere higher. Consequently their momentum is greater, and they are less deflected than before. The second cylinder, therefore, diverges the beam less than the first cylinder converged it, and the beam emerges from the second anode still somewhat convergent. By adjusting the potential of the first anode, the beam can be focused upon the screen, to give a spot a millimetre or less in diameter.

Electron-focusing devices are called electron-lenses, or electron-optical systems. For example, the action of the anodes A_1 and A_2 is roughly analogous to that of a pair of glass lenses on a beam of light, the first glass lens being converging, and the second diverging, but weaker.

C.R.O. Supplies

The necessary voltage supplies or circuits for a cathode-ray oscillograph or oscilloscope are shown in block form in Fig. 38.20.

1. The *power pack* supplies e.h.t.—very high d.c. voltage for the tube electrodes such as the electron lens; h.t.—rectified and smoothed d.c. voltage for the amplifiers, for example; and l.t.—low a.c. voltage such as 6·3 V for the valve heaters.

2. *Tube controls.* The brightness and focus, and correction of astigmatism, are controlled by varying the voltage of the appropriate tube electrode.

3. *Y- or Input-Amplifier.* This is a variable gain linear amplifier which amplifies the signal applied to the Y-plates.

4. *Y-shift.* This is in the Y-amplifier circuit and shifts the signal trace up or down on the screen.

5. *Time-base.* This is a 'sawtooth' oscillator (p. 813). In the oscillator circuit, the feedback can be adjusted with stability control until the oscillator is just not free running, and the trigger is then applied to 'lock' the trace.

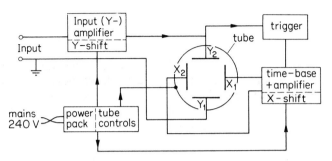

Fig. 38.20 Voltage supplies for oscilloscope

6. *Trigger unit.* This applies a triggering pulse to the time-base oscillator from the Y- or input-voltage. The time-base frequency is then synchronised ('locked') with that of the input signal, so that a stationary trace is obtained on the screen.

7. *Time-base amplifier.* This amplifies the time-base voltage and applies it to the X-plates. The trace can be made to expand or contract horizontally by varying the amplification.

8. *X-shift.* This shifts the time-base horizontally to the left or right.

Uses of Oscilloscope

In addition to displaying waveforms, the oscilloscope can be used for measurement of voltage, frequency and phase.

1. *A.C. voltage*

An unknown a.c. voltage, whose peak value is required, is connected to the Y-plates. With the time-base switched off, the vertical line on the screen is centred and its length then measured, Fig. 38.21 (i). This is proportional to twice the amplitude or peak voltage, V_0. By measuring the length corresponding to a known a.c. voltage V, then V_0 can be found by proportion.

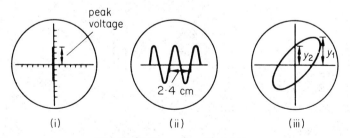

Fig. 38.21 Uses of oscilloscope

Alternatively, using the same gain, the waveforms of the unknown and known voltages, V_0 and V, can be displayed on the screen. The ratio V_0/V is then obtained from measurement of the respective peak-to-peak heights.

2. Comparison of frequency

If a calibrated time-base is available, frequency measurements can be made. In Fig. 38.21 (ii), for example, the trace shown is that of an alternating waveform with the time-base switched to the '5 ms/cm' scale. This means that the time taken for the spot to move 1 cm horizontally across the screen is 5 milliseconds. The horizontal distance on the screen for one cycle is 2·4 cm. This corresponds to a time of $5 \times 2·4$ ms or $12·0$ ms $= 12 \times 10^{-3}$ seconds, which is the period T.

$$\therefore \text{ frequency} = \frac{1}{T} = \frac{1}{12 \times 10^{-3}} = 83 \text{ Hz}.$$

If a comparison of frequencies f_1, f_2 is required, then the corresponding horizontal distances on the screen are measured. Suppose these are d_1, d_2 respectively. Then, since $f \propto 1/T$,

$$\frac{f_1}{f_2} = \frac{T_2}{T_1} = \frac{d_2}{d_1}.$$

3. Measurement of phase

The use of a double beam oscilloscope to measure phase difference is given on p. 415. If only a single beam tube is available, an elliptical trace can be obtained. With the time-base switched off, one input is joined to the X-plates and the other to the Y-plates. We consider only the case when the frequencies of the two signals are the same. An ellipse will then be seen generally on the screen, as shown in Fig. 38.21 (iii) (see p. 428).

The trace is centred, and the peak vertical displacement y_2 at the middle O, and the peak vertical displacement y_1 of the ellipse, are then both measured. Suppose the x-displacement is given by $x = a \sin \omega t$, where a is the amplitude in the x-direction, and the y-displacement by $y = y_1 \sin (\omega t + \varphi)$, where y_1 is the amplitude in the y-direction and φ is the phase angle. When $x = 0$, $\sin \omega t = 0$, so that $\omega t = 0$. In this case, $y = y_2 = y_1 \sin \varphi$. Hence $\sin \varphi = y_2/y_1$, from which φ can be found.

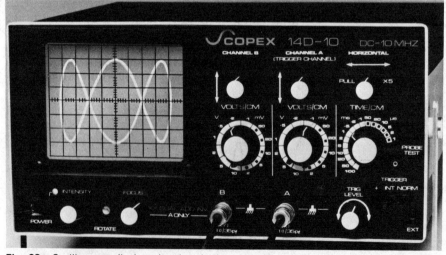

Fig. 38D Oscilloscope display, showing the Lissajous figure obtained when the frequencies on the X and Y-plates are in the ratio 1 : 3. In this modern double-beam instrument, the time-base (X-deflection) ranges from 1 μs cm⁻¹ to 100 ms cm⁻¹ and the voltage sensitivity (Y-deflection) from 2 mV cm⁻¹ to 10 V cm⁻¹. (*Courtesy of Scopex Instruments Ltd*)

Example

A resistor of 1000 Ω carries an alternating current. When an oscilloscope is used to measure the p.d. across it the reading is 20·0 V r.m.s. When an a.c. voltmeter is used the reading is 18 V. What is the resistance of the voltmeter?

Oscilloscope This gives an accurate reading of the p.d. as no current is diverted through the instrument owing to its very high input resistance. So the current in the 1000 Ω resistor is given by

$$I = \frac{20}{1000} = 0.02 \text{ A.}$$

Voltmeter This gives an inaccurate reading of the p.d. as some current is diverted through its resistance R which is in parallel with the 1000 Ω resistor.

The current I in the 1000 Ω resistance

$$= \frac{R}{1000 + R} \times 0.02 \text{ A.}$$

Since $V = I \times 1000$, it follows that

$$\frac{R}{1000 + R} \times 0.02 \times 1000 = 18.$$

So
$$20\,R = 18\,000 + 18R$$

$$R = 9000 \text{ Ω.}$$

Exercises 38

Electrons

1. Electrons are accelerated from rest by a p.d. of 100 V. What is their final velocity?

The electron beam now enters normally a uniform electric field of intensity 10^5 V m^{-1}. Calculate the flux density B of a uniform magnetic field applied perpendicular to the electric field if the path of the beam is unchanged from its original direction. Draw a sketch showing the electron beam and the two fields. (Assume $e/m_e = 1.8 \times 10^{11}\,\text{C}\,\text{kg}^{-1}$.)

2. A beam of protons is accelerated from rest through a potential difference of 2000 V and then enters a uniform magnetic field which is perpendicular to the direction of the proton beam. If the flux density is 0·2 T calculate the radius of the path which the beam describes. (Proton mass = 1.7×10^{-27} kg. Electronic charge = -1.6×10^{-19} C.) (L.)

3. (a) Draw a labelled diagram to show how an electron, of mass m, carrying a charge e, travelling at a constant speed, is deflected on entering a uniform magnetic field of magnetic flux density B at right angles to the direction of the field. Write down an expression for the force acting on the electron whilst it is in the field and then explain, using your knowledge of mechanics, why the path of the electron is the shape it is. Explain carefully how the path would be modified if the electron were slowing down whilst in the magnetic field.

(b) Observations on single electrons are not possible. The path of a beam of electrons from an electron gun inside a spherical glass enclosure can be made visible by introducing into the enclosure a tiny amount of gas, e.g. argon. The path appears as a faint coloured glow through the gas. Explain this.

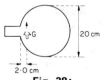

2·0 cm

Fig. 38A

In such a tube, of dimensions shown in Fig. 38A, electrons are ejected in the direction of the arrow from a gun, G, at a speed of 1.0×10^7 m s^{-1}. There is a uniform magnetic field at right angles to the plane of the diagram. Calculate the value of the magnetic flux density that would just confine the path of the beam to the tube without hitting the walls. (Specific charge of electron, $e/m = 1.76 \times 10^{11}$ C kg^{-1}.) (L.)

4. (a) Describe an experiment to determine the ratio of the charge to the mass (e/m) for an electron. Show how the result is derived from the observations.

(b) In an evacuated tube electrons are accelerated from rest through a potential difference of 3600 V and then travel in a narrow beam through a field free space before entering a uniform magnetic field the flux lines of which are perpendicular to the beam. In the magnetic field the electrons describe a circular arc of radius 0·10 m. Calculate (i) the speed of the electrons entering the magnetic field, (ii) the magnitude of the magnetic flux density.

(c) If an electron described a complete revolution in a magnetic field how much energy would it acquire? ($e/m = 1·8 \times 10^{11}$ C kg^{-1}.) (N.)

5. Give an account of a method by which the charge associated with an electron has been measured.

Taking this electronic charge to be $-1·60 \times 10^{-19}$ C, calculate the potential difference in volt necessary to be maintained between two horizontal conducting plates, one 5 mm above the other, so that a small oil drop, of mass $1·31 \times 10^{-14}$ kg with two electrons attached to it, remains in equilibrium between them. Which plate would be at the positive potential? ($g = 9·8$ m s^{-2}.) (L.)

6. Describe an experiment to determine the ratio of the charge to the mass of electrons. Draw labelled diagrams of (a) the apparatus, (b) any necessary electrical circuits, and show how the result is calculated from the observations.

Two plane metal plates 4·0 cm long are held horizontally 3·0 cm apart in a vacuum, one being vertically above the other. The upper plate is at a potential of 300 V and the lower is earthed. Electrons having a velocity of $1·0 \times 10^7$ m s^{-1} are injected horizontally midway between the plates and in a direction parallel to the 4·0 cm edge. Calculate the vertical deflection of the electron beam as it emerges from the plates. (e/m for electron = $1·8 \times 10^{11}$ C kg^{-1}.) (N.)

7. Show that if a free electron moves at right angles to a magnetic field the path is a circle. Show also that the electron suffers no force if it moves parallel to the field. Point out how the steps in your proof are related to fundamental definitions.

If the path of the electron is a circle, prove that the time for a complete revolution is independent of the speed of the electron.

In the ionosphere electrons execute $1·4 \times 10^6$ revolutions in a second. Find the strength of the magnetic flux density B in this region. (Mass of an electron = $9·1 \times 10^{-31}$ kg; electronic charge = $1·6 \times 10^{-19}$ C.) (C.)

8. The electron is stated to have a mass of approximately 10^{-30} kg and a negative charge of approximately $1·6 \times 10^{-19}$ C. Outline the experimental evidence for this statement. Formulae may be quoted without proof. You are not required to justify the actual numerical values quoted.

An oil drop of mass $3·25 \times 10^{-15}$ kg falls vertically, with uniform velocity, through the air between *vertical* parallel plates which are 2 cm apart. When a p.d. of 1000 V is applied to the plates the drop moves towards the negatively charged plate, its path being inclined at 45° to the vertical. Explain why the vertical component of its velocity remains unchanged and calculate the charge on the drop.

If the path of the drop suddenly changes to one at 26° 30′ to the vertical, and subsequently to one at 37° to the vertical, what conclusions can be drawn? (O. & C.)

9. An electron with a velocity of 10^7 m s^{-1} enters a region of uniform magnetic flux density of 0·10 T, the angle between the direction of the field and the initial path of the electron being 25°. By resolving the velocity of the electron find the axial distance between two turns of the helical path. Assume that the motion occurs in a vacuum and illustrate the path with a diagram. ($e/m = 1·8 \times 10^{11}$ C kg^{-1}.) (N.)

10. Give an account of Millikan's experiment for determining the value of the electronic charge e.

In a Millikan-type apparatus the horizontal plates are 1·5 cm apart. With the electric field switched off an oil drop is observed to fall with the steady velocity $2·5 \times 10^{-2}$ cm s^{-1}. When the field is switched on the upper plate being positive, the drop just remains stationary when the p.d. between the two plates is 1500 V.

(a) Calculate the radius of the drop. (b) How many electronic charges does it carry? (c) If the p.d. between the two plates remains unchanged, with what velocity will the drop move when it has collected two more electrons as a result of exposure to ionising radiation? (Oil density $= 900$ kg m^{-3}, viscosity of air $= 1\cdot8 \times 10^{-5}$ N s m^{-2}.) (O. & C.)

11. An electron of charge $-e$ and mass m is initially projected with speed v at right angles to a uniform field of flux density B. Show that the electron moves in a circular path and derive an expression for the radius of the circle. Show also that the time taken to describe one complete circle is independent of the speed of the electron.

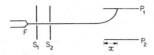

Fig. 38B

Electrons are emitted with negligible speed, *in vacuo*, from a filament F. They are accelerated by a potential difference of 1200 V applied between the plates S_1 and F, as shown in Fig. 38B. The electrons are collimated into a narrow horizontal beam by passing through holes in S_1 and in a second plate S_2 which is at the same potential as S_1. The electron beam subsequently enters the space between two large parallel horizontal plates P_1 and P_2 which are $0\cdot02$ m apart. The point of entry is midway between the plates. The mean potential of P_1 and P_2 is equal to that of S_1 but P_1 is at a positive potential of 150 V with respect of P_2. Neglecting the effect of gravity and of non-uniform fields near the plate boundaries, calculate the distance x travelled by the electrons between P_1 and P_2 before they strike P_1 (O. & C.)

12. Give an account of an experiment to obtain the value of the charge associated with the electron.

An electron beam after being accelerated from rest through a potential difference of 5000 V in vacuo is allowed to impinge normally on a fixed surface. If the incident current is 50 μA determine the force exerted on the surface assuming that it brings the electrons to rest. ($e = 1\cdot6 \times 10^{-19}$ C.) (L.)

13. The value of e/m_e for the electron may be measured using the apparatus shown in Fig. 38c. The electrons are emitted from the cathode K and accelerated by a potential V applied between the first narrow slits S_1 and K. They enter along the axis of a tube bent in the form of a semicircle of radius R centred at O. The tube is placed in a uniform magnetic field of flux density B acting over the shaded region shown. The whole apparatus is evacuated. The electrons which pass through further slits S_2 and S_3 are collected by an electrode E. The current reaching E is adjusted to its maximum value by varying B.

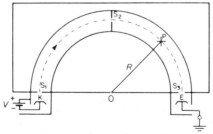

Fig. 38c

(a) Why is the apparatus evacuated? (b) Find an expression for the speed v of the electrons at S_1 in terms of V, e and m_e. (c) What is the magnitude and direction of the magnetic force acting on an electron at a point such as P? (d) Why is the path between S_1, S_2 and S_3 of circular form? (e) What is the direction of B? (f) Derive an expression for e/m_e in terms of R, B and V. (g) The current reaching E is $4\cdot0 \times 10^{-14}$ A. How many electrons are collected by E in one second? (h) The filament is replaced by a source emitting positively-charged ions. Explain what changes would be necessary if their charge-to-mass ratio is to be measured. (O. & C.)

14. Describe and give the theory of the Millikan oil drop experiment for the determination of the electronic charge. What is the importance of the experiment?

In one such experiment a single charged drop was found to fall under gravity at a terminal velocity of 0·0040 cm per second and to rise at 0·0120 cm per second when a field of 2×10^5 V m^{-1} was suitably applied. Calculate the electronic charge given that the radius, a, of the drop was $6·0 \times 10^{-7}$ m and that the viscosity, η, of the gas under the conditions of the experiment was $1·80 \times 10^{-5}$ N s m^{-2}. (N.)

Cathode-Ray Oscilloscope

15. A long uniformly-wound solenoid is placed with its axis vertical and its ends are connected to the Y-plates of a cathode-ray oscilloscope with a suitable time-base. A short bar magnet with its axis vertical is placed well above the solenoid, and is dropped so that it falls through the solenoid and finishes well below it. Sketch the trace you would expect to see on the oscilloscope and explain its shape. (L.)

16. When a sine-form voltage of frequency 1250 Hz is applied to the Y-plates of a cathode-ray oscilloscope the trace on the tube is as shown in Fig. 38D (i).

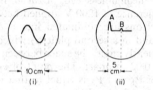

Fig. 38D

If a radar transmitter sends out short pulses, and at the same time gives a voltage to the Y-plates of the oscilloscope, with the time-base setting unchanged, the deflection A is produced as shown in Fig. 38D (ii). An object reflects the radar pulse which, when received at the transmitter and amplified, gives the deflection B. What is the distance of the object from the transmitter? (Speed of radar waves = 3×10^8 m s^{-1}.) (L.)

17. (a) Draw a clear labelled diagram showing the essential features of a single beam cathode-ray oscilloscope tube. Explain, without giving circuit details, how the brightness and focusing of the electron beam are controlled.

(b) What is the *time-base* in an oscilloscope? Sketch a graph showing the variation of time-base voltage with time.

(c) How would you use an oscilloscope, the time-base of which is not calibrated, to measure the frequency of a sinusoidal potential difference which is of the order of 50 Hz?

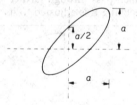

Fig. 38E

(d) With the time-base disconnected, two alternating potential differences of the same frequency are applied to the X- and Y-plates respectively, the gains being equal. Fig. 38E shows the appearance of the trace on the screen. The potentials may be represented by $x = a \sin \omega t$ and $y = a \sin (\omega t + \varphi)$. What is the value of φ, the phase angle between the potentials? Explain your reasoning. (L.)

18. (a) (i) An a.c. power supply is designed to produce output potential differences up to 20 V r.m.s. in 2 V r.m.s. steps using a transformer with suitable tappings. Describe how, with the aid of a cathode-ray oscilloscope, you would check the relative accuracy of each step in voltage (i.e. 4 V, 6 V, 8 V, etc.) using the lowest output voltage (2 V) as your reference. (Your account should include details about electrical connections between the power pack and the oscilloscope and should explain the use of the oscilloscope in ensuring that the best accuracy is achieved.)

(ii) In order to check the accuracy of the 2 V r.m.s. tapping, it is compared using the oscilloscope with a known standard a.c. potential difference of value 1 V peak-to-peak.

If the length of the trace corresponding to the 1 V peak-to-peak standard is 2·1 cm and that corresponding to the 2 V r.m.s. tapping from the power pack is 11·5 cm, what is the actual r.m.s. value of the output potential difference from this tapping?

(b) An a.c. voltmeter and a cathode ray oscilloscope are used in turn to measure the potential difference across a resistance of 10^4 Ω in which there is an alternating current. The readings are 9·0 V and 10·0 V r.m.s. respectively. Calculate the resistance of the a.c. voltmeter. What is the advantage of using a cathode ray oscilloscope for this type of measurement? (L.)

19. Draw a labelled diagram of the structure of a cathode ray tube as used in an oscilloscope. Explain how the electron beam is produced, focused, deflected and detected.

Draw a block diagram showing the essential units of a cathode ray oscilloscope and explain briefly their functions. (Details of the circuitry are *not* required.)

Explain the effect on the sensitivity of the oscilloscope of varying the accelerating potential. (L.)

20. One sinusoidal voltage alternating at 50 Hz is connected across the X plates of a cathode ray oscilloscope and another 50 Hz sinusoidal alternating voltage of approximately the same amplitude is connected across the Y plates. Sketch what you would expect to observe on the screen if the phase difference between the voltages is (a) zero, (b) $\pi/2$, (c) $\pi/4$.

If the voltage on the X plates is replaced by a 100 Hz sinusoidal alternating voltage of similar amplitude sketch what you would observe on the screen.

Explain briefly why figures of this type are useful in the study of alternating voltages. (N.)

21. Explain what is meant by (a) a linear time-base, (b) a sinusoidal time-base, in a cathode ray oscilloscope.

The X and Y deflection sensitivities of a cathode ray oscilloscope are each 5 V cm^{-1}. A sinusoidal potential difference alternating at 50 Hz and of r.m.s. value 20 V is applied to the Y plates of the instrument. A potential difference of the same form and frequency but of r.m.s. value 10 V is simultaneously applied to the X plates. Sketch and explain the pattern seen on the oscilloscope when the potential differences are (a) in phase, (b) 90° out of phase. Indicate the appropriate dimensions on your sketches. (N.)

22. Draw a clear diagram showing the chief features of a cathode-ray oscilloscope tube. Explain the action of the focusing and brilliance controls, and show how the appropriate voltages are applied.

Describe the use of the cathode-ray oscilloscope (a) as a voltmeter, (b) as a means of demonstrating the waveform of a steady musical note of fixed frequency and (c) as a method of measuring short time intervals. (O.)

39 Junction Diode. Transistor and Applications

Energy Bands in Solids

As we see later, the allowed energy levels in a single atom are discrete (separate) and spaced widely apart. In the solid state, however, as in a crystal, large numbers of atoms are packed closely together, and the electrons are influenced strongly by the assembly of nuclei. The allowable energy levels then broaden into *bands* of energy, Fig. 39.1 (i). The bands contain allowable energy levels very close to each other, as at P. There may also be *forbidden bands* of energy, as at Q, which electrons cannot occupy. The lowest available energy band is called the *valence band*. The next available energy band is called the *conduction band*.

In an *insulator*, the valence band energy levels are completely filled by electrons. The conduction band is empty and the two bands are separated by a wide energy gap much greater than kT in magnitude where k is the Boltzmann constant (p. 223), called the 'forbidden' band, Fig. 39.1 (ii). The electrons in the valence band have thermal energy of the order kT but at room temperature they cannot gain sufficient energy from an applied p.d. to move to higher unoccupied energy levels. So the material is an insulator.

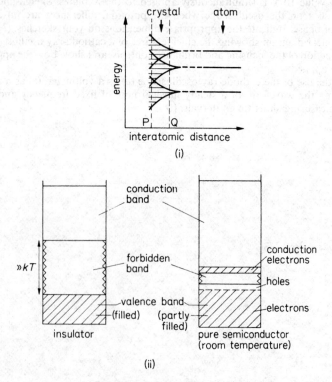

Fig. 39.1 Energy bands in solids

Semiconductors are a class of materials with a narrow forbidden band between the valence and conduction bands; the energy gap is of the order kT. At 0 K, all the energy levels in the valence band are occupied and the material is then an insulator. At normal temperatures, however, the thermal energy of some valence electrons, of the order kT, is sufficient for them to reach the conduction band, where they may become conduction electrons. The gap left in the valence band of energies by the movement of an electron is called a *hole*, Fig. 39.1 (ii). In semiconductor theory, both holes and conduction electrons play an active part, as we soon see.

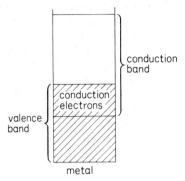

Fig. 39.2 Bands in metal

In *metals*, however, the valence and conduction bands can overlap, as shown diagrammatically in Fig. 39.2. The electrons in the overlapping region of energy are conduction electrons. Since there is a large number of conduction electrons, metals are good conductors.

Semiconductors. Movement of Charge Carriers

Semiconductors are a class of solids with electrical resistivity between that of a conductor and an insulator. For example, the resistivity of a conductor is of the order 10^{-8} Ω m, that of an insulator is 10^4 Ω m and higher, and that of a semiconductor is 10^{-1} Ω m. Silicon and germanium are examples of semiconductor elements widely used in the electronics industry.

Silicon and germanium atoms are tetravalent. They have four electrons in their outermost shell, called *valence electrons*. One valence electron is shared with each of four surrounding atoms in a tetrahedral arrangement, forming 'covalent bonds' which maintain the crystalline solid structure (p. 133). Figure 39.3 (i) is a two-dimensional representation of the structure.

At 0 K, all the valence electrons are firmly bound to the nucleus of their

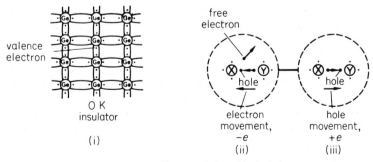

Fig. 39.3 Semiconductor. Electron ($-$) and hole ($+$) movement

particular atoms. At room temperature, however, the thermal energy of a valence electron may become greater than the energy binding to its nucleus. The covalent bond is then broken. The electron leaves the atom, X say, and becomes a free electron. This leaves X with a vacancy or *hole*, Fig. 39.3 (ii). Since X now has a net positive charge, an electron in a neighbouring atom may then be attracted. Thus the hole appears to move to Y.

The hole movement through a semiconductor is random. But if a battery is connected, the valence electrons are urged to move in one direction and to fill the holes. The holes then drift in the direction of the field. Thus the holes move as if they were carriers with a positive charge $+e$, where e is the numerical value of the charge on an electron, Fig. 39.3 (iii). The current in the semiconductor is also carried by the free electrons present. These are equal in number to the holes in a pure semiconductor and drift in the opposite direction since they are negative charges. The mobility of an electron, its average velocity per unit electric field intensity, is usually much greater than that of a hole.

In electrolytes (p. 676), the current is also carried by moving negative and positive charges but the carriers here are *ions*. It should be noted that, in a pure semiconductor, there are equal numbers of electrons and holes, the charge carriers. *Electron-hole pairs* are said to be produced by the movement of an electron from bound state in an atom to a higher energy level, where it becomes a free electron.

Effect of Temperature Rise

In contrast to a semiconductor, the charge carriers in a metal such as copper are only free electrons. Further, as the temperature of the metal rises, the amplitude of vibration of the atoms increases and more 'collisions' with atoms are then made by drifting electrons. Thus, as stated on p. 668, the resistance of a pure metal increases with temperature rise.

In the case of a semiconductor, however, the increase in thermal energy of the valence electrons due to temperature rise enables more of them to break the covalent bonds and become free electrons. Thus more electron-hole pairs are produced which can act as carriers of current. Hence, in contrast to a pure metal, the electrical resistance of a pure semiconductor *decreases* with temperature rise. This is one way of distinguishing between a pure metal and a pure semiconductor. Note that the pure or *intrinsic* semiconductor always has equal numbers of electrons and holes, whatever its temperature.

P- and N-type Semiconductors

A pure or intrinsic semiconductor has charge carriers which are thermally generated. These are relatively few in number. By 'doping' a semiconductor with a tiny amount of impurity such as one part in a million, thus forming a so-called *extrinsic* semiconductor, a considerable increase can be made to the number of charge carriers.

Arsenic atoms, for example, have five electrons in their outermost or valence band. When an atom of arsenic is added to a germanium crystal, the atom settles in a lattice site with four of its electrons shared with neighbouring germanium atoms, Fig. 39.4 (i). The fifth electron may thus become free to wander through the crystal. Since an impurity atom may provide one free electron, an enormous increase occurs in the number of electron carriers. The impure semiconductor is called an 'n-type semiconductor' or *n-semiconductor*, where 'n' represents the negative charge on an electron. Thus the *majority carriers* in an n-semiconductor are electrons. Positive charges or holes are also present in the n-semiconductor.

These are thermally generated, as previously explained, and since they are relatively few they are called the *minority carriers*. The impurity (arsenic) atoms are called *donors* because they donate electrons as carriers.

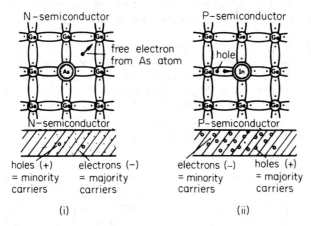

Fig. 39.4 n- and p-semiconductors

P-semiconductors are made by adding foreign atoms which are trivalent to pure germanium or silicon. Examples are boron or indium. In this case the reverse happens to that previously described. Each trivalent atom at a lattice site attracts an electron from a neighbouring atom, thereby completing the four valence bonds and forming a hole in the neighbouring atom, Fig. 39.4 (ii). In this way an enormous increase occurs in the number of holes. Thus in a p-semiconductor, the majority carriers are holes or positive charges. The minority carriers are electrons, negative charges, which are thermally generated. The impurity atoms are called *acceptors* in this case because each 'accepts' an electron when the atom is introduced into the crystal.

Summarising: In a n-semiconductor, conduction is due mainly to negative charges or electrons, with positive charges (holes) as minority carriers. In a p-semiconductor, conduction is due mainly to positive charges or holes, with negative charges (electrons) as minority carriers.

P-N Junction

By a special manufacturing process, p- and n-semiconductors can be melted so that a boundary or *junction* is formed between them. This junction is extremely thin and of the order 10^{-3} mm. It is called a *p-n junction*, Fig. 39.5 (i).

When a scent bottle is opened, the high concentration of scent molecules in the bottle causes the molecules to diffuse into the air. In the same way, the high

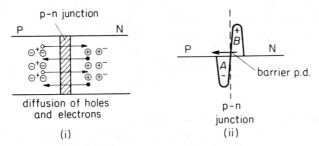

Fig. 39.5 n-p junction and barrier p.d.

concentration of holes (positive charges) on one side of a p-n junction, and the high concentration of electrons on the other side, causes the two carriers to diffuse respectively to the other side of the junction, as shown. The electrons which move to the p-semiconductor side recombine with holes there. These holes therefore disappear, and an excess negative charge A appears on this side, Fig. 39.5 (ii).

In a similar way, an excess positive charge B builds up in the n-semiconductor when holes diffuse across the junction. Together with the negative charge A on the p-side, an e.m.f. or p.d. is produced which opposes more diffusion of charges across the junction. This is called a *barrier p.d.* and when the flow ceases it has a magnitude of a few tenths of a volt. The narrow region or layer at the p-n junction which contains the negative and positive charges is called the *depletion layer*. The width of the depletion layer is of the order 10^{-3} mm.

Junction Diode

When a battery B, with an e.m.f. greater than the barrier p.d., is joined with its positive pole to the p-semiconductor, P, and its negative pole to the n-semi-conductor, N, p-charges (holes) are urged across the p-n junction from P to N and n-charges (electrons) from N to P, Fig. 39.6 (i). Thus an appreciable current is obtained. The p-n junction is now said to be *forward-biased*, and when the applied p.d. is increased, the current increases.

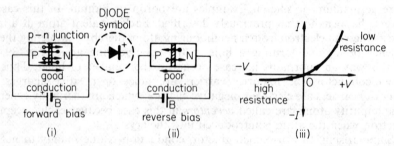

Fig. 39.6 Junction diode characteristic

When the poles of the battery are reversed, only a very small current flows, Fig. 39.6 (ii). In this case the p-n junction is said to be *reverse-biased*. This time only the minority carriers, negative charges in the p-semiconductor and positive charges in the n-semiconductor, are urged across the p-n junction by the battery. Since the minority carriers are thermally-generated, the magnitude of the reverse current depends only on the temperature of the semiconductors. It may also be noted that the reverse-bias p.d. increases the width of the depletion layer, since it urges electrons in the p-semiconductor and holes in the n-semiconductor further away from the junction.

It can now be seen that the p-n junction acts as a *rectifier*. It has a low resistance for one direction of p.d. and a high resistance for the opposite direction, as shown by the characteristic curve in Fig. 39.6 (iii). It is therefore called a *junction diode*. The junction diode has advantages over a diode valve; for example, it needs only a low voltage battery B to function; it does not need time to warm up; it is less bulky, and it is cheaper to manufacture in large numbers.

Full Wave Rectification. Filter Circuit

In a.c. mains transistor receivers, diodes are used to rectify the alternating mains voltage and to produce steady or d.c. voltage for the circuit of the receiver.

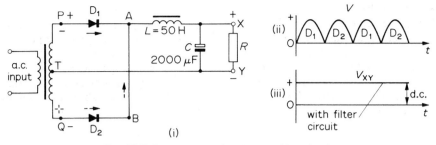

Fig. 39.7 Full wave rectification and filter circuit

Figure 39.7 (i) shows how two diodes, D_1 and D_2, can produce *full-wave rectification*. The secondary PQ of the mains transformer is centre-tapped at T, so that a.c. voltages of *opposite* polarity are applied simultaneously to D_1 and D_2 on one half of the input cycle. Thus in Fig. 39.7 (i), D_1 conducts well but D_2 does not. On the other half of the same cycle D_2 conducts well but D_1 does not. The output voltage V between A and T, with a resistor joined between them, would hence be that shown in Fig. 39.7 (ii). It is a unidirectional voltage. Further, compared with half-wave rectification discussed on p. 811, it has a smaller degree of fluctuation and a greater average voltage.

The output voltage V is equivalent to a steady voltage together with varying voltages. In order to filter off the varying voltages, a *filter circuit* is used. One form of filter circuit consists of a high inductance $L = 50$ H say, in series with a large capacitance $C = 200$ μF say. To a frequency of 50 Hz, the reactance $X_L = 2\pi f L = 2 \times 3.14 \times 50 \times 100 = 31\,400$ Ω; the reactance $X_C = 1/2\pi f C = 1/(2 \times 3.4 \times 50 \times 2000 \times 10^{-6}) = 1.6$ Ω. Since L and C are in series with V, very little of the varying voltage appears across C; practically the whole of it appears across L. So the output voltage V_{XY} across a resistance R shown is a fairly steady or d.c. voltage—it has only a small 'ripple' of a.c. voltage, Fig. 39.7 (iii).

Figure 39.8 shows a *bridge circuit* which produces full-wave rectification without the use of a centre-tapped secondary as in Fig. 39.7 (i). As shown, four diodes are used. On one half of a cycle, when P is +ve relative to Q, only the diodes D_1 conduct. On the other half of the same cycle, only the diodes D_2 conduct. The varying d.c. across A, B is thus similar to that shown in Fig. 39.7 (ii). In this circuit, however, only a 'smoothing' capacitor C is used. Unlike the circuit in Fig. 39.7 (i), C becomes charged to practically the *peak* value of the varying d.c. voltage. The four-diode bridge rectifier thus provides a greater d.c. output voltage than the circuit in Fig. 39.7 (i).

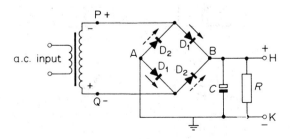

Fig. 39.8 Bridge rectifier circuit

Zener Diode

When the reverse bias or p.d. is increased across a p-n junction, a large increase in current is suddenly obtained at a voltage Z, Fig. 39.9 (i). This is called the *Zener effect*, after the discoverer. It is partly due to the high electric field which exists across the narrow p-n junction at the breakdown or Zener voltage Z, which drags more electrons from their atoms and thus increases considerably the number of electron-hole pairs. Ionisation by collision also contributes to the increase in carriers.

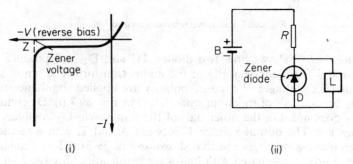

Fig. **39.9** Zener diode and voltage regulation

Zener diodes are used as voltage regulators or stabilisers in circuits. In Fig. 39.9 (ii), a suitable diode D is placed across a circuit L. Although the battery supply B may fluctuate, and produce changes of current in L and D, if R is suitably chosen, the voltage across D remains practically constant over a reverse current range of tens of milliamperes at the Zener voltage shown in Fig. 39.9 (i). The voltage across L thus remains stable.

The Transistor

The junction diode is a component which can only rectify. The *transistor* is a more useful component; it is a *current amplifer*. A transistor is made from three layers of p- and n-semiconductors. They are called respectively the *emitter* (E), *base* (B) and *collector* (C). Figure 39.10 (i) illustrates a *p-n-p transistor*, with electrodes connected to the respective three layers. In a *n-p-n transistor*, the emitter is n-type, the base is p-type and the collector is n-type, Fig. 39.10 (ii). The base is deliberately made very thin in manufacture. The transistor is thus a three-terminal device.

Figure 39.10 shows the circuit symbols for p-n-p and n-p-n transistors. The arrows show the directions of conventional current (+ve charge or hole movement) between the emitter E and base B, so that electrons would flow in the

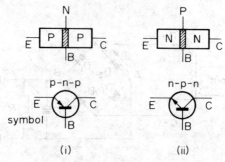

Fig. **39.10** Transistors and symbols

opposite direction. In an actual transistor, the collector terminal is displaced more than the others for recognition or has a dot near it.

Current Flow in Transistors

The transistor may be regarded as two p-n junctions back-to-back. Figure 39.11 (i) shows batteries correctly connected to a p-n-p transistor. The emitter-base is forward-biased; the collector-base is *reverse*-biased; and the base is common. This is called the *common-base* (C-B) mode of using a transistor. Note carefully the polarities of the two batteries. The positive pole of the supply voltage X is joined to the emitter E but the *negative* pole of the supply voltage Y is joined to the collector C. If batteries are connected the wrong way round to a transistor the latter may be seriously damaged. In the case of a n-p-n transistor, therefore, the negative pole of one battery is joined to the emitter and the positive pole of the other is joined to the collector, Fig. 39.11 (ii).

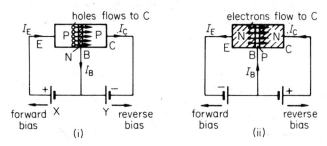

Fig. 39.11 Transistor action

Consider Fig. 39.11 (i). Here the emitter-base is forward biased by X, so that positive charges or holes flow across the junction from E to the base B. The base is so thin, however, that the great majority of the holes are urged across the base to the collector by the battery Y. Thus a current I_C flows in the collector circuit. The remainder of the holes combine with the electrons in the n-base, and this is balanced by electron flow in the base circuit, so that a small current I_B is obtained here. From Kirchhoff's first law, it follows that, if I_E is the emitter current,

$$I_E = I_C + I_B.$$

Typical values for a.f. amplifier transistors are: $I_E = 1.0$ mA, $I_C = 0.98$ mA, $I_B = 0.02$ mA.

Although the action of n-p-n transistors are similar in principle to p-n-p transistors, the carriers of the current in the former case are mainly electrons and in the latter case holes. Electrons are more speedy carriers than holes (p. 824). Thus n-p-n transistors are used in high-frequency and computer circuits, where the carriers are required to respond very quickly to signals.

Common-Emitter (C-E) Characteristics

A transistor can be arranged in one of three ways or modes in circuits. Figure 39.11 shows the *common-base* (C-B) mode, so-called because the base is common to the input (emitter-base) and output (collector-base) circuits. Figure 39.12 shows the *common-emitter* (C-E) mode. A third arrangement is a *common-collector* (C-C) mode. As we explain later, the common-emitter provides satisfactory current amplification and is widely used in a.f. amplifiers.

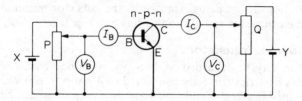

Fig. 39.12 Common-emitter characteristics

Figure 39.12 shows a circuit for obtaining the characteristics of a n-p-n transistor in the common-emitter mode. X and Y may be batteries of 1·5 V and 4·5 V respectively, connected to potentiometers P and Q of 1 kΩ and 5 kΩ. This enables the base-emitter p.d., V_{BE} or V_B, and the collector-emitter p.d., V_{CE} or V_C to be varied. The p.d. is measured by high resistance voltmeters, preferably d.c. solid state voltmeter types capable of measuring p.d. in steps such as 50 mV. The meter for base current, I_B, should be a microammeter and for the collector current, I_C, a milliammeter. Typical results are shown in Fig. 39.13 (i), (ii) and (iii).

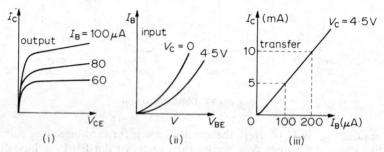

Fig. 39.13 Output and transfer characteristics

Output characteristic ($I_C - V_C$, with I_B constant). The 'knee' of the curves shown in Fig. 39.13 (i) correspond to a low p.d. of the order of 0·2 V. For higher p.d. the output current I_C varies *linearly* with V_C for a given base current. The linear part of the characteristic is used in a.f. amplifier circuits, so that the output voltage variation is then undistorted.

The *output resistance* r_0 is defined as $\Delta V_C / \Delta I_C$, where the changes take place on the straight part of the characteristic. r_0 is an a.c. resistance; it is the effective resistance in the output circuit for an a.c. signal input. It should be distinguished from the d.c. resistance, V_C / I_C, which is not required in amplifier circuit analysis.

The small gradient of the straight part of the characteristic shows that r_0 is high. For example, suppose $\Delta V_C = 2$ V and $\Delta I_C = 0·2$ mA $= 2 \times 10^{-5}$ A. Then $r_0 = 2/(2 \times 10^{-5}) = 100\,000\,\Omega$. If a varying resistance load is used in the output or collector circuit, the high value of r_0 relative to the load shows that the output current is fairly constant. So the output voltage is proportional to the load resistance.

Transfer characteristic ($I_C - I_B$, V_C constant). The output current I_C varies fairly linearly with the input current I_B, Fig. 39.13 (ii). The current *transfer ratio* β, or current gain, is defined as the ratio $\Delta I_C / \Delta I_B$ under a.c. signal conditions. It should be distinguished from the d.c. current gain, I_C / I_B. From Fig. 39.13 (ii),

$$\beta = \frac{(10 - 5)\ \text{mA}}{(200 - 100)\ \mu\text{A}} = 50$$

Input characteristic ($I_B - V_B$, V_C constant). The input resistance r_i is defined as the ratio $\Delta V_B / \Delta I_B$. As the input characteristic in Fig. 39.13 (iii) is non-linear, then r_i varies. At any point of the curve, r_i is equal to the gradient of the tangent to the curve and is of the order of kilohms.

Current Amplification in C-E Mode

In general, the magnitude of β, $\Delta I_C / \Delta I_B$, for the common-emitter circuit is high, from about 20 to 500 for many transistors. Thus the base current is a sensitive control over the collector current.

We can obtain a rough value for β by assuming that when electrons are emitted from the n-emitter towards the p-base, a constant fraction α reaches the n-collector where α is typically 0·98 (see p. 829). Thus $I_C = \alpha I_E$. Now from p. 829,

$$I_E = I_C + I_B.$$

Substituting $I_E = I_C/\alpha$ and simplifying, we obtain

$$I_C = \left(\frac{\alpha}{1-\alpha}\right) I_B$$

$$\therefore \Delta I_C = \left(\frac{\alpha}{1-\alpha}\right) \Delta I_B$$

$$\therefore \frac{\Delta I_C}{\Delta I_B} = \beta = \frac{\alpha}{1-\alpha} = \frac{0·98}{0·02} = 49.$$

Voltage Amplification and Power Gain

As we have just seen, the transistor in the common-emitter mode is a current amplifier. To change the output (a.c.) current to a voltage V_0, a resistance load R can be used in the collector or output circuit. Figure 39.14 shows also diagrammatically the base bias necessary for no distortion of V_0 (p. 830).

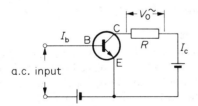

Fig. 39.14 Voltage amplification and power gain (diagrammatic)

We can illustrate the voltage amplification by supposing that $R = 5$ kΩ, the input resistance $r_i = 2$ kΩ, the input (a.c.) voltage is 10 mV or 0·01 V peak value, and the transfer ratio $\beta = 50$.

The peak a.c. current I_b flowing in the base circuit is

$$I_b = \frac{0·01 \text{ V}}{2000 \text{ }\Omega} = 5 \times 10^{-6} \text{ A}$$

$$\therefore I_c = \beta I_b = 50 \times 5 \times 10^{-6} = 2·5 \times 10^{-4} \text{ A}$$

$$\therefore V_0 = I_c R = 2·5 \times 10^{-4} \times 5000$$

$$= 1·25 \text{ V peak.}$$

$$\therefore \text{ voltage amplification } A_V = \frac{V_0}{V_i} = \frac{1\cdot25}{0\cdot01} = 125.$$

Also, power gain = current gain × voltage gain = 50 × 125 = 6250.

Leakage Current and Temperature Rise

When the base current I_B is zero, some current still flows in the collector circuit in the common-emitter arrangement. This is due to the minority carriers present in the collector-base part of the transistor, which is reverse-biased. The collector current when I_B is zero is denoted by I_{CEO} and is called the *leakage current*. Since the minority carriers are thermally generated, the leakage current depends on the temperature of the transistor.

In the common-base arrangement in Fig. 39.11, the leakage current obtained when I_E is zero is denoted by I_{CBO}. This is also due to minority carriers in the collector-base, which is reverse-biased. Thus, more accurately, we should write in place of $I_C = \alpha I_E$ (p. 831)

$$I_C = \alpha I_E + I_{CBO} \qquad . \qquad . \qquad . \qquad . \qquad . \qquad (1)$$

Now from p. 831, $\alpha/(\alpha-1) = \beta$, so that $\alpha = \beta/(\beta+1)$. Further, from previous, $I_E = I_C + I_B$. Substituting in (1) for α and for I_E and simplifying, we obtain

$$I_C = \beta I_B + (\beta+1)I_{CBO} \qquad . \qquad . \qquad . \qquad . \qquad (2)$$

Thus the leakage current I_{CEO} in the common-emitter circuit = $(\beta+1)I_{CBO}$. The current I_{CEO} also flows in the base-emitter circuit when the transistor is operating. A temperature change from 25°C to 45°C, which would increase the current I_{CBO} by 10μA say, would then produce a current $I_{CEO} = (\beta+1)I_{CBO} = 50 \times 10\ \mu\text{A} = 0\cdot5$ mA, assuming $\beta = 49$. This is the increase in collector current if the temperature rose to 45°C from 25°C. The relatively large current change would have an appreciable effect on the output in the collector circuit; for example, it could lead to a distorted output in an a.f. amplifier circuit.

On this account the C-E circuit, which is very sensitive to temperature change, *must* be stabilised for excessive temperature rise. This is explained shortly. Silicon transistors are much less sensitive to temperature change than germanium transistors and are hence used more widely.

C-E Amplifier Circuit

Figure 39.15 shows a n-p-n transistor in a simple C-E amplifier circuit. It uses one battery supply, V_{CC}. A load, R_L, is placed in the collector or output circuit. A resistor R provides the necessary bias, V_{BE}, for the base-emitter circuit. The base-emitter is then forward-biased but the collector-base is reverse-biased, that is, the potential of B is positive relative to E but negative relative to C.

In practice, Fig. 39.15 is unsuitable as an amplifier circuit since there is no arrangement for temperature stabilisation (p. 832). A more reliable C-E a.f. amplifier circuit is shown in Fig. 39.16. Its principal features are:

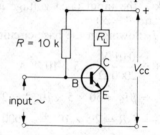

Fig. 39.15 Simple amplifier circuit

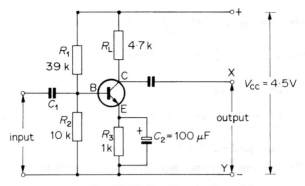

Fig. 39.16 Amplifier circuit

(i) a potential divider arrangement, R_1, R_2, which provides the necessary base-bias;

(ii) a load R_L which produces the output across X, Y;

(iii) a capacitor C_1 which stops the d.c. component in the input signal entering the circuit;

(iv) a large capacitor C_2 across a resistor R_3, which prevents undesirable feedback of the amplified signal to the base-emitter circuit;

(v) an emitter resistance R_3, which stabilises the circuit for excessive temperature rise. Thus if the collector current rises, the current through R_3 increases. This lowers the p.d. between E and B, so that the collector current is automatically lowered.

Example

Figure 39.17 shows a simple form of silicon common-emitter amplifier. When the collector-emitter voltage V_{CE} is between +1 V and +9 V, the collector current is about 100 times the base current and the base-emitter voltage V_{BE} is about 0·7 V.

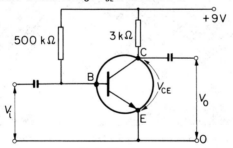

Calculate (i) the base current I_B and the voltage V_{CE} in the circuit, (ii) the voltage gain if the input (base-emitter) a.c. resistance is 2000 Ω, (iii) the largest (limiting) peak value of the input a.c. voltage if V_{CE} varies between 2·2 V and 8·2 V for linear amplification

(i) I_B. Since $V_{BE} = 0·7$ V, p.d. across 500 kΩ resistor $= 9 - 0·7 = 8·3$ V.

So

$$I_B = \frac{8·3}{5 \times 10^5} = 1·66 \times 10^{-5} \text{ A} = 17 \ \mu A \text{ (approx.).}$$

V_{CE}. So $I_C = 100 I_B = 100 \times 1·66 \times 10^{-5}$ A $= 1·66 \times 10^{-3}$ A.

Thus   p.d. across 3 kΩ $= I_c R = 1·66 \times 10^{-3} \times 3000 = 4·98$ V.

So $V_{CE} = 9$ V $- 4·98$ V $= 4$ V (approx).

(ii) *Voltage gain.* If V_i is input, a.c. base current is

$$I_b = \frac{V_i}{R_b} = \frac{V_i}{2000}.$$

So   a.c. output $I_c = 100 I_b = 100 \times \dfrac{V_i}{2000} = \dfrac{V_i}{20}$.

Thus   a.c. output $V_o = I_c R = \dfrac{V_i}{20} \times 3000 = 150 \, V_i$.

So   voltage gain $= \dfrac{V_o}{V_i} = 150$.

(iii) *Largest input voltage.* Since V_{CE} varies between 2·2 V and 8·2 V, peak value of voltage across collector load 3 kΩ is

$$V_o = \tfrac{1}{2}(8\cdot2 - 2\cdot2) = 3 \text{ V}.$$

Since gain = 150, peak value of largest input voltage V_i is given by

$$V_i = \frac{3 \text{ V}}{150} = \frac{1}{50} \text{ V} = 20 \text{ mV}.$$

Fig. 39A An engineer engaged in designing the layout of an integrated circuit. The X and Y co-ordinates at the top of the screen refer to the position of the white tracking cross seen in the centre. The cross is moved by the light pen held by the engineer. (*Courtesy of Mullard Limited*)

Oscillatory Circuit

A coil and capacitor are electrical components which together can produce *oscillations* of current. We can see this by supposing that a capacitor C is discharged through a coil of inductance L and negligible resistance, Fig. 39.18. Then if I is the current flowing at a time t, it follows from p. 753 that

$$\text{p.d. across inductance} = -L\frac{dI}{dt} = \text{p.d. across capacitor} = \frac{Q}{C}$$

$$\therefore -L\frac{dI}{dt} = \frac{Q}{C}.$$

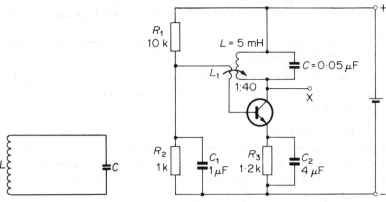

Fig. 39.18 Basic oscillatory circuit   **Fig. 39.19** Transistor oscillator circuit

But

$$I = \frac{dQ}{dt}.$$

$$\therefore -L\frac{d^2Q}{dt^2} = \frac{Q}{C}$$

$$\therefore \frac{d^2Q}{dt^2} = -\frac{1}{LC} \cdot Q \qquad . \qquad . \qquad . \qquad \text{(i)}$$

This is a 'simple harmonic' equation (see p. 65). Thus Q, the charge circulating, varies with time t according to the relation

$$Q = Q_0 \sin \omega t . \qquad . \qquad . \qquad . \qquad . \qquad \text{(ii)}$$

where Q is the maximum value of the varying charge and ω is a constant given by $\omega^2 = 1/LC$, or

$$\omega = \frac{1}{\sqrt{LC}}.$$

The *frequency*, *f*, of the oscillatory charge is given by

$$f = \frac{\omega}{2\pi} = \frac{1}{2\pi\sqrt{LC}} \qquad . \qquad . \qquad . \qquad . \qquad \text{(iii)}$$

A coil-capacitor series circuit is thus a basic oscillatory circuit, and the frequency of the oscillations of charge (or current) depends on the magnitudes of the inductance L and capacitance C.

The physical reason for the oscillations is the constant interchange or feedback of energy between the capacitor and the coil, as explained on p. 408.

Transistor Oscillator Circuit

In order to produce continuous oscillations in the oscillatory (L-C) circuit, energy must be fed back continuously to the circuit to compensate for that lost as heat.

Figure 39.19 shows one form of transistor oscillator circuit. Its main features are:

   (i) a coil-capacitor, L-C, load in the collector circuit;
  (ii) positive feedback through the coil L_1 to maintain the oscillations in the L-C circuit;
 (iii) a potential divider arrangement, R_1, R_2, to provide the necessary base bias;

(iv) an emitter resistor R_3 to stabilise the circuit for excessive temperature rise;

(v) large capacitors C_1 and C_2 across R_2 and R_3 respectively, which prevent undesirable feedback to the base circuit.

Approximately, the frequency of oscillation is given by $f = 1/2\pi\sqrt{LC}$, in this case an audio-frequency of about 10 kHz. Other frequencies may be obtained by changing the magnitude of C.

Transistor Switch

In addition to its use as a current amplifier, the transistor can be used as a *switch* in computer circuits. Millions of switching operations are needed daily in working computers, so swift switches are required. On this account n-p-n transistors are preferred. Here the charge carriers are mainly electrons, which have a much greater speed for a given voltage than holes or p-charges.

The basic circuit is shown in Fig. 39.20 (i). It consists of a n-p-n transistor connected in the common-emitter mode, with a resistance R in the output or collector circuit. Since this is a n-p-n transistor, the positive pole of the supply V_{cc} is connected to the collector C and the negative pole to the emitter E.

A typical output voltage (V_0)-input voltage (V_1) characteristic of the circuit is shown in Fig. 39.20 (ii). At very low input voltages the output voltage is practically $+6$ V, the supply voltage V_{cc} for the circuit. At input voltages of more than a fraction of a volt, however, the output voltage is nearly zero. This is explained shortly on page 837.

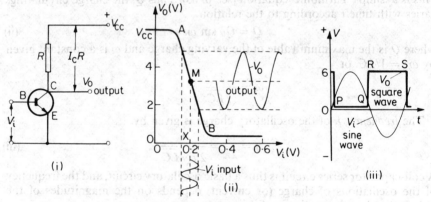

Fig. 39.20 Transistor switch

Sine Wave Input. Amplifier Use

Suppose the input V_i is a *sine wave* voltage of peak value 6 V. Fig. 39.20 (iii). For a large part of the $+$ve half cycle, V_i will be greater than $+0.4$ V. So the output voltage V_0 will be practically zero along PQ. When V_i is less than $+0.4$ V and negative on the $-$ve half of the cycle, V_0 will be practically a constant high voltage RS as shown. So the output V_0 is roughly a *square wave* voltage.

To use the transistor as an *amplifier*, the output voltage V_0 must have the same waveform as the input a.c. voltage V_i. This time the straight inclined line AB of the characteristic in Fig. 39.20 (ii) must be used. So, as shown, (*a*) the base bias should correspond to the midpoint M of AB, a bias equal to OX or about 0.2 V, and (*b*) the input voltage to be amplified must have a peak value not greater than XB (0.1 V), otherwise the output waveform is distorted during part of the input cycle.

States of Transistor

We can explain the characteristic curve by noting that, if I_c is the collector current flowing for a particular input voltage, the output voltage V_0, or V_{ce}, is less than the supply voltage V_{cc} by the potential drop across R, which is I_cR. Thus

$$V_0 = V_{cc} - I_cR$$

In general, I_c depends on the base current I_c and this is governed by the base-emitter or input voltage V_i.

Suppose V_i is very low or practically zero. Then I_c is practically zero, and the transistor is said to be 'cutoff'. From above, we can see that the output voltage V_0 is then practically equal to V_{cc} or high. See Fig. 39.20 (ii). Conversely, suppose V_i is high so that the transistor is 'saturated', that is, any further increase in base current produces no rise in I_c. The p.d. across R is then large and so the output voltage is practically zero from above. See Fig. 39.20 (ii).

Thus depending on the input voltage, the transistor can switch between two states—cutoff or saturation. The output voltage then switches between two levels, $+V_{cc}$ and practically 0. In the special type of computer circuits known as *logic circuits* or *logic gates*, the binary digits '1' and '0' can be represented by $+V_{cc}$ and 0 respectively, or by 0 and $+V_{cc}$, by this switching of states. It should be noted that the transistor acts 'non-linearly', whereas it acts 'linearly' in amplifiers (p. 830).

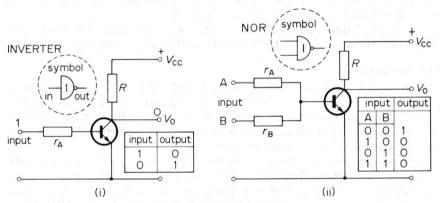

Fig. **39.21** INVERTER and NOR gates

Logic Gates

We can now discuss briefly some useful logic gates.

Figure 30.21 (i) shows the circuit for an INVERTER gate. It consists of a transistor in the common-emitter mode connection, with an appropriate load resistance R and base resistance r_A. Suppose the input is a '1', for example, $+V_{cc}$ volt. A high base current then flows in r_A, and as explained before, the transistor becomes saturated and the output is '0'. Conversely, if the input is '0' (zero volt), the transistor is cutoff and the output is $+V_{cc}$ or '1'. Thus the output is always the *inverse* or opposite of the input. This is shown in a so-called 'truth table' in Fig. 39.21 (i), which also contains the symbol for the INVERTER gate.

Figure 39.21 (ii) shows the circuit for a NOR gate. It is similar to the circuit in Fig. 39.21 (i) except that two inputs and two base resistors, r_A and r_B, are provided. If both inputs are '0' or zero volt, the transistor is cutoff; hence the output is '1' or $+V_{cc}$ volt. If either input or both inputs, A and B, are '1', then, with appropriate vales for r_A and r_B, the transistor saturates and the output is

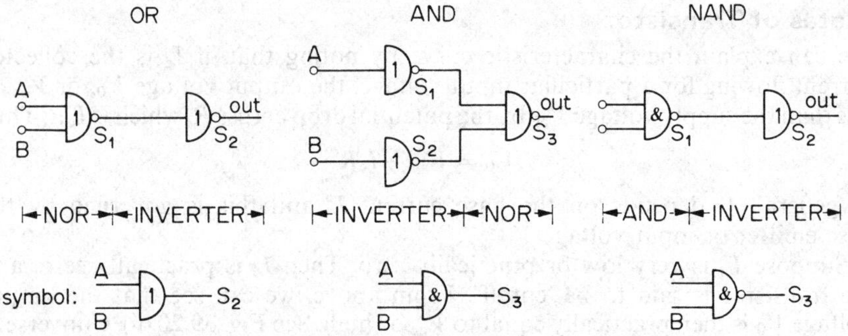

OR   AND   NAND

⎸◄NOR►⎹◄INVERTER►⎹   ⎸◄INVERTER►⎹◄NOR►⎹   ⎸◄AND►⎹◄INVERTER►⎹

symbol:

Fig. 39.22 OR and NAND gates

'0'. These results are shown in the truth table in Fig. 39.21 (ii), together with the symbol for this gate. It is called a NOR gate because the output is '1' if neither A *nor* B is '1'; in all other cases the output is '0'.

Figure 39.22 (i) shows in symbol form an OR gate; it is made of a NOR gate followed by an INVERTER gate, so that the output S_1 of the NOR gate is inverted to produce a final output S_2. By writing the inputs A and B in truth table form as on p. 837, we find that S_2 is a '1' if either A *or* B is a '1'.

Figure 39.22 (ii) shows in symbol form an AND gate; it consists of two IN-VERTER gates followed by a NOR gate. The truth table for S_1, S_2 and S_3 shows that the output S_3 is a '1' only if A *and* B are '1'.

Figure 39.22 (iii) shows the construction of a NAND gate. The truth table shows that the output S_3 is '1' if A *and* B, individually or together, are not '1'.

Phototransistor

A *photodiode* is a junction diode sensitive to light. When the diode is reverse-biased, minority carriers flow in the circuit and constitute a so-called 'dark' current. If the junction of the diode is now illuminated, the light energy produces more electron-hole pairs, which are then swept across the junction. The increased current which flows is the 'light' current.

A *phototransistor* is a transistor sensitive to light in which the base is usually left disconnected. When light falls on the emitter side, more electron-hole pairs are produced in the base. This is amplified by transistor action, and a larger collector current is obtained. In principle the phototransistor is a photodiode plus amplifier.

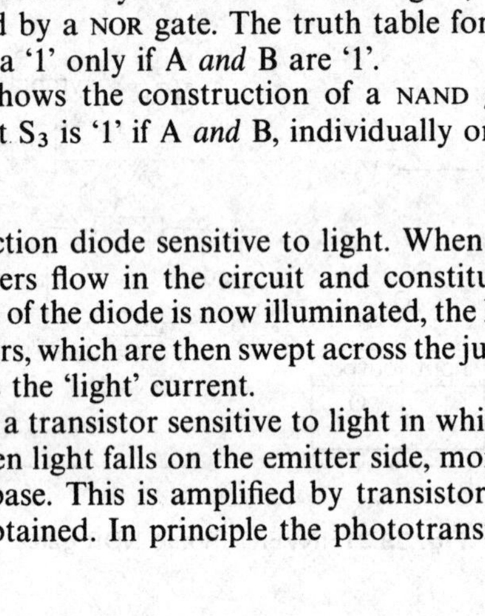

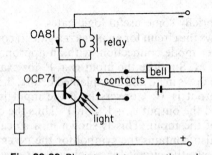

Fig. 39.23 Phototransistor operating relay

Figure 39.23 shows a circuit in which a suitable p-n-p phototransistor is connected in series with a relay coil D and a d.c. supply voltage. When the phototransistor is illuminated, the increase in collector current closes the contacts of a magnetic relay. Current then flows in a circuit connected to the relay, and a bell, for example, may then ring, Fig. 39.23. When the light is switched off, the falling current in the relay coil produced an induced voltage in the same direction as the battery supply. This would raise the collector voltage and prevent the switch-off at the contacts. The diode OA81 across the coil acts as a safeguard. As soon as the rising induced voltage becomes equal to the battery voltage the diode conducts, and prevents any further rise in collector voltage.

Field Effect Transistor (f.e.t.)

The junction transistor operates by the movement of two types of charge carriers, electrons (negative charges) and holes (positive charges). Further, the transistor has two p-n junctions. In this section we give a brief account of the *field effect transistor* (f.e.t.). As we shall see, this type of transistor uses only one type of charge carrier when working and has only one p-n junction.

Figure 39.24 (i) shows the basic construction of the early f.e.t. It consists of a bar of n-type silicon (p-silicon could also be used), with a p.d. V_{DS} applied between one end called the *source* S and the other end called the *drain* D. By diffusion of impurities, a heavily-doped p-region forms a p-n junction with the bar, and the electrode at this p-region is called the *gate* G. The symbol for the n-type f.e.t. is shown in Fig. 39.24 (ii). The arrow is reversed in the p-type f.e.t.

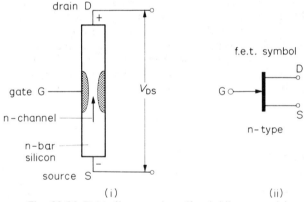

Fig. 39.24 Field effect transistor (f.e.t.) (diagrammatic)

Principle of f.e.t.

Consider the f.e.t. with batteries connected as shown in Fig. 39.25. The potential of the drain D is +ve relative to the source S, so the electrons in the n-bar flow from S to D. This current is called the *drain current* I_D.

The battery between the gate G and S makes the potential of G −ve relative to S. Now the n-bar near the p-junction has a +ve potential relative to S. Thus the p-n junction near the gate is *reverse-biased*. Consequently, as shown diagrammatically in Fig. 39.25, a *depletion region* is obtained at the p-n junction (p. 826). The width of the depletion region depends on the magnitude of the p.d. V_{GS} when the p.d. V_{DS} is constant. If V_{GS}, is made more −ve the wider is the depletion region at the p-n junction. This is shown diagrammatically in Fig. 39.25.

The depletion region, which has no free electrons, narrows the conducting n-channel in the bar. When V_{GS} is made more −ve, the current is reduced. As it provides a sensitive control over the drain current, small changes in gate voltage can produce large amplification of the current. Thus signals, or a.c. voltages, can be amplified by the f.e.t. transistor.

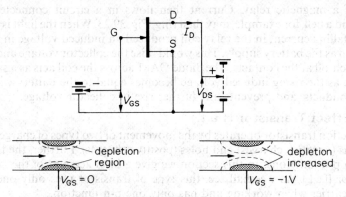

Fig. 39.25 Action of field effect transistor

F.E.T. Characteristics. F.E.T. Amplifier

Some typical *output characteristics* ($I_D - V_{DS}$, V_{GS} constant) and *input characteristics* ($I_D - V_{GS}$, V_{DS} constant) are shown in Fig. 39.26 (i), (ii).

When V_{GS} is kept constant, the drain current I_D first increases linearly as the voltage V_{DS} is increased from zero, since the bar then acts as an ohmic conductor. This corresponds to the line A in Fig. 39.26 (i). When V_{DS} increases further, the p-n junction at the gate becomes more *reverse-biased*, since the positive potential of the n-bar at the gate increases relative to S and hence relative to G. The depletion region then widens, as stated above, and so the current I_D begins to increase at a slower rate. The slow rise of I_D with V_{DS} begins at X in Fig. 39.26 (i) and continues along the straight line B.

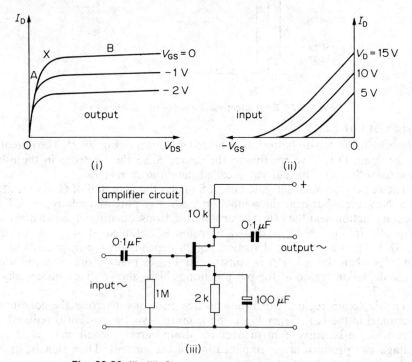

Fig. 39.26 (i), (ii) Characteristics (iii) Amplifier circuit

When the gate voltage V_{GS} is made more negative, the reverse-bias is increased. The current I_D is now decreased. Figure 39.26 (ii) shows the effect on the input characteristics.

An a.f. amplifier circuit which uses the f.e.t. transistor is shown in Fig. 39.26 (iii). The load resistance is 10 kΩ, the temperature stabilisation is provided by the 2 kΩ resistance and the necessary bias for an undistorted output is provided by the 1 MΩ resistor. The 0·1 μF capacitors prevents direct current or voltage reaching the input and output circuits.

Exercises 39

1. A semiconductor diode and a resistor of constant resistance are connected in some way inside a box having two external terminals, as shown in Fig. 39A. When a potential difference V of 1·0 V is applied across the terminals the ammeter reads 25 mA. If the same potential difference is applied in the reverse direction the ammeter reads 50 mA.

What is the most likely arrangement of the diode and the resistor? Explain your deduction. Calculate the resistance of the resistor and the forward resistance of the diode. (*L*.)

2.

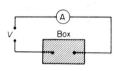

Fig. 39A

Fig. 39B

The circuit in Fig. 39B shows four junction diodes and a resistor R connected to a sinusoidally alternating supply. Sketch graphs showing the variation with time over two cycles of the supply of (*a*) the potential of C with respect to A, (*b*) the potential of B with respect to A, and (*c*) the potential of B with respect to D. (*L*.)

3. Silicon has a valency of four (i.e. its electronic structure is 2:8:4). Explain the effect of doping it with an element of valency three (i.e. of electronic structure 2:8:3). Explain the process by which a current is carried by the doped material.

Describe the structure of a solid-state diode. Draw a circuit diagram showing a reverse-biased diode and explain why very little current will flow. Suggest why a suitable reverse-biased diode could be used to detect alpha particles. (*L*.)

4. What is meant by a *semiconductor*? Explain how the conductivity of such a material changes with (*a*) temperature, and (*b*) the presence of impurities.

Describe the structure of a solid state diode, explaining the nature of the semi-conducting materials from which it is made. Explain the action of the diode in rectifying an alternating current.

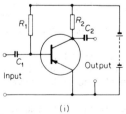

(i)

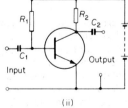

(ii)

Fig. 39c

Figure 39C shows simple forms of transistor voltage amplifiers using (i) a *p-n-p* transistor, and (ii) an *n-p-n* transistor. Choose *one* of these circuits and explain the functions of the components R_1, R_2, C_1 and C_2. (*L*.)

5. Sketch characteristic curves for *either* a junction transistor (relating collector current, base current and collector-emitter voltage) *or* a field-effect transistor (drain current versus drain-source voltage for fixed values of gate-source voltage.) Indicate on the axes of the graphs typical values for each of the quantities involved.

Draw circuit containing *either* a junction transistor *or* a field-effect transistor which would function as a single-stage voltage amplifier and show, with reference to the characteristic curves you have drawn, how the circuit achieves amplification. Indicate typical values for each of the circuit components.

Explain how you would measure the gain (amplification) of the circuit you describe as a function of frequency (*O. & C.*)

6. Describe how the energy levels of electrons in solids differ from those in free atoms. Distinguish, in terms of the filling of such energy levels, between metals, insulators and intrinsic semiconductors.

Explain why the addition of small quantities of suitable impurities to an intrinsic semi-conductor may result in a considerable decrease in its resistivity.

The mobility μ of charge carriers in a conductor is defined by the equation $v = \mu E$, where v is the drift velocity produced by an electric field E. A rod of *p*-type germanium of length 10 mm and cross-section area 1 mm^2 contains 3×10^{21} holes per m^3, the electron density being negligible. Given that the mobility of the holes is 0·35 m^2 V^{-1} s^{-1}, what is the resistance between the ends of the rod? (Electronic charge $e = 1\cdot6 \times 10^{-19}$ C.) (*O. & C.*)

7. What is intrinsic semiconduction? Discuss in terms of band theory the effect of temperature upon the conductivity of intrinsic semiconductors.

What are meant by '*n*-type' and '*p*-type' semiconducting materials? Discuss these materials in terms of valence electrons. Explain the operation of a rectifying device made from *n*-type semiconducting materials.

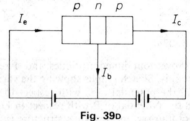

Fig. 39D

The diagram shows a *p-n-p* transistor and its associated power supplies. Explain why the current I_c is considerably greater than the current I_b. (*L.*)

8. A transistor in the common-emitter arrangement provides the following results for I_C, collector current, and V_C, collector voltage, for various constant base current I_B:

$$I_C(mA)$$

V_C (V)	$I_B = 20\ \mu A$	$40\ \mu A$	$60\ \mu A$	$80\ \mu A$
3	0·91	1·60	2·30	3·00
5	0·93	1·70	2·50	3·25
7	0·97	1·85	2·70	3·55
9	1·00	2·05	3·00	4·05

Plot the characteristics, and from them find (i) the current gain at 8 V, (ii) the output resistance for a base current of 40 μA.

9. Explain what is meant by p-type and n-type semiconductors. Describe a p-n junction diode. Draw a graph which shows the variation of the current through such a diode with the potential difference across it, and explain why the diode behaves differently when the potential difference across it is reversed.

Describe the junction transistor. Sketch curves to show the variation of the collector current with the collector-base voltage for various values of the emitter current and explain their form. (*O. & C.*)

10. A common-emitter audio-frequency amplifier circuit needs (i) a load, (ii) temperature stabilisation, (iii) an undistorted output. Explain the reasons for these three items and show in a circuit sketch with a n-p-n transistor how each is provided by suitable components.

40 Photoelectricity.
Energy Levels. X-Rays

Photoelectricity: Particle Nature of Waves

Photoelectricity

In 1888 Hallwachs discovered that an insulated zinc plate, negatively charged, lost its charge if exposed to ultra-violet light. Hertz had previously noticed that a spark passed more easily across the gap of an induction coil when the negative metal terminal was exposed to sunlight. Later investigators such as Lenard and others showed that electrons were liberated from a zinc plate when exposed to ultra-violet light. Light thus gives energy to the electrons in the surface atoms of the metal, and enables them to break through the surface. This is called the *photoelectric effect*.

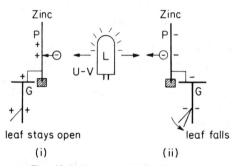

Fig. 40.1 Photoelectric demonstration

Figure 40.1 (i) shows diagrammatically a simple demonstration of the photo-electric effect. The surface of a zinc plate P was rubbed with emery paper until the surface was clean and bright. P was then insulated and connected to the cap of a gold-leaf electroscope G, as shown, and given a *positive* charge by induction. Some of the charge spread to the leaf which then opened.

In a dark room, P was exposed to ultra-violet radiation from a small lamp L placed near it. The leaf stayed open. However, when the experiment was repeated with the plate P charged *negatively*, the leaf slowly collapsed, Fig. 40.1 (ii).

The results are explained as follows. Electrons are usually emitted by the plate P when exposed to ultra-violet light. When P is positively-charged, any electrons (negative charges) liberated would be attracted back to P. When P is negatively charged, however, the electrons emitted by P are now *repelled away from the plate*. So P loses negative charge and the leaf slowly falls.

Velocity or Kinetic Energy of Photo-electron

In 1902 Lenard found that the *velocity* or *kinetic energy* of the electron emitted from an illuminated metal was independent of the intensity of the particular incident monochromatic light. It appeared to vary only with the *wavelength* or *frequency* of the incident light.

Further, for a given metal, no electrons were emitted when it was illuminated by light of wavelength longer than a particular wavelength, no matter how great was the intensity of the light beam. But as soon as the metal was illuminated by light whose wavelength was *lower* than the particular wavelength, electrons were emitted. Even though the light beam was made extremely weak in intensity, it was estimated that the electrons were emitted about 10^{-9} second after exposure to the light, that is, practically simultaneously with exposure to the weak light.

Classical or Wave Theory

On the wave theory of light, the so-called classical theory, these results are very surprising. If we assume light is sent out in waves from a source, the greater the intensity of the light the greater will be the energy per second reaching the illuminated plate. So the classical theory can explain why the number of electrons emitted increases as the light intensity increases.

But it can not explain the result that the velocity or kinetic energy of the emitted electrons is independent of the intensity of the incident light beam. On the classical theory, the greater the intensity of the beam, the greater should be the kinetic energy of the emitted electrons because the energy per second reaching the plate increases with intensity.

Further, on the classical theory electrons should always be emitted by light of *any* wavelength if the incident light beam is strong enough. Experiment, however, shows that however intense the light beam, *no* electrons are emitted if the wavelength is greater than a particular value.

Quantum Theory of Radiation. Planck Constant

In 1902 Planck had shown that the experimental observations in black-body radiation could be explained on the basis that the energy from the body was emitted in separate packets of energy. Each packet was called a *quantum* of energy and the amount of energy E carried was equal to hf, where f is the *frequency* of the radiation and h was a constant called the *Planck constant*. So

$$E = hf \qquad . \qquad . \qquad . \qquad . \qquad (1)$$

This is the *quantum theory of radiation*. Until Planck's quantum theory, it was considered that radiation was emitted continuously and not in separate packets of energy. Since $h = E/f$, the unit of h is joule second or J s. Measurements of radiation showed that, approximately, $h = 6.63 \times 10^{-34}$ J s.

The quantum of energy carried by radiation of wavelength 3×10^{-6} m can be calculated from $E = hf$. Since $f = c/\lambda$, where c is the speed of electromagnetic waves, about 3×10^8 m s^{-1},

$$f = \frac{3 \times 10^8}{3 \times 10^{-6}} = 10^{14} \text{ Hz.}$$

So $E = hf = 6.63 \times 10^{-34} \times 10^{14} = 6.6 \times 10^{-20}$ J (approx.).

Work Function

The minimum amount of work or energy necessary to take a free electron out of a metal against the attractive forces of surrounding positive ions is called the *work function* of the metal, symbol w_0. The work function is related to thermionic emission since this phenomenon is also concerned with electrons breaking away from the metal.

The work functions of caesium, sodium and beryllium are respectively about 1·9 eV, 2·0 eV and 3·9 eV. 1 eV (electronvolt) is a unit of energy equal to 1 electron charge $e \times 1$ volt, which is (p. 581)

$$1\cdot6 \times 10^{-19} \text{ C} \times 1 \text{ V} = 1\cdot6 \times 10^{-19} \text{ J.}$$

So the work function of sodium, $w_0 = 2 \text{ eV} = 2 \times 1\cdot6 \times 10^{-19} = 3\cdot2 \times 10^{-19}$ J.

Einstein's Particle (Photon) Theory

In 1905 Einstein suggested that the experimental results in photoelectricity could be explained by applying a quantum theory of light. He assumed that light of frequency f contains packets or quanta of energy hf. On this basis, light consists of *particles*, and these are called *photons*. The *number* of photons per unit area of cross-section of the beam of light per unit time is proportional to its intensity. But the *energy* of a photon is proportional to its frequency and is independent of the light intensity.

Let us apply the theory to the metal sodium. Its work function w_0 is 2·0 eV or $3\cdot2 \times 10^{-19}$ J. According to the theory, if the quantum energy in the incident light is $3\cdot2 \times 10^{-19}$ J, then electrons are just liberated from the metal. The particular frequency f_0 is called the *threshold frequency* of the metal. The *threshold wavelength* $\lambda_0 = c/f_0$.

We can now calculate f_0 and λ_0 for the metal sodium. Since $E = hf_0 = w_0$,

$$f_0 = \frac{w_0}{h} = \frac{3\cdot2 \times 10^{-19}}{6\cdot6 \times 10^{-34}} = 4\cdot8 \times 10^{14} \text{ Hz.}$$

So

$$\lambda_0 = \frac{c}{f} = \frac{3 \times 10^8}{4\cdot8 \times 10^{14}} = 6\cdot2 \times 10^{-7} \text{ m.}$$

Thus electrons are not liberated from sodium if the incident light has a frequency *less* than $4\cdot8 \times 10^{14}$ Hz or a wavelength *longer* than $6\cdot2 \times 10^{-7}$ m. This agrees with experiment.

Einstein's Photoelectric Equation

Using Einstein's theory, a simple result is obtained for the maximum energy of the electrons liberated from an illuminated metal.

If the quantum of energy hf in the light incident on a metal is say $4\cdot2 \times 10^{-19}$ J, and the work function w_0 of the metal is $3\cdot2 \times 10^{-19}$ J, the *maximum* energy of the liberated electrons is $(4\cdot2 \times 10^{-19} - 3\cdot2 \times 10^{-19})$ or $1\cdot0 \times 10^{-19}$ J because w_0 is the minimum energy to liberate electrons from the metal. The electrons with maximum energy come from the metal surface, as we have already stated. Owing to collisions, others below the surface emerge with a smaller energy.

We now see that the maximum kinetic energy, $\frac{1}{2}m_e v_m{}^2$, of the emitted electrons (photoelectrons) is given generally by

$$\frac{1}{2}m_e v_m{}^2 = hf - w_0 \quad . \quad . \quad . \quad . \quad \text{(i)}$$

So

$$hf = \frac{1}{2}m_e v_m{}^2 + w_0 \quad . \quad . \quad . \quad \text{(ii)}$$

Equation (ii) or (i) is called *Einstein's photoelectric equation*.

Example

Sodium has a work function of 2·0 eV. Calculate the maximum energy and speed of the emitted electrons when sodium is illuminated by radiation of wavelength 150 nm.

What is the least frequency of radiation (threshold frequency) for which electrons are emitted? (Assume $h = 6\cdot6 \times 10^{-34}$ J s, $e = -1\cdot6 \times 10^{-19}$ C, $m_e = 9\cdot1 \times 10^{-31}$ kg, $c = 3 \times 10^8$ m s^{-1}.)

$$\text{Incident photon energy} = hf = h\frac{c}{\lambda} = \frac{6\cdot6 \times 10^{-34} \times 3 \times 10^8}{150 \times 10^{-9}}$$

$$= 13\cdot2 \times 10^{-19} \text{ J}.$$

So   maximum kinetic energy $= hf - w_0 = 13\cdot2 \times 10^{-19} \text{ J} - 2 \times 1\cdot6 \times 10^{-19} \text{ J}$

$$= 10 \times 10^{-19} = 10^{-18} \text{ J} \qquad . \qquad . \qquad . \qquad \text{(i)}$$

Thus $\frac{1}{2}m_e v_m^2 = 10^{-18}.$

So
$$v = \sqrt{\frac{2 \times 10^{-18}}{9\cdot1 \times 10^{-31}}} = \sqrt{\frac{20}{9\cdot1}} \times 10^6$$

$$= 1\cdot5 \times 10^6 \text{ m s}^{-1} \qquad . \qquad . \qquad . \qquad . \qquad \text{(ii)}$$

The threshold frequency is given by $hf_0 = w_0$. So

$$f_0 = \frac{w_0}{h} = \frac{2 \times 1\cdot6 \times 10^{-19}}{6\cdot6 \times 10^{-34}}$$

$$= 4\cdot8 \times 10^{14} \text{ Hz} \qquad . \qquad . \qquad . \qquad . \qquad \text{(iii)}$$

Measuring Maximum Kinetic Energy. Stopping Potential

The maximum kinetic energy of the liberated photoelectrons can be found by a method analogous to finding the kinetic energy of a ball moving horizontally. In Fig. 40.2 (i), the moving ball is made to move up a smooth inclined plane PQ each time the slope of the plane is increased from zero. At one inclination the ball will *just* reach the top of the plane. In this case the kinetic energy at the bottom P is just equal to the potential energy at Q. Knowing the height h of Q above P, the kinetic energy is equal to mgh.

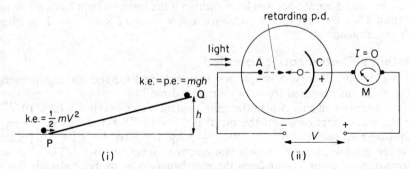

Fig. 40.2 Stopping potential

Figure 40.2 (ii) shows an analogous electrical experiment using a potential 'hill' whose gradient can be varied. Here a varying p.d. V is applied between the plates A and C inside an evacuated glass tube, and the potential of A is *negative* relative to C. This means that when photoelectrons are liberated from C, they are acted on by a retarding force.

When C is illuminated by a suitable beam, the photoelectrons liberated have a varying kinetic energy as we have previously explained. If the negative p.d. V between A and C is very small, many electrons can reach A from C. As V is increased negatively, however, the current I recorded on the meter M decreases because fewer electrons have energies sufficient to overcome the retarding force.

At a particular negative value V_s the current I becomes zero. This is the value of the negative p.d. which just stops the electrons with maximum energy from reaching A. V_s is called the *stopping potential*. Since an electron would lose an amount of energy given by charge $\times$ p.d. in moving from C to A, we see that

$$\frac{1}{2}m_e v_m^2 = eV_s.$$

So eV_s is a measure of the maximum energy of the emitted photoelectrons.

Verifying Einstein's Equation. Measurement of h

Figure 40.3 illustrates the basic features of a laboratory apparatus for investigating photoelectricity. It contains (1) a photoelectric cell X, which has inside it a photosensitive metal C of large area, the cathode, and a collector of the electrons, A, in a vacuum, (2) a potential divider arrangement Y for varying the p.d. V between the anode A and cathode, C, and (3) a d.c. amplifier (p. 598) for measuring a small current.

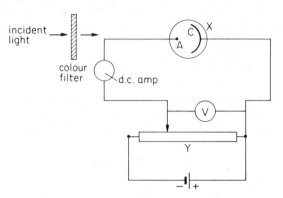

Fig. 40.3 Photoelectricity experiment: $eV_s = hf - w_o$

As shown, A is made *negative* in potential relative to C. The photoelectrons emitted from C then experience a retarding p.d. The p.d. V is increased negatively until the current becomes zero and the 'stopping potential', V_s, is then read from the voltmeter.

Using pure filters, the frequency f of the incident light can be varied and the corresponding value of V_s obtained. Figure 40.4 (i) shows some typical results. For different wavelengths λ_1, λ_2, λ_3, the respective stopping potentials V_s (when $I = 0$) are V_1, V_2, V_3. The shortest wavelength λ_1 (greatest frequency) required the highest stopping potential V_1.

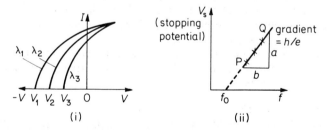

Fig. 40.4 Results of experiment

The values of V_s are now plotted against the corresponding frequency f of the incident light. Figure 40.4 (ii) shows the result. A straight line graph PQ is obtained. Now from Einstein's photoelectric equation, $eV_s = hf - w_0$. So

$$V_s = \frac{h}{e}f - \frac{w_0}{e}$$

Thus from Einstein's theory, the gradient a/b of the line PQ is h/e, or

$$h = e \times \frac{a}{b}$$

Knowing e, the electron charge, h can be calculated. Careful measurements by Millikan (below) gave a result for h of 6.26×10^{-34} J s, which was very close to the value of h found from experiments on black-body radiation. This confirmed Einstein's photoelectric theory that light can be considered to consist of particles with energy hf.

The line PQ in Fig. 40.4 (ii) also enables the theshold frequency f_0 and the work function w_0 to be found. From $eV_s = hf - w_0$, we have $0 = hf - w_0$ when $V_s = 0$. So $f = w_0/h = f_0$. Thus the intercept of PQ with the frequency axis gives the value of f_0. The work function can then be calculated from $w_0 = hf_0$.

Millikan's Experiments

In 1916, Millikan carried out an extensive and accurate investigation of Einstein's photoelectric equation, which in those days had not yet been accepted by all scientists. He used the alkali metals lithium, sodium and potassium. These metals emit electrons when illuminated by visible light, and cylinders of them, A, B, C, were placed round a wheel W (Fig. 40.5). To avoid tarnishing and the formation

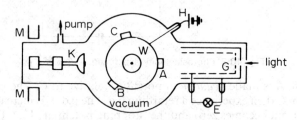

Fig. 40.5 Millikan's photoelectric experiment

of oxide films on the metal surface, which lead to considerable error, the metals were housed in a vacuum. Their surfaces were kept clean by a cutting knife K, which could be moved and turned by means of a magnet M outside.

The metal, A say, was kept at a variable positive potential by a battery H and illuminated by a beam of monochromatic light of wavelength λ_1 from a spectrometer. Any photoelectrons emitted could reach a gauze cylinder G, which was connected to one side of an electrometer E whose other terminal was earthed, and a current I would then flow in E.

The potential V of the collector G relative to G was varied first positively, when all the electrons were collected, and then negatively until the stopping potential V_s was obtained.

Experimental Results

Figure 40.6 (i) shows the results when a metal such as caesium is illuminated by monochromatic light of a given wavelength λ which is below the threshold value for the metal. When the retarding p.d. $-V$ is increased negatively, the stopping potential V_s is the *same* for a beam of low intensity and one of high intensity. This is because $eV_s = hf - w_0 = hc/\lambda - w_0$, and λ and w_0 are constant for a given metal and wavelength.

Figure 40.6 (i) also shows that when V is positive, (i) all the photoelectrons are now collected so that the current is constant, and (ii) a beam of high intensity produces more electrons than one of low intensity. If Q has twice the intensity of P, the current I is twice as much.

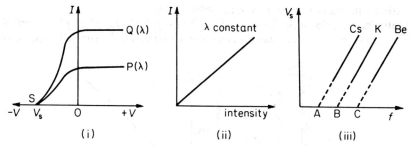

Fig. 40.6 Photoelectricity results

Figure 40.6 (ii) shows how the current I varies with intensity for a given wavelength λ below the threshold value. The number of electrons emitted is proportional to the intensity and a straight line graph is obtained.

Figure 40.6 (iii) shows the results when the stopping potential V_s is plotted against the frequency f for three different metals caesium (Cs), potassium (K) and beryllium (Be). The respective threshold frequencies A, B, C are different because the metals have different work functions; but the *slope of the three straight line graphs is the same* since the slope is given by h/e (see p. 847) and h and e are constants.

Example

Caesium has a work function of 1·9 electronvolts. Find (i) its threshold wavelength, (ii) the maximum energy of the liberated electrons when the metal is illuminated by light of wavelength $4·5 \times 10^{-7}$ m, (iii) the stopping p.d. (1 electronvolt $= 1·6 \times 10^{-19}$ J, $h = 6·6 \times 10^{-34}$ J s, $c = 3·0 \times 10^8$ m s^{-1}).

(i) The threshold frequency f_0 is given by $hf_0 = w_0 = 1·9 \times 1·6 \times 10^{-19}$ J. Now

Now threshold wavelength, $\lambda_0 = c/f_0$

$$\therefore \lambda_0 = \frac{c}{w_0/h} = \frac{ch}{w_0}$$

$$= \frac{3 \times 10^8 \times 6·6 \times 10^{-34}}{1·9 \times 1·6 \times 10^{-19}}$$

$$= 6·5 \times 10^{-7} \text{ m.}$$

(ii) Maximum energy of liberated electrons $= hf - w_0$, where f is the frequency of the incident light. But $f = c/\lambda$.

$$\therefore \text{ max. energy} = \frac{hc}{\lambda} - w_0$$

$$= \frac{6·6 \times 10^{-34} \times 3 \times 10^8}{4·5 \times 10^{-7}} - 1·9 \times 1·6 \times 10^{-19}$$

$$= 1·4 \times 10^{-19} \text{ J.}$$

(iii) The stopping potential V_s is given by $eV_s = 1·4 \times 10^{-19}$ J. Since $e = 1·6 \times 10^{-19}$ C,

$$\therefore V = \frac{1·4 \times 10^{-19}}{1·6 \times 10^{-19}} = 0·9 \text{ V (approx.)}$$

Photo-emissive and Photo-voltaic Cells

Photoelectric cells are used in photometry, in industrial control and counting operations, in television, and in many other ways.

Photoelectric cells of the kind we describe on p. 847 are called *photo-emissive* cells, because in them light causes electrons to be emitted. Another type of cell

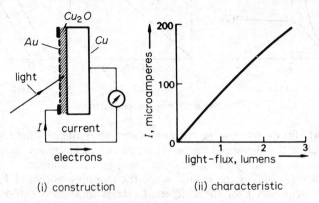

(i) construction   (ii) characteristic

Fig. 40.7 A photo-voltaic cell

is *photo-voltaic*, because it generates an e.m.f. and can therefore provide a current without a battery. One form of such a cell consists of a copper disc, oxidised on one face (Cu_2O/Cu), as shown in Fig. 40.7 (i). Over the exposed surface of the oxide a film of gold (Au) is deposited, by evaporation in a vacuum; the film is so thin that light can pass through it. When it does so it generates an e.m.f. in a way which we cannot describe here.

Photo-voltaic cells are sensitive to visible light. Figure 40.7 (ii) shows how the current from such a cell, through a galvanometer of resistance about 100 ohms, varies with the light-flux falling upon it. The current is not quite proportional to the flux. Photo-voltaic cells are obviously convenient for photographic exposure meters, for measuring illumination in factories, and so on, but as measuring instruments they are less accurate than photo-emissive cells.

Photo-conductive Cells

A photo-conductive cell is one whose resistance changes when it is illuminated. A common form consists of a pair of interlocking comb-like electrodes made of gold (Au) deposited on glass (Fig. 40.8 (i)); over these a thin film of selenium (Se) is deposited. In effect, the selenium forms a large number of strips, electrically in parallel; this construction is necessary because selenium has a very high resistivity (about 700 Ω m in the dark). The resistance between the terminals, XY, falls from about 10^7 ohms in the dark to about 10^6 ohms in bright light.

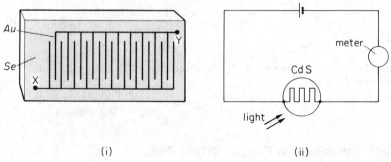

(i)   (ii)

Fig. 40.8 (i) A selenium cell (ii) Circuit for CdS lightmeter

In conjunction with amplifiers, photo-conductive cells were used as fire alarms during the last war.

Cadmium sulphide, CdS, is a photo-conductive material which has a greater light sensitivity than selenium. It is therefore widely used as the light-sensitive cell in exposure meters of cameras. Unlike the photo-voltaic cell, a battery is needed with the cell, Fig. 40.8 (ii).

Quantisation of Energy. Energy Levels

Energy of Atoms

The average energy of a monatomic molecule moving in a gas at room temperature is $\frac{3}{2}kT$ or $\frac{3}{2} \times 1.4 \times 10^{-23} \times 300$ J, which is 6.3×10^{-21} J. Since 1 eV is 1.6×10^{-19} J, this energy corresponds to about 0·04 eV. Thus when collisions between molecules take place, the energy exchange is of the order of 0·04 eV. In these conditions the collisions are *perfectly elastic*, that is, the internal energy of the atoms is not increased by collision and the kinetic energy of the colliding molecules is unchanged.

In 1914, Franck and Hertz bombarded atoms by electrons of much higher energy, of the order of several electronvolts. They used sodium vapour at a very low pressure of about 1 mm of mercury in a tube containing a heated tungsten filament F, a grid plate G, and a plate A, Fig. 40.9 (i). Electrons were emitted from F, and the distance FG was arranged to be much greater than the

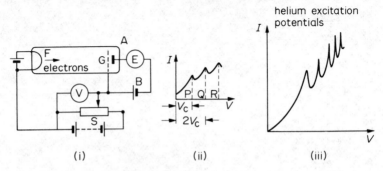

Fig. 40.9 Franck and Hertz experiment

mean free path of the electrons in the gas, in which case the electrons would make collisions with the atoms before reaching G. The p.d. V between F and G could be varied by the potentiometer S. The electrons emitted from F were accelerated to kinetic energies depending on the magnitude of V, which was measured by a voltmeter. A small p.d., less than 1 volt, was applied between A and G so that A was negative in potential relative to G. The plate A was close to G, and electrons reaching G and passing through to A were subjected to a retarding field. The number per second reaching A was measured by a meter E.

When the accelerating p.d. V between F and G was increased from zero, the current in E rose until the p.d. reached a value P, Fig. 49.9 (ii). As V was increased further the current diminished to a minimum, rose again to a new peak at a higher p.d. Q, then diminished again and rose to another peak at a higher p.d. R. The p.d. V_c between successive peaks was found to be constant and equal to 2·10 V for sodium vapour. Similar results were found for other gases. The p.d. V_c was called the *critical p.d.* or *critical potential*.

Energy Levels

From the graph, it can be seen that the current begins to drop at the critical potential V_c. Here the electrons have an energy of V_c electronvolt or eV_c joule. This energy is just sufficient to raise the internal energy of the sodium atom by collision. Energies less than eV_c fail to increase the energy of the atom. After giving up this energy, the electrons then have insufficient energy to overcome the small retarding p.d. between G and A. Thus the current starts to fall. At a p.d. of $2V_c$ a dip again begins to form. This is due to the electrons giving up energy equal to $2eV_c$ to two atoms at separate collisions.

It thus appears that the energy of the atom cannot be increased unless the energy of the colliding particle is greater than V_c electronvolts. In this case an *inelastic* collision takes place. The atom now takes up an energy equal to eV_c joules, which the colliding particle loses. This way of increasing the energy of an atom is called *excitation*. The interval, V_c, between successive peaks of the graph is the *excitation potential* of the atom.

The results of the Franck–Hertz experiment show that the energy of an atom is constant unless the atom is given enough energy to raise this by a *definite* amount. *No intermediate energy change is allowed*. An atom, therefore, exists in one of a set of well defined *energy levels*. If helium, for example, is used in a Franck–Hertz tube, a graph shown in Fig. 40.9 (iii) is obtained. Here each peak corresponds to a different energy level of the atom. Thus a whole sequence of different energy levels can be found in the helium atom.

As we have seen, at ordinary temperatures, the thermal energy of molecules in a gas is insufficient to cause excitation. If the gas can be heated to an enormously high temperature, of the order of 100 000 K, the molecules can gain enough energy to cause excitation.

Bohr's Theory of Energy Levels

A *model* of the hydrogen atom was proposed by Bohr in 1911. This explained satisfactorily the existence of energy levels and the spectrum of the hydrogen atom. Later, however, it was shown that the model could not be applied to other and more complex atoms and a more satisfactory 'quantum theory' of the atom has since been developed.

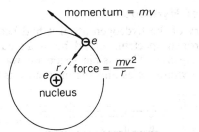

Fig. 40.10 Bohr's theory of hydrogen atom

Bohr considered one electron of charge $-e$ and mass m, moving with speed v and acceleration v^2/r in an orbit round a central hydrogen nucleus of charge $+e$ (Fig. 40.10). In classical physics, charges undergoing acceleration emit radiation and, therefore, they would lose energy. On this basis the electron would spiral towards the nucleus and the atom would collapse. Bohr, therefore, suggested that in those orbits where the angular momentum is a multiple of $h/2\pi$ the energy is constant. Twelve years later, de Broglie proposed that a particle such as the electron may be considered to behave as a *wave* of wavelength $\lambda = h/p$, where h is the Planck constant and p is the *momentum* of the moving particle.

If the electron can behave as a *wave*, it must be possible to fit a whole number of wavelengths around the orbit. In this case a standing wave pattern is set up and the energy in the wave is confined to the atom. A progressive wave would imply that the electron is moving from the atom and is not in a stationary orbit. See p. 421.

If there are n waves in the orbit and λ is the wavelength,

$$n\lambda = 2\pi r \qquad . \qquad . \qquad . \qquad . \qquad . \qquad \text{(i)}$$

$$\therefore \lambda = \frac{2\pi r}{n} = \frac{h}{p} = \frac{h}{mv} \qquad . \qquad . \qquad . \qquad . \qquad \text{(ii)}$$

So $$mvr = \frac{nh}{2\pi} \qquad . \qquad . \qquad . \qquad . \qquad . \qquad \text{(iii)}$$

Now $mv \times r$ is the moment of momentum or *angular momentum* of the electron about the nucleus. So equation (iii) states that the *angular momentum is a multiple of $h/2\pi$*. The quantisation of angular momentum is a key point in atomic theory.

In Fig. 40.10, the electron moving round the nucleus has kinetic energy due to its motion and potential energy in the electrostatic field of the nuclear charge $+e$. Bohr calculated the total energy E of the electron in terms of its charge, mass, orbital radius and the number n which quantises the angular momentum. He then assumed that the electron, or atom, can pass from one energy value or level such as E_2, corresponding to a number $n = n_2$, to another energy level such as E_1 of lower value, corresponding to a number $n = n_1$; and that the difference in energy is released in the form of radiation of energy hf, where h is the Planck constant and f is the frequency of the radiation. So

$$E_2 - E_1 = hf = hc/\lambda,$$

where λ is the wavelength of the radiation, since $c = f\lambda$ for the electromagnetic wave of speed c. From the values of the energy, Bohr was able to calculate the wavelength emitted.

Spectral Series of Hydrogen

Before Bohr's theory of the hydrogen atom it had been found that the wavelengths of the hydrogen spectrum could be arranged in a formula or series named after its discoverer. The visible spectrum was the Balmer series, the ultraviolet was the Lyman series and the infrared was the Paschen series.

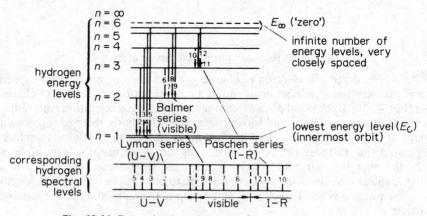

Fig. 40.11 Energy levels and spectra of hydrogen (*not to scale*)

Bohr's theory of energy levels accounted for all the series. As shown in Fig. 40.11, the hydrogen spectrum is obtained simply by using different numbers for n in calculating the energy levels. The ultraviolet series, for example, is obtained when the energy of the atom falls to the lowest energy level E_0 corresponding to $n = 1$; the visible spectrum is obtained for energy falls to a higher level corresponding to $n = 2$; and the infrared spectrum is obtained for energy falls to the higher level when $n = 3$.

From Fig. 40.11, the energy change E for the ultraviolet series is greater than for the visible spectrum. Since $E = hf$, the frequency of ultraviolet radiation is greater than visible radiation. Wavelength, λ, is inversely-proportional to f, since $c = f\lambda$. Hence the ultraviolet wavelengths are *shorter* than those in the visible spectrum. This is discussed further on p. 857.

Energy Levels of the Atom

Bohr's theory of the hydrogen atom was unable to predict the energy levels in complex atoms, which had many electrons. Quantum or wave mechanics, beyond the scope of this book, is used to explain the spectral frequencies of these atoms. The fundamental ideas of Bohr's theory, however, are still retained, for example, the angular momentum of the electron has quantum values and the energy levels of the atom have only allowed separated values characteristic of the atom. We therefore suppose that a given atom has a series of defined and discrete (separated) energy levels, E_0, E_1, E_2, and that no other or intermediate energy level is possible. The lowest energy level E_0 is called the *ground state energy*. All physical systems are in stable equilibrium in the lowest energy state.

If the atom absorbs energy, and the energy of the atom reaches one of its allowed values E_1, the atom is said to be in an *excited state*. Once an atom has been excited to a higher energy level E_n, it will try to reduce its energy. The energy lost if the atom reverts directly to the ground state is $(E_n - E_0)$. *This energy is radiated in the form of electromagnetic radiation.* The energy of the photon emitted is hf where f is the frequency of the radiation, and so

$$hf = E_n - E_0.$$

Sometimes it is possible for the energy to change back to the ground state through an intermediate energy level E_m. In this case *two* different frequencies f_1, f_2 are radiated which are given respectively by the equations:

$$hf_1 = E_n - E_m, \qquad hf_2 = E_m - E_0.$$

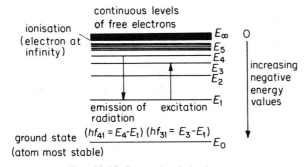

Fig. 40.12 Energy levels in the atom

It is customary to draw the energy levels on a vertical scale and to mark the *transitions* from one energy level to another with an arrow. Figure 40.12 shows roughly the energy levels of an atom, namely, its ground state E_0, its excited states, E_1, E_2, ... E_∞. It will be noted that the energy levels become more closely spaced at the higher excited states.

Excitation and Ionisation Potentials. Spectra

The energy required to raise an atom from its ground state to an excited state is called the *excitation energy* of the atom. If the energy is eV, where e is the electron charge, V is known as the *excitation potential* of the atom (p. 853).

If the atom is in its ground state with energy E_0, and absorbs an amount of energy eV which just removes an electron completely from the atom, then V is said to be the *ionisation potential* of the atom. The potential energy of the atom is here denoted by E_∞, as the ejected electron is so far away from the attractive influence as to be, in effect, at infinity. E_∞ is taken as the 'zero' energy of the atom, and its other values are thus negative. The ionisation potential V is given by $E_\infty - E_0 = eV$, or by $-E_0 = eV$.

If a gas is excited by a high voltage to produce a discharge, and the light is examined in a spectrometer, an emission spectrum is seen.

As they move through the discharge in the gas, some electrons have sufficient energy to excite atoms to a higher energy level. A number of gases such as neon produce a *line spectrum*, that is, the spectrum consists of a number of well defined lines, each having a particular wavelength of frequency. These separated lines are experimental evidence for the existence of separate or 'quantised' energy levels in the atom.

Hydrogen Energy Levels

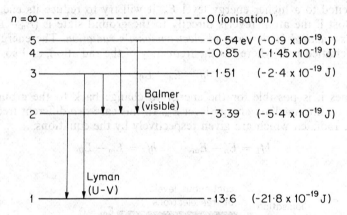

Fig. 40.13 Energy levels of hydrogen

Figure 40.13 shows roughly some of the energy levels of the hydrogen atom in electronvolts (eV) and in joules (J). The quantum numbers $n = 1,2,\ldots,\infty$ on the left correspond to values used to calculate the energy values. The ground state, $n = 1$, is -13.6 eV (-21.8×10^{-19} J) and this is the lowest energy level, where the hydrogen atom is most stable. The higher energy levels are all calculated from the formula -13.6 eV$/n^2$, where $n = 1,2,3,4,5\ldots$ respectively.

The energy levels *increase* to the value 0, which is the ionisation level. When the electron or atom is given an amount of energy equal to the ionisation energy, the electron is just able to become a 'free' electron. If the electron is given a greater amount of energy than the ionisation energy of 21.8×10^{-19} J, such as 22.8×10^{-19} J, the excess energy of 1.0×10^{-19} J is then the kinetic energy of the free electron outside the atom. In general, the free electron can have a continuous range of energies. Inside the atom, however, it can have only one of the energy values characteristic of the atom.

Emission Spectra of Hydrogen

Let us calculate the wavelength of the emitted radiation when the hydrogen atom is excited from its ground state ($n = 1$) where its energy level is -21.8×10^{-19} J to the higher level ($n = 2$) of energy -5.4×10^{-19} J, and then falls back to the ground state.

Since E = energy *change* = $hf = hc/\lambda$, then, using standard values,

$$\lambda = \frac{hc}{E} = \frac{6.6 \times 10^{-34} \times 3 \times 10^8}{(-5.4 \times 10^{-19}) - (-21.8 \times 10^{-19})}$$

$$= \frac{6.6 \times 3}{16.4} \times 10^{-34+8+19}$$

$$= 1.2 \times 10^{-7} \text{ m (approx.).}$$

This wavelength is in the *ultraviolet spectrum* or *Lyman series*.

Suppose we now calculate the wavelength of the radiation emitted from an energy level -2.4×10^{-19} J ($n = 3$) to -5.4×10^{-19} J ($n = 2$). This is an energy change E of 3.0×10^{-19} J. Now from $\lambda = hc/E$ used above, we see that $\lambda \propto 1/E$ since h and c are constants. This means that the smaller the value of E, the *greater* is the wavelength. We have just shown that $\lambda = 1.2 \times 10^{-7}$ m when $E = 16.4 \times 10^{-19}$ J. So, by proportion, for $E = 3.0 \times 10^{-19}$ J, λ is given by

$$\frac{\lambda}{1.2 \times 10^{-7} \text{ m}} = \frac{16.4 \times 10^{-19}}{3.0 \times 10^{-19}}.$$

So
$$\lambda = \frac{1.2 \times 10^{-7} \times 16.4}{3.0} = 6.6 \times 10^{-7} \text{ m.}$$

This wavelength is in the *visible spectrum* or *Balmer series*.

Types of Emission Spectra

Emission spectra are classified into *continuous, line* and *band* spectra. With few exceptions, incandescent solids and liquids produce a *continuous spectrum*, one in which all wavelengths are found over a wide range. *Line spectra* are obtained from *atoms* in gases such as hydrogen in a discharge tube, and the visible spectrum of a sodium salt vaporised in a Bunsen flame consists of two lines close together.

Gases such as carbon dioxide in a discharge tube also produce a *band spectrum*, each band consisting of a series of lines very close together at the sharp edge or head of the band and farther apart at the other end or tail. Band spectra are essentially due to *molecules*. The different band heads in a band system are due to small allowed discrete energy changes in the vibrational states of the molecule. The fine lines in a given band are due to still smaller allowed discrete energy changes in the rotational states of the molecule.

Absorption Spectra

Atoms can absorb energy in a number of ways. In a flame, inelastic collisions with energetic molecules can raise atoms to higher energy levels. In a discharge tube, inelastic collisions with bombarding electrons can raise atoms to higher energy levels.

An atom can also absorb energy from a photon. If the photon energy $E = hf$ is just sufficient to excite an atom to one of its higher energy levels, the photon will be absorbed. When it returns to the ground state the excited atom emits the same wavelength as the photon but equally in all directions. So the intensity of the radiation in the direction of the incident photon is reduced. A *dark* line is thus seen whose wavelength is that of the absorbed photon

Absorption of photons explain the dark lines in the sun's visible spectrum, first observed by Fraunhofer (p. 377). The sun emits a continuous spectrum of photons. Vaporised elements in the outer or cooler parts of the sun's atmosphere absorb those photons which have the same frequency or energy to excite them to higher energy levels. The sun's spectrum is now darker at wavelengths *characteristic of the elements in the sun's atmosphere*. Since absorption spectra are always characteristic of the absorbing elements, these elements can be identified from their absorption spectra. In this way we know that elements such as iron exist in the sun.

In the sun's atmosphere, hot gases such as hydrogen are already excited to a higher energy level by collision with atoms. The absorbed photons are thus able to excite the atoms to still higher energy levels, which are in the sun's visible spectrum or Balmer series. So the Fraunhofer (dark) lines are present in the visible spectrum of the sun.

Normally, however, hydrogen is in the *ground* state, corresponding to the energy level -13.6 eV (p. 856). The next higher energy level is about -3.4 eV. This is a relatively high energy 'jump' of 10.2 eV and corresponds to a photon in the ultraviolet (Lyman) series. Visible light has photons of energy less than 3.4 eV (see p. 857). So photons in visible light are not able to be absorbed by hydrogen atoms in their normal or ground state. Consequently hydrogen is *transparent* to visible light. So if a wide band of wavelengths is passed through hydrogen, only those wavelengths in the ultraviolet would produce absorption lines.

Spontaneous and Stimulated Emission

As we have seen, an atom may undergo a transition between energy states if it emits or absorbs a photon of the appropriate energy. When an atom in the ground state, for example, absorbs a photon and makes a transition to a higher energy state or level, the photon is said to have 'stimulated' the absorption. The atom cannot increase its energy 'spontaneously', that is, in the absence of a photon.

We might expect that the atom in its higher energy state would emit a photon spontaneously, that is, the probability of emission would be independent of the number of photons present in the environment of the atom. In 1917 Einstein proved, however, that this is not the case. The probability per unit time that an atom will decay to a lower energy state and emit a photon is the sum of two terms. One is a spontaneous emission term. The other is a '*stimulated emission*' term, which is proportional to the number of photons of the relevant energy *already present in the environment of the atoms*. Further, the photon produced by stimulated emission is always *in phase* with the stimulating photon.

The Laser Principle

Consider an atom with an energy level E above the ground state for which the spontaneous emission probability is nearly zero but the stimulated emission probability is large. Suppose that, by a method described later, we put a large number of the atoms into this excited state E. So long as there are no photons of energy $E - E_0$ in the system, where E_0 is the ground state energy, the atoms are unable to decay because the stimulated emission term is proportional to the number of photons present.

If, however, a few photons of energy $E - E_0$ are introduced, these will immediately stimulate the emission of a number of photons of the same kind. This increases the number of photons, which in turn stimulate the emission of more photons. Thus a 'chain-reaction' or 'avalanche' effect is produced, with the result that all the atoms present give up their photons very rapidly. This process is called *laser* action (*light amplification by stimulated emission of radiation*) since the original 'pulse' of photons has been amplified into a much more powerful pulse.

Characteristics of Laser Light

The light pulse from a laser is:

(i) *monochromatic* light because all the photons have the same energy $(E - E_0)$ above the ground state and hence the same frequency $(E - E_0)/h$, where h is the Planck constant;

(ii) *coherent* light because all the photons or waves are in phase;

(iii) *intense* light because all the emitted waves are in phase or coherent. If the photons or waves were out of phase or incoherent, the resultant intensity would be the sum of the individual intensities and this is proportional to $n \times a^2$, where n is the number of waves and a is the amplitude of each wave. Since the waves are in phase, however, their total amplitude is na. Hence the resultant intensity is proportional to $(na)^2$ or $n^2 a^2$. This intensity is greater by a factor n than that obtained from incoherent waves. Since n is very large, this is an enormous increase in intensity.

Principle of Ruby Laser

In a system of atoms in thermal equilibrium, the number of atoms in the ground state is much greater than in a higher energy state. This is called a 'normal population' of atoms among the available energy states. In a laser, however, a '*population inversion*' is produced, as we now explain.

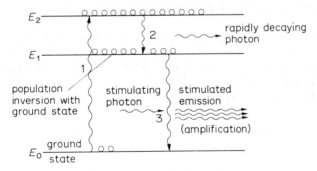

Fig. 40.14 Stages in laser action

Consider an atom which has the three energy states E_0 (ground state), E_1 and E_2, Fig. 40.14. By means of a flash tube, for example, a large number of atoms is first excited or 'pumped' to an energy state E_2 by photons of energy $E_2 - E_0$ (stage 1). The excited atoms decay spontaneously to the lower energy state E_1, emitting photons of energy $E_2 - E_1$ (stage 2).

Now the energy state E_1 has the special property of having a large stimulated emission probability and a low spontaneous emission probability, so that the atoms at this level are unable to decay spontaneously. The energy level E_1 is thus filled with a far greater number of atoms than the ground state E_0. Consequently we have a 'population inversion' between these two states.

Any stray photon with the requisite energy $E_1 - E_0$ will now produce stimulated emission, followed by the 'chain reaction' or laser action described before (stage 3). The action is enhanced by silvering the ends (and sides) of the solid, such as ruby crystal, which produce multiple reflections of photons within the system of excited atoms. Each photon will thus stimulate the emission of many others.

Figure 40.15 (i) illustrates the principle of the ruby laser. A very powerful and narrow beam of red light, wavelength 694·3 nm, emerges through a section at one end which is only partially silvered.

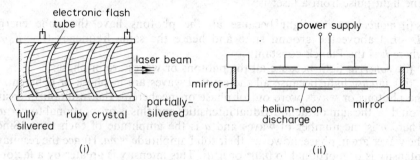

Fig. 40.15 (i) Ruby laser (ii) Gas laser

The Gas Laser

Figure 40.15 (ii) shows the basic features of a gas laser. A helium-neon gas mixture (carbon dioxide gas is also used) is contained inside a long quartz tube with optically plane mirrors at each end. A powerful radio-frequency generator is used to produce a discharge in the gas so that the helium are excited or 'pumped up' to a higher energy level. By collision with these atoms, the neon atoms are excited to a higher energy level so that a population inversion is obtained. Stimulated emission then occurs as the neon atoms decay to a lower energy level.

A gas laser has several advantages over the ruby (solid) laser. An important advantage is that the light is produced as a continuous beam rather than in ultra-short pulses as in the ruby laser. Further, crystal and other imperfections in the solid, which lead to slight beam divergence and slight spread of wavelengths, are avoided in the gas laser.

X-Rays

In 1895, Röntgen found that some photographic plates, kept carefully wrapped in his laboratory, had become fogged. Instead of merely throwing them aside he set out to find the cause of the fogging. He traced it to a gas-discharge tube, which he was using with a low pressure and high voltage. This tube appeared to emit a radiation that could penetrate paper, wood, glass, rubber, and even aluminium a centimetre and a half thick. Röntgen could not find out whether the radiation was a stream of particles or a train of waves—Newton had the same difficulty with light—and he decided to call it X-rays.

Nature and Production of X-rays

We now regard X-rays as waves, similar to light waves, but of much shorter wavelength: about 10^{-10} m, or 0·1 nm. They are produced when fast electrons, or cathode rays, strike a target, such as the walls or anode of a low-pressure discharge tube.

In a modern X-ray tube there is no gas, or as little as high-vacuum technique can achieve: the pressure is about 10^{-5} mmHg. The electrons are provided by thermionic emission from a white-hot tungsten filament. In Fig. 40.16, F is the filament and T is the metal target embedded in a copper anode A. A cylinder C round F, at a negative potential relative to F, focuses the electron beam on T. Because there is so little gas, the electrons on their way to the anode do not

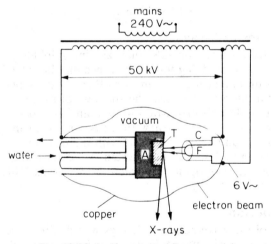

Fig. 40.16 An X-ray tube (diagrammatic)

lose any appreciable amount of their energy in ionising atoms. From the a.c. mains, transformers provide about 6 volts for heating the filament, and about 50 000 volts for accelerating the electrons. On the half-cycles when the target is positive, the electrons bombard it, and generate X-rays. On the half-cycles when the target is negative, nothing happens at all—there is too little gas in the tube for it to break down. Thus the tube acts, in effect, as its own rectifier (p. 811), providing pulses of direct current between target and filament. The heat generated at the target by the electronic bombardment is so great that the metal must be cooled. In large tubes this is done by circulating water or oil, as shown. The target in an X-ray tube may be tungsten, for example, which has a high melting-point.

So X-rays (waves) are produced by bombarding matter with electrons (particles). The production of X-rays is therefore the inverse process to the photo-electric effect, where electrons (particles) are liberated from metals by incident light waves.

Example

An X-ray tube, operated at a d.c. potential difference of 40 kV, produces heat at the target at the rate of 720 W. Assuming 0·5% of the energy of the incident electrons is converted into X-radiation, calculate (i) the number of electrons per second striking the target, (ii) the velocity of the incident electrons. (Assume charge-mass ratio of electron, $e/m_e = 1·8 \times 10^{11}$ C kg^{-1}.)

(i) Heat per second at target = 99·5% × IV, where I is the current flowing and V is the p.d. applied. So

$$0·995 \times I \times 40\,000 = 720.$$

Then
$$I = \frac{720}{40\,000 \times 0·995} = 0·018 \text{ A (approx.).}$$

So number of electrons per second $= \dfrac{I}{e} = \dfrac{0·018}{1·6 \times 10^{-19}}$

$$= 1·1 \times 10^{17}.$$

(ii) Energy of incident electrons $= \frac{1}{2}m_e v^2 = eV$

So
$$v = \sqrt{2V \times \frac{e}{m_e}} = \sqrt{2 \times 40\,000 \times 1·8 \times 10^{11}}$$

$$= 1·2 \times 10^8 \text{ m s}^{-1}.$$

Effects and Uses of X-rays

When X-rays strike many minerals, such as zinc sulphide, they make them fluoresce. (It was while studying this fluorescence that Becquerel discovered the radiations from uranium.) If a human—or other—body is placed between an X-ray tube and a fluorescent screen, the shadows of its bones can be seen on the screen, because they absorb X-rays more than flesh does. Unusual objects, such as swallowed safety-pins, if they are dense enough, can also be located. X-ray photographs can likewise be taken, with the plate in place of the screen. In this way cracks and flaws can be detected in metal castings.

X-rays are used to investigate the structure of crystals. This important use of X-rays is discussed shortly.

Crystal Diffraction

The first proof of the wave-nature of X-rays was due to Laue in 1913, many years after X-rays were discovered. He suggested that the regular small spacing of atoms in crystals might provide a natural diffraction grating if the wavelengths of the rays were too short to be used with an optical line grating. Experiments by Friedrich and Knipping showed that X-rays were indeed diffracted by a thin

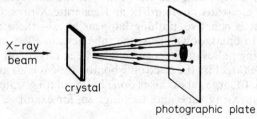

Fig. 40.17 Laue crystal diffraction

crystal, and produced a pattern of intense spots round a central image on a photographic plate placed to receive them, Fig. 40.17. The rays had thus been scattered by interaction with electrons in the atoms of the crystal. The diffraction pattern obtained gave information on the geometrical spacing of the atoms. An X-ray diffraction pattern produced by a crystal is shown on p. 867.

Bragg Law

The study of the atomic structure of crystals by X-ray analysis was initiated in 1914 by Sir William Bragg and his son Sir Lawrence Bragg, with notable achievements. They soon found that a monochromatic beam of X-rays was reflected from a plane in the crystal rich in atoms, a so-called atomic plane, as if this acted like a mirror.

This important effect can be explained by Huygens' wave theory in the same way as the reflection of light by a plane surface. Suppose a monochromatic parallel X-ray beam is incident on a crystal and interacts with atoms such as A, B, C, D in an atomic plane P, Fig. 40.18 (i). Each atom scatters the X-rays. Using Huygens' construction, wavelets can be drawn with the atoms as centres, which all lie on a plane wavefront reflected at an equal angle to the atomic plane P. When the X-ray beam penetrates the crystal to other atomic planes such as Q, R parallel to P, reflection occurs in a similar way, Fig. 40.18 (ii). Usually, the beam or ray reflected from one plane is weak in intensity. If, however, the reflected beams or rays from all planes are *in phase* with each other, an intense reflected beam is produced by the crystal.

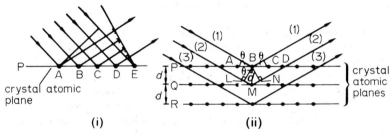

Fig. 40.18 Reflection (diffraction) at crystal atomic planes

Suppose, then, that the glancing angle on an atomic plane in the crystal is θ, and d is the distance apart of consecutive parallel atomic planes, Fig. 40.18 (ii). The path difference between the rays marked (1) and (2) $= LM + MN = 2LM = 2d \sin \theta$. Thus an intense X-ray beam is reflected when

$$2d \sin \theta = n\lambda,$$

where λ is the wavelength and n has integral values. This is known as *Bragg's law*. Hence, as the crystal is rotated so that the glancing angle is increased from zero, and the beam reflected at an equal angle is observed each time, an intense beam is suddenly produced for a glancing angle θ_1 such that $2d \sin \theta_1 = \lambda$. When the crystal is rotated further, an intense reflected beam is next obtained for an angle θ_2 when $2d \sin \theta_2 = 2\lambda$. Thus several orders of diffraction images may be observed. Many orders are obtained if λ is small compared with $2d$. Conversely, no images are obtained if λ is greater than $2d$.

The intense diffraction (reflection) images from an X-ray tube are due to X-ray lines *characteristic of the metal used* as the target, or 'anti-cathode' as it was originally known. The more penetrating or harder X-ray lines emitted from the metal are called the *K lines*; the less penetrating or softer lines are called the *L lines*. The wavelength of the K lines are shorter than those of the L lines, so that the frequencies of the K lines are greater than those of the L lines. The X-radiation from some metals thus consists of characteristic lines in the K series or L series or both.

Moseley's Law

In 1914 Moseley measured the frequency f of the characteristic X-rays from many metals, and found that, for a particular type of emitted X-ray such as K_α, whose origin is explained later, the frequency f varied in a regular way with the *atomic number Z* of the metal. When a graph of Z v. $f^{1/2}$ was plotted, an almost perfect straight line was obtained, Fig. 40.19. Moseley therefore gave an empirical relation, known as *Moseley's law*, between f and Z as

$$f = a(Z - b)^2,$$

where a, b are constants.

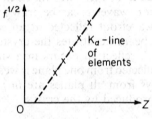

Fig. 40.19 Moseley's law

Since the regularity of the graph was so marked, Moseley predicted the discovery of elements with atomic numbers 43, 61, 72 and 75, which were missing from the graph at that time. These were later discovered. He also found that though the atomic weights of iron, nickel and cobalt increased in this order, their positions from the graph were: iron ($Z = 26$), cobalt ($Z = 27$) and nickel ($Z = 28$). The chemical properties of the three elements agree with the order by atomic number and not by atomic weight. Rutherford's experiments on the scattering of α-particles (p. 888) had shown that the atom contained a central nucleus of charge $+Ze$ where Z is the atomic number, and Moseley's experiments confirmed the importance of Z in atomic theory.

Explanation of Emission Spectra

Since the frequency of the X-ray spectra of elements is related to Ze, the charge on the nucleus, Moseley's results showed that the radiation was due to energy changes of the atom resulting from the movement of *electrons close to the nucleus*.

We now know, in fact, that the electrons in atoms are in groups which have various energy states. These groups are called 'shells'. Electrons nearest the nucleus are in the so-called K shell. The next group, which are further away from the nucleus, are in the so-called L shell. The M shell is farther from the nucleus than the L shell. Electrons farther from the nucleus have greater energy than those nearer, since more work is needed to move them farther away. So electrons in the L shell have greater energy than those in the K shell; electrons in the M shell have greater energy than those in the L shell.

In the X-ray tube, energetic electrons bombard the metal target and may eject an electron from the innermost shell, the K shell. The atom is then raised to an excited state since its energy has increased, and it is unstable. An electron from the L shell may now move into the vacancy in the K shell, thereby decreasing the energy of the atom. Simultaneously, radiation is emitted of frequency f, where E = energy change of atom = hf. Thus $f = E/h$, and as E is very high for metals, the frequency f is very high and the wavelength is correspondingly short. It is commonly of the order of 10^{-9} m (1 nm) or less. The K series of X-ray lines is due to movement of electrons from the L (K_α line) or M (K_β line) shells to the K shell.

X-ray spectra are thus due to energy changes in electrons close to the nucleus of metals. In contrast, the optical spectra of the metals is due to energy changes in the outermost shell of the atom. Here the energy changes are about 1000 times smaller. So the frequencies of the optical lines are about 1000 times smaller, that is, their wavelengths are about 1000 times longer than those of X-rays.

Continuous X-ray Background Radiation

The characteristic X-ray spectrum from a metal is usually superimposed on a background of continuous, or so-called 'white', radiation of small intensity. Figure 40.20 illustrates the characteristic lines, K_α, K_β, of a metal and the continuous background of radiation for two values of p.d., 40 000 and 32 000 volts,

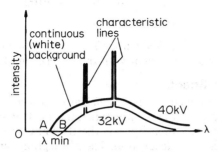

Fig. 40.20 X-ray characteristic lines and background

across an X-ray tube. It should be noted that (i) the wavelengths of the characteristic lines are independent of the p.d.—they are characteristic of the metal, and (ii) the background of continuous radiation has increasing wavelengths which slowly diminish in intensity, but as the wavelengths diminish they are cut off *sharply*, as at A and B.

When the bombarding electrons collide with the metal atoms in the target, most of their energy is lost as heat. A little energy is also lost in the form of electromagnetic radiation. Here the frequencies are given $E = hf$ with the usual notation, and the numerous energy changes produce the background radiation in Fig. 40.20.

The existence of a sharp minimum wavelength at A or B can be explained only by the quantum theory. The energy of an electron before striking the metal atoms of the target is eV, where V is the p.d. across the tube. If a direct collision is made with an atom and *all* the energy is absorbed, then, on quantum theory, the X-ray quantum produced has maximum energy.

$$\therefore hf_{max} = eV$$

$$\therefore f_{max} = \frac{eV}{h} \qquad \qquad \qquad \text{(i)}$$

$$\therefore \lambda_{min} = \frac{c}{f_{max}} = \frac{ch}{eV} \qquad . \qquad . \qquad . \qquad \text{(ii)}$$

Verification of Quantum Theory

These conclusions are borne out by experiment. Thus for a particular metal target, experiment shows that the minimum wavelength is obtained for p.d.s of 40 kV and 32 kV at glancing angles of about 3·0° and 3·8° respectively. The ratio of the minimum wavelengths is hence, from Bragg's law,

$$\frac{\lambda_1}{\lambda_2} = \frac{\sin 3·0°}{\sin 3·8°} = 0·8 \text{ (approx.)}.$$

From (ii), $\lambda_{min} \propto 1/V$.

$$\therefore \frac{\lambda_1}{\lambda_2} = \frac{32}{40} = 0·8.$$

With a tungsten target and a p.d. of 30 kV, experiment shows that a minimum wavelength of $0·42 \times 10^{-10}$ m is obtained, as calculated from values of d and θ. From (ii),

$$\therefore \lambda_{min} = \frac{ch}{eV} = \frac{3·0 \times 10^8 \times 6·6 \times 10^{-34}}{1·6 \times 10^{-19} \times 30\,000} \text{m}$$

$$= 0·41 \times 10^{-10} \text{ m},$$

using $c = 3·0 \times 10^8$ m s^{-1}, $h = 6·6 \times 10^{-34}$ J s, $e = 1·6 \times 10^{-19}$ C, $V = 30\,000$ V. This is in good agreement with the experimental result.

X-Ray Absorption Spectra

X-rays are absorbed by metals. The *coefficient of absorption* can be expressed by a relation of the form $I = I_0 e^{-\lambda x}$, where I_0 is the incident intensity of the X-ray beam, I is the transmitted intensity, x is the thickness of the metal sheet and λ is the linear coefficient of absorption. The absorption can also be expressed by the relation $I = I_0 e^{-\mu m}$, where μ, the mass absorption coefficient $= \lambda/\rho$, ρ is the density of the metal, and m is its mass per unit area.

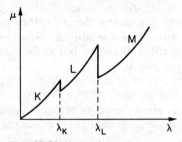

Fig. 40.21 X-ray absorption spectra

Figure 40.21 shows how the coefficient of absorption μ varies with the wavelength of the X-ray beam. Absorption occurs when the X-ray photons possess sufficient energy to eject an electron from the K-shell. In this case the collisions are inelastic. At a particular wavelength λ_K, however, the collisions become elastic and in this case the X-ray photons lose no energy. So the coefficient of absorption now drops sharply, as shown. A similar explanation concerning the L,M and other shells accounts for the discontinuity at the wavelength λ_L and other wavelengths.

Wave Nature of Matter

Electron Diffraction

We have seen how the wave nature of X-rays has been established by X-ray diffraction experiments, see Fig. 40.22 (i). Similar experiments, first performed by Davisson and Germer, show that streams of *electrons* produce diffraction patterns and hence also exhibit wave properties, see Fig. 40.22 (ii). Electron diffraction is now as useful a research tool as X-ray diffraction.

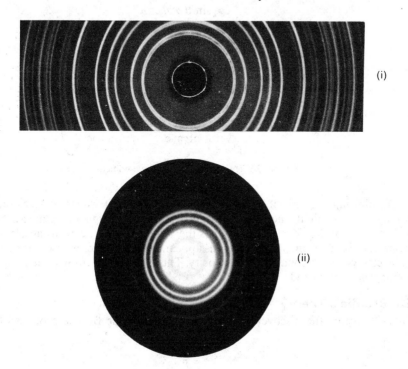

(i)

(ii)

Fig. 40.22 (i) X-ray diffraction rings produced by a crystal (ii) Electron diffraction rings produced by a thin gold film. Similarity of (i) wave and (ii) particle

A TELTRON tube available for demonstrating electron diffraction is shown diagrammatically in Fig. 40.23. A beam of electrons impinges on a layer of graphite which is extremely thin, and a diffraction pattern, consisting of rings, is seen on the tube face. Sir George Thomson first obtained such a diffraction pattern using a very thin gold film, Fig. 40.22 (ii). If the voltage V on the

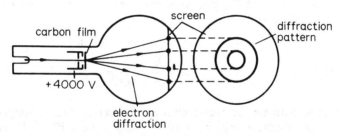

Fig. 40.23 Electron diffraction tube

anode is increased, the velocity, v, of the electrons is increased. The rings are then seen to become narrow, showing that the wavelength λ of the electron waves decreases with increasing v or increasing voltage V.

If a particular ring of radius R is chosen, the angle of deviation φ of the incident beam is given by $\varphi = 2\theta$, where θ is the angle between the incident beam and the crystal planes, Fig. 40.24. Now $\tan \varphi = R/D$, and if φ is small, $\varphi = R/D$ to a good approximation. Hence $\theta = R/2D$. If the Bragg law is true for diffraction as well as X-ray diffraction, then, with the usual notation, $2d \sin \theta = n\lambda$.

$$\therefore \; \lambda \propto \sin \theta \propto \theta \propto R \qquad . \qquad . \qquad . \qquad . \qquad \text{(i)}$$

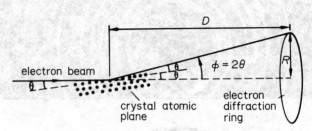

Fig. 40.24 Theory of diffraction experiment

On plotting a graph of R against $1/\sqrt{V}$ for different values of accelerating voltage V, a straight line graph passing through the origin is obtained. Now $\frac{1}{2}m_e v^2 = eV$, or $1/\sqrt{V} \propto 1/v$, where v is the velocity of the electrons accelerated from rest. Hence, from (i), the electrons appear to act as waves whose wavelength is inversely-proportional to their velocity. This is in agreement with de Broglie's theory, now discussed.

De Broglie's Theory

In 1925, before the discovery of electron diffraction, de Broglie proposed that

$$\lambda = \frac{h}{p}, \qquad . \qquad . \qquad . \qquad . \qquad . \qquad \text{(ii)}$$

where λ is the wavelength of waves associated with particles of momentum p, and h is the *Planck constant*, $6\cdot63 \times 10^{-34}$ J s. The quantity h was first used by Planck in his theory of heat radiation and it is a constant which enters into all branches of atomic physics. It is easy to see that de Broglie's relation is consistent with the experimental result obtained with the electron diffraction tube. Here the gain in kinetic energy of the electrons is eV so that

$$\tfrac{1}{2}m_e v^2 = eV,$$

where v is the velocity of the electrons, assuming they start from rest. Thus $v = \sqrt{2eV/m_e}$ and hence

$$p = m_e v = \sqrt{2eVm_e}$$

$$\therefore \; \lambda = \frac{h}{p} = \frac{h}{m_e v} = \frac{h}{\sqrt{2eVm_e}} \propto V^{-1/2}.$$

We can now estimate the wavelength of an electron beam. Suppose $V = 3600$ V. For an electron, $m = 9\cdot1 \times 10^{-31}$ kg, $e = 1\cdot6 \times 10^{-19}$ C, and $h = 6\cdot6 \times 10^{-34}$ J s.

$$\therefore \lambda = \frac{h}{\sqrt{2eVm_e}} = \frac{6 \cdot 6 \times 10^{-34}}{\sqrt{2 \times 1 \cdot 6 \times 10^{-19} \times 3600 \times 9 \cdot 1 \times 10^{-31}}}$$

$$= 2 \times 10^{-11} \text{ m}.$$

This is about 30 000 times smaller than the wavelength of visible light. On this account electron beams are used in *electron microscopes*. These instruments can produce resolving powers far greater than that of an optical microscope.

Wave Nature of Matter

Electrons are not the only particles which behave as waves. The effects are less noticeable with more massive particles because their momenta are generally much higher, and so the wavelength is correspondingly shorter. Since appreciable diffraction is observed only when the wavelength is of the same order as the grating spacing, the heavier particles, such as protons, are diffracted much less. Slow neutrons, however, are used in diffraction experiments, since the low velocity and high mass combine to give a momentum similar to that of electrons used in electron diffraction. The wave nature of α-particles is important in explaining α-decay.

Duality

From what has been said, it is clear that particles can exhibit wave properties, and that waves can sometimes behave as particles. As we saw in the photoelectric effect, electromagnetic waves appear to have a particle nature. Further, γ-rays behave as electromagnetic waves of very short wavelength but on detection by Geiger-Müller tubes, which count *individual* pulses, they behave as particles (see p. 874). It would appear, therefore, that a paradox exists since wave and particle structure appear mutually exclusive.

Scientists gradually realised, however, that the dual aspect of wave-particle properties are completely general in nature. All physical entities can be described either as waves or particles; the description to choose is entirely a matter of convenience. The two aspects, wave and particle, are linked through the two relations

$$E = hf; \qquad p = h/\lambda.$$

On the left of each of these relations, E and p refer to a particle description. On the right, f and λ refer to a wave description. Note that the Planck constant is the constant of proportionality in *both* these equations, a fact which can be predicted by Einstein's Special Theory of Relativity. Further, from these equations, the frequency of the wave is proportional to the *energy* and the wavelength is inversely proportional to the *momentum*.

In the case of a photon of frequency f, we have $\lambda = c/f$. Thus $E = hf$ and $p = h/\lambda = hf/c = E/c$. Hence

$$E = pc = mc^2.$$

So, as a particle, the mass of a photon is E/c^2 or hf/c^2. About 1923, Compton investigated the scattering of X-rays (high-frequency photons) by matter. He showed that the experimental results agreed with the assumption that a particle of mass hf/c^2 collided elastically with an electron in the atom of the material.

Exercises 40

Photoelectricity

1. When a metallic surface is exposed to monochromatic electromagnetic radiation electrons may be emitted. Apparatus is arranged so that (a) the intensity (energy per unit time per unit area) and (b) the frequency of the radiation may be varied. If each of these is varied in turn whilst the other is kept constant, what is the effect on (i) the number of electrons emitted per second, and (ii) their maximum speed? Explain how these results give support to the quantum theory of electromagnetic radiation.

The photoelectric work function of potassium is $2 \cdot 0$ eV. What potential difference would have to be applied between a potassium surface and the collecting electrode in order just to prevent the collection of electrons when the surface is illuminated with radiation of wavelength 350 nm? What would be (iii) the kinetic energy, and (iv) the speed, of the most energetic electrons emitted in this case? (Speed of electromagnetic radiation *in vacuo* $= 3 \cdot 0 \times 10^8$ m s^{-1}, electronic charge $= -1 \cdot 6 \times 10^{-19}$ C, mass of electrons $= 9.1 \times 10^{-31}$ kg, Planck constant $= 6 \cdot 6 \times 10^{-34}$ J s.) (L.)

2. Explain the terms *photoelectric threshold frequency* and *work function* used in connection with the photoelectric effect.

A photoelectric cell consists of a conducting plate coated with photoemissive material and a metal ring as current collector, mounted *in vacuo* in a transparent envelope. Graphs are plotted showing the relation between the collector current I and the potential V of the collector with respect to the emitter, for the following cases (Fig. 40A):

A. Emissive material illuminated with monochromatic light of wavelength λ. B. As A, but the intensity of the incident light has been changed. C. As A, but the wavelength of the light has been changed. D. As A, but the emissive material is different.

Explain the general form of curve A. In what way do curves B,C,D (i) resemble A, (ii) differ from A? What explanation of the differences can be offered? (O. &. C.)

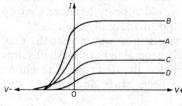

Fig. 40A

3. When light of frequency $5 \cdot 4 \times 10^{14}$ Hz is shone on to a metal surface the maximum energy of the electrons emitted is $1 \cdot 2 \times 10^{-19}$ J. If the same surface is illuminated with light of frequency $6 \cdot 6 \times 10^{14}$ Hz the maximum energy of the electrons emitted is $2 \cdot 0 \times 10^{-19}$ J. Use this data to calculate a value for the Planck constant. (L.)

4. When light is incident on a metal plate electrons are emitted only when the frequency of the light exceeds a certain value. How has this been explained?

The maximum kinetic energy of the electrons emitted from a metallic surface is $1 \cdot 6 \times 10^{-19}$ J when the frequency of the incident radiation is $7 \cdot 5 \times 10^{14}$ Hz. Calculate the minimum frequency of radiation for which electrons will be emitted. Assume that Planck constant $= 6 \cdot 6 \times 10^{-34}$ J s. (N.)

5. Describe and explain one experiment in which light exhibits a wave-like character and one experiment which illustrates the existence of photons.

Light of frequency $5 \cdot 0 \times 10^{14}$ Hz liberates electrons with energy $2 \cdot 31 \times 10^{-19}$ J from a certain metallic surface. What is the wavelength of ultra-violet light which liberates electrons of energy $8 \cdot 93 \times 10^{-19}$ J from the same surface? (Take the velocity of light to be $3 \cdot 0 \times 10^8$ m s^{-1}, and Planck constant (h) to be $6 \cdot 62 \times 10^{-34}$ J s.) (L.)

6. What is meant by a photon?

The Einstein equation for the emission of photoelectrons from a metal due to radiation of frequency v can be written $hv = \frac{1}{2}m_e v^2 + \varphi$, where m_e is the electron mass, v is the greatest speed with which an electron can leave the metal and φ is a constant for the metal called the 'work function'. (*a*) What does this equation imply as to the nature of light? How is the quantity φ interpreted? (*b*) Describe in detail a method for determining experimentally, for a given metal, the value of the greatest kinetic energy of the emitted electrons corresponding to a definite value of *v*.

The following values were obtained for a certain metal:

Frequency in Hz	$7{\cdot}0 \times 10^{14}$	$6{\cdot}3 \times 10^{14}$	$5{\cdot}5 \times 10^{14}$
Greatest k.e. in J	$1{\cdot}65 \times 10^{-19}$	$1{\cdot}17 \times 10^{-19}$	$0{\cdot}65 \times 10^{-19}$

Use these three sets of readings as the basis of a linear graph, and find from the graph the values of *h* and φ. (*O*.)

7. Einstein's equation for the photoelectric emission of electrons from a metal surface can be written $hf = \frac{1}{2}mv^2 + \varphi$, where φ is the work function of the metal, and consistent energy units are used for each term in the equation. Explain briefly the physical process that this equation represents. Outline an experiment by which you could determine the values of *h* (or of *h/e*) and φ.

For caesium the value of φ is $1{\cdot}35$ electrovolts. (*a*) What is the longest wavelength that can cause photo-electric emission from a caesium surface? (*b*) What is the maximum velocity with which photoelectrons will be emitted from a caesium surface illuminated with light of wavelength $4{\cdot}0 \times 10^{-7}$ (*c*) What potential difference will just prevent a current passing through a caesium photocell illuminated with light of wavelength $4{\cdot}0 \times 10^{-7}$ m? ($e = 1{\cdot}6 \times 10^{-19}$ C, $m = 9 \times 10^{-31}$ kg, $h = 6{\cdot}6 \times 10^{-34}$ J s.) (*O*.)

Energy Levels

8. Figure 40B shows three energy levels for the atoms of a particular substance, the energies being in electronvolts (eV). A beam of electrons passing through this substance may excite electrons from the various energy levels. What is the *minimum* potential difference through which the beam of electrons must be accelerated from rest to cause any excitation between two of these levels? At what speed would these electrons be travelling? (Mass of an electron $= 9{\cdot}0 \times 10^{-31}$ kg. 1 eV $= 1{\cdot}6 \times 10^{-19}$ J.) (*L*.)

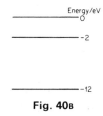

Energy/eV

—————————0

——————————2

——————————12

Fig. 40B

9. Explain why a glowing gas, such as that in a neon tube, gives only certain wavelengths of light and why that gas is capable of absorbing the same wavelengths. (*L*.)

10. Describe an experiment which provides evidence for the belief that a quantity of electric charge is always a multiple of a fundamental unit of charge. Explain how, starting from the measurements made in this experiment, you would calculate the value of this unit.

The ground state of the electron in the hydrogen atom may be represented by the energy $-13{\cdot}6$ eV and the first two excited states by $-3{\cdot}4$ eV and $-1{\cdot}5$ eV respectively, on a scale in which an electron completely free of the atom is at zero energy. Use this data to calculate the ionisation potential of the hydrogen atom and the wavelengths of three lines in the emission spectrum of hydrogen.

(Charge of the electron $= -1{\cdot}6 \times 10^{-19}$ C. Speed of light in a vacuum $= 3{\cdot}0 \times 10^8$ m s^{-1}. The Planck constant $= 6{\cdot}6 \times 10^{-34}$ J s.) (*L*.)

11. What are the chief characteristics of a line spectrum? Explain how line spectra are used in analysis for the identification of elements.

Figure 40c, which represents the lowest energy levels of the electron in the hydrogen

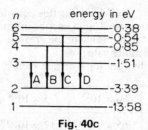

Fig. 40c

atom, specifies the value of the principal quantum number n associated with each state and the corresponding value of the energy of the level, measured in electron volts. Work out the wavelengths of the lines associated with the transitions A, B, C, D marked in the figure. Show that the other transitions that can occur give rise to lines that are either in the ultra-violet or the infra-red regions of the spectrum. (Take 1 eV to be $1\cdot6 \times 10^{-19}$ J; Planck constant h to be $6\cdot5 \times 10^{-34}$ J s; and c, the velocity of light in *vacuo*, to be 3×10^8 m s^{-1}.) (*O*.)

12. Explain what is meant by (*a*) excitation by collision, (*b*) ionisation by collision.

Light or X-radiation may be emitted when fast-moving electrons collide with atoms. Using atomic theory, explain in each case how the radiation is emitted.

Write down *two* differences and *one* similarity between optical emission spectra and X-ray emission spectra produced in this way.

Electrons are accelerated from rest through a p.d. of 5000 V. Calculate the wavelength of the associated electron waves. (Use $h = 6\cdot6 \times 10^{-34}$ J s, $e = 1\cdot6 \times 10^{-19}$ C, $m = 9\cdot1 \times 10^{-31}$ kg.)

13. Give an account of the Rutherford-Bohr model of the atom with special reference to the arrangement of the extra-nuclear electrons in orbits and shells. Outline some experimental evidence in support of your description.

An electron of energy 20 eV comes into collision with a hydrogen atom in its ground state. The atom is excited into a state of higher internal energy and the electron is scattered with reduced velocity. The atom subsequently returns to its ground state with emission of a photon of wavelength $1\cdot216 \times 10^{-7}$ m. Determine the velocity of the scattered electron. ($e = 1\cdot6 \times 10^{-19}$ C; $c = 3\cdot0 \times 10^8$ m s^{-1}; $h = 6\cdot625 \times 10^{-34}$ J s; m(electron) = $9\cdot1 \times 10^{-31}$ kg.) (*L*.)

X-Rays

14. (*a*) Explain briefly why a modern X-ray tube can be operated directly from the output of a step-up transformer.

(*b*) An X-ray tube works at a d.c. potential difference of 50 kV. Only 0·4% of the energy of the cathode rays is converted into X-radiation and heat is generated in the target at a rate of 600 W. Estimate (i) the current passed into the tube, (ii) the velocity of the electrons striking the target. (Electron mass = $9\cdot00 \times 10^{-31}$ kg, electron charge = $-1\cdot60 \times 10^{-19}$ C.) (*N*.)

15. (*a*) When atoms absorb energy by colliding with moving electrons, light or X-radiation may subsequently be emitted. For each type of radiation, state typical values of the energy per atom which must be absorbed and explain in atomic terms how each type of radiation is emitted. (*b*) State *one* similarity and *two* differences between optical atomic emission spectra and X-ray emission spectra produced in this way.

(*c*) Electrons are accelerated from rest through a potential difference of 10 000 V in an X-ray tube. Calculate (i) the resultant energy of the electrons in eV; (ii) the wavelength of the associated electron waves; (iii) the maximum energy and the minimum wavelength of the X-radiation generated. (Charge of electron = $1\cdot6 \times 10^{-19}$ C, mass of electron = $9\cdot11 \times 10^{-31}$ kg, Planck constant = $6\cdot62 \times 10^{-34}$ J s, speed of electromagnetic radiation in vacuo = $3\cdot00 \times 10^8$ m s^{-1}.) (*N*.)

16. Describe the properties of X-rays and compare them with those of ultra-violet radiation. Outline the evidence for the wave nature of X-rays.

The energy of an X-ray photon is hf joule where $h = 6\cdot63 \times 10^{-34}$ J s and f is the frequency in hertz (cycles per second). X-rays are emitted from a target bombarded by electrons which have been accelerated from rest through 10^5 V. Calculate the minimum possible wavelength of the X-rays assuming that the corresponding energy is equal to the whole of the kinetic energy of one electron. (Charge of an electron = 1.60×10^{-19} C; velocity of electromagnetic waves *in vacuo* = 3.00×10^8 m s^{-1}.) (*O. & C.*)

17. Describe the atomic process in the target of an X-ray tube whereby X-ray line spectra are produced. Determine the ratio of the energy of a photon of X-radiation of wavelength 0.1 nm to that of a photon of visible radiation of wavelength 500 nm. Why is the potential difference applied across an X-ray tube very much greater than that applied across a sodium lamp producing visible radiation? (*N.*)

18. Draw a labelled diagram of a modern X-ray tube. How may (*a*) the intensity, (*b*) the penetrating power, of the X-rays be controlled?

An X-ray tube is operated with an anode potential of 10 kV and an anode current of 15.0 mA. (i) Estimate the number of electrons hitting the anode per second. (ii) Calculate the rate of production of heat at the anode, stating any assumptions made. (iii) Describe the characteristics of the emitted X-ray spectrum and account for any special features. (Electron charge $e = 1.60 \times 10^{-19}$ C; Planck constant, $h = 6.63 \times 10^{-34}$ J s; speed of light, $c = 3.00 \times 10^8$ m s^{-1}.) (*C.*)

19. Explain how the minimum wavelength emitted by a hot cathode X-ray tube can be controlled and deduce an expression from which this wavelength can be calculated. (*L.*)

20. Describe a modern form of X-ray tube and explain its action.

Outline the evidence for believing (*a*) that X-rays are an electromagnetic radiation, (*b*) that wavelengths in the X-ray region are of the order of 10^{-3} times those of visible light. (*O.*)

21. Describe a modern form of X-ray tube and explain briefly the energy changes that take place while it is operating.

Calculate the energy in electron-volts of a quantum of X-radiation of wavelength 0.15 nm.

An X-ray tube is operated at 8000 V. Why is there no radiation of shorter wavelength than that calculated above (while there is a great deal of longer wavelength)? When the voltage is increased considerably above this value why does the spectrum of the radiation include a few (*but only a few*) sharp strong lines, the wavelengths of which depend on the material of the target? (Take $e = 1.6 \times 10^{-19}$ C; $h = 6.5 \times 10^{-34}$ J s; $c = 3 \times 10^8$ m s^{-1}.) (*O.*)

Miscellaneous

22. (*a*) Describe a simple experiment by which the photoelectric effect may be demonstrated.

(*b*) Explain what is meant by (i) a photon and (ii) work function. How do these two concepts explain that: (iii) the number of electrons emitted is proportional to the intensity of the radiation incident on a clean metal surface, (iv) these electrons are emitted with a range of velocities from zero up to a maximum value which increases with the frequency of the incident radiation but is independent of the intensity, (v) there is, for each metal, a certain minimum frequency of radiation below which no emission occurs.

(*c*) Give a brief description of a use to which the photoelectric effect is put. (*AEB, 1982*)

23. The accelerating voltage across an X-ray tube is 33.0 kV. Explain why the frequency of the X-radiation cannot exceed a certain value and calculate this maximum frequency. (The Planck constant = 6.6×10^{-34} J s; charge on an electron = 1.6×10^{-19} C. (*AEB, 1981*)

24. In an X-ray tube the current through the tube is 1.0 mA and the accelerating potential is 15 kV. Calculate (i) the number of electrons striking the anode per second, (ii) the speed of the electrons on striking the anode assuming they leave the cathode with zero speed, (iii) the rate at which cooling fluid, entering at 10°C, must circulate through the anode if the anode temperature is to be maintained at 35°C. Neglect any of the kinetic energy of the electrons which is converted to X-radiation. (Electronic charge = 1.6×10^{-19} C; mass of electron = 9.1×10^{-31} kg; specific heat capacity of liquid = 2.0×10^3 J kg^{-1} K^{-1}.) (*AEB, 1982*)

41 Radioactivity. The Nucleus

In 1896 Becquerel found that a uranium compound affected a photographic plate wrapped in light-proof paper. He called the phenomenon *radioactivity*. We shall see later that natural radioactivity is due to one or more of three types of radiation emitted from heavy elements such as uranium whose nuclei are unstable. These were originally called α-, β- and γ-rays but α- and β-'rays' were soon shown to be actually particles.

α- and β-particles and γ-rays all produce ionisation as they move through a gas. On average, α-particles produce about 1000 times as many ions per unit length of their path as β-particles, which in turn produce about 1000 times as many ions as γ-rays. There are numerous detectors of ionising radiations such as α- and β-particles and γ-rays. We begin by describing two detectors used in laboratories.

Geiger-Müller Tube

A Geiger-Müller (GM) tube is widely used for detecting ionising particles or radiation. In one form it contains a central thin wire A, the anode, insulated from a surrounding cylinder C, the cathode, which is metal or graphite-coated. The tube may have a very thin mica end window, Fig. 41.1. A is kept at a positive potential V such as $+400\ V$ relative to C, which may be earthed.

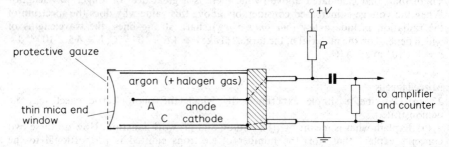

Fig. 41.1 Principle of Geiger-Müller tube

When a single ionising particle enters the tube, a few electrons and ions are produced in the gas. If V is above the breakdown potential of the gas, the number of electrons and ions are multiplied enormously (see p. 798). The electrons are attracted by and move towards A, and the positive ions move towards C. Thus a 'discharge' is suddenly obtained between A and C. The current flowing in the high resistance R produces a p.d. which is amplified and passed to a counter, discussed on p. 875. The counter registers the passage of an ionising particle or radiation passing through the tube.

The discharge persists for a short time, as secondary electrons are emitted from the cathode by the positive ions which arrive there. This would upset the recording of other ionising particles following fast on the first one recorded. The air in the tube is therefore replaced by argon mixed with a halogen vapour, whose molecules absorb the energy of the positive ions on collision. So the discharge is quenched quickly. Electrical methods are also used for quenching.

The anode wire A in the GM tube must be *thin*, so that the charge on it produces an intense electric field E close to its surface (E is inversely-proportional to the radius of the wire). An electron-ion pair, produced by an ionising particle or radiation, is then accelerated to high energies near the wire, thus producing, by collision with gas molecules, an 'avalanche' of more electron-ion pairs.

Solid State Detector

A *solid state detector*, Fig. 41.2, is made from semiconductors. Basically, it has a p-n junction which is given a small bias in the non-conduction direction (p. 826). When an energetic ionising particle such as an α-particle falls on the

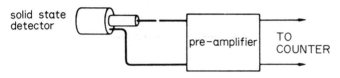

Fig. 41.2 Solid state detector

detector, more electron-hole pairs are created near the junction. These charge carriers move under the influence of the biasing potential and so a pulse of current is produced. The pulse is fed to an amplifier and the output passed to a counter.

The solid state detector is particularly useful for α-particle detection. If the amplifier is specially designed, β-particles and γ-rays of high energy may also be detected. This type of detector can thus be used for all three types of radiation.

Dekatron Counter. Ratemeter

As we have seen, each ionising particle or radiation produces a pulse voltage in the external circuit of a Geiger-Müller or solid state detector. In order to measure the number of pulses from the detectors, some form of counter must be used.

A *dekatron counter* consists of two or more dekatron tubes, each containing a glow or discharge which can move round a circular scale graduated in numbers 0–9, together with a mechanical counter, Fig. 41.3. Each impulse causes the discharge in the first tube, which counts units, to advance one digit. The circuit is designed so that on the tenth pulse, which returns the first counter to zero, a

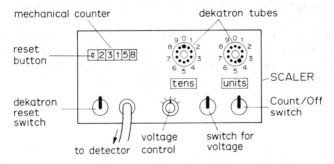

Fig. 41.3 Dekatron counter

pulse is sent to the second tube. The glow here then moves on one place. The second tube thus counts the number of tens of pulses. After ten pulses are sent to the second tube, corresponding to a count of 100, the output pulse from the second tube is fed to the mechanical counter. This, therefore, registers the hundreds, thousands and so on. Dekatron tubes are used in radioactive experiments because they can respond to a rate of about 1000 counts per second. This is far above the count rate possible with a mechanical counter.

In contrast to a scaler, which counts the actual number of pulses, a *ratemeter* is a device which gives directly the average number of pulses per second or *count rate*. The principle is shown in Fig. 41.4. The pulses received are passed to a capacitor C, which then stores the charge. C discharges slowly through a high resistor R and the average discharge current is recorded on a microammeter A. The greater the rate at which the pulses arrive, the greater will be the meter reading. The meter thus records a current which is proportional to the count rate.

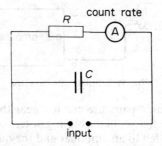

Fig. 41.4 Principle of a ratemeter

A switch marked 'time constant' on most ratemeters allows the magnitude of C to be chosen. If a large value of C is used, the capacitor will take a relatively long time to charge and correspondingly it will be a long time before the average count rate can be taken. The reading obtained, however, will be more accurate since the count rate is then averaged over a longer time (see below). For high accuracy, a small value of C may be used only if the count rate is very high.

Errors in Counting Experiments

Radioactive decay is random in nature (p. 881). If the count rate is high, it is not necessary to wait so long before readings are obtained which vary relatively slightly from each other. If the count rate is low, successive counts will have larger percentage differences from each other, unless a much longer counting time is employed.

The accuracy of a count does not depend on the time involved but on *the total count obtained*. If N counts are received, the statistics of random processes show that this is subject to a statistical error of $\pm\sqrt{N}$. The proof is beyond the scope of this book. The percentage error is thus

$$\frac{\sqrt{N}}{N} \times 100 = \frac{100}{\sqrt{N}}\%.$$

If 10 per cent accuracy is required, $\sqrt{N} = 10$ and hence $N = 100$. Thus 100 counts must be obtained. If the counts are arriving at about 10 every second, it will be necessary to wait for 10 seconds to obtain a count of 100 and so achieve 10 per cent accuracy. Thus a ratemeter circuit must be arranged with a time constant (CR) of 10 seconds, so that an average is obtained over this time. If, however, the counts are arriving at a rate of 1000 per second on average, it will be necessary to wait only 1/10th second to achieve 10 per cent accuracy.

Thus the 1 second time constant scale on the ratemeter will be more than adequate.

Existence of α-, β-particles and γ-rays

The existence of three different types of emission from radioactive substances can be shown by experiments such as those now outlined.

1. (a) When a radium-224 source S_1 is placed above a *spark counter*, which consists of a wire W close to an earthed metal base B, sparks are obtained between W and B, Fig. 40.5 (i). If a sheet of thin paper is now introduced between S_1 and W, the sparks stop.

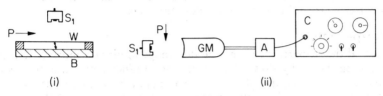

Fig. **41.5** Detection by (i) spark counter (ii) GM tube

(b) The radium source S_1 is now placed in front of a GM tube connected to a counter C (or ratemeter) through a low-noise pre-amplifier A, Fig. 41.5 (ii). The count rate is measured and the paper P is then introduced between S_1 and the GM tube. Practically no difference in the count rate is observed.

We thus conclude that there is a type of radiation from S_1 which is detected by a spark counter but is absorbed by a thin sheet of paper—these are α-particles.

2. There are other types of radiation which do not affect a spark counter but affect a GM tube. We can show this by placing a strontium-90 source S_2 that its radiation is shielded from the GM tube by a lead plate about 1 cm thick, Fig. 41.6. The count rate is then extremely low. A strong magnetic field B, in a direction perpendicular to the radiation, is now introduced beyond the lead as shown. When the field B is directed into the paper, the count rate increases. So the magnetic field has deflected the radiation towards the GM tube, which then detects it.

We conclude that there is a type of radiation which can be detected by a GM tube but is absorbed by lead 1 cm thick—these are β-particles.

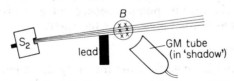

Fig. **41.6** Deflection of β-particles by magnetic field

3. (a) We can repeat the experiment in Fig. 41.6 with the radium source S_1 in place of S_2. This time the lead does not cut out all the radiation; some passes through it to the GM tube. But as before, there is an increase in the count rate when the magnetic field B is used as shown. So the radium source contains β-particles together with a third type of radiation which goes through 1 cm thickness of lead.

(b) Using a cobalt-60 source S directly in front of a GM tube as in Fig. 41.5 (ii), the count rate is high. We can cut out any α-particles emitted by placing a thin sheet of paper between S and the GM tube. But using a fairly thick lead plate between S and the GM tube, the count rate continues to be high. Further, a magnetic field B between the lead and the GM tube makes no difference to the count rate.

Conclusion. There are at least three different types of emission from radioactive sources: (1) A type recorded by a spark counter and cut off by thin paper—α-particles.

(2) A type recorded by a GM tube, cut out by lead, and deflected by a magnetic field—β-particles.

(3) A type recorded by a GM tube, not cut out except by very thick lead, and not affected by a magnetic field—γ-rays.

Further experiments show that the particles or radiation from most other radioactive sources fall into one of these three classes.

Alpha-particles

It is found that α-particles have a limited range in air at atmospheric pressure. This can be shown by slowly increasing the distance between a pure α-source and a detector. The count rate is observed to fall rapidly to zero at a separation greater than a particular value, which is called the 'range' of the α-particles. The range depends on the source and on the air pressure.

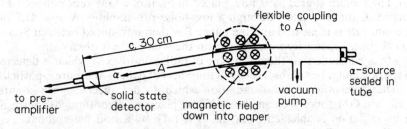

Fig. 41.7 Charge on an α-particle

Using the apparatus of Fig. 41.7, it can be shown α-particles have *positive* charges. When there is no magnetic field, the solid state detector is placed so that the tube A is horizontal in order to get the greatest count. When the magnetic field is applied, the detector has to be moved *downwards* in order to get the greatest count. This shows that the α-particles are deflected by a small amount downwards. By applying Fleming's left-hand rule, we find that particles are *positively* charged. The vacuum pump is needed in the experiment, as the range of α-particles in air at normal pressures is too small.

Nature of α-particle

Lord Rutherford and his collaborators found by deflection experiments that an α-particle had a mass about four times that of a hydrogen atom, and carried a charge $+2e$, where e was the numerical value of the charge on an electron. The relative atomic mass of helium is about four. It was thus fairly certain that an *α-particle was a helium nucleus*, that is, a helium atom which has lost two electrons.

In 1909 Rutherford and Royds showed conclusively that α-particles were helium nuclei. Radon, a gas given off by radium which emits α-particles, was collected above mercury in a thin-walled tube P, Fig. 41.8. After several days some of the α-particles passed through P into a surrounding vacuum Q, and in about a week, the space in Q was reduced in volume by raising mercury reservoirs. A gas was collected in a capillary tube R at the top of Q. A high

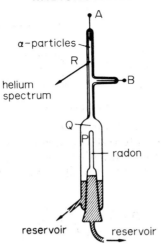

Fig. 41.8 Rutherford and Royd's experiment on α-particles

voltage was then connected to electrodes at A and B, and the spectrum of the discharge was observed to be exactly the same as the characteristic spectrum of helium.

Beta-particles and Gamma-rays

By deflecting β-particles with perpendicular magnetic and electric fields, their charge-mass ratio could be estimated. This is similar to Thomson's experiment, p. 804. These experiments showed that *β-particles are electrons moving at high speeds*. Generally, β-particles have a greater penetrating power of materials than α-particles. They also have a greater range in air than α-particles, since their ionisation of air is relatively smaller, but their path is not so well defined.

Using a strong bar magnet, it can be shown that β-particles are strongly deflected by a magnetic field. The direction of the deflection corresponds to a stream of *negatively*-charged particles, that is, opposite to the deflection of α-particles in the same field. This is consistent with the idea that β-particles are usually fast-moving electrons.

The nature of γ-rays was shown by experiments with crystals. Diffraction phenomena are obtained in this case, which suggest that *γ-rays are electromagnetic waves* (compare X-rays, p. 861). Measurement of their wavelengths, by special techniques with crystals, show they are shorter than the wavelengths of X-rays and of the order 10^{-11} m. γ-rays can penetrate large thicknesses of metals, but they have far less ionising power in gases than β-particles.

If a beam of γ-rays is allowed to pass through a very strong magnetic field no deflection is observed. This is consistent with the fact that γ-rays are electromagnetic waves and carry no charge.

Inverse-square Law for γ-rays

If γ-rays are a form of electromagnetic radiation and undergo negligible absorption in air, then the intensity I should vary inversely as the square of the distance between the source and the detector. The apparatus shown in Fig. 41.9 can be used to investigate if this is the case. A suitable γ-source is placed at a suitable distance from a GM tube connected to a scaler, and I will then be proportional to the count rate C.

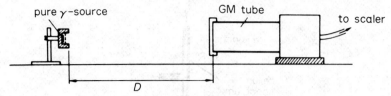

Fig. 41.9 Inverse square law for γ-rays

Suppose D is the measured distance from a fixed point on the γ-source support to the front of the GM tube. To obtain the true distance from the source to the region of gas inside the tube where ionisation occurs, we need to add an unknown but constant distance h to D. Then, assuming an inverse-square law, $I \propto 1/(D + h)^2$. Thus

$$D + h \propto \frac{1}{\sqrt{I}} \propto \frac{1}{\sqrt{C}}.$$

A graph of $1/\sqrt{C}$ is therefore plotted against D for varying values of D. If the inverse-square law is true, a straight line graph is obtained which has an intercept on the D-axis of $-h$. Note that if I is plotted against $1/D^2$ and h is not zero, a straight line graph is *not* obtained from the relation $I \propto 1/(D + h)^2$. Consequently we need to plot D against $1/\sqrt{I}$.

If a pure β-source is substituted for the γ-source and the experiment is repeated, a straight-line graph is not obtained. The absorption of β-particles in air is thus appreciable compared with γ-rays.

Absorption of Radiation by Metals

A GM tube and counter or ratemeter can be used to investigate the absorption of γ-rays or β-particles by metals.

With γ-rays, a pure γ-source is placed at a suitable distance from the window of the GM tube. Successive equal thicknesses of lead sheets, bound tightly together, are then positioned between the source and the GM tube. The count rate C decreases as more plates are used and C is measured for different values of n, the number of plates.

With β-particles, it is convenient to investigate the absorption by aluminium. As before, the count rate C is observed with increasing number n of aluminium plates of equal thickness.

Figure 41.10 (i) shows roughly the results obtained in either case; the count rate C decreases with n along a curve. In the case of the β-particles, however, the count rate C reaches an almost constant low value (not shown) as n becomes larger. When this first occurs, the metal thickness is taken as the *range* of the β-particles in it. This effect is due to the electromagnetic radiation (called

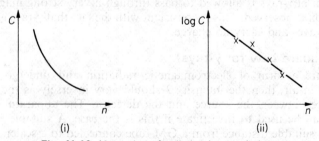

Fig. 41.10 Absorption of radiation by metal plates

'Bremsstrahlung') produced when the electrons slow down, or change direction, as they interact with the aluminium atoms. When $\log C$ is plotted against n, however, a fairly straight line is obtained for the small values of n in Fig. 41.10 (i), as shown in Fig. 41.10 (ii).

The count rate C is a measure of the activity or intensity of radiation I of the radioactive source. The number of plates n is proportional to the thickness t of the metal. Thus, from Fig. 41.10 (ii), the intensity I decreases as t increases according to a law of the form

$$\log I = a - \mu t,$$

where a and μ are constants. When $t = 0$, $I = I_0$ say; so, from the equation, $a = \log I_0$. Substituting for a, then

$$\log I = \log I_0 - \mu t.$$

If we use logs to the base e, then

$$\ln \frac{I}{I_0} = -\mu t$$

$$\therefore I = I_0 e^{-\mu t} \qquad . \qquad . \qquad . \qquad . \qquad . \qquad (1)$$

Thus the intensity I decreases exponentially with the thickness t of the metal. The constant μ is called the *linear absorption coefficient* of the metal. The *mass absorption coefficient* is usually used in comparing absorbing powers of metals. This is defined as the ratio μ/ρ, where ρ is the density of the metal.

Half-life and Decay Constant

Radioactivity, or the emission of α- or β-particles and γ-rays, is due to dis-integrating nuclei of atoms (p. 899). The disintegrations obey the statistical law of chance. Thus although we cannot tell which particular atom is likely to disintegrate next, the number of atoms disintegrating per second, dN/dt, is directly proportional to the number of atoms, N, present at that instant. Hence:

$$\frac{dN}{dt} = -\lambda N,$$

where λ is a constant characteristic of the atom concerned called the *radioactivity decay constant*. The negative sign indicates that N becomes *smaller* when t increases. Thus, if N_0 is the number of radioactive atoms present at a time $t = 0$, and N is the number at the end of a time t, we have, by integration,

$$\int_{N_0}^{N} \frac{dN}{N} = -\lambda \int_0^t dt.$$

$$\therefore \left[\ln N \right]_{N_0}^{N} = -\lambda t.$$

$$\therefore N = N_0 e^{-\lambda t} \qquad . \qquad . \qquad . \qquad . \qquad (i)$$

Thus the number N of undecayed atoms left decreases exponentially with the time t, and this is illustrated in Fig. 41.11.

The *half-life* $T_{1/2}$ of a radioactive element is defined as the time taken for half the atoms to disintegrate (see Fig. 41.11), that is, in a time $T_{1/2}$ the radio-activity of the element diminishes to half its value. Hence, from (i),

$$\frac{N_0}{2} = N_0 e^{-\lambda T_{1/2}}$$

So

$$\tfrac{1}{2} = 2^{-1} = e^{-\lambda T_{1/2}}$$

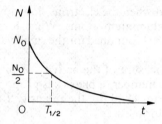

Fig. 41.11 Radioactive decay with time

Taking logs to the base e on both sides of the equation and simplifying,

$$\therefore T_{1/2} = \frac{1}{\lambda}\ln2 = \frac{0\cdot693}{\lambda} \qquad . \qquad . \qquad . \qquad . \qquad \text{(ii)}$$

The half-life varies considerably in a particular radioactive series. In the uranium series shown in the Table on p. 891, for example, uranium I has a half-life of the order of 4500 million years, radium has one of about 1600 years, radium F about 138 days, radium B about 27 minutes, and radium C' about 10^{-4} second.

Measurement of Half-Life

The half-life of radon-220, a radioactive gas with a short half-life, can be measured by the apparatus shown in Fig. 41.12 (i). C is a metal can or *ionisation chamber* containing a metal rod P insulated from C. B is a suitable d.c. supply such as 100 V connected between P and C, with a d.c. amplifier M joined in series as shown. This type of meter can measure very small currents, as explained on p. 598.

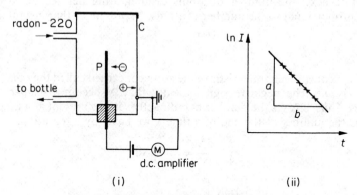

Fig. 41.12 Measurement of half-life

A plastic bottle containing the gas radon-220 is connected to the chamber C. The radioactive gas is passed into C by squeezing the bottle and when the atoms decay the α-particles produced ionise the air in C. The negative and positive ions produced move between P and C due to the supply voltage B and so a small current or ionisation current is registered on the meter M.

Starting with an appreciable deflection in M, the falling current I is noted at equal intervals of time t such as 15 seconds. A graph of $\ln I$ against t is then plotted from the results and a straight line AB is drawn through the points. Figure 41.12 (ii). Now $I = I_0 e^{-\lambda t}$, since the number of disintegrations per second is proportional to the number of ions produced per second and hence to the current I. Taking logs to the base e,

$$\ln I = \ln I_0 - \lambda t.$$

So the gradient a/b of the straight-line AB is λ numerically and this can now be found. The half-life $T_{1/2}$ is given by $T_{1/2} = 0.693/\lambda$ and so it can be calculated.

For a substance with a longer half-life such as days, weeks or years, we can proceed as follows. Firstly, weigh a small mass of the specimen S, m say. If the relative molecular mass of S is M and S is monatomic, the number of atoms N in a mass m is given by

$$N = \frac{m}{M} \times 6.03 \times 10^{23}$$

using the Avogadro constant.

Secondly, determine the rate of emission, dN/dt, all round the specimen S by placing S at a distance r from the end face, area A, of a GM tube connected to a scaler. Suppose the measured count rate through the area A is dN'/dt. Then the count rate all round S, in a sphere of area $4\pi r^2$, is given by

$$\frac{dN}{dt} = \frac{4\pi r^2}{A} \times \frac{dN'}{dt}.$$

Since $dN/dt = -\lambda N$, the decay constant λ can be calculated as N and dN/dt are now known. The half-life is then obtained from $T_{1/2} = 0.693/\lambda$.

Carbon Dating

Carbon has a radioactive isotope ^{14}C. It is formed when neutrons react with nitrogen in the air, as below. The neutrons are produced by cosmic rays which enter the upper atmosphere and interact with air molecules.

$$^{14}_{7}N + ^{1}_{0}n \rightarrow ^{14}_{6}C + ^{1}_{1}H.$$

The radioactive isotope ^{14}C is absorbed by living material such as plants or vegetation in the form of carbon dioxide. The amount of the isotope absorbed is very small and it reaches a maximum concentration as long as the plants are living. Experiment shows that the activity of the radioactive isotope in living materials is about 19 counts per minute per gram. The half-life of the isotope is about 5600 years.

When a plant dies, no more of the isotope is absorbed. Wood formed from dead or decaying plants or vegetation, which were alive thousands of years ago, contain the radioactive isotope but its activity is much less owing to decay of the atoms. Measuring the activity of the isotope in ancient wood or similar carbon materials can provide information about its age, a method known as carbon dating. For example, suppose the measured activity in a piece of ancient wood is 14 counts per minute per gram. Then, from $I = I_0 e^{-\lambda t}$,

$$\frac{I}{I_0} = \frac{14}{19} = e^{-\lambda t}.$$

Taking logs to the base e on both sides we obtain

$$-0.305 = -\lambda t.$$

So

$$t = \frac{0.305}{\lambda}.$$

But
$$\lambda = \frac{0 \cdot 693}{T_{1/2}} = \frac{0 \cdot 693}{5600 \text{ y}}.$$

Hence
$$t = \frac{0 \cdot 305 \times 5600 \text{ y}}{0 \cdot 693} = 2465 \text{ y}.$$

So the ancient wood is about 2500 y old. By carbon dating the ancient material, Stonehenge in England was found to have a date of about 4000 years.

Examples

1. A point source of γ radiation has a half-life of 30 minutes. The initial count rate, recorded by a Geiger counter placed $2 \cdot 0$ m from the source, is 360 s^{-1}. The distance between the counter and the source is altered. After $1 \cdot 5$ hours the count rate recorded is 5 s^{-1}. What is the new distance between the counter and the source? (*L*.)

$1 \cdot 5$ h $= 3 \times 30$ min $= 3 \times$ half-life of source
So at *beginning* of $1 \cdot 5$ h, count rate $= 2 \times 2 \times 2 \times 5$ s$^{-1} = 40$ s^{-1}

$$= \frac{1}{9} \times 360 \text{ s}^{-1} = \frac{1}{9} \times \text{initial rate}$$

$$\therefore \text{ intensity of radiation} = \frac{1}{9} \times \text{initial intensity at } 2 \cdot 0 \text{ m}$$

But
$$\text{intensity} \propto \frac{1}{d^2}, \text{ where } d \text{ is the distance}$$

$$\therefore \text{ new distance} = 3 \times \text{initial distance} = 6 \cdot 0 \text{ m}$$

2. Lanthanum has a stable isotope ^{139}La and radioactive isotope ^{138}La of half-life $1 \cdot 1 \times 10^{10}$ years whose atoms are $0 \cdot 1\%$ of those of the stable isotope. Estimate the rate of decay or activity of ^{138}La with 1 kg of ^{139}La. (Assume the Avogadro constant $= 6 \times 10^{23}$ mol^{-1}.)

$$\text{Decay rate } \frac{dN}{dt} = -\lambda N, \qquad \cdot \qquad \cdot \qquad \cdot \qquad (1)$$

where λ is the decay constant and N is the number of atoms in ^{138}La. Now number of atoms in 1 kg (1000 g) of ^{139}La $= \dfrac{6 \times 10^{23} \times 1000}{139}$.

Since $0 \cdot 1\% = 10^{-3}$, then

$$\text{number of atoms in } {}^{138}\text{La}, \ N = \frac{10^{-3} \times 6 \times 10^{23} \times 1000}{139} = \frac{6 \times 10^{23}}{139}.$$

Also,
$$\lambda = \frac{0 \cdot 693}{T_{1/2}} = \frac{0 \cdot 693}{1 \cdot 1 \times 10^{10} \times 365 \times 24 \times 3600}.$$

From (1),
$$\frac{dN}{dt} = \frac{0 \cdot 693 \times 6 \times 10^{23}}{1 \cdot 1 \times 10^{10} \times 365 \times 24 \times 3600 \times 139}$$

$$= 8600 \text{ s}^{-1}.$$

3. At a certain instant, a piece of radioactive material contains 10^{12} atoms. The half-life of the material is 30 days.

(1) Calculate the number of disintegrations in the first second. (2) How long will elapse before 10^4 atoms remain? (3) What is the count rate at this time?

(1) We have $N = N_0 e^{-\lambda t}$

$$\therefore \frac{dN}{dt} = -N_0 \lambda e^{-\lambda t} = -\lambda N.$$

Hence, when $N = 10^{12}$,
$$\frac{dN}{dt} = -\lambda 10^{12}.$$

Now
$$\lambda = \frac{0\cdot693}{T} = \frac{0\cdot693}{30 \times 24 \times 60 \times 60} s^{-1}$$

$\therefore$ number of disintegrations per second
$$= \frac{10^{12} \times 0\cdot693}{30 \times 24 \times 60 \times 60} = 2\cdot7 \times 10^5.$$

(2) When $N = 10^4$, we have
$$10^4 = 10^{12}e^{-\lambda t}$$
$$\therefore 10^{-8} = e^{-\lambda t}$$

Taking logs to base 10,
$$\therefore -8 = -\lambda t \log e$$
$$\therefore t = \frac{8}{\lambda} \frac{1}{\log e} = \frac{8T}{0\cdot693 \log e}$$
$$= 797 \text{ days (approx.)}.$$

(3) Since
$$\frac{dN}{dt} = -\lambda N$$

$\therefore$ number of disintegrations per hour $= \dfrac{0\cdot693}{30 \times 24} \times 10^4 = 9\cdot6.$

Cloud Chambers

C. T. R. Wilson's *cloud chamber*, invented in 1911, was one of the most useful early inventions for studying radioactivity. It enabled photographs to be obtained of the paths of ionising particles or radiation.

Basically, Wilson's cloud chamber consists of a chamber containing saturated water or alcohol vapour. Figure 41.13 (i) illustrates the basic principle of a cloud chamber C which uses alcohol vapour. An excess amount of liquid alcohol is placed on a dark pad D on a piston P. When the piston is moved down quickly, the air in C undergoes an adiabatic expansion and cools. The dust nuclei are all carried away after a few expansions by drops forming on them, and then the dust-free air in C is subjected to a controlled adiabatic expansion of about 1·31 to 1·38 times its original volume. The air is now supersaturated, that is,

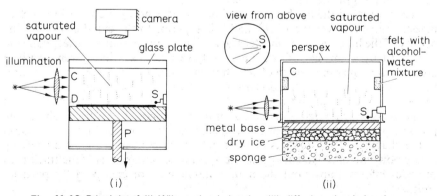

Fig. 41.13 Principle of (i) Wilson cloud chamber (ii) diffusion cloud chamber

the vapour pressure is greater than the saturation vapour pressure at the reduced temperature reached but no vapour condenses. Simultaneously, the air is exposed to α-particles from a radioactive source S, for example. Water droplets immediately collect round the ions produced, which act as centres of formation. The drops are illuminated, and photographed by light scattered from them.

α-particles produce short continuous straight trails, as shown at the top of Fig. 41.13. β-particles, which have much less mass, produce longer but straggly paths owing to collisions with gas molecules. Wilson's cloud chamber has proved of great value in the study of radioactivity and nuclear structure. See Fig. 41.19.

Diffusion cloud chamber. Figure 41.13 (ii) shows the principle of another form of cloud chamber. It has a perspex chamber C, with a strip of felt at the top containing excess of a mixture of water and alcohol. The dark metal base is kept at a low temperature of about −50°C by dry ice (solid carbon dioxide) packed below it. Vapour thus diffuses continously from the top to the bottom of the chamber.

Above the cold metal base there is supersaturated vapour. A radioactive source S near the base, emitting α-particles, for example, produces vapour trails which can be seen on looking down through the top of C. These trails show the straight line paths of the emitted α-particles. See Fig. 41.13. Unlike the Wilson cloud chamber, this type does not require adiabatic expansion of the gas inside. In both cases, the ions formed can be cleared up by applying a suitable p.d. between the top and bottom of the chamber.

The random nature of radioactive decay can be seen by using a cloud chamber. Particles emitted by a radioactive substance do not appear at equal intervals of time but are sporadic or entirely random. The length of the track of an emitted particle is a measure of its initial energy. The tracks of α-particles are nearly all the same, showing that the α-particles were all emitted with the same energy. Sometimes two different lengths of tracks are obtained, showing that the α-particles may have one or two energies on emission.

Glaser's Bubble Chamber

In the same way as air can be supersaturated with water vapour, a liquid under pressure can be heated to a temperature higher than that at which boiling normally takes place and is then said to be *superheated*. If the pressure is suddenly released, bubbles may not form in the liquid for perhaps 30 seconds or more. During this quiet period, if ionising particles or radiation are introduced into the liquid, nuclei are obtained for bubble formation. The liquid quickly evaporates into the bubble, which grows rapidly, and the bubble track when photographed shows the path of the ionising particle.

Glaser invented the bubble chamber in 1951. It is now widely used in nuclear investigations all over the world, and it is superior to the cloud chamber. The density of the liquid ensures shorter tracks than in air, so that a nuclear collision of interest by a particle will be more likely to take place in a given length of liquid than in the same length of air. Photographs of the tracks are much clearer than those taken in the cloud chamber, and they can be taken more rapidly. In 1963 a 1·5 metre liquid hydrogen bubble chamber was constructed for use at the Rutherford High Energy Laboratory, Didcot, England. High energy protons, accelerated by millions of volts are used to bombard hydrogen nuclei in the chamber. The products of the reaction are bent into a curved track by a very powerful magnetic field, and the appearance and radius of the track then provides information about the nature, momentum or energy of the particles emitted.

Scintillations and Photomultiplier

In the early experiments on radioactivity, Rutherford observed the scintillations produced when an α-particle was incident on a material such as zinc sulphide. This is now used in the *scintillation photomultiplier*, whose principle is illustrated in Fig. 41.14.

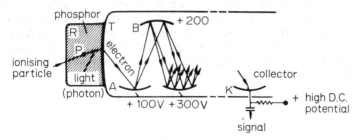

Fig. 41.14 Principle of photomultiplier

When an ionising particle strikes the scintillation material or *phosphor* R, the light falls on a photo-sensitive material T and ejects electrons. In one type of tube, these are now focused towards and accelerated to five times as numerous as those incident on it. The secondary electrons then strike an electrode B after further acceleration, thus multiplying the number of electrons further, and so on along the tube. A single ionising particle can produce a million electrons in a photomultiplier tube, and the pulse of current is amplified further and recorded. By choosing a suitable phosphor, scintillation counters can detect electrons and gamma rays, as well as fast neutrons.

Emulsions

Special photographic emulsions have been designed for investigating nuclear reactions. The emulsions are much thicker than those used in ordinary photography, and in addition the concentration of silver bromide in the gelatine is many times greater than in ordinary photography. α-particles, protons and neutrons can be detected in specially-prepared emulsions by the track of silver granules produced, which has usually a very short range of the order of a millimetre or less. Consequently, after the plate is developed the track is observed under a high power microscope, or a photomicrograph is made. Nuclear emulsions were particularly useful in investigations of cosmic rays at various altitudes.

The Nucleus

Discovery of Nucleus

In 1909 Geiger and Marsden, at Lord Rutherford's suggestion, investigated the scattering of α-particles by thin films of metal of high atomic mass, such as gold foil. They used a radon tube S in a metal block as a source of α-particles, and limited the particles to a narrow pencil, Fig. 41.15. The thin metal foil A was placed in the centre of an evacuated vessel, and the scattering of the particles after passing through A was observed on a fluorescent screen B, placed at the focal plane of a microscope M. Scintillations were seen on B whenever it is struck by α-particles.

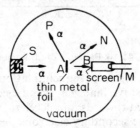

Fig. 41.15 Discovery of nucleus—Geiger and Marsden

Geiger and Marsden found that α-particles struck B not only in the direction SA, but also when the microscope M was moved round to N and even to P. Thus though the majority of α-particles were scattered through small angles, some particles were scattered through very large angles. Rutherford found this very exciting news. It meant that some α-particles had come into the repulsive field of a highly concentrated positive charge at the heart or centre of the atom. As we now explain, Rutherford showed that an atom has a *nucelus*, in which all the positive charge and most of its mass is concentrated.

Paths of Scattered Particles

Rutherford assumed that an atom has a *nucleus*, in which all the positive charge and most of the mass is concentrated. The beam of α-particles incident on a thin metal foil are then scattered through various angles, as shown roughly in

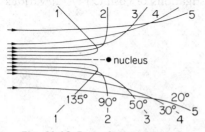

Fig. 41.16 Rutherford scattering law

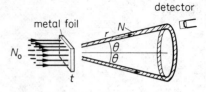

Fig. 41.17 Rutherford scattering law

Fig. 41.16. Those particles very close to the nucleus are deflected through a large angle.

Rurtherford obtained a formula for the number N of α-particles scattered through an angle θ. In a series of experiments using a detector as illustrated diagrammatically in Fig. 41.17, Geiger and Marsden verified the Rutherford formula and thus confirmed the existence of the nucleus.

Atomic Nucleus and Atomic Number

In 1911 Rutherford proposed the basic structure of the atom which is accepted today, and which subsequent experiments by Moseley and others have confirmed. A neutral atom consists of a very tiny nucleus of diameter about 10^{-15} m which contains practically the whole mass of the atom. The atom is largely empty. If a drop of water was magnified until it reached the size of the earth, the atoms inside would then be only a few metres in diameter and the atomic nucleus would have a diameter of only about 10^{-2} millimetre.

The nucleus of hydrogen is called a *proton*, and it carries a charge of $+e$, where e is the numerical value of the charge on an electron. The helium nucleus has a charge of $+2e$. The nucleus of copper has a charge of $+29e$, and the uranium nucleus carries a charge of $+92e$. Generally, the positive charge on a nucleus is $+Ze$, where Z is the *atomic number* of the element and is defined as the number of protons in the nucleus (see also p. 864). Under the attractive influence of the positively-charged nucleus, a number of electrons equal to the atomic number move round the nucleus and surround it like a negatively-charged cloud.

Discovery of Protons in Nucleus. Mass Number

In 1919 Rutherford found that energetic α-particles could penetrate nitrogen atoms and that protons were ejected after the collision. The apparatus used is shown in Fig. 41.18. A source of α-particles, A, was placed in a container D from which all the air had been pumped out and replaced by nitrogen. Silver foil, B, sufficiently thick to stop α-particles, was then placed between A and a

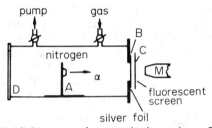

Fig. 41.18 Discovery of protons in the nucleus—Rutherford

Fig. 41.19 Transmutation of nitrogen by collision with α-particle. An oxygen nucleus right-curved track, and a proton, left straight track are produced

fluorescent screen C, and scintillations were observed by a microscope M. The particles which have passed through B were shown to have a similar range, and the same charge, as protons. See also Fig. 41.19.

Protons were also obtained with the gas fluorine, and with other elements such as the metals sodium and aluminium. It thus becomes clear that *the nuclei of all elements contain protons*. The number of protons must equal the number of electrons surrounding the nucleus, so that each is equal to the atomic number, Z, of the element. A proton is represented by the symbol, $_1^1H$; the top number denotes the *mass number* or *nucleon number* A (a nucleon is a particle in the nucleus) and the bottom number is the *atomic* or proton number Z. The helium nucleus such as an α-particle is represented by $_2^4H$; its mass or nucleon number A is 4 and its proton number Z is 2.

One of the heaviest nuclei, uranium, can be represented by $_{92}^{238}U$; it has a nucleon number A of 238 and a proton number Z of 92.

Discovery of Neutron in Nucleus

In 1930 Bothe and Becker found that a very penetrating radiation was produced when α-particles were incident on beryllium. Since the radiation had no charge it was thought to be γ-radiation of very great energy. In 1932 Curie-Joliot placed a block of paraffin-wax in front of the penetrating radiation, and showed that protons of considerable range were ejected from the paraffin-wax. The energy of the radiation could be calculated from the range of the ejected proton, and it was then found to be improbably high.

In 1932 Chadwick measured the velocity of protons and of nitrogen nuclei when they were ejected from materials containing hydrogen and nitrogen by the penetrating radiation. He used polonium, A, as a source of α-particles and the unknown radiation X, obtained by impact with beryllium, B, was then incident on a slab C of paraffin-wax, Fig. 41.20. The velocity of the protons emitted from C could be found from their range in air, which was determined by placing various thicknesses of mica, D, in front of an ionisation chamber E, until no effect was produced here. By previous calibration of the thickness of mica in terms of air thickness, the range in air was found.

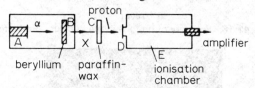

Fig. 41.20 Discovery of the neutron—Chadwick

Chadwick repeated the experiment with a slab of material containing nitrogen in place of paraffin-wax. He then applied the laws of conservation of linear momentum and energy to the respective collisions with the hydrogen and nitrogen atoms, assuming that the unknown radiation was a *particle* carrying no charge and the collisions were elastic. From the equations obtained, he calculated the mass of the particle, and found it to be about the same mass as the proton. Chadwick called the new particle a *neutron*, and it is now considered that *all nuclei contain protons and neutrons*. The neutron is represented by the symbol $_0^1n$ as it has a mass number of 1 and zero charge.

We can now see that a helium nucleus, $_2^4He$, has 2 protons and 2 neutrons, a total mass or nucleon number of 4. The sodium nucleus, $_{11}^{23}Na$, has 11 protons and 12 neutrons. The uranium nucleus, $_{92}^{238}U$, has 92 protons and 146 neutrons. Generally, a nucleus represented by $_Z^AX$ has Z protons and $(A - Z)$ neutrons.

Radioactive Disintegration

Naturally occuring radioactive elements such as uranium, actinium and thorium disintegrate to form new elements, and these in turn are unstable and form other elements. Between 1902 and 1909 Rutherford and Soddy made a study of the elements formed from a particular 'parent' element. The *uranium series* is listed in the table below.

Element	Symbol	Atomic Number	Mass Number	Half-life Period (T)	Particle emitted
Uranium I	U	92	238	4500 million years	α
Uranuium X_1	Th	90	234	24 days	β, γ
Uranium X_2	Pa	91	234	1·2 minutes	β, γ
Uranium II	U	92	234	250 000 years	α
Ionium	Th	90	230	80 000 years	α, γ
Radium	Ra	88	226	1600 years	α, γ
Radon	Rn	86	222	3·8 days	α
Radium A	Po	84	218	3 minutes	α
Radium B	Pb	82	214	27 minutes	β, γ
Radium C*	Bi	83	214	20 minutes	β or α
Radium C'	Po	84	214	1.6×10^{-4} seconds	α
Radium C"	Tl	81	210	1·3 minutes	β
Radium D	Pb	82	210	19 years	β, γ
Radium E	Bi	83	210	5 days	β
Radium F	Po	84	210	138 days	α, γ
Lead	Pb	82	206	(stable)	

* Radium C exhibits branching; it produces Radium C' on β-emission or Radium C" on α-emission. Radium D is then produced from Radium C' by α-emission or from Radium C" by β-emission.

The new element formed after disintegration can be identified by considering the particles emitted from the nucleus of the parent atom. An α-particle, a helium nucleus, has a charge of $+2e$ and a mass number 4. Uranium I, of atomic number 92 and mass number 238, emits an α-particle from its nucleus of charge $+92e$, and hence the new nucleus formed has an atomic number 90 and a mass number 234. This was called uranium X_1, and since the element thorium (Th) has an atomic number 90, uranium X_1 is actually thorium.

A β-particle, an electron, and a γ-ray, an electromagnetic wave, have a negligible effect on the mass of a nucleus when they are emitted. A β-particle has a charge of $-e$. Now uranium X_1 has a nuclear charge of $+90e$ and a mass number 234, and emits β and γ rays. Consequently the mass number is unaltered, but the nuclear charge increases to $+91e$, and hence a new element is formed of atomic number 91. This is uranium X_2 in the series, and is actually the element protactinium. The symbols of the new elements formed are shown in brackets in the column of elements in the Table. The series contains isotopes of uranium (U), lead (Pb), thorium (Th) and bismuth (Bi), that is, elements which have the same atomic numbers but different mass numbers (see p. 894).

Summarising, we can say that:

(i) when the nucleus of an element loses an α-particle, the element is displaced two places to the left in the periodic table of the elements, which follows in the order of its atomic number, and lowers its mass number by four units;

(ii) when the nucleus of an element loses a β-particle, the element is displaced one place to the right in the periodic table and its mass number is unaltered.

This law was stated in 1913 by Soddy, Russell and Fajans.

Nuclear Mass, Nuclear Energy

Positive Rays and Atomic Mass

As we saw in the discharge tube (p. 747), at low pressures electrons (negative charges) will flow from the cathode to the anode. If a hole is made in the cathode, some rays appear to pass through the metal, as shown in Fig. 41.21.

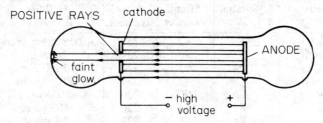

Fig. 41.21 Positive rays

These rays, which move in the opposite direction to the cathode rays (electrons), were called *positive rays*. They were first thought to come from the anode. They are now known to be formed when electrons from the cathode collide with the gas atoms and strip some electrons from the atom. Positive gas ions are then produced. These move slowly towards the cathode on account of the electric field between anode and cathode, and if the cathode is pierced they pass through the hole.

In 1911, Sir J. J. Thomson measured the masses of individual atoms for the first time. The gas concerned was passed slowly through a bulb B a low pressure, and a high voltage was applied. Cathode rays or electrons then flow from the cathode to the anode (not shown), and positive rays due to ionisation flow to the cathode, C, whose axis was pierced by a fine tube, Fig. 41.22. The positive rays are *ions*, that is, atoms which have lost one or more electrons. After flowing through C they are subjected to *parallel magnetic and electric fields* set at right angles to the incident beam, which are applied between the poles N, S of an

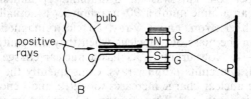

Fig. 41.22 Principle of Thomson's experiment on positive rays

electromagnet. Pieces of mica, G, G are used to insulate N, S from the magnet core. The ions were deflected by the fields and were incident on a photographic plate P. After development, parabolic traces were found on P. See Fig. 44.25.

By considering the vertical (y) and horizontal (x) deflections of the ions due to the two fields, theory now given shows that $x^2 = ky$, where k is a factor which depends on the charge–mass ratio, Q/M, of the ions, the field values and the dimensions of the apparatus.

Theory. The force F on a charge Q moving normally to a vertical magnetic field B is given by $F = BQv$, where v is the velocity, Fig. 41.23. So the horizontal acceleration $a_x = BQv/M$, where M is the mass of the ion. The horizontal distance x travelled as the ion leaves the field is given by $x = \frac{1}{2}at^2$, where t is the time to travel a distance l between the N and S poles in Fig. 41.22, that is, $t = l/v$. So

$$x = \frac{1}{2}\frac{BQv}{M}\left(\frac{l}{v}\right)^2 = \frac{BQ}{2M} \cdot \frac{l^2}{v}. \qquad \cdot \qquad \cdot \qquad \cdot \qquad \cdot \qquad (1)$$

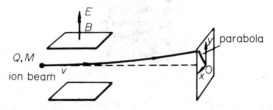

Fig. 41.23 Theory of Thomson's experiment

The force F on a charge Q moving normally to the vertical electric field of intensity E is given by EQ, and so the acceleration vertically, $a_y = EQ/M$. The vertical distance y travelled as the ion leaves the field is given by $y = \frac{1}{2}a_y t^2$, where $t = l/v$. So

$$y = \frac{1}{2}\frac{EQ}{M} \cdot \frac{l^2}{v^2}. \qquad \cdot \qquad \cdot \qquad \cdot \qquad \cdot \qquad \cdot \qquad (2)$$

Eliminating v by squaring (1) and dividing by (2), we obtain

$$x^2 = \frac{B^2 l^2}{2E}\left(\frac{Q}{M}\right)y, \quad \text{or} \quad x^2 = ky,$$

where $$k = (B^2 l^2/2E) \times (Q/M).$$

So ions with the same charge-mass ratio Q/M, although moving with different velocities, all lie on a *parabola* of the form $x^2 = \text{constant} \times y$.

Determination of Masses. Isotopes

As the zero of the parabola was not clear, Thomson reversed the field to obtain a parabolic trace on the other side of the y-axis, as shown in Fig. 41.24. Now

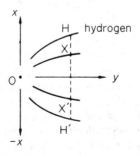

Fig. 41.24 Determination of mass of atom

for a given value of y, $x^2 \propto Q/M$. Consequently any hydrogen ions present would produce the outermost parabola H, since they have the greatest value of Q/M. The masses of ions can thus be measured by comparing the squares of the x-values of the individual parabolas, such as the squares of X'X and H'H, for example. In this way Thomson obtained a *mass spectrometer*, one which gave the masses of individual atoms. See Fig. 41.25.

Fig. 41.25 Mass spectrometer. Positive-ray parabolas due to mercury, carbon monoxide, oxygen and carbon ions

With chlorine gas, two parabolas were obtained which gave atomic masses of 35 and 37 respectively. Thus the atoms of chlorine have different masses but the same chemical properties, and these atoms are said to be *isotopes* of chlorine. In chlorine, there are three times as many atoms of mass 35 as there are of mass 37, so that the average atomic mass is $(3 \times 35 + 1 \times 37)/4$, or 35·5. The element xenon has as many as nine isotopes. One part in 5000 of hydrogen consists of an isotope of mass 2 called deuterium, or heavy hydrogen. An unstable isotope of hydrogen of mass 3 is called tritium.

Bainbridge Mass Spectrometer

Thomson's earliest form of mass spectrometer was followed by more sensitive forms. Bainbridge devised a mass spectrometer in which the ions were incident on a photographic plate after being deflected by a magnetic field.

The principle of the spectrometer is shown in Fig. 41.26. Positive ions were

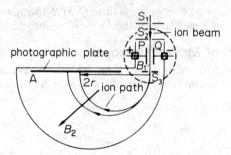

Fig. 41.26 Principle of Bainbridge mass spectrometer

produced in a discharge tube (not shown) and admitted as a fine beam through slits S_1, S_2. The beam then passed between insulated plates P, Q, connected to a battery, which created an electric field of intensity E. A uniform magnetic field B_1, perpendicular to E, was also applied over the region of the plates, and all ions, charge Q, with the *same velocity* v given by $B_1 Qv = EQ$ will then pass undeflected through the plates and through a slit S_3. The selected ions were now deflected in a circular path of radius r by a uniform perpendicular magnetic field B_2, and an image was produced on a photographic plate A, as shown. In this case, if the mass of the ion is M,

$$\frac{Mv^2}{r} = B_2 Qv.$$

$$\therefore \frac{M}{Q} = \frac{rB_2}{v}.$$

But for the selected ions, $v = E/B_1$ from above.

$$\therefore \frac{M}{Q} = \frac{rB_2B_1}{E}$$

$$\therefore \frac{M}{Q} \propto r,$$

for given magnetic and electric fields.

Since the ions strike the photographic plate at a distance $2r$ from the middle of the slit S_3, it follows that the separation of ions carrying the same charge is directly proportional to their mass. Thus a 'linear' mass scale is achieved. A resolution of 1 in 30 000 was obtained with a later type of spectrometer.

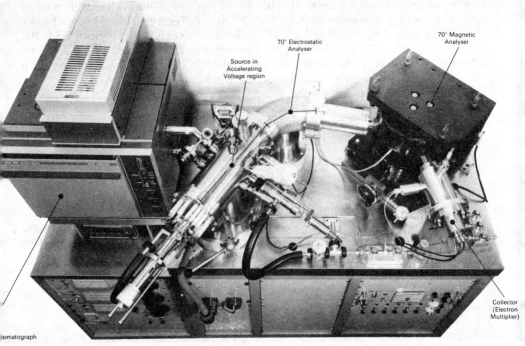

Fig. 41A *Industrial mass spectrometer.* Double-focusing by an electrostatic and a magnetic field is used, which produces mass and resolution superior to single focusing. The gas chromatograph (left) separates mixtures into their components. The ion sources are then passed into the front tube (left) to be accelerated by a high voltage. The ions are now deflected electrostatically by the analyser at the top and then deflected by the magnetic field of the second analyser. The separated ions are collected in the tube at the front. Here they are recorded electronically and their masses and relative abundance in the compound analysed are deduced from the readings. (*Courtesy of VG Micromass Limited*)

Example

In a mass spectrograph, an ion X of mass number 24 and charge $+e$ and an ion Y of mass 22 and charge $+2e$ both enter the magnetic field with the same velocity. The radius of the circular path of X is 0·25 m. Calculate the radius of the circular path of Y.

From above, $$r \propto \frac{M}{Q}$$

For ion Y, $$\frac{M}{Q} = \frac{22}{2e}$$

For ion X, $$\frac{M}{Q} = \frac{24}{e}$$

So
$$\frac{r_Y}{0 \cdot 25} = \frac{22/2e}{24/e} = \frac{11}{24}$$

$$\therefore r_Y = \frac{11}{24} \times 0 \cdot 25 = 0 \cdot 11 \text{ m (approx.)}.$$

Einstein's Mass-Energy Relation

In 1905 Einstein showed from his Theory of Relativity that mass and energy can be changed from one form to the other. The energy E produced by a change of mass m is given by the relation:

$$E = mc^2,$$

where c is the numerical value of the velocity of light. E is in joule when m is in kg and c has the numerical value 3×10^8 (p. 477). Thus a change in mass of 1 kg could theoretically produce 9×10^{13} joules of energy. Now 1 kilowatt-hour of energy is 1000×3600 or $3 \cdot 6 \times 10^6$ joules, and hence 9×10^{13} joules is $2 \cdot 5 \times 10^7$ or 25 million kilowatt-hours. Consequently a change in mass of 1 g could be sufficient to keep the electric lamps in a million houses burning for about a week in winter, on the basis of about seven hours' use per day.

In electronics and in nuclear energy, the unit of energy called an *electron-volt* (eV) is often used. This is defined as the energy gained by a charge equal to that on an electron moving through a p.d. of one volt.

$$1 \text{ eV} = 1 \cdot 6 \times 10^{-19} \text{ J (p. 581)}.$$

The *megelectronvolt* (MeV) is a larger energy unit, and is defined as 1 million eV.

So
$$1 \text{ MeV} = 1 \cdot 6 \times 10^{-13} \text{ J}.$$

Unified Atomic Mass Unit

If another unit of energy is needed, then one may use a unit of mass, since mass and energy are interchangeable. The *unified atomic mass unit* (u) is defined as 1/12th of the mass of the carbon atom $^{12}_6C$. Now the number of molecules in 1 mole of carbon is $6 \cdot 02 \times 10^{23}$, the Avogadro constant, and since carbon is monatomic, there are $6 \cdot 02 \times 10^{23}$ atoms of carbon. These have a mass 12 g.

$$\therefore \text{ mass of 1 atom of carbon} = \frac{12}{6 \cdot 02 \times 10^{23}} \text{ g} = \frac{12}{6 \cdot 02 \times 10^{26}} \text{ kg}$$

$$= 12 \text{ u}.$$

$$\therefore 1 \text{ u} = \frac{12}{12 \times 6 \cdot 02 \times 10^{26}} \text{ kg}$$

$$= 1 \cdot 66 \times 10^{-27} \text{ kg (approx.)}.$$

From our previous calculation, 1 kg change in mass produces 9×10^{16} joules; and we have seen that $1 \text{ MeV} = 1 \cdot 6 \times 10^{-13}$ joule.

$$\therefore 1 \text{ u} = \frac{1 \cdot 66 \times 10^{-27} \times 9 \times 10^{16}}{1 \cdot 6 \times 10^{-13}} \text{ MeV}$$

$$\therefore 1 \text{ u} = 931 \text{ MeV (approx.)} \qquad . \qquad . \qquad . \qquad . \qquad (1)$$

This relation is used to change mass units to MeV, and vice-versa, as we shall see shortly. An electron mass, $9 \cdot 1 \times 10^{-31}$ kg, corresponds to about $0 \cdot 5$ MeV.

Binding Energy

The protons and neutrons in the nucleus of an atom are called *nucleons*. The work or energy needed to take all the nucleons apart so that they are completely separated is called the *binding energy* of the nucleus. Hence, from Einstein's mass-energy relation, it follows that the total mass of all the separated nucleons is greater than that of the nucleus, in which they are together. *The difference in mass is a measure of the binding energy.*

As an example, consider a helium nucleus $^4_2$He. This has 4 nucleons, 2 protons and 2 neutrons. The mass of a proton is 1·0073 and the mass of a neutron is 1·0087 u.

$$\therefore \text{ total mass of 2 protons plus 2 neutrons} = 2 \times 1\cdot0073 + 2 \times 1\cdot0087$$
$$= 4\cdot0320 \text{ u}.$$

But the helium nucleus has a mass of 4·0015 u.

$$\therefore \text{ binding energy} = \text{mass difference of nucleons and nucleus}$$
$$= 4\cdot0320 - 4\cdot0015 = 0\cdot0305 \text{ u}.$$
$$= 0\cdot0305 \times 931 \text{ MeV} = 28\cdot4 \text{ MeV}.$$

The *binding energy per nucleon* of a nucleus is binding energy divided by the total number of nucleons. In the case of the helium nucleus, since there are four nucleons (2 protons and 2 neutrons), the binding energy per nucleon is then 28·4/4 or 7·1 MeV.

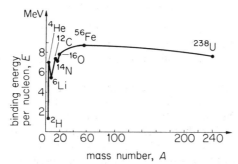

Fig. 41.27 Variation of binding energy per nucleon with mass number

Figure 41.27 shows roughly the variation of the binding energy per nucleon among the elements. Excluding the nuclei lighter than ^{12}C, we can see from Fig. 41.27 that the average binding energy per nucleon, E/A, is fairly constant for the great majority of nuclei. The average value is about 8 MeV per nucleon. The peak occurs at approximately the iron nucleus ^{56}Fe, which is therefore one of the most stable nuclei.

Later we shall use the curve in Fig. 41.27 to show that energy is produced when heavy elements such as uranium undergo *fission* to form two lighter masses or when very light elements such as hydrogen undergo *fusion* to form a heavier element.

Nuclear Forces and Binding Energy

Inside the nucleus, the protons repel each other due to electrostatic repulsion of like charges. So for the nucleus to be stable there must exist other forces between the nucleons. These are called *nuclear forces*. They have short range, shorter than the interatomic distances, and are much stronger than electro-magnetic interactions. They must provide a net attractive force greater than any repulsive electric forces.

A nucleus with a nucleon (mass) number $(N + Z)$ and proton (atomic) number Z has Z protons and N neutrons. When these particles come together in the nucleus there is an increase in potential energy due to the electrostatic forces of the protons but a *greater* decrease in potential energy due to the nuclear

forces of the nucleons. There is therefore a net decrease in the potential energy
of all the nucleons. This decrease expressed per nucleon is the *binding energy
per nucleon*.

So when the nucleons come together in the nucleus, there is a loss of energy
equal to the binding energy. This results in a decrease in mass, from Einstein's
mass-energy relation. As we have seen, the decrease in mass is the difference in the
mass of the individual nucleons when they are completely separated and the mass
of the nucleus when they are together; the so-called 'mass defect'. Expressed in
symbols, the binding energy of the nucleus $^{N+Z}_{Z}X$ of an atom X is given by:
Binding energy

$$= \text{mass of } N \text{ neutrons} + \text{mass of } Z \text{ protons} - \text{mass of nucleus } ^{N+Z}_{Z}X.$$

Energy of Disintegration

It is instructive to consider, from an energy point of view, whether a particular
nucleus is likely to disintegrate with the emission of an α-particle. As an illustration, consider radium F or polonium, $^{210}_{84}Po$. If an α-particle could be emitted
from the nucleus, the reaction products would be the α-particle or helium
nucleus, $^{4}_{2}He$, and a lead nucleus, $^{206}_{82}Pb$, a reaction which could be represented
by:

$$^{210}_{84}Po \rightarrow ^{206}_{82}Pb + ^{4}_{2}He \quad . \quad \quad . \quad \quad . \quad \quad . \quad \quad \text{(i)}$$

Here the sum of the mass numbers, 210, and the sum of the nuclear charges,
$+84e$, of the lead and helium nuclei are respectively equal to the mass number
and nuclear charge of the polonium nucleus, from the law of conservation of
mass and of charge.

If we require to find whether energy has been released or absorbed in the
reaction, we should calculate the total mass of the lead and helium nuclei and
compare this with the mass of the polonium nucleus. It is more convenient to
use atomic masses rather than nuclear masses, and since the total number of
electrons required on each side of (i) to convert the nuclei into atoms is the
same, we may use atomic masses in the reaction. These are as follows:

$$\text{lead, } ^{206}_{82}Pb, = 205 \cdot 969 \text{ u}$$

$$\alpha\text{-particle, } ^{4}_{2}He, = \quad 4 \cdot 004 \text{ u}$$

$$\therefore \text{ total mass} = 209 \cdot 973 \text{ u}$$

Now   polonium, $^{210}_{84}Po, = 209 \cdot 982$ u.

Thus the atomic masses of the products of the reaction are together *less* than
the original polonium nucleus, that is,

$$^{210}_{84}Po \rightarrow ^{206}_{82}Pb + ^{4}_{2}He + Q,$$

where Q is the energy released. It therefore follows that polonium can disintegrate
with the emission of an α-particle and a release of energy (see *uranium series*,
p. 891), that is, the polonium is unstable.

Suppose we now consider the possibility of a lead nucleus, $^{206}_{82}Pb$, disintegrating with the emission of an α-particle, $^{4}_{2}He$. If this were possible, a mercury
nucleus, $^{202}_{80}Hg$, would be formed. The atomic masses are as follows:

$$\text{mercury, } ^{202}_{80}Hg, = 201 \cdot 971 \text{ u}$$

$$\alpha\text{-particle, } ^{4}_{2}He, = \quad 4 \cdot 004 \text{ u}$$

$$\therefore \text{ total mass} = 205 \cdot 975 \text{ u}$$

Now   lead, $^{206}_{82}Pb, = 205 \cdot 969$ u.

Thus, unlike the case previously considered, the atomic masses of the mercury nucleus and α-particle are together *greater* than the lead nucleus, that is,

$$^{206}_{82}\text{Pb} + Q \rightarrow {}^{202}_{80}\text{Hg} + {}^{4}_{2}\text{He},$$

where Q is the energy which must be *given* to the lead nucleus to obtain the reaction products. It follows that the lead nucleus by itself is stable to α-decay.

Generally, then, a nucleus would tend to be unstable and emit an α-particle if the sum of the atomic masses of the products are together *less* than that of the nucleus; and it would be stable if the sum of the atomic masses of the possible reaction products are together *greater* than the atomic mass of the nucleus.

Stable and Unstable Nuclei

Many factors contribute to the binding energy, E, of nuclei and therefore to their stability. The α-particle, ${}^{4}_{2}\text{He}$, appears to be particularly stable. Figure

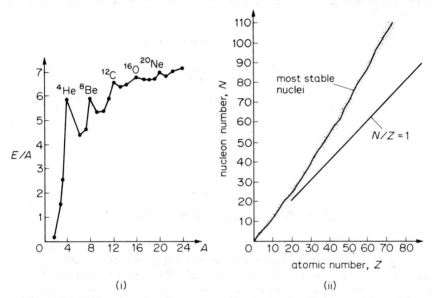

Fig. 41.28 (i) Variation of nuclear energy with atomic mass (ii) Most stable nuclei

41.28 (i) shows the binding energy per nucleon, E/A, for some light nuclei. Peaks occur for nuclei such as ${}^{8}_{4}\text{Be}$, ${}^{12}_{6}\text{C}$, ${}^{16}_{8}\text{O}$, ${}^{20}_{10}\text{Ne}$. Each of these nuclei can be formed by adding an α-particle to the preceding nucleus.

Another factor affecting the stability of nuclei is the *neutron-proton ratio*. Figure 41.28 (ii) shows the number of neutrons, N, plotted against the number of protons (atomic number), Z, for all stable nuclei. It can be seen that for the light stable nuclei, such as ${}^{12}_{6}\text{C}$ and ${}^{16}_{8}\text{O}$, this ratio is 1. For nuclei heavier than ${}^{40}_{20}\text{Ca}$, the ratio N/Z increases slowly towards about 1·6. There are no stable nuclei above about $Z = 92$ (uranium).

Nuclear Emissions and Nuclear Stability

Unstable nuclei are radioactive. Their decay may occur in three main ways:

(1) *α-particle emission.* If the nucleus has excess protons, an α-particle emission would reduce the protons by 2 and the neutrons by two. So, generally, if A is the nucleon (mass) number and Z is the proton (atomic) number of the atom X concerned, then

$$_Z^A X \rightarrow \,_{Z-2}^{A-4} Y + \,_2^4 He.$$

(2) β^- *particle (electron) emission.* If the nucleus has too many neutrons for stability, the neutron-proton ratio is reduced by β-particle (electron) emission. Here a neutron changes to a proton, so

$$_0^1 n \rightarrow \,_1^1 H + \,_{-1}^0 e \text{ (electron).}$$

Hence A remains unchanged but Z increases by 1.

(3) β^+ *particle (positron) emission.* If the nucleus is deficient in neutrons, a decay by β^+ (positron) emission may occur. A proton changes to a neutron:

$$_1^1 H \rightarrow \,_0^1 n + \,_1^0 e \text{ (positron).}$$

So A is unchanged but Z decreases by 1.

The effects of these three types of decays, (1), (2) and (3), on A and Z are summarised in Fig. 41.29; the boxes indicate diagrammatically unit changes in A and Z.

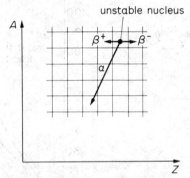

Fig. 41.29 Effect of particle emission on A and Z

In nuclei lighter than the isotope ^{40}Co, the numbers of neutrons and protons are roughly equal. Heavier elements such as thorium or uranium have a greater number of protons which tend to repel each other and so make the nucleus less tightly bound. So most stable heavy nuclei have more neutrons than protons. See also p. 899.

The nucleus formed by a decay may also be unstable. This gives rise to 'decay series or chains', which terminate when a stable nucleus, for example, ^{206}Pb (lead), is reached. A series consisting entirely of α-decays (α-particle emission) would increase the ratio of neutrons to protons as Z decreased. As an example, suppose $^{244}_{94}$Pu decays by α-emission to $^{240}_{92}$U. Now $^{240}_{92}$U has $N = 148$ and $Z = 92$, so the neutron-proton ratio, N/Z, is $1 \cdot 609$. *If* this nucleus were to decay by another α-emission to $^{236}_{90}$Th, the ratio would be increased to 146/90 or $1 \cdot 622$, which is too far from the line of stability shown in Fig. 41.28 (ii). In fact, $^{240}_{92}$U decays to $^{240}_{93}$Np by β^- emission. The ratio N/Z is now 147/93 or $1 \cdot 581$, close to the line of the stability. In a natural decay series, then, there must always be some β^- decays in addition to α-decays.

β^+ (positron) decays are much rarer since they would result in an increase in N and a decrease in Z, thus increasing the ratio N/Z.

Artificial Disintegration

Uranium, thorium and actinium are elements which disintegrate naturally. The artificial disintegration of elements began in 1919, when Rutherford used

α-particles to bombard nitrogen and found that protons were produced (p. 889). Some nuclei of nitrogen had changed into nuclei of oxygen, that is, transmutation had occurred, a reaction which can be represented by:

$$^{14}_{7}\text{N} + ^{4}_{2}\text{He} \rightarrow ^{17}_{8}\text{O} + ^{1}_{1}\text{H}.$$

In 1932 Cockcroft and Walton produced nuclear disintegrations by accelerating protons with a high-voltage machine producing about half a million volts, and then bombarding elements with the high-speed protons. When the light element lithium was used, photographs of the reaction taken in the cloud chamber showed that two α-particles were produced. The particles shot out in opposite direction from the point of impact of the protons, and as their range in air was equal, the α-particles had initially equal energy. This is a consequence of the principle of the conservation of momentum. The nuclear reaction was:

$$^{7}_{3}\text{Li} + ^{1}_{1}\text{H} \rightarrow ^{4}_{2}\text{He} + ^{4}_{2}\text{He} + Q \quad . \quad\quad . \quad\quad . \quad\quad . \quad\quad (i)$$

where Q is the energy released in the reaction.

To calculate Q, we should calculate the total mass of the lithium and hydrogen nuclei and subtract the total mass of the two helium nuclei. As already explained, however, the total number of electrons required to convert the nuclei to neutral atoms is the same on both sides of equation (i). So atomic masses can be used in the calculation in place of nuclear masses. The atomic masses of lithium and hydrogen are 7·016 and 1·008 u respectively, a total of 8·024 u. The atomic mass of the two α-particles is $2 \times 4{\cdot}004$ u or 8·008 u. thus:

$$\text{energy released, } Q, = 8{\cdot}024 - 8{\cdot}008 = 0{\cdot}016 \text{ u}$$

$$= 0{\cdot}016 \times 931 \text{ MeV} = 14{\cdot}9 \text{ MeV}.$$

Each α-particle has therefore an initial energy of 7·4 MeV, and this theoretical value agreed closely with the energy of each α-particle measured from its range in air. The experiment was the earliest verification of Einstein's mass-energy relation.

Cockcroft and Walton were the first scientists to use protons for disrupting atomic nuclei after accelerating them by high voltage. Today, giant high-voltage machines are being built at Atomic Energy centres for accelerating protons to enormously high speeds. The products of the nuclear explosion with light atoms such as hydrogen will yield valuable information on the structure of the nucleus.

Energy from Radioactive Isotopes

Energy from the decay of unstable radioactive isotopes is sometimes used where a continuous powerful but compact energy source is required. Such isotopes have been used to provide power for the batteries of heart 'pacemakers' and for scientific apparatus used in space vehicles.

As an example, consider the isotope Po 210. This has a half-life of about 140 days and emits α-particles each of energy 5·3 MeV.

Since the molar mass is 210 g and the Avogadro constant is about 6×10^{23} mol^{-1}, each gram of the isotope contains $6 \times 10^{23}/210 = 2{\cdot}9 \times 10^{21}$ atoms. Thus in one half-life or 140 days, about $1{\cdot}4 \times 10^{23}$ atoms decay.

These atoms release a total energy (using 1 MeV $= 1{\cdot}6 \times 10^{-13}$ J)

$$= 1{\cdot}4 \times 10^{21} \times 5{\cdot}3 \times 1{\cdot}6 \times 10^{-13} \text{ J}$$

$$= 1{\cdot}2 \times 10^{9} \text{ J (approx.).}$$

Thus the mean output power per gram

$$= \frac{1 \cdot 2 \times 10^9}{140 \times 24 \times 3600} = 100 \text{ W (approx.)}.$$

During the next half-life period, 140 days, the mean power output will only be 50 W, since half the remaining atoms decay in this time.

Energy Released in Fission

In 1934 Fermi began using neutrons to produce nuclear disintegration. These particles are generally more effective than α-particles or protons for this purpose, because they have no charge and are therefore able to penetrate more deeply into the positively-charged nucleus. Usually the atomic nucleus changes only slightly after disintegration, but in 1939 Frisch and Meitner showed that a uranium nucleus had disintegrated into two relatively-heavy nuclei. This is called *nuclear fission*, and as we shall now show, a large amount of energy is released in this case.

Natural uranium consists of about 1 part by mass of uranium atoms $^{235}_{92}U$ and 140 parts by mass of uranium atoms $^{238}_{92}U$. In a nuclear reaction with natural uranium and slow neutrons, it is usually the nucleus $^{235}_{92}U$ which is fissioned. If the resulting nuclei are lanthanum $^{148}_{57}La$ and bromine $^{85}_{35}Br$, together with several neutrons, then:

$$^{235}_{92}U + ^1_0n \rightarrow ^{148}_{57}La + ^{85}_{35}Br + 3^1_0n \qquad . \qquad . \qquad . \qquad (i)$$

Now $^{235}_{92}U$ and ^1_0n together have a mass of $(235 \cdot 1 + 1 \cdot 009)$ or $236 \cdot 1$ u. The lanthanum, bromine and neutrons produced together have a mass

$$= 148 \cdot 0 + 84 \cdot 9 + 3 \times 1 \cdot 009 = 235 \cdot 9 \text{ u}$$

$\therefore$ energy released = mass difference

$$= 0 \cdot 2 \text{ u} = 0 \cdot 2 \times 931 \text{ MeV} = 186 \text{ MeV}$$

$$= 298 \times 10^{-13} \text{ J (approx.)}.$$

This is the energy released per atom of uranium fissioned. In 1 kg of uranium there are about

$$\frac{1000}{235} \times 6 \times 10^{23} \quad \text{or} \quad 26 \times 10^{23} \text{ atoms,}$$

since the Avogadro constant, the number of atoms in a mole of any element, is $6 \cdot 02 \times 10^{23}$. Thus if all the atoms in 1 kg of uranium were fissioned, total energy released

$$= 26 \times 10^{23} \times 298 \times 10^{-13} \text{ joules}$$
$$= 2 \times 10^7 \text{ kilowatt-hours (approx.)},$$

which is the amount of energy given out by burning about 3 million tonnes of coal. The energy released per gram of uranium fissioned $= 8 \times 10^{10}$ J (approx.).

Fast neutrons are absorbed or captured by nuclei of U238 without producing fission. Slow or thermal neutrons in uranium, however, which have the same temperature as that of the uranium, produce fission when incident on nuclei of U235. About 2·5 neutrons per fission are released. When slowed, each of these neutrons produce fission in another U235 nucleus, and so on. Thus a rapid multiplying *chain reaction* can be obtained in the uranium mass, liberating swiftly an enormous amount of energy. This is the basic principle of the nuclear

Fig. 41.30 Nuclear research reactor, ZEUS. This view shows the heart of the reactor, containing a highly enriched uranium central core surrounded by a natural uranium blanket for breeding studies

reactor or pile. As we now describe, graphite can be used to slow down the speed of the fast neutrons released in fission.

Nuclear Reactors

Figure 41.31 shows, in diagrammatic form, the principle of one type of nuclear reactor, used for commercial power generation.

Uranium fuel in the form of thick rods are encased in long aluminium tubes, which are air-tight to contain any gases released and prevent oxidation of the surrounding fuel. The tubes are lowered into hundreds of channels inside

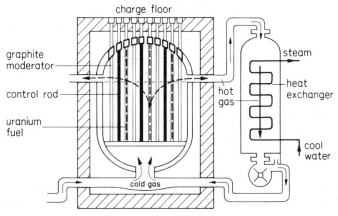

Fig. 41.31 Nuclear reactor principle

blocks. As explained previously, the graphite is needed to moderate or reduce the speed of the neutrons released on fission until they become slow or thermal neutrons; they can then produce fission on collision with other U235 nuclei. The graphite *moderator* is in the form of blocks of pure carbon arranged in a stack.

Heavy water (deuterium oxide) has also been used as a moderator. It reduces the speed of a colliding neutron to about one-half, whereas graphite reduces the speed to about one-seventh. With a graphite moderator the separation of the uranium rods is about 20 cm. In this distance the fast neutron makes about 200 collisions with carbon nuclei and then becomes slow enough to produce fission on collision with a U235 nucleus. The separation of the uranium rods in a water moderator is less than in a graphite moderator as the speed of the colliding neutron is reduced much more with a deuterium nucleus, as stated above.

The power level reached by a reactor is proportional to the *neutron flux density*; this is the number of neutrons per second crossing unit cross-sectional area of the reactor. If the neutrons are produced too fast, the reactor may disintegrate. Ideally, the neutron reproduction factor should be just greater than 1 to make the chain reaction self-sustaining. Boron-coated steel rods are used to control the rate of neutron production, Fig. 41.31. The *control rods* are lowered or raised in channels inside the graphite block by electric motors operated from a control room until the neutron reproduction rate is just greater than 1, and the reactor is then said to go 'critical'. In the event of an electrical failure or other danger, the rods fall and automatically shut off the reactor. Boron atoms have a high absorption cross-section for neutrons and thus capture slow neutrons.

The energy produced by the nuclear reaction would make the reactor too hot. A *coolant* is therefore required. Water and gas such as carbon dioxide have been used as coolants. Molten sodium, which has a high value of (specific heat capacity × density) and a high thermal conductivity, has also been used as a coolant. In a gas-cooled reactor which produces power for the Grid system, the hot gas is led from the reactor into a heat exchanger, Fig. 41.31. Here the heat is transferred to water circulating through pipes so that steam is produced and this is used to drive turbines for electrical power generation.

Finally, it should be noted that if the mass of the uranium is too small, the neutrons will escape and a chain reaction is not produced. The *critical mass* is the least mass to make a self-sustaining chain reaction. The critical mass for a reactor in the shape of a sphere is less than that in the shape of a cube since the surface (neutron escape) area is a minimum for a given volume or mass of material (see p. 139). Fermi, a distinguished pioneer in nuclear research, made one of the first pilot reactors in 1942 in roughly a spherical shape.

Further details of reactors must be obtained from the United Kingdom Atomic Energy Authority or from specialist books.

Energy Released in Fusion

In fission, energy is released when a heavy nucleus is spit into two lighter nuclei. Energy is also released if light nuclei are *fused* together to form heavier nuclei, and fusion reaction, as we shall see, is also a possible source of considerable energy. As an illustration, consider the fusion of the nuclei of deuterium, ^2_1H. Deuterium is an isotope of hydrogen known as 'heavy hydrogen', and its nucleus is called a 'deuteron'. The fusion of two deuterons can result in a helium nucleus, ^3_2He, as follows:

$$^2_1H + ^2_1H \rightarrow ^3_2He + ^1_0n.$$

Now   mass of two deuterons

$$= 2 \times 2{\cdot}015 = 4{\cdot}030 \text{ u},$$

and   mass of helium plus neutron

$$= 3{\cdot}017 + 1{\cdot}009 = 4{\cdot}026 \text{ u}$$

$\therefore$ mass converted to energy by fusion

$$= 4{\cdot}030 - 4{\cdot}026 = 0{\cdot}004 \text{ u}$$

$$= 0{\cdot}004 \times 931 \text{ MeV} = 3{\cdot}7 \text{ MeV}$$

$$= 3{\cdot}7 \times 1{\cdot}6 \times 10^{-13} \text{ J} = 6{\cdot}0 \times 10^{-13} \text{ J}$$

$\therefore$ energy released per deuteron $= 3{\cdot}0 \times 10^{-13}$ J.

6×10^{26} is the number of atoms in a kilomole of deuterium, which has a mass of about 2 kg. Thus if all the atoms could undergo fusion,

energy released per kg

$$= 3{\cdot}0 \times 10^{-13} \times 3 \times 10^{26} \text{ J}$$

$$= 9 \times 10^{13} \text{ J (approx.)}.$$

Other fusion reactions can release much more energy, for example, the fusion of the nuclei of deuterium, ^2_1H, and tritium, ^3_1H, isotopes of hydrogen, releases about 30×10^{13} joules of energy per kg according to the reaction:

$$^2_1H + ^3_1H \rightarrow ^4_2He + ^1_0n.$$

In addition, the temperature required for this fusion reaction is less than that needed for the fusion reaction between two deuterons given above, which is an advantage. Hydrogen contains about 1/5000th by mass of deuterium or heavy hydrogen, needed in fusion reactions, and this can be obtained by electrolysis of sea-water, which is cheap and in plentiful supply.

Binding Energy Curve and Fusion. Thermonuclear Reaction

In Fig. 41.27 (p. 897), the binding energy per nucleon is plotted against the mass number of the nucleons or nucleon number. Since the curve rises from hydrogen, the binding energy per nucleon of the helium nucleus, ^4_2He, is greater than that for deuterium, ^2_1H. Thus the binding energy of the helium nucleus, which has four nucleons, is greater than that of two deuterium nuclei, which also have four nucleons.

Now the binding energy of a nucleus is proprtional to the difference between the mass of the individual nucleons and the mass of the nucleus (the so-called 'mass defect') see p. 898. So the mass of the helium nucleus is *less* than that of the two deuterium nuclei. Hence if two deuterium nuclei can be fused together to form a helium nucleus, the mass lost will be released as energy.

The rising part of the binding energy curve in Fig. 41.27 shows that elements with low mass number can produce energy by fusion. In contrast, the falling part of the curve shows that very heavy elements such as uranium can produce energy by *fission* of their nuclei to nuclei of *smaller* mass number (see p. 897).

For fusion to take place, the nuclei must at least overcome electrostatic repulsion when approaching each other. Consequently, for practical purposes, fusion reactions can best be achieved with the lightest elements such as hydrogen, whose nuclei carry the smallest charges and hence repel each other least.

In attempts to obtain fusion, isotopes of hydrogen such as deuterium, $^2_1$H, and tritium, $^3_1$H, are heated to tens of millions of degrees centigrade. The thermal energy of the nuclei at these high temperatures is sufficient for fusion to occur. One technique of promoting this *thermonuclear reaction* is to pass enormously high currents through the gas, which heat it. A very high percentage of the atoms are then ionised and the name *plasma* is given to the gas. The interstellar space of the *aurora borealis* contains a weak form of plasma, but the interior of stars contains a highly concentrated form of plasma. The gas discharge consists of parallel currents, carried by ions, and the powerful magnetic field round one current due to a neighbouring current draws the discharge together (see p. 712). This is the so-called 'pinch effect'. The plasma, however, wriggles and touches the sides of the containing vessel, thereby losing heat. The main difficulty in thermonuclear experiments in the laboratory is to retain the heat in the gas for a sufficiently long time for a fusion reaction to occur. The stability of plasma is now the subject of considerable research.

It is believed that the energy of the sun is produced by thermonuclear reactions in the heart of the sun, where the temperature is many millions of degrees centigrade. Bethe has proposed a cycle of nuclear reactions in which, basically, protons are converted to helium by fusion in the sun, with the liberation of a considerable amount of energy.

Exercises 41

Radioactivity

1. A Geiger-Müller tube has (i) an anode wire of small radius, (ii) some halogen vapour. Explain the reasons for these features.

Give *two* reasons why the GM tube count rate is less than the rate of disintegration of the radioactive atoms in the specimen investigated.

2. Explain why an α-particle produces a straight-line track in a cloud chamber whereas a β-particle produces a straggly track.

What roughly is the range of an α-particle in air at normal pressure?

3. What is meant by the *half-life* of a radioactive element? Draw a labelled sketch of the relation $N = N_0e^{-\lambda t}$ to illustrate your answer.

The initial number of atoms in a radioactive element is 6.0×10^{20} and its half-life is 10 h. Calculate (a) the number of atoms which have decayed in 30 h, (b) the amount of energy liberated if the energy liberated per atom decay is 4.0×10^{-13} J.

4. What is *γ-radiation*? How, simply, would you distinguish between γ-radiation and a beam of β-particles?

To verify the inverse-square law for γ-radiation, the count rate C is measured by a GM tube for different distances r from the specimen S to the *end window* of the tube. Explain how the measurements are used to verify the law if the distance from S to the place inside the tube where ionisation occurs is $(r + a)$ where a is a constant.

5. The radioactive decay of an element is expressed by the relation $dN/dt = -\lambda N$.

From $N = N_0e^{-\lambda t}$, show that the half-life $T_{1/2} = 0.693/\lambda$.

6. A radioactive isotope X is present as 1% of the atoms of a monatomic element Y, which has a relative atomic mass 80. Assuming the Avogadro constant is 6.0×10^{23} mol^{-1}, calculate (i) the number of atoms of X in 10 g of Y, (ii) the rate of decay (dN/dt) of X after 10 h if its half-life is 5 h. (Use the relation $dN/dt = -\lambda N$.)

7. Give concisely the important facts about mass, charge and velocity associated with α, β and γ-radiations respectively. State the effect, if any, of the emission of each of these radiations on (a) the mass number and (b) the atomic number, of the element concerned.

Describe an experiment *either* to measure the range of α-particles in air *or* to verify that the intensity of γ-radiation varies inversely at the square of the distance from the source, being provided with a suitable radioactive source for the experiment chosen. (L.)

8. (a) A sample initially consisting of N_0 radioactive atoms of a single isotope. After a time t the number N of radioactive atoms of the isotope is given by $N = N_0 e^{-\lambda t}$.

(i) Sketch a graph of this equation and show on the graph the time equal to the half-life of the sample, $T_{1/2}$. (ii) Explain what is meant by the disintegration rate of the sample and represent this on the graph at zero time and at time $T_{1/2}$. State the ratio of these two disintegration rates. (iii) Explain the physical significance of the constant λ in the equation above.

(b) If you were provided with a small gamma ray source of very long half-life, describe the arrangement you would use, and the measurements you would make to investigate the inverse-square law for the gamma rays. Show how you would use your measurements to verify this law. (N.)

9. Describe how you would investigate the absorption of beta particles from a source of long half-life by different thicknesses of aluminium. Sketch a graph of the results you would expect to obtain and comment on any special features of the graph.

A small source of beta particles is placed on the axis of a Geiger-Müller tube and a few centimetres from the window of the tube. State and explain three reasons why the observed count rate is less than the disintegration rate of the source.

A source, of which the half-life is 130 days, contains initially 1.0×10^{20} radioactive atoms, and the energy released per disintegration is 8.0×10^{-13} J. Calculate (a) the activity of the source after 260 days have elapsed and (b) the total energy released during this period. (N.)

10. Describe the structure of a Geiger-Müller tube. Why are some tubes fitted with thin end windows? Why does the anode of a Geiger-Müller tube have to be made of a *thin* wire?

Explain the principle of operation of a cloud chamber. Describe and explain the differences between the tracks formed in such a chamber by alpha and beta particles.

A radioactive source has decayed to 1/128th of its initial activity after 50 days. What is its half-life? (L.)

11. Explain the meaning of the terms *element, nucleus, isotope* and *half-life*.

A nucleus $_Z^A X$ undergoes radioactive decay to a daughter nucleus $_{Z+1}^A Y$. Discuss what emissions may be produced and how one might test for their presence.

The counting rates registered at various times after preparation of a sample of $_Z^A X$ are:

Time (min)	0	10	20	30	40	50	100	200	300	400	500
Counting rate (counts per second)	5000	3710	2670	1980	1530	1095	325	95	102	94	109

By suitably plotting these data, determine the half-life of $_Z^A X$. Estimate the uncertainty in your results. (O. & C.)

12. What do you understand by *radioactivity* and *half-life*? Plot an accurate graph to show how the number of radioactive atoms of a given element (expressed as a percentage of those initially present) varies with time. Use a time scale extending over five half-lives.

The isotope $_{19}^{40}K$, with a half-life of 1.37×10^9 years, decays to $_{18}^{40}Ar$, which is stable. Moon rocks from the Sea of Tranquillity show that the ratio of these potassium atoms to argon atoms is 1/7. Estimate the age of these rocks, stating clearly any assumptions you make. Certain other rocks give a value of 1/4 for this ratio. By means of your graph, or otherwise, estimate their age. (C.)

13. Discuss the assumptions on which the law of radioactive decay is based.

What is meant by the *half-life* of a radioactive substance?

A small volume of a solution which contained a radioactive isotope of sodium had an activity of 12 000 disintegrations per minute when it was injected into the bloodstream of a patient. After 30 hours the activity of 1.0 cm^3 of the blood was found to be 0·50 disintegrations per minute. If the half-life of the sodium isotope is taken as 15 hours, estimate the volume of blood in the patient. (*N.*)

14. List the chief properties of α-radiation, β-radiation and γ-radiation. Describe a type of Geiger-Müller tube that can be used to detect all three types of radiation. If you were supplied with a radioactive preparation that was emitting all three types of radiation, describe and explain how you would use the tube to confirm that each of them was present.

A radon ($^{222}_{86}$Rn) nucleus, of mass $3·6 \times 10^{-25}$ kg, decays by the emission of an α-particle of mass $6·7 \times 10^{-27}$ kg and energy $8·8 \times 10^{-13}$ J. (*a*) Write down the values of the mass number A and the atomic number Z for the resulting nucleus. (*b*) Calculate the momentum of the emitted α-particle. (*c*) Find the velocity of recoil of the resulting nucleus. (*O.*)

15. What is meant by the *half-life period* (*half-life*) of a radioactive material?

Describe how the nature of α-particles has been established experimentally.

The half-life period of the body polonium-210 is about 140 days. During this period the average number of α-emissions per day from a mass of polonium initially equal to 1 microgram is about 12×10^{12}. Assuming that one emission takes place per atom and that the approximate density of polonium is 10 g cm^{-3}, estimate the number of atoms in 1 cm^3 of polonium. (*N.*)

Nucleus. Nuclear Energy

16. (*a*) Describe in terms of nuclear structure the three different stable forms of the element neon, which have nucleon numbers 20, 21 and 22. Why is it impossible to distinguish between these forms chemically?

A beam of singly ionised atoms is passed through a region in which there are an electric field of strength E and a magnetic field of flux density B at right angles to each other and to the path of the atoms. Ions moving at a certain speed v are found to be undeflected in traversing the field region. Show (i) that $v = E/B$ and (ii) that these ions can have any mass.

If the emerging beam contained ions of neon, suggest how it might be possible to show that all three forms of the element were present.

(*b*) Explain what is meant by the *binding energy* of the nucleus. Calculate the binding energy per nucleon for $^4_2$He and $^3_2$He. Comment on the difference in these binding energies and explain its significance in relation to the radioactive decay of heavy nuclei. (Mass of $^1_1$H = 1·007 83 u. Mass of $^1_0$n = 1·008 67 u. Mass of $^3_2$He = 3·016 64 u. Mass of $^4_2$He = 4·003 87 u. 1 u ≡ 931 MeV.) (*L.*)

17. The symbol for one isotope of uranium is $^{238}_{92}$U. What does this tell you about the structure of its nucleus? What is meant by the *binding energy* of the nucleus? Using the values given below calculate its value for $^{238}_{92}$U.

The $^{238}_{92}$U atom can decay, with the emission of an alpha-particle, to give $^{234}_{90}$Th. By considering energies, explain why this decay is possible.

In a certain experiment electrons are emitted by a hot cathode in a valve containing xenon at low pressure. These electrons are accelerated by an applied voltage and are liable to collide with xenon atoms. When the electron speeds are low they are liable to have perfectly elastic collisions with the atoms, but when the speeds reach a certain critical value the collisions are no longer perfectly elastic. Explain this. (The following atomic masses may be used: $^{238}_{92}$U = 238·0508 u, $^{234}_{90}$Th = 234·0436 u, $^1_0$n = 1·008665 u, $^1_1$p = 1·007825 u, $^4_2$He = 4·0026 u. 1 u = 931 MeV.) (*L.*)

18. (*a*) Sketch, on the same diagram, the paths of three alpha particles of the same energy which are directed towards a nucleus so that they are deflected through (i) about 10°, (ii) 90°, (iii) 180° respectively.

(*b*) For the deflection of 180°, describe in qualitative terms how (i) the kinetic energy, (ii) the potential energy of the alpha particle varies during its path, assuming the nucleus remains stationary.

(*c*) If, in (*b*) above, the alpha particle has an initial kinetic energy of $1·60 \times 10^{-13}$ J and the nucleus has a charge of $+50$ e, calculate the nearest distance of approach of the alpha particle to the nucleus. (Magnitude of electronic charge, $e = 1·60 \times 10^{-19}$ C, permittivity of free space, $\varepsilon_0 = 10^{-9}/36\pi$ or $8·85 \times 10^{-12}$ F m^{-1}.) *N.*)

19. Radioactive decay occurs by either α or β emission. Write down the general equation for the decay on an isotope X with nucleon number A and proton number Z by (a) α emission, (b) β emission.

Parts of a radioactive series consists of the following sequence of emissions: α, α, β, α, β, α. Draw this part of the series of a N against Z graph. (N is the number of neutrons in a nucleus.) What are the total changes in N and Z during this part of the series?

How does the ratio $N:Z$ vary for stable nuclei throughout the Period Table? Discuss the relevance of this variation to (i) the fact that in a radioactive decay series the α emissions are interspersed with β emissions, and (ii) the likely radioactivity of the fission fragments in a nuclear fission.

The isotopes with the longest half-lives occurring in the four natural radioactive decay series have half-lifes of $1\cdot4 \times 10^{10}$ years, $4\cdot5 \times 10^9$ years, $7\cdot1 \times 10^8$ years and $2\cdot2 \times 10^6$ years respectively. What is the significance of the fact that only those series containing the three half-lives occur naturally in the earth? (L.)

20. (a) Explain the meaning of the term *mass difference* and state the relationship between the mass difference and the *binding energy* of a nucleus. (b) Sketch a graph of nuclear binding energy per nucleon versus mass number for the naturally occurring isotopes and show how it may be used to account for the possibility of energy release by nuclear fission and nuclear fusion.

(c) The sun obtains its radiant energy from a thermonuclear fusion process. The mass of the sun is 2×10^{30} kg and it radiates 4×10^{23} kW at a constant rate. Estimate the life time of the sun, in years, if $0\cdot7\%$ of its mass is converted into radiation during the fusion process and it loses energy only by radiation. (1 year may be taken as 3×10^7 s.) The speed of light, $c = 3 \times 10^8$ m s^{-1}. (N.)

21. (a) Z protons and N neutrons are combined to form a nucleus $^{Z+N}_{Z}X$. Describe the energy changes which occur as the $(Z + N)$ free particles are combined. Explain the concept of *binding energy per nucleon* which arises in this description. Sketch a graph of binding energy nucleon against nucleon number and use this graph to explain how the process of nuclear fission and nuclear fusion are possible, and the values of nucleon number at which they may occur.

(b) A typical fission reaction is

$$^{235}_{92}U + {}^{1}_{0}n \rightarrow {}^{148}_{57}La + {}^{85}_{35}Br + \text{neutrons.}$$

How many neutrons are released in this reaction? What is the importance of these neutrons in a nuclear reactor? Why are the products of lanthanum (La) and bromine (Br) likely to be radioactive and what type of radioactivity are they likely to exhibit? (L.)

22. Explain why the mass of a nucleus is always less than the combined masses of its consitutent particles.

Naturally occurring chlorine is a mixture of two isotopes, $^{35}_{17}Cl$ with a relative abundance of 75% and $^{37}_{17}Cl$ with a relative abundance of 25%. Explain fully what is meant by this statement, and describe briefly how the relative abundance of two isotopes could be verified experimentally.

When natural chlorine is irradiated with slow neutrons from a reactor, another isotope of chlorine of mass number 38 is produced with decays by β^- emission. What nuclear reaction would you expect to be responsible for producing this isotope, and what are the mass number and atomic number of the nucleus remaining after the β^- decay? (O. & C.)

23. Explain what is meant by *nuclear fission* and show how nuclear chain reactions can arise. Explain why energy is also released in nuclear fission and why it occurs only with heavy nuclei.

Outline an experimental method by which the half-life of a radioactive isotope emitting α-particles can be measured, stating how a result of high accuracy can be achieved. (You may assume that the half-life is of the order of hours.)

The $^{238}_{92}U$ nucleus decays according to $^{238}_{92}U \rightarrow {}^{234}_{90}Th + {}^{4}_{2}He$. Determine the kinetic energy of the emitted α-particle. (Mass of $^{238}_{92}U$ nucleus $= 3\cdot853\,95 \times 10^{-25}$ kg, mass of $^{234}_{90}Th$ nucleus $= 3\cdot787\,37 \times 10^{-25}$ kg, mass of α-particle $= 6\cdot648\,07 \times 10^{-27}$ kg.) (O. & C.)

24. (a) Explain (i) atomic number, (ii) mass number, (iii) isotope. (b) Account for the fact that, although nuclei do not contain electrons, some radioactive emit β-particles. (c) Cobalt has only one stable isotope ^{59}Co. What form of radioactive decay would isotope ^{60}Co expect to undergo?

(d) Radioactive nucleus $^{210}_{84}$Po emits α-particles of single energy, the product nuclei being $^{206}_{82}$Pb. (i) Using the data below, calculate the energy in MeV released in each disintegration. (ii) Explain why this energy does not all appear as kinetic energy, E, of the α-particle. (iii) Calculate E, taking integer values of nuclear masses.

Nucleus: $^{210}_{84}$Po, 		$^{206}_{82}$Pb, 		α-particle

Mass (u): 209·936 70, 205·929 421, 4·001 504 (1 atomic mass unit, u = 931 MeV) (N.)

25. A beam of protons, accelerated through a potential difference of 250 kV and equivalent to a current of 1 μA, strikes a thin target of $^{7}_{3}$Li atoms. Alpha particles are emitted from the target in pairs at the rate of 6 pairs per minute.

(i) Calculate the number of protons striking the target per alpha-pair produced. (ii) Write down the equation for the nuclear reaction which occurs. (iii) Calculate the total kinetic energy (in joules) of a pair of alpha particles. (iv) Calculate the closest distance of approach of a proton to a $^{7}_{3}$Li nucleus, assuming the nucleus to behave as a point charge and taking the electrical potential at a distance r from a point charge Q to be $Q/4\pi\varepsilon_0 r$.

(v) Calculate the ratio of the energy of one alpha-pair to the energy of one proton, and the ratio of the total kinetic energy per minute incident on the target to the total kinetic energy per minute leaving it, and comment on the results.

Mass of $^{1}_{1}$H = 1·673516 × 10^{-27} kg, and mass of $^{7}_{3}$Li = 11·650250 × 10^{-27} kg, mass of $^{4}_{2}$He = 6·646263 × 10^{-27} kg. Electronic charge = 1·6 × 10^{-19} C. (O. & C.)

26. Describe an experiment which justifies the view that an atom contains a small central nucleus. Explain why nuclear reactions can sometimes be used as a source of energy. Distinguish between the processes of nuclear fusion and nuclear fission, giving an example of each process.

The energy liberated in the fission of a single uranium 235 atom is 3·2 × 10^{-11} J. Calculate the power production corresponding to the fission of 1 g of uranium per day. (Assume Avogadro constant = 6·0 × 10^{23} mol^{-1}.) (L.)

27. Explain what is meant by (a) the mass defect, (b) the binding energy of an atomic nucleus.

The binding energy of the isotope of hydrogen $^{3}_{1}$H is greater than that of the isotope of helium $^{3}_{2}$He. Suggest a reason for this. (N.)

28. Define *nucleon number* (*mass number*) and *proton number* (*atomic number*) and explain the term *isotope*. Describe a simple form of mass spectrometer and indicate how it could be used to distinguish between isotopes.

In the naturally occurring radioactive decay series there are several examples in which a nucleus emits an α-particle followed by two β-particles. Show that the final nucleus is an isotope of the original one. What is the change in mass number between the original and final nuclei? (L.)

29. Explain the term *nuclear binding energy*. Sketch a graph showing the variation of binding energy per nucelon with nucleon number (mass number) and show how both nuclear fission and nuclear fusion can be explained from the shape of this curve.

Calculate in MeV the energy liberated when a helium nucleus ($^{4}_{2}$He) is produced (a) by fusing two neutrons and two protons, and (b) by fusing two deuterium nuclei ($^{2}_{1}$H). Why is the quantity of energy different in the two cases?

(The neutron mass is 1·008 98 u, the proton mass is 1·007 59 u, the nuclear masses of deuterium and helium are 2·014 19 u and 4·002 77 u respectively. 1 u is equivalent to 931 MeV.) (L.)

30. Using the information on atomic masses given below, show that a nucleus of uranium 238 can disintegrate with the emission of an alpha particle according to the reaction:

$$^{238}_{92}U \rightarrow \, ^{234}_{90}Th + \, ^{4}_{2}He.$$

Calculate (a) the total energy released in the disintegration, (b) the kinetic energy of the alpha particle, the nucleus being at rest before disintegration.

Mass of ^{238}U = 238·124 92 u. Mass of ^{234}Th = 234.116 50 u. Mass of ^{4}He = 4.003 87 u. 1 u is equivalent to 930 MeV. (N.)

Revision Papers

The questions in these papers cover basic topics in all branches of the subject.

Paper 1

1. A block of wood of mass 1·00 kg is suspended freely by a thread. A bullet of mass 10 g is fired horizontally at the block and becomes embedded in it. The block swings to one side, rising a vertical distance of 50 cm. With what speed did the bullet hit the block? (Acceleration of free fall, $g = 10$ m s^{-2}) (*L.*)

2. A graph of load against extension is plotted from the results obtained in stretching a vertical wire, before the elastic limit is reached. Show how (i) Young modulus, (ii) the energy stored in the wire can be found from the graph.
   A steel wire of length 2 m and cross-sectional area 2×10^{-6} m^2 is stretched by a load of 50 N. Calculate the energy stored in the wire if the elastic limit is not reached. (Young modulus for steel $= 2 \cdot 0 \times 10^{11}$ N m^{-2}.)

3. Derive the expression $p = \frac{1}{3}\rho c^2$ for the pressure p of a gas, stating the assumptions you make.
   If the mass of 1 mole of hydrogen is 2 g, and this occupies a volume of 0·022 m^3 at 273 K and 10^5 N m^{-2} pressure, calculate the root-mean-square velocity of hydrogen molecules at 546 K.

4. When light from the sun is examined using a spectrometer, the spectrum is crossed by dark lines. Explain briefly why this is so. (*L.*)

5. Light consisting of two wavelengths which differ by 160 nm passes through a diffraction grating with $2 \cdot 5 \times 10^5$ lines per metre. In the diffracted light the third order of one wavelength coincides with the fourth order of the other. What are the two wavelengths and at what angle of diffraction does this occur? (*L.*)

6. A battery and parallel plate capacitor C are connected in series, as shown in the diagram. One plate of the capacitor is fixed and the other oscillates, altering the distance between the plates with a small amplitude.
   Explain why there would be an alternating current in the circuit. (*L.*)

Fig. 1

7. A beam of protons is accelerated from rest through a potential difference of 2000 V and then enters a uniform magnetic field which is perpendicular to the direction of the proton beam. If the flux density is 0·2 T, calculate the radius of the path which the beam describes. (Proton mass $= 1 \cdot 7 \times 10^{-27}$ kg; electronic charge $= -1 \cdot 6 \times 10^{-19}$ C.) (*L.*)

8. A point source of γ-radiation has a half-life of 30 minutes. The initial count rate, recorded by a Geiger counter placed 2·0 m from the source, is 360 s^{-1}. The distance between the counter and the source is altered. After 1·5 hour the count rate recorded is 5 s^{-1}. What is the new distance between the counter and the source? (*L.*)

9. If a bar of copper of uniform area of cross-section is well lagged and its ends maintained at different temperatures, the temperature gradient along the bar is uniform. Explain why this is so.
   Discuss the effect on the temperature gradient if the cross-sectional area of the bar had not been uniform but had increased uniformly, being greater at the hot end of the bar. Illustrate your answer with a graph. (*L.*)

10. Give an expression for the electric field strength at a distance d from an isolated point charge Q. Define the term *potential* at any point in the field. Sketch a graph showing how the potential changes with distance from the charge.

One plate of a parallel plate capacitor is earthed and the other plate has a positive charge. For the region between the plates sketch a graph of potential against distance from the earthed plate if the plates are close together.

How would you find the electric field strength at any point from these graphs? Justify your statement by deducing the relationship between field strength and potential gradient.

A parallel plate capacitor with air as a dielectric has plates of area 4.0×10^{-2} m^2 which are 2.0 mm apart. It is charged by connecting it to a 100 V battery. It is then disconnected from the battery and connected in parallel with a similar uncharged capacitor with plates of half the area which are twice the distance apart. Calculate the final charge on each capacitor. Edge effects may be neglected. (Permittivity of air = 8.8×10^{-12} F m^{-1}.) (L.)

Paper 2

1. A body moving with simple harmonic motion has velocity v and acceleration a when the displacement from its mean position is x. Sketch graphs (i) of a against x, and (ii) of v against x. (L.)
2. A glass capillary tube is dipped into a beaker of a liquid which wets the glass. The length of tube above the surface of the liquid is considerably less than the height of capillary rise to be expected for a tube of its diameter. The liquid rises to the top of the capillary tube, but does not overflow.

   Describe and explain the shape of the meniscus at the top of the tube. (L.)
3. Draw a graph showing the distribution of energy in the spectrum of a black body. Explain what quantity is plotted against the wavelength.

   By considering how this energy distribution varies with temperature, explain the colour changes which occur when a piece of iron is heated from cold to near its melting point. (L.)
4. An object is placed 0.15 m in front of a converging lens of focal length 0.10 m. It is required to produce an image on a screen 0.40 m from the lens on the opposite side to the object. This is to be achieved by placing a second lens midway between the first lens and the screen.

   What type and of what focal length should this lens be? Sketch a diagram showing two rays from a non-axial point on the object to the final image. (L.)
5. A source of sound of frequency 2500 Hz is placed in front of a flat wall. If a microphone is moved from the source directly towards the wall, a series of minimum values in its output is observed at equally spaced points.

   Why does this effect occur? Calculate the separation of these points if the speed of sound in air at this temperature is 330 m s^{-1}. (L.)
6. A galvanometer of resistance 40 Ω requires a current of 10 mA to give a full scale deflection. A shunt is put in position to convert it to a meter reading up to 1.0 A full scale deflection. A resistance bobbin, intended to convert the galvanometer to one reading up to 1.0 V full scale deflection, is now attached to the instrument but the shunt is inadvertently left in position.

   By considering the potential difference across the bobbin and the meter, calculate the voltage which would produce full scale deflection of the meter. (L.)
7. Explain why a glowing gas, such as that in a neon tube, gives only certain wavelengths of light, and why that gas is capable of absorbing the same wavelengths. (L.)
8. Explain (i) mass number, (ii) atomic number, (iii) isotope.

   Nuclei do not contain electrons. Account for the fact that β-particles may be emitted from some radioactive nuclei. What effect has an emission of a β-particle on the mass number and atomic number of the atom concerned?
9. Using Huygens' principle of secondary wavelets explain, making use of a diagram, how a refracted wavefront is formed when a beam of light, travelling in glass, crosses

the glass–air boundary. Show how the sines of the angles of incidence and refraction are related to the speeds of light in air and glass (L.)

10. Define the coulomb. Deduce an expression for the current I in a wire in terms of the number of free electrons per unit volume, n, the area of cross-section of the wire, A, the charge on the electron, e, and its drift velocity, v.

A copper wire has 1.0×10^{29} free electrons per cubic metre, a cross-sectional area of 2.0 mm^2 and carries a current of 5.0 A. Calculate the force acting on each electron if the wire is now placed in a magnetic field of flux density 0.15 T which is perpendicular to the wire. Draw a diagram showing the directions of the electron velocity, the magnetic field and this force on an electron.

Why would the force be greater for the same current if a similar wire of semi-conducting material were used? Explain, without experimental detail, how this effect could be used to determine whether a slab of semiconducting material was n-type or p-type. (Charge on an electron $= -1.6 \times 10^{-19}$ C.) (L.)

Paper 3

1. A man is able to jump vertically 1.5 m on earth. What height might he be expected to jump on a planet of which the density is one third that of the earth but of which the radius is one half that of the earth? (L.)

2. A mass of 0.4 kg is attached to the end of a spring of force constant 20 N m^{-1} (force constant = force per unit extension). The mass and spring are placed on a smooth horizontal table and one end of the spring is fixed. If the mass is displaced 0.02 m along the axis of the spring and released, calculate its period of oscillation and the maximum energy of the spring.

3. In an experiment to measure the thermal conductivity of a bad conductor, explain why (i) the material is made thin, (ii) it is sandwiched between thick metal plates.

The heat loss from a room through a glass window of area 2 m^2 and thickness 4 mm is 6×10^5 J min^{-1}. Calculate the air temperature outside the window if the room temperature is $17°$C. (Thermal conductivity of glass $= 1.0$ W m^{-1} K^{-1}.)

4. A lens of focal length 1.00 m and one of focal length 0.05 m are arranged as a simple astronomical telescope so that the final image is formed at infinity.

(i) Draw a ray diagram showing how the final image is formed. (ii) If the objective has a diameter of 0.10 m, calculate the diameter of the eyering.

State the advantages of using an objective of greater diameter than 0.10 m.

5. In Young's experiment on interference, (i) why are coherent sources used, (ii) what is the appearance of the bands if the light source is white, (iii) what effect is produced on the bands in (ii) if one slit is covered with a thin piece of glass?

In a Young's experiment, light of wavelength 6.0×10^{-7} m is used, the slits are 0.6 mm apart and the bright bands formed on a screen S are 0.8 mm apart. Calculate the distance of S from the slits.

6. A metal disc rotates anticlockwise at an angular velocity ω in a uniform magnetic field which is directed into the paper in the diagram (Fig. 2), and covers the whole of the disc. A and B are brushes.

By considering an electron between the brushes show that a potential difference exists between A and B.

Explain what changes, if any, will be needed in the torque required to keep the disc rotating at the same angular velocity ω if A and B are connected by a resistor. (L.)

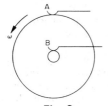

Fig. 2

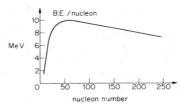

Fig. 3

7. How does Einstein's theory explain the photoelectric effect?

   The frequency of incident radiation on a metal surface is $5\cdot0 \times 10^{14}$ Hz and electrons with maximum energy $2\cdot3 \times 10^{-19}$ J are emitted. What wavelength of incident radiation is required to liberate electrons with maximum energy $5\cdot6 \times 10^{-19}$ J? (Planck constant $h = 6\cdot6 \times 10^{-34}$ J s, $c = 3\cdot0 \times 10^8$ m s^{-1}.)

8. Figure 3 shows roughly how the variation of binding energy (B.E.) per nucleon varies with nucleon (mass) number. See p. 913.

   How does the graph show that nuclear fission and nuclear fusion can both produce energy?

9. A constant voltage a.c. generator, of 20 V r.m.s. and variable frequency, is connected in series with a resistor of resistance $2\cdot0\ \Omega$, a coil of inductance $5\cdot0$ H and a capacitor of capacitance $2\cdot0\ \mu$F. The frequency is adjusted until the current in the circuit has a maximum value of $2\cdot0$ A r.m.s. Calculate the resistance of the wire of the inductor and the value of this frequency. (Assume $\pi^2 = 10$.) (L.)

10. Describe the *principle* of the experiment which established the nuclear model of the atom, explaining how the deduction is made from the observations.

    The emission spectrum of the hydrogen atom consists of a series of lines. Explain why this suggests the existence of definite energy levels for the electron in the atom. By considering the intervals between the energy levels explain the spacing of the lines in the visible hydrogen spectrum.

    The ionisation potential of the hydrogen atom is $13\cdot6$ V. Use the data below to calculate (a) the speed of an electron which could just ionise the hydrogen atom, and (b) the minimum wavelength which the hydrogen atom can emit. (Charge on electron $= -1\cdot60 \times 10^{-19}$ C, mass of electron $= 9\cdot11 \times 10^{-31}$ kg, Planck constant $= 6\cdot63 \times 10^{-34}$ J s, speed of light $= 3\cdot00 \times 10^8$ m s^{-1}.) (L.)

Paper 4

1. Assuming that the human body may be regarded as a cylinder, make estimates of the necessary quantities and calculate an approximate value for the ratio of the angular velocity of an ice skater spinning with his arms extended to the angular velocity when his arms are pulled in close to his body. (The moment of inertia of a cylinder of mass M and radius r about its axis is $Mr^2/2$ and the moment of inertia of a rod of mass m and length $2l$ about an axis through its midpoint at right angles to its length is $ml^2/3$.) (L.)

2. (a) A heavy rigid bar is supported horizontally from a fixed support by two vertical wires, A and B, of the same initial length and which experience the same extension. If the ratio of the diameter of A to that of B is 2 and the ratio of Young's modulus of A to that of B is 2, calculate the ratio of the tension in A to that in B.

   (b) If the distance between the wires is D, calculate the distance of the wire A from the centre of gravity of the bar. (N.)

3. The value of the property X of a certain substance is given by

$$X_t = X_0 + 0\cdot50t + (2\cdot0 \times 10^{-4})t^2,$$

   where t is the temperature in degrees Celsius measured on a gas thermometer scale. What would be the Celsius temperature defined by the property X which corresponds to a temperature of 50°C on this gas thermometer scale? (L.)

4. Explain what is meant by *longitudinal waves, transverse waves, plane-polarized waves.* Can sound waves in a gas be polarized? Explain your answer.

   A stationary source of sound of frequency 1100 Hz is approached by an observer moving at a velocity of 30 m s^{-1}. What frequency is heard by the observer? (Velocity of sound $= 330$ m s^{-1}.)

5. Draw a sketch of a *reflector telescope*, showing how the image is viewed by an observer. What is the advantage of a reflector telescope over a refractor telescope?

   A person has a near point of 50 cm and this requires correction to 25 cm. Calculate The power in dioptres of the lens required, and draw a diagram showing the correction.

6. Explain the purpose of (i) the soft iron cylinder (core) and (ii) the aluminium frame

on which the coil is wound, in the moving-coil meter for measuring current. Draw diagrams to illustrate your answers.

7. Explain how the minimum wavelength emitted by a hot cathode X-ray tube can be controlled.

Calculate the minimum wavelength obtained from an X-ray tube with an anode voltage of 33 000 V. (Planck constant $h = 6.6 \times 10^{-34}$ J s, $e = 1.6 \times 10^{-19}$ C, $c = 3 \times 10^8$ m s^{-1}.)

8. (i) What is meant by *elastic* and *inelastic* collisions between electrons and atoms of a gas? When does an inelastic collision take place?

(ii) Calculate the binding energy per nucleon in MeV of the alpha-particle, given that: Mass of neutron = 1.0090 u, mass of proton = 1.0076 u, mass of alpha-particle = 4.0026 u, and 1 u = 930 MeV.

9. A simple pendulum, suspended from a fixed point, consists of a light cord of length 500 mm and a bob of weight 2.0 N. The bob is made to move in a horizontal circular path. If the maximum tension which the cord can withstand is 5.0 N, show whether or not it is possible for the radius of the path of the bob to be 300 mm. (*L.*)

10. The model of a gas as a large number of elastic bodies moving about in a random manner is the basic idea of the kinetic theory of gases. In terms of this model, explain: (*a*) what is meant by an ideal gas, (*b*) how a gas exerts a pressure when enclosed in a container, (*c*) why the atmospheric pressure decreases with height, (*d*) how the atmosphere, which is not in a container, exerts pressure at all.

Which of the assumptions made to develop a quantitative expression for the pressure of an ideal gas require to be modified to explain the behaviour of a real gas? Illustrate your answer by considering a *p-V* isothermal for an ideal gas and a *p-V* isothermal for a real gas at a temperature below its critical value.

A series of experiments was performed by Andrews to obtain *p-V* isothermals for carbon dioxide. Sketch a set of *p-V* isothermals for water, noting particularly any dissimilarities between the curves for water and carbon dioxide. (*L.*)

Paper 5

1. A mercury barometer, with a scale attached, has a little air above the mercury. The top of the tube is 1.00 m above the level of the mercury in the reservoir. When the tube is vertical the height of the mercury column is 700 mm. When the tube is inclined at 60° to the vertical the reading of the mercury level on the scale is 950 mm. To what height would the mercury have risen in the vertical tube had there not been any air in it? (*L.*)

2. When two moving spheres collide their total kinetic energy is not conserved. Explain this statement.

What conservation law is obeyed in this collision? Give an explanation of this law.

3. A sealed vessel contains a mixture of air and water vapour in contact with water. The total pressures in the vessel at 27°C and 60°C are respectively 1.0×10^5 Pa and 1.3×10^5 Pa. If the saturated vapour pressure of water at 60°C is 2.0×10^4 Pa, what is its value at 27°C? (1 Pa = 1 N m^{-2}.) (*L.*)

4. What is Huygens' principle? Show how Snell's law of refraction is explained by the wave theory.

The wavelength of yellow light in air is 6.0×10^{-7} m. Calculate its wavelength in water of refractive index 4/3, giving an explanation.

5. Two glass plates in contact at one end form a wedge-shaped film of air. Explain (i) why straight interference bands are produced when the film is illuminated by monochromatic light incident normally on the air film, (ii) why a dark band is produced at the line of intersection of the plates.

Interference bands 1.50 mm apart were obtained using monochromatic light of wavelength 6.0×10^{-7} m when the plates were separated by a thin piece of foil at one end 10.00 cm from the line of contact. Calculate (i) the angle in radians of the air wedge and (ii) the thickness of the foil.

6. A coil of wire has a *resistance*, a *reactance* and an *impedance*. Explain the meaning of this statement, and state the relationship between these quantities.

A coil with resistance R and inductance L is in series with a capacitor C. An alternating voltage V is connected to the circuit and its frequency f is varied from low to high values while V is kept constant.

Describe and explain how the current flowing varies with the frequency f.

7. Explain why the capacitance of a parallel-plate capacitor increases when (i) the plates are closer together and (ii) a dielectric fills the air space between the plates.

A p.d. of 100 V is connected between the plates of a parallel-plate air capacitor and the connections then removed. The area of each plate is 0.04 m^2 and their separation is 0.01 m. Calculate (i) the electric intensity between the plates, (ii) the charge on each plate, (iii) the energy stored in the capacitor. $(\varepsilon_0 = 8.85 \times 10^{-12} \text{ F m}^{-1}.)$

8. A moving electric charge enters (i) a uniform electric field perpendicular to its direction, (ii) a uniform magnetic field perpendicular to its direction. Write down in each case (a) the force acting on the charge, explaining any symbols used, (b) the path of the charge in the field.

Why can energy be gained by the charge moving through the electric field but not through the magnetic field?

Describe briefly an experiment to measure the *charge–mass ratio* (specific charge) of electrons.

9. Two parallel wires have currents passed through them which are in the same direction. Draw a diagram showing the directions of the currents and of the forces on the wires.

Alternating currents are now passed through the wires. Explain what forces would act if the currents were (a) in phase, and (b) out of phase by π rad. (*L*.)

10. What do you understand by (a) forced vibrations, (b) free vibrations, and (c) resonance? Illustrate your answer by giving three distinct examples, one each of (a), (b) and (c).

Explain how a stationary sound wave may be set up in a gas column and how you would demonstrate the presence of nodes and antinodes. State what measurements would be required in order to deduce the speed of sound in air from your demonstration and show how you would calculate your result.

The speed of sound, c, in an ideal gas is given by the formula $c = \sqrt{\gamma p/\rho}$ where p is the pressure, ρ is the density of the gas and γ is a constant. By considering this formula explain the effect of a change in (i) temperature and (ii) pressure on the speed of sound. (*L*.)

Multiple Choice Papers

(*Where necessary, assume* $g = 10$ m s^{-2} *or* 10 N kg^{-1}.)

Paper 1 (Time: 1½ hours)
Section I

Questions 1–7
The graphs below represent a quantity y plotted as a function of another quantity x. For each of the Questions 1 to 7, choose one of the graphs A to E which best fits the relationship between y and x.

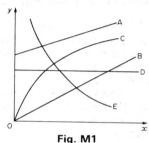

Fig. M1

Quantity y	*Quantity x*
1. pressure of one mole of an ideal gas at constant temperature	volume of one mole
2. velocity of a ball thrown down from a height	time
3. velocity of light in a vacuum	frequency of light
4. electric field strength between parallel plates with given p.d.	distance between plates
5. spacing of fringes in Young's slits experiment	separation of slits
6. current values on closing an inductance-resistance circuit	time
7. energy of photon	frequency

Section II

Questions 8–25
Select the best answer of the five suggested answers following the statement

8. A particle of mass 0·1 kg is rotated at the end of a string in a vertical circle of radius 1·0 m at a constant speed of 5 m s^{-1}. The tension in newton in the string at the highest point of its path is:
   A 0·5   B 1·0   C 1·5   D 3·5   E 15

9. At a distance $2R$ from the centre of the earth, assumed to be a sphere of radius R, the weight of a kilogram mass is 2·5 N. At a distance $3R$ from the centre, the weight of the same mass is
   A 4·75 N   B 3·75 N   C 2·5 N   D 1·1 N   E 0·8 N

10. A simple harmonic oscillator has a period of 0·01 s and an amplitude of 0·2 m. The magnitude of the velocity in m s^{-1} at the centre of oscillation is:
    A 20   B 20π C 100   D 40π E 100π

11. A sphere has a surface area of 0·1 m^2 and a temperature of 400 K, and the power radiated from it is 150 W. Assuming the sphere is a black body radiator, the power in kilowatt radiated when the area expands to 2·0 m^2 and the temperature changes to 800 K is:
    A 144   B 96   C 48   D 16   E 12

12. The molecules of a given mass of gas have a root-mean-square velocity of 200 m s^{-1} at 27°C and $1 \cdot 0 \times 10^5$ N m^{-2} pressure. When the temperature is 127°C and the pressure $0 \cdot 5 \times 10^5$ N m^{-2}, the root-mean-square velocity in m s^{-1} is:

    A $\dfrac{400}{\sqrt{3}}$ B $100\sqrt{2}$ C $\dfrac{100\sqrt{2}}{3}$ D $\dfrac{100}{3}$ E $\dfrac{50\sqrt{2}}{3}$

13. When light travels from air to glass,
    A the frequency is the same and the wavelength in glass is greater than in air
    B the frequency decreases and the wavelength is smaller in glass than in air
    C the frequency is the same and the wavelength is smaller in glass than in air
    D the frequency increases and the speed increases
    E the frequency increases and the speed decreases

14. Glass has a refractive index of 3/2 and water a refractive index of 4/3. If the speed of light in glass is $2 \cdot 00 \times 10^8$ m s^{-1}, the speed in water in m s^{-1} is:
    A $2 \cdot 67 \times 10^8$ B $2 \cdot 25 \times 10^8$ C $1 \cdot 78 \times 10^8$ D $1 \cdot 50 \times 10^8$ E $1 \cdot 33 \times 10^8$

15. A pipe open at both ends has an effective length of $0 \cdot 5$ m. The frequency in Hz of the note one octave above its fundamental frequency, if the velocity of sound in air is 340 m s^{-1}, is:
    A 170   B 340   C 510   D 680   E 1020

16. A small loudspeaker radiates 5 W of power, and the sound intensity 2 m from the speaker is 1 W m^{-2}. If the power of the speaker is doubled, the intensity in W m^{-2} at a distance 4 m from the speaker is:
    A $0 \cdot 5$ B 1   C 2   D 4   E 5

17. In Fig. M2, the amplitude of waves arriving at P from two coherent sources at S and R is a from each source. The speed of the waves is 12 m s^{-1} and the frequency is 6 Hz. The amplitude of the resultant wave at P is
    A 0   B $2a$ C $3a$ D $4a$ E $5a$

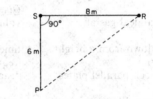

Fig. M2

18. Detergents in hot water enable grease to be removed from plates by
    A changing the angle of contact between grease and plate to an acute angle
    B increasing the temperature of the liquid
    C decreasing the density of the liquid
    D changing the angle of contact between grease and plate to an obtuse angle
    E raising the surface tension of the water

19. In Fig. M3, a current of $1 \cdot 4$ A flows towards the bridge circuit and no current flows in the galvanometer, G. The current in amperes flowing in the 2 Ω resistor is
    A $1 \cdot 4$ B $1 \cdot 2$ C $1 \cdot 0$ D $0 \cdot 6$ E $0 \cdot 3$

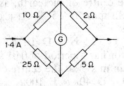

Fig. M3

20. In Fig. M4, the power developed in the 12 Ω resistor is 6 W. The power in W developed in the 8 Ω resistor is:
    A 2   B 4   C 6   D 9   E 16

Fig. M4

21. Two connected charges of $+q$ and $-q$ respectively are a fixed distance AB apart in a non-uniform electric field whose lines of force are shown in Fig. M5.

Fig. M5

The resultant effect on the two connected charges is
   A a couple in the plane of the paper and no resultant force
   B a resultant force in the plane of the paper and no couple
   C a couple normal to the plane of the paper and a resultant force
   D a resultant force normal to the paper and a couple in the plane of the paper
   E a couple in the plane of the paper and a resultant force in the plane of the paper

22. A horizontal coil is connected to a ballistic galvanometer. When the coil is reversed through 180° a deflection θ is obtained in the galvanometer.
   A The induced e.m.f. produced is proportional to the horizontal component of the earth's magnetic field flux density
   B the charge flowing is proportional to θ
   C the current flowing is proportional to θ
   D the induced e.m.f. is independent of the size of the coil
   E the charge flowing depends on the time taken to reverse the coil

23. In Fig. M6, the coils X and Y have the same number of turns and length. Each has a flux density B in the middle, and a flux density $B/2$ at the ends, when carrying the same current I.

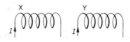

Fig. M6

When the coils are joined together to make a long coil of twice the length of X or Y and a current I now flows through the coil, the flux density in the middle is given by
   A 0   B $B/2$ C B D $3B/2$ E $2B$

24. Figure M7 shows the energy levels P, Q, R, S, T of an atom where T represents the ground state.

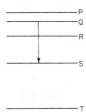

Fig. M7

A red line in the emission spectrum of the atom can be obtained by an energy level change from Q to S as shown. A blue line can be obtained by an energy level change from

A P to Q   B Q to R   C R to S   D T to R   E R to T

25. In 6 days, the number of disintegrated atoms of a radioactive substance is 7/8ths of its original (undisintegrated) atoms. In 10 days, the fraction of the original number of atoms which have disintegrated is

A 31/32   B 77/80   C 15/16   D 71/80   E 29/32

Section III

Questions 26-38

In each of the questions 26-38, one or more of the responses 1, 2, 3 are correct. Decide which of the responses is/are correct and then choose

A if 1, 2, 3 are all correct
B if 1 and 2 only are correct
C if 2 and 3 only are correct
D if 1 only is correct
E if 3 only is correct

Directions summarised				
A	B	C	D	E
1, 2, 3	1, 2	2, 3	1	3
correct	only	only	only	only

26. Figure M8 shows a ray refracted through a glass prism

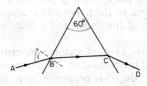

Fig. M8

1 The deviation by the prism is the angle between AB and BC
2 The deviation always decreases when the angle of incidence i increases
3 The deviation is a maximum when angle i is $90°$

27. Some equations in physics are true only under certain conditions. Which of the following is/are true under *all* conditions?
1 Equation for force: $F = ma$
2 Gravitation: $F = GMm/r^2$
3 Wave motion: wavelength $\lambda = c/f$, where c is the wave velocity and f is the frequency

28. The First Law of Thermodynamics may be expressed by the relation $\Delta Q = \Delta U + \Delta W$, where Q is the quantity of heat given to a system of internal energy U and W is the external work done.
1 ΔU depends only on the temperature change.
2 $\Delta W = V.\Delta p$, where V is the volume of the gas and p is the pressure.
3 $\Delta U = \Delta W$ at constant temperature.

29. Figure M9 shows the variation of alternating current I with time t in a pure inductance L.
1 The voltage across L is a minimum at X
2 The average power in L is zero
3 The magnetic field in the inductor is a maximum at Y

Fig. M9

30 A suspended long metal wire is stretched a small distance x in metres by a load W in newtons suspended at the other end.

1 The energy stored in the wire can always be found from the area between the force-extension graph and the extension-axis

2 The energy per unit volume stored in the wire $= \frac{1}{2}$ stress $\times$ strain

3 The loss in potential energy of the load W is equal to the gain in energy of the wire in stretching a length x

31. A sounding tuning fork is placed with its bottom end on the surface of a table so that the prongs continue to vibrate.

   1 The material of the table is not set into vibration

   2 The loudness of the sound is then increased

   3 The vibrations of the prongs then die away more quickly

32. When light is incident normally on a diffraction grating

   1 the central diffraction image is coloured

   2 several coloured spectra are formed on one side only

   3 the colour in each spectrum nearest the central image is blue

33. When an X-ray tube is working

   1 a diode is not necessary to rectify the a.c. mains voltage to d.c. voltage needed for the anode

   2 the emitted X-rays have a maximum frequency

   3 the X-rays characteristic of the metal target have a wavelength which depend on the speed of the incident electrons

34. An air-wedge film is formed by two microscope slides X and Y and illuminated normally by monochromatic blue and red light.

   1 The separation of the blue bands is less than that of the red bands

   2 A coloured band is formed at the end where the slides touch

   3 The bands are formed by interference between waves reflected at the top of X and the bottom of Y

35. A metal disc is rotated about its central axis O in a uniform magnetic field normal to its plane and a galvanometer G is connected to contacts at O and a point P on the circumference.

   1 The induced current in G is due to the cutting of flux by the radius OP

   2 When the disc is rotated twice as fast, the current is doubled

   3 The electrical energy is produced by the work done against an opposing force while the disc is turned

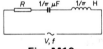

Fig. M10

36. In the a.c. circuit shown in Fig. M10, the supply voltage has a constant r.m.s. value V but variable frequency f. At resonance, the circuit

   1 has a current I given by $I = V/R$

   2 has a resonant frequency of 500 Hz

   3 has a voltage across the capacitor which is 180° out of phase with that across the inductor

37. In the subject of Waves,

   1 Sound waves can sometimes be polarized

   2 Light waves undergo a phase change of π when reflected at a denser medium

   3 X-rays are transverse waves because they can be polarized

38. In a pure (intrinsic) semiconductor

   1 the Hall voltage is zero if the n- and p-velocities are equal

   2 the n- and p-charges are equal in number before and after the magnetic field is applied

   3 The electrical resistance rises when its temperature rises

Section IV

Questions 39–45

In the two groups of questions which follow, each situation is followed by a set of statements. Select the best answer.

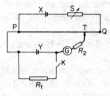

Fig. M11

Figure M11 shows a potentiometer circuit for measuring the internal resistance r of a cell Y. When the cell is on open circuit T is the balance-point on the wire AB.

39. The resistor R_2 is used
  A to make the galvanometer G more sensitive
  B to produce a long balance length
  C because G is a sensitive galvanometer
  D to increase the terminal p.d. of the cell
  E to protect the potentiometer wire
40. When the experiment is repeated with the same values of R_1 and S, the balance length on closing K is found nearer to the end Q than before. A possible cause is
  A the cell X may have run down slightly
  B the cell Y may have run down slightly
  C the resistance R_2 may have been changed
  D the resistance of PQ may have been increased by the current flow
  E the zero in G may have altered
41. If the balance length on PQ is l_1 when a resistance R_1 is used and K is closed, and l_2 when K is not closed, the internal resistance r is calculated from
  A $R_1 l_1/l_2$ B $l_1/R_1 l_2$ C $(l_2-l_1)R_1/l_2$ D $(l_2-l_1)R_1/l_1$ E $l_2 R_1/l_1$
42. The resistance R_1 is now considerably increased and the new balance length l_3 is found when K is closed.
  A l_3 is greater than l_2
  B l_1 is greater than l_3
  C $l_2 = (l_1 + l_3)/2$
  D $r = l_3 R_1/2$
  E l_3 is not much less than l_2

Questions 43–45

Figure M12 shows a beam of electrons with a velocity v entering an electric field due to a potential difference V applied to two parallel plates X, Y of length l which are distance d apart.

Fig. M12

43. When the speed v is increased, the beam
  A follows a circular path of smaller radius
  B is deflected more than before
  C follows a parabolic path between the plates
  D is deflected along its original path
  E follows a more helical path
44. When the applied voltage V is increased
  A the beam takes longer to travel between the plates
  B the vertical acceleration of the electrons is decreased

C the deflection of the beam is decreased

D the vertical acceleration of the electrons is increased

E the horizontal velocity is increased

45. The beam is subjected to a magnetic field perpendicular to the electric field and the beam passes through the plates along its original direction. In this case

A the velocity is given by $v = e/m$, where e is the electronic charge and m is the mass of the electron

B the velocity v is given by $v = V/Bd$

C the force on the electrons due to B is given by $l/v = Bev$

D the momentum gained by the electrons in the electric field is cancelled out by the momentum gained in the magnetic field

E the energy gained by the electrons in the electric field is cancelled out by the energy gained in the magnetic field

Paper 2 (Time: $1\frac{1}{2}$ hours)

Section I

Questions 1–7

The effects A to F below can occur with waves. For each of the Questions 1–7 state the effect necessary to explain the phenomenon concerned.

A reflection   B interference   C diffraction   D polarisation
E refraction   F scattering

1. Beats

2. A phase change of π may be produced

3. Stationary waves of current are formed in an aerial

4. On a clear night sound waves can be heard over a long distance

5. Colour of the sky

6. The resolving power of an optical telescope depends on the diameter of the objective

7. A radio receiver aerial is rotated to obtain a maximum signal

Section II

Questions 8–25

8. In Fig. M13, the direction in which an extra force must act in order to maintain equilibrium is:

A E 60°S   B E 60°N   C E 30°N   D W 30°S   E W 60°S

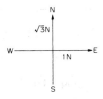

Fig. M13

9. A ball is released from a height and bounces up and down until coming to rest. Which of the following graphs best represents the acceleration a of the ball during the time t of its motion:

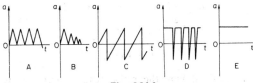

Fig. M14

10. A mass on the end of a spring undergoes simple harmonic motion with a frequency of 0·5 Hz. If the attached mass is reduced to one-quarter of its value, the new frequency in Hz is:

    A 0·25   B 1·0   C 2·0   D 4·5   E 5·0

11. If the distance of the earth to the moon is about 60 earth radii, then the gravitational field at the moon in N kg^{-1} due to the earth is about:

    A 6   B 0·16   C 3·2 × 10^{-2} D 2·8 × 10^{-3} E 5·4 × 10^{-3}

12. The density of air at a pressure of 10^5 N m^{-2} is 1·2 kg m^{-3}. Under these conditions, the root-mean-square velocity of the air molecules in m s^{-1} is:

    A 500   B 1000   C 1500   D 3000   E 5000

13. An astronomical telescope has an eyepiece of focal length 5 cm. If the angular magnitification in *normal* adjustment is 10, the distance between the objective and eyepiece in cm is:

    A 110   B 55   C 50   D 45   E 2

14. An audio oscillator is tuned to a tuning fork of nominal value 512 Hz. The oscillator is accurately calibrated and when it reads 514 Hz a beat frequency of 2 Hz is heard. When it reads 510 Hz, another beat frequency of 6 Hz is heard. The actual frequency in Hz of the fork is:

    A 506   B 510   C 512   D 516   E 518

15. Figure M15 shows the cooling curve of a pure wax material after heating. It cools from P to Q and solidifies along QR. If *l* and *c* are the respective values of the specific latent heat of fusion and specific heat capacity of the liquid wax, the ratio *l*/*c* in K is:

    A 100   B 80   C 40   D 20   E 2

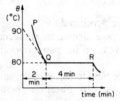

Fig. M15

16. In the potentiometer circuit shown, an e.m.f. E is balanced at C, 40 cm from the end B of wire AB, which has a length of 100 cm and a resistance of 4 Ω. If the p.d. across the resistor R of 200 Ω is 1·00 V, the e.m.f. E in mV is:

    A 12   B 10   C 8   D 6   E 4

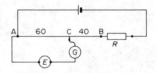

Fig. M16

17. In the circuit shown, the galvanometer G reads zero. If battery H has negligible internal resistance, the value of the resistor X in Ω is:

    A 10   B 100   C 200   D 500   E 1000

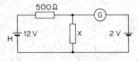

Fig. M17

18. In the a.c. series circuit shown, the r.m.s. voltage across C is 3 V. The r.m.s. voltage across the resistor R in V is:

    A 1   B 2   C 3   D 4   E 8

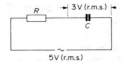

Fig. M18

19. In the diagram shown, a stationary oil-drop P between two parallel plates X and Y has a charge of $3\cdot2 \times 10^{-19}$ C and a weight of $1\cdot6 \times 10^{-14}$ N. The p.d. V between the plates in V is:

    A 500   B 700   C 1000   D 1500   E 2000

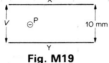

Fig. M19

20. A $0\cdot01$ μF capacitor is charged to a potential of 500 V. It is then connected to an instrument of input capacitance $1\cdot00$ μF. The p.d. across the instrument in V is now:

    A $1\cdot00$ B $4\cdot95$ C $5\cdot00$ D $49\cdot5$ E $50\cdot0$

21. A copper disc of radius $0\cdot1$ m is spun about its centre at a steady rate of 10 rev s^{-1} in a uniform magnetic field of $0\cdot1$ T with its plane perpendicular to the field. The e.m.f. in V induced across a radius of the disc is roughly:

    A $3\cdot1$ B $1\cdot0$ C $0\cdot8$ D $6\cdot2 \times 10^{-1}$ E $3\cdot1 \times 10^{-2}$

22. A steady p.d. of 10 V produces heat at a constant rate in a resistor. The peak value of the alternating voltage in V which will produce half the heating effect in the same resistor is:

    A 2   B 8   C 10   D 20   E 38

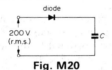

Fig. M20

23. In the circuit shown, a 200 V (r.m.s.) mains supply is connected to a diode and a capacitor C in series. The p.d. in V across C finally is:

    A $70\cdot7$ B 100   C 141   D 283   E 350

24. Electrons with energy 5 eV are incident on a cathode C in a photoelectric cell. The maximum energy of the emitted photoelectrons is 2 eV. When electrons of energy 6 eV are incident on C, no photoelectrons reach the anode A if the stopping potential of A relative to C is:

    A $+9$ V   B $+3$V   C -1 V   D -2 V   E -3 V

25. The rate of decay of atoms in a radioactive sample is proportional to

    A the half-life period
    B the number of undecayed atoms
    C the number of decayed atoms
    D the mass (nucleon) number of the atom
    E the atomic (proton) number of the atom

Section III

Questions 26–38
In each of the questions 26–38, one or more of the responses 1, 2, 3 are correct. Decide which of the responses is/are correct and then answer

    A if 1, 2, 3 are all correct
    B if 1 and 2 only are correct
    C if 2 and 3 only are correct
    D if 1 only is correct
    E if 3 only is correct

Directions summarised				
A	B	C	D	E
1, 2, 3	1, 2	2, 3	1	3
correct	only	only	only	only

26. Which of the following laws is/are valid under *all* conditions?
    1 Ohm's law for a conductor.
    2 Boyle's law for gases.
    3 Conservation of linear momentum.

27. A beam of electrons, initially straight, is deflected in a circular arc by a uniform magnetic field B which is everywhere perpendicular to the beam.
    1 No energy is gained by the electrons when the field is applied.
    2 The momentum of the electrons is proportional to the radius of the circular arc.
    3 If the field is increased sufficiently the path of the beam may become parabolic.

28. Figure M21 shows the variation of potential energy V with separation r of two molecules.

Fig. M21

    1 The molecules can be in equilibrium at P.
    2 At distances farther than Q the force is repulsive.
    3 The molecules cannot be easily compressed because the portion XY is negative.

29. The equation $E = \sigma AT^4$ relates to a perfect black body radiator.
    1 σ is independent of the nature of the surface of the black body.
    2 The equation is true for all shapes of black bodies.
    3 E is measured in joule.

30. In the kinetic theory of gases,
    1 The pressure of a gas is proportional to the mean speed of the molecules.
    2 The root-mean-square speed of the molecules is proportional to the pressure.
    3 The rate of diffusion is proportional to the mean speed of the molecules.

31. The following statements refer to a diffraction grating.
    1 When white light is incident normally on a grating, the red light is deviated least.
    2 The greater the number of lines per millimetre, the smaller is the number of the highest order obtainable.
    3 The waves passing through the clear spaces interfere with each other.

32. A Skylab is moving in a circular orbit of radius R_0 round the centre of the earth of radius R. An astronaut inside feels weightless because
    1 the reaction of the floor on him is less than his weight
    2 the reaction of the floor is zero
    3 the centripetal force on him is mg' where m is his mass and g' is given by $g'/g = R^2/R_0^2$ where g is the acceleration due to gravity at the earth's surface

33. A battery is connected to a variable multiplate capacitor C when it has a value of 1 μF and then removed. When C is altered to 10 μF
    1 the charge on C is unaltered
    2 the common area of the plates has increased ten times
    3 the energy stored is decreased to one-tenth of its original value

34. In the potentiometer experiment shown in Fig. M22, the galvanometer deflection in G is always in the same direction when the slider C is moved along the wire from A to B. In this case
    1 the resistance R may be too high
    2 the resistance S may be too high
    3 the cell Y may have run down

Fig. M22

35. Figure M23 shows a long vertical current-carrying wire XY parallel to the side AD of a vertical rectangular current-carrying coil ABCD suspended near it from the middle of the upper side AB.

1 The coil will move sideways towards XY
2 There are no resultant forces on the sides AB and DC
3 The coil will rotate about its suspension

Fig. M23

36. In Fig. M24, a coil X, joined to a battery, is placed close to a coil AB connected to a resistor R. When the key K is pressed to complete the circuit of X,
   1 the induced current value in AB depends on the magnitude of R
   2 the induced current flows from Q to P through R
   3 the induced e.m.f. in AB depends on the magnitude of R

Fig. M24

37. A beam of electrons, each moving with a momentum U, enters a uniform magnetic field of flux density B, perpendicular to its motion. The charge on an electron is e and its mass is m.
   1 The energy gained is $U^2/2m$
   2 The centripetal force on an electron is Bem/U
   3 The radius of the electron path is U/Be
38. In the nuclear reaction
$$_1^1\text{H} + _3^7\text{Li} \rightarrow 2\,_2^4\text{He}$$
the atomic masses of hydrogen and lithium are 1·008 and 7·016 u respectively and the atomic mass of the helium atom is 4·004 u.
   1 Energy is released by nuclear fission
   2 The energy released can be calculated from $E = mc^2$, where m is the mass change produced and c is the speed of light numerically
   3 The relation $E = mc^2$ can be verified by measuring the initial energy of the α-particles

Section IV

Questions 39–45
In the two groups of questions which follow, each situation is followed by a set of statements. Select the best answer.

Questions 39–42

Fig. M25

In Fig. M25 a lamp L with a straight filament producing white light is placed behind a narrow slit S. S_1 and S_2 are two close narrow slits and Young's fringes are seen through a low-power eyepiece M.
39. The fringes are seen brightest when
   A M is moved closer to S_1, S_2
   B slits S_1, S_2 are closer together
   C slit S is parallel to S_1, S_2

       D the filament is $90°$ to S

       E slit S is $90°$ to S_1, S_2

40. The bright fringes are $1\cdot0$ mm apart when the distance D *to the front lens* of the eyepiece M is $0\cdot8$ m and $S_1S_2 = 0\cdot4$ mm. The average wavelength λ of the light from L is then

       A greater than $5\cdot0 \times 10^{-7}$ m

       B $5\cdot0 \times 10^{-7}$ m

       C less than $5\cdot0 \times 10^{-7}$ m

       D 2×10^{-6} m

       E greater than 2×10^{-6} m

41. The bright fringes seen in M are

       A centrally white with several coloured spectra on one side

       B centrally white with several coloured spectra on both sides

       C coloured in the centre with blue nearer the middle than red

       D coloured in the centre with red nearer the middle than the blue

       E white in the centre with no colours either side

42. When a very thin piece of glass is placed in front of S_1 only, the bright fringes then

       A appear clearer

       B appear much darker

       C turn through a small angle

       D are displaced to one side of the middle

       E disappear

Questions 43–45

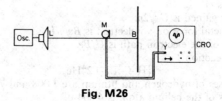

Fig. M26

In Fig. M26 a loudspeaker L, connected to an oscillator, produces sound waves of frequency 2000 Hz. These waves are incident on a large plane board B, and a microphone M, joined to a cathode-ray oscilloscope C.R.O., is moved from B to L along a straight line BL normal to the board.

43. When M is moved through a total distance of $33\cdot0$ cm, five successive positions of maximum wave amplitude are detected by means of the oscilloscope. The speed of sound in m s^{-1} is

       A 340   B 330   C 320   D 264   E 132

44. When the frequency is changed to 2200 Hz, the number of successive positions of maximum wave amplitude in a distance of $37\cdot5$ cm is

       A 6   B 5   C 4   D 3   E 2

45. The board B is now completely removed and the time-base voltage is switched off from the oscilloscope. The loudspeaker is joined to the X-plates and the microphone M is left connected to the Y-plates.

     As M is now moved towards L, a similar straight line trace is seen on the oscilloscope screen in two successive positions P and Q of the microphone. Then

       A the distance PQ is half a wavelength

       B the vibrations at P and Q are $90°$ out of phase

       C the distance PQ is twice the wavelength

       D the straight line is due to a stationary wave between M and L

       E the distance PQ is one wavelength

Answers to Exercises and Papers

Mechanics

Exercise 1A (p. 13)
1. (i)5 s (ii) 62·5 m (iii) 17 m s^{-1}
2. (i) 4 s (ii) 20 m
4. 3 s
5. (i) 3 s (ii) 30 m (iii) 72°
7. (i) 16 s (ii) 8868 m; 10 240 m
8. (i) 5 m s^{-2} (ii) 1·4 s (iii) 3 s (iv) 22·5 m
9. (i) 20 m s^{-1} (ii) 14·1 m s^{-1} at 45° to OA
10. (i) 10 km h^{-1} (ii) 2·4 km
11. 21 675 m, 3125 m, 25 s, 465 m s^{-1}
12. B

Exercise 1B (p. 25)
1. 2000 N, 3000 N
2. 550 N, 500 N
3. (i) 600 N (ii) 400 N
4. (i) 10 N s (ii) 100 N
5. 2·0 N
6. 2·5 m s^{-2}
7. 2 m s^{-1}, 4 m s^{-1}
8. 10 m s^{-1}, 2·0 s
9. 200 N
10. 0·9 v, 30°
11. (a) 1·25 N (b) 0·48 N m^{-2}
    (c) 0·96 N m^{-2}
12. 1·45 × 10^6 m s^{-1}, 9 × 10^5 m s^{-1}
13. 2·25 × 10^5 N m^{-2}
14. 2 m s^{-2}
15. 8/3, −4/3 m s^{-1}
17. 254 m s^{-1}, 23° to orig. dir.

Exercise 1C (p. 34)
1. 180 J, 60 N s; 3 s, 9 m
2. 600 J, 6 m
3. (i) 4 s (ii) 20 m (iii) 10 J, 10 J
4. (i) 100 J (ii) 500 J
5. (i) 5·2 m s^{-1}, 58 J (ii) 1·2 m s^{-1},
   314 J
6. 20 J, 15 J
7. 1000 J
8. 1·25 m
9. 4 kW, 14 kW
10. (i) B, C, E, G (ii) A, D, F, H
12. $x = 4$
13. $x = \frac{3}{2}, y = \frac{1}{2}, z = -\frac{1}{2}; t = k\sqrt{a^3\rho/\gamma}$
15. (a) 0·3 N (b) 0·6 W (c) 0·3 W
16. 2/3
17. $x = 2, k = 2·2$
18. $m^2 Hg/(M + m)d$
19. 56·5:1, 0·082 MeV
20. (i) 400 kW (ii) 656 kW

21. 2E (3E less E)
22. (a) 4 m s^{-1} (b) 0, 4 m s^{-1}
24. (a) 10 m s^{-1} (b) 24 J, 12·65 m s^{-1}
25. 2E/103
26. (a) 10/3 N (b) 5/9 W (c) 5/18 W
27. 1667 N, 83 330 J
28. 1/3
29. 8000 N, 22 kW

Exercise 2A (p. 45)
1. (i) 2 rad s^{-1} (ii) 96 N
2. (i) 118 N (ii) 32°
3. 42°, 13 450 N
4. 224 N, 64 N
5. 15·8 N
6. 7·3 × 10^{-5} rad s^{-1}, 0·068 N
7. 0·5%
8. (i) 2·3 N (ii) 1·7 m s^{-1}
9. 1·6 rev s^{-1}
10. (i) 4·2 m s^{-1} (ii) 11 N (iii) 7 N
11. 90 N
12. 0·675 rev s^{-1}
13. (a) mgl (b) $\sqrt{2}\,gl$ (c) 2g up (d) 3mg
14. 7·7 rad s^{-1}, 122 cm away
15. (a) 9·2 N (b) 8·6 N

Exercise 2B (p. 58)
1. 250 N
2. 0·25 g
4. $M^{-1}L^3T^{-2}$
5. Y
6. 9910 s
7. 6 × 10^{24} kg
8. 9·9 m s^{-2}
9. 4·5 × 10^{-9} rad s^{-1}
10. (a) 7852 m s^{-1} (b) 5250 s
11. 889 N
12. 0·0646 m s^{-2}
13. (b) $T = k(GD)^{-\frac{1}{2}}$
    (c) $T = 2\pi\sqrt{D^3/G(M_1 + M_2)}$
14. 24 h
15. 9·77 m s^{-2}
16. 3 m, 6:1
17. 27·3 days
19. 42 600 km
20. 7·1 × 10^{10} J

Exercise 2C (p. 74)
1. (i) 0·05 s (ii) 0, 3200 π^2 cm s^{-2}
   (iii) 80 π cm s^{-1}

2. (a) 0.08 m s^{-1} (b) 1.57 s
3. (a) 2.5 m s^{-1} (b) 790 m s^{-2}
4. A, D, E, F
5. (i) 1.26 s (ii) 0.1 m s^{-2} (iii) 4×10^{-5} J
   (iv) 4×10^{-5} J
6. 10.0 m s^{-2}, 4.5 m; 2.45: 1
7. 2.7 s (i) 0.24 J (ii) 0.47 m s^{-1}
8. 0.20 s
9. (a) (i) 20 mm (ii) 0.2 s (iii) 628 mm s^{-1}
   (b) 7.9×10^{-2} J
10. (i) 0.4 s (ii) 0.5 N (iii) 0.005 J
11. 0.2 J
12. 6.3 cm
13. (i) 1.0 N (ii) 0.8 s, 0.04 J
14. 1.6 Hz
15. 5×10^{-2} m
16. (a) 2.43 s (b) 0.150 J, 0.387 m s^{-1}
    (c) 0.56 (d) 0.067 m
17. (i) 0.0034 J (ii) 3.45 N, 2.55 N

Exercise 3 (p. 92)
1. (i) 2000 J (ii) 200 kg m^2 s^{-1}
   (iii) 3.2 rev s^{-1}
2. (i) 8 rad s^{-1} (ii) 25 133 J
3. 6.4 rev, 8 s
4. 50 rad s^{-1}, 25 000 J
5. 4 rev s^{-1}
6. (i) 2 rad s^{-1} (ii) 15 J
7. (i) 6 rad s^{-1} (ii) 12 N m (iii) 14.3 rev
8. (i) 2.2 rad s^{-1} (ii) 3.2 rad s^{-1}
9. 1.3×10^{-4} kg m^2
10. 18.3 rad s^{-1}
11. (a) 20 rad s^{-2} (b) 0.32 N m
12. 7.3×10^{-4} kg m^2
13. 6.3×10^{12} rad s^{-1}
14. (i) 2×10^7 J (ii) 10 km
15. (i) 2304 J, 38.4 kg m^2 s^{-1} (ii) 0.42 N m
    (iii) 92.2 s
16. (a) 4ω (b) $4\omega/5$
17. 6.2×10^{-4} kg m^2
18. 21.6 rad s^{-1}

Exercise 4 (p. 110)
1. E
2. (i) 0.8 N m (ii) 0.4 N m; 10.1 J
3. 690 N, 1390 N
4. 500 N m; 167 N
5. 22.6 cm
6. (i) 0.65 m (ii) 0.60 m
7. 41.7 N, 41.7 N
8. 180 N, 159 N
9. 0.0063 nm from N atom
10. 1710 N
11. (i) 3:2 (ii) 40:9
12. 1.62 N
13. 5.1 cm
14. 38 cm
15. 773.8 mmHg
16. Wρ/σ (a) 0.95 (b) 1.19
17. (a) 750 kg (b) 270 kg (c) 0.42 m s^{-2}

18. 1.5 kg s^{-1}
19. 9 m s^{-1}
21. 1 m s^{-1}, 9.25×10^4 Pa
22. 4 m s^{-1}, 12 kg s^{-1}
23. (i) 2.8 m s^{-1} (ii) 5.7×10^{-3} m^3 s^{-1}

Properties of Matter

Exercise 5 (p. 134)
1. (i) 1/2000 (ii) 10^{11} N m^{-2} (iii) 0.025 J
2. (i) Y
3. 6.4×10^6 N m^{-2}, 6×10^{-5},
   1.1×10^{11} N m^{-2}
4. 67 N
5. 0.08 mm
6. 0.04 J, 0.08 J
7. 1.2×10^8 N m^{-2}, 29.6 N
8. 8.3×10^{-3} J
9. (a) C:1.25×10^{-3}, S:0.75×10^{-3};
   (b) 471 N
10. (a) 2.0×10^{11} N m^{-2} (b) 4.8×10^{-2} J
11. 20 m s^{-1}
12. 1.47×10^8 N m^{-2}, 2.2×10^{-3} m, 0.16 J
13. 1.0×10^{10} N m^{-2}
14. 1.2 N
15. 1.1×10^4 J
16. (a) 2.3×10^{-6} m (b) 5.7×10^{-5} J
17. (c) (i) 4.02×10^{-3} (ii) 8.03×10^8 N m^{-2}
    (iii) 2.08 J
18. 40 N, 74 N
19. 115 g
25. 3×10^6 N m^{-2}

Exercise 6 (p. 157)
1. 1.4×10^{-2} N, 8.7×10^{-3} N
2. -3×10^7 N m^{-2} s^{-1}
3. 1.2×10^{-3} J
4. (i) 6.5 cm (ii) 5.4 cm
5. angle of contact 60° at top
6. (i) 1.00056×10^5 N m^{-2} (ii) 9×10^{-3} J
11. 7.2 cm, angle of contact 46°
12. -7.2×10^9 N m^{-2} s^{-1}
13. 0.75 cm
14. $8 \pi T (b^2 - a^2)$
15. 133 mm
17. (a) 3.2 cm (b) 7.1×10^{-2} N m^{-1}
18. (i) 3×10^{19} (ii) 6×10^{-21} J mol^{-1}
    (iii) 2×10^{28} (iv) 10^8 J

Exercise 7 (p. 169)
1. 5.07 cm
2. (a) 17° (b) 0.36 N
3. 0.6
6. (a) $(\rho - \sigma)g/\sigma$ (b) $2(\rho - \sigma)gr^2/9\eta$
7. 15 N
8. 0.238 m s^{-1}
9. (a) $x = 2$, $y = 1$, $z = 2$ (b) $k = 2.3$
10. 0.5 cm, 13.5 cm
11. 0.025 m s^{-1}
12. 17 s

Heat

Exercise 9 (p. 191)
1. (i) 12 W (ii) 2 W (iii) 600 J kg^{-1} K^{-1}
2. (i) 100 s (ii) 0·2 kg
3. 6000 J kg^{-1} K^{-1}, 10%
4. 6·76 × 10^{-20} J molecule^{-1}
5. 5·7 V
6. 0·45 W
7. 4200 J kg^{-1} K^{-1}
8. (1) 1·3 W (2) 960 J kg^{-1} K^{-1}
9. 2·6 × 10^3 J kg^{-1} K^{-1}, 1·5 W
10. 3·78 × 10^5 J kg^{-1}
11. 0·0935 kg
12. 360 J K^{-1}, 0·035 kg
13. 2·23 × 10^6 J kg^{-1}
14. 1·6 × 10^5 J kg^{-1}

Exercise 10 (p. 216)
1. (i) 2·4 × 10^5 N m^{-2} (ii) 0·24 m^3 (iii) 0·13 m^3
3. 1·1 kg
4. 100°C
5. (i) 2080 J kg^{-1} K^{-1} (ii) 842 mmHg
6. 0·014 m^3
7. 1·3 × 10^5, 2·3·10^5 N m^{-2}
8. (ii) 8·3 J mol^{-1} K^{-1}
9. (i) 576 J (ii) 410 J (iii) 166 J
10. (i) 2000 J (ii) 450 K (iii) 60 J
12. 1·44 × 10^4, 1·03 × 10^4 J kg^{-1} K^{-1}
13. 8·3 J mol^{-1} K^{-1}, 25·9 J
14. 164 J
15. $EA\alpha T$, $EA\alpha^2 lT^2/2$
16. (a) 1·5 × 10^5 N m^{-2}, 435 K (b) 30·9 J
17. 1·67
18. 10·48 × 10^5 N m^{-2}, 666 K
19. (a) 586 K (b) 101·2 J (c) 355 J
20. 222 K, 384 mmHg
22. 4·40 × 10^5 Pa, 15°C; 7·66 × 10^5 Pa, 228°C

Exercise 11 (p. 233)
2. (i) 1732 m s^{-1} (ii) 433 m s^{-1}
3. (i) 1039 m s^{-1} (ii) 900 m s^{-1}
5. 597 m s^{-1}
7. 4·35 × 10^{16} m^{-3}
9. (a) 1·07 (b) 4/1 (c) 0·87
10. 0·21 mmHg
11. 461 m s^{-1}, 64 rad s^{-1}
12. 1305 m s^{-1}, 0°C
13. Kr
14. 164 J, 30 J
15. (b) 311, 519 J kg^{-1} K^{-1} (c) 249 J K^{-1}, 1·15 × 10^5 J

Exercise 12 (p. 254)
2. 12·5°C. (a) 4·66 (b) 5·27 kN m^{-2}
5. 707 mm
6. 1·078 × 10^5, 1·018 × 10^5 N m^{-2}

7. 91·7 torr
8. 44·5 mmHg
15. 3·35 × 10^{-29} m^3

Exercise 13 (p. 285)
1. (i) 4·5 × 10^7 J (ii) 1·8 × 10^8 J
5. (a) 90 W m^{-2} (b) 3 m; 19·5°C, 5·5°C; 1/32
6. 0·017 kg s^{-1}, 45·9°C, 29·2°C
7. 354 J min^{-1}
8. 1·47 × 10^{-4} kg s^{-1}
9. (a) 90°C (b) 57·6 W
10. 89°C
11. 94%
12. 12 kW. (a) 3·75 K (b) 11·25 K (c) 0·18 mm
14. 1070 K
17. 2·44, 0·82 W
18. 4·55 × 10^{26} W m^{-2}, 1600 W m^{-2}
21. 5·7 × 10^{-8} W m^{-2} K^{-4}
22. 5490°C
23. 1105 K
24. 0·068 K s^{-1}
25. 1300 K
26. 2140 K
27. 5·6 × 10^7 J
28. 5450°C

Exercise 14 (p. 296)
1. 434 K
2. 4 s
3. 300 N
4. 270°C
5. 79°C
6. 762·4 mmHg
7. 21·7°C

Exercise 15 (p. 306)
1. 16·6°C, 17·0°C
3. 1·43°C
5. 68°C, −272°C
7. 385°C
8. (a) −274°C (b) 99·45 cm
9. 57·8°C

Geometrical Optics

Exercise 16 (p. 325)
2. 4a, 2na
4. (i) 15 cm, 1·5 (ii) 12 cm, 3
5. 6 cm, 0·4
6. u = 10 cm, r = 40 cm, concave
7. 6·35 × 10^{-3} m
8. 2 rad or 114°
9. (a) 240 cm (b) 1·3 cm

Exercise 17 (p. 337)
1. 35·3°
2. 41·8°

3. (i) 26·3° (ii) 56·4°
4. 3 cm from bottom
5. (i) 41·8° (ii) 48·8° (iii) 62·5°
7. 1·47
8. (b) 12 cm from mirror
9. (b) 1·60
10. 1·41
11. (i) 60°04′ (ii) 48°27′

Exercise 18 (p. 347)
1. 42°
2. (i) 1·52 (ii) 52·2°
3. 49°
4. 60°, 55°30′, 1·648
6. 4°48′
9. 43°35′

Exercise 19A (p. 368)
1. (i) 12 cm, $m = 1$ (ii) 12 cm, $m = 3$
2. $6\frac{2}{3}$ cm, 0·6
3. $13\frac{1}{3}$, 40 cm
4. (i) 5·1 cm (ii) 22·5 cm
5. 1·82 mm, 20 cm from 2nd lens, virtual, 9·1 mm
6. 12·0 cm, 18·7 cm
7. 4 cm
8. (a) 120 cm from converging lens (b) 92·2 cm (c) 2·2
9. (a) beside object O (b) 72·5 cm from O
10. 1·4
11. 1·35
12. 4 cm

Exercise 19B (p. 372)
1. diverging, $f = 200$ cm, 22·2 cm
2. converging, $f = 28·6$ cm, 35·3 to 25 cm
3. near point 50 cm, far point 200 cm

Exercise 19C (p. 380)
2. (a) 49°12′ (b) 50°38′ (c) 1°26′

Exercise 20 (p. 401)
1. (i) infinity (ii) 20
3. 11·25 cm
6. 0·0091 rad
7. 7×10^{-6} rad
8. (a) 4 (b) 4·8
9. 5°
11. $f_e = 4$ cm, dia. = 0·5 cm
12. (a) 22·4 (b) 4·9 cm dia.
13. $6·25 \times 10^{-3}$ rad
14. (i) 4·2 cm (ii) 6; 5
16. 20·2 cm sepn.
17. 16 cm from second lens
18. 3·2, 8·8 cm
19. 8·7 cm, 46·7

20. 0·55 cm, 2:1
21. 89° or 91°
22. 1/15 s
23. $6·8 \times 10^{-2}$ s

Waves

Exercise 21 (p. 437)
1. 0·42 m s^{-1}
3. (i) 1·33 m (ii) 400 Hz
4. (i) $5\pi/3$
 (ii) $y = 0·03 \sin 2\pi(250t - 25x/3)$
 (iii) 6 cm
7. 3400 Hz
8. 10^{10} Hz
9. 4 max intensity per 3 s
11. 3:1
12. (i) 1·73 A (ii) 1·5:1
13. (a) 1/9th (b) 2·9 cm, $1·03 \times 10^{10}$ Hz
14. (i) 100 Hz (ii) 1·7 m (iii) 170 m s^{-1}
 (iv) π (v) $0·2 \sin (400\pi t + 20\pi x)/7$
15. (a) 330 m s^{-1} (b) $6·6 \times 10^{-4}$ m s^{-1}
16. 330 m s^{-1}, 579 mm
17. 5·9 s
19. 349 m s^{-1}
20. 332 m s^{-1}
22. 680 Hz

Wave Optics

Exercise 22 (p. 454)
3. 18·6°
4. 47°10′, 41°48′ to vertical at oil surface
9. $4·0 \times 10^{-7}$ m
11. $1·9 \times 10^{-3}$ deg.
12. 31·25 km, 300 rev s^{-1}
14. 6250 m
15. 250, 6×10^4 cycles

Exercise 23 (p. 477)
2. $1·8 \times 10^{-3}$ m; $1·5 \times 10^{-3}$ rad
4. $1·8 \times 10^{-3}$ rad, 0·064 mm
6. $2·27 \times 10^{-5}$ m
7. 643 nm
9. 0·34 mm
10. 1·5
11. $6·25 \times 10^{-7}$ m
12. $2·4 \times 10^{-3}$ rad
14. 1·0 m
16. $1·5 \times 10^{-5}$ m
17. $7·1 \times 10^{-6}$ m, recede

Exercise 24 (p. 496)
1. (a) $2·5 \times 10^{-3}$ mm (b) 13·9° (c) 4
4. 2×10^{-6} m
6. 1376 m s^{-1}
7. 600 nm, 285 000 m^{-1}, 43·2°

8. 3; 5.895×10^{-7} m
9. (i) 1.78×10^{-6} m (ii) 7.52×10^{-7} m
 (iii) $38.2°$, $57.6°$
10. 7; $0°$, $\pm 17.1°$, $\pm 36.1°$, $\pm 62.1°$
11. 3; 1.2 cm
12. 5 m s^{-1}
13. 2.32×10^{-7} m
14. 2.35 mm, 0.0785 rad
15. 2.7 mm

Exercise 25 (p. 506)
 3. $67.5°$
 6. 1.0 mm

Sound

Exercise 26 (p. 526)
 3. 34°C
 6. (a) 4:1, 2500 Hz (b) 9:1 (c) 1:1
  (e) 12.4 cm
 8. 25 W
 9. 1.83 s
 11. 625.9 Hz
 12. 1091, 909 Hz
 13. 638, 567 Hz
 14. (a) away (b) 4×10^5 m s^{-1}
 15. 425 Hz
 16. 514, 545 Hz
 17. 12 Hz; 1007, 993 Hz
 18. (i) 66 cm (ii) (1)550 (2)545 (3)600 Hz
 19. 1115 Hz
 20. 1174, 852 Hz

Exercise 27 (p. 556)
 1. (i) $\lambda/2$ (ii) $\lambda/4$ (iii) $\lambda/2$, 567 Hz
 2. 20 cm
 3. (i) 320 cm s^{-1} (ii) 10 cm (iii) 40 cm
  (iv) 1200 Hz
 4. (a) 0.322 m (b) 0.645 m
 5. (i) max (ii) 0 (iii) max (iv) half-max
 6. 267 Hz
 9. $+5.2°C$
 12. 300 Hz
 13. 170 Hz
 14. resonance at 115 Hz
 15. $v = \sqrt{(F/\rho)}/r$
 16. 239 Hz
 17. -3.2%, $+6.8\%$
 19. (a) 2 m (b) 100 Hz
 20. pluck 1/6th from end
 22. 20.5 N

Electricity

Exercise 28 (p. 591)
 1. 2×10^4 V m^{-1}, 1.6×10^{-14} N
 3. 1.25×10^{-18} C

 5. 1.6×10^{-19} C
 7. 1.5×10^6 V
 9. 3 electrons, 19.6 m s^{-2} (2 g)
 11. 100 V
 12. (a) 9×10^4 N C^{-1} (V m^{-1}) (b) 9000 V
 13. 3.45×10^{-20} C too high
 14. 1.33×10^{-5} C, 6×10^5 V, yes

Exercise 29 (p. 615)
 1. 0.2 μF, 2.5×10^{-4} J
 2. (a) 9×10^{-4} C, 18×10^{-4} C; 0.135 J,
  0.27 J (b) 6×10^{-4} C, 6×10^{-4} C;
  0.06 J, 0.03 J
 3. (a) 6×10^{-8} C (b) 3×10^{-6} J
  (c) 6×10^{-6} J; 1×10^{-6} J
 4. 133.3 μC, 133.3 μC
 5. (i) 160 V (ii) 0.14 J (iii) 0.128 J
 6. 33.9 m
 7. 4
 8. 9×10^{-9} F, 20 times
 11. 3.6×10^{-5} J, 2.4×10^{-5} J
  (i) 3.6×10^{-5} J, 3.84×10^{-4} J
  (ii) 3.6×10^{-5} J, 5.4×10^{-5} J
 12. 75 V
 13. 1.35×10^{-3} C, 3 times, 84
 14. (a) 2×10^{-11} F, (b) 1.2×10^{-6} J. 300 V
 15. 1.13×10^5 V m^{-1} (a) 226 V
  (b) 2.2×10^{-10} F (c) 5.65×10^{-6} J
 16. 4.25×10^{-5} A
 17. 4.6×10^{-6}:1

Exercise 30 (p. 652)
 1. 3.2 V, 10 Ω
 2. (i) 0.4 A (ii) 10 Ω (iii) 12 V
 3. 5 V, 0; 5.8 V, 4.8 V
 4. 14.4 V, 9.6 V; 8 V, 5.3 V
 5. 0.5 A, 6 Ω
 6. parallel 2.17 Ω
 7. (a) series 725 Ω (b) parallel 0.10 Ω
 8. (i) (a) 1.5 V (b) 9/16 V (c) 1.5 V (ii) 9.6 V
 9. 2 V, 4 Ω parallel
 10. (i) 12.5 mA (ii) 6.25 mA (iii) 0.25 V
  (iv) 0.25 V
 11. (a) 25 V (b) 50 V
 12. $64n/(n^2 + 16)$, 8 A, 3/8 Ω
 13. $X = 2000$ Ω, 1000 Ω
 14. -1%
 15. 1.54 V
 16. 0.01 A
 17. 2:1
 18. 4 m, 1 m
 19. (a) R_1 (b) R_2
 20. (a) 1 Ω (b) 20 Ω or 0.05 Ω
 21. 18 Ω
 22. 2220 J kg^{-1} K^{-1}, 72 J K^{-1}
 23. (1) 1:5 (2) 1:7.5
 24. $\frac{1}{2}$, 1, 2 kW
 25. 36, 39%

26. 18.75 kW

Exercise 31 (p. 669)
1. 1.53 V, 60.0 cm
2. 0.12 A
3. 1197 Ω, 3 mV
6. 15.8 mV
7. 0.825 m, 1.29 V
9. 12, less
11. 9 mV, $101.6\,\Omega$
12. (a) 1.5 V (b) 1.6 V (c) 3 (d) 2 V
14. 5×10^{-4} A, 2.5×10^{-3} V, $1962\,\Omega$
15. 0.004 K^{-1}, 50°C
16. $14.9\,\Omega$
17. 0.367 m
18. 1.4:1, 1500 s
19. 0.0037 K^{-1}
20. 8.8×10^{-5} K^{-1}
21. 5.3×10^{-4} K^{-1}
22. -5.1×10^{-4} K^{-1}

Exercise 32 (p. 689)
1. (i) 19 300 C (ii) 7720 s (iii) 0.2 g
2. (a) 3.0×10^{23}, 6.0×10^{23} (b) 0.098 g, 0.34 g
3. (a) 96 320 C mol^{-1} (b) 5.06×10^{22}
6. 4.2×10^{-4} m^3
7. 18.6 cm^3
8. 1.5 V
9. $23\,\Omega$, 1.6 p, 50%

Exercise 33 (p. 706)
1. (i) 0.2 N (ii) 60°
2. 5 A
3. 5×10^{-3} N m, 2.5×10^{-3} N m
4. 1.0 T, PQ
5. 1.6×10^{-14} N
6. 2.6×10^{22} m^{-3}
7. $21.8°$
8. 69 μA
9. (a) 10 (b) 1/4
10. 3.75×10^{-24} N
11. 13/12
12. 0.4 T
13. 3.5×10^{-4} m s^{-1}, BI/Net
14. (a) shunt $0.0125\,\Omega$ (b) series $9995\,\Omega$
15. 2.1×10^{22}

Exercise 34 (p. 724)
1. (b) 1×10^{-5} T (c) 3×10^{-5} T, greater
2. 0.8 A
3. (b) 0.04 m (c) 1.33×10^{-5} N m^{-1}
4. 6×10^{-5} N away from X
6. 3.7×10^{-3} V
7. 6 cm from 12 A wire
8. 3.64 A
9. (i) 1.6×10^{-4} T (ii) 0

10. 1.5×10^{-4} N

Exercise 35 (p. 757)
3. (a) 0 (b) 0.6 V. P
4. (a) 0 (b) 0.1 V; 10^{-3} C
5. 3.1 V. (i) 3.1 V (ii) 1.6 V (iii) 0
10. 7.5×10^{-4} V
11. $\mu_0 n I A f$, 100 $\mu_0 n A f$
13. (i) Bvd (ii) 0 (iii) Bvd; (i), (iii) $I = Bvd/R$, $F = B^2 vd^2/R$
14. (a) 2.5×10^{-3} Wb (b) 2.5×10^{-3} C
15. (b) 15 A, 6×10^{-5} T
16. (a) (i) 4.0 A (ii) 1.0 A s^{-1} (iii) 0.5 A s^{-1}
    (b) 26.8 div.
17. 4 μH
18. (a) $\mu_0 N^2 \pi r^2 I/Lt$ (b) $\mu_0 N^2 \pi r^2 I^2/L$
19. 44.2 rev s^{-1}
21. 1.05 V
23. 10^{-4} C
26. (a) 25 000 V (b) 2×10^{-2} A
27. (a) 4 A (b) 2 A s^{-1}
28. 0.53 V
29. (a) 2.51×10^{-2} T (b) 4.02×10^{-2} Wb
    (c) 8.04 mH (d) 2.41 mH (e) 0.72 mH
    (f) 28.9 mV
30. 25 μF

Exercise 36 (p. 771)
2. 2000
3. 0.22 A
4. 2500

Exercise 37 (p. 790)
1. (i) 2.1 A (ii) 2.1 A (iii) 3 A
2. (i) 2 A (r.m.s.) (ii) 0.016 A (r.m.s.)
   (iii) 0.0031 A (r.m.s.)
3. (i) R, 4 V; L, 3 V; C, 4 V (ii) 5 V
   (iii) 1 V (iv) 4.1 V; 0.8 W, 0, 0
4. (a) 28.3 V (b) 64 Hz (c) 20 W
6. 0.5 A
7. (b) 7.8 V, 6.2 V
8. (b) 0.11 H (c) 2.56:1
10. $37.5\,\Omega$
13. (a) $306\,\Omega$ (b) 0.19 H
14. 3.46 A
15. (a) 0.1 A (b) 3 W
16. $60.6\,\Omega$
17. $R = 190.0\,\Omega$, $C = 50.6$ μF
19. $31.4\,\Omega$, $1592\,\Omega$, 356 Hz

Atomic Physics

Exercise 38 (p. 817)
1. 6×10^6 m s^{-1}, 1.7×10^{-2} T
2. 33 mm
3. 6.3×10^{-4} T
4. (b) (i) 3.6×10^7 m s^{-1} (ii) 2×10^{-3} T
   (c) 0

5. 2006 V
6. 1.44×10^{-2} m
7. 0.5×10^{-4} T
8. $4e$, changes to $2e$ and $3e$
9. 3.2 mm
10. (a) 1.5×10^{-6} m (b) 8
(c) 6.25×10^{-5} m s^{-1}
11. 0.08 m
12. 1.2×10^{-8} N
13. (g) 2.5×10^5
14. 1.6×10^{-19} C
16. 60 km
17. (d) $30°$
18. (a) (ii) 1.94 V (b) 9×10^4 Ω

Exercise 39 (p. 841)
1. 40 Ω, 40 Ω
6. 59.5 Ω
8. (i) 45 (ii) 13 k Ω

Exercise 40 (p. 870)
1. 1.5 V (iii) 2.5×10^{-19} J
(iv) 7.3×10^5 m s^{-1}
3. 6.7×10^{-34} J s
4. 5.1×10^{14} Hz
5. 2.0×10^{-7} m
6. (c) 6.6×10^{-34} J s, 3.0×10^{-19} J
7. (a) 9×10^{-7} m (b) 7.9×10^5 m s^{-1}
(c) 1.7 V
8. 2 V, 8.4×10^5 m s^{-1}
10. 13.6 V
11. A, B, C, D = 6.5, 4.8, 4.3, 4.0×10^{-7} m
12. 1.7×10^{-11} m
13. 1.86×10^6 m s^{-1}
14. (i) 0.012 A (ii) 1.33×10^8 m s^{-1}
15. (i) 10^4 eV (ii) 1.23×10^{-11} m
(iii) 1.6×10^{-15} J, 1.24×10^{-10} m
16. 1.24×10^{-11} m
17. 5000:1
18. (i) 9.375×10^{16} (ii) 150 W
21. 8125 eV
23. 8×10^{18} Hz
24. (i) 6.25×10^{15} (ii) 7.3×10^7 m s^{-1}
(iii) 3×10^{-4} kg s^{-1}

Exercise 41 (p. 906)
3. (a) 5.25×10^{20} (b) 2.1×10^8 J
6. (i) 7.5×10^{20} (ii) 7.2×10^{15} s^{-1}
9. (a) 1.54×10^{12} s^{-1} (b) 6×10^7 J
10. 7.14 days
11. 23 s
12. 4.1×10^9 y, 3.2×10^9 y
13. 6000 cm^3
14. (a) 218, 84 (b) 1.08×10^{-19} N s
(c) 3.1×10^5 m s^{-1}
15. 3.36×10^{22}
16. 6.78, 2.39 MeV/nucleon

17. 1.8×10^3 MeV
18. 1.44×10^{-13} m
20. 1×10^{11} y
21. (b) 3
23. 8.9×10^{-13} J
24. (c) $_0^1$n (d) (ii) 5.38 MeV (iii) 5.27 MeV
25. (i) 6.25×10^{13} (iii) 2.8×10^{-12} J
(iv) 1.7×10^{-14} m (v) 70:1, 8.9×10^{11}:1
26. 0.95 MW
29. (a) 28.3 MeV (b) 23.8 MeV
30. (a) 4.23 MeV (b) 4.16 MeV

Revision Papers (p. 911–916)

1

1. 319 m s^{-1}
2. 6.25×10^{-3} J
3. 2570 m s^{-1}
5. 480 nm, 640 nm, $28.7°$
7. 32.6 mm
8. 6.0 m
10. 14.1×10^{-9} C,
3.5×10^{-9} C

2

4. -0.2 m
5. 0.066 m
6. 60.4 V
10. 3.75×10^{-24} N

3

1. 9 m
2. 0.89 s, 4×10^{-3} J
3. $-3°$C
4. 5×10^{-3} m
5. 0.8m
7. 3.0×10^{-7} m
9. 8 Ω, 50 Hz
10. (a) 2.19×10^6 m s^{-1}
(b) 0.91×10^{-7} m

4

2. (a) 8:1, (b) $D/9$
3. $49.05°$C
4. 1200 Hz
5. $+2D$
7. 3.8×10^{-11} m
8. 7.1 MeV
9. Possible

5

1. 745 mm
3. 10^3 Pa
4. 4.5×10^{-7} m
5. (i) 2×10^{-4} rad,
(ii) 2×10^{-5} m
7. (i) 10^4 V m^{-1}; (ii) 3.5×10^{-9} C;
(iii) 1.8×10^{-9} J

Multiple Choice Paper 1 (p. 917)

1. E	**2.** A	**3.** D	**4.** E	**5.** E
6. C	**7.** B	**8.** C	**9.** D	**10.** D
11. C	**12.** A	**13.** C	**14.** B	**15.** D
16. A	**17.** B	**18.** D	**19.** C	**20.** D
21. E	**22.** B	**23.** C	**24.** E	**25.** A
26. E	**27.** C	**28.** D	**29.** A	**30.** B
31. C	**32.** E	**33.** B	**34.** D	**35.** A
36. A	**37.** C	**38.** B	**39.** C	**40.** A
41. D	**42.** E	**43.** C	**44.** D	**45.** B

Multiple Choice Paper 2 (p. 923)

1. B	**2.** A	**3.** B	**4.** E	**5.** F
6. C	**7.** D	**8.** E	**9.** D	**10.** B
11. D	**12.** A	**13.** B	**14.** D	**15.** D
16. A	**17.** B	**18.** D	**19.** A	**20.** B
21. E	**22.** C	**23.** D	**24.** E	**25.** B
26. E	**27.** B	**28.** D	**29.** B	**30.** E
31. C	**32.** C	**33.** A	**34.** D	**35.** B
36. D	**37.** E	**38.** C	**39.** C	**40.** A
41. B	**42.** D	**43.** B	**44.** A	**45.** E

Index

Absolute expansivity, 293
  measurement, current, 713
  temperature, 198
  zero, 198
Absorption of radiation, 272
  radioactivity, 880
  spectra, 283, 376, 857
A.C. circuits, 772
Acceleration, 5, 88
  and force, 16
  angular, 82
  in circle, 39
  of charge, 853
  of gravity, 8, 42, 50
Accommodation (eye), 370
Accumulator, 686
Acetate rod, 564
Achromatic doublet, 379
Adhesion, 142
Adiabatic change, 210–4
  , sound waves, 435
  curves, 211
  modulus, 123
A.f. amplifier (transistor), 832
Aerial, 421
Air breakdown, 579
Air-cell method, 336
Air wedge fringes, 463
Alpha-particle, 877
  , range of, 878
  , scattering of, 888
Alternating current, 737, 772
Alternators, 739
Ammeter, a.c., 772, 812
  , d.c., 627, 699
Ampere, 713
  balance, 713
  , circuit law of, 722
Amplification, transistor, 832
Amplitude, 63, 407, 521
  , a.c., 738, 774
ANDREWS' experiments, 247
Angle of contact, 141
Angular acceleration, 82
  magnification, 383, 392
  momentum, 84
  velocity, 39
Anode (electrolysis), 674
  of valve, 810
Antinode, 419–21

Aperture, 319, 400
Apparent depth, 330
APPLETON layer, 337
ARCHIMEDES' principle, 104
Armature, 744
ARRHENIUS' theory, 677
Artificial disintegration, 900
Astronomical telescope, 383
Atmospheric pressure, 103
  refraction, 337
Atomic mass, 890
  nucleus, 889
  number, 889
  spacing, 863
  structure, 889, 891
  unit, 896
Audio frequency, 511
AVOGADRO constant, 129, 201, 681

B (flux density), 695, 718–24
  , measurement of, 711, 749
Back-e.m.f. of induction, 754
  of motor, 745
  of polarization, 683
BAINBRIDGE spectrometer, 894
Balancing column method, 294
Ballistic galvanometer, 596, 748
Balmer series, 854
Band spectra, 375
Banking of track, 43
Bar, the, 102
Barometer, 103
  correction, 103, 295
Barrier p.d., 826
Base (transistor), 828
Beats, 512
BECQUEREL, 874
BERNOULLI's principle, 107
Beta-particles, 879
Binding energy, 898
Binoculars, 391
BIOT and SAVART law, 718
Biprism experiment, 469
Black body, 274
  radiation, 275
Blooming, lens, 470
BOHR's theory, 853
Boiling-point, 243, 245
Bolometer, 269
BOLTZMANN constant, 223

Bonds, 133
Boyle's law, 194, 251
Boys, G, 49
Bragg's law, 863
Breakdown potential, 579
Break strain, 132
  stress, 115
Brewster's law, 501
Bridge rectifier, 812, 827
Brittle, 116, 128
Brownian motion, 129
Bubble chamber, 886
  raft, 127
Bulk modulus, 122, 434

Calibration of thermometer, 174
  voltmeter, 661
Calorimeter, 184
Camera lens, 399
Candela, 313
Capacitance, 596, 598
  , measurement of large, 596
  , measurement of small, 596
Capacitor, charging of, 593, 612
  , discharging of, 593, 611
Capacitor, electrolytic, 599, 681
  mica, 599
  paper, 599
  parallel-plate, 600
  variable, 599
Capacitors in parallel, 606
  in series, 607
Capillarity, 141, 149
Carbon dating, 883
Carnot cycle, 216
Cassegrain, 390
Cathode (valve), 810
Cathode-ray oscilloscope, 812–6
Cathode rays, 796–8
Cells, secondary, 685
  , series and parallel, 643
Celsius temperature scale, 175
Centre of gravity, 98
  of mass, 98
Centripetal force, 41
Chadwick, 890
Chain reaction, 902
Characteristics of sound, 511
Charge carriers, 705
Charge on conductor, 571, 597
  on electron, 564, 795
Charles' law, 196
Chromatic aberration, 378
Circle, motion in, 38
Circular coil, 715, 718
Closed pipe, 534
Cloud chamber, 885

Cokcroft-Walton, 901
Coefficient of friction, 160
  of viscosity, 163
Coercive force, 766
Coherent sources, 456
Cohesion, 142
Collector (transistor), 828
Collimator, 342
Colours of sunlight, 373
  of thin films, 471
Combination of thin lenses, 362
Common-base circuit, 829
  -emitter amplifier, 831
    characteristics, 831
    power gain, 831
    voltage gain, 831
Comparator method, 291
Comparison of e.m.f.s., 657
  of resistance, 662
Components of force, 94
  of velocity, 9
  resolved, 9
Compound microscope, 394
  pendulum, 90
Concave mirror, 319
Concentric spherical capacitor, 601
Condensing lens, 398
Conductance, 622
Conduction in metals, 257, 265, 620
  through gases (electrical), 796
Conduction, thermal, 257
  of bad conductor, 264
  of good conductor, 263, 265
Conductivity, electrical, 632
Conductors, 563, 620
Conical pendulum, 43
Conservation of energy, 31, 177
  of momentum, 22, 84
Conservative forces, 30
Constant pressure experiment, 198
  volume gas scale, 298
    thermometer, 298
Constructive interference, 456
Continuous flow calorimeter, 181
  spectra, 376
Convex mirror, 319, 321, 324
Cooling, correction for, 186
  Newton's law of, 185
Copper oxide rectifier, 811
Corkscrew rule, 691
Corona discharge, 579
Corpuscular theory, 444
Cosmic rays, 798
Coudé, 390
Coulomb, the, 574
Couple on coil, 82, 696
Couples and work, 82

Covalent bond, 133
Critical angle, 334
  mass, 904
  potential, 852
  temperature (gas), 248
  velocity (viscosity), 165
CROOKES' dark space, 796
Crystal structure, 863
Cubic expansivity, gas, 197
  liquid, 293
CURIE-JOLIOT, 890
CURIE temperature, 771
Current balance, 713
  density, 632
  , electron, 620
Current unit, 713
Current, potentiometer, 660
Curvature, 447
Curved mirror, 319

DALTON'S law of partial pressures, 196
Damped oscillation, 409
Damping of galvanometer, 410, 743
D.c. amplifier, 597
DE BROGLIE'S law, 853, 868
Decay constant, 881
  series, 891
Deceleration, 5
Declination, 717
Decibel, 522
Defects of vision, 370
Degrees of freedom, 228
Dekatron counter, 875
Demagnetization, 768
Density, 104
  of earth, 54
  of liquid, and temperature, 293
Depletion layer, 826
Depth of field, 401
Destructive interference, 457
Deuterium, 904
Deviation by mirror, 315
  by prism, 340, 353
  by thin lens, 361
  by thin prism, 352
Deviations from gas laws, 250
Diamagnetism, 770
Dielectric, 599, 603
  constant, 602
  polarization, 603
  strength, 602
Diffraction at lens, 484
  at slit, 425, 480–3
  , electron, 867
  , microwave, 429
  , X-ray, 862
Diffraction grating, 489–92

Diffuse reflection, 314
Diffusion, 223
  cloud chamber, 886
Dimensions, 32
  applications of, 33, 124, 165
Diode valve, 810
Dioptre, 447
Dip, 716, 751
Dipole (electric), 603
Disc generator, 734
Discharge, capacitor, 593, 611
  tube, 796
Disintegration, nuclear, 898, 900
Dislocation, 125
Dispersion by prism, 373, 445
Dispersive power, 377
Displacement (waves), 413, 416
  of lens, 358
  by rectangular block, 330, 332
Dissociation, electrolytic, 676
Distance-time graph, 4
Diverging lens, 349, 351, 360
Division of amplitude, 463
  wavefront, 460
Domains, magnetic, 768
DOPPLER effect, 513–20
Double refraction, 502
Drift velocity, 621
Driving mirror, 322
Drops, formation of, 885
Ductile, 116
DULONG and PETIT, 294
Dust-tube experiment, 542, 552
Dynamo, 736

Earth, density of, 54
  , escape velocity, 56
  mass of, 54
  potential of, 55, 582
Earth's horizontal component, 716, 751
  magnetic field, 716
  vertical component, 716, 751
Echo, 423
Eddy currents, 742
Efficiency (electrical), 646
EINSTEIN'S mass law, 896
  photon theory, 845
  relativity theory, 477
Elastic collisions, 24, 852
  deformation, 115
  limit, 115
Elasticity, 114
  , adiabatic, 123, 435
  , isothermal, 123, 435
  , modulus of, 117, 434
Electric field, 574, 801
  intensity, 575, 584

potential, 580–9
Electrical calorimetry, 180, 182, 188
  symbols, 622
Electrodes, 674
Electrolysis, 674
  back e.m.f., 683
  , FARADAY's laws of, 674
Electrolyte, 674
Electromagnetic induction, 726
  waves, 428, 503
Electromagnetism, 691, 709
Electrometer, 597
Electron charge, 564, 793, 795
  diffraction, 867
  , e/m_e, 804, 806
  lens, 814
  mass, 805
  motion, 620, 801–8
  orbit, 421, 853
  shells, 864
  -volt, 581
Electrons, 563, 632, 799
Electrophorus, 567
Electroscope, 565, 588
Electrostatic fields, 574
  intensity, 575, 584
  shielding, 589
Emitter (transistor), 828
E.M.F., 640, 645
  , determination, 657, 662
Emission spectra, 375, 857
Emissivity, 281
Emulsion detector, 887
End-correction, 538
Energy and matter, 896
Energy bands in solids, 822
Energy, electrical, 408, 637
  exchanges, 67, 408
  in capacitor, 608, 756
  in coil, 755
  levels (atom), 852–7
  luminous, 312
  mechanical, 28
  nuclear, 892
  shells, 864
  sound, 521
Energy in wire, 120
  , kinetic, 29
  , potential, 29
  rotational, 77, 87, 229
  translational, 222
  vibrational, 67
Equation of state, 200
Equilibrium, conditions of, 95
Equipotential, 586
Errors, counting 876
Escape velocity, 56

Ether, 477
Evaporation, 239
Excess pressure in bubble, 147
Excitation potential, 856
Expansion, cubic, 293
  , linear, 290
  of gas, 197
  of scale, 295
Explosive forces, 25
External work, 204, 210
Extraordinary ray, 500, 502
Eye, 370
Eyepiece, RAMSDEN's, 398
Eye-ring, 384, 395

Falling sphere, viscosity by, 167
Far point of eye, 370
Farad, 596
Faraday constant, 675
FARADAY's dark space, 796
  ice-pail experiment, 658
  laws of electrolysis, 674–5
  law of induction, 729
Feedback, energy, 835
Ferromagnetism, 690, 768
Field effect transistor, 839
Field of view of microscope, 395
  of mirror, 322
  of telescope, 385
Filter circuit, 826
  pump, 108
Fine beam tube, 804
Fission, nuclear, 902
Fixed points, 174, 298
Flame, sensitive, 538
FLEMING's left-hand rule, 693
  right-hand rule, 731
Flotation, 106
Flow, laminar, 166
  turbulent, 165
Fluid, 101
  motion, 106
Flux, electric, 576
  , magnetic, 729
Flux density, B, 749, 764
  -linkages, 729, 748
f-number, 399
Focal length of lenses, 365–8
  of mirrors, 320
Focusing, electron, 814
Force, 15, 16, 19
  between currents, 712
  due to expansion, 119
  on charges, 702, 733
  on conductor, 693
Forced oscillation, 410
FOUCAULT's method for c, 450

FRANCK-HERTZ, 852
FRAUNHOFER's lines, 377
Free oscillation, 411
Freezing, 236
Frequency, 511
  by beats, 513
  by resonance tube, 541
  by sonometer, 548
  fundamental, 534, 536, 546
FRESNEL's biprism, 469
Friction, 160–2
Fusion, nuclear, 904
  (solid), 236

GABOR, 494
Galvanometer, 699
  , sensitivity of, 700
Gamma-rays, 879
  , inverse-square law, 879
  , range, 879
Gas, 194, 219, 250
  constant, 200
  equation, 200
  , ideal, 200, 247
  laws, 194–201
  , mole of, 201
  , real, 250
  thermometer, 298
  velocity of sound in, 553
Gases, diffusion of, 223
Gaseous state, 219, 239, 247
GAUSS's theorem, 577
GEIGER and MARSDEN, 888
  -MÜLLER (GM) tube, 874
Generators, 732, 741
Germanium, 823
Glancing angle, 315
Gold-leaf electroscope, 565, 588
GRAHAM's law, 223
Gravitation, law of, 48
Gravitational constant, 48
  intensity, 51
  mass, 54
  orbits, 52, 56
  potential, 30, 55
Gravity, acceleration of, 8, 42, 50
  motion under, 8
Grazing incidence, 346
Ground state, 855
Gyration, radius of, 78

HALE telescope, 388
Half-life, 881
Half-wave rectification, 811
Hall voltage, 704
Harmonics of pipes, 537
  of strings, 547

Heat capacity, 179, 207
  energy, 176
  engine, 215
Heating effect of current, 636
Heavy hydrogen, 904
Helium, 878, 905
HELMHOLTZ coils, 720, 807
Henry, the, 753
High temperature measurement, 304
  tension transmission, 638
Holes, 823
Holography, 493–6
Hollow conductor, 570, 588
Homopolar generator, 734
HOOKE's law, 115
Horizontal component, 716, 751
Hot-wire ammeter, 772
HUYGENS' construction, 441
Hydrogen atom, 853, 854, 856
  isotopes, 904
Hydrogen gas thermometer, 299
Hysteresis, 765

Ice pail experiment, 569
  point, 174
Ideal gas, 200, 247
Illuminance, 313
Images in curved mirrors, 320–4
  in lenses, 350–9
  in plane mirror, 316
Impedance, 782
Impulse, 18
Impurity, metal, 632
  , semiconductor, 824
Induced charge, 566–73
  current, 726
  e.m.f., 728, 730, 737
Inductance, 752–6, 778
Induction, electrostatic, 565
Inelastic collisions, 24
Inertia, 15
  , moments of, 78
Infrared radiation, 268
  spectrometer, 271
Insulators, 563, 822
Intensity, electric, 575, 584
Intensity of magnetization, 764
  of sound, 521
Interaction, fields, 696
Interference (light), 456
  in thin film, 471–3
  in wedge films, 463
  of waves, 425, 512
Intermolecular forces, 130, 251
  energy, 130
Internal energy of gas, 205, 227, 229
  work, 204, 251

Internal resistance of cell, 640, 659
International temperature scale, 306
Inverse-square law (electrostatics), 577
  (radiation), 270, 879
Ionic bond, 133
Ionization current, 598
  of gases, 796–8
  potential, 856
Ions, 676, 797
Isothermal bulk modulus, 123
  change, 209
  curves, 209, 248, 253
  sound waves, 435
  work, 210
Isotopes, 894, 901

JAEGER's method, 153
JOULE-KELVIN effect, 251
Joule, the, 27
JOULE heating, 636
Junction diode, 826

KELVIN scale, 174, 298
KEPLER's laws, 47, 87
Kilowatt, 638
Kilowatt-hour, 639
Kinetic energy, 29, 87
  theory of gases, 219
  of liquids, 242
KIRCHHOFF's laws (electricity), 648
  (radiation), 282, 376
KUNDT's tube, 552
  determination of γ, 554
  velocity in gas, 553
  velocity in rod, 552

Laminar flow, 166
Laminations, 739
LAPLACE's correction, 434
Laser, 858–60
Latent heat, 188, 245
  of evaporation, 188, 245
  of fusion, 190
Lateral inversion, 317
  magnification, 322, 356
Laue diffraction, 862
Leakage current, 832
LEES' disc method, 264
Lens, 239
  formula, 354, 356
  of eye, 370
LESLIE cube, 273
LENZ's law, 727
Limiting angle of prism, 346
Line spectra, 375
Linear flow, 163, 258
  expansivity, 290

Lines of force, electric, 575–9
  , magnetic, 709–17
Liquefaction of gases, 249
Liquid state, 236
LISSAJOUS' figures, 428
LLOYD's mirror, 470
Logic gates, 837
Long sight, 371
Longitudinal waves, 413
LORENZ method, resistance, 667
Loudness, 522
Loudspeaker, 524, 749
Lumen, 312
Luminance, 313
Luminous flux, 312
  intensity, 312
Lux, 313
Lyman series, 854

Magnetic cycle, 765
  fields, 690
  flux, 729
  flux-density, 695, 708, 715, 718
  intensity, 764
    solenoid, 692, 709
    straight wire, 691, 710
  materials, 763
  meridian, 717
  moment, 698
  susceptibility, 765
Magnification, angular, 383, 392
  lateral (linear), 322, 356
  of microscope, 392, 395
  of telescope, 383–7
Magnifying glass, 392
Mains frequency, 549
Majority carriers, 824
Mass-energy relation, 896
  number, 890
  spectrometers, 894
Maximum power, 646
MAXWELL, 227, 691
  distribution law, 227
Mean free path, 231
Mean square velocity, 220, 222
  velocity, 222, 225
Megelectron-volt, 896
Melting-point, 237
  effect of pressure, 238
Mercury, absolute expansion, 294
  thermometer, 173
Metal-glass seal, 292
Metal conductor, 263, 619
  rectifier, 811
Metallic bond, 133
Metre bridge, 666
Mica capacitor, 599

MICHELSON's interferometer, 473
method (light), 453
Morley, 474
Microfarad, 596
Microphone, ribbon, 524
Microscope, compound, 394
, simple, 392
Microwaves, 428, 431
MILLIKAN's experiment, e, 794
, photoelectric, 848
Minimum deviation by prism, 446
distance, object-image, 357
Mirage, 337
Mirror galvanometer, 699
Modulus of elasticity, adiabatic, 123
, bulk, 122, 434
, isothermal, 123, 434
of gases, 123, 434
of rigidity, 124
YOUNG modulus, 116, 433
Molar gas constant, 200
heat capacity, 207, 230
Mole, 201
Molecular forces, 130, 131, 251
speeds, 226
sphere of attraction, 138
Molecules, 128, 228
Moment, magnetic, 698
Moment of couple, 98, 697
of force, 96
Moments of inertia, 78, 82
, measurement, 89
theorems of, 79
Momentum, angular, 84
linear, 18
conservation of angular, 84
of linear, 22
Moon, motion of, 48
MOSELEY's law, 864
Motion, NEWTON's laws of, 15
in circle, 38
in straight line, 7
of projectile, 10
simple harmonic, 60
under gravity, 8
Motors, 744
Mount Palomar telescope, 388
Moving-coil ammeter, 699
galvanometer, 700
loudspeaker, 524, 749
voltmeter, 625, 629, 661
Moving-iron meter, 772
Multimeters, 628
Multiple images, 329
Multiplier, 627
Musical interval, 511
Mutual induction, 756

Natural frequency, 411
Near point of eye, 370, 382
Negative charge, 564
NEUMANN's law, 729
Neutral temperature, 651
Neutron, 890
-proton ratio, 899
NEWTON's law of cooling, 185
law of gravitation, 48
laws of motion, 15
rings, 466
velocity (sound) formula, 434
Newton, the, 16
NICOL prism, 503
NIFE cell, 688
Nodes, 418
in air, 531, 543
in pipes, 534–7
in rod, 552
in strings, 418, 546
Non-conservative forces, 31
Non-inductive resistance, 753
Non-ohmic conductors, 635, 753
N-p-n transistor, 828
N-semiconductor, 824
Nuclear charge, 678, 864, 889
forces, 857
mass, 892
reactions, 898–905
reactor, 903
stability, 899
structure, 890
Nucleon, 896
Nucleus, 888

Objective, microscope, 394
, telescope, 383, 388, 485
Octave, 511
OERSTED, 691
Ohmic conductor, 634
Ohm-meter, 628
OHM's law (electricity), 633
electrolytes, 682
Open pipe, 536
Optical fibre, 335
instruments, 382
lever, 316
path, 458
pyrometer, 305
spectra, 375, 493, 856
Ordinary ray, 500, 502
Oscillation, damped, 409
, electric, 408, 834
, energy exchanges, 67
of liquid, 73
of rigid body, 90
of spring, 67–8

, undamped, 749
Oscillator, transistor, 835
Oscillatory current, 835
Overtones, 523
    of pipes, 535, 537
    of strings, 546

Paper capacitor, 599
Parallel-plate capacitor, 600
Parallel a.c. circuits, 785
Parallel axes, theorem of, 79
    forces, 97
Parallelogram of forces, 94
    of velocities, 9
Paramagnetism, 770
Parking orbit, 53
Particle, nature, 845
    -wave duality, 869
Pascal, the, 102
Peak value, 774
PELTIER effect, 651
Pendulum, simple, 71
Period, 63
Permeability, 710
    relative, 602
    , vacuum, 574, 604
Permittivity, 574, 604
    relative, 602
    , vacuum, 574, 604
PERRIN tube, 799
Phase angle, 409, 414, 458, 780–8
Phasor, 780
Phon, 522
Phonon, 266
Photon, 845
Photo-cells, 849–51
Photoelasticity, 505
Photoelectricity, 843
Phototransistor, 838
Pipe, flow through, 164
Pitch, 511
Pitot-static tube, 109
PLANCK constant, 844, 853
Plane conductor, 578
    mirror, 314
    polarization, 500
Plane-progressive wave, 416
Planetary motion, 47, 87
Plasma, 519, 906
Plastic deformation, 125
PLATEAU's spherule, 139
Platinum resistance thermometer, 301
P-n junction, 825
P-n-p transistor, 828
Points, action at, 568, 579
POISEUILLE's formula, 164
Polar molecules, 604

Polarization (e.m. wave), 431, 499
    (light), 499
    and electric vector, 504
    by double refraction, 501
    by Polaroid, 499, 504
    by reflection, 500
    by tourmaline, 500
Polarization of dielectric, 603
Polarizing angle, 501
Pole of mirror, 319
Polythene rod, 564
Positive charge, 564
    column, 796
    ion, 676, 892
    rays, 892
Potential difference, 580, 637
    divider, 624
Potential, electric, 580, 854
Potential energy (mechanical), 30
    (molecular), 130
    gradient, 584
    gravitational, 55
Potentiometer, 656–63
Power (electrical), 638, 646
    (a.c.) 787
    , mechanical, 28
    of lens, 447
Pressure, atmospheric, 103
    coefficient, 198
    curves (sound), 413, 420
    , gas, 219
    liquid, 101
    standard, 103
PRÉVOST's theory, 279
Principal focus, 349
    section, 502
Principle of Superposition, 417, 456
Prism, 339
    binoculars, 391
    deviation by, 340, 347
    limiting angle of, 346
    minimum deviation by, 340
    total reflecting, 347
Progressive wave, 414
Projectiles, 9
Projection lantern, 398
Proton, 889
    number, 890
    -neutron ratio, 899
P-semiconductor, 825
Pulse voltage, 613
Pure spectrum, 373
Pyrometry, 304

Quality of sound, 522
Quantity of charge, 574, 675, 793
    of heat, 179

Quantization of energy, 852
Quantum theory, 844, 852, 866
  of light, 844
Quenching agent, 874

Radial field, 70
Radiation (thermal), 267
  laws, 278
  pyrometer, 304
  wavelengths, 269
Radioactivity, 874
Radio telescope, 486
Radium, 891
Radius of curvature, lens, 361
  of gyration, 78
  of mirrors, 320
RAMSDEN's eyepiece, 398
Range (radioactivity), 878–9
Rarefaction, 413
Ratemeter, 876
Ratio of heat capacities, 211, 230
Reactance of capacitor, 777
  coil, 779
Reaction, 21
Reactor, 903
Real gases, 247
  image, 321, 324, 351
Rectification, 811, 826
Rectilear propagation, 483
Reflection of light, 442
  of sound, 423
Reflector telescope, 388
Refraction at plane surface, 326, 443
  at spherical surface, 447
  through prism, 339, 446
  wave theory of, 443
Refraction of sound, 424
Refractive index, 326, 444
  absolute, 326
  of glass, 331
  of glass prism, 345
  of lens, 361
  of liquid, 336, 363
Refractor telescope, 383, 390
Relative permeability, 710, 764
  permittivity, 602
  velocity, 12
Remanence, 765
Resistance, 621
  absolute method, 667
  low, 662
Resistance box, 623
Resistance thermometer, 301
Resistances in parallel, 632
  in series, 622
Resistivity, 631
  electron theory, 632

Resolution of forces, 9, 94
Resolving power, radio telescope, 486
  , telescope, 484
Resonance (sound), 539
  tube, 541
Resonance, series, 784
Resonant frequency, 411, 540, 785
Retina, 370
Reversibility of light, 312
Reversible change, 210
REYNOLDS, 165
Rigid body, motion of, 77–90
Rigidity, modulus of, 124
Rod, longitudinal vibration in, 552
ROENTGEN, 861
Rolling object, 87
Root-mean-square current, 773
  velocity, 221
Rotational dynamics, 77
Rotational energy, 77, 88, 229
Rubber molecules, 127
RUTHERFORD, 878, 888
  and ROYDS, 878

Saccharimetry, 506
Satellites, 47, 52
Saturated vapours, 239–44
Saturation current, diode, 810
  , gas, 797
Scalars, 3
Scaler, 875
Scales of temperature, 174, 298
Scattering law, 888
Scintillation photomultiplier, 887
Screening, 589
Search coil, 749, 750
SEARLE's method, 263
Secondary cells, 685
  coil, 740
  electrons, 887
SEEBECK effect, 650
Selenium cell, 850
Self-induction, 752–6
Semiconductors, 823
Sensitivity of meter, 700
Series a.c. circuits, 782
Series-wound motors, 746
Shear modulus, 125
  strain, 125
  stress, 125
Shells, energy, 864
Short sight, 370
Shunt, 627
  -wound motors, 747
Siemens, the, 622
Sign convention, 322, 354
Silicon, 823

Simple harmonic motion, 60, 407
  pendulum, 71
Slide-wire bridge, 666
Slip in metals, 126
Soap film, 145, 147, 473
Solar constant, 279
Solenoid, 692, 709, 720
Solid angle, 312
  friction, 160
  state, 236
    detector, 875
Solid, velocity in, 432, 553
Sonometer, 548
Sound waves, 413, 434, 511
Spark counter, 877
Specific heat capacity, 179–83
  , constant pressure, 207
  , constant volume, 207
  of liquid, 181
  of solid, 180, 184
  of water, 183
Spectra, 375, 493, 856
  , hydrogen, 375, 854, 856
Spectrometer, 342
Spectrum, hydrogen, 375
  pure, 373
  sunlight, 373
Speed in circle, 38
Sphere capacitance, 601
  , field due to, 577
Spherical aberration, lenses, 397
  mirrors, 319
Spin, electron, 763, 768
Spiral, electron, 806
Spontaneous emission, 858
Spring, helical, 68
Stability, nuclear, 899
Standard cell, 659, 684
Starting resistance, motor, 744
States of matter, 130, 236, 239
Static bodies, 94
Stationary (standing) waves, 418, 422, 530
  aerial, 421
  , electron orbit, 421
  , light, 420
  pipe, 419, 534–42
  rod, 551
  sound, 531
  string, 418, 420
Steam point, 173
STEFAN constant, 278
  law, 278
Stimulated emission, 858
STOKES' law, 167
Stopping potential, 846
Straight conductor, 691, 710, 721
Strain, bulk, 122

, shear, 124
, tensile, 117
, torsional, 125
Streamlines, 106, 163
Stress, bulk, 122
, shear, 125
, tensile, 116
, torsional, 125
Stretched wire energy, 120
Strings, harmonics in, 547
  resonance in, 547
  , waves in, 545
Sun, mass of, 55
  , radiation of, 267, 279
Supercooling, 237
Surface density charge, 572, 579
Surface energy, 139, 154–7
Surface tension, 138
  and excess pressure, 146
  measurement of, 143, 153
  variation of, 153
Susceptibility, 765

Tape recorder, 525
Telescope, astronomical, 383
  , radio-, 486
  , reflector, 388
  , terrestrial, 390
Temperature, 173, 222, 298
  coefficient, 668
  gradient, 258–62
Tensile strain, 116
  stress, 116
Terminal p.d., 640
  velocity, 166
Terrestrial magnetism, 716
Tesla (T), the, 695
Thermal capacity, 179
  conductivity, 258, 263
  expansion, 290
  insulators, 261
Thermistor, 668
Thermocouple, 301, 303, 650
Thermodynamic scale, 174
Thermodynamics, first law, 205
  second law, 215
Thermoelectric thermometer, 303
Thermoelectricity, 650
Thermometers, 173, 298
Thermonuclear reaction, 905
Thermopile, 270
THOMSON effect, 651
  experiment, e/m_e, 804
Threshold of hearing, 552
Timbre, 522
Time constant, 612
Time-base (C.R.O.), 813

Toroid, 709
Torque, 80, 97
 on coil, 696–8
 rotating body, 81
Torricelli theorem, 109
Torsion, 125
Torsional oscillation, 90
Total internal reflection, 333
 radiation pyrometer, 304
 reflecting prism, 347
Transformers, 739
Transistor, 828
 amplifier, 831–3
 , field effect, 839
 oscillator, 835
 switch, 836
Translational energy, 222
Transmutation, 889
Transverse waves, 34, 413
Triangle of forces, 95
Triple point, 174, 298
Tritium, 905
Tuning, 785
Turbulent flow, 165

Ultraviolet rays, 268
Ultrasonics, 511
Uniform acceleration, 7
 velocity, 3
Unsaturated vapour, 239
Upthrust, fluid, 105
Uranium, 874
 fission, 902
 series, 891

Valence electron, 822
Valency, 675
Valve, thermionic, 810
VAN DE GRAAF generator, 568
VAN DER WAALS' bond, 133
 equation, 251
Vapour pressure, 239
Variable capacitor, 599
Variation of g, 44, 50
Vectors, 3, 12, 781–8
Velocities, addition of, 12
 , subtraction of, 12
Velocity, angular, 38
 critical, 165
 , relative, 12
 , terminal, 166
 -time graph, 5
 , uniform, 3
Velocity of light, 449
 FOUCAULT, 450
 MICHELSON, 452
 ROEMER, 449

Velocity of sound, 432
 in air, 123, 415, 427, 542
 in gas, 432, 434, 553
 in pipe, 541
 in rod, 433, 553
 in string, 33, 545
Vibrating reed switch, 596, 598
Vibrational energy, 521
Vibrations, forced, 410, 539
 in pipes, 534
 in strings, 545
 longitudinal, 417
 resonant, 539
 transverse, 33, 412, 545
Virtual image, 317, 322, 351
 object, 324, 360, 367
Viscosity, 163
 falling sphere and, 166
 , gas, 168, 232
Visual angle, 382
Volt, 580, 637
Voltage amplification, 831
Voltameter, 674
Voltmeter, moving coil, 625, 629
Volume expansivity, gas, 197
 , liquid, 293

Watt, 28
Wattmeter, 701
Wave, 412
 equation, 416
 nature (particle), 867
 properties, 412–7
 theory of light, 440
 velocity, 432
Waveform analysis, 523
Wavefront, 440
 , curvature of, 447
 , lens, 447
Wavelength of light, 462, 491
 of sound, 425, 531–6
Weber, the, 729
Weight, 16, 42
Weightlessness, 51
WESTON cadmium cell, 684
Wetting surfaces, 142
WHEATSTONE bridge, 665
WIEDEMANN-FRANZ law, 260, 265
WIEN's law of radiation, 278
WILSON cloud chamber, 885
Work, 28
 done by torque, 82
 done by gas, 204, 215
 hardening, 127
 in stretching wire, 120
Work function, 844

X-rays, 861
, diffraction of, 862
spectra, 864

Yield point, 115
YOUNG's fringes, 459–62
, modulus, 117

, determination of, 118, 553

Zener diode, 828
Zero potential (atom), 856
(electrical), 582
(gravitational), 55
Zeroth law, 176